EXPERIMENTS IN FOUR DIMENSIONS

BY DAVID L. HEISERMAN

TAB BOOKS Inc.
BLUE RIDGE SUMMIT, PA 17214

To Judy, my wife: A promise of seventeen years fulfilled.

FIRST EDITION
FIRST PRINTING

Library of Congress Cataloging in Publication Data

Heiserman, David L., 1940-
Experiments in four dimensions.

Bibliography: p.
Includes index.
1. Physics—Experiments. 2. Hyperspace. I. Title.
QC33.H38 1983 530.1'1 82-19328
ISBN 0-8306-0141-4
ISBN 0-8306-1541-5 (pbk.)

Contents

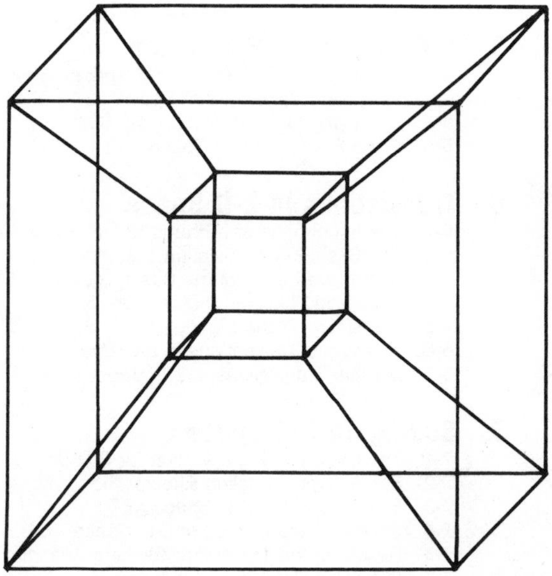

Introduction

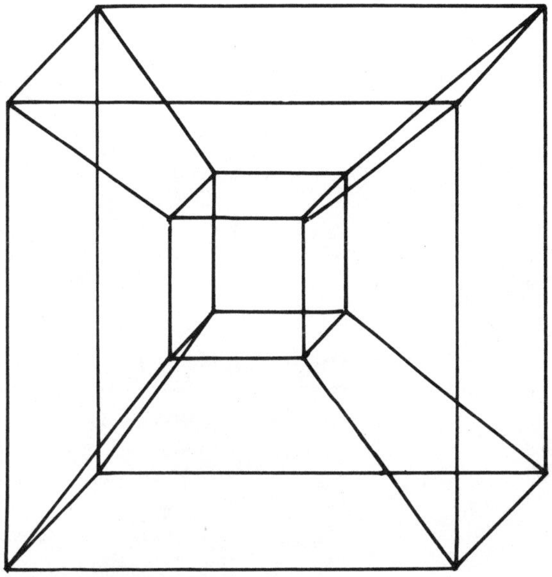

Some plain paper, a pad of graph paper, and a pencil: that's all you need in the way of equipment to engage actively in this adventure in four dimensions. An active imagination and a willingness to work hard and learn new things is all you need in the way of personal traits to follow my lead, and, eventually, wander on your own into an imaginary world of four dimensions.

You will find that this book is quite different from most others that deal with the esoteric concepts of four dimensions. Rather than describing, it explains; rather than offering mere information, it propounds ideas. It offers the tools and points the way. It is wholly up to you to make of it what you will.

Incidentally, neither I nor the publisher assume any liability if you use this material carelessly and invent a 4-dimensional object that draws you or any of your personal belongings away into hyperspace.

Special Note: The English letters A, B, C, and D, used in the *tables* correspond to the Greek letters α, β, γ, and δ, used in the *figures* and *text*.

Other TAB books by the author:

No. 841 *Build Your Own Working Robot*
No. 1101 *How to Design & Build Your Own Custom TV Games*
No. 1191 *Robot Intelligence . . . with experiments*
No. 1205 *PASCAL*
No. 1241 *How To Build Your Own Self-Programming Robot*
No. 1341 *How To Design and Build Your Own Custom Robot*
No. 1391 *Projects in Machine Intelligence for Your Home Computer*

How to Use This Book

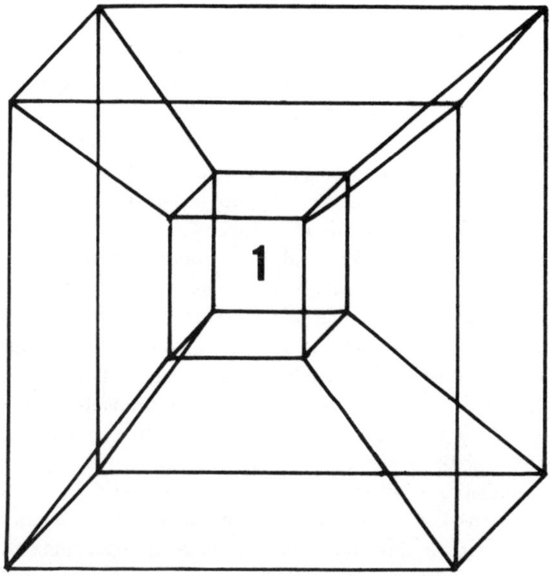

Everyone takes space for granted most of the time. I don't necessarily mean outer space, but space in the most general sense of the word. I mean the space that surrounds us and plays a powerful role in everything we think and do. And I mean the kind of space that characterizes the entire universe as we commonly perceive it.

Every commonly understood attribute of space is built around the notion of 3-dimensional space, and it seems most natural to characterize space as having qualities of width, height, and depth. But that doesn't have to be the case. In fact, thinking only in terms of 3-D space can be considered quite provincial in the light of expanding knowledge.

This book is about alternative spaces. It's about spaces of dimensions lower than three, spaces of three dimensions, and spaces of four dimensions. And the book does more than describe alternative spaces; it explains them and gives you an opportunity to manipulate them.

1-1 MORE ABOUT ALTERNATIVE SPACES

People can, and often do, work with a world having something other than three dimensions. That is the world of just two dimensions; it is the world characterized by pictures drawn onto the plane of a page or canvas. One of the big differences between the work of a painter and a sculptor is that the latter works with an artistic medium of three dimensions, while the painter works with a medium of just two dimensions.

The sculptor attempts to portray objects of the ordinary, familiar 3-D world by means of a medium of the same number of dimensions. In a manner of speaking, a painter works from something of a disadvantage; he or she is attempting to portray the 3-dimensional universe in a 2-D format. Having to drop down from three to two dimensions, the painter must use some tricks to fool the viewer of the art into thinking that a quality of depth exists when, in reality, it does not. (A great deal of abstract art produced during the first half of this century reflects the painters' dissatisfaction with the usual style of portraying 3-D space in a 2-D medium. Many broke the conventional bonds and, in fact, attempted to show more on a 2-D canvas that even a 3-D perspective can show.)

The problems of portraying 3-D space on the 2-D plane of a page aren't limited to artists. Anyone who has attempted to make perspective drawings of 3-D objects encounter the same difficulties all along the way. Can you imagine how it is possible to portray a 2-dimensional figure in 1-dimensional space—on a single line? Or even more intriguing, how about portraying imaginary 4-

dimensional objects in spaces of three and two dimensions? The experiments suggested in this book give you an opportunity to do that sort of work to your heart's content. But there is more.

Not only will you see how to portray objects, but manipulate them as well. It is entirely possible, and quite often exciting, to apply some mathematical procedures for moving objects within the space that surrounds them. You will learn how to rotate 3-dimensional objects, for instance, and plot the results on the 2-D plane of the page. Plotting the figure is one thing, but observing the overall effect of the rotation is even more fascinating. And the manipulations aren't limited to rotations—there are other ways to manipulate objects mathematically.

The idea of being able to portray and manipulate geometric objects extends to the mysterious realm of four dimensions, hyperspace. Imagine being able to portray 4-D hyperspace objects in a more humanly understandable form of 3-D and 2-D spaces. Then to top that, manipulate the hyperspace object and see how the operation can be interpreted from ordinary space.

Provided you follow the work and experiments suggested in this book, you will have an opportunity to explore a strange and wonderful world of the imagination. And given sufficient experience and imagination, you will be able to do something that few have ever done successfully and in a responsible fashion: create custom 4-D objects.

1-2 DEALING WITH THE MATHEMATICS

Mathematics is the most appropriate tool for specifying and manipulating geometric spaces. In many instances, it is the only tool for doing the job in a complete, fully meaningful and responsible way. Even a casual glance through the pages of this book will show that it isn't short on mathematics; you are not going to have to suffer the adverse consequences of relying on a book that pulls its mathematical punches. Too many "popularized" books on the geometry of higher dimensions beg off on the mathematics and leave the reader with descriptions, but no explanations (let alone the tools for doing anything on your own).

Omitting the mathematics in this book, presumably for the sake of making it more appealing to a wider audience, would be like sending a carpenter to build a house without a hammer, presumably so he wouldn't risk hitting himself on the thumb. The intent is perfectly understandable, but the results would amount to little of significance to you.

Now I know that this need to face the necessary tools of mathematics is going to cause some readers a bit of difficulty. Generally, most people aren't fully prepared to deal with mathematics of the level required here. To get the most from this book, you will need, or have to acquire, basic skills in algebra, trigonometry and analytic geometry at a level taught at most high schools. If you do not possess those skills already, then you now have some motivation to do so. If you've had such courses in the past, but sense that they left you with no feeling for the subject, you, too, have some motivation for brushing up on the subjects.

The character and scope of the mathematical presentations in this book are designed with two assumptions in mind. First, the reader will be willing to bolster his or her gaps in basic mathematical operations from another source—from good text books on algebra, trigonometry, and analytic geometry. (Descriptive geometry of the type usually described as Euclidian plane and solid geometry will be of little use here.) You must be able to solve elementary algebraic equations, being comfortable with operations such as substituting terms, factoring, and solving square roots. As far as the trigonometry is concerned, you should be able to interpret trigonometric expressions, use standard trig tables, apply elementary trig identities, and solve trigonometric equations. The discussions do not require very much in terms of skills in analytic geometry. In fact, I have written most of that material in such a way that you can learn the necessary geometry as you go along; perhaps even being unaware that you are doing so.

The second major assumption running through the discussions is that most readers do not have an intuitive feeling for the creative use of mathematics. This book represents a creative application of mathematics in one particular context—in the context of exploring four different dimensions of space in an analytic fashion. I suspect that a good many readers will ask this question many times through the course of the work: How did that author know he was supposed to do that? As you develop an all-important intuitive feeling for mathematics, you will find yourself asking that question less frequently and spending far less time searching through procedures decribed in previous discussions. You will simply know what has to be done next and why it must be done.

If you presently lack that intuitive grasp of mathematics, you will certainly develop it as you work carefully and consciencously through the discussions, examples, and suggested experiments. The harder you work at it, and try to see how and why things are done the way they are, the better your chances of having some fun and doing some things that few have ever done before.

You might not be much of a mathematician going into this book, but you will be a tolerably good one when you are through with the final chapter. And more than anything else, I hope you come away with a great sense of confidence, enthusiasm, and a determination to do some original mathematical work.

No doubt about it, there is a price to pay for success—success in mathematics or any other human endeavor that is worth anything. That price is hard work and, in the case of intellectual endeavors such as this, careful and responsible thinking as well. Stick with it; it's worth it.

1-3 TAKE THINGS ONE STEP AT A TIME

Things aren't going to stand up to my promises if you don't work your way through this book from the very beginning and one step at a time. After having a chance to read this chapter, begin your real work from the beginning of Chapter 2. Whatever you might already know about the mathematics of spaces can be of great help to you, but you must follow the presentations as I offer them here. Skipping over material with the thought in mind that you already know all about the subject at hand will be a big mistake.

This is not an ordinary text book on geometry. It is an exploration of four dimensions; and as such, the material isn't presented in the same way one would find it in a text book. For instance, you will find some important suggestions and mini-dissertations imbedded in a number of the discussions. Such things do not fit well into a conventional text book; and if you miss them, you run the risk of losing some special insight that can be of real interest. Even some of the exercises include ideas that aren't mentioned anywhere else.

Study each chapter thoroughly with pencil and paper at hand. Study every equation, example, and proof carefully, expecting to gain some new knowledge and insight at any moment. The higher your expectations, the more you will achieve. Work the examples, even if they seem trivial at the moment. Every bit of experience enhances your ability to deal with the less trivial and more abstract ideas that are crucial in the later chapters of the book. Walk into those later chapters without going through all of the material that precedes them, and you will be lost, disappointed and no better off than before.

1-4 REQUIRED TOOLS AND MATERIALS

All that you need to work through this book is a supply of pencils, some ordinary notebook paper, graph paper and a table of trigonometric sines and cosines. Those are the minimal requirements. You can, however, save yourself a lot of time and eliminate many calculation errors by using a scientific or engineering calculator. A calculator, especially a programmable one, can be of great help. Or if you are fortunate enough to have access to a small computer, you can save yourself even more time.

Please understand, however, that such electronic aids are just that—aids. They cannot do your thinking for you, and they certainly lack the imagination required for interpreting the facts. There is nothing you can do with a calculator or computer that you cannot do with an ordinary pencil, a sheet of paper, and a trig table.

1-5 A NOTE ABOUT ERRORS

Anytime you deal with mathematics, and particularly a long series of mathematical operations, you are bound to make mistakes now and then. And that applies to us, author and publisher, as well as to you, reader and experimenter. Fortunately, mathematics, like science in general, is self-correcting. An error cannot pass unnoticed for long, because anything coming from it eventually turns out to be self-contradictory. There are so many safeguards built into the schemes suggested in this book that you will eventually spot your own errors in arithmetic or general procedure; and by the same token, you will eventually uncover any errors that might exist in your edition of this book. I am fully aware of the frustration that can result from having to deal with the errors of others; but if you are fully aware of what you are reading, you will be able to detect and correct any errors quite readily.

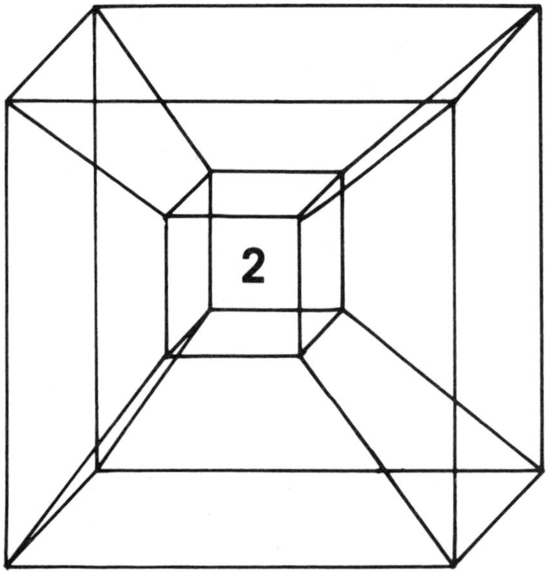

A First Look at 1-D Space

One-dimensional space is a simple kind of space that can be characterized as an indefinitely long and straight line. There can be limits on its length, of course, and there is no reason in principle why it must be straight; but viewing a 1-D world as an indefinitely long and straight line suits the needs of our experiments quite nicely.

An imaginary 1-dimensional creature would have its perception of the world limited to the character of that line-world space. Such a creature would be able to move to and fro along the confines of its 1-D space, and it would most likely possess the ability to perceive and manipulate other kinds of objects existing in that 1-dimensional world.

But the creature would have no perception of anything outside the realm of its own line-world. Other lines and other spaces of higher dimensions would be beyond the creature's natural mechanisms of perception and, perhaps, its understanding as well. Given a bit of imagination and the tools of analytic geometry, though, the creature might be able to think about worlds other than his own and dimensions higher than 1; but its real sense of reality is limited to its 1-D line-world. The creature is, in a manner of speaking, a prisoner of its own geometry.

Through the next few chapters, it might be helpful for you to bear in mind the character of the 1-D creature and its line-world space. Put yourself in that creature's place, and pretend that these are the analytic tools you have available for understanding that somewhat limited universe.

2-1 A COORDINATE SYSTEM FOR 1-D SPACE

Dealing with a world of any number of dimensions requires establishing a *coordinate system*—a standard of measurement—that is suitably designed for it. Figure 2-1 shows four legitimate kinds of coordinate systems for 1-D space. First notice that they are all straight-line coordinate systems. That feature matches the straight-line quality that we have already assigned to 1-D space in general.

Next notice that all four coordinate systems are limited in length. While 1-D space, itself, might have an unlimited length, it is wholly impractical to work with the entire space at one time. The measuring sticks that we use in our own 3-D world are also limited in length, but that doesn't affect their usefulness over normal distances. So there is nothing wrong with limiting the length of a coordinate system to some practical size.

Now notice that all four coordinate systems have a point of origin. That isn't a necessary quality of 1-D space, itself, but it is a practical requirement for measuring

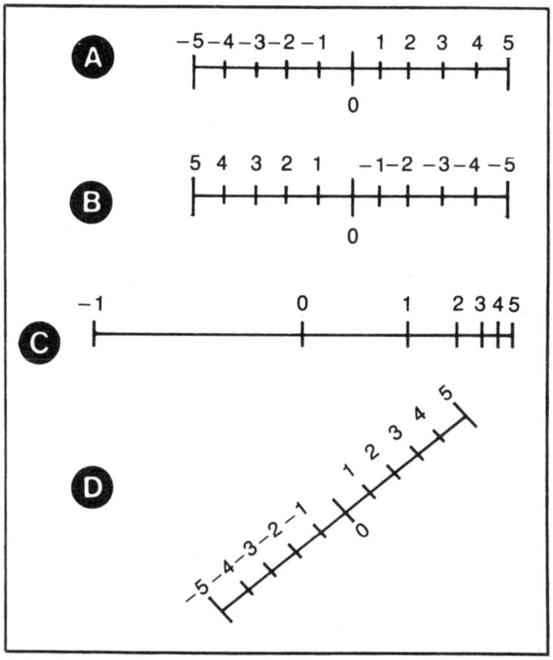

Fig. 2-1. Four kinds of coordinate systems for characterizing 1-D space.

things in space. There has to be a point of reference for all measurements, and the origin of the coordinate system—usually designated with a Ø or O—is such a point of reference. And we have not yet eliminated any of the coordinate systems shown in Fig. 2-1.

For our purposes here, and indeed through the remainder of this book, a coordinate system must be *linear*. That is, the distance between any two consecutive measuring units must be the same everywhere on the coordinate system. The distance between measuring units −3 and −2 in Fig. 2-1A, for instance, is the same as the distance between measuring units 4 and 5 in that same coordinate system. That is also true for the coordinate systems illustrated in Figs. 2-1B and 2-1D. The spacing between consecutive measuring units on one coordinate system need not be the same on another coordinate system, but they must be the same on any given coordinate system. That need for using a linear coordinate system thus eliminates Fig. 2-1C as one that is useful for our purposes. It is an example of a *non-linear* coordinate system for 1-D space.

By convention—and by convention only—we usually draw a 1-D coordinate system parallel to one of the edges of the page. And when working with 1-D coordinate systems, in particular, the custom is to draw the scale parallel to the top and bottom edges of the page. So the 1-D coordinate system in Fig. 2-1D is a perfectly legitimate one in principle, but it does not suit the conventions being established here. That leaves Figs. 2-1A and 2-1B as the remaining possibilities.

Finally, you should notice that the remaining coordinate systems in Figs. 2-1A and 2-1B are identical in all respects but one: the units of measure are specified in opposite directions. In Fig. 2-1A, the units of measure increase in value in a left-to-right fashion, while those in Fig. 2-1B increase from right to left. Again, both are legitimate methods of measurement, but we ought to establish a convention that will apply unless circumstances dictate otherwise.

Unless it is important to do otherwise, our standard 1-D coordinate system will be labeled in such a way that measuring-unit values increase from left to right. That leaves only Fig. 2-1A as the conventional 1-D coordinate system for our immediate purposes. A brief summary of the nature of 1-D coordinates and conventions is in order.

Principles:

1. A coordinate system for 1-D space is a straight line.
2. A coordinate system has an origin that serves as a point of reference.

Conventions:

1. The 1-D coordinate system ought to be drawn parallel to the top and bottom edges of the page.
2. The measuring-unit values ought to increase in value in a left-to-right fashion.

2-2 SPECIFYING POINTS IN 1-D SPACE

Points are the elementary components of 1-D space; and being able to specify the positions of points according to a 1-D coordinate system is a natural part of working with 1-D space. Figure 2-2 shows six different points plotted in 1-D space and specified with regard to their positions according to a standard 1-D coordinate system. Each point on the plot is assigned a particular label that distinguishes its name from all other points. Those are the designations P_1, P_2, P_3, and so on through P_6. There is nothing particularly important about which name, or index, is assigned to which point; but once you make the assignments you are committed to them from then on.

There's more to the matter than simply assigning somewhat arbitrary names to some points that are plotted in 2-D space. In order to be meaningful in a specific situation, you must also assign a coordinate value to each point. That point is situated at some specific position on

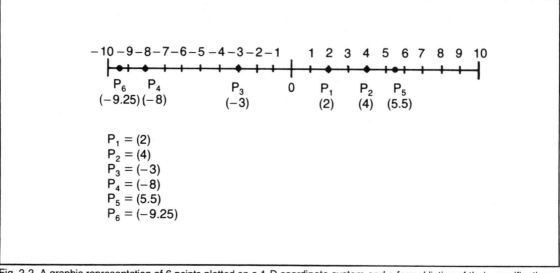

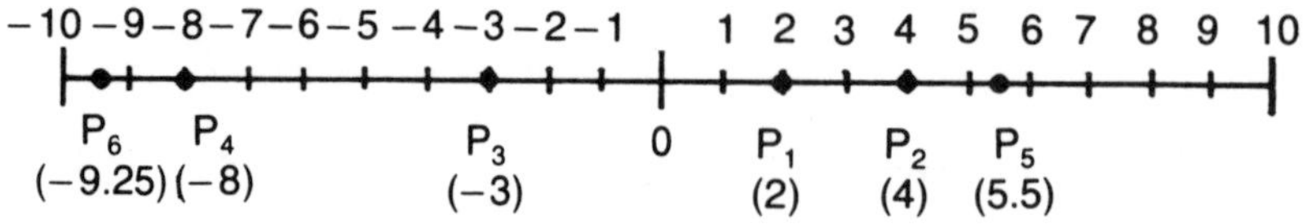

$P_1 = (2)$
$P_2 = (4)$
$P_3 = (-3)$
$P_4 = (-8)$
$P_5 = (5.5)$
$P_6 = (-9.25)$

Fig. 2-2. A graphic representation of 6 points plotted on a 1-D coordinate system and a formal listing of their specifications.

the coordinate system, and you should designate that position in some fashion. That is the purpose of the numbers enclosed in parentheses.

So each of the six points plotted in that figure carry a point name as well as a specific coordinate value, or measuring-unit position on the coordinate system. Thus a point designation such as $P_2 = (4)$ literally means: A point called "2" is located at coordinate position 4 on a 1-D coordinate system. I think you can appreciate how simple the shorthand notation, $P_2 = (4)$, seems compared to saying, "A point labeled number 2 is located at 1-D coordinate position 4." By way of further examples, notice the point designations $P_1 = (2)$, $P_2 = (4)$, $P_3 = (-3)$, and so on. Can you state their literal meanings?

In a very general sense, a point in 1-D space is specified with this sort of notation:

where: $P_n=(x_n)$

P_n is simply a convenient label for any point *n*

x_n is the coordinate of that same point *n*

Literally, such an expression refers to the name of some point *n* in 1-D space and assigns a coordinate value to it. I am using x as the general coordinate for any point in 1-D space in order to be consistent with later work in this book.

This sort of shorthand notation saves a great deal of time and space—mathematicians use such notation almost exclusively. They don't use it in order to confuse neophytes or to make themselves feel especially smart; they use special notation and symbols in order to make things simpler, not more difficult. Understanding the so-called special "language of mathematics" is most often a matter of understanding the notation. Understand the notation and you are well on your way to understanding any topic of a mathematical nature.

So Fig. 2-2 is a graphic representation of a situation that specifies six points in 1-D space. And to make sure there is no confusion about the point specifications, the presentation also includes a listing of the six point designations. Both presentations can be important for a reliable description of the situation at hand: the graphic representation offers an "eyeball view" that can be especially meaningful in a visual sense, while the list of point designations offers the most objective representation (there is no possibility of measuring errors, for instance). Take a crack at interpreting and plotting points by working Exercise 2-1.

Exercise 2-1

1. Construct a 1-D coordinate system (Fig. 2-3) having a range of −10 to 10, then plot the following points on it:

$P_1 = (5)$ $P_2 = (-4)$ $P_3 = (6.5)$
$P_4 = (-8.2)$ $P_5 = (\sqrt{9})$ $P_6 = (-2\cos 45°)$

2. Determine the coordinates of the points that are plotted on the 1-D coordinate system in Fig. 2-3:

P_1 = _____ P_2 = _____ P_3 = _____
P_4 = _____ P_5 = _____ P_6 = _____

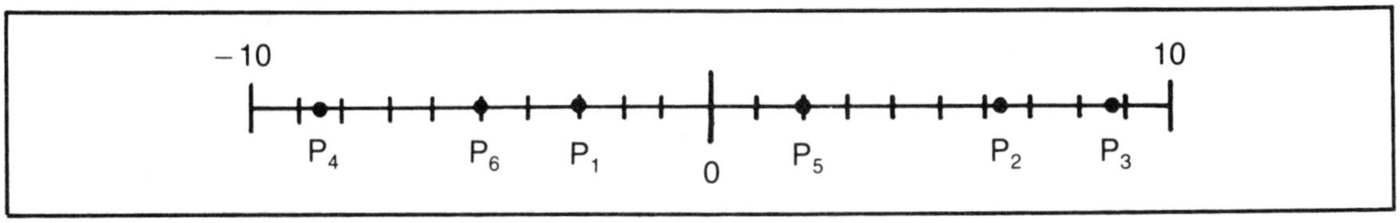

Fig. 2-3. Six points plotted on a 1-D coordinate system as required for part 2 of Exercise 2-1.

2-3 SPECIFYING LINES IN 1-D SPACE

Points are the basic elements of 1-D space; and it so happens that two different points in 1-D space define a second important element—a line. It is possible to draw a straight line between any two different points in 1-D space, and it is likewise possible to define a line in terms of its two endpoints.

Figure 2-4A shows a line drawn in 1-D space. Its two endpoints are labeled P_u and P_v. Those two points completely specify line n as symbolized by:

$$L_n = \overline{P_u P_v}$$

where:

L_n designates any line *n* in 1-D space
P_u designates one of the two endpoints, *u*
P_v designates the second of two endpoints, *v*

The bar across the two point designations indicates a connection, or enclosure of the elements under it. In this case, the bar implies that line L_n is the element "enclosed" by points P_u and P_v.

Those expressions use some very general notation that might puzzle you if you are not yet accustomed to working with ideas in the most productive and general forms. The idea is that there are two points specified just about anywhere you want in 2-D space. In order to avoid citing any particular points and thereby risk conveying some limitations I certainly don't intend to convey, I am letting P_u and P_v stand for any two points. Substitute numerals 1 and 2 for subscripts u and v if you like, but doing that will suggest that the idea is limited to two particular points, P_1 and P_2. I want to be able to substitute numerals such as 4, 5, 6 and so on in order to cover any

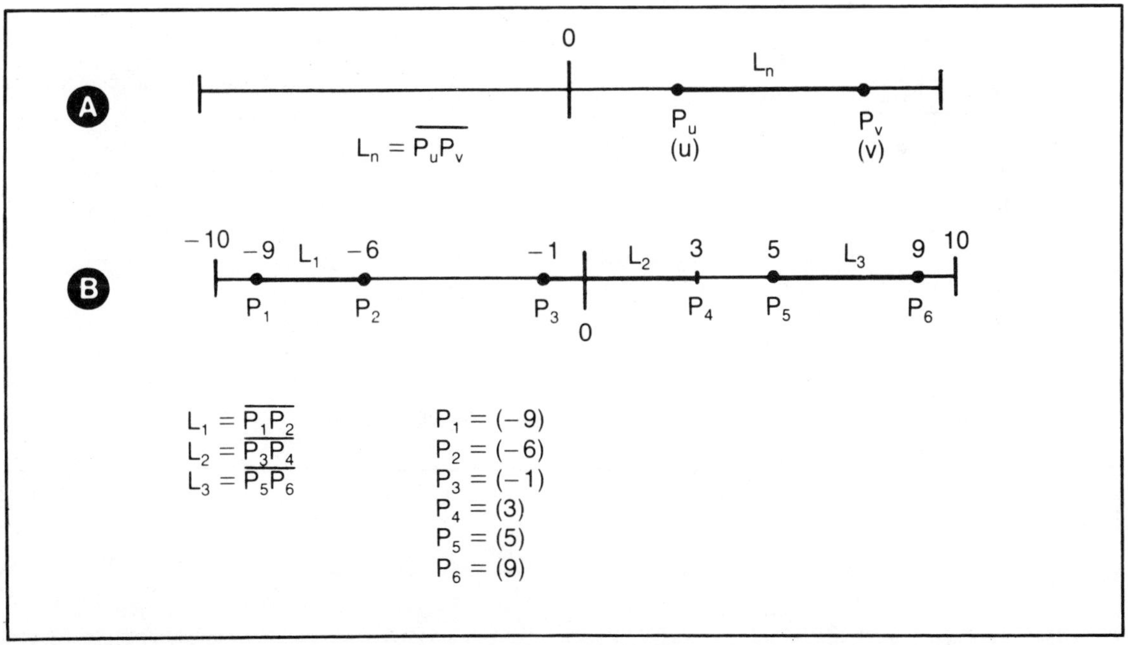

Fig. 2-4. Plotting and specifying lines in 1-D space. (A) The general case where any line L_n is defined in terms of the endpoints that bound it, P_u and P_v. (B) A specific case for three different lines in 1-D space. Notice especially that line L_1 is bounded by endpoints P_1 and P_2, line L_2 is bounded by endpoints P_3 and P_4, and that line L_3 is bounded by points P_5 and P_6.

pair of points ever encountered, and I don't want to prejudice the situation by naming two specific points. That is the essence of generality; and generality is of paramount importance to serious and responsible scientific investigation.

Figure 2-4B shows three specific lines plotted in 1-D space. Each has a pair of endpoints that define it, and line L_1 is defined by endpoints P_1 and P_2, line L_2 is defined by endpoints P_3 and P_4, and line L_3 is defined by endpoints P_5 and P_6.

It is possible to get a complete and reasonably precise view of the three lines from the graphic representation, alone. But the analytic representation—the list of point-and-line (P-L) specifications—is important, too. The analytic representation of the situation in Fig. 2-4B first specifies the three lines in terms of their respective endpoint designations, and then it defines the endpoints, themselves, in terms of their individual coordinates. The graphic representation is nice for visualizing the situation, but the analytic version is absolutely precise and, indeed, more useful for further work. See what you can do with these ideas by working through Exercise 2-2.

Exercise 2-2

1. Construct a 1-D coordinate system having a range of – 10 to 10, and then plot the lines specified here:

$L_1 = \overline{P_1P_2}$ $\quad P_1 = (-9)$
$L_2 = \overline{P_3P_4}$ $\quad P_2 = (-7)$
$L_3 = \overline{P_5P_6}$ $\quad P_3 = (3)$
$P_4 = (-1)$
$P_5 = (5)$
$P_6 = (8)$

2. Determine the point-line (P-L) specifications for the lines plotted in Fig. 2-5:

$L_1 = \overline{P_{\_}P_{\_}}$ $\quad P_1 = (\_\_\_)$ $\quad P_2 = (\_\_\_)$
$L_2 = \overline{P_{\_}P_{\_}}$ $\quad P_3 = (\_\_\_)$ $\quad P_4 = (\_\_\_)$
$L_3 = \overline{P_{\_}P_{\_}}$ $\quad P_5 = (\_\_\_)$ $\quad P_6 = (\_\_\_)$

2-4 LENGTHS OF LINES IN 1-D SPACE

A line is defined in terms of the coordinates of its two endpoints; and knowing the exact coordinates of the endpoints makes it possible to generate a value that represents the length of the line relative to the given coordinate system. Length, as far as we are concerned here, is defined as the unsigned distance between the two endpoints of a line. By "unsigned" I mean that distance, or the length of a line, always has a positive value. And there is a simple equation for determining the length of a line in 1-D space, given the exact coordinates of its two endpoints.

$$s_n = \sqrt{(x_u - x_v)^2} \qquad \textbf{Equation 2-1}$$

where:

s_n = the length of line n in 1-D space
x_u = the coordinate of one endpoint of line n
x_v = the coordinate of the second endpoint of line n

Example 2-1 spells out the details for using that length equation.

Example 2-1

A line in 2-D space has the following specifications:

$L = \overline{P_1P_2}$ $\quad P_1 = (2)$
$P_2 = (6)$

What is the length of the line?

1. Gather the coordinates into a convenient form:

$x_1 = 2 \qquad x_2 = 6$

2. From Equation 2-1:

$$s = \sqrt{(x_2 - x_1)^2}$$

3. Substituting the data from Step 1 into the Equation in Step 2:

$$s = \sqrt{(2-6)^2}$$
$$s = \sqrt{(-4)^2}$$
$$s = \sqrt{16}$$
$$s = 4$$

The line is 4 units long.

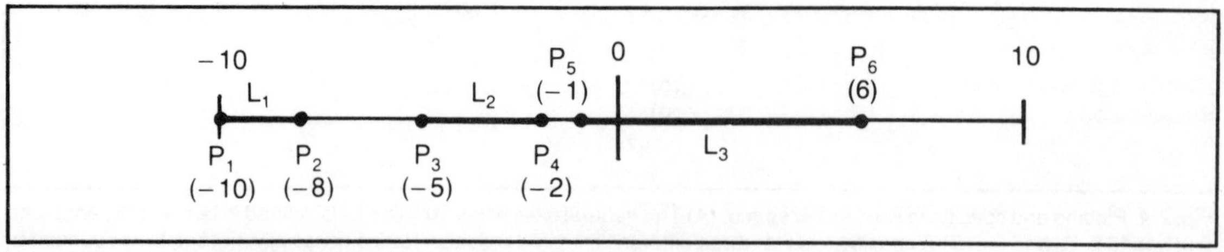

Fig. 2-5. Three lines plotted in 1-D space as required for part 2 of Exercise 2-2.

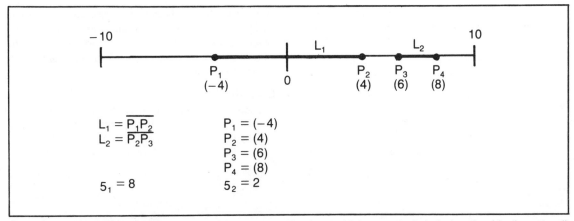

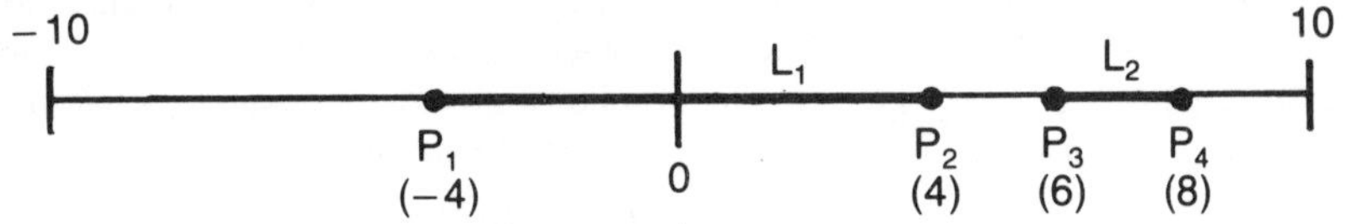

$L_1 = \overline{P_1P_2}$

$L_2 = \overline{P_2P_3}$

$P_1 = (-4)$

$P_2 = (4)$

$P_3 = (6)$

$P_4 = (8)$

$5_1 = 8$ $\quad$ $5_2 = 2$

Fig. 2-6 Two lines plotted and specified in 1-D space. The specifications in this instance include the lengths of the lines as well as the coordinates of the endpoints.

The example begins by specifying a particular line in 2-D space. The L specification shows that the line is bounded by endpoints P_1 and P_2, and then the P specifications show the coordinates of those two points—2 and 6, respectively.

Step 1 in the example is one that will become very familiar to you as you advance through the experiments suggested in this book. The idea is to express the coordinates of the points using a form of notation that corresponds with the equations that use them. So the coordinate of point P_1, according to the original specifications, is 2; while the coordinate of point P_2 is 6. Step 1 summarizes that vital information in a different form: x_1 (specified by point P_1) is equal to 2, and x_2 (specified by point P_2) is equal to 6.

Step 2 restates the general form of the length equation, Equation 2-1, in a form specifically tailored for the situation at hand, and Step 3 solves the equation on the basis of the original specifications and the information cited in Step 1.

It might come as no big surprise to you that the line is exactly 4 units long. You could have "eyeballed" that value by plotting the points on a good 1-D coordinate system and counting the number of measuring units from one point to the other. Doesn't this seem to be a lot of trouble to do something that is done by some simpler graphic methods, or even "common sense?" Well, that might be true in this case, but you won't have to progress much farther into the book before eyeball reckoning and "common sense" will begin to fail you. When that happens, you'll need all the experience and analytical insight you can muster—that is the reason for dealing with an "easy" idea in such a round-about manner.

Figure 2-6 shows two lines specified in 1-D space. Line L_1 is defined by endpoints P_2 and P_3, while line L_2 is defined by the coordinates of endpoints P_3 and P_4. Applying the length equation to line L_1, its length, s_1, turns out to be 8 measuring units; then applying the same equation to the endpoint specifications for line L_2, its length, given as s_2, is just 2 units. Test your confidence with Exercise 2-3.

Exercise 2-3

1. Given the following specifications for lines in 1-D space, use Equation 2-1 to calculate their lengths:

$L_1 = \overline{P_1P_2}$ $\qquad$ $P_1 = (2)$

$L_2 = \overline{P_3P_4}$ $\qquad$ $P_2 = (6)$

$L_3 = \overline{P_5P_6}$ $\qquad$ $P_3 = (0)$

$P_4 = (1)$

$P_5 = (-6)$

$P_6 = (-8)$

s_1 ——— $\quad$ s_2 ——— $\quad$ s_3 ———

2. Draw a carefully measured 1-D coordinate system having a range of -10 to 10. Plot the points and lines specified in Part 1 of this exercise, then check your length calculations by actually measuring the lengths of the lines on the coordinate system.

2-5 ANALYSIS IN 1-D SPACE

Analysis is a vital part of experiments in any number of dimensions. In a manner of speaking, an analysis is a full-blown experiment; it describes with great precision the nature of a point or line in 1-dimensional space, and it includes a graphic representation of the situation as well.

Here are the essential components of a complete analysis, or summary of an experiment, in 1-D space:

☐ A graphic representation of the situation, including plots of all points and lines.

☐ A complete set of specifications for all of the lines and their endpoints.

☐ A summary of the calculated lengths for each of the lines.

There is really nothing more of relevance concerning the 1-D situation, because such an analysis represents a 1-D situation completely and quite precisely. Test your understanding of this matter by working through the following exercise. It suggests your first of many experiments in four dimensions.

Exercise 2-4

Make up a set of P-L specifications for several lines of your own choosing; then complete the analysis by plotting the points and lines on a 1-D coordinate system, and calculating the length of each line.

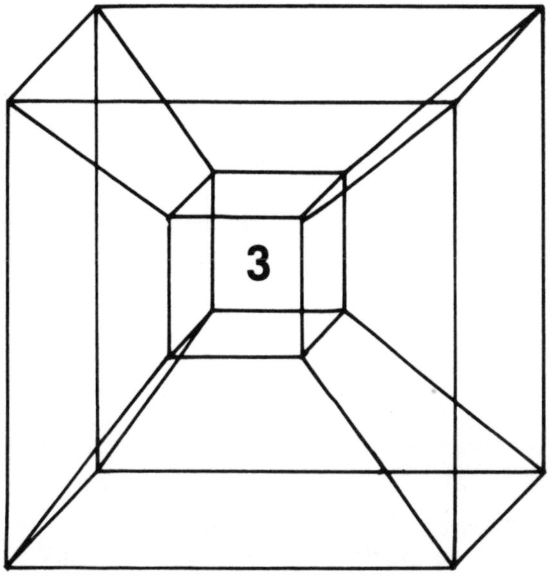

Translations in 1-D Space

The previous chapter characterizes 1-D space as a simple kind of straight-line space, and it introduces an appropriate coordinate system for setting up points and lines in that space. Now it is time to do something meaningful with that space, that coordinate system, and those points and lines.

A translation is a type of geometric transformation that expresses a displacement—a distance and a direction—between coordinate systems that are sharing the same 1-D world. Figure 3-1 shows a pair of coordinate systems, systems A and B, plotted onto a 1-D line-world. The two coordinate systems happen to be identical in all respects but one: their relative position in 1-D space. It can be said that a 1-D translation exists between them.

Looking at the diagram from the viewpoint of coordinate system A, it appears that system B is displaced by some distance to the right. Adopting the viewpoint of coordinate system B, it seems that system A is displaced by the same amount, but in the opposite direction—to the left. Perhaps you are already getting the important idea that a translation is characterized by both a distance and a direction.

A *geometric translation*, as opposed to a relativistic translation described later in this chapter, is most often used for displacing a coordinate system from one place to another in its 1-D line-world. If there happen to be any points or lines associated with the original coordinate system, they are also translated through the same distance and direction. So in effect, geometric 1-D translations can be used for moving coordinate systems, points and lines from one place to another in the 1-D space that they share.

Figure 3-2 shows two coordinate systems, one labeled X and the second labeled X′. The X′ version represents a translated version of coordinate system X. A translation has duplicated the reference coordinate system in a different position in the 1-D world. The coordinate systems should be superimposed on the 1-D line-world, but I have moved them off that line for the sake of clarity. Imagine in your mind's eye that they are superimposed on the 1-D line-world they represent.

Notice especially how a point $P = (x)$ specified in the original X coordinate system is carried by the translation to point $P' = (x')$ in the translated version, X′. Point $P = (x)$ has the same coordinate system as point $P' = (x')$ has to the origin of the X′ coordinate system. Translations do not affect the coordinates of points with respect to their own coordinate systems, but the coordinates are certainly different as regarded from one coordinate system or the other. From the perspective of an

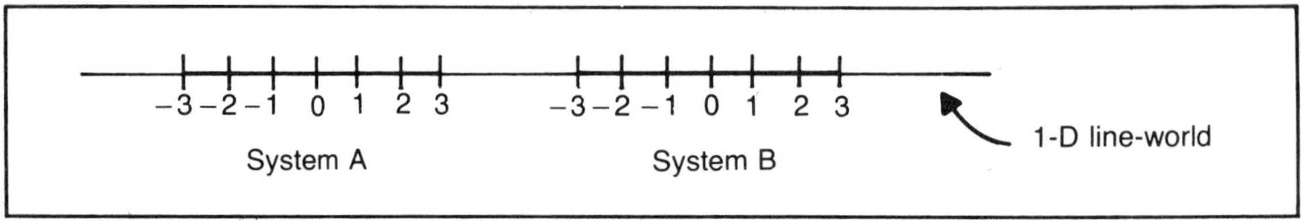

Fig. 3-1. Two identical 1-D coordinate systems that are displaced in 1-dimension by a translation effect.

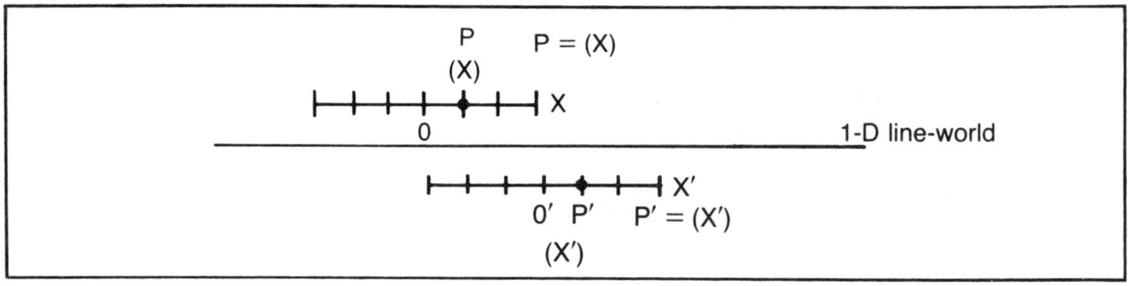

Fig. 3-2. A 1-D translation displaces coordinate system X and its point (P = (x) to a new position in 1-D space characterized by coordinate system X′ and point P′ = (x′). The two coordinate systems should be superimposed on the 1-D line-world, but they are separated here for clarity.

intelligent creature living in coordinate system X, for instance, the coordinate of point P′ is different from that of point P.

If I had specified a line in the X coordinate system, it, too, would have moved along with the X′ coordinate system. A translation literally carries a coordinate system and any points and lines specified by it to some other place in the 1-D line-world. By convention, the original coordinate system and its points are designated with X and P = (x) expressions, while the translated version of the same system used X′ and P′ = (x′) expressions.

One of the most compelling kinds of experiments with translations in 1-D space involves setting up a coordinate system and specifying some points or lines with respect to it. The experimenter then translates that coordinate system by some amount and in one direction or another, and finally calculates the coordinates of the translated versions of the points or lines as regarded from the original coordinate system.

Stated in a somewhat more formal fashion: Given the coordinates of a point or line in 1-D space, apply a translation to them and determine the coordinates of the translated versions as regarded from the original coordinate system.

3-1 THE IMPORTANCE OF A FRAME OF REFERENCE

Figure 3-3A shows a pair of identical coordinate systems that are displaced with respect to one another in 1-D space. (Please imagine that they are actually superimposed on the 1-D line-world.) It is an ambiguous translation situation because it does not show which coordinate system is shifted with respect to the other. From the viewpoint of coordinate system A, for instance, system B is displaced 3 measuring-unit values to the right. The same translation situation appears somewhat different from the viewpoint of coordinate system B, however. In that case, the translation distance of system A is 3 measuring units, but the direction is to the left. So there is room for legitimate dispute about the direction of translation.

Resolving that dispute, or ambiguity, is a matter of selecting one of the two coordinate systems to be the *frame of reference* for the experiment. Once you've selected the frame-of-reference coordinate system, all translation effects are reckoned with respect to it.

Figure 3-3B duplicates the original translation situation, but show system A as the chosen frame of reference, X. Selecting that coordinate system as the frame of reference resolves any ambiguity regarding the direction of translation: System B, now shown as system X′, is clearly translated 3 units to the right relative to the frame of reference. Three units to the right—that's the nature of this version of the translation.

Just to give you a chance to look at the same situation from a different viewpoint, Fig. 3-3C duplicates the original situation once again. Here, however, system B plays the role of the chosen frame of reference, X, and the other system is said to be translated with respect to it. The distance of translation is still 3 units, but the direction is to the left relative to the frame of reference.

A translation is truly meaningful only when first specifying which of two coordinate systems will serve as the frame of reference. Once you make that choice, there is no longer any legitimate argument regarding the direction of translation—the direction will always be taken with respect to the chosen frame of reference.

Initially, it makes no difference which coordinate system you select as the frame of reference, but you must stick with the choice once it is made. And by convention, you will assign X to the frame-of-reference coordinate system and P = (x) terms to its points. The second coordinate system—the one regarded as translated with

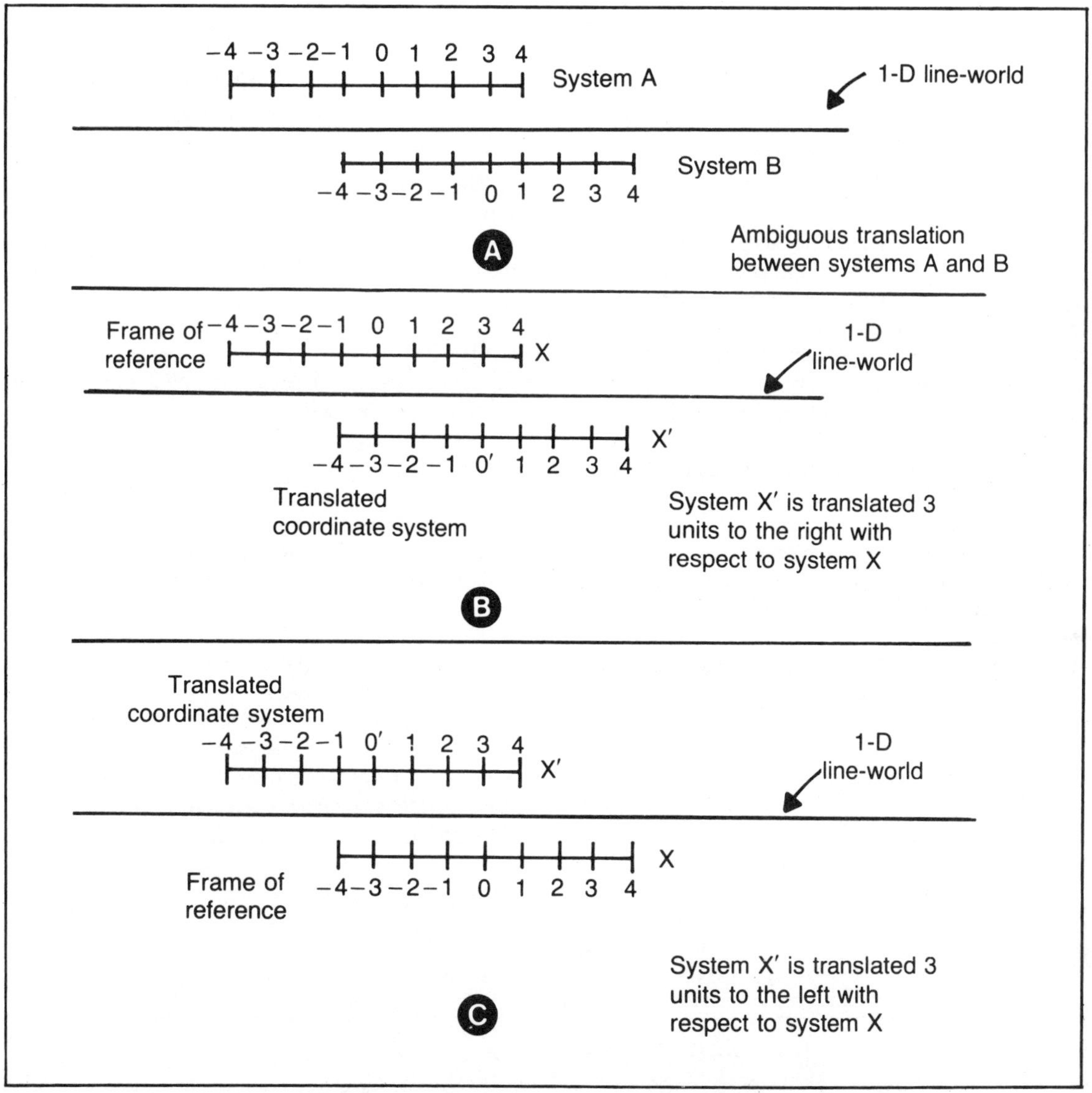

Fig. 3-3. Translation effects are largely meaningless without a frame of reference. (A) A translation exists between systems A and B, but the direction of translation is ambiguous because neither is cited as the frame of reference. (B) Establishing system X as the frame of reference, system X′ is translated 3 units to the right relative to X. (C) Establishing system X as the frame of reference, system X′ is translated 3 units to the left relative to X.

respect to the frame of reference—will then carry X′ and P′ = (x′) expressions.

3-2 THE BASIC EQUATION FOR TRANSLATIONS IN 1-D SPACE

Equation 3-1 includes just three terms, but it is fully adequate for conducting most experiments with translations in 1-D space.

$$x_n' = x_n + T_x \qquad \textbf{Equation 3-1}$$

where:

x_n' = the coordinate of the translated version of point *n* relative to the chosen frame of reference

x_n = the coordinate of the reference point *n* relative to the chosen frame of reference

T_X = the translation term

The equation deals directly with this sort of situation: I have specified the coordinate of a point in the chosen frame of reference, and I have applied a translation to the coordinate frame of reference. What is the coordinate of the translated version of the point as reckoned from the frame of reference?

The x_n term is the coordinate of the original reference point in the frame of reference. Term T_x is the *translation term*—a term that expresses both the distance and direction of translation with respect to the chosen frame of reference. Finally, term x_n' is the coordinate of the translated version of the reference point as reckoned from the frame of reference. Take note of the fact that all three terms in the equation apply directly to the chosen frame of reference.

See how the equation works by studying Example 3-1.

Example 3-1

A point in a frame-of-reference coordinate system is located at P = (2). What is the coordinate of that point, relative to the frame of reference, after applying a translation of $T_x = -3$? Plot the situation.

1. Gather the essential data:

$$x = 2 \qquad T_x = -3$$

2. From Equation 3-1:

$$x' = x + T_x$$

3. Substituting values from Step 1 into the equation in Step 2:

$$x' = 2 + (-3)$$
$$x' = -1$$

The translation thus displaces reference point P = (2) to P′ = (−1). See the situation plotted in Fig. 3-4.

The example begins by citing the coordinates of a reference point and a translation term to be applied to it. Step 1 summarizes that information, using notation that is most suitable for the experiment at hand. In this instance, the coordinate of point P is specified as a particular component of the X coordinate system.

Step 2 suggests a form of the translation equation specially tailored for a 1-point translation operation. The n subscripts shown in Equation 3-1 aren't necessary here because the experiment deals with just one reference point, x.

Step 3 represents a straightforward substitution of the available data into the translation equation. Working out the arithmetic, it shows that the coordinate of the translated version of the reference point is at (−1) relative to the frame of reference.

The overall conclusion is that a translation of −3 moves a point from P = (2) to P′ = (−1) relative to the chosen frame of reference.

The translation term in the equation is important enough to warrant some special consideration. The distance of the translation term—its absolute value—expresses the amount of translation. The amount of

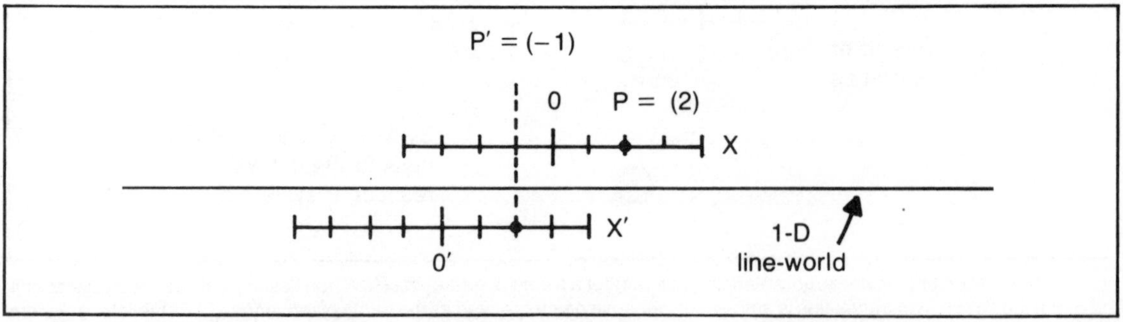

Fig. 3-4. A plot of the result of Example 3-1. In reality, the two coordinate systems should be superimposed on the 1-D line-world.

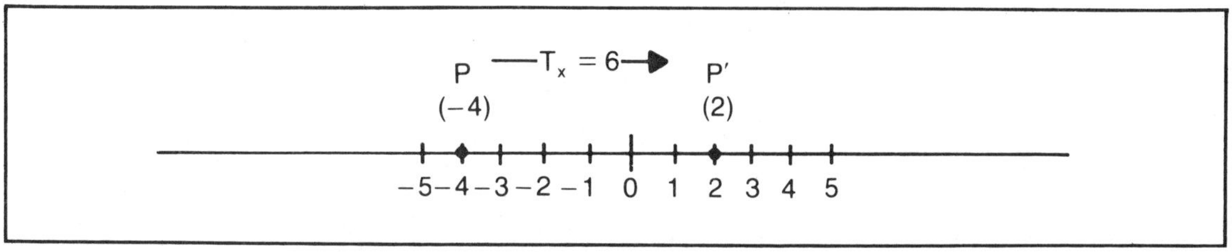

Fig. 3-5. A translation term of 6 effectively displaces point P by 6 measuring units to the right, thus generating the translated version of that point, point P′. The matter is clarified by showing only the frame-of-reference coordinate system.

translation will always be the same, no matter which coordinate system is chosen to serve as the frame of reference for the experiment. The sign of that term, positive or negative, indicates the direction of the translation; and that is sensitive to which coordinate system is chosen to the frame of reference.

By convention, a positive-valued translation term displaces the translated coordinate system in the direction of increasing measuring-unit values, relative to the frame of reference. Negative-valued translation terms, on the other hand, displace the translated coordinate system in the direction of decreasing measuring-unit values on the frame of reference.

You have already seen that all three terms in the equation refer directly to the chosen frame of reference. In a manner of speaking, the translated version of the coordinate system just goes along for the ride. Since all of the expression refer to the frame-of-reference coordinate system, it is often more convenient and less confusing to omit the drawing of the translated version. It is sufficient to show just the reference coordinate system and the points or lines involved in the experiment. Check out this example.

Example 3-2

A reference point is located at coordinate $P = (-4)$ in the frame of reference. What is the coordinate of the translated version of that point after applying a translation of 6? Plot the results, showing only the frame-of-reference coordinate system and the two points involved.

1. Gathering the relevant information:

$$x = -4 \quad T_x = 6$$

2. From Equation 3-1:

$$x' = x + T_x$$

3. Substituting the data from Step 1 into the equation in Step 2:

$$x' = -4 + 6$$
$$x' = 2$$

The translation displaces point $P = (-4)$ to $P3 = (2)$. See the diagram in Fig. 3-5.

The example isn't much different from the previous one; the statement of the experiment simply calls for omitting the translated frame of reference from the final plot. Follow the procedure, check the math if you wish, but notice how the plot in Fig. 3-5 more clearly shows that point P is translated to point P′. Test your own ability to work with these ideas by working Exercise 3-1.

Exercise 3-1

Each of the following experiments cites the coordinate of a reference point and a translation term to be applied to its frame-of-reference coordinate system. Carry out the transformation and plot the results in each case, showing only the reference coordinate system and the two points involved.

1. $P = (2)$, $T_x = 4$
2. $P = (-2)$, $T = 4$
3. $P = (2)$, $T_x = -4$
4. $P = (-2)$, $T_x = -4$

Try some translation experiments using reference points and translation terms of your own choosing.

Bear in mind that it is not enough to be competent at solving equations by simply plugging in specific numbers and cranking out the results. That is important, of course, but it is even more important to understand what is going on. Understanding equations and being able to intercept their meaning is actually more important than the mechanics of solving them for specific situations.

3-3 ALTERNATIVE FORMS OF THE BASIC EQUATION

Equation 3-2 represents a simple algebraic rearrangement of the basic equation for translations in 1-D space. This version solves directly for a translation term, given the coordinate of the reference point and the translated version of it. All terms still refer to the chosen frame of reference.

$$T_x = x_n' - x_n \qquad \textbf{Equation 3-2}$$

where:

T_x = the translation term

x_n' = the coordinate of the translated version of point *n* relative to the chosen frame of reference

x_n = the coordinate of reference point *n* relative to the chosen frame of reference

The equation is especially useful when dealing with a pair of identical coordinate systems, and the idea is to determine the translation that exists between them. You can do that, given the coordinate of at least one reference point and the coordinate of a translated version of that point relative to the frame of reference. See Example 3-3.

Example 3-3

A certain translation experiment shows that a reference point P is located at coordinate 3 with respect to the frame of reference. A translated version of that point, P′, is located at coordinate 7. What translation exists between those two points?

1. Gathering the information provided:

$$x = 3 \quad x' = 7$$

2. From Equation 3-2:

$$T_x = x' - x$$

3. Substituting the data from Step 1 into the equation in Step 2:

$$T_x = 7 - 3$$
$$T_x = 4$$

The translation that exists in this case is 4—point P′ is located 4 units to the right of point P.

The example begins by citing two different points in 1-D space: a reference point P and a translated version of it, P′. Step 1 then puts that information into a form that is more appropriate for Equation 3-2, and then Step 2 restates that equation for a single-point case. Step 3 concludes the experiment by substituting the given information and carrying out the necessary arithmetic. The final conclusion is that a translation of 4 must exist between the two coordinate systems. In other words, the translated coordinate system is 4 units to the right of the frame-of-reference coordinate system. Try to picture the situation in your mind, or else draw out a plot of the situation.

The basic translation equation can also be rearranged as shown in Equation 3-3. This is technically called the *inverse* of the basic translation equation. It solves for the coordinate of the reference point, given a translation term and the coordinate of a translated version of that point.

$$x_n = x_n' = -T_x \qquad \textbf{Equation 3-3}$$

where:

x_n = the coordinate of reference point *n* relative to the chosen frame of reference

x_n' = the coordinate of the translated version of point *n* relative to the chosen frame of reference

T_x = the translation term

Note: This is the inverse version of Equation 3-1.

You can test your understanding of the two alternative forms of the translation equation by working through Exercise 3-2.

Exercise 3-2

1. Use Equations 3-1, 3-2 and 3-3 to complete the following lists of translation information:

A. P = (3)	$T_x = -5$	P′ = ______
B. P = (−3)	$T_x = 6$	P′ = ______
C. P = (2)	T_x = ____	P′ = (−4)
D. P = (−5)	T_x = ____	P′ = (0)
E. P = ____	$T_x = 2$	P′ (5)
F. P = ____	$T_x = -4$	P′ = (−2.5)

2. Given two points, P = (9) and P′ = (4), what is the translation that exists between them, relative to the coordinate sytem for P?

3. Given a translation of $T_x = -5$ and the translated version of a point, P′ = (−4), what is the coordinate of the reference point, P?

TRANSLATING LINES IN 1-D SPACE

A line in 1-D space can be defined in terms of the coordinates of its endpoints, and translating a line in 1-D space is a matter of translating its individual endpoints and reconstructing the line between them. See Example 3-4.

Example 3-4

A line in 1-D space has these specifications relative to a chosen frame of reference:

$$L = \overline{P_1P_2} \qquad P_1 = (4)$$
$$P_2 = (2)$$

What are the specifications for that line after applying a translation of $T_x = 5$ relative to the frame of reference? Plot the overall experiment.

1. Gathering the relevant information into a useful form:

$x_1 = 4 \quad x_2 = 2 \quad T_x = 5$

2. Translating endpoint P_1 according to Equation 3-1:

$$x_i = x_1 + T_x$$
$$x_i = 4 + 5$$
$$x_i = 9$$

The translation thus displaces endpoint $P_1 = (4)$ to $P_1 = (9)$.

3. Translating endpoint P_2 according to Equation 3-1:

$$x_2' = x_2 + T_x$$
$$x_2' = 2 + 5$$
$$x_2' = 7$$

So the translation displaces endpoint $P_2 = (2)$ to $P_2' = (7)$.

4. Gathering the specifications for the translated version of the line:

$$L' = P_1' P_2' \qquad P_1' = (9)$$
$$P_2' = (7)$$

See the diagram of the experiment in Fig. 3-6.

The example begins with a statement of the experiment to be performed. It specifies a point in 1-D space and a translation that is to be applied to it. The general idea is to come up with a set of specifications for the translated version of that line.

Step 1 gathers the information supplied by the statement of the experiment, and puts it into a form that matches up with the needs of the equation that will use it. From the original specifications, you can see that the X component of endpoint P_1 is 4; or in other words, $x_1 = 4$. By the same token, the X component of endpoint P_2 is 2, and that is shown in Step 2 as $x_2 = 2$.

Step 2 solves for the coordinate of the translated version of endpoint P_1. It states the translation equation in terms of that point, substitutes the values from Step 1, and then solves for the value of the translated version of that same endpoint.

Step 3 performs an identical series of steps, but this time the variables refer to endpoint P_2.

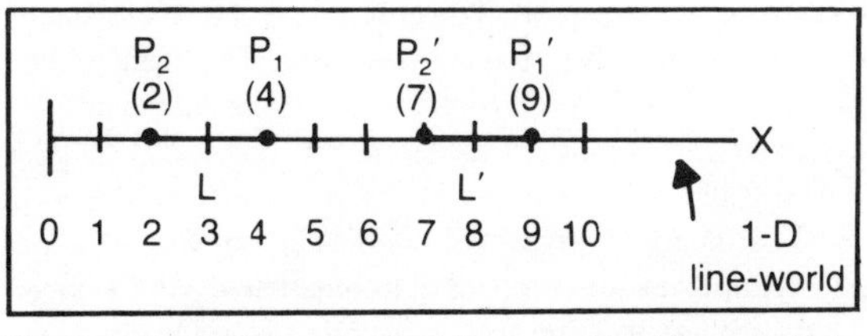

Fig. 3-6. Translating a line in 2-D space. See Example 3-4.

Step 4 completes the project by gathering the new specifications for the line and plotting the experiment.

Build your confidence by working through Exercise 3-3.

Exercise 3-3

The following experiments cite specifications for a reference line and a translation term relative to a chosen frame of reference. Perform the translation, generate the specifications for the translated version of the line, and plot the overall results. (Note: some of the reference and translated lines will overlap to some extent. Get accustomed to dealing with that situation.)

1. $L = \overline{P_1P_2}$ $\quad P_1 = (2)$, $P_2 = (6)$
   $T_x = -4$

2. $L = \overline{P_1P_2}$ $\quad P_1 = (2)$, $P_2 = (-2)$
   $T_x = 2$

3. $L = \overline{P_1P_2}$ $\quad P_1 = (-8)$, $P_2 = (-1)$
   $T_x = 6$

4. $L = \overline{P_1P_2}$ $\quad P_1 = (0)$, $P_2 = (-2)$
   $T_x = -2$

Discussions in Chapter 2 describe a line in terms of the coordinates of its endpoints and analyze it in terms of its length. Length of lines, or the distance between two points, is a vital quality of 1-D space, and Equation 2-1 expresses the length of a line in terms of the coordinates of its endpoints.

That's all by way of review of previous discussions. What is important to consider now is what effect a translation might have upon the length of a line. How do the lengths of the lines—both the reference and translated versions—compare? See the previous examples and exercises.

It turns out that *a translation in 1-D spaces does not affect the length of a line*. A technical expression used many times through this book describes such a quality as *invariant*. The lengths of lines, or the distances between pairs of points, is said to be an invariant under translations in 1-D space.

You can calculate the length of any reference line, apply some translation to it, calculate the length of the translated version, and you will always find that the values are identical. That is a demonstration of the idea, however; and mathematicians cannot be satisfied with demonstrations—the notion must be proven in a formal sense.

Proof 3-1 shows that length is an invariant under translations in 1-D space. Study the proof carefully. Not only is the result important, but so is the method. There will come a time in your experiments in four dimensions where the spaces and geometric entities are so abstract and esoteric that it is virtually impossible to use a common-sense view of the events. When common sense fails, you will have to rely on purely abstract, mathematical methods to formulate even the most basic ideas; and that includes proving notions that you can only suspect are true.

Proof 3-1

1. The length of any line in 1-D space is:

$$s_n = \sqrt{(x_u - x_v)^2} \qquad \textbf{Equation 2-1}$$

2. Adjusting the notation in Step 1 to relate to a translated line in 1-D space:

$$s_n' = \sqrt{(x_u' - x_v')^2}$$

3. From Equation 3-1:

$$x_u' = x_u + T_x$$
$$x_v' = x_v + T_x$$

4. Substituting the expressions of Step 3 into the equation in Step 2:

$$s_n' = \sqrt{[(x_u + T_x) - (x_v + T_x)]^2}$$

5. Simplifying the equation in Step 4:

$$s_n' = \sqrt{(x_u - x_v)^2}$$

6. Since:

$$s_n = \sqrt{(x_u - x_v)^2} \qquad \text{(Step 1)}$$

and

$$s_n' = \sqrt{(x_u - x_v)^2} \qquad \text{(Step 5)}$$

it follows that

$$s_n = s_n'$$

and the proposition is proven.

The overall plan of this proof is to show that the length of any reference line, s_n, in 1-D space is equal in length to a version of it that is translated through any distance and in either direction. The length of the translated version will carry the notation s′ in order to be consistent with previous notation convention. In short, the goal of the proof is to show that $s_n = s_n'$. Do that, and you can be certain that length is an invariant under 1-D translations.

Thus the proof begins by specifying the standard length equation for lines in 1-D space. To preserve the generality of the proof, I am designating the endpoints in their most general forms. That is the formal and most general expression for sn—the length of the reference line relative to the chosen frame of reference.

Step 2 takes the same equation and adjusts the notation so that it deals with variables related to a translated line. Variables x_u' and x_v' refer to the translated versions of endpoints x_u and x_v, respectively. The result is an expression for the length of a translated version of line *n* in 1-D space.

Step 3 simply states the basic geometric translation equation two times—once for each of the endpoints of the line, and then Step 4 substitutes the expressions into the equation in Step 2. Applying some elementary algebra, the rather cumbersome equation in Step 4 reduces to that of Step 5. That step actually represents the length of a translated line in terms of the endpoint coordinates of the reference line.

Finally, comparing the right side of the equation in Step 5 with the right side of the equation in Step 1, you can rightly conclude that $s_n = s_n'$, and the proposition is proven. For any reference line in 1-D space, you can apply any translation you wish, and the length of the line will remain unchanged.

Certainly that is a lot of work to go through just to prove something that might be intuitively obvious or at least shown by measuring and translating lines a great many times. But this is just the beginning. The time will soon arrive when intuition is little more than a guide, and you should always be aware of the old idea that one exception negates the rule. Keep things general and don't trust unproven ideas.

3-5 RELATIVISTIC TRANSLATION EFFECTS

I have been making a big point of the fact that the three terms in the geometric translation equations refer directly to the chosen frame of reference. You are certainly free to choose which one of the two coordinate systems will serve as the frame of reference for the experiment at hand, but making that choice commits you to refering all terms to it. The coordinate of the reference point refers directly to the chosen frame of reference, the translation term refers directly to it, and the coordinate of the translated version of the point refers directly to the chosen frame of reference.

That is how things have been so far anyway. One of the adventurous elements of mathematics is coming face to face with a situation that is usually taken for granted;

and then playing around with it. Until now, the discussions take for granted that there is just one legitimate frame of reference. You can select either of two possible coordinate systems as the frame of reference; but your selection automatically eliminates the usefulness of the translated coordinate system. In fact, I went so far as to suggest that there is no need to draw the translated coordinate system on the graphs of 1-D translation experiments.

Is it possible to work with two different frames of reference under a single translation? It certainly is. In fact transformations between two different frames of reference are the mathematical cornerstones of relativistic physics. In the context of translations in 1-D space, there are two frames of reference: the *observer's frame of reference* and the *translated frame of reference*.

The observer's frame of reference is the one we consciously select as our own frame of reference. A physicist might say that it is the fixed frame of reference; we can say that it is the coordinate system that is not translated.

The second frame of reference, the one being introduced here, is the translated frame of reference. None of the previous discussions dealt with this frame of reference.

Working with these two frames of reference simultaneously allows us to view a translation situation from two distinctly different points of view: from our own point of view (the observer's frame of reference) and from the viewpoint of the translated coordinate system (the translated frame of reference).

The basic geometric translation equation (Equation 3-1) does, in effect, move a coordinate system and any points and lines to a new position in 1-D space. The *relativistic* translation equation, however, does a different sort of job. It does not move anything. It adheres more directly to the notion that a translation is a mechanism for expressing the nature of a displacement that already exists between a pair of otherwise identical coordinate systems.

A typical relativistic scenario goes something like this: You are an intelligent 1-dimensional being using your own coordinate system to determine the coordinate of some point in 1-D space. Another equally intelligent being located elsewhere in your 1-D world is using an identical coordinate system to determine the coordinate of that same point. What is the coordinate of the point as reckoned from the translated frame of reference?

There are two coordinate systems and one point in question. If the coordinate systems are displaced by some translation effect, you and your counterpart using the other coordinate system are bound to see the point having different coordinates.

See Figure 3-7. In that example, the point in question at coordinate 8 from the observer's frame of reference, but it takes on coordinate −6 from the translated frame of reference. There are two different frames of reference and two different ways of looking at the same point. That is the essence of relativistic transformations and, indeed classical relativistic physics as well.

Equation 3-4 deals with the relativistic translations. Given the coordinate of some reference point, x_n, as reckoned from the observer's frame of reference and a translation, T_x, that exists between the two frames of reference, the equation solves for the coordinate of the point, x_n*, as viewed from the translated frame of reference. Putting it simply: the equation lets you figure out how things look from the other guy's point of view.

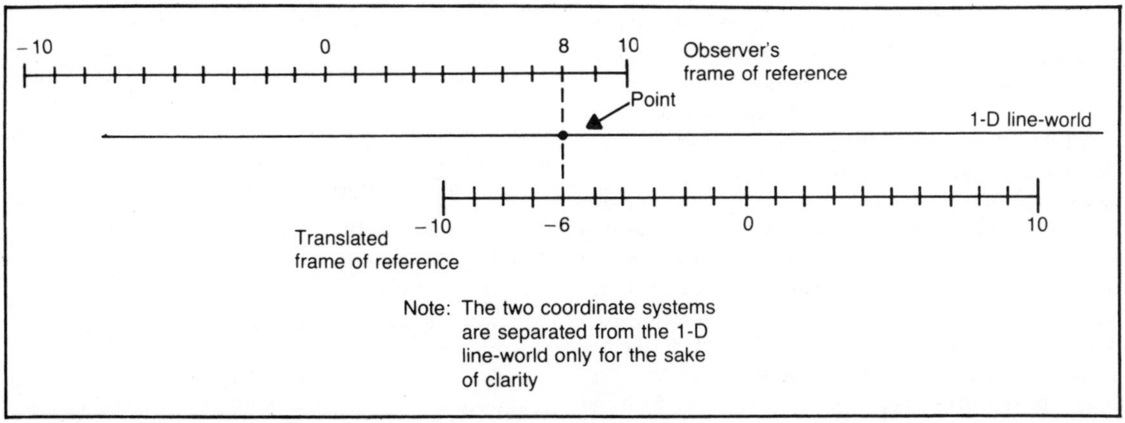

Fig. 3-7. Relativistic view of a point in 1-D space.

$$x_n^* = x_n - T_x \qquad \textbf{Equation 3-4}$$

where:

x_n^* = the coordinate of point *n* relative to the translated frame of reference

x_n = the coordinate of point *n* relative to the observer's frame of reference.

T_x = the translation term relative to the observer's frame of reference

Example 3-5 illustrates the application of the relativistic equation.

Example 3-5

You are a 1-D creature conducting an experiment with the help of a 1-D companion who is located elsewhere in your line-world. Using your own coordinate system, the observer's frame of reference, you find the coordinate of a point to be P = (4). Your companion, however, is using a coordinate system that is translated by $T_x = -5$ relative to your own. Calculate the coordinate of the point in question as regarded from your companion's point of view, P*.

1. Gathering the relevant data:

$$x = 4 \qquad T_x = -5$$

2. From Equation 3-4:

$$x^* = x - T_x$$

3. Substituting the information in Step 1 into the equation in Step 2:

$$x^* = 4-(-5)$$
$$x^* = 9$$

The reference point views at coordinate P = (4) in the observer's frame of reference thus appears at P* = (9) as viewed from the translated frame of reference.

The example deals with a situation whereby you and a companion sharing a 1-D line-world are attempting to locate the coordinate of a particular point that also exists in that world. The catch is that the two of you are using measuring sticks (coordinate systems) that are situated (translated) at slightly different positions. With those measuring sticks thus located at different places, the two of you are bound to get a different reading for the position of the point. You are fully aware of that, and you are particularly interested in finding out what your companion's reading might be. You know your reading from direct observation of your own measuring stick (the observer's frame of reference) and you've somehow determined the translation between your measuring stick and that of your companion. All that remains, then, is to substitute those values into the relativistic translation equation and crank out the results.

Step 1 in the example cites the relevant data as terms that are compatible with the form of the equation, Step 2 restates the equation for this fairly simple, 1-point case, and Step 3 works out the solution. Doing all of that, you discover that the point you find at measuring unit 4 in your frame of reference will be found at measuring unit 9 in your companion's frame of reference.

See what you can do on your own with Exercise 3-4.

Exercise 3-4

1. Use Equation 3-4 to complete the following series of relativistic terms. Plot the situations using Fig. 3-7 as a model.

A. P = (2)	$T_x = 4$	P* = ______
B. P = (3)	$T_x = -6$	P* = ______
C. P = (−4)	$T_x = 5$	P* = ______
D. P = (−5)	$T_x = -2$	P* = ______

2. Using the observer's frame of reference, you find that a point in 1-D space is located at coordinate (8). Sliding that frame of reference by $T_x = -6$, you end up with a translated frame of reference. What is the coordinate of the original point as regarded from that frame of reference? (Hint: This is the typical relativistic scenario stated in a somewhat different fashion.)

The relativistic translation equation applies equally well when working with a line in 1-D space. The general idea is to solve the equation for both of the endpoints specified for that line. See Example 3-6.

Example 3-6

Relative to an observer's frame of reference, a line in 1-D space has the following specifications:

$$L = \overline{P_1P_2} \qquad P_1 = (3)$$
$$P_2 = (7)$$

There is a second frame of reference that is translated by $T_x = -5$ relative to your own. What are the specifications of that line as regarded from that second frame of reference? How does the length of the line compare as viewed from both frames of reference?

1. Gathering the given information into a more useful form:

$$x_1 = 3 \quad x_2 = 7 \quad T_x = -5$$

2. Applying Equation 3-4 to endpoint P_1 of that line:

$$x_1^* = x_1 - T_x$$
$$x_1^* = 3-(-5)$$
$$x_1^* = 8$$

3. Applying Equation 3-4 to endpoint P_2 of the line:

$$x_2^* = x_2 - T_x$$
$$x_2^* = 7-(-5)$$
$$x_2^* = 12$$

4. Gathering the information to specify the line as viewed from the translated frame of reference:

$$L^* = \overline{P_1^*P_2^*} \qquad P_1^* = (8)$$
$$P_2^* = (12)$$

5. Using Equation 2-1 to determine the length of the line relative to the observer's frame of reference:

$$s = \sqrt{(X_2-x_1)^2}$$
$$s = \sqrt{(7-3)^2}$$
$$s = \sqrt{4^2}$$
$$s = 4$$

The line is 4 units long from that frame of reference.

6. Using Equation 2-1 to determine the length of the line relative to the translated frame of reference:

$$s^* = \sqrt{(x_2^*-x_1^*)^2}$$
$$s^* = \sqrt{(12-8)^2}$$
$$s^* = \sqrt{4^2}$$
$$s^* = 4$$

The line is also 4 units long as viewed from the translated frame of reference.

7. Plot the situation. See Fig. 3-8.

That example begins by citing the specifications for a line as reckoned from the observer's frame of reference and a translation that exists between the observer's and a translated frame of reference. That information is adequate for determining the specifications of that same line as viewed from the translated frame of reference.

Step 1 gathers the important information and assigns the numerical data to variables that suit the translation equation. Then Step 2 states the relativistic translation equation for endpoint P_1 and substitutes the relevant information to determine the coordinate of that same endpoint as seen from the translated frame of reference, point P_1'. Step 3 performs an identical series of operations for the second endpoint of the line.

Step 4 uses the calculated endpoint coordinates to summarize the specifications for the line as viewed from the translated frame of reference.

The original statement of the experiment also calls for finding the length of the line measured from both frames of reference. Step 5 deals with the length of the line, 5, as seen from the observer's frame of reference. Finally, Step 6 adjusts the notation of the length equation to suit that of the translated frame of reference and then works out the length from that point of view.

It should come as no big surprise to you that the line has the same length as regarded from both frames of reference; length, you recall, is an invariant under 1-D translations—including translations of a relativistic form. See if you can prove that principle in a formal fashion; use Proof 3-1 as a guide, but build it around Equation 3-4 instead of 3-1.

Equations 3-5 and 3-6 are alternative versions of the basic relativistic equation for 1-D translations. The first solves for the translation that exists between the two frames of reference, given the coordinate of a reference point as viewed from both systems. The second equation is the inverse form of the basic relativistic equation, and it solves for the coordinate of a point as viewed from the observer's frame of reference. See Example 3-7 and Exercise 3-5.

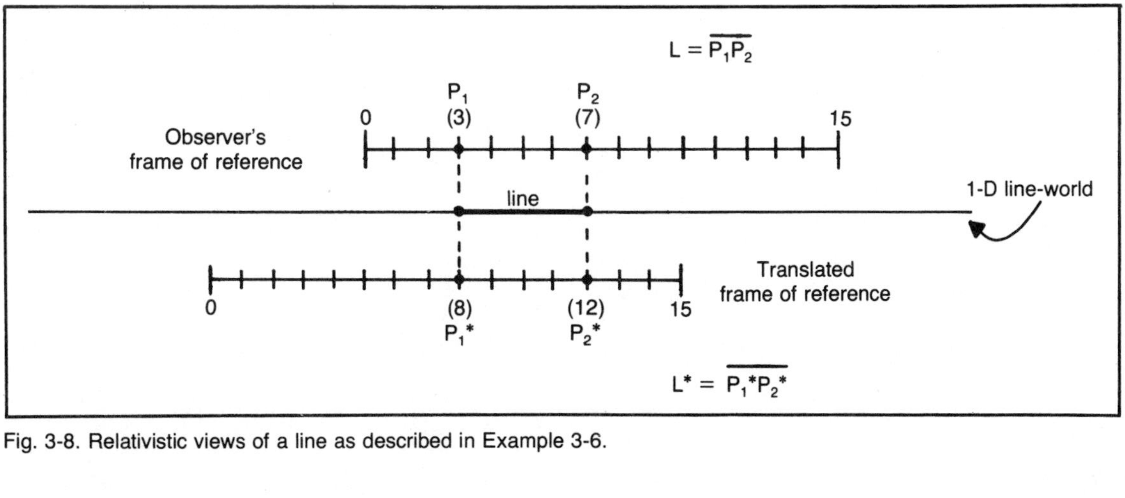

Fig. 3-8. Relativistic views of a line as described in Example 3-6.

$$T_x = x_n - x_n^* \qquad \textbf{Equation 3-5}$$

where:

T_x = the translation that exists between the observer's and translated frames of reference

x_n = the coordinate of point *n* relative to the observer's frame of reference

x_n^* = the coordinate of point *n* relative to the translated frame of reference

$$x_n = x_n^* + T_x \qquad \textbf{Equation 3-6}$$

where:

x_n = the coordinate of point *n* relative to the observer's frame of reference

x_n^* = the coordinate of point *n* relative to the translated frame of reference

T_x = the translation that exists between the observer's and translated frames of reference

Example 3-7

You and a companion are working according to different frames of reference in a 1-D world. Using the observer's frame of reference, you find the coordinate of a 1-dimensional "star" to be (2.25). Communicating with your companion in the experiment, you discover that he or she sees that same star at coordinate (−4.5). What translation, relative to your frame of reference, exists between your two coordinate systems?

1. Gathering the relevant information:

$$x = 2.25 \qquad x^* = -4.5$$

2. Applying Equation 3-5:

$$T_x = x - x^*$$
$$T_x = 2.25 - (-4.5)$$
$$T_x = 6.75$$

Using conventional terminology, your companion's frame of reference is located 6.75 measuring units to the right of your own.

Exercise 3-5

1. What translation exists between two coordinate systems if a reference point is at coordinate (−5) as viewed from the observer's frame of reference and that same point is at (2) relative to the translated frame of reference?

2. A line in 1-D space is found to be 3 units long as reckoned from the observer's frame of reference. How long is that same line as noted from a translated frame of reference? (Note: This question can be answered without citing any specific coordinates nor a translation.)

3-6 SUCCESSIVE TRANSLATIONS IN 1-D SPACE

Translations in 1-D space are *commutative*. That is to say, you can apply a succession of translations in any sequence and end up with the same overall translation. Suppose you want to apply this succession of translation-term values: 2, −4, 5. All together, that figures out to a translation of 2 + (−4) + 5 3 = 3. Scramble the order of that same sequence, and you will get the same overall translation; for example, the sequence −4, 2, 5 adds up to an overall translation of (−4) + 2 + 5 = 3. The idea applies equally well to geometric and relativistic translation situations.

Scalings in 1-D Space

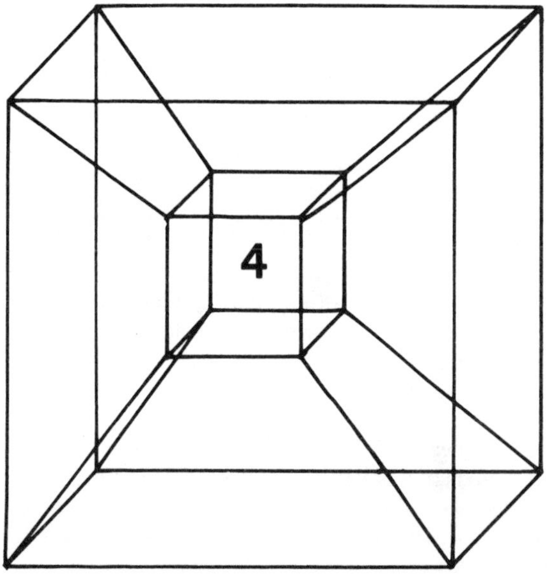

The previous chapter introduces a simple kind of 1-D transformation that is known as a translation. A translation expresses a displacement that exists between a pair of coordinate systems; or, alternatively, it serves as a mathematical mechanism for moving coordinate systems, points and lines along 1-dimensional world.

One-dimensional space allows a second kind of transformation known as a *scaling.* You are about to discover that it can portray some interesting effects.

Instead of simply displacing a coordinate system, a 1-D scaling alters the spacing between the measuring units of a coordinate system by some amount, often reducing or expanding its size. Suppose that you lay a wide rubber band flat on a table and draw a 1-D coordinate system along its length. Stretching that rubber band increases the spacing between the original measuring units equally along its length. That would represent an expansion-type scaling of a 1-D coordinate system. Or you could stretch out the rubber band first, draw the scaling on it, and then release the tension. The result, a smaller-sized coordinate system, would then represent a reduction-type scaling in one dimension.

Figure 4-1 shows two 1-D coordinate systems representing a 1-D line-world. The coordinate systems should be superimposed on the line-world but, as done in the previous chapter, I've separated them for the sake of visual clarity. Just imagine that both coordinate systems are actually laying on the 1-D line-world.

The important notion illustrated here is that the two coordinate systems differ only in the scaling of their measuring-unit values. Both are legitimte, conventional 1-D coordinate systems as defined in Chapter 2. Notice, particularly, that they are both linear systems. Looking at system A, for instance, you can see that the spacing between any two consecutive measuring-unit values is the same everywhere on that system. Likewise, you can see that the spacing of the measuring-units values on system B is everywhere the same. Indeed, both systems are linear in their own right. The only discernable difference is the relative scaling of the two systems.

Indeed, the two coordinate systems are scaled with respect to one another by some amount. By what amount? You cannot answer that question properly until establishing which of the two will serve as the frame of reference. Just as an expression of a 1-D translation is ambiguous without defining a frame of reference, so is an expression of a scaling transformation. From one point of view, coordinate system B appears to be expanded in scale compared to system A. It is possible to look at the situation from the other way around, however: coordinate

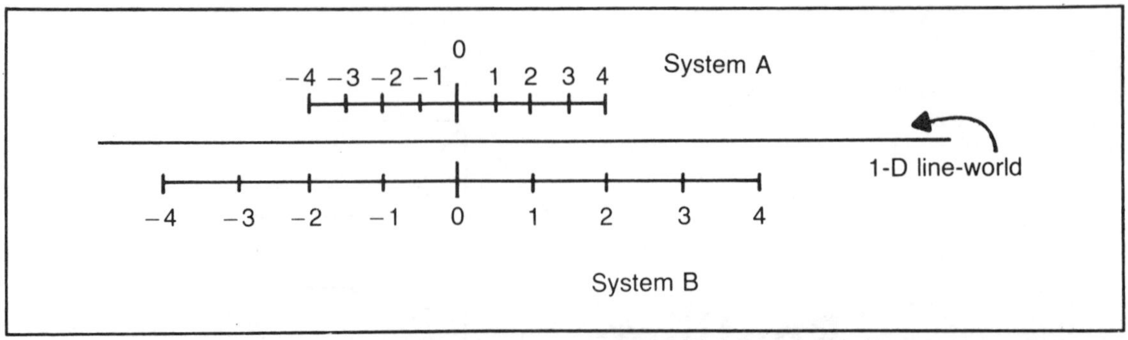

Fig. 4-1. A 1-D line-world and a pair of 1-D coordinate systems, systems A and B, that are scaled with respect to one another.

system A is reduced with respect to system B. Resolving this sort of ambiguity is a matter of defining which system will be regarded as the frame-of-reference coordinate system from the outset.

Figure 4-2 illustrates the same scaling situation, but it also defines the frame of reference. What was called system A in the previous illustration is taken as the frame of reference and renamed the X coordinate system. By default, then, the other coordinate systems becomes the one that is scaled relative to X. Using the same notation introduced earlier, that coordinate system is labeled X′. Clearly, the scaling is now characterized as an expansion of coordinate system X to X′.

Now notice that the origins of those coordinate systems in are lined up with one another. Pure scalings, unlike translations, do not affect the position of the origin of a coordinate system. But then see how a point that is not at the origin—point (P = (x) specified in the frame-of-reference coordinate system—is carried by this particular expansion-type scaling to point P′ = (x′) in the scaled version of the coordinate system.

One of the most common kinds of experiments with scaling transformations in 1-D space involves setting up a frame-of-reference coordinate system, specifying some points and lines relative to it, scaling the system by some amount, and then calculating the coordinates of the scaled version of the scheme.

Finally, the subject of scaling coordinate systems includes a special effect called *geometric reflection*. Certain kinds of scaling transformations generate mirror images of the original coordinate system without necessarily changing its size. Working with reflection-type scalings can be quite interesting and informative.

Thus, a scaling can be viewed as a transformation that literally expands, contracts or changes the direction of scaling of a coordinate system and any points and lines specified relative to it.

The notation convention is the same as that for translations in 1-D space: the original coordinate system and its points are designated with X and P = (x) expressions, while the scaled version of the same system uses X′ and P′ = (x′) expressions.

4-1 THE BASIC EQUATION FOR SCALINGS IN 1-D SPACE

Equation 4-1 represents the basic scaling equation for 1-D space.

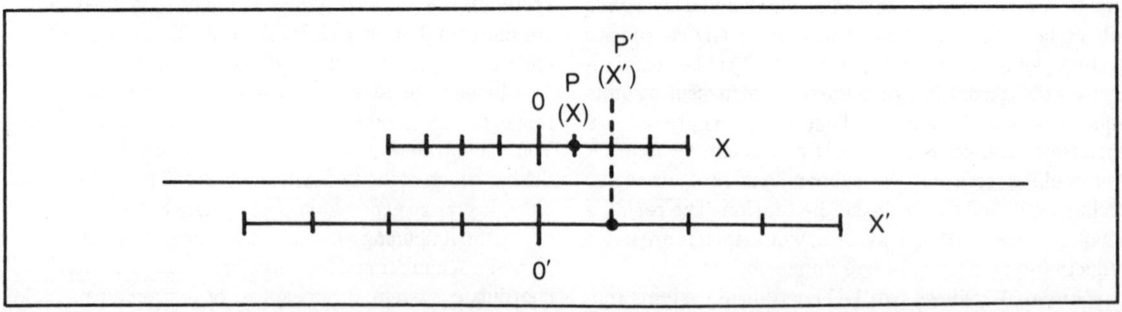

Fig. 4-2. A 1-D line-world and a pair of 1-D coordinate systems, X and X′. Using the notation established earlier, X is taken as the frame of reference and X′ represents the scaled frame of reference.

$$x_n' = x_n K_x \qquad \textbf{Equation 4-1}$$

where:

x_n' = the coordinate of scaled point *n* relative to the chosen frame of reference

x_n = the coordinate of reference point *n* relative to the chosen frame of reference

K_x = the scaling factor

That equation applies directly to experiments where the experiment knows a scaling factor, K_x, and the coordinate of some off-origin reference point, P(x), relative to the chosen frame of reference. Working the equation leads to the coordinate of the point in the scaled frame of reference, but also reckoned with respect to the frame of reference.

See how the equation works by studying Example 4-1.

Example 4-1

Let P = (2) be a point specified in a frame-of-reference coordinate system. Apply a scaling of $K_x = 0.5$ to the system in order to generate a scaled coordinate system. What is the coordinate of the scaled version of the reference point as reckoned from the frame of reference? Plot the scaling situation.

1. Gathering the relevant data:

$$x = 2 \qquad K_x = 0.5$$

2. From Equation 4-1:

$$x' = xK_x$$

3. Substituting the values from Step 1 into the equation in Step 2:

$$x' = 2(0.5) \qquad x' = 1$$

See Fig. 4-3.

The scaling factor in the equation is an all-important part of the operation—important enough to justify some specials. The magnitude of the scaling factor (its absolute value) expresses the amount of scaling in this way:

1. If the scaling factor is greater than 1, the scaled coordinate system will be larger than the coordinate frame of reference (an expansion-type scaling).
2. If the scaling factor is equal to 1, the scaled frame of reference will be identical to the coordinate frame of reference (no scaling effect at all).
3. If the scaling factor is less than 1 but greater than zero, the scaled frame of reference will be smaller than the coordinate frame of reference (reduction-type scaling).
4. If the scaling factor is zero, the scaled frame of reference will be reduced to a Ø– dimensional point.

The foregoing summary of scaling effects takes into account only positive values for the scaling factor. Some interesting things take place when the scaling factor happens to have a negative value—a value less than zero. Such a situation implies a reflective-type scaling.

Figure 4-4 shows the effect of applying a scaling factor of 1 and a scaling factor of −1. In both instances, system A is taken as the frame of reference, but you can see that there is a dramatic difference in the direction of scaling in the scaled coordinate system in those examples.

Applying a scaling factor of 1 as illustrated in Fig. 4-4A shows that it has no affect at all on the coordinate system—the "scaled" version is identical to the reference version. That confirms the information included in

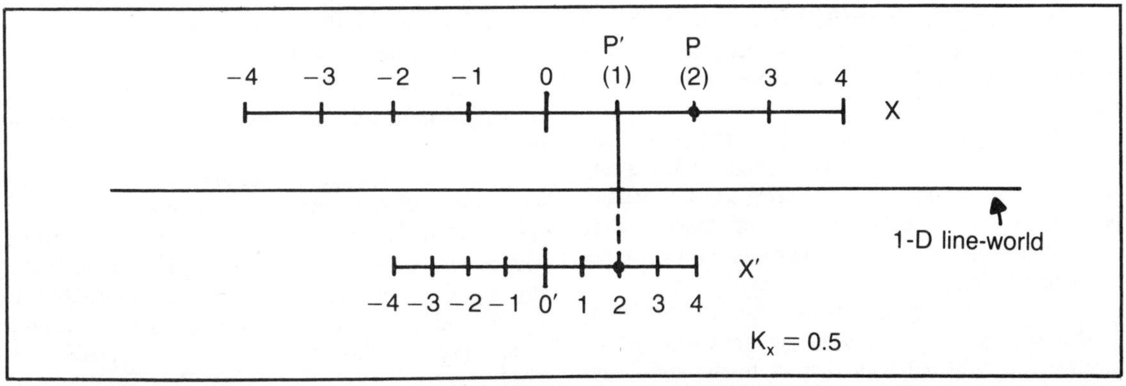

Fig. 4-3. A point at coordinate 2 in the scaled frame of reference appears at coordinate 1 in the reference coordinate system. See Example 4-1.

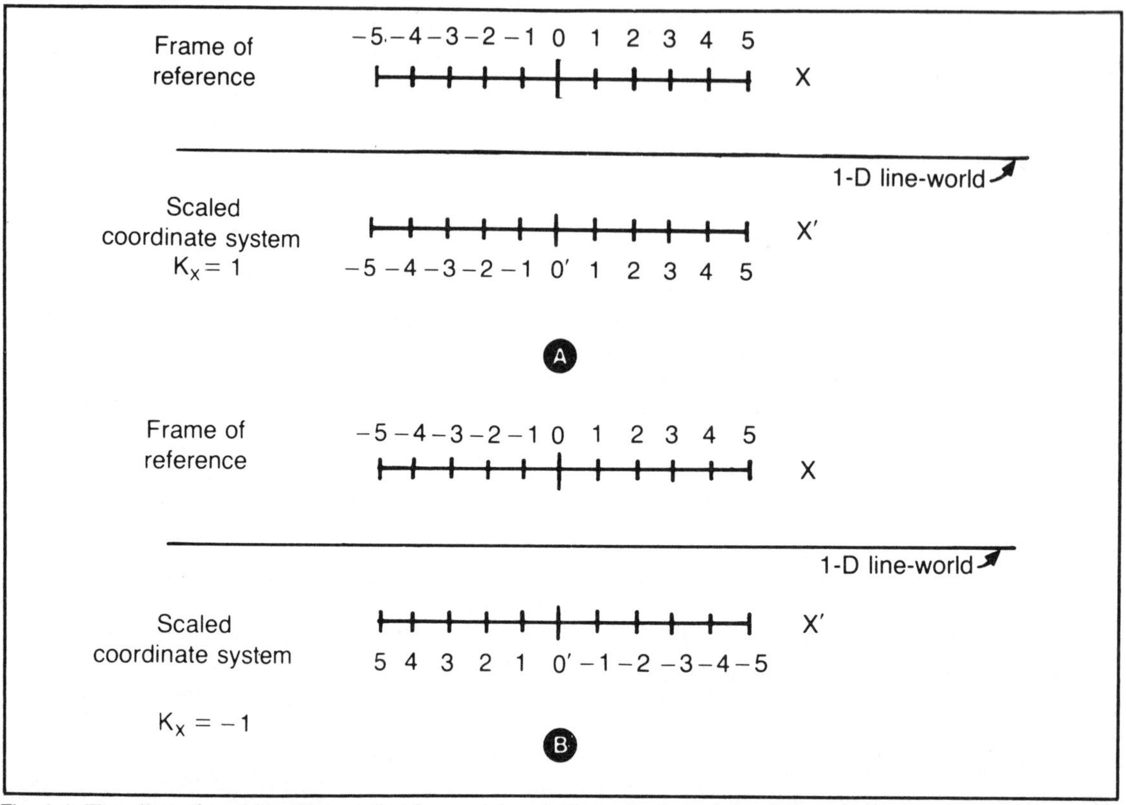

Fig. 4-4. The effect of applying (A) a scaling factor of 1 and (B) a reflective scaling factor of −1.

the previous summary; that scaling factors of 1 cause no change in a coordinate system. The scaling of −1 portrayed in Fig. 4-4B, however, shows a significant difference between the frame of reference and the scaled coordinate system. There is no change in the spacing between measuring units, but the direction of measurement is reversed. That is a pure reflection in 1-D space; the coordinate systems are mirror images of one another. In a general sense, negative-valued scaling factors always imply a reflection effect.

That example happens to show a scaling of −1; the coordinate system is reversed without showing any change in the scaling of the measuring units. It is quite possible to set up reflection-type scalings that also alter the scaling of the system. Consider this continuation of the summary initiated earlier:

5. If the scaling factor is less than zero but greater than −1, the scaled frame of reference will be both smaller and reflected with respect to the coordinate frame of reference (reduction-type scaling with a reflection).

6. If the scaling factor is equal to −1, the scaled frame of reference will be reflected, but not changed in size, with respect to the frame of reference (pure reflection).

7. If the scaling factor is less than −1, the scaled frame of reference will be both reflected and increased in size with respect to the frame of reference (expansion-type scaling with a reflection).

Getting down to the workaday experimental level, you can begin simplifying matters by noting that all three terms in the equation refer directly to the chosen frame of reference. So as suggested for such experiments with translations, there is nothing lost by omitting the scaled version of the coordinate systems from any drawings your experiments by generate. It is sufficient, and actually less confusing, to show just the reference coordinate system and the points or lines involved in the experiment. Check out this example.

Example 4-2

Let P = (3) be a point specified in a frame-of-reference coordinate system, then apply a scaling of $K_x = -2$ to it. Determine the coordinate of the scaled version of the point, relative to the frame of reference, and plot the situation.

1. Gathering the given data:

$$x = 3 \qquad K_x = -2$$

2. Applying Equation 4-1:

$$x' = xK_x$$
$$x' = 3(-2)$$
$$x' = -6$$

See the plot in Fig. 4-5.

Test your ability to work with these ideas by working Exercise 4-1.

Exercise 4-1

1. Use Equation 4-1 to complete the following series of data:

A. $x = 4$	$K_x = 2$	$x' =$ ____
B. $x = 2$	$K_x = -3$	$x' =$ ____
C. $x = -1.5$	$K_x = 1.5$	$x' =$ ____
D. $x = -2$	$K_x = -3$	$x' =$ ____
E. $x = 4$	$K_x = 0$	$x' =$ ____

2. A reference point is specified at coordinate (3) in the frame of reference. What is the coordinate of that same point, relative to the frame of reference, upon applying a scaling of -1.5? Plot the result.

Being able to plug specific numbers into the equation and cranking out a numerical result is important, but it ought not be regarded as the most important procedure anywhere in this book. Understanding is far, far more vital and mind expanding than technique. Please try avoiding any confusion between genuine ideas and the simple mathematics that happens to represent them in some way.

4-2 ALTERNATIVE FORMS OF THE BASIC EQUATION

Equation 4-2 represents a simple algebraic rearrangement of the basic equation for scalings in 1-D space; is solved for the scaling factor, given the coordinate of an off-origin reference point and the scaled version of it. All terms still refer to the chosen frame of reference.

$$K_x = x_n'/x_n \qquad \textbf{Equation 4-2}$$

where:

K_x = the scaling factor relative to the chosen frame of reference

x_n' = the coordinate of the scaled version of point *n* relative to the chosen frame of reference

x_n = the coordinate of reference point *n* relative to the chosen frame of reference

The equation is especially useful when dealing with a pair of scaled coordinate systems and you want to determine the scaling that exists between them. You can do that, given the coordinate of at least one off-origin reference point and the coordinate of a scaled version of that point relative to the frame of reference. See Example 4-3.

Example 4-3

You have found that a certain scaling transformation changes the coordinate of a reference point from P = (4) to P′ = (6). What is the nature of that scaling?

1. Gathering the relevant data:

$$x = 4 \qquad x' = 6$$

2. Expressing Equation 4-2 in terms of the problem at hand:

$$K_x = x'/x$$

3. Substituting the data from Step 1 into the equation in Step 2:

$$K_x = 6/4$$
$$K_x = 1.5$$

The scaling that exists is thus characterized by a factor of 1.5.

Incidentally, you have probably noticed that I am making continual reference to "off-origin" points. That is

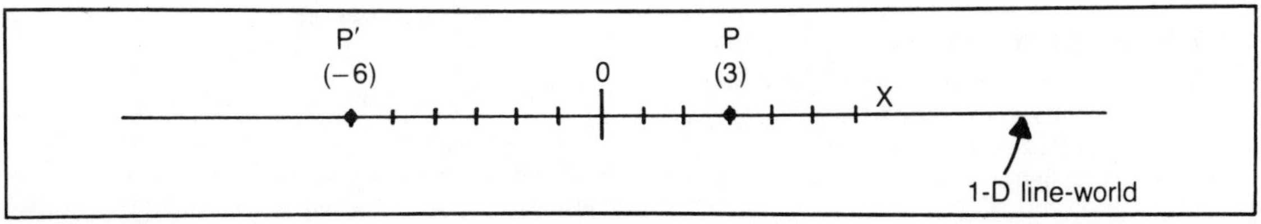

Fig. 4-5. Applying a scaling and plotting the results such that it shows only the frame-of-reference coordinate system. See Example 4-2.

necessary because it is impossible to work with the scaling that exists between two coordinate systems on the basis of the origins—pure scalings do not affect the origins.

The basic scaling equation can also be rearranged to provide the inverse of the basic scaling equation. See Equation 4-3. It solves for the coordinate of the reference point, given a scaling factor and the coordinate of a scaled version of that point.

$$x_n = x_n'/K_x \qquad \textbf{Equation 4-3}$$

where:

x_n = the coordinate of reference point *n* relative to the chosen frame of reference

x_n' = the coordinate of the scaled version of point *n* relative to the chosen frame of reference

K_x = the scaling factor

Note: This is the inverse version of Equation 4-1.

You can test your understanding of the alternative forms of the scaling equation by working Exercise 4-2.

Exercise 4-2

1. Use Equations 4-1, 4-2 or 4-3 as necessary for completing the following lists of data:

A. $x = 2$	$K_x = -1$	$x' =$ ____
B. $x = 0$	$K_x = 100$	$x' =$ ____
C. $x =$ ____	$K_x = 2$	$x' = 4$
D. $x = 1$	$K_x =$ ____	$x' = 3$
E. $x =$ ____	$K_x = -0.5$	$x' = -4$

2. Given two points relative to a frame of reference, determine the scaling that exists if the reference point has a coordinate of (–0.5) and its scaled version has a coordinate of (–4). Plot the situation, showing only the frame-of-reference coordinate system and the two points in question.

3. You know that a scaling of 2 exists between two coordinate system, and that the coordinate of the scaled version of a point has coordinate (–4) compared to the frame of reference. What is the coordinate of the non-scaled version of that point, also reckoned from the frame of reference.

4-3 SCALING LINES IN 1-D SPACE

A line in 1-D space, defined in terms of the coordinates of its endpoints, can be scaled by applying the same scaling factor to those endpoints separately, and then reconstructing the line between them. See Example 4-4.

Example 4-4

A line in 1-D space has the following specifications:

$$L = \overline{P_1P_2} \qquad P_1 = (1)$$
$$P_2 = (3)$$

Apply a scaling of $K_x = -1.5$ to the coordinate system and its line. Determine the specifications for the scaled version of the line and plot the experiment. Calculate the lengths of both the reference and scaled versions of the line.

1. Gathering the data from the given information:

$$x_1 = 1 \qquad x_2 = 3 \qquad K_x = -1.5$$

2. Using Equation 4-1 for point P_1:

$$x_1' = x_1K_x$$
$$x_1' = 1(-1.5)$$
$$x_1' = -1.5$$

The scaling thus places endpoint $P_1 = (1)$ to $P_1 = (-1.5)$.

3. Applying Equation 4-1 to endpoint P_2:

$$x_2' = x_2K_x$$
$$x_2' = 3(-1.5)$$
$$x_2' = -4.5$$

The scaling places endpoint $P_2 = (3)$ to P_2' (–4.5).

4. Compiling the specifications for the scaled version of the line:

$$L' = \overline{P_1'P_2'} \qquad P_1' = (-1.5)$$
$$P_2' = (-4.5)$$

5. Plot the results of the experiment. See Fig. 4-6.

6. Adjust the notation of the basic length equation (Equation 2-1) to conform to that of the reference line:

$$s = \sqrt{(x_2-x_1)^2} \qquad s = \sqrt{2^2}$$
$$s = \sqrt{(3-1)^2} \qquad s = 2$$

So the reference version of the line is 2 units long relative to the frame of reference.

7. Adjust the notation of the length equation to conform to that of the scaled version of a line:

$$s' = \sqrt{(x_2'-x_1')^2} \qquad s' = \sqrt{(-3)^2}$$
$$s' = \sqrt{[-4.5-(-1.5)]^2} \qquad s' = 3$$

And you can see that the scaled version of the line is 3 units long relative to the frame of reference.

The example begins with the statement of an experiment; it specifies a line in 1-D space in terms of the coordinates of its two endpoints, and it indicates a scaling factor that is to be applied.

Step 1 gathers the relevant information into a form

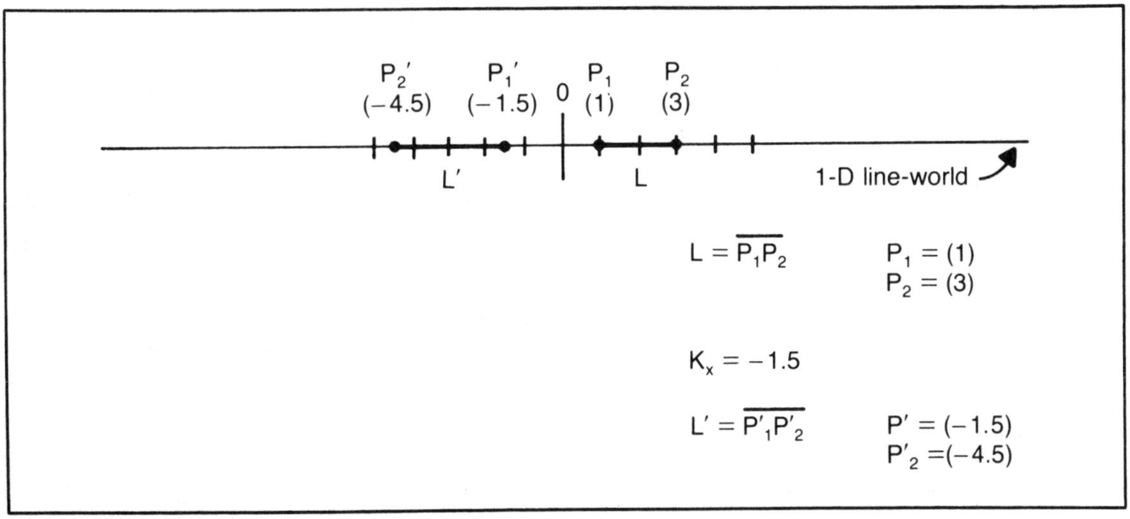

Fig. 4-6. Scaling a line L to get line L′. See Example 4-4.

that is notationally more consistent with the terms in the scaling equation. Step 2 then restates the scaling equation, using a notation that applies directly to the variables for point 1 in the line. Solving the equation turns up a scaled coordinate of (−1.5). Step 3 does the same thing, but uses the terminology and variables for endpoint 2 of the line.

Step 4 gathers the scaled coordinates and fits them into a formal set of specifications for the scaled version of the line. Step 5 then simply refers to the diagram of the situation in Fig. 4-6.

Step 6 sets up the basic length equation for 1-D space as applied to the endpoints of the reference line. Running through the calculations, based upon the original data, the line is found to be 2 units long.

Step 7 also uses the basic length equation, but this time the notation is adjusted so that it applies to the terminology of a transformed line in 1-D space. Using the data generated for that scaled line, the length turns out to be 3 units.

Try the scheme yourself by working Exercise 4-3.

Exercise 4-3

Each of the following experiments cites the specifications for a line in 1-D space and a scaling factor. Apply that scaling and determine the specifications of the scaled version of the line, relative to the frame of reference, of course. Plot the experiment, calculate and compare the lengths of the two lines.

1. $L = \overline{P_1P_2}$ $P_1 = (0)$ $P_2 = (2)$

$K_x = 4$

2. $L = \overline{P_1P_2}$ $P_1 = (0)$ $P_2 = (-3)$

$K_x = 2$

3. $L = \overline{P_1P_2}$ $P_1 = (-1)$ $P_2 = (2)$

$K_x = -2$

4. $L = \overline{P_1P_2}$ $P_1 = (-4)$ $P_2 = (4)$

$K_x = 0.5$

Whereas translations in 1-D space have no effect on the length of a reference line, the previous example and exercise clearly demonstrate that scalings can have a significant effect upon the lengths of lines. Unless the scaling factor happens to be 1 or −1, the length of a line specified in the scaled frame of reference will be quite different from that of the reference version.

It turns out that a scaling in 1-D space affects the length of a line by a factor equal to the scaling factor, itself. Thus, length is not necessarily preserved by scaling operations. *The lengths of lines, or the distances between pairs of points, is said to be a variant under scalings in 1-D space.*

Proof 4-1 shows that length is a variant under scalings in 1-D space. The proof is a negative one; it proves that something is not true. But in the process of working through it, we get an equation that expressed the length of a scaled line in terms of its reference length and the scaling factor.

Proof 4-1

1. The general equation for the length of a line in 1-D space is:

$$s_n = \sqrt{(x_u - x_v)^2} \qquad \textbf{(Equation 2-1)}$$

2. Adjusting the notation in Step 1 to account for any transformation:

$$s_n' = \sqrt{(x_u' - x_v')^2}$$

3. From Equation 4-1:

$$x_u' = x_u K_x$$
$$x_v' = x_v K_x$$

4. Substituting the expressions in Step 3 into the equation in Step 2:

$$s_n' = \sqrt{x_u K_x - x_v K_x)^2}$$

5. Simplifying the result in Step 4:

$$s_n' = K_x \sqrt{(x_u - x_v)^2}$$

6. Substituting the expression in Step 1 into that of Step 5:

$$s_n' = K_x s_n$$

The length of a scaled version of a line, s_n', can be quite different from that of the reference version, s_n. There are only two situations where the lengths of the lines will be unchanged by a scaling in 1-D space; where $s_n' = s_n$:

1. When the scaling factor is 1 or −1 (the −1 case is covered by the fact that length is an unsigned term)
2. When the reference line is 0 units long.

4-4 RELATIVISTIC SCALING EFFECTS

You have been seeing that the three terms in the geometric scaling equations refer directly to the chosen frame of reference. You can choose which one of the two coordinate systems will serve as the frame of reference for the experiment at hand, but making that choice commits you to refering all terms to it: the coordinate of the reference point refers directly to the chosen frame of reference, the scaling factor refers directly to it, and the coordinate of the scaled version of the point refers directly to the chosen frame of reference.

Is it possible to work with two different frames of reference under a single scaling? You saw in Chapter 3 that it could be done for translations in 1-D space. Well, it can be done here, too. And the interpretation is, I think, an intriguing one.

The idea is to work with two frames of reference; again, the observer's frame of reference and the scaled frame of reference. The observer's frame of reference is the one we choose to make our own, while the second frame of reference is the one that is scaled with respect to ours.

Dealing with these two frames of reference at the same time permits us to view a scaling transformation from two different points of view: from our own point of view (the observer's frame of reference) and from the viewpoint of the scaled coordinate system (the scaled frame of reference).

The basic *geometric scaling equation* (Equation 4-1) does, in effect, rescale a coordinate system and any off-origin points and lines to a new system of measurement 1-D space. The scaling equations express that change of scale.

The *relativistic scaling equation,* on the other hand, adheres more directly to the notion that a scaling transformation is a mechanism for expressing the nature of a change-of-scale or reflection that already exists between a pair of otherwise identical coordinate systems. And as in the case of relativistic translations, we deal here with a single reference point and two distinctly different frames of reference. In the relativistic case, the observer has an opportunity to determine the coordinate of some reference point in a common 1-D world as viewed from the scaled frame of reference.

Imagine that you are an intelligent 1-dimensional being using your own coordinate system to determine the coordinate of some point in 1-D space. Another equally intelligent being located elsewhere in your 1-D world is using his or her own version of a coordinate system to determine the coordinate of that same point. There are two coordinate systems and one point in question. If the coordinate systems differ only by some pure scaling effect, you and your counterpart using the other coordinate system are bound to see the point having different coordinates.

See Figure 4-7. In that example, the point in question at coordinate 9 from the observer's frame of reference, but it takes on coordinate 3 from the scaled frame of reference—two different frames of reference, two different views of the same point. That is the essence of relativistic transformations, whether they are translations, scalings, or a combination of the two.

Equation 4-4 deals with the relativistic scalings in this way: Given the coordinate of some reference point, xn, as reckoned from the observer's frame of reference and a scaling, Kx, that exists betweeen the two frames of reference, the equation solves for the coordinate of the point, x*, as viewed from the scaled frame of reference. In a manner of speaking, the equation lets you see how

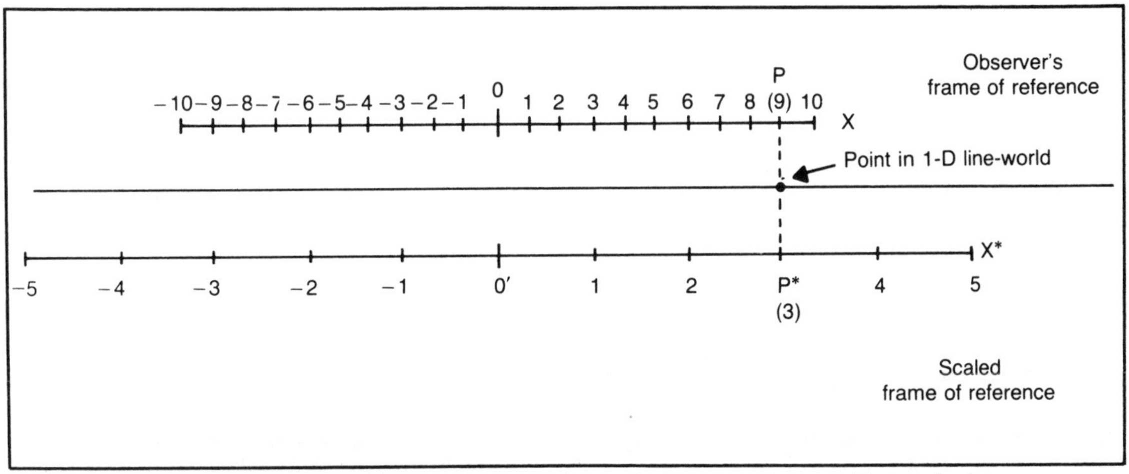

Fig. 4-7. A reference point as reckoned from an observer's frame of reference and a scaled frame of reference.

things look from the other system's point of view.

$$x_n^* = x_n/K_x \qquad \textbf{Equation 4-4}$$

where:

x_n^* = the coordinate of reference point *n* relative to the scaled frame of reference

x_n = the coordinate of reference point *n* relative to the observer's frame of reference

K_x = the scaling factor that exists between the scaled and observer's frames of reference, relative to the observer's frame of reference

Example 4-5 illustrates the application of the relativistic equation.

Example 4-5

Two experimenters are playing around with measuring sticks (frames of reference) that have different scalings. Using your own measuring stick (the observer's frame of reference), you find that a spot on a table is 5 measuring units from the left-hand edge of the table. If your partner's measuring stick is scaled by a factor of 2.5 relative to yours, where will he or she determine the position of that spot?

1. Gathering the information into a precise form:

$$x = 5 \qquad K_x = 2.5$$

2. Using Equation 4-3:

$$x^* = x/K_x$$
$$x^* = 5/2.5$$
$$x^* = 2$$

So your partner's measuring stick will show the spot on the table to be 2 units, instead of 5, from the left-hand edge of the table.

More interesting experiments come to mind when dealing with the relativistic scaling effects of lines in 1-D space. Figure 4-8 shows two coordinate systems that differ by some scaling factor; the X′ system appears to be scaled by 0.5 relative to the frame of reference, X. Both coordinate systems can be used for determining the position and length of a line given in 1-D space, but given a significant scaling effect between the two, they are bound to come up with different coordinate values and lengths. The relativistic version of the scaling equation yields the endpoint coordinates of the line as viewed from the scaled frame of reference, X′.

I think you will find that Example 4-6 cites an interesting, if altogether imaginary, experiment in 1-D space. The summary of the results at the end of the experiment amounts to a complete representation of the spacing between two stars in 1-D space as reckoned from two different scales of measurement. Ponder the experiment for a while, and you might come up with some interesting variations of the scenario.

Example 4-6

You are a scientist living and working in a 1-D universe. Using your system of measurement, you find that two stars are located at coordinates $P_1 = (9)$ and $P_2 = (-4)$. Another investigator living in that universe uses a system of measurement that is scaled by a factor of 2.5 relative to your own. What will that other investigator find in terms of the coordinates of those same two stars?

1. Gathering the given information into a more precise form:

$$x_1 = 9 \qquad x_2 = -4 \qquad K_x = 2.5$$

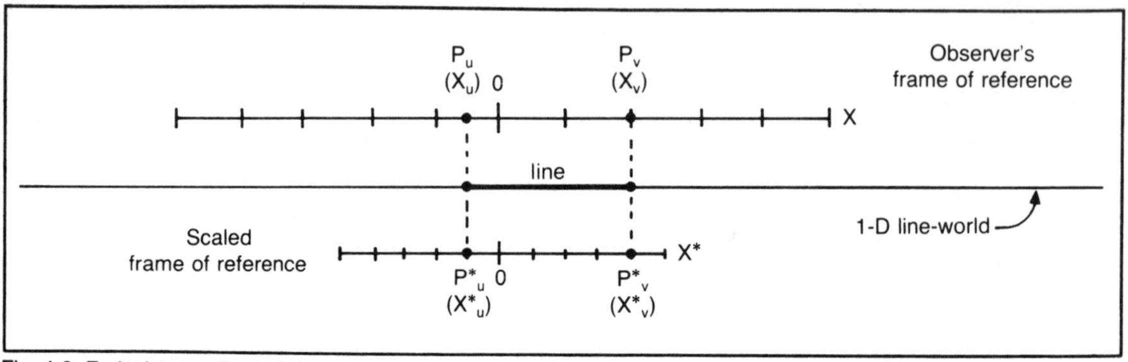

Fig. 4-8. Endpoint coordinates of a reference line as viewed from the observer's and scaled frames of reference.

2. Adjusting the notation in Equation 4-4 and solving it:

$$x_1^* = x_1/K_x$$
$$x_1^* = 9/2.5$$
$$x_1^* = 3.6$$

Point $P_1 = (9)$ thus scales to $P_1^* = (3.6)$.

3. Repeating that step after adjusting the notation of the equation so that it refers to point P_2:

$$x_2^* = x_2/K_x$$
$$x_2^* = -4/2.5$$
$$x_2^* = -1.6$$

The scaling thus places $P_2 = (-4)$ in your system of measurement to $P_2^* = (-1.6)$ in the other investigator's system of measurement.

Regarding the distance between the two stars, you will find that it is 13 measuring units; but using the other investigator's frame of reference, the spacing is 5.2 measuring units. Perform the calculations to support those figures. Describing the situation in terms of an imaginary line between the two stars, you will see this:

$$L = \overline{P_1P_2} \qquad P_1 = (9)$$
$$P_2 = (-4)$$

The other investigator will see:

$$L^* = \overline{P_1^*P_2^*} \qquad P_1^* = (3.6)$$
$$P_2^* = (-1.6)$$

Equations 4-5 and 4-6 are alternative versions of the basic relativistic equation for 1-D scalings. The first solves for the scaling that exists between the two frames of reference, given the coordinate of a reference point as viewed from both systems. The second equation is the inverse form of the basic relativistic equation, and it solves for the coordinate of a point as viewed from the observer's frame of reference. See Example 4-7 and Exercise 4-4.

$$K_x = x_n/x_n^* \qquad \textbf{Equation 4-5}$$

where:

K_x = the scaling factor that exists between the observer's and the scaled frames of reference, relative to the observer's

x_n = the coordinate of a reference point *n* relative to the observer's frame of reference

x_n^* = the coordinate of reference point *n* relative to the scaled frame of reference

$$x_n = x_n^*K_x \qquad \textbf{Equation 4-6}$$

where:

x_n = the coordinate of some reference point *n* relative to the observer's frame of reference

x_n^* = the coordinate of reference point *n* relative to the scaled frame of reference

K_x = the scaling factor that exists between the observer's and the scaled frames of reference, relative to the observer's

Note: This is the inverse form of Equation 4-4.

Example 4-7

Two scientists working in a 1-D universe are aware that a scaling of some magnitude exists between their respective systems of measurement, but they don't know the exact nature of it. In order to resolve the matter, they agree to cite the positions of two stars according to each other's system of measurement and exchange the data. Here is a summary of their data:

Investigator X finds the coordinates of the stars to be $P_1 = (-1)$ and $P_2 = (2)$.

Investigator X* finds the coordinates to be $P_1^* = (3)$ and $P_2^* = (-6)$.

What will Investigator X find to be the scaling, relative to his or her own system of measurement?

1. Gathering the given information into a form more suitable for the available equations:

$$x_1 = -1 \qquad x_2 = 2 \qquad x_1^* = (3) \qquad x_2^* = (-6)$$

2. Working first with points P_1 and P_1^*, Equation 4-5:

$$K_x = x_1/x_1^*$$
$$K_x = -1/3$$

3. Doing it again with points P_2 and P_2^*:

$$K_x = x_2/x_2^*$$
$$K_x = 2/-6$$
$$K_x = -1/3$$

The scaling that exists between the two investigator's measuring systems is clearly $-1/3$ relative to that of Investigator X. Not only is the scaling of the other's system smaller, but it is reversed in direction.

Example 4-7 is especially interesting in the light of some recent experiments that human scientists are actually performing with regard to the shape of our real universe. There is some question regarding the shape and scale of the universe, and investigators are currently using a more sophisticated version of the experiment in this example in an attempt to resolve the question.

4-5 AMBIGUOUS 1-POINT TRANSFORMATIONS

You might have noticed that I often cite two different points when setting up experiments that call for determining the scaling factor that exists between a pair of coordinate systems. Why would I do that when the same thing can be accomplished by studying the reference and scaled version of a single point?

There has been a certain ambiguity lurking within many of the previous examples and exercises. Set up a reference coordinate system and specify the coordinate of some point that is not located on the origin. Then apply a scaling of your choice and plot the overall on the frame-of-reference coordinate system.

Now look at those two points—the reference point and the scaled version of it. Can you see that you could accomplish the very same effect by means of a translation?

Looking at the same situation from a somewhat different viewpoint, suppose that you are given the coordinates of two different points in 1-D space. One is shown to be a reference point (x) and the other is shown to be a transformed version of it, point (x′). What is the nature of the transformation that exists between them? It could be a translation or a scaling; there's no telling which.

If the reference point happens to be at coordinate 2 with respect to the frame of reference and the transformed version turns up at coordinate 6 relative to the frame of reference, you can legitimately characterize the transformation as either a translation of +4 or a scaling of +3. A serious investigator ought to be intolerant of an ambiguity of this kind.

Equations, alone, are inadequate for uncovering the nature of the transformation that offset the two points in the first place. This is where the design of an experiment becomes more important.

The way around the ambiguity is to redesign the experiment so that it takes into account the transformation of at least two different points. The rationale behind that notion rests upon two important principles cited in earlier discussions:

1. Length, or the distance between two points in an invariant under 1-D translations.

2. Length, or the distance between two points, is a variant under 1-D scalings.

Bearing in mind those principles, you can set up a 2-point experiment for determining whether a displacement is caused by a translation or a scaling and, of course, the associated translation term or scaling factor.

* If the length of the transformed line is equal to that of its reference version, then the transformation is characterized by a pure translation.
* If the length of the transformed version of the line is different from its reference version, then a scaling transformation is at work.

4-6 SUCCESSIVE SCALINGS IN 1-D SPACE

Successive scalings in 1-D space are commutative. That is to say, you can apply a combination of two or more scalings in any sequence and end up with the same overall scaling. Suppose you want to apply this succession of scaling-factor values: 2, -1, 0.5. Taken together, that figures out to a scaling of $(2)(-1)(0.5) = -1$. Mix up the order of that same combination of scaling factors, and you will get the same overall scaling; for example, the sequence $-1,2,0.5$ multiplies out to an overall scaling of $(-1)(2)(0.6) = -1$. The idea applies equally well to geometric and relativistic scaling operations.

4-7 COMBINATIONS OF TRANSLATIONS AND SCALINGS

Translations are communitative and scalings are commutative, but combinations of the two are not. You can execute a given series of translations in any desired sequence and come up with the same overall result every time. Likewise, you can execute a series of scaling factors in any sequence and come up with the same overall result. Mix translations and scalings, however, and the order of execution becomes important.

Suppose that you want to apply some translation and a scaling to a point or line in 1-D space. If you apply the translation equation first, it is quite likely that you will get a different result than you would by applying the scaling first.

The only conditions that will yield the same result, regardless which transformation is performed first, are the trivial ones: those where the translation term is zero or the scaling factor is 1. Can you use some methods of analytic geometry to prove that idea?

A First Look at 2-D Space

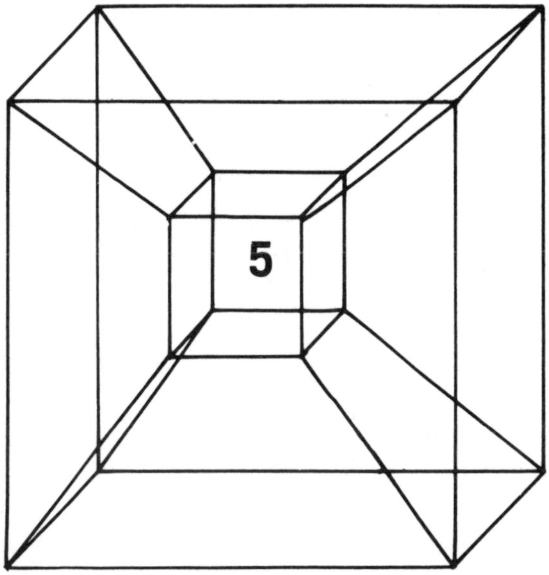

Two-dimensional space can be characterized as a plane that extends indefinitely in all directions from a given reference point. Think of it as an infinitely large, but infinitely thin sheet of paper. It is something quite different from the straight-line view of the simpler 1-D space.

The similarities between 1-D and 2-D spaces, however, can be as striking as the differences. Much that you already know about 1-D space applies directly to matters of 2-D space, and many of the procedures you have been practicing for 1-D spaces carry over to 2-D spaces as well. But there are significant differences.

Two-dimensional space is far more interesting and flexible. In 1-D space, for instance, the range of possible transformations is limited to just two: translations and scalings. In 2-D space, however, there is a third possible transformation: rotations. Furthermore, the elements of 2-D space include points, lines, and plane figures (triangles, rectangles, and the like). One-dimensional space supports only points and lines.

And finally, the analysis of lines in 1-D space deals only with the endpoint specifications and lengths of lines, while analyses in 2-D space deals with the endpoint specifications, lengths, orientation of lines, and the angle between two lines. Indeed, there is a lot more involved in working with 2-D space, but I think you will agree that it's worth the additional work.

5-1 A COORDINATE SYSTEM FOR 2-D SPACE

A suitable coordinate system for 2-D space must match the essential qualities of that space. It is not enough to specify points in 2-D space by means of a simple 1-D coordinate system. This matter calls for a 2-dimensional coordinate system—one that can express units of height as well as width, for instance.

Figure 5-1 illustrates several legitimate 2-D coordinate systems. Having already done some considerable work with 1-D coordinates, you might recognize the fact that these 2-D coordinate systems appear to be composed of a pair of intersecting 1-D coordinate systems. That is the case. In Fig. 5-2A, for example, a 1D coordinate system that is labeled −X and X can indicate the position of a point in a horizontal direction. The second 1-D coordinate system, labeled −Y and Y, can indicate the position of a point in a vertical direction. Put the two together, and the coordinate system can completely describe the position of any point in the 2-dimensional plane-world.

The same general idea applies to Figs. 5-2B and 5-2C, although there are differences in the directions of

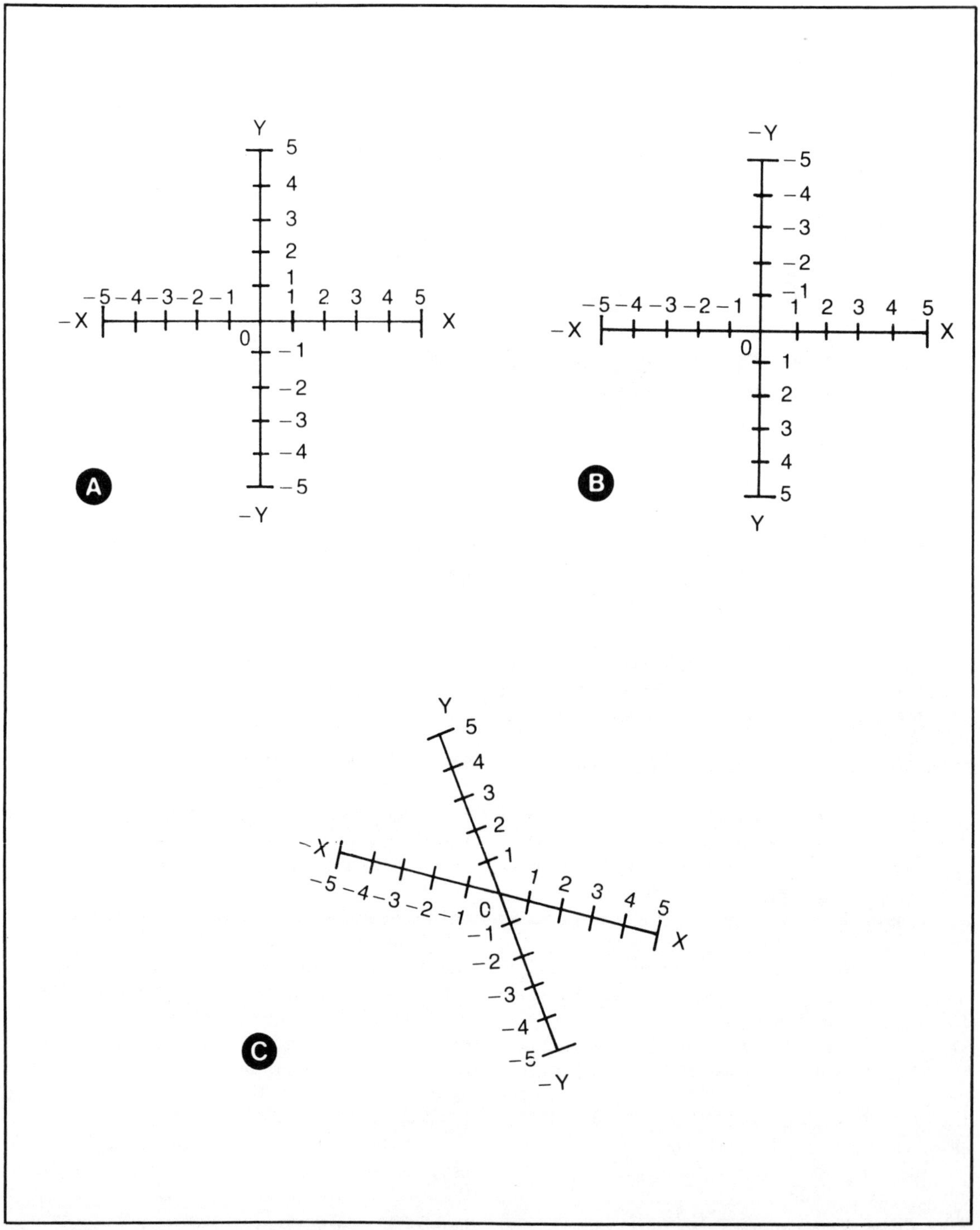

Fig. 5-1. Three possible coordinate systems for portraying 2-D space. (A) The accepted conventional version. (B) The Y axis shows measuring-unit values increasing in a top-to-bottom fashion. (C) The X and Y axes are not perpendicular.

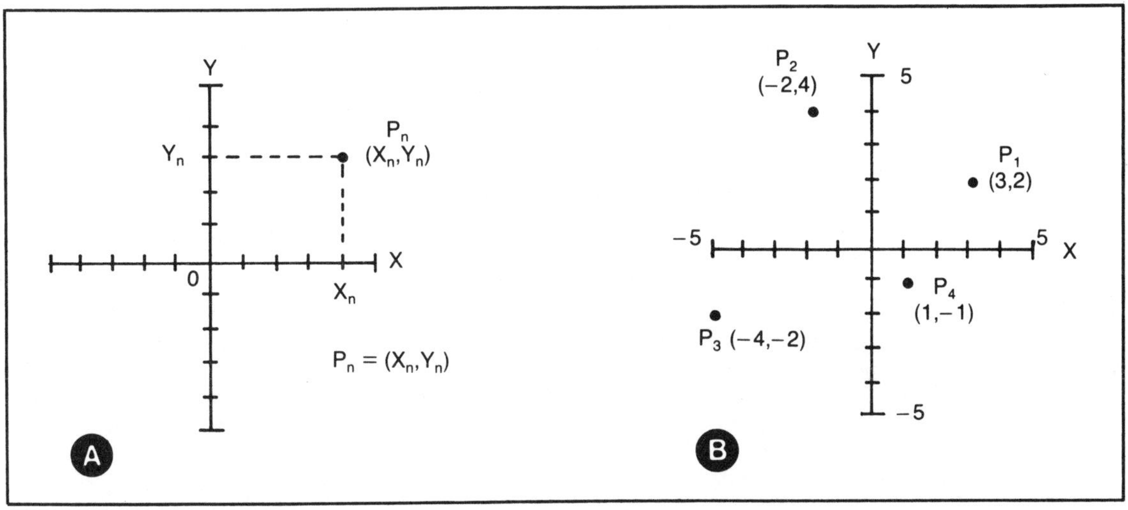

Fig. 5-2. Specifying the coordinates of points in a 2-D coordinate system. (A) A single point. (B) Four different points.

increasing measuring-unit values and the orientation of the coordinate axes relative to one another.

From the discussion so far you can rightly conclude that a suitable coordinate system for 2-D space is composed of a pair of 1-D coordinate systems, or axes, that intersect at a common point of origin. All three shown here meet that criterion.

But convention dictates that the two coordinate axes that define a 2-D coordinate system ought to be oriented at right angles to one another. That is simply a convention; not a necessity. In fact there is a whole branch of modern geometry, called non-Euclidian geometry, that allows the axes to intersect at any desired angle. For our purposes, it is necessary to adopt the right-angle-axis convention; and that eliminates the 2-D coordinate system in Fig. 5-1C from the list of possibilities.

Looking at the two remaining coordinate systems, those in Figs. 5-1A and 5-1B, you can see that they differ only by the direction of increasing measuring-unit values on the Y axis. The arrangement of measuring-unit values on the horizontal X axis are identical in both instances and, indeed, follow the format adopted earlier for 1-D coordinate systems. So it is time to establish a convention for designating measuring-unit values on the Y axis of a 2-D coordinate system.

Notice in Fig. 5-1A that the measuring-unit values on the Y axis increase from bottom to top, while in Fig. 5-1B they increase in a top-to-bottom fashion. The most popular convention favors the former: increasing Y-axis measuring-unit values running upward. There is absolutely nothing wrong with the coordinate system in Fig. 5-1B; it simply doesn't fit the convention used in this book nor that in many other places either.

Unless clearly stated otherwise, Fig. 5-1A represents the standard coordinate system for work in 2-D space. Take note of its main characteristics:

☐ It is composed of two axes that are labeled X and Y.

☐ The axes are arranged in such a way that they are perpendicular to one another and intersect at their common origin.

☐ Both axes are linear.

☐ The X axis shows increasing measuring unit values in a left-to-right fashion, while the Y axis shows increasing measuring-unit values in a bottom-to-top fashion.

5-2 SPECIFYING POINTS IN 2-D SPACE

A point in 1-D space always falls directly on the 1-D coordinate system, thus leaving little room for error in terms of citing its coordinate position. While there are two such axes in a 2-D coordinate system, there is no reason why a point in 2-D space should fall directly onto either of them.

Specifying the coordinate of a point in 2-D space calls for two different readings from the coordinate system—one from each of the axes. A position reading from the X axis, for instance, shows the horizontal component of a point's position; and a reading from the Y axis indicates the vertical component of the position. See the general example in Fig. 5-2A.

That drawing specifies point P in 2-D space. It is located at measuring unit x_n along the X axis and at measuring unit y_n along the Y axis. Notice the dashed lines drawn between the point and perpendicular to the X and Y axes. Those lines are shown here only for purposes of explanation, and they will not appear in any future drawings of this type. It is sufficient to indicate the X and Y components of the point's position, or its coordinate, in 2-D space.

It is absolutely necessary to specify the position of a point in 2-D space using two different coordinate readings: one from the X axis and one from the Y axis. The conventional form of notation looks like this:

$$P_n = (x_n, y_n)$$

where:

P_n = the label designating point n

x_n = the X-axis component of the coordinate of point *n*

y_n = the Y-axis component of the coordinate of point *n*

Another convention (and this is a truly firm one) is that the X and Y components of a coordinate are expressed within parentheses, separated by a comma, and arranged such that the X component appears first. Looking at some of the points plotted for you in Fig. 5-2B, then, a designation such as $P_1 = (3,2)$ literally means: Point number 1 is located in 2-D space in such a way that it is at unit 3 along the X axis and at unit 2 along the vertical Y axis.

Or you can describe the four points in the figure this way:

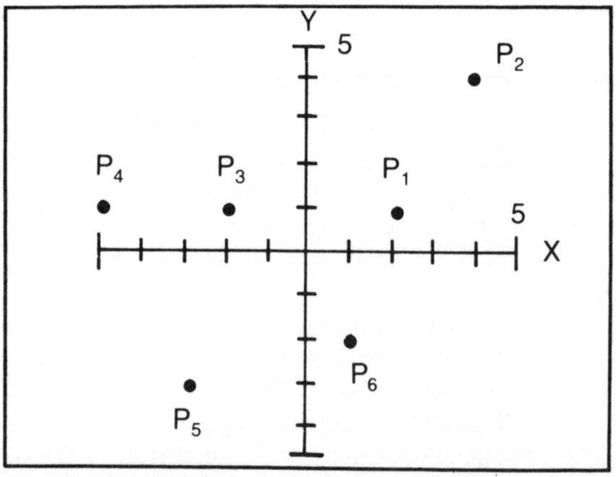

Fig. 5-3. Determine the coordinates of these six points plotted in 2-D space. See Exercise 5-1.

Point 1 — $P_1 = (3,2)$ $x_1 = 3$ $y_1 = 2$
Point 2 — $P_2 = (-2,4)$ $x_2 = -2$ $y_2 = 4$
Point 3 — $P_3 = (-4, -2)$ $x_3 = -4$ $y_3 = -2$
Point 4 — $P_4 = (1, -1)$ $x_4 = 1$ $y_4 = -1$

Exercise 5-1

1. Construct a standard 2-D coordinate system that has X and Y ranges of −5 to 5. Then plot the following points on it:

$P_1 = (1,2)$ $P_2 = (-1,4)$
$P_3 = (-3,6)$ $P_4 = (-3,-4)$
$P_5 = (-4,3.5)$ $P_6 = (3,-5.5)$

2. Cite the coordinates for the six points plotted in Fig. 5-3.

$P_1 = (\_\_\_,\_\_\_)$ $P_2 = (\_\_\_,\_\_\_)$
$P_3 = (\_\_\_,\_\_\_)$ $P_4 = (\_\_\_,\_\_\_)$
$P_5 = (\_\_\_,\_\_\_)$ $P_5 = (\_\_\_,\_\_\_)$

5-3 SPECIFYING LINES IN 2-D SPACE

As in 1-D space, lines in 2-D space are defined according to the coordinates of their endpoints. Line L_n in Fig. 5-4A, for example, is defined by endpoints P_u and P_v; then those endpoints are defined, in turn, in terms of their 2-D coordinates. In this general case, endpoint $P_u = (x_u, y_u)$ and endpoint $P_v = (x_v, y_v)$.

The complete specification for that line looks like this:

$$L = \overline{P_u P_v} \qquad P_u = (x_u, y_u)$$
$$P_v = (x_v, y_v)$$

where the bar across the P terms in the L expression indicates enclosure. I will be referring to those expressions as the P-L specifications for a line.

Figure 5-4B shows the complete specifications and plots three different lines in 2-D space. Notice that the three lines; L_1, L_2 and L_3; are first defined in terms of their endpoints, and then the endpoints, themselves, are defined in terms of their 2-D coordinates. Study the drawing and specifications carefully to make sure you understand the notation.

Exercise 5-2

1. Construct a standard 2-D coordinate system having X and Y ranges of −5 to 5. Plot the four lines specified here:

$L_1 = \overline{P_1P_2}$ $P_1 = (-4,4)$
$L_2 = \overline{P_3P_4}$ $P_2 = (4,4)$

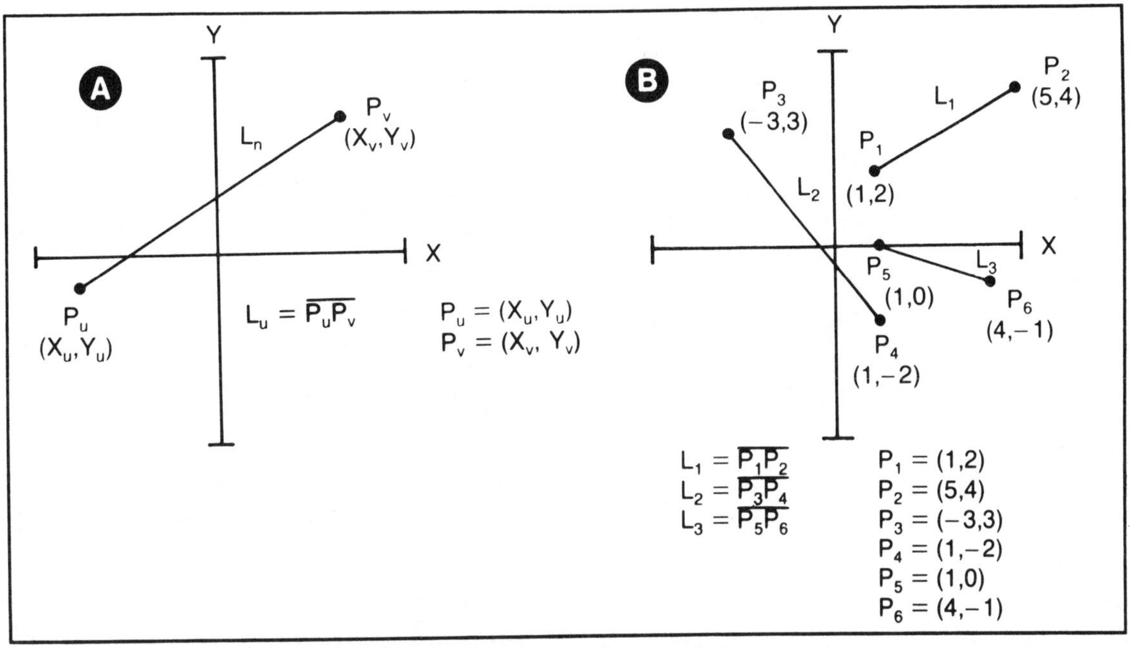

Fig. 5-4. Plotting lines in a 2-D coordinate system. (A) A single line. (B) Three different lines.

$L_3 = \overline{P_5P_6}$ $\quad P_3 = (-2,-3)$

$P_4 = (1,1)$

$P_5 = (-1,1)$

$P_6 = (4,-4)$

2. Generate the P-L specifications for the three lines that are plotted in Fig. 5-5.

$L_1 = \overline{P\_\ P\_}$ $\quad P_1 = (\_\_\_,\_\_\_)$ $\quad P_4 = (\_\_\_,\_\_\_)$

$L_2 = \overline{P\_\ P\_}$ $\quad P_2 = (\_\_\_,\_\_\_)$ $\quad P_5 = (\_\_\_,\_\_\_)$

$L_3 = \overline{P\_\ P\_}$ $\quad P_3 = (\_\_\_,\_\_\_)$ $\quad P_6 = (\_\_\_,\_\_\_)$

5-3.1. Lengths of Lines in 2-D Space

A line in 2-D space is defined in terms of its endpoints, and if you happen to know the coordinates of those endpoints, Equation 5-1 lets you calculate the length of that line.

$$s_n = \sqrt{(x_u - x_v)^2 + (y_u - y_v)^2} \qquad \textbf{Equation 5-1}$$

where:

s_n = the length of line *n* in 2-D space

x_u,y_u = the coordinate of endpoint *u* of line *n*

x_v,y_v = the coordinate of endpoint *v* of line *n*

There is one line, two points and one length involved here. The line, L_n, is defined in terms of endpoints $P_u = (x_u,y_u)$ and $P_v = (x_v,y_v)$. Getting the length of the

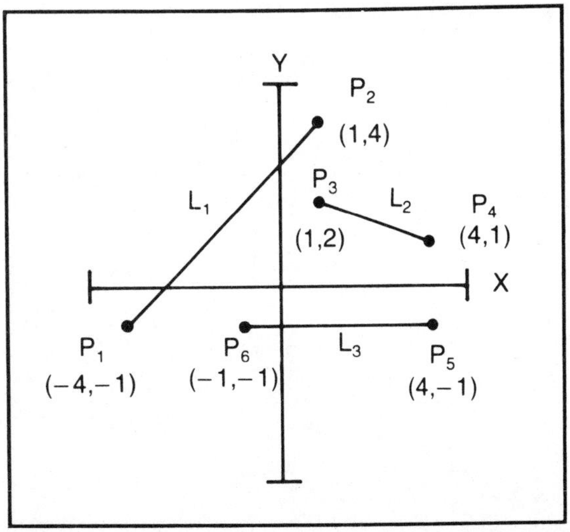

Fig. 5-5. Generate the P-L specifications for these three lines plotted in 2-D space. See Part 2 of Exercise 5-2.

line, S_n, is a matter of summing the squares of the differences between the X and Y components of the endpoints, and then taking the square root of the result. Figure 5-6 defines a line in this fashion and shows the equation for finding its length.

Example 5-1 illustrates the procedure for finding the lengths of two different lines, L_1 and L_2. The example begins, quite necessarily, with the complete P-L specifications for the two lines. There is no need to plot them for the sake of this example—the necessary information is available.

Example 5-1

Suppose that two lines in 2-D space have the following specifications:

$$L_1 = \overline{P_1P_2} \quad P_1 = (-2,1)$$
$$L_2 = \overline{P_3P_4} \quad P_2 = (4,-1)$$
$$P_3 = (-4,-3)$$
$$P_4 = (5,4)$$

Calculate the lengths of those lines.

1. Gathering the coordinates into a more useful format:

$x_1 = -2$	$x_2 = 4$	$x_3 = -4$	$x_4 = 5$
$y_1 = 1$	$y_2 = -1$	$y_3 = -3$	$y_4 = 4$

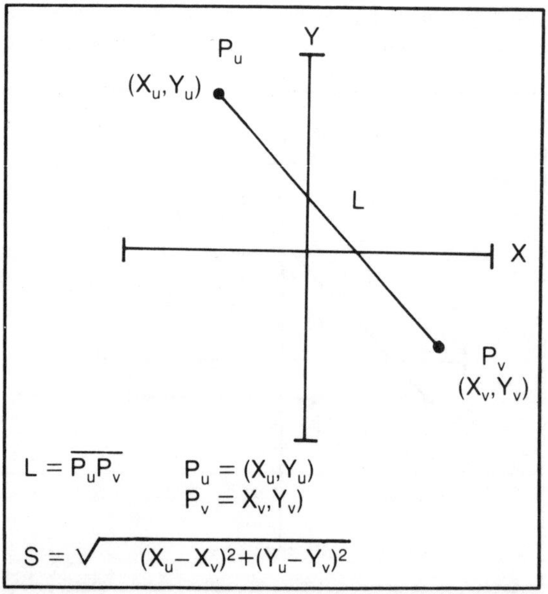

Fig. 5-6. Calculating the length of line specified in a 2-D coordinate system.

2. Applying Equation 5-1 to line L_1:

$$s_1 = \sqrt{(x_2-x_1)^2 + (y_2-y_1)^2} \qquad s_1 = \sqrt{36 + 4}$$
$$s_1 = \sqrt{(4-(-2))^2 + (-1-1)^2} \qquad s_1 = \sqrt{40}$$
$$s_1 = \sqrt{(6)^2 + (-2)^2} \qquad s_1 = 6.32$$

Thus line L_1 is about 6.32 units long.

3. Applying Equation 5-1 to line L_2:

$$s_2 = \sqrt{(x_4-x_3)^2 + (y_4-y_3)^2} \qquad s_2 = \sqrt{81 + 49}$$
$$s_2 = \sqrt{(5-(-4))^2 + (4-(-3))^2} \qquad s_2 = \sqrt{130}$$
$$s_2 = \sqrt{(9)^2 + (7)^2} \qquad s_2 = 11.4$$

So line L_2 is about 11.4 units long.

Step 1 is optional, but it can save some confusion during the later steps. The general idea is to specify the components of each endpoint coordinate. Point P_1 is composed of coordinate elements $x_1 = -2$ and $y_1 = 1$, point P_2 is composed of elements $x_2 = 4$ and $y_2 = -1$, and so on for all four points specified in that particular situation. It is nothing more than a summary of the components of the four points.

Step 2 uses Equation 5-1 to solve for the length of line L_1. The length of that line, designated s_1, is composed of coordinate elements x_1, y_1, x_2 and y_2. You know that is the case from the original specifications for the line. Once you have stated the length equation in terms of specific coordinate elements, simply substitute the values in the component summary (Step 1) and solve the equation.

See if you can follow the solution for the length of line L_2 in Step 3 of the example. Here the length of line s_2, is expressed in terms of the components of the endpoints of that line: x_3, y_3, x_4 and y_4.

Exercise 5-3

1. Given the following specifications for three lines in 2-D space, use Equation 5-1 to calculate their lengths:

$$L_1 = \overline{P_1P_2} \qquad P_1 = (1,2)$$
$$L_2 = \overline{P_3P_4} \qquad P_2 = (5,5)$$
$$L_3 = \overline{P_5P_6} \qquad P_3 = (5,2)$$
$$P_4 = (4-4)$$
$$P_5 = (-5,1)$$
$$P_6 = (1,-4)$$

$s_1 =$ ____ $\qquad s_2 =$ ____ $\qquad s_3 =$ ____

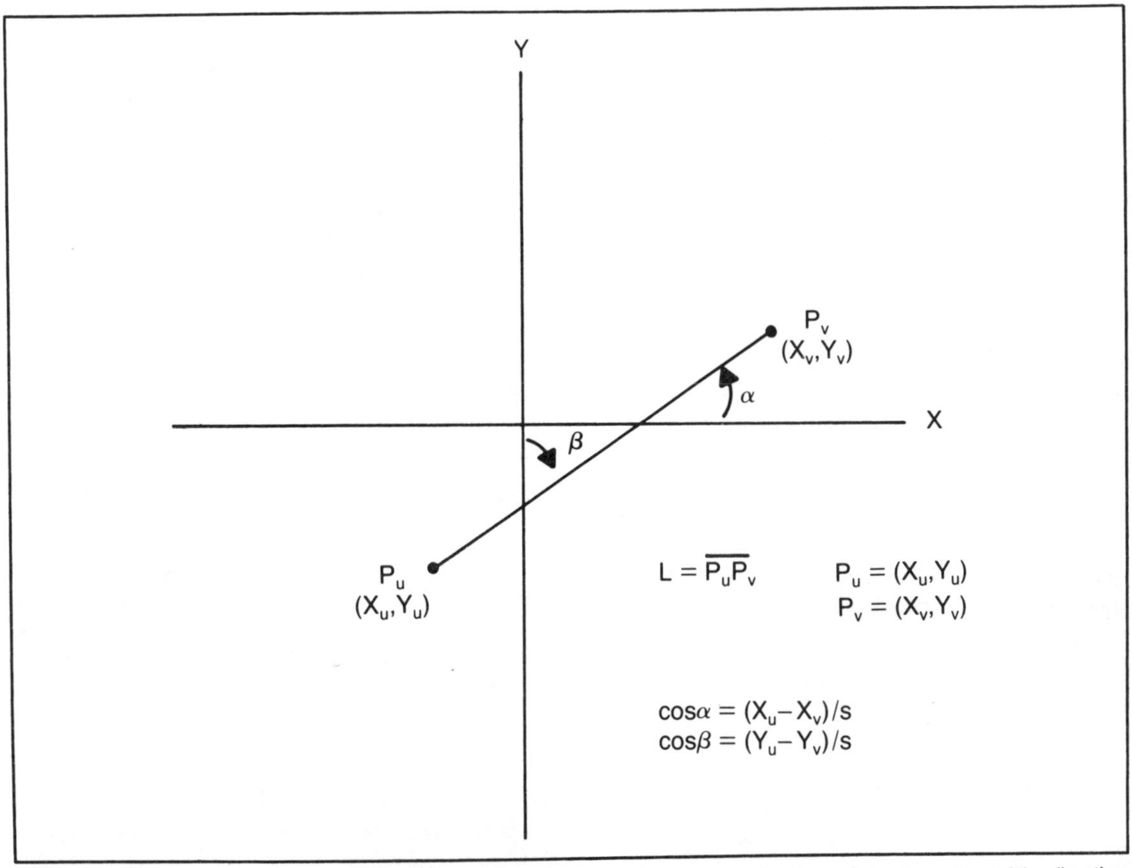

Fig. 5-7. A line plotted in 2-D space, showing the direction angles and the equations for determining the values of the direction cosines.

2. Construct a 2-D coordinate system, plot the lines specified in Part 1 of this exercise, and check your calculated lengths by comparing them with actual measurements on the coordinate system.

5-3.2 Direction Cosines in 2-D Space

Lines in 2-D space possess a quality that is totally foreign to their 1-D counterparts—their orientation with respect to the coordinate axes. Every line in 1-D space is oriented the same way with respect to its simple, single-line axis; they must fall directly on the axis. There is, however, an infinite number of different ways to orient lines in 2-D space. The lines may be drawn directly along the axes in 2-D space, too; but they may also be drawn off the axes, across the axes, or anywhere else at any desired slope on the page. The orientation of a line in 2-D space—its orientation with respect to the coordinate axes—is a vital part of its overall character.

This book specifies the orientation of lines in terms of direction cosines and their close relatives, direction numbers. Since direction numbers are usually derived from direction cosines, we have to deal with the latter topic first.

The line in Fig. 5-7 is plotted at some angle with respect to both the X and Y axes. The angle between the X axis and the plotted line is called angle α (alpha), while the angle between the Y axis and the plotted line is called angle β (beta). These two angles are known as the *direction angles* of the line. In 2-D space, a line has two direction angles.

Actually, there is a choice of four different angles for each of the two direction angles; but a convenient convention allows the direction angles to be taken as the acute angles in each case—as the angles that are less than 90 degrees. Angle α is taken as one of the acute angles between the line and the X axis, and angle β is taken as one of the acute angles between the lines and the Y axis.

Given the coordinates of the endpoints of a line in 2-D space, it is quite possible to use a bit of trigonometry to calculate the two direction angles. But it is easier, and usually more useful as well, to calculate the cosines of those direction angles instead. The cosine values of the direction angles of a line are called its *direction cosines.*

$$\cos\alpha_n = (x_u - x_v)/s_n$$
$$\cos\beta_n = (y_u - y_v)/s_n \qquad \textbf{Equation 5-2}$$

where:

$\cos\alpha_n$ = X-axis direction cosine for line *n*
$\cos\beta_n$ = Y-axis direction cosine for line *n*
x_u, y_u = the coordinate of endpoint *u* of line *n*
x_v, y_v = the coordinate of endpoint *v* of line *n*
s_n = the length of line *n* (see Equation 5-1)

Equation 5-2 leads directly to the values of the direction cosines of a line in 2-D space. The only information required for solving the equation is the coordinates of the two endpoints of the line in question. Example 5-2 illustrates the use of this equation.

Example 5-2

Calculate the direction cosines for the following lines:

$L_1 = \overline{P_1P_2}$ $\qquad P_1 = (-1,2)$
$L_2 = \overline{P_3P_4}$ $\qquad P_2 = (1,0)$
$P_3 = (1,2)$
$P_4 = (-3,-2)$

1. Gathering the coordinates into a more useful format

$x_1 = -2 \qquad x_2 = 1 \qquad x_3 = 1 \qquad x_4 = -3$
$y_1 = 2 \qquad y_2 = 0 \qquad y_3 = 2 \qquad y_4 = -2$

2. Calculating the lengths of the two lines:

$$s_1 = \sqrt{(x_2-x_1)^2 + (y_2-y_1)^2}$$
$$s_1 = \sqrt{[1-(-1)]^2 + (0-2)^2}$$
$$s_1 = \sqrt{2^2 + (-2)^2}$$
$$s_1 = \sqrt{4+4}$$
$$s_1 = \sqrt{8}$$
$$s_1 = 3\sqrt{2}$$
$$s_2 = \sqrt{(x_4-x_3)^2 + (y_4-y_3)^2}$$
$$s_2 = \sqrt{(-3-1)^2 + (-2-2)^2}$$
$$s_2 = \sqrt{(-4)^2 + (-4)^2}$$
$$s_2 = \sqrt{16 + 16}$$
$$s_2 = \sqrt{32}$$
$$s_2 = 5\sqrt{2}$$

3. Applying Equation 5-2 to calculate the direction cosines for line L_2:

$\cos\alpha_1 = (x_2-x_1)/s_1 \qquad \cos\beta_1 = (y_2-y_1)/s_1$
$\cos\alpha_1 = (1-(-1))/3\sqrt{2} \qquad \cos\beta_1 = (0-2)/3\sqrt{2}$
$\cos\alpha_1 = 2/3\sqrt{2} \qquad \cos\beta_1 = -2/3\sqrt{2}$
$\cos\alpha_1 = \sqrt{2/3} \qquad \cos\beta_1 = -\sqrt{2/3}$

4. Applying Equation 5-2 to calculate the direction cosines for line L_2:

$\cos\alpha_2 = (x_4-x_3)/s_2 \qquad \cos\beta_2 = (y_4-y_3)/s_2$
$\cos\alpha_2 = (-3-1)/5\sqrt{2} \qquad \cos\beta_2 = (-2-2)/5\sqrt{2}$
$\cos\alpha_2 = -4/5\sqrt{2} \qquad \cos\beta_2 = -4/5\sqrt{2}$
$\cos\alpha_2 = -2\sqrt{2/5} \qquad \cos\beta_2 = -2\sqrt{2/5}$

5. Summarizing the direction cosines for the two lines:

$\cos\alpha_1 = \sqrt{2/3} \qquad \cos\beta_1 = -\sqrt{2/3}$
$\cos\alpha_2 = -2\sqrt{2/5} \qquad \cos\beta_2 = -2\sqrt{2/5}$

The example begins by completely specifying two different lines in 2-D space, then Step 1 simply summarizes the values of the components of the endpoints. Check those values carefully against the original specifications. Make sure you understand where those values come from and what they mean.

Step 2 calculates the lengths of the two lines. It is a necessary procedure because the lengths of the lines is one of the variables in Equation 5-2. I stated earlier that you need only the endpoint coordinates in order to solve for the direction cosines. Am I backing away from that statement? Certainly not. The length of the line is, indeed, one of the variables in the equation; but you know that you can calculate the length of the line from the endpoint data. So it's true that you need only the endpoint coordinates to solve for the direction cosines of a line; but it's a 2-step process. You must first calculate the lengths of the lines. That is the purpose of Step 2 in the example. And it isn't just wasted work, because a complete analysis of a line in 2-D space calls for showing its length anyhow. Thus Step 2 in the example turns up the lengths of the two lines, where s_1 is the length of line L_1, and s_2 is the length of line L_2.

Step 3 begins by expressing Equation 5-2 as it applies to Line L_1. Using the components of endpoints P_1 and P_2 and the length of line L_1 calculated in the previous step, the procedure generates the direction cosines for that line, $\cos\alpha_1$ and $\cos\beta_1$.

Notice in the conclusion of Step 3 that I converted the preliminary solutions to a form having a rational denominator. In other words, I cleared the square-root term out of the denominator by multiplying both the numerator and denominator by the square-root of 2. It isn't absolutely necessary to rationalize the denominator in this fashion, but it is a commonly accepted practice to do so. It's one of those little features that distinguish an experienced and knowledgeable experimenter from a novice.

Of course this whole matter of rationalizing denominators would not have arisen if I had converted the square root of 2 to its decimal value, 1.414; but you are going to find more precise final results when leaving a value in its rational form as long as possible through an analysis.

Step 4 generates the direction cosines for line L_2 in a similar way. Here you can see Equation 5-2 stated in terms of that line—in terms of the endpoint coordinates for line L_2 (points P_3 and P_4) and the length of that same line as calculated in Step 2. Again, I rationalized the final results so that there are no square-root terms in the denominators of the fractions.

Step 5 merely summarizes the direction cosines for the two lines specified originally. Incidentally, if you ever wish to calculate the angles, themselves, just determine the arc cosines of those values. You can do that by reading a printed trig table "backwards," or by taking advantage of the calculating power of an electronic calculator or home computer.

Exercise 5-4

1. Calculate the direction cosines for the lines that are specified in Exercise 5-3.

$\cos\alpha_1 =$ ____ $\cos\beta_1 =$ ____
$\cos\alpha_2 =$ ____ $\cos\beta_2 =$ ____
$\cos\alpha_3 =$ ____ $\cos\beta_3 =$ ____

2. Use the information supplied in Fig. 5-8 to:

a. Generate a complete set of P-L specifications

$L = \overline{P\_P\_}$ $P_1 =$ (____,____)
$P_2 =$ (____,____)

b. Calculate the length of the line

s = ______

c. Calculate the direction cosines of the line

$\cos\alpha =$ ____ $\cos\beta =$ ____

3. Make up some P-L specifications of your own, and then calculate the lengths of the lines and their respective direction cosines. Plot the lines on a standard 2-D coordinate system.

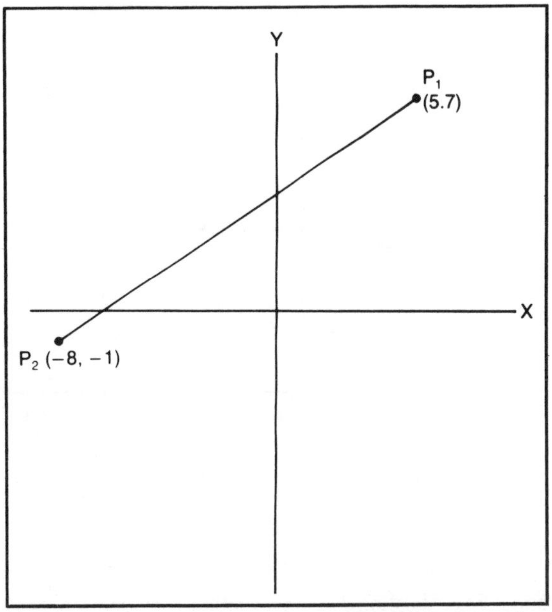

Fig. 5-8. A reference line in 2-D space. See Part 2 of Exercise 5-4.

5-3.3 Angle Between Lines in 2-D Space

You have just seen that lines in 2-D space can be oriented in an infinitely large number of ways with respect to the coordinate axes. Those orientations are expressed quite precisely in terms of the direction cosines and direction numbers for each line. There is, however, one more quality of lines in 2-D space that is missing altogether in 1-D space: Two different lines in 2-D space have a certain angle between them.

Figure 5-9 shows two lines, L_m and L_n, in 2-D space. They happen to meet at a common coordinate point, P_2. It is the angle that exists where those two lines meet that is of concern now.

The angle, itself, is labeled $\phi_{m,n}$ implying that it is the angle created by the interception of lines L_m and L_n. You will find that particular notation for the angle between two lines used throughout this book. If the lines happened to be labeled L_3 and L_4, the angle designation would be $\phi_{3,4}$.

The trick is to determine the size of that angle in degrees. I suggest a 2-step procedure that begins by solving for the cosine of the angle and then using a trig table or electronic calculator to determine the arc cosine of that cosine value. See Equations 5-3 and 5-4.

$$\cos\phi_{m,n} = \cos\alpha_m \cos\alpha_n + \cos\beta_m \cos\beta_n \quad \textbf{Equation 5-3}$$

where:

$\cos\phi_{m,n}$ = the cosine of the angle between lines *m* and *n*

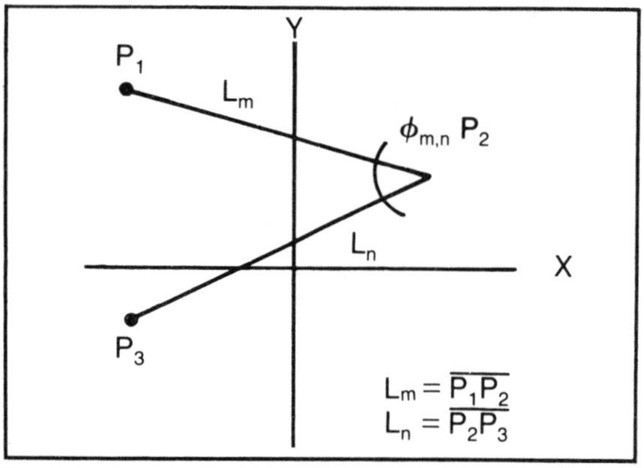

Fig. 5-9. The angle between two lines plotted in 2-D space.

$\cos\alpha_m$ = the X-axis direction cosine for line *m*
$\cos\beta_m$ = the Y-axis direction cosine for line *m*
$\cos\alpha_n$.= the X-axis direction cosine for line *n*
$\cos\beta_n$ = the X-axis direction cosine for line *n*

Note: See Equation 5-2 for direction cosines

Equation 5-4 solves for the cosine of the angle between any two lines in terms of the direction cosines of those lines. Simply multiply together the cosα elements of the two lines, multiply together the cosβ elements of the same two lines, and then sum the results. You end up with the cosine of the angle between them.

Quite often the result is a negative value, but that isn't relevant in this case. Direction cosines are frequently negative, and it is important to retain that sign. Not as far as the cosine of the angle between two lines is concerned, though. Ignore a minus sign that will occasionally turn up in front of the result of solving Equation 5-3.

$$\phi_{m,n} = \text{Arc}\cos(\cos\phi_{m,n}) \quad \textbf{Equation 5-4}$$

where:

$\phi_{m,n}$ = the angle between lines *m* and *n*

$\cos\phi_{m,n}$ = the cosine of the angle between lines *m* and *n* (see Equation 5-4)

Equation 5-4 simply shows that you can put the final touch on the procedure by determining the arc cosine of the value. Doing that, either by consulting trig tables or using a calculator or home computer, you have the actual angle in degrees.

So whenever you want to find the angle between any two lines in 2-D space, Equation 5-4 will finish up the job for you. But in order to solve that equation, you must have the results of solving for the cosine of that angle—the results of applying Equation 5-3. What must you know in order to solve Equation 5-3? Well, you have to know the direction cosines of both lines involved in the analysis. That takes you back to Equation 5-2. But you can't solve for the direction cosines without knowing the lengths of the two lines, and that means applying Equation 5-1. And all you need for solving that equation is the P-L specifications for the two lines under analysis.

Putting these ideas together in a practical fashion, it means that you can solve for the angle between any two lines that are completely specified in 2-D space. Starting from the P-L specifications, the procedure goes like this:

☐ Calculate the lengths of the two lines (Equation 5-1).

☐ Calculate the direction cosines of the two lines (Equation 5-2).

☐ Calculate the cosine of the angle between the two lines (Equation 5-3).

☐ Determine the angle, itself (Equation 5-4).

Example 5-3 runs through the procedure for you in a step-by-step fashion.

Example 5-3

Figure 5-10 shows two lines that meet at point P_2 and the specifications for those lines. Determine the angle between them.

1. Restating the given specifications:

$$L_1 = \overline{P_1P_2} \qquad P_1 = (-3,-2)$$
$$L_2 = \overline{P_2P_3} \qquad P_2 = (3,1)$$
$$P_3 = (-2,2)$$

2. Gathering the components of the coordinates into a useful format:

$$x_1 = -3 \qquad x_2 = 3 \qquad x_3 = -2$$
$$y_1 = -2 \qquad y_2 = 1 \qquad y_3 = 2$$

3. Calculating the lengths of the lines:

$$s_1 = \sqrt{(x_2-x_1)^2 + (y_2-y_1)^2}$$
$$s_1 = \sqrt{[3-(3)]^2 + [1-(-2)]^2}$$
$$s_1 = \sqrt{45} = 6.7$$
$$s_2 = \sqrt{(x_3-x_2)^2 + (y_3-y_2)^2}$$
$$s_2 = \sqrt{(-2-3)^2 + (2-1)^2}$$
$$s_2 = \sqrt{26} = 5.1$$

4. Calculating the direction cosines:

$$\cos\alpha_1 = (x_2-x_1)/s_1 \qquad \cos\beta_1 = (y_2-y_1)/s_1$$
$$\cos\alpha_1 = [-3-(-3)]/\sqrt{45} \qquad \cos\beta_1 = ([1-(-2)]/\sqrt{45}$$
$$\cos\alpha_1 = 2\sqrt{45}/15 \qquad \cos\beta_1 = \sqrt{45}/15$$

$\cos\alpha_2 = (x_3 - x_2)/s_2 \qquad \cos\beta_2 = (y_3 - y_2)/s_2$

$\cos\alpha_2 = (-2-3)/\sqrt{26} \qquad \cos\beta_2 = (2-1)/\sqrt{26}$

$\cos\alpha_2 = 5\sqrt{26}/26 \qquad \cos\beta_2 = \sqrt{26}/26$

5. Calculating the angle between L_1 and L_2:

$\cos\phi_{1,2} = \cos\alpha_1\cos\alpha_2 + \cos\beta_1\cos\beta_2$

$\cos\phi_{1,2} = (2\sqrt{45}/15)(-5\sqrt{26}/26) + (\sqrt{45}/15)$
$(\sqrt{26}/26$

$\cos\phi_{1,2} = -9\sqrt{1170/390}$ or about -0.79

Since the minus sign isn't relevant, the cosine of the angle is about 0.79.

6. Determining the angle, itself:

$$\phi_{1,2} = \text{arc cos}(\cos\phi_{1,2})$$
$$\phi_{1,2} = \text{arc cos}(0.79$$
$$\phi_{1,2} = 52°$$

The main objective of Example 5-3 is to calculate the angle $\phi_{1,2}$ between lines L_1 and L_2 in Fig. 5-10. The procedure in this case begins with a restatement of the lines' specifications. That isn't really necessary in this case because the specifications also appear in the drawing. But I decided to document the specifications again anyway.

Step 2 merely gathers the relevant components of the endpoint coordinates, assigning them the point indices as dictated by the specifications. You are probably well aware of the fact that I run through this procedure in every case. It is a good idea, I think. Try setting up an analysis without summarizing the components one time, and you will see the potential for confusion and errors.

Step 3 restates the length equation according to the lines' specifications and then solves it for both lines. In this particular instance, L_1 turns out to be about 6.7 units long, and line L_2 is about 5.1 units long. Those are only approximations, however, and you should leave the values in their more precise form—in a square-root form in this example.

Step 4 works out the direction cosines for both lines. Notice that I write the direction-cosine equation as it applies to the specifications, then solve it.

Step 5 begins by stating Equation 5-3 as it applies to the angle between lines that are designated L_1 and L_2. The result is an approximate decimal value of -0.79. The minus sign isn't relevant, however; so there is no need to give it any special attention, a value of 0.79 is just as good.

Finally, Step 6 determines the angle in degrees. It is an approximated angle in this case, but it is close enough for the purpose at hand. Come up with a value with a precision to the nearest minute of arc if you like. Going any farther is a waste of time, though.

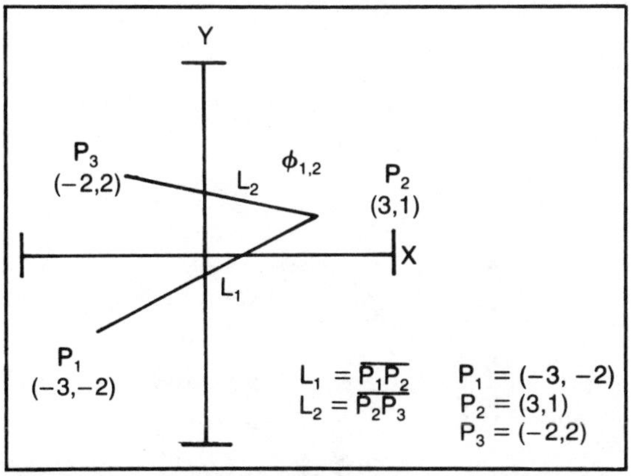

Fig. 5-10. Determine the acute angle between these two lines plotted in 2-D space. See Example 5-3.

Exercise 5-5

Use the following P-L specifications for completing the analysis of a pair of lines in 2-D space:

$$L_1 = \overline{P_1P_2} \qquad P_1 = (-5,2)$$
$$L_2 = \overline{P_2P_3} \qquad P_2 = (3,2)$$
$$P_3 = (5,-1)$$

1. Plot the lines on a 2-D coordinate system.
2. Gather the components of the coordinates:

$x_1 =$ ______ $\quad x_2 =$ ______ $\quad x_3 =$ ______
$y_1 =$ ______ $\quad y_2 =$ ______ $\quad y_3 =$ ______

3. Calculate the lengths of the two lines:

$s_1 =$ ______ $\quad s_2 =$ ________

4. Calculate the direction cosines for the two lines:

$\cos\alpha_1 =$ ______ $\quad \cos\beta_1 =$ ________
$\cos\alpha_2 =$ ______ $\quad \cos\beta_2 =$ ________

5. Calculate the cosine of angle $\phi_{1,2}$:

$\cos\phi_{1,2} =$ ________

6. Determine the angle between the two lines:

$\phi_{1,2} =$ ________

7. Does your calculated angle come close to fitting an actual measurement between the lines? It should.

5-4 PLANE FIGURES IN 2-D SPACE

Thus far in this chapter you have found several qualities of 2-D space that are not included in 1-D space—qualities that are actually meaningless in 1-D space. Those qualities include the orientation of lines with respect to the coordinate axes and the notion of an angle between two lines. There is one more feature of special importance, and that is the possibility of working with plane figures in 2-D space.

Points in 2-D space are defined in terms of their coordinates, and lines in 2-D space are defined in terms of the points that mark their endpoints. And now we are ready to deal with the notion that a plane figure can be defined in terms of the lines that bound it. Points bound lines, and lines bound plane figures.

5-4.1 Specifying Plane Figures

Lines in 1-D and 2-D space can be defined by the point-line (P-L) specifications. Introducing plane figures makes it necessary to extend the specifications to include a bounded plane. The specifications now include definitions for points, lines and plane figures—they are P-L-F specifications.

The point portion of the specifications, as before, specifies the individual points in the figure according to their exact coordinates in 2-D space. Then the line portion specifies the lines according to their designated endpoints. Again, there is nothing new. But finally, the figure, itself, ought to be specified in terms of the lines that bound it. That is the F part of the specifications.

Looking at Fig. 5-11A, the specifications begin by defining figure F in terms of the lines that bound it. In this instance, the lines are L_1, L_2 and L_3. Then the specifications show the three lines as bounded by their endpoints—certain combinations of points P_1, P_2 and P_3. Finally, the specifications show the coordinates of each of the points.

Compare those specifications with the ones cited in Fig. 5-11B. One of the essential differences is that the F specification for a quadralateral must cite four lines instead of three.

Test your understanding by working through Exercise 5-6.

Exercise 5-6

1. Develop a set of P-L-F specifications for the plane figure shown in Fig. 5-12.

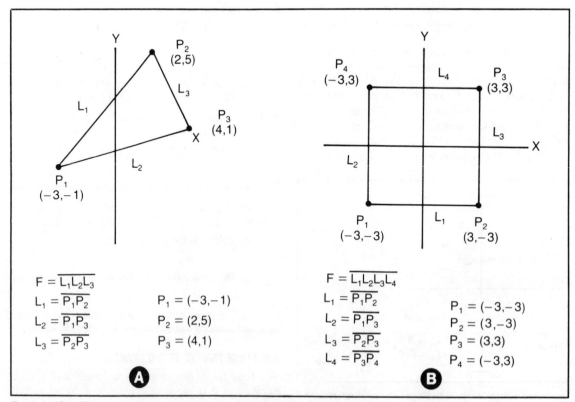

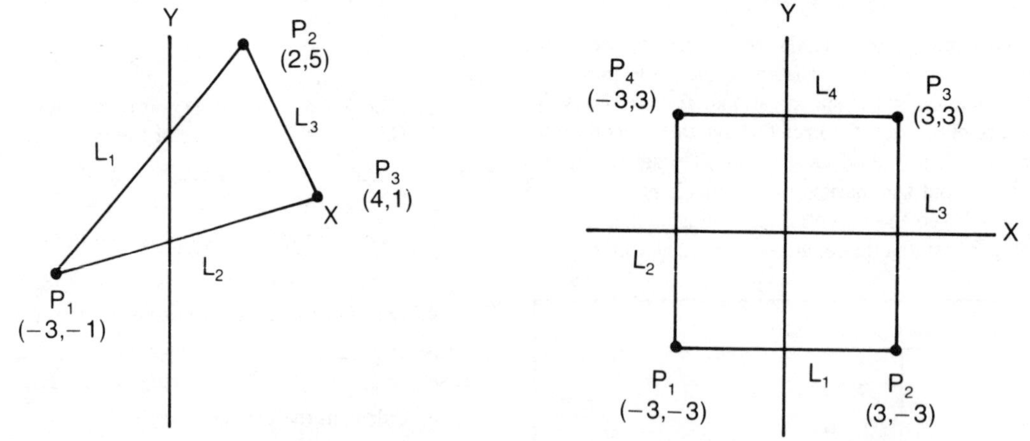

$F = \overline{L_1L_2L_3}$

$L_1 = \overline{P_1P_2}$ $P_1 = (-3,-1)$

$L_2 = \overline{P_1P_3}$ $P_2 = (2,5)$

$L_3 = \overline{P_2P_3}$ $P_3 = (4,1)$

A

$F = \overline{L_1L_2L_3L_4}$

$L_1 = \overline{P_1P_2}$ $P_1 = (-3,-3)$

$L_2 = \overline{P_1P_3}$ $P_2 = (3,-3)$

$L_3 = \overline{P_2P_3}$ $P_3 = (3,3)$

$L_4 = \overline{P_3P_4}$ $P_4 = (-3,3)$

B

Fig. 5-11. Specifying plane figures in 2-D space. (A) A triangle. (B) A quadralateral.

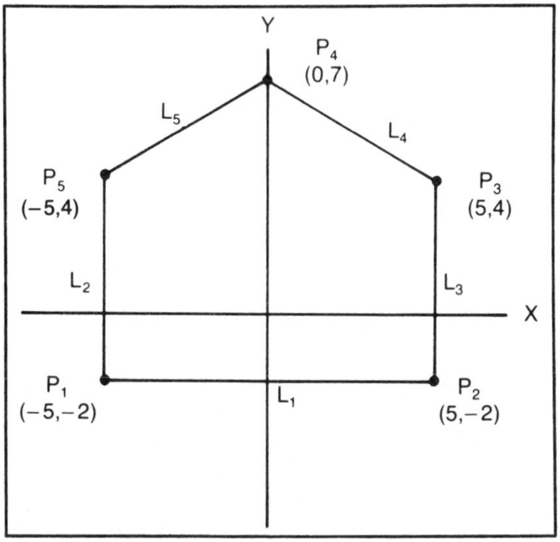

Fig. 5-12. One type of pentagon plotted as a plane figure in 2-D space. See Part 1 of Exercise 5-6 and Part 2 of Exercise 5-7.

$F = \overline{L_{\_}\ L_{\_}\ L_{\_}\ L_{\_}\ L_{\_}}$ $L_1 = \overline{P_{\_}P_{\_}}$ $P_1 = (\_\_\_, \_\_\_)$

$L_2 = \overline{P_{\_}P_{\_}}$ $P_2 = (\_\_\_, \_\_\_)$

$L_3 = \overline{P_{\_}P_{\_}}$ $P_3 = (\_\_\_, \_\_\_)$

$L_4 = \overline{P_{\_}P_{\_}}$ $P_4 = (\_\_\_, \_\_\_)$

$L_5 = \overline{P_{\_}P_{\_}}$ $P_5 = (\_\_\_, \_\_\_)$

2. Plot the plane figure that is defined by the following specifications. It should turn out to be a rather familiar figure.

$F = \overline{L_1L_2L_3L_4L_5}$ $L_1 = \overline{P_1P_3}$ $P_1 = (5,2)$

$L_2 = \overline{P_3P_5}$ $P_2 = (0,5)$

$L_3 = \overline{P_5P_2}$ $P_3 = (-5,2)$

$L_4 = \overline{P_2P_4}$ $P_4 = (-4,-4)$

$L_5 = \overline{P_4P_1}$ $P_5 = (4,-4)$

5-4.2 Analyzing Plane Figures in 2-D Space

A complete analysis of plane figures in 2-D space should include the following:

1. A 2-D plot of the figure—a graphic representation.
2. A complete set of P-L-F specifications.
3. A summary of the lengths of the lines in the figure.
4. A summary of the direction cosines of all lines.
5. A summary of all angles in the figure.

See all of this worked out for you in Example 5-4.

Example 5-4

Run a complete analysis of the triangle specified in Fig. 5-11A.

1. Gathering the components of the coordinates:

$x_1 = -3$ $x_2 = 2$ $x_3 = 4$
$y_1 = -1$ $y_2 = 5$ $y_3 = 1$

2. Calculating the lengths of the lines:

$s_1 = \sqrt{(x_2-x_1)^2 + (y_2-y_1)^2}$
$s_1 = \sqrt{[2-(-3)]^2 + [5-(-1)]^2}$
$s_1 = \sqrt{61}$
$s_2 = \sqrt{(x_3-x_1)^2 + (y_3-y_1)^2}$
$s_2 = \sqrt{[4-(-3)]^2 + [1-(-1)]^2}$
$s_2 = \sqrt{53}$
$s_3 = \sqrt{(x_3-x_2)^2 + (y_3-y_2)^2}$
$s_3 = \sqrt{(4-2)^2 + (1-5)^2}$
$s_3 = 2\sqrt{5}$

Summary of lengths:

$s_1 = \sqrt{61}$ $s_2 = \sqrt{53}$ $s_3 = 2\sqrt{5}$

3. Calculating the direction cosines:

$\cos\alpha_1 = (x_2-x_1)/s_1$ $\cos\beta_1 = (y_2-y_1)/s_1$
$\cos\alpha_1 = [2-(-3)]/\sqrt{61}$ $\cos\beta_1 = [5-(-1)]/\sqrt{61}$
$\cos\alpha_1 = 5\sqrt{61}/61$ $\cos\beta_1 = 6\sqrt{61}/61$

$\cos\alpha_2 = (x_3-x_1)/s_2$ $\cos\beta_2 = (y_3-y_1)/s_2$
$\cos\alpha_2 = [4-(-3)]/\sqrt{53}$ $\cos\beta_2 = [1-(-1)]/\sqrt{53}$
$\cos\alpha_2 = 7\sqrt{53}/53$ $\cos\beta_2 = 2\sqrt{53}/53$

$\cos\alpha_3 = (x_3-x_2)/s_3$ $\cos\beta_3 = (y_3-y_2)/s_3$
$\cos\alpha_3 = [(4-2)]/(2\sqrt{5})$ $\cos\beta_3 = (1-5)/(2\sqrt{5})$
$\cos\alpha_3 = \sqrt{5}/5$ $\cos\beta_3 = -2\sqrt{5}/5$

Summary of direction cosines:

$\cos\alpha_1 = 5\sqrt{61}/61 \quad \cos\beta_1 = 6\sqrt{61}/61$

$\cos\alpha_2 = 7\sqrt{53}/53 \quad \cos\beta_2 = 2\sqrt{53}/53$

$\cos\alpha_3 = \sqrt{5}/5 \quad \cos\beta_3 = -2\sqrt{5}/5$

4. Calculating the cosines of the angles between lines:

$\cos\phi_{1,2} = \cos\alpha_1\cos\alpha_2 + \cos\beta_1\cos\beta_2$

$\cos\phi_{1,2} = (5\sqrt{61}/61)7\sqrt{53}/53) + (6\sqrt{61}/61)$
$(2\sqrt{53}/53)$

$\cos\phi_{1,2} = 0.827$

$\cos\phi_{1,3} = \cos\alpha_1\cos\alpha_3 + \cos\beta_1\cos\beta_3$

$\cos\phi_{1,3} = (5\sqrt{61}/61)\ (\sqrt{5}/5)-(6\sqrt{61}/61)$
$(2\sqrt{5}/5)$

$\cos\phi_{1,3} = 0.401$

$\cos\phi_{2,3} = \cos\alpha_2\cos\alpha_3 + \cos\beta_2\cos\beta_3$

$\cos\phi_{2,3} = (7\sqrt{53}/53)\ (\sqrt{5}/5)-(2\sqrt{53}/53)$
$(2\sqrt{5}/5)$

$\cos\phi_{2,3} = 0.184$

Summarizing the cosines of the angles:

$\cos\phi_{1,2} = 0.827 \quad \cos\phi_{1,3} = 0.401 \quad \cos\phi_{2,3} = 0.184$

5. Determining the angles, themselves:

$\phi_{1,2} = \text{arc} \cos(0.827) = 34°10'$
$\phi_{1,3} = \text{arc} \cos(0.401) = 66°\ 22'$
$\phi_{2,3} = \text{arc} \cos(0.184) = 79°\ 23'$

Sum of the angles in this triangle:
179° 55′

That is close enough to 180° to provide the confidence that the job is done correctly.

The figure being analyzed is already plotted and specified in Fig. 5-11A, so the mathematical part of the analysis can begin by calculating the lengths of the three lines. Step 1 in the example sets up the components of the coordinates, and Step 2 works out the lengths of the lines according to Equation 5-1.

Step 3 uses the direction-cosine equation to calculate the direction cosines for all three lines; and it concludes with a summary of them.

Step 4 sets up the equations for the cosine of each of the three angles in the plane figure. $\text{Cos}\phi_{1,2}$ is the cosine of the angle between lines 1 and 2, $\cos\phi_{1,3}$ is the cosine of the angle between lines 1 and 3, and $\cos\phi_{2,3}$ represents the cosine of the angle between lines 2 and 3.

Step 5 completes the analysis by specifying the three angles, themselves. Notice how I checked the final results by summing the three angles. According to elementary plane geometry, the sums of the three angles in any triangle should be 180 degrees. This analysis came in close enough—179 degrees, 55 minutes.

Test your understanding by working through Exercise 5-7.

Exercise 5-7

1. Generate a complete analysis for the figure in Fig. 5-11B. If you are handling it properly, you should find that all lines are 6 units long and all angles are 90°. Remember that the sum of the four angles in a quadralateral should be 360°.

2. Generate a complete analysis for the 5-sided plane figure in Fig. 5-12.

Translations in 2-D Space

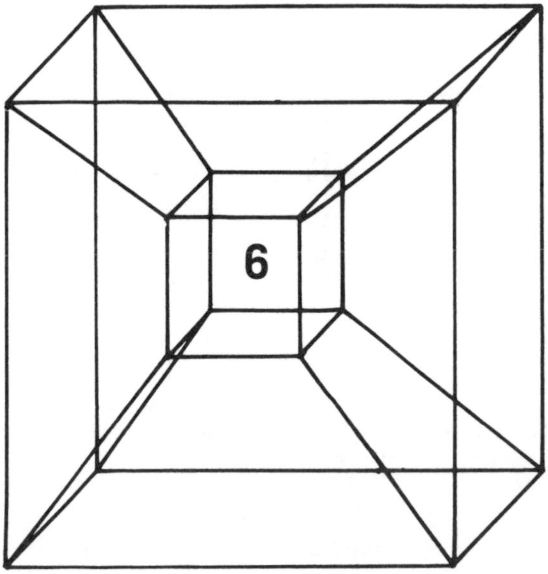

Recall from Chapter 3 that translations effectively displace a coordinate system and all of its points and lines along the line-world of 1-D space. A single translation term determines both the direction and distance of those 1-dimensional translations.

The same general idea applies to translations in the plane world of 2-D space, but of course there is now a second direction of translation—along the Y axis. And what's more, working in 2-D space brings up the possibility of translating plane figures as well as simple points and lines.

6-1 THE IMPORTANCE OF A FRAME OF REFERENCE

Figure 6-1 illustrates a translation between a pair of 2-D coordinate systems. The two systems are identical in every respect but one: their relative positions on the page.

As in the case of 1-D translations, you must select one of the two systems to serve as the frame of reference. Initially, it makes no difference which one you choose; but once you make that selection, you are committed to it.

Suppose that you select system A to be the frame of reference. That being the case, you can say that system B is translated relative to system A. This translation happens to show system B translated about 3.5 units to the right along the X axis of the reference system, and about 2.5 units downward along the Y axis. Relative to system A, the translation displaces system B to the right and downward.

But if you decide that coordinate system B is the frame of reference, the translation displaces system A relative to it. The translation in that case shifts system A about 3.5 units to the left along the X axis and about 2.5 upward along the Y axis of system B.

Two different distances are involved here—one along the X axis of the selected frame of reference and another along the Y axis of that frame of reference. The distances along the X axes are the same, no matter which coordinate system happens to be serving as the frame of reference. That applies to the distances along the Y axes, too.

The sign of the translation—the direction—depends on which coordinate system is serving as the frame of reference. If system A is taken as the frame of reference, the X-axis displacement of system B is in a positive direction and the Y-axis displacement is in a negative direction. Using system B as the frame of reference, however, reverses the directions of translation: the X-axis translation of system A, relative to B, is in a negative

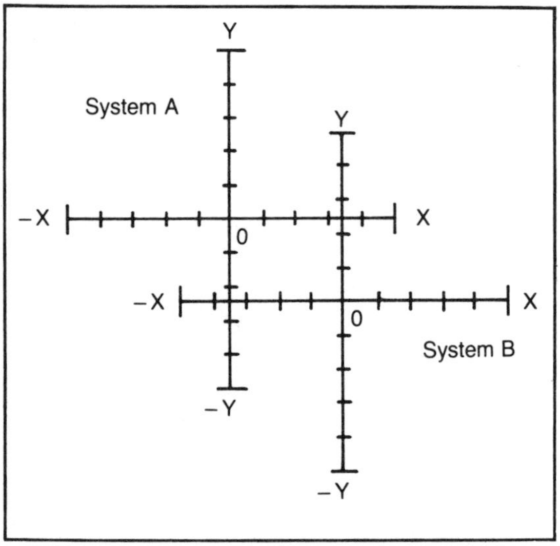

Fig. 6-1. A pair of 2-D coordinate systems having a 2-D translation between them.

direction, and the Y-axis translation is in a positive direction. Indeed, selecting the coordinate system that is to serve as the frame of reference is an all-important first step in dealing with the translation of coordinates.

6-2 THE BASIC 2-D TRANSLATION EQUATION

The basic 2-D translation equation, like its 1-D counterpart, refers directly to a point in the chosen frame of reference. The coordinate of that point is taken relative to the frame of reference, the translation is applied with respect to the frame of reference, and the coordinate of the translated coordinate is likewise reckoned from the frame of reference. The only real difference between translations in 1-D and 2-D spaces is that the latter deals with two perpendicular axes instead of one. See Equation 6-1.

$$x_n' = x_n + T_x \qquad y_n' = y_n + T_y \qquad \textbf{Equation 6-1}$$

where:

x_n', y_n' = the coordinate of the translated version of point *n* relative to the frame of reference

x_n, y_n = the coordinate of point *n* relative to the frame of reference

T_x = the X-axis translation term

T_y = the Y-axis translation term

Just as it requires two components to specify the coordinate of a point in 2-D space, it requires two translation terms to specify the full nature of a 2-D translation—one translation term for each component of a coordinate. So the equation deals with two components of a reference point, two translation terms, and two components of the translated version of a point. Using some notation established earlier, the 2-D translation equation transforms a reference point *n* from:

$$P_n = (x_n, y_n)$$

to:

$$P_n' = (x_n', y_n').$$

In that equation, the absolute values of the translation terms determine the distance of translation along the X and Y axes, while the signs of those two terms fix the direction of translation (+ in the direction of increasing measuring-unit values and – in the direction of decreasing measuring-unit values).

Notice that if you ignore the Y-axis elements of the 2-D translation, the situation degenerates to the basic 1-D translation equation (Equation 3-1). Likewise, you can ignore the X-axis elements and end up with a 1-D translation equation that works with the vertical Y axis. In a manner of speaking, then, a 1-D translation is simply a special case of a 2-D translation.

Example 6-1 demonstrates the general procedure for executing the translation of a point in 2-D space. I hope you can appreciate its simplicity and still understand its significance.

Example 6-1

A point in 2-D space happens to have a coordinate of (1,2). What is the coordinate of the translated version of that point if the translation terms are $T_x = 3$, $T_y = -5$? Plot the experiment.

1. Gathering the relevant data:

$$x = 1 \qquad y = 2 \qquad T_x = 3 \qquad T_y = -5$$

2. Applying Equation 6-1 to the situation:

$$x' = x + T_x \qquad y' = y + T_y$$
$$x' = 1 + 3 \qquad y' = 2 + (-5)$$
$$x' = 4 \qquad y' = -3$$

That translation thus carries P = (1,2) to P′ = (4, –3). See the result plotted in Fig. 6-2.

The example first cites a 2-D coordinate for a point in the chosen frame of reference. You should be able to sketch a 2-D coordinate system and plot that point with no difficulty at all. But then the example specifies a 2-dimensional translation. You know it is a 2-D translation by the fact that it spells out translation terms for two different axes—the x and y axes.

Step 1 then breaks the situation down into the individual components of the point's coordinate and summarizes the translation terms to be applied to those components. Step 2 simply restates the basic geometric translation equation of two dimensions (Equation 6-1) for the situation at hand, substitutes the values gleaned from the original information, and solves the equation. In this simple 1-point experiment, the translation displaces point P = (1,2) to point P′ = (4, −3).

Notice that the plot of that example (Fig. 6-2) does not show the translated coordinate system. It is sufficient, and generally less confusing, to show only the frame-of-reference coordinate system and the reference and translated version of the points involved.

Test your understanding of the procedure by working through Exercise 6-1. See if you can anticipate the translation effect in your mind's eye before you actually plot it.

Exercise 6-1

Each of the following experiments suggests the coordinate of a reference point P and a set of 2-dimensional translation terms T_x and T_y. Calculate the coordinate of the translated version of the point, P, and then plot the situation.

1. P = (1,2) $T_x = 1$ $T_y = 2$ P′ = (____, ____)
2. P = (−2,3) $T_x = 1$ $T_y = 1$ P′ = (____, ____)
3. P = (2,2) $T_x = -1$ $T_y = 2$ P′ = (____, ____)
4. P = (1,−1) $T_x = 0$ $T_y = 4$ P′ = (____, ____)
5. P = (2,3) $T_x = -2$ $T_y = -1$ P′ = (____, ____)
6. P = (−2,2) $T_x = 3$ $T_y = 0$ P′ = (____, ____)
7. P = (3,3) $T_x = 0$ $T_y = 0$ P′ = (____, ____)

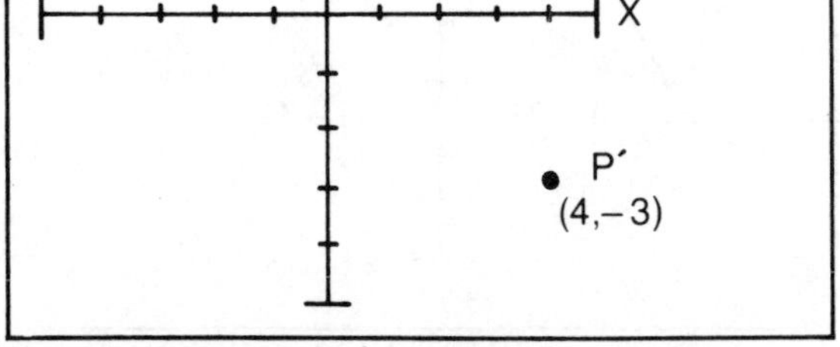

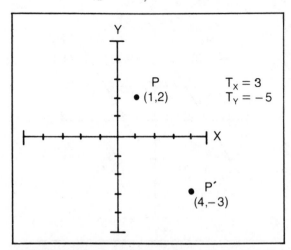

Fig. 6-2. The 2-D translation of a point from P = (1,2) to P′ = (4,−3). See Example 6-1.

6-3 ALTERNATIVE FORMS OF THE BASIC EQUATION

There are two alternative forms of the basic equation for translations in 2-D space. The first, Equation 6-2, represents a rearranged version that directly solves for the translation terms that exist between two coordinate systems, given the coordinates of a reference point and a translated version of it.

$$T_x = x_n' - x_n \qquad T_y = y_n' - y_n \qquad \textbf{Equation 6-2}$$

where:

T_x = the X-axis translation term

T_y = the Y-axis translation term

x_n',y_n' = the coordinate of the translated version of point *n* relative to the frame of reference

x_n,y_n = the coordinate of point *n* relative to the frame of reference

Check out Example 6-2.

Example 6-2

You have a situation where a reference point is located at coordinate P = (1,2), but you want to apply a 2-D translation such that the point turns up at coordinate P′ = (−3,−3). What are the necessary translation terms?

1. Gathering the components of the given coordinates:

$$x = 1 \qquad y = 2 \qquad x' = -3 \qquad y' = -3$$

2. Applying Equation 6-2:

$$T_x = x' - x \qquad T_y = y' - y$$
$$T_x = -3 - 1 \qquad T_y = -3 - 2$$
$$T_x = -4 \qquad T_y = -5$$

Thus a translation of $T_x = -4$, $T_y = -5$ translates P = (1,2) to P′ = (−3,−3).

The example cites two different points in 2-D space: a reference point P and a translated version of it, P′. The general idea is to determine the nature of the translation that separates those points.

Step 1 gathers the information in a form that is more meaningful for plugging the data into the appropriate equation; the equation in this case being Equation 6-2. The remaining work is rather straightforward, soon leading to a set of translation terms. You can check the results by applying those translation terms to the coordinate of the reference point via Equation 6-1. If the result is the coordinate of the given translated version of the point, you can be quite sure that you made no errors in arithmetic.

The alternative version in Equation 6-3 represents the inverse of the basic translation equation. You can use it to solve for the coordinate of a reference point, P, given the translation terms and the coordinate of the translated point, P′. Its real usefulness, however, rests with the fact that it leads directly to the relativistic forms of the equations for 2-D translations.

$$x_n = x_n' - T_x \qquad \textbf{Equation 6-3}$$
$$y_n = y_n' - T_y$$

where:

x_n, y_n = the coordinate of point *n* relative to the frame of reference

x_n', y_n' = the coordinate of the translated version of point *n* relative to the frame of reference

T_x = the X-axis translation term

T_y = the Y-axis translation term

Note: This is the inverse version of Equation 6-1

Use the following exercise to test your understanding of the basic notions of translations in 2-D space.

Exercise 6-2

Use Equations 6-1, 6-2 or 6-3 to complete the following lists of 2-D translation data:

1. $P = (2,2) \qquad T_x = 1$
2. $P = (2,-1) \qquad T_x = \_\_\_$
3. $P = (3,\_\_\_) \qquad T_x = -4$
4. $P = (\_\_\_, \_\_\_) \qquad T_x = 0$

$T_y = -1 \qquad P' = (\_\_\_, \_\_\_)$
$T_y = \_\_\_ \qquad P' = (0,0)$
$T_y = 0 \qquad P' = (\_\_\_, -1)$
$T_y = 2 \qquad P' = (0,0)$

6-4 TRANSLATING LINES IN 2-D SPACE

Lines in 2-D space can be defined in terms of their endpoint coordinates. Discussions in Chapter 3 show how to translate a line in 1-D space by translating the individual endpoints. The same idea holds here as well. Translating a line in 2-D space is a simple matter of applying the same translation terms to the endpoint coordinates of the reference line, and then constructing a straight line between the translated points. See the following example.

Example 6-3

Given the following specifications for a line in 2-D space, apply the prescribed 2-D translation and plot the results.

$$L = \overline{P_1P_2} \qquad P_1 = (-4,-2)$$
$$P_2 = (-1,3)$$
$$T_x = 5 \qquad T_y = 1$$

1. Gathering the data:

$$x_1 = -4 \quad y_1 = -2 \quad T_x = 5$$
$$x_2 = -1 \quad y_2 = 3 \quad T_y = 1$$

2. Applying Equation 6-1 to endpoint P_1:

$$x_1' = x_1 + T_x \qquad y_1' = y_1 + T_y$$
$$x_1' = =4 + 5 \qquad y_1' = -2 + 1$$
$$x_1' = 1 \qquad y_1' = -1$$

The translation transforms endpoint $P_1 = (-4,-2)$ to $P_i = (1,-1)$.

3. Applying Equation 6-1 to endpoint P_2:

$$x_2' = x_2 + T_x \qquad y_2 = y_2 + T_y$$
$$x_2' = -1 + 5 \qquad y_2' = 3 + 1$$
$$x_2' = 4 \qquad y_2' = 4$$

So the translation carries endpoint $P_2 = (-1,3)$ to $P_2' = (4,4)$

4. Gathering the specifications for the translated version of the line:

$$L' = \overline{P_1'P_2'} \qquad P_1' = (1, -1)$$
$$P_2' = (4,4)$$

See the experiment plotted in Fig. 6-3.

The example specifies a line in terms of its endpoints, and then it specifies the endpoints in terms of their 2-D coordinates. This is a typical P-L specification for a line that is situated in 2-D space. Since this is an experiment with translation effects in 2 dimensions, the original statement includes a set of translation terms as

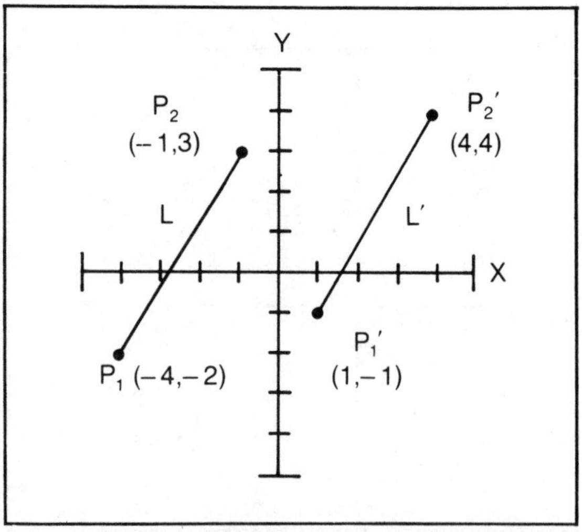

Fig. 6-3. The 2-D translation of a line L to line L′. See Example 6-3.

well. The idea is to apply those translation terms to the individual endpoints of that reference line, L, in order to generate the point coordinates for the translated version of that same line, L′.

Step 1 establishes the values of the significant variables in the experiment, and Step 2 uses Equation 6-1 to determine the coordinate of the translated version of point 1. The same equation, adjusted to show the variables for point 2, generates the coordinate for the translated version of that second endpoint. See Step 3. The example concludes with Step 4, specifying the translated version of the line in terms of its P′-L′ specifications.

One of the most significant applications of the translation equations is that of displacing a given line from one point to another. Suppose that you have the specifications for a line as specified for a certain position in 2-D space, but for the sake of conducting a particular experiment, you want to line situated elsewhere in that space. Example 6-4 deals with such a situation.

Example 6-4

A line in 2-D space has the following specifications:

$$L = \overline{P_1P_2} \qquad P_1 = (-3,5) \quad P_2 = (2,4)$$

What 2-D translation is necessary for adjusting the line in such a way that point P_2 is at (0,0) instead of (2,4)? What is the coordinate of point P_1 after making the adjustment?

1. Setting up the information provided:

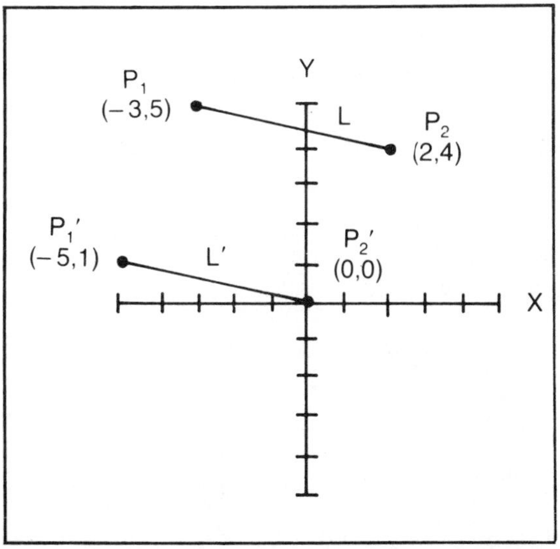

Fig. 6-4. Translations in 2-D space can be used for adjusting the position of line L to the position shown by line L′. See Example 6-4.

$$x_1 = -3 \quad y_1 = 5 \quad x_1' = ? \quad y_1' = ?$$
$$x_2 = 2 \quad y_2 = 4 \quad x_2' = 0 \quad y_2' = 0$$

2. Using the coordinates for point P_2 and P_2' as applied to Equation 6-2:

$$T_x = x_2' - x_2 \qquad T_y = y_2' - y_2$$
$$T_x = 0-2 \qquad T_y = 0-4$$
$$T_x = -2 \qquad T_y = -4$$

So a translation of T_x −2, T_y −4 executes the necessary transformation.

3. Using the calculated translation terms and applying them to the coordinate of P_1, Equation 6-1 shows that:

$$x_1' = x_1' + T_x \qquad y_1' = y_1 + T_y$$
$$x_1' = -3 + (-2) \qquad y_1' = 5 + (-4)$$
$$x_1' = -5 \qquad y_1' = 1$$

So the same translation that displaces point P_2 = (2,4) to P_2' = (0,0) will translate endpoint P_1 = (−3,5) to P_1' = (−5,1)

4. Summarizing the specifications for the translated version of the line:

$$L' = \overline{P_1'P_2'} \qquad P_1' = (-5,1) \quad P_2' = (0,0)$$

See the experiment as plotted in Fig. 6-4.

The purpose of the example is to fix the position of the specified line in such a way that point 2 has coordinate (0,0) instead of (2,4). You want to translate the line to that point without disturbing its length, direction cosines nor any angles it might make with other lines.

The general approach is to solve first for the translation terms that are necessary for moving the reference point to the coordinate position you want it to have. That requires an application of Equation 6-2 (see Step 2 in the example).

Once you know the translation that is necessary for translating the key point to its desired location, what remains is to apply the calculated translation terms to the remaining point. In a manner of speaking, you want that second endpoint to go along for the ride. See Step 3. With the endpoints thus translated to a desired point, all that remains is to respecify the line in its new position (Step 2). See a plot of the experiment in Fig. 6-4.

Specify some lines of your own choosing, pick a key endpoint, then calculate the translation terms required for placing that point elsewhere in the 2-D coordinate system. Further test your understanding of the translation of the lines in 2-D space by working through Exercise 6-3.

Exercise 6-3

1. Given the following specifications for a line in 2-D

space and a set of translation terms, calculate the endpoint coordinates of the translated version of the line and plot the results.

$L = \overline{P_1P_2}$ $P_1 = (3,3)$ $T_x = -4$
$P_2 = (-1,0)$ $T_y = 2$

2. Conduct the following experiment for the translation of a line in 2-D space. Plot both the reference and translated version of the line.

$L = \overline{P_1P_2}$ $P_1 = (-3,4)$ $T_x = 6$
$P_2 = (-3,4)$ $T_y = 0$

3. A line in 2-D space has the following specifications:

$L = \overline{P_1P_2}$ $P_1 = (-1,-1)$
$P_2 = (4,4)$

What 2-D translation is necessary for displacing point P_1 to coordinate $P_1' = (-2,1)$? What is the resulting coordinate of endpoint P_2? Plot the entire experiment.

6-4.1 Length is an Invariant

The length of a line in 2-D space is not affected by a simple translation in 2 dimensions. In other words, length (or distance) is an invariant under 2-D translations.

You can demonstrate that fact yourself by applying some translation to a line specified in 2-D space, calculating the lengths of both the reference and translated versions (Equation 5-1), and then comparing the two values. If you have carried out the experiment properly, you should find that the lengths are exactly the same. Try it.

A demonstration of that sort, no matter how convincing it might seem, is not adequate proof of the matter, however. You will find a formal proof of the principle of invariance of length, or distance, under 2-D translations shown here as Proof 6-1. Study the proof carefully, because you will be asked to follow its general outline for proving the same principle in higher dimensions. And besides, analytic proofs of this sort are part of the real substance of mathematics. The greater your ability to prove an idea, the better your chances of dealing with abstract matters in a responsible fashion.

Proof 6-1

1. The length of any line in 2-D space is given by:

$$s_n = \sqrt{(x_u - x_v)^2 + (y_u - y_v)^2} \quad \textbf{(Equation 5-1)}$$

2. A version of the 2-D length equation that applies to a line that is transformed in 2-D space is:

$$s_n' = \sqrt{(x_u' - x_v')^2 + (y_u' - y_v')^2}$$

3. From Equation 6-1:

$$x_u' = x_u + T_x \qquad y_u' = y_u + T_y$$
$$x_v' = x_v + T_x \qquad y_v' = y_v + T_y$$

4. Substituting the expressions in Step 3 into the equation in Step 3:

$$s_n' = \sqrt{[(x_u + T_x) - (x_v + T_x)]^2 + [(y_u + T_y) - (y_v + T_y)]^2}$$
$$s_n' = \sqrt{(x_u + T_x - x_v - T_x)^2 + (y_u + T_y - y_v - T_y)^2}$$
$$s_n' = \sqrt{(x_u - x_v)^2 + (y_u - y_v)^2}$$

The proposition that length is an invariant under 2-D translations is thus proven by $s_n' = s_n$

6-4.2 Direction Cosines are Invariant

Translations in two dimensions do not affect the length of a line, nor do they change the direction cosines. Recall from the previous chapter that direction cosines indicate the angular orientation of a line with respect to its coordinate system. Every line has a well defined set of direction cosines.

The purpose of this discussion is to show that the direction cosines, or angular orientation, of a line is not affected by simple translations in 2 dimensions. Look back at some of the previous examples and exercises that deal with the translation of lines in 2-D space. You will see that the reference and translated versions of the lines are always parallel to one another; the position of a line can be drastically changed by a 2-D translation, but its angle with respect to the coordinate axes is not.

Example 6-5 demonstrates the principle for one specific case.

Example 6-5

An experiment calls for translating the following line by $T_x = -1$, $T_y = 2$:

$L = \overline{P_1P_2}$ $P_1 = (-2,1)$
$P_2 = (4,-3)$

Perform the prescribed translation and calculate the direction cosines for both the reference and translated version. Compare the results.

1. Gathering the original data:

$x_1 = -2$ $y_1 = 1$ $T_x = -1$
$x_2 = 4$ $y_2 = -3$ $T_y = 2$

2. Applying Equation 6-1 to endpoint P_1:

$$x_1' = x_1 + T_x \qquad y_1' = y_1 + T_y$$
$$x_1' = -2 + (-1) \qquad y_1' = 1 + 2$$
$$x_1' = -3 \qquad y_1' = 3$$

Thus $P_1' = (-3,3)$

3. Applying Equation 6-1 to endpoint P_2:

$$x_2' = x_2 + T_x \qquad y_2' = y_2 + T_y$$
$$x_2' = 4 + (-1) \qquad y_2' = -3 + 2$$
$$x_2' = 3 \qquad y_2' = -1$$

Thus $P_2' = (3,-1)$

4. Specifying the translated version of the original line:

$$L' = \overline{P_1'P_2'} \qquad P_1' = (-3,3)$$
$$P_2' = (3,-1)$$

5. Using Equation 5-1 to calculate the length of the reference line, L:

$$s = \sqrt{(x_2-x_1)^2 + (y_2-y_1)^2}$$
$$s = \sqrt{[4-(-2)]^2 + (-3-1)^2}$$
$$s = \sqrt{6^2 + 4^2}$$
$$s = 52$$

6. Using Equation 5-1 to calculate the length of translated line L′:

$$s' = \sqrt{(x_2'-x_1')^2 + (y_2'-y_1')^2}$$
$$s' = \sqrt{[3-(-3)]^2 + (-1-3)^2}$$
$$s' = \sqrt{6^2 + 4^2}$$
$$s' = \sqrt{52}$$

7. Using Equation 5-2 to calculate the direction cosines for line L:

$$\cos\alpha = (x_2-x_1)/s \qquad \cos\beta = (y_2-y_1)/s$$
$$\cos\alpha = [4-(-2)]/\sqrt{52} \qquad \cos\beta = (-3-1)/\sqrt{52}$$
$$\cos\alpha = 3\sqrt{52}/26 \qquad \cos\beta = -\sqrt{52}/13$$

8. Using Equation 5-2 to calculate the direction cosines for translated line L′:

$$\cos\alpha' = (x_2'-x_1')/s' \qquad \cos\beta' = (y_2'-y_1')/s'$$
$$\cos\alpha' = [3-(-3)]/\sqrt{52} \qquad \cos\beta' = (-1-3)/\sqrt{52}$$
$$\cos\alpha' = 3\sqrt{52}/26 \qquad \cos\beta' = -\sqrt{52}/13$$

A comparison of the results of Steps 7 and 8 shows that the direction cosines are identical for L and L′.

The experiment begins by specifying a particular line in 2-D space and then applying a certain translation to its endpoints. Applying the same translation terms to the two endpoints of a line amounts to the same thing as translating the line, itself. That part of the experiment

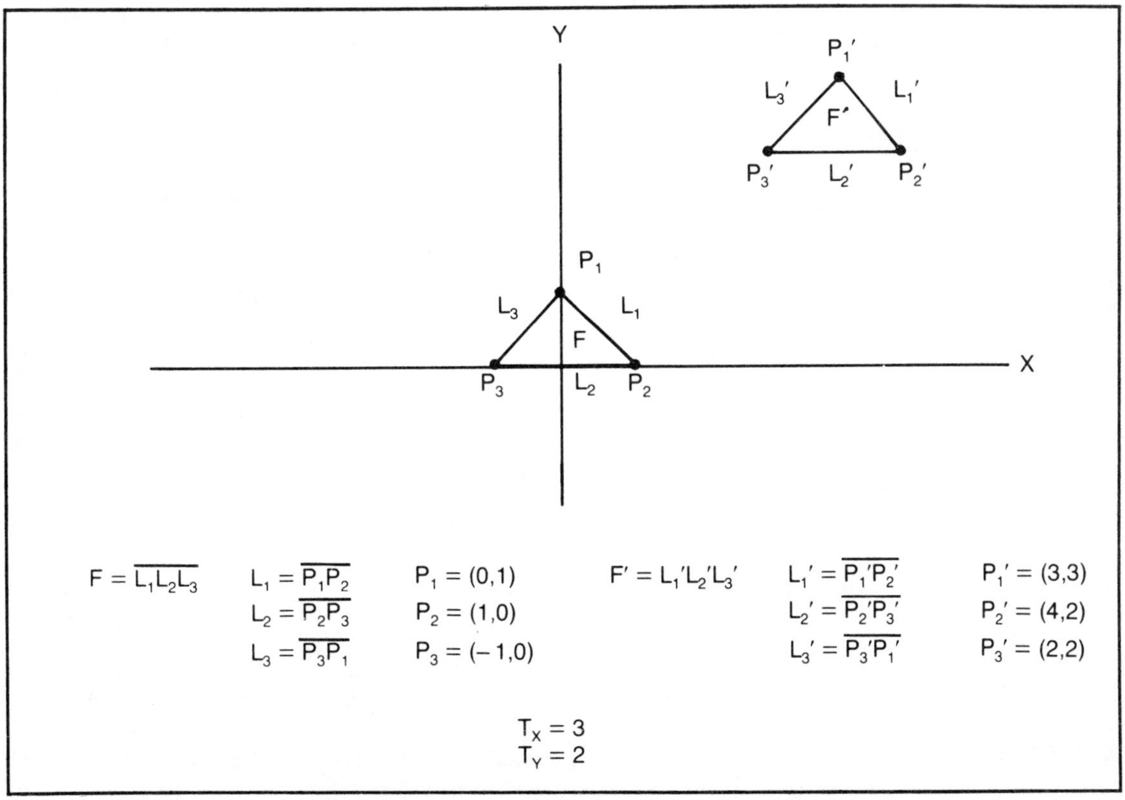

Fig. 6-5. The 2-D translation of a plane figure, F to F′. See Example 6-6.

concludes with the specification of the translated version of the line. See Steps 1 through 4.

Step 5 then calculates the length of the reference line and Step 6 uses the same method to find the length of the translated version of that same line. The fact that the two lines have the same length should come as no big surprise—you saw the proof of the invariance of length in the previous section of this chapter.

Step 7 calls upon Equation 5-2 to calculate the direction cosines of the reference line. The data is, of course, based upon that supplied originally for that line. Then Step 8 works out the direction cosines for the translated version of the line. Don't let the notation puzzle you; I have simply adjusted it to use the primed notation that suits the notation convention we've been using all along for designating transformed points in an experiment.

The results of Steps 7 and 8 clearly demonstrate that the direction cosines of the lines are identical. But again, a single demonstration (or a thousand different ones, for that matter) do not constitute a proof. A formal, analytic proof is in order.

Proof 6-2

1. The general expressions for the direction cosines of any line in 2-D space are:

$$\cos\alpha_n = (x_u - x_v)/s_n$$
$$\cos\beta_n = (y_u - y_v)/s_n$$

2. Adjusting the notation in Step 1 to account for any transformed line in 2-D space:

$$\cos\alpha_n' = (x_u' - x_v')/s_n'$$
$$\cos\beta_n' = (y_u' - y_v')/s_n'$$

3. From Equation 6-1:

$$x_u' = x_u + T_x \qquad y_u' = y_u + T_y$$
$$x_v' = x_v + T_x \qquad y_v' = y_u + T_y$$

4. Substituting the expressions in Step 3 into the equations in Step 2:

$$\cos\alpha_n' = [(x_u + T_x) - (x_v + T_x)]/s_n' = (x_u - x_v)/s_n'$$
$$\cos\beta_n' = [(y_u + T_y) - (y_v + T_y)]/s' = (y_u - y_v)/s_n'$$

5. We have already proven that $s_n' = s_n$ (Proof 6-1), so substituting s_n for s_n' in Step 4:

$$\cos\alpha_n' = (x_u - x_v)/s_n$$
$$\cos\beta_n' = (y_u - y_v)/s_n$$

Clearly, $\cos\alpha_n' = \cos\alpha_n$ and $\cos\beta_n' = \cos\beta_n$. Thus the direction cosines of a line in 2-D space are invariants under 2-D translations.

The goal of the proof is to show that the direction cosines of any reference line are equal to the corresponding direction cosines of the translated version of it—under any possible translation of 2 dimensions.

Step 1 in the proof merely restates the basic equation for the direction cosines of a line oriented in 2-D space; that is Equation 5-2. Step 2 then states the same equation, but uses the notation that is appropriate for a transformed version of the line. If you understand the real meaning of equations, as opposed to simply using them as guides for cranking out specific values for specific experiments, you ought to have no difficulty with this matter of adjusting the notation to suit special needs.

Step 2 is a restatement of the basic translation equation for a line in 2-D space. Don't be put off by the notation; I am simply substituting very general labels, u and v, for the specific labels 1 and 2 that have been used in specific examples and experiments in the past.

Step 4 substitutes the expressions from Step 3 into the equations in Step 2, and then reduces them by means of elementary algebra. Knowing in advance that the lengths of the two lines are necessarily equal under a 2-D translation (Proof 6-1), matters proceed to the results in Step 6.

Finally, comparing the results in Step 6 with the expressions in Step 1, you should be able to see that the direction cosines are equal. The proposition is thus proven, and you can bet from this time forward that no simple translation is going to affect the direction cosines of a line in 2-D space.

6-4.3 Angle Between Lines is Invariant

Finally, it is safe to say that the angle between two lines in 2-D space is not affected by a translation of 2 dimensions. A descriptive proof of the principle is adequate in this instance.

Equation 5-3, the main equation for calculating the cosine of the angle between a pair of lines in 2-D space, clearly shows that the angle depends only upon the values of the lines' direction cosines. Proof 6-2 in the previous section of this chapter showed that the direction cosines of lines are not affected by a translation, therefore it is safe to conclude that the angle between the lines is not affected.

That, incidentally, is a descriptive proof, as opposed to analytic proofs offered earlier as Proofs 6-1 and 6-2. Descriptive proofs are generally less technical and often less troublesome than their analytic counterparts; but they have to be used with great care. It is often too easy to toss around a few descriptive proofs when, in actuality, an experimenter is merely hoping to avoid some careful analytic work. An irresponsible application of descriptive proofs can lead to a lot of misinterpretations and outright errors in logic. This one happens to be well-founded. Conduct a formal analytic proof if you have the slightest doubt in your mind that the angle between two lines is an invariant under 2-D translations.

6-4.4 Using the Principles of Invariance

In the foregoing sections, you have seen some proofs of the following principles:

- ☐ The length of a line is invariant under a 2-D translation.
- ☐ The direction cosines of a line is an invariant under 2-D translations.
- ☐ The angle between two lines is an invariant under 2-D translations.

What practical use does that knowledge serve? Well, it can serve at least two purposes. First, it simplifies the analysis of a translation situation in 2-D space. If you have calculated the lengths, direction cosines and angles between reference lines, there is no real need to run through all of the calculations for analyzing the same values for translated versions of those lines. A 2-D translation affects only the positions of points—nothing else.

A wise experimenter will use the principles in a slightly different way, though. Suppose that you have just conducted a simple translation of some lines, and then you want to calculate the lengths, direction cosines and angles between lines for both the reference and translated versions. How can you be certain that you've made no errors in those calculations? One good way is to compare the results. The lengths, direction cosines and angles between pairs of lines are not supposed to be affected by a translation. If, upon comparing the values, you find that some differences exist, you can bet that you made an error in arithmetic somewhere along the way. Knowing that you have made an error is just as important as finding and remedying it.

6-5 TRANSLATING PLANE FIGURES IN 2-D SPACE

Just as a line can be translated by translating its individual endpoints, so can a plane figure be translated by translating its points. All of the principles that apply to translating lines apply equally well in the plane-figure case: the invariance of the lengths of lines, direction cosines and the angle between lines. Nothing changes but the position of the figure relative to the frame of reference.

Check out these ideas by studying the following example and the complete analysis of the experiment in Tables 6-1 and 6-2.

Example 6-6

Apply a 2-D translation of $T_x = 3$, $T_y = 2$ to the following plane figure specified in 2-D space:

$$F = \overline{L_1L_2L_3} \qquad L_1 = \overline{P_1P_2} \qquad P_1 = (0,1)$$
$$L_2 = \overline{P_2P_3} \qquad P_2 = (1,0)$$
$$L_3 = \overline{P_3P_1} \qquad P_3 = (-1,0)$$

1. Gathering the relevant data:

$$x_1 = 0 \quad y_1 = 1 \quad T_x = 3$$
$$x_2 = 1 \quad y_2 = 0 \quad T_y = 2$$
$$x_3 = -1 \quad y_3 = 0$$

2. Apply the 2-D translation equation, Equation 6-1, to point P_1:

$$x_1' = x_1 + T_x \qquad y_1' = y_1 + T_y$$
$$x_1' = 0 + 3 \qquad y_1' = 1 + 2$$
$$x_1' = 3 \qquad y_1' = 3$$

Thus $P_1 = (0,1)$ translates to $P_1' = (3,3)$

3. Applying Equation 6-1 to point P_2:

Table 6-1. Analysis of the Reference Triangle in Fig. 6-5.

(A, B, C, and D correspond to α, β, γ, and δ.)

SPECIFICATIONS

$F=\overline{L_1L_2L_3}$	$L_1=\overline{P_1P_2}$	$P_1=(0,1)$
	$L_2=\overline{P_2P_3}$	$P_2=(1,0)$
	$L_3=\overline{P_3P_1}$	$P_3=(-1,0)$

LENGTHS OF LINES

$s_1=\sqrt{2}$	$s_2=2$	$s_3=\sqrt{2}$

DIRECTION COSINES

$\cos A_1=\sqrt{2}/2$	$\cos B_1=-\sqrt{2}/2$
$\cos A_2=-1$	$\cos B_2=0$
$\cos A_3=\sqrt{2}/2$	$\cos B_3=\sqrt{2}/2$

ANGLES

$\phi_{1,2}=45^\circ$	$\phi_{1,3}=90^\circ$	$\phi_{2,3}=45^\circ$

Table 6-2. Analysis of the Translated Version of the Triangle in Fig. 6-5.

(A, B, C, and D correspond to α, β, γ, and δ.)

SPECIFICATIONS

$F' = \overline{L_1'L_2'L_3'}$	$L_1' = \overline{P_1'P_2'}$	$P_1' = (3,3)$
	$L_2' = \overline{P_2'P_3'}$	$P_2' = (4,2)$
	$L_3' = \overline{P_3'P_1'}$	$P_3' = (2,2)$

LENGTHS OF LINES

$s_1' = \sqrt{2}$	$s_2' = 2$	$s_3' = \sqrt{2}$

DIRECTION COSINES

$\cos A_1' = \sqrt{2}/2$	$\cos B_1' = -\sqrt{2}/2$
$\cos A_2' = -1$	$\cos B_2' = 0$
$\cos A_3' = \sqrt{2}/2$	$\cos B_3' = \sqrt{2}/2$

ANGLES

$\phi_{1,2}' = 45^\circ$	$\phi_{1,3}' = 90^\circ$	$\phi_{2,3}' = 45^\circ$

$$x_2' = x_2 + T_x \qquad y_2' = y_2 + T_y$$
$$x_2' = 1 + 3 \qquad y_2' = 0+2$$
$$x_2' = 4 \qquad y_2' = 2$$

So the given translation displaces $P_2 = (1,0)$ to $P_2' = (4,2)$

4. Applying the translation to point P_3:

$$x_3' = x_3 + T_x \qquad y_3' = y_3 + T_y$$
$$x_3' = -1 + 3 \qquad y_3' = 0 + 2$$
$$x_3' = 2 \qquad y_3' = 2$$

And point $P_3 = (-1,0)$ is translated to $P_3' = (2,2)$

5. Composing the specifications for the translated version of the reference plane figure:

$F' = \overline{L_1'L_2'L_3'}$	$L_1' = \overline{P_1'P_2'}$	$P_1' = (3,3)$
	$L_2' = \overline{P_2'P_3'}$	$P_2' = (4,2)$
	$L_3' = \overline{P_3'P_1'}$	$P_3' = (2,2)$

See a plot of the translation experiment in Fig. 6-5.

The example cites a triangular plane figure in 2-D space; that is a fact that you can discern by studying the P-L-F specifications. (The F specifications, for instance, cite three bounding lines.) After gathering the relevant information in Step 1, the project goes on to calculate the coordinates of the translated version of the three points. See Steps 2, 3 and 4. Step 5 summarizes the specifications for the translated version of the line and points to Fig. 6-5 as the overall graphic representation of the experiment.

Tables 6-1 and 6-2 represent a complete and formal analysis of the experiment. Table 6-1 shows the analysis of the reference line, while Table 6-2 shows the analysis of the translated version of that line. Notice, especially that the two analyses are virtually identical except for the coordinates of the points and the notation. The coordinates of the points are different because the experiment represents a translation in 2-D space; unless the project happens to feature translation terms of zero, the coordinates of the points are bound to change. The notation is different, too; but only to separate the reference version of the line from its translated version.

The invariant qualities—lengths of the lines, the direction cosines of the lines, and the angles between the lines—are clearly identical in both analyses.

Try your hand at translating plane figures in 2-D space and analyzing the results by working through this exercise.

Exercise 6-4

Use the following specifications for the reference figure in all of the experiments cited here:

$F = \overline{L_1L_2L_3L_4}$	$L_1 = \overline{P_1P_2}$	$P_1 = (-1,-1)$
	$L_2 = \overline{P_2P_3}$	$P_2 = (1,-1)$
	$L_3 = \overline{P_3P_4}$	$P_3 = (1,1)$
	$L_4 = \overline{P_4P_1}$	$P_4 = (1,-1)$

1. Plot the reference figure and conduct a complete analysis of it. Include the lengths of all four lines, the direction cosines for all four lines, and the relevant angles in the figure.

2. Apply the following translation to the reference figure:

$$T_x = 2 \qquad T_y = 2$$

Conduct a complete analysis of the result.

3. Apply the following translation to the reference figure:

$$T_x = 1 \qquad T_y = -1$$

4. What translation must be applied to the reference figure in order to displace point P_2 to the origin of its

coordinate system? (Hint: Use Equation 6-2.) Cite the complete specifications for the figure after determining and carrying out that translation.

6-6 RELATIVISTIC TRANSLATIONS IN 2-D SPACE

The fundamental geometric translations of points in 2-D space is embodied in Equation 6-1. The equation is most useful for displacing points, lines and plane figures in 2-D spaces. All of the terms in the equation—the components of the reference coordinate, the translation terms, and the components of the translated version of the point—refer directly to the chosen frame of reference. From one useful point of view, translating a coordinate system is little more than a vehicle for moving points from one place to another in a systematic fashion.

Another way to view a geometric translation is by regarding the scheme two coordinate systems separated by some translation. That view is especially meaningful for the relativistic forms of 2-D translations. The general idea is to set up two coordinate frames of reference that differ from one another only by a displacement caused by an existing translation—one selected to be the observer's frame of reference and the other to be the translated frame of reference. The relativistic versions of the translation equations then deal with the coordinates of point as they appear from those two coordinate systems.

Equation 6-4, for example, expresses the components of the coordinates of some point in 2-D space relative to the translated frame of reference, given the coordinate of that same point relative to the observer's frame of reference and some translation terms. The translation terms in this instance represent the translation that exists between the two frames of reference, relative to the observer's.

$$x_n^* = x_n - T_x \qquad y_n^* = y_n - T_y \qquad \textbf{Equation 6-4}$$

where:

x_n^*, y_n^* = the coordinate of reference point *n* relative to the translated frame of reference

x_n, y_n = the coordinate of reference point *n* relative to the observer's frame of reference

T_x = The X-axis translation term relative to the observer's frame of reference

T_y = the Y-axis translation term relative to the observer's frame of reference

Maybe that is a lot to take in at once. In essence, the scheme involves two coordinate systems and a single point in 2-D space. The coordinate systems are displaced by the translation terms, relative to the system chosen to be the observer's frame of reference. The second coordinate system is then considered to be the translated coordinate frame of reference. The one point that is presumably observed simultaneously from both frames of reference. The coordinate of that one point, relative to the observer's frame of reference is given by P. The coordinate of that same point, but taken relative to the translated frame of reference is point P*. The idea applies equally well to lines and plane figures. Those notions simply call for calculating the relativistic coordinates for a greater number of points. See Example 6-7.

Example 6-7

You and a companion are standing in a room looking at an interesting pattern of triangles drawn on the floor. From your point of view (the observer's frame of reference), you see one particular triangle as having the following specifications:

$$F = \overline{L_1L_2L_3} \qquad L_1 = \overline{P_1P_2} \qquad P_1 = (-1,-1)$$
$$L_2 = \overline{P_2P_3} \qquad P_2 = (0,1)$$
$$L_3 = \overline{P_3P_1} \qquad P_3 = (1,-1)$$

Your companion is observing the same triangle, but at a slightly different position in the room. You figure that your companion's position (the translated frame of reference) is characterized by a displacement of $T_x = 4$, $T_y = -2$—perhaps you can say that he or she is 4 steps to your right and 2 steps behind. At any rate, determine the specifications for that triangle as viewed from your companion's position in the room.

1. Gathering the given information:

$$x_1 = -1 \qquad y_1 = -1 \qquad T_x = 4$$
$$x_2 = 0 \qquad y_2 = 1 \qquad T_y = -2$$
$$x_3 = 1 \qquad y_3 = -1$$

2. Applying Equation 6-4 to your point P_1:

$$x_1^* = x_1 - T_x \qquad y_1^* = y_1 - T_y$$
$$x_1^* = -1-4 \qquad y_1^* = -1-(-2)$$
$$x_1^* = -5 \qquad y_1^* = 1$$

The point that you see as $P_1 = (-1,-1)$ appears as $P_1^* = (-5,1)$ from your companion's point of view.

3. Applying Equation 6-4 to your point P_2:

$$x_2^* = x_2 - T_x \qquad y_2^* = y_2 - T_y$$
$$x_2^* = 0-4 \qquad y_2^* = 1-(-2)$$
$$x_2^* = -4 \qquad y_2^* = 3$$

The point you see as $P_2 = (0,1)$ appears as $P_2^* = (-4,3)$ to your companion.

4. Applying Equation 6-4 to your point P_3:

$$x_3^* = x_3 - T_x \qquad y_3^* = y_3 - T_y$$
$$x_3^* = 1-4 \qquad y_3^* = -1-(-2)$$
$$x_3^* = -3 \qquad y_3^* = 1$$

So the one that you see as $P_3 = (1,-1)$ appears at $P_3^* = (-3,1)$.

5. Constructing the specifications of the figure as reckoned from the translated frame of reference:

$$F = L_1^*L_2^*L_3^* \quad L_1^* = \overline{P_1^*P_2^*} \quad P_1^* = (-5,1)$$
$$L_2^* = \overline{P_2^*P_3^*} \quad P_2^* = (-4,3)$$
$$L_3^* = \overline{P_3^*P_1^*} \quad P_3^* = (-3,1)$$

6. Doublecheck the results by calculating the lengths of all three lines as viewed from both frames of reference. If the corresponding lines are equal, you can be reasonably certain that the arithmetic for the translation is correct.

For the lines as viewed from the observer's frame of reference: $s_1 = \sqrt{5} \quad s_2 = \sqrt{5} \quad s_3 = 2$

For the lines as viewed from the translated frame of reference: $s_1^* = \sqrt{5} \quad s_2^* = \sqrt{5} \quad s_3^* = 2$

Since the lengths of corresponding lines are equal, the translation is most likely free of errors.

That is an example of an application of *classical, or Galilean, relativity*. There is really nothing new here in the context of traditional physics, and it should not be confused in any way with the terminology of Einstein's modern relativity. That modern approach to experiments in four dimensions must be set aside for another book in the future. Nevertheless, this is a legitimate form of relativity at work, and I hope you can appreciate how intriguing it can be. You see, equations, in themselves, tend to be rather sterile and uninteresting; it is the interpretation of those equations and the situations that take advantage of them that stirs real enthusiasm and challenges one's sense of creativity and imagination.

Study the statement of the experiment in that example very carefully, making sure you can picture the situation in your mind. Then study the steps to see how to handle the experiment in a rational way. I encourage you to take this seriously, because the time is coming when you won't be able to picture the situation at all in your mind—common sense and ordinary experience will not have any meaning at all. Test your understanding by making up some relativistic situations of your own.

Relativistic translations become particularly interesting when regarding them as a means for determining the translating that exists between two coordinate systems in 2-D space. Suppose that there are two coordinate systems in 2-D space that are identical in all respects but one—they are displaced from one another. You, the observer living in the observer's frame of reference, do not know the translation terms for the displacement between the two coordinate systems, but you have the ability to determine the coordinate of some point in space and, equally important, find out the coordinate of that same point from the view of someone else living in the displaced coordinate system. Given that information, Equation 6-5 lets you discover the translation that exists between your coordinate system and the other one.

$$T_x = x_n - x_n^* \qquad T_y = y_n - y_n^* \qquad \textbf{Equation 6-5}$$

where:

T_x = The X-axis translation term that exists between an observer's and translated frame of reference, relative to the observer's frame of reference.

T_y = the Y-axis translation term that exists between an observer's and translated frame of reference, relative to the observer's frame of reference

x_n,y_n = the coordinate of point *n* relative to the observer's frame of reference

x_n^*,y_n^* = the coordinate of point *n* relative to the translated frame of reference

The next example illustrates the use of that relativistic translation equation.

Example 6-8

You, an observer in 2-dimensional Flatland, find that a star has coordinate (4,–2.2). An observer elsewhere in Flatland observes the same star at coordinate (5.5, 1.8). What translation characterizes the other's viewing position relative to your own?

1. Sorting out the data and putting it into a meaningful form:

$x = 4 \quad y = -2.2 \quad x^* = 5.5 \quad y^* = 1.8$

2. Using Equation 6-5:

$$T_x = x - x^* \qquad T_y = y - y^*$$
$$T_x = 4 - 5.5 \qquad T_y = -2.2 - 1.8$$
$$T_x = -1.5 \qquad T_y = -4$$

Using a conventional arrangement of 2-D coordinates, the other observer must be located 1.5 measuring units to your left and 4 units behind you.

The inverse form of the relativistic translation equation for 2-D space is shown here as Equation 6-6. In its most literal sense, it says that you, an observer, can determine the coordinate of some point in 2-D space, given the coordinate as reckoned from a translated frame of reference and the translation that exists between the two frames of reference.

$$x_n = T_x + x_n^* \qquad y_n = T_y + y_n^* \qquad \textbf{Equation 6-6}$$

where:

x_p,y_p = the coordinate of point *n* relative to the observer's frame of reference

x_n^*, y_n^* = the coordinate of point *n* relative to the translated frame of reference

T_x and T_y = the X- and Y-axis translation terms that exist between the observer's and translated frames of reference, relative to the observer's frame of reference

Note: This is the inverse form of Equation 6-4.

6-7 MULTIPLE TRANSLATIONS IN 2-D SPACE

Translations in 2-D space, like those in 1 dimension, are perfectly commutative. Any prescribed set of translations result in the same overall translation, regardless of the order in which they are applied. You can, for example, apply three or four X-axis translations and then a corresponding number of Y-axis translation, and end up with the same coordinate points you would get by applying X- and Y-axis translations alternately. Try it for yourself, just making certain that the families of translations are the same.

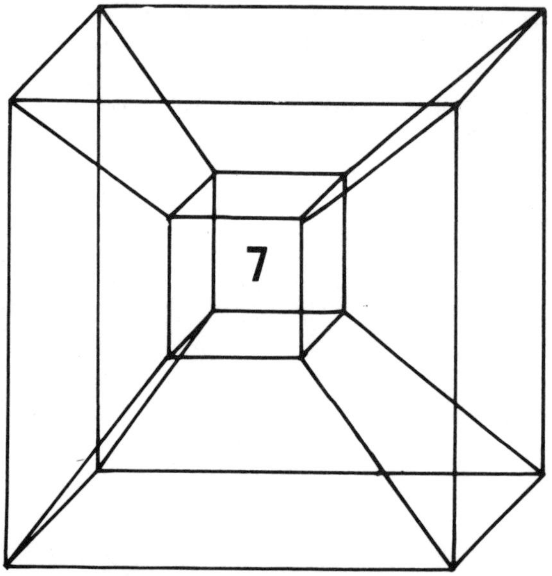

Scaling in 2-D Space

When I was a youngster, I was taken with the idea of making poster-sized reproductions of some of my favorite comic book characters. I managed the job by drawing relatively small grid of horizontal and vertical lines on the picture I wanted to enlarge, and then drawing a larger, but otherwise identical, grid on a sheet of poster board. That was just the start. The next step was to copy the figure I wanted to reproduce from the smaller to the larger gridwork, attempting to make proportionally larger versions of the picture enclosed in each grid block. See an example in Fig. 7-1.

These pictures are examples of scaling transformation in 2-D space. The axes are somewhat different from the convention established for most of the experiments in this book, but the main principles of 2-D scaling still show through.

There is, for example, a need for establishing a frame-of-reference coordinate system. In Fig. 7-1, the scaling is regarded as an expansion-type scaling if system A is taken as the frame of reference. But if system B is taken as the frame of reference, it follows that the scaling to system A is a reduction-type scaling.

The scalings are also linear. Looking at just one picture or the other, you can see that the spacing between the grid lines is everywhere the same; the horizontal grids are equally spaced, and the vertical grid lines are equally spaced. Scaling transformations, as described throughout this book, have that special linear quality.

It is also possible to perform reflection-type scalings in two dimensions. Compare system A in Fig. 7-1 with the three different reflections of it in Fig. 7-2.

Finally, this introduction to 2-D scalings brings up the possibility of scaling the horizontal and vertical axes by different amounts. See Fig. 7-3. The previous examples, whether they are reflections or not, represent symmetrical scalings in 2-D space. That is, the amounts of horizontal and vertical scaling are equal. But it is possible to scale them differently, carrying out something called an asymmetrical scaling in two dimensions.

Using system A in Fig. 7-1 as the frame of reference, you can see that Fig. 7-3A is expanded in the horizontal direction only. On the other hand, you can expand the system in the vertical direction as illustrated in Fig. 7-3B.

These represent just a few possible scalings in two dimensions. You can, for instance, reflect and change the size of the original, doing something such as reflecting it horizontally and stretching it vertically. Or you can stretch and reflect one dimension at the same time. There

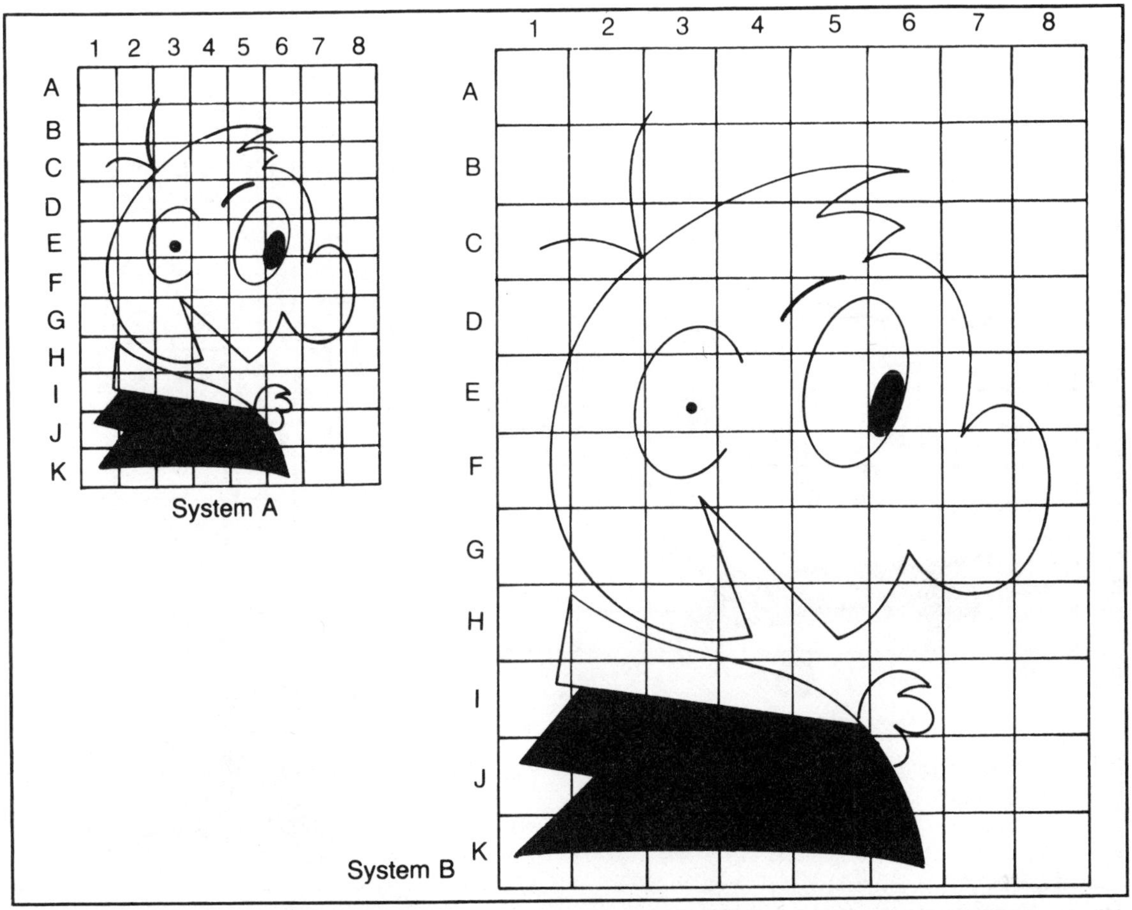

Fig. 7-1. Scaling a 2-dimensional drawing.

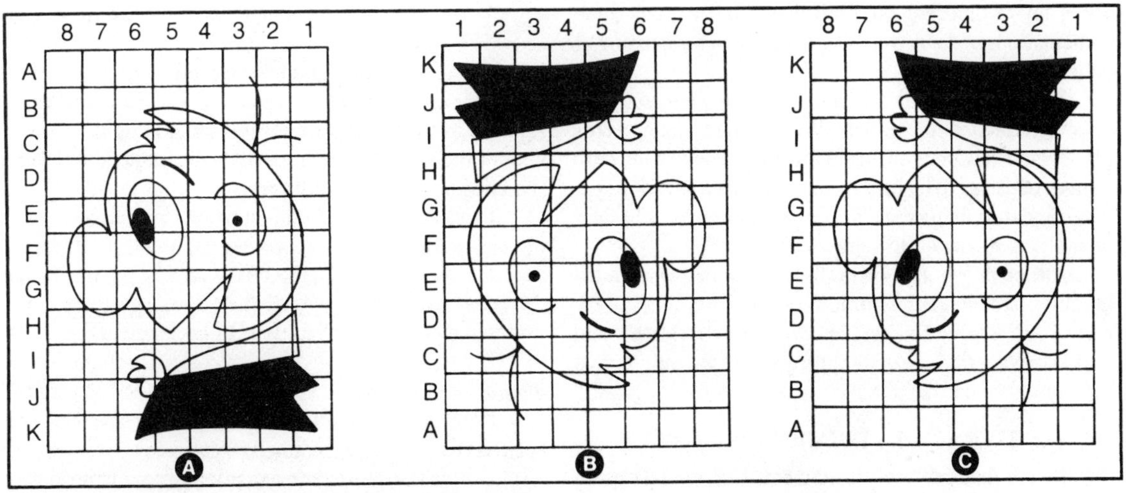

Fig. 7-2. Reflection-type scalings of a 2-D drawing.

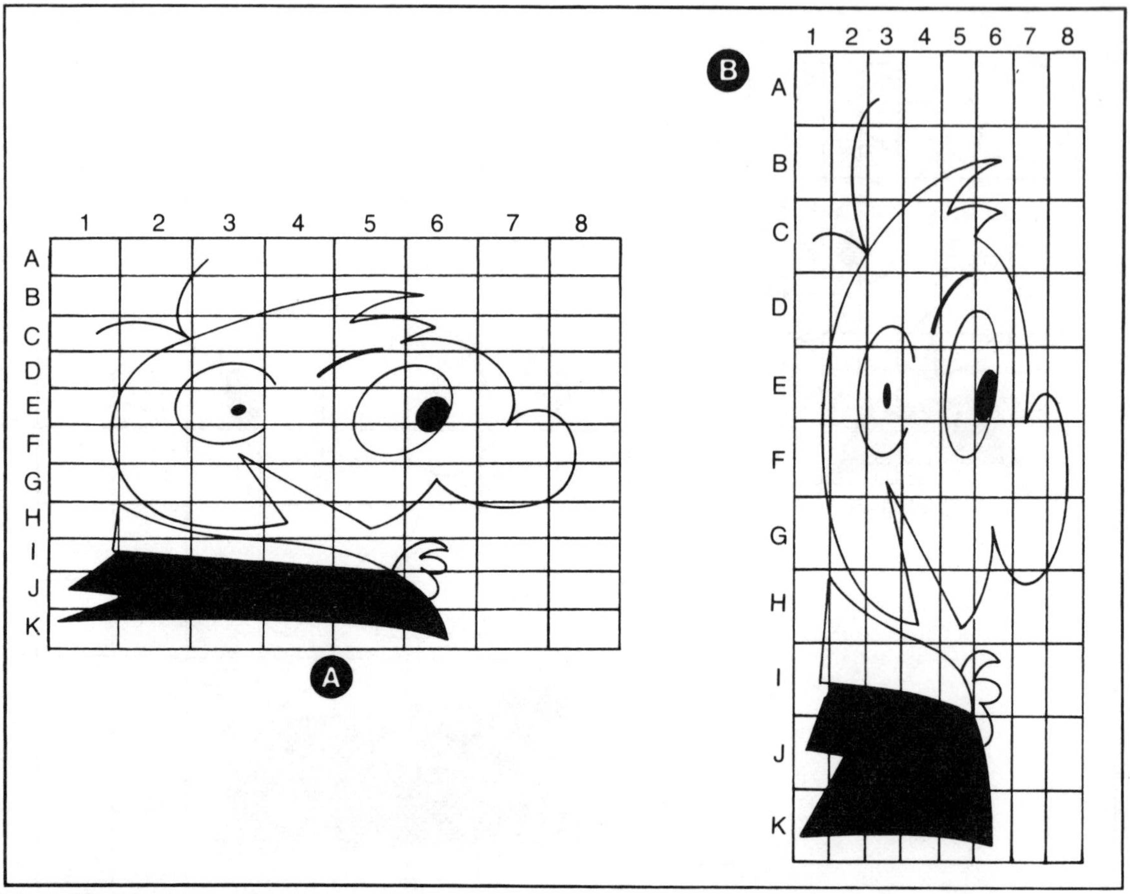

Fig. 7-3. Asymmetrical scalings of a 2-D drawing.

are a lot of different possibilities. Why not try some combinations of your own? The results can be amusing as well as enlightening.

What kind of 2-D scaling takes place when you look at your reflection in a mirror? What kind of scaling takes place when you focus an image through a convex and onto a sheet of paper? What is the conventional frame of reference for determining the difference between left and right on a standard anatomical drawing?

Although we live in a 3-dimensional world, we quickly become accustomed to dealing with 2-dimensional things; things that are often reflected and scaled in ways that are quite interesting if we care to give the matter some thought.

7-1 THE BASIC SCALING EQUATION FOR 2-D SPACE

Recall that the basic scaling equation for 1-D space has the general form:

$$x' = K_x x_n \qquad \textbf{(Equation 4-1)}$$

where x_n is the coordinate of a reference point *n* relative to the chosen frame of reference, K_x is the scaling factor relative to the frame of reference, and x_n' is the coordinate of the scaled version of the point as reckoned from the chosen frame of reference.

The 2-D version of that equation simply adds a scaling factor for the Y axis. See Equation 7-1.

$$x_n' = x_n K_x$$
$$y_n' = y_n K_y \qquad \textbf{Equation 7-1}$$

where:

x_n', y_n' = the coordinate of a scaled version of point *n* relative to the chosen frame of reference

x_n, y_n = the coordinate of a reference point *n* relative to the frame of reference

K_x = the X-axis scaling factor relative to the frame of reference

K_y = the Y-axis scaling factor relative to the frame of reference

Given the 2-D coordinate of a point in 2-D space, the general idea is to multiply the components of that coordinate by their respective scaling factors. Multiply the X component of the coordinate by the K_x factor, and the Y component by the K_y factor. Check out the following example.

Example 7-1

Two different reference points in 2-D space have the following coordinates:

$$P_1 = (2,1) \qquad P_2 = (-1,-0.5)$$

Determine the coordinates of their scaled versions, P_1' and P_2', where the scaling factors are $K_x = 2$ and $K_y = 2$.

1. Gathering the relevant information:

$$x_1 = 2 \qquad y_1 = 1 \qquad K_x = 2$$
$$x_2 = -1 \qquad y_2 = -0.5 \qquad K_y = 2$$

2. Applying Equation 7-1 to the coordinate and scaling factors for point P_1:

$$x_1' = x_1K_x \qquad y_1' = Y_1K_y$$
$$x_1' = 2(2) \qquad y_1' = 1(2)$$
$$x_1' = 4 \qquad y_1' = 2$$

So the scaling transforms $P_1 = (2,1)$ to $P_1' = (4,2)$

3. Applying Equation 7-1 to the coordinate and scaling factors for point P_2:

$$x_2' = x_2K_x \qquad y_2' = y_2K_y$$
$$x_2' = -1(2) \qquad y_2' = (-0.5)\,(2)$$
$$x_2' = -2 \qquad y_2' = -1$$

The scaling transforms $P_2 = (1-0.5)$ to $P_2' = (-2,-1)$. See the project plotted in Fig. 7-4.

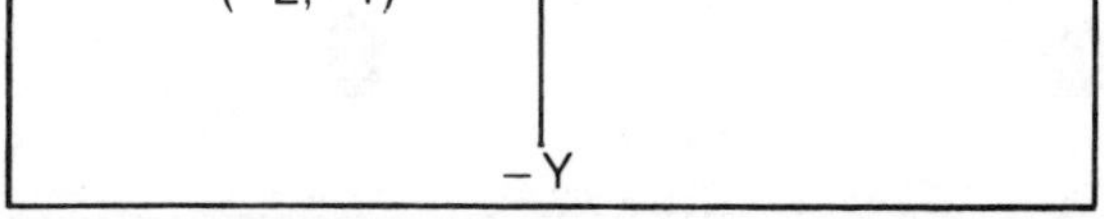

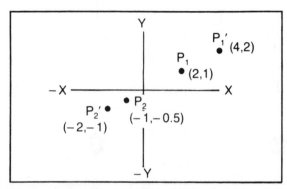

Fig. 7-4. Scaling two reference points in 2-D space. See Example 7-1.

The example cites two different points in 2-D space, and asks you to scale the coordinate system. The objective is to determine the coordinates of the scaled versions of the points, relative to the chosen frame of reference.

It is a rather straightforward project. First break down the information into a notational form that suits the scaling equation, and then apply the scaling factors to the X and Y components of the reference points. The results are shown at the conclusion of Steps 2 and 3. Now you try it with these exercises.

Exercise 7-1

Use Equation 7-1 to complete the following series of data:

1. $P = (1,2) \quad K_x = 2 \quad K_y = 2$
   P' = (_____,_____)
2. $P = (-2,1) \quad K_x = 0.5 \quad K_y = 0.5$
   P' = (_____,_____)
3. $P = (0.5, 1) \quad K_x = -1 \quad K_y = -1$
   P' = (_____,_____)
4. $P = (-2,-4) \quad K_x = -0.5 \quad K_y = 0.5$
   P' = (_____,_____)

7-2 SOME OVERALL SCALING EFFECTS

You have already seen some scaling effects in the figures at the beginning of this chapter; but now having an opportunity to see the basic scaling equation for 2-D space, it is time to put the ideas onto firmer analytic grounds.

Recall the following effects described in Chapter 4 generated by changing around the value and sign of a scaling factor:

☐ A scaling factor greater than 1 increases the scaling.

☐ A scaling factor equal to 1 changes nothing.

☐ A scaling factor that is greater than zero but less than one reduces the scaling.

☐ A scaling factor equal to zero reduces the axis to a 0-dimensional point.

☐ A scaling factor less than zero but greater than −1 both reduces the scaling and reflects the axis about the origin.

☐ A scaling factor equal to −1 causes a reflection about the origin.

☐ A scaling factor less than −1 both reduces the scaling and reflects the axis about the origin.

Those same principles hold for scalings in 2-D space, applying equally well to both the X- and Y-axis scalings. If the X- and Y-axis scaling factors happen to be equal, the scaling is considered a symmetrical one. If the scaling factors are not equal, the transformation is an asymmetrical one.

Figure 7-5 illustrates some combinations of reflection-type scalings in 2-D space. Figure 7-5A is taken as the frame of reference in all instances; it is the one that shows the directions of increasing measuring-unit values as established in Chapter 5.

A symmetrical scaling of 1, as shown in Fig. 7-5B, causes no reflection nor, for that matter, any change at all in the coordinate system.

Figure 7-5C illustrates a reflection about the X axis; any points above the X axis of the reference system end up below that axis, and any points originally below the X axis switch to a corresponding place above it. That is accomplished by reversing the sign of all Y-axis components; by using a negative-valued Y-axis scaling factor and a positive-valued X-axis factor.

Figure 7-5D represents a reflection about the Y axis. Relative to the frame-of-reference coordinate system, a Y-axis reflection reverses the sign, or sense, of the X axis.

Finally, Fig. 7-5E shows a reflection of both the X and Y axes. The actual axis of reflection isn't defined from the perspective of a true 2-D world, but the general idea is to reverse the signs of both axes.

Set up some experiments of your own. Specify a coordinate frame of reference and a point or two in it. Apply some scalings and plot the results—both the reference and scaled versions of the points. See if you can

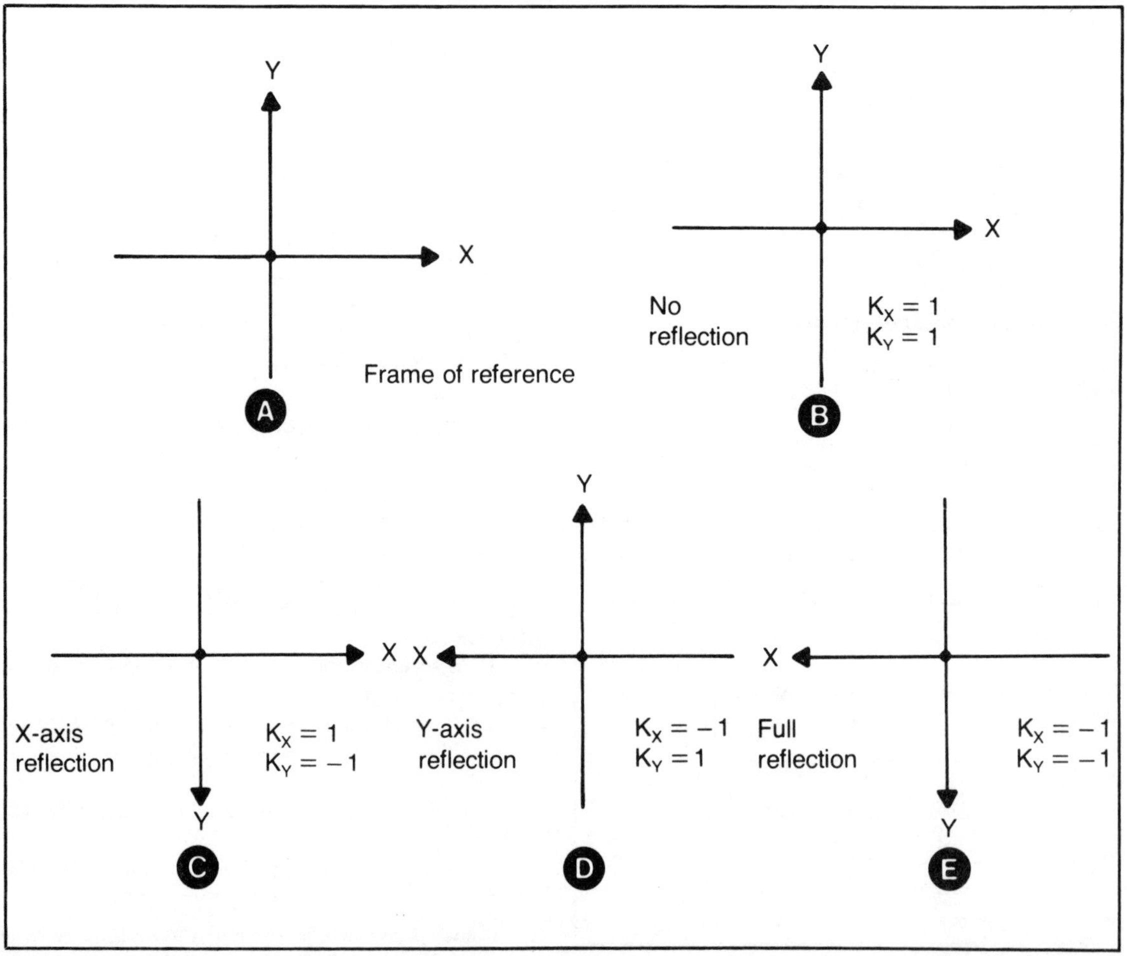

Fig. 7-5. Some classes of 2-D scalings. (A) The reference coordinate system. (B) A symmetrical scaling of 1. (C) A reflection about the X axis. (D) A reflection about the Y axis. (E) A full reflection in 2-D space.

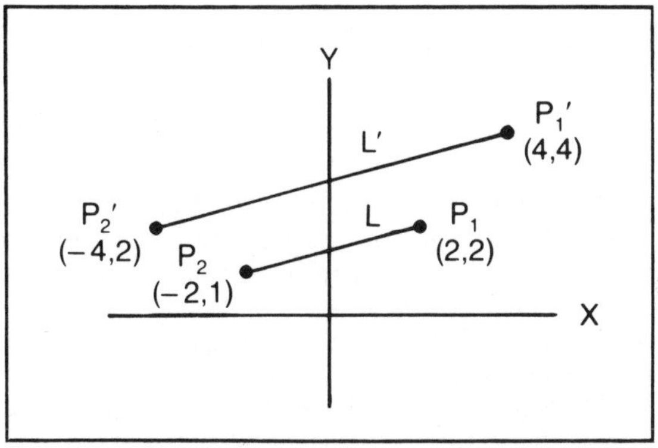

Fig. 7-6. A scaling of line L to L′ in 2-D space. See Example 7-2.

come up with sets of 2-D scaling factors for doing the following kinds of operations:

- ☐ A symmetrical, expansion-type scaling.
- ☐ A symmetrical, reduction-type scaling.
- ☐ An expansion-type scaling with a reflection about the X or Y axis.
- ☐ An asymmetrical scaling that reduces the system in one dimension and expands it in the other.
- ☐ A full-reflection scaling.

7-3 SCALING LINES IN 2-D SPACE

A line in 2-D space is defined in terms of the coordinates of its endpoints. So scaling a line in 2-D space is a matter of scaling the two endpoints individually, then constructing the line between them. Take a look at Example 7-2.

Example 7-2

A reference line in the frame-of-reference coordinate system has the following specifications:

$$L = \overline{P_1P_2} \qquad P_1 = (2,2) \quad P_2 = (-2,1)$$

Apply a symmetrical scaling of $K_x = 2$, $K_y = 2$, and plot the two lines.

1. Gathering the data:

$$x_1 = 2 \qquad y_1 = 2 \qquad K_x = 2$$
$$x_2 = -2 \qquad y_2 = 1 \qquad K_y = 2$$

2. Applying the scaling equation to point P_1:

$$x_1' = x_1K_x \qquad y_1' = y_1K_y$$
$$x_1' = 2(2) \qquad y_1' = 2(2)$$
$$x_1' = 4 \qquad y_1' = 4$$

Thus $P_1 = (2,2)$ scales to $P_1' = (4,4)$

3. Applying the scaling equation to point P_2:

$$x_2' = x_2K_x \qquad y_2' = y_2K_y$$
$$x_2' = -2(2) \qquad y_2' = 1(2)$$
$$x_2' = -4 \qquad y_2' = 2$$

So $P_1 = (-2,1)$ scales to $P_2' = (-4,2)$

4. Specifying the scaled version of the line:

$$L' = \overline{P_1'P_2'} \qquad P_1' = (4,4) \quad P_2' = (-4,2)$$

See the experiment plotted for you in Fig. 7-6.

The example cites a simple line in 2-D space, then asks you to perform a symmetrical scaling of 2. Step 1 sets up the values according to the customary notation, then Step 2 solves Equation 7-1 for the two endpoints. Compare the reference line, L, with its scaled version, L′, in Fig. 7-6.

Build some confidence in your own understanding of the procedure by working through Exercise 7-2.

Exercise 7-2

Use the following line specifications for all of the experiments cited here:

$$L = \overline{P_1P_2} \qquad P_1 = (-4,-2) \quad P_2 = (4,3)$$

Perform the indicated scaling and plot the results in each instance.

1. Carry out a symmetrical expansion of $K_x = 2$, $K_y = 2$.
2. Carry out an asymmetrical scaling of $K_x = 0.5$, $K_y = 2$.
3. Carry out an X-axis reflection of $K_x = 1$, $K_y = -1$.
4. Carry out a Y-axis reflection of $K_x = -1$, $K_y = 1$.
5. Carry out a full reflection/expansion of $K_x = -2$, $K_y = -2.5$.

7-3.1 Effects on the Lengths of Lines

Scalings in 2-D space clearly have some effect on the length of a line. You can see that demonstrated by calculating the length of a reference line, performing a scaling, calculating the length of the scaled version, and comparing the results. Only symmetrical scalings of 1 or −1 will show the two lines having the same length. Generally speaking, scalings affect the length of a line in 2-D space. Length, or distance, is a variant under scaling transformations in 2-D space.

7-3.2 Effects on Direction Cosines

Scalings in 2-D space can also affect the direction cosines of a line. That isn't always the case, but just one instance is sufficient to conclude that direction cosines are variant under 2-D scalings. More often than not, the direction cosines of a scaled line are different from those for its reference version.

Any transformation that affects the orientation of a line with respect to the coordinate axes is bound to affect the angles between the line and the axes; and anything that changes those angles changes the direction cosines.

To clarify the idea in your mind, specify a line or two in 2-D space, apply some 2-D scalings, then compare the direction cosines of the reference and scaled versions. You will be able to find some particular instances where the direction cosines are unaffected, but such instances are special ones.

7-3.3 Effect on the Angle Between Two Lines

If scalings in 2-D space can affect the direction cosines of a line under many different circumstances, and since the angle between two lines is a function of their direction cosines (Equation 4-3), it follows that the angle between two lines is likewise affected.

Again there are a number of special situations where a scaling of two dimensions does not affect the angle, but just one instance is sufficient to make a general rule: angles between lines are variant under scaling transformations in 2-D space.

As suggested in the previous sections, you can test the principle yourself by specifying some reference lines, applying a scaling to them, and comparing the angles between the reference and scaled versions.

7-3.4 Summary of Scaling Effects On Lines

The foregoing discussions can be summarized this way. Bear these principles in mind when you are attempting to set up scaling-type experiments of your own.

☐ Length, or distance, is a variant under 2-D scalings.

☐ Direction cosines are variant under 2-D scalings.

☐ The angle between lines is a variant under 2-D scalings.

In short, every feature of lines can be affected by a scaling in two dimensions.

7-4 SCALING PLANE FIGURES IN 2-D SPACE

Scaling plane figures in 2-D space is, in principle, no different from scaling lines nor, for that matter, scaling points. Scale the lines in a plane figure, and you end up scaling the figure, itself. Scale the endpoints of a line, and you end up scaling the line. Putting that all together: scaling a plane figure in 2-D space is a matter of scaling the coordinates of the points that comprise the endpoints of its lines. Scale every specified point in the figure, and you scale the figure, itself.

The practical applications of scalings include changing the relative size of a figure, or changing its shape by stretching or squashing it down in one direction. You can, for example, use a scaling factor greater than 1 in the X dimension to transform a square into a rectangle. That one would end up wider than it is tall. Or you could do the same thing by applying a scaling factor less than 1 in the Y direction.

Scalings cannot change the basic nature of a plane figure—they cannot add or delete lines and angles from it as long as you avoid scaling factors of zero—but they can change its general appearance and point coordinates. (Scaling factors of 0 can cause a dramatic change in the character of a figure, but that is considered a special case that has some applications of questionable value.) It all goes back to the notion that none of the features of a line is invariant under 2-D scalings. Example 7-3 illustrates the idea.

Example 7-3

Suppose that you have the specifications for a small triangle in 2-D space:

$$F = \overline{L_1L_2L_3} \qquad L_1 = \overline{P_1P_2} \qquad P_1 = (-1,0)$$
$$L_2 = \overline{P_2P_3} \qquad P_2 = (1,0)$$
$$L_3 = \overline{P_3P_1} \qquad P_3 = (0,1)$$

You want to make it larger, stretching it more in the vertical direction than in the horizontal. The following scaling factors will do that:

$$K_x = 2 \qquad K_y = 4$$

Carry out that scaling, plot both the reference and scaled version of the triangle, then conduct a complete analysis for both of them.

1. Gather the data for the reference version of the triangle:

$$x_1 = -1 \qquad y_1 = 0$$
$$x_2 = 1 \qquad y_2 = 0$$
$$x_3 = 0 \qquad y_3 = 1$$

2. Scale point P_1:

$$x_1' = x_1K_x \qquad y_1' = y_1K_y$$
$$x_1' = -1(2) \qquad y_1' = 0(4)$$
$$x_1' = -2 \qquad y_1' = 0$$

Thus $P_1' = (-2,0)$

3. Scale point P_2:

$$x_2' = x_2K_x \qquad y_2' = y_2K_y$$

$$x_2' = 1(2) \qquad y_2' = 0(4)$$
$$x_2' = 2 \qquad y_2' = 0$$

Thus $P_2' = (2,0)$

4. Scale point P_3:

$$x_3' = x_3K_x \qquad y_3' = y_3K_y$$
$$x_3' = 0(2) \qquad y_3' = 1(4)$$
$$x_3' = 0 \qquad y_3' = 4$$

Thus $P_3' = (0,4)$

5. Cite the specifications for the scaled version of the figure and plot the results:

$$F' = \overline{L_1'L_2'L_3'} \quad L_1' = \overline{P_1'P_2'} \quad P_1' = (-2,0)$$
$$L_2' = \overline{P_2'P_3'} \quad P_2' = (2,0)$$
$$L_3' = \overline{P_3'P_1'} \quad P_3' = (0,4)$$

See the figures plotted in Fig. 7-7.

6. Conduct a complete analysis of the reference version of the figure:

(A) Length of line L_1:

$$s_1 = \sqrt{(x_2-x_1)^2 + (y_2-y_1)^2}$$
$$s_1 = \sqrt{[1-(-1)]^2 + (0-0)^2}$$
$$s_1 = \sqrt{2^2}$$
$$s_1 = 2$$

(B) Length of line L_2:

$$s_2 = \sqrt{(x_3-x_2)^2 + (y_3-y_2)^2}$$
$$s_2 = \sqrt{(0-1)^2 + (1-0)^2}$$
$$s_2 = \sqrt{(-1)^2 + (1)^2}$$
$$s_2 = \sqrt{2}$$

(C) Length of line L_3:

$$s_3 = \sqrt{(x_1-x_3)^2 + (y_1-y_3)^2}$$
$$s_3 = \sqrt{(-1-0)^2 + (0-1)^2}$$
$$s_3 = \sqrt{(-1)^2 + (-1)^2}$$
$$s_3 = \sqrt{2}$$

(D) Direction cosines for line L_1:

$$\cos\alpha_1 = (x_2-x_1)/s_1 \qquad \cos\beta_1 = (y_2-y_1)/s_1$$
$$\cos\alpha_1 = 2/2 \qquad \cos\beta_1 = 0/2$$
$$\cos\alpha_1 = 1 \qquad \cos\beta_1 = 0$$

(E) Direction cosines for line L_2:

$$\cos\alpha_2 = (x_3-x_2)/s_2 \qquad \cos\beta_2 = (y_3-y_2)/s_2$$
$$\cos\alpha_2 = -1/\sqrt{2} \qquad \cos\beta_2 = 1/\sqrt{2}$$
$$\cos\alpha_2 = -\sqrt{2}/2 \qquad \cos\beta_2 = \sqrt{2}/2$$

(F) Direction cosines for line L_3:

$$\cos\alpha_3 = (x_1-x_3)/s_3 \qquad \cos\beta_3 = (y_1-y_3)/s_3$$
$$\cos\alpha_3 = -1/\sqrt{2} \qquad \cos\beta_3 = -1/\sqrt{2}$$
$$\cos\alpha_3 = -\sqrt{2}/2 \qquad \cos\beta_3 = -\sqrt{2}/2$$

(G) Angle between lines L_1 and L_2:

$$\cos\phi_{1,2} = \cos\alpha_1\cos\alpha_2 + \cos\beta_1\cos\beta_2$$
$$\cos\phi_{1,2} = 1(-\sqrt{2}/2) + 0(\sqrt{2}/2)$$
$$\cos\phi_{1,2} = -\sqrt{2}/2$$
$$\phi_{1,2} = 45°$$

(H) Angle between lines L_1 and L_3:

$$\cos\phi_{1,3} = \cos\alpha_1\cos\alpha_3 + \cos\beta_1\cos\beta_3$$
$$\cos\phi_{1,3} = 1(-\sqrt{2}/2) + 0(1\sqrt{2}/2)$$
$$\cos\phi_{1,3} = -\sqrt{2}/2$$
$$\phi_{1,3} = 45°$$

(I) Angle between lines L_2 and L_3:

$$\cos\phi_{2,3} = \cos\alpha_2\cos\alpha_3 + \cos\beta_2\cos\beta_3$$
$$\cos\phi_{2,3} = -\sqrt{2}/2(-\sqrt{2}/2) + \sqrt{2}/2(-\sqrt{2}/2)$$
$$\cos\phi_{2,3} = 1/2 - 1/2 = 0$$
$$\phi_{2,3} = 90°$$

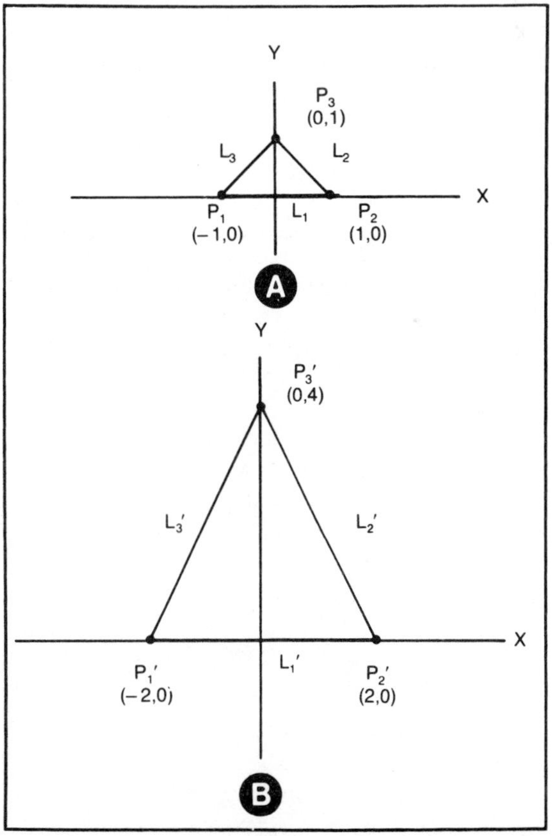

Fig. 7-7. Scaling a plane figure in 2-D space. (A) The reference figure. (B) The scaled version of the same figure. See Example 7-3.

Table 7-1. Summary of an Analysis of the Reference Triangle in Fig. 7-7A. Also See Example 7-3.

(A, B, C, and D correspond to α, β, γ, and δ.)

SPECIFICATIONS		
$F=\overline{L_1L_2L_3}$	$L_1=\overline{P_1P_2}$	$P_1=(-1,0)$
	$L_2=\overline{P_2P_3}$	$P_2=(1,0)$
	$L_3=\overline{P_3P_1}$	$P_3=(0,1)$
LENGTHS OF LINES		
$s_1=2$	$s_2=\sqrt{2}$	$s_3=\sqrt{2}$
DIRECTION COSINES		
$\cos A_1=1$	$\cos B_1=0$	
$\cos A_2=-\sqrt{2}/2$	$\cos B_2=\sqrt{2}/2$	
$\cos A_3=-\sqrt{2}/2$	$\cos B_3=-\sqrt{2}/2$	
ANGLES		
$\phi_{1,2}=45^\circ$	$\phi_{1,3}=45^\circ$	$\phi_{2,3}=90^\circ$

See the analysis of the reference figure summarized in Table 7-1.

7. Conduct a complete analysis of the scaled version of the figure:

(A) Length of line L_1':

$$s_1' = \sqrt{(x_2'-x_1')^2 + (y_2'-y_1')^2}$$
$$s_1' = \sqrt{[2-(-2)]^2 + (0-0)^2}$$
$$s_1' = \sqrt{(4)^2 + (0)^2}$$
$$s_1' = 4$$

(B) Length of line L_2':

$$s_2' = \sqrt{(x_3'-x_2')^2 + (y_3'-y_2')^2}$$
$$s_2' = \sqrt{(0-2)^2 + (4-0)^2}$$
$$s_2' = \sqrt{(-2)^2 + (4)^2}$$
$$s_2' = 2\sqrt{5}$$

(C) Length of line L_3':

$$s_3' = \sqrt{(x_1'-x_3')^2 + (y_1'-y_3')^2}$$
$$s_3' = \sqrt{(-2-0)^2 + (0-4)^2}$$
$$s_3' = \sqrt{(-2)^2 + (-4)^2}$$
$$s_3' = 2\sqrt{5}$$

(D) Direction cosines for line L_1':

$$\cos\alpha_1' = (x_2'-x_1')/s_1' \qquad \cos\beta_1' = (y_2'-y_1')/s_1'$$
$$\cos\alpha_1' = 4/4 \qquad \cos\beta_1' = 0/4$$
$$\cos\alpha_1' = 1 \qquad \cos\beta_1' = 0$$

(E) Direction cosines for line L_2':

$$\cos\alpha_2' = (x_3'-x_2')/s_2' \qquad \cos\beta_2' = (y_3'-y_2'/s_2'$$
$$\cos\alpha_2' = -2/2\sqrt{5} \qquad \cos\beta_2' = 4/2\sqrt{5}$$
$$\cos\alpha_2' = -\sqrt{5}/5 \qquad \cos\beta_2' = 2\sqrt{5}/5$$

(F) Direction cosines for line L_3':

$$\cos\alpha_3' = (x_1'-x_3')/s_3' \qquad \cos\beta_3' = (y_1'-y_3')/s_3'$$
$$\cos\alpha_3' = -2/2\sqrt{5} \qquad \cos\beta_3' = -4/2\sqrt{5}$$
$$\cos\alpha_3' = -\sqrt{5}/5 \qquad \cos\beta_3' = -2\sqrt{5}/5$$

(G) Angle between lines L_1' and L_2':

$$\cos\phi_{1,2}' = \cos\alpha_1'\cos\alpha_2' + \cos\beta_1'\cos\beta_2'$$
$$\cos\phi_{1,2}' = 1(-\sqrt{5}/5) + 0(2\sqrt{5}/5)$$
$$\cos\phi_{1,2}' = -\sqrt{5}/5$$
$$\phi_{1,2}' = 63^\circ\ 26'$$

(H) Angle between lines L_1' and L_3':

$$\cos\phi_{1,3}' = \cos\alpha_1'\cos\alpha_3' + \cos\beta_1'\cos\beta_3'$$
$$\cos\phi_{1,3}' = 1(-\sqrt{5}/5) + 0(-2\sqrt{5}/5)$$
$$\cos\phi_{1,3}' = -\sqrt{5}/5$$
$$\phi_{1,3}' = 63^\circ\ 26'$$

(I) Angle between lines L_2' and L_3':

$$\cos\phi_{2,3}' = \cos\alpha_2'\cos\alpha_3' + \cos\beta_2'\cos\beta_3',$$
$$\cos\phi_{2,3}' = -\sqrt{5}/5)-\sqrt{5}/5 + 2\sqrt{5}/5)-2\sqrt{5}/5$$
$$\cos\phi_{2,3}' = 3/5$$
$$\phi_{2,3}' = 53^\circ\ 6'$$

See the summary of this analysis in Table 7-2.

Hopefully, the example appears quite straightforward to you at the start. It specifies a plane figure in 2-D space and asks you to apply a certain asymmetrical scaling. The general idea is to apply the scaling factors to their corresponding components of all three points, and then connect the resulting coordinates with lines as prescribed in the specifications. Those operations conclude with Step 5. But the example doesn't end there. It also asks for a complete analysis of both the reference and scaled versions of the figure.

An analysis—a complete analysis—of a plane figure in 2-D space must include the basic specifications for the figure, a summary of the lengths of all lines that bound the figure, a summary of direction cosines for all of the lines, and the relevant angles between pairs of lines.

Step 6 details the procedure for finding the lengths of lines, direction cosines and angles for the reference version of the figure, Fig. 7-7A. Step 7 goes through the very same procedure, but applies it to the specifications for the scaled version of the figure, Fig. 7-7B. Study those procedures carefully, and make certain you understand the notation and how to use it.

I have summarized the analyses in Tables 7-1 and 7-2. Those tables, along with the diagrams in Fig. 7-7, comprise a complete analysis of a scaling experiment in 2-D space. It is the sort of thing you should be doing on your own by now.

As a point of interest, notice from the analyses that the prescribed scaling transformation affected the length of all three lines, changed most of the direction cosines, and changed all three angles.

Make up some experiments of your own. Specify some plane figures that happen to interest you and tinker with various combinations of scaling factors, including some reflections. It's a lot of work, but you will need this sort of experience behind you when you get into spaces of higher dimensions.

7-5 RESOLVING THE TRANSLATION/SCALING AMBIGUITY

Equation 7-2 represents a straightforward algebraic rearrangement of the basic scaling equation for 2-D space. In this instance, it solves for the two scaling factors, given the components of the reference and scaled version of a point.

$$K_x = x_n'/x_n \qquad K_y = y_n'/y_n \qquad \textbf{Equation 7-2}$$

where:

K_x and K_y = the scaling factors, relative to the frame of reference, for the X and Y axes, respectively

x_n', y_n' = the coordinate of the scaled version of point n relative to the chosen frame of reference

x_n', y_n' = the coordinate of reference point n relative to the frame of reference

The following example illustrates the use of that equation.

Example 7-4

A transformation that is known to be a scaling transforms a reference point P = (2,5) to P′ = (−4,1). What are the scaling factors behind that transformation?

1. Gathering the relevant data:

   $x = 2 \quad y = 5 \quad x' = -4 \quad y' = 1$

2. From Equation 7-2:

   $K_x = x'/x \qquad K_y = y'/y$
   $K_x = -4/2 \qquad K_y = 1/5$
   $K_x = -2 \qquad K_y = 0.2$

The scaling is thus characterized by a reflection and doubling of the X-axis components, and a reduction of a factor of 0.2 along the Y axis.

As in 1 dimension, the results of scaling single points in 2-D space is indistinguishable from translating points in 2-D space. You have already seen that ambiguity in Chapter 4.

Try scaling the coordinate of a single point in 2-D space, then solve for a set of translation terms, using Equation 5-2 as a guide. You will find that you can, indeed, come up with a set of translation terms for describing an effect actually generated by applying a set of scaling factors. Or return to Example 7-4 and see if you can calculate a set of translation terms that achieve exactly the same end result.

Table 7-2. Summary of an Analysis of the Scaled Version of Triangle in Fig. 7-7B. Also See Example 7-3.

(A, B, C, and D correspond to α, β, γ, and δ.)

SPECIFICATIONS

$F' = \overline{L'_1 L'_2 L'_3}$	$L'_1 = \overline{P'_1 P'_2}$	$P'_1 = (-2,0)$
	$L'_2 = \overline{P'_2 P'_3}$	$P'_2 = (2,0)$
	$L'_3 = \overline{P'_3 P'_1}$	$P'_3 = (0,4)$

LENGTHS OF LINES

$s'_1 = 2$	$s'_2 = \sqrt{2}$	$s'_3 = \sqrt{2}$

DIRECTION COSINES

$\cos A'_1 = 1$	$\cos B'_1 = 0$
$\cos A'_2 = -\sqrt{5}/5$	$\cos B'_2 = 2\sqrt{5}/5$
$\cos A'_3 = -\sqrt{5}/5$	$\cos B'_3 = -2\sqrt{5}/5$

ANGLES

$\phi'_{1,2} = 63^\circ\ 26'$

$\phi'_{1,3} = 63^\circ\ 26'$

$\phi'_{2,3} = 53^\circ\ 6'$

It is important to deal with that ambiguity in order to get a reliable idea of what is happening; and resolving it is a matter of working out a transformation that exists with regard to the length of a line, or the distance between two points. A single point does not offer enough information to resolve the matter.

The procedure for testing for a prevailing translation or scaling is rather simple: use Equation 4-2 to solve for the translation terms that might exist between the points as represented in both the reference and transformed versions. If you end up with the same set of translation terms for both points, then the transformation can be characterized by a translation. But if the translation terms are different for the points, the transformation has to include a scaling. (A translation might be included as well, but that is the topic that is reserved for special treatment in Chapter 9.)

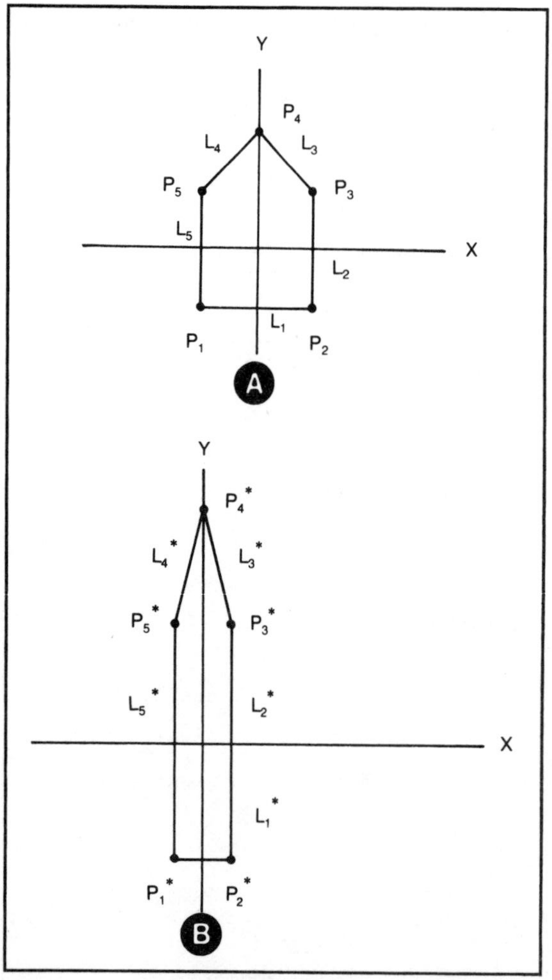

Fig. 7-8. A relativistic scaling effect in 2-D space. (A) A reference figure as it appears from the observer's frame of reference. (B) The same plane figure as it appears from a scaled frame of reference.

7-6 RELATIVISTIC VERSIONS OF 2-D SCALINGS

Equation 7-3 represents the inverse form of the basic scaling equation shown earlier as Equation 7-1. In the most literal sense, it solves for the coordinate of a reference point, given the coordinate of a scaled point and the scaling factors that exist between two coordinate systems. Everything is regarded with respect to the chosen frame of reference.

$$x_n = x_n'/K_x \qquad y_n = y_n'/K_y \qquad \textbf{Equation 7-3}$$

where:

x_n,y_n = the coordinate of a reference point n as reckoned from the chosen frame of reference

x_n',y_n' = the coordinate of the scaled version of point n as reckoned from the frame of reference

K_x,K_y = the scaling factors for the X and Y axes

Note: This is the inverse form of Equation 7-1

That equation leads directly to the expression for relativistic scalings in 2-D space. A simple redefinition of terms yields Equation 7-4.

$$x_n^* = x_n/K_x \qquad y_n^* = y_n/K_y \qquad \textbf{Equation 7-4}$$

where:

x_n^*,y_n^* = the coordinate of a reference point n relative to the scaled frame of reference

x_n,y_n = the coordinate of reference point n relative to the observer's frame of reference

K_x and K_y = the scaling factors that exist between the observer's and scaled frames of reference, relative to the observer's frame of reference

The relativistic equation solves directly for the components of some reference point relative to a scaled frame of reference. The scaling factors and coordinate of the reference point are both reckoned relative to the observer's frame of reference.

The most useful point of view to adopt in this case goes something like this. A certain known scaling exists between two different frames of reference—the observer's frame of reference and a scaled frame of reference. That scaling is specified relative to the observer's frame of reference. What's more, there is a point of reference somewhere in 2-D space, and its coordinate is known relative to the observer's frame of reference. The idea is to use that equation to determine the coordinate of that same reference point, but relative to the scaled frame of reference. In a matter of speaking, the equation lets you know how someone living on the scaled frame of reference sees the position of some point you are observing from your own frame of reference. That is the essence of relativity.

The notion is just as interesting when dealing with a reference line, and even more intriguing when working with plane figures in 2-D space. How is that line or plane figure oriented in space as viewed from that other coordinate system? Keeping in mind that the lengths of lines, direction cosines and angle beween lines are variant under 2-D scalings (including their relativistic forms), you might be able to appreciate how different a space object can appear from the two frames of reference. Check out this example.

Example 7-5

You, an observer in 2-dimensional Flatland, use your own coordinate system to describe an object this way:

$$F = \overline{L_1L_2L_3L_4L_5} \qquad L_1 = \overline{P_1P_2} \qquad P_1 = (-1,-1)$$
$$L_2 = \overline{P_2P_3} \qquad P_2 = (1,-1)$$
$$L_3 = \overline{P_3P_4} \qquad P_3 = (1,1)$$
$$L_4 = \overline{P_4P_5} \qquad P_4 = (0,2)$$
$$L_5 = \overline{P_5P_1} \qquad P_5 = (-1,1)$$

Another investigator is using a different coordinate scale that, compared to yours, is characterized by the scaling factors $K_x = 0.5$, $K_y = 2$. How does that object appear from the other investigator's scaling system?

1. Gathering the data:

$x_1 = -1$	$y_1 = -1$	$K_x = 0.5$
$x_2 = 1$	$y_2 = -1$	$K_y = 2$
$x_3 = 1$	$y_3 = 1$	
$x_4 = 0$	$y_4 = 2$	
$x_5 = -1$	$y_5 = 1$	

2. Applying Equation 7-4 to all of the points:

$x_1^* = -0.5$	$y_1^* = -2$
$x_2^* = 0.5$	$y_2^* = -2$
$x_3^* = 0.5$	$y_3^* = 2$
$x_4^* = 0$	$y_4^* = 4$
$x_5^* = -0.5$	$y_5^* = 2$

Table 7-3. Summary of an Analysis of a Reference Figure as it Appears From an Observer's Frame of Reference. See Fig. 7-8A and Example 7-5.

(A, B, C, and D correspond to α, β, γ, and δ.)

SPECIFICATIONS

$$F=\overline{L_1L_2L_3L_4L_5}$$

$L_1=\overline{P_1P_2}$	$P_1=(-1,-1)$
$L_2=\overline{P_2P_3}$	$P_2=(1,-1)$
$L_3=\overline{P_3P_4}$	$P_3=(1,1)$
$L_4=\overline{P_4P_5}$	$P_4=(0,2)$
$L_5=\overline{P_5P_1}$	$P_5=(-1,1)$

LENGTHS OF LINES

$s_1=2$	$s_2=2$	$s_3=\sqrt{2}$
$s_4=\sqrt{2}$	$s_5=2$	

DIRECTION COSINES

$\cos A_1=1$	$\cos B_1=0$
$\cos A_2=0$	$\cos B_2=1$
$\cos A_3=-\sqrt{2}/2$	$\cos B_3=\sqrt{2}/2$
$\cos A_4=-\sqrt{2}/2$	$\cos B_4=-\sqrt{2}/2$
$\cos A_5=0$	$\cos B_5=-1$

ANGLES

$\phi_{1,2}=90^\circ$	$\phi_{1,5}=90^\circ$	$\phi_{2,3}=135^\circ$
$\phi_{3,4}=90^\circ$	$\phi_{4,5}=135^\circ$	

3. Specifying the object as viewed from the scaled frame of reference:

$$F = \overline{L_1^*L_2^*L_3^*}\ \overline{L_4^*L_5^*} \qquad L_1^* = \overline{P_1^*P_2^*} \qquad P_1^* = (-0.5,-2)$$
$$L_2^* = P_2^*P_3^* \qquad P_2^* = (0.5,-2)$$
$$L_3^* = \overline{P_3^*P_4^*} \qquad P_3^* = (0.5,2)$$
$$L_4^* = P_4^*P_5^* \qquad P_4^* = (0,4)$$
$$L_5^* = P_5^*P_1^* \qquad P_5^* = (-0.5,2)$$

Table 7-4. Summary of an Analysis of a Reference Figure as it Appears From a Scaled Frame of Reference. See Fig. 7-8B and Example 7-5.

(A, B, C, and D correspond to α, β, γ, and δ.)

SPECIFICATIONS

$F' = \overline{L'_1 L'_2 L'_3 L'_4 L'_5}$

$L'_1 = \overline{P'_1 P'_2}$	$P'_1 = (-1,-1)$
$L'_2 = \overline{P'_2 P'_3}$	$P'_2 = (1,-1)$
$L'_3 = \overline{P'_3 P'_4}$	$P'_3 = (1,1)$
$L'_4 = \overline{P'_4 P'_5}$	$P'_4 = (0,2)$
$L'_5 = \overline{P'_5 P'_1}$	$P'_5 = (-1,1)$

LENGTHS OF LINES

$s'_1 = 1$	$s'_2 = 4$	$s'_3 = \sqrt{17}/2$
$s'_4 = \sqrt{17}/2$	$s'_5 = 4$	

DIRECTION COSINES

$\cos A'_1 = 1$	$\cos B'_1 = 0$
$\cos A'_2 = 0$	$\cos B'_2 = 1$
$\cos A'_3 = -\sqrt{17}/17$	$\cos B'_3 = 4\sqrt{17}/17$
$\cos A'_4 = -\sqrt{17}/17$	$\cos B'_4 = -4\sqrt{17}/17$
$\cos A'_5 = 0$	$\cos B'_5 = 1$

ANGLES

$\phi'_{1,2} = 90^\circ$	$\phi'_{1,5} = 90^\circ$	$\phi'_{2,3} = 165^\circ\ 58'$
$\phi'_{3,4} = 28^\circ\ 56'$	$\phi'_{4,5} = 165^\circ\ 58'$	

See the plots in Fig. 7-8 and a complete analysis of the reference and scaled views in Tables 7-3 and 7-4.

The reference figure that is specified in the example is shown in Fig. 7-8A. Maybe you would like to think of it as a typical house in Flatland. At any rate, the example shows how you can compress and stretch such a figure by means of an asymmetrical scaling.

The example, itself, deals only with the basic scaling operations. What is of equal importance is the analysis of the experiment as portrayed in Tables 7-3 and 7-4.

I am leaving it up to you to confirm my results of the analysis by doing the calculations on your own. You will benefit from the work, especially when it comes to working out the angles between lines 2 and 3, and lines 4 and 5. For the reference version of that figure, you will find that the calculations show angles of 45 degrees for both of them. Looking at the figure, however, you can see that the angles are much greater than 45 degrees. What's wrong? It is time to write to the publisher and complain about gross errors in the book? No, it is time to do something that few people know how to do—think a little in terms of something besides plug-'em-in-and-crank-'em out arithmetic.

When I originally described the equation for determining the cosine of the angle between two lines in 2-D space, I said that the equation solves for the acute angle between the lines. Did I say that just to make impressive sounds? No, I said it because it really means something; and in this particular example, it makes a big difference in the results and, indeed, the credibility of the experimenter.

In this case, the two angles in question are internal angles. The acute angle between the lines lays outside the figure. Whenever that happens—when you can see from the drawings that the angle is obviously much greater than the equations say they should be—adjust the matter by subtracting the result of the calculations from 180 degrees. Whereas the equation for the angle will turn up 45 degrees (obviously too small), fix the situation by 180 – 45 = 135 degrees. The same situation occurs when calculating the corresponding angles for the scaled version of the figure—Fig. 7-8B.

Equation 7-5 is a variation of the basic relativistic equation. This one solves for the scaling that exists between two coordinate frames of reference, given the coordinates of a point as views from both of the coordinate systems. In principle at least, the equation lets you determine the nature of the scaling that exists between your frame of reference and that of another individual. It assumes, of course, that you can find out how the other individual is viewing the point in question.

$$K_x = x_n/x_n^*$$
$$K_y = y_n/y_n^*$$

Equation 7-5

where:

K_x and K_y = the scaling factors that exist between the observer's and scaled frames of reference, relative to the observer's frame of reference

x_n, y_n = the coordinate of reference point *n* relative to the observer's frame of reference

x_n^*, y_n^* = the coordinate of reference point *n* relative to the scaled frame of reference

As with the geometric scaling operations, there is a translation/scaling ambiguity inherent in the idea as long as you and the individual in the other frame of reference are working with a single reference point. Alter the experiment to deal with two reference points—say, the endpoints of a line—and you can resolve it. At least you can come up with a clear-cut set of translation terms or scaling factors that represent the transformation that exists between the frames of reference.

7-7 MULTIPLE 2-D SCALINGS

As in 1-D space, a succession of scalings applied in 2-D space is cummutative—it makes no difference which order you apply them. And, in fact, you can multiply all the X-axis scaling factors to come up with one overall X-axis scaling factor; and you can multiply together all of the Y-axis scaling factors to come up with a composite Y-axis scaling factor. Just don't mix X- and Y-axis scaling factors—keep them separated.

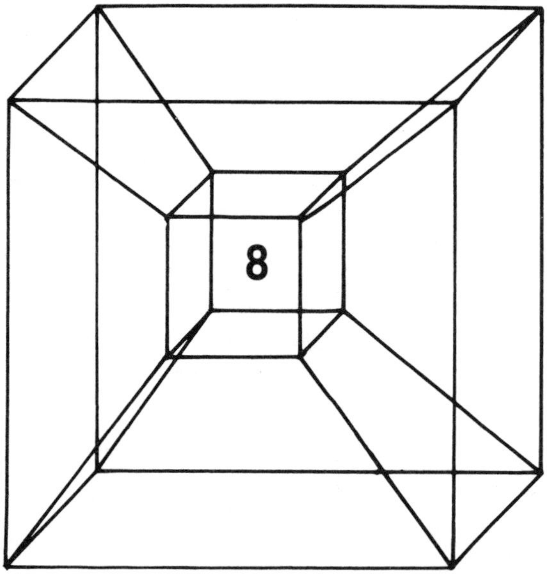

Rotations in 2-D Space

The line-world of 1-D space permits the use of two kinds of geometric elements and two kinds of transformations. The geometric elements are points and lines, and the transformations are translations and scalings. That fairly well represents experiments in 1-D space.

Moving up a dimension, the plane-world of 2-D space permits the use of three different kinds of geometric elements: point, lines and plane figures. And you have just studied two possible kinds of 2-D transformations: translations and scalings. There is, however, a third kind of transformation that is possible in 2-D space—one that is totally foreign to 1-D space.

You might guess from the title of this chapter that the third possible transformation in 2-D space is a *rotation*. That is indeed the case. Specifically, it is time to introduce the notion of rotating a coordinate system about a point.

Figure 8-1 summarizes the three kinds of transformations for 2-D space. You can see that a translation (Fig. 8-1A) simply displaces a coordinate system, but keeps the corresponding axes parallel to one another and maintains the same scales of measurement. A scaling transformation (Fig. 8-1B) affects the scaling of the measuring units on a coordinate system, yet keeps the cooresponding axes parallel to one another. Neither a translation nor a scaling in 2-D space alters the angular orientation of a transformed coordinate system relative to its reference version. A rotation transformation, however, literally turns a coordinate system about the origin. See Fig. 8-1C.

8-1 THE GENERAL CHARACTER OF 2-D ROTATIONS

A rotation in two dimensions always rotates a coordinate frame of reference about the origin and through some angle. It is possible to use a bit of mathematical trickery to "fool" the system into doing a rotation about some off-origin point; and that is a valuable trick that is described later in this chapter. But the general rule is that a rotation in 2-D space always occurs around the origin of the frame of reference. The origin remains fixed, and everything else rotates around it. You can see that effect in Fig. 8-1C.

The amount and direction of rotation are both determined by a signed term, the angle of rotation. The amount of rotation is specified in degrees throughout this book, but units of radians work equally well. So you will be expected to have some understanding of the meaning of angles of rotation such as 30 degrees, 180 degrees, 270 degrees and so on.

It is not enough to specify just the amount of rotation, however. It is equally important to specify a direction of

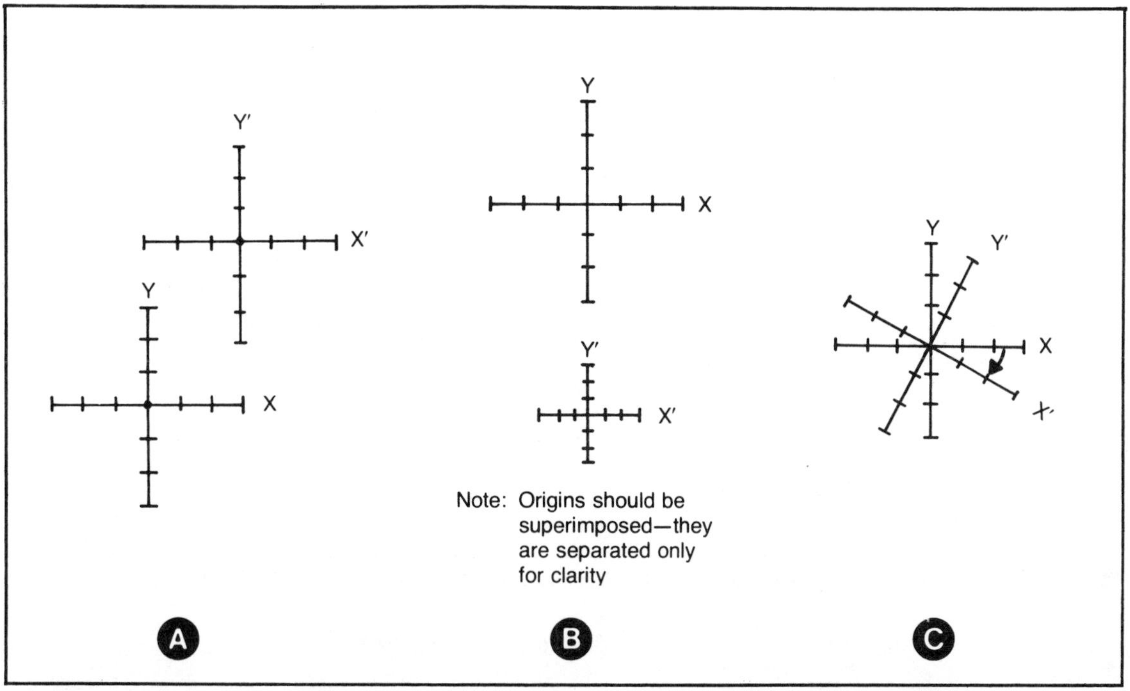

Fig. 8-1. The three types of translations that are possible in 2-D space. (A) Translation of coordinates. (B) Scaling of coordinates [the origins should be superimposed, but they are separated here for the sake of clarity]. (C) Rotation of coordinates about the origin.

rotation. That is handled by the sign of the angle of rotation—positive or negative.

There are two possible directions of rotation about the origin of a coordinate system in 2-D space: clockwise and counterclockwise. It makes no difference in principle whether we assign a positive value to a clockwise or counterclockwise direction or rotation. But to remain consistent with some work introduced later in this book, we are forced to establish a convention whereby a positive direction of rotation is one that rotates the coordinate system in a counterclockwise direction. It follows, then, that a negative direction of rotation is one that turns the coordinate system in a clockwise direction.

☐ A positive angle of rotation represents a rotation in the counterclockwise direction.

☐ A negative angle of rotation represents a rotation in the clockwise direction.

See the examples in Fig. 8-2.

8-2 THE BASIC 2-D ROTATION EQUATION

Equation 8-1 shows the basic equation for geometric rotations in 2-D space. It suggests a rotation of angle θ (theta) about the origin of the coordinate system.

$$x_n' = x_n \cos\theta - y_n \sin\theta$$
$$y_n' = x_n \sin\theta + y_n \cos\theta$$

Equation 8-1

where:

x_n', y_n' = the coordinate of the rotated version of point n relative to the frame of reference

x_n, y_n = the coordinate of point n relative to the frame of reference

θ = the angle of rotation relative to the frame of reference

See an application of the equation in Example 8-1.

Example 8-1

A point in 2-D space has coordinate P = (4,5) relative to its frame of reference. What is the coordinate of that point, relative to the frame of reference, upon applying a rotation of 45 degrees?

1. Gathering the information:

$x = 4 \qquad y = 5 \qquad \theta = 45°$

$\cos\theta = 0.71$

$\sin\theta = 0.71$

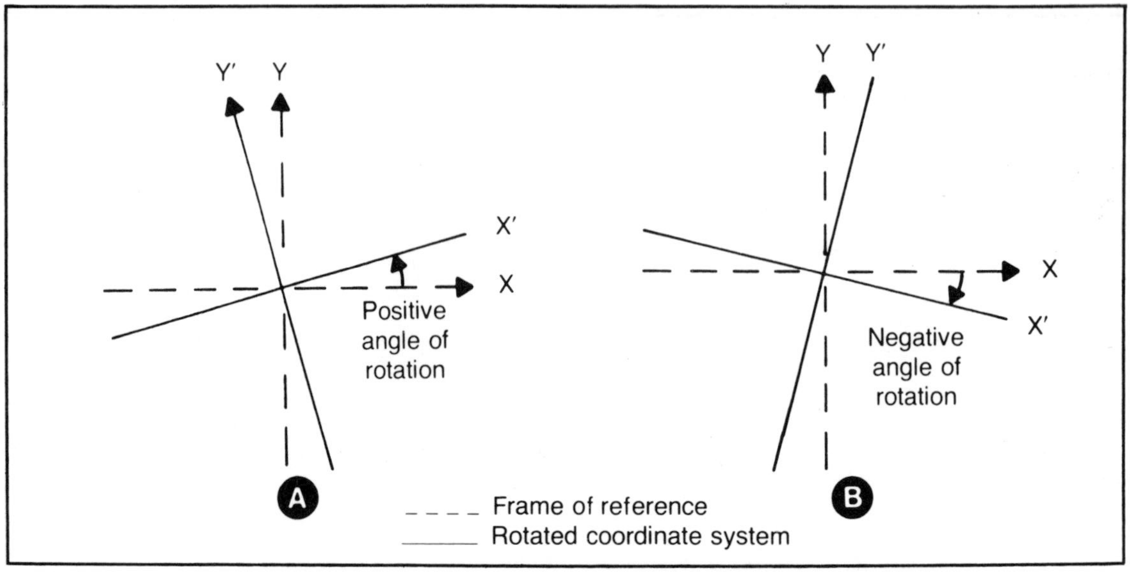

Fig. 8-2. The two possible directions of rotation in 2-D space. (A) Positive angle of rotation. (B) Negative angle of rotation.

2. Applying Equation 8-1:

$x' = x\cos\theta - y\sin\theta$	$y' = x\sin\theta + y\cos\theta$
$x' = 4(0.71) - 5(0.71)$	$y' = 4(0.71) + 5(0.71)$
$x' = 2.8 - 3.6$	$y' = 2.8 + 3.6$
$x' = -0.8$	$y' = 6.4$

The rotation of 45 degrees thus transforms reference point P = (4,5) to P′ = (−0.8,6.4). See the two points plotted against a background of the frame of reference in Fig. 8-3.

Step 1 in the example gathers the given information into a notational form that is compatible with the equations, and it shows the cosine and sine values of the angle of rotation. Step 2 completes the arithmetic part of the job by applying the basic 2-D rotation equation to the given information.

Using the previous example as a guide, see if you can rotate the coordinate systems and points specified in Exercise 8-1.

Exercise 8-1

Execute the prescribed rotations and plot the results in each of the following cases:

1. P = (4,5) $\theta = 30°$ P′ = (_____,_____)
2. P = (−4,5) $\theta = 60°$ P′ = (_____,_____)
3. P = (3,2) $\theta = -45°$ P′ = (_____,_____)
4. P = (−3,4) $\theta = -30°$ P′ = (_____,_____)
5. P = (−4,−4) $\theta = 60°$ P′ = (_____,_____)
6. P = (−2,−3) $\theta = -30°$ P′ = (_____,_____)

8-3 ROTATING LINES IN 2-D SPACE

Rotating a line that is specified in 2-D space is a matter of applying the rotation equation, using the same angle of rotation, to the two endpoints of that line. See Example 8-2.

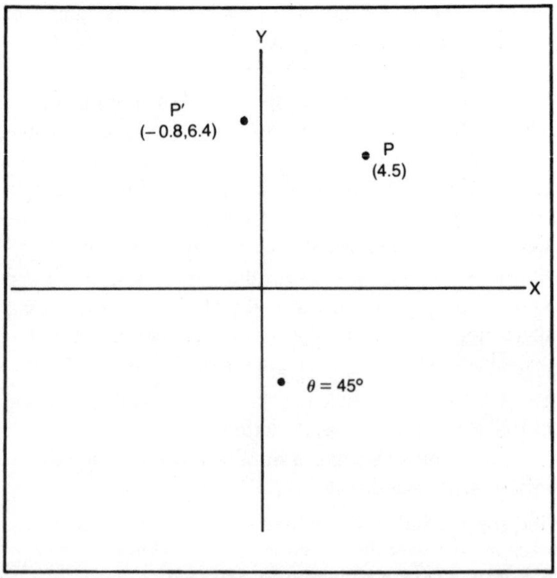

Fig. 8-3. A rotation about the origin of a coordinate system carries reference point P to its rotated position, P′. See Example 8-1.

Example 8-2

A line situated in 2-D space has the following specifications:

$$L = \overline{P_1P_2} \qquad P_1 = (4,1)$$
$$P_2 = (-3,-2)$$

Rotate that line through an angle of 30 degrees, specify the rotated version of it, and plot both onto a frame-of-reference coordinate system.

1. Gathering the information into a more symbolic form:

$$x_1 = 4 \qquad y_1 = 1 \qquad \theta = 30^\circ$$
$$x_2 = -3 \qquad y_2 = -2 \qquad \sin\theta = 1/2$$
$$\cos\theta = \sqrt{3}/2$$

2. Set up and solve Equation 8-1 for endpoint P_1:

$$x_1' = x_1 \cos\phi - y_1 \sin\phi \qquad y_1' = x_1 \sin\phi + y_1 \cos\phi$$
$$x_1' = 4(\sqrt{3}/2) - 1(1/2) \qquad y_1' = 4(1/2) + 1(\sqrt{3}/2)$$
$$x_1' = 2.7 \qquad y_1' = 2.9$$

So point $P_1 = (4,1)$ is rotated to $P_1' = (2.7,2.9)$

3. Set up and solve Equation 8-1 for endpoint P_2:

$$x_2' = x_2\cos\theta - y_s \sin\theta$$
$$x_2' = -3(\sqrt{3}/2) - (-2)\,(1/2)$$
$$x_2' = -1.6$$

$$y_2' = x_2 \sin\theta + y_2 \cos\theta$$
$$y_2' = 0 - 3(1/2) + (-2)\,(\sqrt{3}/2)$$
$$y_2' = -2.4$$

And point $P_2 = (-3, -2)$ is rotated to $P_2' = (-1.6, -2.4)$.

4. Specify the rotated version of the line:

$$L' = \overline{P_1'P_2'} \qquad P_1' = (2.7,2.9)$$
$$P_2' = (-1.6,-2.4)$$

See the plot of this rotation in Fig. 8-4.

The example cites the specifications for a simple line in 2-D space and suggests an angle of rotation of 30 degrees (presumably about the origin of the coordinate system). The first step, as usual, is to gather the given data into a more mathematical form and show the sine and cosine values of the angle.

Steps 2 and 3 show that rotating a line in 2-D space is a matter of applying the same angle of rotation to its endpoint coordinates. Step 4 completes the job by summarizing the specifications for the rotated version of the line. Specify some lines of your own in 2-D space, select some angles of rotation, perform the rotation and plot the results.

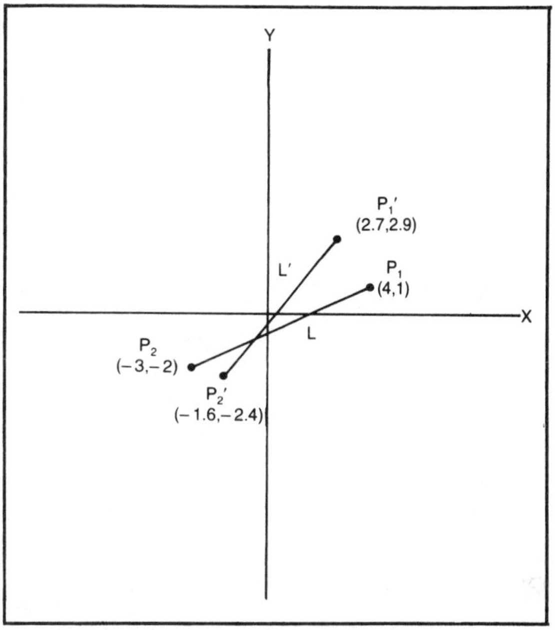

Fig. 8-4. A rotation of line L to L'. See Example 8-2.

8-3.1 Length is Invariant

The preceding example and suggested experiments ought to show you that rotations in 2-D space do not affect the lengths of lines, or the distance between two points. Length thus appears to be an invariant under 2-D rotations. You can demonstrate the principle by specifying a line in 2-D space, calculating its length (Equation 5-1), rotating the line through some angle, and then calculating the length of the rotated version of the line. If you've made no errors in method or calculation, you will always find that the line has the same length before and after applying the rotation transformation. Try that experiment a couple of times with different lines and different angles of rotation.

These demonstrations are not adequate proof of the matter, however. Mathematicians, and anyone else attempting to carry out some responsible work of this kind, demand formal proofs. The formal proof of the invariance of length under 2-D rotations is a lengthy one, so I am not going to interrupt the flow of new ideas by asking you to study it now. You will find this important proof featured in Section 8-9 at the conclusion of this chapter.

8-3.2 Direction Cosines are Variant Under 2-D Rotations

The direction cosines of a line carry information regarding the orientation of the line with respect to the axes of a coordinate system. Alter the orientation of the line, and you change its direction cosines. You saw in Chapter 6 that simple translations in 2-D space do not affect the direction cosines of a line—the line might change position, but its angle with respect to the axes of the coordinate system remains unchanged.

In Chapter 7, on the other hand, you found that certain kinds of asymmetrical scalings in 2-D space can dramatically change the orientation of a line with respect to the frame-of-reference coordinate system; in other words, direction cosines are variant under 2-D scalings. The same is true here for rotation-type transformations.

A rotation in 2-D space turns a coordinate system through some angle and in a direction that is specified by the angle of rotation. It figures, then, that most rotations will indeed alter the angle a line makes with respect to the axes of the frame of reference. Hence it is safe to say that direction cosines are variant under rotations in 2-D space.

Calculate and compare the direction cosines for some of the lines rotated in previous examples and exercises. You will find that, except under some very special circumstances, the direction cosines for the reference and rotated versions of a line are quite different.

8-3.3 Angle Between Lines is an Invariant

Because the equation for calculating the angle between two lines (Equation 5-3) is expressed entirely in terms of the direction cosines of those lines, and because the direction cosines of a line are variant under 2-D rotations, one might conclude that the angle between two lines is likewise a variant under 2-D rotations. It seems logical, but it isn't true. Surprisingly enough, the angle between two lines is an invariant under 2-D rotations, and I'd like you to develop a formal analytic proof of that fact.

8-4 ROTATING PLANE FIGURES IN 2-D SPACE

If you can rotate a single point in 2-D space, you can rotate two of them; and if you can rotate two points, you can rotate a line. If you can rotate a line in 2-D space, you can rotate any number of them; and if you can rotate any number of lines, you can certainly rotate a plane figure in 2-D space.

That line of reasoning rests on the fact that a plane figure can be defined in terms of the lines that bound it, the lines can be defined in terms of the endpoints that bound them, and the endpoints can be defined in terms of their cordinates. It all builds up and fits together quite nicely; you can effectively rotate a plane figure in 2-D space by rotating the points specified for it. Take a look at Example 8-3.

Example 8-3

A plane figure in 2-D space had the following specifications:

$$F = \overline{L_1L_2L_3} \qquad L_1 = \overline{P_1P_2} \qquad P_1 = (0,0)$$
$$L_2 = \overline{P_2P_3} \qquad P_2 = (0,1)$$
$$L_3 = \overline{P_3P_1} \qquad P_3 = (1,0)$$

Rotate that figure through an angle of 45 degrees, and then conduct a complete analysis of both the reference and rotated versions.

1. Gathering the relevant information for performing the rotation:

$$x_1 = 0 \qquad y_1 = 0 \qquad \theta = 45°$$
$$x_2 = 0 \qquad y_2 = 1 \qquad \sin\theta = \sqrt{2}/2$$
$$x_3 = 1 \qquad y_3 = 0 \qquad \cos\theta = \sqrt{2}/2$$

2. Applying the rotation equation to the three endpoints in succession:

$$x_1' = x_1\cos\theta - y_1\sin\theta \qquad y_1' = x_1\sin\theta + y_1\cos\theta$$
$$x_1' = 0(\sqrt{2}/2) - 0(\sqrt{2}/2) \qquad y_1' = 0(\sqrt{2}/2) + 0(\sqrt{2}/2)$$
$$x_1' = 0 \qquad y'_1 = 0$$

So point $P_1 = (0,0)$ remains at $P_1' = (0,0)$.

$$x_2' = x_2\cos\theta - y_2\sin\theta \qquad y_2' = x_2\sin\theta + y_2\cos\theta$$
$$x_2' = 0(\sqrt{2}/2) - 1(\sqrt{2}/2 \qquad y_2' = 0(\sqrt{2}/2) + 1(\sqrt{2}/2)$$
$$x_2' = -\sqrt{2}/2 \qquad y_2' = \sqrt{2}/2$$

Thus point $P_2 = (0,1)$ rotates to $P_2' = (-\sqrt{2}/2, \sqrt{2}/2)$

$$x_3' = x_3\cos\theta - y_3\sin\theta \qquad y_3' = x_3\sin\theta + y_3\cos\theta$$
$$x_3' = 1(\sqrt{2}/2) - 0(\sqrt{2}/2) \qquad y_3' = 1(\sqrt{2}/2) + 0(\sqrt{2}/2$$
$$x_3' = \sqrt{2}/2 \qquad y_3' = \sqrt{2}/2$$

And point $P_3 = (1,0)$ rotates to $P_3' = (\sqrt{2}/2, \sqrt{2}/2)$

3. Specifying the rotated version of the figure:

$$F' = \overline{L_1'L_2'L_3'} \qquad L_1' = \overline{P_1'P_2'} \qquad P_1' = (0,0)$$
$$L_2' = \overline{P_2'P_3'} \qquad P_2' = (-\sqrt{2}/2, \sqrt{2}/2)$$
$$L' = \overline{P'P'} \qquad P' = (\sqrt{2}/2, \sqrt{2}/2)$$

See the two versions of the figure plotted in Fig. 8-5.

4. Conducting a complete analysis of the reference figure, F:

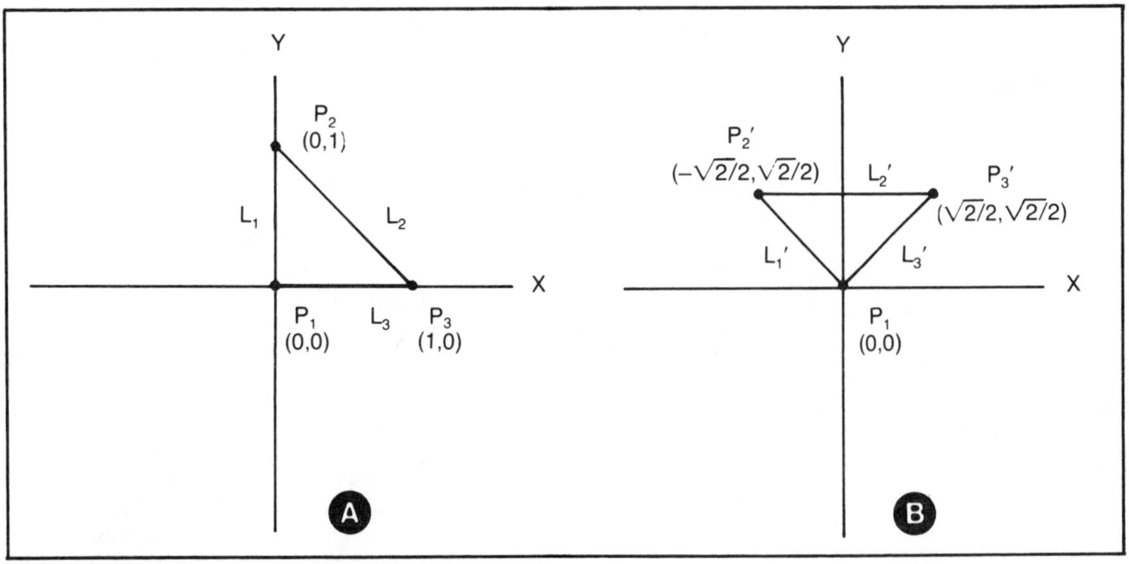

Fig. 8-5. Rotation of a triangular figure about the origin of the coordinate system. See Example 8-3.

(A) Lengths of lines

$$s_1 = \sqrt{(x_2-x_1)^2 + (y_2-y_1)^2}$$
$$s_1 = \sqrt{(0-0)^2+(1-0)^2}$$
$$s_1 = 1$$
$$s_2 = \sqrt{(x_3-x_2)^2+(y_3-y_2)^2}$$
$$s_2 = \sqrt{(1-0)^2+(0-1)^2}$$
$$s_2 = \sqrt{2}$$
$$s_3 = \sqrt{(x_1-x_3)^2+(y_1-y_3)^2}$$
$$s_3 = \sqrt{(0-1)^2+(0-0)^2}$$
$$s_3 = 1$$

(B) Direction cosines

$$\cos\alpha_1 = (x_2-x_1)/s_1 \qquad \cos\beta_1 = (y_2-y_1)/s_1$$

$$\cos\alpha_1 = 0/1 \qquad \cos\beta_1 = 1/1$$
$$\cos\alpha_1 = 0 \qquad \cos\beta_1 = 1$$

$$\cos\alpha_2 = (x_3-x_2)/s_2 \qquad \cos\beta_2 = (y_3-y_2)/s_2$$

$$\cos\alpha_2 = 1/\sqrt{2} \qquad \cos\beta_2 = -1/\sqrt{2}$$
$$\cos\alpha_2 = \sqrt{2}/2 \qquad \cos\beta_2 = -\sqrt{2}/2$$

$$\cos\alpha_3 = (x_1-x_3)/s_3 \qquad \cos\beta_3 = (y_1-y_3)/s_3$$

$$\cos\alpha_3 = -1/1 \qquad \cos\beta_3 = 0/1$$
$$\cos\alpha_3 = -1 \qquad \cos\beta_3 = 0$$

(C) Angles between lines

$$\cos\phi_{1,2} = \cos\alpha_1\cos\alpha_2+\cos_1\cos\beta_2$$
$$\cos\phi_{1,2} = (0)(\sqrt{2}/2)+1(-\sqrt{2}/2$$
$$\cos\phi_{1,2} = -\sqrt{2}/2$$
$$\phi_{1,2} = 45°$$
$$\cos\phi_{1,3} = \cos\alpha_1\cos\alpha_3+\cos\beta_1\cos\beta_3$$
$$\cos\phi_{1,3} = 0(-1)+1(0)$$
$$\cos\phi_{1,3} = 0$$
$$\phi_{1,3} = 90°$$
$$\cos\phi_{2,3} = \cos\alpha_2\cos\alpha_3+\cos\beta_2\cos\beta_3$$
$$\cos\phi_{2,3} = \sqrt{2}/2(-1)+(-\sqrt{2}/2)(0)$$
$$\cos\phi_{2,3} = -\sqrt{2}/2$$
$$\phi_{2,3} = 45°$$

See the analysis of the reference plane figure summarized in Table 8-1.

5. Conducting a complete analysis of the rotated version of the figure, F′.

(A) Assemble the relevant data

$$x_1' = 0 \qquad y_1' = 0$$
$$x_2' = -\sqrt{2}/2 \qquad y_2' = \sqrt{2}/2$$
$$x_3' = \sqrt{2}/2 \qquad y_3' = \sqrt{2}/2$$

(B) Lengths of lines

$$s_1' = \sqrt{(x_2'-x_1')^2+(y_2'-y_1')^2}$$
$$s_1' = \sqrt{(-\sqrt{2}/2-0)^2+(\sqrt{2}/2-0)^2}$$

Table 8-1. A Summary of the Analysis of the Reference Version of the Triangle in Fig. 8-5A. Also See Example 8-3.

(A, B, C, and D correspond to α, β, γ, and δ.)		
SPECIFICATIONS		
$F=\overline{L_1L_2L_3}$	$L_1=\overline{P_1P_2}$	$P_1=(0,0)$
	$L_2=\overline{P_2P_3}$	$P_2=(0,1)$
	$L_3=\overline{P_1P_3}$	$P_3=(1,0)$
LENGTHS OF LINES		
$s_1=1$	$s_2=\sqrt{2}$	$s_3=1$
DIRECTION COSINES		
$\cos A_1=0$		$\cos B_1=1$
$\cos A_2=\sqrt{2}/2$		$\cos B_2=-\sqrt{2}/2$
$\cos A_3=-1$		$\cos B_3=0$
ANGLES		
$\phi_{1,2}=45^\circ$	$\phi_{1,3}=90^\circ$	$\phi_{2,3}=45^\circ$

$$s_1' = 1$$
$$s_2' = \sqrt{(x_3'-x_2')^2+(y_3'-y_2')^2}$$
$$s_2' = \sqrt{[\sqrt{2}/2-(-\sqrt{2}/2)]^2+(\sqrt{2}/2-\sqrt{2}/2)^2}$$
$$s_2' = \sqrt{(\sqrt{2})^2+(0)^2}$$
$$s_2' = \sqrt{2}$$
$$s_3' = \sqrt{(x_1'-x_3')^2+(y_1'-y_3')^2}$$
$$s_3' = \sqrt{(0-\sqrt{2}/2)^2+(0-\sqrt{2}/2)^2}$$
$$s_3' = 1$$

(C) Direction cosines

$$\cos\alpha_1' = (x_2'-x_1')/s_1' \qquad \cos\beta_1' = (y_2'-y_1')/s_1'$$
$$\cos\alpha_1' = -\sqrt{2}/2 \qquad \cos\beta_1' = \sqrt{2}/2$$
$$\cos\alpha_2' = (x_3'-x_2')/s_2' \qquad \cos\beta_2' = (y_3'-y_2')/s_2'$$
$$\cos\alpha_2' = 1 \qquad \cos\beta_2' = 0$$
$$\cos\alpha_3' = (x_1'-x_3')/s_3' \qquad \cos\beta_3' = (y_1'-y_3')/s_3'$$
$$\cos\alpha_3' = -\sqrt{2}/2 \qquad \cos\beta_3' = -\sqrt{2}/2$$

(D) Angles between lines

$$\cos\phi_{1,2}' = \cos\alpha_1'\cos\alpha_2'+\cos\beta_1'\cos\beta_2'$$
$$\cos\phi_{1,2}' = (-\sqrt{2}/2)(1)+(\sqrt{2}/2)(0)$$
$$\cos\phi_{1,2}' = -\sqrt{2}/2$$
$$\phi_{1,2}' = 45^\circ$$
$$\cos\phi_{1,3}' = \cos\alpha_1'\cos\alpha_3'+\cos\beta_1'\cos\beta_3'$$
$$\cos\phi_{1,3}' = (-\sqrt{2}/2)(-\sqrt{2}/2)+(\sqrt{2}/2)(-\sqrt{2}/2)$$
$$\cos\phi_{1,3}' = 1/2-1/2 = 0$$
$$\phi_{1,3}' = 90^\circ$$
$$\cos\phi_{2,3}' = \cos\alpha_2'\cos\alpha_3'+\cos\beta_2'\cos\beta_3'$$
$$\cos\phi_{2,3}' = (1)(-\sqrt{2}/2)+(0)(-\sqrt{2}/2)$$
$$\cos\phi_{2,3}' = -\sqrt{2}/2$$
$$\phi_{2,3}' = 45^\circ$$

See the analysis of this rotated figure summarized in Table 8-2.

The example specifies a triangular figure in 2-D space and asks you to rotate it through an angle of 45 degrees. Step 1 is the information-gathering step, and Step 2 applies the rotation equation to the three individual points in the figure. Step 3 concludes the rotation phase of the experiment by citing the specifications for the rotated version of the triangle.

The experiment also calls for conducting complete analyses of both the reference and rotated versions. So Step 4 carries out the analysis for the reference figure, and Step 5 does the same for the rotated version of it. See the plots in Fig. 8-5 and summaries of the analyses in Tables 8-1 and 8-2.

Try your hand at a complete analysis of a rotation of plane figures by working through Exercise 8-2.

Exercise 8-2

1. Given the following specifications for a plane figure in 2-D space, rotate it 30 degrees and plot the two figures. Conduct a complete analysis of both.

$$F = \overline{L_1L_2L_3} \qquad L_1 = \overline{P_1P_2} \qquad P_1 = (-2,-2)$$
$$L_2 = \overline{P_2P_3} \qquad P_2 = (2,-2)$$
$$L_3 = \overline{P_3P_1} \qquad P_2 = (0,2)$$

2. Given the following plane figure in a frame of reference, rotate it −45 degrees, plot the results and conduct a complete analysis of both versions.

$$F = \overline{L_1L_2L_3L_4} \qquad L_1 = \overline{P_1P_2} \qquad P_1 = (0,0)$$
$$L_2 = \overline{P_2P_3} \qquad P_2 = (1,0)$$
$$L_3 = \overline{P_3P_4} \qquad P_3 = (1,1)$$
$$L_4 = \overline{P_4P_1} \qquad P_4 = (0,1)$$

3. Specify some plane figures of your own choosing, apply some angles of rotation to each of them and plot the experiments. Conduct complete analyses in all cases.

8-5 ALTERNATIVE FORMS OF THE ROTATION EQUATION

Equation 8-2 is a variation of the basic rotation equation. In this instance, it solves directly for the cosine of the angle of rotation, given the coordinate of both a reference point and the rotated version of it. You can then determine the angle, itself, by taking the arc cosine of the result. Such an operation is useful for determining an angle of rotation that happens to exist between two otherwise identical 2-D coordinate systems.

$$\cos\theta(x_n x_n'+y_n y_n')/(x_n^2+y_n^2) \qquad \textbf{Equation 8-2}$$

where:

θ = the angle of rotation relative to the frame of reference

x_n,y_n = the coordinate of point *n* relative to the frame of reference

x'_n,y'_n = the coordinate of the rotated version of point *n* relative to the frame of reference

Equation 8-3 is the result of rearranging the basic rotation equation in such a way that it is solved directly for the coordinate of a reference point, given an angle of rotation that exits between two coordinate systems and the coordinate of the rotated version of that point. It is the inverse form of the basic 2-D rotation equation, and its most important role is that of leading us to the notion of relativistic rotations.

$$x_n = x_n'\cos\theta+y_n\sin\theta$$
$$y_n = -x_n'\sin\theta+y_n\cos\theta \qquad \textbf{Equation 8-3}$$

where:

x_n,y_n = the coordinate of point *n* relative to the frame of reference

x_n',y_n' = the coordinate of the rotated version of point *n* relative to the frame of reference

θ = the angle of rotation relative to the frame of reference

Note: These expressions apply only to a rotation about the origin of the frame of reference

8-6 RELATIVISTIC ROTATION IN 2-D SPACE

Suppose that you are aware of a pair of 2-D coordinate systems that are identical in every respect but one—they are rotated by some angle and direction with respect to one another. You have chosen one to be the observer's frame of reference; that's usually your own point of view. The second coordinate system, rotated with respect to your own, is then the rotated frame of reference. That sets the stage for the important notion of relativistic rotations in 2-D space.

Further suppose that there is a point somewhere in that 2-D space that can be observed by intelligent creatures living on the two coordinate systems. Having a rotation existing between the two frames of reference, it is quite likely that the creatures will disagree on the exact coordinate of that one reference point. Why shouldn't they disagree? The coordinate systems that they are using for reckoning the coordinate are not lined up the same way in space.

At any rate, it is possible for you, the observer, to calculate the coordinate of that reference point as viewed from the rotated frame of reference. That is the function of Equation 8-4. Then see the idea demonstrated in Example 8-4.

$$x_n^* = x_n\cos\phi+y_n\sin\theta$$
$$y_n^* = -x_n\sin\theta+y_n\cos\theta \qquad \textbf{Equation 8-4}$$

where:

x_n^*,y_n^* = the coordinate of point *n* relative to the rotated frame of reference

Table 8-2. A Summary of the Analysis of the Rotated Version of a Triangle. See Fig. 8-5B and Example 8-3.

(A, B, C, and D correspond to α, β, γ, and δ.)

SPECIFICATIONS

$F' = \overline{L'_1 L'_2 L'_3}$

$L'_1 = \overline{P'_1 P'_2}$	$P'_1 = (0,0)$
$L'_2 = \overline{P'_2 P'_3}$	$P'_2 = (-\ 2/2,\ 2/2)$
$L'_3 = \overline{P'_1 P'_3}$	$P'_3 = (\ 2/2,\ 2/2)$

LENGTHS OF LINES

$s'_1=1$	$s'_2=\sqrt{2}$	$s'_3=1$

DIRECTION COSINES

$\cos A'_1=-\sqrt{2}/2$	$\cos B'_1=\sqrt{2}/2$
$\cos A'_2=1$	$\cos B'_2=0$
$\cos A'_3=-\sqrt{2}/2$	$\cos B'_3=-\sqrt{2}/2$

ANGLES

$\phi'_{1,2}=45^\circ$	$\phi'_{1,3}=90^\circ$	$\phi'_{2,3}=45^\circ$

x_n, y_n = the coordinate of point *n* relative to the observer's frame of reference

θ = the angle of rotation that exists between the two frames of reference, relative to the observer's frame of reference

Note: These expressions apply only when the origins of the two frames of reference are superimposed

Example 8-4

You and a friend are standing in a room observing a pattern drawn on the floor. From your viewing angle, the pattern has the following specifications:

$$F = \overline{L_1 L_2 L_3 L_4 L_5}$$

$L_1 = \overline{P_1 P_2}$	$P_1 = (-1,-1)$
$L_2 = \overline{P_2 P_3}$	$P_2 = (1,-1)$
$L_3 = \overline{P_3 P_4}$	$P_3 = (1,1)$
$L_4 = \overline{P_4 P_5}$	$P_4 = 0,2)$
$L_5 = \overline{P_5 P_1}$	$P_5 = (-1,1)$

See the figure from your point of view in Fig. 8-6A. Your friend, however, is situated such that he or she is rotated 90 degrees with respect to your position. How does your friend see the figure?

1. Gathering the relevant information:

$x_1 = -1$	$y_1 = -1$	$0 = 90°$
$x_2 = 1$	$y_2 = -1$	$\cos\theta = 0$
$x_3 = 1$	$y_3 = 1$	$\sin\theta = 1$
$x_4 = 0$	$y_4 = 2$	
$x_5 = -1$	$y_5 = 1$	

2. Applying the relativistic rotation equation to those five sets of points, they become:

$$P_1' = (1,-1)$$
$$P_2' = (1,1)$$
$$P_3' = (-1,1)$$
$$P_4' = (-2,0)$$
$$P_5' = (-1,-1)$$

See Fig. 8-6B.

Get into the fun yourself by making up and conducting some experiments of your own along these lines.

8-7 OFF-ORIGIN ROTATIONS

All of the rotations described thus far take place about the origin of the frame-of-reference coordinate system. Figure 8-7A illustrates the rotation of a line about the origin through a series of angles. Clearly, the line has the origin of its coordinate system as the central point of rotation.

Figure 8-7B, however, shows something quite different. It is the same line and the same reference coordinate system, but the rotation is taking place about one endpoint of the line, rather than the origin of its coordinate system. The rotation still takes place about a single point—an off-origin point.

The ability to rotate a point, line or plane figure about some point other than the origin brings up some

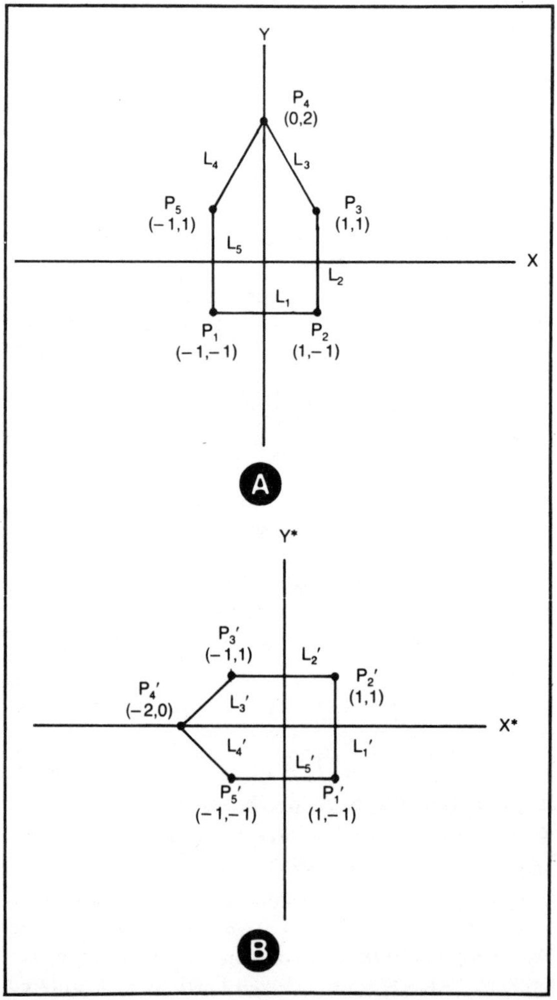

Fig. 8-6. Relativistic views of a reference figure in 2-D space. (A) The view from the observer's frame of reference. (B) The view from the rotated frame of reference. See Example 8-4.

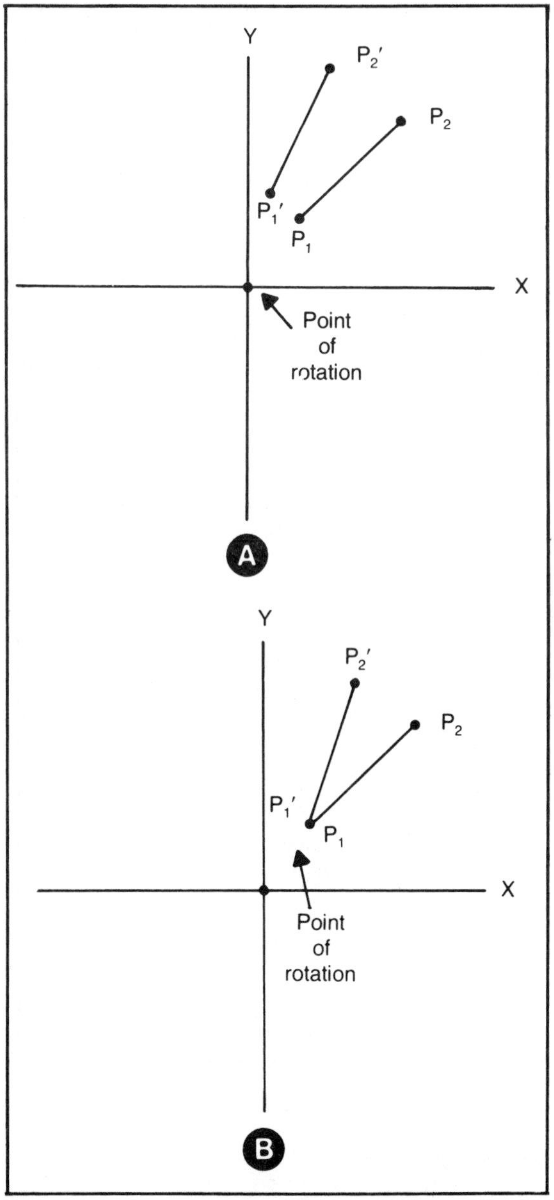

Fig. 8-7. Rotating a line in 2-D space. (A) Rotation about the origin of the coordinate system. (B) Rotation about an off-origin point.

interesting and useful experiments. Equation 8-5 leads the way.

$$x_n' = (x_n - d_x)\cos\theta - (y_n - d_y)\sin\theta + d_x$$
$$y_n' = (x_n - d_x)\sin\theta + (y_n - d_y)\cos\theta + d_y$$
Equation 8-5

where:

x_n',y_n' = the coordinate of the rotated version of point *n* relative to the frame of reference.

x_n,y_n = the coordinate of reference point *n* relative to the frame of reference.

d_y,d_n = the coordinate of the point of rotation

θ = the angle of rotation

Most of the variables in that equation serve the same functions as in the basic equation for origin-centered rotations. The only difference here is the addition of the dx and dy terms. Those two terms handle the displacement of the desired point of rotation from the origin of the frame-of-reference coordinate system. Substitute zeros for those displacement terms, and you end up with the usual sort of rotation about the origin. But specify an off-origin point of rotation such as (2,4), substitute $d_x = 2$ and $d_y = 4$ into this equation, and you will find the rotation taking place about point (2,4) in the coordinate system rather than origin point (0,0).

Here is an example.

Example 8-5

Given the following triangle in 2-D space, rotate it 45 degrees about point P_2:

$$F = \overline{L_1L_2L_3}$$
$$L_1 = \overline{P_1P_2} \qquad P_1 = (0,0)$$
$$L_2 = \overline{P_2P_3} \qquad P_2 = (1,0)$$
$$L_3 = \overline{P_3P_1} \qquad P_3 = (0,1)$$

1. Setting up the data:

$x_1 = 0 \quad y_1 = 0 \quad \theta = 45° \qquad d_x = x_2 = 1$
$x_2 = 1 \quad y_2 = 0 \quad \cos\theta = \sqrt{2}/2 \quad d_y = y_2 = 0$
$x_3 = 0 \quad y_3 = 1 \quad \sin\theta = \sqrt{2}/2$

2. Applying the off-origin equation to point P_1:

$$x_1' = (x_1 - d_x\cos\theta - (y_1 - d_y)\sin\theta + d_x$$
$$x_1' = (0-1)(\sqrt{2}/2) - (0-0)(\sqrt{2}/2) + 1$$
$$x_1' = -\sqrt{2}/2 + 1 = 0.3$$
$$y_1' = (x_1 - d_x)\sin\theta + (y_1 - d_y)\cos\theta + d_y$$
$$y_1' = (-1)(\sqrt{2}/2) + (0)(\sqrt{2}/2) + 0$$
$$y_1' = -\sqrt{2}/2 = -0.7$$

So point $P_1 = (0,0)$ rotates to $P_1' = (0.3,-0.7)$

3. Applying the equation to point P_2:

$$x_2' = (x_2 - d_x)\cos\theta - (y_2 - d_y)\sin\theta + d_x$$
$$x_2' = (1-1)(\sqrt{2}/2) - (0-0)(\sqrt{2}/2) + 1$$
$$x_2' = (0)(\sqrt{2}/2) - 0 + 1$$

$x_2' = 1$

$y_2' = (x_2-d_x)\sin\theta+(y_2-d_y)\cos\theta+d_y$
$y_2' = (0)(\sqrt{2}/2)+(0)(\sqrt{2}/2)+0$
$y_2' = 0$

So the rotation keeps $P_2 = (1,0)$ at $P_2' = (1,0)$—as it should, because that happens to be the point of rotation.

4. Applying the equation to point P_3:

$x_3' = (x_3-d_x)\cos\theta-(y_3-d_y)\sin\theta+d_x$
$x_3' = (0-1)(\sqrt{2}/2)-(1-0)(\sqrt{2}/2)+1$
$x_3' = -\sqrt{2}/2-\sqrt{2}/2+1 = -0.4$

$y_3' = (x_3-d_x)\sin\theta-(y_3-d_y)\cos\theta+d_y$
$y_3' = (-1)(\sqrt{2}/2)+(1)(\sqrt{2}/2)+0$
$y_3' = 0$

Thus point $P_3 = (0,1)$ is rotated to $P_3' = (-0.4,0)$. See the situation plotted for you in Fig. 8-8.

Specify some lines and plane figures of your own. Make up some off-origin points of rotation, set up some angles of rotation, and use Equation 8-5 to determine the coordinates of the rotated versions of the points. Be sure to plot the "before-and-after" results.

Off-axis rotations affect the lines in a coordinate system in the same way that origin-centered rotations do: length is an invariant, direction cosines are variant, and angles between lines are invariant.

8-8 COMBINATIONS OF SUCCESSIVE ROTATIONS

Rotations in 2-D space, whether of the on- or off-origin type, are commutative. You can execute a series of rotations in any sequence and end up with the same overall result every time. Try it for yousełf.

8-9 PROOF OF THE INVARIANCE OF LENGTH UNDER 2-D ROTATIONS

The following proof shows that the length of a line remains unchanged by any sort of origin-centered rotation in 2-D space. In the light of the short discussion in the previous section, you should be able to appreciate the fact that the proof applies to sequences of rotations as well.

Proof 8-1

The objective of this proof is to show that the length of any line in 2-D space remains unchanged by a rotation about the origin of the coordinate system. To make things run a bit simpler, I am going to define some special notation:

$$\Delta x_n = x_{n,u} - x_{n,v}$$
$$\Delta y_n = y_{n,u} - y_{n,v}$$

where Δ is read "delta," and literally means "difference between." So Δx_n means the difference between the X components of endpoints *u* and *v* of line *n;* and Δy_n stands for the difference between the Y components of endpoints *u* and *v* of the same line.

1. The general expression for the length of a reference line in 2-D space is:

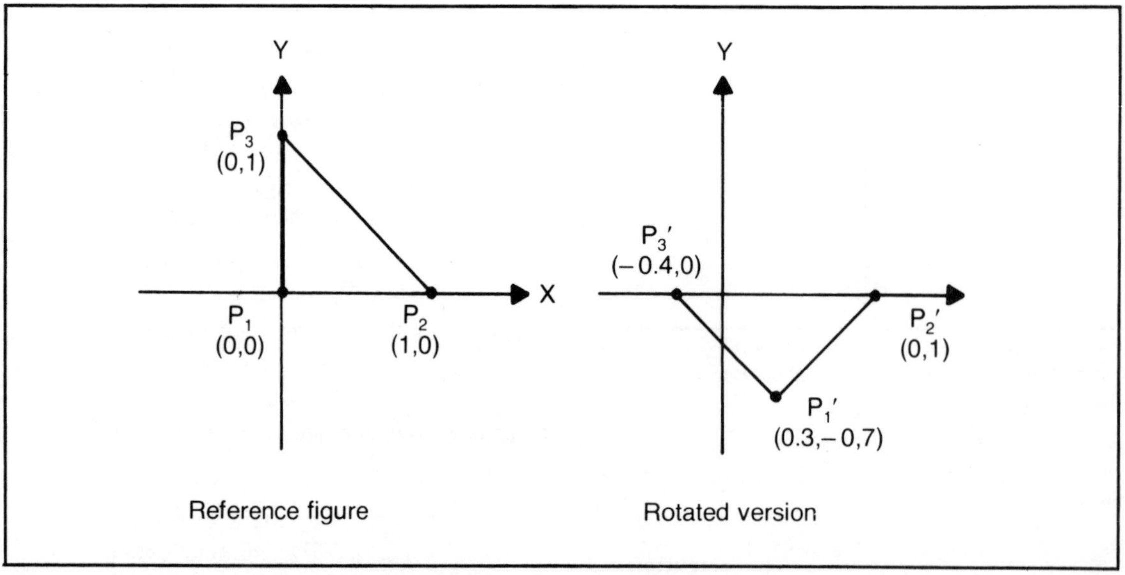

Fig. 8-8. Rotating a triangle about an off-origin point in 2-D space. See Example 8-5.

$$s_n = \sqrt{(\Delta x_n)^2+(\Delta y_n)^2}$$

2. The general expression for the length of any transformed line in 2-D space is:

$$s_n' = \sqrt{(\Delta x_n')^2+(\Delta y_n')^2}$$

3. Applying the equation for a 2-D rotation about the axis of the coordinate system, the components shown on the right-hand side of the equation in Step 2 become:

$$x_n' = \Delta x_n \cos\theta - \Delta y_n \sin\theta$$
$$y_n' = \Delta x_n \sin\theta + \Delta y_n \cos\theta$$

4. Squaring both expressions in Step 3:

$$(\Delta x_n')^2 = \Delta x_n^{\,2}\cos^2\theta - 2\Delta x_n \Delta y_n \cos\theta\sin\theta + \Delta y_n^{\,2}\sin^2\theta$$
$$(\Delta y_n')^2 = \Delta x_n^{\,2}\sin^2\theta + 2\Delta x_n \Delta y_n \cos\theta\sin\theta + \Delta y_n^{\,2}\cos^2\theta$$

5. Summing the expressions in Step 4:

$$(\Delta x_n')^2+(\Delta y_n')^2 = \Delta x_n^{\,2}(\sin^2\theta+\cos^2\theta)+\Delta y_n^{\,2}(\sin^2\theta+\cos^2\theta)$$
$$(\Delta x_n')^2+(\Delta y_n')^2 = \Delta x_n^{\,2}+\Delta y_n^{\,2}$$

6. From Steps 2 and 5;

$$s_n' = \sqrt{(\Delta x_n')^2+(\Delta y_n')^2} = \sqrt{\Delta x_n^{\,2}+\Delta y_n^{\,2}} = s_n$$

That is a lengthy and somewhat tedious proof, but it deserves your time and full attention for a couple of different reasons. First, it represents a formal analytic proof of the type a responsible experimenter must conduct in order to keep his or her ideas on a firm footing. If you can follow a proof of that extent and, better yet, conduct such a proof on your own, you know that you are maturing in this business.

A second reason I have offered this proof in such painstaking detail is beause I want to make certain that you know how to deal with abstract ideas in a responsible way—how to avoid the traps that so often snare those who make a habit of thinking only in terms of simple analogies. In fact, one of the real wonders of mathematics is the surprises it holds for anyone who cares to approach it on its own terms. Logical thinking, apart from the algebra and trigonometry of pure analysis, has its place in mathematics. But if you have ever played around with logical puzzles, you are aware of the fallibility of descriptive thinking. Were it not for the pitfalls of common logic, such interesting puzzles would not exist.

It is unfortunate that so few people take the time to learn the methods and master the kind of thinking that goes into pure mathematical analysis. What is perhaps even more unfortunate is the existence of so many books on scientific subjects that make a self-conscious effort to eliminate any sort of mathematical analysis. Doing that, the reader is left with nothing but analogies and so-called "logical thinking" that is loaded with pitfalls: and that adds up to a lot of misconceptions, superficial ideas and an impotent grasp of things. Bypassing mathematical analysis doesn't open any new doors to knowledge and understanding—it closes them. We will not allow that to happen here.

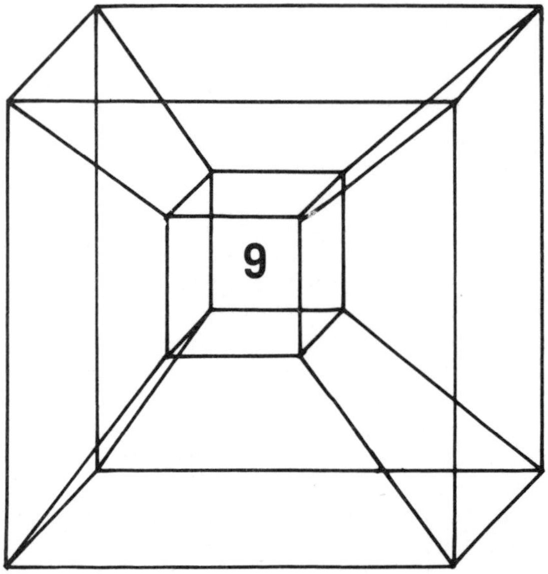

Experiments in Two Dimensions

This chapter represent both a summary and an extension of what you've learned thus far about 2-D space. If you can understand the summaries and grasp the essence of these suggested experiments, you are ready for even bigger and better things.

9-1 A DIRECTORY OF GEOMETRIC ENTITIES FOR 2-D SPACE

Two-dimensional space, as defined in this book, allows three different kinds of geometric entities: points, lines and plane figures. Any you have seen that points represent the most elementary, or primitive, entities; they are defined in terms of their coordinates in 2-D space.

Lines represent a somewhat more complicated entity—but not much more complicated. A line, as defined in this book, is specified in terms of the points that mark its ends. Thus a line is ultimately defined in terms of elementary points.

Plane figures represent the next step in the evolution of geometric entities, so it follows that they should be directly specified in terms of the lines that bound them. Since the lines, themselves, are defined in terms of points, the conclusion is that even plane figures are ultimately defined by elementary points in space.

You have also seen this notion of the evolution of geometric entities reflected in the F-L-P specifications in the past few chapters. The P specifications cite the coordinates of points, the L specifications cite the points that bound lines, and the F specifications cite the lines that bound a plane figure.

If you have been especially observant, you will have noticed that the L-P specifications for lines differ only at the P, or point-specification, level. The L specification for one line has the same form as that of any other line. The specifications for one line differs from that of any other line only by the coordinates of its endpoints.

In a similar way, the F specifications for a triangular plane figures are virtually identical in form; they show three lines, or L designations. Furthermore, the L designations for triangles look like L specifications for lines in any other sort of figure—a pair of point designations. So as far as the F-L specifications are concerned, one triangle looks exactly like any other. It is at the point-coordinate level that triangles differ from one another. The point coordinates—the P specifications—determine the shape, size and position of the triangle.

And that same general idea applies to any figure in 2-D space. All quadralaterals are identical as far as their F-L specifications are concerned, and differ only at the

P-coordinate level. Pentagons are identical at the F-L level, and so are plane figures of six and more sides.

The overall conclusion is that the P specifications—the coordinates of points that comprise a plane figure—dictate its size, general shape and position in a 2-D coordinate system. Perhaps you have noticed that all of the transformation equations apply directly to the points in a figure; there is no need to deal with anything else. Translate, scale or rotate the points in a figure, and you translate, scale or rotate the entire figure. The F and L specifications simply distinguish one kind of plane figure from another.

The following discussions emphasize these ideas, and extend them to include the formal analysis of the entities. The primary purpose is to demonstrate the logical simplicity and consistency of both the specifications and analyses of geometric entities in 2-D space. A secondary purpose is to help you build up a useful file of lines and space objects for future study and experiments.

9-1.1 Basic Lines in 2-D Space

Figure 9-1 and Table 9-1 summarize the specifications and analysis of any line in 2-D space. The line is defined in terms of two general endpoints, and the endpoints are defined in terms of their 2-D coordinates. That completes the specifications, the L-P specifications for any line.

The analysis of a line in 2-D space ought to include its length and its pair of direction cosines. Nothing else is necessary for a complete analysis.

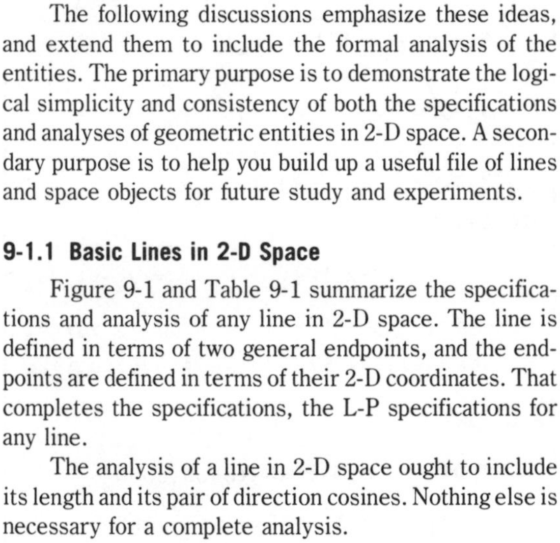

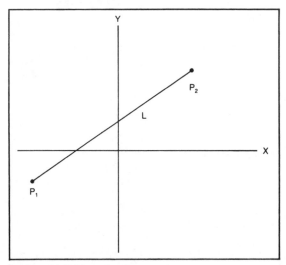

Fig. 9-1. A general line in 2-D space.

Table 9-1. Guide to the Analysis of a General Line in 2-D Space. See Fig. 9-1.

(A, B, C, and D correspond to α, β, γ, and δ.)

SPECIFICATIONS

$L=\overline{P_1P_2}$ $\qquad P_1=(x_1,y_1)$

$P_2=(x_2,y_2)$

LENGTH

$$s=\sqrt{(x_2-x_1)^2+(y_2-y_1)^2}$$

DIRECTION COSINES

$$\cos A=(x_2-x_1)/s$$

$$\cos B=(y_2-y_1)/s$$

Now notice in the analysis that the length of the line and its direction cosines are all calculated from the most primitive elements of the line—the endpoint coordinates. So when it comes to specifying and analyzing a line in 2-D space, all you have to do is supply some coordinates in the original specifications. Complete those two P specifications by substituting coordinate values for the X- and Y-axis components, and you have a specific line that you can analyze.

You are free to substitute any coordinate numbers you choose, just as long as the two points end up with different coordinates. Supply the same coordinates for both points, and you will end up with a single point that is not sufficient for defining a line.

Make up some line specifications of your own. Just plug in some numbers for the coordinates (making certain that the coordinates are different), plot the line on a 2-D coordinate system, and conduct a formal analysis by working out the length and direction cosines. That amounts to a complete experiment with lines in 2-D space.

Let me stress once again that lines in 2-D space ultimately differ only by their endpoint coordinates. Of course the lengths and direction cosines can be quite different from one line to another, but the forms of the equations remain unchanged.

If you think about the direction cosines for lines in 2-D space a little bit, or play around with some of your own for a while, you ought to come up with a couple of potentially useful principles:

☐ A line is parallel to the X axis whenever its alpha direction cosine is equal to 1 or – 1, and it is perpendicular to that axis whenever its alpha direction cosine is equal to zero. That means there are angles of zero or 90 degrees, respectively, between the line and the X axis.

☐ A line is parallel to the Y axis whenever its beta direction cosine is equal to 1 or – 1, and it is perpendicular to the Y axis whenever its beta direction cosine is equal to zero.

Those two principles represent just a couple of special cases. Generally speaking, you can always determine the angle between a line and the X axis of its coordinate system by taking the arc cosine of its alpha direction cosine. Likewise, you can determine the angle between the line and its Y axis by taking the arc cosine of its beta direction cosine. Please don't accept these ideas outright; think about them and convince yourself that they make sense.

9-1.2 Triangles in 2-D Space

Figure 9-2 and Table 9-2A outline the specifications and procedure of analysis for any triangle in 2-D space. The F specification in the table shows that every triangle is composed of three lines. That specification never changes for triangles. Then the specifications for the three lines show that each is composed of a pair of endpoints, but that there are only three different endpoints

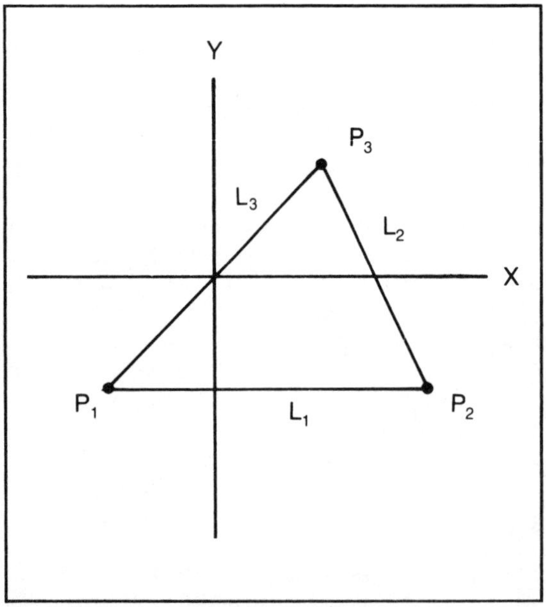

Fig. 9-2. A general triangle in 2-D space.

Table 9-2. A General Triangle in 2-D Space, Fig. 9-2. (A) Guide to Analysis. (B) Suggested Coordinates.

(A, B, C, and D correspond to α, β, γ, and δ.)

SPECIFICATIONS

A

$F=L_1L_2L_3$

$L_1=\overline{P_1P_2}$ $\qquad P_1=(x_1,y_1)$

$L_2=\overline{P_2P_3}$ $\qquad P_2=(x_2,y_2)$

$L_3=\overline{P_3P_1}$ $\qquad P_3=(x_3,y_3)$

LENGTH

$$s_1=\sqrt{(x_2-x_1)^2+(y_2-y_1)^2}$$

$$s_2=\sqrt{(x_3-x_2)^2+(y_3-y_2)^2}$$

$$s_3=\sqrt{(x_1-x_3)^2+(y_1-y_3)^2}$$

DIRECTION COSINES

$\cos A_1=(x_2-x_1)/s_1$ $\qquad \cos B_1=(y_2-y_1)/s_1$

$\cos A_2=(x_3-x_2)/s_2$ $\qquad \cos B_2=(y_3-y_2)/s_2$

$\cos A_3=(x_1-x_3)/s_3$ $\qquad \cos B_3=(y_1-y_3)/s_3$

ANGLES

$$\cos\phi_{1,2}=\cos A_1\cos A_2+\cos B_1\cos B_2$$

$$\cos\phi_{1,3}=\cos A_1\cos A_3+\cos B_1\cos B_3$$

$$\cos\phi_{2,3}=\cos A_2\cos A_3+\cos B_2\cos B_3$$

CHECK

$$\phi_{1,2}+\phi_{1,3}+\phi_{2,3}=180^\circ$$

B

FIGURE VERSION 1:

$P_1=(0,0)$ $\qquad P_2=(1,0)$ $\qquad P_3=(0,1)$

FIGURE VERSION 2:

$P_1=(-1,0)$ $\qquad P_2=(1,0)$ $\qquad P_3=(0,1)$

FIGURE VERSION 3:

$P_1=(-1,-1)$ $\qquad P_2=(2,1)$ $\qquad P_3=(3,3)$

involved. That particular combination of lines and point specifications ensures a complete closure of the three lines—a closure that is necessary for making a complete, unbroken, triangular figure in 2-D space.

You can see that the analysis of a triangle is somewhat more involved than that of a single line. You must, for example, calculate the lengths and direction cosines for three different lines. Then additionally, you should figure the angles between the lines—the internal angles of the triangle. The equations actually solve for the cosines of the angles, but of course you know that you can get the angles, themselves, by taking the arc cosine of the results.

But again, everything in the specifications and analysis rests ultimately with your choice of point coordinates; the form of the specifications and the equations for working out the analysis are vitually identical for every triangular figure in 2-D space.

You can generate any number of different triangles in 2-D space by assigning specific and carefully selected numbers for the component of the coordinates in the P specifications. See the suggest coordinates for some basic triangles in Table 9-2B. While you are conducting formal analyses of triangles in 2-D space, bear in mind three important ideas:

☐ The sums of the three angles in a triangle must be 180 degrees. Allow yourself a degree of error one way or the other, because it is difficult to achieve absolute precision when working with trigonometric quantities. But when you complete the analysis of a triangle and find that the sum of the three internal angles is very close to 180 degrees, you can be reasonably certain that you've made no serious errors in specifying and analyzing it. On the other hand, a failure to get close to 180 degrees worth of angles in a triangle clearly indicates that something is wrong somewhere in your work.

☐ If one of the internal angles of the triangle is greater than 90 degrees, the analysis will turn up an angle that is much smaller than it should be. Remedy that situation by subtracting the calculated angle from 180 degrees.

☐ If the angle between any two lines happens to be 90 degrees, you have specified a right triangle. If an angle happens to be zero degrees, you've made a mistake either when specifying the original coordinates or carrying out the arithmetic.

Plot the triangles suggested in Table 9-2B and conduct a complete analysis of them. Begin keeping a file of your more interesting and successful experiments of this kind.

9-1.3 Quadralaterals in 2-D Space

A quadralateral plane figure differs from a triangle only by having four points, lines and angles instead of three (Fig. 9-3). The F specification in Table 9-3A shows the three lines, and the L specifications define those lines in terms of their endpoints. The specifications are designed here to ensure a complete closure of the figure. Plug in coordinate values, such as those suggested in Table 9-3B, and the specifications will lead you to a valid 4-sided plane figure every time.

The formal analysis includes the lengths of the four lines, their direction cosines, and a family of four different angles between them. In principle, it is possible to find 16 different angles—16 different combinations of two lines each—but only four are necessary. Those are the four obvious internal angles of the figure.

While you are making up some coordinates, plotting the figures and conducting complete analyses of them, it will be helpful to bear in mind some special ideas:

☐ The sum of the four angles in a quadralateral is 360 degrees. Use that fact to doublecheck the results of an analysis, but allow a degree one way or the other to account for round-off errors in your calculations.

☐ Two lines in the figure are parallel to one another if the sum of the products of their corresponding direction cosines is 1 or −1. (That amounts to the same thing as

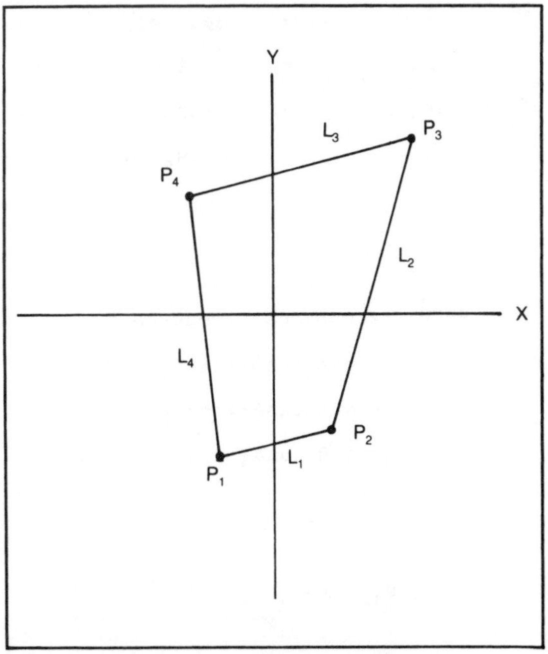

Fig. 9-3. A general quadralateral in 2-D space.

Table 9-3. A General Quadralateral in 2-D Space, Fig. 9-3. (A) Guide to Analysis. (B) Suggested Coordinates.

(A, B, C, and D correspond to α, β, γ, and δ.)

A

SPECIFICATIONS

$F=\overline{L_1L_2L_3L_4}$

$L_1=\overline{P_1P_2}$	$P_1=(x_1,y_1)$
$L_2=\overline{P_2P_3}$	$P_2=(x_2,y_2)$
$L_3=\overline{P_3P_4}$	$P_3=(x_3,y_3)$
$L_4=\overline{P_4P_1}$	$P_4=(x_4,y_4)$

LENGTH

$$s_1=\sqrt{(x_2-x_1)^2+(y_2-y_1)^2}$$

$$s_2=\sqrt{(x_3-x_2)^2+(y_3-y_2)^2}$$

$$s_3=\sqrt{(x_4-x_3)^2+(y_4-y_3)^2}$$

$$s_4=\sqrt{(x_1-x_4)^2+(y_1-y_4)^2}$$

DIRECTION COSINES

$\cos A_1=(x_2-x_1)/s_1$	$\cos B_1=(y_2-y_1)/s_1$
$\cos A_2=(x_3-x_2)/s_2$	$\cos B_2=(y_3-y_2)/s_2$
$\cos A_3=(x_4-x_3)/s_3$	$\cos B_3=(y_4-y_3)/s_3$
$\cos A_4=(x_1-x_4)/s_4$	$\cos B_4=(y_1-y_4)/s_4$

ANGLES

$$\cos\phi_{1,2}=\cos A_1\cos A_2+\cos B_1\cos B_2$$

$$\cos\phi_{1,4}=\cos A_1\cos A_4+\cos B_1\cos B_4$$

$$\cos\phi_{2,3}=\cos A_2\cos A_3+\cos B_2\cos B_3$$

$$\cos\phi_{3,4}=\cos A_3\cos A_4+\cos B_3\cos B_4$$

CHECK

$$\phi_{1,2}+\phi_{1,4}+\phi_{2,3}+\phi_{3,4}=360^\circ$$

B

FIGURE VERSION 1:

$P_1=(-1,-1)$	$P_2=(1,-1)$
$P_3=(1,1)$	$P_4=(-1,1)$

FIGURE VERSION 2:

$P_1=(-2,-1)$	$P_2=(2,-1)$
$P_3=(2,1)$	$P_4=(-2,1)$

FIGURE VERSION 3:

$P_1=(-3,-2)$	$P_2=(1,-2)$
$P_3=(1,2)$	$P_4=(-1,2)$

FIGURE VERSION 4:

$P_1=(-1,0)$	$P_2=(0,-1)$
$P_3=(1,0)$	$P_4=(0,1)$

saying that the angle between the two lines is zero degrees.)

☐ Two lines in the figure are perpendicular to one another if the sum of the products of their corresponding direction cosines is zero. (That is the same as saying that the angle between them is 90 degrees.)

Also try working such experiments by plotting a desired quadralateral onto the coordinate system, and then determining the initial coordinates from your drawing.

9-1.4 Pentagons in 2-D Space

Obviously, we can carry this progression of plane figures up to any number of sides, but I am going to conclude with the general 5-sided figure shown in Fig. 9-4. You can carry matters to 6-sided figures and more on your own.

The specifications in Table 9-4A show that a general pentagon is composed of five bounding lines and five different points. The analysis concludes with equations

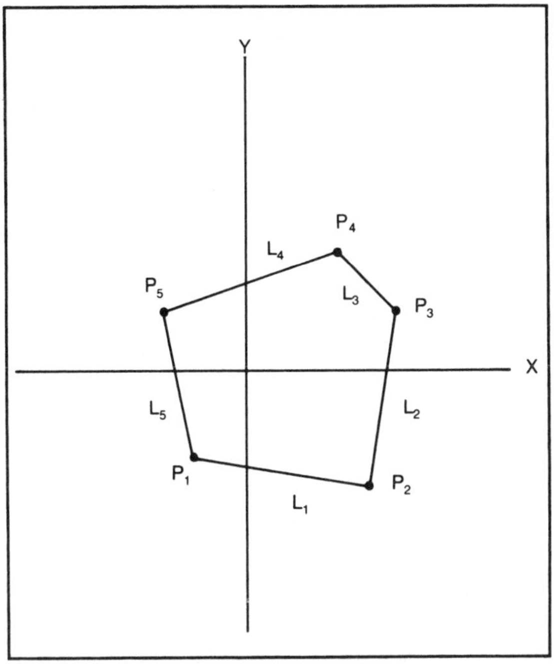

Fig. 9-4. A general pentagon in 2-D space.

for determining the cosines of five different internal angles and an equation for checking the final results.

Having presumably studied the general analysis for the 3- and 4-sided plane figures already described in this chapter, you should have no trouble following this one. The main idea is that all closed, 5-sided figures have basically the same specifications and call for the same general analysis. The uniqueness of a given pentagon rests with the coordinate values assigned to its points. Assign carefully selected sets of five 2-D coordinates to those points, and you are sure to end up with a legitimate pentagon. Try those suggested in Table 9-4B.

There is one point of special interest regarding such figures: The sum of the angles in a pentagon is 540 degrees. Use that fact to doublecheck all your work as suggested at the conclusion of Table 9-4A. (By way of a special project, see if you can find a method for determining what the sum of internal angles should be for closed figures of six or more sides.)

9-2 A SUMMARY OF SIMPLE TRANSFORMATIONS IN 2-D SPACE

Simple transformations are those that do not combine two or more different kinds of transformations. The simple transformations in 2-D space include:

- ☐ Translations or combinations of translations.
- ☐ Scalings or combinations of scalings.
- ☐ Rotations or combinations of rotations.

Combinations of translations and scalings do not fall into the category of simple transformations, nor do combinations of translations and rotations, and combinations of scalings and rotations. The closing sections of this chapter deal with such combinations of transformations in some detail.

Here, however, the subject is simple transformations; and the nice thing about simple transformations is that you can adhere to some general principles that can be of great help when conducting an analysis of a transformed line or plane figure. Consider this brief summary.

1. Under simple translations:
   - A. Length is an invariant
   - B. Direction cosines are invariant
   - C. Angles between lines are invariant
   - D. Simple translations are commutative
2. Under simple scalings:
   - A. Length is a variant
   - B. Direction cosines are variant
   - C. Angles between lines are variant
   - D. Simple scalings are commutative
3. Under simple rotations:
   - A. Length is an invariant
   - B. Direction cosines are variant
   - C. Angles between lines are invariant
   - D. Simple rotations are commutative

The next objective of this discussion is to elaborate on those special ideas and suggest some experiments.

9-3 SIMPLE TRANSLATIONS AND SUGGESTED EXPERIMENTS

Simple translations, you recall, generally carry out or express a spatial displacement between a reference figure and a translated version of it. One of the special features of simple translations is that the two versions are identical in every respect but one—their position with respect to the frame-of-reference coordinate system.

9-3.1 Taking Advantage of the Invariants

Knowing that lengths, direction cosines and angles are invariant under simple translations lets you do a couple of things. First, you can count on those parameters remaining unchanged upon translating any line or plane figure by any amount and in any direction in 2-D space. The practical conclusion is that you have to do the

Table 9-4. A General Pentagon in 2-D Space, Fig. 9-4. (A) Guide to Analysis. (B) Suggested Coordinates.

A

(A, B, C, and D correspond to α, β, γ, and δ.)

SPECIFICATIONS

$$F=\overline{L_1L_2L_3L_4L_5}$$

$L_1=\overline{P_1P_2}$	$P_1=(x_1,y_1)$
$L_2=\overline{P_2P_3}$	$P_2=(x_2,y_2)$
$L_3=\overline{P_3P_4}$	$P_3=(x_3,y_3)$
$L_4=\overline{P_4P_5}$	$P_4=(x_4,y_4)$
$L_5=\overline{P_5P_1}$	$P_5=(x_5,y_5)$

LENGTH

$$s_1=\sqrt{(x_2-x_1)^2+(y_2-y_1)^2}$$

$$s_2=\sqrt{(x_3-x_2)^2+(y_3-y_2)^2}$$

$$s_3=\sqrt{(x_4-x_3)^2+(y_4-y_3)^2}$$

$$s_4=\sqrt{(x_5-x_4)^2+(y_5-y_4)^2}$$

$$s_5=\sqrt{(x_1-x_5)^2+(y_1-y_5)^2}$$

DIRECTION COSINES

$\cos A_1=(x_2-x_1)/s_1$	$\cos B_1=(y_2-y_1)/s_1$
$\cos A_2=(x_3-x_2)/s_2$	$\cos B_2=(y_3-y_2)/s_2$
$\cos A_3=(x_4-x_3)/s_3$	$\cos B_3=(y_4-y_3)/s_3$
$\cos A_4=(x_5-x_4)/s_4$	$\cos B_4=(y_5-y_4)/s_4$
$\cos A_5=(x_1-x_5)/s_5$	$\cos B_5=(y_1-y_5)/s_5$

ANGLES

$$\cos\phi_{1,2}=\cos A_1\cos A_2+\cos B_1\cos B_2$$

$$\cos\phi_{2,3}=\cos A_2\cos A_3+\cos B_2\cos B_3$$

$$\cos\phi_{3,4}=\cos A_3\cos A_4+\cos B_3\cos B_4$$

$$\cos\phi_{4,5}=\cos A_4\cos A_5+\cos B_4\cos B_5$$

$$\cos\phi_{5,1}=\cos A_5\cos A_1+\cos B_5\cos B_1$$

CHECK

$$\phi_{1,2}+\phi_{2,3}+\phi_{3,4}+\phi_{4,5}+\phi_{5,1}=440^\circ$$

B

FIGURE VERSION 1:

$P_1=(-1,0)$	$P_2=(1,0)$
$P_3=(1,1)$	$P_4=(0,2)$
$P_5=(-1,1)$	

FIGURE VERSION 2:

$P_1=(-2,0)$	$P_2=(2,0)$
$P_3=(3,4)$	$P_4=(0,6)$
$P_5=(-3,4)$	

mathematics of the formal analysis only one time. Do it for the reference version of the line or figure, and then assume that there will be no change in length, direction cosines and angles for the analysis of the translated version. That can save you some time when it comes to conducting a formal experiment with simple translation effects.

A more careful experimenter, however, might use the invariant features of translations for another purpose. Since the length, direction cosines and angles are not supposed to change under a simple translation, taking the trouble to run a full analysis of both the reference and translated figure provides a good means for double-checking your calculations. Compare the results of the two analyses. If the lengths, direction cosines and angles are identical, you can be sure that you have made no errors in method or calculation anywhere along the way. If they differ in any respect, however, you know it is time to begin looking for a mistake.

When it comes to conducting your own experiments, it is just as important to know that you've made a mistake as it is to correct it. There is probably no one else around to check your work for you.

The fact that simple translations are commutative means that you can apply any combination of transformations in any sequence and end up with the same overall result. That might seem to be a trivial or self-evident notion, but its importance becomes clear when studying combinations of transformations.

9-3.2 Suggested Experiments and Studies

As simple as translation effects often appear to be, there is still room for conducting some interesting and meaningful experiments and special studies—especially for an experimenter who has little previous experience with the sort of thinking that goes into real mathematics or, in particular, the methods of analytic geometry.

There are several classes of experiments that beginners ought to find of special interest. One such class of experiment runs like this:

□ Specify a custom reference figure in 2-D space (see Section 9-1).

□ Conduct a complete analysis of the reference figure.

□ Specify and carry out a chosen simple 2-D translation.

□ Set up the formal specifications for the translated version of the figure and conduct a complete analysis of it.

□ Plot both the reference and translated version of the figure.

Conduct that sort of experiment for a variety of plane figures, including right, oblique and equilateral triangles; square, rectangular and trapezoidal quadralaterals; and several different pentagons, including a regular one. If the experiments seem a bit tedious or uninspiring after you've done a couple of them, at least notice that you are building a nice file of fully specified and analyzed plane figures.

A second class of simple translation experiments has more practical value. Check this outline:

□ Specify a reference plane figure in 2-D space and plot it onto a standard 2-D coordinate system.

□ Select a point in the coordinate system where you want a certain point in the figure to appear.

□ Solve for the translation terms that are necessary for displacing that selected point in your reference figure to the desired place in the coordinate system (Equation 6-2).

□ Translate the entire figure to that new position by applying the translation terms to its remaining points (Equation 6-1).

□ Specify the figure in its new position and plot it on the coordinate system.

That kind of experiment lets you adjust the position of a plane figure, presumably in order to set up the specifications for a more involved experiment.

A third kind of study involves applying successive translations to some reference figure, using the specifications for a previous version as the starting point for translating it again. The idea is to generate a series of displaced plane figures. Suppose for instance that you translate a reference figure with X- and Y-axis translations terms of 1. Use the specifications for that translated figure as the reference figure for a performing the same translation again. Execute the same translation effect any number of times, and you will find yourself plotting an array of figures.

That kind of experiment can be especially interesting and, indeed, somewhat artistic in nature, if you select certain progressions of translation terms. It can be a time-consuming process, but it pays off when you can show someone else that your "strange" experiments in geometry can be justified on the grounds of artistic endeavor.

Getting into an entirely different phase of experimentation, try working with the relativistic versions of the simple translation equations. Basically, the idea is to view a figure from two different points of view; from points of view that are identical except for a spatial displacement between them. Here is an outline for an elementary version of such an experiment:

□ Specify a reference figure in the 2-D space of the observer's frame of reference.

□ If you haven't done so in past experiments, conduct a complete analysis of that figure as viewed from the observer's frame of reference.

□ Specify some translation terms and use the relativistic forms of the translation equation (Equation 6-4) to determine the specifications for the figure as it appears from the translated frame of reference.

□ Conduct a complete analysis of the figure as it appears from the translated frame of reference and compare the results with the analysis of the figure in the observer's frame of reference.

A variation of that sort of experiment combines the simple geometric and relativistic versions of the translation equations. The general idea is to set up a figure and specify it in both the observer's and translated frames of reference. That part of the procedure is shown in the previously outlined experiment. Then apply a geometric translation to the figure as it is specified in the observer's frame of reference. That, in effect, moves the figure. Then use the relativistic equations to determine the new position of the figure as viewed from the observer's frame of reference. Do a long series of those geometric translations, and you will get a strobe-like impression of a moving figure as it appears from two different frames of reference.

Along the lines of special studies, there is a lot of work for investigators who are more interested in general principles than simple translation effects. Consider the following suggestions:

☐ Without referring to previous discussions in this book, see if you can conduct a formal proof of the invariance of length, direction cosines and angles under both geometric and relativistic translations in 2-D space.

☐ Study the effect of specifying two or more identical coordinates for lines and plane figures.

☐ See if you can derive an equation for calculating the length of the perimeter of a plane figure as a function of its endpoint coordinates.

☐ See if you can prove that either the alpha or the beta direction cosine is adequate for expressing the orientation of a line in 2-D space with respect to the coordinate axes.

☐ Use the results of the paragraph cited above to derive an equation for the angle between two lines, given only the alpha or beta direction cosine.

☐ Look up the definition of the slope of a line in 2-D space in another source. Derive an equation that expresses the slope of a line in terms of the notation suggested throughout this book.

☐ Based on the results of the paragraph cited above, derive an equation for calculating the angle between two lines in 2-D space as a function of the slopes of the lines.

9-4 SIMPLE SCALINGS AND SUGGESTED EXPERIMENTS

Simple scalings in 2-D space can expand, contract and reflect (reverse) the scaling of a coordinate system. What's more, the scalings can be symmetrical or asymmetrical; symmetrical scalings apply the same scaling factor to both the X and Y axis, while assymetrical scalings do not.

For the purposes of the present discussion, it is convenient to classify 2-dimensional scalings as symmetrical or asymmetrical, and as reflective or non-reflective.

☐ Under symmetrical, non-reflective scalings:

A. Length is a variant
B. Direction cosines are invariant
C. Angles are invariant

Those ideas make sense when you understand that a symmetrical scaling expands or reduces a line or plane figure evenly in both the X and Y directions. The lengths of lines (and the size of a figure) can certainly change under those circumstances, but the angles between lines remain fixed. And as long as the symmetrical scaling is non-reflective as well, the lines maintain the same orientation (direction cosines) with respect to their coordinate axes. Using the nomenclature of classical descriptive geometry: Plane figures are similar under symmetrical, non-reflective scalings.

☐ Under symmetrical, reflective scalings:

A. Length is a variant
B. Direction cosines are variant
C. Angles are invariant

You can see that the only difference between scalings that are symmetrical and non-reflective and those that are symmetrical and reflective is the variance of the direction cosines. Under symmetrical and reflective scalings, the direction cosines simply change sign; the direction cosines for the reference and scaled version of a figure maintain the same absolute values, but show opposite signs (directions).

☐ Under assymetrical scalings, both non-reflective and reflective:

A. Length is a variant
B. Direction cosines are variant
C. Angles are variant

Notice that asymmetrical scalings are the only ones that affect the angles between lines. Just picture a triangle, for instance, being scaled by one amount along the X axis and a different amount along the Y axis. Clearly such a transformation will affect the angles that comprise the figure; and the greater the difference between the two scaling factors, the greater the change in the angles.

In short, you cannot count on any parameter of a line or figure in 2-D space remaining unchanged under scaling transformations. The practical significance of that idea is that you are often forced to conduct complete analyses of both the reference and scaled versions of the figure.

9-4.1 Some Suggested Experiments

Experiments with simple scaling transformations can be a bit more dramatic than those for translations. Scalings offer an opportunity to mold a figure; expanding, compressing and reflecting it. Consider this type of experiment:

☐ Specify a reference figure in 2-D space according to the methods outlined in Section 9-1.

☐ Conduct a complete analysis of that reference figure.

☐ Specify some desired scaling factors and carry out the scaling operation (Equation 7-1).

☐ Write out the specifications for the scaled version of the figure and conduct a complete analysis of it.

☐ Plot both the reference and scaled versions of the figure.

Conduct that sort of experiment for a variety of figures specified in 2-D space. If you've done some of the translation experiments suggested earlier, you should already have a file of reference figures already specified and analyzed; all you have to do is carry out some scalings and analyze the results. Be sure to do some work with reflective scalings as well as the usual types of symmetrical and asymmetrical scalings.

Play around with this sort of experiment for a while, and you will get a feeling for picking scaling factors that accomplish the effects that you want. The idea is to mold the shape and size of some plane figure you have already specified in a "standard" form.

Applying successive scalings can create some interesting visual effects when the results are plotted onto the same coordinate system. It can be a tricky and time-consuming process, but the results can be visually striking. Try specifying and plotting a relatively small plane figure, then apply a symmetrical scaling of 1.5 to it. Plot that scaled figure, and then use it as a reference figure for doing another scaling of 1.5. Continue that procedure until the figure grows too large to fit onto the paper. Try enhancing the visual effect by plotting the successively larger figure with different-colored pencil or ink. Then try progressions of scaling factors, including some experiments with assymetrical and reflection-type scalings.

An entirely different sort of scaling experiment suggests using the relativistic scaling equation (Equation 7-4). The general idea is to plot a plane figure as it appears from two different frames of reference—frames of reference that differ only in the scaling that exists between them. Assymetrical and reflection-type relativistic scaling effects are particularly intriguing.

Combine simple, geometric scalings with relativistic scalings, and you have the most sophisticated type of experiment you can perform in this context. It isn't easy, but it comes close to representing some of the work that currently interests physicists with regard to the nature of our seemingly expanding universe.

9-4.2 Some Suggested Special Studies

The special studies suggested here are often less dramatic than the experiments just cited, but they lead to some special insight that can be of great help when it comes to setting up and conducting your own experiments. Knowing how to handle special ideas with some confidence leads one to a more responsible and productive frame of mind. Consider these:

☐ Derive a general equation for the length of a scaled line as a function of the coordinates of the points in the reference version of the figure and the scaling factors.

☐ Derive a general equation for the length of a scaled line as a function of its reference length and the scaling factors.

☐ Conduct a formal, analytic proof of the fact that a line that is parallel to one of the coordinate axes remains parallel to that axis under any sort of scaling in 2-D space.

☐ Derive a general equation for calculating the angle between two lines as a function of the corresponding angle in the reference figure and the given scaling factors.

☐ Derive a set of equations for calculating the coordinates of points relative to a scaled frame of reference, given the corresponding coordinates of the figure relative to the observer's frame of reference, a geometric scaling with respect to the observer's frame of reference, and the relativistic scaling that exists between the observer's and scaled frames of reference.

9-5 SIMPLE ROTATIONS AND SUGGESTED EXPERIMENTS

Simple rotations turn a coordinate system about its own origin. In effect, they carry out angular displacements of points, lines and plane figures specified in that coordinate system. The reference (non-rotated) and rotated versions of those entities are identical except for their position and angular orientation (direction cosines) with respect to the coordinate axes. The lengths of lines and angles between them are invariant.

The same principles apply to rotating a coordinate system about some off-origin point. As described later, such off-origin rotations are actually combinations of rotations and translations; but nature of the off-origin rotation equation allows us to treat that particular combination as though it were a simple transformation.

9-5.1 Taking Advantage of the Invariants

Only the direction cosines of lines are variant under simple rotations in 2-D space—the lengths of corresponding lines and the corresponding angles are invariant. So you can doublecheck your work by comparing the analyses of the lengths of lines and the figures' angles; especially the angles. Why is a comparison of corresponding angles important? First because they are invariant, and second because the calculation of the angles for the rotated version of the figure uses the direction cosines for that figure. If you mess up the lengths of lines or the direction cosines, the trouble will show up when comparing the corresponding angles of the two versions of the figure.

9-5.2 Suggested Experiments and Studies

The most straightforward kinds of experiments with simple rotations in 2-D space take this general form:

☐ Specify a custom reference figure in 2-D space (Section 9-1).

☐ Conduct a complete analysis of the reference figure.

☐ Carry out a desired rotation, either an on-origin or off-origin version (Equations 8-1 or 8-5).

☐ Establish the formal specifications for the rotated version of the figure and conduct a complete analysis of it.

☐ Plot both the reference and rotated versions of the figure.

Assuming that you have already generated a file of 2-D figures in some previous experiments, this sort of general experiment suggests extending the file to show the figures in different positions and with different orientations with respect to the coordinate axes. You can then use the specifications, analyses and plots as reference material for initiating other kinds of experiments.

It is often tempting at this time to consider experiments whereby you set up a reference figure, locate a point where you want one of the coordinates of the figure to appear, and then attempt to solve for an angle of rotation that will do the job for you (Equation 8-2). I suggested that sort of procedure for simple translations, so it ought to work for rotations, too. But there's a problem—you must be very careful about selecting the target coordinate.

Suppose that you want to rotate a figure such that its point P rotates to point P′ elsewhere on the coordinate system. The tricky part is that the straight-line distance from the point of rotation to point P′ must be the same as that from the point of origin to point P. Break that rule, and you won't be able to get a real solution to the equation that solves for the necessary angle of rotation. That sort of situation is better handled by a special technique from linear algebra that is really beyond the scope of this book.

But what about the idea of applying a rotation of some relatively small angle such as 15 degrees to some reference figure, and then applying another 15-degree rotation to the result? Do that several times—apply a rotation to the result of a previous rotation—and you will end up with a visually interesting plot. If you have the patience and precision of a good draftsperson, you can turn out some fascinating drawings that way.

Then, of course, there is the matter of relativistic rotations. The objective is to specify a relativistic rotation that exists between an observer's and rotated coordinate system, specify a figure as it would appear from the observer's point of view, and then use the relativistic rotation equations to find out how that figure appears from the rotated frame of reference (Equation 8-3).

Combining relativistic and geometric rotations into the same experiment can be a worthwhile effort, too. Consider this general experimental procedure:

☐ Specify a fixed relativistic rotation that is to exist between the observer's and the rotated frames of reference.

☐ Specify a reference plane figure relative to the observer's frame of reference.

☐ Apply the relativistic rotation equation to see how the reference figure appears from the rotated frame of reference.

☐ Apply some geometric angle of rotation to the reference figure.

☐ Repeat Steps 3 and 4 any number of times, plotting the results in each case.

There are a number of special studies with simple rotations in 2-D space that can go a long way toward bolstering your confidence in designing and conducting such experiments. Here is a list of some of them.

☐ Conduct a formal analytic proof of the fact that length is an invariant under 2-D rotations, including off-origin rotations.

☐ Conduct a formal analytic proof that the internal angles of a plane figure in 2-D space are invariant under simple 2-D rotations, including off-origin rotations.

☐ Derive a set of equations that determine the direction cosines of a line that is being rotated in 2-D space as a function of the angle of rotation.

9-6 COMBINATIONS OF TRANSLATIONS AND SCALINGS

Combinations of translation and scaling transformations are perhaps the simplest to execute, and they are certainly easy to perceive in a 2-D environment. The situation is this: Given some reference line or plane figure specified in 2-D space, apply a combination of translation and scaling operations to generate the final specifications for a transformed—both translated and scaled—version. There is a special feature inherent in the idea, and a careful experimenter ought to be fully aware of it.

You know from previous work that translations, in themselves, are commutative. Given two or more translations to be performed in succession, their sequence of execution is not relevant. The entity always ends up with the same final specifications. And you have seen that is also true for scaling operations, alone.

But it turns out that combinations of translation and scaling operations are not commutative. Given both a translation and a scaling, you are likely to end up with different results, depending upon which order you execute them. The following discussions demonstrate that fact.

9-6.1 Translation Followed by Scaling

Example 9-1 suggest a particular translation that is to be applied to a reference plane figure. Further, the example cites a certain scaling that is to be applied to the result. Setting up some new notation to make the discussions of this sort run more smoothly, designate that combination of two different transformations this way:

$$T{:}K$$

That implies a translation T followed by a scaling K.

Example 9-1

Given the following specifications for a plane figure in 2-D space, apply a translation of $T_x = 1$, $T_y = 0$ followed by a scaling of $K_x = 2$, $K_y = 0.5$.

$F = \overline{L_1L_2L_3}$

$L_1 = \overline{P_1P_2}$ $\quad P_1 = (-1,0)$

$L_2 = \overline{P_2P_3}$ $\quad P_2 = (1,0)$

$L_3 = \overline{P_3P_1}$ $\quad P_3 = (0,1)$

See that reference figure plotted in Fig. 9-5A.

1. Applying the given translation as the first of the two translations, the points become:

$P_{1,1} = (0,0)$ $\quad P_{2,1} = (2,0)$ $\quad P_{3,1} = (1,1)$

See the result in Fig. 9-5B.

2. Applying the given scaling to the results of Step 1, points become:

$P_{1,2} = (0,0)$ $\quad P_{2,2} = (4,0)$ $\quad P_{3,2} = (2,0.5)$

See the result plotted in Fig. 9-5C.

3. Drawing up the specifications for the final result:

$F' = \overline{L_1'L_2'L_3'}$

$L_1' = \overline{P_1'P_2'}$ $\quad P_1' = (0,0)$

$L_2' = \overline{P_2'P_3'}$ $\quad P_2' = (4,0)$

$L_3' = \overline{P_3'P_1'}$ $\quad P_3' = (2,0.5)$

Step 1 in the example shows the effect of the initial translation upon the coordinates of the points in the original figure. Point P_1 is translated to point $P_{1,1}$; point P_2 is translated to $P_{2,1}$; and point P_3 is translated to $P_{3,1}$.

Before going any farther with this discussion of the example, notice some additional new notation. The result of the initial transformation is shown generally as one that transforms any point P_n in the reference entity to point $P_{n,1}$; where n indicates the point-number label, and the 1 designates the result of the first-applied transformation. You will see shortly that a designation such as $P_{n,2}$ will designate the coordinate of point *n* following the second transformation. That notation is a bit awkward, but you will find with experience that it represents a great way to keep things straight when combining two or more kinds of transformations.

Returning now to the example, Step 2 applies the given scaling factors to the results of Step 1. Overall, point P_1 in the original reference figure is transformed to

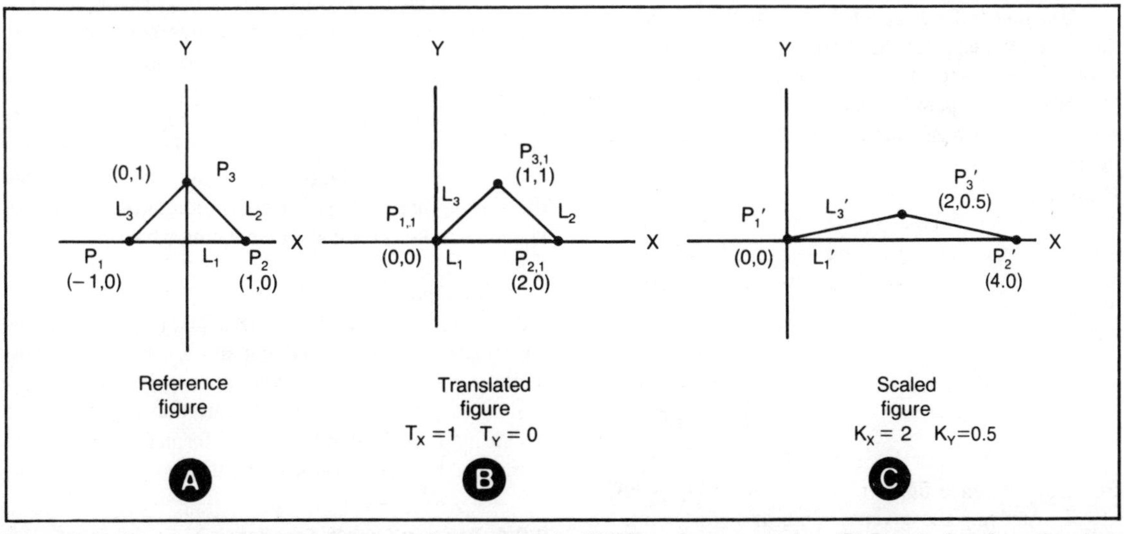

Fig. 9-5. Applying a T:K transformation to a reference triangle. (A) The reference version. (B) Result of applying the given translation. (C) The result of applying the scaling. See Example 9-1.

point $P_{1,2}$; P_2 is transformed to $P_{2,2}$; and P_3 is tranformed to $P_{3,2}$. The transformation—the specified T:K transformation—is done, and you can see the sequence plotted in Fig. 9-5.

Step 3 simply specifies the transformed version of the figure, using notation that conforms to that used in all previous studies.

The idea is a simple one, and so is the procedure. I just hope that you aren't tripping over the notation. Apply the given translation to the original reference figure, apply the given scaling to the resulting points, and then respecify the figure. That's all there is to it. The special notation simply helps organize the analytic procedures.

Now conduct analyses of the original reference figure and the final transformed version. It should come as no great surprise to you that the line parameters are rather different for the two versions. And why not? The overall transformation includes a scaling, and you know that scaling operations quite often change everything about lines and angles between lines. Specify several other T:K transformations of your own, and execute them, using Example 9-1 as a guide.

9-6.2 Scaling Followed by Translation

Whenever you are dealing with two different kinds of transformations, there are always two different ways to execute them. You have just seen the effects of doing a translation followed by a scaling—a T:K transformation. The alternative is to conduct a scaling followed by a translation. That would be designated a K:T transformation.

Example 9-2 sets up a K:T transformation that cites the same original reference figure, translation terms and scaling factors as the previous example. Study that example and the accompanying diagrams (Fig. 9-6) and you will see how a K:T transformation affects the same reference figure.

Example 9-2

Given the specifications for a reference triangle in 2-D space (the same one cited in Example 9-1), apply a scaling of $K_x = 2$, $K_y = 0.5$ followed by a translation of $T_x = 1$, $T_y = 0$.

$$F = \overline{L_1L_2L_3}$$
$$L_1 = \overline{P_1P_2} \qquad P_1 = (-1,0)$$
$$L_2 = \overline{P_2P_3} \qquad P_2 = (1,0)$$
$$L_3 = \overline{P_3P_1} \qquad P_3 = (0,1)$$

See that reference figure plotted again in Fig. 9-6A.

1. Applying the given scaling first, the points become:

$$P_{1,1} = (-2,0) \qquad P_{2,1} = (2,0) \qquad P_{3,1} = (0,0.5)$$

2. Applying the given translation to the result in Step 1, the points become:

$$P_{1,2} = (-1,0) \qquad P_{2,2} = (3,0) \qquad P_{3,3} = (1,0.5)$$

See the plot in Fig. 9-6C.

3. Specifying the final transformed version of the figure:

$$F' = \overline{L_1'L_2'L_3'}$$
$$L_1 = \overline{P_1'P_2'} \qquad P_1' = (-1,0)$$
$$L_2' = \overline{P_2'P_3'} \qquad P_2' = (3,0)$$
$$L_3' = \overline{P_3'P_1} \qquad P_3' = (1,0.5)$$

Using the example as a guide, execute some K:T transformations of your own. Be sure to plot and fully analyze at least the original reference figure and its final transformed version.

9-6.3 Drawing Some Conclusions

I mentioned earlier that combinations of translation and scaling transformations are not commutative. The idea is that a T:K transformation is something different from a K:T transformation. You can clearly see the difference in this case by comparing the final results of the two previous examples. The triangles in Figs. 9-5C and 9-6C are congruent (in our analytic terminology, that means that corresponding lines have the same lengths and direction cosines, and corresponding angles are equal). The congruent figures, however, are located at different places in the frame-of-reference coordinate systems. It makes a big difference in the final result whether you do a T:K or a K:T transformation.

That principle takes on some practical significance in the context of setting up a custom reference figure from the specifications for a "standard" version of it; a version perhaps selected from Section 9-1 of this chapter. The scaling operation molds the figure with respect to its general configuration and size, while the translation adjusts its position in the coordinate system. The fact that translations and scalings are not commutative means that you have to be careful about selecting which is executed first.

Trial-and-error works in your favor in this case. Take some "standard" figure and adjust its position in the coordinate system by applying an appropriate transformation. If you are satisfied with the results, then apply a scaling to get the size and configuration that you want. Tinkering around with those transformations will eventually get you the specifications that you want.

9-6.4 Deriving Custom Equations

The final comments in the previous section might

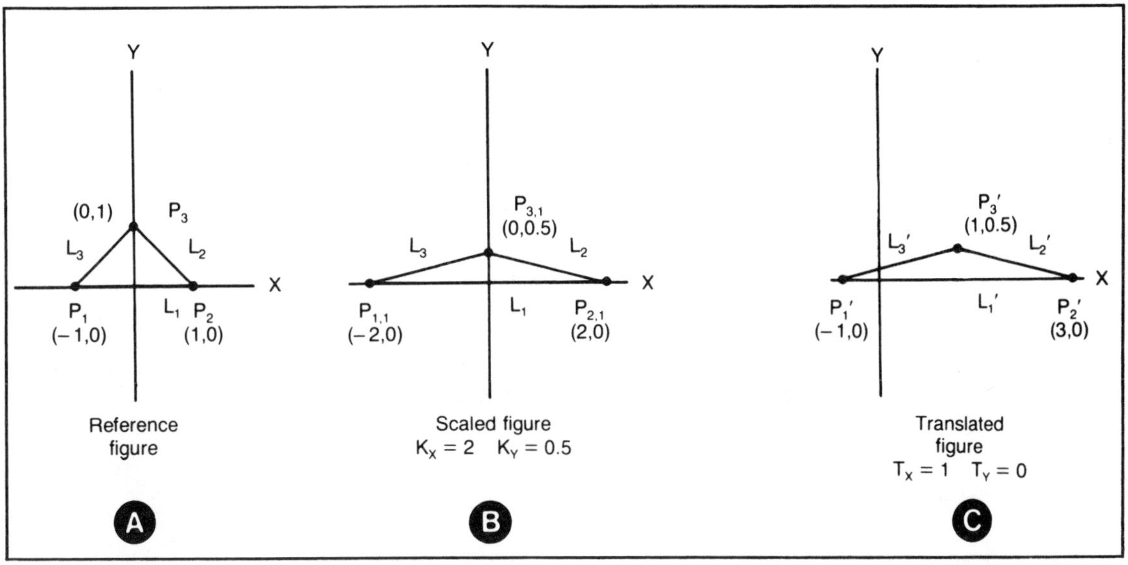

Fig. 9-6. Applying a K:T transformation to a reference triangle. (A) The reference version of the figure. (B) Result of applying the given scaling. (C) The result of applying the translation. See Example 9-2.

leave you with the impression that this entire matter of the non-commutative nature of translations and scalings is purely academic. To be sure, you can conduct a lot of useful experiments on a step-by-step, trial-and-error basis without being aware of that principle. But the idea serves another purpose of greater interest to experimenters who like to work with such ideas in a more generally and fully analytic way.

Suppose, for example, that you want to develop a formal proof of the fact that T:K and K:T transformations are not commutative; and perhaps you want to take the notion a step farther and generate some equations that lead directly to the coordinates, lengths of lines, direction cosines and angles under those two kinds of transformations. This is where the notation introduced earlier becomes a powerful tool.

Derivation 9-1 illustrates a procedure for determining the coordinate of any point upon subjecting it to a T:K transformation.

Derivation 9-1

The objective of this derivation is to obtain a transformation-of-points equation for a T:K transformation in 2-D space.

1. Expressing the translation as the first transformation:

$$x_{n,1} = x_n + T_x \qquad y_{n,1} = y_n + T_y$$

2. Expressing the scaling as the second transformation:

$$x_{n,2} = x_{n,1}K_x \qquad y_{n,2} = y_{n'}{,}1^{K}{}_{y}$$

3. Substituting the expressions in Step 1 into those of Step 2:

$$x_{n,2} = (x_n + T_x)K_x \qquad y_{n,2} = (y_n + T_y)K_y$$

4. And adjusting Step 3 for using our standard notation:

$$x_n' = (x_n + T_x)K_x \qquad y_n' = (y_n + T_y)K_y$$

Derivation 9-2 repeats the procedure, but applies it to a K:T transformation. The result is clearly different from that obtained for the T:K case.

Derivation 9-2

The objective of this derivation is to generate an equation for a basic K:T transformation of points in 2-D space.

1. Expressing the scaling as the first transformation:

$$x_{n,1} = x_nK_x \qquad y_{n,1} = y_nK_y$$

2. Expressing the translation as the second transformation:

$$x_{n,2} = x_{n,1} + T_x \qquad y_{n,2} = y_{n,1} + T_y$$

3. Substituting the expressions in Step 1 into those of Step 2:

$$x_{n,2} = x_nK_x + T_x \qquad y_{n,2} = y_nK_y + T_y$$

4. Using our standard notation:

$$x_n' = x_nK_x + T_x \qquad y_n' = y_nK_y + T_y$$

Derivation 9-3 uses the result of the T:K transformation in Derivation 9-1 to come up with an equation that directly yields the length of any line that has undergone that sort of transformation.

Derivation 9-3

Given the basic equation for the transformation of points under a T:K transformation of 2 dimensions, derive an equation for the length of a line as transformed in that fashion.

1. Let endpoints *u* and *v* of a transformed line in 2-D space be defined this way:

$$P_u' = (x_u', y_u') \qquad P_v' = (x_v', y_v')$$

2. The length of such a line is then determined by the general length equation for 2-D space:

$$s' = \sqrt{(x_u' - x_v')^2 + (y_u' - y_v')^2}$$

3. From the general T:K transformation equation:

$$x_u' = (x_u + T_x)K_x \qquad y_u' = (y_u + T_y)K_y$$
$$x_v' = (x_v + T_x)K_x \qquad y_v' = (y_v + T_y)K_y$$

4. Substituting the expressions of Step 3 into the equation in Step 2, and then simplifying the result:

$$s' = \sqrt{K_x^2(x_u - x_v)^2 + k_y^2(y_u - y_v)^2}$$

Use Derivation 9-3 as a guide for deriving an equation that expresses the length of any line under a K:T transformation. Compare the results, and you should see that they are identical. That, itself, constitutes a formal analytic proof of the fact that length of a line is the same under T:K and K:T transformations.

See if you can work out derivations for equations that lead directly to the direction cosines and angles between lines under both T:K and K:T transformations. Comparing those results ultimately leads to the conclusion demonstrated earlier—that the same reference figure subjected to T:K and K:T transformations will be congruent.

Combinations of translations and scalings are not commutative with respect to the position of a figure, but they are commutative with respect to the final lengths of lines, direction cosines and angles between lines. Mathematical life is interesting, isn't it? Things seem complicated only when you are attempting to work from rote memory; try working from careful thinking and sound analytic procedures, and you'll find yourself really understanding good ideas rather than memorizing them.

9-7 COMBINATIONS OF TRANSLATIONS AND ROTATIONS

You know that sequences of translations are commutative and so are sequences of rotations; but combinations of translations and rotations are not. Using the same approach and, indeed, the same reference figure cited earlier, the final results of Examples 9-3 and 9-4 demonstrate the principle that T:R and R:T transformations are not the same kinds of transformations.

Example 9-3 considers the T:R case—a translation followed by a rotation. Study it carefully, making certain that you understand every step; especially the notation and the method of substituting the result of a previous step into the set-up for the next. Conduct complete analyses for both the initial reference figure and the final transformed version of it.

Example 9-3

Given the following triangle specified in 2-D space, apply a translation of $T_x = -1$, $T_y = -1$ followed by a rotation of 45 degrees about the origin of the coordinate system.

$$F = \overline{L_1L_2L_3}$$
$$L_1 = \overline{P_1P_2} \qquad P_1 = (-1,0)$$
$$L_2 = \overline{P_2P_3} \qquad P_2 = (1,0)$$
$$L_3 = \overline{P_3P_1} \qquad P_3 = (0,1)$$

See Fig. 9-7A.

1. Applying the given translation to the reference figure, the points become:

$$P_{1,1} = (-1,-1) \qquad P_{2,1} = (0,-1) \qquad P_{3,1} = (-1,0)$$

See the plot in Fig. 9-7B.

2. Applying the given rotation to the results of Step 1, the points become:

$$P_{1,2} = (0,-\sqrt{2}/2 \qquad P_{2,2} = (\sqrt{2}/2,-\sqrt{2}/2$$
$$P_{3,2} = (-\sqrt{2}/2,-\sqrt{2}/2)$$

See the plot in Fig. 9-7C.

3. Gathering the specifications for the overall transformation:

$$F' = \overline{L_1'L_2'L_3'}$$
$$L_1' = \overline{P_1'P_2'} \qquad P_1' = (0,-1.4)$$
$$L_2' = \overline{P_2'P_3} \qquad P_2 = (0.7,-0.7)$$
$$L_3' = \overline{P_3'P_1'} \qquad P_3' = (-0.7,-0.7)$$

Example 9-4 runs through the same general procedure, but for the R:T case—rotation followed by translation. Be sure to analyze the final transformed version.

Example 9-4

Given the following reference triangle in 2-D space (the same one cited in Example 9-3), apply a rotation of 45 degrees about the origin followed by a translation of $T_x = -1$, $T_y = -1$.

$$F = \overline{L_1L_2L_3}$$
$$L_1 = \overline{P_1P_2} \qquad P_1 = (-1,0)$$

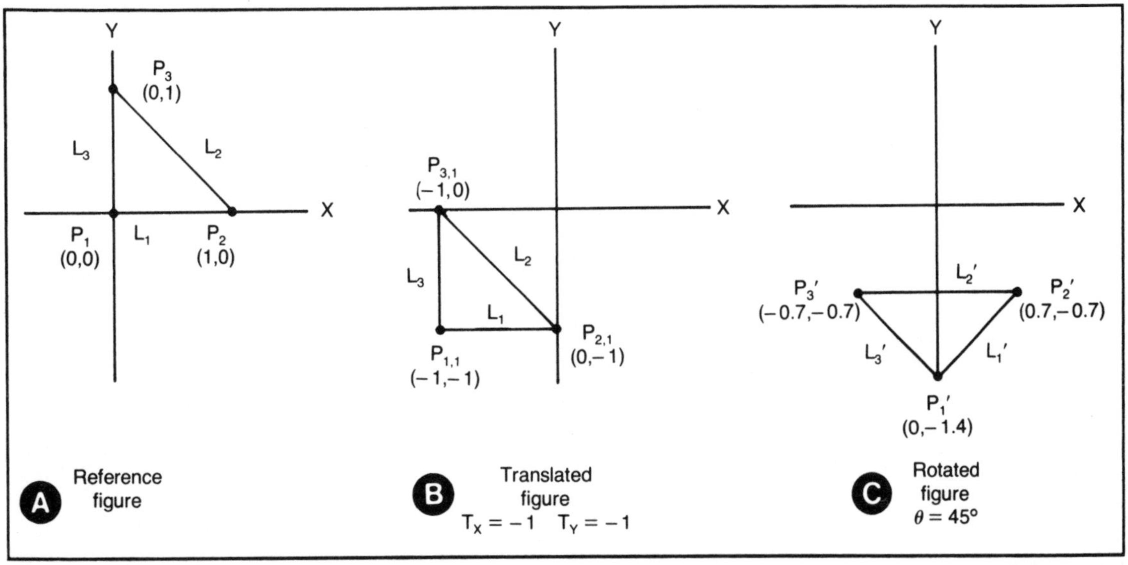

Fig. 9-7. Applying a T:R transformation to a reference triangle. (A) The reference version of the figure. (B) Result of applying the given translation scaling. (C) The result of applying the rotation. See Example 9-3.

$$L_2 = \overline{P_2P_3} \qquad P_2 = (1,0)$$
$$L_3 = \overline{P_3P_1} \qquad P_3 = (0,1)$$

See the plot in Fig. 9-8A.

1. Applying the given rotation first, the points become:

$$P_{1,1} = (0,0) \qquad P_{2,1} = (\sqrt{2}/2,\sqrt{2}/2)$$
$$P_{3,1} = (-\sqrt{2}/2,\sqrt{2}/2)$$

See Fig. 9-8B.

2. Applying the given translation to the results in Step 1, the points become:

$$P_{1,2} = (-1,-1) \qquad P_{2,2} = ([\sqrt{2}-2]/2,[\sqrt{2}-2]/2)$$
$$P_{3,2} = (-[\sqrt{2}+2]/2,[\sqrt{2}-2]/2)$$

See Fig. 9-8C.

3. Assembling the specifications for the overall transformation:

$$F' = \overline{L_1'L_2'L_3'}$$
$$L_1' = \overline{P_1'P_2'} \qquad P_1' = (-1,-1)$$
$$L_2' = \overline{P_2'P_3'} \qquad P_2' = (-0.3,-0.3)$$
$$L_3' = \overline{P_3'P_1'} \qquad P_3' = (1.7,-0.3)$$

Comparing the overall results in Figs. 9-7C and 9-8C, you can see that the triangles end up in different positions for the T:R and R:T cases. That is an adequate negative proof that combinations of translations and rotations are not commutative.

But what about the lengths of corresponding lines, values of corresponding direction cosines and angles? To phrase the question differently: Are the figures resulting from T:R and R:T transformations congruent? I leave the proof, one way or the other, to you. Just apply the techniques of analytic derivation shown in the previous section of this chapter.

And if you haven't the heart, mind and spirit for deriving equations and conducting formal proofs, at least keep in mind that it makes a big difference whether you combine translations and rotations by doing one or the other first.

Back in Chapter 8 you found an equation that is useful for rotating a 2-D coordinate system about any desired off-origin point. (The basic rotation equation, you recall, always assumes a rotation about the origin.) You might also remember that I suggested that the off-origin rotation equation represents a combination of two different transformations—a rotation and a translation. It is, in fact, a very special combination of two translations and a single rotation.

The idea behind an off-origin rotation is to translate the desired point of rotation to the origin of the coordinate system, perform the rotation, and then translate the point of rotation back to its original position. Simple rotations always take place about the origin; so why not slip a desired point of rotation to the origin, do the rotation, and then put it back. It is a case of doing a translation-

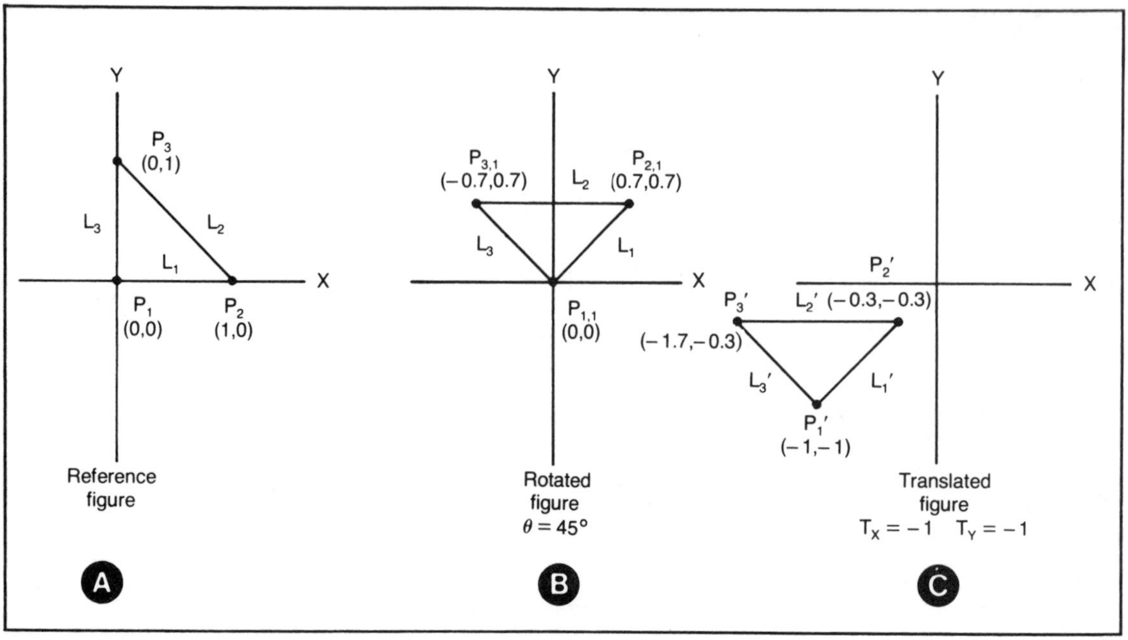

Fig. 9-8. Applying an R:T transformation to a reference triangle. (A) The reference version of the figure. (B) Result of applying the given rotation. (C) The result of applying the translation. See Example 9-4.

rotation-translation transformation. The appropriate notation is:

T1:R:T2

where T1 is the first translation, R is the desired rotation, and T2 is the second translation.

Derivation 9-4 uses that notation, some ideas offered earlier in this chapter, and a few equations that should be quite familiar by now to generate that equation for off-origin rotations in 2-D space.

Derivation 9-4

The general plan is to execute a series of transformation as follows:

☐ Translate the desired point of rotation to the origin of the frame-of-reference coordinate system—translation T_1.

☐ Rotate the coordinate system through a desired angle about the origin—rotation R.

☐ Translate the origin back to the point of rotation—translation T_2.

So the sequence of transformations is T_1:R:T_2.

Here is the derivation, itself:

1. Given a desired point of rotation (d_x,d_y), the first step calls for determining translation T_1 terms that translate the point to the origin of the coordinate system (0,0).

$$T_{x},1 = 0-d_x$$
$$T_{y},1 = 0-d_y$$

2. As a result of applying those T_1 translation terms to the reference figure:

$$x_{n,1} = x_n + T_{x,1} \qquad y_{n,1} = y_n + T_{y,1}$$
$$x_{n,1} = x_n - d_x \qquad y_{n,1} = y_n - d_y$$

3. Applying rotation R to the results in Step 2:

$$x_{n,2} = x_{n,1}\cos\theta - y_{n,1}\sin\theta$$
$$= (x_n - d_x)\cos\theta - (y_n - d_y)\sin\theta$$
$$y_{n,2} = x_{n,1}\sin\theta + y_{n,1}\cos\theta$$
$$= (x_n - d_x)\sin\theta + (y_n - d_y)\cos\theta$$

4. To return the origin back to coordinate (dx,dy):

$$T_{x,2} = d_x - 0 \qquad T_{y,2} = d_y - 0$$

5. Applying translation T_2 to the results in Step 4:

$$x_{n,3} = x_{n,2} + T_{x,2} = (x_n - d_x)\cos\theta - (y_n - d_y)\sin\theta + d_x$$
$$y_{n,3} = y_{n,2} + T_{y,2} = (x_n - d_x)\sin\theta + y_n - d_y)\cos\theta + d_y$$

6. Adjusting the notation in Step 5 to make it conform to our standard form:

$$x_n' = (x_n - d_x)\cos\theta - (y_n - d_y)\sin\theta + d_x$$
$$y_n' = (x_n - d_x)\sin\theta + (y_n - dy_y)\cos\theta + d_y$$

Step 1 uses Equation 6-3 to solve for the translation terms necessary for displacing the desired point of rotation to the origin, coordinate (0,0). And then Step 2 begins by expressing the first translation as the first in a series of transformations, and it concludes by substituting the results of the previous step. What we have at this point are expressions for the coordinate of any point in the reference figure as a function of its original coordinate and the coordinate of the desired point of rotation.

Step 3 first shows a general expression for applying a rotation as transformation 2, then it shows the result of substituting the expressions from the previous step. Step 4 suggests the translation terms that are necessary for returning the desired point of rotation from the origin (where it was set by Step 2) to its original place in the coordinate system. Step 5 then shows that translation as transformation number 3, and substitutes the expressions from the previous step. As far as the analytic work is concerned, the derivation is complete. Step 4, however, restates the results in terms of our more standard notation.

That, incidentally, is an example of a derivation that deals with one particular kind of transformation that is bracketed by two versions of a different transformation. When you consider all possible combinations of translations, rotations and scalings, you can come up with a rather larger number of possibilities; especially when you let yourself use two or more of each one. There is enough potential here to keep math technicians busy for a long, long time.

Why not make up some tricky little combinations of transformations of your own, then see if you can derive equations for calculating the overall results? Include lengths and direction cosines for lines, and the angles between lines in your studies.

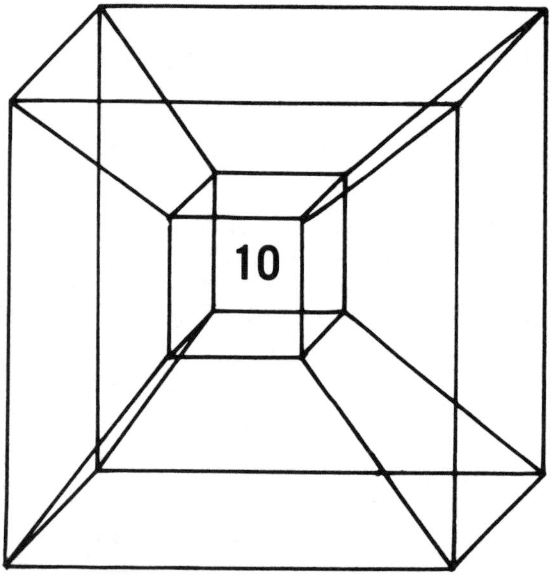

1-D Views of 2-D Space

A common problem in the graphic arts is that of portraying 3-D images in 2-D space. It is a matter of attempting to transmit meaningful spatial information about a 3-dimensional figure onto the 2-dimensional plane of a sheet of paper or canvas. There are a lot of tricks available for doing that sort of task, but it can never be done without making some compromises—something is always lost in a transformation from a space of some dimension to one of a lower dimensions.

We are not yet prepared to deal with the mathematical side of transforming images from 3-D to 2-D space. That comes later. But we are now in a position to consider the techniques and problems associated with portraying 2-D space in a 1-D format. The exercise can be interesting and informative; and just about everything you learn here can be extended to more ambitious projects later on. Specifically, these ideas set the stage for studying 4-dimensional objects and spaces from our more familiar environments of two and three dimensions.

Figure 10-1 shows a point P in 2-D space that is being translated from one place to another. Using the standard equation for translating points in 2-D space, it is possible to move the point to any desired place in its 2-D world.

The dashed line leading from the point to the X axis points to a mapping of point P to point P_{gx}. As point P moves around in its 2-dimensional world, its mapped version, P_g, follows along on the X axis—much as a shadow follows the object that produces it. It is important to realize that point P is free to roam in two dimensions, while point P_{gx} is confined to the 1-dimensional world of the X axis.

Lines in 2-D space can also show mappings onto a 1-D axis. Figure 10-2 shows the endpoints of a line L being mapped onto the X axis, thereby creating a 1-D "shadow" composed of line L_{gx} and its endpoints, P_{gx1} and P_{gx2}. The line can be translated freely through its own 2-D world, but its mapping to the X axis is confined to that 1-D space.

Figure 10-3 shows a line in 2-D space that is rotated about an off-origin point, P_ϕ. Its mapping to the 1-dimensional X axis is shown as line L_{gx}. Note especially how the length of line L_{gx} changes with the rotation of the original line in 2-D space. The length of line L in 2-D space does not change under a normal rotation, but the length of L_{gx} certainly does. And that little observation clearly indicates that at least one kind of distortion can occur when representing geometric elements in one particular space in a space of a lower number of dimensions.

Figure 10-4 illustrates yet another kind of distortion

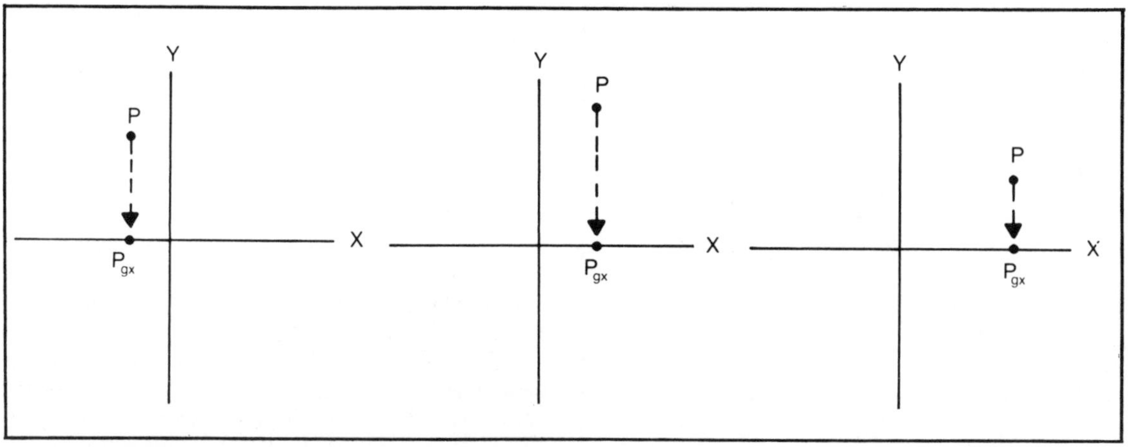

Fig. 10-1. Mapping a translating point to the X axis of a 2-D coordinate system.

that can occur when mapping down from two dimensions to one. The triangle in Fig. 10-4A, for instance, maps down to the X axis as a series of three points, two lines that lay end-to-end, and a third line that is superimposed on the two shorter ones. Indeed the essential character of the 2-D triangle is lost under this mapping to the 1-D lineworld of the X axis.

The same sort of distortion is even greater in the case of the square in Fig. 10-4B. Points 1 and 4 are mapped down to a single point on the X axis—a point that can be labeled Pgx1, Pgx4 or preferably both. From the viewpoint of a creature whose perception is limited to that of the 1-dimensional X axis, there is no distinction between those points; they appear to be one and the same point. The same distortion occurs where points 2 and 3 map down to a single point, Pgx2 and Pgx3.

Lines 1 and 3 are mapped down to a single line on the X axis, but matters are even worse in the case of lines 4 and 2. In 2-D space, lines 4 and 2 map down to points because they are perpendicular to the X axis.

From the viewpoint of the X axis, there is no difference between points 1 and 2, and line 4. Those points are spatially separated as viewed in two dimensions or more, but they are superimposed from the 1-D point of view. And the fact that a line in 2-D space can be reduced to a point when mapped to one dimension represents the most serious kind of distortion.

Without knowing the complete specifications for

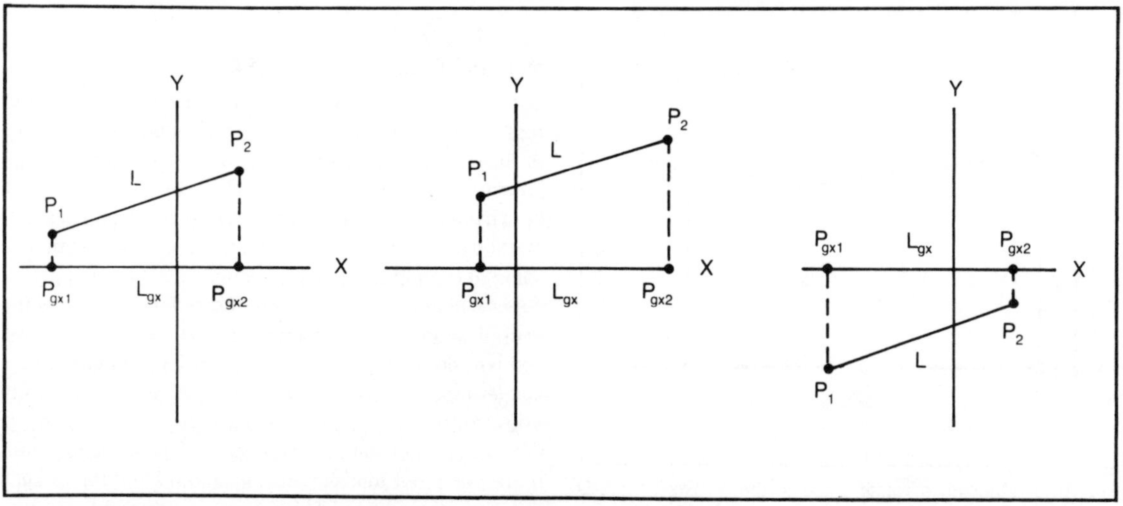

Fig. 10-2. Mapping a translating line to the X axis of a 2-D coordinate system.

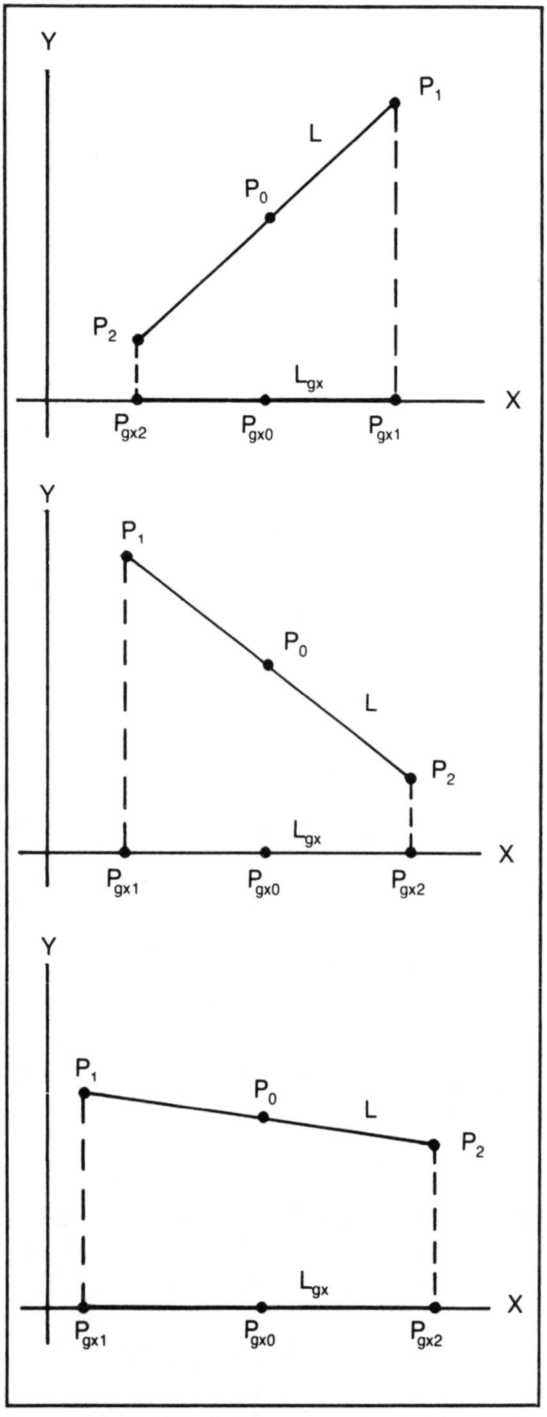

Fig. 10-3. Mapping a rotating line to the X axis of a 2-D coordinate system.

2-dimensional version of that square, a creature confined to the 1-dimensional X-axis world would have no way of knowing that the line appearing in his or her world is actually a "shadow" of a square.

All of the foregoing examples of mapping down from 2-D to 1-D space use mappings to the X axis of a coordinate system. The same principles apply equally well to mappings from 2-D space to the 1-dimensional Y axis. In the case of Y-axis mappings, the points are mapped perpendicular to the Y axis. See if you can plot the Y-axis mappings for Figs. 10-1 through 10-4. Use notation such as Pgy1, Pgy2, Lgy1, and so on.

It is possible to learn a lot about such mappings through purely graphic and descriptive operations such as those described so far, and you have the advantage of being able to portray 2-D coordinates and F-L-P specifications very precisely and completely in the 2-dimensional coordinate system. You have the freedom to apply translations, scalings and rotations to the points, and you possess the ability to observe what is happening from the best possible vantage point. The fun of the matter becomes apparent when you give up some of that advantage and pretend that your senses are limited to the line-world of one of the coordinate axes. You are then forced to perceive the entities of 2-D space in the limited context of 1-D space.

The experiments devised from this point of view can be interesting, enlightening and, hopefully, rather fun as well. The real value will become obvious later on when you attempt to grasp the essence of 4-dimensional objects from our limited perspective of two and three dimensions.

10-1 THE MECHANICS OF MAPPING FROM 2-D TO 1-D SPACE

The figures shown in the first part of this chapter represent a descriptive view of mappings from 2-D space to that of one dimension. Although such descriptive approaches can be useful and visually interesting, they will be virtually useless when working with entities immersed in a higher number of dimensions. So even though some of the following mathematical analyses and procedures will appear unduely complicated in the light of the overall simplicity of descriptive 2-D to 1-D mappings, you will do well to follow them carefully. In fact, it is a single-minded insistence upon working exclusively from a descriptive viewpoint, as opposed to an analytic one, is the sort of thing that generates creates so much impotent literature about four dimensions and related topics such as relativity and particle physics.

Here is a preview of the general analytic procedure:

1. Define the points in 2-D space and establish a set of specifications. You have been doing quite a bit of that sort of work through the past few chapters, so there should be no difficulties in store for you in this first step.

2. Select which of the two coordinate axes will serve as the 1-D space for the experiment. It makes no difference which axis you choose, just as long as you define it from the outset and stay with it throughout the analysis.

3. Calculate the 1-D mapped coordinates for each of the points from the entity in 2-D space. You will find the necessary equations in the next part of this discussion.

4. Plot the mapped version of the 2-dimensional entity onto the chosen 1-dimensional axis.

Step 3 in the procedure calls for an equation for mapping points down from 2-D to 1-D space. See the most general form in Equation 10-1.

$$x_{gn} = x_n K_{gx}$$
$$y_{gn} = y_n K_{gy}$$

Equation 10-1

where:

x_n, y_n = the coordinate of point *n* specified in 2-D space

x_{gn} = the X coordinate for point *n* as mapped to the X axis of the coordinate system

y_{gn} = the Y coordinate for point *n* as mapped to the Y axis of the coordinate system

K_{gx} = a mapping-scaling factor for the X dimension; it is equal to 1 for an X-axis mapping and equal to 0 for a Y-axis mapping

K_{gy} = a mapping-scaling factor for the Y dimension; it is equal to 0 for an X-axis mapping and equal to 1 for a Y-axis mapping

The idea of the equation is to map the coordinate of point n in 2-D space—specified as (xn,yn)—to a point on either the X axis (xgn) or Y axis (ygn). You must "program" the equation to map to either the X or Y axis by adjusting the values of the mapping-scaling factors, Kgx and Kgy.

If you "program" the equation for an X-axis mapping, you must let Kgx = 1 and Kgy = 0. Substituting those values into the general form of the equation, Equation 10-1, you end up with a special X-axis mapping equation, Equation 10-2.

$$x_{gn} = x_n$$
$$y_{gn} = 0$$

Equation 10-2

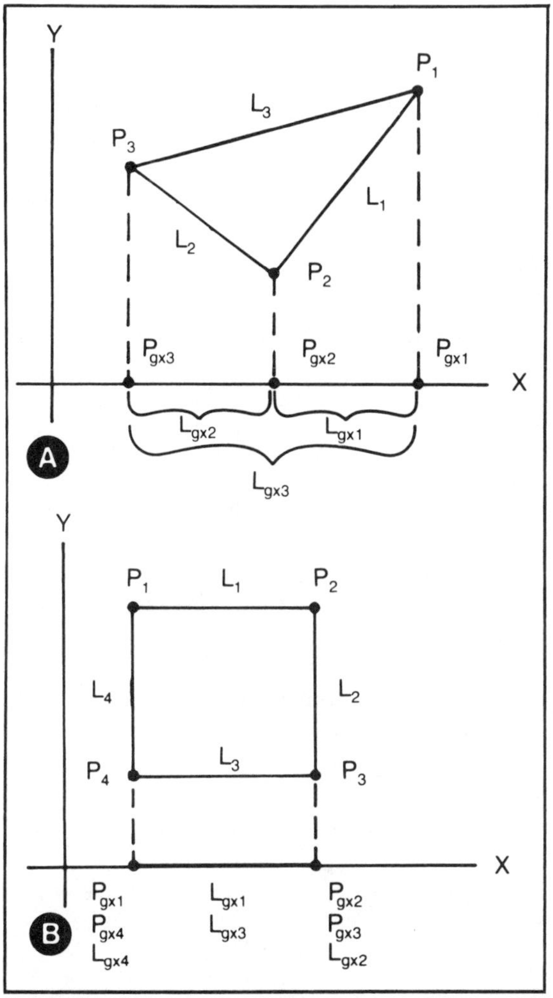

Fig. 10-4. Mapping plane figures to the X axis of a 2-D coordinate system. (A) This triangle maps to a line and three points. (B) This square maps to the X axis as a line and two points.

where:

x_n = the X-axis component of point *n* in 2-D space

x_{gn} = the X coordinate for point *n* as mapped to the X axis

y_{gn} = always zero; the Y coordinate for point *n* as mapped to the X axis

Note: This is a special form of Equation 10-1; it applies only to mappings to the X axis of the coordinate system.

But if you "program" a Y-axis mapping by letting Kgx = 0 and Kgy = 1, you get the special Y-axis equation, Equation 10-3.

$$x_{gn} = 0$$
$$Y_{gn} = y_n \qquad \textbf{Equation 10-3}$$

where:

y_n = the Y-axis component of point *n* in 2-D space

x_{gn} = always zero; the X coordinate for point *n* as mapped to the Y axis

y_{gn} = the Y coordinate of point *n* as mapped to the Y axis

Note: This is a special form of Equation 10-1; it applies only to mappings to the Y axis of the coordinate system.

Check out a basic application of these equations in Example 10-1.

Example 10-1

Let a reference point in 2-D space have this coordinate:

$$P = (2,3)$$

Cite the X-axis mapping of that point.
Cite the Y-axis mapping of that point.

1. Setting the given information into a component form:

$$x = 2,\ y = 3$$

2. Using Equation 10-2 to determine the coordinate of the X-axis mapping:

$$x_g = x$$
$$y_g = 0$$

So the mapped version of the point is at (2,0) or, more simply, $x_g = (2)$

3. Using Equation 10-3 to determine the coordinate of the Y-axis mapping of the original point:

$$x_g = 0$$
$$y_g = 3$$

The point thus maps to (0,3), or $y_g = (3)$
See Fig. 10-5.

Test your understanding of the matter by working Exercise 10-1.

Exercise 10-1

Map the following points first to the X axis (point P_{gx}) and then to the Y axis (point P_{gy}).

1. P = (4,5) P_{gx} = _____ P_{gy} = _____
2. P = (−1,4) P_{gx} = _____ P_{gy} = _____
3. P = (−1,−4) P_{gx} = _____ P_{gy} = _____
4. P = (4,−1) P_{gx} = _____ P_{gy} = _____

There really is a simpler, albeit less formal, way to treat this idea of mapping geometric entities from 2-D to

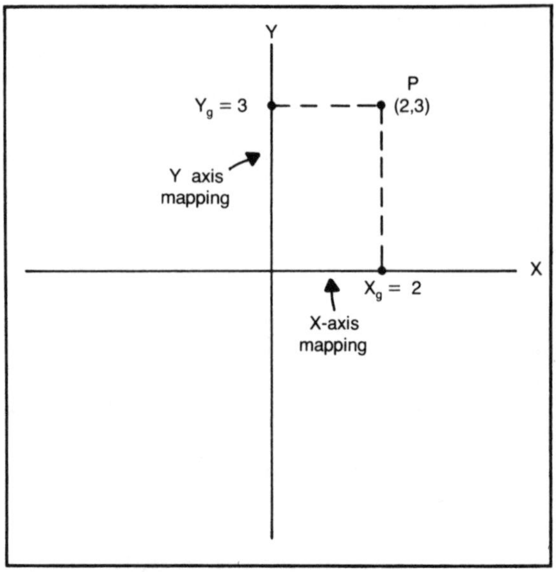

Fig. 10-5. Mapping a point in 2-D space to the 1-D line-worlds of the X and Y axes. See Example 10-1.

one of the 1-dimensional axes of the coordinate system: If you want to map to the X axis, simply set all Y-axis components of the specified points in two dimensions to 0; if you want to map to the Y axis, simply set all X-axis components to zero. As far as any creature living on one axis is concerned, the other axis does not exist.

10-2 MAPPING LINES FROM 2-D TO 1-D SPACE

As you might suspect, the procedure for mapping lines to 1-D space is a matter of mapping the individual endpoints and then drawing a line between them as prescribed in the 2-D specifications. See that done for you in Example 10-2.

Example 10-2

Suppose that a line has these specifications:

$$L = \overline{P_1P_2} \qquad P_1 = (4,3)$$
$$P_2 = (-2,5)$$

Map that line to the X axis of its coordinate system.

1. Gathering the information:

$$x_1 = 4 \qquad y_1 = 3$$
$$x_2 = -2 \qquad y_2 = 5$$

2. Using Equation 10-2 to map point P_1 to the X axis:

$$x_{g1} = x_1 \qquad y_{g1} = 0$$
$$x_{g1} = 4$$
$$p_{g1} = (4,0)$$

3. Using the same equation for point P_2:

$$x_{g2} = x_2 \qquad y_{g2} = 0$$
$$x_{g2} = -2$$
$$p_{g2} = (-2,0)$$

See the mapping in Fig. 10-6.

Make up some specifications for lines in 2-D space, then map them to either the X or Y axis.

10-2.1 Length of a Mapped Line

The examples of mapping lines from 2-D to 1-D clearly show that the mapped version can have a length that is different from its true, 2-D counterpart. Equation 10-4 is the general form of an expression for the length of a line that is mapped from 2-D space to either the X axis or the Y axis. (This is another of those equations that you must "program" for a particular case.)

$$s_{gn} = \sqrt{K_{gx}(x_u - x_v)^2 + K_{gy}(y_u - y_v)^2} \qquad \textbf{Equation 10-4}$$

where:

s_{gn} = the length of line *n* as mapped to either the X or Y axis of its 2-D coordinate system

x_u, y_u = the 2-D coordinate of endpoint *u* of the line

x_v, y_v = the 2-D coordinate of endpoint *v* of the line

K_{gx} = a mapping-scaling factor for the X dimension; it is equal to 1 for an X-axis mapping and equal to 0 for a Y-axis mapping

K_{gy} = a mapping-scaling factor for the Y dimension; it is equal to 0 for an X-axis mapping and equal to 1 for a Y-axis mapping

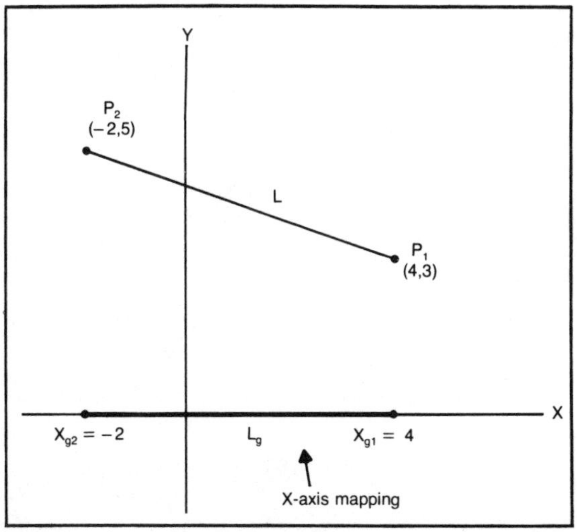

Fig. 10-6. Given this particular orientation of line L in 2-D space, it maps to the 1-dimensional X axis as a line that is somewhat shorter than the version in 2-D space. See Example 10-2.

Letting $K_{gx} = 1$ and $K_{gy} = 0$ sets up the equation for an X-axis mapping, while letting $K_{gx} = 0$ and $K_{gy} = 1$ sets it up for finding the length of a line mapped to the Y axis. See that done for you in the special forms of the equation, Equations 10-5 and 10-6. Then see how they are applied in Example 10-3.

$$s_{gn} = \sqrt{(x_u - x_v)^2} \qquad \textbf{Equation 10-5}$$

where:

s_{gn} = the length of line *n* in 2-D space as mapped to the X axis of its coordinate system

x_u = the X component of one endpoint of the line

x_v = the X component of endpoint *v* of the line

Note: This is a special version of Equation 10-4 that applies only to X-axis mappings.

$$s_{gn} = \sqrt{(y_u - y_v)^2} \qquad \textbf{Equation 10-6}$$

where:

s_{gn} = the length of line *n* in 2-D space as mapped to the Y axis of its coordinate system

y_u = the Y component of endpoint *u* of the line

y_v = the Y component of endpoint *v* of the line

Note: This is a special case of Equation 10-4 that applies only to Y-axis mappings.

Example 10-3

A line in 2-D space has the following specifications:

$$L = \overline{P_1P_2} \qquad P_1 = (4,5)$$
$$P_2 = (1,1)$$

Calculate the true, 2-D length of the line.
Calculate the length of the line as mapped to the X axis.
Calculate the length of the line as mapped to the Y axis.

1. Gathering the relevant information into an equation-oriented form:

$$x_1 = 4 \qquad y_1 = 5$$
$$x_2 = 1 \qquad y_2 = 1$$

2. Calculating the true length:

$$s = \sqrt{(x_2 - x_1)^2 + (y_2 - y_1)^2}$$
$$s = \sqrt{(1-4)^2 + (1-5)^2}$$
$$s = \sqrt{(-3)^2 + (4)^2}$$
$$s = \sqrt{25}$$
$$s = 5$$

So from a 2-D perspective, the line is 5 units long.

3. Using Equation 10-5 to determine the length of the line as mapped to the X axis:

$$s_{gx} = \sqrt{x_2 - x_1)^2}$$
$$s_{gx} = \sqrt{(1-4)^2}$$
$$s_{gx} = 3$$

From the perspective of the X axis, the line is 3 units long.

4. Using Equation 10-6 to determine the length of the line as mapped to the Y axis:

$$s_{gy} = \sqrt{(y_2 - y_1)^2}$$
$$s_{gy} = \sqrt{(1-5)^2}$$
$$s_{gy} = 4$$

So from the perspective of the Y axis, the line is 4 units long.

If you work out enough examples, you might even conclude that the mapped version of a line is never longer than its 2-dimensional counterpart. The lengths can be equal under certain circumstances, the length can be reduced to zero under some conditions, and it is entirely possible to have a mapped line that is anywhere in between 0 and 1 times the length of the 2-dimensional version. But the mapping, or shadow, can never by longer than the true 2-dimensional line it represents. Can you prove that principle in a formal, analytic fashion?

10-2.2 Direction Cosines and Angles

Any line immersed in 2-D space has a set of direction cosines associated with it. Those expressions indicate the angular orientation of the line with respect to the axes of the coordinate system. Then, too, there is always a well-defined angle between any two lines that reside in 2-D space.

But it doesn't take much in the way of a descriptive or mathematical analysis to see that direction cosines and angles between lines are virtually meaningless in 1-D space. Those terms are completely lost when mapping lines from two dimensions to one dimension.

10-3 MAPPING PLANE FIGURES FROM 2-D TO 1-D

Given the F-L-P specifications for a plane figure in 2-D space, mapping that figure to the 1-dimensional X or Y axis is a matter of mapping its individual points and constructing the specified lines between them. See Example 10-4.

Example10-4

A triangular figure in 2-D space has the following specifications:

$F = \overline{L_1L_2L_3}$ $L_1 = \overline{P_1P_2}$ $P_1 = (5,-1)$

$L_2 = \overline{P_2P_3}$ $P_2 = (4,2)$

$L_3 = \overline{P_3P_2}$ $P_3 = (1,5)$

1. Conduct a complete analysis of the figure as it appears in 2-D space.

2. Map the figure to the X axis and compile an analysis of the result.

3. Map the original figure of the Y axis and compile an analysis of the result.

See the 2-D version of the figure in Fig. 10-7 and its analysis in Table 10-1. See the mapped versions in Fig. 10-8 and the analyses in Table 10-2.

The example cites a triangle in 2-D space and then suggest mapping it, in turn, to the X and Y axes. See the analyses of all three versions in Tables 10-1 and 10-2. The analyses of the mapped versions show only the specifications and the lengths of lines; direction cosines and angles between lines have no meaning in 1-D space.

Much of the information concerning the shape and position of a plane figure in 2-D space is lost by mapping it to a 1-D axis. The plane figure, no matter how simple or complex it might be as viewed in its true 2-D form, is reduced to a series of points and end-to-end or superim-

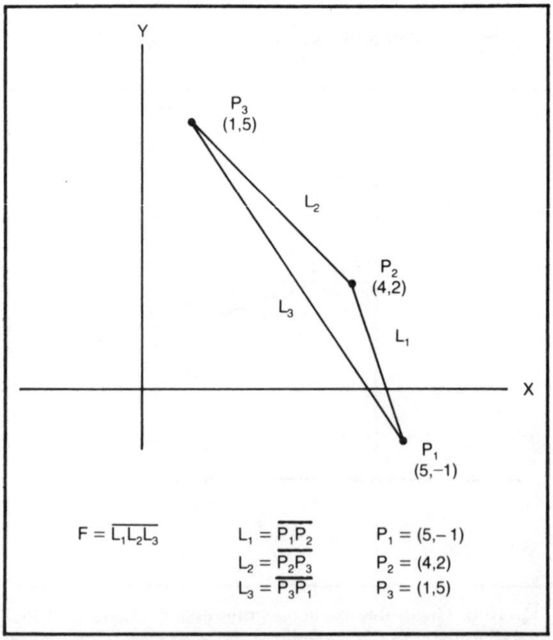

Fig. 10-7. A fully specified triangle in 2-D space. See Example 10-4 and a summary of its analysis in Table 10-1.

Table 10-1. Summary of Analysis of the Plane Figure in Fig. 10-7.

(A, B, C, and D correspond to α, β, γ, and δ.)

SPECIFICATIONS

$F=\overline{L_1L_2L_3}$ $L_1=\overline{P_1P_2}$ $P_1=(5,-1)$

$L_2=\overline{P_2P_3}$ $P_2=(4,2)$

$L_3=\overline{P_3P_1}$ $P_3=(1,5)$

LENGTHS OF LINES

$s_1=\sqrt{10}$ $s_2=3\sqrt{2}$ $s_3=2\sqrt{13}$

DIRECTION COSINES

$\cos A_1=-\sqrt{10}/10$ $\cos B_1=3\sqrt{10}/10$

$\cos A_2=-\sqrt{2}/2$ $\cos B_2=\sqrt{2}/2$

$\cos A_3=2\sqrt{13}/13$ $\cos B_3=-3\sqrt{13}/13$

ANGLES

$\phi_{1,2}=135^\circ\ 26'$ $\phi_{1,3}=15^\circ\ 45'$

$\phi_{2,3}=11^\circ\ 20'$

posed lines. Without the aid of some analytic tools to express the nature of the figure in its 2-dimensional form, it is virtually impossible to grasp its essence from the 1-D view.

10-4 1-D MAPPINGS OF 2-D TRANSLATIONS

Chapter 6 deals with the mechanics and general characteristics of translations in 2-D space. Translating points, lines and plane figures in 2-D space creates a corresponding, translation-type mapping to the 1-dimensional worlds of the X and Y axes. As you might suspect, however, there is always a possibility of a distorted rendition of those 2-D translations (as viewed from 1-D space). Equation 10-7 is a general mathematical expression for the mapping of translations from two dimensions to one.

$$X'_{gn} = K_{gx}(x_n + T_x)$$
$$y'_{gn} = K_{gy}(y_n + T_y) \qquad \textbf{Equation 10-7}$$

where:

x'_{gn} = the coordinate of the translated version of reference point *n* as mapped to the X axis

y'_{gn} = the coordinate of the translated version of reference point *n* as mapped to the Y axis

x_n,y_n = the 2-D coordinate of point *n* relative to the chosen frame of reference

T_x and T_y = the translation terms relative to the frame of reference

K_{gx} = mapping-scaling factor for the X axis; it is equal to 1 for an X-axis mapping and 0 for a Y-axis mapping

K_{gy} = mapping-scaling factor for the Y axis; it is equal to 0 for an X-axis mapping and 0 for a Y-axis mapping

This is a very general equation that I am offering here only for the use of experimenters who wish to pursue the matter in a purely analytic fashion. I do not recommend using it to crank out specific values for specific mappings of translation effects. It is much easier, and equally effective to study mapped translations in this way:

1. Specify a reference line in 2-D space.
2. Select one of the coordinate axes to serve as the 1-D line-world for the experiment.
3. Generate the mapped version of the line by applying Equation 10-2 or 10-3, depending on the choice of an axis in Step 2.
4. Calculate the length of the true, 2-D version of the reference line (Equation 5-1), and the length of its mapped counterpart (Equation 10-5 or 10-6, depending on the choice of an axis in Step 2).
5. Apply some sort of 2-D translation to the true reference line specified in Step 1.
6. Generate the mapped version of the translated line by applying Equation 10-2 and 10-3 to it (again, depending on the choice of a coordinate axis in Step 2).
7. Calculate the length of the 2-D version of the translated line and the length of its mapped version.

Studying the results of such an experiment, you will find that the lengths of the lines in 2-D space might well be quite different from their counterparts mapped to the chosen 1-D axis. But the translation does not affect the lengths as viewed from one space or the other.

10-4.1 Some Important Principles

There are a couple of important principles lurking within these equations and the whole idea of mapping translation effects from 2-D to 1-D space. You will have a better opportunity to use these principles in a later discussion in this chapter, but it is important to point them out at this time.

Notice that a Y-axis translation term in 2-D space

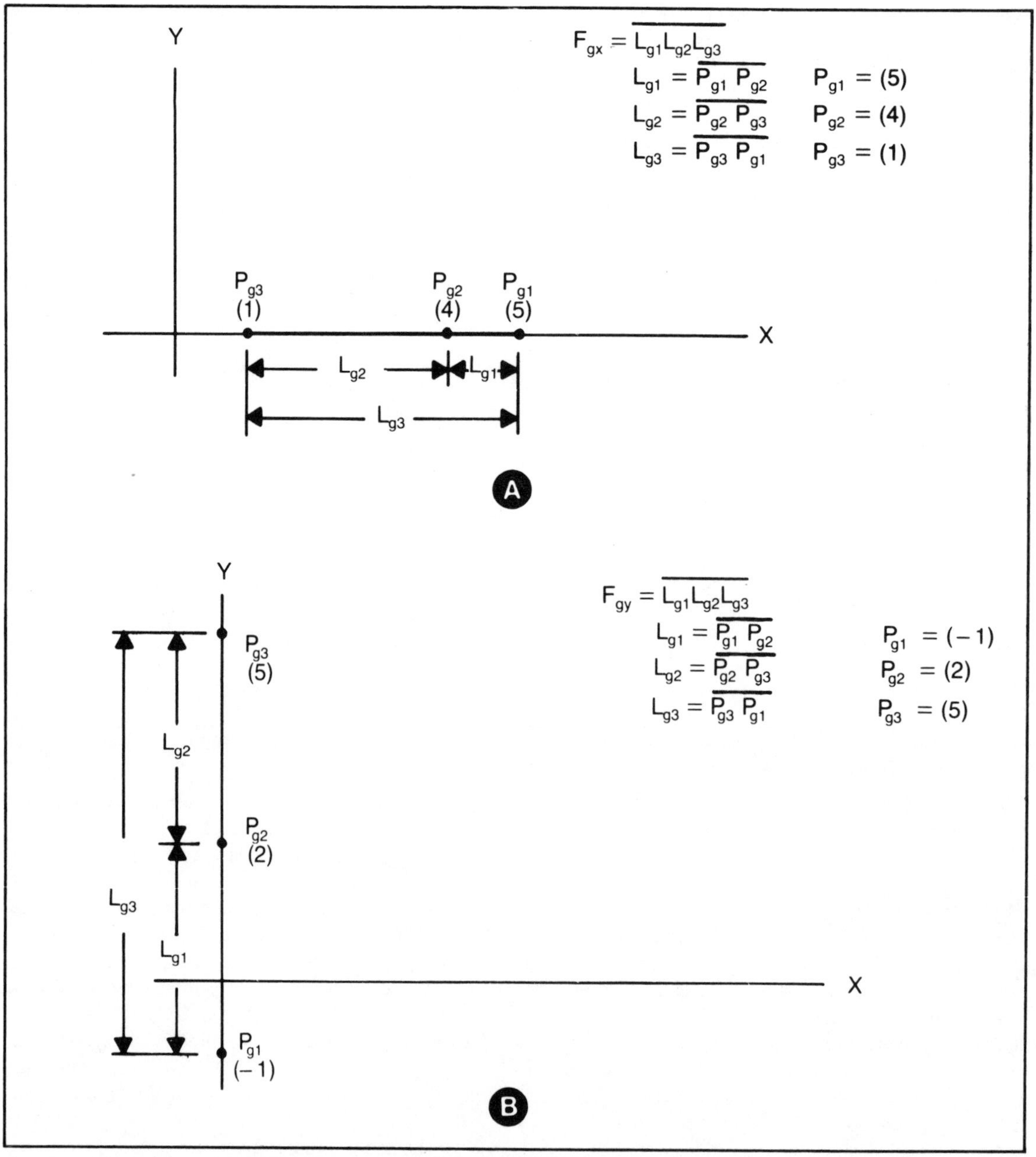

Fig. 10-8. The triangle in Fig. 10-7 is mapped here to the 1-D worlds of the axes of a coordinate system. (A) An X-axis mapping. (B) A Y-axis mapping. See Example 10-4 and a summary of its analysis in Table 10-2.

has no affect at all on the mapping of geometric entities to the 1-dimensional X-axis world. You can translate a point from here to eternity along the Y axis without affecting its mapping position to the X axis in the slightest. A creature whose perception is limited to the X-axis line-world has absolutely no impression of translations in the Y-axis dimension. X-axis translations certainly do affect the positions of points as mapped to the X axis, but Y-axis translations do not. You can demonstrate that principle in a graphic, descriptive fashion; plot a point in 2-D space, map it to the X axis, then apply any Y-axis translation you wish. You will find that the mapped version of the point does not move.

To illustrate that same principle in a more formal way, set up Equation 10-7 for an X-axis mapping and you will find that the Y-axis translation term disappears—it is no longer relevant. Set up that equation for a Y-axis mapping, and the X-axis translation term disappears. Do you see how you can learn things from equations?

Summarizing those principles:

Table 10-2. Summary of the Analysis of the Figure Mapped in Fig. 10-8.

SPECIFICATIONS

$F=\overline{L_{g1}L_{g2}L_{g3}}$	$L_{g1}=\overline{P_{g1}P_{g2}}$	$P_{g1}=(5)$
	$L_{g2}=\overline{P_{g2}P_{g3}}$	$P_{g2}=(4)$
	$L_{g3}=\overline{P_{g3}P_{g1}}$	$P_{g3}=(1)$

LENGTHS OF LINES

$s_{g1}=1$	$s_{g2}=3$	$s_{g3}=4$

See Fig. 10-8A

SPECIFICATIONS

$F=\overline{L_{g1}L_{g2}L_{g3}}$	$L_{g1}=\overline{P_{g1}P_{g2}}$	$P_{g1}=(-1)$
	$L_{g2}=\overline{P_{g2}P_{g3}}$	$P_{g2}=(2)$
	$L_{g3}=\overline{P_{g3}P_{g1}}$	$P_{g3}=(5)$

LENGTHS OF LINES

$s_{g1}=3$	$s_{g2}=3$	$s_{g3}=6$

See Fig.10-8B

☐ Mappings of 2-D translations onto the X axis are totally immune to any Y-axis component of translation.

☐ Mappings of 2-D translations onto the Y axis are totally immune to any X-axis component of translation.

10-4.2 Mapped Lengths of Translated Lines

Discussions in Chapter 6 demonstrate and prove that the length of a line in 2-D space is an invariant—remains unchanged by—a 2-D translation. That principle still holds true, but only as long as viewing the situation from a 2-dimensional perspective.

You have seen earlier in this chapter that the length of a line that is mapped from 2-D space to the 1-D line-world of a coordinate axis will be less than or equal to its 2-dimensional counterpart. A line in a given number of dimensions can never appear longer when mapped to a space of a lower number of dimensions. The length of a line is not an invariant under 2-D to 1-D mappings.

So the length of a line is an invariant under 2-D translations in 2-D space, but it is not an invariant under 2-D to 1-D mappings. And that should bring up a question: Is the length of a line that is mapped from 2-D to 1-D space an invariant under translations in 2-D space?

Try to picture a line L immersed in 2-D space and its mapped counterpart, L_g, on the X axis. Then imagine applying any sort of 2-D translation to line L, thereby generating a second line, L′, in 2-D space. That should map another line to the X axis, line L_g'. You know that the lengths of L and L′ are the same (the length of a line is an invariant under 2-D translations), but how to the lengths of their mapped versions, L_g and L_g', compare? Well, I can give you a quick answer: those lines also have the same length. The length of mapped lines is an invariant under 2-D translation; not under the mapping transformation, itself, but under any subsequent translations.

You can demonstrate the principle by means of the experimental method described in the previous section of this chapter. Such a demonstration, or even a thousand of them, does not constitute a proof of the principle, though; so I am including a formal, analytic proof as Proof 10-1.

Proof 10-1

1. If the length of a line mapped to 1-D space is unchanged by a translation of its 2-D counterpart, then:

$$s_g' = s_g$$

2. Substituting Equation 10-4, using appropriate notation:

$$\sqrt{K_{gx}(x_u'-x_v')^2+K_{gy}(y_u'-y_v')^2}$$
$$=\sqrt{K_{gx}(x_u-x_v)^2+K_{gy}(y_u-y_v)^2}$$

3. Substituting Equation 10-7 into the left side of the equation in Step 2 and reducing the result:

$$\sqrt{K^3_{gx}(x_u - x_v)^2 + K^3_{gy}(y_u - y_v)^2}$$
$$= \sqrt{K_{gx}(x_u - x_v)^2 + K_{gy}(y_u - y_v)^2}$$

4. Since the mapping-scaling factors are either 1 or 0, it is possible to replace the cubed factors on the left side of the equation in Step 3 with K_{gx} and K_{gy}. Thus:

$$\sqrt{K_{gx}(x_u - x_v)^2 + K_{gy}(y_u - y_v)^2}$$
$$= \sqrt{K_{gx}(x_u - x_v)^2 + K_{gy}(y_u - y_v)^2}$$

And the proposition $s_g' = s_g$ is proven.

Step 1 sets up the proof by stating that the lengths of the mapped versions of the reference and translated lines ought to be equal. The idea behind the proof is then to substitute known expressions until the operations come to a point where they show an obvious equality. That finally happens here in Step 4. Indeed, the length of a line mapped from 2-D to 1-D space is an invariant.

10-5 1-D MAPPINGS OF 2-D SCALINGS

Imagine some simple plane figure specified in 2-D space and mapped down to one of the coordinate axes. Then scale the 2-dimensional version of the figure in some fashion, and map the scaled version to the same axis as the original. The result is a representation of a mapping of a scaling transformation from 2-D to 1-D space. Equation 10-8 represents the most general equation for conducting that sort of experiment.

$$x'_{gn} = K_{gx}K_x y_n \qquad y'_{gn} = K_{gy}K_y Y_n \qquad \textbf{Equation 10-8}$$

where:

x_{gn}' = the coordinate of scaled point *n* as mapped to the X axis.

y_{gn}' = the coordinate of scaled point *n* as mapped to the Y axis.

K_{gx} = the mapping-scaling factor; it is equal to 1 for an X-axis mapping and 0 for a Y-axis mapping

K_{gy} = the mapping-scaling factor; it is equal to 0 for an X-axis mapping and 1 for a Y-axis mapping

K_x and K_y = 2-D scaling factors for the X and Y axes, respectively

x_n, y_n = the coordinate of reference point *n* in the 2-D frame of reference

The equation, being a very general one, has to be "programmed" for the mapping you want to perform. Adjusting the mapping-scaling factors as in the previous discussions in this chapter, you get the two special versions of the equation: one for a mapping to the X axis and another for a mapping to the Y axis. Do that for your own benefit. The real reason for showing you the general equation at all is to set up a model for drawing some conclusions later.

You can see that a scaling factor in the X dimension affects only the mappings to the X axis, while scaling factors for the Y dimension affect only the mappings to the Y axis. Viewing the principles in a more formal way, adjust Equation 10-8 for an X-axis mapping, and you will find that the Y-axis scaling factor disappears. Likewise setting up the equation for a Y-axis mapping makes the X-axis scaling factor irrelevant. In summary:

- ☐ A Y-axis scaling in 2-D space does not affect the mapping of a point or line to the X axis.
- ☐ An X-axis scaling in 2-D space does not affect the mapping of a point or line to the Y axis.

And when scaling lines that are immersed in 2-D space, any change in the length of their mapped counterparts is proportional to the scaling factor. Length is not an invariant under 2-D scalings in 2-D space, and it is not an invariant when viewing 1-D mappings of those lines.

A simple negative proof is adequate for establishing the variance of length in this case. Just run a couple of experiments involving the scaling of a line in 2-D space. Map the results to one of the coordinate axes and compare their mapped lengths. Under all but a few special conditions, you will find that their lengths are different. That constitutes a responsible negative proof. (Generally it is easier to prove that something is not true—one instance makes the case.)

10-6 MAPPINGS OF 2-D ROTATIONS

Rotating lines and plane figures in a 2-D space, and then mapping the results to one coordinate axis or the other can be an intriguing sort of experiment. Imagine how a creature living on one of the 1-D axes perceives the rotation of a line in 2-D space.

Figure 10-9 shows a series of rotations of a line about the origin, P_ϕ, in 2-D space. There is nothing new about that; you were introduced to the idea in Chapter 8. What is new, however, is the mapping of that rotation to the X axis of the coordinate system.

In Fig. 10-9A, the reference line is parallel to the X axis—it has no Y-axis component. Thus it maps to the X axis with an equal length. A 1-D creature living in that X-axis world sees the line with its true, 2-dimensional length.

The line in 2-D space is then rotated 45 degrees as shown in Fig. 10-9B. You can see that its mapped version

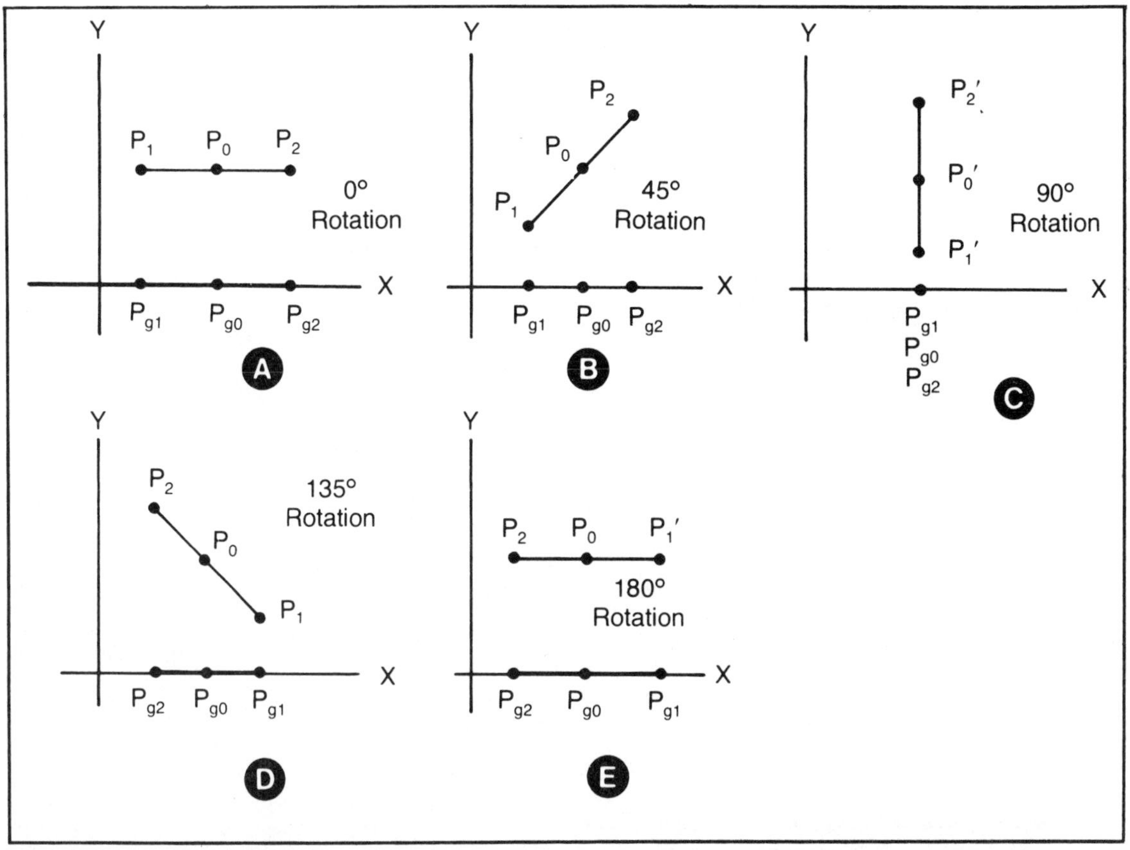

Fig. 10-9. X-axis mappings of a line being rotated about its center in 2-D space. (A) The reference line. (B) At a rotation of 45 degrees. (C) At a rotation of 90 degrees. (D) At a rotation of 135 degrees. (E) At a rotation of 180 degrees.

is substantially shorter. The creature confined to the X-axis world would see only the shorter rendition; he or she would think the line is shorter than it really is in 2-dimensional space.

Then in Fig. 10-9C, the real line is rotated in such a way that it is perpendicular to the X axis. It has no X-axis component, so it maps to the X axis as a single point. Now the creature in the 1-D, X-axis world sees the real line in 2-D space as a single point. That amounts to a complete change in the character of the entity; from a line to a point.

The real line is rotated 135 degrees in 2-D space in Fig. 10-9D. The creature confined to the X-axis world sees it a bit longer again, but with its endpoints reversed.

Then the length of the mapped version of the line in Fig. 10-9E is equal to its true length once again. The positions of the endpoints are reversed from the situation in Fig. 10-9A, however.

So as the real line is rotated around and around in 2-D space, its true length remains fixed (length is an invariant under 2-D rotations), but a creature living in the 1-D line-world of the X axis views the continuous rotation as a line that oscillates in length and sense, or direction. From that creature's point of view, the essence of the transformation is quite different from the reality of the matter. Unless he or she happened to be well-versed in the logic and mathematics of higher dimensioned spaces and the transformation of geometric entities, the situation would be impossible to describe in its true form. Does that bring to mind some questions about modern physics and the philosophy of science in our own, real, 3-D world? Think about it.

Equation 10-9 completely characterizes the mapping of rotating points from 2-D to 1-D space. Apply it to specific instances, if you wish, by adjusting the values of the mapping-scaling terms to suit either an X- or a Y-axis mapping.

$$x_{gn}' = K_{gx}\,[(x_n - x_r)\cos\theta - (y_n - y_r)\sin\theta + x_r]$$
$$y_{gn}' = K_{gy}\,[(x_n - x_r)\sin\theta + (y_n - y_r)\cos\theta + y_r]$$

Equation 10-9

where:

x_{gn}' = the X-axis mapping of rotated version of point

y_{gn}' = the Y-axis mapping of the rotated version of point n

K_{gx} = the mapping-scaling factor; it is equal to 1 for an X-axis mapping and 0 for a Y-axis mapping

K_{gy} = the mapping-scaling factor for the Y axis; it is equal to 0 for an X-axis mapping and 1 for a Y-axis mapping

x_n,y_n = the coordinate of reference point n in the frame of reference

x_r,y_r = the point of rotation relative to the frame of reference

θ = the angle of rotation relative to the frame of reference

I would rather you use the equation as a foundation for uncovering underlying principles, proving some ideas and deriving other equations that might be of interest to you. As far as conducting experiments of the kind illustrated in Fig. 10-9 is concerned, however, you will find a more practical approach taking this general form:

1. Specify a reference point, line or plane figure in 2-D space.
2. Select an axis of the coordinate system that is to serve as the 1-D world for the mapping operations.
3. Map the 2-D version of the reference entity to the chosen 1-D space.
4. Rotate the 2-D version of the entity about a chosen point in 2-D space, using the equations introduced in Chapter 8.
5. Map the rotated version to the chosen axis of the coordinate system.

That is, I think, the simplest way to go about conducting mapping experiments of this sort. You can complete the work by running a complete analysis of both the 2-D version of the project and its mapped version.

10-7 SOME SPECULATIONS ABOUT 1-D AND 2-D WORLDS

With a bit of sound knowledge and experience behind you, you are prepared to speculate in a responsible fashion about the way events in 2-D space must appear to a creature whose entire perception of the universe is limited to a 1-D space. Several writers have suggested such scenarios in the past, but most of them suggested only descriptive explanations and "common-sense" analogies as a basic frame work. You don't have to handle the matter in such a tenuous way now. While the situations I am about to describe are altogether fanciful, the essential ideas are sound.

You have seen from the previous discussions in this chapter that a mapping, or shadow, of some geometric entities in 2-D space are most often distored from that lower-dimensioned mapping. What's more, transformations applied to the 2-D entities create some different kinds of effects as mapped to a 1-D space.

Placing those ideas into a fictional context, we can say that a 2-dimensional creature has some definite advantages over a 1-dimensional creature that might share a part of the 2-D universe. The 2-D creature, for instance, is free to roam its world in two different kinds of directions—let's say north and south, and east and west. A 1-D creature living in the same universe, but is restricted to a 1-D line-world of a coordinate axis, can move only in one dimension. Let's call that east and west. The notion of being able to move in a north-south direction is entirely foreign to the 1-D creature.

Our fictional 2-D and 1-D creatures can both recognize and specify points and lines in their spaces, but the 1-D creature cannot possibly recognize anything approaching a plane figure. Both creatures know about translations and scalings in their respective dimensions, but the 1-D creature is completely lost with the idea of rotations. There is no way to rotate a line that is restricted to one certain 1-D space.

Please notice that the only idea that I have introduced here in an off-the-wall fashion is that of intelligent creatures living in these spaces. Everything else I have said about them—about their special capabilities and limitations—are special interpretations of facts established earlier in a sound and responsible way. I have taken the mathematical principles and interpreted them in the light of a single bit of conjecture. There is nothing wrong with doing that as long as you bear in mind that you are playing with ideas and not attempting to suggest that any such things as 1-D and 2-D creatures or their worlds actually exist.

Also, I want to avoid building ideas purely by analogy. Analogies can serve useful and very real purposes, but they must never become the foundations for building theories—even theories of imaginary worlds.

So getting back to an interpretation of mappings as viewed by creatures in 1 and 2 dimensions, suppose that both creatures can be aware of each other's existence. The 2-D creature, however, has the advantage of being able to see the 1-D creature in its true form: as a line in the 1-D line-world, for example. The 1-D creature, however, can only perceive the 2-D creature as a vague shadow; as a mapping of the creature onto the 1-D world. The 2-D creature might have any sort of complicated plane shape, but you have seen that any line or plane figure in 2-D space, no matter how simple or complex, maps to a 1-D world as a series of points, end-to-end lines and superimposed lines.

From the vantage of the 2-D creature, the 1-D creature and its line-world exist in a clear and unambiguous way; and the 2-D creature can freely manipulate any point or line in that simpler space. But from the viewpoint of the 1-D creature, the 2-D creature and its world exist in a fashion, but really aren't there in a common-sense way. The 1-D creature has absolutely no ability to manipulate anything in the 2-D world beyond its own axis of existence.

Given that sort of background, make up some experiments of your own. Use the analytical foundation built up in the first parts of this chapter, but interpret the work in terms of the special creatures and their perceptions of things.

10-8 PERPENDICULAR 1-D WORLDS

If you have spent much time reading science-fiction stories, you have most likely come across the notion of parallel worlds—worlds that exist in the same universe and are quite similar in most respects. The problem is that each world is supposed to be inaccessable from the other, but of course it makes a good story when something happens that forces a hero to slip from one to the other.

A slight variation of the fantasy offered in the previous section reminds me of the parallel-worlds idea. The worlds aren't parallel in this case, but rather perpendicular to one another. Conjecture? Yes. Consistent with some sound principles? That, too.

The scenario in this instance concerns two different 1-D worlds. They are situated in 2-D space, however. You need a 2-D space to make the idea work. Further suppose that the 1-D worlds can be represented as the two axes of a 2-D coordinate system. Each is populated with 1-D creatures whose perception of the universe is limited to the qualities of their respective 1-D, axis line-worlds.

Given that situation, it is possible to imagine the two worlds existing quite independent of one another. The creatures are almost always unaware of each other's existence. They do share a common 2-D space, of which they are probably unaware, and you have seen from previous analytical work that they have entirely different views of "shadowy" events in 2-D space. You can experiment with those ideas by mapping 2-D space entities onto both the X and Y axes, pretending that each axis represents a different 1-D world.

There exists, however, an unique point in this scheme where the creatures of one axis world can be aware, in a very limited way, of events in the other world. That point is the origin of this hypothetical coordinate-system universe. At that one point, it is possible for creatures from the two different worlds to sense one another's presence as a point. I suppose that a popular myth in those worlds concerns the existence of such a point, and their people tell stories of brave searches for it and encounters with it. Why not make up some such stories of that sort for yourself?

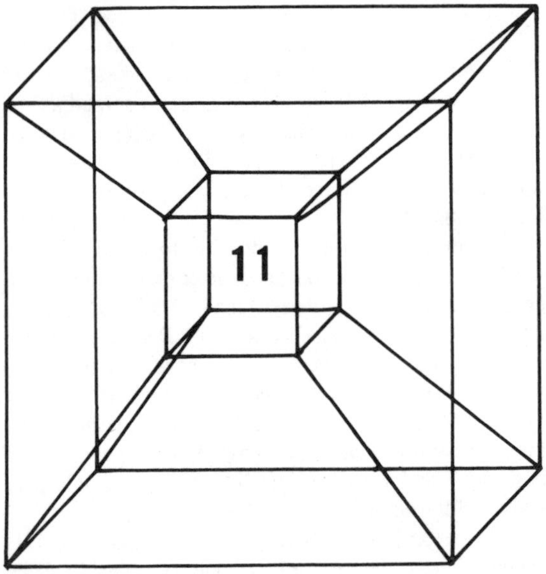

A First Look at 3-D Space

Three-dimensional space is the most familiar of all spaces as far as human beings are concerned. We are born in it, we live in it, and we die in it. At least that is the usual perception of things. Our own world and, indeed, our entire universe is generally regarded as a 3-D space. In this chapter there will be no need to speculate about the existence of imaginary intelligent—we are the beings of 3-D space.

Three-dimensional space is a logical extension of 1-D and 2-D spaces. Whereas 1-D space is characterized only by a quality of width, 2-D space possesses the qualities of both width and height. Moving into the world of 3-D space adds the element of depth. Thus, 3-D space has the qualities of width, height and depth.

There is one catch, however, as far as plotting the visual results of experiments in 3-D space is concerned. Recall from the work in the previous chapter that mapping down from 2-D to 1-D space creates some ambiguous visual results. Mappings that reduce the dimension of a line or plane figure almost always result in the loss of some essential visual information. And because of that, plotting the results of experiments in 3-D space is going to cause some difficulties because the plots are done on the 2-D plane of a sheet of paper.

So it is going to be difficult to achieve true visual plots of experiments in 3-D space. The 2-D renderings on a sheet of paper will inevitably include some spatial and angular distortion; and more than ever before, we will have to rely on the numerical analyses to convey the true qualities of the work.

11-1 A COORDINATE SYSTEM FOR 3-D SPACE

Just as 3-D space, itself, is a logical extension of 1-D and 2-D spaces, so is the coordinate system that characterizes it. A 2-D coordinate system is made up of two 1-D coordinate systems that intersect at right angles at a common point of origin. A 3-D coordinate system, then, is one composed of three 1-D coordinate axes that likewise intersect at right angles at a common point of origin. It is usually said that the axes are mutually perpendicular.

There are three axes now, and in order to be consistent with previous work, they ought to be labeled the X, Y and Z axes. The X and Y axes will represent the "width" and "height" axes of 2-D space. The third axis, the Z axis, will represent the new property of "depth." See two examples in Fig. 11-1.

It is impossible to portray the situation perfectly on the 2-D plane of the page, because at least one of the axes extends in and out of the page. The angles at the origin are

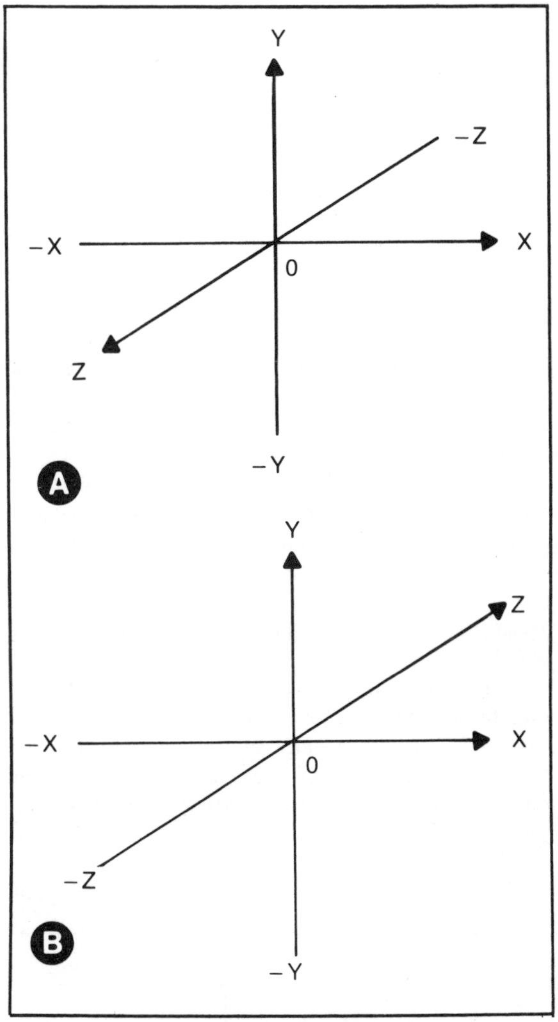

Fig. 11-1. Two legitimate coordinate systems for portraying 3-D space. (A) A right-handed system. (B) A left-handed system.

supposed to be mutually perpendicular, but we have to be content simply defining them that way. In both instances, I have plotted the X and Y axes as though they belong to a 2-D coordinate system. Notice that they are shown at right angles to one another. The Z axis, however, is shown at an angle of about 30 degrees from the X axis. That represents a big compromise—it, too, is supposed to be drawn at right angles to both the X and Y axes.

If you want to get an accurate mental image of the 3-D coordinate system, imagine the X and Y axes drawn as they are here, but the Z axis extending straight in and out of the plane of the page.

11-1.1 The Right-Handed 3-D Coordinate System

The only difference between these two 3-D coordinate systems is the sense of the Z axis. In Fig. 11-1A, the values of the measuring units on the Z axis increase in an out-of-the-page fashion. In the second example, however, the positive end of the Z axis is portrayed as extending into the plane of the page.

They are both perfectly valid 3-D coordinate systems, but in order to save a lot of confusion through the experiments, we must establish one or the other as our standard. The choice is especially important because it dictates the position of plus and minus signs in the rotation equations for 3-D space.

The first coordinate system in Fig. 11-1—the one where the positive end of the Z axis supposedly extends out of the page—is technically known as a right-hand coordinate system for 3-D space. The second is a left-hand system. Although both are quite valid, I have used the right-hand system for generating certain vital equations later in this book. So we must agree to use the right-hand coordinate system in Fig. 11-1A. (No slight is intended, but southpaws who are offended by this selection will have to adjust their equations accordingly. Perhaps some ambitious lefty will write a book of experiments in four dimensions for left-handed individuals!)

Incidentally, you can always test a 3-D system for a right-hand quality this way: Imagine that the Z axis is threaded like a common right-hand screw. If the system is a right-hand system, turning the positive end of the X axis in the direction of the positive end of the Y axis will tighten the screw. Or if the X axis is threaded in a right-hand fashion, turning the positive end of the Y axis toward the positive end of the Z axis will tighten the screw. Or, finally, if the Y axis is threaded like a right-hand screw, turning the positive end of the Z axis toward the positive end of the X axis will tighten it. That isn't a very scientific definition of a right-hand coordinate system, but it is certainly a workable description.

At any rate, a brief summary of the essential characteristics of our choice of 3-D coordinate system is in order:

- ☐ It is composed of three axes labeled X, Y and Z.
- ☐ The axes are mutually perpendicular and intersect at a common point of origin.
- ☐ All three axes are linear (see Section 2-1).
- ☐ It is a right-hand coordinate system.

11-1.2 Perspective Views of 3-D Coordinates

One way to cope with the visual distortion that is inevitably present when attempting to plot 3-D coordinates onto the 2-D plane of a sheet of paper is by using a perspective plot. Figure 11-1 described earlier is an example of perspective plots of 3-D coordinates. Two of the three axes are drawn at right angles and without any distortion on the plane of the page, while the third is drawn at some angle relative to the other two. It is a compromise that deliberately introduces angular distortion, but it is a necessary trade-off.

It is possible to live with this sort of angular distortion, and yet get a more complete and truer visual impression of 3-D space, by plotting at least three different perspective views of this variety. See Fig. 11-2.

The three views show two of the coordinate axes drawn in an undistorted fashion, while the third is offset from the horizontal axis by some angle ψ (psi). The X-Y perspective view shows the X and Y axes drawn flat on the plane of the page and at a true right angle to one another. The Z axis, actually running out of the plane of the page, is shown at angle ψz with respect to the X axis. That is the same perspective view used in Fig. 11-1.

The X-Z perspective view shows the X and Z axes drawn at right angles to one another, and the Y axis offset from the X axis by some angle ψy. Take special note of the fact that the positive end of the Y axis is supposedly extending directly into the plane of the page. That is a necessary feature in order to preserve the right-hand quality of the coordinate system.

Finally, the Y-Z perspective view shows the Y and Z axes drawn perpendicular to one another, and the X axis apparently extending out of the plane of the page. The X axis is offset from the horizontal Y axis by angle ψx. Give all three of those perspective views the "screw test" for a right-handed coordinate system.

Any one of the perspective views can convey a somewhat distorted, but meaningful, visual impression of 3-D space. More from habit than anything else, I tend to use the X-Y perspective view more often than the other two. But if there is any suspicion that some important quality of an object in 3-D space is lost from that perspective, I do not hesitate to look at it from one or both of the others.

11-1.3 Planar Views of 3-D Space

The perspective views of 3-D space introduce some spatial and angular distortion. It is a controlled amount of distortion, but it is nevertheless present. An alternative to perspective views is planar views. See Fig. 11-3.

Planar views of 3-D space show only two of the three axes; and the two that are shown are drawn without a bit of distortion. They are both linear and fixed at right angles to one another. The presence of the third axis is implied, and not actually shown on the drawing. The planar views look and work just like 2-D coordinate systems.

The X-Y planar view, for instance, shows the X and Y axes in an undistorted fashion. Using such a view for plotting 3-D spaces will not introduce any distortion with regard to the X and Y elements of the space. The implied Z axis runs straight in and out of the page at the origin; the positive end extending out of the page for a right-hand coordinate system.

The X-Z planar view looks at the coordinate system

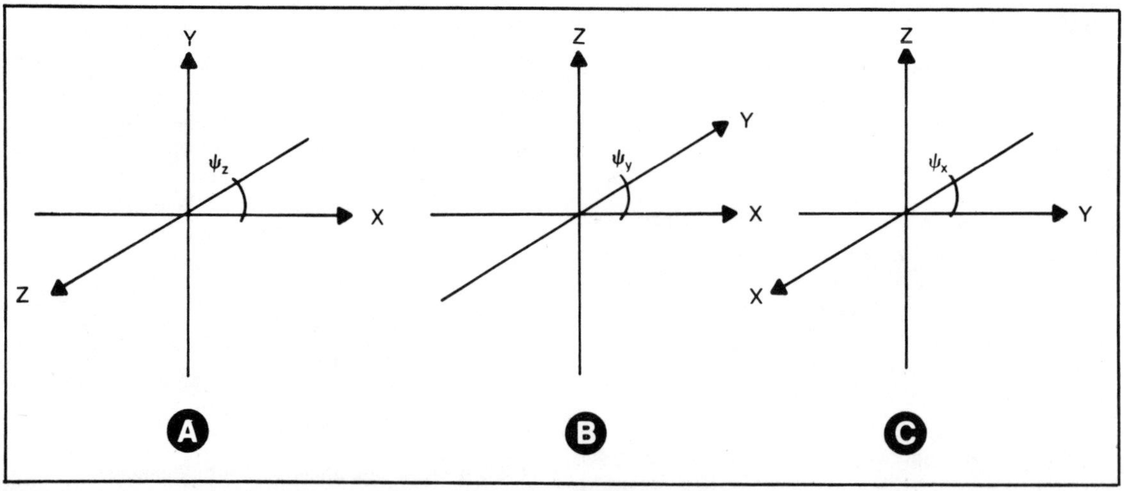

Fig. 11-2. A series of three perspective views for portraying a 3-D coordinate system onto the 2-D plane of a page. (A) The X-Y perspective view. (B) The X-Z perspective view. (C) The Y-Z perspective view.

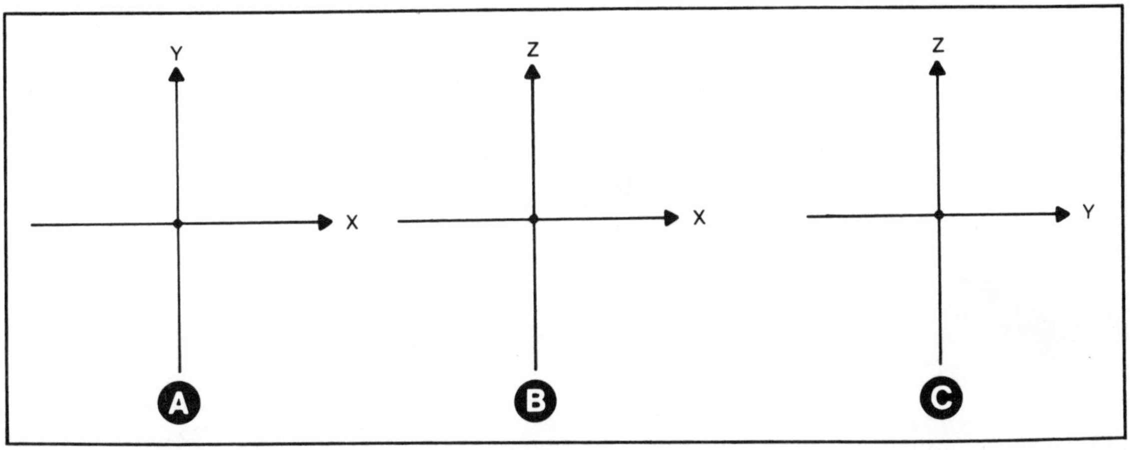

Fig. 11-3. A series of three planar views for portraying 3-D coordinate systems onto the plane of a page. (A) The X-Y planar view. (B) The X-Z planar view. (C) The Y-Z planar view.

from a different angle. Here the X and Z axes are drawn at right angles to one another and without any distortion in that plane. The implied Y axis extends straight in and out of the page at the origin, with its positive end running into the page.

The remaining view, the Y-Z planar view, shows the Y and Z axes without any sort of distortion. The positive end of the implied X axis runs straight out of the page from the origin.

Of course such views of 3-D space cannot convey any visual information about the axis that is not shown. The information for the axes that are shown, however, comes across without any spatial or angular distortion. An experimenter can get a fairly good impression of an object in 3-D space by viewing it from all three of these planar views.

Whether to use perspective or planar views of 3-D space is sometimes a purely arbitrary choice. Often one kind of view is better than the other; but generally there's no telling which is better until you try one of each. Just bear in mind that virtually all spatial and angular qualities of a 3-D object will be distorted by a perspective view; but at the same time, such a view can convey a great deal of important visual information from a single viewing angle. Planar views, on the other hand, always lose information about the axis that isn't shown; but the information related to the two axes that are shown will not include any spatial or angular distortion.

11-2 SPECIFYING AND PLOTTING POINTS IN 3-D SPACE

Three-dimensional space is a "solid" sort of space, and its coordinate system is characterized by three mutually perpendicular axes. By convention, we label those axes X, Y and Z. Now recall that you can specify the position of a point in 1-D space with a single x component; and you can specify the position of a point in 2-D space by means of x and y components—one for each of the two axes. Following this logical progression, it seems that you might be able to specify the position of a point in 3-D space by means of X, Y and Z components. Look at the matter this way:

- ☐ In 1-D space, $P_n = (x_n)$
- ☐ In 2-D space, $P_n = (x_n,y_n)$
- ☐ In 3-D space, $P_n = (x_n,y_n,z_n)$

Thus specifying the position of some point *n* in 3-D space is a matter of citing its X-, Y- and Z-axis components in that order. See Fig. 11-4.

The figure shows the plotting of a single point, P_n, in 3-D space in a couple of different ways. The first drawing illustrates the situation from an X-Y perspective view. Note that the point is located in space by means of its X-, Y- and Z-axis displacements from the origin. The remaining drawings show the same point plotted from the three planar views. In each of those cases, the point's position is reckoned by its displacement from the origin of the two axes that are shown. (Although the planar views show only two coordinate axes at a time, notice that I've indicated all three elements of the point's coordinate on each drawing. Somewhere there should be a record of the information contained in third component, even if you cannot visually appreciate its effect.)

Figure 11-5 shows four specific points in 3-D space as plotted to an X-Y perspective view and all three planar

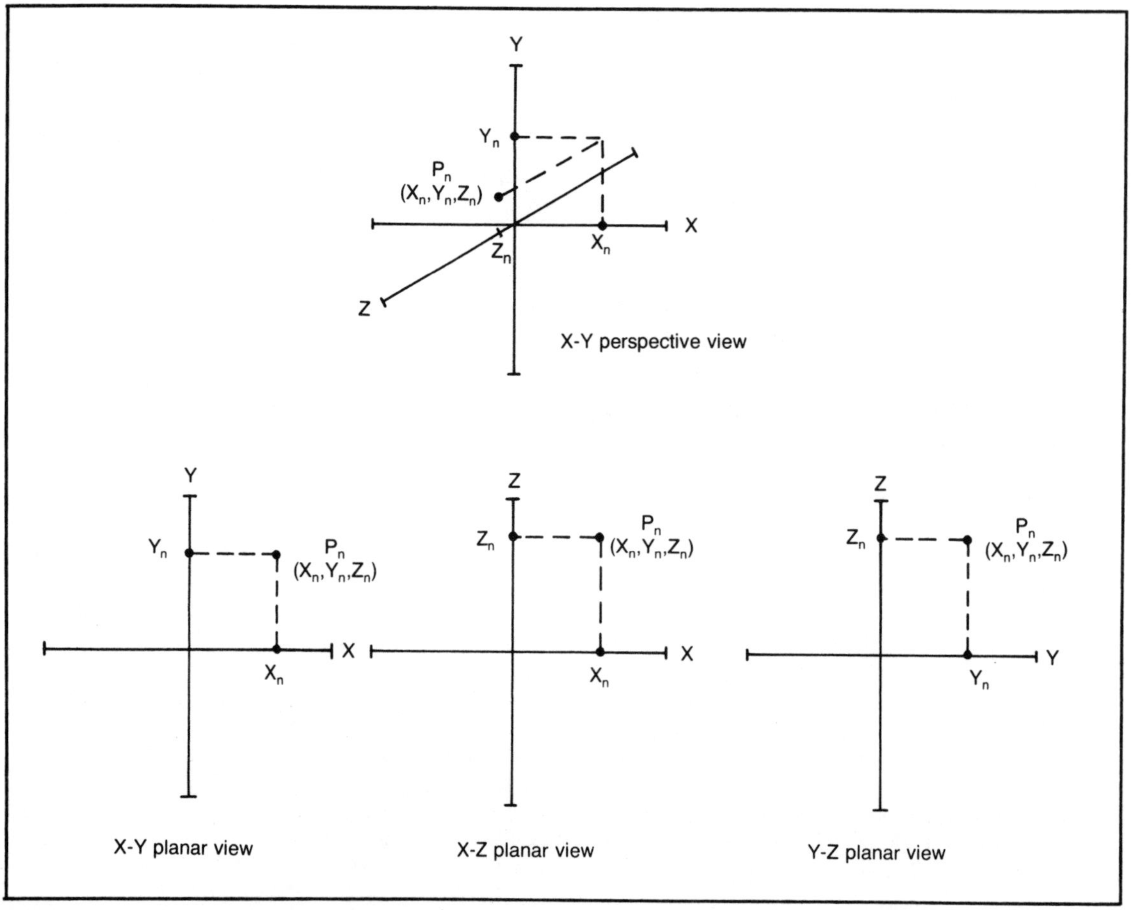

Fig. 11-4. An X-Y perspective view and the three planar views of a single point plotted in space.

views. It is important to realize that each point has the same true coordinate in each case. Notice, however, that the points sometimes appear in different positions relative to one another on the planar views. Points 2 and 3 are especially interesting in the Y-Z planar view because they appear superimposed; they happen to have the same Y- and Z-axis components.

11-2.1 Calculating Precise Coordinates for Perspective Views

Technically speaking, a precision perspective view of 3-D space represents a special sort of geometric transformation; one that transforms a set of true 3-D coordinates to 2-D coordinates. And since it is a formal sort of geometric transformation, there has to be some equations for executing it in a precise fashion. The following equations represent such a transformation.

$$x_{pn} = x_n - z_n \cos\psi_z$$
$$y_{pn} = y_n - z_n \sin\psi_z$$

Equation 11-1

where:

x_{pn}, y_{pn} = the perspective-mapped coordinate of point *n*

x_n, y_n, z_n = the true 3-D coordinate of point *n*

ψ_z = the angle between the X and Z axes

Note: This equation applies only to the X-Y perspective view of 3-D space

$$x_{pn} = x_n + y_n \cos\psi_y$$
$$z_{pn} = z_n + y_n \sin\psi_y$$

Equation 11-2

where:

x_{pn}, z_{pn} = the perspective-mapped coordinate of point *n*

x_n, y_n, z_n = the true 3-D coordinate of point *n*

ψ_y = the angle between the X and Y axes

Note: This equation applies only to the X-Z perspective view of 3-D space

$$y_{pn} = y_n - x_n \cos\psi_x$$
$$z_{pn} = z_n - x_n \sin\psi_x$$ **Equation 11-3**

where:

y_{pn}, z_{pn} = the perspective-mapped coordinate of point n

x_n, y_n, z_n = the true 3-D coordinate of point n

ψ_x = the angle between the Y and Z axes

Note: This equation applies only to the Y-Z perspective view of 3-D space

To use these equations, first select the perspective view you want to generate from set of 3-D coordinates, select an angle for the offset between the horizontal axis and the angled axis, then apply the appropriate equation. If you want to generate a precision X-Y perspective view of four different points that are specified for 3-D space, for example, then you should apply Equation 11-1 with some chosen angle and the components of the coordinates. See Example 11-1.

Example 11-1

Make a precision X-Y perspective plot of the following points specified in 3-D space.

$$P_1 = (0,0,0)$$
$$P_2 = (4,4,4)$$
$$P_3 = (-4,4,4)$$
$$P_4 = (4,-4,-4)$$

1. Specify the components of the coordinates into a more convenient format:

$x_1 = 0$	$x_2 = 4$	$x_3 = -4$	$x_4 = 4$
$y_1 = 0$	$y_2 = 4$	$y_3 = 4$	$y_4 = -4$
$z_1 = 0$	$z_2 = 4$	$z_3 = 4$	$z_4 = -4$

2. Solve Equation 11-1 using an angle of 30 degrees:

$x_{pl} = x_1 - z_1\cos 30°$ $\quad y_{pl} = y_1 - z_1\sin 30°$
$x_{pl} = 0 - (0)(0.87)$ $\quad y_{pl} = 0 - (0)(0.5)$
$x_{pl} = 0$ $\quad y_{pl} = 0$

$x_{p2} = x_2 - z_2\cos 30°$ $\quad y_{p2} = y_2 - z_2\sin 30°$
$x_{p2} = 4 - (4)(0.87)$ $\quad y_{p2} = 4 - (4)(0.5)$
$y_{p2} = 0.02$ $\quad y_{p2} = 2$

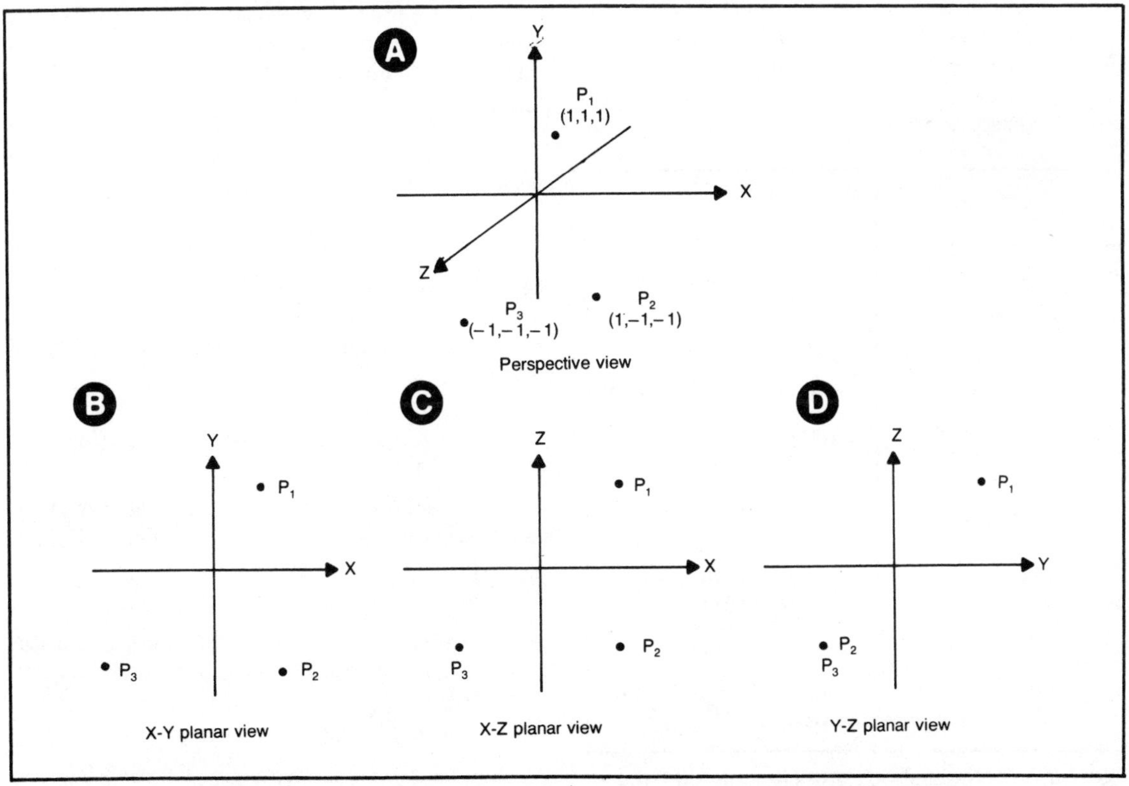

Fig. 11-5. An X-Y perspective view and the three planar views of four different points plotted in 3-D space.

$x_{p3} = x_3 - z_3\cos30^\circ$ $\quad y_{p3} = y_3 - z_3\sin30^\circ$
$x_{p3} = -4-(4)\ (0.87)$ $\quad y_{p3} = 4-(4)\ (0.5)$
$x_{p3} = -7.48$ $\quad y_{p3} = 2$

$x_{p4} = x_4 - z_4\cos30^\circ$ $\quad y_{p4} = y_4 - z_4\sin30^\circ$
$x_{p4} = 4-(-4)\ (0.87)$ $\quad y_{p4} = -4-(-4)\ (0.5)$
$x_{p4} = 7.48$ $\quad y_{p4} = -2$

3. Summarizing the X-Y perspective coordinates:

$P_{pl} = (0,0)$ $\quad P_{p2} = (0.02,2)$
$P_{p3} = (-7.48,2)$ $\quad P_{p4} = (7.48,-2)$

4. Plot the points on an X-Y perspective coordinate system. See Fig. 11-6.

Step 1 in the example simply restates the coordinates of the points in a convenient form. Steps 2, 3 and 4 use the appropriate equations to generate the coordinates for the X-Y, X-Z and Y-Z planar views, respectively.

I selected an offset angle of 30 degrees for all three perspective views in that example. I think it is a good choice because it tends to create the most satisfying visual results for most plotting situations.

I must confess, however, that I do not always make precision perspective views, but rather "eyeball" perspective sketches. The drawings in Fig. 11-6 are precision perspective drawings, and you should go to the

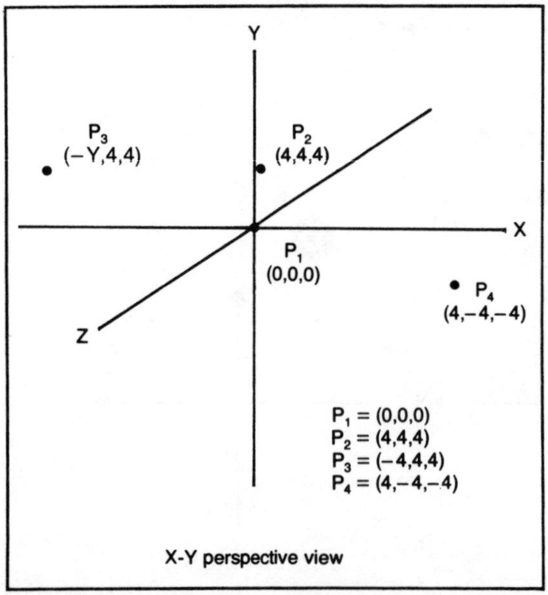

Fig. 11-6. An X-Y perspective plot of four points in 3-D space. See Example 11-1.

trouble of doing them that way for the final documentation of an experiment; but there is no need for such a high level of visual precision during the course of setting up an experiment and running through it in a preliminary sort of way.

11-2.2 Making Planar Views of 3-D Space

It is far easier to generate planar views of 3-D space than to built perspective views. In fact, the work required for making precision plots of all three planar views is less troublesome than setting up a single perspective drawing. Perhaps that, alone, justifies the generous use of planar views.

The procedure is simple: just ignore the component of the coordinates that does not appear on the chosen planar view. When plotting an X-Y planar view from 3-D coordinates, for instance, simply plot the X and Y components—ignore the Z components. And when making an X-Z perspective plot, ignore the Y components of the coordinates. Finally, ignore the X components of the coordinates when setting up a Y-Z planar view of points specified in 3-D space.

Remember to include the unused components in the point specifications, however. See, for example, the planar views in Fig. 11-5 and some others in Fig. 11-7.

Example 11-2

Plot the three planar views of the following points specified in 3-D space:

$P_1 = (0,0,0)$
$P_2 = (4,4,4)$
$P_3 = (-4,4,4)$
$P_4 = (4,-4,-4)$

1. Break down the coordinates into their individual components:

$x_1 = 0$ $\quad x_2 = 4$ $\quad x_3 = -4$ $\quad x_4 = 4$
$y_1 = 0$ $\quad y_2 = 4$ $\quad y_3 = 4$ $\quad y_4 = -4$
$z_1 = 0$ $\quad z_2 = 4$ $\quad z_3 = 4$ $\quad z_4 = -4$

2. Generate the plotted points for the X-Y planar view by suppressing the Z-axis component in each case:

Plot P_1 at (0,0) $\quad$ Plot P_2 at (4,4)
Plot P_3 at (−4,4) $\quad$ Plot P_4 at (4,−4)

3. Generate the plotted points for the X-Z planar view by suppressing the Y-axis component in each case:

Plot P_1 at (0,0) $\quad$ Plot P_2 at (4,4)
Plot P_3 at (−4,4) $\quad$ Plot P_4 at (4,−4)

4. Generate the plotted points for the Y-Z planar view by suppressing the X-axis component in each case:

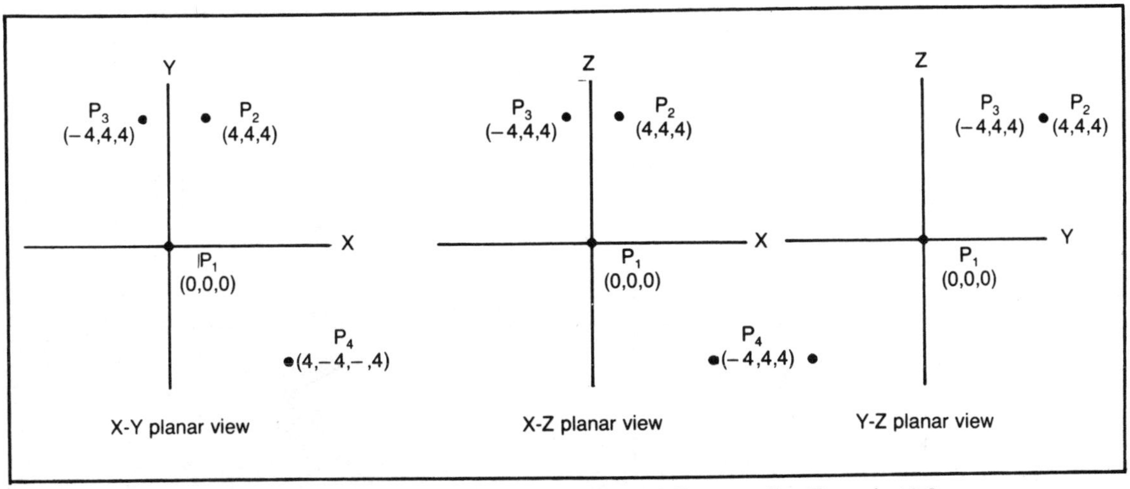

Fig. 11-7. The three planar views of three different points specified in 3-D space. See Example 11-2.

Plot P_1 at (0,0) Plot P_2 at (4,4)
Plot P_3 at (4,4) Plot P_4 at (−4,−4)

See the results in Fig. 11-7.

Test your understanding of making perspective and planar mappings of 3-D space by working through Exericse 11-1.

Exercise 11-1

Use the following coordinates for all of the work recommended in this exercise:

$P_1 = (0,0,0)$ $P_2 = (1,0,0)$

$P_3 = (1,0,1)$ $P_4 = (0,0,1)$

$P_5 = (0,1,0)$ $P_6 = (1,1,0)$

$P_7 = (1,1,1)$ $P_8 = (0,1,1)$

1. Make a precision X-Y perspective plot of points P_1, P_2, P_3 and P_4. Use a ψ_z of 30°.
2. Make a precision X-Z perspective plot of points P_3, P_4, P_5 and P_6. Use a ψ_y of 30°.
3. Plot the three planar views of all eight points.

11-3 SPECIFYING LINES IN 3-D SPACE

You have seen that it is possible to define and specify a line in terms of the coordinates of its two endpoints. The only difference between the specifications for lines in 3-D space is that the endpoint coordinates are composed of three components instead of just one or two.

Thus the general specifications for a line in 3-D space look like this:

$$L_n = \overline{P_u P_v} \qquad P_u = (x_u, y_u) \quad P_v = (x_v, y_v)$$

See an X-Y perspective view of this general situation in Fig. 11-8.

Example 11-3 illustrates the procedure for a specific line.

Example 11-3

Generate an X-Y perspective view and all three planar views for this line specified in 3-D space:

$$L = \overline{P_1 P_2} \qquad P_1 = (4,4,-4) \quad P_2 = (-4,-4,4)$$

1. Citing the components of the two endpoints:

$x_1 = 4$ $x_2 = -4$
$y_1 = 4$ $y_2 = -4$
$z_1 = -4$ $z_2 = 4$

2. Using Equation 11-1 to generate the X-Y perspective coordinates:

$x_{p1} = x_1 - z_1\cos 30°$ $y_{p1} = y_1 - z_1\sin 30°$
$x_{p1} = 4-(-4)\ (0.87)$ $y_{p1} = 4-(-4)\ (0.5)$
$x_{p1} = 7.48$ $y_{p1} = 6$
$x_{p2} = x_2 - z_2\cos 30°$ $y_{p2} = y_2 - z_2\sin 30°$
$x_{p2} = -4-4\ (0.87)$ $y_{p2} = -4-4\ (0.5)$
$x_{p2} = -7.48$ $y_{p2} = -6$

3. Suppressing the Z components to determine the coodinates for an X-Y planar view:

Plot P_1 at (4,4) Plot P_2 at (−4,−4)

4. Suppressing the Y components of the original specifications to determine the coordinates for an X-Z planar view:

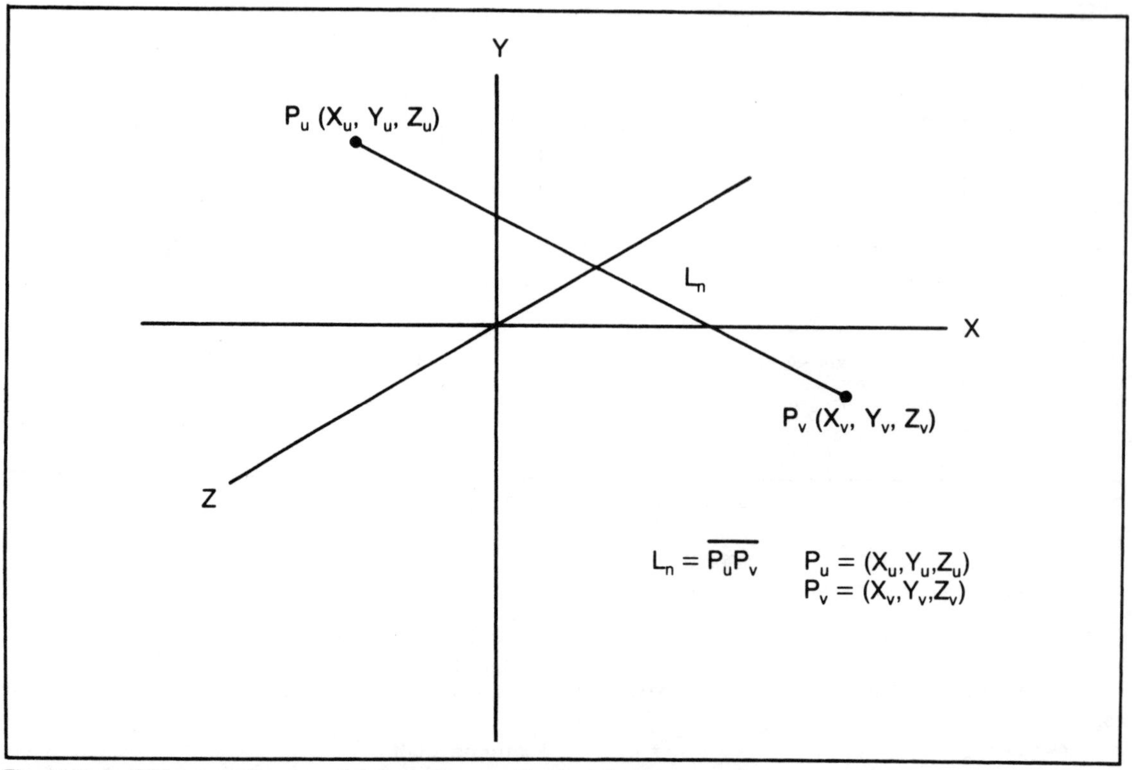

Fig. 11-8. An X-Y perspective view of a single line specified in 3-D space.

Plot P_1 at (4,−4)   Plot P_2 at (−4,4)

5. Suppressing the X components to determine the coordinates for the Y-Z planar view:

Plot P_1 at (4,−4)   Plot P_2 at (−4,4)

See the results in Fig. 11-9.

Try plotting the lines as specified in Exericse 11-2.

Exercise 11-2

Let two lines in 3-D space be defined this way:

$L_1 = \overline{P_1P_2}$ $P_1 = (-1,-1,1)$
$L_2 = \overline{P_3P_4}$ $P_2 = (1,1,-1)$
$P_3 = (2,-1,0)$
$P_4 = (-,0,-2)$

1. Make an X-Y perspective view of both lines.
3. Make a Y-Z perspective view of both lines.
4. Make planar plots of both lines from all three viewing planes.

11-3.1 Lengths of Lines in 3-D Space

Perhaps it comes as no surprise that the equation for the length of a line in 3-D space differs from that of 2-D space only by the addition of the Z-axis components of the endpoint coordinates. See Equation 11-4.

$$s_n = \sqrt{(x_u - x_v)^2 + (y_u - y_v)^2 + (z_u - z_v)^2} \quad \textbf{Equation 11-4}$$

where:

s_n = the length of line *n* in 3-D space
x_u, y_u, z_u = the coordinate of one endpoint of line *n*
x_v, y_v, z_v = the coordinate of the second endpoint of line *n*

The general idea is to square the difference between the X components of the two endpoints, square the difference between the Y components, square the difference between the Z components of the endpoints, then sum the result and find the square root of the whole thing. That yields the length of a line that is immersed in 3-D space.

Example 11-4

Two lines in 3-D space have the following specifications:

$L_1 = \overline{P_1P_2}$ $P_1 = (-2,3,-4)$
$L_2 = \overline{P_3P_4}$ $P_2 = (3,-1,2)$
$P_3 = (2,-3,4)$
$P_4 = (-1,-1,1)$

Calculate the lengths of the lines and make a precision X-Y perspective plot of them.

1. Gathering the components of the coordinates:

$$\begin{array}{llll} x_1 = -2 & x_2 = 3 & x_3 = 2 & x_4 = -1 \\ y_1 = 3 & y_2 = -1 & y_3 = -3 & y_4 = -1 \\ z_1 = -4 & z_2 = 2 & z_3 = 4 & z_4 = 1 \end{array}$$

2. Calculating the length of line L_1:

$$s_1 = \sqrt{(x_2-x_1)^2 + (y_2-y_1)^2 + (z_2-z_1)^2}$$
$$s_1 = \sqrt{[3-(-2)]^2 + (-1-3)^2 + [2-(-4)]^2}$$
$$s_1 = \sqrt{5^2+(-4)^2+(6)^2}$$
$$s_1 = \sqrt{25+16+36}$$
$$s_1 = \sqrt{77}$$
$$s_1 = 8.8$$

3. Calculating the length of line L_2:

$$s_2 = \sqrt{(x_4-x_3)^2 + (y_4-y_3)^2 + (z_4-z_3)^2}$$
$$s_2 = \sqrt{(-1-2)^2 + [-1-(-3)]^2 + (1-4)^2}$$
$$s_2 = \sqrt{(-3)^2+2^2+(-3)^2}$$
$$s_2 = \sqrt{9+4+9}$$
$$s_2 = \sqrt{22}$$
$$s_2 = 4.7$$

4. Make the X-Y perspective plot, using an angle of 30° and Equation 11-1. See Fig. 11-10.

That example specifies two different lines in 3-D space. The general idea is to apply Equation 11-4 to find the length of both lines. Getting the procedure started in a tidy way is a matter of breaking down the four endpoint

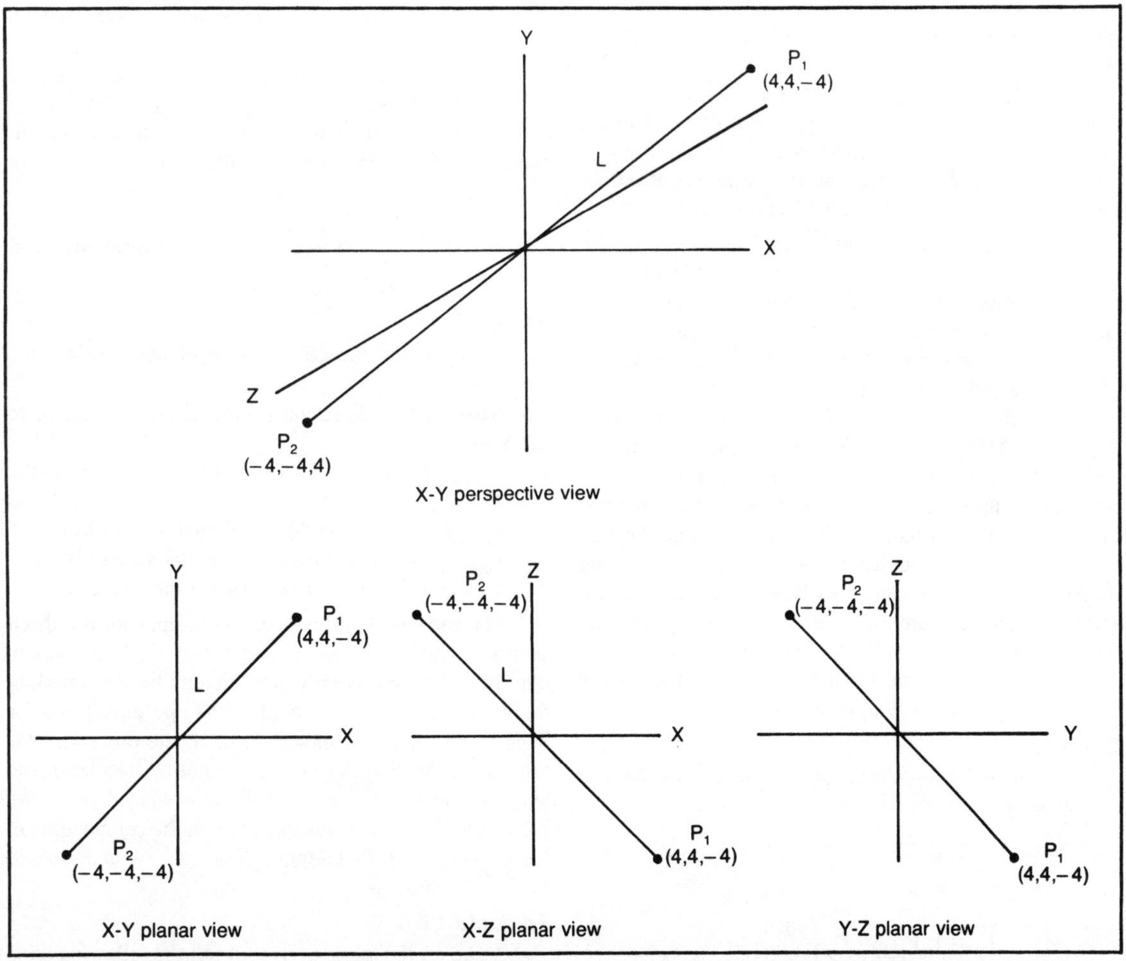

Fig. 11-9. An X-Y perspective view and the three planar views of a line specified in 3-D space. See Example 11-3.

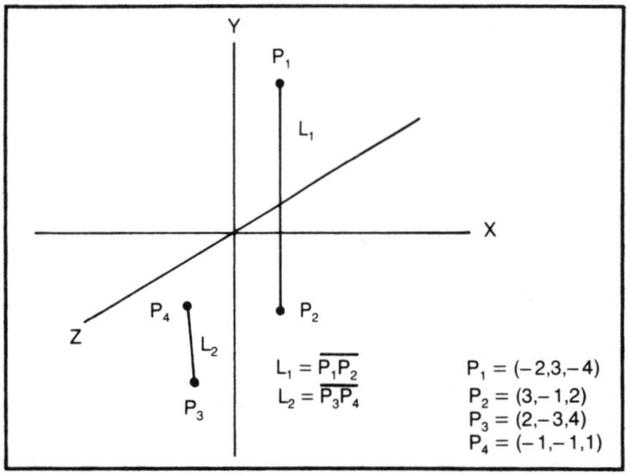

Fig. 11-10. An X-Y perspective view of two different lines that are specified in 3-D space. See Exercise 11-4.

coordinates into their individual components. That is done for you in Step 1.

Step 2 adjusts the notation for Equation 11-4 so that it applies specifically to the endpoints of line 1. Solving the equation with those terms shows that the line is about 8.8 units long. Step 3 carries out the same operations as they apply to the endpoint specifications for line 2.

It is especially important to see that the calculated lengths of those two lines does not match up with actual measurements of the lines as plotted in Fig. 11-10. The discrepancy is caused by the nature of the perspective drawing; such drawings almost always introduce spatial (length) distortion. It should be clear, then, that the numerical results of this length-of-lines analysis are quite important. The numerical analysis provides the truest and most reliable results; the graphic representation—the plotted lines—serve as little more than a general description of the situation. In fact, we could go through this entire book without having to plot a single, counting on the abstract mathematical analyses for conducting the experiments. The numerical result is all-important; the drawings are simply guides to form our thinking.

Try your hand at calculating some lengths of lines in 3-D space by working Exercise 11-3.

Exercise 11-3

Calculate the lengths of the following lines that are specified in 3-D space:

$\overline{L_1 = P_1P_2}$ $\quad P_1 = (-2,2,2)$

$\overline{L_2 = P_1P_3}$ $\quad P_2 = (2,2,-2)$

$\overline{L_3 = P_2P_3}$ $\quad P_3 = (2,-2,-2)$

$P_4 = (-2,-2,2)$

$s_1 =$ ______ $\quad s_2 =$ ______ $\quad s_3 =$ ______

11-3.2 Direction Cosines in 3-D Space

The direction cosines for lines in 3-D space include three direction angles—one relative to the X axis, one relative to the Y axis, and one relative to the Z axis. See Fig. 11-11.

The scheme defines angles alpha and beta as in the 2-D case; they are the acute angles between the line and the X and Y axes, respectively. Moving into 3-D space simply add the third angle γ (gamma). Angle gamma is the acute angle between the line and the Z axis.

Figure 11-11A shows the three direction angles by means of an X-Y perspective drawing, while the remaining illustrations show the same line and direction angles from the three planar views. Notice that each of the planar views includes just two direction angles. Why not all three?

So that is a descriptive definition of direction angles for a line in 3-D space. The direction cosines for a line in 3-D space are defined as the cosines of these three angles. See the mathematical definition in Equation 11-5.

$$\cos\alpha_n = (x_u - x_v)/s_n$$
$$\cos\beta_n = (y_u - y_v)/s_n \qquad \textbf{Equation 11-5}$$
$$\cos\gamma_n = (z_u - z_v)/s_n$$

where:

$\cos\alpha_n$ = the direction cosine of line *n* relative to the X axis

$\cos\beta_n$ = the direction cosine of line *n* relative to the Y axis

$\cos\gamma_n$ = the direction cosine of line *n* relative to the Z axis

x_u, y_u, z_u = the coordinate of endpoint *u* of line *n*

x_v, y_v, z_v = the coordinate of endpoint *v* of line *n*

s_n = the length of line *n* (see Equation 11-4)

You can see that the direction cosines for the three axes are equal to the X-, Y- and Z-axis displacements of the line's endpoint coordinates divided by the length of the line. The idea is identical to the one introduced for lines in 2-D space. In practice, then, the procedure for calculating the direction cosines begins by calculating the length of the line (Equation 11-4) followed by dividing the X-, Y- and Z-axis differences between the components of the endpoints by that length. Take a look at Example 11-5.

Example 11-5

Make an X-Y perspective plot of the following lines and calculate their direction cosines:

$$L_1 = \overline{P_1P_4} \qquad P_1 = (0,0,0)$$
$$L_2 = \overline{P_2P_3} \qquad P_2 = (1,0,1)$$
$$P_3 = (1,1,0)$$
$$P_4 = (1,1,1)$$

1. Use Equation 11-1 to generate the X-Y perspective coordinates of the four points, then join them with lines as specified. See Fig. 11-12.

2. Gathering the components of the coordinates:

$$x_1 = 0 \quad x_2 = 1 \quad x_3 = 1 \quad x_4 = 1$$
$$y_1 = 0 \quad y_2 = 0 \quad y_3 = 1 \quad y_4 = 1$$
$$z_1 = 0 \quad z_2 = 1 \quad z_3 = 0 \quad z_4 = 1$$

3. Calculating the lengths of the lines:

$$s_1 = \sqrt{(x_4-x_1)^2+(y_4-y_1)^2+(z_4-z_1)^2}$$
$$s_1 = \sqrt{(1-0)^2+(1-0)^2+(1-0)^2}$$
$$s_1 = \sqrt{3}$$
$$s_2 = \sqrt{(x_3-x_2)^2+(y_3-y_2)^2+(z_3-z_2)^2}$$
$$s_2 = \sqrt{(1-1)^2+(1-0)^2+(0-1)^2}$$
$$s_2 = \sqrt{2}$$

4. Calculating the direction cosines for the two lines:

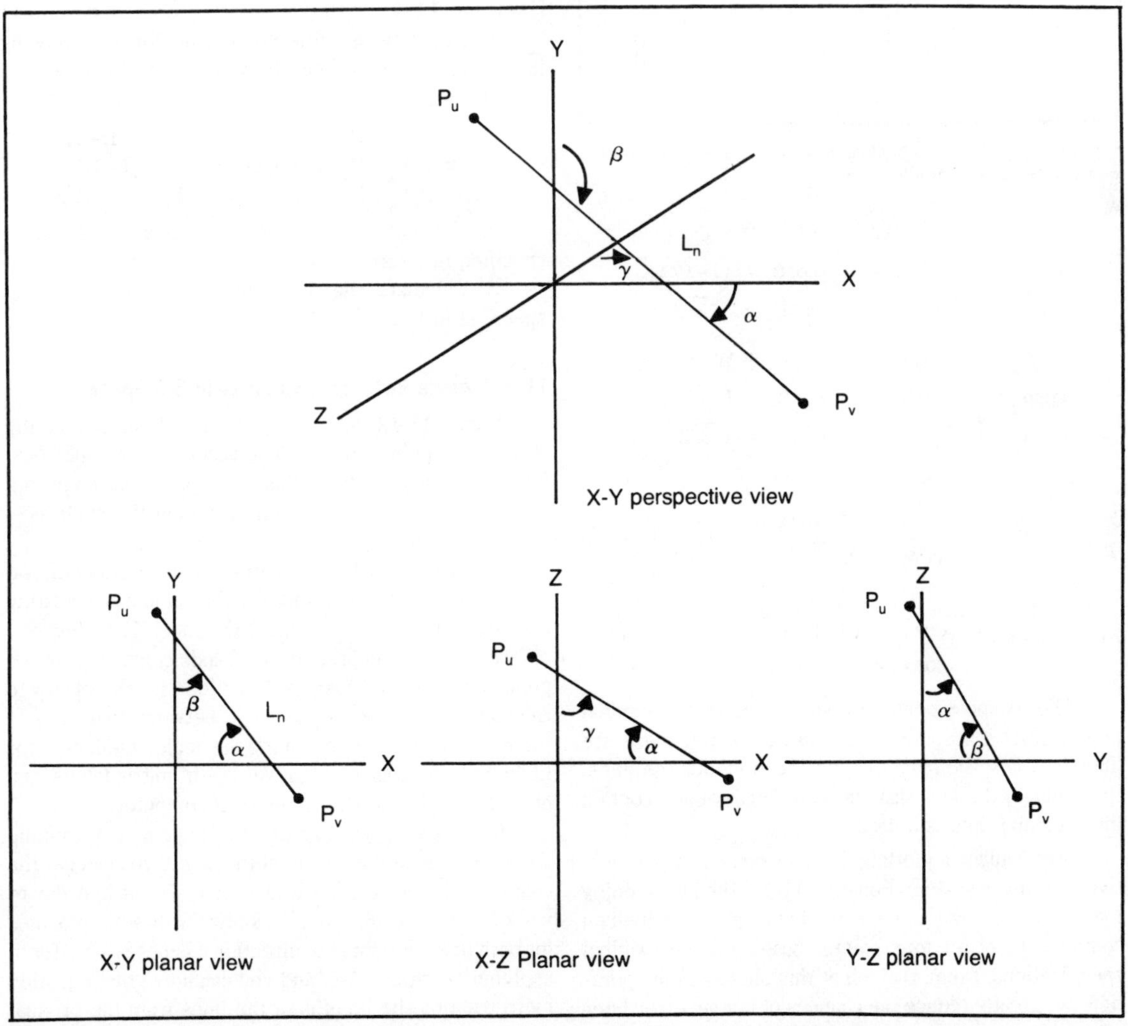

Fig. 11-11. The three direction angles for a single line specified in 3-D space as seen from an X-Y perspective view and the three planar views.

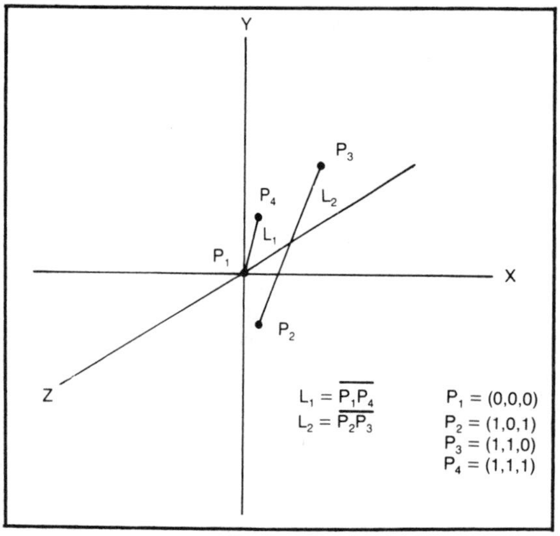

Fig. 11-12. An X-Y perspective view of two lines specified in 3-D space. See Example 11-5.

$\cos\alpha_1 = (x_4 - x_1)/s_1$ $\qquad$ $\cos\beta_1 = (y_4 - y_1)/s_1$

$\cos\alpha_1 = (1-0)/\sqrt{3}$ $\qquad$ $\cos\beta_1 = (1-0/\sqrt{3}$

$\cos\alpha_1 = \sqrt{3}/3$ $\qquad$ $\cos\beta_1 = \sqrt{3}/3$

$\cos\alpha_2 = (x_3 - x_2)/s_2$ $\qquad$ $\cos\beta_2 = (y_3 - y_2)/s_2$

$\cos\alpha_2 = (1-1)/\sqrt{2}$ $\qquad$ $\cos\beta_2 = 1-0)/\sqrt{2}$

$\cos\alpha_2 = 0$ $\qquad$ $\cos\beta_2 = \sqrt{2}/2$

$$\cos\gamma_1 = (z_4 - z_1)/s_1$$
$$\cos\gamma_1 = (1-0)/\sqrt{3}$$
$$\cos\gamma_1 = \sqrt{3}/3$$
$$\cos\gamma_2 = (z_3 - z_2)/s_2$$
$$\cos\gamma_2 = (0-1)/\sqrt{2}$$
$$\cos\gamma_2 = -\sqrt{2}/2$$

The example begins by specifying two individual lines in 3-D space. Although you cannot tell these are lines in 3-D space from the specifications for the lines, themselves, the fact that the coordinates each contain three components is a dead giveaway.

Step 1 suggest plotting the lines on an X-Y perspective coordinate system. Equation 11-1 is the key to doing that. Step 2 shows the assignments for the individual components of all four points, based on the original specifications. Doing that job at this phase of the operation can greatly reduce the chances of human error later on.

The next step, Step 3, is to calculate the lengths of the two lines. You can see the length equation presented with terms that are adjusted to the specifications of the lines, an intermediate step, and finally the calculated lengths. I have left them in their rational form (square-root form in this instance) in order to preserve absolute precision all the way through.

Step 4 completes the task at hand by calculating the direction cosines for both lines. Doing so is a matter of adjusting the notation in Equation 11-5 to suit the line-number designations. Again, I have left the final values in a rational form.

If you think you are ready to test your understanding, try Exercise 11-4.

Exercise 11-4

1. Calculate the direction cosines of the following lines specified in 3-D space. Make an X-Y perspective plot of the results.

$L_1 = P_1P_2$ $\quad$ $P_1 = (-1,-1,2)$ $\quad$ $P_2 = (1,-1,-2)$
$L_2 = P_3P_4$ $\quad$ $P_3 = (1,1,2)$ $\quad$ $P_4 = (-1,1,-2)$
$L_3 = P_4P_6$ $\quad$ $P_5 = (1,-1,2)$ $\quad$ $P_6 = (-1,1,2)$

2. Calculate the direction cosines for the lines specified in Exercise 11-2.

3. Calculate the direction cosines for the lines specified in Exercise 11-3.

11-3.3 Angle Between Two Lines in 3-D Space

Figure 11-13 shows two lines in 3-D space meeting at a certain point, point 2. The accompanying specifications show that the lines share that particular point, but the important matter at hand is finding the angle $\phi_{m,n}$ between them.

The procedure for calculating the angle between two lines in 3-D space is essentially the same as calculating that angle between lines in 2-D space. The only real difference is the addition of the Z-axis component of the points and direction cosines. Equation 11-6 shows how to calculate the cosine of the angle between two lines in three dimensions. Determining the angle, itself, calls for finding the arc cosine of that value—from trig tables, and electronic calculator or a personal computer.

In any event, you can find the angle by determining the arc cosine of the cosine of the angle; you can get the cosine of the angle from Equation 11-6; but in order to work Equation 11-6, you must know the direction cosines for both lines. Finding the direction cosines is a matter of applying Equation 11-5, and you can solve that equation by calculating the lengths of the lines from the original specifications. Putting all of this into a step-by-step procedure:

1. Calculate the lengths of the two lines from the original specifications (Equation 11-4).

2. Calculate the direction cosines of both lines (Equation 11-5).

3. Calculate the cosine of the angle between the two lines (Equation 11-6).

4. Determine the angle by taking the arc cosine of the result of Step 3.

Example 11-6 demonstrates the procedure. The steps are numbered differently, because some of the operations just outlined require more than one basic operation. The general flow of events are to be rather clear, however.

Example 11-6

Calculate the angle between lines L_1 and L_2:

$$L_1 = \overline{P_1P_2} \qquad P_1 = (0,0,0)$$
$$L_2 = \overline{P_1P_3} \qquad P_2 = (1,2,4)$$
$$P_3 = (0,-3,-5)$$

1. Gathering the component parts of the coordinates:

$$x_1 = 0 \quad x_2 = 1 \quad x_3 = 0$$
$$y_1 = 0 \quad y_2 = 2 \quad y_3 = -3$$
$$z_1 = 0 \quad z_2 = 4 \quad z_3 = -5$$

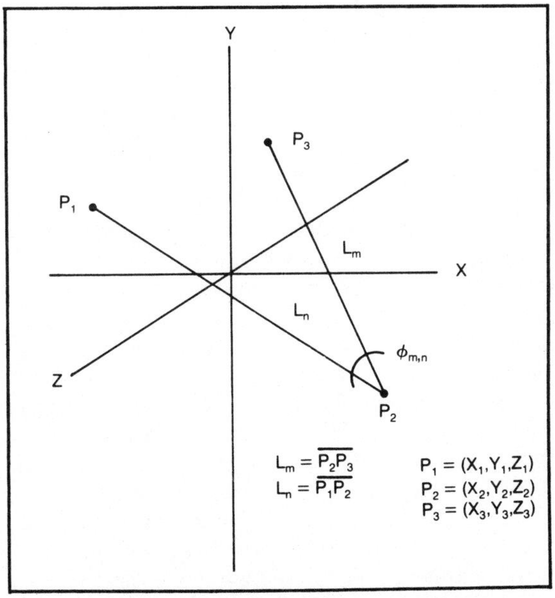

Fig. 11-13. The angle between two lines specified in 3-D space as seen from an X-Y perspective view.

2. Calculating the lengths of the lines:

$$s_1 = \sqrt{(x_2-x_1)^2+(y_2-y_1)^2+(z_2-z_1)^2}$$
$$s_1 = \sqrt{(1-0)^2+(2-0)^2+(4-0)^2}$$
$$s_1 = \sqrt{21}$$
$$s_2 = \sqrt{(x_3-x_1)^2+(y_3-y_1)^2+(z_3-z_1)^2}$$
$$s_2 = \sqrt{(0-0)^2+(-3-0)^2+(-5-0)^2}$$
$$s_2 = \sqrt{34}$$

3. Calculating the direction cosines for both lines:

$$\cos\alpha_1 = (x_2-x_1)/s_1 \qquad \cos\beta_1 = (y_2-y_1)/s_1$$
$$\cos\alpha_1 = (1-0)/\sqrt{21} \qquad \cos\beta_1 = (2-0)/\sqrt{21}$$
$$\cos\alpha_1 = \sqrt{21}/21 \qquad \cos\beta_1 = 2\sqrt{21}/21$$
$$\cos\gamma_1 = (z_2-z_1)/s_1$$
$$\cos\gamma_1 = (4-0)/\sqrt{21}$$
$$\cos\gamma_1 = 2\sqrt{21}/21$$
$$\cos\alpha_2 = (x_3-x_1)/s_2 \qquad \cos\beta_2 = (y_3-y_1)/s_2$$
$$\cos\alpha_2 = (0-0)/\sqrt{34} \qquad \cos\beta_2 = (-3-0)/\sqrt{34}$$
$$\cos\alpha_2 = 0 \qquad \cos\beta_2 = -3\sqrt{34}/34$$
$$\cos\gamma_2 = (z_3-z_1)/s_2$$
$$\cos\gamma_2 = (-5-0)/\sqrt{34}$$
$$\cos\gamma_2 = -5\sqrt{34}/34$$

4. Calculating the cosine of angle $\phi_{1,2}$:

$$\cos\phi_{1,2} = \cos\alpha_1\cos\alpha_2+\cos\beta_1\cos\beta_2+\cos\gamma_1\cos\gamma_2$$
$$\cos\phi_{1,2} = 0+0.224+0.748$$
$$\cos\phi_{1,2} = 0.972$$

5. Determining the angle from its cosine value:

$$\phi_{1,2} = 13°20'$$

The two lines specified in that example have point P1 in common, so the angle between the lines will be most apparent at that place. Sketch an X-Y perspective view of the lines if you wish. The first step, as is most often the case, is to make a table of the components of the coordinates. Step 2 then uses that information in conjunction with the line specifications and the 3-D length equation to figure out the lengths of the lines. This, you should recall, is vital to determining the direction cosines.

Step 3 works out the direction cosines, and Step 4 puts together the direction-cosine information to come up with the cosine of the angle in question. Determining the arc cosine of the value is the final step in the operation.

Incidentally, you are going to find it virtually impossible to doublecheck the results by measuring the angle as it appears on any sort of 3-D coordinate system that is drawn onto a sheet of paper. Just as plotting 3-D lines distorts the true length of those lines, such plots also distort angles. You simply must work the procedure for calculating the angle very carefully and doublecheck the work by running through it two or three more times.

Use Exercise 11-5 to test your understanding of the procedure.

Exercise 11-5

Using Example 11-6 as a guide, determine the angle between these two lines in 3-D space:

$$L_1 = \overline{P_1P_2} \quad P_1 = (1-1,-1)$$
$$L_2 = \overline{P_2P_3} \quad P_2 = (1,1,1)$$
$$P_3 = (-2,1,2)$$

Make an X-Y perspective view of the lines, noting the apparent distortion of the graphic version of the angle between them.

11-4 PLANE FIGURES IN 3-D SPACE

Plane figures can fit quite nicely into a 3-D space. If you imagine that a sheet of paper represents an infinitely thin plane figure, you should have no trouble realizing that it can be described quite adequately in our 3-D universe. As long as the dimension of a geometric element is less than or equal to the number of dimensions in the space surrounding it, there is no real problem with specifying it in that space.

The specifications for a plane figure that is immersed in 3-D space are practically identical to its 2-D specifications. In fact the F and L specifications are the same in either dimension. The difference is that the points specified for a plane figure in 3-D space can include three components—one for each of the 3-D axes.

Example 11-7 demonstrates these ideas and shows how to go about plotting a plane figure that is specified in 3-D space.

Example 11-7

Make an X-Y perspective plot and the three planar plots of the following figure:

$$F = \overline{L_1L_2L_3L_4} \quad L_1 = \overline{P_1P_2} \quad P_1 = (-1,1,2)$$
$$L_2 = \overline{P_1P_4} \quad P_2 = (1,2,0)$$
$$L_3 = \overline{P_2P_3} \quad P_3 = (2,0,-1)$$
$$L_4 = \overline{P_3P_4} \quad P_4 = (1,-1,1)$$

1. Gathering the components of the coordinates:

$x_1 = -1$	$x_2 = 1$	$x_3 = 2$	$x_4 = 1$
$y_1 = 1$	$y_2 = 2$	$y_3 = 0$	$y_4 = -1$
$z_1 = 2$	$z_2 = 0$	$z_3 = -1$	$z_4 = 1$

2. Calculating the X-Y perspective coordinates for each point (Equation 11-1):

$x_{p1} = x_1 - z_1\cos30°$	$yx_{p1} = y_1 - z_1\sin30°$
$x_{p1} = -1-2\ (0.87)$	$y_{p1} = 1-(2)\ (0.5)$
$x_{p1} = -2.74$	$y_{p1} = 0$
$x_{p2} = x_2 - z_2\cos30°$	$y_{p2} = y_2 - z_2\sin30°$
$x_{p2} = 1-(0)\ (0.87)$	$y_{p2} = 2-(0)\ (0.5)$
$x_{p2} = 1$	$y_{p2} = 2$
$x_{p3} = x_3 - z_3\cos30°$	$y_{p3} = y_3 - z_3\sin30°$
$x_{p3} = 2-(-1)\ (0.87)$	$y_{p3} = (0)-(-1)\ (0.5)$
$x_{p3} = 2.87$	$y_{p3} = 0.5$
$x_{p4} = x_4 - z_4\cos30°$	$y_{p4} = y_4 - z_4\sin30°$
$x_{p4} = 1-(1)\ (0.87)$	$y_{p4} = -1-(1)\ (0.5)$
$x_{p4} = 0.13$	$y_{p4} = -1.5$

See the figure plotted in Fig. 11-14.

3. Generate the coordinates for the three planar views:

For the X/Y planar view: P_1 plots as (−1,1)
P_2 plots as (1,2)
P_3 plots as (2,0)
P_4 plots as (1,−1)

For the X-Z planar view: P_1 plots as (−1,2)
P_2 plots as (1,0)
P_3 plots as (2,−1)
P_4 plots as (1,1)

For the Y-Z planar view: P_1 plots as (1,2)
P_2 plots as (2,0)
P_3 plots as (0,−1)
P_4 plots as (−1,1)

See Fig. 11-14.

The specifications introduced in the example show a plane figure—a quadralateral—immersed in 3-D space. The F and L specifications could apply equally well to the same figure in 2-D space, but the fact that the point coordinates have 3 components is a clear indication that the figure is in 3-D space.

The purpose of the example is to plot that figure from an X-Y perspective view and the three planar views, but a good first step is to set up the assignments for the individual components of all four points. See Step 1.

Step 2 uses Equation 11-1 to generate the coordinates for making the X-Y perspective view, and Step 3 works out the coordinates for the three planar views.

Indeed, this is a plane figure, but since it is oriented in 3-D space, the usual 2-D plotting procedures are not

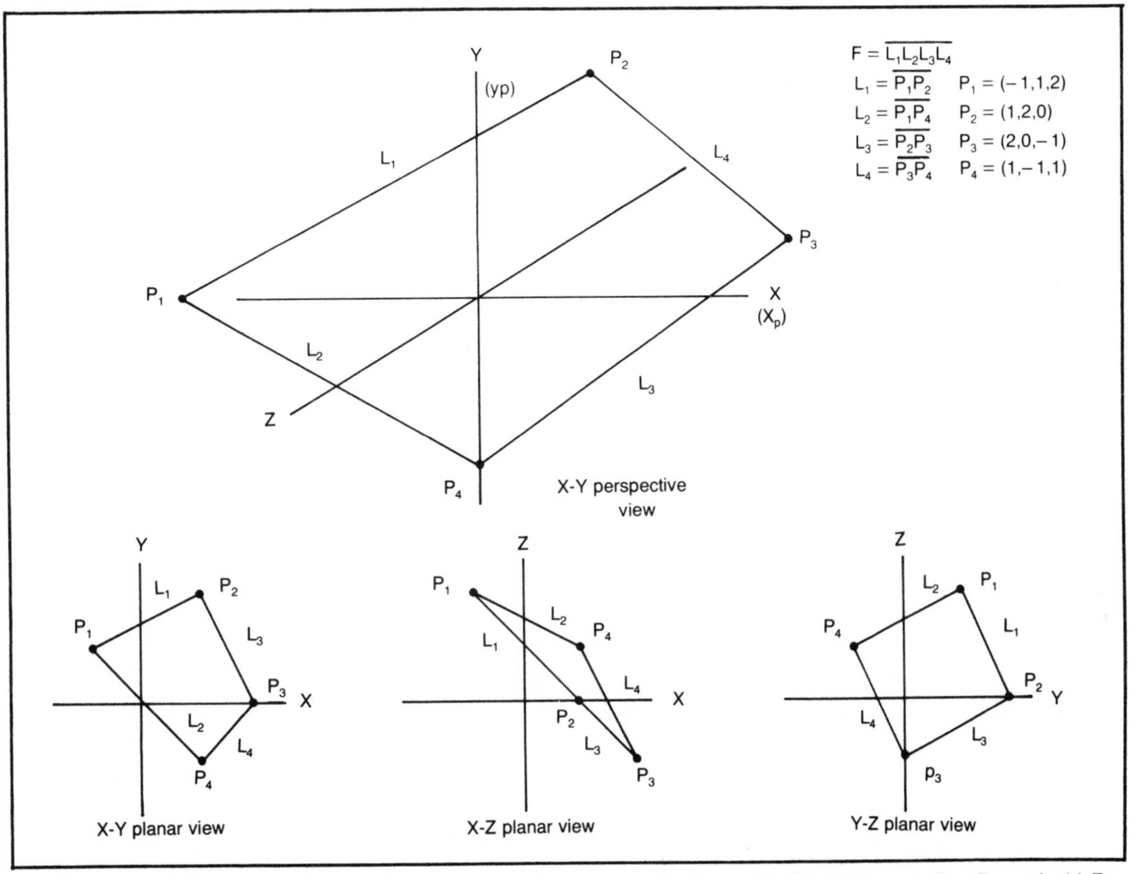

Fig. 11-14. An X-Y perspective and the three planar views of a plane figure specified in 3-D space. See Example 11-7.

adequate. The 3-D plotting procedures are quite necessary.

Try your hand at plotting plane figures immersed in a 3-D environment by working Exercise 11-6.

Exercise 11-6

Make these plots of the given plane figures in 3-D space:

a. X-Y perspective view (Equation 11-1)
b. X-Z perspective view (Equation 11-2)
c. Y-Z perspective view (Equation 11-3)
d. X-Y planar view
e. X-Z planar view
f. Y-Z planar view

1. $F = \overline{L_1L_2L_3}$ $\quad L_1 = \overline{P_1P_2}$ $\quad P_1 = (0,0,0)$
   $L_2 = \overline{P_1P_3}$ $\quad P_2 = (1,-1,1)$
   $L_3 = \overline{P_2P_3}$ $\quad P_3 = (0,1,0)$

2. $F = \overline{L_1L_2L_3L_4}$ $\quad L_1 = \overline{P_1P_2}$ $\quad P_1 = (-1,1,-1)$
   $L_2 = \overline{P_1P_4}$ $\quad P_2 = (1,-1,1)$
   $L_3 = \overline{P_2P_3}$ $\quad P_3 = (1,1,1)$
   $L_4 = \overline{P_3P_4}$ $\quad P_4 = (-1,2,-1)$

A complete analysis of a plane figure in 3-D space follows the same general pattern we have been using all along:

1. Determine the F-L-P specifications for the figure.
2. Establish the components of all points.
3. Plot the figure as represented in 3-D space.
4. Calculate the lengths of the lines.
5. Calculate the direction cosines of all the lines.
6. Calculate the cosine of the relevant angles in the figure.
7. Determine the value of the relevant angles in the figure.

Example 11-8 demonstrates a complete analysis for a triangle in 3-D space.

Example 11-8

Conduct a complete analysis of the following plane figure suspended in 3-D space:

$F = \overline{L_1L_2L_3}$ $\quad L_1 = \overline{P_1P_2}$ $\quad P_1 = (0,1,2)$

$L_2 = \overline{P_2P_3}$ $\quad P_2 = (4,2,3)$

$L_3 = \overline{P_3P_1}$ $\quad P_3 = (3,4,5)$

1. Collect the components of the coordinates:

$x_1 = 0$	$x_2 = 4$	$x_3 = 3$
$y_1 = 1$	$y_2 = 2$	$y_3 = 4$
$z_1 = 2$	$z_2 = 3$	$z_3 = 5$

2. Use Equation 11-1 to generate the coordinates for an X-Y perspective view of the figure:

$x_{p1} = x_1 - z_1\cos 30°$ $\quad y_{p1} = y_1 - z_1\cos 30°$

$x_{p1} = -1.74$ $\quad y_{p1} = 0$

$x_{p2} = x_2 - z_2\cos 30°$ $\quad y_{p2} = y_2 - z_2\cos 30°$

$x_{p2} = 1.39$ $\quad y_{p2} = 0.5$

$x_{p3} = x_3 - z_3\cos 30°$ $\quad y_{p3} = y_3 - z_3\cos 30°$

$x_{p3} = -1.35$ $\quad y_{p3} = 1.5$

See the X-Y perspective view in Fig. 11-15.

3. Generate the coordinates for the three planar views from the original specifications:

For the X-Y view: Plot P_1 at (0,1)
Plot P_2 at (4,2)
Plot P_3 at (3,4)

For the X-Z view: Plot P_1 at (0,2)
Plot P_2 at (4,3)
Plot P_3 at (3,5)

For the Y-Z view: Plot P_1 at (1,2)
Plot P_2 at (2,3)
Plot P_3 at (4,5)

See the planar views in Fig. 11-15.

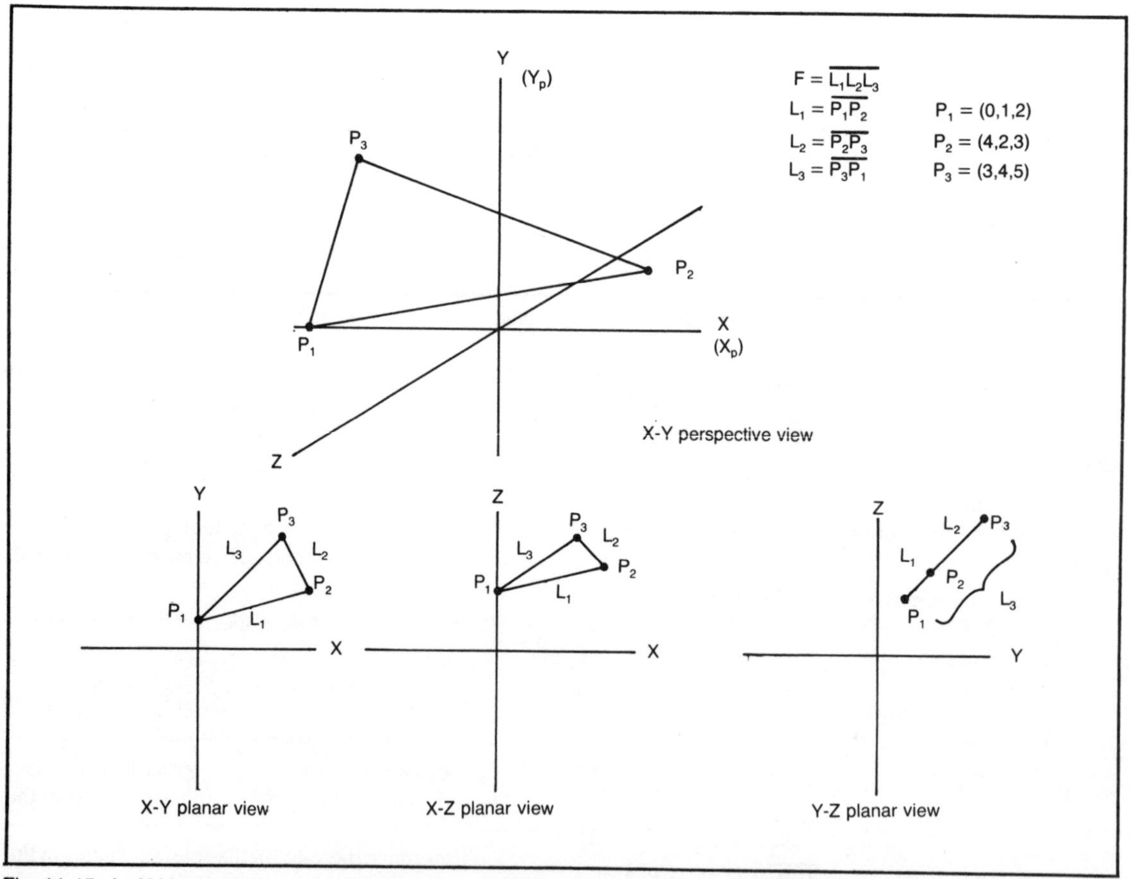

Fig. 11-15. An X-Y perspective and the three planar views of a triangle that is specified in 3-D space. See Example 11-8.

4. Finding the lengths of the lines:

$$s_1 = \sqrt{(x_2-x_1)^2+(y_2-y_1)^2+(z_2-z_1)^2}$$
$$s_1 = \sqrt{(4-0)^2+(2-1)^2+(3-2)^2}$$
$$s_1 = \sqrt{16+1+1}$$
$$s_1 = 3\sqrt{2}$$

Thus line L_1 is 4.24 units long.

$$s_2 = \sqrt{(x_3-x_2)^2+(y_3-y_2)^2+(z_3-z_2)^2}$$
$$s_2 = \sqrt{(3-4)^2+(4-2)^2+(5-3)^2}$$
$$s_2 = \sqrt{1+4+4}$$
$$s_2 = 3$$

Line L_2 is 3 units long.

$$s_3 = \sqrt{(x_1-x_3)^2+(y_1-y_3)^2+(z_1-z_3)^2}$$
$$s_3 = \sqrt{(0-3)^2+(1-4)^2+(2-5)^2}$$
$$s_3 = \sqrt{9+9+9}$$
$$s_3 = 3\sqrt{3}$$

Or line L_3 is about 5.2 units long.

Summarizing the lengths for future reference:

$$s_1 = 3\sqrt{2} \qquad s_2 = 3 \qquad s_3 = 3\sqrt{3}$$

5. Calculating the direction cosines for the three lines:

$$\cos\alpha_1 = (x_2-x_1)/s_1 \qquad \cos\beta_1 = (y_2-y_1)/s_1$$
$$\cos\alpha_1 = 4/3\sqrt{2} \qquad \cos\beta_1 = 1/3\sqrt{2}$$
$$\cos\alpha_1 = 2\sqrt{2}/3 \qquad \cos\beta_1 = \sqrt{2}/6$$

$$\cos\gamma_1 = (z_2-z_1)/s_1$$
$$\cos\gamma_1 = 1/3\sqrt{2}$$
$$\cos\gamma_1 = \sqrt{2}/6$$

$$\cos\alpha_2 = (x_3-x_2)/s_2 \qquad \cos\beta_2 = (y_3-y_2)/s_2$$
$$\cos\alpha_2 = -1/3 \qquad \cos\beta_2 = 2/3$$

$$\cos\gamma_2 = (z_3-z_2)/s_2$$
$$\cos\gamma_2 = 2/3$$

$$\cos\alpha_3 = (x_1-x_3)/s_3 \qquad \cos\beta_3 = (y_1-y_3)/s_3$$
$$\cos\alpha_3 = -3/3\sqrt{3} \qquad \cos\beta_3 = -3/3\sqrt{3}$$
$$\cos\alpha_3 = -\sqrt{3}/3) \qquad \cos\beta_3 = -\sqrt{3}/3$$

$$\cos\gamma_3 = (z_1-z_3)/s_3$$
$$\cos\gamma_3 = -3/3\sqrt{3}$$
$$\cos\gamma_3 = -\sqrt{3}/3$$

Summarizing the direction cosines:

$$\cos\alpha_1 = 2\sqrt{2}/3 \qquad \cos\beta_1 = \sqrt{2}/6$$
$$\cos\alpha_2 = -1/3 \qquad \cos\beta_2 = 2/3$$
$$\cos\alpha_3 = -\sqrt{3}/3 \qquad \cos\beta_3 = -\sqrt{3}/3$$

$$\cos\gamma_1 = \sqrt{2}/6$$
$$\cos\gamma_2 = 2/3$$
$$\cos\gamma_3 = -\sqrt{3}/3$$

6. Calculating the cosine of the three angles in the figure:

$$\cos\phi_{1,2} = \cos\alpha_1\cos\alpha_2+\cos\beta_1\cos\beta_2+\cos\gamma_1\cos\gamma_2$$
$$\cos\phi_{1,2} = (2\sqrt{2}/3)(-/3)+(\sqrt{2}/6)(2/3)+(\sqrt{2}/6)(2/3)$$
$$\cos\phi_{1,2} = 0$$

$$\cos\phi_{1,3} = \cos\alpha_1\cos\alpha_3+\cos\beta_1\cos\beta_3+\cos\gamma_1\cos\gamma_3$$
$$\cos\phi_{1,3} = (2\sqrt{2}/3)(-\sqrt{3}/3)+(\sqrt{2}/6)(-\sqrt{3}/3)+(\sqrt{2}/6)(-\sqrt{3}/3)$$
$$\cos\phi_{1,3} = -\sqrt{6}/3$$

$$\cos\phi_{2,3} = \cos\alpha_2\cos\alpha_3+\cos\beta_2\cos\beta_3+\cos\gamma_2\cos\gamma_3$$
$$\cos\phi_{2,3} = (-1/3)(-\sqrt{3}/3)+(2/3)(-\sqrt{3}/3)+(2/3)(-\sqrt{3}/3)$$
$$\cos\phi_{2,3} = -\sqrt{3}/3$$

Summarizing the cosines of the angles:

$$\cos\phi_{1,2} = 0 \quad \cos\phi_{1,3} = -\sqrt{6}/3 \quad \cos\phi_{2,3} = -\sqrt{3}/3$$

7. Determining the three angles, themselves, by finding the arc cosine of the previous results:

$$\phi_{1,2} = 90° \qquad \phi_{1,3} = 35°\ 44' \qquad \phi_{2,3} = 54°\ 44'$$

8. Specify some general conclusions:

a. The figure is a triangle—three sides.
b. The triangle is a right triangle—the angle between lines L_1 and L_2 is a right angle.

Now you try it with Exercise 11-7.

Exercise 11-7

1. Complete the analysis of the figure specified in Example 11-7.
2. Complete the analysis of the two figures cited in Exercise 11-6.
3. Specify some plane figures in 3-D space of your own design, then carry out a complete analysis of them.

11-5 SPACE OBJECTS IN 3-D SPACE

So far in this chapter we have dealt only with geometric elements (points, lines and plane figures) that can be specified in 2-D space as well as 3-D space. Now it is time to introduce something truly new: three-dimensional geometric entities, or space objects.

A line is bounded by its endpoints and a plane figure is bounded by lines. In 3-D space, it is possible to generate another entity that is bounded by plane figures. That is, of course, a 3-dimensional space object.

The specifications for a space object include an S element; it is an element defined in terms of the plane figures, F elements, that bound it. See, for instance, the drawings and specifications in Fig. 11-16. The figure in that case is a prism. The specifications show that it is a space object bounded by 5 plane figures. Some of the plane figures are then bounded by 3 lines, and others are bounded by 4-lines—some of the bounding figures are triangles and some are quadralaterals. Then the specifications define each line in terms of its endpoint designations; and finally, the endpoints are specified according to their coordinates in 3-D space. That space object cannot be properly presented in any space of less than 3 dimensions.

Example 11-9 deals with the space object in particular.

Example 11-9

Generate an X-Y perspective view and the three planar views of this object:

$$S = \overline{F_1F_2F_3F_4F_5}$$

$F_1 = \overline{L_1L_2L_4}$ $\quad L_1 = \overline{P_1P_2}$ $\quad P_1 = (-1,0,1)$
$F_2 = \overline{L_1L_3L_5L_7}$ $\quad L_2 = \overline{P_1P_3}$ $\quad P_2 = (1,0,1)$
$F_3 = \overline{L_2L_3L_6L_8}$ $\quad L_3 = \overline{P_1P_4}$ $\quad P_3 = (0,1,1)$
$F_4 = \overline{L_4L_5L_6L_9}$ $\quad L_4 = \overline{P_2P_3}$ $\quad P_4 = (-1,0,-1)$
$F_5 = \overline{L_7L_8L_9}$ $\quad L_5 = \overline{P_2P_5}$ $\quad P_5 = (1,0,-1)$
$L_6 = \overline{P_3P_6}$ $\quad P_6 = (0,1,-1)$
$L_7 = \overline{P_4P_5}$
$L_8 = \overline{P_4P_6}$
$L_9 = \overline{P_5P_6}$

1. Gathering the components of the point coordinates:

$x_1 = -1 \quad x_2 = 1 \quad x_3 = 0$
$y_1 = 0 \quad y_2 = 0 \quad y_3 = 1$
$z_1 = 1 \quad z_2 = 1 \quad z_3 = 1$

$x_4 = -1 \quad x_5 = 1 \quad x_6 = 0$
$y_4 = 0 \quad y_5 = 0 \quad y_6 = 1$
$z_4 = -1 \quad z_5 = -1 \quad z_6 = -1$

2. Determine the x_{pn} and y_{pn} components of the X-Y perspective coordinates for each point (Equation 11-1):

$x_{pl} = x_1 - z_1\cos 30° \quad y_{pl} = y_1 - z_1\sin 30°$
$x_{pl} = -1-(1)\ (0.87) \quad y_{pl} = 0-(1)\ (0.5)$
$x_{pl} = -1.87 \quad y_{pl} = -0.5$

$x_{p2} = x_2 - z_2\cos 30° \quad y_{p2} = y_2 - z_2\sin 30°$
$x_{p2} = 1-(1)\ (0.87) \quad y_{p2} = 0-(1)\ (0.5)$
$x_{p2} = 0.13 \quad y_{p2} = -0.5$

$x_{p3} = x_3 - z_3\cos 30° \quad y_{p3} = y_3 - z_3\sin 30°$
$x_{p3} = 0-(1)\ (0.87) \quad y_{p3} = 1-(1)\ (0.5)$
$x_{p3} = -0.87 \quad y_{p3} = 0.5$

$x_{p4} = x_4 - z_4\cos 30° \quad y_{p4} = y_4 - z_4\sin 30°$
$x_{p4} = -1-(-1)\ (0.87) \quad y_{p4} = 0-(-1)\ (0.5)$
$x_{p4} = -0.13 \quad y_{p4} = 0.5$

$x_{p5} = x_5 - z_5\cos 30° \quad y_{p5} = y_5 - z_5\sin 30°$
$x_{p5} = 1-(-1)\ (0.87) \quad y_{p5} = 0-(-1)\ (0.5)$
$x_{p5} = 1.87 \quad y_{p5} = 0.5$

$x_{p6} = x_6 - z_6\cos 30° \quad y_{p6} = y_6 - z_6\sin 30°$
$x_{p6} = 0-(-1)\ (0.87) \quad y_{p6} = 1-(-1)\ (0.5)$
$x_{p6} = 0.87 \quad y_{p6} = 1.5$

See the results in the X-Y perspective view in Fig. 11-16.

3. Determining the coordinates for the planar views from the original specifications:

For the X-Y planar view: P_1 plots as (−1,0)
P_2 plots as (1,0)
P_3 plots as (0,1)
P_4 plots as (−1,0)
P_5 plots as (1,0)
P_6 plots as (0,1)

For the X-Z planar view: P_1 plots as (−1,1)
P_2 plots as (1,1)
P_3 plots as (0,1)
P_4 plots as (−1,−1)
P_5 plots as (1,−1)
P_6 plots as (0,−1)

For the Y-X planar view: P_1 plots as (0,1)
P_2 plots as (0,1)
P_3 plots as (1,1)
P_4 plots as (0,−1)
P_5 plots as (0,−1)
P_6 plots as (1,−1)

See Fig. 11-16.

The specifications at the beginning of that example are quite extensive compared to those suggested in earlier discussions. That is a natural consequence of having to specify more than a single plane figure. This particular space object is bounded by five plane figures, each having their own point and line specifications. (If you think the specifications are getting lengthy, wait until you reach the 4-D objects introduced in Chapter 17!)

$$S = \overline{F_1F_2F_3F_4F_5}$$

$F_1 = \overline{L_1L_2L_4}$	$L_1 = \overline{P_1P_2}$	$P_1 = (-1,0,1)$
$F_2 = \overline{L_1L_3L_5L_7}$	$L_2 = \overline{P_1P_3}$	$P_2 = (1,0,1)$
$F_3 = \overline{L_2L_3L_6L_8}$	$L_3 = \overline{P_1P_4}$	$P_3 = (0,1,1)$
$F_4 = \overline{L_4L_5L_6L_9}$	$L_4 = \overline{P_2P_3}$	$P_4 = (-1,0,-1)$
$F_5 = \overline{L_7L_8L_9}$	$L_5 = \overline{P_2P_5}$	$P_5 = (1,0,-1)$
	$L_6 = \overline{P_3P_6}$	$P_6 = (0,1,-1)$
	$L_7 = \overline{P_4P_5}$	
	$L_8 = \overline{P_4P_6}$	
	$L_9 = \overline{P_5P_6}$	

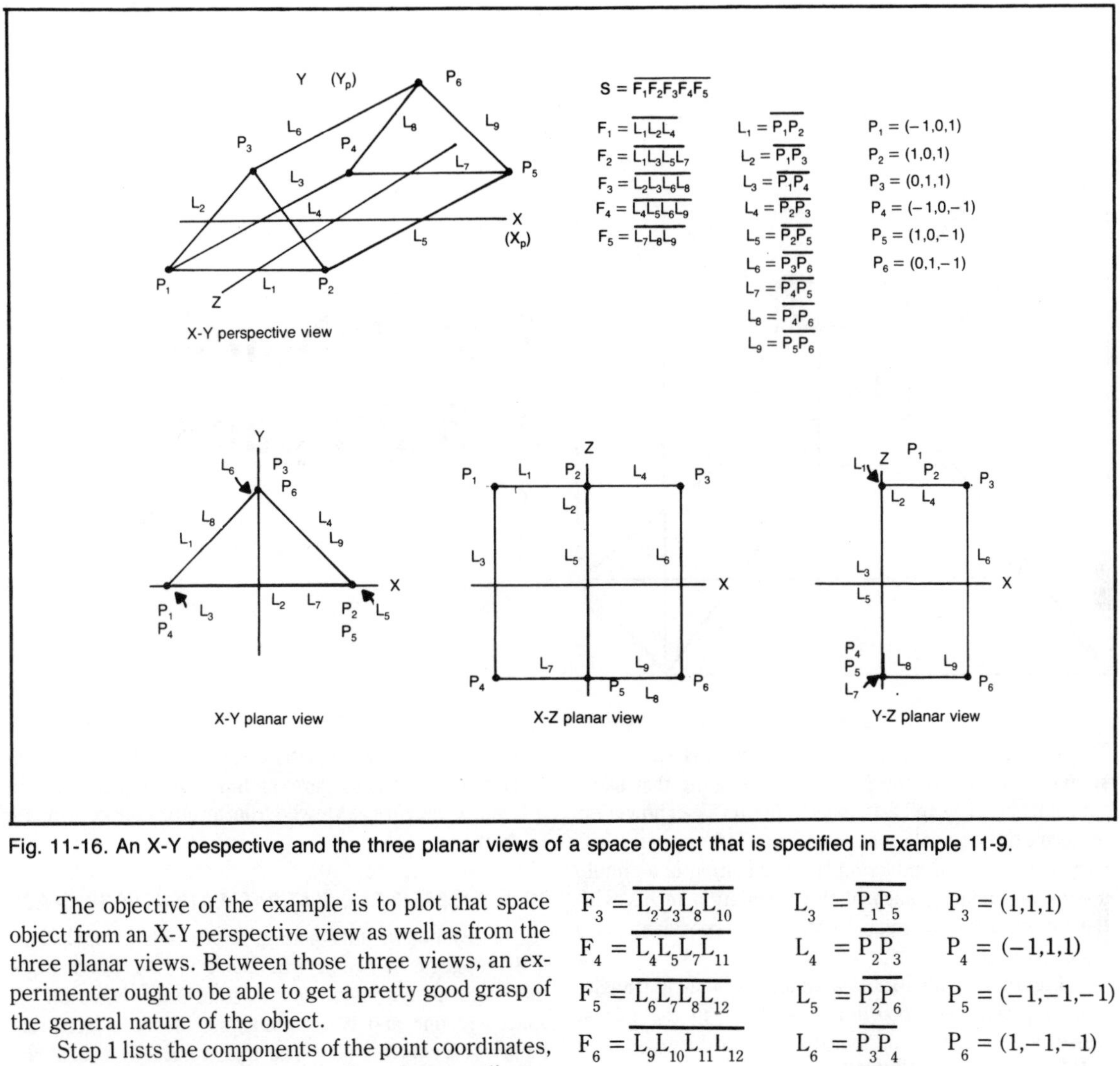

Fig. 11-16. An X-Y pespective and the three planar views of a space object that is specified in Example 11-9.

The objective of the example is to plot that space object from an X-Y perspective view as well as from the three planar views. Between those three views, an experimenter ought to be able to get a pretty good grasp of the general nature of the object.

Step 1 lists the components of the point coordinates, while Step 2 generates the perspective-view coordinates for all six points in the object. Finally, Step 3 generates the coordinates for plotting the space object from the three planar views.

Study the example carefully, then see what you can do with Exercise 11-8.

Exercise 11-8

Generate all three perspective and all three planar views of the following object in 3-D space:

$$S = \overline{F_1F_2F_3F_4F_5F_6}$$

$F_1 = \overline{L_1L_2L_4L_6}$	$L_1 = \overline{P_1P_2}$	$P_1 = (-1,-1,1)$
$F_2 = \overline{L_1L_3L_5L_9}$	$L_2 = \overline{P_1P_4}$	$P_2 = (1,-1,1)$

$F_3 = \overline{L_2L_3L_8L_{10}}$	$L_3 = \overline{P_1P_5}$	$P_3 = (1,1,1)$
$F_4 = \overline{L_4L_5L_7L_{11}}$	$L_4 = \overline{P_2P_3}$	$P_4 = (-1,1,1)$
$F_5 = \overline{L_6L_7L_8L_{12}}$	$L_5 = \overline{P_2P_6}$	$P_5 = (-1,-1,-1)$
$F_6 = \overline{L_9L_{10}L_{11}L_{12}}$	$L_6 = \overline{P_3P_4}$	$P_6 = (1,-1,-1)$
	$L_7 = \overline{P_3P_7}$	$P_7 = (1,1,-1)$
	$L_8 = \overline{P_4P_8}$	$P_8 = (-1,1,-1)$
	$L_9 = \overline{P_5P_6}$	
	$L_{10} = \overline{P_5P_8}$	
	$L_{11} = \overline{P_6P_7}$	
	$L_{12} = \overline{P_7P_8}$	

The analysis of space objects follows the same general path as already described for plane figures in 3-D space. The procedure is more involved, however, in the sense that there are more points, lines and angles to deal with.

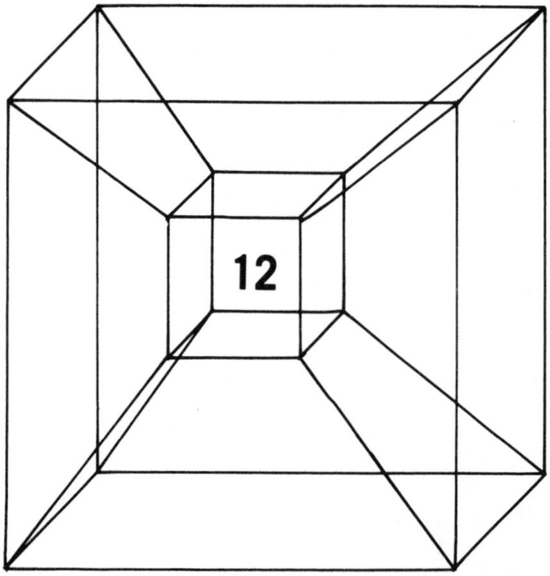

Translations in 3-D Space

Translating a point that is specified in 3-D space is a simple extension of the procedure for doing that same sort of task in 1-D and 2-D spaces. In this case, however, there are three translation terms—one for each of the 3-D coordinate axes. Conducting the translation is a simple matter of summing each of the translation terms with their corresponding components of the coordinate of a point in space.

Figure 12-1 illustrates a simple geometric translation of a reference coordinate system. In the purest sense, a 3-D translation transforms coordinate system X-Y-Z to system X′-Y′-Z′. The translation terms represent the translation-type displacement between those two coordinate systems.

Three-dimensional translations have the same general characteristics as those already described for lower dimensions: the position of a point is a variant, but the lengths of lines, their direction cosines and angle between them are invariant. Taking advantage of those characteristics can save some work when it comes to carrying out the complete analyses of a 3-D translation.

Furthermore, there is a handy form of the geometric translation equation of three dimensions that allows you to solve for the translation terms that are necessary for moving an object to any desired place in 3-D space. And finally there are some relativistic forms of the translation equation that let you figure out how a geometric entity in 3-D space appears as viewed from an alternate coordinate system.

12-1 THE BASIC TRANSLATION EQUATION FOR 3-D SPACE

The fundamental equation for geometric translations in 3-D space is shown here as Equation 12-1. You can see that it is an extension of the translation equation for spaces of one and two dimensions; it has a form that includes terms for all three axes of a 3-D coordinate system.

$$x_n' = x_n + T_x$$
$$y_n' = y_n + T_y$$
$$z_n' = z_n + T_z$$

Equation 12-1

where:

x_n', Y_n', z_n' = the coordinate of the translated version of point *n* relative to the frame of reference

x_n, y_n, z_n = the coordinate of point *n* relative to the frame of reference

T_x = the X-axis translation term

T_y = the Y-axis translation term

T_z = the Z-axis translation term

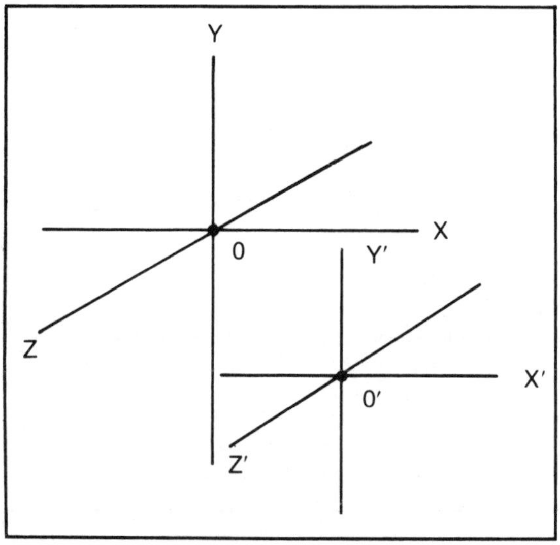

Fig. 12-1. A 3-D translation represents the displacement between reference coordinate system X-Y-Z and system X′-Y′-Z′.

The following example demonstrates the procedure for translating a point within a 3-dimensional coordinate system. Perhaps you can appreciate the fact that it is going to be more difficult to plot the reference and translated versions of the point than it is to carry out the arithmetic of the translation.

Example 12-1

Let a reference point in 3-D space have the following coordinate:

$$P = (1,2,3)$$

Apply a 3-D translations of $T_x = 2$, $T_y = -1$, $T_z = -2$.

1. Gathering the relevant data:

$$x = 1 \quad y = 2 \quad z = 3$$

2. Applying Equation 12-1:

$$x' = x+T_x \quad y' = y+T_y \quad z' = z+T_z$$
$$x' = 1+3 \quad y' = 2-1 \quad z' = 3-2$$
$$x' = 3 \quad y' = 1 \quad z' = 1$$

The translation thus displaces P = (1,2,3) to P′ = (3,1,1). See the plot in Fig. 12-2.

By way of an exercise, make up some coordinates for points in three dimensions and apply some 3-D translation terms of your own.

12-2 TRANSLATING LINES IN 3-D SPACE

You found in Chapter 11 that you can define a line in 3-D space in terms of the coordinates of its two endpoints, and you have just seen the equation for translating a point in 3 dimensions. It logically follows, then, that you should also be able to translate lines in 3-D space by means of that same equation: just translate the endpoints individually, and then connect them with a line as prescribed in the original specifications.

An analysis of both lines—the reference line, and the translated version of it—is in order for completing such experiments. You can save yourself some time and work in that regard by taking advantage of these general principles for the translation of lines in any number of dimensions:

☐ The length of a line is an invariant under translations of any number of dimensions.

☐ The direction cosines of a line are invariant under translations of any dimension.

☐ The angle between two lines is an invariant under translations of any dimension.

As far as the translated version of a line is concerned, recalling those principles can save you the trouble of having to do a complete analysis from scratch. Of course it is still necessary to do the complete analysis of the reference line, and you will find that it is necessary to apply Equation 12-1 for finding the endpoint coordinates for the translated version of the line. But there is no need to repeat the work as far as the length, direction cosines and angles between lines in the translated version are concerned. Those values will be the same as their counterparts for the reference line.

The following Exercise (2-1) represents an experiment whereby a reference point specified in 3-D space is

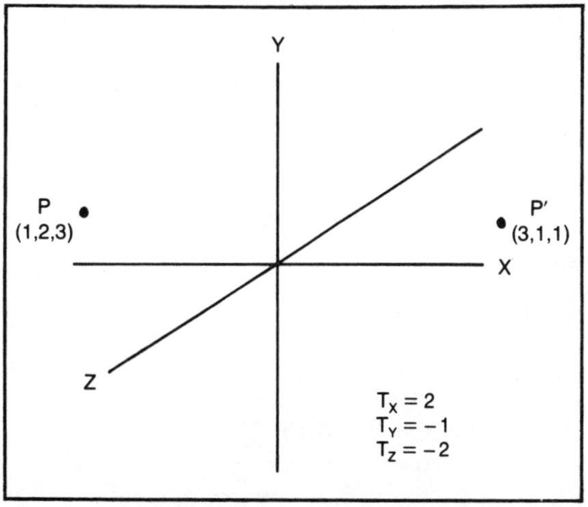

Fig. 12-2. An X-Y perspective plot of a 3-D translation of reference point P to point P′. See Example 12-1.

subject to a 3-D translation. You will find plots and a complete analysis of the reference and translated versions of the line in Fig. 12-3.

Exercise 12-1

Given this line in 3-D space:

$$L = \overline{P_1P_2} \qquad P_1 = (2,4,-3) \quad P_2 = (-3,-1,4)$$

apply this translation:

$$T_x = 4 \qquad T_y = -3 \qquad T_z = 1$$

1. Gathering the relevant data:

$$x_1 = 2 \qquad y_1 = 4 \qquad z_1 = -3$$
$$x_2 = -3 \qquad y_2 = -1 \qquad z_2 = 4$$

2. Applying the translation to endpoint P_1:

$$x_1' = x_1+T_x \qquad y_1' = y_1+T_y \qquad z_1' = z_1+T_z$$
$$x_1' = 2+4 \qquad y_1' = 4-3 \qquad z_1' = -3+1$$
$$x_1' = 6 \qquad y_1' = 1 \qquad z_1' = -2$$

So endpoint $P_1 = (2,4,-3)$ translates to $P_1' = (6,1,-2)$

3. Applying the translation to endpoint P_2:

$$x_2' = x_2 + T_x \qquad y_2' = Y_2 + T_y \qquad z_2' = z_2 + T_2$$
$$x_2' = -3+4 \qquad y_2' = -1-3 \qquad z_2' = 4+1$$
$$x_2' = 1 \qquad y_2' = -4 \qquad z_2' = 5$$

And endpoint $P_2 = (-3,-1,4)$ translates to $P_2' = (1,-4,5)$.

4. Summarizing the specifications for the translated version of the line:

$$L' \; P_1'P_2' \qquad P_1' \; (6,1,-2) \quad P' \; (1,-4,5)$$

See the plot and analyses in Fig. 12-3.

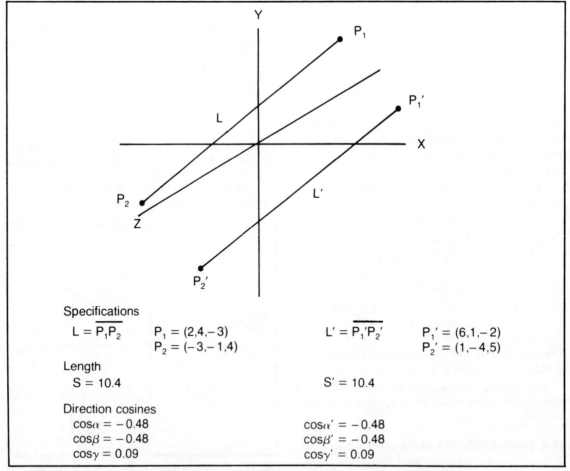

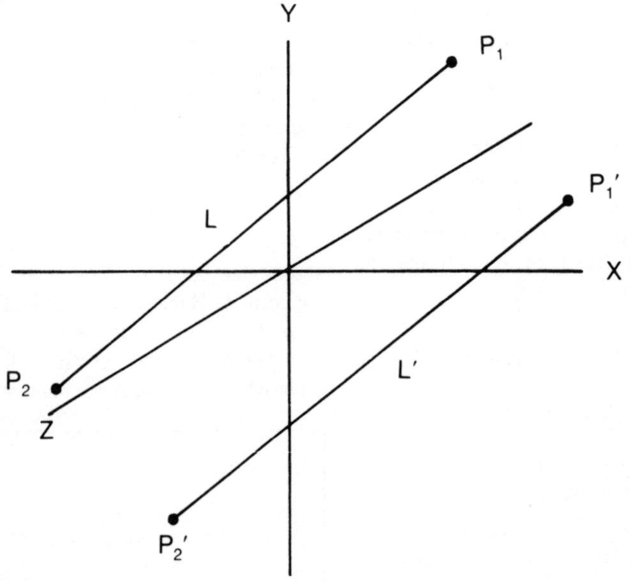

Specifications

$L = \overline{P_1P_2}$ $\quad P_1 = (2,4,-3)$, $P_2 = (-3,-1,4)$

$L' = \overline{P_1'P_2'}$ $\quad P_1' = (6,1,-2)$, $P_2' = (1,-4,5)$

Length

$S = 10.4$ $\qquad S' = 10.4$

Direction cosines

$\cos\alpha = -0.48$ $\qquad \cos\alpha' = -0.48$

$\cos\beta = -0.48$ $\qquad \cos\beta' = -0.48$

$\cos\gamma = 0.09$ $\qquad \cos\gamma' = 0.09$

Fig. 12-3. An X-Y perspective plot of a 3-D translation of reference line L to line L'. See Example 12-2.

Table 12-1. A Summary of the Analysis of the Reference Plane Figure Described in Example 12-2 and Portrayed as Figure F in Fig. 12-4.

(A, B, C, and D correspond to α, β, γ, and δ.)

SPECIFICATIONS

$F=\overline{L_1L_2L_3L_4}$

$L_1=\overline{P_1P_2}$	$P_1=(2,2,-2)$
$L_2=\overline{P_2P_3}$	$P_2=(-2,2,-2)$
$L_3=\overline{P_3P_4}$	$P_3=(-2,-2,2)$
$L_4=\overline{P_4P_1}$	$P_4=(2,-2,2)$

LENGTHS OF LINES

$s_1=4$	$s_2=4\sqrt{2}$	$s_3=4$	$s_4=4\sqrt{2}$

DIRECTION COSINES

$\cos A_1=-1$	$\cos B_1=0$	$\cos C_1=0$
$\cos A_2=0$	$\cos B_2=-\sqrt{2}/2$	$\cos C_2=\sqrt{2}/2$
$\cos A_3=1$	$\cos B_3=0$	$\cos C_3=0$
$\cos A_4=0$	$\cos B_4=\sqrt{2}/2$	$\cos C_4=\sqrt{2}/2$

ANGLES

$\phi_{1,2}=90^\circ$	$\phi_{1,4}=90^\circ$
$\phi_{2,3}=90^\circ$	$\phi_{3,4}=90^\circ$

12-3 TRANSLATING PLANE FIGURES IN 3-D SPACE

Just as you can translate a line in 3-D space by translating its individual endpoint, so you can translate a plane figure that is immersed in 3-D space by translating its individual endpoints. And, again, nothing changes in the analysis but the position of the figure.

Example 12-2 suggests doing a complete analysis of a quadralateral being translated in 3-D space. The work done there supports the general principles for translating any sort of geometric entity in 3-D space. Compare, for instance, the analyses in Tables 12-1 and 12-2.

Example 12-2

Translate the following reference figure by $T_x = -1$, $T_y = 3$, $T_z = 2$. Plot the translation and conduct a complete analysis of both versions of the figure.

$F = \overline{L_1L_2L_3L_4}$ $L_1 = \overline{P_1P_2}$ $P_1 = (2,2,-2)$

$L_2 = \overline{P_2P_3}$ $P_2 = (-2,2,-2)$

$L_3 = \overline{P_3P_4}$ $P_3 = (-2,-2,2)$

$L_4 = \overline{P_4P_1}$ $P_4 = (2,-2,2)$

1. Gathering the data:

$x_1 = 2$	$y_1 = 2$	$z_1 = -2$
$x_2 = -2$	$y_2 = 2$	$z_2 = -2$
$x_3 = -2$	$y_3 = -2$	$z_3 = 2$
$x_4 = 2$	$y_4 = -2$	$z_4 = 2$

2. After applying the translation:

$F' = \overline{L_1'L_2'L_3'L_4'}$ $L_1' = \overline{P_1'P_2'}$ $P_1' = (1,5,0)$

Table 12-2. A Summary of the Analysis of the Translated Version of the Plane Figure, F′, Portrayed in Fig. 12-4.

(A, B, C, and D correspond to α, β, γ, and δ.)

SPECIFICATIONS

$F' = \overline{L'_1 L'_2 L'_3 L'_4}$

$L'_1 = \overline{P'_1 P'_2}$	$P'_1 = (2,2,-2)$
$L'_2 = \overline{P'_2 P'_3}$	$P'_2 = (-2,2,-2)$
$L'_3 = \overline{P'_3 P'_4}$	$P'_3 = (-2,-2,2)$
$L'_4 = \overline{P'_4 P'_1}$	$P'_4 = (2,-2,2)$

LENGTHS OF LINES

$s'_1 = 4$	$s'_2 = 4\sqrt{2}$	$s'_3 = 4$	$s'_4 = 4\sqrt{2}$

DIRECTION COSINES

$\cos A'_1 = -1$	$\cos B'_1 = 0$	$\cos C'_1 = 0$
$\cos A'_2 = 0$	$\cos B'_2 = -\sqrt{2}/2$	$\cos C'_2 = \sqrt{2}/2$
$\cos A'_3 = 1$	$\cos B'_3 = 0$	$\cos C'_3 = 0$
$\cos A'_4 = 0$	$\cos B'_4 = \sqrt{2}/2$	$\cos C'_4 = \sqrt{2}/2$

ANGLES

$\phi'_{1,2} = 90^\circ$	$\phi'_{1,4} = 90^\circ$
$\phi'_{2,3} = 90^\circ$	$\phi'_{3,4} = 90^\circ$

$L_2' = \overline{P_2'P_3'}$ $\quad P_2' = (-3,5,0)$

$L_3' = \overline{P_3'P_4'}$ $\quad P_3' = (-3,1,4)$

$L_4' = \overline{P_4'P_1'}$ $\quad P_4' = (1,1,4)$

See the plot of this translation in Fig. 12-4, the analysis of the reference figure in Table 12-1, and the analysis of the translated version in Table 12-2.

12-4 TRANSLATING SPACE OBJECTS IN 3-D SPACE

There is little more to be said about translating space objects than has already been said about translating lines and plane figures in 3-D space. Generally speaking, the transformation, itself, is a simple one: apply the translation equation to each of the points in the space object. That provides the new point coordinates. What remains, then, is to rewrite the line, plane and space specifications with the appropriate notation. Nothing changes significantly but the coordinates of the points.

When carrying out complete analyses of both the reference and translated versions of the space object, you will occasionally encounter situations where the direction cosines for a given 3-D perspective view or a given planar view appear different for the reference and translated version. Don't worry about it. Assuming that you have made no errors, calculating the angle between the lines in question turns up an angle of zero degrees—the lines are, in fact, parallel to one another and, therefore, unchanged with respect to their orientation in space.

The following example cites a reference space object and a translation in 3-D space.

Example 12-3

Apply a translation of $T_x = -5$, $T_y = -5$, $T_z = 5$ to this space object:

$S = \overline{F_1F_2F_3F_4}$

$F_1 = \overline{L_1L_2L_3}$ $\quad L_1 = \overline{P_1P_2}$ $\quad P_1 = (0,0,0)$

$F_2 = \overline{L_2L_4L_5}$ $\quad L_2 = \overline{P_2P_3}$ $\quad P_2 = (4,0,0)$

$F_3 = \overline{L_3L_5L_6}$ $\quad L_3 = \overline{P_3P_1}$ $\quad P_3 = (0,4,0)$

$F_4 = \overline{L_1L_4L_6}$ $\quad L_4 = \overline{P_2P_4}$ $\quad P_4 = (0,0,-4)$

$L_5 = \overline{P_4P_3}$

$L_6 = \overline{P_4P_1}$

Applying the given translation transforms the specifications to:

$S' = \overline{F_1'F_2'F_3'F_4'}$

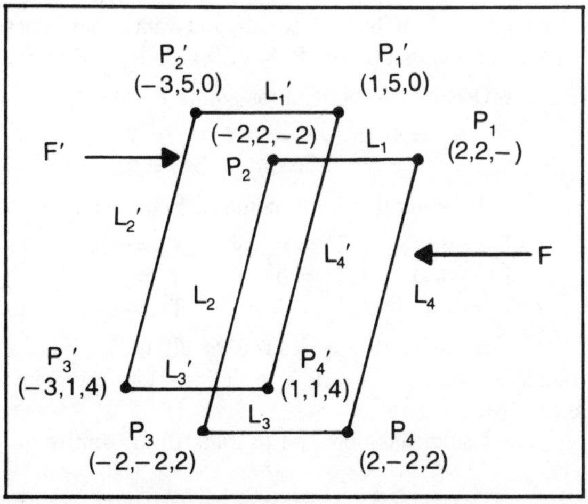

Fig. 12-4. An X-Y perspective plot of a 3-D translation of a reference plane figure, F, to its translated version, F'. See Example 12-2 and a complete analysis of the situation in Tables 12-1 and 12-2.

$F_1' = \overline{L_1'L_2'L_3'}$ $\quad L_1' = \overline{P_1'P_2'}$ $\quad P_1' = (-5,-5,5)$

$F_2' = \overline{L_2'L_4'L_5'}$ $\quad L_2' = \overline{P_2'P_3'}$ $\quad P_2' = (-1,-5,5)$

$F_3' = \overline{L_3'L_5'L_6'}$ $\quad L_3' = \overline{P_3'P_1'}$ $\quad P_3' = (-5,-1,5)$

$F_4' = \overline{L_1'L_4'L_6'}$ $\quad L_4' = \overline{P_2'P_4'}$ $\quad P_4' = (-5,-5,1)$

$L_5' = \overline{P_4'P_3'}$

$L_6' = \overline{P_4'P_1'}$

See this translation plotted in Fig. 12-5.

Work out the analysis of both the reference and translated objects on your own.

Make up some specifications for simple reference space objects and translate them in some fashion. See if you can alter the position of a space object from its reference position to some other clearly defined position in the frame-of-reference coordinate system.

12-5 ALTERNATIVE VERSIONS OF THE GEOMETRIC EQUATION

Equation 12-2 shows the basic equation for geometric translation as reorganized to solve for the translation terms. Given the coordinate of some reference point and a translated version of that same point, the equation leads directly to the translation terms that exists between them.

$$T_x = x_n' - x_n \qquad T_y = y_n' - y_n \qquad T_z = z_n' - z_n \qquad \textbf{Equation 12-2}$$

where:

T_x = the X-axis translation term

T_y = the Y-axis translation term

T_z = the Z-axis translation term

x_n', y_n', z_n' = the coordinate of translated point n relative to the frame of reference

x_n, y_n, z_n = the coordinate of point n to the frame of reference

Earlier discussions cited 1-D and 2-D versions of this equation as a means for determining the translation that exists between two coordinate systems. That can be done in the case of geometric entities that are separated by a translation in 3-D space as well, but I am leaving such applications to you. What is of more practical consequence for an experimenter working in a 3-dimensional environment is the ability to solve for the translation terms that are necessary for adjusting the position of a point, line, figure or space object in 3-D space.

You have seen from previous work that it is far easier to derive the specifications for entities in 3-D space when they are situated in a particular way with respect to the origin of the coordinate system. Many interesting kinds of experiments, on the other hand, call for working with the entities as situated elsewhere on the coordinate system. Equation 12-2 provides a means for relocating the points to some other desired position.

To use that technique, begin with the specifications that locate the entity in its simplest location on the coordinate system. Then cite a new coordinate for one of the critical points in the entity. Regarding the original coordinate as the reference coordinate and the desired coordinate as the translated version of the point, Equation 12-2 provides the necessary translation terms for moving that point and all others in the figure or object at hand. Calculate the translation terms for the critical point you have chosen, then use those terms as a basic for translating the remaining point by applying them in Equation 12-1.

Example 12-4

Suppose that you have the basic specifications and a complete analysis of a certain 5-sided pyramid. Here are the specifications:

$S = \overline{F_1F_2F_3F_4F_5}$	$F_1 = \overline{L_1L_2L_4L_6}$	$L_1 = \overline{P_1P_2}$	$P_1 = (-1,0,1)$
	$F_2 = \overline{L_1L_3L_5}$	$L_2 = \overline{P_1P_4}$	$P_2 = (1,0,1)$
	$F_3 = \overline{L_2L_3L_8}$	$L_3 = \overline{P_1P_5}$	$P_3 = (1,0,-1)$
	$F_4 = \overline{L_4L_5L_7}$	$L_4 = \overline{P_2P_3}$	$P_4 = (-1,0,-1)$
	$F_5 = \overline{L_6L_7L_8}$	$L_5 = \overline{P_2P_5}$	$P_5 = (0,1,0)$
		$L_6 = \overline{P_3P_4}$	
		$L_7 = \overline{P_3P_5}$	
		$L_8 = \overline{P_4P_5}$	

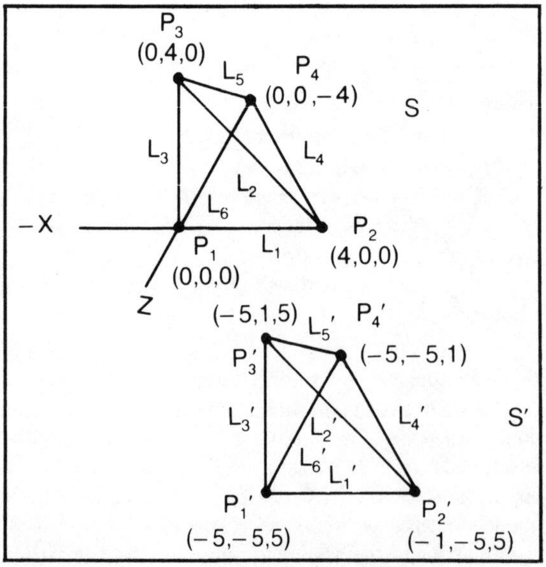

Fig. 12-5. An X-Y perspective plot of the 3-D translation of space object S to S′. See Example 12-3.

For the sake of a new experiment, you want to translate the object in such a way that point P_2 is at the origin of the coordinate system. In other words, you want to perform a transformation that places $P_2 = (1,0,1)$ to $P_2' = (0,0,0)$.

1. Organizing the data for points P_2 and P_2'

$$x_2 = 1 \quad y_2 = 0 \quad z_2 = 1$$
$$x_2' = 0 \quad y_2' = 0 \quad z_2' = 0$$

2. Applying that information to Equation 12-2:

$$T_x = x_2' - x_2 \quad T_y = y_2' - y_2 \quad T_z = z_2' - z_2$$
$$T_x = 0 - 1 \quad T_y = 0 - 0 \quad T_z = 0 - 1$$
$$T_x = -1 \quad T_y = 0 \quad T_z = -1$$

So you can achieve the desired result by applying a translation of $T_x = -1$, $T_y = 0$, $T_z = -1$ to the reference object.

3. Using Equation 12-1 to translate the entire object:

$$P_1' = (-2,0,0) \qquad P_4' = (-2,0,-2)$$
$$P_2' = (0,0,0) \qquad P_5' = (-1,1,-1)$$
$$P_3' = (0,0,-2)$$

Plot and analyze the translation situation in that example.

Equation 12-3 represents the inverse form of the basic equation for geometric translations in 3-D space. Given the translation terms and the coordinate of at least one point in the translated version of the figure, it lets you calculate the reference-point coordinates. You can use it that way if you choose, but I am showing it here only as a means for leading into the relativistic version of 3-D translations.

$$\begin{aligned} x_n &= x_n' - T_x \\ y_n &= y_n' - T_y \\ z_n &= z_n' - T_z \end{aligned} \qquad \textbf{Equation 12-3}$$

where:

x_n, y_n, z_n = the coordinate of point *n* relative to the frame of reference

x_n', y_n', z_n' = the coordinate of the translated version of point *n* relative to the frame of reference

T_x = the X-axis translation term

T_y = the Y-axis translation term

T_z = the Z-axis translation term

Note: This is the inverse version of Equation 12-1

12-6 RELATIVISTIC TRANSLATIONS IN 3-D

From the tone of the foregoing discussions, it is easy to lose sight of the fact that a translation actually represents a translation-type transformation that happens to exist between two coordinate systems. Most of the discussions to this point have regarded a 3-D translation as a particular means for moving geometric entities from one place to another in 3-D space. Both views serve particular purposes.

In the context of relativistic translations, the idea of moving a geometric entity from place to place is not very meaningful; it is better to view a translation as an expression of the displacement between two otherwise identical 3-D coordinate systems.

Equation 12-4 represents the basic relativistic version of a 3-dimensional translation that exists between two 3-D coordinate systems. Used in its most direct fashion, the equation yields the coordinate of some point in 3-D space as regarded from a translated frame of reference. Generally the project begins by specifying some geometric entity as it appears from the viewpoint of the observer's 3-D frame of reference. Then given a set of translation terms for the displacement of the translated frame of reference, relative to the observer's frame of reference, the equation shows how the entity appears relative to that second frame of reference.

$$\begin{aligned} x_n^* &= x_n - T_x \\ y_n^* &= y_n - T_y \\ z_n^* &= z_n - T_z \end{aligned} \qquad \textbf{Equation 12-4}$$

where:

x_n^*, y_n^*, z_n^* = coordinate of reference point *n* relative to the translated frame of reference

x_n, y_n, z_n = the coordinate of reference point *n* relative to the observer's frame of reference

T_x = the X-axis translation term relative to the observer's frame of reference

T_y = the Y-axis translation term relative to the observer's frame of reference

T_z = the Z-axis translation term relative to the observer's frame of reference

Example 12-5 illustrates the situation.

Example 12-5

Suppose that you are using the observer's frame of reference and see a house-shaped object this way:

$S = \overline{F_1F_2F_3F_4F_5F_6F_7}$

$F_1 = \overline{L_1L_2L_4L_6L_8}$	$L_1 = \overline{P_1P_2}$	$P_1 = (0,0,0)$
$F_2 = \overline{L_1L_3L_5L_{11}}$	$L_2 = \overline{P_1P_4}$	$P_2 = (1,0,0)$
$F_3 = \overline{L_2L_3L_9L_{12}}$	$L_3 = \overline{P_1P_6}$	$P_3 = (1,1,0)$
$F_4 = \overline{L_4L_5L_7L_{13}}$	$L_4 = \overline{P_2P_3}$	$P_4 = (0,1,0)$
$F_5 = \overline{L_6L_7L_{10}L_{14}}$	$L_5 = \overline{P_2P_7}$	$P_5 = (0,2,0)$
$F_6 = \overline{L_8L_9L_{10}L_{15}}$	$L_6 = \overline{P_3P_5}$	$P_6 = (0,0,1)$
$F_7 = \overline{L_{11}L_{12}L_{13}L_{14}L_{15}}$	$L_7 = \overline{P_3P_8}$	$P_7 = (1,0,1)$
	$L_8 = \overline{P_4P_5}$	$P_8 = (1,1,1)$
	$L_9 = \overline{P_4P_9}$	$P_9 = (0,1,1)$
	$L_{10} = \overline{P_5P_{10}}$	$P_{10} = (0,2,1)$
	$L_{11} = \overline{P_6P_7}$	
	$L_{12} = \overline{P_6P_9}$	
	$L_{13} = \overline{P_7P_8}$	
	$L_{14} = \overline{P_8P_{10}}$	
	$L_{15} = \overline{P_9P_{10}}$	

A friend is using a frame of reference that is translated this way relative to your own:

$$T_x = -5,\ T_y = 4,\ T_z = -2$$

Using conventional XYZ orientation, you can say that your friend is located 5 units to your left, 4 units above, and 2 units in front of you. How does that house-shaped object appear from your friend's frame of reference?

Applying Equation 12-4 transforms the points as follows:

$P_1 = (0,0,0) \rightarrow P_1* = (5,-4,2)$
$P_2 = (1,0,0) \rightarrow P_2* = (6,-4,2)$
$P_3 = (1,1,0) \rightarrow P_3* = (6,-3,2)$
$P_4 = (0,1,0) \rightarrow P_4* = (5,-3,2)$
$P_5 = (0,2,0) \rightarrow P_5* = (5,-2,2)$
$P_6 = (0,0,1) \rightarrow P_6* = (5,-4,3)$
$P_7 = (1,0,1) \rightarrow P_7* = (6,-4,3)$
$P_8 = (1,1,1) \rightarrow P_8* = (6,-3,3)$
$P_9 = (0,1,1) \rightarrow P_9* = (5,-3,3)$
$P_{10} = (0,2,1) \rightarrow P_{10}* = (5,-2,3)$

Complete the S-F-L-P specifications for the object as viewed from the translated frame of reference, plot both versions, and generate a complete analysis for both.

12-7 COMBINATIONS OF 3-D TRANSLATIONS

Multiple translations in 3-D space are communitative. That is, you can apply a given series of translations in any desired order and still end up with the same overall translation. The idea applies to geometric translations as well as their relativistic versions.

Scalings in 3-D Space

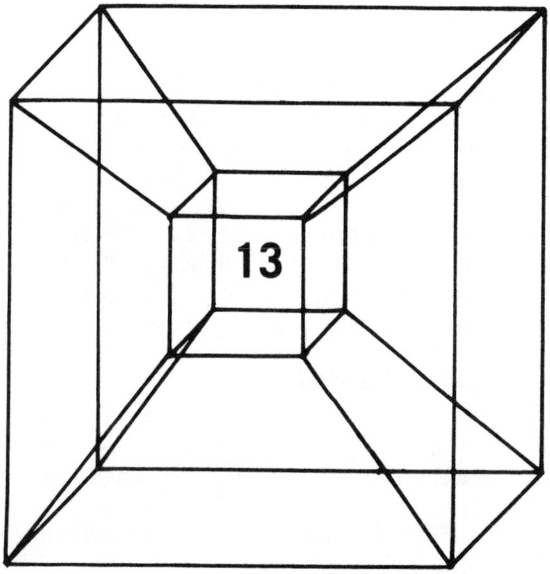

Scaling points that are specified in 3-D space is an extension of the procedure for scaling them in 2-D space. The only difference is that, in the 3-D case, you are working with three scaling factors instead of two. The mathematics is the same: multiply each of the scaling factors by their corresponding component in the coordinate of a point.

Three-dimensional scalings have the same general characteristics as those already described for one and two dimensions; namely that all parameters of a line can change, and that there are useful alternative versions of the geometric scaling equation of three dimensions as well as the usual family of relativistic forms. The only truly new feature here is that reflection-type scalings in 3-D space can take place about planes as well as lines.

13-1 THE BASIC SCALING EQUATION FOR 3-D SPACE

Equation 13-1 represents the fundamental equation for geometric scalings in 3-D space. You ought to be able to appreciate the fact that it is an extension of the scaling equation for 2-D space—it prescribes the same multiplication tasks, but for three axes instead of just two.

$$x_n' = x_n K_x$$
$$y_n' = y_n K_y$$
$$z_n' = z_n K_z$$

Equation 13-1

where:

x_n', y_n', z_n' = the coordinate of the scaled version of point *n* relative to the frame of reference.

x_n, y_n, z_n = the coordinate of a reference point *n* relative to the frame of reference.

K_x, K_y and K_z = the scaling factors for the X, Y and Z axes, relative to the frame of reference.

Example 13-1 demonstrates the procedure for scaling a point within a 3-dimensional coordinate system.

Example 13-1

Given the reference point P=(0,1,2) in 3-D space, apply a scaling of K_x=2, K_y=3, K_z=−2:

1. Scaling the X component:

$$x_n' = x_n K_x$$
$$x_n' = (0)\ (2)$$
$$x_n' = 0$$

2. Scaling the Y component:

$$y_n' = y_n K_y$$
$$y_n' = (1)(3)$$
$$y_n' = 3$$

3. Scaling the Z component:

$$z_n' = z_n K_z$$
$$z_n' = (2)(-2)$$
$$z_n' = -4$$

Point P=(0,1,2) is thus scaled to point P′=(0,3,4).

You can see that the procedure is quite simple. In fact the scaling procedure, itself, is simpler than the task of plotting the points on a 3-D perspective coordinate system or mapping them to the X-Y, X-Z and Y-Z planes. Do those plots for the scaling just illustrated, and then make up some coordinates for points in 3 dimensions and apply some 3-D scaling terms of your own.

13-2 TYPES OF 3-D SCALINGS

Like 2-D scalings, 3-D scalings can be classified as symmetrical (equal scaling along all axes) or asymmetrical (unequal scaling along the axes), and as reflective or non-reflective.

The following summary is a review of some scaling effects that apply to any number of dimensions. Term K in this particular summary applies to any or all of the axes of scaling.

1. When K is greater than 1, the scaling is an expansion-type scaling.
2. When K is equal to 1, there is no change of scale.
3. When K is less than 1, but greater than zero, the scaling is a reduction-type scaling.
4. When K is equal to zero, all components of the given axis are reduced to nothing.
5. When K is less than zero, but greater than −1, the scaling is a reduction-type scaling with a reflection effect.
6. When K is equal to −1, there is a reflection effect but no scaling of size.
7. When K is less than −1, there is an expansion-type scaling with a reflection effect.

Now consider this progression of ideas about scaling transformations:

1. In 1-D space, reflections take place about a point (0-D space).
2. In 2-D space, reflections take place about a line (1-D space).
3. In 3-D space, reflections take place about a plane (2-D space).

Table 13-1 relates those ideas to all possible combinations of three scaling factors in 3-D space. Look it over carefully, and test your understanding by trying some of those combinations for yourself. Try using values other than 1 and −1 for more interesting effects.

13-3 SCALING LINES IN 3-D SPACE

A line in 3-D space is defined in terms of the coordinates of its two endpoints, so scaling a line is a matter of scaling the coordinates of its endpoints and constructing the specified line between them. There is nothing especially difficult about that.

Conducting a complete analysis of a scaling in 3 dimensions can be a moderately complex task because

Table 13-1. A Summary of all Possible Combinations of 1 and −1 Scaling Effects.

K_x	K_y	K_z	Effect
1	1	1	no reflection
1	1	-1	reflection about the X-Y plane
1	-1	1	reflection about the X-Z plane
1	-1	-1	reflection about the X axis
-1	1	1	reflection about the Y-Z plane
-1	1	-1	reflection about the Y axis
-1	-1	1	reflection about the Z axis
-1	-1	-1	full 3-D reflection

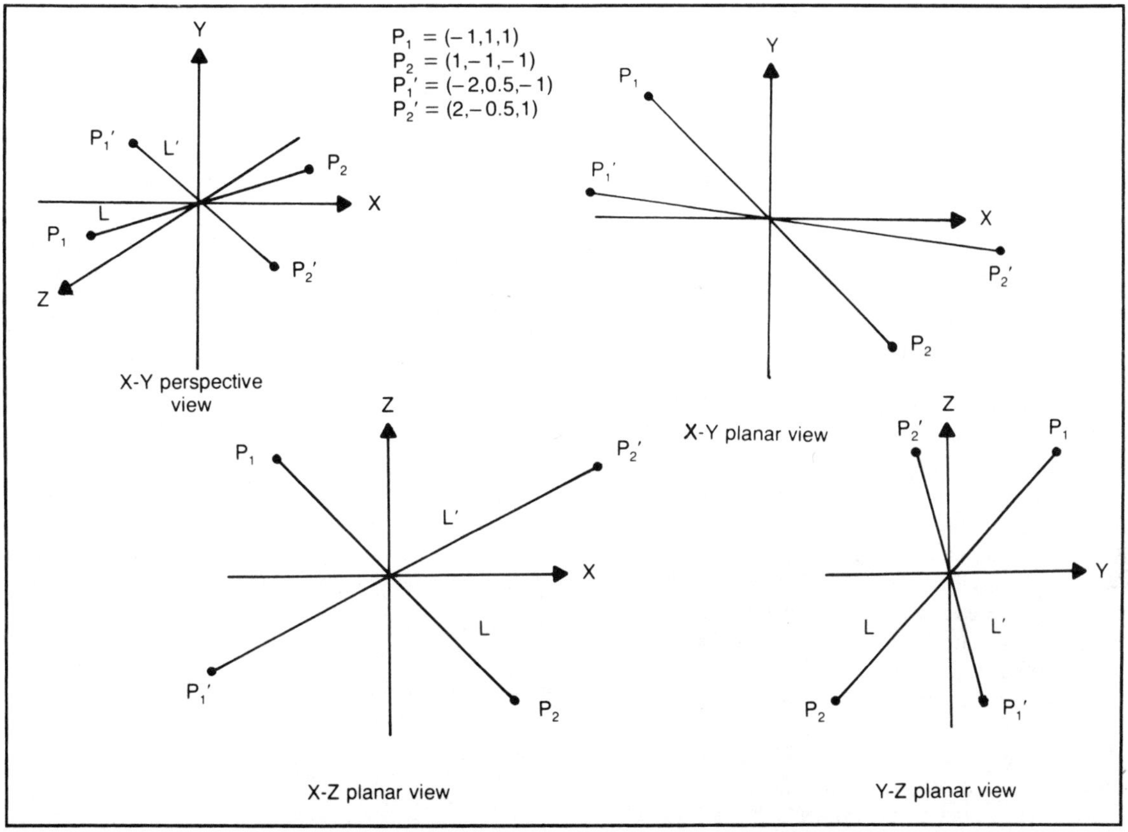

Fig. 13-1. An X-Y perspective and the three planar views of line L being transformed to line L′ by a 3-D scaling.

nothing is invariant; recall that every line parameter can change under a scaling—point positions, lengths of lines, direction cosines, and the angle between lines.

Example 13-2 illustrates the procedure for scaling a line in 3-D space and conducting a primary analysis of the situation. See the plotted results in Fig. 13-1.

Example 13-2

Given the line in 3-D space as specified below, apply a 3-D scaling of $K_x=2$, $K_y=0.5$, $K_z=-1$. Then conduct primary analyses of both the reference line and scaled versions, and make an X-Y perspective plot of the situation.

$L = \overline{P_1P_2}$ $\qquad$ $P_1 = (-1,1,1)$
$\qquad\qquad\qquad$ $P_2 = (1,-1,-1)$

1. Gathering the available data:

$x_1 = -1 \quad y_1 = 1 \quad z_1 = 1 \quad K_x = 2$
$x_2 = 1 \quad y_2 = -1 \quad z_2 = -1 \quad K_y = 0.5$
$K_z = -1$

2. Scaling point P_1:

$x_1' = x_1K_x \qquad y_1' = y_1K_y \qquad z_1' = z_1K_z$
$x_1' = (-1)(2) \qquad y_1' = (1)(0.5) \qquad z_1' = (1)(-1)$
$x_1' = -2 \qquad y_1' = 0.5 \qquad z_1' = -1$

Thus point P_1=(−1,1 1) is scaled to point P_1'= (−2.0.5,−1)

3. Scaling point P_2:

$x_2' = x_2K_x \qquad y_2' = y_2K_y \qquad z_2' = z_2K_z$
$x_2' = (1)(2) \qquad y_2' = (-1)(0.5) \qquad z_2' = (-1)(-1)$
$x_2' = 2 \qquad Y_2' = -0.5 \qquad z_2' = 1$

Thus point P_2 = (1,−1,−1) is scaled to P_2' = (2,−0.5.1)

4. Conducting the analysis of the reference version of the line:

$$s = \sqrt{(x_2-x_1)^2 + (y_2-y_1)^2 + (z_2-z_1)^2}$$
$$s = \sqrt{(1+1)^2 + (-1-1)^2 + (-1-1)^2}$$
$$s = 2\sqrt{3}$$

$\cos = (x_2 - x_1)/s \quad \cos = (y_2 - y_1)/s \quad \cos = (z_2 - z_1)/s$

$\cos = \sqrt{3}/3 \quad \cos = -\sqrt{3}/3 \quad \cos = -\sqrt{3}/3$

5. Conducting the analysis of the scaled version of the line:

$$s' = \sqrt{(x_2' - x_1')^2 + (y_2' - y'_1)^2 + (z_2' - z_1')^2}$$

$$s = \sqrt{(2+2)^2 + (-0.5-0.5)^2 + (1+1)^2}$$

$$s = \sqrt{21}$$

$\cos\alpha' = (x_2' - x_1')/s' \quad \cos\beta' = (y_2' - y_1')/s'$

$\cos\alpha' = 4\sqrt{21}/21 \quad \cos\beta' = -\sqrt{21}/21$

$\cos\gamma' = (z_2' - z_1')/s' \quad \cos\gamma' = 2\sqrt{21}/21$

After citing the specifications for the line, the example summarizes the point components and scaling factors in a symbolic fashion. Step 2 then carries out the prescribed scaling on point 1 in the reference version of the line. Step 3 does the same for point 2, and that completes the scaling operation, itself.

But the example calls for an analysis of the whole operation, so Step 4 calculates the length of the reference line and its direction cosines. Step 5 then does the same for the scaled version of the line. Comparing the results of those two analyses clearly shows that the scaling operation has a significant effect on the line.

Make up some reference-line specifications and scaling factors of your own. Do the scaling, plot the results and conduct an analysis of both versions of your lines.

13-4 SCALING PLANE FIGURES IN 3-D SPACE

The procedure for scaling plane figures in 3-D space is much the same sort of process as scaling lines: scale the individual points, and then reconstruct the lines according to the original L-P specifications.

Try Exercise 13-1. It asks for an analysis of a triangle in 3-D space before and after applying a certain 3-dimensional scaling.

Exercise 13-1

Use the following figure specified in 3-D space for conducting the given scalings. Plot and analyze the situation in each case:

$F = \overline{L_1L_2L_3} \quad L_1 = \overline{P_1P_2} \quad P_1 = (0,0,-1)$

$L_2 = \overline{P_2P_3} \quad P_2 = (1,0,1)$

$L_3 = \overline{P_3P_1} \quad P_3 = (0,1,0)$

1.	$K_x = 1$	$K_y = -1$	$K_z = 2$
2.	$K_x = 0.5$	$K_y = -2$	$K_z = 1$
3.	$K_x = -1$	$K_y = -1$	$K_z = -1$
4.	$K_x = 2$	$K_y = 2$	$K_z = 2$
5.	$K_x = 0$	$K_y = 1$	$K_z = 1$
6.	$K_x = 0.5$	$K_y = 0.5$	$K_z = 2$
7.	$K_x = 2$	$K_y = -1$	$K_z = 0.25$

13-5 SCALING SPACE OBJECTS IN 3-D SPACE

The procedure for scaling a space object that is immersed in 3-D space is really no different that just described for lines and plane figures. The specifications are a bit more involved and the analysis calls for a few more steps; but the scaling operation, itself, remains a rather simple one.

The following example shows a space object being subjected to a 3-D scaling. Table 13-2 summarizes the primary analysis of the reference version of the object, and you can see a 3-D perspective view of it in Fig. 13-2. The primary analysis of the scaled version appears in Table 13-3, and its graphic version is in Fig. 13-3.

Example 13-3

Using the reference space object specified below, apply a 3-D scaling of $K_x=0.5$, $K_y=-2$, $K_z=0.5$. Plot and analyze both versions of the object.

$S = \overline{F_1F_2F_3F_4F_5F_6F_7}$

$F_1 = \overline{L_1L_2L_4L_6L_8}$	$L_1 = \overline{P_1P_2}$	$P_1 = (-1,-1,1)$
$F_2 = \overline{L_1L_3L_5L_{11}}$	$L_2 = \overline{P_1P_5}$	$P_2 = (1,-1,1)$
$F_3 = \overline{L_2L_3L_{10}L_{11}}$	$L_3 = \overline{P_1P_6}$	$P_3 = (1,1,1)$
$F_4 = \overline{L_4L_5L_7L_{13}}$	$L_4 = \overline{P_2P_3}$	$P_4 = (0,2,1)$
$F_5 = \overline{L_6L_7L_9L_{14}}$	$L_5 = \overline{L_6G_7L_9L_{14}}$	$P_5 = (-1,1,1)$
$F_6 = \overline{L_3L_8L_{10}L_{15}}$	$L_6 = \overline{P_3P_4}$	$P_6 = (-1,-1,-1)$
$F_7 = \overline{L_{11}L_{12}L_{13}L_{14}L_{15}}$	$L_7 = \overline{P_3P_8}$	$P_7 = (1,-1,-1)$
	$L_8 = \overline{P_4P_5}$	$P_8 = (1,1,-1$
	$L_9 = \overline{P_4P_9}$	$P_9 = (0,2,-1)$
	$L_{10} = \overline{P_5P_{10}}$	$P_{10} = (-1,1,-1)$
	$L_{11} = \overline{P_6P_7}$	
	$L_{12} = \overline{P_6P_{10}}$	
	$L_{13} = \overline{P_7P_8}$	
	$L_{14} = \overline{P_8P_9}$	
	$L_{15} = \overline{P_9P_{10}}$	

Table 13-2. Specifications and Analysis of the Reference Space Object Portrayed in Fig. 13-2.

(A, B, C, and D correspond to α, β, γ, and δ.)

SPECIFICATIONS

$S=\overline{F_1F_2F_3F_4F_5F_6F_7}$

$F_1=\overline{L_1L_2L_4L_6L_8}$	$L_1=\overline{P_1P_2}$	$P_1=(-1,-1,1)$
$F_2=\overline{L_1L_3L_5L_{11}}$	$L_2=\overline{P_1P_5}$	$P_2=(1,-1,1)$
$F_3=\overline{L_2L_3L_{10}L_{11}}$	$L_3=\overline{P_1P_6}$	$P_3=(1,1,1)$
$F_4=\overline{L_4L_5L_7L_{13}}$	$L_4=\overline{P_2P_3}$	$P_4=(0,2,1)$
$F_5=\overline{L_6L_7L_9L_{14}}$	$L_5=\overline{L_6L_7L_9L_{14}}$	$P_5=(-1,1,1)$
$F_6=\overline{L_3L_8L_{10}L_{15}}$	$L_6=\overline{P_3P_4}$	$P_6=(-1,-1,-1)$
$F_7=\overline{L_{11}L_{12}L_{13}L_{14}L_{15}}$	$L_7=\overline{P_3P_8}$	$P_7=(1,-1,-1)$
	$L_8=\overline{P_4P_5}$	$P_8=(1,1,-1)$
	$L_9=\overline{P_4P_9}$	$P_9=(0,2,-1)$
	$L_{10}=\overline{P_5P_{10}}$	$P_{10}=(-1,1,-1)$
	$L_{11}=\overline{P_6P_7}$	
	$L_{12}=\overline{P_6P_{10}}$	
	$L_{13}=\overline{P_7P_8}$	
	$L_{14}=\overline{P_8P_9}$	
	$L_{15}=\overline{P_9P_{10}}$	

LENGTHS

$s_1=2$	$s_2=2$	$s_3=3$	$s_4=2$	$s_5=2$
$s_6=\sqrt{2}$	$s_7=2$	$s_8=\sqrt{2}$	$s_9=2$	$s_{10}=2$
$s_{11}=2$	$s_{12}=2$	$s_{13}=2$	$s_{14}=\sqrt{2}$	$s_{15}=\sqrt{2}$

DIRECTION COSINES

$\cos A_1=1$	$\cos B_1=0$	$\cos C_1=0$
$\cos A_2=0$	$\cos B_2=1$	$\cos C_2=0$
$\cos A_3=0$	$\cos B_3=0$	$\cos C_3=-1$
$\cos A_4=0$	$\cos B_4=1$	$\cos C_4=0$
$\cos A_5=0$	$\cos B_5=0$	$\cos C_5=-1$

Table 13-2. Specifications and Analysis of the Reference Space Object Portrayed in Fig. 13-2. **(Continued from page 153.)**

(A, B, C, and D correspond to α, β, γ, and δ.)

DIRECTION COSINES

$\cos A_6=-\sqrt{2}/2$	$\cos B_6=\sqrt{2}/2$	$\cos C_6=0$
$\cos A_7=0$	$\cos B_7=0$	$\cos C_7=-1$
$\cos A_8=-\sqrt{2}/2$	$\cos B_8=-\sqrt{2}/2$	$\cos C_8=0$
$\cos A_9=0$	$\cos B_9=0$	$\cos C_9=-1$
$\cos A_{10}=0$	$\cos B_{10}=0$	$\cos C_{10}=-1$
$\cos A_{11}=1$	$\cos B_{11}=0$	$\cos C_{11}=0$
$\cos A_{12}=0$	$\cos B_{12}=1$	$\cos C_{12}=0$
$\cos A_{13}=0$	$\cos B_{13}=1$	$\cos C_{13}=0$
$\cos A_{14}=-\sqrt{2}/2$	$\cos B_{14}=\sqrt{2}/2$	$\cos C_{14}=0$
$\cos A_{15}=-\sqrt{2}/2$	$\cos B_{15}=-\sqrt{2}/2$	$\cos C_{15}=0$

ANGLES

$\phi_{1,2}=90^\circ$	$\phi_{1,3}=90^\circ$	$\phi_{1,4}=90^\circ$	$\phi_{1,5}=90^\circ\phi$
$\phi_{2,3}=90^\circ$	$\phi_{2,8}=135^\circ$	$\phi_{2,10}=90^\circ$	$\phi_{3,11}=90^\circ\phi$
$\phi_{3,12}=90^\circ$	$\phi_{4,5}=90^\circ$	$\phi_{4,6}=135^\circ$	$\phi_{4,7}=90^\circ$
$\phi_{5,11}=90^\circ$	$\phi_{5,13}=90^\circ$	$\phi_{6,7}=90^\circ$	$\phi_{6,8}=90^\circ$
$\phi_{6,9}=90^\circ$	$\phi_{7,13}=90^\circ$	$\phi_{7,14}=90^\circ$	$\phi_{8,9}=90^\circ$
$\phi_{8,10}=90^\circ$	$\phi_{9,14}=90^\circ$	$\phi_{9,15}=90^\circ$	$\phi_{10,12}=90^\circ$
$\phi_{10,15}=90^\circ$	$\phi_{11,12}=90^\circ$	$\phi_{11,13}=90^\circ$	$\phi_{12,15}=135^\circ$
$\phi_{13,14}=135^\circ$	$\phi_{14,15}=90^\circ$		

1. Applying the given scaling, the points become:

$P_1 = (-0.5,2,0.5)$
$P_2 = (0.5,2,0.5)$
$P_3 = (0.5,-2,0.5)$
$P_4 = (0,-4,0.5)$
$P_5 = (-0.5,-2,0.5)$
$P_6 = (-0.5,2,-0.5)$
$P_7 = (0.5,2,-0.5)$
$P_8 = (0.5,-2,-0.5)$
$P_9 = (0,-4,-0.5)$
$P_{10} = (-0.5,-2,-0.5)$

2. Running an analysis of the reference version of the figure:

A. Calculate the lengths of the lines
B. Determine the direction cosines
C. Determine the relevant angles in the object

See the results summarized in Table 13-2.

3. Running an analysis of the scaled version of the figure:

A. Calculate the lengths of the scaled lines
B. Determine the direction cosines of the scaled lines
C. Determine the angles in the scaled version

See the results summarized in Table 13-3.

Of course it is necessary to begin such an experiment by showing the complete specifications for the

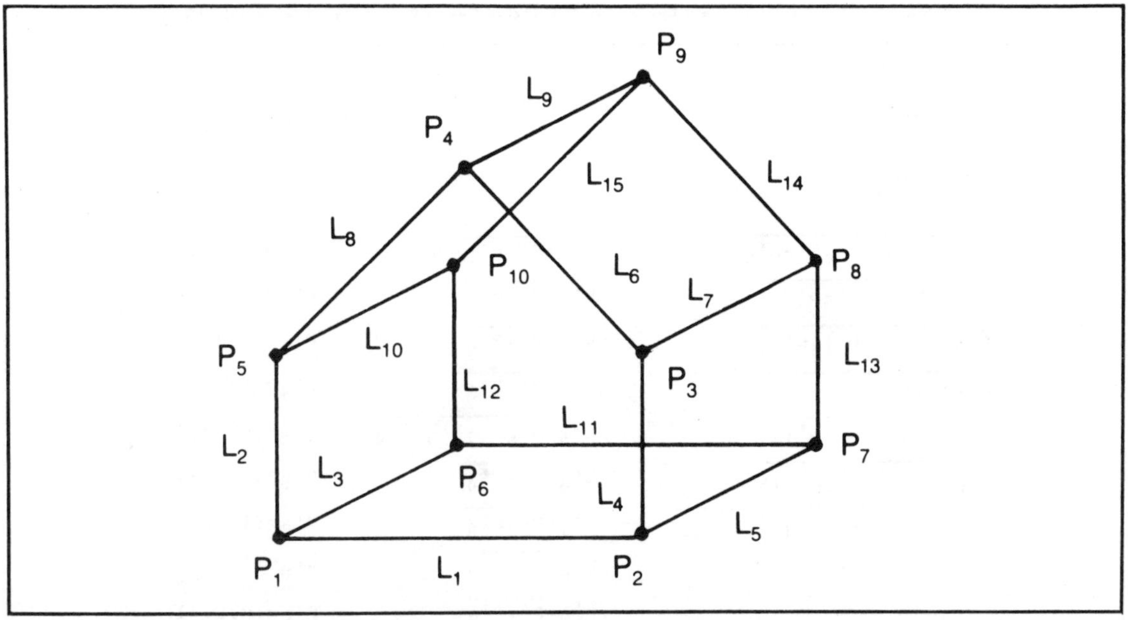

Fig. 13-2. A reference space object specified in 3-D space. See the analysis of this object in Table 13-2.

reference space object. The object in this instance turns out to look something like a little house. Notice from the S specifications that it is composed for seven plane figures (or sides), fifteen lines (or edges) and ten points. Plot the three planar views if you wish.

Step 1 in this case summarizes the results of applying the prescribed scaling factors to the components of the points in the reference version of the object. Check my work by running through that 3-D scaling procedure yourself.

Step 2 in the example summarizes the steps involved in conducting a primary analysis of the reference object, and then the final step outlines the analysis of the scaled version. Those two analyses are summarized in detail in Tables 13-2 and 13-3.

Notice that the scaling in this instance changes the lengths of all the lines, alters the direction cosines for many of them, and changes a few of the critical angles in the object. Again, doublecheck my results by working through the analyses in detail on your own.

The plots of the reference and scaled versions clearly show a change in the object. The scaled version is compressed, or reduced, along the X and Z axes, but stretched and reflected along the Y axis. See Fig. 13-3.

13-6 ALTERNATIVE FORMS OF THE GEOMETRIC EQUATION

Equation 13-2 is the basic equation for geometric scaling as reorganized to solve for the scaling terms. Given the coordinate of a reference point and a scaled version of that point, the expressions lead directly to the scaling that exists between them.

$$K_x = x_n/x_n' \qquad K_y = y_n/y_n' \qquad K_z = z_n/z_n' \qquad \textbf{Equation 13-2}$$

where:

x_n',y_n',z_n' = the coordinate of the scaled version of point n relative to the frame of reference.

x_n,y_nz_n = the coordinate of a reference point n relative to the frame of reference.

K_x, K_y and K_z = the scaling factors for the X, Y and Z axes, relative to the frame of reference.

Earlier descriptions of scalings in one and two dimensions cited appropriate versions of this equation as a means for determining the scaling that exists between two coordinate systems. That can be done in the case of coordinate systems that are separated by a scaling in 3-D space as well.

The equation can be of great practical importance when using it to change the scaling of one or more dimensions of a geometric entity that happens to be specified in some convenient, but less desirable, form. The general idea is to make a triangle taller, for example, by applying a

Table 13-3. Specifications and Analysis of the Scaled Space Object Portrayed in Fig. 13-3.

(A, B, C, and D correspond to α, β, γ, and δ.)

SPECIFICATIONS

$S' = \overline{F'_1 F'_2 F'_3 F'_4 F'_5 F'_6 F'_7}$

$F'_1 = \overline{L'_1 L'_2 L'_4 L'_6 L'_8}$	$L'_1 = \overline{P'_1 P'_2}$	$P'_1 = (-1,-1,1)$
$F'_2 = \overline{L'_1 L'_3 L'_5 L'_{11}}$	$L'_2 = \overline{P'_1 P'_5}$	$P'_2 = (1,-1,1)$
$F'_3 = \overline{L'_2 L'_3 L'_{10} L'_{11}}$	$L'_3 = \overline{P'_1 P'_6}$	$P'_3 = (1,1,1)$
$F'_4 = \overline{L'_4 L'_5 L'_7 L'_{13}}$	$L'_4 = \overline{P'_2 P'_3}$	$P'_4 = (0,2,1)$
$F'_5 = \overline{L'_6 L'_7 L'_9 L'_{14}}$	$L'_5 = \overline{L'_6 L'_7 L'_9 L'_{14}}$	$P'_5 = (-1,1,1)$
$F'_6 = \overline{L'_3 L'_8 L'_{10} L'_{15}}$	$L'_6 = \overline{P'_3 P'_4}$	$P'_6 = (-1,-1,-1)$
$F'_7 = \overline{L'_{11} L'_{12} L'_{13} L'_{14} L'_{15}}$	$L'_7 = \overline{P'_3 P'_8}$	$P'_7 = (1,-1,-1)$
	$L'_8 = \overline{P'_4 P'_5}$	$P'_8 = (1,1,-1)$
	$L'_9 = \overline{P'_4 P'_9}$	$P'_9 = (0,2,-1)$
	$L'_{10} = \overline{P'_5 P'_{10}}$	$P'_{10} = (-1,1,-1)$
	$L'_{11} = \overline{P'_6 P'_7}$	
	$L'_{12} = \overline{P'_6 P'_{10}}$	
	$L'_{13} = \overline{P'_7 P'_8}$	
	$L'_{14} = \overline{P'_8 P'_9}$	
	$L'_{15} = \overline{P'_9 P'_{10}}$	

LENGTHS

$s'_1=1$	$s'_2=4$	$s'_3=1$	$s'_4=4$
$s'_5=1$	$s'_6=\sqrt{17}/17$	$s'_7=1$	$s'_8=\sqrt{17}/17$
$s'_9=1$	$s'_{10}=1$	$s'_{11}=1$	$s'_{12}=4$
$s'_{13}=4$	$s'_{14}=\sqrt{17}/2$	$s'_{15}=\sqrt{17}/2$	

DIRECTION COSINES

$\cos' A_1=1$	$\cos' B_1=0$	$\cos' C_1=0$
$\cos' A_2=0$	$\cos' B_2=-1$	$\cos' C_2=0$
$\cos' A_3=0$	$\cos' B_3=0$	$\cos' C_3=-1$
$\cos' A_4=0$	$\cos' B_4=-1$	$\cos' C_4=0$

DIRECTION COSINES

$\cos'A_5=0$	$\cos'B_5=0$	$\cos'C_5=-1$
$\cos'A_6=-\sqrt{17}/17$	$\cos'B_6=-4\sqrt{17}/17$	$\cos'C_6=0$
$\cos'A_7=0$	$\cos'B_7=0$	$\cos'C_7=-1$
$\cos'A_8=-\sqrt{17}/17$	$\cos'B_8=4\sqrt{17}/17$	$\cos'C_8=0$
$\cos'A_9=0$	$\cos'B_9=0$	$\cos'C_9=-1$
$\cos'A_{10}=0$	$\cos'B_{10}=0$	$\cos'C_{10}=-1$
$\cos'A_{11}=1$	$\cos'B_{11}=0$	$\cos'C_{11}=0$
$\cos'A_{12}=0$	$\cos'B_{12}=-1$	$\cos'C_{12}=0$
$\cos'A_{13}=0$	$\cos'B_{13}=-1$	$\cos'C_{13}=0$
$\cos'A_{14}=-\sqrt{17}/17$	$\cos'B_{14}=-4\sqrt{17}/17$	$\cos'C_{14}=0$
$\cos'A_{15}=-\sqrt{17}/17$	$\cos'B_{15}=4\sqrt{17}/17$	$\cos'C_{15}=0$

ANGLES

$\phi'_{1,2}=90^\circ$	$\phi'_{1,3}=90^\circ$	$\phi'_{1,4}=90^\circ$
$\phi'_{1,5}=90^\circ$	$\phi'_{2,3}=90^\circ$	$\phi'_{2,8}=166^\circ 50'$
$\phi'_{2,10}=90^\circ$	$\phi'_{3,11}=90^\circ$	$\phi'_{3,12}=90^\circ$
$\phi'_{4,5}=90^\circ$	$\phi'_{4,6}=166^\circ 50'$	$\phi'_{4,7}=90^\circ$
$\phi'_{5,11}=90^\circ$	$\phi'_{5,13}=90^\circ$	$\phi'_{6,7}=90^\circ$
$\phi'_{6,8}=28^\circ 55'$	$\phi'_{6,9}=90^\circ$	$\phi'_{7,13}=90^\circ$
$\phi'_{7,14}=90^\circ$	$\phi'_{8,9}=90^\circ$	$\phi'_{8,10}=90^\circ$
$\phi'_{9,14}=90^\circ$	$\phi'_{9,15}=90^\circ$	$\phi'_{10,12}=90^\circ$
$\phi'_{10,15}=90^\circ$	$\phi'_{11,12}=90^\circ$	$\phi'_{11,13}=90^\circ$
$\phi'_{12,15}=166^\circ 50'$	$\phi'_{13,14}=166^\circ 50$	$\phi'_{14,15}=28^\circ 55'$

positive scaling greater than 1 to the vertical axis of its coordinate system. The scaled coordinates can then become the basis for a newly specified pyramid.

Equation 13-3 is the inverse form of the basic equation for geometric scalings in 3-D space. Taken as directed, it lets you calculate the reference-point coordinates, given the scaled version of the coordinates and the scaling terms. It also leads quite naturally to the equation for relativistic scalings.

$$x_n = x_n'/K_x \quad y_n = y_n'/K_y \quad z_n = z_n'/K_z$$

Equation 13-3

where:

x_n', y_n', z_n' = the coordinate of the scaled version of point *n* relative to the frame of reference

x_n, y_n, z_n = the coordinate of a reference point *n* relative to the frame of reference.

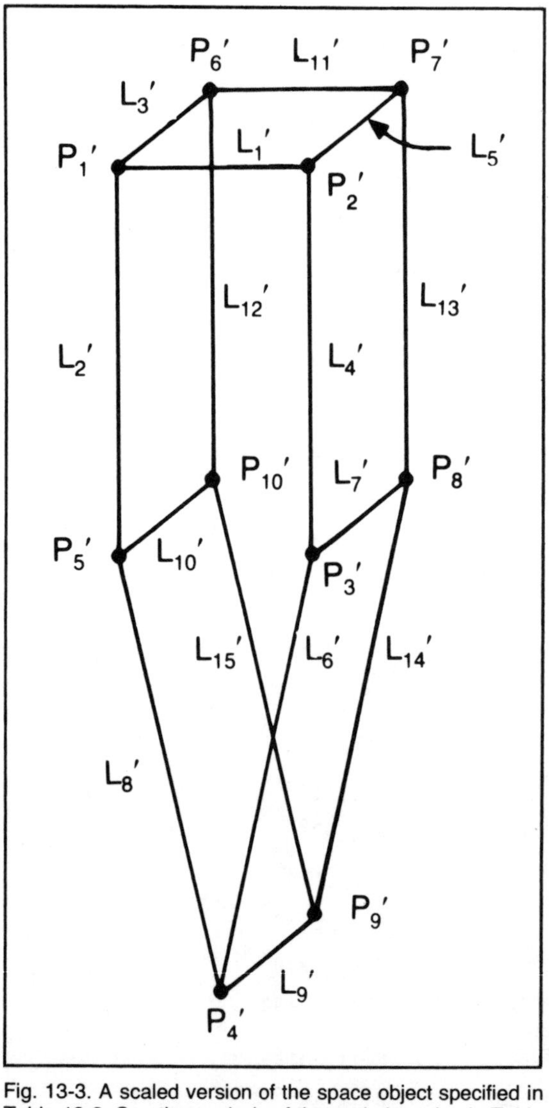

Fig. 13-3. A scaled version of the space object specified in Table 13-2. See the analysis of the scaled version in Table 13-3.

K_x, K_y and K_z = the scaling factors for the X, Y and Z axes, relative to the frame of reference.

13-7 RELATIVISTIC SCALINGS IN 3-D

A scaling in any number of dimensions actually represents a scaling that exists between two coordinate systems. Specifically, a geometric scaling in 3 dimensions represents a scaling-type offset between a 3-D reference coordinate system and another 3-D system that is otherwise identical to it.

In the context of relativistic scalings, however, the idea of changing the size of an object is not very meaningful. It is more useful to view a relativistic scaling as an expression of the size difference, or a difference between the plus-and minus orientation of the axes, between two otherwise identical 3-D coordinate systems.

Equation 13-4 is the basic relativistic version of a 3-dimensional scaling that exists between a pair of 3-D coordinate systems. Given the coordinate of some point in a reference figure—reckoned with respect to the observer's frame of reference—it leads to the coordinate of that point as regarded from the scaled frame of reference. The equation ultimately shows how a geometric entity appears relative to the scaled frame of reference.

$$x_n^* = x_n/K_x$$
$$y_n^* = y_n/K_y$$
$$z_n^* = z_n/K_z$$

Equation 13-4

where:

x_n^*, y_n^*, z_n^* = the coordinate of reference point *n* relative to the scaled frame of reference.

x_n, y_n, z_n = the coordinate of reference point *n* relative to the observer's frame of reference.

K_x, K_y and K_z = the scaling factors that exist between the observer's and scaled frames of reference, relative to the observer's.

Variations of such experiments include running complete analyses of the reference object as regarded from both frames of reference. Such projects show how the object appears in one particular 3-D space from both points of view when the two coordinate systems are scaled in 3-D space.

Equation 13-5 lets you determine the 3-D scaling that exists between two 4-dimensional frames of reference, given the coordinates of a pair of points from both the observer's and scaled frames of reference.

$$K_x = x_n/x_n^*$$
$$K_y = y_n/y_n^*$$
$$K_z = z_n/z_n^*$$

Equation 13-5

where:

x_n^*, y_n^*, z_n^* = the coordinate of reference point *n* relative to the scaled frame of reference.

x_n, y_n, z_n = the coordinate of reference point *n* relative to the observer's frame of reference.

K_x, K_y and K_z = the scaling factors that exist between the observer's and scaled frames of reference, relative to the observer's.

The inverse form of the relativistic 3-D scaling equation is included here only as a matter of record. It is a simple variation of the basic relativistic scaling equation that follows the same general pattern as in 2-D counterpart.

$$x_n = x_n*K_x$$
$$y_n = y_n*K_y$$
$$z_n = z_n*K_z$$

Equation 13-6

where:

x_n*, y_n*, z_n* = the coordinate of reference point *n* relative to the scaled frame of reference.

x_n, y_n, z_n = the coordinate of reference point *n* relative to the observer's frame of reference.

K_x, K_y and K_z = the scaling factors that exist between the observer's and scaled frames of reference, relative to the observer's

13-8 COMBINATIONS OF 3-D SCALINGS

Multiple scaling in 3-D space are communitative. That is, you can apply a given series of scalings in any desired order and still end up with the same overall scaling. This, of course, applies to geometric scalings as well as their relativistic counterparts.

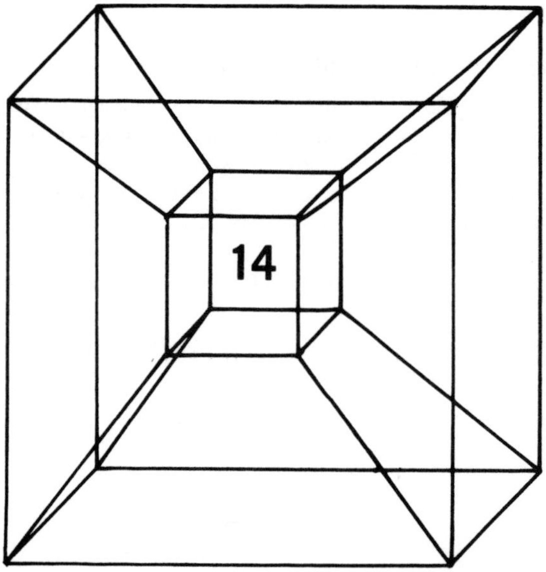

Rotations in 3-D Space

The mathematics necessary for rotating points in 3-D space is a bit more complicated than that of 2-D space. The big difference is that 3-D rotations take place about an axis of rotation as opposed to a simple point of rotation; and that suggests the need for three separate sets of equations. You will first select an *axis of rotation*, and then use the appropriate rotation equations—those for an X-axis rotation, a Y-axis rotation or a Z-axis rotation. Although there are three separate sets of 3-D rotation equations, you will find that their forms are almost identical to the point-of-rotation equations for 2-D space.

Here is a brief introductory account of the evolution of rotations:

☐ Rotations are not defined for 1-D space.

☐ Rotations in 2-D space take place about a point of rotation.

☐ Rotations in 3-D space take place about a line, or axis, of rotation.

Perhaps you can see a logical progression of ideas in that summary. In 2-D space, it is most appropriate to work with a point of rotation; in 3-D space, we work with an axis of rotation. And since there are three different axes in a 3-D coordinate system, it follows that you must have access to three different equations.

Figure 14-1 shows a 3-D coordinate system and illustrates the directions of rotation for each of the three axes. If you have a good sense of spatial perception, you can imagine how that coordinate system responds as you twist it about each of the three axes. Under a pure rotation (no translation or scaling included), the chosen axis of rotation retains its orientation in space while the other two undergo a big change in position. The axes remain at right angles to one another throughout the entire operation, though.

14-1 THE EQUATIONS OF 3-D ROTATION

Recall from Chapter 8 that simple rotations in 2-D space take place about a single point—the origin of the coordinate system, unless you specify otherwise. Setting up an experiment using a 2-D rotation is a matter of specifying a reference point, a point of rotation (if something other than the origin of the coordinate system), selecting an angle of rotation, and then plugging the data into just one of two equations—one for a rotation about the origin or one for a rotation about some off-origin point. Actually, a single equation—the 2-D, off-origin rotation equation—is adequate for any such experiment.

Setting up a rotation in three dimensions requires an

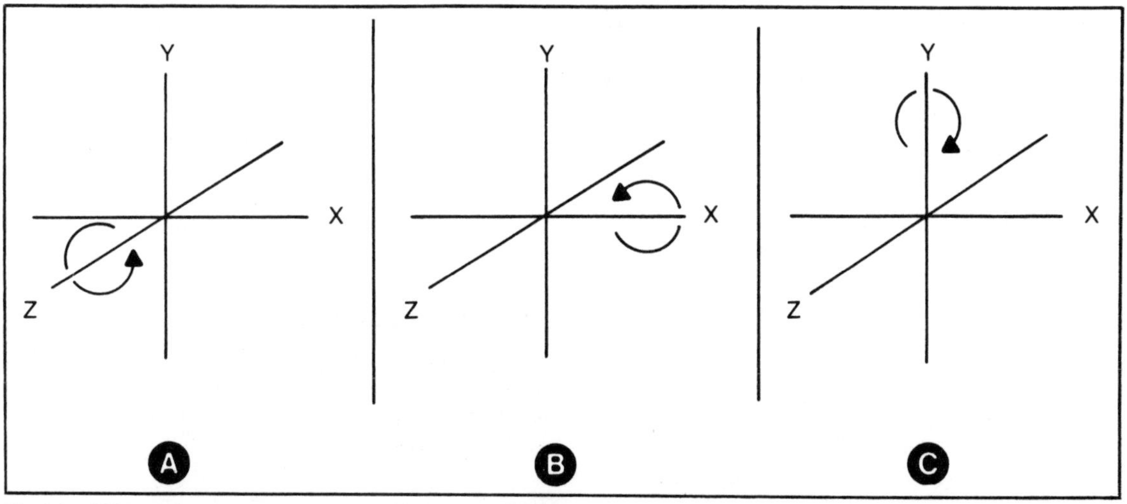

Fig. 14-1. X-Y perspective views of the three axes of 3-D rotation. (A) Positive angle of rotation about the Z axis. (B) Positive angle of rotation about the X axis. (C) Positive angle of rotation about the Y axis.

additional step: selecting the axis of rotation. That step, in turn, dictates the rotation equation that is appropriate for the task at hand. The three equations; shown here as Equations 14-1, 14-2 and 14-3; appear quite similar in two respects. First, they use the same pattern of sine and cosine functions of the angle of rotation; and second, the components of points corresponding to the chosen axis of rotation are not affected by the rotation equations. Doing an X-axis rotation, for example, does not affect the X-axis components of any point in the space.

$$x_n' = x_n\cos\theta - y_n\sin\theta$$
$$y_n' = x_n\sin\theta + y_n\cos\theta \qquad \textbf{Equation 14-1}$$
$$z_n' = z_n$$

where:

x_n',y_n',z_n' = the coordinate of the rotated point n relative to the frame of reference

x_n,y_n,z_n = the coordinate of reference point n relative to the frame of reference

θ = the angle of rotation about the Z axis

Note: This equation applies only to Z-axis rotations about the origin of the frame of reference.

$$x_n' = x_n\cos\theta - z_n\sin\theta$$
$$y_n' = y_n \qquad \textbf{Equation 14-2}$$
$$z_n' = x_n\sin\theta + z_n\cos\theta$$

where:

x_n',y_n',z_n' = the coordinate of the rotated version of point n relative to the frame of reference

x_n,y_n,z_n = the coordinate of the reference point n relative to the frame of reference

θ = the angle of rotation about the Y axis

Note: This equation applies only to Y-axis rotations about the origin of the frame of reference.

$$x_n' = x_n$$
$$y_n' = y_n\cos\theta - z_n\sin\theta \qquad \textbf{Equation 14-3}$$
$$z_n = y_n\sin\theta + z_n\cos\theta$$

where:

x_n',y_n',z_n' = the coordinate of the rotated version of point n relative to the frame of reference

x_n,y_n,z_n = the coordinate of point n relative to the frame of reference

θ = the angle of rotation about the X axis

Note: This equation applies only to X-axis rotations about the origin of the frame of reference.

Use Equation 14-1 for a rotation about the Z axis, Equation 14-2 for a rotation about the Y axis, and Equation 14-3 whenever you want to do a rotation about the X axis.

Compare those three equations with the 2-D versions in Chapter 8, and you will see the same pattern of sine and cosine functions of the angle of rotation. The only real difference is that each 3-D rotation equation includes an axis that is unaffected by the transformation—the axis of rotation. Generally speaking, 3-D rotations are simple extensions of the same transformation in 2-D space.

Example 14-1 demonstrates how easy it is to rotate a point in 3-D space about an axis of the coordinate system.

Example 14-1

Given a reference point P = (1,2,3):

A. Rotate the reference point 30 degrees about the Z axis.
B. Rotate the reference point 30 degrees about the Y axis.
C. Rotate the reference point 30 degrees about the X axis.

1. Gathering the data for all three phases of the experiment:

$$x = 1 \quad y = 2 \quad z = 3 \qquad \theta = 30^\circ$$
$$\sin\theta = 1/2$$
$$\cos\theta = \sqrt{3}/2$$

2. Applying Equation 14-1 to rotate the point about the Z axis:

$$x' = x\cos\theta - y\sin\theta \qquad y' = x\sin\theta + y\cos\theta$$
$$x' = (1)\,(\sqrt{3}/2) - (2)\,(1/2) \quad y' = (1)\,(1/2) + (2)\,(\sqrt{3}/2)$$
$$x' = -0.13 \qquad y' = 2.2$$

$$z' = z$$
$$z' = 3$$

So the 30-degree rotation about the Z axis transforms P = (1,2,3) to P′ = (0.13,2.2,3).

3. Applying Equation 14-2 to rotate the point about the Y axis:

$$x' = x\cos\theta - z\sin\theta \qquad y' = y \quad z' = x\sin\theta + z\cos\theta$$
$$x' = 1(\sqrt{3}/2) - 3\,(1/2) \quad y' = 2 \quad z' = 1(1/2) + 3(\sqrt{3}/2)$$
$$x' = -0.63 \qquad y' = 2 \quad z' = 3.1$$

So the 30-degree rotation about the Y axis transforms P = (1,2,3) to P′ = (−0.63,2,3.1).

4. Applying Equation 14-3 to rotate the point about the X axis:

$$x' = x$$
$$x' = 1$$

$$y' = y\cos\theta - z\sin\theta \qquad z' = y\sin\theta + z\cos\theta$$
$$y' = 2\,(\sqrt{3}/2) - 3\,(1/2) \qquad z' = 2\,(1/2) + 3\,(\sqrt{3}/2)$$
$$y' = 0.23 \qquad z' = 3.6$$

So the 30-degree rotation about the X axis transforms P = (1,2,3) to P′ = (1,0,23,3.6).

Try specifying some points in 3-D space as suggested in Exercise 14-1.

Exercise 14-1

1. Rotate P = (1,0,0) 45 degrees about the Z axis.
2. Rotate P = (1,0,0) −30 degrees about the Z axis.
3. Rotate P = (1,0,0) 45 degrees about the Y axis.
4. Rotate P = (1,0,0) 60 degrees about the Y axis.
5. Rotate P = (1,0,0) −45 degrees about the X axis.
6. Rotate P = (−1,−1,1) 15 degrees about the Z axis.
7. Rotate P = (−1,−1,1) 270 degrees about the Y axis.
8. Rotate P = (−1 −1,1) 15 degrees about the X axis.
9. Make up some point coordinates of your own, then rotate them through some selected combinations of angles and axes.

14-2 ROTATING LINES IN 3-D SPACE

Rotating a line that is immersed in 3-D space is a matter of rotating its two endpoints. It is a straightforward procedure that is demonstrated for you in the following example.

Example 14-2

Given the specifications for a line in 3-D space, rotate its coordinate system 45 degrees about the Z axis. Compare the analysis of the reference and rotated version of the lines.

$$L = \overline{P_1P_2} \qquad P_1 = (-1,1,1)$$
$$P_2 = (1,-1,-1)$$

1. Gathering the relevant data into an equation-oriented form:

$$x_1 = -1 \quad y_1 = 1 \quad z_1 = 1 \quad \theta = 45^\circ$$
$$x_2 = 1 \quad y_2 = -1 \quad z_2 = -1 \quad \sin\theta = \sqrt{2}/2$$
$$\cos\phi = \sqrt{2}/2$$

2. Solving for the rotation of endpoint P_1:

$$x_1' = x_1\cos\theta - y_1\sin\theta$$
$$x_1' = -1(\sqrt{2}/2) - 1\,(\sqrt{2}/2)$$
$$x_1' = -\sqrt{2}$$
$$y_1' = x_1\sin\theta + y_1\cos\theta$$
$$y_1' = -1(\sqrt{2}/2) + 1\,(\sqrt{2}/2)$$
$$y_1' = 0$$

$$z_1' = z_1$$
$$z_1' = 1$$

So the transformation rotates P_1 = (−1,1,1) to P_1' = (− $\sqrt{2}$,0,1).

3. Solving for the rotation of endpoint P_2:

$$x_2' = x_2\cos\theta - y_2\sin\theta$$
$$x_2' = 1\,(\sqrt{2}/2) - (-1)\,(\sqrt{2}/2)$$
$$x_2' = \sqrt{2}$$
$$y_2' = x_2\sin\theta + y_2\cos\theta$$
$$y_2' = 1\,(\sqrt{2}/2) + (-1)\,(\sqrt{2}/2)$$
$$y_2' = 0$$

$$z_2' = z_2$$
$$z_2' = -1$$

And the transformation rotates $P_2 = (1,-1,-1)$ to $P_2' = (\sqrt{2},0,1)$.

4. Analyzing the reference version of the line:

$$s = 2\sqrt{3}$$

$$\cos\alpha = \sqrt{3}/3 \qquad \cos\beta = -\sqrt{3}/3 \quad \cos\gamma = -\sqrt{3}/3$$

5. Analyzing the rotated version of the line:

$$s' = 2\sqrt{3}$$

$$\cos\alpha' = \sqrt{6}/3 \qquad \cos\beta' = 0 \qquad \cos\gamma' = -\sqrt{3}/3$$

The example cites a simple line in 3-D space and calls for a rotation of 45 degrees about the Z axis. Step 1 then sets up the relevant variables, including the sine and cosine values of that particular angle.

Step 2 uses the appropriate equation for a Z-axis rotation, Equation 14-1, to rotate one of the endpoints of the line. Step 3 does the same thing for the second endpoint. The rotation procedure, itself, is completed at the conclusion of that step.

Steps 4 and 5 analyze the reference and rotated versions of the line. You can see that the line has the same length in either case, but that its direction cosines are somewhat different. You are about to find that those results fit a general principle that says the lengths of lines are invariant under simple 3-D rotations, but the direction cosines are not.

Try rotating some lines as suggested in the following exercise.

Exercise 14-2

In each of the following instances, perform the given rotations on the specified reference line. Plot and analyze the entire operation.

1. Rotate the line defined defined by endpoints (0,0,0) and (1,1,1):

   A. 45 degrees about the X axis
   B. 30 degrees about the Y axis
   C. −45 degrees about the Z axis

2. Rotate the line defined by endpoints (−1,0,0) and (1,0,0):

   A. 60 degrees about the X axis
   B. 90 degrees about the Y axis
   C. 180 degrees about the Z axis

3. Rotate the line defined by endpoints (0,−1,0) and (0,1,0):

   A. 45 degrees about the X axis
   B. 90 degrees about the Y axis
   C. −30 degrees about the Z axis

14-2.1 Effects on the Lengths of Lines

As in 2-D space, a rotation in three dimensions does not affect the length of a line. You have already seen that fact demonstrated in the previous example and exercise, and you have seen a formal analytic proof for the 2-D case in Chapter 8 (Proof 8-1). There is thus a great deal of evidence to support the notion that length, or the distance between points, is preserved under 3-D rotations. To get further experience and build your confidence, use Proof 8-1 as a guide for working out a formal analytic proof for the invariance of length under 3-D rotations.

14-2.2 Direction Cosines are Variant

In all but a few trivial instances, rotations in two and three dimensions alter the orientation of a line with respect to the axes of its coordinate system. And since the direction cosines of a line indicate its orientation, it follows that those line parameters are variant under rotations. Comparing the direction cosines for the reference and rotated lines in Example 14-2 illustrates that principle.

14-2.3 Angle Between Lines is Invariant

Although the direction cosines of a line are variant under 3-D rotations, rotations do not affect the angle between lines. That might appear to be a questionable statement in view of the fact that the equation for the angle between two lines is built wholly upon the direction cosines of the lines. But as described earlier for the case of rotations in 2-D space (Chapter 8), the angle is indeed invariant.

14-2.4 Summary of the Principles

You have seen these principles suggested in the foregoing discussions of the rotation of lines in 3-D space:

☐ Length, or distance, is an invariant under 3-D rotations.

☐ Direction cosines of lines are variant under 3-D rotations.

☐ The angle between two lines is invariant under 3-D rotations.

Those principles, demonstrated to some extent in the analyses offered in previous examples and, hopefully, proven in a formal fashion through your own effort, support the notion that rotations in 3-D space amount to little more than a straightforward extension of the same sort of transformation in 2-D space.

14-3 ROTATING PLANE FIGURES IN 3-D SPACE

There is nothing new here in principle: The general procedure for rotating plane figures immersed in 3-D space is not significantly different from rotating such figures in 2-D space. Rotations in the 3-dimensional case take place about a chosen axis of rotation; but the essential mathematical ideas are almost identical. See if you agree after working the experiment suggested in Example 14-3.

Example 14-3

Rotate the triangular figure specified below through an angle of 60 degrees about the Y axis; plot and conduct analyses of both the reference and rotated versions.

$$F = \overline{L_1L_2L_3} \quad L_1 = \overline{P_1P_2} \quad P_1 = (0,0,0)$$
$$L_2 = \overline{P_2P_3} \quad P_2 = (1,0,0)$$
$$L_3 = \overline{P_3P_1} \quad P_3 = (0,1,0)$$

1. Gathering the data:

$$x_1 = 0 \quad y_1 = 0 \quad z_1 = 0 \qquad \theta = 60°$$
$$x_2 = 1 \quad y_2 = 0 \quad z_2 = 0 \qquad \sin\theta = \sqrt{3}/2$$
$$x_3 = 0 \quad y_3 = 1 \quad z_3 = 0 \qquad \cos\theta = 1/2$$

2. Rotating point P_1:

$$x' = x_1\cos\theta - z_1\sin\theta$$
$$x_1' = 0\ (1/2) - 0\ (\sqrt{3}/2)$$
$$x_1' = 0$$
$$z_1' = x_1\sin\theta + z_1\cos\theta$$
$$z_1' = 0\ (\sqrt{3}/2) + 0\ (1/2)$$
$$z_1' = 0$$
$$y_1' = y_1$$
$$y_1' = 0$$
$$y_1' = 0$$

So end point $P_1 = (0,0,0)$ transforms to $P_1' = (0,0,0)$.

3. Rotating point P_2:

$$x_2' = x_2\cos\theta - z_2\sin\theta$$
$$x_2' = 1\ (1/2) - 0\ (\sqrt{3}/2)$$
$$x_2' = 1/2$$
$$z_2' = x_2\sin\theta + z_2\cos\theta$$
$$z_2' = 1\ (\sqrt{3}/2) + 0\ (1/2)$$
$$z_2' = \sqrt{3}/2$$
$$y_2' = y_2$$
$$y_2' = 0$$
$$y_2' = 0$$

So endpoint $P_2 = (1,0,0)$ transforms to $P_2' = (1/2,\ 0,\ \sqrt{3}/2)$.

4. Rotating point P_3:

$$x_3' = x_3\cos\theta - z_3\sin\theta$$
$$x_3' = 0\ (1/2) - 0\ (\sqrt{3}/2)$$
$$x_3' = 0$$
$$z_3' = x_3\sin\theta + z_3\cos\theta$$
$$z_3' = 0\ (\sqrt{3}/2) + 0\ (1/2)$$
$$z_3' = 0$$
$$y_3' = y_3$$
$$y_3' = 1$$
$$y_3' = 1$$

So endpoint $P_3 = (0,1,0)$ transforms to $P_3' = (0,1,0)$.

5. Analyzing the reference version of the figure:

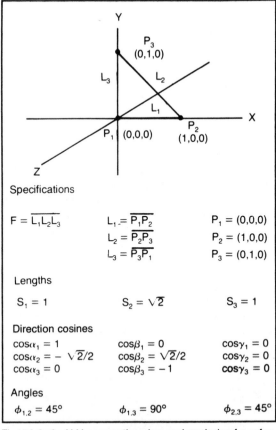

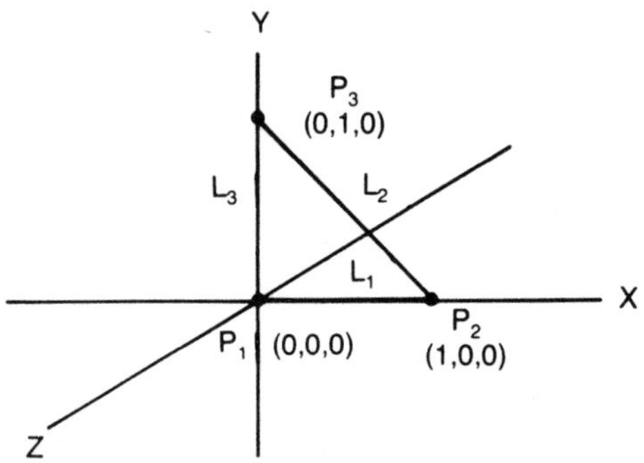

Specifications

$F = \overline{L_1L_2L_3}$	$L_1 = \overline{P_1P_2}$	$P_1 = (0,0,0)$
	$L_2 = \overline{P_2P_3}$	$P_2 = (1,0,0)$
	$L_3 = \overline{P_3P_1}$	$P_3 = (0,1,0)$

Lengths

$S_1 = 1$	$S_2 = \sqrt{2}$	$S_3 = 1$

Direction cosines

$\cos\alpha_1 = 1$	$\cos\beta_1 = 0$	$\cos\gamma_1 = 0$
$\cos\alpha_2 = -\sqrt{2}/2$	$\cos\beta_2 = \sqrt{2}/2$	$\cos\gamma_2 = 0$
$\cos\alpha_3 = 0$	$\cos\beta_3 = -1$	$\cos\gamma_3 = 0$

Angles

$\phi_{1,2} = 45°$	$\phi_{1,3} = 90°$	$\phi_{2,3} = 45°$

Fig. 14-2. An X-Y perspective view and analysis of a reference triangle in 3-D space. See Example 14-3.

$s_1 = 1$	$s_2 = 2$	$s_3 = 1$
$\cos\alpha_1 = 1$	$\cos\beta_1 = 0$	$\cos\gamma_1 = 0$
$\cos\alpha_2 = -\sqrt{2}/2$	$\cos\beta_2 = \sqrt{2}/2$	$\cos\gamma_2 = 0$
$\cos\alpha_3 = 0$	$\cos\beta_3 = -1$	$\cos\gamma_3 = 0$
$\phi_{1,2} = 45°$	$\phi_{1,3} = 90°$	$\phi_{2,3} = 45°$

See Fig. 14-2.

6. Analyzing the rotated version of the figure:

$s_1' = 1$	$s_2' = 2$
$\cos\alpha_1' = 1/2$	$\cos\beta_1' = 0$
$\cos\alpha_2' = -\sqrt{2}/4$	$\cos\beta_2' = \sqrt{2}/2$
$\cos\alpha_3' = 0$	$\cos\beta_3' = -1$
$\phi_{1,2}' = 45°$	$\phi_{1,3}' = 90°$

$s_3' = 1$	$\cos\gamma_3' = 0$
$\cos\gamma_1' = \sqrt{3}/2$	$\phi_{2,3}' = 45°$
$\cos\gamma_2' = -\sqrt{6}/4$	

See Fig. 14-3.

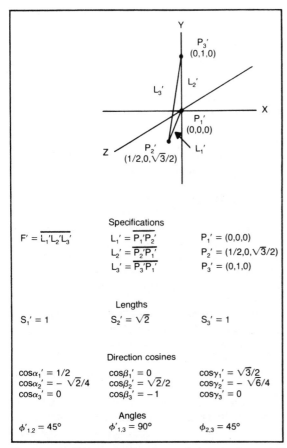

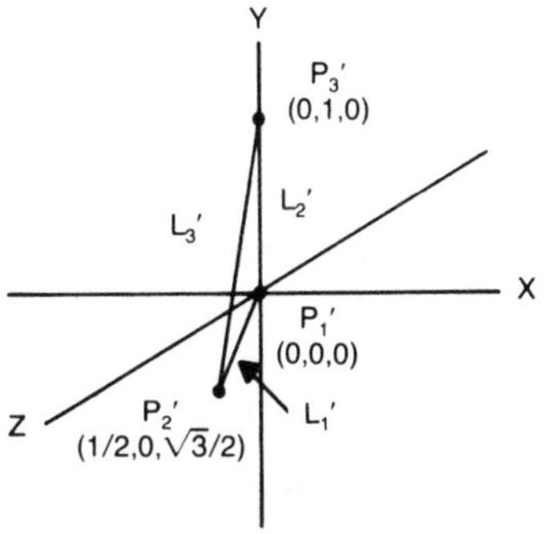

Specifications

$F' = \overline{L_1'L_2'L_3'}$	$L_1' = \overline{P_1'P_2'}$	$P_1' = (0,0,0)$
	$L_2' = \overline{P_2'P_1'}$	$P_2' = (1/2,0,\sqrt{3}/2)$
	$L_3' = \overline{P_3'P_1'}$	$P_3' = (0,1,0)$

Lengths

$S_1' = 1$	$S_2' = \sqrt{2}$	$S_3' = 1$

Direction cosines

$\cos\alpha_1' = 1/2$	$\cos\beta_1' = 0$	$\cos\gamma_1' = \sqrt{3}/2$
$\cos\alpha_2' = -\sqrt{2}/4$	$\cos\beta_2' = \sqrt{2}/2$	$\cos\gamma_2' = -\sqrt{6}/4$
$\cos\alpha_3' = 0$	$\cos\beta_3' = -1$	$\cos\gamma_3' = 0$

Angles

$\phi'_{1,2} = 45°$	$\phi'_{1,3} = 90°$	$\phi_{2,3} = 45°$

Fig. 14-3. An X-Y perspective view and analysis of a rotated version of the triangle specified in Fig. 14-2.

The general idea of that example is to rotate a triangular plane figure 60 degrees about the Y axis. The experiment opens with the formal specifications for the reference version of the object. If you study those coordinates carefully, you can see that it is a typical triangle from your earlier work in 2-D space; I've set it into 3-D space by simply adding a zero-valued Z component to each of the coordinates. The rotation then commits the figure to a 3-D space.

Steps 2, 3 and 4 apply the Y-axis rotation equation, Equation 14-2, to each of the three points in the figure. Step 5 represents an analysis of the reference version shown in Fig. 14-2, and Step 6 carries out a similar analysis for the rotated version of the line. See the rotated version plotted for you in Fig. 14-3.

A comparison of the direction cosines shows that they are quite different in some instances. Nevertheless, the angles between lines remain unchanged by the rotation.

See what you can learn by working through Exercise 14-3.

Exercise 4-3

1. Rotate the reference figure cited in Example 4-3 through an angle of 45 degrees about the X axis; about the Z axis. Plot and analyze the rotated versions, comparing the information with that of the reference version.

2. Rotate some plane figure 45 degrees about the X axis, then take the result of that rotation and use it as a reference figure for a rotation of 45 degrees about the Y axis. Plot and analyze the figure at each step along the way.

3. Consult your personal file of plane figures specified in 2-D space (or generate some by consulting Chapter 9), and then embed them in 3-D space by simply adding a Z component of zero to each coordinate. Rotate the figures into 3-D space by executing some angle of rotation about the X or Y axis. Conduct analyses in each instance and file the results for future reference.

14-4 ROTATING SPACE OBJECTS IN 3-D SPACE

You have seen that virtually every principle from 2-D rotations carries directly over to the 3-D situations. Confirm your understanding by studying the analysis offered here as part of Example 14-4.

Example 14-4

Here are the specifications for a 4-sided pyramid in 3-D space:

$$S = \overline{F_1F_2F_3F_4}$$

$F_1 = \overline{L_1L_2L_3}$	$L_1 = \overline{P_1P_2}$	$P_1 = (0,0,0)$
$F_2 = \overline{L_1L_4L_5}$	$L_2 = \overline{P_2P_3}$	$P_2 = (4,0,0)$
$F_3 = \overline{L_2L_5L_6}$	$L_3 = \overline{P_3P_1}$	$P_3 = (0,0,4)$
$F_4 = \overline{L_3L_4L_6}$	$L_4 = \overline{P_1P_4}$	$P_4 = (0,4,0)$
	$L_5 = \overline{P_2P_4}$	
	$L_6 = \overline{P_3P_4}$	

Rotate it 45 degrees about the X axis.

1. Gathering the given information:

$x_1 = 0$	$y_1 = 0$	$z_1 = 0$	$\theta = 45°$
$x_2 = 4$	$y_2 = 0$	$z_2 = 0$	$\sin\theta = \sqrt{2}/2$
$x_3 = 0$	$y_3 = 0$	$z_3 = 4$	$\cos\theta = \sqrt{2}/2$
$x_4 = 0$	$y_4 = 4$	$z_4 = 0$	

See the analysis of this reference object in Table 14-1.

2. Applying the 45-degree, X-axis rotation, the coordinates for the points in the object become:

Table 14-1. The Summary of the Analysis of the Reference Space Object in Fig. 14-4.

(A, B, C, and D correspond to α, β, γ, and δ.)

SPECIFICATIONS

$S=\overline{F_1F_2F_3F_4}$

$F_1=\overline{L_1L_2L_3}$	$L_1=\overline{P_1P_2}$	$P_1=(0,0,0)$
$F_2=\overline{L_1L_4L_5}$	$L_2=\overline{P_2P_3}$	$P_2=(4,0,0)$
$F_3=\overline{L_2L_5L_6}$	$L_3=\overline{P_3P_1}$	$P_3=(0,0,4)$
$F_4=\overline{L_3L_4L_6}$	$L_4=\overline{P_1P_4}$	$P_4=(0,4,0)$
	$L_5=\overline{P_2P_4}$	
	$L_6=\overline{P_3P_4}$	

LENGTHS

$s_1=4$	$s_2=4\sqrt{2}$	$s_3=4$
$s_4=4$	$s_5=4\sqrt{2}$	$s_6=4\sqrt{2}$

DIRECTION COSINES

$\cos A_1=1$	$\cos B_1=0$	$\cos C_1=0$
$\cos A_2=-\sqrt{2}/2$	$\cos B_2=0$	$\cos C_2=\sqrt{2}/2$
$\cos A_3=0$	$\cos B_3=0$	$\cos C_3=-1$
$\cos A_4=0$	$\cos B_4=-1$	$\cos C_4=0$
$\cos A_5=-\sqrt{2}/2$	$\cos B_5=\sqrt{2}/2$	$\cos C_5=0$
$\cos A_6=0$	$\cos B_6=\sqrt{2}/2$	$\cos C_6=-\sqrt{2}/2$

ANGLES

$\phi_{1,2}=45^\circ$	$\phi_{1,3}=90^\circ$	$\phi_{1,4}=90^\circ$
$\phi_{1,5}=45^\circ$	$\phi_{2,3}=45^\circ$	$\phi_{2,5}=60^\circ$
$\phi_{2,6}=60^\circ$	$\phi_{3,4}=90^\circ$	$\phi_{3,6}=45^\circ$
$\phi_{4,5}=45^\circ$	$\phi_{4,6}=45^\circ$	$\phi_{5,6}=60^\circ$

$$P_1{}' = (0,0,0)$$
$$P_2{}' = (4,0,0)$$
$$P_3{}' = (0,\sqrt{2}/2,\sqrt{2}/2)$$
$$P_4{}' = (0,-\sqrt{2}/2,\sqrt{2}/2)$$

3. Conduct an analysis of the reference version of the object by calculating the lengths and direction cosines of the six lines and the relevant angles between them. See that information summarized in Table 14-1 and its X-Y perspective plot in Fig. 14-4.

4. Conduct an analysis of the rotated version of the object: lengths of lines, direction cosines and angles. See Table 14-2 and Fig. 14-5.

The example cites the specifications for a reference space object—a 4-sided pyramid in this case. Step 1 gathers the relevant information and shows the sine and cosine values for the 45-degree angle of rotation.

Step 2 summarizes the result of applying the X-axis rotation equation (Equation 14-3) to all four points in the object. Check my work by carrying out the mathematics for yourself.

You can see the reference object plotted in Fig. 14-4 and a summary of its analysis in Table 14-1. The plot and summary of analysis for the rotated version appear in Fig. 14-5 and Table 14-2.

Try your hand at the experiments in Exercise 14-4.

Exercise 14-4

1. Apply the following combinations of angles and axes of rotations to the reference space object already

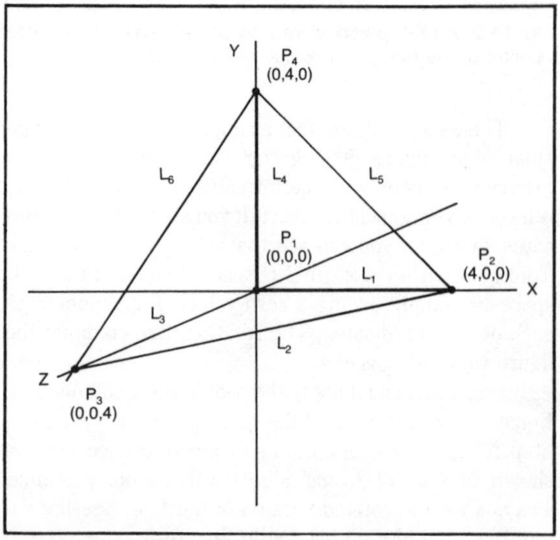

Fig. 14-4. A perspective view of a reference pyramid specified in Example 14-4. See a summary of its analysis in Table 14-1.

Table 14-2. A Summary of the Analysis of the Rotated Space Object in Fig. 14-5.

(A, B, C, and D correspond to α, β, γ, and δ.)

SPECIFICATIONS

$S'=\overline{F'_1F'_2F'_3F'_4}$

$F'_1=\overline{L'_1L'_2L'_3}$	$L'_1=\overline{P'_1P'_2}$	$P'_1=(0,0,0)$
$F'_2=\overline{L'_1L'_4L'_5}$	$L'_2=\overline{P'_2P'_3}$	$P'_2=(4,0,0)$
$F'_3=\overline{L'_2L'_5L'_6}$	$L'_3=\overline{P'_3P'_1}$	$P'_3=(0,0,4)$
$F'_4=\overline{L'_3L'_4L'_6}$	$L'_4=\overline{P'_1P'_4}$	$P'_4=(0,4,0)$
	$L'_5=\overline{P'_2P'_4}$	
	$L'_6=\overline{P'_3P'_4}$	

LENGTHS

$s'_1=4$	$s'_2=4\ 2$	$s'_3=4$
$s'_4=4$	$s'_5=4\ 2$	$s'_6=4\ 2$

DIRECTION COSINES

$\cos A'_1=1$	$\cos B'_1=0$	$\cos C'_1=0$
$\cos A'_2=-\sqrt{2}/2$	$\cos B'_2=-1/2$	$\cos C'_2=1/2$
$\cos A'_3=0$	$\cos B'_3=\sqrt{2}/2$	$\cos C'_3=-\sqrt{2}/2$
$\cos A'_4=0$	$\cos B'_4=\sqrt{2}/2$	$\cos C'_4=\sqrt{2}/2$
$\cos A'_5=-\ 2/2$	$\cos B'_5=1/2$	$\cos C'_5=1/2$
$\cos A'_6=0$	$\cos B'_6=1$	$\cos C'_6=0$

ANGLES

$\phi'_{1,2}=45^\circ$	$\phi'_{1,3}=90^\circ$	$\phi'_{1,4}=90^\circ$
$\phi'_{1,5}=45^\circ$	$\phi'_{2,3}=45^\circ$	$\phi'_{2,5}=60^\circ$
$\phi'_{2,6}=60^\circ$	$\phi'_{3,4}=90^\circ$	$\phi'_{3,6}=45^\circ$
$\phi'_{4,5}=45^\circ$	$\phi'_{4,6}=45^\circ$	$\phi'_{5,6}=60^\circ$

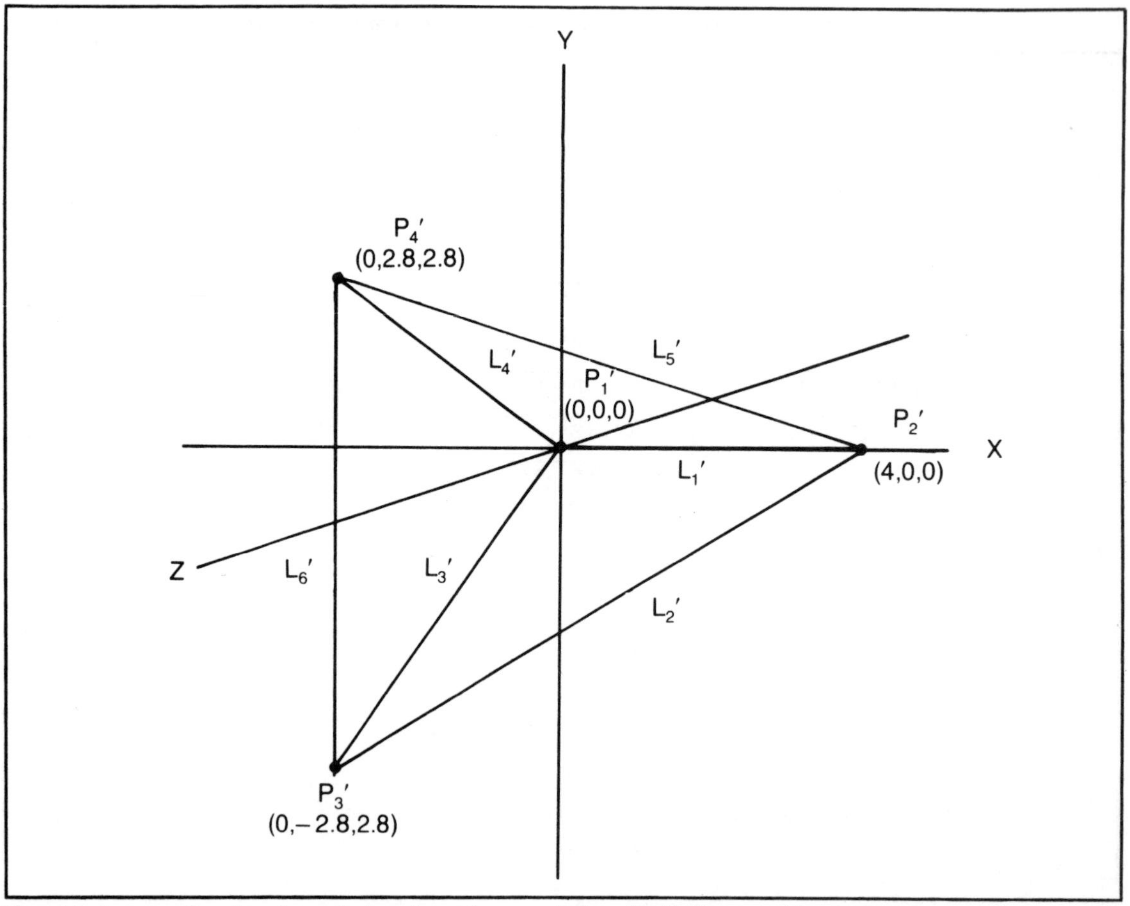

Fig. 14-5. A perspective view of a rotated version of the pyramid described in Example 14-4. See a summary of its analysis in Table 14-2.

specified for you in Example 14-4. Compare the X-Y perspective plots.

A. 45 degrees about the Y axis
B. 45 degrees about the Z axis
C. 90 degrees about the X axis
D. 30 degrees about the Y axis

2. Apply various combinations of angles and axes of rotation to the cube specified here:

$S = \overline{F_1F_2F_3F_4F_5F_6}$

$F_1 = \overline{L_1L_2L_3L_4}$	$L_1 = \overline{P_1P_2}$	$P_1 = (–,1–1,1)$
$F_2 = \overline{L_5L_6L_7L_8}$	$L_2 = \overline{P_2P_3}$	$P_2 = (1,–1,1)$
$F_3 = \overline{L_1L_5L_9L_{10}}$	$L_3 = \overline{P_3P_4}$	$P_3 = (1,1,1)$
$F_4 = \overline{L_2L_6L_8L_{10}}$	$L_4 = \overline{P_4P_1}$	$P_4 = (–1,1,1)$
$F_5 = \overline{L_3L_7L_{11}L_{12}}$	$L_5 = \overline{P_5P_6}$	$P_5 = (–1,–1,–1)$
$F_6 = \overline{L_4L_8L_9L_{12}}$	$L_6 = \overline{P_6P_7}$	$P_6 = (1,–1,–1)$
	$L_7 = \overline{P_7P_8}$	$P_8 = (1,1–1)$
	$L_8 = \overline{P_5P_8}$	
	$L_9 = \overline{P_1P_5}$	
	$L_{10} = \overline{P_2P_6}$	
	$L_{11} = \overline{P_3P_7}$	
	$L_{12} = \overline{P_4P_8}$	

Hint: For the purposes of analysis, the relevant angles in this particular space object are:

$\phi_{1,2}$	$\phi_{1,4}$	$\phi_{1,9}$	$\phi_{1,10}$
$\phi_{2,3}$	$\phi_{2,10}$	$\phi_{2,11}$	$\phi_{3,4}$
$\phi_{3,11}$	$\phi_{3,12}$	$\phi_{4,9}$	$\phi_{4,12}$
$\phi_{5,6}$	$\phi_{5,8}$	$\phi_{5,9}$	$\phi_{5,10}$
$\phi_{6,7}$	$\phi_{6,10}$	$\phi_{6,11}$	$\phi_{7,8}$
$\phi_{7,11}$	$\phi_{7,12}$	$\phi_{8,9}$	$\phi_{8,12}$

14-5. ALTERNATIVE VERSIONS OF THE GEOMETRIC ROTATIONS

Equations 14-4 through 14-6 are the inverse forms of the rotation equations for 3-D space. You can use them for determining the coordinate of reference points relative to the frame of reference, given the angle and axis of rotation, and the coordinate of the rotated version of the point.

$$x_n = x_n'\cos\theta + y_n'\sin\theta$$
$$y_n = -x_n'\sin\theta + y_n'\cos\theta \quad \textbf{Equation 14-4}$$
$$z_n = z_n'$$

where:

x_n,y_n,z_n = the coordinate of point *n* relative to the frame of reference

x_n',y_n',z_n' =the coordinate of the rotated version of point *n* relative to the frame of reference

θ = the angle of rotation about the Z axis relative to the frame of reference.

Note: This is the inverse version of Equation 14-1.

$$x_n = x_n'\cos\theta + z_n'\sin\theta$$
$$y_n = y_n' \quad \textbf{Equation 14-5}$$
$$z_n = -x_n'\sin\theta + z_n'\cos\theta$$

where:

x_n,y_n,z_n = the coordinate of point *n* relative to the frame of reference

x_n',y_n',z_n' = the coordinate of the rotated version of point *n* relative to the frame of reference

θ = the angle of rotation about the Y axis relative to the frame of reference

Note: This is the inverse version of Equation 14-2.

$$x_n = x_n'$$
$$y_n = y_n'\cos\theta + z_n'\sin\theta \quad \textbf{Equation 14-6}$$
$$z_n = -y_n'\sin\theta + z_n'\cos\theta$$

where:

x_n,y_n,z_n = the coordinate of point *n* relative to the frame of reference

x_n',y_n',z_n' = the coordinate of the rotated version of point *n* relative to the frame of reference

θ = the angle of rotation about the X axis relative to the frame of reference

Note: This is the inverse version of Equation 14-3.

One of the important features of these inverse rotation equations is that they lead directly to those expressing relativistic rotation experiments in 3-D space.

14-6 RELATIVISTIC 3-D ROTATIONS

Suppose that you have set up a pair of 3-D coordinate systems in such a way that some rotation exists between them. Furthermore, you have specified a space object relative to the coordinate system you have chosen to be the observer's frame of reference. How does that space object appear from the rotated frame of reference? Answering such questions is the purpose of the relativistic forms of the 3-D rotation equations.

Equations 14-7 through 14-9 assume that a known angle and axis of rotation exists between the observer's frame of reference and the rotated frame of reference. The choice of the angle determines the value of theta, and the choice of an axis of rotation determines which of the three equations must be applied: Equation 14-7 for a Z-axis rotation, Equation 14-8 for a Y-axis rotation, and Equation 14-9 for an X-axis rotation. Then given the coordinate of at least one off-origin point, relative to the observer's frame of reference, the equation cranks out the coordinate relative to the rotated frame of reference.

$$x_n* = x_n\cos\theta + y_n\sin\theta$$
$$y_n* = -x_n\sin\theta + y_n\cos\theta \quad \textbf{Equation 14-7}$$
$$z_n* = z_n$$

where:

x_n*,y_n*,z_n* = the coordinate of point *n* relative to the rotated frame of reference

x_n,y_n,z_n = the coordinate of point *n* relative to the observer's frame of reference

θ = the Z-axis rotation that exists between the observer's and rotated frames of reference, relative to the observer's frame of reference

Note: These expressions apply only when the origins of the two frames of reference are identical and the rotation is about the observer's Z axis.

$$x_n* = x_n\cos\theta + z_n\sin\theta$$
$$y_n* = y_n \quad \textbf{Equation 14-8}$$
$$z_n* = -x_n\sin\theta + z_n\cos\theta$$

where:

x_n*,y_n*,z_n* = the coordinate of point *n* relative to the rotated frame of reference

x_n,y_n,z_n = the coordinate of point *n* relative to the observer's frame of reference

θ = the Y-axis rotation that exists between the observer's and rotated frames of reference, relative to the observer's frame of reference

Note: These expressions apply only when the origins of the two frames of reference are superimposed and the rotation is about the observer's Y axis.

$$x_n* = x_n$$
$$y_n* = y_n\cos\theta + z_n\sin\theta \quad \textbf{Equation 14-9}$$
$$z_n* = -y_n\sin\theta + z_n\cos\theta$$

where:

x_n*, y_n*, z_n* = the coordinate of point *n* relative to the rotated frame of reference

x_n, y_n, z_n = the coordinate of point *n* relative to the observer's frame of reference

θ = the X-axis rotation that exists between the observer's and rotated frames of reference, relative to the observer's frame of reference

Note: These expressions apply only when the origins of the two frames of reference are superimposed and the rotation is about the observer's X axis.

Study Example 14-5 thoroughly, first concentrating on the technique, then again paying special attention to the interpretation of the ideas.

Example 14-5

You are observing a space object as specified below. What are the specifications for that same object as viewed from a frame of reference that is rotated 30 degrees about your X axis?

$$S = \overline{F_1F_2F_3F_4F_5}$$

$$F_1 = \overline{L_1L_2L_3L_4} \quad L_1 = \overline{P_1P_2} \quad P_1 = (-1,0,1)$$
$$F_2 = \overline{L_1L_5L_6} \quad L_2 = \overline{P_2P_3} \quad P_2 = (1,0,1)$$
$$F_3 = \overline{L_2L_6L_7} \quad L_3 = \overline{P_3P_4} \quad P_3 = (1,0,-1)$$
$$F_4 = \overline{L_3L_7L_8} \quad L_4 = \overline{P_4P_1} \quad P_4 = (-1,0,-1)$$
$$F_5 = \overline{L_4L_5L_8} \quad L_5 = \overline{P_1P_5} \quad P_5 = (0,1,0)$$
$$L_6 = \overline{P_2P_5}$$
$$L_7 = \overline{P_3P_5}$$
$$L_8 = \overline{P_4P_5}$$

Using Equation 14-9 to determine the coordinates of the points as viewed from the rotated frame of reference, the specifications become:

$$S* = \overline{F_1*F_2*F_3*F_4*F_5*}$$

$$F_1* = \overline{L_1*L_2*L_3*L_4*} \quad L_1* = \overline{P_1*P_2*}$$
$$F_2* = \overline{L_1*L_5*L_6*} \quad L_2* = \overline{P_2*P_3*}$$
$$F_3* = \overline{L_2*L_6*L_7*} \quad L_3* = \overline{P_3*P_4*}$$
$$F_4* = \overline{L_3*L_7*L_8*} \quad L_4 = \overline{P_4*P_1*}$$
$$F_5* = \overline{L_4*L_5*L_8*} \quad L_5* = \overline{P_1*P_5*}$$
$$L_6* = \overline{P_2*P_5*}$$
$$L_7* = \overline{P_3*P_5*}$$
$$L_8* = \overline{P_4*P_5*}$$

$$P_1* = (-1, 1/2, \sqrt{3}/2)$$
$$P_2* = (1, 1/2, \sqrt{3}/2)$$
$$P_3* = (1, -1/2, -\sqrt{3}/2)$$
$$P_4* = (-1, -1/2, -\sqrt{3}/2)$$
$$P_5* = (0, \sqrt{3}/2, -1/2)$$

Work some experiments of your own.

14-7 SUCCESSIVE 3-D ROTATIONS

Successive rotations in 3-D space about the same axis are commutative. You can perform a certain series of rotations about the X plane, for instance, and come up with the same overall rotation, regardless of the order to choose to apply those rotations.

Successive rotations about different axes, however, are not commutative. Suppose that you execute a rotation about the X axis, followed by one about the Z axis. That will yield a particular result; but you are likely to get an entirely different result if you perform the Z-axis rotation first.

Try an experiment that shows that successive rotations about different axes are not commutative. Specify some point in 3-D space (other than the origin), then apply a rotation of 45 degrees about the X axis, followed by a rotation of 45 degrees about the Y axis. Keep track of the results, and then rotate that same reference point 45 degrees, first about the Y axis and then about the X axis. The results of the two phases of the experiment ought to be different.

That experiment amounts to an adequate negative proof—that combinations of rotations about different axes are not commutative. Such a proof requires only one particular instance. A more sophisticated approach to the idea is to use some mathematical work to determine the special conditions of angles and axes of rotation that produce identical results, regardless of the order of execution.

Experiments in Three Dimensions

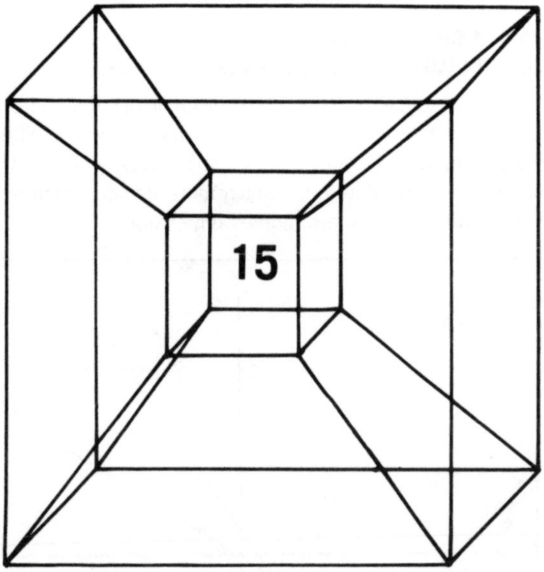

This chapter is both a summary and an extension of what you've learned so far about 3-D space. It serves the same purpose for studies in 3-D space that Chapter 9 does for 3-D space.

Three-dimensional space, as defined in this book, allows four different kinds of geometric entities: points, lines, plane figures, and space objects. And you should understand by now that that sequence of entities represents an evolutionary development from the most primitive to the most sophisticated kinds of entities in 3-D space.

Points are the most elementary entities in space of any number of dimensions, and they are defined in the most elementary terms—in terms of their coordinates in a coordinate system.

Lines are defined and specified in terms of the points that mark their ends. A line is defined in a general way in terms of its endpoints, and specifically in terms of the coordinates of its endpoints.

Plane figures are next up on the evolutionary ladder, because they are directly specified in terms of the lines that bound them. But since the lines, themselves, are defined in terms of points, it follows that plane figures are also defined ultimately in terms of coordinates.

Finally, space objects—the natural products of a 3-D space—are first defined in terms of plane figures that bound them. Then, of course, the planes are defined in terms of their bounding lines, and the lines are defined in terms of their endpoint coordinates.

The essential character of every geometric entity ultimately rests with the specifications for the points that comprise it. You have seen in past experiments that translating, scaling or rotating a geometric entity is a matter of dealing with the coordinates of its points.

The S-F-L-P specifications for objects in 3-D space also reflect the evolution of geometric entities. The P specifications cite the coordinate of points, the L specifications cite the points that bound lines, the F specifications cite the lines that bound a plane figure, and the S specifications cite the planes that bound the space object at hand.

And in keeping with the idea that the character of an entity, especially when it is considered the object of a geometric or relativistic transformation, rests solely with the manipulation of the coordinates of its points; the S, F and L specifications remain unchanged.

The following discussions emphasize these ideas, and extend them to include the formal analysis of the entities. The main purpose is to demonstrate the logical simplicity and consistency of both the specifications and analyses of geometric entities in 3-D space. A secondary purpose is to provide convenient guides for composing a family of useful geometric entities and conducting complete analyses of them.

15-1 A DIRECTORY OF LOWER-DIMENSIONED ENTITIES IN 3-D SPACE

Three-dimensional space is the home of points, lines and plane figures as well as 3-D space objects. The lower-dimensioned entities can be immersed in a 3-D space and manipulated in a much more dynamic fashion than is possible in their "home" dimensions.

15-1.1 Basic Lines in 3-D Space

Figure 15-1 and Table 15-1 plot and summarize the specifications and both the primary and the secondary analyses of any line in 3-D space. The line is defined in terms of two general endpoints, and the endpoints are defined in terms of their 3-D coordinates. That completes the specifications—the L-P specifications—for any line that is embedded in a 3-D space.

The analysis of a line in 3-D space ought to include its length and the cosines of its direction angles alpha, beta and gamma. As far as the primary analysis is concerned, those line parameters should appear in their true,

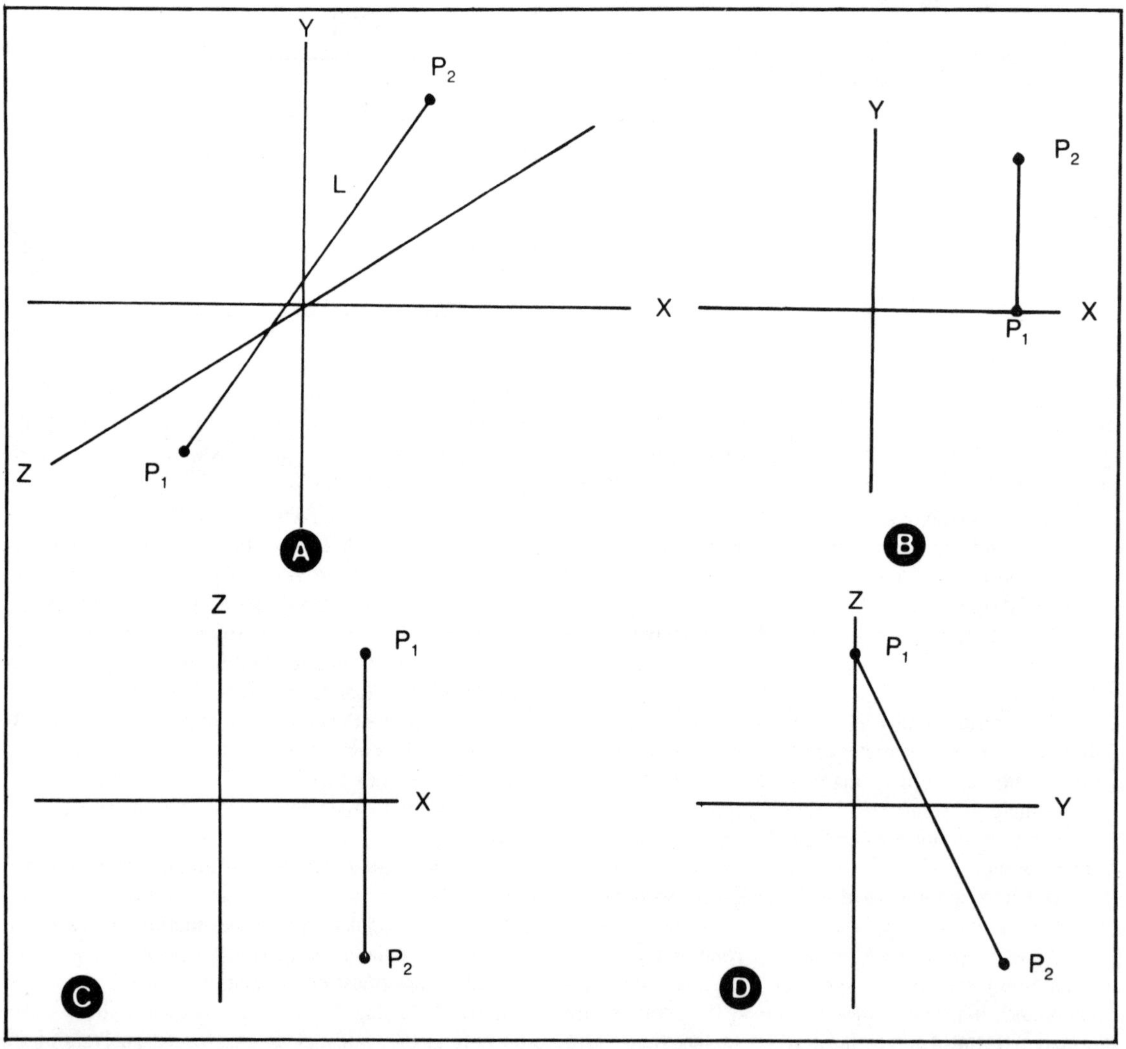

Fig. 15-1. A line specified in 3-D space. (A) X-Y perspective view. (B) X-Y planar view. (C) X-Z planar view. (D) Y-Z planar view. See the outlines of analyses in Table 15-1.

Table 15-1. Specifications and Analyses of a Line in 3-D Space. See Fig. 15-1. (A) Specifications and Analysis in XYZ Space. (B) Secondary Analysis in the X-Y Plane. (C) Secondary Analysis in the X-Z Plane. (D) Secondary Analysis in the Y-Z Plane.

(A, B, C, and D correspond to α, β, γ, and δ.)

SPECIFICATIONS

$L=\overline{P_1P_2}$

$P_1=(x_1,y_1,z_1)$

$P_2=(x_2,y_2,z_1)$

LENGTH

$$s=\sqrt{(x_2-x_1)^2+(y_2-y_1)^2+(z_2-z_1)^2}$$

DIRECTION COSINES

$\cos A=(x_2-x_1)/s$

$\cos B=(y_2-y_1)/s$

$\cos C=(z_2-z_1)/s$

A

SECONDARY ANALYSIS -- XY PLANE

PLOTTED POINTS

$P_1=(x_1,y_1)$

$P_2=(x_2,y_2)$

LENGTH

$$s=\sqrt{(x_2-x_1)^2+(y_2-y_1)^2}$$

DIRECTION COSINES

$\cos A=(x_2-x_1)/s$

$\cos B=(y_2-y_1)/s$

B

SECONDARY ANALYSIS -- XZ PLANE

PLOTTED POINTS

$P_1=(x_1,z_1)$

$P_2=(x_2,z_2)$

LENGTH

$$s=\sqrt{(x_2-x_1)^2+(z_2-z_1)^2}$$

DIRECTION COSINES

$\cos A=(x_2-x_1)/s$

$\cos C=(z_2-z_1)/s$

C

SECONDARY ANALYSIS -- YZ PLANE

PLOTTED POINTS

$P_1=(y_1,z_1)$

$P_2=(y_2,z_2)$

LENGTH

$$s=\sqrt{(y_2-y_1)^2+(z_2-z_1)^2}$$

DIRECTION COSINES

$\cos B=(y_2-y_1)/s$

$\cos C=(z_2-z_1)/s$

D

non-distorted forms. The three secondary analyses, however, cite those parameters as mapped (and generally distorted) to the three planar views; the XY, XZ and YZ views.

In all of the analyses, the length of the line its direction cosines are all calculated from the most primitive elements of the line—the endpoint coordinates. The endpoint coordinates for the primary analysis are the 3-D versions of the coordinates stated in the original specifications. For the XY planar analysis, however, the endpoints are specified in terms of their X and Y components; for the XZ planar analysis, the endpoints are taken from the X and Z components; and for the YZ planar analysis, the components of the endpoints are taken from the original Y and Z components.

So when it comes to specifying and analyzing a line in 3-D space, all you have to do is supply some specific values for the coordinates in the original specifications, plot the line on a 3-D coordinate system, and complete the primary analysis by working out the length and direction cosines. Repeat the procedure for the three planar views to complete the secondary analyses—to see how the line would appear to creatures that are confined to those plane-worlds. And don't be surprised if the line reduces to a single point as plotted and analyzed from one of the planar views. That can happen under certain circumstances.

If you think about the direction cosines for lines in 3-D space a little bit, or carefully compare the results of some of your own experiments, you ought to come up with a couple of potentially useful principles that apply to both the primary and secondary views:

1. A line is parallel to the X axis whenever its alpha direction cosine is equal to 1 or −1, and it is perpendicular to that axis whenever its alpha direction cosine is equal to zero. That means there are angles of zero or 90 degrees, respectively, between the line and the X axis.

2. A line is parallel to the Y axis whenever its beta direction cosine is equal to 1 or −1 and it is perpendicular to the Y axis whenever its beta direction cosine is equal to zero.

3. A line is parallel to the Z axis whenever its gamma direction cosine is equal to 1 or −1 zero, and it is perpendicular to that axis whenever its gamma direction cosine is equal to zero.

Those principles represent just a couple of special cases. Generally speaking, you can always determine the angle between a line and the X axis of its coordinate system by taking the arc cosine of its alpha direction cosine. Likewise, you can figure the angle between the line and its Y axis by taking the arc cosine of its beta direction cosine. What's more, the principles apply to the secondary planar analyses as well as the true primary analysis.

15-1.2 Triangles in 3-D Space

Figure 15-2 and Table 15-2 outline the specifications and procedure of analysis for any triangle in 3-D space. The F specification in the table shows that every triangle is composed of three lines. That specification never changes for triangles. Then the specifications for the three lines show that the figure is composed of just three points. That particular configuration of lines and point specifications ensures the closure of the three lines—a closure that is necessary for making a complete, unbroken, triangular figure in 3-D space.

You can see that the analysis of a triangle is somewhat more involved than that of a single line. You must, for example, calculate the lengths and direction cosines for three different lines. Then additionally, you should figure the angles between the lines—the internal angles of the triangle. The equations actually solve for the cosines of the angles, but of course you know that you can get the angles, themselves, by taking the arc cosine of the results.

Everything that is expressed in the specifications and both levels of analysis rests ultimately with your choice of point coordinates; the form of the specifications and the equations for working out the analyses are virtually identical for every triangular figure in 3-D space.

You can generate any number of different triangles in 3-D space by assigning some carefully selected, specific numbers to the components of the coordinates in the P specifications. See some examples in Table 15-2E.

Even when specifying a set of valid 3-D coordinates for a triangle, the figure quite often reduces to a single line as plotted and analyzed from one of the planar views. That is simply one of the distortion effects that such mappings can create.

While you are conducting formal analyses of some triangles in 3-D space, bear in mind a couple of important ideas that have been outlined in detail in Chapter 9 (ideas that apply equally well to triangles in 2-D and 3-D space): (1) The sum of the three angles in a triangle is 180 degrees, (2) if one of the angles is supposed to be greater than 90 degrees, subtract the calculated value from 180 degrees to determine that angle, (3) if the angle between any two lines happens to be 90 degrees, you have specified a right triangle, and (4) if an angle between two lines calculates to be zero degrees, you've made a mis-

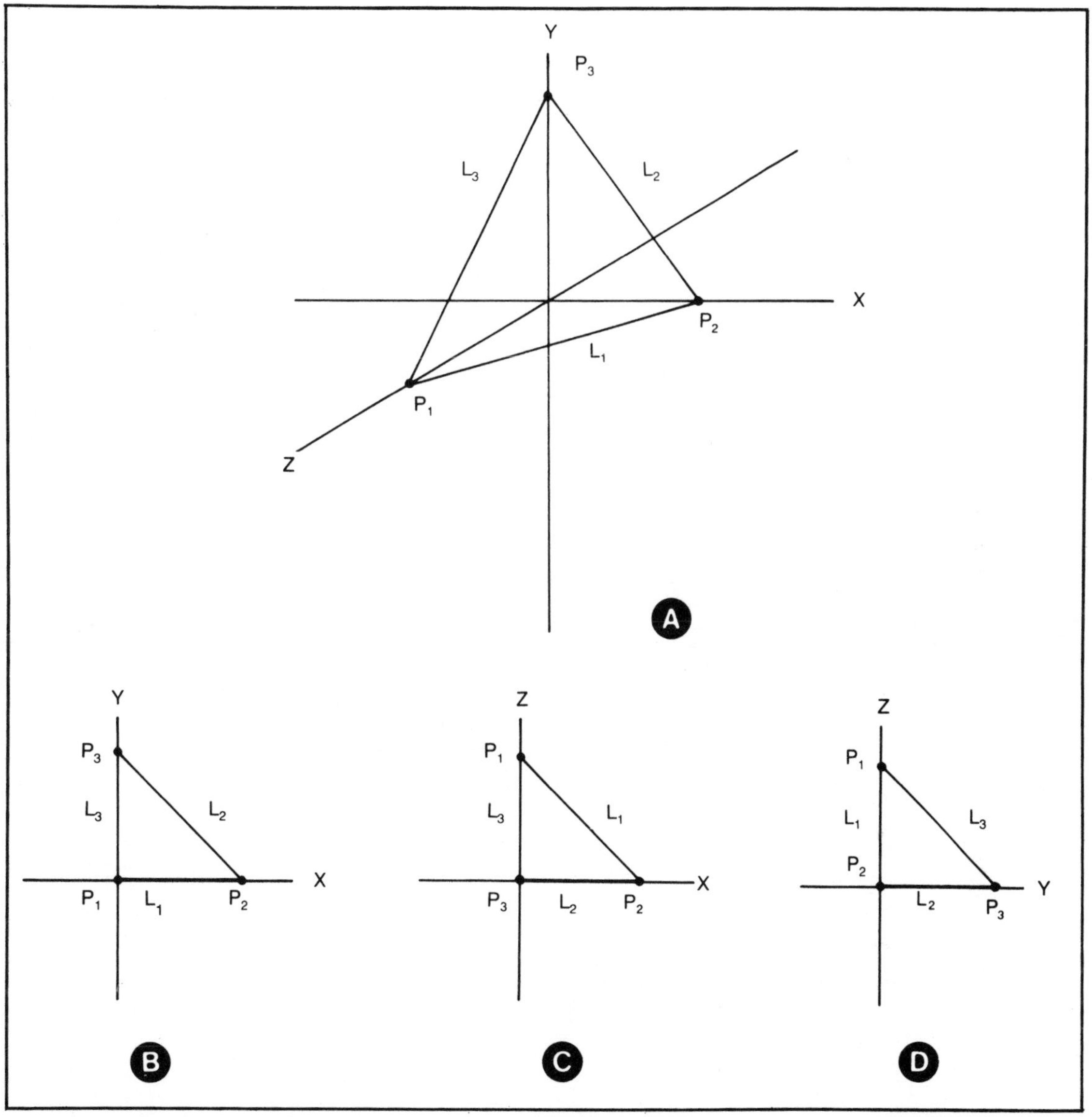

Fig. 15-2. A triangle specified in 3-D space. (A) X-Y perspective view. (B) X-Y planar view. (C) X-Z planar view. (D) Y-Z planar view. See the outlines of analyses in Table 15-2.

take in specifying the original coordinates or carrying out the arithmetic. Add the results of your more interesting studies to your file of figures.

15-1.3 Quadralaterals in 3-D Space

A quadralateral plane figure differs from a triangle only by having four points, lines and angles instead of three (Fig. 15-3). The F specification in Table 15-3 shows the three lines, and the L specifications define those lines in terms of their 3-D endpoints. This particular set of specifications are designed to ensure complete closure of the figure. Refer to some of the sets of coordinates suggested in Table 15-3E, and the general specifications will always lead you to a valid 4-sided plane figure immersed in 3-D space.

The primary and secondary analyses includes the

Table 15-2. Specifications and Analyses of a Triangle in 3-D Space. See Fig. 15-2. (A) Specifications and Analysis in XYZ Space. (B) Secondary Analysis in the X-Y Plane. (C) Secondary Analysis in the X-Z Plane. (D) Secondary Analysis in the Y-Z Plane. (E) Suggested Coordinates.

(A, B, C, and D correspond to α, β, γ, and δ.)

A

SPECIFICATIONS

$F=\overline{L_1L_2L_3}$

$L_1=\overline{P_1P_2}$	$P_1=(x_1,y_1,z_1)$
$L_2=\overline{P_2P_3}$	$P_2=(x_2,y_2,z_2)$
$L_3=\overline{P_3P_1}$	$P_3=(x_3,y_3,z_3)$

LENGTH

$$s_1=\sqrt{(x_2-x_1)^2+(y_2-y_1)^2+(z_2-z_1)^2}$$

$$s_2=\sqrt{(x_3-x_2)^2+(y_3-y_2)^2+(z_3-z_2)^2}$$

$$s_3=\sqrt{(x_1-x_3)^2+(y_1-y_3)^2+(z_1-z_3)^2}$$

DIRECTION COSINES

$\cos A_1=(x_2-x_1)/s_1$	$\cos B_1=(y_2-y_1)/s_1$	$\cos C_1=(z_2-z_1)/s_1$
$\cos A_2=(x_3-x_2)/s_2$	$\cos B_2=(y_3-y_2)/s_2$	$\cos C_2=(z_3-z_2)/s_2$
$\cos A_3=(x_1-x_3)/s_3$	$\cos B_3=(y_1-y_3)/s_3$	$\cos C_3=(z_1-z_3)/s_3$

ANGLES

$$\cos\phi_{1,2}=\cos A_1\cos A_2+\cos B_1\cos B_2+\cos C_1\cos C_2$$

$$\cos\phi_{1,3}=\cos A_1\cos A_3+\cos B_1\cos B_3+\cos C_1\cos C_3$$

$$\cos\phi_{2,3}=\cos A_2\cos A_3+\cos B_2\cos B_3+\cos C_2\cos C_3$$

CHECK

$$\phi_{1,2}+\phi_{1,3}+\phi_{2,3}=180^\circ$$

B

SECONDARY ANALYSIS -- XY PLANE

PLOTTED POINTS

$P_1=(x_1,y_1)$

$P_2=(x_2,y_2)$

$P_3=(x_3,y_3)$

LENGTH

$$s_1=\sqrt{(x_2-x_1)^2+(y_2-y_1)^2}$$

$$s_2=\sqrt{(x_3-x_2)^2+(y_3-y_2)^2}$$

$$s_3=\sqrt{(x_1-x_3)^2+(y_1-y_3)^2}$$

DIRECTION COSINES

B

$\cos A_1=(x_2-x_1)/s_1$ $\cos B_1=(y_2-y_1)/s_1$

$\cos A_2=(x_3-x_2)/s_2$ $\cos B_2=(y_3-y_2)/s_2$

$\cos A_3=(x_1-x_3)/s_3$ $\cos B_3=(y_1-y_3)/s_3$

ANGLES

$$\cos\phi_{1,2}=\cos A_1\cos A_2+\cos B_1\cos B_2$$

$$\cos\phi_{1,3}=\cos A_1\cos A_3+\cos B_1\cos B_3$$

$$\cos\phi_{2,3}=\cos A_2\cos A_3+\cos B_2\cos B_3$$

CHECK

$$\phi_{1,2}+\phi_{1,3}+\phi_{2,3}=180^\circ$$

SECONDARY ANALYSIS -- XZ PLANE

PLOTTED POINTS

$P_1=(x_1,z_1)$

$P_2=(x_2,z_2)$

C

$P_3=(x_3,z_3)$

LENGTH

$$s_1=\sqrt{(x_2-x_1)^2+(z_2-z_1)^2}$$

$$s_2=\sqrt{(x_3-x_2)^2+(z_3-z_2)^2}$$

$$s_3=\sqrt{(x_1-x_3)^2+(z_1-z_3)^2}$$

Table 15-2. Specifications and Analyses of a Triangle in 3-D Space. See Fig. 15-2. (A) Specifications and Analysis in XYZ Space. (B) Secondary Analysis in the X-Y Plane. (C) Secondary Analysis in X-Z Plane. (D) Secondary Analysis in the Y-Z Plane. (E) Suggested Coordinates. (Continued from page 177.)

(A, B, C, and D correspond to α, β, γ, and δ.)

C

DIRECTION COSINES

$\cos A_1=(x_2-x_1)/s_1$ $\quad$ $\cos C_1=(z_2-z_1)/s_1$

$\cos A_2=(x_3-x_2)/s_2$ $\quad$ $\cos C_2=(z_3-z_2)/s_2$

$\cos A_3=(x_1-x_3)/s_3$ $\quad$ $\cos C_3=(z_1-z_3)/s_3$

ANGLES

$\cos\phi_{1,2}=\cos A_1\cos A_2+\cos C_1\cos C_2$

$\cos\phi_{1,3}=\cos A_1\cos A_3+\cos C_1\cos C_3$

$\cos\phi_{2,3}=\cos A_2\cos A_3+\cos C_2\cos C_3$

CHECK

$\phi_{1,2}+\phi_{1,3}+\phi_{2,3}=180^\circ$

D

SECONDARY ANALYSIS -- YZ PLANE

PLOTTED POINTS

$P_1=(y_1,z_1)$

$P_2=(y_2,z_2)$

$P_3=(y_3,z_3)$

LENGTH

$s_1=\sqrt{(y_2-y_1)^2+(z_2-z_1)^2}$

$s_2=\sqrt{(y_3-y_2)^2+(z_3-z_2)^2}$

$s_3=\sqrt{(y_1-y_3)^2+(z_1-z_3)^2}$

DIRECTION COSINES

$\cos B_1=(y_2-y_1)/s_1$ $\quad$ $\cos C_1=(z_2-z_1)/s_1$

$\cos B_2=(y_3-y_2)/s_2$ $\quad$ $\cos C_2=(z_3-z_2)/s_2$

$\cos B_3=(y_1-y_3)/s_3$ $\quad$ $\cos C_3=(z_1-z_3)/s_3$

D

ANGLES

$\cos\phi_{1,2}=\cos B_1\cos B_2+\cos C_1\cos C_2$

$\cos\phi_{1,3}=\cos B_1\cos B_3+\cos C_1\cos C_3$

$\cos\phi_{2,3}=\cos B_2\cos B_3+\cos C_2\cos C_3$

CHECK

$\phi_{1,2}+\phi_{1,3}+\phi_{2,3}=180^\circ$

E

FIGURE VERSION 1:

$P_1=(0,0,0)$ $P_2=(1,0,1)$ $P_3=(0,1,0)$

FIGURE VERSION 2:

$P_1=(-1,0,-1)$ $P_2=(1,0,1)$ $P_3=(0,1,0)$

FIGURE VERSION 3:

$P_1=(-1,-1,-1)$ $P_2=(2,1,2)$ $P_3=(3,3,3)$

lengths of the four lines, their direction cosines, and values for the four angles between them. The essential difference between them is the components for the coordinates of the points.

While you are working with the suggested coordinates, plotting the figures and conducting complete analyses of them, it will be helpful to bear in mind some special ideas that have been outlined for this sort of figure in Chapter 19: (1) The sum of the four angles in a quadralateral is 360 degrees, (2) two lines in the figure are parallel to one another if the sum of the products of their corresponding direction cosines is 1 or -1, and (3) two lines in the figure are perpendicular to one another if the sum of the products of their corresponding direction cosines is 0.

Like triangles that are mapped to 2-D space, quadralaterals often map as simple lines. Whenever that happens, you ought to regard it as a positive feature of that particular experiment, and not as a disaster.

15-1.4 Pentagons in 3-D Space

It is possible to carry this progression of plane figures immersed in 3-D space to any number of sides, but I am leaving that sort of project to you. The present discussion concludes here with the general 5-sided figure shown in Fig. 15-4.

The specifications in Table 15-4 show that a general pentagon is composed of five bounding lines and five different points. The specifications are identical to those for a pentagon in 2-D space except at one level—the points here require three components instead of just two.

Use the general specifications and the suggested sets of coordinates in Table 15-4E to generate a file of pentagons. Check your work by using the rule that says the sum of the angles in a pentagon is 540 degrees.

15-2 A DIRECTORY OF SPACE OBJECTS IN THREE DIMENSIONS

The following summary deals with a 4-sided space

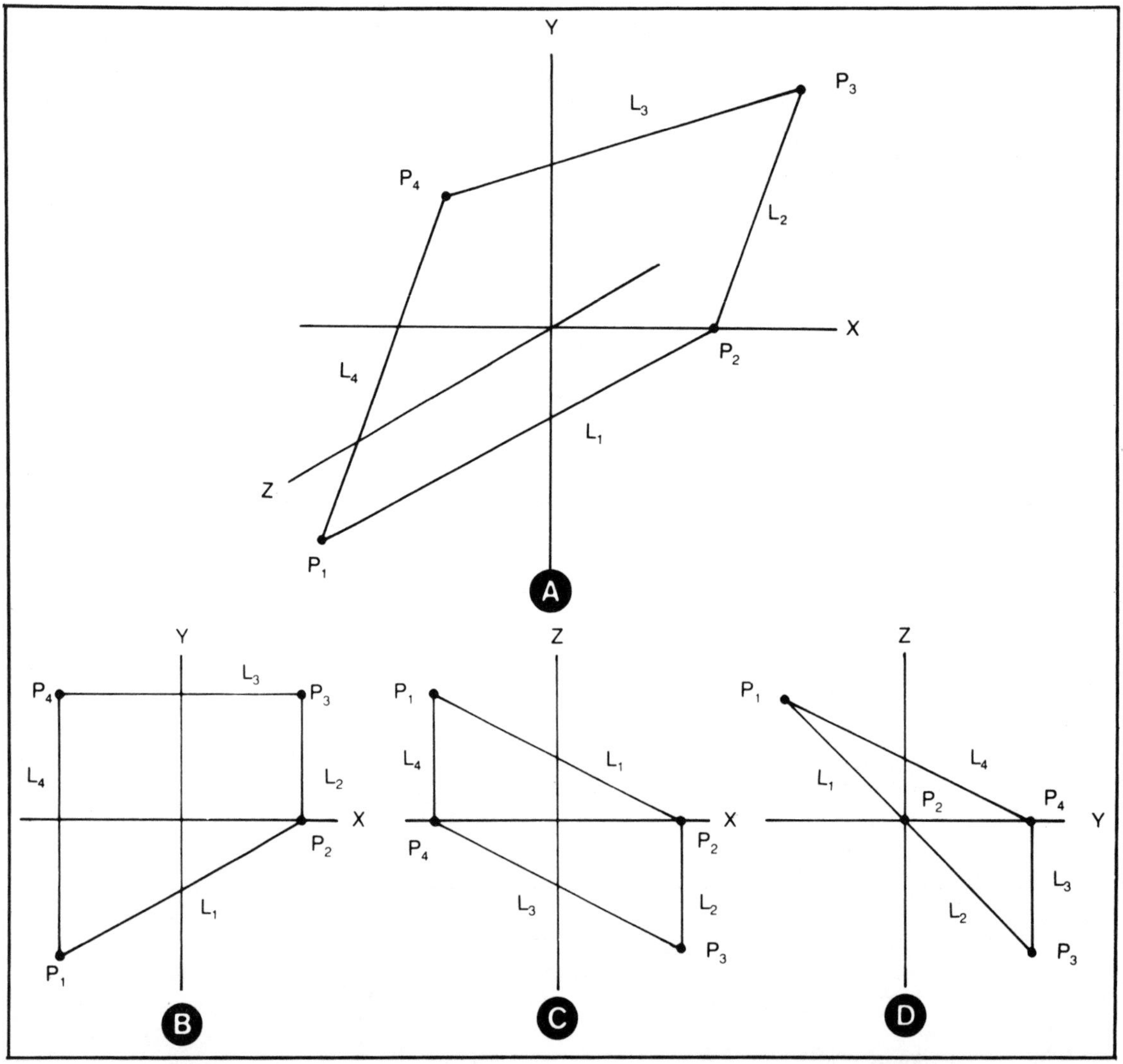

Fig. 15-3. A quadralateral specified in 3-D space. (A) X-Y perspective view. (B) X-Y planar view. (C) X-Z planar view. (D) Y-Z planar view. See the outlines of analyses in Table 15-3.

object, two different kinds of 5-sided objects, a 6-sided object, and a 7-sided object. Given the general specifications and complete guides to full primary and secondary analyses of them, you will be in a position to generate a file of objects of those types. Chapter 16 will then provide the tools for generating any number of different, customized space objects.

15-2.1 The 4-Sided Pyramid

Figure 15-5 shows the 3-D perspective and the three planar views of the simplest possible 3-dimensional space object—a 4-sided pyramid. Table 15-5 shows the general specifications and the outlines for conducting both the primary and secondary levels of analysis.

Notice first that the S specification shows that the object is bounded by 4 plane-figure sides. Then the F specifications show that all four sides are triangular; that they are bounded by three lines apiece. The L specifications indicate the endpoints of the lines, and the P specifications cite the 3-D coordinate for each point.

You can use those specifications to generate any number of variations of this 4-sided pyramid. All you have

Table 15-3. Specifications and Analyses of a Quadralateral in 3-D Space. See Fig. 15-3. (A) Specifications and Analyses in XYZ Space. (B) Secondary Analysis in the X-Y Plane. (C) Secondary Analysis in the X-Z Plane. (D) Secondary Analysis in the Y-Z Plane. (E) Suggested Coordinates.

(A, B, C, and D correspond to α, β, γ, and δ.)

SPECIFICATIONS

$$F=\overline{L_1L_2L_3L_4}$$

$L_1=\overline{P_1P_2}$	$P_1=(x_1,y_1,z_1)$
$L_2=\overline{P_2P_3}$	$P_2=(x_2,y_2,z_2)$
$L_3=\overline{P_3P_4}$	$P_3=(x_3,y_3,z_3)$
$L_4=\overline{P_4P_1}$	$P_4=(x_4,y_4,z_4)$

LENGTH

$$s_1=\sqrt{(x_2-x_1)^2+(y_2-y_1)^2+(z_2-z_1)^2}$$

$$s_2=\sqrt{(x_3-x_2)^2+(y_3-y_2)^2+(z_3-z_2)^2}$$

$$s_3=\sqrt{(x_4-x_3)^2+(y_4-y_3)^2+(z_4-z_3)^2}$$

$$s_4=\sqrt{(x_1-x_4)^2+(y_1-y_4)^2+(z_1-z_4)^2}$$

DIRECTION COSINES

$\cos A_1=(x_2-x_1)/s_1$	$\cos B_1=(y_2-y_1)/s_1$	$\cos C_1=(z_2-z_1)/s_1$
$\cos A_2=(x_3-x_2)/s_2$	$\cos B_2=(y_3-y_2)/s_2$	$\cos C_2=(z_3-z_2)/s_2$
$\cos A_3=(x_4-x_3)/s_3$	$\cos B_3=(y_4-y_3)/s_3$	$\cos C_3=(z_4-z_3)/s_3$
$\cos A_4=(x_1-x_4)/s_4$	$\cos B_4=(y_1-y_4)/s_4$	$\cos C_4=(z_1-z_4)/s_4$

ANGLES

$$\cos\phi_{1,2}=\cos A_1\cos A_2+\cos B_1\cos B_2+\cos C_1\cos C_2$$

$$\cos\phi_{1,4}=\cos A_1\cos A_4+\cos B_1\cos B_4+\cos C_1\cos C_4$$

$$\cos\phi_{2,3}=\cos A_2\cos A_3+\cos B_2\cos B_3+\cos C_2\cos C_3$$

$$\cos\phi_{3,4}=\cos A_3\cos A_4+\cos B_3\cos B_4+\cos C_3\cos C_4$$

CHECK

$$\phi_{1,2}+\phi_{1,4}+\phi_{2,3}+\phi_{3,4}=360^\circ$$

A

Table 15-3. Specifications and Analyses of a Quadralateral in 3-D Space. See Fig. 15-3. (A) Specifications and Analyses in XYZ Space. (B) Secondary Analysis in the X-Y Plane. (C) Secondary Analysis in the X-Z Plane. (D) Secondary Analysis in the Y-Z Plane. (E) Suggested Coordinates. (Continued from page 181.)

SECONDARY ANALYSIS -- XY PLANE

PLOTTED POINTS

$P_1=(x_1,y_1)$

$P_2=(x_2,y_2)$

$P_3=(x_3,y_3)$

$P_4=(x_4,y_4)$

LENGTH

$s_1=\sqrt{(x_2-x_1)^2+(y_2-y_1)^2}$

$s_2=\sqrt{(x_3-x_2)^2+(y_3-y_2)^2}$

$s_3=\sqrt{(x_4-x_3)^2+(y_4-y_3)^2}$

$s_4=\sqrt{(x_1-x_4)^2+(y_1-y_4)^2}$

DIRECTION COSINES

$\cos A_1=(x_2-x_1)/s_1$ $\cos B_1=(y_2-y_1)/s_1$

$\cos A_2=(x_3-x_2)/s_2$ $\cos B_2=(y_3-y_2)/s_2$

$\cos A_3=(x_4-x_3)/s_3$ $\cos B_3=(y_4-y_3)/s_3$

$\cos A_4=(x_1-x_4)/s_4$ $\cos B_4=(y_1-y_4)/s_4$

ANGLES

$\cos\phi_{1,2}=\cos A_1\cos A_2+\cos B_1\cos B_2$

$\cos\phi_{1,4}=\cos A_1\cos A_4+\cos B_1\cos B_4$

$\cos\phi_{2,3}=\cos A_2\cos A_3+\cos B_2\cos B_3$

$\cos\phi_{3,4}=\cos A_3\cos A_4+\cos B_3\cos B_4$

CHECK

$\phi_{1,2}+\phi_{1,4}+\phi_{2,3}+\phi_{3,4}=360^\circ$

B

SECONDARY ANALYSIS -- XZ PLANE

PLOTTED POINTS

$P_1=(x_1,z_1)$

$P_2=(x_2,z_2)$

$P_3=(x_3,z_3)$

$P_4=(x_4,z_4)$

LENGTH

$s_1=\sqrt{(x_2-x_1)^2+(z_2-z_1)^2}$

$s_2=\sqrt{(x_3-x_2)^2+(z_3-z_2)^2}$

$s_3=\sqrt{(x_4-x_3)^2+(z_4-z_3)^2}$

$s_4=\sqrt{(x_1-x_4)^2+(z_1-z_4)^2}$

DIRECTION COSINES

$\cos A_1=(x_2-x_1)/s_1$ $\cos C_1=(z_2-z_1)/s_1$

$\cos A_2=(x_3-x_2)/s_2$ $\cos C_2=(z_3-z_2)/s_2$

$\cos A_3=(x_4-x_3)/s_3$ $\cos C_3=(z_4-z_3)/s_3$

$\cos A_4=(x_1-x_4)/s_4$ $\cos C_4=(z_1-z_4)/s_4$

ANGLES

$\cos\phi_{1,2}=\cos A_1\cos A_2+\cos C_1\cos C_2$

$\cos\phi_{1,4}=\cos A_1\cos A_4+\cos C_1\cos C_4$

$\cos\phi_{2,3}=\cos A_2\cos A_3+\cos C_2\cos C_3$

$\cos\phi_{3,4}=\cos A_3\cos A_4+\cos C_3\cos C_4$

CHECK

$\phi_{1,2}+\phi_{1,4}+\phi_{2,3}+\phi_{3,4}=360^\circ$

C

SECONDARY ANALYSIS YZ PLANE

PLOTTED POINTS

$P_1=(y_1,z_1)$

$P_2=(y_2,z_2)$

$P_3=(y_3,z_3)$

$P_4=(y_4,z_4)$

LENGTH

$s_1=\sqrt{(y_2-y_1)^2+(z_2-z_1)^2}$

$s_2=\sqrt{(y_3-y_2)^2+(z_3-z_2)^2}$

$s_3=\sqrt{(y_4-y_3)^2+(z_4-z_3)^2}$

$s_4=\sqrt{(y_1-y_4)^2+(z_1-z_4)^2}$

DIRECTION COSINES

$\cos B_1=(y_2-y_1)/s_1$ $\quad \cos C_1=(z_2-z_1)/s_1$

$\cos B_2=(y_3-y_2)/s_2$ $\quad \cos C_2=(z_3-z_2)/s_2$

$\cos B_3=(y_4-y_3)/s_3$ $\quad \cos C_3=(z_4-z_3)/s_3$

$\cos B_4=(y_1-y_4)/s_4$ $\quad \cos C_4=(z_1-z_4)/s_4$

ANGLES

$\cos\phi_{1,2}=\cos B_1\cos B_2+\cos C_1\cos C_2$

$\cos\phi_{1,4}=\cos B_1\cos B_4+\cos C_1\cos C_4$

$\cos\phi_{2,3}=\cos B_2\cos B_3+\cos C_2\cos C_3$

$\cos\phi_{3,4}=\cos B_3\cos B_4+\cos C_3\cos C_4$

CHECK

$\phi_{1,2}+\phi_{1,4}+\phi_{2,3}+\phi_{3,4}=360^\circ$

D

FIGURE VERSION 1:

$P_1=(-1,-1,-1)$ $\quad P_2=(1,-1,1)$

$P_3=(1,1,1)$ $\quad P_4=(-1,1,-1)$

FIGURE VERSION 2:

$P_1=(-2,-1,-2)$ $\quad P_2=(2,-1,2)$

$P_3=(2,1,2)$ $\quad P_4=(-2,1,-2)$

FIGURE VERSION 3:

$P_1=(-3,-2,-3)$ $\quad P_2=(1,-2,1)$

$P_3=(1,2,1)$ $\quad P_4=(-1,2,-1)$

FIGURE VERSION 4:

$P_1=(-1,0,-1)$ $\quad P_2=(0,-1,0)$

$P_3=(1,0,1)$ $\quad P_4=(0,1,0)$

E

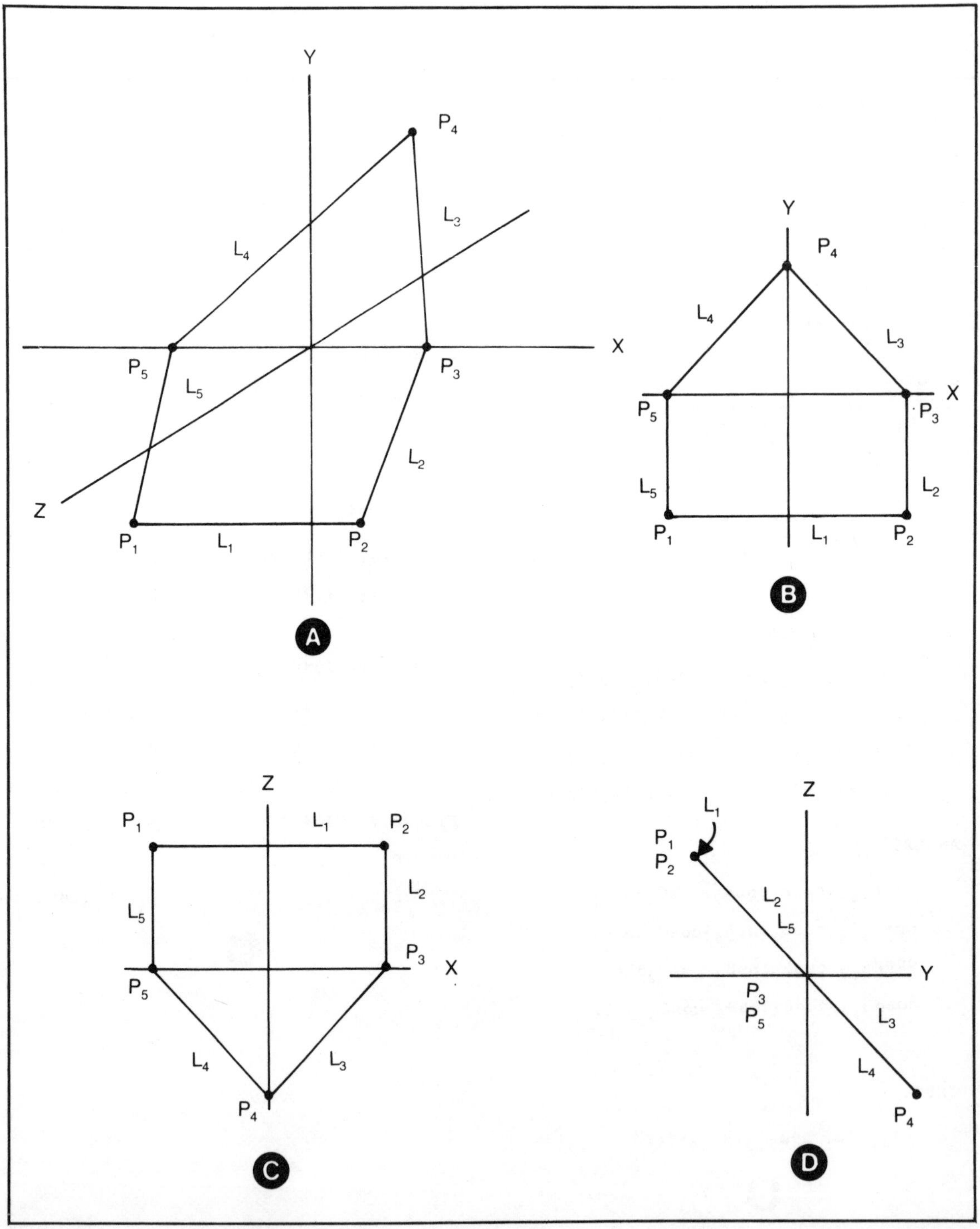

Fig. 15-4. A pentagon specified in 3-D space. (A) X-Y perspective view. (B) X-Y planar view. (C) X-Z planar view. (D) Y-Z planar view. See the outlines of analyses in Table 15-4.

Table 15-4. Specifications and Analyses of a Pentagon in 3-D Space. See Fig. 15-4. (A) Specifications and Analysis in XYZ Space. (B) Secondary Analysis in the X-Y Plane. (C) Secondary Analysis in the X-Z Plane. (D) Secondary Analysis in the Y-Z Plane. (E) Suggested Coordinates.

(A, B, C, and D correspond to α, β, γ, and δ.)

A

SPECIFICATIONS

$$F=\overline{L_1L_2L_3L_4L_5}$$

$L_1=\overline{P_1P_2}$	$P_1=(x_1,y_1,z_1)$
$L_2=\overline{P_2P_3}$	$P_2=(x_2,y_2,z_2)$
$L_3=\overline{P_3P_4}$	$P_3=(x_3,y_3,z_3)$
$L_4=\overline{P_4P_5}$	$P_4=(x_4,y_4,z_4)$
$L_5=\overline{P_5P_1}$	$P_5=(x_5,y_5,z_5)$

LENGTH

$$s_1=\sqrt{(x_2-x_1)^2+(y_2-y_1)^2+(z_2-z_1)^2}$$
$$s_2=\sqrt{(x_3-x_2)^2+(y_3-y_2)^2+(z_3-z_2)^2}$$
$$s_3=\sqrt{(x_4-x_3)^2+(y_4-y_3)^2+(z_4-z_3)^2}$$
$$s_4=\sqrt{(x_5-x_4)^2+(y_5-y_4)^2+(z_5-z_4)^2}$$
$$s_5=\sqrt{(x_1-x_5)^2+(y_1-y_5)^2+(z_1-z_5)^2}$$

DIRECTION COSINES

$\cos A_1=(x_2-x_1)/s_1$	$\cos B_1=(y_2-y_1)/s_1$	$\cos C_1=(z_2-z_1)/s_1$
$\cos A_2=(x_3-x_2)/s_2$	$\cos B_2=(y_3-y_2)/s_2$	$\cos C_2=(z_3-z_2)/s_2$
$\cos A_3=(x_4-x_3)/s_3$	$\cos B_3=(y_4-y_3)/s_3$	$\cos C_3=(z_4-z_3)/s_3$
$\cos A_4=(x_5-x_4)/s_4$	$\cos B_4=(y_5-y_4)/s_4$	$\cos C_4=(z_5-z_4)/s_4$
$\cos A_5=(x_1-x_5)/s_5$	$\cos B_5=(y_1-y_5)/s_5$	$\cos C_5=(z_1-z_5)/s_5$

ANGLES

$$\cos\phi_{1,2}=\cos A_1\cos A_2+\cos B_1\cos B_2+\cos C_1\cos C_2$$
$$\cos\phi_{2,3}=\cos A_2\cos A_3+\cos B_2\cos B_3+\cos C_2\cos C_3$$
$$\cos\phi_{3,4}=\cos A_3\cos A_4+\cos B_3\cos B_4+\cos C_3\cos C_4$$
$$\cos\phi_{4,5}=\cos A_4\cos A_5+\cos B_4\cos B_5+\cos C_4\cos C_5$$
$$\cos\phi_{5,1}=\cos A_5\cos A_1+\cos B_5\cos B_1+\cos C_5\cos C_1$$

CHECK

$$\phi_{1,2}+\phi_{2,3}+\phi_{3,4}+\phi_{4,5}+\phi_{5,1}=540^\circ$$

Table 15-4. Specifications and Analyses of a Pentagon in 3-D Space. See Fig. 15-4. (A) Specifications and Analysis in XYZ Space. (B) Secondary Analysis in the X-Y Plane. (C) Secondary Analysis in the X-Z Plane. (D) Secondary Analysis in the Y-Z Plane. (E) Suggested Coordinates. (Continued from page 185.)

SECONDARY ANALYSIS -- XY PLANE

PLOTTED POINTS

$P_1=(x_1,y_1)$

$P_2=(x_2,y_2)$

$P_3=(x_3,y_3)$

$P_4=(x_4,y_4)$

$P_5=(x_5,y_5)$

LENGTH

$s_1=\sqrt{(x_2-x_1)^2+(y_2-y_1)^2}$

$s_2=\sqrt{(x_3-x_2)^2+(y_3-y_2)^2}$

$s_3=\sqrt{(x_4-x_3)^2+(y_4-y_3)^2}$

$s_4=\sqrt{(x_5-x_4)^2+(y_5-y_4)^2}$

$s_5=\sqrt{(x_1-x_5)^2+(y_1-y_5)^2}$

DIRECTION COSINES

$\cos A_1=(x_2-x_1)/s_1$ $\cos B_1=(y_2-y_1)/s_1$

$\cos A_2=(x_3-x_2)/s_2$ $\cos B_2=(y_3-y_2)/s_2$

$\cos A_3=(x_4-x_3)/s_3$ $\cos B_3=(y_4-y_3)/s_3$

$\cos A_4=(x_5-x_4)/s_4$ $\cos B_4=(y_5-y_4)/s_4$

$\cos A_5=(x_1-x_5)/s_5$ $\cos B_5=(y_1-y_5)/s_5$

ANGLES

$\cos\phi_{1,2}=\cos A_1\cos A_2+\cos B_1\cos B_2$

$\cos\phi_{2,3}=\cos A_2\cos A_3+\cos B_2\cos B_3$

$\cos\phi_{3,4}=\cos A_3\cos A_4+\cos B_3\cos B_4$

$\cos\phi_{4,5}=\cos A_4\cos A_5+\cos B_4\cos B_5$

$\cos\phi_{5,1}=\cos A_5\cos A_1+\cos B_5\cos B_1$

CHECK

$\phi_{1,2}+\phi_{2,3}+\phi_{3,4}+\phi_{4,5}+\phi_{5,1}=540^\circ$

B

SECONDARY ANALYSIS -- XZ PLANE

PLOTTED POINTS

$P_1=(x_1,z_1)$

$P_2=(x_2,z_2)$

$P_3=(x_3,z_3)$

$P_4=(x_4,z_4)$

$P_5=(x_5,z_5)$

LENGTH

$s_1=\sqrt{(x_2-x_1)^2+(z_2-z_1)^2}$

$s_2=\sqrt{(x_3-x_2)^2+(z_3-z_2)^2}$

$s_3=\sqrt{(x_4-x_3)^2+(z_4-z_3)^2}$

$s_4=\sqrt{(x_5-x_4)^2+(z_5-z_4)^2}$

$s_5=\sqrt{(x_1-x_5)^2+(z_1-z_5)^2}$

DIRECTION COSINES

$\cos A_1=(x_2-x_1)/s_1$ $\cos C_1=(z_2-z_1)/s_1$

$\cos A_2=(x_3-x_2)/s_2$ $\cos C_2=(z_3-z_2)/s_2$

$\cos A_3=(x_4-x_3)/s_3$ $\cos C_3=(z_4-z_3)/s_3$

$\cos A_4=(x_5-x_4)/s_4$ $\cos C_4=(z_5-z_4)/s_4$

$\cos A_5=(x_1-x_5)/s_5$ $\cos C_5=(z_1-z_5)/s_5$

ANGLES

$\cos\phi_{1,2}=\cos A_1\cos A_2+\cos C_1\cos C_2$

$\cos\phi_{2,3}=\cos A_2\cos A_3+\cos C_2\cos C_3$

$\cos\phi_{3,4}=\cos A_3\cos A_4+\cos C_3\cos C_4$

$\cos\phi_{4,5}=\cos A_4\cos A_5+\cos C_4\cos C_5$

$\cos\phi_{5,1}=\cos A_5\cos A_1+\cos C_5\cos C_1$

CHECK

$\phi_{1,2}+\phi_{2,3}+\phi_{3,4}+\phi_{4,5}+\phi_{5,1}=540^\circ$

C

SECONDARY ANALYSIS -- YZ PLANE

PLOTTED POINTS

$P_1=(y_1,z_1)$

$P_2=(y_2,z_2)$

$P_3=(y_3,z_3)$

$P_4=(y_4,z_4)$

$P_5=(y_5,z_5)$

LENGTH

$s_1=\sqrt{(y_2-y_1)^2+(z_2-z_1)^2}$

$s_2=\sqrt{(y_3-y_2)^2+(z_3-z_2)^2}$

$s_3=\sqrt{(y_4-y_3)^2+(z_4-z_3)^2}$

$s_4=\sqrt{(y_5-y_4)^2+(z_5-z_4)^2}$

$s_5=\sqrt{(y_1-y_5)^2+(z_1-z_5)^2}$

DIRECTION COSINES

$\cos B_1=(y_2-y_1)/s_1$ $\cos C_1=(z_2-z_1)/s_1$

$\cos B_2=(y_3-y_2)/s_2$ $\cos C_2=(z_3-z_2)/s_2$

$\cos B_3=(y_4-y_3)/s_3$ $\cos C_3=(z_4-z_3)/s_3$

$\cos B_4=(y_5-y_4)/s_4$ $\cos C_4=(z_5-z_4)/s_4$

$\cos B_5=(y_1-y_5)/s_5$ $\cos C_5=(z_1-z_5)/s_5$

ANGLES

$\cos\phi_{1,2}=\cos B_1\cos B_2+\cos C_1\cos C_2$

$\cos\phi_{2,3}=\cos B_2\cos B_3+\cos C_2\cos C_3$

$\cos\phi_{3,4}=\cos B_3\cos B_4+\cos C_3\cos C_4$

$\cos\phi_{4,5}=\cos B_4\cos B_5+\cos C_4\cos C_5$

$\cos\phi_{5,1}=\cos B_5\cos B_1+\cos C_5\cos C_1$

CHECK

$\phi_{1,2}+\phi_{2,3}+\phi_{3,4}+\phi_{4,5}+\phi_{5,1}=540^\circ$

D

FIGURE VERSION 1:

$P_1=(-1,0,-1)$ $P_2=(1,0,1)$

$P_3=(1,1,1)$ $P_4=(0,2,0)$

$P_5=(-1,1,-1)$

FIGURE VERSION 2:

$P_1=(-2,0,-2)$ $P_2=(2,0,2)$

$P_3=(3,4,3)$ $P_4=(0,6,0)$

$P_5=(-3,4,-3)$

E

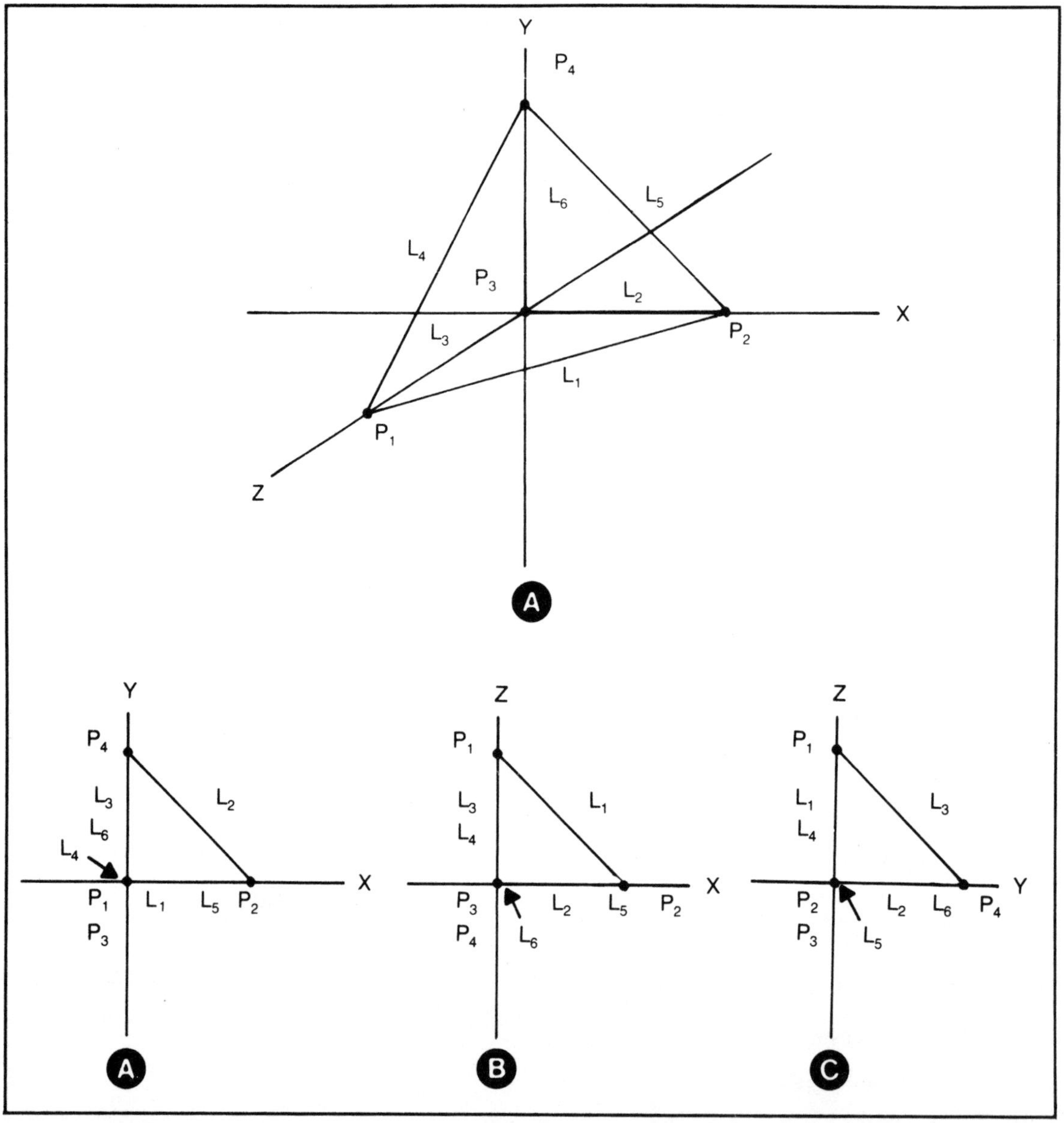

Fig. 15-5. A 4-sided pyramid specified in 3-D space. (A) X-Y perspective view. (B) X-Y planar view. (C) X-Z planar view. (D) Y-Z planar view. See the outlines of analyses in Table 15-5.

to do is substitute valid and specific component values for the point coordinates (see some examples in Table 15-5E).

After specifying some coordinates and conducting the primary analysis, you can check the results by working with the angles. Recalling that the sum of the angles in a triangle is 180 degrees, you can determine which angles appear in one of the sides of the space object and add them together. If the sum is close to 180 degrees, you can be reasonably sure that your analysis of that side is correct. Doing the same for the three remaining sides completes your check of the work.

Don't be surprised, though, if one or more of your planar views and analyses shows the space object re-

Table 15-5. Specifications and Analyses of a 4-Sided Pyramid in 3-D Space. See Fig. 15-5. (A) Specifications and Analysis in XYZ Space. (B) Secondary Analysis in the X-Y Plane. (C) Secondary Analysis in the X-Z Plane. (D) Secondary Analysis in the Y-Z Plane. (E) Suggested Coordinates.

(A, B, C, and D correspond to α, β, γ, and δ.)

SPECIFICATIONS

$S=\overline{F_1F_2F_3F_4}$

$F_1=\overline{L_1L_2L_3}$

$F_2=\overline{L_1L_4L_5}$

$F_3=\overline{L_2L_5L_6}$

$F_4=\overline{L_3L_4L_6}$

$L_1=\overline{P_1P_2}$	$P_1=(x_1,y_1,z_1)$
$L_2=\overline{P_2P_3}$	$P_2=(x_2,y_2,z_2)$
$L_3=\overline{P_3P_1}$	$P_3=(x_3,y_3,z_3)$
$L_4=\overline{P_1P_4}$	$P_4=(x_4,y_4,z_4)$
$L_5=\overline{P_2P_4}$	
$L_6=\overline{P_3P_4}$	

LENGTHS

$$s_1=\sqrt{(x_2-x_1)^2+(y_2-y_1)^2+(z_2-z_1)^2}$$

$$s_2=\sqrt{(x_3-x_2)^2+(y_3-y_2)^2+(z_3-z_2)^2}$$

$$s_3=\sqrt{(x_1-x_3)^2+(y_1-y_3)^2+(z_1-z_3)^2}$$

$$s_4=\sqrt{(x_4-x_1)^2+(y_4-y_1)^2+(z_4-z_1)^2}$$

$$s_5=\sqrt{(x_4-x_2)^2+(y_4-y_2)^2+(z_4-z_2)^2}$$

$$s_6=\sqrt{(x_4-x_3)^2+(y_4-y_3)^2+(z_4-z_3)^2}$$

DIRECTION COSINES

$\cos A_1=(x_2-x_1)/s_1$ $\cos B_1=(y_2-y_1)/s_1$

$\cos C_1=(z_2-z_1)/s_1$

$\cos A_2=(x_3-x_2)/s_2$ $\cos B_2=(y_3-y_2)/s_2$

$\cos C_2=(z_3-z_2)/s_2$

$\cos A_3=(x_1-x_3)/s_3$ $\cos B_3=(y_1-y_3)/s_3$

$\cos C_3=(z_1-z_3)/s_3$

DIRECTION COSINES

$\cos A_4=(x_4-x_1)/s_4$ $\cos B_4=(y_4-y_1)/s_4$

$\cos C_4=(z_4-z_1)/s_4$

$\cos A_5=(x_4-x_2)/s_5$ $\cos B_5=(y_4-y_2)/s_5$

$\cos C_5=(z_4-z_2)/s_5$

$\cos A_6=(x_4-x_3)/s_6$ $\cos B_6=(y_4-y_3)/s_6$

$\cos C_6=(z_4-z_3)/s_6$

ANGLES

$$\cos\phi_{1,2}=\cos A_1\cos A_2+\cos B_1\cos B_2+\cos C_1\cos C_2$$

$$\cos\phi_{1,3}=\cos A_1\cos A_3+\cos B_1\cos B_3+\cos C_1\cos C_3$$

$$\cos\phi_{1,4}=\cos A_1\cos A_4+\cos B_1\cos B_4+\cos C_1\cos C_4$$

$$\cos\phi_{2,3}=\cos A_2\cos A_3+\cos B_2\cos B_3+\cos C_2\cos C_3$$

$$\cos\phi_{2,5}=\cos A_2\cos A_5+\cos B_2\cos B_5+\cos C_2\cos C_5$$

$$\cos\phi_{2,6}=\cos A_2\cos A_6+\cos B_2\cos B_6+\cos C_2\cos C_6$$

$$\cos\phi_{3,4}=\cos A_3\cos A_4+\cos B_3\cos B_4+\cos C_3\cos C_4$$

$$\cos\phi_{3,6}=\cos A_3\cos A_6+\cos B_3\cos B_6+\cos C_3\cos C_6$$

$$\cos\phi_{4,5}=\cos A_4\cos A_5+\cos B_4\cos B_5+\cos C_4\cos C_5$$

$$\cos\phi_{4,6}=\cos A_4\cos A_6+\cos B_4\cos B_6+\cos C_4\cos C_6$$

$$\cos\phi_{5,6}=\cos A_5\cos A_6+\cos B_5\cos B_6+\cos C_5\cos C_6$$

CHECK

F_1:	$\phi_{1,2}+\phi_{1,3}+\phi_{2,3}=180^\circ$
F_2:	$\phi_{1,4}+\phi_{1,5}+\phi_{4,5}=180^\circ$
F_3:	$\phi_{2,5}+\phi_{2,6}+\phi_{5,6}=180^\circ$
F_4:	$\phi_{3,4}+\phi_{3,6}+\phi_{4,6}=180^\circ$

A

Table 15-5. Specifications and Analyses of a 4-Sided Pyramid in 3-D Space. See Fig. 15-5. (A) Specifications and Analysis in XYZ Space. (B) Secondary Analysis in the X-Y Plane. (C) Secondary Analysis in the X-Z Plane. (D) Secondary Analysis in the Y-Z Plane. (E) Suggested Coordinates. **(Continued from page 189.)**

SECONDARY ANALYSIS -- XY SPACE

PLOTTED POINTS

$P_1=(x_1,y_1)$

$P_2=(x_2,y_2)$

$P_3=(x_3,y_3)$

$P_4=(x_4,y_4)$

LENGTHS

$s_1=\sqrt{(x_2-x_1)^2+(y_2-y_1)^2}$

$s_2=\sqrt{(x_3-x_2)^2+(y_3-y_2)^2}$

$s_3=\sqrt{(x_1-x_3)^2+(y_1-y_3)^2}$

$s_4=\sqrt{(x_4-x_1)^2+(y_4-y_1)^2}$

$s_5=\sqrt{(x_4-x_2)^2+(y_4-y_2)^2}$

$s_6=\sqrt{(x_4-x_3)^2+(y_4-y_3)^2}$

DIRECTION COSINES

$\cos A_1=(x_2-x_1)/s_1$ $\qquad$ $\cos B_1=(y_2-y_1)/s_1$

$\cos A_2=(x_3-x_2)/s_2$ $\qquad$ $\cos B_2=(y_3-y_2)/s_2$

$\cos A_3=(x_1-x_3)/s_3$ $\qquad$ $\cos B_3=(y_1-y_3)/s_3$

DIRECTION COSINES

$\cos A_4=(x_4-x_1)/s_4$ $\qquad$ $\cos B_4=(y_4-y_1)/s_4$

$\cos A_5=(x_4-x_2)/s_5$ $\qquad$ $\cos B_5=(y_4-y_2)/s_5$

$\cos A_6=(x_4-x_3)/s_6$ $\qquad$ $\cos B_6=(y_4-y_3)/s_6$

ANGLES

$\cos\phi_{1,2}=\cos A_1\cos A_2+\cos B_1\cos B_2$

$\cos\phi_{1,3}=\cos A_1\cos A_3+\cos B_1\cos B_3$

$\cos\phi_{1,4}=\cos A_1\cos A_4+\cos B_1\cos B_4$

$\cos\phi_{2,3}=\cos A_2\cos A_3+\cos B_2\cos B_3$

$\cos\phi_{2,5}=\cos A_2\cos A_5+\cos B_2\cos B_5$

$\cos\phi_{2,6}=\cos A_2\cos A_6+\cos B_2\cos B_6$

$\cos\phi_{3,4}=\cos A_3\cos A_4+\cos B_3\cos B_4$

$\cos\phi_{3,6}=\cos A_3\cos A_6+\cos B_3\cos B_6$

$\cos\phi_{4,5}=\cos A_4\cos A_5+\cos B_4\cos B_5$

$\cos\phi_{4,6}=\cos A_4\cos A_6+\cos B_4\cos B_6$

$\cos\phi_{5,6}=\cos A_5\cos A_6+\cos B_5\cos B_6$

CHECK

F_1: $\qquad$ $\phi_{1,2}+\phi_{1,3}+\phi_{2,3}=180^\circ$

F_2: $\qquad$ $\phi_{1,4}+\phi_{1,5}+\phi_{4,5}=180^\circ$

F_3: $\qquad$ $\phi_{2,5}+\phi_{2,6}+\phi_{5,6}=180^\circ$

F_4: $\qquad$ $\phi_{3,4}+\phi_{3,6}+\phi_{4,6}=180^\circ$

B

SECONDARY ANALYSIS -- XZ PLANE

PLOTTED POINTS

$P_1=(x_1,z_1)$

$P_2=(x_2,z_2)$

$P_3=(x_3,z_3)$

$P_4=(x_4,z_4)$

LENGTHS

$s_1=\sqrt{(x_2-x_1)^2+(z_2-z_1)^2}$

$s_2=\sqrt{(x_3-x_2)^2+(z_3-z_2)^2}$

$s_3=\sqrt{(x_1-x_3)^2+(z_1-z_3)^2}$

$s_4=\sqrt{(x_4-x_1)^2(z_4-z_1)^2}$

C

$$s_5=\sqrt{(x_4-x_2)^2+(z_4-z_2)^2}$$

$$s_6=\sqrt{(x_4-x_3)^2+(z_4-z_3)^2}$$

DIRECTION COSINES

$\cos A_1=(x_2-x_1)/s_1$	$\cos C_1=(z_2-z_1)/s_1$
$\cos A_2=(x_3-x_2)/s_2$	$\cos C_2=(z_3-z_2)/s_2$
$\cos A_3=(x_1-x_3)/s_3$	$\cos C_3=(z_1-z_3)/s_3$
$\cos A_4=(x_4-x_1)/s_4$	$\cos C_4=(z_4-z_1)/s_4$
$\cos A_5=(x_4-x_2)/s_5$	$\cos C_5=(z_4-z_2)/s_5$
$\cos A_6=(x_4-x_3)/s_6$	$\cos C_6=(z_4-z_3)/s_6$

ANGLES

$\cos\phi_{1,2}=\cos A_1\cos A_2+\cos C_1\cos C_2$

$\cos\phi_{1,3}=\cos A_1\cos A_3+\cos C_1\cos C_3$

ANGLES

$\cos\phi_{1,4}=\cos A_1\cos A_4+\cos C_1\cos C_4$

$\cos\phi_{2,3}=\cos A_2\cos A_3+\cos C_2\cos C_3$

$\cos\phi_{2,5}=\cos A_2\cos A_5+\cos C_2\cos C_5$

$\cos\phi_{2,6}=\cos A_2\cos A_6+\cos C_2\cos C_6$

$\cos\phi_{3,4}=\cos A_3\cos A_4+\cos C_3\cos C_4$

$\cos\phi_{3,6}=\cos A_3\cos A_6+\cos C_3\cos C_6$

$\cos\phi_{4,5}=\cos A_4\cos A_5+\cos C_4\cos C_5$

$\cos\phi_{4,6}=\cos A_4\cos A_6+\cos C_4\cos C_6$

$\cos\phi_{5,6}=\cos A_5\cos A_6+\cos C_5\cos C_6$

CHECK

F_1:	$\phi_{1,2}+\phi_{1,3}+\phi_{2,3}=180^\circ$
F_2:	$\phi_{1,4}+\phi_{1,5}+\phi_{4,5}=180^\circ$
F_3:	$\phi_{2,5}+\phi_{2,6}+\phi_{5,6}=180^\circ$
F_4:	$\phi_{3,4}+\phi_{3,6}+\phi_{4,6}=180^\circ$

C

SECONDARY ANALYSIS -- YZ PLANE

PLOTTED POINTS

$P_1=(y_1,z_1)$

$P_2=(y_2,z_2)$

$P_3=(y_3,z_3)$

$P_4=(y_4,z_4)$

LENGTHS

$$s_1=\sqrt{(y_2-y_1)^2+(z_2-z_1)^2}$$

$$s_2=\sqrt{(y_3-y_2)^2+(z_3-z_2)^2}$$

$$s_3=\sqrt{(y_1-y_3)^2+(z_1-z_3)^2}$$

$$s_4=\sqrt{(y_4-y_1)^2+(z_4-z_1)^2}$$

$$s_5=\sqrt{(y_4-y_2)^2+(z_4-z_2)^2}$$

$$s_6=\sqrt{(y_4-y_3)^2+(z_4-z_3)^2}$$

DIRECTION COSINES

$\cos B_1=(y_2-y_1)/s_1$	$\cos C_1=(z_2-z_1)/s_1$
$\cos B_2=(y_3-y_2)/s_2$	$\cos C_2=(z_3-z_2)/s_2$
$\cos B_3=(y_1-y_3)/s_3$	$\cos C_3=(z_1-z_3)/s_3$
$\cos B_4=(y_4-y_1)/s_4$	$\cos C_4=(z_4-z_1)/s_4$
$\cos B_5=(y_4-y_2)/s_5$	$\cos C_5=(z_4-z_2)/s_5$
$\cos B_6=(y_4-y_3)/s_6$	$\cos C_6=(z_4-z_3)/s_6$

ANGLES

$\cos\phi_{1,2}=\cos B_1\cos B_2+\cos C_1\cos C_2$

D

Table 15-5. Specifications and Analyses of a 4-Sided Pyramid in 3-D Space. See Fig. 15-5. (A) Specifications and Analysis in XYZ Space. (B) Secondary Analysis in the X-Y Plane. (C) Secondary Analysis in the X-Z Plane. (D) Secondry Analysis in the Y-Z Plane. (E) Suggested Coordinates. **(Continued from page 191.)**

$\cos\phi_{1,3}=\cos B_1\cos B_3+\cos C_1\cos C_3$

$\cos\phi_{1,4}=\cos B_1\cos B_4+\cos C_1\cos C_4$

$\cos\phi_{2,3}=\cos B_2\cos B_3+\cos C_2\cos C_3$

$\cos\phi_{2,5}=\cos B_2\cos B_5+\cos C_2\cos C_5$

$\cos\phi_{2,6}=\cos B_2\cos B_6+\cos C_2\cos C_6$

$\cos\phi_{3,4}=\cos B_3\cos B_4+\cos C_3\cos C_4$

$\cos\phi_{3,6}=\cos B_3\cos B_6+\cos C_3\cos C_6$

$\cos\phi_{4,5}=\cos B_4\cos B_5+\cos C_4\cos C_5$

$\cos\phi_{4,6}=\cos B_4\cos B_6+\cos C_4\cos C_6$

$\cos\phi_{5,6}=\cos B_5\cos B_6+\cos C_5\cos C_6$

CHECK

F_1: $\phi_{1,2}+\phi_{1,3}+\phi_{2,3}=180^\circ$

F_2: $\phi_{1,4}+\phi_{1,5}+\phi_{4,5}=180^\circ$

F_3: $\phi_{2,5}+\phi_{2,6}+\phi_{5,6}=180^\circ$

F_4: $\phi_{3,4}+\phi_{3,6}+\phi_{4,6}=180^\circ$

D

E

OBJECT VERSION 1

$P_1=(0,0,1)$ $P_2=(1,0,0)$ $P_3=(0,0,0)$

$P_4=(0,1,0)$

OBJECT VERSION 2

$P_1=(0,0,1)$ $P_2=(1,0,0)$ $P_3=(-1,0,-1)$

$P_4=(0,1,0)$

OBJECT VERSION 3

$P_1=(0,0,1)$ $P_2=(1,1,1)$ $P_3=(0,-2,-1)$

$P_4=(2,2,0)$

duced to a plane figure. This matter of portraying 3-dimensional objects in 2-D space always generates some distortion—distortion that is often so great that the object loses a dimension.

15-2.2 An Object Bounded by 4 Triangles and 1 Rectangle

Figure 15-6 and Table 15-6 represent a 5-sided space object that is bounded by four triangles and a single rectangle. The 5-sided pyramid cited in earlier projects in 3-D space is one version of it. The specifications, however, are written in a very general form, thereby giving you the opportunity to generate any number of different versions by simply plugging in a set of five carefully selected coordinates (see Table 15-6E for some examples).

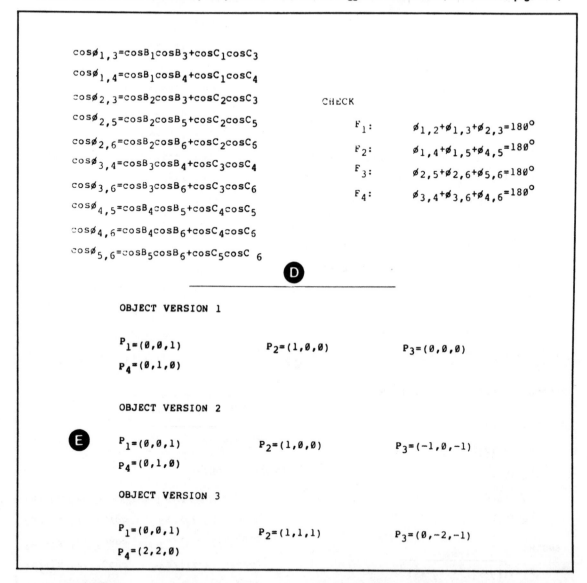

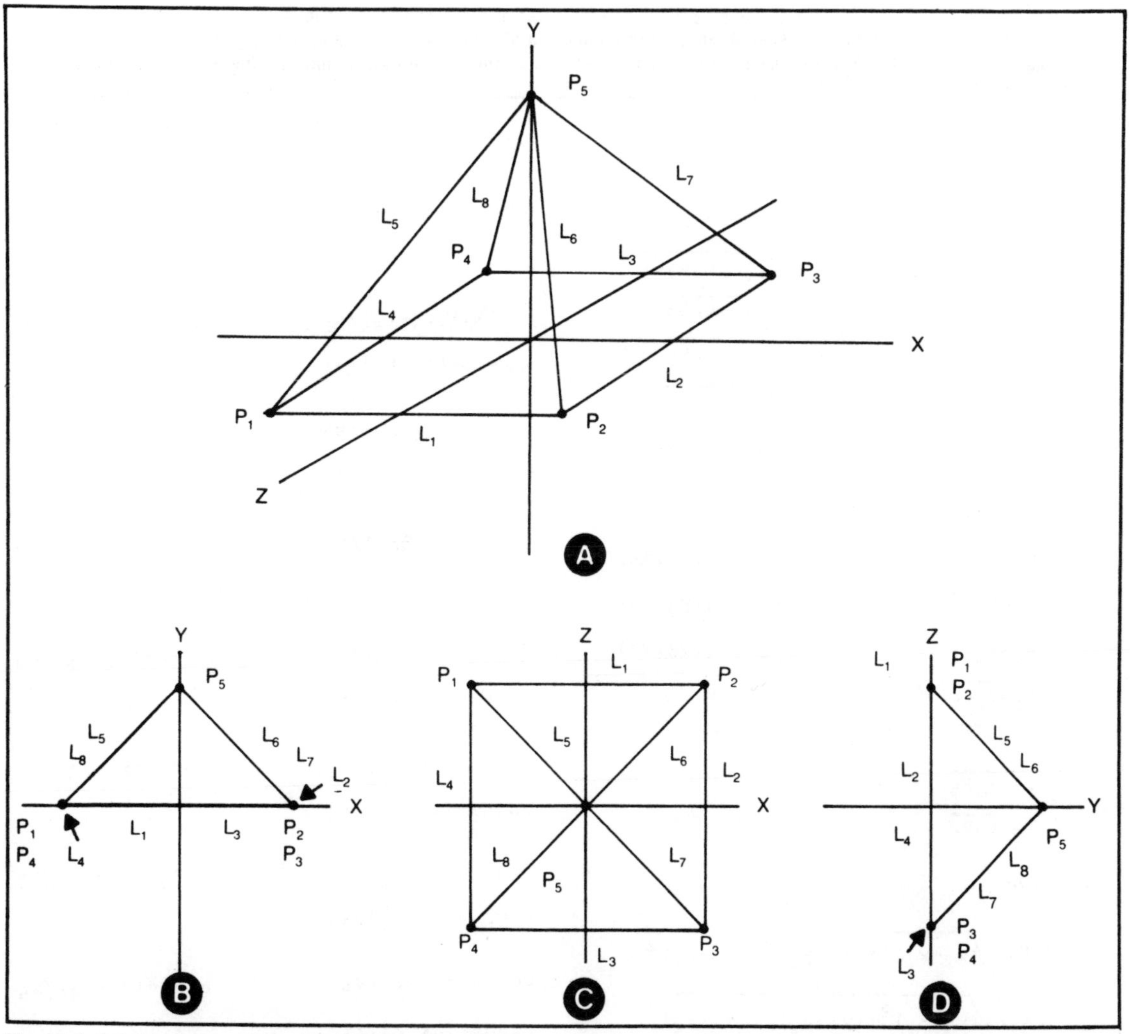

Fig. 15-6. A 5-sided pyramid specified in 3-D space, (A) X-Y perspective view. (B) X-Y planar view. (C) X-Z planar view. (D) Y-Z planar view. See the outlines of analyses in Table 15-6.

The F specifications indicate the lines that are involved in the four triangular and single rectangular side. So when you've completed the primary analysis of your version of this object, you can doublecheck the results by summing the three angles in the triangular sides and the four angles in the rectangular side. The triangles should show a sum of 180 degrees, and the rectangle should have a sum of 360 degrees (allowing a degree either way for round-off errors, of course).

15-2.3 A Space Object of 2 Triangles and 3 Rectangles

It is possible to generate more than one kind of 5-sided space object. The 5-sided pyramid described in the previous discussion is bounded by four triangles and a single rectangle. The object featured here is also a 5-sided space object, but it is bounded by two triangles and three rectangles. It can be generally described as a prism object. See Fig. 15-7 and Table 15-7.

Substitute the suggested specific coordinates (Table 15-7E) into the original specifications, and you will find a family of 5-sided objects of this type. Conduct complete primary and secondary analyses of several of the suggested versions in order to build up your file of experimental data.

Table 15-6. Specifications and Analyses of a 5-Sided Pyramid in 3-D Space. See Fig. 15-6. (A) Specifications and Analysis in XYZ Space. (B) Secondary Analysis in the X-Y Plane. (C) Secondary Analysis in the X-Z Plane. (D) Secondary Analysis in the Y-Z Plane. (E) Suggested Coordinates.

(A, B, C, and D correspond to α, β, γ, and δ.)

SPECIFICATIONS

$S=\overline{F_1F_2F_3F_4F_5}$

$F_1=\overline{L_1L_2L_3L_4}$

$F_2=\overline{L_1L_5L_6}$

$F_3=\overline{L_2L_6L_7}$

$F_4=\overline{L_3L_7L_8}$

$F_5=\overline{L_4L_5L_8}$

$L_1=\overline{P_1P_2}$

$L_2=\overline{P_2P_3}$

$L_3=\overline{P_3P_4}$

$L_4=\overline{P_4P_1}$

$L_5=\overline{P_1P_5}$

$L_6=\overline{P_2P_5}$

$L_7=\overline{P_3P_5}$

$L_8=\overline{P_4P_5}$

$P_1=(x_1,y_1,z_1)$

$P_2=(x_2,y_2,z_2)$

$P_3=(x_3,y_3,z_3)$

$P_4=(x_4,y_4,z_4)$

$P_5=(x_5,y_5,z_5)$

LENGTHS

$$s_1=\sqrt{(x_2-x_1)^2+(y_2-y_1)^2+(z_2-z_1)^2}$$

$$s_2=\sqrt{(x_3-x_2)^2+(y_3-y_2)^2+(z_3-z_2)^2}$$

$$s_3=\sqrt{(x_4-x_3)^2+(y_4-y_3)^2+(z_4-z_3)^2}$$

$$s_4=\sqrt{(x_1-x_4)^2+(y_1-y_4)^2+(z_1-z_4)^2}$$

$$s_5=\sqrt{(x_1-x_5)^2+(y_1-y_5)^2+(z_1-z_5)^2}$$

$$s_6=\sqrt{(x_5-x_2)^2+(y_5-y_2)^2+(z_5-z_2)^2}$$

$$s_7=\sqrt{(x_5-x_3)^2+(y_5-y_3)^2+(z_5-z_1)^2}$$

$$s_8=\sqrt{(x_5-x_4)^2+(y_5-y_4)^2+(z_5-z_4)^2}$$

DIRECTION COSINES

$\cos A_1=(x_2-x_1)/s_1$ $\quad\cos B_1=(y_2-y_1)/s_1$

$\cos C_1=(z_2-z_1)/s_1$

$\cos A_2=(x_3-x_2)/s_2$ $\quad\cos B_2=(y_3-y_2)/s_2$

$\cos C_2=(z_3-z_2)/s_2$

$\cos A_3=(x_4-x_3)/s_3$ $\quad\cos B_3=(y_4-y_3)/s_3$

$\cos C_3=(z_4-z_3)/s_3$

$\cos A_4=(x_1-x_4)/s_4$ $\quad\cos B_4=(y_1-y_4)/s_4$

$\cos C_4=(z_1-z_4)/s_4$

$\cos A_5=(x_1-x_5)/s_5$ $\quad\cos B_5=(y_1-y_5)/s_5$

$\cos C_5=(z_1-z_5)/s_5$

$\cos A_6=(x_5-x_2)/s_6$ $\quad\cos B_6=(y_5-y_2)/s_6$

$\cos C_6=(z_5-z_2)/s_6$

$\cos A_7=(x_5-x_3)/s_7$ $\quad\cos B_7=(y_5-y_3)/s_7$

$\cos C_7=(z_5-z_1)/s_7$

$\cos A_8=(x_5-x_4)/s_8$ $\quad\cos B_8=(y_5-y_4)/s_8$

$\cos C_8=(z_5-z_4)/s_8$

ANGLES

$$\cos\phi_{1,2}=\cos A_1\cos A_2+\cos B_1\cos B_2+\cos C_1\cos C_2$$

$$\cos\phi_{1,4}=\cos A_1\cos A_4+\cos B_1\cos B_4+\cos C_1\cos C_4$$

$$\cos\phi_{1,5}=\cos A_1\cos A_5+\cos B_1\cos B_5+\cos C_1\cos C_5$$

$$\cos\phi_{1,6}=\cos A_1\cos A_6+\cos B_1\cos B_6+\cos C_1\cos C_6$$

$$\cos\phi_{2,3}=\cos A_2\cos A_3+\cos B_2\cos B_3+\cos C_2\cos C_3$$

$$\cos\phi_{2,6}=\cos A_2\cos A_6+\cos B_2\cos B_6+\cos C_2\cos C_6$$

(A)

$\cos\phi_{2,7}=\cos A_2\cos A_7+\cos B_2\cos B_7+\cos C_2\cos C_7$

$\cos\phi_{3,7}=\cos A_3\cos A_7+\cos B_3\cos B_7+\cos C_3\cos C_7$

$\cos\phi_{3,8}=\cos A_3\cos A_8+\cos B_3\cos B_8+\cos C_3\cos C_8$

$\cos\phi_{4,5}=\cos A_4\cos A_5+\cos B_4\cos B_5+\cos C_4\cos C_5$

$\cos\phi_{4,8}=\cos A_4\cos A_8+\cos B_4\cos B_8+\cos C_4\cos C_8$

$\cos\phi_{5,6}=\cos A_5\cos A_6+\cos B_5\cos B_6+\cos C_5\cos C_6$

$\cos\phi_{5,7}=\cos A_5\cos A_7+\cos B_5\cos B_7+\cos C_5\cos C_7$

$\cos\phi_{5,8}=\cos A_5\cos A_8+\cos B_5\cos B_8+\cos B_5\cos C_8$

$\cos\phi_{6,7}=\cos A_6\cos A_7+\cos B_6\cos B_7+\cos B_6\cos C_7$

$\cos\phi_{7,8}=\cos A_7\cos A_8+\cos B_7\cos B_8+\cos C_7\cos C_8$

CHECK

F_1: $\phi_{1,2}+\phi_{2,3}+\phi_{3,4}+\phi_{1,4}=360^\circ$

F_2: $\phi_{1,5}+\phi_{1,6}+\phi_{5,6}=180^\circ$

F_3: $\phi_{2,6}+\phi_{2,7}+\phi_{6,7}=180^\circ$

F_4: $\phi_{3,7}+\phi_{3,8}+\phi_{7,8}=180^\circ$

F_5: $\phi_{4,5}+\phi_{4,8}+\phi_{5,8}=180^\circ$

A

SECONDARY ANALYSIS -- XY PLANE

PLOTTED POINTS

$P_1=(x_1,y_1)$

$P_2=(x_2,y_2)$

$P_3=(x_3,y_3)$

$P_4=(x_4,y_4)$

$P_5=(x_5,y_5)$

LENGTHS

$s_1=\sqrt{(x_2-x_1)^2+(y_2-y_1)^2}$

$s_2=\sqrt{(x_3-x_2)^2+(y_3-y_2)^2}$

$s_3=\sqrt{(x_4-x_3)^2+(y_4-y_3)^2}$

$s_4=\sqrt{(x_1-x_4)^2+(y_1-y_4)^2}$

$s_5=\sqrt{(x_1-x_5)^2+(y_1-y_5)^2}$

$s_6=\sqrt{(x_5-x_2)^2+(y_5-y_2)^2}$

$s_7=\sqrt{(x_5-x_3)^2+(y_5-y_3)^2}$

$s_8=\sqrt{(x_5-x_4)^2+(y_5-y_4)^2}$

DIRECTION COSINES

$\cos A_1=(x_2-x_1)/s_1$ $\cos B_1=(y_2-y_1)/s_1$

$\cos A_2=(x_3-x_2)/s_2$ $\cos B_2=(y_3-y_2)/s_2$

$\cos A_3=(x_4-x_3)/s_3$ $\cos B_3=(y_4-y_3)/s_3$

$\cos A_4=(x_1-x_4)/s_4$ $\cos B_4=(y_1-y_4)/s_4$

$\cos A_5=(x_1-x_5)/s_5$ $\cos B_5=(y_1-y_5)/s_5$

$\cos A_6=(x_5-x_2)/s_6$ $\cos B_6=(y_5-y_2)/s_6$

$\cos A_7=(x_5-x_3)/s_7$ $\cos B_7=(y_5-y_3)/s_7$

$\cos A_8=(x_5-x_4)/s_8$ $\cos B_8=(y_5-y_4)/s_8$

ANGLES

$\cos\phi_{1,2}=\cos A_1\cos A_2+\cos B_1\cos B_2$

$\cos\phi_{1,4}=\cos A_1\cos A_4+\cos B_1\cos B_4$

$\cos\phi_{1,5}=\cos A_1\cos A_5+\cos B_1\cos B_5$

$\cos\phi_{1,6}=\cos A_1\cos A_6+\cos B_1\cos B_6$

$\cos\phi_{2,3}=\cos A_2\cos A_3+\cos B_2\cos B_3$

$\cos\phi_{2,6}=\cos A_2\cos A_6+\cos B_2\cos B_6$

$\cos\phi_{2,7}=\cos A_2\cos A_7+\cos B_2\cos B_7$

$\cos\phi_{3,7}=\cos A_3\cos A_7+\cos B_3\cos B_7$

$\cos\phi_{3,8}=\cos A_3\cos A_8+\cos B_3\cos B_8$

$\cos\phi_{4,5}=\cos A_4\cos A_5+\cos B_4\cos B_5$

$\cos\phi_{4,8}=\cos A_4\cos A_8+\cos B_4\cos B_8$

B

Table 15-6. Specifications and Analyses of a 5-Sided Pyramid in 3-D Space. See Fig. 15-6. (A) Specifications and Analysis in XYZ Space. (B) Secondary Analysis in the X-Y Plane. (C) Secondary Analysis in the X-Z Plane. (D) Secondary Analysis in the Y-Z Plane. (E) Suggested Coordinates. (Continued from page 195.)

ANGLES (Cont'd)

$\cos\phi_{5,6}=\cos A_5\cos A_6+\cos B_5\cos B_6$

$\cos\phi_{5,7}=\cos A_5\cos A_7+\cos B_5\cos B_7$

$\cos\phi_{5,8}=\cos A_5\cos A_8+\cos B_5\cos B_8$

$\cos\phi_{6,7}=\cos A_6\cos A_7+\cos B_6\cos B_7$

$\cos\phi_{7,8}=\cos A_7\cos A_8+\cos B_7\cos B_8$

B

CHECK

F_1: $\phi_{1,2}+\phi_{2,3}+\phi_{3,4}+\phi_{1,4}=360^\circ$

F_2: $\phi_{1,5}+\phi_{1,6}+\phi_{5,6}=180^\circ$

F_3: $\phi_{2,6}+\phi_{2,7}+\phi_{6,7}=180^\circ$

F_4: $\phi_{3,7}+\phi_{3,8}+\phi_{7,8}=180^\circ$

F_5: $\phi_{4,5}+\phi_{4,8}+\phi_{5,8}=180^\circ$

SECONDARY ANALYSIS -- XZ PLANE

PLOTTED POINTS

$P_1=(x_1,z_1)$

$P_2=(x_2,z_2)$

$P_3=(x_3,z_3)$

$P_4=(x_4,z_4)$

$P_5=(x_5,z_5)$

LENGTHS

$s_1=\sqrt{(x_2-x_1)^2+(z_2-z_1)^2}$

$s_2=\sqrt{(x_3-x_2)^2+(z_3-z_2)^2}$

$s_3=\sqrt{(x_4-x_3)^2+(z_4-z_3)^2}$

$s_4=\sqrt{(x_1-x_4)^2+(z_1-z_4)^2}$

$s_5=\sqrt{(x_1-x_5)^2+(z_1-z_5)^2}$

$s_6=\sqrt{(x_5-x_2)^2+(z_5-z_2)^2}$

$s_7=\sqrt{(x_5-x_3)^2+(z_5-z_1)^2}$

$s_8=\sqrt{(x_5-x_4)^2+(z_5-z_4)^2}$

DIRECTION COSINES

$\cos A_1=(x_2-x_1)/s_1$ $\quad$ $\cos C_1=(z_2-z_1)/s_1$

$\cos A_2=(x_3-x_2)/s_2$ $\quad$ $\cos C_2=(z_3-z_2)/s_2$

$\cos A_3=(x_4-x_3)/s_3$ $\quad$ $\cos C_3=(z_4-z_3)/s_3$

$\cos A_4=(x_1-x_4)/s_4$ $\quad$ $\cos C_4=(z_1-z_4)/s_4$

$\cos A_5=(x_1-x_5)/s_5$ $\quad$ $\cos C_5=(z_1-z_5)/s_5$

$\cos A_6=(x_5-x_2)/s_6$ $\quad$ $\cos C_6=(z_5-z_2)/s_6$

$\cos A_7=(x_5-x_3)/s_7$ $\quad$ $\cos C_7=(z_5-z_1)/s_7$

$\cos A_8=(x_5-x_4)/s_8$ $\quad$ $\cos C_8=(z_5-z_4)/s_8$

ANGLES

$\cos\phi_{1,2}=\cos A_1\cos A_2+\cos C_1\cos C_2$

$\cos\phi_{1,4}=\cos A_1\cos A_4+\cos C_1\cos C_4$

$\cos\phi_{1,5}=\cos A_1\cos A_5+\cos C_1\cos C_5$

$\cos\phi_{1,6}=\cos A_1\cos A_6+\cos C_1\cos C_6$

$\cos\phi_{2,3}=\cos A_2\cos A_3+\cos C_2\cos C_3$

$\cos\phi_{2,6}=\cos A_2\cos A_6+\cos C_2\cos C_6$

$\cos\phi_{2,7}=\cos A_2\cos A_7+\cos C_2\cos C_7$

$\cos\phi_{3,7}=\cos A_3\cos A_7+\cos C_3\cos C_7$

$\cos\phi_{3,8}=\cos A_3\cos A_8+\cos C_3\cos C_8$

$\cos\phi_{4,5}=\cos A_4\cos A_5+\cos C_4\cos C_5$

$\cos\phi_{4,8}=\cos A_4\cos A_8+\cos C_4\cos C_8$

C

ANGLES (Cont'd)

$\cos\phi_{5,6}=\cos A_5\cos A_6+\cos C_5\cos C_6$

$\cos\phi_{5,7}=\cos A_5\cos A_7+\cos C_5\cos C_7$

$\cos\phi_{5,8}=\cos A_5\cos A_8+\cos C_5\cos C_8$

$\cos\phi_{6,7}=\cos A_6\cos A_7+\cos C_6\cos C_7$

$\cos\phi_{7,8}=\cos A_7\cos A_8+\cos C_7\cos C_8$

C

CHECK

F_1: $\phi_{1,2}+\phi_{2,3}+\phi_{3,4}+\phi_{1,4}=360^\circ$

F_2: $\phi_{1,5}+\phi_{1,6}+\phi_{5,6}=180^\circ$

F_3: $\phi_{2,6}+\phi_{2,7}+\phi_{6,7}=180^\circ$

F_4: $\phi_{3,7}+\phi_{3,8}+\phi_{7,8}=180^\circ$

F_5: $\phi_{4,5}+\phi_{4,8}+\phi_{5,8}=180^\circ$

SECONDARY ANALYSIS -- YZ PLANE

PLOTTED POINTS

$P_1=(y_1,z_1)$

$P_2=(y_2,z_2)$

$P_3=(y_3,z_3)$

$P_4=(y_4,z_4)$

$P_5=(y_5,z_5)$

LENGTHS

$s_1=\sqrt{(y_2-y_1)^2+(z_2-z_1)^2}$

$s_2=\sqrt{(y_3-y_2)^2+(z_3-z_2)^2}$

$s_3=\sqrt{(y_4-y_3)^2+(z_4-z_3)^2}$

$s_4=\sqrt{(y_1-y_4)^2+(z_1-z_4)^2}$

$s_5=\sqrt{(y_1-y_5)^2+(z_1-z_5)^2}$

$s_6=\sqrt{(y_5-y_2)^2+(z_5-z_2)^2}$

$s_7=\sqrt{(y_5-y_3)^2+(z_5-z_1)^2}$

$s_8=\sqrt{(y_5-y_4)^2+(z_5-z_4)^2}$

DIRECTION COSINES

$\cos B_1=(y_2-y_1)/s_1$ $\cos C_1=(z_2-z_1)/s_1$

$\cos B_2=(y_3-y_2)/s_2$ $\cos C_2=(z_3-z_2)/s_2$

$\cos B_3=(y_4-y_3)/s_3$ $\cos C_3=(z_4-z_3)/s_3$

$\cos B_4=(y_1-y_4)/s_4$ $\cos C_4=(z_1-z_4)/s_4$

$\cos B_5=(y_1-y_5)/s_5$ $\cos C_5=(z_1-z_5)/s_5$

$\cos B_6=(y_5-y_2)/s_6$ $\cos C_6=(z_5-z_2)/s_6$

$\cos B_7=(y_5-y_3)/s_7$ $\cos C_7=(z_5-z_1)/s_7$

$\cos B_8=(y_5-y_4)/s_8$ $\cos C_8=(z_5-z_4)/s_8$

ANGLES

$\cos\phi_{1,2}=\cos B_1\cos B_2+\cos C_1\cos C_2$

$\cos\phi_{1,4}=\cos B_1\cos B_4+\cos C_1\cos C_4$

$\cos\phi_{1,5}=\cos B_1\cos B_5+\cos C_1\cos C_5$

$\cos\phi_{1,6}=\cos B_1\cos B_6+\cos C_1\cos C_6$

$\cos\phi_{2,3}=\cos B_2\cos B_3+\cos C_2\cos C_3$

$\cos\phi_{2,6}=\cos B_2\cos B_6+\cos C_2\cos C_6$

$\cos\phi_{2,7}=\cos B_2\cos B_7+\cos C_2\cos C_7$

$\cos\phi_{3,7}=\cos B_3\cos B_7+\cos C_3\cos C_7$

$\cos\phi_{3,8}=\cos B_3\cos B_8+\cos C_3\cos C_8$

$\cos\phi_{4,5}=\cos B_4\cos B_5+\cos C_4\cos C_5$

$\cos\phi_{4,8}=\cos B_4\cos B_8+\cos C_4\cos C_8$

D

Table 15-6. Specifications and Analyses of a 5-Sided Pyramid in 3-D Space. See Fig. 15-6 (A) Specifications and Analysis in XYZ Space. (B) Secondary Analysis in the X-Y Plane. (C) Secondary Analysis in the X-Z Plane. (D) Secondary Analysis in the Y-Z Plane. (E) Suggested Coordinates. (Continued from page 197.)

ANGLES (Cont'd)	CHECK	
	F_1:	$\phi_{1,2}+\phi_{2,3}+\phi_{3,4}+\phi_{1,4}=360^\circ$
$\cos\phi_{5,6}=\cos B_5\cos B_6+\cos C_5\cos C_6$	F_2:	$\phi_{1,5}+\phi_{1,6}+\phi_{5,6}=180^\circ$
$\cos\phi_{5,7}=\cos B_5\cos B_7+\cos C_5\cos C_7$	F_3:	$\phi_{2,6}+\phi_{2,7}+\phi_{6,7}=180^\circ$
$\cos\phi_{5,8}=\cos B_5\cos B_8+\cos C_5\cos C_8$	F_4:	$\phi_{3,7}+\phi_{3,8}+\phi_{7,8}=180^\circ$
$\cos\phi_{6,7}=\cos B_6\cos B_7+\cos C_6\cos C_7$	F_5:	$\phi_{4,5}+\phi_{4,8}+\phi_{5,8}=180^\circ$
$\cos\phi_{7,8}=\cos B_7\cos B_8+\cos C_7\cos C_8$		

D

OBJECT VERSION 1

$P_1=(-1,0,1)$	$P_2=(1,0,1)$	$P_3=(0,1,-1)$
$P_4=(-1,0,-1)$	$P_5=(0,1,0)$	

OBJECT VERSION 2

$P_1=(-1,0,1)$	$P_2=(0,0,1)$	$P_3=(1,0,-1)$
$P_4=(0,0,-1)$	$P_5=(0,1,0)$	

OBJECT VERSION 3

$P_1=(-1,0,1)$	$P_2=(0,0,1)$	$P_3=(0,0,-1)$
$P_4=(-1,0,-1)$	$P_5=(1,1,0)$	

E

15-2.4 A Space Object of 6 Rectangles

A cube is perhaps the most widely recognized 6-sided object. There are other kinds of 6-sided objects and it happens that a cube is a figure that is bounded by six squares. The sides need not be squares in order to qualify as a member of this family—rectangles, trapezoids and other such 4-sided figures fit the specifications just as well.

Figure 15-8 and Table 15-8 show the plots and analyses of 6-sided objects where all sides happen to be quadralateral figures. The sets of coordinates in Table 15-8E are intended to provide you with a selection of these space objects. Conduct complete analyses of them, including the plots, and save them for future work.

Check your work at the primary-analysis level by summing the four angles in each side. Since they are quadralaterals, the sum in each case should come out to about 180 degrees. Pairs of lines are parallel, incidentally, when the sum of the products of their corresponding direction cosines is 1 or -1; and they are perpendicular when the sum of the products of their corresponding direction is 0.

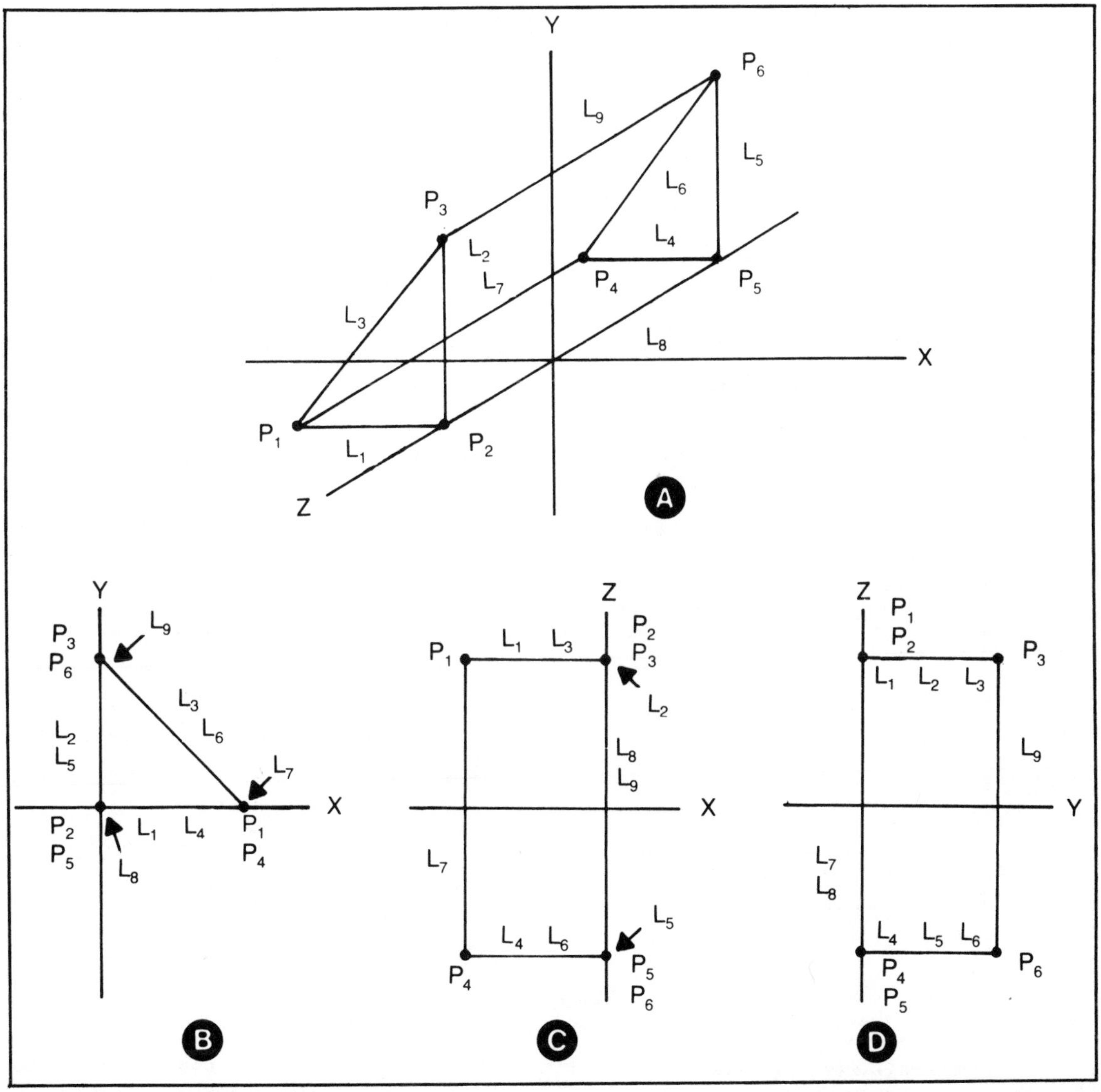

Fig. 15-7. A prism object specified in 3-D space. (A) X-Y perspective view. (B) X-Y planar view. (C) X-Z planar view. (D) Y-Z planar view. See the outlines of analyses in Table 15-7.

15-2.5 A 7-Sided Space Object

Figure 15-9 and Table 15-9 represents a 7-sided space object that is bounded by two pentagons and five rectangles. In its standard form, it looks much like a house object for a Monopoly game. There are a number of variations, however, as indicated by the suggested coordinates in Table 15-9E.

Conduct complete analyses of the object, using some of the suggested coordinates. Remember that the sum of the angles in the pentagons should be close to 540 degrees, while the sum of the angles in the rectangular sides has to be close to 360 degrees.

15-3 A SUMMARY OF SIMPLE TRANSFORMATIONS IN 3-D SPACE

The simple transformations—those based on a single type of transformation—in 3-D space are the same as those in 2-D space:

Table 15-7. Specifications and Analyses of a Prism in 3-D Space. See Fig. 15-7. (A) Specifications and Analysis in XYZ Space. (B) Secondary Analysis in the X-Y Plane. (C) Secondary Analysis in the X-Z Plane. (D) Secondary Analysis in the Y-Z Plane. (E) Suggested Coordinates.

(A, B, C, and D correspond to α, β, γ, and δ.)

SPECIFICATIONS

$S=\overline{F_1F_2F_3F_4F_5}$

$F_1=\overline{L_1L_2L_3}$

$F_2=\overline{L_1L_4L_7L_8}$

$F_3=\overline{L_2L_5L_8L_9}$

$F_4=\overline{L_3L_6L_7L_9}$

$F_5=\overline{L_4L_5L_6}$

$L_1=\overline{P_1P_2}$

$L_2=\overline{P_2P_3}$

$L_3=\overline{P_3P_1}$

$L_4=\overline{P_4P_5}$

$L_5=\overline{P_5P_6}$

$L_6=\overline{P_6P_4}$

$L_7=\overline{P_1P_4}$

$L_8=\overline{P_2P_5}$

$L_9=\overline{P_3P_6}$

$P_1=(x_1,y_1,z_1)$

$P_2=(x_2,y_2,z_2)$

$P_3=(x_3,y_3,z_3)$

$P_4=(x_4,y_4,z_4)$

$P_5=(x_5,y_5,z_5)$

LENGTHS

$s_1=\sqrt{(x_2-x_1)^2+(y_2-y_1)^2+(z_2-z_1)^2}$

$s_2=\sqrt{(x_3-x_2)^2+(y_3-y_2)^2+(z_3-z_2)^2}$

$s_3=\sqrt{(x_1-x_3)^2+(y_1-y_3)^2+(z_1-z_3)^2}$

$s_4=\sqrt{(x_5-x_4)^2+(y_5-y_4)^2+(z_5-z_4)^2}$

$s_5=\sqrt{(x_6-x_5)^2+(y_6-y_5)^2+(z_6-z_5)^2}$

$s_6=\sqrt{(x_4-x_6)^2+(y_4-y_6)^2+(z_4-z_6)^2}$

$s_7=\sqrt{(x_4-x_1)^2+(y_4-y_1)^2+(z_4-z_1)^2}$

$s_8=\sqrt{(x_5-x_2)^2+(y_5-y_2)^2+(z_5-z_2)^2}$

$s_9=\sqrt{(x_6-x_3)^2+(y_6-y_3)^2+(z_6-z_3)^2}$

DIRECTION COSINES

$\cos A_1=(x_2-x_1)/s_1$ $\cos B_1=(y_2-y_1)/s_1$

$\cos C_1=(z_2-z_1)/s_1$

$\cos A_2=(x_3-x_2)/s_2$ $\cos B_2=(y_3-y_2)/s_2$

$\cos C_2=(z_3-z_2)/s_2$

$\cos A_3=(x_1-x_3)/s_3$ $\cos B_3=(y_1-y_3)/s_3$

$\cos C_3=(z_1-z_3)/s_3$

$\cos A_4=(x_5-x_4)/s_4$ $\cos B_4=(y_5-y_4)/s_4$

$\cos C_4=(z_5-z_4)/s_4$

$\cos A_5=(x_6-x_5)/s_5$ $\cos B_5=(y_6-y_5)/s_5$

$\cos C_5=(z_6-z_5)/s_5$

$\cos A_6=(x_4-x_6)/s_6$ $\cos B_6=(y_4-y_6)/s_6$

$\cos C_6=(z_4-z_6)/s_6$

$\cos A_7=(x_4-x_1)/s_7$ $\cos B_7=(y_4-y_1)/s_7$

$\cos C_7=(z_4-z_1)/s_7$

$\cos A_8=(x_5-x_2)/s_8$ $\cos B_8=(y_5-y_2)/s_8$

$\cos C_8=(z_5-z_2)/s_8$

$\cos A_9=(x_6-x_3)/s_9$ $\cos B_9=(y_6-y_3)/s_9$

$\cos C_9=(z_6-z_3)/s_9$

ANGLES

$\cos\phi_{1,2}=\cos A_1\cos A_2+\cos B_1\cos B_2+\cos C_1\cos C_2$

$\cos\phi_{1,3}=\cos A_1\cos A_3+\cos B_1\cos B_3+\cos C_1\cos C_3$

$\cos\phi_{1,7}=\cos A_1\cos A_7+\cos B_1\cos B_7+\cos C_1\cos C_7$

$\cos\phi_{1,8}=\cos A_1\cos A_8+\cos B_1\cos B_8+\cos C_1\cos C_8$

$\cos\phi_{2,3}=\cos A_2\cos A_3+\cos B_2\cos B_3+\cos C_2\cos C_3$

(A)

ANGLES

$\cos\phi_{2,8}=\cos A_2\cos A_8+\cos B_2\cos B_8+\cos C_2\cos C_8$

$\cos\phi_{2,9}=\cos A_2\cos A_9+\cos B_2\cos B_9+\cos C_2\cos C_9$

$\cos\phi_{3,7}=\cos A_3\cos A_7+\cos B_3\cos B_7+\cos C_3\cos C_7$

$\cos\phi_{3,9}=\cos A_3\cos A_9+\cos B_3\cos B_9+\cos C_3\cos C_9$

$\cos\phi_{4,5}=\cos A_4\cos A_5+\cos B_4\cos B_5+\cos C_4\cos C_5$

$\cos\phi_{4,6}=\cos A_4\cos A_6+\cos B_4\cos B_6+\cos C_4\cos C_6$

$\cos\phi_{4,7}=\cos A_4\cos A_7+\cos B_4\cos B_7+\cos C_4\cos C_7$

$\cos\phi_{4,8}=\cos A_4\cos A_8+\cos B_4\cos B_8+\cos C_4\cos C_8$

$\cos\phi_{5,6}=\cos A_5\cos A_6+\cos B_5\cos B_6+\cos C_5\cos C_6$

$\cos\phi_{5,8}=\cos A_5\cos A_8+\cos B_5\cos B_8+\cos C_5\cos C_8$

$\cos\phi_{5,9}=\cos A_5\cos A_9+\cos B_5\cos B_9+\cos C_5\cos C_9$

$\cos\phi_{6,7}=\cos A_6\cos A_7+\cos B_6\cos B_7+\cos C_6\cos C_7$

$\cos\phi_{6,9}=\cos A_6\cos A_9+\cos B_6\cos B_9+\cos C_6\cos C_9$

CHECK

F_1: $\phi_{1,2}+\phi_{1,3}+\phi_{2,3}=180^\circ$

F_2: $\phi_{1,7}+\phi_{1,8}+\phi_{4,7}+\phi_{4,8}=360^\circ$

F_3: $\phi_{2,8}+\phi_{2,9}+\phi_{5,8}+\phi_{5,9}=360^\circ$

F_4: $\phi_{3,7}+\phi_{3,9}+\phi_{6,7}+\phi_{6,9}=360^\circ$

F_5: $\phi_{4,5}+\phi_{4,6}+\phi_{5,6}=180^\circ$

A

SECONDARY ANALYSIS -- XY PLANE

B

PLOTTED POINTS

$P_1=(x_1,y_1)$

$P_2=(x_2,y_2)$

$P_3=(x_3,y_3)$

$P_4=(x_4,y_4)$

$P_5=(x_5,y_5)$

LENGTHS

$s_1=\sqrt{(x_2-x_1)^2+(y_2-y_1)^2}$

$s_2=\sqrt{(x_3-x_2)^2+(y_3-y_2)^2}$

$s_3=\sqrt{(x_1-x_3)^2+(y_1-y_3)^2}$

$s_4=\sqrt{(x_5-x_4)^2+(y_5-y_4)^2}$

$s_5=\sqrt{(x_6-x_5)^2+(y_6-y_5)^2}$

$s_6=\sqrt{(x_4-x_6)^2+(y_4-y_6)^2}$

$s_7=\sqrt{(x_4-x_1)^2+(y_4-y_1)^2}$

$s_8=\sqrt{(x_5-x_2)^2+(y_5-y_2)^2}$

$s_9=\sqrt{(x_6-x_3)^2+(y_6-y_3)^2}$

DIRECTION COSINES

$\cos A_1=(x_2-x_1)/s_1$ $\cos B_1=(y_2-y_1)/s_1$

$\cos A_2=(x_3-x_2)/s_2$ $\cos B_2=(y_3-y_2)/s_2$

$\cos A_3=(x_4-x_3)/s_3$ $\cos B_3=(y_4-y_3)/s_3$

$\cos A_4=(x_1-x_4)/s_4$ $\cos B_4=(y_1-y_4)/s_4$

$\cos A_5=(x_1-x_5)/s_5$ $\cos B_5=(y_1-y_5)/s_5$

$\cos A_6=(x_5-x_2)/s_6$ $\cos B_6=(y_5-y_2)/s_6$

$\cos A_7=(x_5-x_3)/s_7$ $\cos B_7=(y_5-y_3)/s_7$

$\cos A_8=(x_5-x_4)/s_8$ $\cos B_8=(y_5-y_4)/s_8$

Table 15-7. Specifications and Analyses of a Prism in 3-D Space. See Fig. 15-7. (A) Specifications and Analysis in XYZ Space. (B) Secondary Analysis in the X-Y Plane. (C) Secondary Analysis in the X-Z Plane. (D) Secondary Analysis in the Y-Z Plane. (E) Suggested Coordinates. (Continued from page 201.)

ANGLES

$\cos\phi_{1,2}=\cos A_1\cos A_2+\cos B_1\cos B_2$

$\cos\phi_{1,3}=\cos A_1\cos A_3+\cos B_1\cos B_3$

$\cos\phi_{1,7}=\cos A_1\cos A_7+\cos B_1\cos B_7$

$\cos\phi_{1,8}=\cos A_1\cos A_8+\cos B_1\cos B_8$

$\cos\phi_{2,3}=\cos A_2\cos A_3+\cos B_2\cos B_3$

$\cos\phi_{2,8}=\cos A_2\cos A_8+\cos B_2\cos B_8$

$\cos\phi_{2,9}=\cos A_2\cos A_9+\cos B_2\cos B_9$

$\cos\phi_{3,7}=\cos A_3\cos A_7+\cos B_3\cos B_7$

$\cos\phi_{3,9}=\cos A_3\cos A_9+\cos B_3\cos B_9$

$\cos\phi_{4,5}=\cos A_4\cos A_5+\cos B_4\cos B_5$

$\cos\phi_{4,6}=\cos A_4\cos A_6+\cos B_4\cos B_6$

$\cos\phi_{4,7}=\cos A_4\cos A_7+\cos B_4\cos B_7$

B

ANGLES (Cont'd)

$\cos\phi_{4,8}=\cos A_4\cos A_8+\cos B_4\cos B_8$

$\cos\phi_{5,6}=\cos A_5\cos A_6+\cos B_5\cos B_6$

$\cos\phi_{5,8}=\cos A_5\cos A_8+\cos B_5\cos B_8$

$\cos\phi_{5,9}=\cos A_5\cos A_9+\cos B_5\cos B_9$

$\cos\phi_{6,7}=\cos A_6\cos A_7+\cos B_6\cos B_7$

$\cos\phi_{6,9}=\cos A_6\cos A_9+\cos B_6\cos B_9$

CHECK

F_1:	$\phi_{1,2}+\phi_{1,3}+\phi_{2,3}=180^\circ$
F_2:	$\phi_{1,7}+\phi_{1,8}+\phi_{4,7}+\phi_{4,8}=360^\circ$
F_3:	$\phi_{2,8}+\phi_{2,9}+\phi_{5,8}+\phi_{5,9}=360^\circ$
F_4:	$\phi_{3,7}+\phi_{3,9}+\phi_{6,7}+\phi_{6,9}=360^\circ$
F_5:	$\phi_{4,5}+\phi_{4,6}+\phi_{5,6}=180^\circ$

C

SECONDARY ANALYSIS -- XZ PLANE

PLOTTED POINTS

$P_1=(x_1,z_1)$

$P_2=(x_2,z_2)$

$P_3=(x_3,z_3)$

$P_4=(x_4,z_4)$

$P_5=(x_5,z_5)$

LENGTHS

$s_1=\sqrt{(x_2-x_1)^2+(z_2-z_1)^2}$

$s_2=\sqrt{(x_3-x_2)^2+(z_3-z_2)^2}$

$s_3=\sqrt{(x_1-x_3)^2+(z_1-z_3)^2}$

$s_4=\sqrt{(x_5-x_4)^2+(z_5-z_4)^2}$

$s_5=\sqrt{(x_6-x_5)^2+(z_6-z_5)^2}$

$s_6=\sqrt{(x_4-x_6)^2+(z_4-z_6)^2}$

$s_7=\sqrt{(x_4-x_1)^2+(z_4-z_1)^2}$

$s_8=\sqrt{(x_5-x_2)^2+(z_5-z_2)^2}$

$s_9=\sqrt{(x_6-x_3)^2+(z_6-z_3)^2}$

DIRECTION COSINES

$\cos A_1=(x_2-x_1)/s_1$	$\cos C_1=(z_2-z_1)/s_1$
$\cos A_2=(x_3-x_2)/s_2$	$\cos C_2=(z_3-z_2)/s_2$
$\cos A_3=(x_1-x_3)/s_3$	$\cos C_3=(z_1-z_3)/s_3$

DIRECTION COSINES

$\cos A_4=(x_5-x_4)/s_4$

$\cos A_5=(x_6-x_5)/s_5$

$\cos A_6=(x_4-x_6)/s_6$

$\cos A_7=(x_4-x_1)/s_7$

$\cos A_8=(x_5-x_2)/s_8$

$\cos A_9=(x_6-x_3)/s_9$

$\cos C_4=(z_5-z_4)/s_4$

$\cos C_5=(z_6-z_5)/s_5$

$\cos C_6=(z_4-z_6)/s_6$

$\cos C_7=(z_4-z_1)/s_7$

$\cos C_8=(z_5-z_2)/s_8$

$\cos C_9=(z_6-z_3)/s_9$

ANGLES

$\cos\phi_{1,2}=\cos A_1\cos A_2+\cos C_1\cos C_2$

$\cos\phi_{1,3}=\cos A_1\cos A_3+\cos C_1\cos C_3$

$\cos\phi_{1,7}=\cos A_1\cos A_7+\cos C_1\cos C_7$

$\cos\phi_{1,8}=\cos A_1\cos A_8+\cos C_1\cos C_8$

$\cos\phi_{2,3}=\cos A_2\cos A_3+\cos C_2\cos C_3$

$\cos\phi_{2,8}=\cos A_2\cos A_8+\cos C_2\cos C_8$

$\cos\phi_{2,9}=\cos A_2\cos A_9+\cos C_2\cos C_9$

$\cos\phi_{3,7}=\cos A_3\cos A_7+\cos C_3\cos C_7$

$\cos\phi_{3,9}=\cos A_3\cos A_9+\cos C_3\cos C_9$

C

ANGLES (Cont'd)

$\cos\phi_{4,5}=\cos A_4\cos A_5+\cos C_4\cos C_5$

$\cos\phi_{4,6}=\cos A_4\cos A_6+\cos C_4\cos C_6$

$\cos\phi_{4,7}=\cos A_4\cos A_7+\cos C_4\cos C_7$

$\cos\phi_{4,8}=\cos A_4\cos A_8+\cos C_4\cos C_8$

$\cos\phi_{5,6}=\cos A_5\cos A_6+\cos C_5\cos C_6$

$\cos\phi_{5,8}=\cos A_5\cos A_8+\cos C_5\cos C_8$

$\cos\phi_{5,9}=\cos A_5\cos A_9+\cos C_5\cos C_9$

$\cos\phi_{6,7}=\cos A_6\cos A_7+\cos C_6\cos C_7$

$\cos\phi_{6,9}=\cos A_6\cos A_9+\cos C_6\cos C_9$

CHECK

F_1: $\phi_{1,2}+\phi_{1,3}+\phi_{2,3}=180^\circ$

F_2: $\phi_{1,7}+\phi_{1,8}+\phi_{4,7}+\phi_{4,8}=360^\circ$

F_3: $\phi_{2,8}+\phi_{2,9}+\phi_{5,8}+\phi_{5,9}=360^\circ$

F_4: $\phi_{3,7}+\phi_{3,9}+\phi_{6,7}+\phi_{6,9}=360^\circ$

F_5: $\phi_{4,5}+\phi_{4,6}+\phi_{5,6}=180^\circ$

SECONDARY ANALYSIS -- YZ PLANE

D

PLOTTED POINTS

$P_1=(y_1,z_1)$

$P_2=(y_2,z_2)$

$P_3=(y_3,z_3)$

$P_4=(y_4,z_4)$

$P_5=(y_5,z_5)$

LENGTHS

$s_1=\sqrt{(y_2-y_1)^2+(z_2-z_1)^2}$

$s_2=\sqrt{(y_3-y_2)^2+(z_3-z_2)^2}$

$s_3=\sqrt{(y_1-y_3)^2+(z_1-z_3)^2}$

$s_4=\sqrt{(y_5-y_4)^2+(z_5-z_4)^2}$

$s_5=\sqrt{(y_6-y_5)^2+(z_6-z_5)^2}$

$s_6=\sqrt{(y_4-y_6)^2+(z_4-z_6)^2}$

$s_7=\sqrt{(y_4-y_1)^2+(z_4-z_1)^2}$

$s_8=\sqrt{(y_5-y_2)^2+(z_5-z_2)^2}$

$s_9=\sqrt{(y_6-y_3)^2+(z_6-z_3)^2}$

Table 15-7. Specifications and Analyses of a Prism in 3-D Space. See Fig. 15-7. (A) Specifications and Analysis in XYZ Space. (B) Secondary Analysis in the X-Y Plane. (C) Secondary Analysis in the X-Z Plane. (D) Secondary Analysis in the Y-Z Plane. (E) Suggested Coordinates. (Continued from page 203).

(D)

DIRECTION COSINES

$\cos B_1=(y_2-y_1)/s_1$ | $\cos C_1=(z_2-z_1)/s_1$
$\cos B_2=(y_3-y_2)/s_2$ | $\cos C_2=(z_3-z_2)/s_2$
$\cos B_3=(y_1-y_3)/s_3$ | $\cos C_3=(z_1-z_3)/s_3$
$\cos B_4=(y_5-y_4)/s_4$ | $\cos C_4=(z_5-z_4)/s_4$
$\cos B_5=(y_6-y_5)/s_5$ | $\cos C_5=(z_6-z_5)/s_5$
$\cos B_6=(y_4-y_6)/s_6$ | $\cos C_6=(z_4-z_6)/s_6$
$\cos B_7=(y_4-y_1)/s_7$ | $\cos C_7=(z_4-z_1)/s_7$
$\cos B_8=(y_5-y_2)/s_8$ | $\cos C_8=(z_5-z_2)/s_8$
$\cos B_9=(y_6-y_3)/s_9$ | $\cos C_9=(z_6-z_3)/s_9$

ANGLES

$\cos\phi_{1,2}=\cos B_1\cos B_2+\cos C_1\cos C_2$
$\cos\phi_{1,3}=\cos B_1\cos B_3+\cos C_1\cos C_3$
$\cos\phi_{1,7}=\cos B_1\cos B_7+\cos C_1\cos C_7$
$\cos\phi_{1,8}=\cos B_1\cos B_8+\cos C_1\cos C_8$
$\cos\phi_{2,3}=\cos B_2\cos B_3+\cos C_2\cos C_3$
$\cos\phi_{2,8}=\cos B_2\cos B_8+\cos C_2\cos C_8$
$\cos\phi_{2,9}=\cos B_2\cos B_9+\cos C_2\cos C_9$

ANGLES (Cont'd)

$\cos\phi_{3,7}=\cos B_3\cos B_7+\cos C_3\cos C_7$
$\cos\phi_{3,9}=\cos B_3\cos B_9+\cos C_3\cos C_9$
$\cos\phi_{4,5}=\cos B_4\cos B_5+\cos C_4\cos C_5$
$\cos\phi_{4,6}=\cos B_4\cos B_6+\cos C_4\cos C_6$
$\cos\phi_{4,7}=\cos B_4\cos B_7+\cos C_4\cos C_7$
$\cos\phi_{4,8}=\cos B_4\cos B_8+\cos C_4\cos C_8$
$\cos\phi_{5,6}=\cos B_5\cos B_6+\cos C_5\cos C_6$
$\cos\phi_{5,8}=\cos B_5\cos B_8+\cos C_5\cos C_8$
$\cos\phi_{5,9}=\cos B_5\cos B_9+\cos C_5\cos C_9$
$\cos\phi_{6,7}=\cos B_6\cos B_7+\cos C_6\cos C_7$
$\cos\phi_{6,9}=\cos B_6\cos B_9+\cos C_6\cos C_9$

CHECK

F_1: $\phi_{1,2}+\phi_{1,3}+\phi_{2,3}=180^\circ$
F_2: $\phi_{1,7}+\phi_{1,8}+\phi_{4,7}+\phi_{4,8}=360^\circ$
F_3: $\phi_{2,8}+\phi_{2,9}+\phi_{5,8}+\phi_{5,9}=360^\circ$
F_4: $\phi_{3,7}+\phi_{3,9}+\phi_{6,7}+\phi_{6,9}=360^\circ$
F_5: $\phi_{4,5}+\phi_{4,6}+\phi_{5,6}=180^\circ$

(E)

OBJECT VERSION 1

$P_1=(-1,0,1)$	$P_2=(0,0,1)$	$P_3=(0,1,1)$
$P_4=(-1,0,-1)$	$P_5=(0,0,-1)$	$P_6=(0,1,-1)$

OBJECT VERSION 2

$P_1=(-1,1,1)$	$P_2=(0,1,1)$	$P_3=(0,2,1)$
$P_4=(0,-1,-1)$	$P_5=(1,-1,-1)$	$P_6=(1,-1,-1)$

OBJECT VERSION 3

$P_1=(-1,0,1)$	$P_2=(0,0,1)$	$P_3=(1,1,1)$
$P_4=(-1,0,-1)$	$P_5=(0,0,-1)$	$P_6=(-1,1,0)$

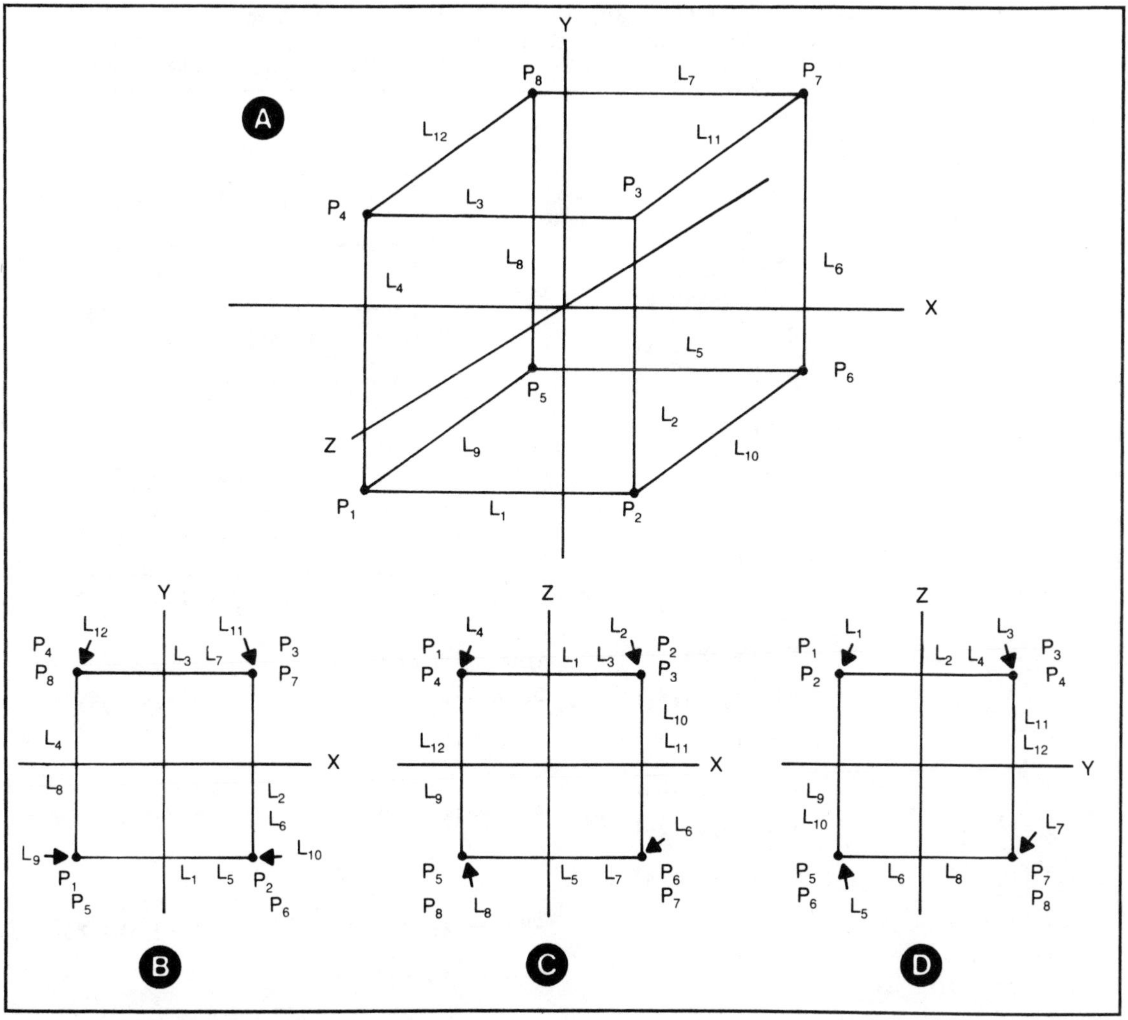

Fig. 15-8. A 6-sided object specified in 3-D space. (A) X-Y perspective view. (B) X-Y planar view. (C) X-Z planar view. (D) Y-Z planar view. See the outlines of analyses in Table 15-8.

- ☐ Translations or combinations of translations.
- ☐ Scalings or combinations of scalings.
- ☐ Rotations or combinations of rotations.

Combinations of translations and scalings do not fall into the category of simple transformations, nor do combinations of translations and rotations, and combinations of scalings and rotations.

The topic at hand, however, is that of simple transformations. You will begin dealing with combinations of different types of transformations in the next section of this chapter.

An important practical feature of simple transformations is that you can use their general principles to simplify or doublecheck the analysis of a transformed line, plane figure or space object. Here is a brief summary of those principles:

☐ Under simple translations: length is an invariant, direction cosines are invariant, angles between lines are invariant, and all simple translations are commutative.

☐ Under simple scalings: length is a variant, direction cosines are variant, angles between lines are variant, and combinations of simple scalings are commutative.

☐ Under simple rotations: length is an invariant, direction cosines are variant, angles between lines are

Table 15-8. Specifications and Analyses of a 6-Sided Object in 3-D Space. See Fig. 15-8. (A) Specifications and Analysis in XYZ Space. (B) Secondary Analysis in the X-Y Plane. (C) Secondary Analysis in the X-Z Plane. (D) Secondary Analysis in the Y-Z Plane. (E) Suggested Coordinates.

(A, B, C, and D correspond to α, β, γ, and δ.)

A

SPECIFICATIONS

$S=\overline{F_1F_2F_3F_4F_5F_6}$

$F_1=\overline{L_1L_2L_3L_4}$

$F_2=\overline{L_1L_5L_9L_{10}}$

$F_3=\overline{L_2L_6L_{10}L_{11}}$

$F_4=\overline{L_3L_7L_{11}L_{12}}$

$F_5=\overline{L_4L_8L_9L_{12}}$

$F_6=\overline{L_5L_6L_7L_8}$

$L_1=\overline{P_1P_2}$

$L_2=\overline{P_2P_3}$

$L_3=\overline{P_3P_4}$

$L_4=\overline{P_4P_1}$

$L_5=\overline{P_5P_6}$

$L_6=\overline{P_6P_7}$

$L_7=\overline{P_7P_8}$

$L_8=\overline{P_8P_5}$

$L_9=\overline{P_1P_5}$

$L_{10}=\overline{P_2P_6}$

$L_{11}=\overline{P_3P_7}$

$L_{12}=\overline{P_4P_8}$

$P_1=(x_1,y_1,z_1)$

$P_2=(x_2,y_2,z_2)$

$P_3=(x_3,y_3,z_3)$

$P_4=(x_4,y_4,z_4)$

$P_5=(x_5,y_5,z_5)$

$P_6=(x_6,y_6,z_6)$

$P_7=(x_7,y_7,z_7)$

$P_8=(x_8,y_8,z_8)$

LENGTHS

$s_1=\sqrt{(x_2-x_1)^2+(y_2-y_1)^2+(z_2-z_1)^2}$

$s_2=\sqrt{(x_3-x_2)^2+(y_3-y_2)^2+(z_3-z_2)^2}$

$s_3=\sqrt{(x_4-x_3)^2+(y_4-y_3)^2+(z_4-z_3)^2}$

$s_4=\sqrt{(x_1-x_4)^2+(y_1-y_4)^2+(z_1-z_4)^2}$

$s_5=\sqrt{(x_6-x_5)^2+(y_6-y_5)^2+(z_6-z_5)^2}$

$s_6=\sqrt{(x_7-x_6)^2+(y_7-y_6)^2+(z_7-z_6)^2}$

$s_7=\sqrt{(x_8-x_7)^2+(y_8-y_7)^2+(z_8-z_7)^2}$

$s_8=\sqrt{(x_5-x_8)^2+(y_5-y_8)^2+(z_5-z_8)^2}$

$s_9=\sqrt{(x_5-x_1)^2+(y_5-y_1)^2+(z_5-z_1)^2}$

$s_{10}=\sqrt{(x_6-x_2)^2+(y_6-y_2)^2+(z_6-z_2)^2}$

$s_{11}=\sqrt{(x_7-x_3)^2+(y_7-y_3)^2+(z_7-z_3)^2}$

$s_{12}=\sqrt{(x_8-x_4)^2+(y_8-y_4)^2+(z_8-z_4)^2}$

DIRECTION COSINES

$\cos A_1=(x_2-x_1)/s_1$ $\cos B_1=(y_2-y_1)/s_1$

$\cos C_1=(z_2-z_1)/s_1$

$\cos A_2=(x_3-x_2)/s_2$ $\cos B_2=(y_3-y_2)/s_2$

$\cos C_2=(z_3-z_2)/s_2$

$\cos A_3=(x_4-x_3)/s_3$ $\cos B_3=(y_4-y_3)/s_3$

$\cos C_3=(z_4-z_3)/s_3$

$\cos A_4=(x_1-x_4)/s_4$ $\cos B_4=(y_1-y_4)/s_4$

$\cos C_4=(z_1-z_4)/s_4$

$\cos A_5=(x_6-x_5)/s_5$ $\cos B_5=(y_6-y_5)/s_5$

$\cos C_5=(z_6-z_5)/s_5$

$\cos A_6=(x_7-x_6)/s_6$ $\cos B_6=(y_7-y_6)/s_6$

$\cos C_6=(z_7-z_6)/s_6$

$\cos A_7=(x_8-x_7)/s_7$ $\cos B_7=(y_8-y_7)/s_7$

$\cos C_7=(z_8-z_7)/s_7$

DIRECTION COSINES (Cont'd) **A** ANGLES (Cont'd)

$\cos A_8=(x_5-x_8)/s_8$ $\cos B_8=(y_5-y_8)/s_8$

$\cos C_8=(z_5-z_8)/s_8$

$\cos A_9=(x_5-x_1)/s_9$ $\cos B_9=(y_5-y_1)/s_9$

$\cos C_9=(z_5-z_1)/s_9$

$\cos A_{10}=(x_6-x_2)/s_{10}$ $\cos B_{10}=(y_6-y_2)/s_{10}$

$\cos C_{10}=(z_6-z_2)/s_{10}$

$\cos A_{11}=(x_7-x_3)/s_{11}$ $\cos B_{11}=(y_7-y_3)/s_{11}$

$\cos C_{11}=(z_7-z_3)/s_{11}$

$\cos A_{12}=(x_8-x_4)/s_{12}$ $\cos B_{12}=(y_8-y_4)/s_{12}$

$\cos C_{12}=(z_8-z_4)/s_{12}$

$\cos\phi_{5,6}=\cos A_5\cos A_6+\cos B_5\cos B_6+\cos C_5\cos C_6$

$\cos\phi_{5,8}=\cos A_5\cos A_8+\cos B_5\cos B_8+\cos C_5\cos C_8$

$\cos\phi_{5,9}=\cos A_5\cos A_9+\cos B_5\cos B_9+\cos C_5\cos C_9$

$\cos\phi_{5,10}=\cos A_5\cos A_{10}+\cos B_5\cos B_{10}+\cos C_5\cos C_{10}$

$\cos\phi_{6,7}=\cos A_6\cos A_7+\cos B_6\cos B_7+\cos C_6\cos C_7$

$\cos\phi_{6,10}=\cos A_6\cos A_{10}+\cos B_6\cos B_{10}+\cos C_6\cos C_{10}$

$\cos\phi_{6,11}=\cos A_6\cos A_{11}+\cos B_6\cos B_{11}+\cos C_6\cos C_{11}$

$\cos\phi_{7,8}=\cos A_7\cos A_8+\cos B_7\cos B_8+\cos C_7\cos C_8$

$\cos\phi_{7,11}=\cos A_7\cos A_{11}+\cos B_7\cos B_{11}+\cos C_7\cos C_{11}$

$\cos\phi_{7,12}=\cos A_7\cos A_{12}+\cos B_7\cos B_{12}+\cos C_7\cos C_{12}$

$\cos\phi_{8,9}=\cos A_8\cos A_9+\cos B_8\cos B_9+\cos C_8\cos C_9$

$\cos\phi_{8,12}=\cos A_8\cos A_{12}+\cos B_8\cos B_{12}+\cos C_8\cos C_{12}$

ANGLES

$\cos\phi_{1,2}=\cos A_1\cos A_2+\cos B_1\cos B_2+\cos C_1\cos C_2$

$\cos\phi_{1,4}=\cos A_1\cos A_4+\cos B_1\cos B_4+\cos C_1\cos C_4$

$\cos\phi_{1,9}=\cos A_1\cos A_9+\cos B_1\cos B_9+\cos C_1\cos C_9$

$\cos\phi_{1,10}=\cos A_1\cos A_{10}+\cos B_1\cos B_{10}+\cos C_1\cos C_{10}$

$\cos\phi_{2,3}=\cos A_2\cos A_3+\cos B_2\cos B_3+\cos C_2\cos C_3$

$\cos\phi_{2,10}=\cos A_2\cos A_{10}+\cos B_2\cos B_{10}+\cos C_2\cos C_{10}$

$\cos\phi_{2,11}=\cos A_2\cos A_{11}+\cos B_2\cos B_{11}+\cos C_2\cos C_{11}$

$\cos\phi_{3,4}=\cos A_3\cos A_4+\cos B_3\cos B_4+\cos C_3\cos C_4$

$\cos\phi_{3,11}=\cos A_3\cos A_{11}+\cos B_3\cos B_{11}+\cos C_3\cos C_{11}$

$\cos\phi_{3,11}=\cos A_3\cos A_{11}+\cos B_3\cos B_{11}+\cos C_3\cos C_{11}$

$\cos\phi_{4,9}=\cos A_4\cos A_9+\cos B_4\cos B_9+\cos C_4\cos C_9$

$\cos\phi_{4,12}=\cos A_4\cos A_{12}+\cos B_4\cos B_{12}+\cos C_4\cos C_{12}$

CHECK

F_1: $\phi_{1,2}+\phi_{1,4}+\phi_{2,3}+\phi_{3,4}=360^\circ$

F_2: $\phi_{1,9}+\phi_{1,10}+\phi_{5,9}+\phi_{5,10}=360^\circ$

F_3: $\phi_{2,10}+\phi_{2,11}+\phi_{6,10}+\phi_{6,11}=360^\circ$

F_4: $\phi_{3,11}+\phi_{3,12}+\phi_{7,11}+\phi_{7,12}=360^\circ$

F_5: $\phi_{4,9}+\phi_{4,12}+\phi_{8,9}+\phi_{8,12}=360^\circ$

F_6: $\phi_{5,6}+\phi_{5,8}+\phi_{6,7}+\phi_{7,8}=360^\circ$

Table 15-8. Specifications and Analyses of a 6-Sided Object in 3-D Space. See Fig. 15-8. (A) Specifications and Analysis in XYZ Space. (B) Secondary Analysis in the X-Y Plane. (C) Secondary Analysis in the X-Z Plane. (D) Secondary Analysis in the X-Z Plane. (E) Suggested Coordinates. (Continued from page 207.)

B

SECONDARY ANALYSIS -- XY PLANE

PLOTTED POINTS

$P_1=(x_1,y_1)$ $P_2=(x_2,y_2)$

$P_3=(x_3,y_3)$ $P_4=(x_4,y_4)$

$P_5=(x_5,y_5)$ $P_6=(x_6,y_6)$

$P_7=(x_7,y_7)$ $P_8=(x_8,y_8)$

LENGTHS

$s_1=\sqrt{(x_2-x_1)^2+(y_2-y_1)^2}$ $s_2=\sqrt{(x_3-x_2)^2+(y_3-y_2)^2}$

$s_3=\sqrt{(x_4-x_3)^2+(y_4-y_3)^2}$ $s_4=\sqrt{(x_1-x_4)^2+(y_1-y_4)^2}$

$s_5=\sqrt{(x_6-x_5)^2+(y_6-y_5)^2}$ $s_6=\sqrt{(x_7-x_6)^2+(y_7-y_6)^2}$

$s_7=\sqrt{(x_8-x_7)^2+(y_8-y_7)^2}$ $s_8=\sqrt{(x_5-x_8)^2+(y_5-y_8)^2}$

$s_9=\sqrt{(x_5-x_1)^2+(y_5-y_1)^2}$ $s_{10}=\sqrt{(x_6-x_2)^2+(y_6-y_2)^2}$

$s_{11}=\sqrt{(x_7-x_3)^2+(y_7-y_3)^2}$ $s_{12}=\sqrt{(x_8-x_4)^2+(y_8-y_4)^2}$

DIRECTION COSINES

$\cos A_1=(x_2-x_1)/s_1$ $\cos B_1=(y_2-y_1)/s_1$

$\cos A_2=(x_3-x_2)/s_2$ $\cos B_2=(y_3-y_2)/s_2$

$\cos A_3=(x_4-x_3)/s_3$ $\cos B_3=(y_4-y_3)/s_3$

$\cos A_4=(x_1-x_4)/s_4$ $\cos B_4=(y_1-y_4)/s_4$

$\cos A_5=(x_6-x_5)/s_5$ $\cos B_5=(y_6-y_5)/s_5$

$\cos A_6=(x_7-x_6)/s_6$ $\cos B_6=(y_7-y_6)/s_6$

$\cos A_7=(x_8-x_7)/s_7$ $\cos B_7=(y_8-y_7)/s_7$

$\cos A_8=(x_5-x_8)/s_8$ $\cos B_8=(y_5-y_8)/s_8$

$\cos A_9=(x_5-x_1)/s_9$ $\cos B_9=(y_5-y_1)/s_9$

$\cos A_{10}=(x_6-x_2)/s_{10}$ $\cos B_{10}=(y_6-y_2)/s_{10}$

$\cos A_{11}=(x_7-x_3)/s_{11}$ $\cos B_{11}=(y_7-y_3)/s_{11}$

$\cos A_{12}=(x_8-x_4)/s_{12}$ $\cos B_{12}=(y_8-y_4)/s_{12}$

ANGLES

$\cos\phi_{1,2}=\cos A_1\cos A_2+\cos B_1\cos B_2$

$\cos\phi_{1,4}=\cos A_1\cos A_4+\cos B_1\cos B_4$

$\cos\phi_{1,9}=\cos A_1\cos A_9+\cos B_1\cos B_9$

$\cos\phi_{1,10}=\cos A_1\cos A_{10}+\cos B_1\cos B_{10}$

$\cos\phi_{2,3}=\cos A_2\cos A_3+\cos B_2\cos B_3$

$\cos\phi_{2,10}=\cos A_2\cos A_{10}+\cos B_2\cos B_{10}$

$\cos\phi_{2,11}=\cos A_2\cos A_{11}+\cos B_2\cos B_{11}$

$\cos\phi_{3,4}=\cos A_3\cos A_4+\cos B_3\cos B_4$

$\cos\phi_{3,11}=\cos A_3\cos A_{11}+\cos B_3\cos B_{11}$

$\cos\phi_{3,11}=\cos A_3\cos A_{11}+\cos B_3\cos B_{11}$

$\cos\phi_{4,9}=\cos A_4\cos A_9+\cos B_4\cos B_9$

$\cos\phi_{4,12}=\cos A_4\cos A_{12}+\cos B_4\cos B_{12}$

$\cos\phi_{5,6}=\cos A_5\cos A_6+\cos B_5\cos B_6$

$\cos\phi_{5,8}=\cos A_5\cos A_8+\cos B_5\cos B_8$

$\cos\phi_{5,9}=\cos A_5\cos A_9+\cos B_5\cos B_9$

$\cos\phi_{5,10}=\cos A_5\cos A_{10}+\cos B_5\cos B_{10}$

B

ANGLES (Cont'd)

$\cos\phi_{6,7}=\cos A_6\cos A_7+\cos B_6\cos B_7$

$\cos\phi_{6,10}=\cos A_6\cos A_{10}+\cos B_6\cos B_{10}$

$\cos\phi_{6,11}=\cos A_6\cos A_{11}+\cos B_6\cos B_{11}$

$\cos\phi_{7,8}=\cos A_7\cos A_8+\cos B_7\cos B_8$

$\cos\phi_{7,11}=\cos A_7\cos A_{11}+\cos B_7\cos B_{11}$

$\cos\phi_{7,12}=\cos A_7\cos A_{12}+\cos B_7\cos B_{12}$

$\cos\phi_{8,9}=\cos A_8\cos A_9+\cos B_8\cos B_9$

$\cos\phi_{8,12}=\cos A_8\cos A_{12}+\cos B_8\cos B_{12}$

CHECK

F_1: $\phi_{1,2}+\phi_{1,4}+\phi_{2,3}+\phi_{3,4}=360^\circ$

F_2: $\phi_{1,9}+\phi_{1,10}+\phi_{5,9}+\phi_{5,10}=360^\circ$

F_3: $\phi_{2,10}+\phi_{2,11}+\phi_{6,10}+\phi_{6,11}=360^\circ$

F_4: $\phi_{3,11}+\phi_{3,12}+\phi_{7,11}+\phi_{7,12}=360^\circ$

F_5: $\phi_{4,9}+\phi_{4,12}+\phi_{8,9}+\phi_{8,12}=360^\circ$

F_6: $\phi_{5,6}+\phi_{5,8}+\phi_{6,7}+\phi_{7,8}=360^\circ$

SECONDARY ANALYSIS -- XZ PLANE

C

PLOTTED POINTS

$P_1=(x_1,z_1)$ $P_2=(x_2,z_2)$

$P_3=(x_3,z_3)$ $P_4=(x_4,z_4)$

$P_5=(x_5,z_5)$ $P_6=(x_6,z_6)$

$P_7=(x_7,z_7)$ $P_8=(x_8,z_8)$

LENGTHS

$s_1=\sqrt{(x_2-x_1)^2+(z_2-z_1)^2}$ $s_2=\sqrt{(x_3-x_2)^2+(z_3-z_2)^2}$

$s_3=\sqrt{(x_4-x_3)^2+(z_4-z_3)^2}$ $s_4=\sqrt{(x_1-x_4)^2+(z_1-z_4)^2}$

Table 15-8. Specifications and Analyses of a 6-Sided Object in 3-D Space. See Fig. 15-8. (A) Specifications and Analysis in XYZ Space. (B) Secondary Analysis in the X-Y Plane. (C) Secondary Analysis in the X-Z Plane. (D) Secondary Analysis in the X-Z Plane. (E) Suggested Coordinates. **(Continued from page 209.)**

C

LENGTHS

$s_5=\sqrt{(x_6-x_5)^2+(z_6-z_5)^2}$

$s_6=\sqrt{(x_7-x_6)^2+(z_7-z_6)^2}$

$s_7=\sqrt{(x_8-x_7)^2+(z_8-z_7)^2}$

$s_8=\sqrt{(x_5-x_8)^2+(z_5-z_8)^2}$

$s_9=\sqrt{(x_5-x_1)^2+(z_5-z_1)^2}$

$s_{10}=\sqrt{(x_6-x_2)^2+(z_6-z_2)^2}$

$s_{11}=\sqrt{(x_7-x_3)^2+(z_7-z_3)^2}$

$s_{12}=\sqrt{(x_8-x_4)^2+(z_8-z_4)^2}$

DIRECTION COSINES

$\cos A_1=(x_2-x_1)/s_1$ $\cos C_1=(z_2-z_1)/s_1$

$\cos A_2=(x_3-x_2)/s_2$ $\cos C_2=(z_3-z_2)/s_2$

$\cos A_3=(x_4-x_3)/s_3$ $\cos C_3=(z_4-z_3)/s_3$

$\cos A_4=(x_1-x_4)/s_4$ $\cos C_4=(z_1-z_4)/s_4$

$\cos A_5=(x_6-x_5)/s_5$ $\cos C_5=(z_6-z_5)/s_5$

$\cos A_6=(x_7-x_6)/s_6$ $\cos C_6=(z_7-z_6)/s_6$

$\cos A_7=(x_8-x_7)/s_7$ $\cos C_7=(z_8-z_7)/s_7$

$\cos A_8=(x_5-x_8)/s_8$ $\cos C_8=(z_5-z_8)/s_8$

$\cos A_9=(x_5-x_1)/s_9$ $\cos C_9=(z_5-z_1)/s_9$

ANGLES

$\cos\phi_{1,2}=\cos A_1\cos A_2+\cos C_1\cos C_2$

$\cos\phi_{1,4}=\cos A_1\cos A_4+\cos C_1\cos C_4$

$\cos\phi_{1,9}=\cos A_1\cos A_9+\cos C_1\cos C_9$

$\cos\phi_{1,10}=\cos A_1\cos A_{10}+\cos C_1\cos C_{10}$

$\cos\phi_{2,3}=\cos A_2\cos A_3+\cos C_2\cos C_3$

$\cos\phi_{2,10}=\cos A_2\cos A_{10}+\cos C_2\cos C_{10}$

$\cos\phi_{2,11}=\cos A_2\cos A_{11}+\cos C_2\cos C_{11}$

$\cos\phi_{3,4}=\cos A_3\cos A_4+\cos C_3\cos C_4$

$\cos\phi_{3,11}=\cos A_3\cos A_{11}+\cos C_3\cos C_{11}$

$\cos\phi_{3,11}=\cos A_3\cos A_{11}+\cos C_3\cos C_{11}$

ANGLES (Cont'd)

$\cos\phi_{4,9}=\cos A_4\cos A_9+\cos C_4\cos C_9$

$\cos\phi_{4,12}=\cos A_4\cos A_{12}+\cos C_4\cos C_{12}$

$\cos\phi_{5,6}=\cos A_5\cos A_6+\cos C_5\cos C_6$

$\cos\phi_{5,8}=\cos A_5\cos A_8+\cos C_5\cos C_8$

$\cos\phi_{5,9}=\cos A_5\cos A_9+\cos C_5\cos C_9$

$\cos\phi_{5,10}=\cos A_5\cos A_{10}+\cos C_5\cos C_{10}$

$\cos\phi_{6,7}=\cos A_6\cos A_7+\cos C_6\cos C_7$

$\cos\phi_{6,10}=\cos A_6\cos A_{10}+\cos C_6\cos C_{10}$

$\cos\phi_{6,11}=\cos A_6\cos A_{11}+\cos C_6\cos C_{11}$

$\cos\phi_{7,8}=\cos A_7\cos A_8+\cos C_7\cos C_8$

$\cos\phi_{7,11}=\cos A_7\cos A_{11}+\cos C_7\cos C_{11}$

$\cos\phi_{7,12}=\cos A_7\cos A_{12}+\cos C_7\cos C_{12}$

$\cos\phi_{8,9}=\cos A_8\cos A_9+\cos C_8\cos C_9$

$\cos\phi_{8,12}=\cos A_8\cos A_{12}+\cos C_8\cos C_{12}$

CHECK

F_1: $\phi_{1,2}+\phi_{1,4}+\phi_{2,3}+\phi_{3,4}=360^\circ$

F_2: $\phi_{1,9}+\phi_{1,10}+\phi_{5,9}+\phi_{5,10}=360^\circ$

F_3: $\phi_{2,10}+\phi_{2,11}+\phi_{6,10}+\phi_{6,11}=360^\circ$

F_4: $\phi_{3,11}+\phi_{3,12}+\phi_{7,11}+\phi_{7,12}=360^\circ$

F_5: $\phi_{4,9}+\phi_{4,12}+\phi_{8,9}+\phi_{8,12}=360^\circ$

F_6: $\phi_{5,6}+\phi_{5,8}+\phi_{6,7}+\phi_{7,8}=360^\circ$

D

SECONDARY ANALYSIS -- YZ PLANE

PLOTTED POINTS

$P_1=(y_1,z_1)$ $\quad$ $P_2=(y_2,z_2)$

$P_3=(y_3,z_3)$ $\quad$ $P_4=(y_4,z_4)$

$P_5=(y_5,z_5)$ $\quad$ $P_6=(y_6,z_6)$

$P_7=(y_7,z_7)$ $\quad$ $P_8=(y_8,z_8)$

LENGTHS

$$s_1=\sqrt{(y_2-y_1)^2+(z_2-z_1)^2} \qquad s_2=\sqrt{(y_3-y_2)^2+(z_3-z_2)^2}$$

$$s_3=\sqrt{(y_4-y_3)^2+(z_4-z_3)^2} \qquad s_4=\sqrt{(y_1-y_4)^2+(z_1-z_4)^2}$$

$$s_5=\sqrt{(y_6-y_5)^2+(z_6-z_5)^2} \qquad s_6=\sqrt{(y_7-y_6)^2+(z_7-z_6)^2}$$

$$s_7=\sqrt{(y_8-y_7)^2+(z_8-z_7)^2} \qquad s_8=\sqrt{(y_5-y_8)^2+(z_5-z_8)^2}$$

$$s_9=\sqrt{(y_5-y_1)^2+(z_5-z_1)^2} \qquad s_{10}=\sqrt{(y_6-y_2)^2+(z_6-z_2)^2}$$

$$s_{11}=\sqrt{(y_7-y_3)^2+(z_7-z_3)^2} \qquad s_{12}=\sqrt{(y_8-y_4)^2+(z_8-z_4)^2}$$

DIRECTION COSINES

$\cos B_1=(y_2-y_1)/s_1$ $\quad$ $\cos C_1=(z_2-z_1)/s_1$

$\cos B_2=(y_3-y_2)/s_2$ $\quad$ $\cos C_2=(z_3-z_2)/s_2$

$\cos B_3=(y_4-y_3)/s_3$ $\quad$ $\cos C_3=(z_4-z_3)/s_3$

$\cos B_4=(y_1-y_4)/s_4$ $\quad$ $\cos C_4=(z_1-z_4)/s_4$

$\cos B_5=(y_6-y_5)/s_5$ $\quad$ $\cos C_5=(z_6-z_5)/s_5$

$\cos B_6=(y_7-y_6)/s_6$ $\quad$ $\cos C_6=(z_7-z_6)/s_6$

$\cos B_7=(y_8-y_7)/s_7$ $\quad$ $\cos C_7=(z_8-z_7)/s_7$

$\cos B_8=(y_5-y_8)/s_8$ $\quad$ $\cos C_8=(z_5-z_8)/s_8$

$\cos B_9=(y_5-y_1)/s_9$ $\quad$ $\cos C_9=(z_5-z_1)/s_9$

$\cos B_{10}=(y_6-y_2)/s_{10}$ $\quad$ $\cos C_{10}=(z_6-z_2)/s_{10}$

$\cos B_{11}=(y_7-y_3)/s_{11}$ $\quad$ $\cos C_{11}=(z_7-z_3)/s_{11}$

$\cos B_{12}=(y_8-y_4)/s_{12}$ $\quad$ $\cos C_{12}=(z_8-z_4)/s_{12}$

ANGLES

$\cos\phi_{1,2}=\cos B_1\cos B_2+\cos C_1\cos_2$

$\cos\phi_{1,4}=\cos B_1\cos B_4+\cos C_1\cos C_4$

$\cos\phi_{1,9}=\cos B_1\cos B_9+\cos C_1\cos C_9$

$\cos\phi_{1,10}=\cos B_1\cos B_{10}+\cos C_1\cos C_{10}$

$\cos\phi_{2,3}=\cos B_2\cos B_3+\cos C_2\cos C_3$

$\cos\phi_{2,10}=\cos B_2\cos B_{10}+\cos C_2\cos C_{10}$

$\cos\phi_{2,11}=\cos B_2\cos B_{11}+\cos C_2\cos C_{11}$

$\cos\phi_{3,4}=\cos B_3\cos B_4+\cos C_3\cos C_4$

$\cos\phi_{3,11}=\cos B_3\cos B_{11}+\cos C_3\cos C_{11}$

$\cos\phi_{3,11}=\cos B_3\cos B_{11}+\cos C_3\cos C_{11}$

$\cos\phi_{4,9}=\cos B_4\cos B_9+\cos C_4\cos C_9$

$\cos\phi_{4,12}=\cos B_4\cos B_{12}+\cos C_4\cos C_{12}$

Table 15-8. Specifications and Analyses of a 6-Sided Object in 3-D Space. See Fig. 15-8. (A) Specifications and Analysis in XYZ Space. (B) Secondary Analysis in the X-Y Plane. (C) Secondary Analysis in the X-Z Plane. (D) Secondary Analysis in the X-Z Plane. (E) Suggested Coordinates. (Continued from page 211.)

ANGLES (Cont'd)

$\cos\phi_{5,6}=\cos B_5\cos B_6+\cos C_5\cos C_6$

$\cos\phi_{5,8}=\cos B_5\cos B_8+\cos C_5\cos C_8$

$\cos\phi_{5,9}=\cos B_5\cos B_9+\cos C_5\cos C_9$

$\cos\phi_{5,10}=\cos B_5\cos B_{10}+\cos C_5\cos C_{10}$

$\cos\phi_{6,7}=\cos B_6\cos B_7+\cos C_6\cos C_7$

$\cos\phi_{6,10}=\cos B_6\cos B_{10}+\cos C_6\cos C_{10}$

$\cos\phi_{6,11}=\cos B_6\cos B_{11}+\cos C_6\cos C_{11}$

$\cos\phi_{7,8}=\cos B_7\cos B_8+\cos C_7\cos C_8$

$\cos\phi_{7,11}=\cos B_7\cos B_{11}+\cos C_7\cos C_{11}$

$\cos\phi_{7,12}=\cos B_7\cos B_{12}+\cos C_7\cos C_{12}$

$\cos\phi_{8,9}=\cos B_8\cos B_9+\cos C_8\cos C_9$

$\cos\phi_{8,12}=\cos B_8\cos B_{12}+\cos C_8\ \phi_{12}$

D

CHECK

F_1: $\phi_{1,2}+\phi_{1,4}+\phi_{2,3}+\phi_{3,4}=360^\circ$

F_2: $\phi_{1,9}+\phi_{1,10}+\phi_{5,9}+\phi_{5,10}=360^\circ$

F_3: $\phi_{2,10}+\phi_{2,11}+\phi_{6,10}+\phi_{6,11}=360^\circ$

F_4: $\phi_{3,11}+\phi_{3,12}+\phi_{7,11}+\phi_{7,12}=360^\circ$

F_5: $\phi_{4,9}+\phi_{4,12}+\phi_{8,9}+\phi_{8,12}=360^\circ$

F_6: $\phi_{5,6}+\phi_{5,8}+\phi_{6,7}+\phi_{7,8}=360^\circ$

OBJECT VERSION 1

$P_1=(-1,-1,1)$ $P_2=(1,-1,1)$ $P_3=(1,1,1)$

$P_4=(-1,1,1)$ $P_5=(-1,-1,-1)$ $P_6=(1,-1,-1)$

$P_7=(1,1,-1)$ $P_8=(-1,1,-1)$

OBJECT VERSION 2

$P_1=(-1,-1,1)$ $P_2=(1,-1,1)$ $P_3=(2,1,1)$

$P_4=(0,1,1)$ $P_5=(-1,-1,-1)$ $P_6=(1,-1,-1)$

$P_7=(2,1,-1)$ $P_8=(0,1,-1)$

OBJECT VERSION 3

$P_1=(-1,-2,0)$ $P_2=(1,-2,1)$ $P_3=(2,0,1)$

$P_4=(0,0,1)$ $P_5=(-1,0,-1)$ $P_6=(1,0,-1)$

$P_7=(2,2,0)$ $P_8=(0,2,-1)$

E

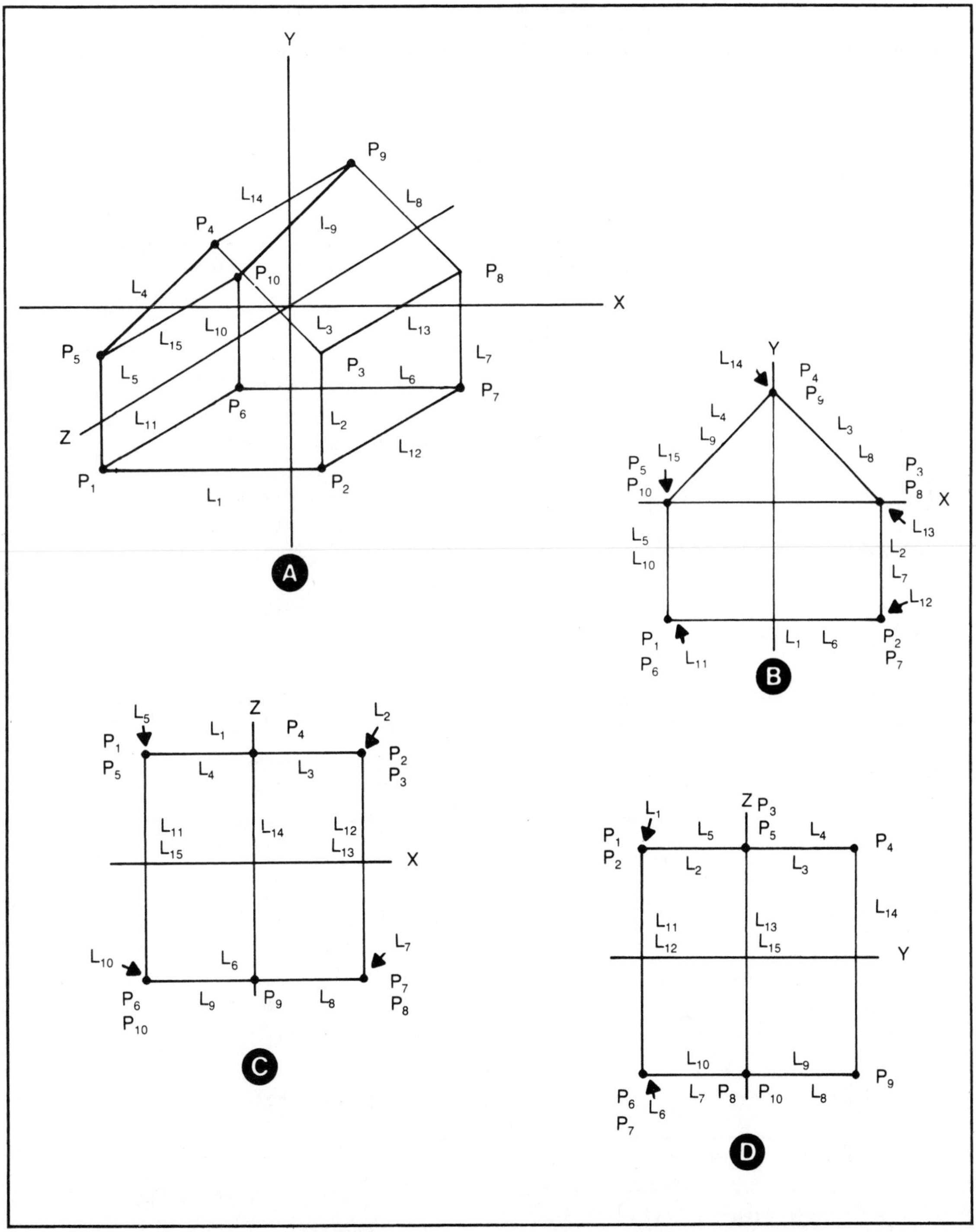

Fig. 15-9. A 7-sided object specified in 3-D space. (A) X-Y perspective view. (B) X-Y planar view. (C) X-Z planar view. (D) Y-Z planar view. See the outlines of analyses in Table 15-9.

Table 15-9. Specifications and Analyses of a 7-Sided Object in 3-D Space. See Fig. 15-9. (A) Specifications and Analysis in XYZ Space. (B) Secondary Analysis in the X-Y Plane. (C) Secondary Analysis in the X-Z Plane. (D) Secondary Analysis in the Y-Z Plane. (E) Suggested Coordinates.

(A, B, C, and D correspond to α, β, γ, and δ.)

A

SPECIFICATIONS

$S=\overline{F_1F_2F_3F_4F_5F_6F_7}$

$F_1=\overline{L_1L_2L_3L_4L_5}$

$F_2=\overline{L_1L_6L_{11}L_{12}}$

$F_3=\overline{L_2L_7L_{12}L_{13}}$

$F_4=\overline{L_3L_8L_{13}L_{14}}$

$F_5=\overline{L_4L_9L_{14}L_{15}}$

$F_6=\overline{L_5L_{10}L_{11}L_{15}}$

$F_7=\overline{L_6L_7L_8L_9L_{10}}$

$L_1=\overline{P_1P_2}$	$P_1=(x_1,y_1,z_1)$
$L_2=\overline{P_2P_3}$	$P_2=(x_2,y_2,z_2)$
$L_3=\overline{P_3P_4}$	$P_3=(x_3,y_3,z_3)$
$L_4=\overline{P_4P_5}$	$P_4=(x_4,y_4,z_4)$
$L_5=\overline{P_5P_1}$	$P_5=(x_5,y_5,z_5)$
$L_6=\overline{P_6P_7}$	$P_6=(x_6,y_6,z_6)$
$L_7=\overline{P_7P_8}$	$P_7=(x_7,y_7,z_7)$
$L_8=\overline{P_8P_9}$	$P_8=(x_8,y_8,z_8)$
$L_9=\overline{P_9P_{10}}$	$P_9=(x_9,y_9,z_9)$
$L_{10}=\overline{P_{10}P_6}$	$P_{10}=(x_{10},y_{10},z_{10})$
$L_{11}=\overline{P_1P_6}$	
$L_{12}=\overline{P_2P_7}$	
$L_{13}=\overline{P_3P_8}$	
$L_{14}=\overline{P_4P_9}$	
$L_{15}=\overline{P_5P_{10}}$	

LENGTHS

$s_1=\sqrt{(x_2-x_1)^2+(y_2-y_1)^2+(z_2-z_1)^2}$

$s_2=\sqrt{(x_3-x_2)^2+(y_3-y_2)^2+(z_3-z_2)^2}$

$s_3=\sqrt{(x_4-x_3)^2+(y_4-y_3)^2+(z_4-z_3)^2}$

$s_4=\sqrt{(x_5-x_4)^2+(y_5-y_4)^2+(z_5-z_4)^2}$

$s_5=\sqrt{(x_1-x_5)^2+(y_1-y_5)^2+(z_1-z_5)^2}$

$s_6=\sqrt{(x_7-x_6)^2+(y_7-y_6)^2+(z_7-z_6)^2}$

$s_7=\sqrt{(x_8-x_7)^2+(y_8-y_7)^2+(z_8-z_7)^2}$

$s_8=\sqrt{(x_9-x_8)^2+(y_9-y_8)^2+(z_9-z_8)^2}$

$s_9=\sqrt{(x_{10}-x_9)^2+(y_{10}-y_9)^2+(z_{10}-z_9)^2}$

$s_{10}=\sqrt{(x_6-x_{10})^2+(y_6-y_{10})^2+(z_6-z_{10})^2}$

$s_{11}=\sqrt{(x_6-x_1)^2+(y_6-y_1)^2+(z_6-z_1)^2}$

$s_{12}=\sqrt{(x_7-x_2)^2+(y_7-y_2)^2+(z_7-z_2)^2}$

$s_{13}=\sqrt{(x_8-x_3)^2+(y_8-y_3)^2+(z_8-z_3)^2}$

$s_{14}=\sqrt{(x_9-x_4)^2+(y_9-y_4)^2+(z_9-z_4)^2}$

$s_{15}=\sqrt{(x_{10}-x_5)^2+(y_{10}-y_5)^2+(z_{10}-z_5)^2}$

DIRECTION COSINES

$\cos A_1=(x_2-x_1)/s_1$ $\cos B_1=(y_2-y_1)/s_1$

$\cos C_1=(z_2-z_1)/s_1$

$\cos A_2=(x_3-x_2)/s_2$ $\cos B_2=(y_3-y_2)/s_2$

$\cos C_2=(z_3-z_2)/s_2$

$\cos A_3=(x_4-x_3)/s_3$ $\cos B_3=(y_4-y_3)/s_3$

$\cos C_3=(z_4-z_3)/s_3$

DIRECTION COSINES (Cont'd)

$\cos A_4=(x_5-x_4)/s_4$ $\cos B_4=(y_5-y_4)/s_4$

$\cos C_4=(z_5-z_4)/s_4$

$\cos A_5=(x_1-x_5)/s_5$ $\cos B_5=(y_1-y_5)/s_5$

$\cos C_5=(z_1-z_5)/s_5$

$\cos A_6=(x_7-x_6)/s_6$ $\cos B_6=(y_7-y_6)/s_6$

$\cos C_6=(z_7-z_6)/s_6$

$\cos A_7=(x_8-x_7)/s_7$ $\cos B_7=(y_8-y_7)/s_7$

$\cos C_7=(z_8-z_7)/s_7$

$\cos A_8=(x_9-x_8)/s_8$ $\cos B_8=(y_9-y_8)/s_8$

$\cos C_8=(z_9-z_8)/s_8$

$\cos A_9=(x_{10}-x_9)/s_9$ $\cos B_9=(y_{10}-y_9)/s_9$

$\cos C_9=(z_{10}-z_9)/s_9$

$\cos A_{10}=(x_6-x_{10})/s_{10}$ $\cos B_{10}=(y_6-y_{10})/s_{10}$

$\cos C_{10}=(z_6-z_{10})/s_{10}$

$\cos A_{11}=(x_6-x_1)/s_{11}$ $\cos B_{11}=(y_6-y_1)/s_{11}$

$\cos C_{11}=(z_6-z_1)/s_{11}$

$\cos A_{12}=(x_7-x_2)/s_{12}$ $\cos B_{12}=(y_7-y_2)/s_{12}$

$\cos C_{12}=(z_7-z_2)/s_{12}$

$\cos A_{13}=(x_8-x_3)/s_{13}$ $\cos B_{13}=(y_8-y_3)/s_{13}$

$\cos C_{13}=(z_8-z_3)/s_{13}$

$\cos A_{14}=(x_9-x_4)/s_{14}$ $\cos B_{14}=(y_9-y_4)/s_{14}$

$\cos C_{14}=(z_9-z_4)/s_{14}$

$\cos A_{15}=(x_{10}-x_5)/s_{15}$ $\cos B_{15}=(y_{10}-y_5)/s_{15}$

$\cos C_{15}=(z_{10}-z_5)/s_{15}$

A

ANGLES

$\cos\phi_{1,2}=\cos A_1\cos A_2+\cos B_1\cos B_2+\cos C_1\cos C_2$

$\cos\phi_{1,5}=\cos A_1\cos A_5+\cos B_1\cos B_5+\cos C_1\cos C_5$

$\cos\phi_{1,11}=\cos A_1\cos A_{11}+\cos B_1\cos B_{11}+\cos C_1\cos C_{11}$

$\cos\phi_{1,12}=\cos A_1\cos A_{12}+\cos B_1\cos B_{12}+\cos C_1\cos C_{12}$

$\cos\phi_{2,3}=\cos A_2\cos A_3+\cos B_2\cos B_3+\cos C_2\cos C_3$

$\cos\phi_{2,12}=\cos A_2\cos A_{12}+\cos B_2\cos B_{12}+\cos C_2\cos C_{12}$

$\cos\phi_{2,13}=\cos A_2\cos A_{13}+\cos B_2\cos B_{13}+\cos C_2\cos C_{13}$

$\cos\phi_{3,4}=\cos A_3\cos A_4+\cos B_3\cos B_4+\cos C_3\cos C_4$

$\cos\phi_{3,13}=\cos A_3\cos A_{13}+\cos B_3\cos B_{13}+\cos C_3\cos C_{13}$

$\cos\phi_{3,14}=\cos A_3\cos A_{14}+\cos A_3\cos B_{14}+\cos B_3\cos C_{14}$

$\cos\phi_{4,5}=\cos A_4\cos A_5+\cos B_4\cos B_5+\cos C_4\cos C_5$

Table 15-9. Specifications and Analyses of a 7-Sided Object in 3-D Space. See Fig. 15-9. (A) Specifications and Analysis in XYZ Space. (B) Secondary Analysis in the X-Y Plane. (C) Secondary Analysis in the X-Z Plane. (D) Secondary Analysis in the Y-Z Plane. (E) Suggested Coordinates. (Continued from page 215.)

ANGLES (Cont'd)

A

$\cos\phi_{4,14}=\cos A_4\cos A_{14}+\cos B_4\cos B_{14}+\cos C_4\cos C_{14}$

$\cos\phi_{4,15}=\cos A_4\cos A_{15}+\cos B_4\cos B_{15}+\cos C_4\cos C_{15}$

$\cos\phi_{5,11}=\cos A_5\cos A_{11}+\cos B_5\cos B_{11}+\cos C_5\cos C_{11}$

$\cos\phi_{5,15}=\cos A_5\cos A_{15}+\cos B_5\cos B_{15}+\cos C_5\cos C_{15}$

$\cos\phi_{6,7}=\cos A_6\cos A_7+\cos B_6\cos B_7+\cos C_6\cos C_7$

$\cos\phi_{6,10}=\cos A_6\cos A_{10}+\cos B_6\cos B_{10}+\cos C_6\cos C_{10}$

$\cos\phi_{6,11}=\cos A_6\cos A_{11}+\cos B_6\cos B_{11}+\cos C_6\cos C_{11}$

$\cos\phi_{6,12}=\cos A_6\cos A_{12}+\cos B_6\cos B_{12}+\cos C_6\cos C_{12}$

$\cos\phi_{7,8}=\cos A_7\cos A_8+\cos B_7\cos B_8+\cos C_7\cos C_8$

$\cos\phi_{7,12}=\cos A_7\cos A_{12}+\cos B_7\cos B_{12}+\cos C_7\cos C_{12}$

$\cos\phi_{7,13}=\cos A_7\cos A_{13}+\cos B_7\cos B_{13}+\cos C_7\cos C_{13}$

$\cos\phi_{8,9}=\cos A_8\cos A_9+\cos B_8\cos B_9+\cos C_8\cos C_9$

$\cos\phi_{8,13}=\cos A_8\cos A_{13}+\cos B_8\cos B_{13}+\cos C_8\cos C_{13}$

$\cos\phi_{8,14}=\cos A_8\cos A_{14}+\cos B_8\cos B_{14}+\cos C_8\cos C_{14}$

$\cos\phi_{9,10}=\cos A_9\cos A_{10}+\cos B_9\cos B_{10}+\cos C_9\cos C_{10}$

$\cos\phi_{9,14}=\cos A_9\cos A_{14}+\cos B_9\cos B_{14}+\cos C_9\cos C_{14}$

$\cos\phi_{9,15}=\cos A_9\cos A_{15}+\cos B_9\cos B_{15}+\cos C_9\cos C_{15}$

$\cos\phi_{10,11}=\cos A_{10}\cos A_{11}+\cos B_{10}\cos B_{11}+\cos C_{10}\cos C_{11}$

$\cos\phi_{10,15}=\cos A_{10}\cos A_{15}+\cos B_{10}\cos B_{15}+\cos C_{10}\cos C_{15}$

CHECK

F_1: $\phi_{1,2}+\phi_{1,5}+\phi_{2,3}+\phi_{3,4}+\phi_{4,5}=540^\circ$

F_2: $\phi_{1,12}+\phi_{1,11}+\phi_{6,11}+\phi_{6,12}=360^\circ$

F_3: $\phi_{2,12}+\phi_{2,13}+\phi_{7,12}+\phi_{7,13}=360^\circ$

F_4: $\phi_{3,13}+\phi_{3,14}+\phi_{8,13}+\phi_{8,14}=360^\circ$

F_5: $\phi_{4,14}+\phi_{4,15}+\phi_{9,14}+\phi_{9,15}=360^\circ$

F_6: $\phi_{5,11}+\phi_{5,15}+\phi_{10,11}+\phi_{10,15}=360^\circ$

F_7: $\phi_{6,7}+\phi_{6,10}+\phi_{7,8}+\phi_{8,9}+\phi_{9,10}=540^\circ$

SECONDARY ANALYSIS -- XY PLANE

B

PLOTTED POINTS

$P_1=(x_1,y_1)$	$P_2=(x_2,y_2)$
$P_3=(x_3,y_3)$	$P_4=(x_4,y_4)$
$P_5=(x_5,y_5)$	$P_6=(x_6,y_6)$
$P_7=(x_7,y_7)$	$P_8=(x_8,y_8)$
$P_9=(x_9,y_9)$	$P_{10}=(x_{10},y_{10})$

LENGTHS

$$s_1=\sqrt{(x_2-x_1)^2+(y_2-y_1)^2} \qquad s_2=\sqrt{(x_3-x_2)^2+(y_3-y_2)^2}$$

$$s_3=\sqrt{(x_4-x_3)^2+(y_4-y_3)^2} \qquad s_4=\sqrt{(x_5-x_4)^2+(y_5-y_4)^2}$$

$$s_5=\sqrt{(x_1-x_5)^2+(y_1-y_5)^2} \qquad s_6=\sqrt{(x_7-x_6)^2+(y_7-y_6)^2}$$

$$s_7=\sqrt{(x_8-x_7)^2+(y_8-y_7)^2} \qquad s_8=\sqrt{(x_9-x_8)^2+(y_9-y_8)^2}$$

$$s_9=\sqrt{(x_{10}-x_9)^2+(y_{10}-y_9)^2} \qquad s_{10}=\sqrt{(x_6-x_{10})^2+(y_6-y_{10})^2}$$

$$s_{11}=\sqrt{(x_6-x_1)^2+(y_6-y_1)^2} \qquad s_{12}=\sqrt{(x_7-x_2)^2+(y_7-y_2)^2}$$

$$s_{13}=\sqrt{(x_8-x_3)^2+(y_8-y_3)^2} \qquad s_{14}=\sqrt{(x_9-x_4)^2+(y_9-y_4)^2}$$

$$s_{15}=\sqrt{(x_{10}-x_5)^2+(y_{10}-y_5)^2}$$

DIRECTION COSINES

$\cos A_1=(x_2-x_1)/s_1$	$\cos B_1=(y_2-y_1)/s_1$
$\cos A_2=(x_3-x_2)/s_2$	$\cos B_2=(y_3-y_2)/s_2$
$\cos A_3=(x_4-x_3)/s_3$	$\cos B_3=(y_4-y_3)/s_3$
$\cos A_4=(x_5-x_4)/s_4$	$\cos B_4=(y_5-y_4)/s_4$
$\cos A_5=(x_1-x_5)/s_5$	$\cos B_5=(y_1-y_5)/s_5$
$\cos A_6=(x_7-x_6)/s_6$	$\cos B_6=(y_7-y_6)/s_6$
$\cos A_7=(x_8-x_7)/s_7$	$\cos B_7=(y_8-y_7)/s_7$
$\cos A_8=(x_9-x_8)/s_8$	$\cos B_8=(y_9-y_8)/s_8$

Table 15-9. Specifications and Analyses of a 7-Sided Object in 3-D Space. See Fig. 15-9. (A) Specifications and Analysis in XYZ Space. (B) Secondary Analysis in the X-Y Plane. (C) Secondary Analysis in the X-Z Plane. (D) Secondary Analysis in the Y-Z Plane. (E) Suggested Coordinates. (Continued from page 217.)

B

DIRECTION COSINES

$\cos A_9=(x_{10}-x_9)/s_9$ $\cos B_9=(y_{10}-y_9)/s_9$

$\cos A_{10}=(x_6-x_{10})/s_{10}$ $\cos B_{10}=(y_6-y_{10})/s_{10}$

$\cos A_{11}=(x_6-x_1)/s_{11}$ $\cos B_{11}=(y_6-y_1)/s_{11}$

$\cos A_{12}=(x_7-x_2)/s_{12}$ $\cos B_{12}=(y_7-y_2)/s_{12}$

$\cos A_{13}=(x_8-x_3)/s_{13}$ $\cos B_{13}=(y_8-y_3)/s_{13}$

$\cos A_{14}=(x_9-x_4)/s_{14}$ $\cos B_{14}=(y_9-y_4)/s_{14}$

$\cos A_{15}=(x_{10}-x_5)/s_{15}$ $\cos B_{15}=(y_{10}-y_5)/s_{15}$

ANGLES

$\cos\phi_{1,2}=\cos A_1\cos A_2+\cos B_1\cos B_2$

$\cos\phi_{1,5}=\cos A_1\cos A_5+\cos B_1\cos B_5$

$\cos\phi_{1,11}=\cos A_1\cos A_{11}+\cos B_1\cos B_{11}$

$\cos\phi_{1,12}=\cos A_1\cos A_{12}+\cos B_1\cos B_{12}$

$\cos\phi_{2,3}=\cos A_2\cos A_3+\cos B_2\cos B_3$

$\cos\phi_{2,12}=\cos A_2\cos A_{12}+\cos B_2\cos B_{12}$

$\cos\phi_{2,13}=\cos A_2\cos A_{13}+\cos B_2\cos B_{13}$

$\cos\phi_{3,4}=\cos A_3\cos A_4+\cos B_3\cos B_4$

$\cos\phi_{3,13}=\cos A_3\cos A_{13}+\cos B_3\cos B_{13}$

$\cos\phi_{3,14}=\cos A_3\cos A_{14}+\cos B_3\cos B_{14}$

$\cos\phi_{4,5}=\cos A_4\cos A_5+\cos B_4\cos B_5$

$\cos\phi_{4,14}=\cos A_4\cos A_{14}+\cos B_4\cos B_{14}$

$\cos\phi_{4,15}=\cos A_4\cos A_{15}+\cos B_4\cos B_{15}$

$\cos\phi_{5,11}=\cos A_5\cos A_{11}+\cos B_5\cos B_{11}$

$\cos\phi_{5,15}=\cos A_5\cos A_{15}+\cos B_5\cos B_{15}$

$\cos\phi_{6,7}=\cos A_6\cos A_7+\cos B_6\cos B_7$

$\cos\phi_{6,10}=\cos A_6\cos A_{10}+\cos B_6\cos B_{10}$

$\cos\phi_{6,11}=\cos A_6\cos A_{11}+\cos B_6\cos B_{11}$

$\cos\phi_{6,12}=\cos A_6\cos A_{12}+\cos B_6\cos B_{12}$

$\cos\phi_{7,8}=\cos A_7\cos A_8+\cos B_7\cos B_8$

$\cos\phi_{7,12}=\cos A_7\cos A_{12}+\cos B_7\cos B_{12}$

$\cos\phi_{7,13}=\cos A_7\cos A_{13}+\cos B_7\cos B_{13}$

$\cos\phi_{8,9}=\cos A_8\cos A_9+\cos B_8\cos B_9$

$\cos\phi_{8,13}=\cos A_8\cos A_{13}+\cos B_8\cos B_{13}$

$\cos\phi_{8,14}=\cos A_8\cos A_{14}+\cos B_8\cos B_{14}$

$\cos\phi_{9,10}=\cos A_9\cos A_{10}+\cos B_9\cos B_{10}$

$\cos\phi_{9,14}=\cos A_9\cos A_{14}+\cos B_9\cos B_{14}$

$\cos\phi_{9,15}=\cos A_9\cos A_{15}+\cos B_9\cos B_{15}$

$\cos\phi_{10,11}=\cos A_{10}\cos A_{11}+\cos B_{10}\cos B_{11}$

$\cos\phi_{10,15}=\cos A_{10}\cos A_{15}+\cos B_{10}\cos B_{15}$

CHECK

F_1: $\phi_{1,2}+\phi_{1,5}+\phi_{2,3}+\phi_{3,4}+\phi_{4,5}=540^\circ$

F_2: $\phi_{1,12}+\phi_{1,11}+\phi_{6,11}+\phi_{6,12}=360^\circ$

F_3: $\phi_{2,12}+\phi_{2,13}+\phi_{7,12}+\phi_{7,13}=360^\circ$

F_4: $\phi_{3,13}+\phi_{3,14}+\phi_{8,13}+\phi_{8,14}=360^\circ$

F_5: $\phi_{4,14}+\phi_{4,15}+\phi_{9,14}+\phi_{9,15}=360^\circ$

F_6: $\phi_{5,11}+\phi_{5,15}+\phi_{10,11}+\phi_{10,15}=360^\circ$

F_7: $\phi_{6,7}+\phi_{6,10}+\phi_{7,8}+\phi_{8,9}+\phi_{9,10}=540^\circ$

SECONDARY ANALYSIS -- XZ PLANE

PLOTTED POINTS

$P_1=(x_1,z_1)$ $P_2=(x_2,z_2)$

$P_3=(x_3,z_3)$ $P_4=(x_4,z_4)$

$P_5=(x_5,z_5)$ $P_6=(x_6,z_6)$

$P_7=(x_7,z_7)$ $P_8=(x_8,z_8)$

$P_9=(x_9,z_9)$ $P_{10}=(x_{10},z_{10})$

LENGTHS

$$s_1=\sqrt{(x_2-x_1)^2+(z_2-z_1)^2} \qquad s_2=\sqrt{(x_3-x_2)^2+(z_3-z_2)^2}$$

$$s_3=\sqrt{(x_4-x_3)^2+(z_4-z_3)^2} \qquad s_4=\sqrt{(x_5-x_4)^2+(z_5-z_4)^2}$$

$$s_5=\sqrt{(x_1-x_5)^2+(z_1-z_5)^2} \qquad s_6=\sqrt{(x_7-x_6)^2+(z_7-z_6)^2}$$

$$s_7=\sqrt{(x_8-x_7)^2+(z_8-z_7)^2} \qquad s_8=\sqrt{(x_9-x_8)^2+(z_9-z_8)^2}$$

$$s_9=\sqrt{(x_{10}-x_9)^2+(z_{10}-z_9)^2} \qquad s_{10}=\sqrt{(x_6-x_{10})^2+(z_6-z_{10})^2}$$

$$s_{11}=\sqrt{(x_6-x_1)^2+(z_6-z_1)^2} \qquad s_{12}=\sqrt{(x_7-x_2)^2+(z_7-z_2)^2}$$

$$s_{13}=\sqrt{(x_8-x_3)^2+(z_8-z_3)^2} \qquad s_{14}=\sqrt{(x_9-x_4)^2+(z_9-z_4)^2}$$

$$s_{15}=\sqrt{(x_{10}-x_5)^2+(z_{10}-z_5)^2}$$

DIRECTION COSINES

$\cos A_1=(x_2-x_1)/s_1$ $\cos C_1=(z_2-z_1)/s_1$

$\cos A_2=(x_3-x_2)/s_2$ $\cos C_2=(z_3-z_2)/s_2$

$\cos A_3=(x_4-x_3)/s_3$ $\cos C_3=(z_4-z_3)/s_3$

$\cos A_4=(x_5-x_4)/s_4$ $\cos C_4=(z_5-z_4)/s_4$

$\cos A_5=(x_1-x_5)/s_5$ $\cos C_5=(z_1-z_5)/s_5$

$\cos A_6=(x_7-x_6)/s_6$ $\cos C_6=(z_7-z_6)/s_6$

$\cos A_7=(x_8-x_7)/s_7$ $\cos C_7=(z_8-z_7)/s_7$

Table 15-9. Specifications and Analyses of a 7-Sided Object in 3-D Space. See Fig. 15-9. (A) Specifications and Analysis in XYZ Space. (B) Secondary Analysis in the X-Y Plane. (C) Secondary Analysis in the X-Z Plane. (D) Secondary Analysis in the Y-Z Plane. (E) Suggested Coordinates. (Continued from page 219.)

C

DIRECTION COSINES

$\cos A_8=(x_9-x_8)/s_8$ | $\cos C_8=(z_9-z_8)/s_8$

$\cos A_9=(x_{10}-x_9)/s_9$ | $\cos C_9=(z_{10}-z_9)/s_9$

$\cos A_{10}=(x_6-x_{10})/s_{10}$ | $\cos C_{10}=(z_6-z_{10})/s_{10}$

$\cos A_{11}=(x_6-x_1)/s_{11}$ | $\cos C_{11}=(z_6-z_1)/s_{11}$

$\cos A_{12}=(x_7-x_2)/s_{12}$ | $\cos C_{12}=(z_7-z_2)/s_{12}$

$\cos A_{13}=(x_8-x_3)/s_{13}$ | $\cos C_{13}=(z_8-z_3)/s_{13}$

$\cos A_{14}=(x_9-x_4)/s_{14}$ | $\cos C_{14}=(z_9-z_4)/s_{14}$

$\cos A_{15}=(x_{10}-x_5)/s_{15}$ | $\cos C_{15}=(z_{10}-z_5)/s_{15}$

ANGLES

$\cos\phi_{1,2}=\cos A_1\cos A_2+\cos C_1\cos C_2$

$\cos\phi_{1,5}=\cos A_1\cos A_5+\cos C_1\cos C_5$

$\cos\phi_{1,11}=\cos A_1\cos A_{11}+\cos C_1\cos C_{11}$

$\cos\phi_{1,12}=\cos A_1\cos A_{12}+\cos C_1\cos C_{12}$

$\cos\phi_{2,3}=\cos A_2\cos A_3+\cos C_2\cos C_3$

$\cos\phi_{2,12}=\cos A_2\cos A_{12}+\cos C_2\cos C_{12}$

$\cos\phi_{2,13}=\cos A_2\cos A_{13}+\cos C_2\cos C_{13}$

$\cos\phi_{3,4}=\cos A_3\cos A_4+\cos C_3\cos C_4$

$\cos\phi_{3,13}=\cos A_3\cos A_{13}+\cos C_3\cos C_{13}$

$\cos\phi_{3,14}=\cos A_3\cos A_{14}+\cos C_3\cos C_{14}$

$\cos\phi_{4,5}=\cos A_4\cos A_5+\cos C_4\cos C_5$

$\cos\phi_{4,14}=\cos A_4\cos A_{14}+\cos C_4\cos C_{14}$

$\cos\phi_{4,15}=\cos A_4\cos A_{15}+\cos C_4\cos C_{15}$

$\cos\phi_{5,11}=\cos A_5\cos A_{11}+\cos C_5\cos C_{11}$

$\cos\phi_{5,15}=\cos A_5\cos A_{15}+\cos C_5\cos C_{15}$

$\cos\phi_{6,7}=\cos A_6\cos A_7+\cos C_6\cos C_7$

$\cos\phi_{6,10}=\cos A_6\cos A_{10}+\cos C_6\cos C_{10}$

$\cos\phi_{6,11}=\cos A_6\cos A_{11}+\cos C_6\cos C_{11}$

$\cos\phi_{6,12}=\cos A_6\cos A_{12}+\cos C_6\cos C_{12}$

$\cos\phi_{7,8}=\cos A_7\cos A_8+\cos C_7\cos C_8$

$\cos\phi_{7,12}=\cos A_7\cos A_{12}+\cos C_7\cos C_{12}$

$\cos\phi_{7,13}=\cos A_7\cos A_{13}+\cos C_7\cos C_{13}$

$\cos\phi_{8,9}=\cos A_8\cos A_9+\cos C_8\cos C_9$

$\cos\phi_{8,13}=\cos A_8\cos A_{13}+\cos C_8\cos C_{13}$

$\cos\phi_{8,14}=\cos A_8\cos A_{14}+\cos C_8\cos C_{14}$

$\cos\phi_{9,10}=\cos A_9\cos A_{10}+\cos C_9\cos C_{10}$

$\cos\phi_{9,14}=\cos A_9\cos A_{14}+\cos C_9\cos C_{14}$

$\cos\phi_{9,15}=\cos A_9\cos A_{15}+\cos C_9\cos C_{15}$

$\cos\phi_{10,11}=\cos A_{10}\cos A_{11}+\cos C_{10}\cos C_{11}$

$\cos\phi_{10,15}=\cos A_{10}\cos A_{15}+\cos C_{10}\cos C_{15}$

CHECK

F_1: $\phi_{1,2}+\phi_{1,5}+\phi_{2,3}+\phi_{3,4}+\phi_{4,5}=540^\circ$

F_2: $\phi_{1,12}+\phi_{1,11}+\phi_{6,11}+\phi_{6,12}=360^\circ$

F_3: $\phi_{2,12}+\phi_{2,13}+\phi_{7,12}+\phi_{7,13}=360^\circ$

F_4: $\phi_{3,13}+\phi_{3,14}+\phi_{8,13}+\phi_{8,14}=360^\circ$

F_5: $\phi_{4,14}+\phi_{4,15}+\phi_{9,14}+\phi_{9,15}=360^\circ$

F_6: $\phi_{5,11}+\phi_{5,15}+\phi_{10,11}+\phi_{10,15}=360^\circ$

F_7: $\phi_{6,7}+\phi_{6,10}+\phi_{7,8}+\phi_{8,9}+\phi_{9,10}=540^\circ$

D

SECONDARY ANALYSIS -- YZ PLANE

PLOTTED POINTS

$P_1=(y_1,z_1)$ $P_2=(y_2,z_2)$

$P_3=(y_3,z_3)$ $P_4=(y_4,z_4)$

$P_5=(y_5,z_5)$ $P_6=(y_6,z_6)$

$P_7=(y_7,z_7)$ $P_8=(y_8,z_8)$

$P_9=(y_9,z_9)$ $P_{10}=(y_{10},z_{10})$

LENGTHS

$s_1=\sqrt{(y_2-y_1)^2+(z_2-z_1)^2}$ $s_2=\sqrt{(y_3-y_2)^2+(z_3-z_2)^2}$

$s_3=\sqrt{(y_4-y_3)^2+(z_4-z_3)^2}$ $s_4=\sqrt{(y_5-y_4)^2+(z_5-z_4)^2}$

$s_5=\sqrt{(y_1-y_5)^2+(z_1-z_5)^2}$ $s_6=\sqrt{(y_7-y_6)^2+(z_7-z_6)^2}$

$s_7=\sqrt{(y_8-y_7)^2+(z_8-z_7)^2}$ $s_8=\sqrt{(y_9-y_8)^2+(z_9-z_8)^2}$

$s_9=\sqrt{(y_{10}-y_9)^2+(z_{10}-z_9)^2}$ $s_{10}=\sqrt{(y_6-y_{10})^2+(z_6-z_{10})^2}$

$s_{11}=\sqrt{(y_6-y_1)^2+(z_6-z_1)^2}$ $s_{12}=\sqrt{(y_7-y_2)^2+(z_7-z_2)^2}$

$s_{13}=\sqrt{(y_8-y_3)^2+(z_8-z_3)^2}$ $s_{14}=\sqrt{(y_9-y_4)^2+(z_9-z_4)^2}$

$s_{15}=\sqrt{(y_{10}-y_5)^2+(z_{10}-z_5)^2}$

DIRECTION COSINES

$\cos B_1=(y_2-y_1)/s_1$ $\cos C_1=(z_2-z_1)/s_1$

$\cos B_2=(y_3-y_2)/s_2$ $\cos C_2=(z_3-z_2)/s_2$

$\cos B_3=(y_4-y_3)/s_3$ $\cos C_3=(z_4-z_3)/s_3$

$\cos B_4=(y_5-y_4)/s_4$ $\cos C_4=(z_5-z_4)/s_4$

$\cos B_5=(y_1-y_5)/s_5$ $\cos C_5=(z_1-z_5)/s_5$

$\cos B_6=(y_7-y_6)/s_6$ $\cos C_6=(z_7-z_6)/s_6$

$\cos B_7=(y_8-y_7)/s_7$ $\cos C_7=(z_8-z_7)/s_7$

Table 15-9. Specifications and Analyses of a 7-Sided Object in 3-D Space. See Fig. 15-9. (A) Specifications and Analysis in XYZ Space. (B) Secondary Analysis in the X-Y Plane. (C) Secondary Analysis in the X-Z Plane. (D) Secondary Analysis in the Y-Z Plane. (E) Suggested Coordinates. (Continued from page 221.)

D

DIRECTION COSINES

$\cos B_8=(y_9-y_8)/s_8$ $\cos C_8=(z_9-z_8)/s_8$

$\cos B_9=(y_{10}-y_9)/s_9$ $\cos C_9=(z_{10}-z_9)/s_9$

$\cos B_{10}=(y_6-y_{10})/s_{10}$ $\cos C_{10}=(z_6-z_{10})/s_{10}$

$\cos B_{11}=(y_6-y_1)/s_{11}$ $\cos C_{11}=(z_6-z_1)/s_{11}$

$\cos B_{12}=(y_7-y_2)/s_{12}$ $\cos C_{12}=(z_7-z_2)/s_{12}$

$\cos B_{13}=(y_8-y_3)/s_{13}$ $\cos C_{13}=(z_8-z_3)/s_{13}$

$\cos B_{14}=(y_9-y_4)/s_{14}$ $\cos C_{14}=(z_9-z_4)/s_{14}$

$\cos B_{15}=(y_{10}-y_5)/s_{15}$ $\cos C_{15}=(z_{10}-z_5)/s_{15}$

ANGLES

$\cos\phi_{1,2}=\cos B_1\cos B_2+\cos C_1\cos C_2$

$\cos\phi_{1,5}=\cos B_1\cos B_5+\cos C_1\cos C_5$

$\cos\phi_{1,11}=\cos B_1\cos B_{11}+\cos C_1\cos C_{11}$

$\cos\phi_{1,12}=\cos B_1\cos B_{12}+\cos C_1\cos C_{12}$

$\cos\phi_{2,3}=\cos B_2\cos B_3+\cos C_2\cos C_3$

$\cos\phi_{2,12}=\cos B_2\cos B_{12}+\cos C_2\cos C_{12}$

$\cos\phi_{2,13}=\cos B_2\cos B_{13}+\cos C_2\cos C_{13}$

$\cos\phi_{3,4}=\cos B_3\cos B_4+\cos C_3\cos C_4$

$\cos\phi_{3,13}=\cos B_3\cos B_{13}+\cos C_3\cos C_{13}$

$\cos\phi_{3,14}=\cos B_3\cos B_{14}+\cos C_3\cos C_{14}$

$\cos\phi_{4,5}=\cos B_4\cos B_5+\cos C_4\cos C_5$

$\cos\phi_{4,14}=\cos B_4\cos B_{14}+\cos C_4\cos C_{14}$

$\cos\phi_{4,15}=\cos B_4\cos B_{15}+\cos C_4\cos C_{15}$

$\cos\phi_{5,11}=\cos B_5\cos B_{11}+\cos C_5\cos C_{11}$

$\cos\phi_{5,15}=\cos B_5\cos B_{15}+\cos C_5\cos C_{15}$

$\cos\phi_{6,7}=\cos B_6\cos B_7+\cos C_6\cos C_7$

$\cos\phi_{6,10}=\cos B_6\cos B_{10}+\cos C_6\cos C_{10}$

$\cos\phi_{6,11}=\cos B_6\cos B_{11}+\cos C_6\cos C_{11}$

$\cos\phi_{6,12}=\cos B_6\cos B_{12}+\cos C_6\cos C_{12}$

$\cos\phi_{7,8}=\cos B_7\cos B_8+\cos C_7\cos C_8$

$\cos\phi_{7,12}=\cos B_7\cos B_{12}+\cos C_7\cos C_{12}$

$\cos\phi_{7,13}=\cos B_7\cos B_{13}+\cos C_7\cos C_{13}$

$\cos\phi_{8,9}=\cos B_8\cos B_9+\cos C_8\cos C_9$

$\cos\phi_{8,13}=\cos B_8\cos B_{13}+\cos C_8\cos C_{13}$

$\cos\phi_{8,14}=\cos B_8\cos B_{14}+\cos C_8\cos C_{14}$

$\cos\phi_{9,10}=\cos B_9\cos B_{10}+\cos C_9\cos C_{10}$

$\cos\phi_{9,14}=\cos B_9\cos B_{14}+\cos C_9\cos C_{14}$

$\cos\phi_{9,15}=\cos B_9\cos B_{15}+\cos C_9\cos C_{15}$

$\cos\phi_{10,11}=\cos B_{10}\cos B_{11}+\cos C_{10}\cos C_{11}$

$\cos\phi_{10,15}=\cos B_{10}\cos B_{15}+\cos C_{10}\cos C_{15}$

CHECK

F_1: $\phi_{1,2}+\phi_{1,5}+\phi_{2,3}+\phi_{3,4}+\phi_{4,5}=540^\circ$

F_2: $\phi_{1,12}+\phi_{1,11}+\phi_{6,11}+\phi_{6,12}=360^\circ$

F_3: $\phi_{2,12}+\phi_{2,13}+\phi_{7,12}+\phi_{7,13}=360^\circ$

F_4: $\phi_{3,13}+\phi_{3,14}+\phi_{8,13}+\phi_{8,14}=360^\circ$

F_5: $\phi_{4,14}+\phi_{4,15}+\phi_{9,14}+\phi_{9,15}=360^\circ$

F_6: $\phi_{5,11}+\phi_{5,15}+\phi_{10,11}+\phi_{10,15}=360^\circ$

F_7: $\phi_{6,7}+\phi_{6,10}+\phi_{7,8}+\phi_{8,9}+\phi_{9,10}=540^\circ$

```
OBJECT VERSION 1

P1=(-1,-1,1)        P2=(1,-1,1)        P3=(1,1,1)
P4=(Ø,2,1)          P5=(-1,1,1)        P6=(-1,-1,-1)
P7=(1,-1,-1)        P8=(1,1,-1)        P9=(Ø,2,-1)
P1Ø=(-1,1,-1)

OBJECT VERSION 2

P1=(-2,-2,1)        P2=(Ø,-2,1)        P3=(Ø,Ø,1)
P4=(-1,1,1)         P5=(-2,Ø,1)        P6=(Ø,Ø,-1)
P7=(2,Ø,-1)         P8=(2,2,-1)        P9=(1,4,-1)
P1Ø=(Ø,2,-1)

OBJECT VERSION 3

P1=(-1,-1,1)        P2=(1,-1,1)        P3=(1,1,1)
P4=(Ø,Ø.5,1)        P5=(-1,1,1)        P6=(-1,-1,-1)
P7=(1,-1,-1)        P8=(1,1,-1)        P9=(Ø,Ø.5,-1)
P1Ø=(-1,1,-1)
```

invariant, and simple rotations about one particular axis are commutative. (Combinations of simple rotations about different axes are not commutative.)

The next objective of this discussion is to elaborate on those special ideas and suggest some experiments.

15-4 SIMPLE TRANSLATIONS AND SUGGESTED EXPERIMENTS

Simple translations in any number of dimensions express a spatial displacement between a reference figure and a translated version of it. One of the special features of simple translations is that the two versions are identical in every respect but one: their position with respect to the frame-of-reference coordinate system.

15-4.1 Suggested Experiments

There are several classes of procedures and that takes advantage of the characteristics of simple translations in 3-D space. The most straightforward sort of translation experiment goes something like this:

1. Specify a reference space object in 3-D space (see Section 15-1).
2. Conduct complete primary and secondary analyses of the reference space object.
3. Specify and carry out a chosen simple 3-D translation.
4. Set up the formal specifications for the translated version of the space object, and conduct complete primary and secondary analyses of it.

5. Plot both the reference and translated versions of the space object.

Run that sort of experiment for a variety of space objects, and you will be building up a file of specifications, analyses and figures that can be invaluable for later experiments. The second general class of simple translation experiments follows this general outline:

1. Specify a reference space object in 3-D space and plot its 3-D perspective view and the three planar views.
2. Select a point in the space object that you want to displace to another position in the 3-D coordinate system. Specify the coordinate of that new position, too.
3. Solve for the translation terms that are necessary for displacing that selected point in your reference space object to the new place in the coordinate system.
4. Translate the entire space object to that new position by applying the translation terms to the remaining points in that space object.
5. Specify the figure in its new position and plot it on the coordinate system.

A third kind of experiment involves applying successive translations to some reference space object, using the specifications for a previous version as the starting point for translating it again. The idea is to generate a series of displaced space objects that can create both a meaningful and visually interesting set of plots.

Experiments with relativistic 3-D translations can be especially meaningful because they relate directly to the universe as we perceive it. The basic idea is to view a reference object from two different viewing spaces. Here is an outline for such an experiment:

1. Specify a reference space object in the 3-D space of the observer's frame of reference.
2. Conduct complete primary and secondary analyses of the object as viewed from the observer's frame of reference.
3. Specify a translation that is to exist between the observer's and translated frames of reference, then use the relativistic forms of the translation equations to determine the specifications for the space object as it appears from the translated frame of reference.
4. Conduct complete primary and secondary analyses of the figure as it appears from the translated frame of reference and compare the results with the analysis of the space object in the observer's frame of reference.

An interesting variation of that experiment combines the simple geometric and relativistic versions of the translation equations. The objective is to deal with the geometric translation of a space object as viewed from two different coordinate systems which are, themselves, separated by a relativistic translation. Here is a general procedure:

1. Establish a fixed, relativistic translation that is to exist between the two frames of reference.
2. Specify the initial space object as it is to appear from the observer's frame of reference, and then conduct at least a primary analysis of it.
3. Apply the relativistic translation equation to get the specifications for the object as it appears from the translated frame of reference. Do at least a primary analysis of the results.
4. Apply some geometric translation to the initial object in the observer's frame of reference, and then apply the relativistic equation to find out the specifications of that translated version as taken from the translated frame of reference.
5. Repeat Step 4 as desired.

That particular experiment represents the motion, or succesive translation, of a space object as viewed from two different frames of reference in 3-D space. Imagine yourself and a companion observing the position of a flying aircraft from two different places. You are observing the aircraft from the observer's frame of reference, and your companion is observing it from the fixed, translated frame of reference. The relativistic translation in Step 1 represents the difference between you and your companion's positions in 3-D space. The series of geometric translations then represent the motion of the aircraft relative to your position, and the results of working those equations show how the motion appears from your companion's position.

The ultimate translation experiment is one where both the relativistic and geometric translation terms change. The general outline is the same as the one just described except that the relativistic translation terms change, too. Instead of you and your companion observing a moving aircraft from fixed frames of reference, the frames of reference are moving, too. (If we were working with non-linear translations, this would be a very natural place to slip quite easily into a discussion of Einstein's Special Theory of Relativity.)

15-4.2 Some Special Studies

There are some important lines of study for investigators who are interested in the most general views of geometric and relativistic translations. Consider these studies:

1. Conduct formal analytic proofs of the invariance of length, direction cosines and angles under both geometric and relativistic translations in 3-D space.

2. See if you can discover a general relationship between the direction cosines of line in 3-D space, its possible reduction to a point as mapped to a planar view.

3. See if you can discover a general relationship between the direction cosines of the lines of a side in a space object and its reduction to a line as mapped to a given planar view.

4. Research other text sources for discussions of the volumes of space objects from an analytic viewpoint. See if you can develop a technique for calculating the volume of any space object, given its primary specifications.

5. Consult a basic text on physics, and generate an equation that expresses a change in translation terms as a function of velocity and time.

15-5 SIMPLE SCALINGS AND SUGGESTED EXPERIMENTS

Simple scaling transformations in 3-D space can expand, contract and reflect (reverse) the scaling of a coordinate system. And of course the scalings can be symmetrical or asymmetrical as well.

Summarizing the essential qualities of simple 3-D scalings, recall that under symmetrical, non-reflective scalings, length is a variant, direction cosines are invariant, and angles are invariant. Considering symmetrical scalings, both reflective and non-reflective, however, length is a variant, direction cosines are variant, and angles are invariant. Finally, you should know by now that under asymmetrical scalings, both non-reflective and reflective, length is a variant, direction cosines are variant, and angles are variant. That might seem confusing to you if you haven't done a thorough job of studying the basic scaling effects in earlier chapters. If you find yourself confused (or attempting to memorize such principles as opposed to understanding them), I suggest that you have hurried your work a bit too much.

15-5.1 Some Suggested Experiments

You have already seen that experiments with simple scaling transformation can be more dramatic than those for translations. Scalings offer an opportunity to mold a figure or space object; expanding, compressing and reflecting it relative to its "standard" form. Consider this type of experiment:

1. Specify a reference figure or space object according to the methods outlined in Sections 15-1 and 15-2.

2. Conduct complete primary and secondary analyses of that reference figure or object.

3. Specify some desired scaling factors and carry out the scaling operation.

4. Write out the specifications for the scaled version of the figure or object, and conduct complete analyses of it.

5. Plot both the reference and scaled versions of the figure or space object.

Conduct such experiments for a number of different figures and space objects specified in 3-D space. Be sure to select both symmetrical and asymmetrical scalings at one time or another. The truly useful application of the work is to mold the shape and size of space objects and plane figure immersed in 3-D space that you have already specified in a "standard" form.

Then try applying successive scalings to a space object of some sort, making it expand, shrink or reflect about a plane or axis. That can be a time-consuming process, but the results are usually quite striking in a visual sense.

An entirely different sort of scaling experiment can be structured around the 3-D relativistic scaling equations, plotting a space object as it appears from two different frames of reference—frames of reference that differ only in the scaling that exists between them. A-symmetrical and reflection-type relativistic scaling effects are particularly intriguing, because they can represent different, and sometimes highly unusual, scaling system that might exist elsewhere in our known universe.

There is a great deal of evidence these days that suggest the notion of an expanding universe. You can get some idea about the mathematics involved in such research by combining simple, 3-D geometric scalings with relativistic scalings.

15-5.2 Some Suggested Special Studies

Experimenters who feel prepared to deal with 3-D scalings in a more general fashion, the following special studies can point the way to some fascinating work.

1. Derive a general equation for the length of a scaled line in three dimensions as a function of the coordinates of the points in the reference version of the figure and the scaling factors.

2. Derive a general equation for the length of a scaled line in 3-D space as a function of its reference length and the scaling factors.

3. Derive a general equation for calculating the angle between two lines in 3-D space as a function of the

corresponding angle in the reference figure and the given scaling factors.

15-6 SIMPLE ROTATIONS AND SUGGESTED EXPERIMENTS

Simple rotations in 3-D space turn a coordinate system about a selected axis of rotation. The reference (non-rotated) and rotated versions of any geometric entities specified within the coordinate systems are identical except for their position and angular orientation (direction cosines of any lines) with respect to the coordinate axes. The lengths of lines and angles between them are invariant.

15-6.1 Taking Advantage of the Invariants

Only the direction cosines of lines are variant under simple rotations in 3-D space—the lengths of corresponding lines and the corresponding angles are invariant. So you can doublecheck your work by comparing the analyses of the lengths of lines and the angles that comprise the figure or space object—especially the angles.

15-6.2 Suggested Experiments and Studies

The most straightforward kinds of experiments with simple rotations in 3-D space take this general form:

1. Specify a space object or reference figure in 3-D space.
2. Conduct complete analyses of the reference entity.
3. Carry out a desired rotation about a selected axis.
4. Establish the formal specifications for the rotated version of the entity and conduct complete primary and secondary analyses of it.
5. Plot both the reference and rotated versions of the entity.

Assuming that you have already generated a file of 3-D figures in some previous experiments, that sort of experiment suggests extending the file to show the figures having different orientations with respect to the coordinate axes. You can then use the specifications, analyses and plots as reference material for initiating other kinds of experiments.

Try applying some 3-D rotations of relatively small angle such as 15 degrees to some reference figure, and then applying another 15-degree rotation to the result. Vary the axis of rotation to get some other kinds of effects.

Then, of course, there is the matter of relativistic rotations. The objective is to specify a relativistic rotation that exists between an observer's and rotated coordinate system, specify a space object as it would appear from the observer's point of view, and then use the relativistic rotation equations to find out how that figure appears from the rotated frame of reference. Combining relativistic and geometric rotations into the same experiment can also be a worthwhile effort:

1. Specify a fixed relativistic rotation that is to exist between the observer's and the rotated frames of reference.
2. Specify a reference space object relative to the observer's frame of reference.
3. Apply the relativistic rotation equation to see how the reference object appears from the rotated frame of reference.
4. Apply some geometric angle of rotation to the space object.
5. Repeat Steps 3 and 4 any number of times, plotting the results in each case.

As far as special studies are concerned, consider these:

1. Conduct a formal analytic proof of the fact that length is an invariant under 3-D rotations, regardless of the chosen axis of rotation.
2. Conduct a formal analytic proof that the internal angles of a space object in 3-D space are invariant under simple 3-D rotations, regardless of the chosen axis of rotation.
3. Derive a set of equations that determine the direction cosines of a line that is being rotated in 3-D space as a function of the angle of rotation.

15-7 COMBINATIONS OF TRANSLATIONS, SCALINGS AND ROTATIONS

You saw in Chapter 9 that combinations of translations and scalings in 2-D space are not commutative; and as you might suspect, the same is true in 3-D space. It makes a big difference whether you do the translation or scaling first.

You can adequately prove the principle by first translating some specified space object and then scaling it. Then beginning with the same reference object, to the scaling followed by the rotation. Unless you have happened to pick some rare combinations of scalings and transformations, the results will be different.

Use the technique outlined in Chapter 9 to derive equations that lead directly to the coordinates of points under 3-D versions of T:K and K:T transformations. See Derivations 9-1 and 9-2.

If you are interested in purely mathematical matters, use the results of your derivations to determine the relatively rare combinations of scaling factors and translation terms that let a T:K transformation equal the results of a K:T sequence. (Hint: How about setting the scaling factors to 1?)

Just as combinations of translations and scalings are not commutative, neither are combinations of translations and scalings. A T:R transformation is equal to an R:T transformation only under certain combinations of angles and axes of rotation. You can demonstrate that idea for yourself by specifying some simple space object, rotating and then translating it. Starting over with the same reference object, execute the same translation terms first, followed by the same angle and axis of rotation used in the first part of the experiment. Chances are quite good that you will end up with different results.

Make up some combinations of transformations of your own; including translations, scalings and rotations; then see if you can derive equations for calculating the overall results. Extend your derivations to include lengths and direction cosines for lines, and the angles between lines.

Before leaving the work suggested in this chapter, give some special consideration to combinations of rotations about different axes. Are rotations about the X and Y axes commutative, for instance? Do you get the same overall result by rotating a point in 3-D space say 45 degrees about the X axis, then 45 degrees about the Y axis as you would be executing that combination the other way around? Does the angle of rotation in such combinations of different axes of rotation make a significant difference in the principle that should emerge from your study of these particular questions? Demonstrations, proofs and derivations of equations are in order, but I am leaving that to your own sense of curiosity and ambition.

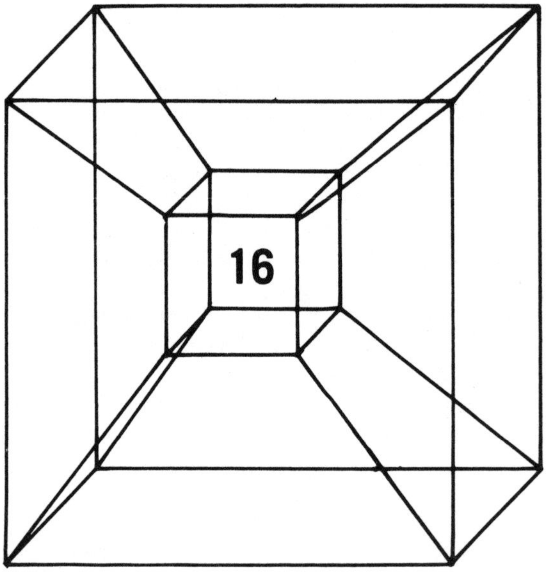

Constructing Space Objects from 2-D Data

It is possible to generate reasonably meaningful 2-dimensional information from 3-D specifications. You do that, for instance, every time you generate the planar views of a 3-D space or object that is immersed in that space. There is really nothing new about that; we have been doing it in this book, and a lot of others have been doing it for many generations.

Here is something new, though: Given the proper sort of 2-D data, it is possible to figure out the information that is necessary for developing the full set of 3-D specifications for a space object. It can be a lengthly and sometimes cumbersome procedure, but it can be done.

Why go to the trouble of generating 3-D figures from 2-D data when we can construct the object outright? With a bit of imagination, combined with some ability to see 3-D object with the mind's eye, it is usually easier to dream up the specifications for 3-D objects outright. You have probably done a lot of that sort of thing already.

The reason for getting into the subject of creating 3-D objects from 2-D data is that the principles apply equally well to the procedure for generating custom 4-D object from 3-D data. When that time arrives, you will have nothing but analytic methods at your disposal; ordinary spatial perception won't be worth very much. So master the procedure now while you can visualize both the raw information at the beginning and the finished object at the end. Prepare yourself for the time when you can properly visualize only the beginning of the experiment.

There is a step-by-step procedure for generating specifications for higher-dimensioned object from lower-dimensioned data; but the purpose of each step will not be clear until you've had a chance to see the procedure at work. So rather than describing the procedure in detail at the outset, I am going to offer a specific example first. After going through that example, you will be in a better position to appreciate a generalized list of the operations involved.

16-1 A FIRST EXAMPLE

Suppose that some hypothetical 3-D space object shows the three planar views illustrated in Fig. 16-1. There is sufficient information available there to cite 2 of the 3 necessary components for the coordinate of each point. The X-Y planar view, for instance, provides the X and Y components for three different points, the X-Z planar view shows the X and Z components for three points, and the Y-Z planar views shows the Y and Z components for three points.

Notice how I have assigned X and Y components to

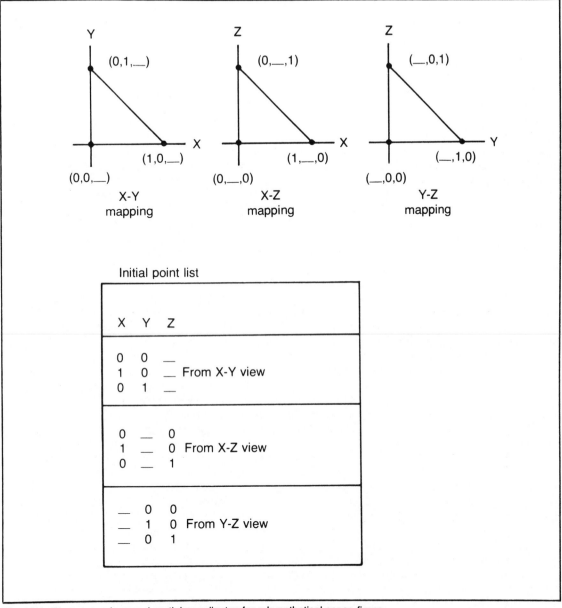

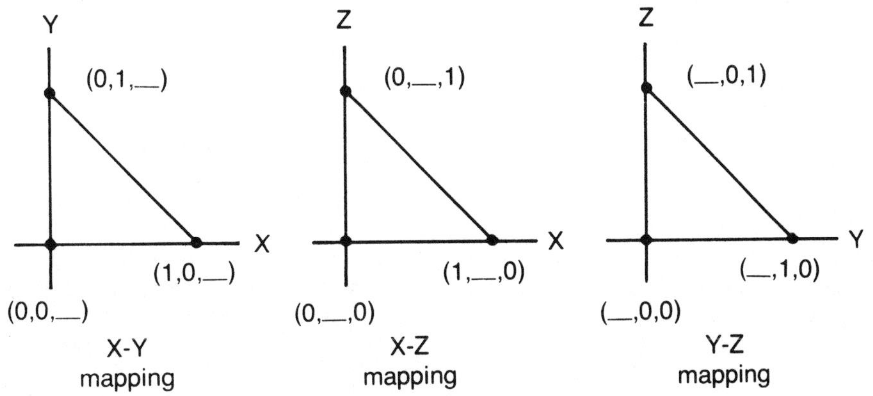

Initial point list

X	Y	Z	
0	0	_	
1	0	_	From X-Y view
0	1	_	
0	_	0	
1	_	0	From X-Z view
0	_	1	
_	0	0	
_	1	0	From Y-Z view
_	0	1	

Fig. 16-1. Planar mappings and partial coordinates for a hypothetical space figure.

each point in the X-Y planar view, but left the Z-component position blank. The same idea applies to the two remaining planar views. The existence of those blanks in the point coordinates imply that we do not know what those values might be—not yet, anyway.

It turns out that there is enough information distributed among those three planar views and partial coordinate specifications to build up the full specifications for a 3-D space object that those views represent. That is the goal of the whole project, and the next major step is to supply some values for the blanks.

The *initial point list* in that figure summarizes the information that is available at this time. Check the list against the planar views, and you will see where the information in the list come from. You can see that the Z-axis components are missing from the first part of the list, the Y-axis components are missing from the second, and that the X-axis components are missing from the last part of the list.

Now, looking down through that list, you can see that the Z-axis components have values of either 0 or 1. That means that the blanks in the first part of the list can be filled with Z-axis components of both 0 and 1—nothing else. So the first point in the list, shown as (0,0,___) can be respecified as (0,0,0), and (0,01). The 0 and 1 values for Z that are specified later in the list can be substituted for the blank space. The same idea holds for the second and thrid partial coordinates in the list: partial coordinate (1,0,___) can be both (1,0,0) and 1,0,1), while the partial coordinate (0,1___) can be completed as (0,1,0) and (0,1,1). Just substituting the Z-axis components shown elsewhere on the list can generate six complete, 3-component coordinates for the first part of the list.

Using a similar line of reasoning, you can look through the list of partial coordinates and find that the Y-axis components can also have values of 0 or 1. Thus is is possible to generate six complete 3-D coordinates from the X-Z view. Simply substitute first a 0 and then a 1 for the blank in each of the three partial coordinates. The idea works the same in this case for completing the blanks for the partial coordinates of the Y-Z plane.

When you have completed that part of the job, you actually end up with a list of all possible points generated from the sets of partial coordinates. See the first column in the listing in Fig. 16-2.

Looking over that list of *all possible points*, you will find a number of duplicates. Eliminate the duplicates to get a listing such as that in the *eliminate duplicates* column. In this particular example, you are left with seven different points generated from the original list of partial coordinates.

The next step is to compare that list of coordinates with the points plotted in the original planar views. Doing that, you will find that there is no point plotted at (1,0,1), (0,1,1) nor (1,1,0). Eliminate those three points from the working list of coordinates. The final result is listed for you in the *plotted points* column in Fig. 16-2.

Assign those remaining points some point labels and plot the results as planar views of the object. See the drawings in Fig. 16-2 and the *final point specifications*. There happens to be just four viable points in the space object being created here.

The fact that points 1 and 2 are shown as the same point in the X-Y planar view implies that there is a Z-dimension line running between them. Likewise, having points 1 and 4 plotted at the same place in the X-Z view suggests that there is a Y-axis line between them. Finally, there must be an X-axis line running between points 1 and 3 in the Y-Z view.

Each of the three mappings in Fig. 16-2 can thus represent two different faces of a space object. From the X-Y view, for example, one face can be composed of the lines from points 1 to 3, from 3 to 4 and from 4 to 1. The second possible figure from the same view is one that starts from point 1 and goes to point 3, then from point 3 to point 4, and finally from point 4 back to point 2. The same general idea happens to apply to the X-Z and Y-Z planar views as well. All of the possible mappings are shown in Fig. 16-3.

As you might suspect, we are getting ready to generate a complete set of line specifications for the 3-D space object. Working from the mappings in Fig. 16-3, you can generate a set of all possible lines. See the list in Table 16-1.

It so happens that each of the six views of the figure generates three possible lines; and that works out to a total of 18 possible lines. But you can see from the list of *all possible lines* that there are a number of duplicates. Eliminating the duplicate lines leaves just six unique ones. See the *eliminate duplicates* list in that table.

The final part of the line-generating procedure is to eliminate lines that don't really plot to the planar views of the space object. In this instance, there are none in that category—all of the lines in the *eliminate duplicates* list appear at one place or another on the plots in Fig. 16-3. So the *plotted lines* list is identical to the former list—in this particular example.

Figure 16-4 updates the possible views of the object to include the newly generated line specifications as well as the point labels generated earlier in the project. With the points and lines thus fully specified, the next step is to define the faces for the space object.

The previous figure suggests that there might be as many as six faces in the hypothetical space object. Considering each of those views one at a time and in order, we can generate a *tenative face list* such as the one shown in Table 16-2.

After generating that *tentative face list* (based on the drawings and specifications in Fig. 16-4), you can eliminate some duplicates. See the list of faces specified in the *eliminate duplicates* list. It would appear that the space object in this project has just four faces.

The *final face list* is the result of eliminating faces that do not really exist. Generally that is done by making

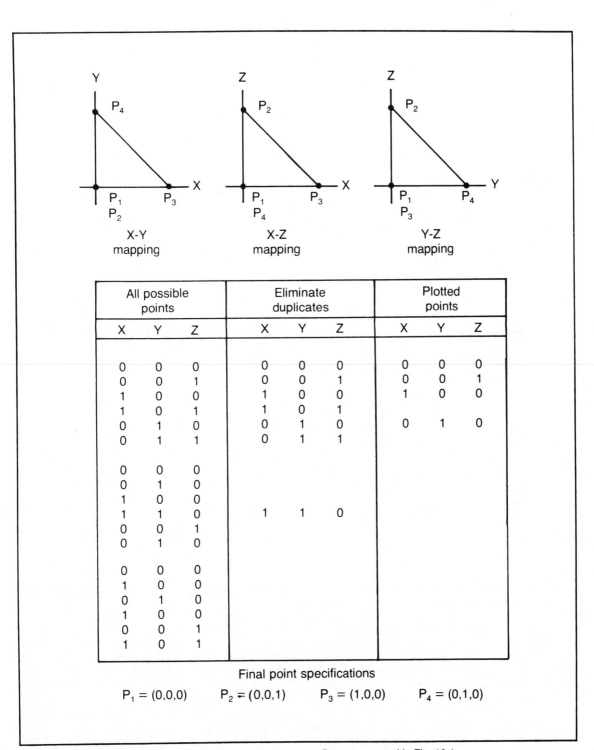

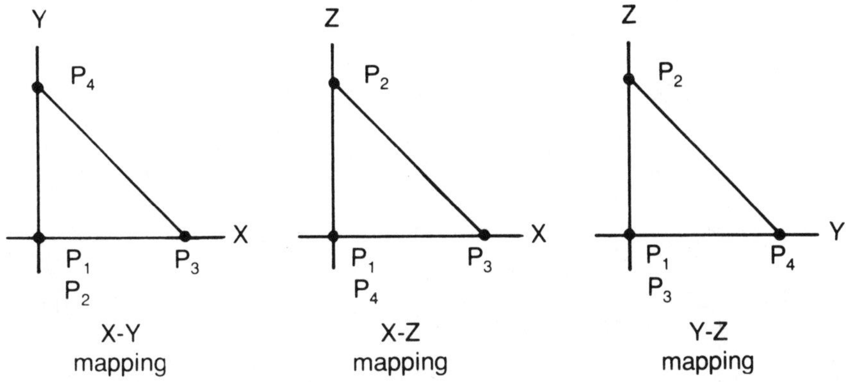

All possible points			Eliminate duplicates			Plotted points		
X	Y	Z	X	Y	Z	X	Y	Z
0	0	0	0	0	0	0	0	0
0	0	1	0	0	1	0	0	1
1	0	0	1	0	0	1	0	0
1	0	1	1	0	1			
0	1	0	0	1	0	0	1	0
0	1	1	0	1	1			
0	0	0						
0	1	0						
1	0	0						
1	1	0	1	1	0			
0	0	1						
0	1	0						
0	0	0						
1	0	0						
0	1	0						
1	0	0						
0	0	1						
1	0	1						

Final point specifications

$P_1 = (0,0,0)$ $P_2 = (0,0,1)$ $P_3 = (1,0,0)$ $P_4 = (0,1,0)$

Fig. 16-2. Evolution of the final point specifications for the space figure suggested in Fig. 16-1.

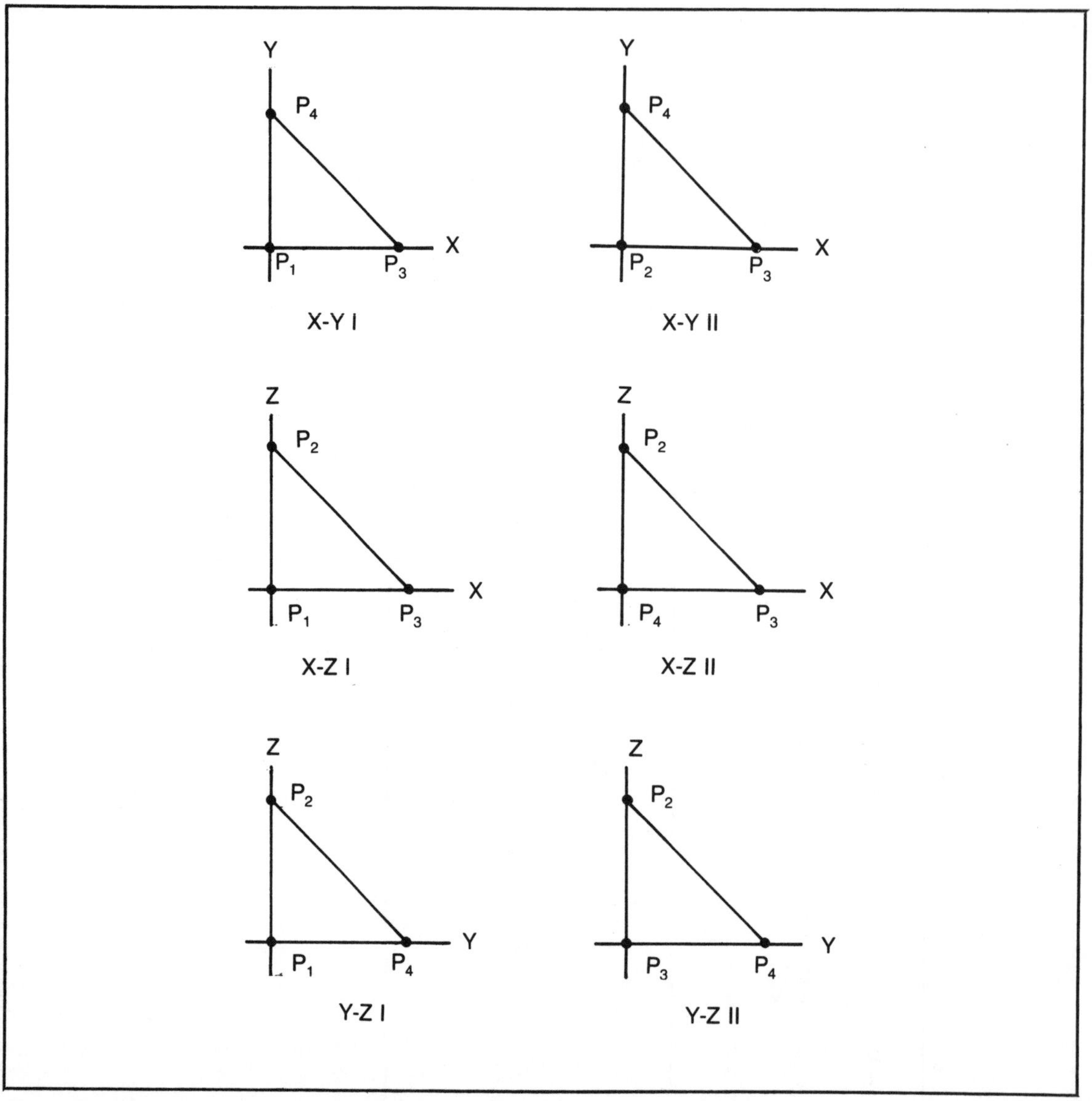

Fig. 16-3. Separation of planar views to show all possible lines.

sure that each line appears only twice in the list of *eliminate duplicates*. Here there are no invalid faces on the list; the *final face list* is thus the same as the *eliminate duplicates* list. That isn't always the case, though. The *final face list* represents the face specifications for the space object, and they are shown in the list of *final face specifications*.

The job is practically done. You've started with some simple planar views of a hypothetical space object and generated final point, line and face specifications. What more could you want? You have generated a 3-D space object from purely 2-D data. You have created a complete geometric entity from data of a lower dimension. Can you appreciate the power you will have at your disposal when it comes to generating the full specifications for strange 4-dimensional objects based on more humanly compatible 3-D information? You will be asked to do that when the time is right—in Chapter 18.

The project is completed by summarizing the P-L-F specifications you have generated and conducting a complete analysis of the result. What you do with the object after that is up to you. You can translate, scale or rotate it any way you choose.

16-2 A FIRST SUMMARY OF THE PROCEDURE

The procedure for generating a complete set of specifications for a space object, given only the 2-dimensional planar views of it, follows this general outline:

1. Generate the point specifications from the original data
   A. Make a list of all possible points
   B. Eliminate duplicate points
   C. Eliminate invalid points
2. Generate the line specifications from the points
   A. Make a list of all possible lines
   B. Eliminate the duplicate lines
   C. Eliminate any invalid lines

Table 16-1. Working Point List for Generating the Final Point Specifications for the Space Object Suggested in Fig. 6-1.

ALL POSSIBLE LINES	ELIMINATE DUPLICATES	PLOTTED LINES
P_1P_3	P_1P_3	P_1P_3
P_3P_4	P_3P_4	P_3P_4
P_1P_4	P_1P_4	P_1P_4
P_2P_3	P_2P_3	P_2P_3
P_3P_4		
P_2P_4	P_2P_4	P_2P_4
P_1P_3		
P_2P_3		
P_1P_2	P_1P_2	P_1P_2
P_3P_4		
P_2P_3		
P_2P_4		
P_1P_4		
P_2P_4		
P_1P_2		
P_3P_4		
P_2P_4		
P_2P_3		

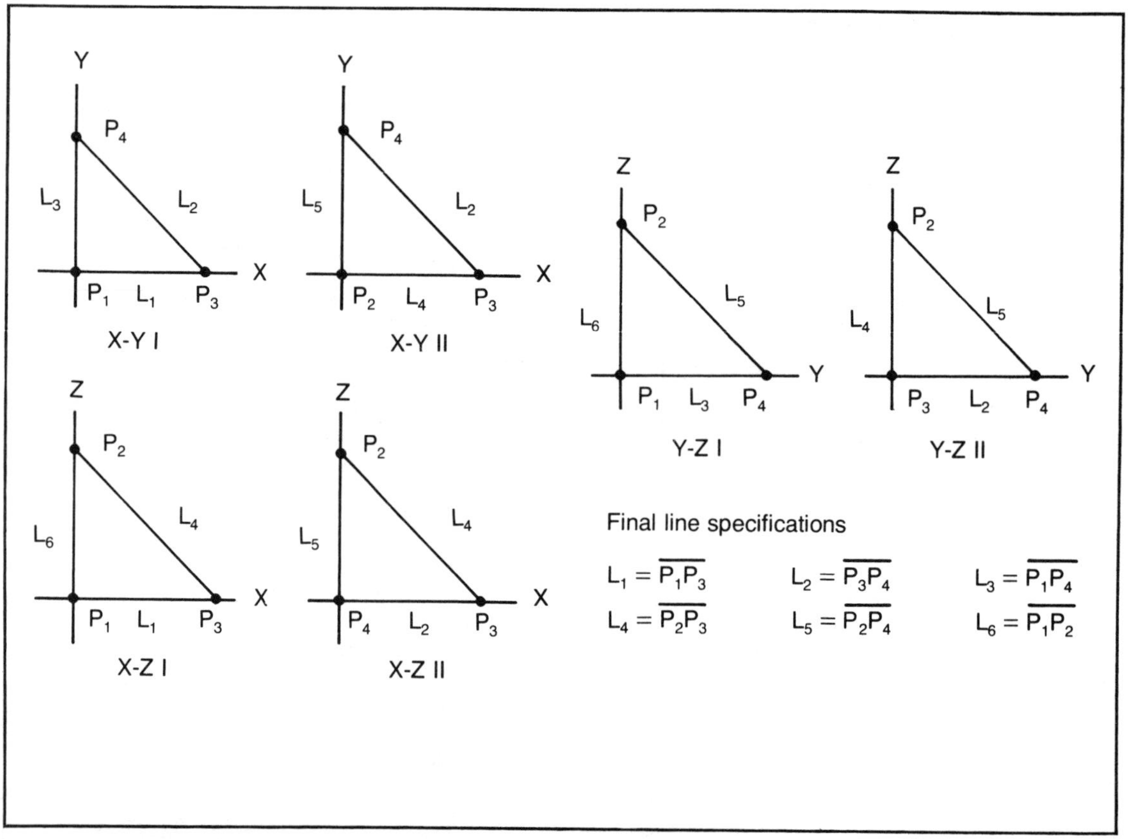

Fig. 16-4. Final line specifications for the space figure suggested in Fig. 16-1.

Final line specifications

$$L_1 = \overline{P_1P_3} \qquad L_2 = \overline{P_3P_4} \qquad L_3 = \overline{P_1P_4}$$

$$L_4 = \overline{P_2P_3} \qquad L_5 = \overline{P_2P_4} \qquad L_6 = \overline{P_1P_2}$$

3. Generate the face specifications from the lines
   A. Make a list of all possible faces
   B. Eliminate duplicate faces
   C. Eliminate any invalid faces
4. Generate the space-object specifications from the faces
5. Conduct a complete analysis of the object

The three steps involved in the first three major phases of the procedure serve similar purposes: look at all possible combinations, eliminate the duplicates, and eliminate any elements that do not appear on the planar views. The more carefully and systematically you approach each step, the less the chances of making a time-consuming and frustrating error. Since the work in any particular phase of the task is based upon work done in a previous phase, I think you can appreciate how devastating an error can be—it carries through all the work until you eventually find that there is something going wrong. Here is another example that will help build your understanding of this important procedure.

16-3 A SECOND EXAMPLE

The initial planar views of some sort of 3-D space object are shown in Fig. 16-5. You might be able to guess the kind of space object those views represent, but pretend you cannot. So make up the *initial point list*.

The *initial point list* shows that the missing Z components for the X-Y mapping can take on values of −1 or 1—those are the Z components that are available from the information about the other two views. Likewise, the Y component for the X-Z mapping can take on values of −1, 0 and 1—all three are available from the other two mappings. The same goes for the X component for the Y-Z mapping—values of −1,0 and 1.

Working from that information, you are ready to generate the *all possible points* list. See Table 16-3. And after doing that, eliminate the duplicates.

Next, attempt to plot the points that remain on the *eliminate duplicates* list. A coordinate is valid if it plots onto a point appearing on all three initial planar views of the object. A coordinate is invalid if it fails to plot onto one of the original points in any one of the initial views. Coordinate (0,0,1) has to be discarded, for instance, because it doesn't match up with any one of the given points in the X-Z and Y-Z mappings. It fits onto the X-Y mapping but not onto the other two.

The *plotted points* list represents the coordinates that remain after eliminating duplicates and invalid coordinates. If you've made no errors or omissions, that list represents the final point specifications for the true 3-D version of the object. This particular space object has exactly five points in it.

Figure 16-6 shows the final point specifications and a set of mappings that use those assigned point coordinates. That figure is also the starting point for the next major phase of the job.

How many different lines are represented by the planar views in Fig. 16-6? There appears to be eight of them in the X-Y mapping, alone: four around the outside of the square, and four more leading from the corners of the square and into its middle.

There is a possibility that the triangle in the X-Z planar view, or mapping, represents up to four different triangles. There can be lines between points 1 and 2, between points 1 and 3, between points 4 and 2, and between points 4 and 3. That view, in other words, might represent as many as four lines along its base. And there are four possible lines leading up to the apex at point 5. That totals up to eight possible lines for the X-Z mapping. And the same happens to be true for the Y-Z mapping. Can you specify all eight possible lines from that Y-Z planar view? Table 16-4 lists *all possible lines* from those three views of the object we are creating here.

After eliminating the duplicates, attempt to plot the remaining lines. In this case, two lines fail to plot prop-

Table 16-2. Working Out the Final Face and Space-Object Specifications for the 3-D Object Suggested in Fig. 16-1.

TENTATIVE FACE LIST	ELIMINATE DUPLICATES	FINAL FACE LIST
$L_1L_2L_3$	$L_1L_2L_3$	$L_1L_2L_3$
$L_2L_4L_5$	$L_2L_4L_5$	$L_2L_4L_5$
$L_1L_4L_6$	$L_1L_4L_6$	$L_1L_4L_6$
$L_2L_4L_5$		
$L_3L_5L_6$	$L_3L_5L_6$	$L_3L_5L_6$
$L_2L_4L_4$		

FINAL FACE SPECIFICATIONS:

$F_1=\overline{L_1L_2L_3}$ $F_2=\overline{L_2L_4L_5}$

$F_3=\overline{L_1L_4L_6}$ $F_4=\overline{L_3L_5L_6}$

FINAL SPACE OBJECT SPECIFICATIONS:

$S=\overline{F_1F_2F_3F_4}$

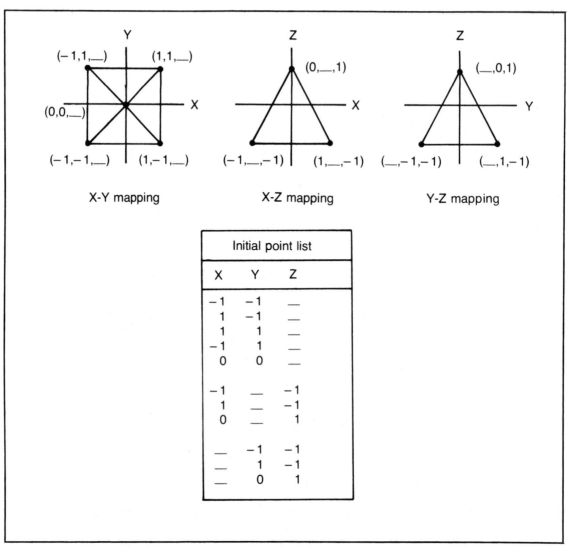

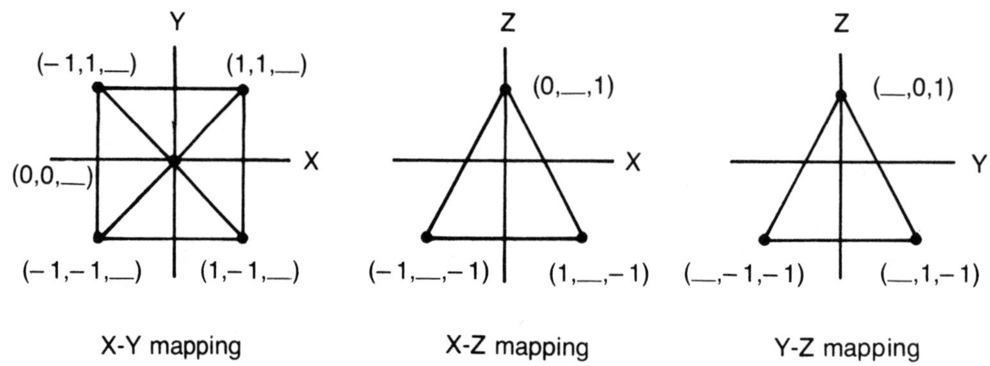

Initial point list		
X	Y	Z
−1	−1	—
1	−1	—
1	1	—
−1	1	—
0	0	—
−1	—	−1
1	—	−1
0	—	1
—	−1	−1
—	1	−1
—	0	1

Fig. 16-5. Initial planar mappings and point list for a space figure.

erly; the line suggested by endpoints 2 and 4, for instance, calls for a single line extending diagonally across the square in the X-Y mapping. We earlier defined that distance as being composed of two lines: those defined by endpoints 4 and 5, and 2 and 5. Thus line P2P4 out to be eliminated from the line list because it isn't necessary for completing the figure. The same idea applies to suggested line P1P3—it is already handled in a more satisfactory fashion by two other lines, P1P5 and P3P4. The *plotted lines* remaining on the list are the actual lines for the space object. Apparently the object is composed of five points and eight lines.

Those final line specifications are formalized for you in Fig. 16-7. The drawings represent the faces that can be generated from those lines. It turns out that there are two possible faces represented by the X-Z mapping, and two more represented by the Y-Z mapping. Take a moment to appreciate the consistency of the information presented thus far.

Those figures are the starting point for generating a *tenative face list* in Table 16-5. A list of five faces (defined in terms of bounding lines) remains after eliminating the duplicates.

There are no invalid faces in this instance. Notice

that each line appears exactly two times in the list. If one of the lines doesn't appear on the list, you've made a mistake somewhere along the way; the same is true if the line appears just one time. If a line designation appears more than twice, it means that one of the faces should be eliminated—it would represent a needless face. But everything is "clean" in this particular presentation, and it turns out that the object has five faces.

The final step, specifying the space object in terms of its grouping of faces, is a simple matter of writing an S specification for it. See the final entry in Table 16-5.

It's time for you to take over the job. Run through the experiment to gather the final S-F-L-P specifications, and then doublecheck the results by mapping the information to X-Y, X-Z and Y-X planar views. Plot an X-Y perspective view to get some idea of how the object looks suspended in 3-D space.

Now try to generate a complete set of specifications for a 3-D space object that is represented by the three planar views offered in Fig. 16-8. Use the previous project as a general guide.

16-4 HOW TO KNOW WHEN YOU'VE MADE A MISTAKE

I am sure that you can appreciate that a mistake anywhere along the way in this procedure can grow into a significant problem later on. Errors in the procedure rarely go away on their own accord (although that luckily happens sometimes), so you ought to be prepared to deal with them as soon as possible.

There are some certain signs that indicate that you have made a mistake in a previous step in the procedure. I will describe a few of those signs with the thought in mind that you will use them to pick up a problem, retrace your steps, find the error and fix it before things get out of hand.

During the point-specification phase of the job, the

Table 16-3. The Working Point List for Generating the Final Point Specifications for the Space Object Started in Fig. 16-5.

ALL POSSIBLE POINTS			ELIMINATE DUPLICATES			PLOTTED POINTS		
X	Y	Z	X	Y	Z	X	Y	Z
-1	-1	-1	-1	-1	-1	-1	-1	-1
-1	-1	1	-1	-1	1			
1	-1	-1	1	-1	-1	1	-1	-1
1	-1	1	1	-1	1			
1	1	-1	1	1	-1	1	1	-1
1	1	1	1	1	1			
-1	1	-1	-1	1	-1	-1	1	-1
Ø	Ø	-1	Ø	Ø	-1			
Ø	Ø	1	Ø	Ø	1	Ø	Ø	1
-1	-1	-1	-1	-1	-1	-1	-1	-1
-1	-1	-1						
-1	Ø	-1	-1	Ø	-1			
-1	1	-1						
1	-1	-1						
1	Ø	-1	1	Ø	-1			
1	1	-1						
Ø	-1	1	Ø	-1	1			
Ø	Ø	1						
Ø	1	1						
-1	-1	-1						
Ø	-1	-1	Ø	-1	-1			
1	-1	-1						
-1	1	-1						
Ø	1	-1	Ø	1	-1			
1	1	-1						
-1	Ø	1	-1	Ø	1			
Ø	Ø	1						
1	Ø	1	1	Ø	1			

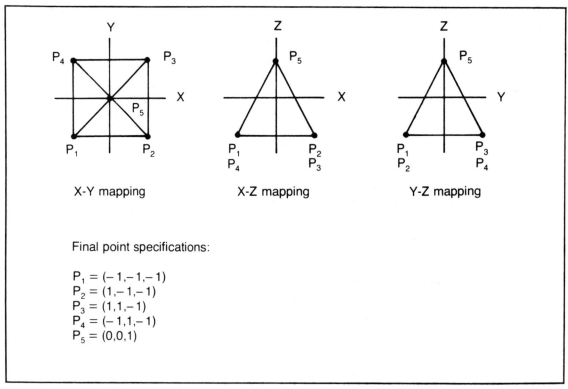

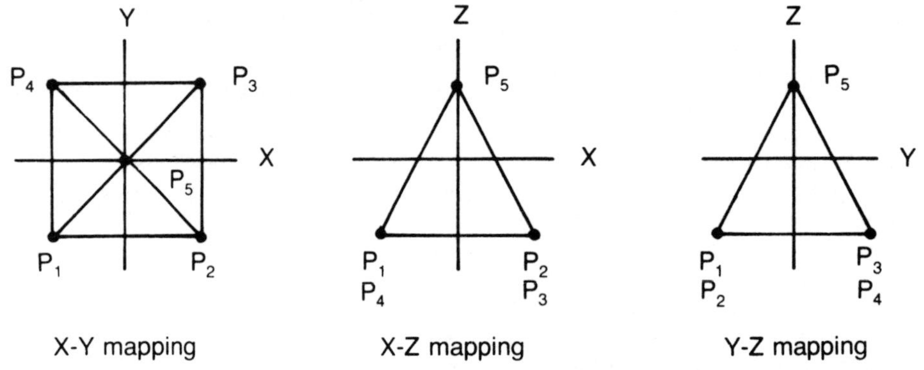

Final point specifications:

$P_1 = (-1,-1,-1)$
$P_2 = (1,-1,-1)$
$P_3 = (1,1,-1)$
$P_4 = (-1,1,-1)$
$P_5 = (0,0,1)$

Fig. 16-6. Final point specifications and revised mappings for the space figure suggested in Fig. 16-5.

most common signal for an error is that of coming up with a point that doesn't plot properly onto one view, but happens to be a critical point for completing one of the other planar views. That is a logical inconsistency that says a point cannot be used in one view, but must be used in another. Assuming that the initial drawings, themselves, are self-consistent, you have most likely made an error in assigning the partial coordinates or copying the coordinates from one list to the next.

The same sort of signal can appear in the line-specification phase. You might find a line that does not plot properly onto one planar view, but is an absolutely necessary line for one of the other views. Whenever that happens, doublecheck your current drawings and line listings for copying errors. And if you've failed to detect an earlier error in the point specifications, it will show up here as either an inconsistent line (as just described) or a line that has an endpoint that isn't shared with any other line. Here are some signs of errors at the face-specifications phase of the procedure:

1. A line fails to appear at all in the face specifications.

2. A line appears only once in the face specifications.

3. A line appears more than twice in the face specifications.

Of course you can take that list and digest it to the point where it says, in effect: a line should appear exactly two times in a set of face specifications. That is logically correct, but it doesn't help much when it comes to troubleshooting an error.

Assuming that you have made no errors in the earlier point and line specifications, the occurrence of troubles signaled by items 1 and 2 above indicate a need for defining more faces—check your current drawings to make sure you have found all possible faces. The occurrence of the trouble in item 3 indicates that you have probably included a face that isn't needed to complete the boundary of the figure.

Here are the planar views of a somewhat more complicated space object. Generate the true 3-D specifications, keeping an eye open for possible errors along the way. Begin from the partial coordinates offered in Fig. 16-9.

Table 16-4. Working Out the Line Specifications From the Previously Determined Point Specifications.

ALL POSSIBLE LINES	ELIMINATE DUPLICATES	PLOTTED LINES
P_1P_2	P_1P_2	P_1P_2
P_2P_3	P_2P_3	P_2P_3
P_3P_4	P_3P_4	P_3P_4
P_1P_4	P_1P_4	P_1P_4
P_1P_5	P_1P_5	P_1P_5
P_2P_5	P_2P_5	P_2P_5
P_3P_5	P_3P_5	P_3P_5
P_4P_5	P_4P_5	P_4P_5
P_1P_1		
P_1P_3	P_1P_3	
P_2P_4	P_2P_4	
P_3P_4		
P_2P_5		
P_3P_5		
P_1P_5		
P_4P_5		
P_1P_3		
P_1P_4		
P_2P_3		
P_2P_4		
P_3P_5		
P_4P_5		
P_1P_5		
P_2P_5		

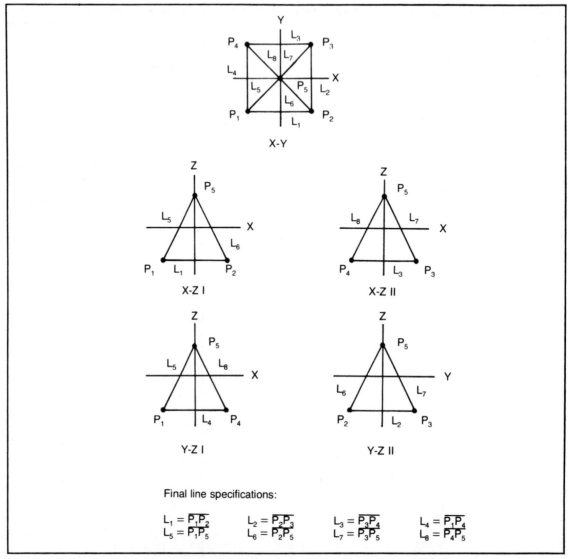

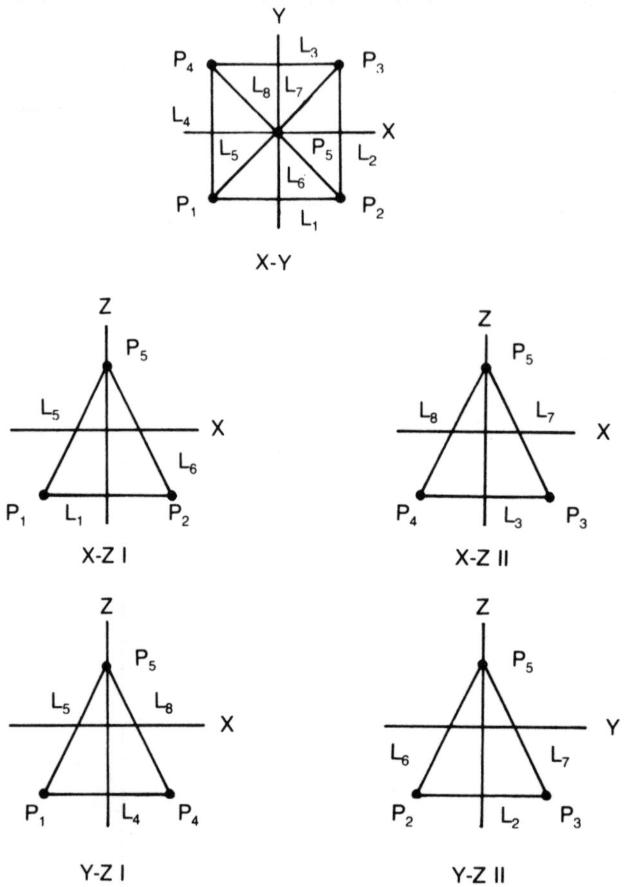

Final line specifications:

$L_1 = \overline{P_1P_2}$ $L_2 = \overline{P_2P_3}$ $L_3 = \overline{P_3P_4}$ $L_4 = \overline{P_1P_4}$
$L_5 = \overline{P_1P_5}$ $L_6 = \overline{P_2P_5}$ $L_7 = \overline{P_3P_5}$ $L_8 = \overline{P_4P_5}$

Fig. 16-7. Evolution of the final line specifications for the space figure suggested in Fig. 16-5.

16-5 GENERATING THE INITIAL INFORMATION

Throughout all of the discussions so far in this chapter, I have been supplying the initial planar views and partial coordinates for you. Now it is time for you to begin thinking in terms of your own custom space objects.

Of course you can probably sketch out the specifications for just about any 3-D object you choose without having to go through all of this work; but remember that the real purpose of this chapter is to prepare the way for doing the same sort of construction procedure for 4-dimensional objects—objects that you cannot visualize from the outset.

Custom initial planar views and partial coordinates are not always easy to establish from scratch. It takes some experience, imagination, educated guesswork and some trial-and-error tinkering. In fact, by the time you have set up some self-consistent initial planar views and partial coordinates, you have already gone a long way toward establishing the final 3-D specifications.

Suppose that you want to set up an object that looks

something like a little house figure—a 5-sided object that is shown as an X-Y planar view in Fig. 16-10. That is how you start: sketch an X-Y planar view and assign some simple partial coordinates to the points. It is easy to set up that initial view, because there are not restrictions on how it should look. Getting the other two views is a bit more difficult.

One thing you know for certain from the initial X-Y mapping is that the X components of the next mapping—the X-Z view—must have values limited to those prescribed for the first view. Those locked-in X-component values in this particular example are 1,0, and −1. If you use any other X components in the X-Z mapping, they will not be consistent with the first drawing. Self-consistency is the rule here.

Since the X components of the X-Y view are limited to −1,0 and 1 in this example, it follows that the X components of points on the X-Z view are limited to those

Table 16-5. Evolution of the Final Face and Space-Object Specifications for the Space Object Suggested in Fig. 16-5.

TENTATIVE FACE LIST	ELIMINATE DUPLICATES	FINAL FACE LIST
$L_1L_2L_3L_4$	$L_1L_2L_3L_4$	$L_1L_2L_3L_4$
$L_1L_5L_6$	$L_1L_5L_6$	$L_1L_5L_6$
$L_2L_6L_7$	$L_2L_6L_7$	$L_2L_6L_7$
$L_3L_7L_8$	$L_3L_7L_8$	$L_3L_7L_8$
$L_4L_5L_8$	$L_4L_5L_8$	$L_4L_5L_8$
$L_1L_5L_6$		
$L_3L_7L_8$		
$L_4L_5L_8$		
$L_2L_6L_7$		

FINAL FACE SPECIFICATIONS:

$F_1=\overline{L_1L_2L_3L_4}$

$F_2=\overline{L_1L_5L_6}$

$F_3=\overline{L_2L_6L_7}$

$F_4=\overline{L_3L_7L_8}$

$F_5=\overline{L_4L_5L_8}$

SPACE OBJECT SPECIFICATIONS:

$S=\overline{F_1F_2F_3F_4F_5}$

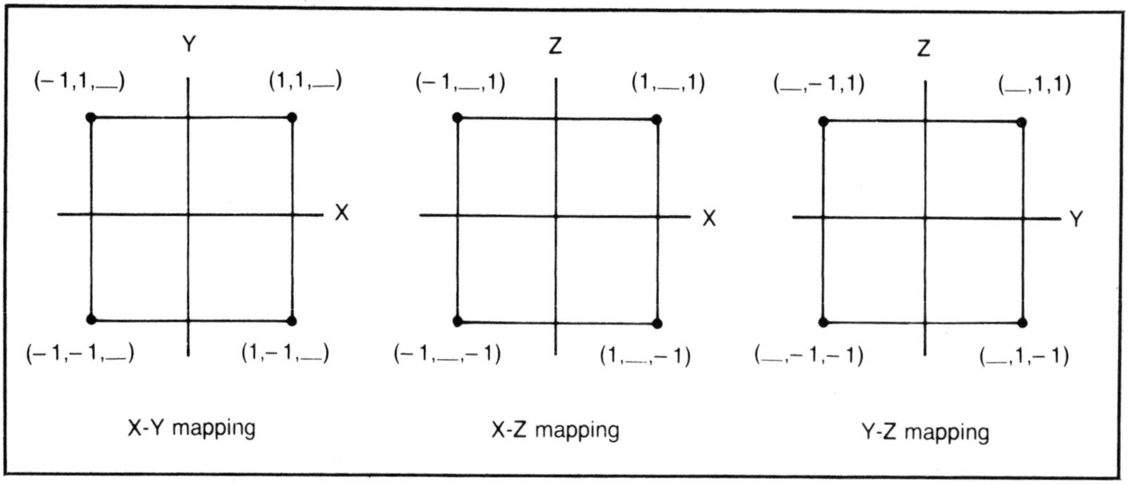

Fig. 16-8. Initial planar mappings and partial coordinates for a space figure you can transform into full 3-D specifications.

same values. See the fixed X-component values drawn as vertical lines in Fig. 16-11A.

There are a lot of different ways to add Z-axis components to those lines; but since the X-Z view represents a "top view" of the object, I am going to keep things simple and make two horizontal lines that make the figure look something like the top view of a house figure. See the result and partial coordinates for the X-Z planar view in Fig. 16-11B.

Now we have a self-consistent set of partial coordinates for the X-Y and X-Z planar views. The initial setup of the X-Y view is purely arbitrary. But once you have established its partial coordinates, you are a bit more restricted when it comes to devising that second view—the X-Z planar view. And as you might imagine by now, the matter of generating the third view—the Y-Z view—is even more restricted in choices. In fact, you have no real choices in the matter.

Examining the partial coordinates for the X-Y planar view, you have available in this example Y-axis components of −1, 1 and 2. And looking at the partial coordinates for the X-Z view, you are left with possible Z components of −1 and 1. Given that information, you can generate the ALL POSSIBLE Y-Z partical coordinates

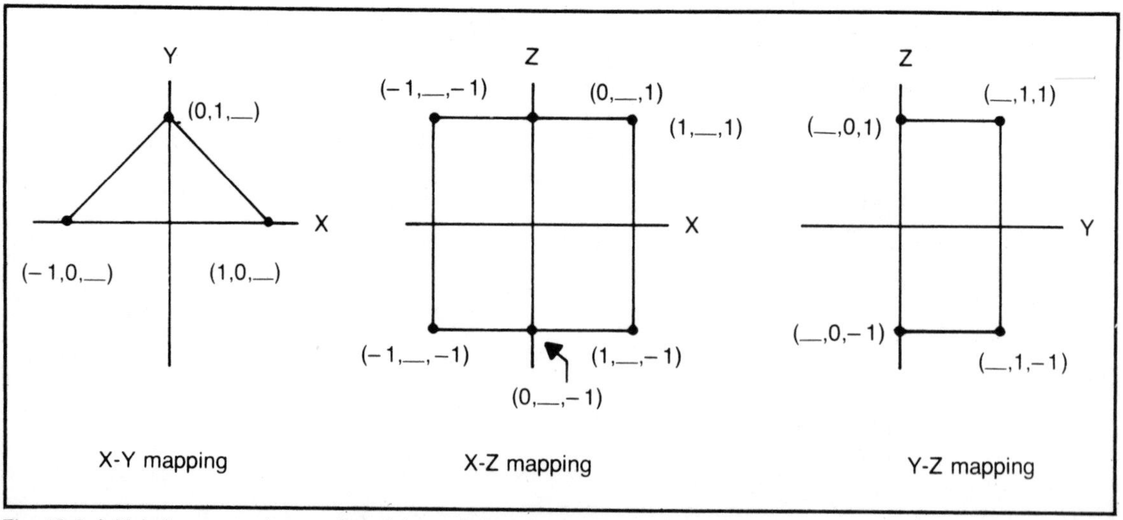

Fig. 16-9. Initial planar mappings and partial coordinates for a different space figure.

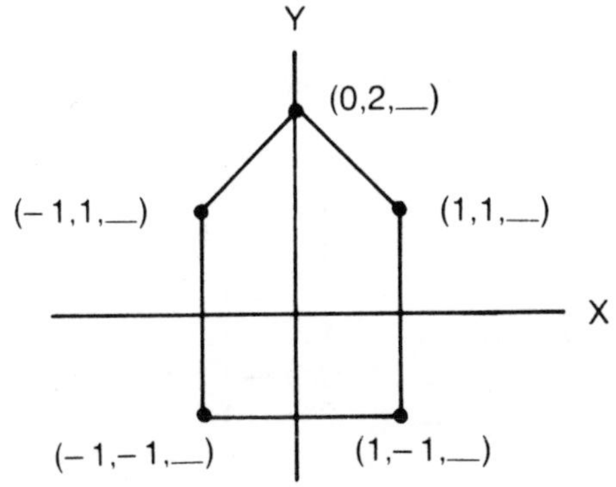

Partial coordinates for the X-Y mapping

X	Y	Z
1	−1	__
1	1	__
0	2	__
−1	1	__
−1	−1	__

Fig. 16-10. X-Y mapping of a space figure to be generated.

shown in Fig. 16-12. Eliminate the duplicates (there happen to be none in this example), and you are left with a set of partial coordinates for the Y-Z mapping. Using some educated guesswork and maybe a bit of inspiration, the points can be plotted and connected with lines as shown in that diagram.

Doublecheck the lists of partial coordinates for possible inconsistencies and points that fail to plot on one or more view, and you are ready to begin generating the full 3-D specifications. I will leave that task for this house-shaped object to you.

This example happens to work out rather nicely. If there were any serious problems with the initial planar drawings or their partial 3-D coordinates, the errors would show up as clear inconsistencies somewhere through the construction procedure.

Suppose, now, that you have studied the technique for a while and have mastered its basic principles. You

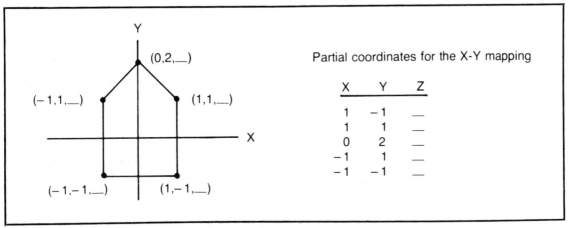

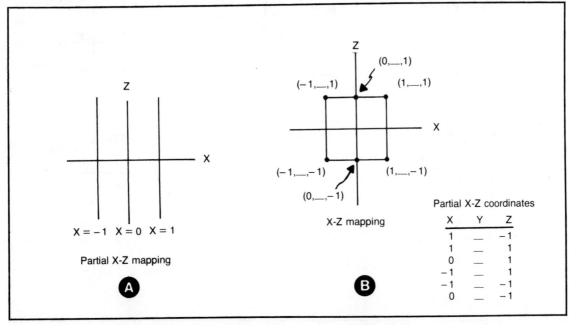

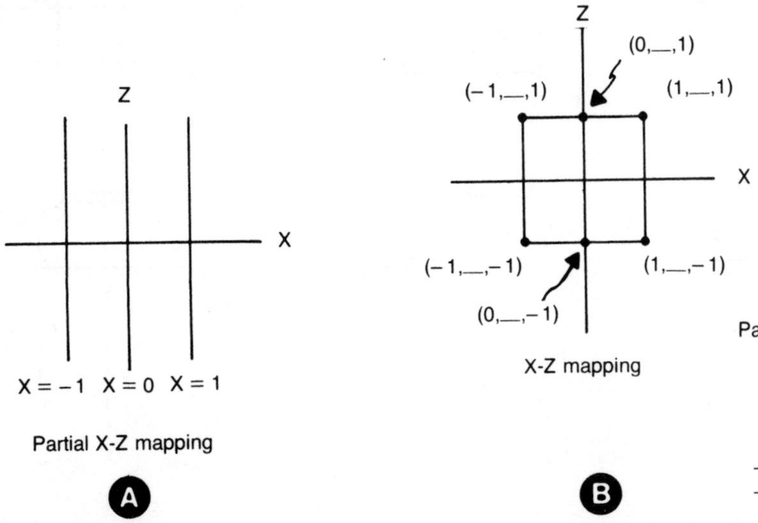

Partial X-Z coordinates

X	Y	Z
1	__	−1
1	__	1
0	__	1
−1	__	1
−1	__	−1
0	__	−1

Fig. 16-11. Evolution of an X-Z mapping based on the partial coordinates in Fig. 16-10.

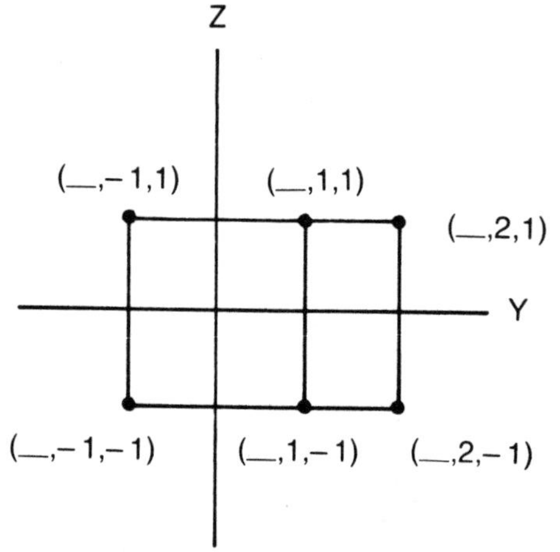

All possible Y-Z

X	Y	Z
—	−1	−1
—	−1	1
—	1	−1
—	1	1
—	2	−1
—	2	1

Fig. 16-12. A Y-Z mapping dictated by the partial coordinates in Figs. 16-10 and 16-11.

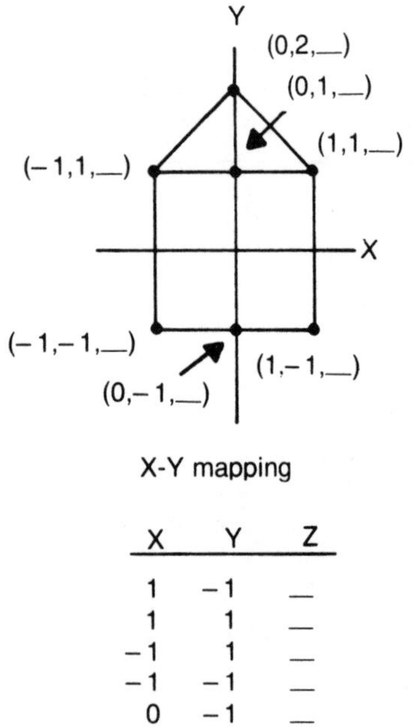

X-Y mapping

X	Y	Z
1	−1	—
1	1	—
−1	1	—
−1	−1	—
0	−1	—
0	1	—
0	2	—

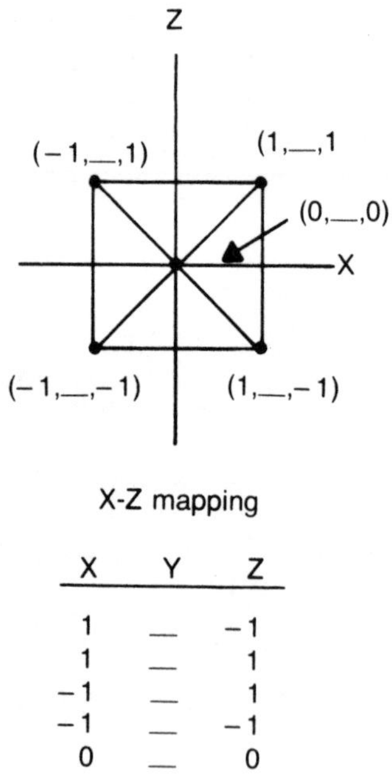

X-Z mapping

X	Y	Z
1	—	−1
1	—	1
−1	—	1
−1	—	−1
0	—	0

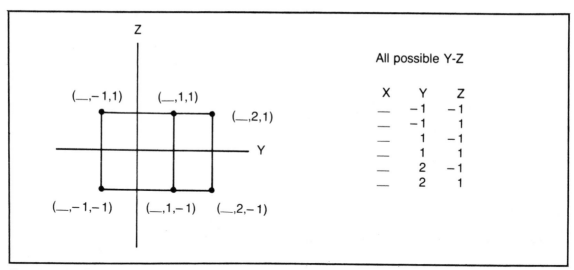

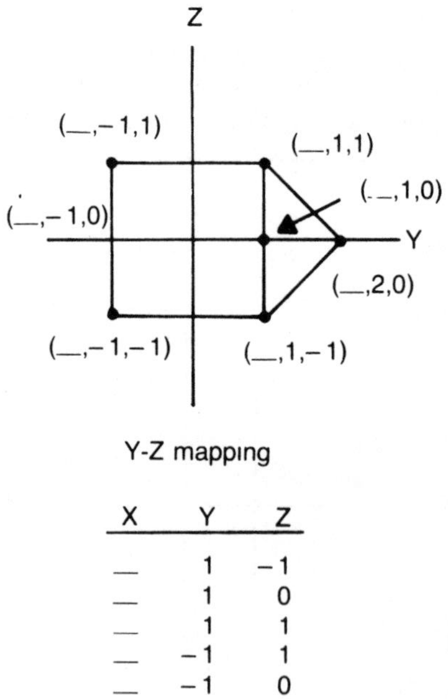

Y-Z mapping

X	Y	Z
—	1	−1
—	1	0
—	1	1
—	−1	1
—	−1	0
—	−1	−1
—	2	0

Fig. 16-13. Setting up the initial mappings and partial coordinates for a somewhat more complex space figure.

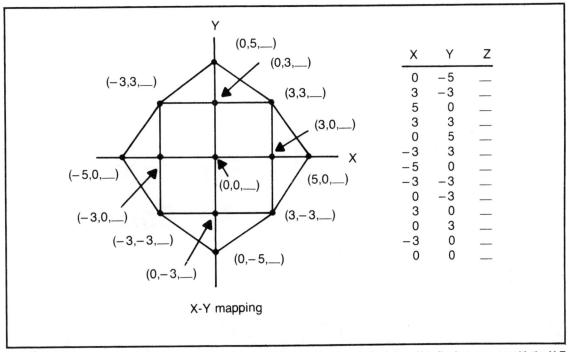

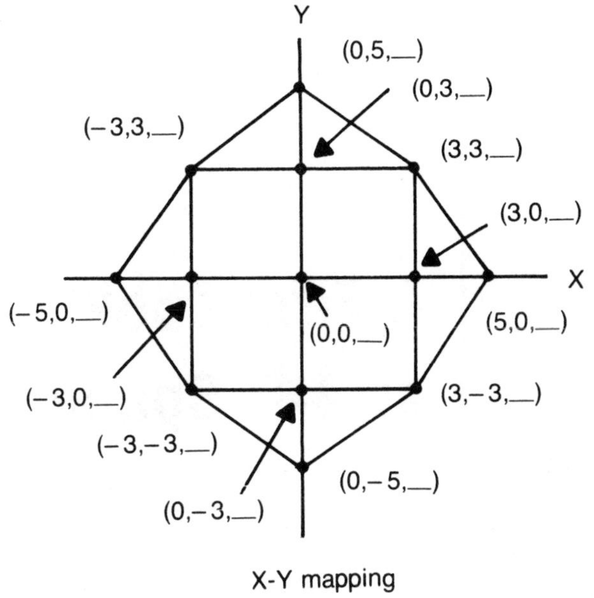

X	Y	Z
0	−5	—
3	−3	—
5	0	—
3	3	—
0	5	—
−3	3	—
−5	0	—
−3	−3	—
0	−3	—
3	0	—
0	3	—
−3	0	—
0	0	—

Fig. 16-14. The procedure can evolve some complex space figures. Use the methods just described to come up with the X-Z and Y-Z mappings.

have even set up some initial planar views and partial coordinates that you ultimately transformed into full 3-D specifications for custom space objects. With that sort of success under your belt, you will find it rather easy to generate views of slightly more complex object, using some of your simpler views as starting points.

For instance, you can modify the X-Y mapping of the house-shaped object in the previous example to look like the version in Fig. 16-13. That is a matter of adding two more points and four more lines. The X-Z mapping is made consistent with the revised X-Y view; and given those two new views, the nature of the Y-Z mapping ought to be quite apparent.

Can you see how the planar view of the 3-D space object in Fig. 16-14 evolves from the earlier ones? If not, you need more study in this chapter.

Once you get rolling with this idea of creating increasingly complex planar views and partial coordinates, I suggest that you begin keeping a permanent file of them for future experiments.

And, incidentally, there is no point in going through the entire construction procedure if all you want to do is change the size, position or orientation of an object that you have already constructed. Simply apply the appropriate 3-D translation, scaling or rotation to do that sort of task.

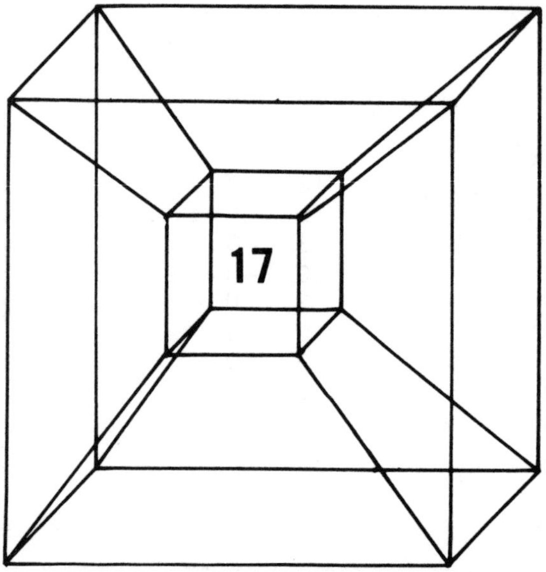

A First Look at 4-D Space

Here it is at last! Here is the introduction to something entirely new to most people living on this planet today—a workable vision of 4-dimensional space. All of the previous work; virtually every equation, example, exercise, proof, and derivation cited this far; has prepared you for this moment. If you have been conscientious about studying the previous work, you are fully prepared for the things to come; and if you have been doing the exercises and thinking for yourself, you can have a lot of fun and excitement with the principles of 4-dimensional space.

You are in for some trouble, however, if you have skipped over a significant amount of the earlier examples and exercises. You are going to be lost here, because you lack the firm foundation that is required for thinking and experimenting successfully in terms of an abstract space of 4 dimensions. There are no satisfactory short cuts.

Four-dimensional space is a strange place; a very strange place. Human beings have no direct perception of it, and most find it difficult to conjure even a mental impression of it. But let's try, anyway.

Recall the characterization of 1-D space as a simple straight line of indefinite length. A suitable coordinate system for 1-D space is likewise a straight line that includes some linear scaling and a point of origin. Any imaginary beings that are confined to that space can move about within it, but have no direct perception of anything beyond that 1-dimensional line-world.

Two-dimensional space is characterized by an infinitely thin flat plane that can extend indefinitely in all directions. Its coordinate system is composed of a pair of perpendicular 1-D coordinate systems that intersect at a common point of origin. Imaginary 2-D creatures can move about at an infinite number of different angles in their plane-world, but they cannot directly perceive anything that isn't on the plane, itself.

Spaces of 3 dimensions are the most familiar to humans. It is the character of the real universe as we perceive it at the present time. It has qualities that can be roughly, but generally meaningfully, described as width, height and depth. A suitable coordinate system for 3-D space is composed of three mutually perpendicular coordinate systems that intersect at their common point of origin. Creatures living in 3-D space have no difficulty thinking in terms of the lower spaces of 1 and 2 dimensions, but the notion of anything outside the bounds of 3-D space are impossible to perceive in a direct fashion.

Four-dimensional space represents a logical next step in the progression of spaces—it follows quite naturally from the qualities just described for spaces of 1, 2 and 3 dimensions. It has the properties of width, height and depth of 3-D space, plus one more that we can call about anything we want. The coordinate system is composed of four 1-D coordinates that are mutually perpendicular and intersect at a common origin. Yes, I know it is difficult to envision a fourth coordinate axis that is perpendicular to three others that are likewise perpendicular to one another. If you can picture four mutually perpendicular axes in your mind, you either have an extraordinary high level of spatial perception or you don't really understand the situation. Human perception is not structured to deal with 4-D space in a direct way, so the only avenue for experimenting with it at all is that of abstract geometry—the subject and purpose of this book.

17-1 A COORDINATE SYSTEM FOR 4-D SPACE

If you think it has been tough to represent 3-D spaces on the 2-D plane of a sheet of paper, wait until you try representing 4-D spaces on a sheet of paper. That represents a drop of two dimensions, and you are going to have to do a lot of work to get any real meaning through the resulting spatial and angular distortion.

Figure 17-1 is a rough representation of a 4-D coordinate system. The purpose is simply to show that it includes four different axes: axes labeled X, Y, Z and W. They are supposed to be perpendicular to one another; but on the plane of a page, it isn't possible to show any more than two perpendicular lines. In this case, I've shown the X and Y axes perpendicular to one another. The Z and W axes are sketched in at some arbitrary angles just for the sake of indicating their presence.

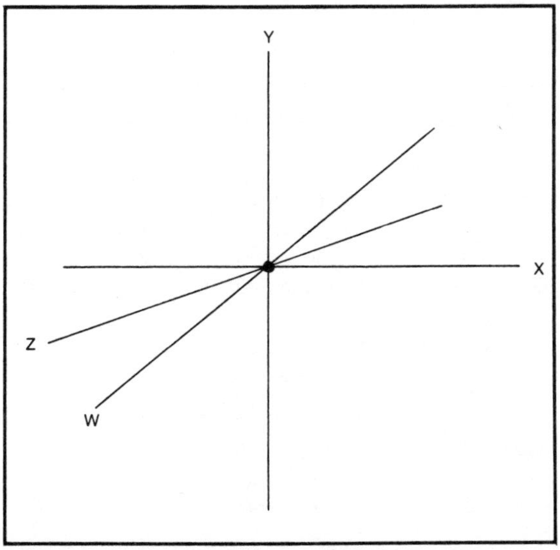

Fig. 17-1. A general perspective view of a coordinate system portraying 4-D space, or hyperspace.

Recall that one way to portray 3-D spaces on the plane of a page of paper is to plot the 3-D coordinate system from three different planar views. Generally speaking, a series of lower-dimensional views can adequately portray the qualities of a higher-dimensional space. Applying that notion to the problem of drawing objects in 4-D space, we can represent 4-D space by means of a series of 3-D perspective views.

Figure 17-2 shows four 3-D perspective views of a suitable 4-D coordinate system. It is, in fact a right-handed 4-D coordinate system that will serve as the standard through the remainder of this book.

The first view—the XYZ perspective view—is identical to the 3-D version used through all of the previous work in 3-D space. The X and Y axes are shown perpendicular to one another, while the Z axis is offset from the X axis by some angle fz. In reality, the positive end of the X axis extends straight out of the page through the origin of the coordinate system, but it cannot be drawn that way on a sheet of paper. Hopefully, you learned to live with that trade-off through earlier work.

The three remaining views follow the same general format, but are new in the sense that they portray a fourth axis—the W axis of 4-D space. In the case of the XYW perspective view, the X and Y axes are drawn perpendicular to each other, while W axis is offset from the X axis by some angle fw. Again, this is a 3-D representation of three axes of 4-D space, and the positive end of the W axis is really supposed to be sticking straight out of the page at you.

The XZW perspective view shows the X and Z axes of the coordinate system as perpendicular to one another; and in this instance, it is the W axis that is extending out of the page. The YZW perspective view completes the series by representing the coordinate system's Y and Z axes as being perpendicular and the W axis extending out of the page, but shown as offset from the Y axis at some angle fw.

That is an adequate representation of a 4-D coordinate system. It is an attempt to portray 4-D space by lowering it just one dimension—down to 3-D representations. But since it is equally difficult to portray 3-D space on the plane of a page, it is necessary to resort to the 3-D perspective views of the situation. Using these four views usually exposes the essential qualities of anything plotted in 4-D space. The numerical analysis, however, still stands as the most accurate and reliable way to deal with the fourth dimension.

In summary, this book uses two techniques for plotting 4-D spaces to the plane of a sheet of paper: a single,

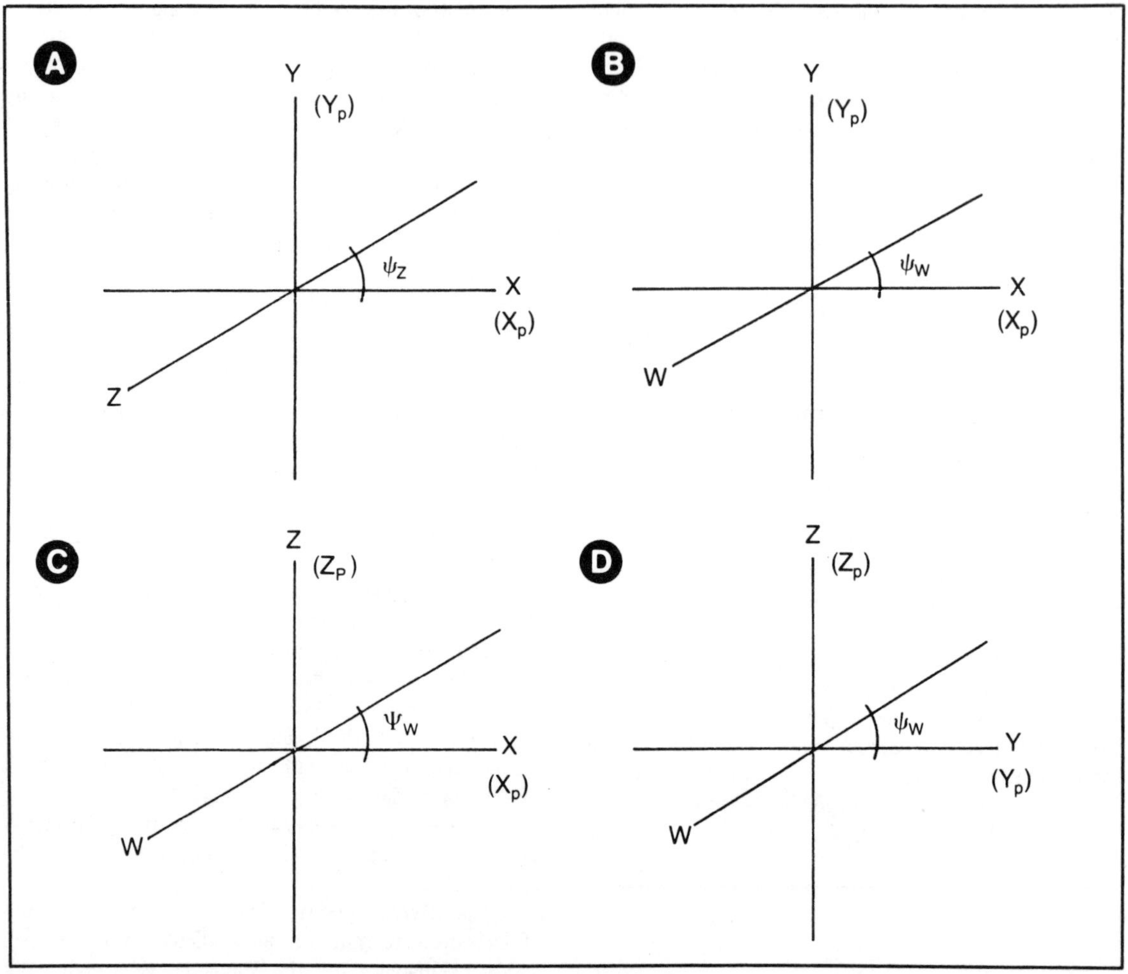

Fig. 17-2. 3-D perspective views of a 4-D coordinate system. (A) XYZ perspective view. (B) XYW perspective view. (C) XZW perspective view. (D) YZW perspective view.

general perspective view, and a set of four 3-D perspective views. The general perspective view is one that shows the four axes of the coordinate system meeting at the origin; but the orientation of the axes with respect to one another is irrelevant in a practical sense. The angular and spatial distortion will be so bad in any case that there is no need to be fussy about the precisions of this sort of 2-dimensional representation of 4-D space. The purpose of this view is to show some of the most important characteristics of a hyperspace object—number of points, number of lines, number of lines comprising plane figures, and the number and type of plane figures that bound space objects.

The sets of 3-D perspective views convey information in a somewhat more precise form. Each, in itself, is an inadequate portrayal; but taken together, they do a pretty good job of showing what a hyperspace object looks like. The perspective views look much like perspective drawings for ordinary 3-D space. It just so happens that 4-D space can be broken down into four separate and different 3-D spaces. Those spaces are:

XYZ space
XYW space
XZW space
YZW space

In a manner of speaking, those spaces represent four different, alternative 3-D spaces. It's fun to suppose that we inhabit one of them—say, the XYZ space—and that other creatures such as ourselves inhabit the other spaces. You will soon find that our XYZ rendition of some hyperspace object can be quite different from that of a person living in one of those other space universes.

17-2 SPECIFYING AND PLOTTING POINTS IN 4-D SPACE

A point in 1-D space can be completely specified by means of a 1-component coordinate—an X component as far as our own convention is concerned. The a point in 2-D space is completely specified by a 2-component coordinate; namely the X and Y components. Then a point in 3-D space can be properly specified by 3 components—X, Y and Z components. So now it should come as no great surprise that a point in 4-D space must be specified in terms of a 4-component coordinate. Check out this summary:

- ☐ In 1-D space: $P_n = (x_n)$
- ☐ In 2-D space: $P_n = (x_n, y_n)$
- ☐ In 3-D space: $P_n = (x_n, y_n, z_n)$
- ☐ In 4-D space: $P_n = (x_n, y_n, z_n, w_n)$

Plotting 4-D point to the four perspective views defined in Fig. 17-2 is a matter of combining the plotting techniques for 3-D space. Notice in the perspective views that one axis is missing in each case. The W axis is not portrayed in any way at all in the XYZ perspective view, the Z axis is absent from the XYW view, the Y axis is not shown in the XZW view, and finally the X axis is missing from the YZW perspective view.

So the first step in plotting a point specified in 4-D space is to ignore the component of the axis that isn't portrayed in a given perspective view. That is analogous to mapping 3-D spaces directly to its three planar views (a notion introduced in Chapter 11). Thus:

- ☐ When generating an XYZ perspective view, ignore the W-axis component.
- ☐ When generating an XYW perspective view, ignore the Z-axis component.
- ☐ When generating an XZW perspective view, ignore the Y-axis component.
- ☐ When generating a YZW perspective view, ignore the X-axis component.

Doing that, you are left with just three components in each case. Then treat the three components as a coordinate of a point in 3 dimensions; use some perspective-generating equation to transform the three components to a 2-D format that is suitable for plotting on a sheet of paper.

Eliminate one component and then treat the remaining three as components of a point in 3 dimensions. That's the general idea. Equations 17-1 through 17-4 are appropriate for that latter step.

$$x_{pn} = x_n - z_n \cos\psi_z \qquad y_{pn} = y_n - z_n \sin\psi_z$$ **Equation 17-1**

where:

x_n, y_n, z_n = components of the true coordinate of points *n*

x_{pn}, y_{pn} = perspective-view components to be plotted for point *n*

ψ_z = perspective angle between the X and Z axis

Note: These expressions generate the plotted coordinate for an XYZ perspective view of 4-D space

$$x_{pn} = x_n - w_n \cos\psi_w \qquad y_{pn} = y_n - w_n \sin\psi_w$$ **Equation 17-2**

where:

x_n, y_n, w_n = components of the true coordinate of point *n* in 4-D space

x_{pn}, y_{pn} = perspective-view components to be plotted for point *n*

ψ_w = the perspective angle between the X and W axes

Note: These expressions generate the plotted coordinate for an XYW perspective view of 4-D space.

$$x_{pn} = x_n - w_n \cos\psi_w \qquad z_{pn} = z_n - w_n \sin\psi_w$$ **Equation 17-3**

where:

x_n, z_n, w_n = components of the true coordinate of point *n* in 4-D space

x_{pn}, z_{pn} = perspective-view components to be plotted for point *n*

ψ_w = the perspective angle between the X and W axes

Note: These expressions generate the plotted coordinate for an XZW perspective view of 4-D space

$$y_{pn} = y_n - w_n \cos\psi_w \qquad z_{pn} = z_n - w_n \sin\psi_w$$ **Equation 17-4**

where:

y_n, z_n, w_n = components of the true coordinate of point *n* in 4-D space

y_{pn}, z_{pn} = perspective-view components to be plotted for point *n*

ψ_w = the perspective angle between the Y and W axes

Note: The expressions generate the plotted coordinate for a YZW perspective view of 4-D space

Equation 17-1 generates the perspective-view coordinates for the XYZ perspective view of 4-D space. Notice that the W-axis component is missing and that it turns up the plottable point, (x_{pn},y_{pn}). The equation provides a coordinate that can be plotted on the perpendicular X and Y axes. The Z axis is shown only to indicate its presence.

In a similar fashion, Equation 17-2 generates a 2-dimensional coordinate, (x_{pn},y_{pn}). The result is affected by the W-axis component of the 4-D version of the coordinate, but the point is plotted without reference to it. The Z-axis component of the 4-D coordinate is not relevant in this case.

Use Equation 17-3 to get the perspective-view coordinate for the XZW view of 4-D space. The result is a 2-D coordinate of (x_{pn},z_{pn}) which can be plotted on the X-Z axes. Here the W-axis component affects the point's position on the drawing, but the Y-axis component is not relevant.

Finally, Equation 17-4 turns up the plottable point (y_{pn},z_{pn}) for the YZW perspective view. The W-axis component of the 4-D coordinate affects the result, but the point, itself, is plotted only with regard to the Y and Z axes. The X component of the original 4-D specification plays no part here. See how this all works by studying Example 17-1.

Example 17-1

Plot the four points specified in 4-D space onto the four 3-D perspective views, using a perspective angle of 30 degrees in each instance.

$$P_1 = (0,0,0,0) \quad P_2 = (1,1,1,0)$$
$$P_3 = (0,-1,-1,-1) \quad P_4 = (-1,0,0,1)$$

1. Gathering the components of those points:

$$\begin{array}{llll} x_1 = 0 & x_2 = 1 & x_3 = 0 & x_4 = -1 \\ y_1 = 0 & y_2 = 1 & y_3 = -1 & y_4 = 0 \\ z_1 = 0 & z_2 = 1 & z_3 = -1 & z_4 = 0 \\ w_1 = 0 & w_2 = 0 & w_3 = -1 & w_4 = 1 \end{array}$$

2. Applying Equation 17-1 to transform the true 4-D coordinates to XYZ perspective-view coordinates:

$$\begin{array}{ll} x_{pl} = x_1 - z_1\cos 30^\circ & y_{pl} = y_1 - z_1\sin 30^\circ \\ x_{pl} = 0-(0)(0.87) & y_{pl} = 0-(0)(0.5) \\ x_{pl} = 0 & y_{pl} = 0 \end{array}$$
$$P_{pl} = (0,0)$$

Using the same equation and procedure for points P_2, P_3 and P_4:

$$P_{p2} = (0.13,0.5)$$
$$P_{p3} = (0.87,-0.5)$$
$$P_{p4} = (-1,0)$$

See those four points plotted on the XYZ perspective view in Fig. 17-3.

3. Applying Equation 17-2 to transform the true 4-D coordinates to XYW perspective-view coordinates:

$$\begin{array}{ll} x_{pl} = x_1 - w_2\cos 30^\circ & y_{pl} = y_1 - w_1\sin 30^\circ \\ x_{pl} = 0-(0)(0.87) & y_{pl} = 0-(0)(0.5) \end{array}$$
$$P_{pl} = (0,0)$$

Using the same equation and procedure for points P_2, P_3 and P_4:

$$P_{p2} = (1,1)$$
$$P_{p3} = (0.87,-0.5)$$
$$P_{p4} = (-1.87,-0.5)$$

See the points plotted on the XYW perspective view in Fig. 17-3.

4. Applying Equation 17-3 to transform the true 4-D coordinates to XZW perspective-view coordinates:

$$\begin{array}{ll} x_{pl} = x_1 - w_1\cos 30^\circ & z_{pl} = z_1 - w_1\sin 30^\circ \\ x_{pl} = 0-(0)(0.87) & z_{pl} = 0-(0)(0.5) \\ x_{pl} = 0 & z_{pl} = 0 \end{array}$$
$$P_{pl} = (0, 0)$$

Using the same equation and procedure for points P_2, P_3 and P_4:

$$P_{p2} = (1,1)$$
$$P_{p3} = (0.87,-0.5)$$
$$P_{p4} = (-1.87,-0.5)$$

See these points plotted on the XZW perspective view in Fig. 17-3.

5. Applying Equation 17-4 to transform the true 4-D coordinates to YZW perspective-view coordinates:

$$\begin{array}{ll} y_{pl} = y_1 - w_1\cos 30^\circ & z_{pl} = z_1 - w_1\sin 30^\circ \\ y_{pl} = 0-(0)(0.87) & z_{pl} = 0-(0)(0.5) \\ y_{pl} = 0 & z_{pl} = 0 \end{array}$$
$$P_{pl} = (0,0)$$

Using the same equation and procedure for points P_2, P_3 and P_4:

$$P_{p2} = (1,1)$$
$$P_{p3} = (-0.13,-0.5)$$
$$P_{p4} = (-0.87,-0.5)$$

See these points plotted on the YZW perspective view in Fig. 17-3.

The example cites four different points specified in 4-D space, then Step 1 shows the individual components

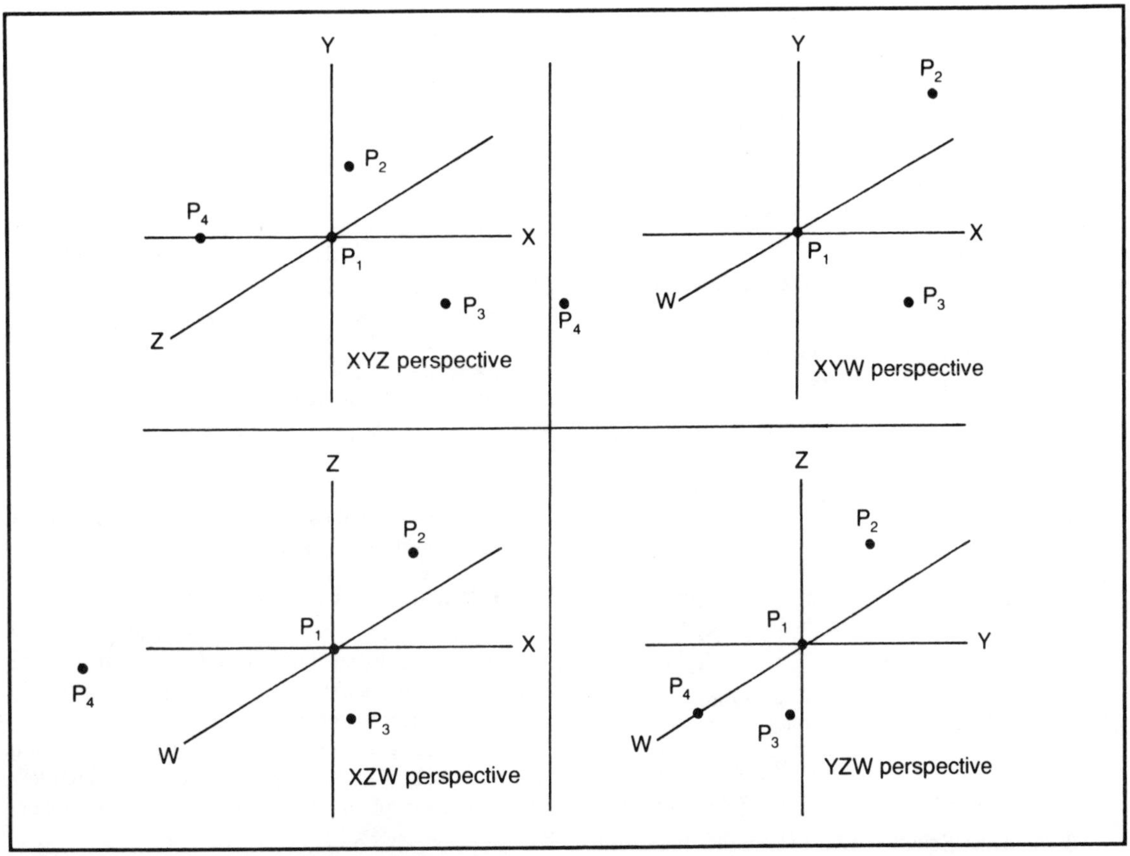

Fig. 17-3. Four points specified in hyperspace and plotted onto the four 3-D perspective views. See Example 17-1.

along with their point designations. Step 2 uses the given information and Equation 17-1 to generate the perspective-view coordinates for the XYZ view of the situation. I've shown the procedure in some detail for point P_1, but I simply summarized the results for the three other points. You can doublecheck my results by applying the equation to the appropriate components in the original data. Steps 3, 4 and 5 use the same procedure to get the perspective-view coordinates for the four points as plotted to the XYW, XZW and YZW views, respectively

The final results shown in Fig. 17-3 might not appear very exciting nor especially meaningful to a casual observer, but at least you and I know it represents some points that are specified in 4-D space. And that, I think, is both exciting and meaningful. Work Exercise 17-1 to test your understanding and build some confidence.

Exercise 17-1

Generate the four perspective views of the following points specified in 4-D space. Use Example 17-1 as a guide.

$P_1 = (1,1,1,1)$ $\quad$ $P_2 = (-1,1,1,1)$
$P_3 = (-1,-1,1,1)$ $\quad$ $P_4 = (1,-1,1,1)$
$P_5 = (1,1,1,-2)$ $\quad$ $P_6 = (-1,1,1,-2)$
$P_7 = (-1,-1,1,-2)$ $\quad$ $P_8 = (1,-1,1,2)$

17-3 SPECIFYING AND PLOTTING LINES IN 4-D SPACE

A line in 4-D space can be defined in terms of the coordinates of its two endpoints. There is nothing new about that. The only difference is that the points in this case are specified by 4 components, rather than 2 or 3. So generally speaking, the specification for a line in 4-D space looks like this:

$$L_n = P_uP_v \qquad P_u = (x_u,y_u,z_u,w_u)$$
$$P_v = (x_v,y_v,z_v,w_v)$$

Where P_u and P_v are the bounding endpoints of line L_n.

Notice that you cannot tell whether or not the line is in 1-D, 2-D, 3-D or 4-D space by looking at the line specification alone. It is the point specifications that signal the dimension of the space surrounding the line. Example 17-2 illustrates a situation for plotting two different lines that are specified in 4-D space.

Example 17-2

Make all four perspective plots of these two lines specified in 4-D space:

$$L_1 = \overline{P_1P_2} \qquad P_1 = (1,1,1,1)$$
$$L_2 = \overline{P_3P_4} \qquad P_2 = (-1,-1,-1,-1)$$
$$P_3 = (-1,-1,1,1)$$
$$P_4 = (1,1,-1,-1)$$

1. Tabulating the components of the four coordinates:

$$\begin{array}{llll} x_1 = 1 & x_2 = -1 & x_3 = -1 & x_4 = 1 \\ y_1 = 1 & y_2 = -1 & y_3 = -1 & y_4 = -1 \\ z_1 = 1 & z_2 = -1 & z_3 = 1 & z_4 = -1 \\ w_1 = 1 & w_2 = -1 & w_3 = 1 & w_4 = -1 \end{array}$$

2. Applying Equation 17-1 to generate the coordinates for the XYZ perspective view:

$$P_{pl} = (0.13,0.5) \qquad P_{p2} = (-0.13,-0.5)$$
$$P_{p3} = (-1.87,-1.5) \qquad P_{p4} = (1.87,-0.5)$$

3. Applying Equation 17-2 to generate the coordinates for the XYW perspective view:

$$P_{pl} = (0.13,0.5) \qquad P_{p2} = (-0.13,-0.5)$$
$$P_{p3} = (1.87,-1.5) \qquad P_{p4} = (1.87,-0.5)$$

4. Applying Equation 17-3 to generate the coordinates for the XZW perspective view:

$$P_{pl} = (0.13,0.5) \qquad P_{p2} = (-0.13,-0.5)$$
$$P_{p3} = (-1.87,0.5) \qquad P_{p4} = (1.87,-0.5)$$

5. Applying Equation 17-4 to generate the coordinates for the YZW perspective view:

$$P_{p1} = (0.13,0.5) \qquad P_{p2} = (-0.13,-0.5)$$
$$P_{p3} = (-1.87,0.5) \qquad P_{p4} = (-0.13,-0.5)$$

6. Plot the perspective-view coordinates on their respective graphs, then draw lines between the points as specified in the original information. See Fig. 17-4.

The specifications at the beginning of that example specify the two endpoints of the lines. Line 1 is bounded by points 1 and 2, while line 2 is bounded by points 3 and 4. The coordinate assignments for the points completes the specifications.

Step 1 draws upon the specifications to develop a complete table of components, then Steps 2 through 5 generate the perspective-view coordinates for the four standard 3-D perspective views of 4-D space. This, of course, represents the work first described in the opening section of the present chapter.

All that remains, then, is to draw lines between the points on the drawings, using the original line specifications as a guide to getting the lines between the proper points. See Fig. 17-4.

Make up some line specifications of your own, being certain to indicate four components for each of the endpoints. Then use the procedure demonstrated in Example 17-2 to plot your lines. Not many people in this world have attempted such a thing; but if you have been doing all the work prescribed thus far, you will find it a relatively simple and quite meaningful task.

17-3.1 Lengths of Lines in 4-D Space

Length, or distance, is just as meaningful in 4-D space as it is in lower-dimensioned spaces. It is more difficult, if not impossible, to visualize the true length of a line in 4-D; but it nevertheless exists.

The basic length equation is expanded in Equation 17-5 to include W-axis components. It is a straightforward extension of the length equations for all lower-dimensioned spaces.

Equation 17-5

$$s_n = \sqrt{(x_u-x_v)^2 + (y_u-y_v)^2 + (z_u-z_v)^2 + (w_u-w_v)^2}$$

where:

s_n = the length of line *n* in 4-D space
x_u,y_u,z_u,w_u = the coordinate of endpoint *u* of line *n*
x_v,y_v,z_v,w_v = the coordinate of endpoint *v* of line *n*

All you have to do is set up the components of the endpoints of the line, determine the displacement of those endpoints for each of the four axes, square the differences, sum them, and take the square root of the result. There's nothing new but the presence of the W-axis components. Take a look at Example 17-3.

Example 17-3

Calculate the length of these two lines specified in 4-D space:

$$L_1 = \overline{P_1P_2} \qquad P_1 = (0,1,3,5)$$
$$L_2 = \overline{P_3P_4} \qquad P_2 = (-2,4,6,8)$$
$$P_3 = (3,2,-1,0)$$
$$P_4 = (4,-6,8,-2)$$

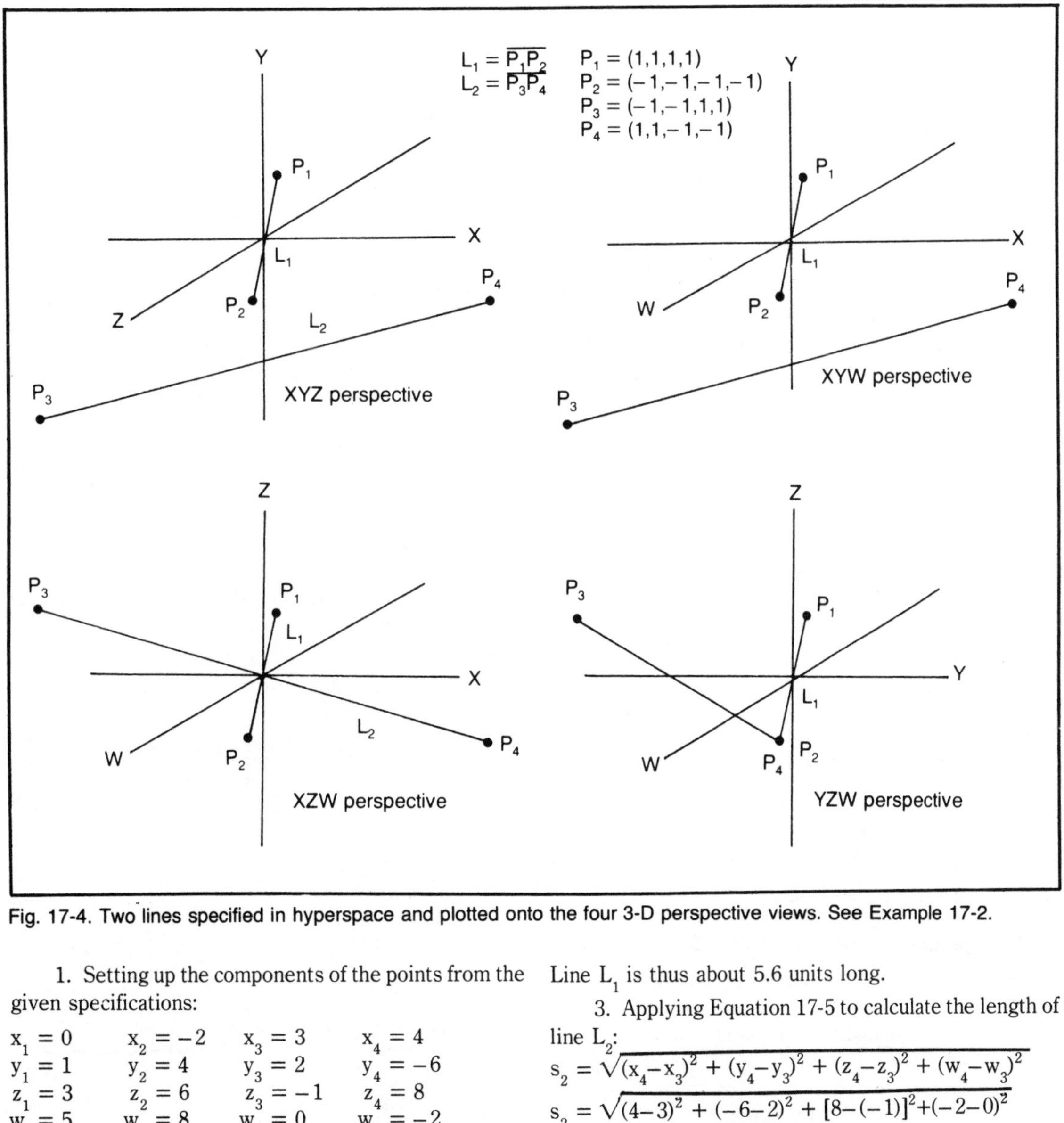

Fig. 17-4. Two lines specified in hyperspace and plotted onto the four 3-D perspective views. See Example 17-2.

1. Setting up the components of the points from the given specifications:

$x_1 = 0$	$x_2 = -2$	$x_3 = 3$	$x_4 = 4$
$y_1 = 1$	$y_2 = 4$	$y_3 = 2$	$y_4 = -6$
$z_1 = 3$	$z_2 = 6$	$z_3 = -1$	$z_4 = 8$
$w_1 = 5$	$w_2 = 8$	$w_3 = 0$	$w_4 = -2$

2. Applying Equation 17-5 to calculate the length of line L_1:

$$s_1 = \sqrt{(x_2-x_1)^2 + (y_2-y_1)^2(z_2-z_1)^2 + (w_2-w_1)^2}$$

$$s_1 = \sqrt{(-2-0)^2 + (4-1)^2 + (6-3)^2 + (8-5)^2}$$

$$s_1 = \sqrt{(-2)^2 + (3)^2 + (3)^2+(3)^2}$$

$$s_1 = \sqrt{4 + 9 + 9 + 9}$$

$$s_1 = \sqrt{31} = 5.6$$

Line L_1 is thus about 5.6 units long.

3. Applying Equation 17-5 to calculate the length of line L_2:

$$s_2 = \sqrt{(x_4-x_3)^2 + (y_4-y_3)^2 + (z_4-z_3)^2 + (w_4-w_3)^2}$$

$$s_2 = \sqrt{(4-3)^2 + (-6-2)^2 + [8-(-1)]^2+(-2-0)^2}$$

$$s_2 = \sqrt{(1)^2 + (-8)^2 + (9)^2 + (-2)^2}$$

$$s_2 = \sqrt{1 + 64 + 81 + 4}$$

$$s_2 = \sqrt{150} = 12.2$$

So line L_2 is about 12.2 units long.

The example specifies two lines that are defined in terms of four different points in 4-D space. Step 1, as usual, summarizes the assignment of components. The real work begins in Step 2 where I assign the endpoint components for line 1 to the variables in the length equation. Working through the arithmetic shows that the line is about 5.6 units long.

Then in Step 3, I adjust the assignment of endpoint components for line 2 to the basic length equation for 4 dimensions and work through the arithmetic again. That line turns out to be about 12.2 units long.

Plot those two lines on the four perspective views of 4-D space if you wish. Chances are quite good, however, that the actual measurement of the lines on any of the views will not match up with the calculated values for length. Why not? The perspective views represent a drop of two dimensions, each one inserting some spatial distortion. There is no reliable way to check the results of calculating the length of a line in 4-D space by direct measurement. Just be careful with the arithmetic to ensure accurate results. Try your hand at calculating the lengths of some lines immersed in 4-D space by working Exercise 17-2.

Exercise 17-2

1. Calculate the lengths of these two lines in 4-D space:

$L_1 = \overline{P_1P_2}$ $\quad P_1 = (0,0,0,0)$
$L_2 = \overline{P_2P_3}$ $\quad P_2 = (1,1,1,1)$
$\quad P_3 = (-1,-1,1,1)$

s_1 ______ $\quad s_2$ ______

2. Calculate the lengths of the lines specified in Example 17-2:

s_1 ______ $\quad s_2$ ______

3. Devise some P-L specifications for lines in 4-D space and calculate their lengths.

17-3.2 Direction Cosines of Lines in 4-D Space

The equation for the length of a line in 4-D space differs in form from that of 3-D space only by the addition of the W-axis components. The same notion holds for direction cosines; the general forms are identical, but the 4-D case has four sets of expressions instead of three.

$$\cos\alpha_n = (x_u - x_v)/s_n$$
$$\cos\beta_n = (y_u - y_v)/s_n$$
$$\cos\gamma_n = (z_u - z_v)/s_n$$
$$\cos\delta_n = (w_u - w_v)/s_n$$

Equation 17-6

where:

$\cos\alpha_n$ = the direction cosine of line n relative to the X axis

$\cos\beta_n$ = the direction cosine of line n relative to the Y axis

$\cos\gamma_n$ = the direction cosine of line n relative to the Z axis

$\cos\delta_n$ = the direction cosine of line n relative to the W axis

x_u,y_u,z_u,w_u = the coordinate of endpoint u of line n

x_v,y_v,z_v,w_v = the coordinate of endpoint v of line n

The expressions show that you must have access to information regarding the displacements of the individual components of the endpoint coordinates and the length of the line. The general specifications for a line in 4-D space include the components of the endpoint coordinates, and you have just seen how to use those values for calculating the length of the line. The only new feature is the cosine of an acute angle between the line and the W axis—angle d (delta). See Fig. 17-5.

That figure shows the four perspective views of a 4-D coordinate system. Between the four views, you can see all four direction angles for the line. Example 17-4 demonstrates the procedure for calculating the four direction cosines for a line that is imbedded in 4-D space. The example considers two different lines, just to make certain you understand how to deal with the notation for two or more lines.

Example 17-4

Calculate the direction cosines for the two lines specified here in 4-D space:

$L_1 = \overline{P_1P_2}$ $\quad P_1 = (1,1,1,1)$
$L_2 = \overline{P_3P_4}$ $\quad P_2 = (1,1,1,-1)$
$\quad P_3 = (0,0,0,0)$
$\quad P_4 = (-1,2,2,-2)$

1. Setting up the components of those coordinates:

$x_1 = 1$	$x_2 = 1$	$x_3 = 0$	$x_4 = -1$
$y_1 = 1$	$y_2 = 1$	$y_3 = 0$	$y_4 = 2$
$z_1 = 1$	$z_2 = 1$	$z_3 = 0$	$z_4 = 2$
$w_1 = 1$	$w_2 = -1$	$w_3 = 0$	$w_4 = -2$

2. Calculating the length of line L_1:

$$s_1 = \sqrt{(x_2-x_1)^2 + (y_2-y_1)^2 + (z_2-z_1)^2 + (w_2-w_1)^2}$$
$$s_1 = 2$$

3. Calculating the length of line L_2:

$$s_2 = \sqrt{(x_4-x_3)^2 + (y_4-y_3)^2 + (z_4-z_3)^2 + (w_4-w_3)^2}$$
$$s_2 = \sqrt{13}$$

4. Calculating the direction cosines for line L_1:

$$\cos\alpha_1 = (x_2-x_1)/s_1 = (1-1)/2 = 0$$
$$\cos\beta_1 = (y_2-y_1)/s_1 = (1-1)/2 = 0$$
$$\cos\gamma_1 = (z_2-z_1)/s_1 = (1-1)/2 = 0$$
$$\cos\delta_1 = (w_2-w_1)/s_1 = (-1-1)/2 = -1$$

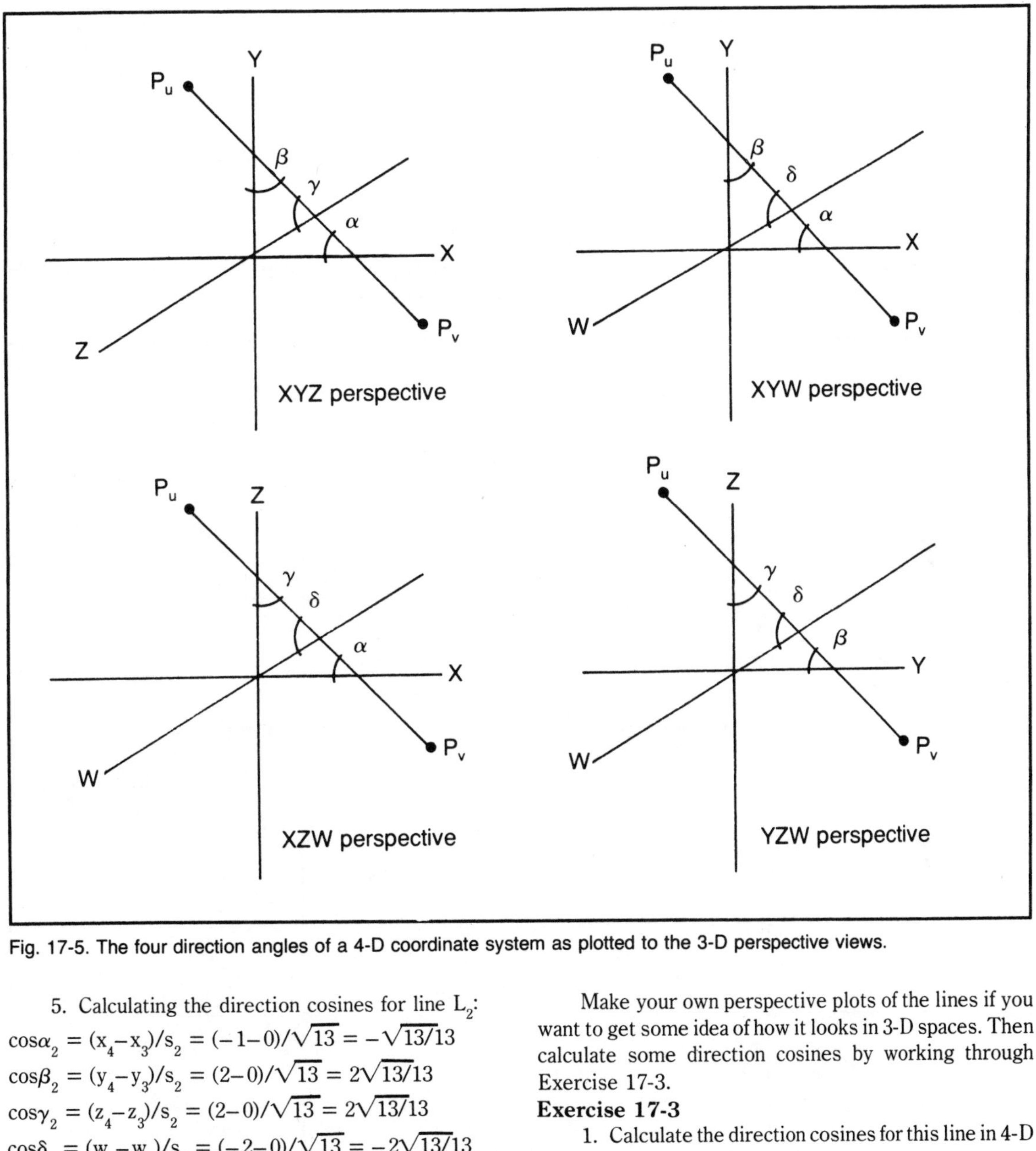

Fig. 17-5. The four direction angles of a 4-D coordinate system as plotted to the 3-D perspective views.

5. Calculating the direction cosines for line L_2:

$\cos\alpha_2 = (x_4 - x_3)/s_2 = (-1-0)/\sqrt{13} = -\sqrt{13}/13$

$\cos\beta_2 = (y_4 - y_3)/s_2 = (2-0)/\sqrt{13} = 2\sqrt{13}/13$

$\cos\gamma_2 = (z_4 - z_3)/s_2 = (2-0)/\sqrt{13} = 2\sqrt{13}/13$

$\cos\delta_2 = (w_4 - w_3)/s_2 = (-2-0)/\sqrt{13} = -2\sqrt{13}/13$

The example begins by specifying the two lines in terms of their endpoints and endpoint coordinates, and then much of the work that remains relies on the component table in Step 1. Steps 2 and 3 calculate the lengths of the specified lines, leaving the results in a rational form. Then Steps 4 and 5 use Equation 15-6 to determine the direction cosines for the two lines.

Make your own perspective plots of the lines if you want to get some idea of how it looks in 3-D spaces. Then calculate some direction cosines by working through Exercise 17-3.

Exercise 17-3

1. Calculate the direction cosines for this line in 4-D space:

$$L = \overline{P_1P_2} \qquad P_1 = (1,1,1,1) \quad P_2 = (0,1,0,1)$$

2. Calculate the direction cosines for the lines specified in part 1 of Exercise 17-2.

3. Calculate the direction cosines for the lines specified in Example 17-2.

17-3.3 Angle Between Lines in 4-D Space

Figure 17-6 represents two lines immersed in 4-D space, but they are actually shown from the four perspective views. These lines happen to share a common point, point 2, and the angle between them is shown as $\psi_{1,2}$.

You ought to be fully prepared by now to deal with the equation for finding the cosine of the acute angle between any two lines immersed in 4-D space. It looks very much like the equation for the cosine of an angle between lines in 3-D space. It is simply extended to include the W-axis terms.

Equation 17-7

$$\cos\phi_{m,n} = \cos\alpha_m \cos\alpha_n + \cos\beta_m \cos\beta_n + \cos\gamma_m \cos\gamma_n + \cos\delta_m \cos\delta_n$$

where:

$\cos\alpha_m$, $\cos\beta_m$, $\cos\gamma_m$ and $\cos\delta_m$ = the direction cosines for line *n*

$\cos\alpha_n$, $\cos\beta_n$, $\cos\gamma_n$ and $\cos\delta_n$ = the direction cosines for line *n*

Note: Taking the arc cosine of the result yields the angle between lines *m* and *n*

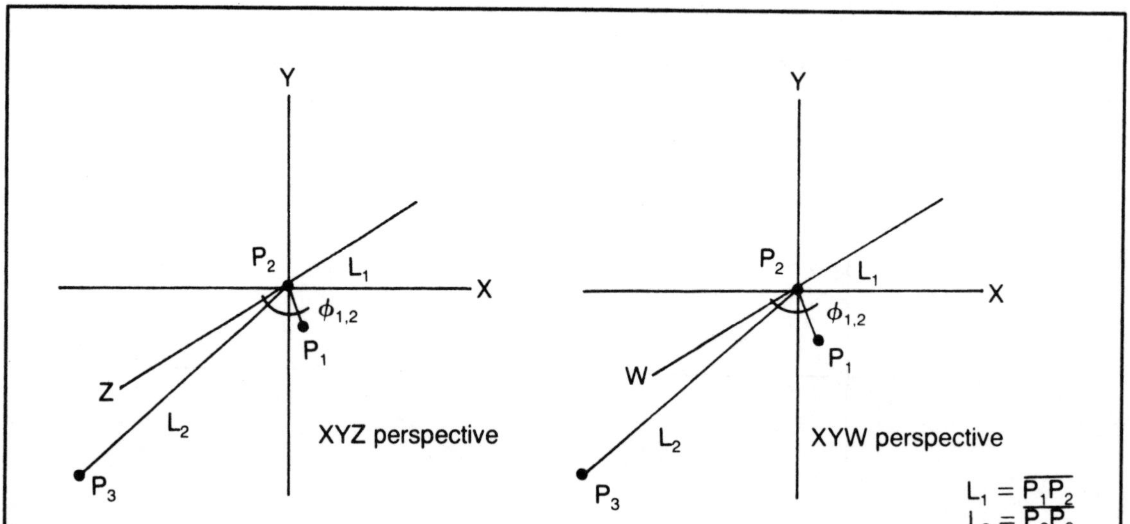

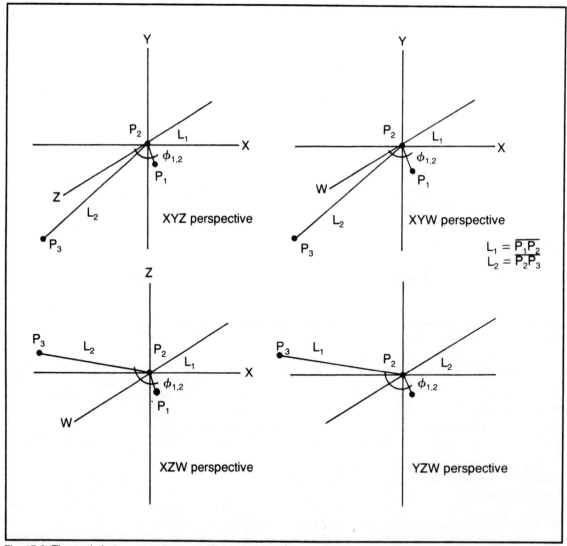

Fig. 17-6. The angle between two lines specified in hyperspace and plotted to the four 3-D perspective views.

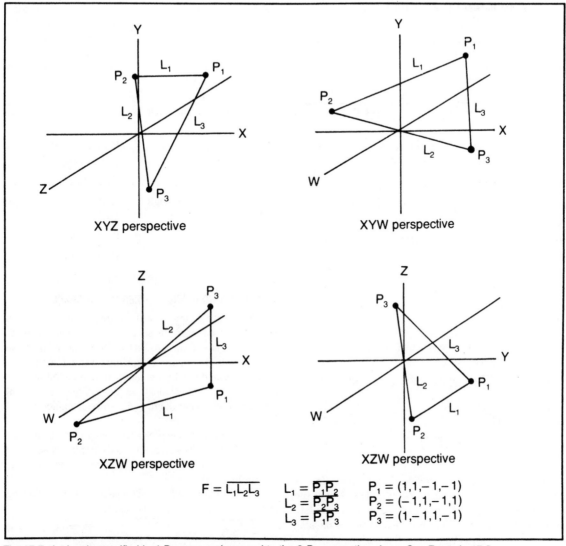

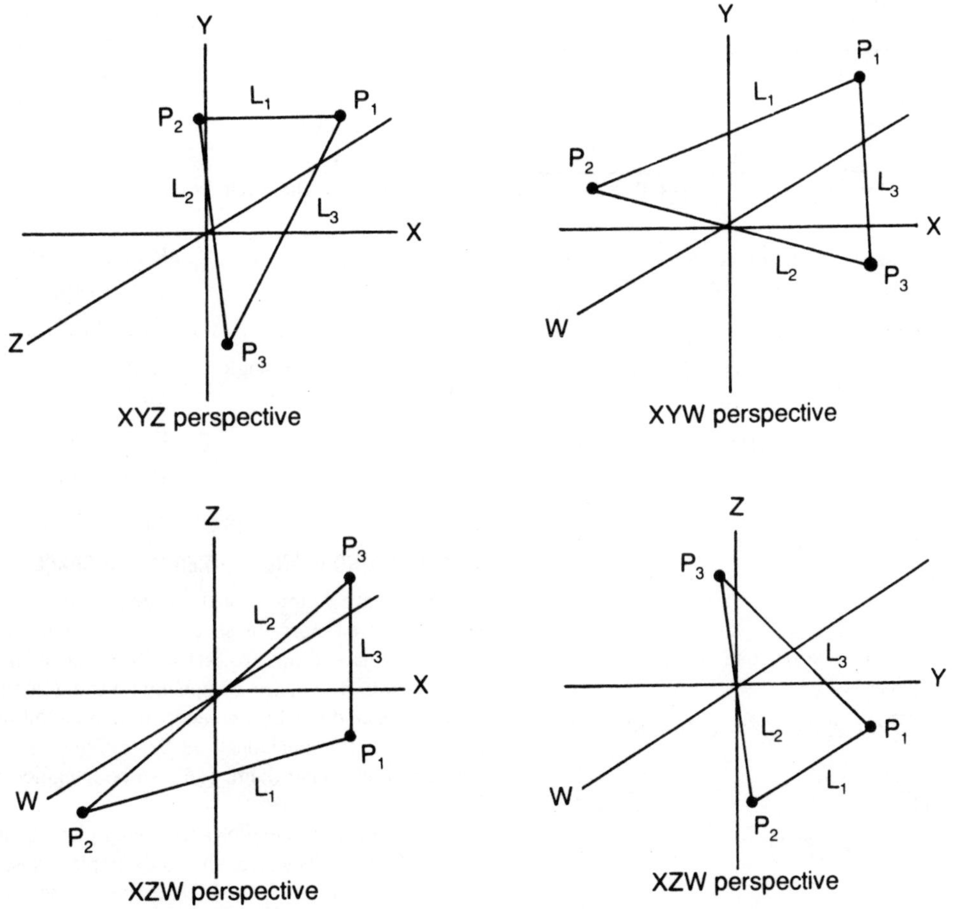

$F = \overline{L_1L_2L_3}$ $L_1 = \overline{P_1P_2}$ $P_1 = (1,1,-1,-1)$
$L_2 = \overline{P_2P_3}$ $P_2 = (-1,1,-1,1)$
$L_3 = \overline{P_1P_3}$ $P_3 = (1,-1,1,-1)$

Fig. 17-7. A triangle specified in 4-D space and mapped to the 3-D perspective views. See Example 17-6.

As before, determining the angle, itself, is a multistep procedure. Given the L-P specifications for two lines in 4-D space, use the information to generate the lengths of the lines and the direction cosines. Then use the direction cosines to figure the cosine of the angle as prescribed by Equation 17-7. Finally, determine the arc cosine of that value to get the angle in degrees. See Example 17-5.

Example 17-5

Calculate the angle between these two lines specified in 4-D space:

$L_1 = \overline{P_1P_2}$ $P_1 = (1,1,-1,-1)$
$L_2 = \overline{P_2P_3}$ $P_2 = (-1,1,-1,1)$
$P_3 = (1,-1,1,-1)$

1. Establish the components of the coordinates cited in the specifications:

$x_1 = 1$	$x_2 = -1$	$x_3 = 1$
$y_1 = 1$	$y_2 = 1$	$y_3 = -1$
$z_1 = -1$	$z_2 = -1$	$z_3 = 1$
$w_1 = -1$	$w_2 = 1$	$w_3 = -1$

2. Calculate the lengths of the two lines:

$s_1 = \sqrt{(x_2-x_1)^2 + (y_2-y_1)^2 + (z_2-z_1)^2 + (w_2-w_2)^2}$

$s_1 = \sqrt{(-1-1)^2 + (1-1)^2 (-1-(-1)^2 + (1-(-1))^2}$

$s_1 = 2\sqrt{2}$

$s_2 = \sqrt{(x_3-x_2)^2 + (y_3-y_2)^2 + (z_3-z_2)^2 + (w_3-w_2)^2}$

$s_2 = \sqrt{(1-(-1))^2 + (-1-1)^2 + (1-(-1))^2 + (-1-1)^2}$

$s_2 = 4$

3. Calculating the direction cosines:

$\cos\alpha_1 = (x_2-x_1)/s_1 = (-1-1)/2\sqrt{2} = -\sqrt{2}/2$

$\cos\beta_1 = (y_2-y_1)/s_1 = (1-1)/2\sqrt{2} = 0$

$\cos\gamma_1 = (z_2-z_1)/s_1 = [-1-(-1)]/2\sqrt{2} = 0$

$\cos\delta_1 = (w_2-w_1)/s_1 = [1-(-1]/2\sqrt{2} = \sqrt{2}/2$

$\cos\alpha_2 = (x_3-x_2)/s_2 = [1-(-1)]/4 = 1/2$

$\cos\beta_2 = (y_3-y_2)^2/s_2 = (-1-1)/4 = -1/2$

$\cos\gamma_2 = (z_3-z_2)/s_2 = [1-(-1)]/4 = 1/2$

$\cos\delta_2 = (w_3-w_2)/s_2 = (-1-1)/4 = -1/2$

4. Calculating the cosine of the angle between lines L_1 and L_2:

$\cos\phi_{1,2} = \cos\alpha_1\cos\alpha_2 + \cos\beta_1\cos\beta_2 + \cos\gamma_1\cos\gamma_2 + \cos\delta_1\cos\delta_2$

$\cos\phi_{1,2} = (-\sqrt{2}/2)(1/2) + (0)(-1/2) + (0)(1/2) + (\sqrt{2}/2)(-1/2)$

$\cos\phi_{1,2} = -\sqrt{2}/2$

5. Determining the angle:

$\phi_{1,2} = 45°$

The example specifies two lines in 4-D space. In this particular instance, they share point 2. The flow of the procedure is virtually identical to that of doing a complete analysis of those lines. The only element that is missing is a plotting of them. You can do that for yourself.

Step 1 sets up the components of the endpoint coordinates, Step 2 works out the lengths of the lines, Step 3 calculates the direction cosines, and Step 4 uses the most recent equation to determine the cosine of the angle between those two lines. The final step establishes the angle, itself.

If you have followed by suggestion and plotted these lines as represented by the four perspective views, you aren't likely to measure the angle and find it to be the calculated value of 45 degrees. This apparent discrepancy merely demonstrates the fact that perspective views of 4-D space will include some amount of angular distortion as well as spatial distortion. Or to put it very simply: making 3-D perspective views of 4-D space screws up the lengths of lines and angles between them. Only the mathematical analyses can be absolutely accurate and reliable—the perspective drawings are supposed to guide you to your thinking. Test your understanding by working the next exercise.

Exercise 17-4

1. Calculate the angle between these two lines in 4-D space:

$L_1 = \overline{P_1P_2}$ $\quad P_1 = (0,0,1,1)$

$L_2 = \overline{P_2P_3}$ $\quad P_2 = (0,1,0,1)$

$P_3 = (0,0,0,1)$

2. Calculate the angle between these two lines in 4-D space:

$L_1 = \overline{P_1P_2}$ $\quad P_1 = (1,0,0,1)$

$L_2 = \overline{P_2P_3}$ $\quad P_2 = (0,0,1,1)$

$P_3 = (0,1,0,1)$

17-4 PLOTTING PLANE FIGURES IN 4-D SPACE

Points are 0-dimensional geometric entities that can be plotted in 4-D space by specifying four components for each coordinate. Lines are 1-D entities that can likewise be plotted in 4-D space by specifying the 4-component coordinates of the lines' two endpoints. So it follows that plane figure, 2-dimensional entities defined in terms of the lines that bound them, can also be specified in 4-D space.

Example 17-6 specifies a triangular plane figure that is immersed in 4-D space, shows a complete analysis of the situation, and offers the 3-D perspective views of it.

Example 17-6

Conduct a complete 4-D analysis of the following plane figure specified in 4-D space.

$F = \overline{L_1L_2L_3}$ $\quad L_1 = \overline{P_1P_2}$ $\quad P_1 = (1,1,-1,-1)$

$L_2 = \overline{P_2P_3}$ $\quad P_2 = (-1,1,-1,1)$

$L_3 = \overline{P_1P_3}$ $\quad P_3 = (1,-1,1,-1)$

1. Establish the components of the coordinates cited in the specifications:

$x_1 = 1$	$x_2 = -1$	$x_3 = 1$
$y_1 = 1$	$y_2 = 1$	$y_3 = -1$
$z_1 = -1$	$z_2 = -1$	$z_3 = 1$
$w_1 = -1$	$w_2 = 1$	$w_3 = -1$

2. Calculate the perspective coordinates for all three 3-D perspective views, using a 30-degree perspective angle:

For the XYZ view: $P_{p1} = (1.87,1.5)$

$P_{p2} = (-0.13,1.5)$
$P_{p3} = (0.13,-1.5)$

For the XYW view: $P_{p1} = (1.87,1.87)$
$P_{p2} = (-1.87,0.5)$
$P_{p3} = (1.87,-0.5)$

For the XZW view: $P_{p1} = (1.87,-0.5)$
$P_{p2} = (-1.87,-1.5)$
$P_{p3} = (1.87,1.5)$

For the YZW view P_{p1} ='(1.87,−0.5)
$P_{p2} = (0.13,-1.5)$
$P_{p3} = (-0.13,1.5)$

3. Plot the three perspective views. See Fig. 17-7.

4. Calculate the lengths of the three lines:

$s_1 = \sqrt{(x_2-x_1)^2 + (y_2-y_1)^2 + (z_2-z_1)^2 + (w_2-w_1)^2}$

$s_1 = 2\sqrt{2}$

$s_2 = \sqrt{(x_3-x_2)^2 + (y_3-y_2)^2 + (z_3-z_2)^2 + (w_3-w_2)^2}$

$s_2 = 4$

$s_3 = \sqrt{(x_3-x_1)^2 + (y_3-y_1)^2 + (z_3-z_1)^2 + (w_3-w_1)^2}$

$s_3 = 2\sqrt{2}$

5. Calculate the direction cosines for all three lines:

$\cos\alpha_1 = (x_2-x_1)/s_1 = -\sqrt{2}/2$
$\cos\beta_1 = (y_2-y_1)/s_1 = 0$
$\cos\gamma_1 = (z_2-z_1)/s_1 = 0$
$\cos\delta_1 = (w_2-w_1)/s_1 = 2/2$

$\cos\alpha_2 = (x_3-x_2)/s_2 = 1/2$
$\cos\beta_2 = (y_3-y_2)/s_2 = -1/2$
$\cos\gamma_2 = (z_3-z_2)/s_2 = 1/2$
$\cos\delta_2 = (w_3-w_2)/s_2 = -1/2$

$\cos\alpha_3 = (x_3-x_1)/s_3 = 0$
$\cos\beta_3 = (y_3-y_1)/s_3 = -\sqrt{2}/2$
$\cos\gamma_3 = (z_3-z_1)/s_3 = \sqrt{2}/2$
$\cos\delta = (w_3 = w_1)/s_3 = 0$

6. Calculate the cosines of the three angles:

$\cos\phi_{1,2} = \cos d_1 \cos d_2 + \cos\beta_1\cos\beta_2 + \cos\gamma_1\cos\gamma_2 + \cos\delta_1\cos\delta_2$

$\cos\phi_{1,2} = (-\sqrt{2}/2)\ (1/2) + (0)\ (-1/2) + (0)\ (1/2) + (\sqrt{1}/2)\ (-1/2)$

$\cos\phi_{1,2} = -\sqrt{2}/2$

$\cos\phi_{1,3} = \cos\gamma_1\cos\gamma_3 + \cos\beta_1 + \cos\beta_3 + \cos\gamma_1\cos\gamma_3 + \cos\delta_1\ \cos\delta_3$

$\cos\phi_{1,3} = (-\sqrt{2}/2)\ (0) + (0)\ (-\sqrt{2}) + (0)\ (\sqrt{2}/2) + (\sqrt{2}/2)\ (0)$

$\cos\phi_{1,3} = 0$

$\cos\phi_{2,3} = \cos\alpha_2\cos\alpha_3 + \cos\beta_2\cos\beta_3 + \cos\gamma_2\cos\gamma_3 + \cos\delta_2\cos\delta_3$

$\cos\phi_{2,3} = (1/2)\ (0) = (-1/2)\ (-\sqrt{2}/2) + (1/2)\ (\sqrt{2}/2) + (-1/2)\ (0)$

$\cos\phi_{2,3} = \sqrt{2}/2$

7. Determine the angles:

$\phi_{1,2} = 45°$ $\quad \phi_{1,3} = 90°$ $\quad \phi_{2,3} = 45°$

The plane figure in this instance happens to be a right triangle immersed in 4-D space.

Notice that the F and L specifications in that example could apply to a triangular plane figure in two, three or four dimensions. It is the point specifications, each showing four components per coordinate, that indicates that the figure is in a 4-D space.

The analysis is an all-important part of the experiment. You can see that it includes a summary of the lengths of all the lines, the direction cosines of all the lines, and the angles between pairs of lines.

The graphic representation (Fig. 17-7 in this case) can be important, too. Although it necessarily exhibits a great deal of spatial and angular distortion, the drawing represents and important visual bridge between purely abstract mathematical analysis and human perception. Use Exercise 17-5 to test your understanding.

Exercise 17-5

1. Using Example 17-6 as a guide, plot and conduct a complete analysis of the following quadralateral that is immersed in 4-D space:

$F = \overline{L_1L_2L_3L_4}$ $\quad L_1 = \overline{P_1P_2}$ $\quad P_1 = (-1-1,1,-1)$
$L_2 = \overline{P_2P_3}$ $\quad P_2 = (1,-1,1,-1)$
$L_3 = \overline{P_3P_4}$ $\quad P_3 = (1,1,-1,1)$
$L_4 = \overline{P_4P_1}$ $\quad P_4 = (-1,1,-1,1)$

2. You can fix an endless variety of triangles in 4-D space by using the specifications provided here and completing the coordinates in such a way that no two are alike:

$F = \overline{L_1L_2L_3}$ $\quad L_1 = \overline{P_1P_2}$ $\quad P_1$ = (____,____,____,____)
$L_2 = \overline{P_2P_3}$ $\quad P_2$ = (____,____,____,____)
$L_3 = \overline{P_3P_1}$ $\quad P_3$ = (____,____,____,____)

See if you can use that format to generate at least three or four different triangles. Plot them and conduct a complete analysis in each case.

17-5 PLOTTING SPACE OBJECTS IN 4-D SPACE

The only real difference between plotting plane figures and space objects in 4-D space is the relative magnitude of the task. Space objects, usually composed of a larger number of points and lines, force the experimenter to work longer and harder at generating a complete analysis and graphic representations. Beyond that, the transition for working with plane figures in 4-D space to space objects in that same space is a very natural one.

Example 17-7 treats a particular space object that is imbedded in 4-D space. An entire experiment is represented there: specifications, analysis of all lines and relevant angles, and a graphic representation.

Example 17-7

Conduct an analysis of the following space object that is immersed in 4-D space:

$S = \overline{F_1F_2F_3F_4F_5}$

$F_1 = \overline{L_1L_2L_3L_4}$	$L_1 = \overline{P_1P_2}$	$P_1 = (-1,0,1,1)$
$F_2 = \overline{L_1L_5L_8}$	$L_2 = \overline{P_2P_3}$	$P_2 = (1,0,1,1)$
$F_3 = \overline{L_2L_6L_7}$	$L_3 = \overline{P_3P_4}$	$P_3 = (1,0,-1,1)$
$F_4 = \overline{L_3L_7L_8}$	$L_4 = \overline{P_1P_4}$	$P_4 = (-1,0,-1,1)$
$F_5 = \overline{L_4L_5L_8}$	$L_5 = \overline{P_1P_5}$	$P_5 = (0,2,0,0)$
	$L_6 = \overline{P_2P_5}$	
	$L_7 = \overline{P_3P_5}$	
	$L_8 = \overline{P_4P_5}$	

See Fig. 17-8.

LENGTHS OF LINES

$s_1 = 2$	$s_2 = 2$	$s_3 = 2$	$s_4 = 2$
$s_5 = \sqrt{7}$	$s_6 = \sqrt{7}$	$s_7 = \sqrt{7}$	$s_8 = \sqrt{7}$

DIRECTION COSINES

$\cos\alpha_1 = 1$	$\cos\beta_1 = 0$
$\cos\alpha_2 = 0$	$\cos\beta_2 = 0$
$\cos\alpha_3 = -1$	$\cos\beta_3 = 0$
$\cos\alpha_4 = 0$	$\cos\beta_4 = 0$
$\cos\alpha_5 = \sqrt{7}/7$	$\cos\beta_5 = 2\sqrt{7}/7$
$\cos\alpha_6 = -\sqrt{7}/7$	$\cos\beta_6 = 2\sqrt{7}/7$
$\cos\alpha_7 = -\sqrt{7}/7$	$\cos\beta_7 = 2\sqrt{7}/7$
$\cos\alpha_8 = \sqrt{7}/7$	$\cos\beta_8 = 2\sqrt{7}/7$

$\cos\gamma_1 = 0$	$\cos\delta_1 = 0$
$\cos\gamma_2 = -1$	$\cos\delta_2 = 0$
$\cos\gamma_3 = 0$	$\cos\delta_3 = 0$
$\cos\gamma_4 = -1$	$\cos\delta_4 = 0$
$\cos\gamma_5 = -\sqrt{7}/7$	$\cos\delta_5 = -\sqrt{7}/7$
$\cos\gamma_6 = -\sqrt{7}/7$	$\cos\delta_6 = -\sqrt{7}/7$
$\cos\gamma_7 = \sqrt{7}/7$	$\cos\delta_7 = -\sqrt{7}/7$
$\cos\gamma_8 = \sqrt{7}/7$	$\cos\delta_8 = -\sqrt{7}/7$

ANGLES

$\phi_{1,2} = 90°$	$\phi_{1,4} = 90°$	$\phi_{1,5} = 67° \, 40'$
$\phi_{1,6} = 67° \, 40'$	$\phi_{2,3} = 90°$	$\phi_{2,6} = 67° \, 40'$
$\phi_{2,7} = 67° \, 40'$	$\phi_{3,4} = 90°$	$\phi_{3,7} = 67° \, 40'$
$\phi_{3,8} = 67° \, 40'$	$\phi_{4,5} = 67° \, 40'$	$\phi_{4,8} = 67° \, 40'$
$\phi_{5,6} = 44° \, 35'$	$\phi_{5,8} = 44° \, 35'$	$\phi_{6,7} = 44° \, 35°$
$\phi_{7,8} = 44° \, 35'$		

Conduct some experiments of your own, and try the suggested experiments in Exercise 17-6.

Exercise 17-6

Conduct a complete analysis of the following space object that is specified in 4-D space:

$s = \overline{F_1F_2F_3F_4F_5}$

$F_1 = \overline{L_1L_2L_4}$	$L_1 = \overline{P_1P_2}$	$P_1 = (-1,0,1,0)$
$F_2 = \overline{L_1L_3L_5L_7}$	$L_2 = \overline{P_1P_3}$	$P_2 = (1,0,1,0)$
$F_3 = \overline{L_2L_3L_6L_8}$	$L_3 = \overline{P_1P_4}$	$P_3 = (0,1,1,0)$
$F_4 = \overline{L_4L_5L_6L_9}$	$L_4 = \overline{P_2P_3}$	$P_4 = (-1,0,1,1)$
$F_5 = \overline{L_7L_8L_9}$	$L_5 = \overline{P_2P_5}$	$P_5 = (1,0,1,1)$
	$L_6 = \overline{P_3P_6}$	$P_6 = (0,1,1,1)$
	$L_7 = \overline{P_4P_5}$	
	$L_8 = \overline{P_4P_6}$	
	$L_9 = \overline{P_5P_6}$	

17-6 WORKING WITH HYPERSPACE OBJECTS IN 4-D SPACE

Here we enter yet another realm that is not often explored by people with the proper analytical tools. Up to this point in the discussion, you have been working with lower-dimensioned geometric entities in 4-D space. You have seen how to plot 0-D points, 1-D lines, 2-D plane figures, and 3-D space objects in a 4-dimensional environment. Now it is time to introduce the premere geometric entity of 4-dimensional space—hyperspace objects.

Anyone planning to do some experiments with hyperspace objects ought to have at least an intuitive notion of what those objects are like. This is not to say that an intuitive notion is adequate for setting up and conducting such experiments (and it's a good thing, too, because hyperspace objects are virtually impossible to picture, even in a purely mental form). At any rate, there are a couple of different ways to introduce hyperspace objects through analogies. Here is one such line of reasoning:

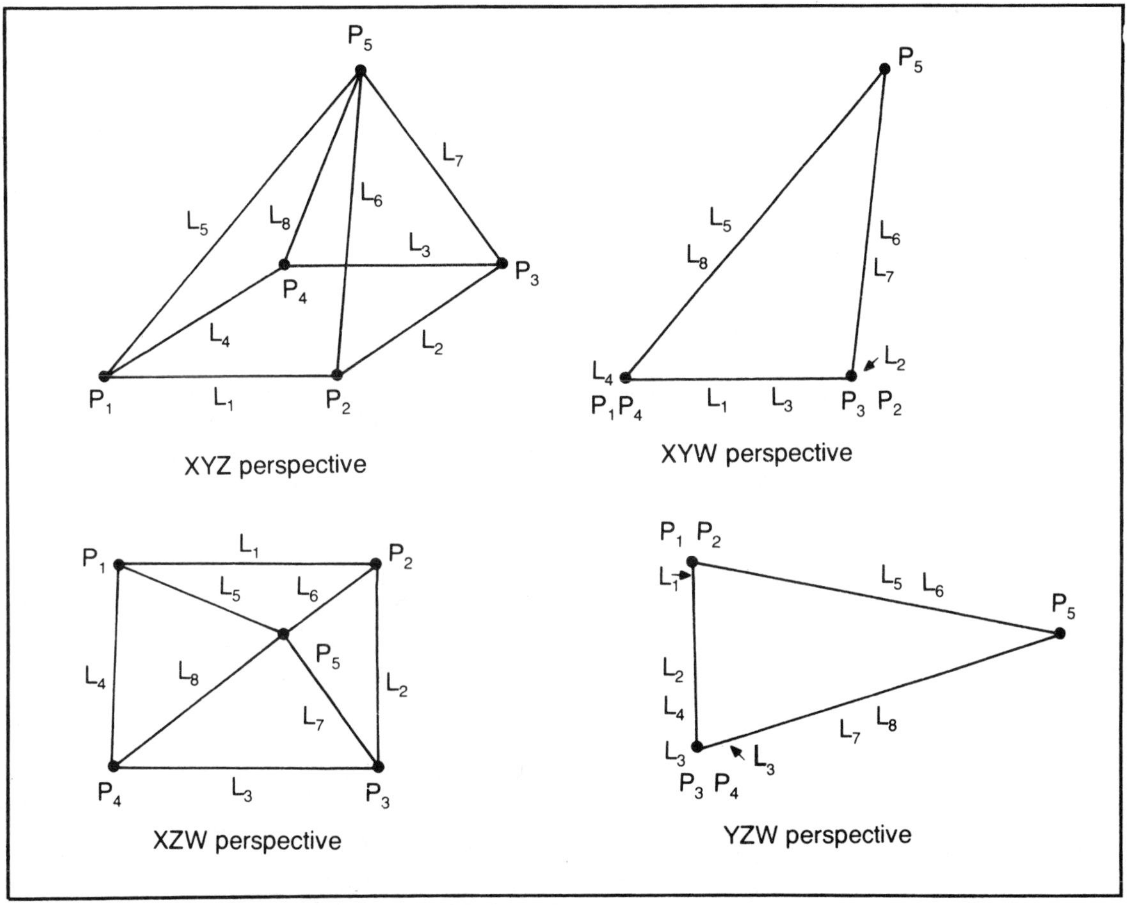

Fig. 17-8. The four 3-D perspective views of the space object specified in Example 17-7.

☐ A line is the most natural product of 1-D spaces, and it is the most complex geometric entity that can be properly specified in that space.

☐ A plane figure arises naturally from a 2-D space, and it is the most complex entity that can be fully specified in 2 dimensions.

☐ A space object is a natural product of 3-D space, and it is the most complex sort of object that can be completely specified in 3 dimensions.

☐ A hyperspace object is the natural product of 4-D space, and it is the most complex sort of object that can be fully specified in that space.

Lines come out of a 1-D environment, plane figures come out of a 2-D environment, space objects come out of a 3-D environment and hyperspace objects come out of a 4-D environment. And here is another way to approach the idea of hyperspace objects:

☐ Lines are bounded by points.

☐ Plane figures are bounded by lines.

☐ Space objects are bounded by plane figures.

☐ Hyperspace objects are bounded by space objects.

It shouldn't be too difficult for you to envision a line being defined in terms of the coordinates of its endpoints (L-P specifications); and you know quite well by now that plane figures are defined in terms of the lines that bound it and, in turn, the coordinates of the endpoints of those lines (F-L-P specifications).

Now, as far as space objects are concerned, you have seen that they can be defined in terms of the plane figures that bound them, the plane figures can be defined in terms of the lines that bound them and, finally, the lines, themselves, are defined in terms of the coordinates of their endpoints (S-F-L-P specifications).

There is nothing new to all of that. But the next step is the critical one for the present discussion: Hyperspace objects are defined in terms of the space objects that bound them, the space objects are defined in terms of the plane figures that bound them, the plane figures are defined in terms of the lines that bound them, and then those lines are defined in terms of the coordinates of their endpoints (H-S-F-L-P specifications). Yes, it is necessary to introduce H notation for specifying the space objects that bound a hyperspace object. You will see that notation used in the next example.

That is a lengthy, but meaningful, line of reasoning. If nothing else, it demonstrates the evolution of spaces and geometric entities from one through four dimensions. It's all very straightforward, clean and logical. One notion simply builds upon the previous one. The same line of reasoning can be used for setting up spaces of any number of dimensions.

In a manner of speaking, then, there is really nothing mysterious about 4-D space and hyperspace objects that it can engender. It is entirely possible to carry the line of reasoning upward to any number of dimensions of space without getting lost.

Matters become difficult in principle only when attempting to come up with a visual portrayal of phenomena of hyperspace. It is difficult at best, and most often virtually impossible, for the human mind to grasp the true essence of hyperspace objects in a visual way. Generally the best we can do is visualize the objects in bits and pieces as they appear from various 3-dimensional perspectives. Mathematical analysis is the only tool available for dealing with hyperspace objects in their true forms.

As difficult as it might be to perceive hyperspace objects in their entirety, the mathematical analysis of them is a simple extension of the analysis of lower-dimensioned entities in 4-D space. The task is a lengthy one, but assuming that you approach the experiment in a systematic and careful fashion, the results can be as clear and meaningful as they are for simpler, lower-dimensioned objects.

Actually, the most critical part of an experiment with

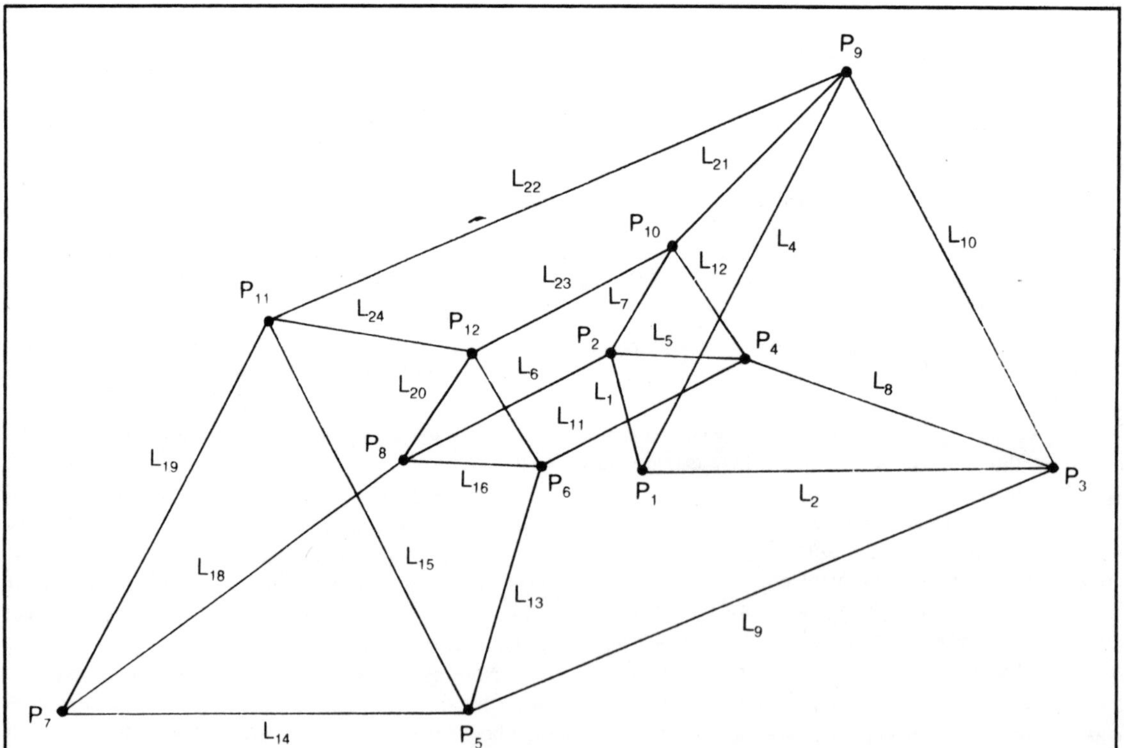

Fig. 17-9. General 4-D plot of the hyperspace object specified in Table 17-1.

Table 17-1. Specifications and a Summary of the Primary Analysis of a Hyperspace Object.

(A, B, C, and D correspond to α, β, γ, and δ.)

SPECIFICATIONS

$H=\overline{S_1S_2S_3S_4S_5S_6S_7}$

$S_1=\overline{F_4F_5F_6F_{12}F_{16}}$

$S_2=\overline{F_7F_8F_9F_{13}F_{17}}$

$S_3=\overline{F_1F_3F_5F_8F_{11}}$

$S_4=\overline{F_{14}F_{15}F_{16}F_{17}F_{18}}$

$S_5=\overline{F_2F_3F_6F_9F_{18}F_{19}}$

$S_6=\overline{F_{10}F_{11}F_{12}F_{13}F_{15}F_{19}}$

$S_7=\overline{F_1F_2F_4F_7F_{10}F_{14}}$

$F_1=\overline{L_1L_2L_5L_8}$	$F_2=\overline{L_1L_3L_6L_{18}}$
$F_3=\overline{L_1L_4L_7L_{21}}$	$F_4=\overline{L_2L_3L_9L_{14}}$
$F_5=\overline{L_2L_4L_{10}}$	$F_6=\overline{L_3L_4L_{19}L_{22}}$
$F_7=\overline{L_5L_6L_{11}L_{16}}$	$F_8=\overline{L_5L_7L_{12}}$
$F_9=\overline{L_6L_7L_{20}L_{23}}$	$F_{10}=\overline{L_8L_9L_{11}L_{13}}$
$F_{11}=\overline{L_8L_{10}L_{12}L_{21}}$	$F_{12}=\overline{L_9L_{10}L_{15}L_{22}}$
$F_{13}=\overline{L_{11}L_{12}L_{17}L_{23}}$	$F_{14}=\overline{L_{13}L_{14}L_{16}L_{18}}$
$F_{15}=\overline{L_{13}L_{15}L_{17}L_{24}}$	$F_{16}=\overline{L_{14}L_{15}L_{19}}$
$F_{17}=\overline{L_{16}L_{17}L_{20}}$	$F_{18}=\overline{L_{18}L_{19}L_{20}L_{24}}$
	$F_{19}=\overline{L_{21}L_{22}L_{23}L_{24}}$

$L_1=\overline{P_1P_2}$	$L_2=\overline{P_1P_3}$	$P_1=(-1,0,0,0)$
$L_3=\overline{P_1P_7}$	$L_4=\overline{P_1P_9}$	$P_2=(-1,0,0,1)$
$L_5=\overline{P_2P_4}$	$L_6=\overline{P_2P_8}$	$P_3=(1,0,0,0)$
$L_7=\overline{P_2P_{10}}$	$L_8=\overline{P_3P_4}$	$P_4=(1,0,0,1)$
$L_9=\overline{P_3P_5}$	$L_{10}=\overline{P_3P_9}$	$P_5=(1,0,1,0)$
$L_{11}=\overline{P_4P_6}$	$L_{12}=\overline{P_4P_{10}}$	$P_6=(1,0,1,1)$
$L_{13}=\overline{P_5P_6}$	$L_{14}=\overline{P_5P_7}$	$P_7=(-1,0,1,0)$
$L_{15}=\overline{P_5P_{11}}$	$L_{16}=\overline{P_6P_8}$	$P_8=(-1,0,1,1)$
$L_{17}=\overline{P_2P_{12}}$	$L_{18}=\overline{P_7P_8}$	$P_9=(0,1,0,0)$
$L_{19}=\overline{P_7P_{11}}$	$L_{20}=\overline{P_8P_{12}}$	$P_{10}=(0,1,0,1)$
$L_{21}=\overline{P_9P_{10}}$	$L_{22}=\overline{P_9P_{11}}$	$P_{11}=(0,1,1,0)$
$L_{23}=\overline{P_{10}P_{12}}$	$L_{24}=\overline{P_{11}P_{12}}$	$P_{12}=(0,1,1,1)$

Table 17-1. Specifications and a Summary of the Primary Analysis of a Hyperspace Object. **(Continued from page 263.)**

LENGTHS OF LINES

$s_1=1$	$s_2=2$	$s_3=1$	$s_4=\sqrt{2}$
$s_5=2$	$s_6=1$	$s_7=\sqrt{2}$	$s_8=1$
$s_9=1$	$s_{10}=\sqrt{2}$	$s_{11}=1$	$s_{12}=\sqrt{2}$
$s_{13}=1$	$s_{14}=2$	$s_{15}=\sqrt{2}$	$s_{16}=2$
$s_{17}=\sqrt{2}$	$s_{18}=1$	$s_{19}=\sqrt{2}$	$s_{20}=\sqrt{2}$
$s_{21}=1$	$s_{22}=1$	$s_{23}=1$	$s_{24}=1$

DIRECTION COSINES

$\cos A_1=0$	$\cos B_1=0$	$\cos C_1=0$	$\cos D_1=1$
$\cos A_2=1$	$\cos B_2=0$	$\cos C_2=0$	$\cos D_2=0$
$\cos A_3=0$	$\cos B_3=0$	$\cos C_3=1$	$\cos D_3=0$
$\cos A_4=\sqrt{2}/2$	$\cos B_4=\sqrt{2}/3$	$\cos C_4=0$	$\cos D_4=0$
$\cos A_5=1$	$\cos B_5=0$	$\cos C_5=0$	$\cos D_5=0$
$\cos A_6=0$	$\cos B_6=0$	$\cos C_6=1$	$\cos D_6=0$
$\cos A_7=\sqrt{2}/3$	$\cos B_7=\sqrt{2}/2$	$\cos C_7=0$	$\cos D_7=0$
$\cos A_8=0$	$\cos B_8=0$	$\cos C_8=0$	$\cos D_8=1$
$\cos A_9=0$	$\cos B_9=0$	$\cos C_9=1$	$\cos D_9=0$
$\cos A_{10}=-\sqrt{2}/2$	$\cos B_{10}=\sqrt{2}/2$	$\cos C_{10}=0$	$\cos D_{10}=0$
$\cos A_{11}=0$	$\cos B_{11}=0$	$\cos C_{11}=1$	$\cos D_{11}=0$
$\cos A_{12}=-\sqrt{2}/2$	$\cos B_{12}=\sqrt{2}/2$	$\cos C_{12}=0$	$\cos D_{12}=0$
$\cos A_{13}=0$	$\cos B_{13}=0$	$\cos C_{13}=0$	$\cos D_{13}=1$
$\cos A_{14}=-1$	$\cos B_{14}=0$	$\cos C_{14}=0$	$\cos D_{14}=0$
$\cos A_{15}=-\sqrt{2}/2$	$\cos B_{15}=\sqrt{2}/2$	$\cos C_{15}=0$	$\cos D_{15}=0$
$\cos A_{16}=-1$	$\cos B_{16}=0$	$\cos C_{16}=0$	$\cos D_{16}=0$
$\cos A_{17}=-\sqrt{2}/2$	$\cos B_{17}=\sqrt{2}/2$	$\cos C_{17}=0$	$\cos D_{17}=0$
$\cos A_{18}=0$	$\cos B_{18}=0$	$\cos C_{18}=0$	$\cos D_{18}=1$
$\cos A_{19}=\sqrt{2}/2$	$\cos B_{19}=\sqrt{2}/2$	$\cos C_{19}=0$	$\cos D_{19}=0$
$\cos A_{20}=\sqrt{2}/2$	$\cos B_{20}=\sqrt{2}/2$	$\cos C_{20}=0$	$\cos D_{20}=0$
$\cos A_{21}=0$	$\cos B_{21}=0$	$\cos C_{21}=0$	$\cos D_{21}=1$

DIRECTION COSINES

$\cos A_{22}=0$	$\cos B_{22}=0$	$\cos C_{22}=1$	$\cos D_{22}=0$
$\cos A_{23}=0$	$\cos B_{23}=0$	$\cos C_{23}=1$	$\cos D_{23}=0$
$\cos A_{24}=0$	$\cos B_{24}=0$	$\cos C_{24}=0$	$\cos D_{24}=1$

ANGLES

$\phi_{1,2}=90^\circ$	$\phi_{1,3}=90^\circ$	$\phi_{1,4}=90^\circ$
$\phi_{1,5}=90^\circ$	$\phi_{1,6}=90^\circ$	$\phi_{1,7}=90^\circ$
$\phi_{2,3}=90^\circ$	$\phi_{2,4}=45^\circ$	$\phi_{2,8}=90^\circ$
$\phi_{2,9}=90^\circ$	$\phi_{2,10}=45^\circ$	$\phi_{3,4}=90^\circ$
$\phi_{3,14}=90^\circ$	$\phi_{3,18}=90^\circ$	$\phi_{3,19}=90^\circ$
$\phi_{4,10}=90^\circ$	$\phi_{4,21}=90^\circ$	$\phi_{4,22}=90^\circ$
$\phi_{5,6}=90^\circ$	$\phi_{5,7}=45^\circ$	$\phi_{5,8}=90^\circ$
$\phi_{5,11}=90^\circ$	$\phi_{5,12}=45^\circ$	$\phi_{6,7}=90^\circ$
$\phi_{6,16}=90^\circ$	$\phi_{6,17}=90^\circ$	$\phi_{6,18}=90^\circ$
$\phi_{7,12}=90^\circ$	$\phi_{7,21}=90^\circ$	$\phi_{7,23}=90^\circ$
$\phi_{8,9}=90^\circ$	$\phi_{8,10}=90^\circ$	$\phi_{8,11}=90^\circ$
$\phi_{8,12}=90^\circ$	$\phi_{9,10}=90^\circ$	$\phi_{9,13}=90^\circ$
$\phi_{9,14}=90^\circ$	$\phi_{9,15}=90^\circ$	$\phi_{10,21}=90^\circ$
$\phi_{10,22}=90^\circ$	$\phi_{11,12}=90^\circ$	$\phi_{11,13}=90^\circ$
$\phi_{11,16}=90^\circ$	$\phi_{11,17}=90^\circ$	$\phi_{12,21}=90^\circ$
$\phi_{12,23}=90^\circ$	$\phi_{13,14}=90^\circ$	$\phi_{13,15}=90^\circ$
$\phi_{13,16}=90^\circ$	$\phi_{13,17}=90^\circ$	$\phi_{14,15}=45^\circ$
$\phi_{14,18}=90^\circ$	$\phi_{14,19}=45^\circ$	$\phi_{15,19}=90^\circ$
$\phi_{15,22}=90^\circ$	$\phi_{15,24}=90^\circ$	$\phi_{16,17}=45^\circ$
$\phi_{16,18}=90^\circ$	$\phi_{16,20}=45^\circ$	$\phi_{17,20}=90^\circ$
$\phi_{17,23}=90^\circ$	$\phi_{17,24}=90^\circ$	$\phi_{18,19}=90^\circ$
$\phi_{18,20}=90^\circ$	$\phi_{19,22}=90^\circ$	$\phi_{19,24}=90^\circ$
$\phi_{20,23}=90^\circ$	$\phi_{20,24}=90^\circ$	$\phi_{21,22}=90^\circ$
$\phi_{21,23}=90^\circ$	$\phi_{22,24}=90^\circ$	$\phi_{23,24}=90^\circ$

a hyperspace object is coming up with a reliable set of initial specifications. The difficulty stems from the fact that the human mind is wholly unaccustomed to dealing with hyperspace objects. One cannot simply dream up a reliable set of specifications based upon common experience and intuition. That can be done for lower-dimensioned geometric entities; but common experience with 4-D objects is virtually nonexistent. Thus I am devoting the entire next chapter to a formal procedure for generating the specifications for custom hyperspace objects. Generating custom specifications for objects of lower dimensions isn't all that difficult because ordinary experiences of a lifetime breed a sense of familiarity with them. That is hardly the case for hyperspace objects. It is difficult to believe that any human being grew up thinking, working and playing in terms of bonafide hyperspace.

The examples of hyperspace objects offered through the remainder of this book were generated using the technique described in the next chapter. So until you've had an opportunity to study and master that technique, you will have little choice but to work from specifications I have to offer.

Figure 17-9 illustrates a hyperspace version of a prism. The specifications and a complete analysis appear in Table 17-1. It is, indeed, a complex figure compared to its 3-dimensional counterpart.

Try your hand at plotting and analyzing the hyperspace object in the following exercise.

Exercise 17-7

Plot the following hyperspace object to the four 3-D perspective views and run a complete analysis.

$$H = \overline{S_1S_2S_3S_4S_5}$$

$$S_1 = \overline{F_1F_2F_3F_4}$$
$$S_2 = \overline{F_4F_5F_6F_7}$$
$$S_3 = \overline{F_1F_5F_8F_9}$$
$$S_4 = \overline{F_2F_6F_8F_{10}}$$
$$S_5 = \overline{F_3F_7F_9F_{10}}$$

$$F_1 = \overline{L_1L_2L_7}$$
$$F_2 = \overline{L_1L_3L_8}$$
$$F_3 = \overline{L_2L_3L_9}$$
$$F_4 = \overline{L_7L_8L_9}$$
$$F_5 = \overline{L_4L_5L_7}$$
$$F_6 = \overline{L_4L_6L_8}$$
$$F_7 = \overline{L_5L_6L_9}$$
$$F_8 = \overline{L_1L_4L_{10}}$$
$$F_9 = \overline{L_2L_5L_{10}}$$
$$F_{10} = \overline{L_3L_6L_{10}}$$

$$L_1 = \overline{P_1P_3}$$
$$L_2 = \overline{P_1P_4}$$
$$L_3 = \overline{P_1P_5}$$
$$L_4 = \overline{P_2P_3}$$
$$L_5 = \overline{P_2P_4}$$
$$L_6 = \overline{P_2P_5}$$
$$L_7 = \overline{P_3P_4}$$
$$L_8 = \overline{P_3P_5}$$
$$L_9 = \overline{P_4P_5}$$
$$L_{10} = \overline{P_1P_2}$$

$$P_1 = (0,0,0,0)$$
$$P_2 = (0,0,0,1)$$
$$P_3 = (1,0,0,0)$$
$$P_4 = (0,1,0,0)$$
$$P_5 = (0,0,1,0)$$

Constructing Custom Hyperspace Objects

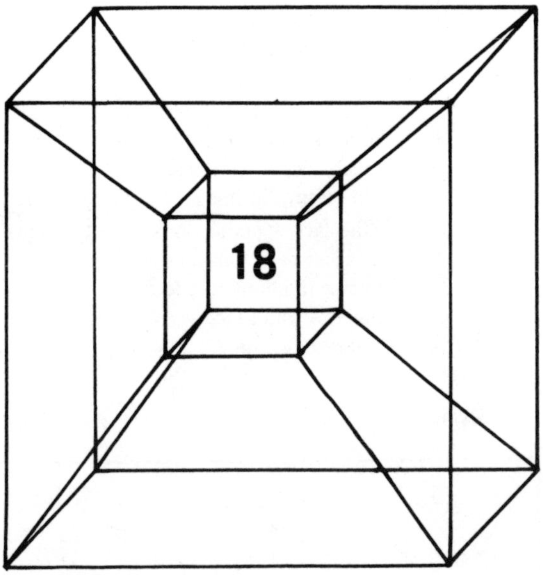

Chapter 16 dealt with a procedure for generating the full 3-D specifications for space objects that are originally specified in terms of 2-D data, 2-D planar views and partial 3-D coordinates. If you possess an average level of 3-dimensional spatial perception, it is, admittedly, simpler to generate complete 3-D specifications for just about any space object by playing around a bit with perspective drawings and 3-D coordinates. But if you treated that chapter seriously and developed a feeling for generating 3-D space-object specifications from partial information, you are prepared for what is about to take place. (I hate to think about what is going to happen if you aren't fully prepared.)

This is something new—quite new. You are about to engage in a workable procedure for generating customized 4-dimensional hyperspace objects. Not many people have done that; and even those who have constructed such objects from scratch generally work from a limited, analogy-like approach. The problem with that approach is that it restricts the variety of possible objects and all but eliminates the possibility of discovering new ones.

Assuming that you want to experiment with a variety of hyperspace objects, you will have to generate more than is already available in the literature and, indeed, more than the number offered in this book. Study and work through this chapter carefully, because it describes a workable technique for generating objects of endless variety. This work is so critical to any future experiments that a brief review of the procedure as presented in Chapter 16 is in order first.

You saw in that chapter that the construction procedure begins with a set of self-consistent 2-D mappings of the 3-D object to be specified. You have total freedom when it comes to sketching the initial X-Y planar view and selecting the partial coordinates for its points. The second view, the X-Z view, permits a lesser degree of freedom because some of the points will have been determined from the initial X-Y planar view. Finally, the third view—the Y-Z planar view—allows practically no freedom regarding the choice of coordinates. Those coordinates are almost entirely determined by the partial coordinates specified for the first two views. At that part of the task, the important thing is not so much the shape of the object as it is the consistency of the partial coordinates. There should be no points on any one of the three planar views that cannot be mapped onto points on the other two views.

After developing a satisfactory, self-consistent set of three planar views and partial 3-D coordinates, the general 3-D construction procedure follows this sort of plan:

1. Generate the point specifications from the original Data
   - A. Make a list of all possible points
   - B. Eliminate duplicate points
   - C. Eliminate invalid points
2. Generate the line specifications from the points
   - A. Make a list of all possible lines
   - B. Eliminate duplicate lines
   - C. Eliminate invalid lines
3. Generate the face specifications from the lines
   - A. Make a list of all possible faces
   - B. Eliminate the duplicate faces
   - C. Eliminate invalid faces
4. Generate the space-object specifications from the faces

Given some insight into the procedure (insight that can come only from first-hand experience), it is possible to come up with some shortcuts that relieve some of the more tedious and ultimately nonproductive elements in this general procedure. You can, for instance, eliminate duplicate and invalid points while you are generating the first list of all possible points. That saves some time. What's more, you can save yourself a great deal of work by listing only the lines that appear on the drawings, rather than all possible lines; and you can do the same for generating the face specifications. Care, insight and experience can go a long way toward making the whole procedure simpler and more attractive in terms of sheer human effort.

The same general procedure and possible shortcuts apply to the matter of generating full 4-D hyperspace specifications from 3-D data. In principles at least, the only difference is that the extended, 4-D version of the job includes one additional step: generating the 4-dimensional H specifications from the 3-D space data.

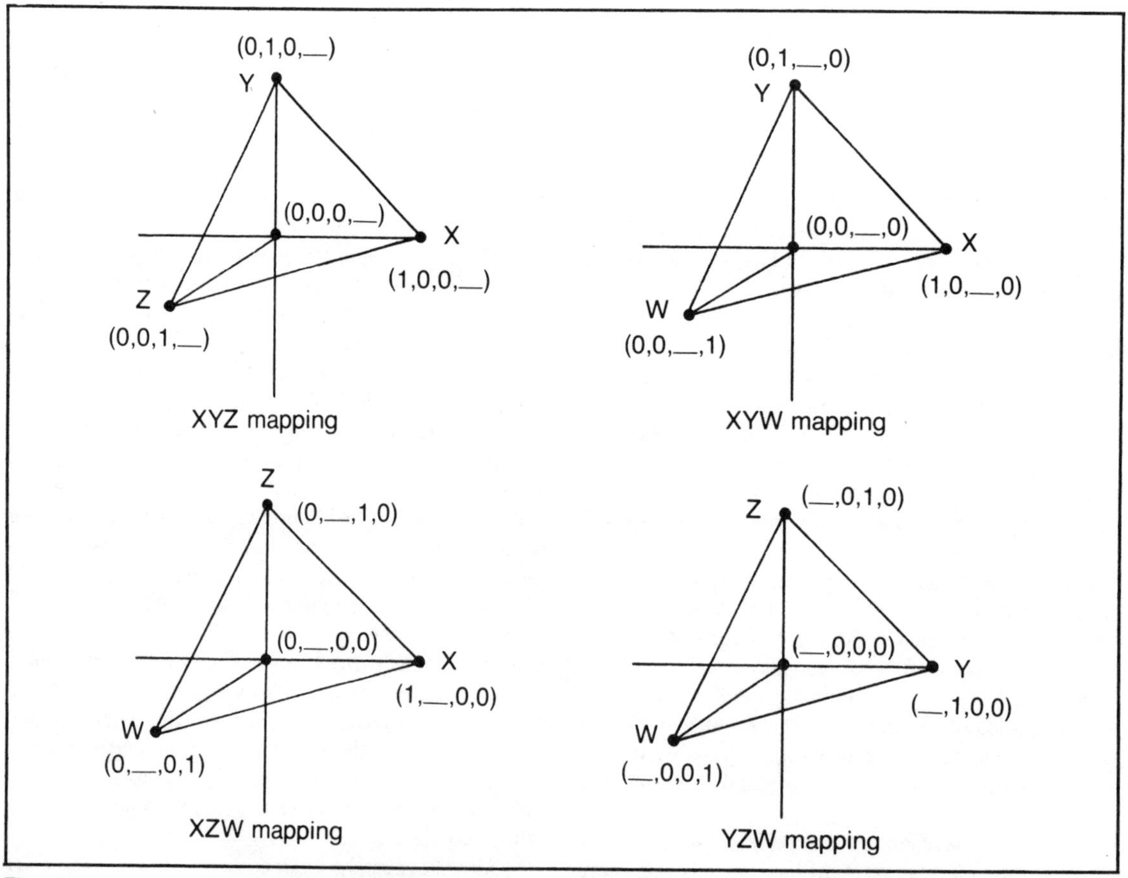

Fig. 18-1. A proposed hyperspace pyramid showing the partial coordinates under the four 3-D perspective mappings.

Table 18-1. Summary of Partial Point Coordinates for the Hyperspace Pyramid Suggested by the Plots in Fig. 18-1.

INITIAL POINT SPECIFICATIONS

X	Y	Z	W	
Ø	Ø	Ø	_	
1	Ø	Ø	_	FROM XYZ MAPPING
Ø	1	Ø	_	
Ø	Ø	1	_	
Ø	Ø	_	Ø	
1	Ø	_	Ø	FROM XYW MAPPING
Ø	1	_	Ø	
Ø	Ø	_	1	
Ø	_	Ø	Ø	
1	_	Ø	Ø	FROM XZW MAPPING
Ø	_	1	Ø	
Ø	_	Ø	1	
_	Ø	Ø	Ø	
_	1	Ø	Ø	FROM YZW MAPPING
_	Ø	1	Ø	
_	Ø	Ø	1	

The original mappings are now perspective drawings of the hyperspace object as viewed from the XYZ, XYW, XZW and YZW spaces. There are thus four different 3-D perspective views rather than just three 2-D planar views. And as far as the partial coordinates are concerned, they are now partial 4-D coordinates. The four perspective views of the hyperspace object and the set of partial coordinates must for a self-consistent, albeit partial, view of the proposed object.

From there, the technique follows that of the 3-D construction procedure: generating point specifications from the original information, generating line specifications from the point coordinates, generating face specifications from the lines, generating space-object specifications from the faces and, finally, generating the hyperspace specifications from the space-object information.

In this chapter, I will develop a couple of hyperspace objects in great detail and show just the final specifications for a couple more. You will have to take matters into your own hands after that.

18-1 GENERATING HYPERSPACE OBJECT NUMBER 1

The objective of this example is to generate a set of specifications for a hyperspace object that is analogous to a 4-sided pyramid of 3-D space. Its special importance is that it represents the simplest possible hyperspace object.

Figure 18-1 shows the proposed hyperspace pyramid from the four 3-E perspective views. It is a regular object in this case, appearing the same from all four views. Note the consistency of the tentative point coordinates.

Table 18-1 summarizes all of the tentative point coordinates. That information is then extended to include all possible points in Table 18-2.

It thus seems that the hyperspace pyramid is composed of just 5 points. Figure 18-2 shows the final point specifications and the four 3-D perspective views having those point designations.

Each of the four views in that figure happens to include 9 possible lines, and they are summarized for you in the possible lines column in Table 18-3. Eliminating the duplicate lines, we end up with a list of *plotted lines*—a series of line specifications that will soon become the final line specifications for this hyperspace object.

Figure 18-3 uses the previous views and line designations to develop a series of two versions for each

Table 18-2. Summary of the Evolution of Actual Point Coordinates for the Hyperspace Pyramid Suggested in Fig. 18-1.

ALL POSSIBLE POINTS				DUPLICATES ELIMINATED				PLOTTED POINTS			
X	Y	X	W	X	Y	X	W	X	Y	X	W
Ø	Ø	Ø	Ø	Ø	Ø	Ø	Ø	Ø	Ø	Ø	Ø
Ø	Ø	Ø	1	Ø	Ø	Ø	1	Ø	Ø	Ø	1
1	Ø	Ø	Ø	1	Ø	Ø	Ø	1	Ø	Ø	Ø
1	Ø	Ø	1	1	Ø	Ø	1				
Ø	1	Ø	Ø	Ø	1	Ø	Ø	Ø	1	Ø	Ø
Ø	1	Ø	1	Ø	1	Ø	1				
Ø	Ø	1	Ø	Ø	Ø	1	Ø	Ø	Ø	1	Ø
Ø	Ø	1	1	Ø	Ø	1	1				
Ø	Ø	Ø	Ø								
Ø	Ø	1	Ø								
1	Ø	Ø	Ø								
1	Ø	1	Ø	1	Ø	1	Ø				
Ø	1	Ø	Ø								
Ø	1	1	Ø	Ø	1	1	Ø				
Ø	Ø	Ø	1								
Ø	Ø	1	1	Ø	Ø	1	1				
Ø	Ø	Ø	Ø								
Ø	1	Ø	Ø								
1	Ø	Ø	Ø								
1	1	Ø	Ø	1	1	Ø	Ø				
Ø	Ø	1	Ø								
Ø	1	1	Ø								
Ø	Ø	Ø	1								
Ø	1	Ø	1	Ø	1	Ø	1				
Ø	Ø	Ø	Ø								
1	Ø	Ø	Ø								
Ø	1	Ø	Ø								
1	1	Ø	Ø								
Ø	Ø	1	Ø								
1	Ø	1	Ø								
Ø	Ø	Ø	1								
1	Ø	Ø	1								

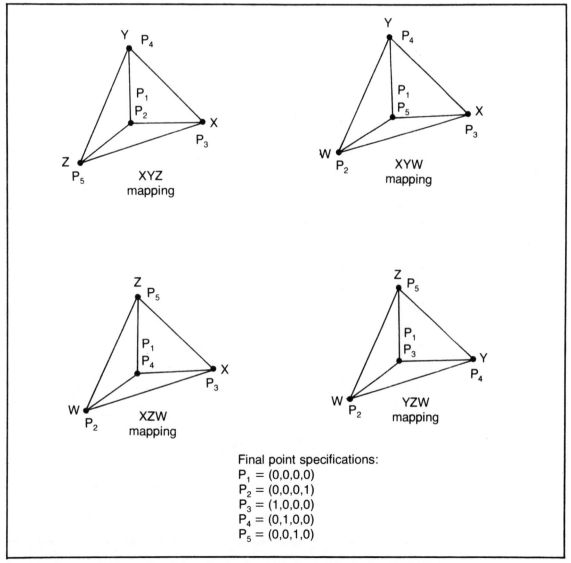

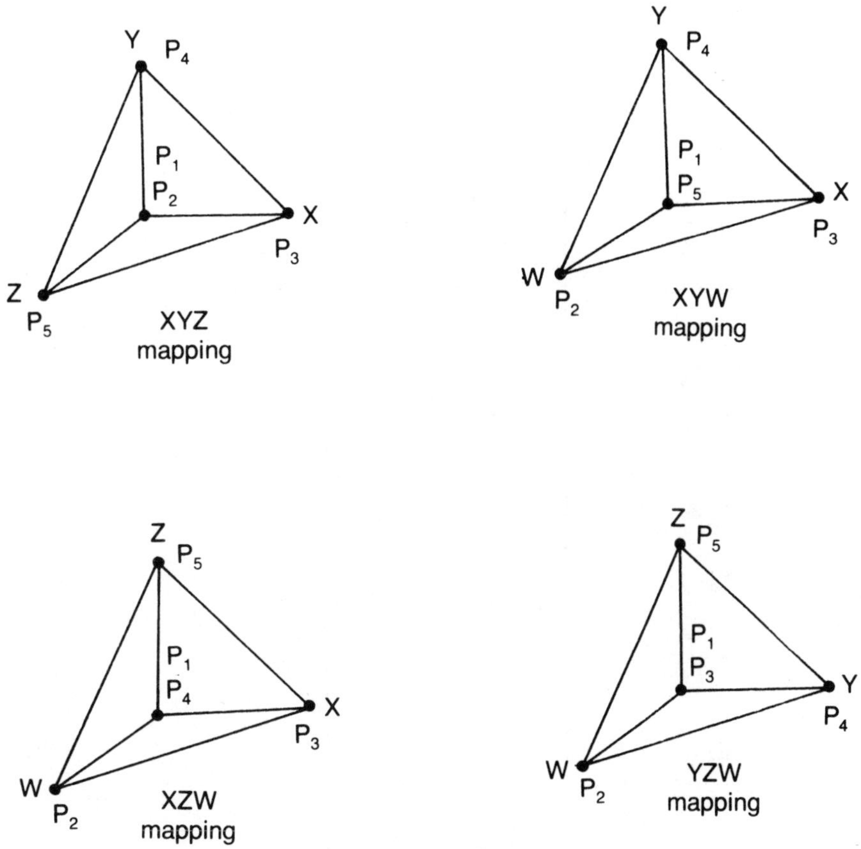

Final point specifications:
$P_1 = (0,0,0,0)$
$P_2 = (0,0,0,1)$
$P_3 = (1,0,0,0)$
$P_4 = (0,1,0,0)$
$P_5 = (0,0,1,0)$

Fig. 18-2. The evolving hyperspace figure showing the final point coordinates from Table 18-2.

viewing space. The line designations on those drawings come from the final line specifications that accompany the figures.

Table 18-4 shows the series of lines that can define faces of the hyperspace object. After eliminating the duplications, ten lines remain; and they are specified for you in Fig. 18-4.

The views in Fig. 18-4 do not include views XYW-II, XZW-II and YZW-II from the preceding drawings because they are exact duplicates of other views. The views in Fig. 18-4, in other words, represent the unique 3-D perspective views of the hyperspace object.

Those five remaining views lead directly to the final space-object specifications; it seems that there are five space objects bounding this hyperspace pyramid. And getting the final hyperspace specifications is a simple matter of gathering the five space-object designations into one group.

Again, if you have worked the detailed examples for constructing 3-D space objects from 2-D data (Chapter

Table 18-3. The Evolution of Line Specifications Based on the Results Plotted in Fig. 18-2.

POSSIBLE LINES	DUPLICATES ELIMINATED	PLOTTED LINES
P_1P_3	P_1P_3	P_1P_3
P_1P_4	P_1P_4	P_1P_4
P_1P_5	P_1P_5	P_1P_5
P_2P_3	P_2P_3	P_2P_3
P_2P_4	P_2P_4	P_2P_4
P_2P_5	P_2P_5	P_2P_5
P_3P_4	P_3P_4	P_3P_4
P_3P_5	P_3P_5	P_3P_5
P_4P_5	P_4P_5	P_4P_5
P_1P_2	P_1P_2	P_1P_2
P_1P_3		
P_1P_4		
P_2P_3		
P_2P_4		
P_2P_5		
P_3P_4		
P_3P_5		
P_4P_5		

16), you should have little trouble following this particular 4-D construction. If you don't feel comfortable with it, take some time to study it more carefully.

Figure 18-5 is a general space view of this hyperspace object. It is a free-form sketch in the sense that the points and lines aren't plotted to precision; the drawing is intended to illustrate the main qualities of the hyperspace object. The specifications carry the burden of precision for you.

Given the specifications just derived, you should be able to see that this hyperspace object composed of 5 points, 10 lines and 10 faces. If you care to decompose the object further, using the specifications and Fig. 18-4 as guides, you will be able to find the five 3-D space objects that bound it.

Table 18-5 summarizes the general specifications and full analysis of the object for future reference.

18-2 CONSTRUCTING HYPERSPACE OBJECT NUMBER 2

This hyperspace object is the result of an initial attempt to create a 4-D version of a right prism. You can see the basic idea in the XYZ and XYW perspective views in Fig. 18-6. Unfortunately for my original plan, the same prism-looking object could not be carried to the XZW and

POSSIBLE LINES	DUPLICATES ELIMINATED	PLOTTED LINES
P_1P_2		
P_1P_3		
P_1P_5		
P_2P_3		
P_2P_4		
P_2P_5		
P_3P_4		
P_3P_5		
P_4P_5		
P_1P_2		
P_1P_4		
P_1P_5		
P_2P_3		
P_2P_4		
P_2P_5		
P_3P_4		
P_3P_5		
P_4P_5		

YZW perspective mappings—the necessary points created inconsistencies that could not be resolved. So it was necessary to generate the cube object shown in the XZW and YZW mappings. The partial coordinates and drawings are now fully self-consistent.

Table 18-6 shows the evolution of the final point specifications, and Fig. 18-7 shows the results. This hyperspace object thus has 12 points that are distributed between two prism-looking objects and two cubes.

Table 18-7 then shows the evolution of the line specifications. The shortcut technique that I used for generating the list of possible lines was one that did, indeed, generate all possible lines; and it did so without coming up with any duplicates that had to be eliminated. Thus the list leads directly to the plotted lines list.

The final point specifications and list of plotted lines lead from the ambiguous point designations in Fig. 18-7 to the more precise renditions of the objects in Fig. 18-8. That set of drawings includes the final line specifications as well.

At this point in the construction procedure, we know that the hyperspace object is composed of 10 points and 24 individual lines. Furthermore, there are no duplicates in the sequence of drawings in Fig. 18-8, so we can be

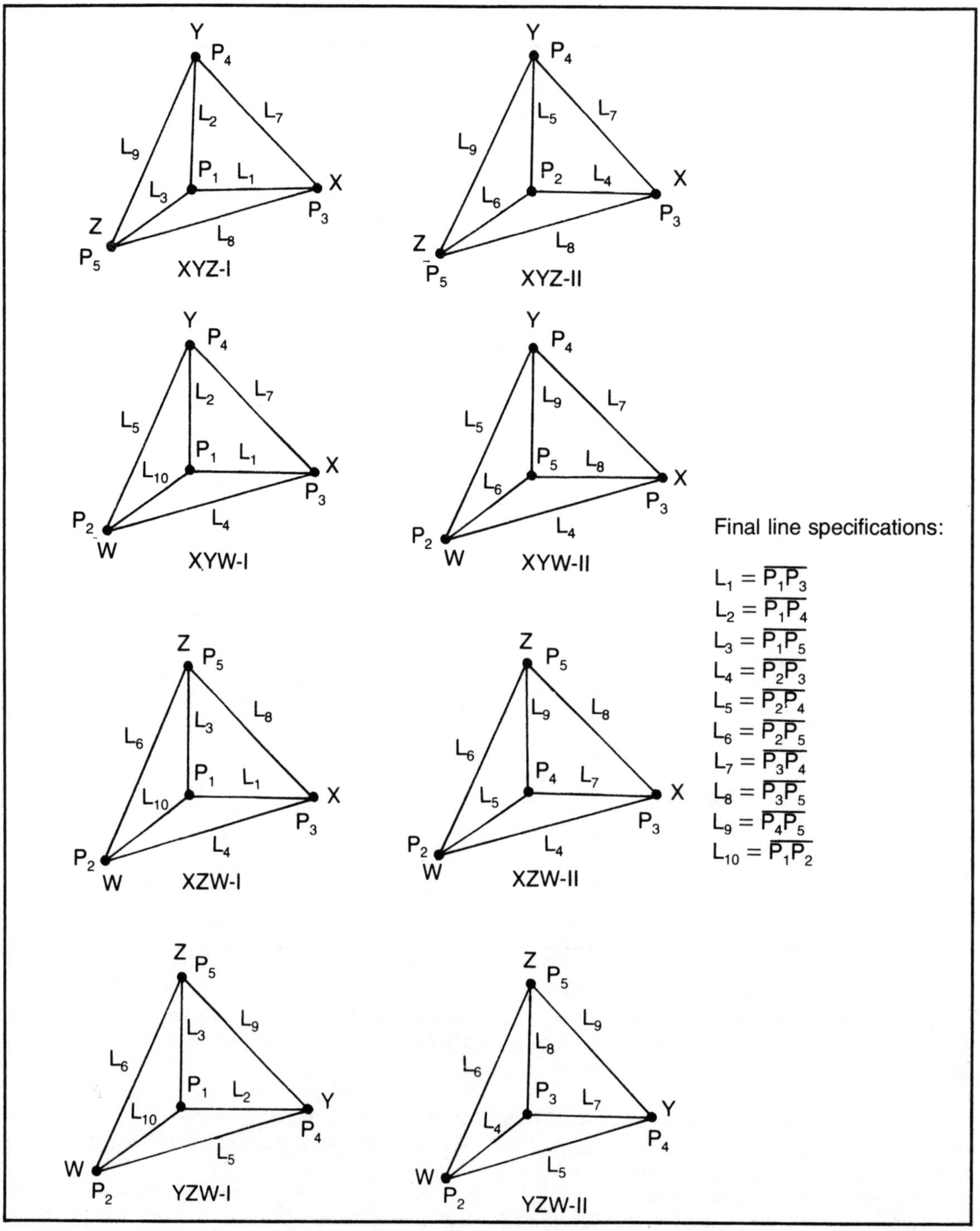

Fig. 18-3. The lines derived from Table 18-3 lead to two different objects as viewed from the four different 3-D perspective views.

reasonably confident that the hyperspace object will be bounded by 4 prism-like and 4 cube-like space objects.

An analysis of those eight bounding space objects leads to the evolution of the face specifications (see Table 18-8), the final space specifications and, at last, the rather long, but straightforward, hyperspace specifications (see Table 18-9).

Table 18-10 shows a complete analysis of the hyperspace object, and Fig. 18-9 represents a general perspective view of it.

The final space specifications indicate that the hyperspace object is bounded by four 5-sided space objects and four 6-sided objects. Those are the prism-like and cube-like objects mentioned earlier. You can locate them, using those specifications and the views in Fig. 18-8 as guides, and separating them out one at a time.

Table 18-4. The Evolution of the Final Figure, or "Sides," Specifications for the Proposed Hyperspace Object.

POSSIBLE FACES	DUPLICATES ELIMINATED
$L_1L_2L_7$	$L_1L_2L_7$
$L_1L_3L_8$	$L_1L_3L_8$
$L_2L_3L_9$	$L_2L_3L_9$
$L_7L_8L_9$	$L_7L_8L_9$
$L_4L_5L_7$	$L_4L_5L_7$
$L_4L_6L_8$	$L_4L_6L_8$
$L_5L_6L_9$	$L_5L_6L_9$
$L_7L_8L_9$	
$L_1L_2L_7$	
$L_1L_4L_{10}$	$L_1L_4L_{10}$
$L_2L_5L_{10}$	$L_2L_5L_{10}$
$L_4L_5L_7$	
$L_4L_5L_7$	
$L_4L_6L_8$	
$L_5L_6L_9$	
$L_7L_8L_9$	
$L_1L_3L_8$	
$L_1L_4L_{10}$	
$L_3L_6L_{10}$	$L_3L_6L_{10}$
$L_4L_6L_8$	

POSSIBLE FACES	DUPLICATES ELIMINATED
$L_4L_5L_7$	
$L_4L_6L_8$	
$L_5L_6L_9$	
$L_7L_8L_9$	
$L_2L_3L_9$	
$L_2L_5L_{10}$	
$L_3L_6L_{10}$	
$L_5L_6L_9$	
$L_4L_5L_7$	
$L_4L_6L_8$	
$L_5L_6L_9$	
$L_7L_8L_9$	

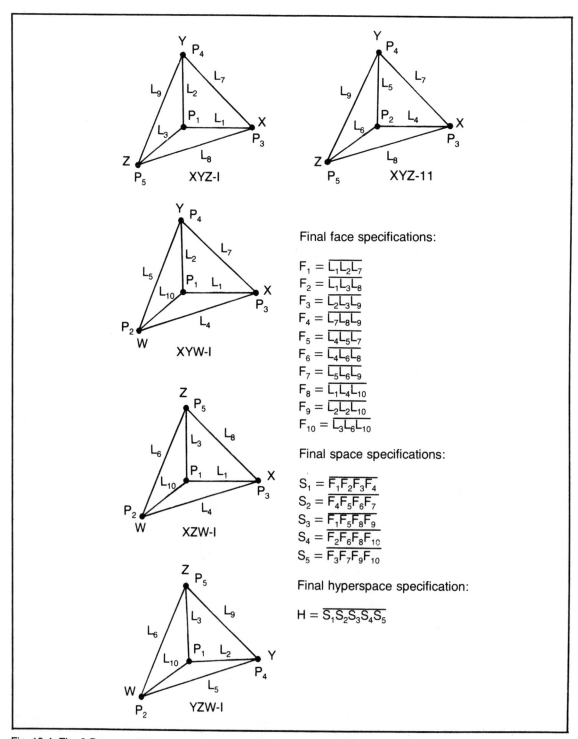

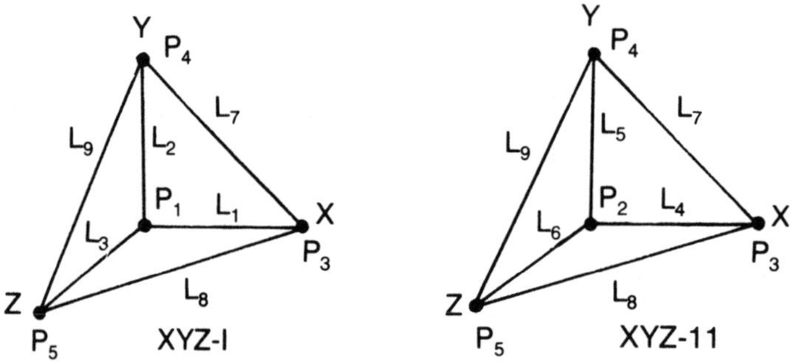

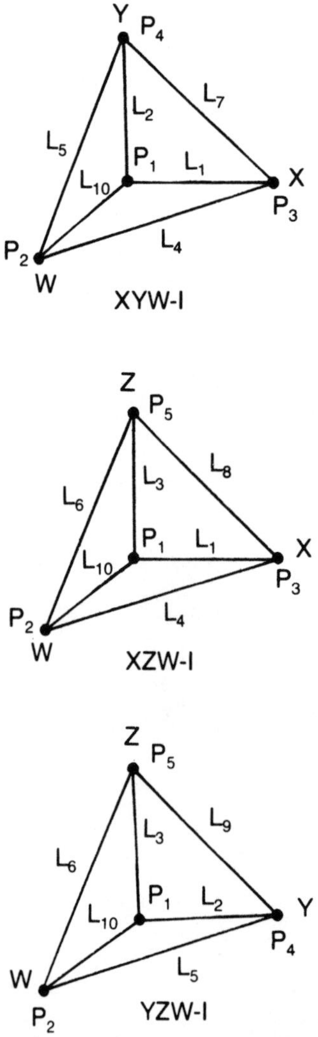

Final face specifications:

$F_1 = \overline{L_1L_2L_7}$
$F_2 = \overline{L_1L_3L_8}$
$F_3 = \overline{L_2L_3L_9}$
$F_4 = \overline{L_7L_8L_9}$
$F_5 = \overline{L_4L_5L_7}$
$F_6 = \overline{L_4L_6L_8}$
$F_7 = \overline{L_5L_6L_9}$
$F_8 = \overline{L_1L_4L_{10}}$
$F_9 = \overline{L_2L_2L_{10}}$
$F_{10} = \overline{L_3L_6L_{10}}$

Final space specifications:

$S_1 = \overline{F_1F_2F_3F_4}$
$S_2 = \overline{F_4F_5F_6F_7}$
$S_3 = \overline{F_1F_5F_8F_9}$
$S_4 = \overline{F_2F_6F_8F_{10}}$
$S_5 = \overline{F_3F_7F_9F_{10}}$

Final hyperspace specification:

$H = \overline{S_1S_2S_3S_4S_5}$

Fig. 18-4. The 3-D perspective objects that remain after eliminating the duplicates.

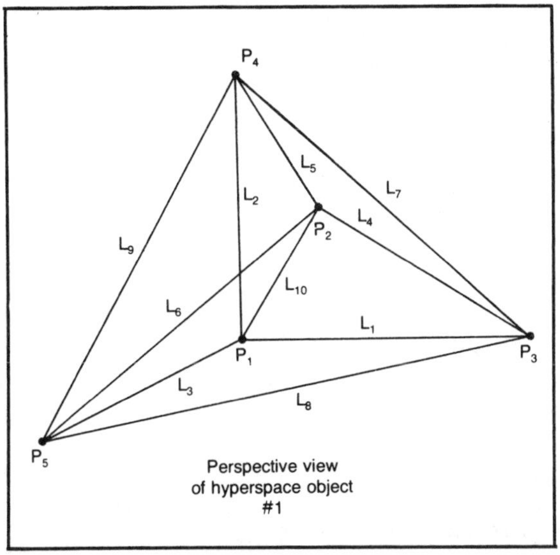

Fig. 18-5. A general 4-D perspective view of the finished hyperspace pyramid. See the specifications and primary analysis in Table 18-5.

Incidentally, I sketched that final space view in such a way that it emphasizes the prism-like nature of the object. It is entirely possible—given the freedom that is allowed for drawing such views—to redraw the object in a fashion that emphasizes its cube-like elements. Give it a try on your own.

Right now you probably know more about a 4-D prism than 99.99 percent of the people in the world do.

18-3 A HYPERSPACE VERSION OF A 5-SIDED PYRAMID

Figure 18-10 shows the four perspective views of a hyperspace version of our 5-sided pyramid. It appears as that familiar 3-D object from three different 3-D spaces, but as a cube from XZW space. Figure 18-11 is a general perspective illustration of it. See Table 18-11 for a complete analysis. I generated this object using the procedures already outlined and demonstrated for you.

18-4 A HYPERSPACE CUBE

The hyperspace versions of a cube is illustrated in a general perspective fashion in Fig. 18-12, and you will find its specifications in Table 18-12. It is the most com-

Table 18-5. Specifications and Primary Analysis of the Hyperspace Object Illustrated in Figs. 18-4 and 18-5.

(A, B, C, and D correspond to α, β, γ, and δ.)

SPECIFICATIONS

$H=\overline{S_1S_2S_3S_4S_5}$

$S_1=\overline{F_1F_2F_3F_4}$

$S_2=\overline{F_4F_5F_6F_7}$

$S_3=\overline{F_1F_5F_8F_9}$

$S_4=\overline{F_2F_6F_8F_{10}}$

$S_5=\overline{F_3F_7F_9F_{10}}$

$F_1=\overline{L_1L_2L_7}$	$L_1=\overline{P_1P_3}$	$P_1=(0,0,0,0)$
$F_2=\overline{L_1L_3L_8}$	$L_2=\overline{P_1P_4}$	$P_2=(0,0,0,1)$
$F_3=\overline{L_2L_3L_9}$	$L_3=\overline{P_1P_5}$	$P_3=(1,0,0,0)$
$F_4=\overline{L_7L_8L_9}$	$L_4=\overline{P_2P_3}$	$P_4=(0,1,0,0)$
$F_5=\overline{L_4L_5L_7}$	$L_5=\overline{P_2P_4}$	$P_5=(0,0,1,0)$
$F_6=\overline{L_4L_6L_8}$	$L_6=\overline{P_2P_5}$	
$F_7=\overline{L_5L_6L_9}$	$L_7=\overline{P_3P_4}$	

Table 18-5. Specifications and Primary Analysis of the Hyperspace Object Illustrated in Figs. 18-4 and 18-5. (Continued from page 277.)

$F_8=\overline{L_1L_4L_{10}}$	$L_8=\overline{P_3P_5}$
$F_9=\overline{L_2L_5L_{10}}$	$L_9=\overline{P_4P_5}$
$F_{10}=\overline{L_3L_6L_{10}}$	$L_{10}=\overline{P_1P_2}$

LENGTHS

$s_1=1$	$s_2=1$	$s_3=1$	$s_4=\sqrt{2}$
$s_5=\sqrt{2}$	$s_6=\sqrt{2}$	$s_7=\sqrt{2}$	$s_8=\sqrt{2}$
$s_9=\sqrt{2}$	$s_{10}=1$		

DIRECTION COSINES

$\cos A_1=1$	$\cos B_1=0$	$\cos C_1=0$	$\cos D_1=0$
$\cos A_2=0$	$\cos B_2=1$	$\cos C_2=0$	$\cos D_2=0$
$\cos A_3=0$	$\cos B_3=0$	$\cos C_3=1$	$\cos D_3=0$
$\cos A_4=\sqrt{2}/2$	$\cos B_4=0$	$\cos C_4=0$	$\cos D_4=-\sqrt{2}/2$
$\cos A_5=0$	$\cos B_5=\sqrt{2}/2$	$\cos C_5=0$	$\cos D_5=-\sqrt{2}/2$
$\cos A_6=0$	$\cos B_6=0$	$\cos C_6=\ 2/2$	$\cos D_6=-\sqrt{2}/2$
$\cos A_7=-\sqrt{2}/2$	$\cos B_7=\sqrt{2}/2$	$\cos C_7=0$	$\cos D_7=0$
$\cos A_8=-\sqrt{2}/2$	$\cos B_8=0$	$\cos C_8=\sqrt{2}/2$	$\cos D_8=0$
$\cos A_9=0$	$\cos B_9=-\sqrt{2}/2$	$\cos C_9=\ 2/2$	$\cos D_9=0$
$\cos A_{10}=0$	$\cos B_{10}=0$	$\cos C_{10}=0$	$\cos D_{10}=1$

ANGLES

$\phi_{1,2}=90^\circ$	$\phi_{1,3}=90^\circ$	$\phi_{1,4}=45^\circ$
$\phi_{1,7}=45^\circ$	$\phi_{1,8}=45^\circ$	$\phi_{1,10}=90^\circ$
$\phi_{2,3}=90^\circ$	$\phi_{2,5}=45^\circ$	$\phi_{2,7}=45^\circ$
$\phi_{2,10}=90^\circ$	$\phi_{3,6}=45^\circ$	$\phi_{3,8}=45^\circ$
$\phi_{3,9}=45^\circ$	$\phi_{3,10}=90^\circ$	$\phi_{4,5}=60^\circ$
$\phi_{4,6}=60^\circ$	$\phi_{4,7}=60^\circ$	$\phi_{4,8}=60^\circ$
$\phi_{4,10}=45^\circ$	$\phi_{5,6}=60^\circ$	$\phi_{5,7}=60^\circ$
$\phi_{5,9}=60^\circ$	$\phi_{5,10}=45^\circ$	$\phi_{6,8}=60^\circ$
$\phi_{6,9}=60^\circ$	$\phi_{6,10}=45^\circ$	$\phi_{7,8}=60^\circ$
$\phi_{7,9}=60^\circ$	$\phi_{8,9}=60^\circ$	

Initial point list

X	Y	Z	W	
0	0	0	__	
1	0	0	__	
0	1	0	__	From XYZ mapping
0	0	1	__	
1	0	1	__	
0	1	1	__	
0	0	__	0	
1	0	__	0	From XYW mapping
0	1	__	0	
0	0	__	1	
1	0	__	1	
0	1	__	1	

X	Y	Z	W	
0	__	0	0	
1	__	0	0	
0	__	1	0	From XZW mapping
1	__	1	0	
0	__	0	1	
1	__	0	1	
0	__	1	1	
1	__	1	1	

X	Y	Z	W	
__	0	0	0	
__	1	0	0	
__	0	1	0	From YZW mapping
__	1	1	0	
__	0	0	1	
__	1	0	1	
__	0	1	1	
__	1	1	1	

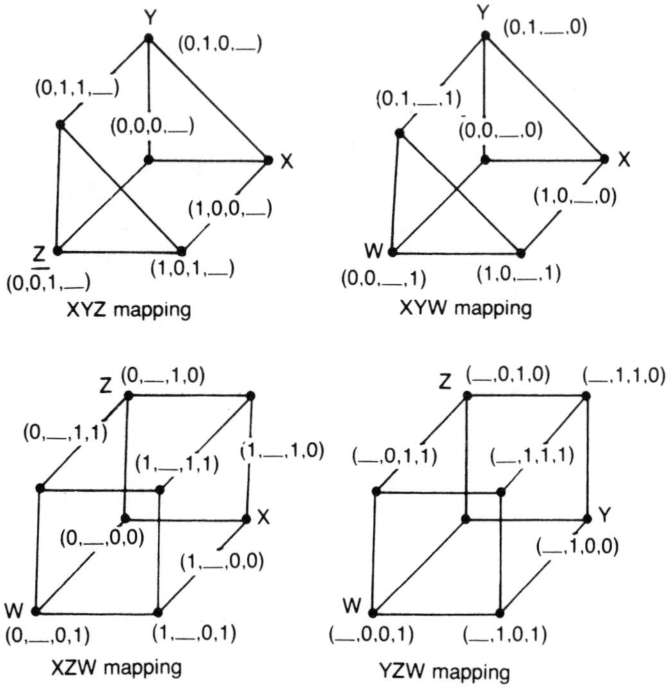

Fig. 18-6. The 3-D perspective views and partial coordinates for hyperspace object number 2.

Table 18-6. Evolution of the Final Point Specifications for Hyperspace Object Number 2.

ALL POSSIBLE POINTS				DUPLICATES ELIMINATED				PLOTTED POINTS			
X	Y	Z	W	X	Y	Z	W	X	Y	Z	W
Ø	Ø	Ø	Ø	Ø	Ø	Ø	Ø	Ø	Ø	Ø	Ø
Ø	Ø	Ø	1	Ø	Ø	Ø	1	Ø	Ø	Ø	1
1	Ø	Ø	Ø	1	Ø	Ø	Ø	1	Ø	Ø	Ø
1	Ø	Ø	1	1	Ø	Ø	1	1	Ø	Ø	1
Ø	1	Ø	Ø	Ø	1	Ø	Ø	Ø	1	Ø	Ø
Ø	1	Ø	1	Ø	1	Ø	1	Ø	1	Ø	1
Ø	Ø	1	Ø	Ø	Ø	1	Ø	Ø	Ø	1	Ø
Ø	Ø	1	1	Ø	Ø	1	1	Ø	Ø	1	1
Ø	1	1	Ø	Ø	1	1	Ø	Ø	1	1	Ø
Ø	1	1	1	Ø	1	1	1	Ø	1	1	1
Ø	Ø	Ø	Ø								
Ø	Ø	1	Ø								
1	Ø	Ø	Ø								
1	Ø	1	Ø	1	Ø	1	Ø	1	Ø	1	Ø
Ø	1	Ø	Ø								
Ø	1	1	Ø								
Ø	Ø	Ø	1								
Ø	Ø	1	1								
1	Ø	Ø	1								
1	Ø	1	1	1	Ø	1	1	1	Ø	1	1
Ø	1	Ø	1								
Ø	1	1	1								
Ø	Ø	Ø	Ø								
Ø	1	Ø	Ø								
1	Ø	Ø	Ø								
1	1	Ø	Ø	1	1	Ø	Ø				
Ø	Ø	1	Ø								
Ø	1	1	Ø								
1	Ø	1	Ø								
1	1	1	Ø	1	1	1	Ø				
Ø	Ø	Ø	1								
Ø	1	Ø	1								
1	Ø	Ø	1								
1	1	Ø	1	1	1	Ø	1				
Ø	Ø	1	1								
Ø	1	1	1								
1	Ø	1	1								
1	1	1	1	1	1	1	1				
Ø	Ø	Ø	Ø								
1	Ø	Ø	Ø								
Ø	1	Ø	Ø								
1	1	Ø	Ø								
Ø	Ø	1	Ø								
1	Ø	1	Ø								
Ø	1	1	Ø								
1	1	1	Ø								
Ø	Ø	Ø	1								
1	Ø	Ø	1								
Ø	1	Ø	1								
1	1	Ø	1								
Ø	Ø	1	1								
1	Ø	1	1								
Ø	1	1	1								
1	1	1	1								

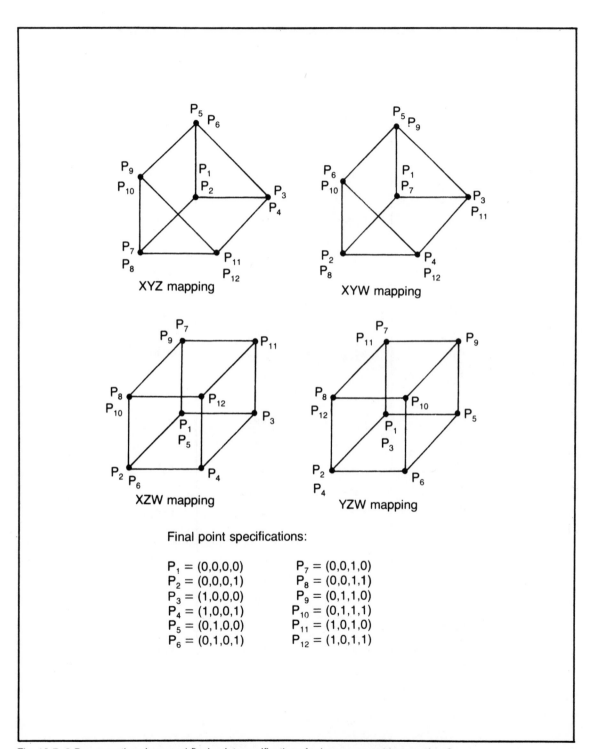

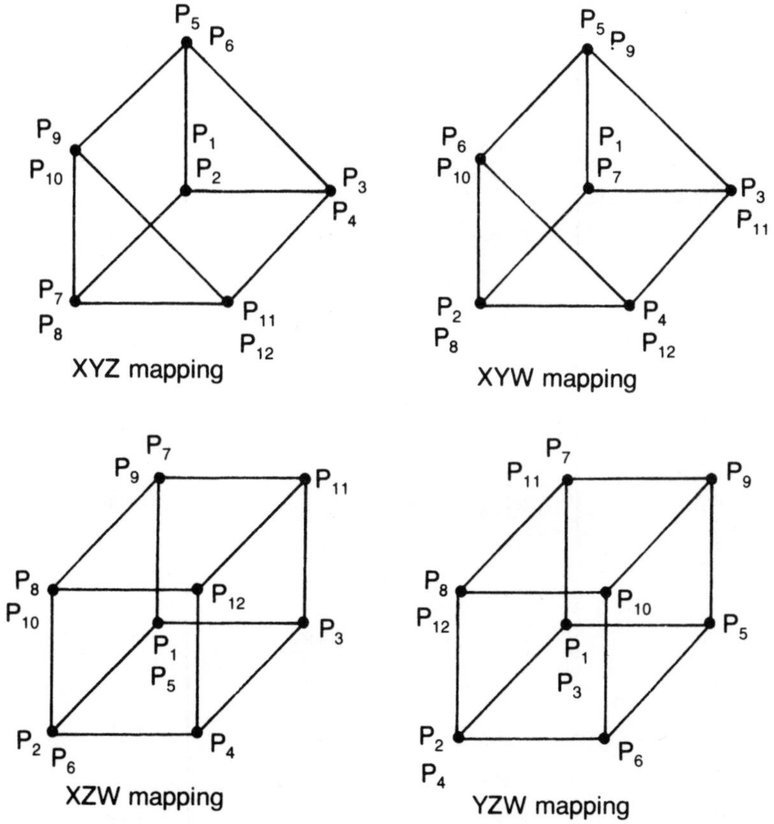

Final point specifications:

$P_1 = (0,0,0,0)$	$P_7 = (0,0,1,0)$
$P_2 = (0,0,0,1)$	$P_8 = (0,0,1,1)$
$P_3 = (1,0,0,0)$	$P_9 = (0,1,1,0)$
$P_4 = (1,0,0,1)$	$P_{10} = (0,1,1,1)$
$P_5 = (0,1,0,0)$	$P_{11} = (1,0,1,0)$
$P_6 = (0,1,0,1)$	$P_{12} = (1,0,1,1)$

Fig. 18-7. 3-D perspective views and final point specifications for hyperspace object number 2.

Table 18-7. Evolution of the Line Specifications for Hyperspace Object Number 2.

POSSIBLE LINES	PLOTTED LINES	POSSIBLE LINES	PLOTTED LINES
P_1P_2	P_1P_2	P_4P_{11}	
P_1P_3	P_1P_3	P_4P_{12}	P_4P_{12}
P_1P_4		P_5P_6	P_5P_6
P_1P_5	P_1P_5	P_5P_9	P_5P_9
P_1P_6		P_5P_{10}	
P_1P_7	P_1P_7	P_6P_9	
P_1P_8		P_6P_{10}	P_6P_{10}
P_2P_3		P_7P_8	P_7P_8
P_2P_4	P_2P_4	P_7P_9	P_7P_9
P_2P_5		P_7P_{10}	
P_2P_6	P_2P_6	P_7P_{11}	P_7P_{11}
P_2P_7		P_8P_9	
P_2P_8	P_2P_8	P_8P_{10}	P_8P_{10}
P_3P_4	P_3P_4	P_8P_{11}	
P_3P_5	P_3P_5	P_8P_{12}	P_8P_{12}
P_3P_6		P_9P_{10}	P_9P_{10}
P_3P_{11}	P_3P_{11}	P_9P_{11}	P_9P_{11}
P_3P_{12}		P_9P_{12}	
P_4P_5		$P_{10}P_{11}$	
P_4P_6	P_4P_6	$P_{11}P_{12}$	$P_{11}P_{12}$

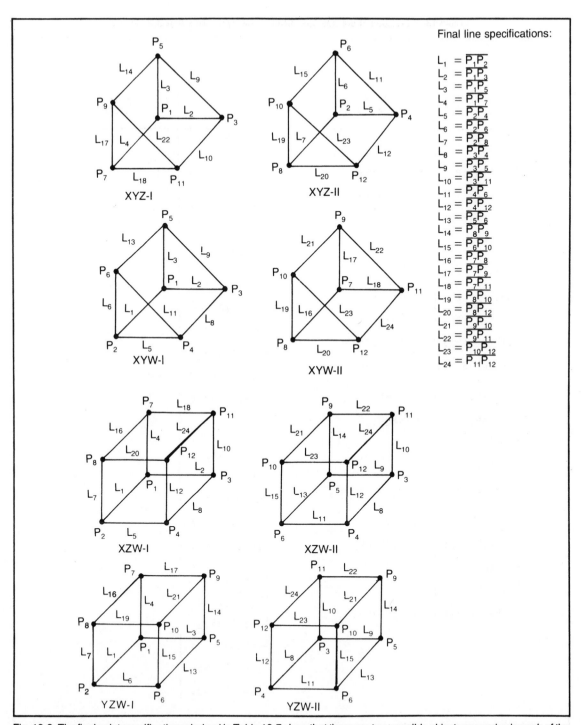

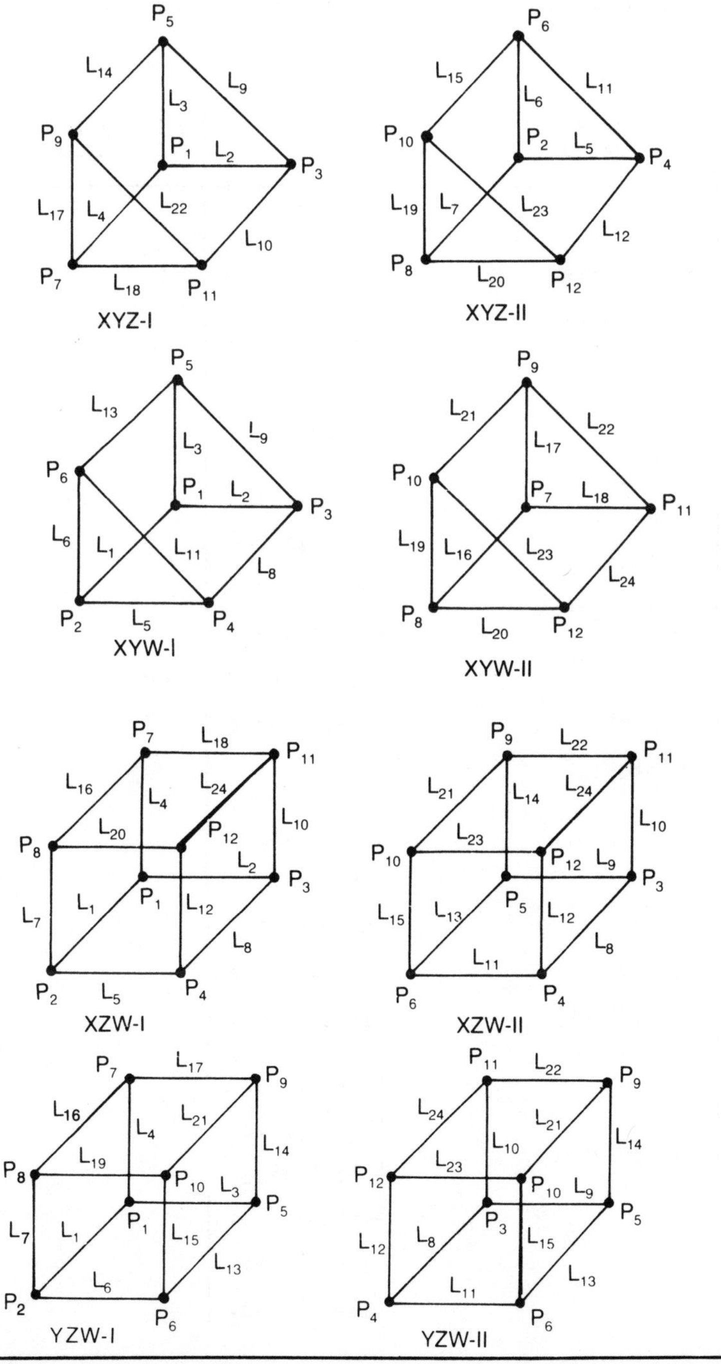

Final line specifications:

$L_1 = \overline{P_1P_2}$
$L_2 = \overline{P_1P_3}$
$L_3 = \overline{P_1P_5}$
$L_4 = \overline{P_1P_7}$
$L_5 = \overline{P_2P_4}$
$L_6 = \overline{P_2P_6}$
$L_7 = \overline{P_2P_8}$
$L_8 = \overline{P_3P_4}$
$L_9 = \overline{P_3P_5}$
$L_{10} = \overline{P_3P_{11}}$
$L_{11} = \overline{P_4P_6}$
$L_{12} = \overline{P_4P_{12}}$
$L_{13} = \overline{P_5P_6}$
$L_{14} = \overline{P_8P_9}$
$L_{15} = \overline{P_6P_{10}}$
$L_{16} = \overline{P_7P_8}$
$L_{17} = \overline{P_7P_9}$
$L_{18} = \overline{P_7P_{11}}$
$L_{19} = \overline{P_8P_{10}}$
$L_{20} = \overline{P_8P_{12}}$
$L_{21} = \overline{P_9P_{10}}$
$L_{22} = \overline{P_9P_{11}}$
$L_{23} = \overline{P_{10}P_{12}}$
$L_{24} = \overline{P_{11}P_{12}}$

Fig. 18-8. The final point specifications derived in Table 18-7 show that there are two possible objects appearing in each of the 3-D perspective views.

Table 18-8. Evolution of the F Specifications for Hyperspace Object Number 2.

POSSIBLE FACES	DUPLICATES ELIMINATED
$L_2L_3L_9$	$L_2L_3L_9$
$L_2L_4L_{10}L_{18}$	$L_2L_4L_{10}L_{18}$
$L_3L_4L_{14}L_{17}$	$L_3L_4L_{14}L_{17}$
$L_9L_{10}L_{14}L_{22}$	$L_9L_{10}L_{14}L_{22}$
$L_{17}L_{18}L_{22}$	$L_{17}L_{18}L_{22}$
$L_5L_6L_{11}$	$L_5L_6L_{11}$
$L_5L_7L_{12}L_{20}$	$L_5L_7L_{12}L_{20}$
$L_6L_7L_{15}L_{19}$	$L_6L_7L_{15}L_{19}$
$L_{11}L_{12}L_{15}L_{23}$	$L_{11}L_{12}L_{15}L_{23}$
$L_{19}L_{20}L_{23}$	$L_{19}L_{20}L_{23}$
$L_1L_2L_5L_8$	$L_1L_2L_5L_8$
$L_1L_3L_6L_{13}$	$L_1L_3L_6L_{13}$
$L_2L_3L_9$	
$L_5L_6L_{11}$	
$L_8L_9L_{11}L_{13}$	$L_8L_9L_{11}L_{13}$
$L_{16}L_{17}L_{19}L_{21}$	$L_{16}L_{17}L_{19}L_{21}$
$L_{16}L_{18}L_{20}L_{24}$	$L_{16}L_{18}L_{20}L_{24}$
$L_{17}L_{18}L_{22}$	
$L_{19}L_{20}L_{23}$	
$L_{21}L_{22}L_{23}L_{24}$	$L_{21}L_{22}L_{23}L_{24}$

POSSIBLE FACES	DUPLICATES ELIMINATED
$L_1L_2L_5L_8$	
$L_1L_4L_7L_{16}$	$L_1L_4L_7L_{16}$
$L_2L_4L_{10}L_{18}$	
$L_5L_7L_{12}L_{20}$	
$L_8L_{10}L_{12}L_{24}$	$L_8L_{10}L_{12}L_{24}$
$L_{16}L_{18}L_{20}L_{24}$	
$L_8L_9L_{11}L_{13}$	
$L_8L_{10}L_{12}L_{24}$	
$L_9L_{10}L_{14}L_{22}$	
$L_{11}L_{12}L_{15}L_{23}$	
$L_{13}L_{14}L_{15}L_{21}$	$L_{13}L_{14}L_{15}L_{21}$
$L_{21}L_{22}L_{23}L_{24}$	
$L_1L_3L_6L_{13}$	
$L_1L_4L_7L_{16}$	
$L_3L_4L_{14}L_{17}$	
$L_6L_7L_{15}L_{19}$	
$L_{13}L_{14}L_{15}L_{21}$	
$L_{16}L_{17}L_{19}L_{21}$	
$L_8L_9L_{11}L_{13}$	
$L_8L_{10}L_{12}L_{24}$	
$L_9L_{10}L_{14}L_{22}$	
$L_{11}L_{12}L_{15}L_{23}$	
$L_{13}L_{14}L_{15}L_{21}$	
$L_{21}L_{22}L_{23}L_{24}$	

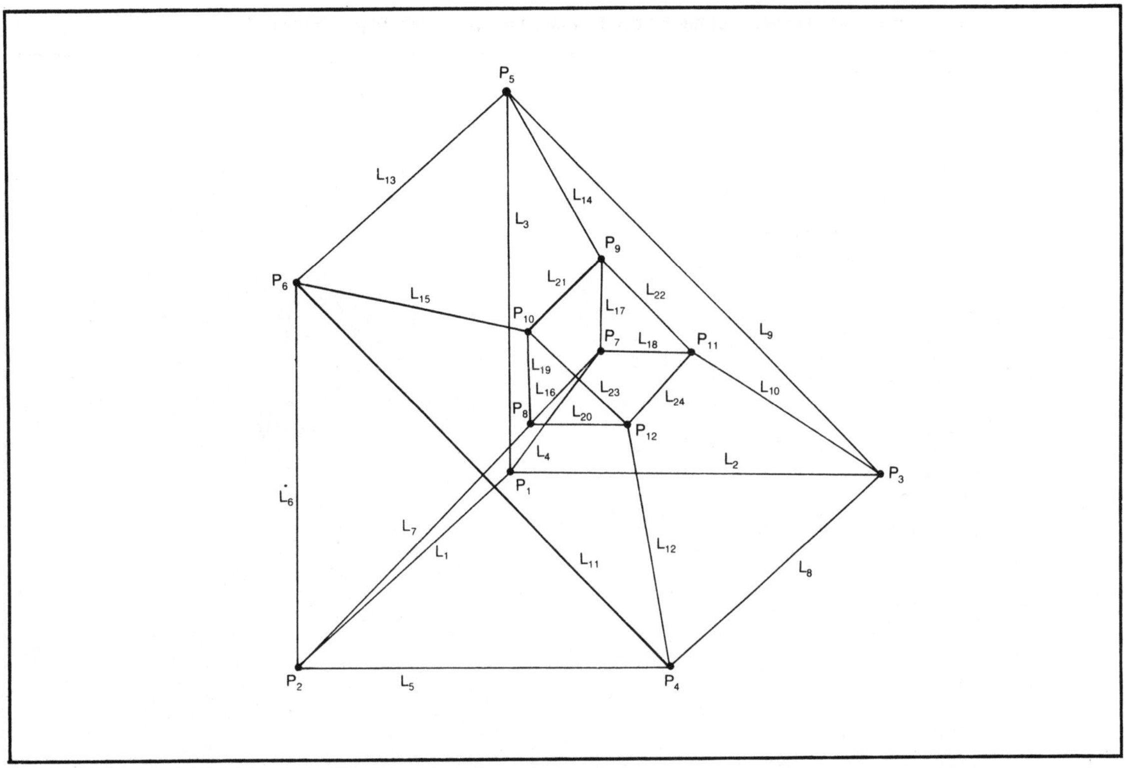

Fig. 18-9. General 4-D view of hyperspace object number 2. See the formal specifications and primary analysis in Table 18-10.

Table 18-9. Final Specifications for Hyperspace Object Number 2. See Fig. 18-9.

FINAL FACE SPECIFICATIONS

$F_1=\overline{L_2L_3L_9}$

$F_2=\overline{L_2L_4L_{10}L_{18}}$

$F_3=\overline{L_3L_4L_{14}L_{17}}$

$F_4=\overline{L_9L_{10}L_{14}L_{22}}$

$F_5=\overline{L_{17}L_{18}L_{22}}$

$F_6=\overline{L_5L_6L_{11}}$

$F_7=\overline{L_5L_7L_{12}L_{20}}$

$F_8=\overline{L_6L_7L_{15}L_{19}}$

$F_9=\overline{L_{11}L_{12}L_{15}L_{23}}$

$F_{10}=\overline{L_{19}L_{20}L_{23}}$

$F_{11}=\overline{L_1L_2L_5L_8}$

$F_{12}=\overline{L_1L_3L_6L_{13}}$

$F_{13}=\overline{L_8L_9L_{11}L_{13}}$

$F_{14}=\overline{L_{16}L_{17}L_{19}L_{21}}$

$F_{15}=\overline{L_{16}L_{18}L_{20}L_{24}}$

$F_{16}=\overline{L_{21}L_{22}L_{23}L_{24}}$

$F_{17}=\overline{L_1L_4L_7L_{16}}$

$F_{18}=\overline{L_8L_{10}L_{12}L_{24}}$

$F_{19}=\overline{L_{13}L_{14}L_{15}L_{21}}$

FINAL SPACE SPECIFICATIONS

$S_1=\overline{F_1F_2F_3F_4F_5}$

$S_2=\overline{F_6F_7F_8F_9F_{10}}$

$S_3=\overline{F_1F_6F_{11}F_{12}F_{13}}$

$S_4=\overline{F_{14}F_{15}F_{16}F_{17}F_{18}}$

$S_5=\overline{F_2F_7F_{11}F_{15}F_{19}F_{20}}$

$S_6=\overline{F_4F_9F_{13}F_{18}F_{20}F_{21}}$

$S_7=\overline{F_3F_8F_{12}F_{14}F_{19}F_{21}}$

$S_8=\overline{F_4F_9F_{13}F_{18}F_{20}F_{21}}$

FINAL HYPERSPACE SPECIFICATION

$H=\overline{S_1S_2S_3S_4S_5S_6S_7S_8}$

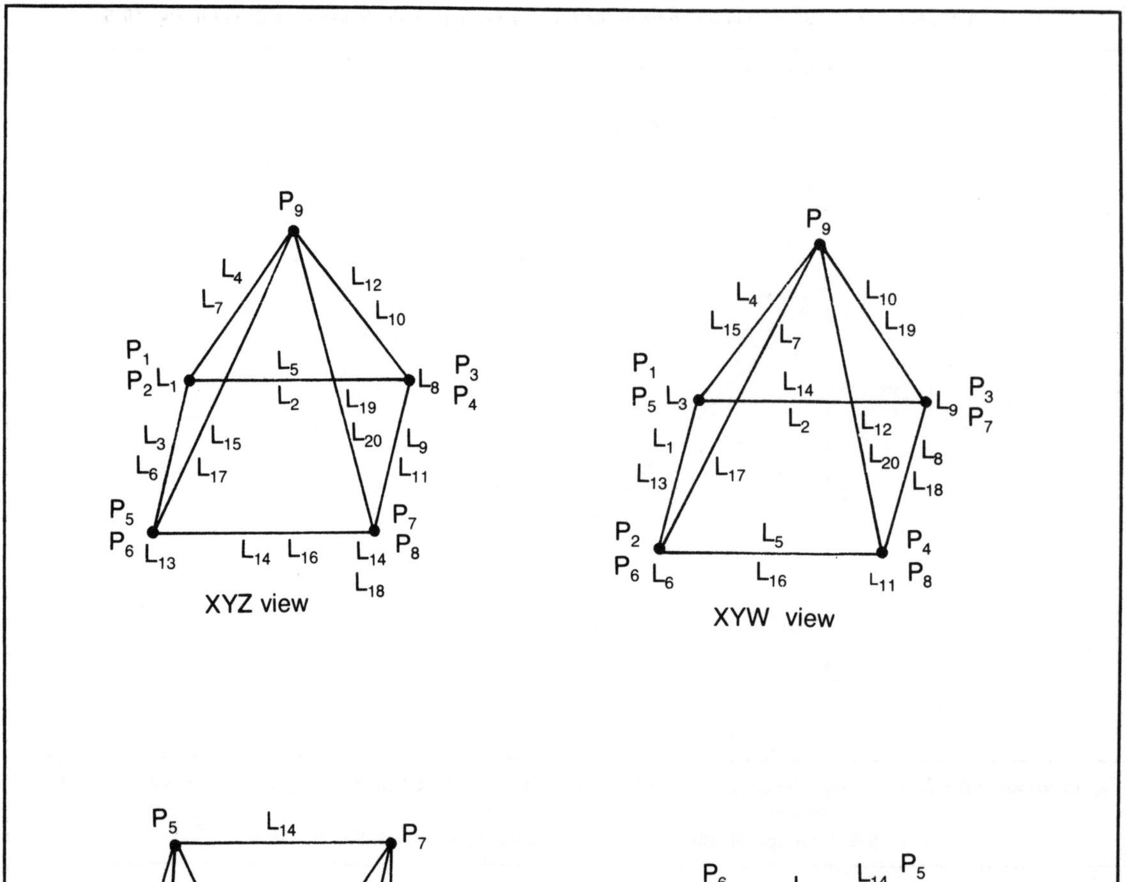

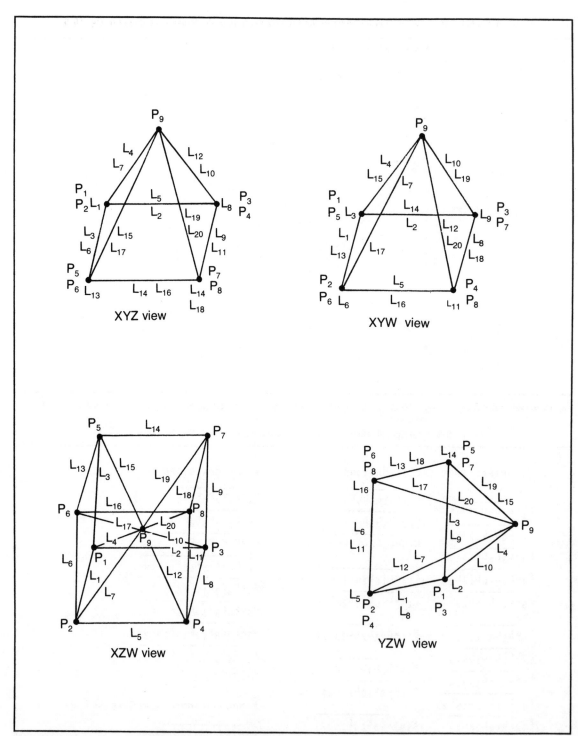

Fig. 18-10. Initial 3-D perspective views for a hyperspace version of a 5-sided pyramid.

Table 18-10. Full Specifications and Summary of the Primary Analysis of the Hyperspace Object in Fig. 18-9.

(A, B, C, and D correspond to α, β, γ, and δ.)

SPECIFICATIONS

$H=\overline{S_1S_2S_3S_4S_5S_6S_7S_8}$

$S_1=\overline{F_1F_2F_3F_4F_5}$

$S_2=\overline{F_6F_7F_8F_9F_{10}}$

$S_3=\overline{F_1F_6F_{11}F_{12}F_{13}}$

$S_4=\overline{F_{14}F_{15}F_{16}F_{17}F_{18}}$

$S_5=\overline{F_2F_7F_{11}F_{15}F_{19}F_{20}}$

$S_6=\overline{F_4F_9F_{13}F_{18}F_{20}F_{21}}$

$S_7=\overline{F_3F_8F_{12}F_{14}F_{19}F_{21}}$

$S_8=\overline{F_4F_9F_{13}F_{18}F_{20}F_{21}}$

$F_1=\overline{L_2L_3L_9}$	$F_2=\overline{L_2L_4L_{10}L_{18}}$
$F_3=\overline{L_3L_4L_{14}L_{17}}$	$F_4=\overline{L_9L_{10}L_{14}L_{22}}$
$F_5=\overline{L_{17}L_{18}L_{22}}$	$F_6=\overline{L_5L_6L_{11}}$
$F_7=\overline{L_5L_7L_{12}L_{20}}$	$F_8=\overline{L_6L_7L_{15}L_{19}}$
$F_9=\overline{L_{11}L_{12}L_{15}L_{23}}$	$F_{10}=\overline{L_{19}L_{20}L_{23}}$
$F_{11}=\overline{L_1L_2L_5L_8}$	$F_{12}=\overline{L_1L_3L_6L_{13}}$
$F_{13}=\overline{L_8L_9L_{11}L_{13}}$	$F_{14}=\overline{L_{16}L_{17}L_{19}L_{21}}$
$F_{15}=\overline{L_{16}L_{18}L_{20}L_{24}}$	$F_{16}=\overline{L_{21}L_{22}L_{23}L_{24}}$
$F_{17}=\overline{L_1L_4L_7L_{16}}$	$F_{18}=\overline{L_8L_{10}L_{12}L_{24}}$
	$F_{19}=\overline{L_{13}L_{14}L_{15}L_{21}}$

$L_1=\overline{P_1P_2}$	$L_2=\overline{P_1P_3}$	$L_3=\overline{P_1P_5}$
$L_4=\overline{P_1P_7}$	$L_5=\overline{P_2P_4}$	$L_6=\overline{P_2P_6}$
$L_7=\overline{P_2P_8}$	$L_8=\overline{P_3P_4}$	$L_9=\overline{P_3P_5}$
$L_{10}=\overline{P_4P_{11}}$	$L_{11}=\overline{P_4P_6}$	$L_{12}=\overline{P_4P_{12}}$
$L_{13}=\overline{P_5P_6}$	$L_{14}=\overline{P_5P_9}$	$L_{15}=\overline{P_6P_{10}}$
$L_{16}=\overline{P_7P_8}$	$L_{17}=\overline{P_7P_9}$	$L_{18}=\overline{P_7P_{11}}$
$L_{19}=\overline{P_8P_{10}}$	$L_{20}=\overline{P_8P_{12}}$	$L_{21}=\overline{P_9P_{10}}$
$L_{22}=\overline{P_9P_{11}}$	$L_{23}=\overline{P_{10}P_{12}}$	$L_{24}=\overline{P_{11}P_{12}}$

Table 18-10. Full Specifications and Summary of the Primary Analysis of the Hyperspace Object in Fig. 18-9. **(Continued from page 287.)**

$P_1=(0,0,0,0)$	$P_2=(0,0,0,1)$	$P_3=(1,0,0,0)$
$P_4=(1,0,0,1)$	$P_5=(0,1,0,0)$	$P_6=(0,1,0,1)$
$P_7=(0,0,1,0)$	$P_8=(0,0,1,1)$	$P_9=(0,1,1,0)$
$P_{10}=(0,1,1,1)$	$P_{11}=(1,0,1,0)$	$P_{12}=(1,0,1,1)$

LENGTHS

$s_1=1$	$s_2=1$	$s_3=1$	$s_4=1$
$s_5=1$	$s_6=1$	$s_7=1$	$s_8=1$
$s_9=\ 2$	$s_{10}=1$	$s_{11}=\ 2$	$s_{12}=1$
$s_{13}=1$	$s_{14}=1$	$s_{15}=1$	$s_{16}=1$
$s_{17}=1$	$s_{18}=1$	$s_{19}=1$	$s_{20}=1$
$s_{21}=1$	$s_{22}=\ 2$	$s_{23}=\ 2$	$s_{24}=1$

DIRECTION COSINES

$\cos A_1=0$	$\cos B_1=0$	$\cos C_1=0$	$\cos D_1=1$
$\cos A_2=1$	$\cos B_2=0$	$\cos C_2=0$	$\cos D_2=0$
$\cos A_3=0$	$\cos B_3=1$	$\cos C_3=0$	$\cos D_3=0$
$\cos A_4=0$	$\cos B_4=0$	$\cos C_4=1$	$\cos D_4=0$
$\cos A_5=1$	$\cos B_5=0$	$\cos C_5=0$	$\cos D_5=0$
$\cos A_6=0$	$\cos B_6=1$	$\cos C_6=0$	$\cos D_6=0$
$\cos A_7=0$	$\cos B_7=0$	$\cos C_7=1$	$\cos D_7=0$
$\cos A_8=0$	$\cos B_8=0$	$\cos C_8=0$	$\cos D_8=1$
$\cos A_9=-\sqrt{2}/2$	$\cos B_9=\sqrt{2}/2$	$\cos C_9=0$	$\cos D_9=0$
$\cos A_{10}=0$	$\cos B_{10}=0$	$\cos C_{10}=1$	$\cos D_{10}=0$
$\cos A_{11}=-\sqrt{2}/2$	$\cos B_{11}=\sqrt{2}/2$	$\cos C_{11}=0$	$\cos D_{11}=0$
$\cos A_{12}=0$	$\cos B_{12}=0$	$\cos C_{12}=1$	$\cos D_{12}=0$
$\cos A_{13}=0$	$\cos B_{13}=0$	$\cos C_{13}=0$	$\cos D_{13}=1$
$\cos A_{14}=0$	$\cos B_{14}=0$	$\cos C_{14}=1$	$\cos D_{14}=0$
$\cos A_{15}=0$	$\cos B_{15}=0$	$\cos C_{15}=1$	$\cos D_{15}=0$

DIRECTION COSINES

$\cos A_{16}=0$	$\cos B_{16}=0$	$\cos C_{16}=0$	$\cos D_{16}=1$
$\cos A_{17}=0$	$\cos B_{17}=1$	$\cos C_{17}=0$	$\cos D_{17}=0$
$\cos A_{18}=1$	$\cos B_{18}=0$	$\cos C_{18}=0$	$\cos D_{18}=0$
$\cos A_{19}=0$	$\cos B_{19}=1$	$\cos C_{19}=0$	$\cos D_{19}=0$
$\cos A_{20}=1$	$\cos B_{20}=0$	$\cos C_{20}=0$	$\cos D_{20}=0$
$\cos A_{21}=0$	$\cos B_{21}=0$	$\cos C_{21}=0$	$\cos D_{21}=1$
$\cos A_{22}=\sqrt{2}/2$	$\cos B_{22}=-\sqrt{2}/2$	$\cos C_{22}=0$	$\cos D_{22}=0$
$\cos A_{23}=\sqrt{2}/2$	$\cos B_{23}=-\sqrt{2}/2$	$\cos C_{23}=0$	$\cos D_{23}=0$
$\cos A_{24}=0$	$\cos B_{24}=0$	$\cos C_{24}=0$	$\cos D_{24}=1$

ANGLES

$\phi_{1,2}=90^\circ$	$\phi_{1,3}=90^\circ$	$\phi_{1,4}=90^\circ$	$\phi_{1,5}=90^\circ$
$\phi_{1,6}=90^\circ$	$\phi_{1,7}=90^\circ$	$\phi_{2,3}=90^\circ$	$\phi_{2,4}=90^\circ$
$\phi_{2,8}=90^\circ$	$\phi_{2,9}=45^\circ$	$\phi_{2,10}=90^\circ$	$\phi_{3,4}=90^\circ$
$\phi_{3,9}=45^\circ$	$\phi_{3,13}=90^\circ$	$\phi_{3,14}=90^\circ$	$\phi_{4,16}=90^\circ$
$\phi_{4,17}=90^\circ$	$\phi_{4,18}=90^\circ$	$\phi_{5,6}=90^\circ$	$\phi_{5,7}=90^\circ$
$\phi_{5,8}=90^\circ$	$\phi_{5,11}=90^\circ$	$\phi_{5,12}=90^\circ$	$\phi_{6,7}=90^\circ$
$\phi_{6,11}=45^\circ$	$\phi_{6,13}=90^\circ$	$\phi_{6,15}=90^\circ$	$\phi_{7,16}=90^\circ$
$\phi_{7,19}=90^\circ$	$\phi_{7,20}=90^\circ$	$\phi_{8,9}=90^\circ$	$\phi_{8,10}=90^\circ$
$\phi_{8,11}=90^\circ$	$\phi_{8,12}=90^\circ$	$\phi_{9,10}=90^\circ$	$\phi_{9,13}=90^\circ$
$\phi_{9,14}=90^\circ$	$\phi_{10,18}=90^\circ$	$\phi_{10,22}=90^\circ$	$\phi_{10,24}=90^\circ$
$\phi_{11,12}=90^\circ$	$\phi_{11,13}=90^\circ$	$\phi_{11,15}=90^\circ$	$\phi_{12,20}=90^\circ$
$\phi_{12,23}=90^\circ$	$\phi_{12,24}=90^\circ$	$\phi_{13,14}=90^\circ$	$\phi_{13,15}=90^\circ$
$\phi_{14,17}=90^\circ$	$\phi_{14,21}=90^\circ$	$\phi_{14,22}=90^\circ$	$\phi_{15,19}=90^\circ$
$\phi_{15,21}=90^\circ$	$\phi_{15,23}=90^\circ$	$\phi_{16,17}=90^\circ$	$\phi_{16,18}=90^\circ$
$\phi_{16,19}=90^\circ$	$\phi_{16,20}=90^\circ$	$\phi_{17,18}=90^\circ$	$\phi_{17,21}=90^\circ$
$\phi_{17,22}=45^\circ$	$\phi_{18,22}=45^\circ$	$\phi_{18,24}=90^\circ$	$\phi_{19,20}=90^\circ$
$\phi_{19,21}=90^\circ$	$\phi_{19,23}=45^\circ$	$\phi_{20,23}=45^\circ$	$\phi_{20,24}=90^\circ$
$\phi_{21,22}=90^\circ$	$\phi_{21,23}=90^\circ$	$\phi_{22,24}=90^\circ$	$\phi_{23,24}=^\circ$

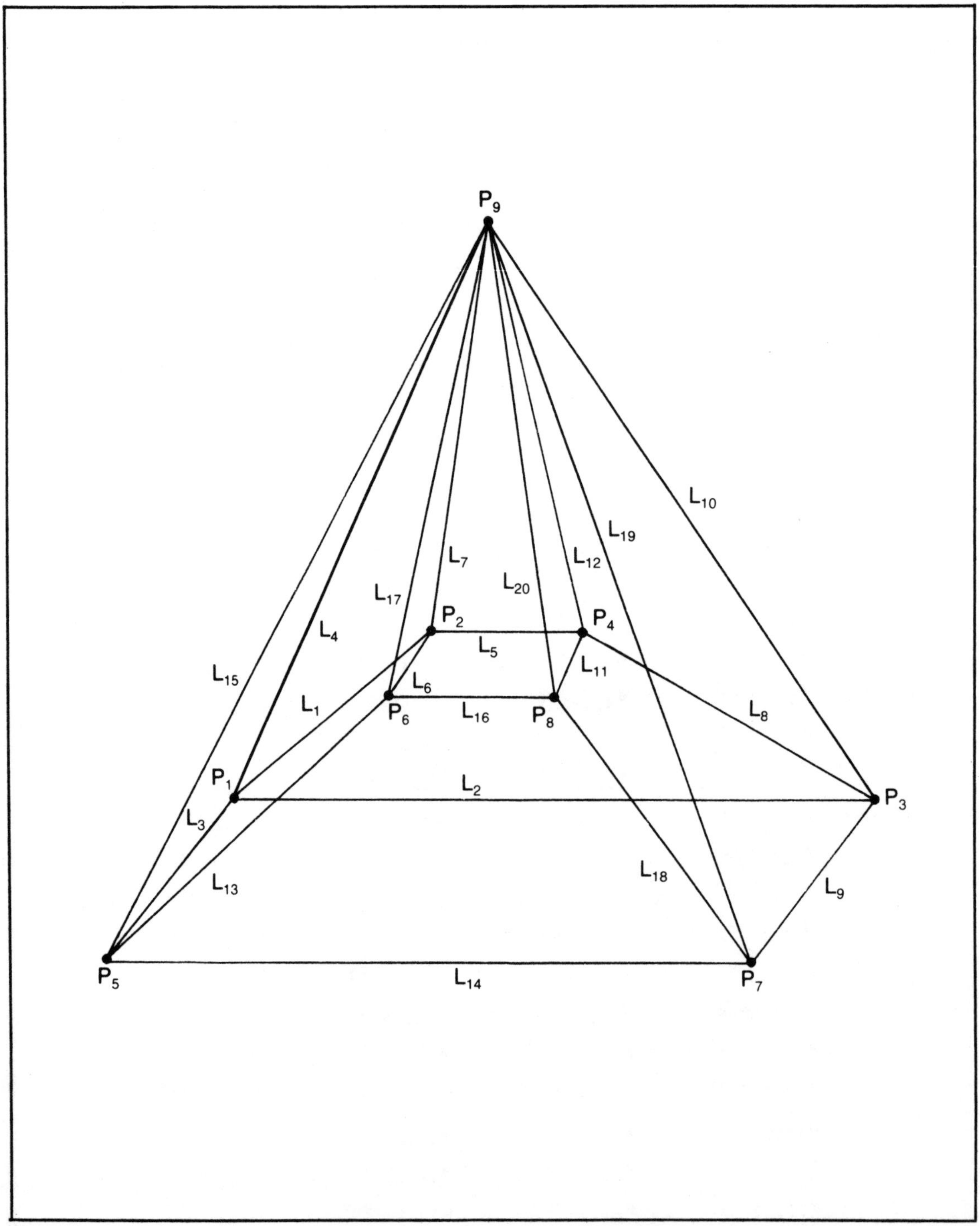

Fig. 18-11. General 4-D view of the hyperpyramid. See the formal specifications and a summary of the primary analysis in Table 18-11.

Table 18-11. Specifications and a Primary Analysis of the Hyperpyramid Shown in Fig. 18-11.

(A, B, C, and D correspond to α, β, γ, and δ.)

SPECIFICATIONS

$H=\overline{S_1S_2S_3S_4S_5S_6S_7}$

$S_1=\overline{F_1F_2F_4F_7F_{10}F_{14}}$

$S_2=\overline{F_1F_3F_5F_8F_{11}}$

$S_3=\overline{F_2F_3F_6F_9F_{15}}$

$S_4=\overline{F_4F_5F_6F_{12}F_{16}}$

$S_5=\overline{F_7F_8F_9F_{13}F_{17}}$

$S_6=\overline{F_{10}F_{11}F_{12}F_{13}F_{18}}$

$S_7=\overline{F_{14}F_{15}F_{16}F_{17}F_{18}}$

$F_1=\overline{L_1L_2L_5L_8}$	$F_2=\overline{L_1L_3L_6L_{13}}$
$F_3=\overline{L_1L_4L_7}$	$F_4=\overline{L_2L_3L_9L_{14}}$
$F_5=\overline{L_2L_4L_{10}}$	$F_6=\overline{L_3L_4L_{15}}$
$F_7=\overline{L_5L_6L_{11}L_{16}}$	$F_8=\overline{L_5L_7L_{12}}$
$F_9=\overline{L_6L_7L_{17}}$	$F_{10}=\overline{L_8L_9L_{11}L_{18}}$
$F_{11}=\overline{L_8L_{10}L_{12}}$	$F_{12}=\overline{L_9L_{10}L_{19}}$
$F_{13}=\overline{L_{11}L_{12}L_{20}}$	$F_{14}=\overline{L_{13}L_{14}L_{16}L_{18}}$
$F_{15}=\overline{L_{13}L_{15}L_{17}}$	$F_{16}=\overline{L_{14}L_{15}L_{19}}$
$F_{17}=\overline{L_{16}L_{17}L_{20}}$	$F_{18}=\overline{L_{18}L_{19}L_{20}}$

SPECIFICATIONS (Cont'd)

$L_1=\overline{P_1P_2}$	$P_1=(-1,0,-1,-1)$
$L_2=\overline{P_1P_3}$	$P_2=(-1,0,-1,1)$
$L_3=\overline{P_1P_5}$	$P_3=(1,0,-1,-1)$
$L_4=\overline{P_1P_9}$	$P_4=(1,0,-1,1)$
$L_5=\overline{P_2P_4}$	$P_5=(-1,0,1,-1)$
$L_6=\overline{P_2P_6}$	$P_6=(-1,0,1,1)$
$L_7=\overline{P_2P_9}$	$P_7=(1,0,1,-1)$
$L_8=\overline{P_3P_4}$	$P_8=(1,0,1,1)$
$L_9=\overline{P_3P_7}$	
$L_{10}=\overline{P_3P_9}$	
$L_{11}=\overline{P_4P_8}$	
$L_{12}=\overline{P_4P_9}$	
$L_{13}=\overline{P_5P_6}$	
$L_{14}=\overline{P_5P_7}$	
$L_{15}=\overline{P_5P_9}$	
$L_{16}=\overline{P_6P_8}$	
$L_{17}=\overline{P_6P_9}$	
$L_{18}=\overline{P_7P_8}$	
$L_{19}=\overline{P_7P_9}$	
$L_{20}=\overline{P_8P_9}$	

LENGTHS

$s_1=2$	$s_2=2$	$s_3=2$	$s_4=2$	$s_5=2$
$s_6=2$	$s_7=2$	$s_8=2$	$s_9=2$	$s_{10}=2$
$s_{11}=2$	$s_{12}=2$	$s_3=2$	$s_{14}=2$	$s_{15}=2$
$s_{16}=2$	$s_{17}=2$	$s_{18}=2$	$s_{19}=2$	$s_{20}=2$

DIRECTION COSINES

$\cos A_1=0$	$\cos B_1=0$	$\cos C_1=0$	$\cos D_1=1$
$\cos A_2=1$	$\cos B_2=0$	$\cos C_2=0$	$\cos D_2=0$
$\cos A_3=0$	$\cos B_3=0$	$\cos C_3=1$	$\cos D_3=0$
$\cos A_4=1/2$	$\cos B_4=1/2$	$\cos C_4=1/2$	$\cos D_4=1/2$
$\cos A_5=1$	$\cos B_5=0$	$\cos C_5=0$	$\cos D_5=0$

Table 18-11. Specifications and a Primary Analysis of the Hyperpyramid Shown in Fig. 18-11. **(Continued from page 291.)**

DIRECTION COSINES

$\cos A_6=0$	$\cos B_6=0$	$\cos C_6=1$	$\cos D_6=0$
$\cos A_7=1/2$	$\cos B_7=1/2$	$\cos C_7=1/2$	$\cos D_7=-1/2$
$\cos A_8=0$	$\cos B_8=0$	$\cos C_8=0$	$\cos D_8=1$
$\cos A_9=0$	$\cos B_9=0$	$\cos C_9=1$	$\cos D_9=0$
$\cos A_{10}=-1/2$	$\cos B_{10}=1/2$	$\cos C_{10}=1/2$	$\cos D_{10}=1/2$
$\cos A_{11}=0$	$\cos B_{11}=0$	$\cos C_{11}=1$	$\cos D_{11}=0$
$\cos A_{12}=-1/2$	$\cos B_{12}=1/2$	$\cos C_{12}=1/2$	$\cos D_{12}=-1/2$
$\cos A_{13}=0$	$\cos B_{13}=0$	$\cos C_{13}=0$	$\cos D_{13}=1$
$\cos A_{14}=1$	$\cos B_{14}=0$	$\cos C_{14}=0$	$\cos D_{14}=0$
$\cos A_{15}=1/2$	$\cos B_{15}=1/2$	$\cos C_{15}=-1/2$	$\cos D_{15}=1/2$
$\cos A_{16}=1$	$\cos B_{16}=0$	$\cos C_{16}=0$	$\cos D_{16}=0$
$\cos A_{17}=1/2$	$\cos B_{17}=1/2$	$\cos C_{17}=-1/2$	$\cos D_{17}=-1/2$
$\cos A_{18}=0$	$\cos B_{18}=0$	$\cos C_{18}=0$	$\cos D_{18}=1$
$\cos A_{19}=-1/2$	$\cos B_{19}=1/2$	$\cos C_{19}=-1/2$	$\cos D_{19}=1/2$
$\cos A_{20}=-1/2$	$\cos B_{20}=1/2$	$\cos C_{20}=-1/2$	$\cos D_{20}=-1/2$

ANGLES

$\phi_{1,2}=90^\circ$	$\phi_{1,3}=90^\circ$	$\phi_{1,4}=60^\circ$	$\phi_{1,5}=90^\circ$
$\phi_{1,6}=90^\circ$	$\phi_{1,7}=60^\circ$	$\phi_{2,3}=90^\circ$	$\phi_{2,4}=60^\circ$
$\phi_{2,8}=90^\circ$	$\phi_{2,9}=90^\circ$	$\phi_{2,10}=60^\circ$	$\phi_{3,4}=60^\circ$
$\phi_{3,13}=90^\circ$	$\phi_{3,14}=90^\circ$	$\phi_{3,15}=60^\circ$	$\phi_{4,7}=60^\circ$
$\phi_{4,10}=60^\circ$	$\phi_{4,15}=60^\circ$	$\phi_{5,6}=90^\circ$	$\phi_{5,7}=60^\circ$
$\phi_{5,8}=90^\circ$	$\phi_{5,11}=90^\circ$	$\phi_{5,12}=60^\circ$	$\phi_{6,7}=60^\circ$
$\phi_{6,13}=90^\circ$	$\phi_{6,16}=90^\circ$	$\phi_{6,17}=60^\circ$	$\phi_{7,12}=60^\circ$
$\phi_{7,17}=60^\circ$	$\phi_{8,9}=90^\circ$	$\phi_{8,10}=60^\circ$	$\phi_{8,11}=90^\circ$
$\phi_{8,12}=60^\circ$	$\phi_{9,10}=60^\circ$	$\phi_{9,14}=90^\circ$	$\phi_{9,18}=90^\circ$
$\phi_{9,19}=60^\circ$	$\phi_{10,12}=60^\circ$	$\phi_{10,19}=60^\circ$	$\phi_{11,12}=60^\circ$
$\phi_{11,16}=90^\circ$	$\phi_{11,18}=90^\circ$	$\phi_{11,20}=60^\circ$	$\phi_{12,20}=60^\circ$
$\phi_{13,14}=90^\circ$	$\phi_{13,15}=60^\circ$	$\phi_{13,16}=90^\circ$	$\phi_{13,17}=60^\circ$
$\phi_{14,15}=60^\circ$	$\phi_{14,18}=90^\circ$	$\phi_{14,19}=60^\circ$	$\phi_{15,17}=60^\circ$
$\phi_{15,19}=60^\circ$	$\phi_{16,17}=60^\circ$	$\phi_{16,18}=90^\circ$	$\phi_{16,20}=60^\circ$
$\phi_{17,20}=60^\circ$	$\phi_{18,19}=60^\circ$	$\phi_{18,20}=60^\circ$	$\phi_{19,20}=60^\circ$

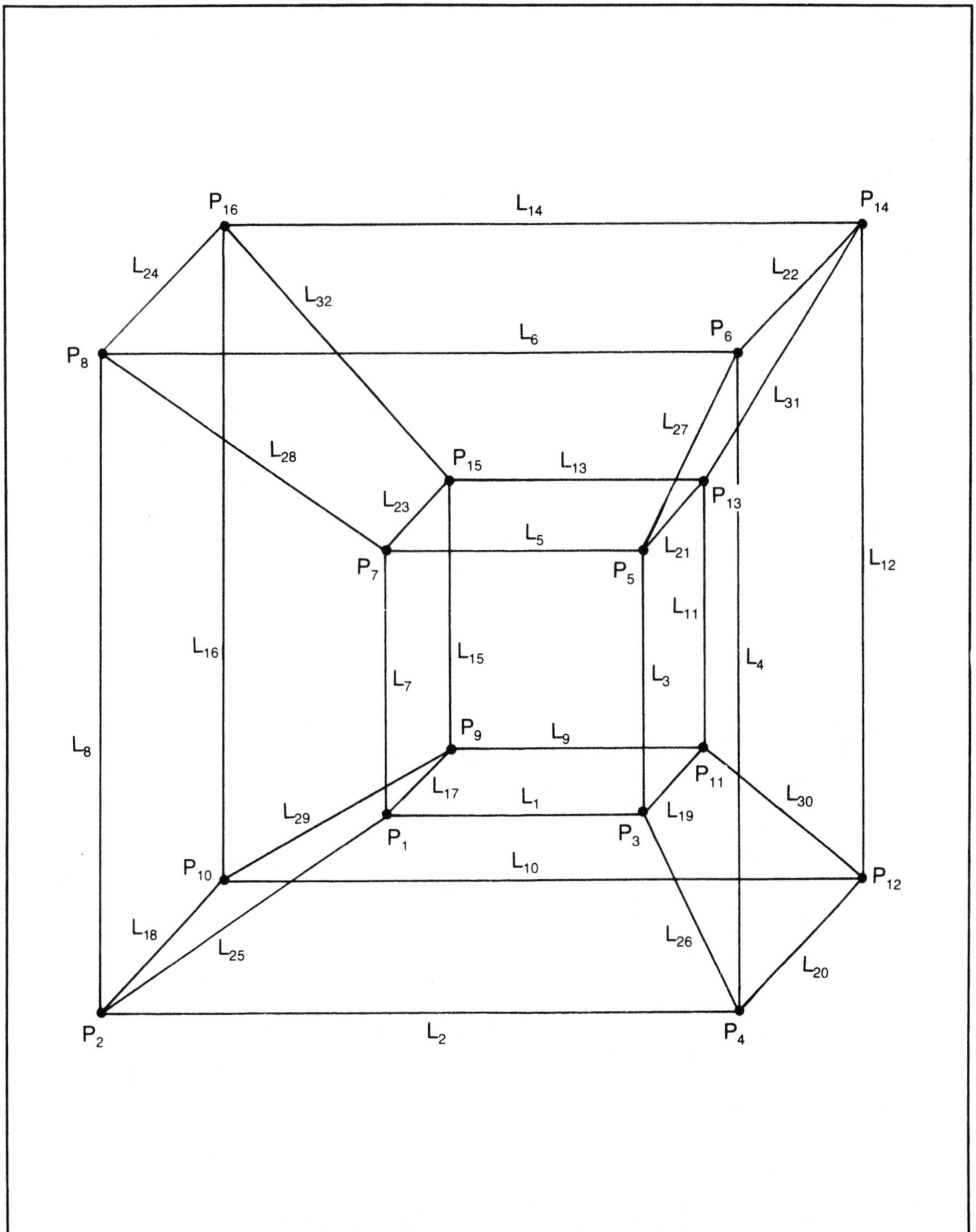

Fig. 18-12. General 4-D view of a hypercube. See the specifications and summary of its primary analysis in Table 18-12.

Table 18-12. Specifications and a Primary Analysis of the Hypercube Illustrated in Fig. 18-12.

SPECIFICATIONS

$H=\overline{S_1S_2S_3S_4S_5S_6S_7S_8}$

$S_1=\overline{F_1F_2F_3F_4F_5F_6}$

$S_2=\overline{F_7F_6F_9F_{10}F_{11}F_{12}}$

$S_3=\overline{F_1F_7F_{13}F_{14}F_{15}F_{16}}$

$S_4=\overline{F_3F_9F_{17}F_{18}F_{19}F_{20}}$

$S_5=\overline{F_5F_{11}F_{15}F_{19}F_{21}F_{22}}$

$S_6=\overline{F_6F_{12}F_{16}F_{20}F_{23}F_{24}}$

$S_7=\overline{F_4F_{10}F_{14}F_{18}F_{22}F_{24}}$

$S_8=\overline{F_2F_8F_{13}F_{17}F_{21}F_{23}}$

$F_1=\overline{L_1L_3L_5L_7}$	$F_2=\overline{L_3L_{11}L_{19}L_{21}}$
$F_3=\overline{L_9L_{11}L_{13}L_{15}}$	$F_4=\overline{L_7L_{15}L_{17}L_{23}}$
$F_5=\overline{L_1L_9L_{17}L_{19}}$	$F_6=\overline{L_5L_{13}L_{21}L_{23}}$
$F_7=\overline{L_2L_4L_6L_8}$	$F_8=\overline{L_4L_{12}L_{20}L_{22}}$
$F_9=\overline{L_{10}L_{12}L_{14}L_{16}}$	$F_{10}=\overline{L_8L_{16}L_{18}L_{24}}$
$F_{11}=\overline{L_2L_{10}L_{18}L_{20}}$	$F_{12}=\overline{L_6L_{14}L_{22}L_{24}}$
$F_{13}=\overline{L_3L_4L_{26}L_{27}}$	$F_{14}=\overline{L_7L_8L_{25}L_{28}}$
$F_{15}=\overline{L_1L_2L_{25}L_{26}}$	$F_{16}=\overline{L_5L_6L_{27}L_{28}}$
$F_{17}=\overline{L_{11}L_{12}L_{30}L_{31}}$	$F_{18}=\overline{L_{15}L_{16}L_{29}L_{32}}$
$F_{19}=\overline{L_9L_{10}L_{26}L_{30}}$	$F_{20}=\overline{L_{13}L_{14}L_{31}L_{32}}$
$F_{21}=\overline{L_{19}L_{20}L_{26}L_{30}}$	$F_{22}=\overline{L_{17}L_{18}L_{25}L_{29}}$
$F_{23}=\overline{L_{21}L_{22}L_{27}L_{31}}$	$F_{24}=\overline{L_{23}L_{24}L_{28}L_{32}}$

SPECIFICATIONS (Cont'd)

$L_1=\overline{P_1P_3}$	$L_2=\overline{P_2P_4}$	$L_3=\overline{P_3P_5}$
$L_4=\overline{P_4P_6}$	$L_5=\overline{P_5P_7}$	$L_6=\overline{P_6P_8}$
$L_7=\overline{P_1P_7}$	$L_8=\overline{P_2P_8}$	$L_9=\overline{P_9P_{11}}$
$L_0=\overline{P_{10}P_{12}}$	$L_{11}=\overline{P_{11}P_{13}}$	$L_{12}=\overline{P_{12}P_{14}}$
$L_{13}=\overline{P_{13}P_{15}}$	$L_{14}=\overline{P_{14}P_{16}}$	$L_{15}=\overline{P_9P_{15}}$
$L_{16}=\overline{P_{10}P_{16}}$	$L_{17}=\overline{P_1P_9}$	$L_{18}=\overline{P_2P_{10}}$
$L_{19}=\overline{P_3P_{11}}$	$L_{20}=\overline{P_4P_{12}}$	$L_{21}=\overline{P_5P_{13}}$
$L_{22}=\overline{P_6P_{14}}$	$L_{23}=\overline{P_7P_{15}}$	$L_{24}=\overline{P_8P_{16}}$
$L_{25}=\overline{P_1P_2}$	$L_{26}=\overline{P_3P_4}$	$L_{27}=\overline{P_5P_6}$
$L_{28}=\overline{P_7P_8}$	$L_{29}=\overline{P_9P_{10}}$	$L_{30}=\overline{P_{11}P_{12}}$
$L_{31}=\overline{P_{13}P_{14}}$	$L_{32}=\overline{P_{15}P_{16}}$	

$P_1=(-1,-1,1,-1)$	$P_2=(-1,-1,1,1)$
$P_3=(1,-1,1,-1)$	$P_4=(1,-1,1,1)$
$P_5=(1,1,1,-1)$	$P_6=(1,1,1,1)$
$P_7=(-1,1,1,-1)$	$P_8=(-1,1,1,1)$
$P_9=(-1,-1,-1,-1)$	$P_{10}=(-1,-1,-1,1)$
$P_{11}=(1,-1,-1,-1)$	$P_{12}=(1,-1,-1,1)$
$P_{13}=(1,1,-1,-1)$	$P_{14}=(1,1,-1,1)$
$P_{15}=(-1,1,-1,-1)$	$P_{16}=(-1,1,-1,1)$

monly cited object in serious hyperspace literature, perhaps because it is the simplest to generate. It is hardly the simplest possible object, though.

From the specifications alone, you can see that the object is composed of 16 points, 32 lines, 24 faces and 8 space objects. Confirm the specifications by constructing such an object on your own, then complete the work with a full analysis of it.

Translations in 4-D Space

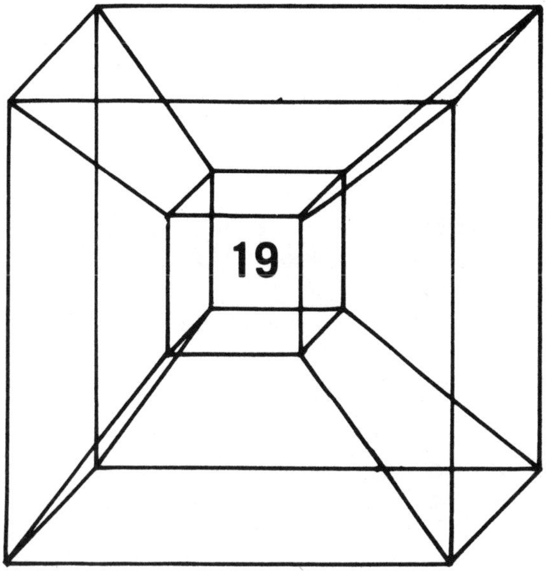

Translating a point that is specified in 4-D space is a simple extension of the procedure for doing the same sort of task in 2-D and 3-D spaces. In this case there are four translation terms—one for each of the 4-D coordinate axes. And summing each of those translation terms with their corresponding component of the coordinate of a point carries out the translation.

The 4-D translations have the same general characteristics as those you have already explored in lower dimensions. For instance, the position of a point is not preserved under a translation, but the lengths of lines, direction cosines and angle between lines are, indeed, preserved.

Furthermore, there is a handy inverse form of the geometric translation equation of four dimensions, and you will find there is also a useful form that allows you to solve for the translation terms that are necessary for moving an object to any desired place in 4-D space. There are some relativistic versions of the translation equation described in this chapter, too, but their glamour is somewhat diminished by the exciting experiments involving 3-D views of objects being translated in 4-D space.

19-1 THE BASIC TRANSLATION EQUATION FOR 4-D SPACE

The fundamental equation for geometric translations in 4-D space is shown here as Equation 19-1. You can see that it is a simple extension of the translation equation for lower spaces. It enables you to translate the components of a reference point to some other point in 4-D space by summing the individual reference components with its respective translation terms. It's that easy (even if it is often terribly difficult to picture the whole thing in your mind).

$$\begin{aligned} x_n' &= x_n + T_x \\ y_n' &= y_n + T_y \\ z_n' &= z_n + T_z \\ w_n' &= w_n + T_w \end{aligned} \qquad \textbf{Equation 19-1}$$

where:

x_n', y_n', z_n', w_n' = the coordinate of the translated version of point *n* relative to the frame of reference

x_n, y_n, z_n, w_n = the coordinate of reference point *n* relative to the frame of reference

T_x = the X-axis translation term

T_y = the Y-axis translation term
T_z = the Z-axis translation term
T_w = the W-axis translation term

Example 19-1 demonstrates the procedure for translating a point within a 4-dimensional coordinate system.

Example 19-1

Given P =(1,2,3,4) in hyperspace, subject it to this 4-D translation:

$$T_x = 4,\ T_y = -3,\ T_z = 3,\ T_w = -4$$

After carrying out the translation, plot and specify the points as viewed in hyperspace, XYZ space, XYW space, XZW space, and YZW space.

1. Applying Equation 19-1 to do the translation:

$$x' = x + T_x \qquad y' = y + T_y$$
$$x' = 1 + 4 \qquad y' = 2 - 3$$
$$x' = 5 \qquad y' = -1$$

$$z' = z + T_z \qquad w' = w + T_w$$
$$z' = 3 + 3 \qquad w' = 4 - 4$$
$$z' = 6 \qquad w' = 0$$

The translation thus carries P=(1,2,3,4) to P′ =(5,−1,6,0). See the plot of that primary analysis in Fig. 19-1A.

2. The plotted points for the XYZ-space secondary analysis are P = (1,2,3) for the reference version and P′ = 5,−1,6) for the translated version. See a precision XYZ-space plot in Fig. 19-1B.

3. The plotted points for the XYW-space secondary analysis are P = (1,2,4) for the reference version and P′ = (5,−1,0) for the translated version. See a precision XYW-space plot in Fig. 19-1C.

4. The plotted points for the XZW-space secondary analysis are P = (1,3,4) for the reference version and P′ = (5,6,0) for the translated version. See a precision XZW-space plot in Fig. 19-1D.

5. The plotted points for the YZW-space secondary analysis are P = (2,3,4) for the reference version and P′ = (−1,6,0) for the translated version. See a precision YZW-space plot in Fig. 19-1E.

After specifying a particular point in hyperspace and setting up a specific 4-D translation, Step 1 in the example carries out the arithmetic that is involved in the translation. The result is a transformation of hyperspace point P to hyperspace point P′.

Human beings have difficulty visualizing hyperspace or, in this case, the position of a point in hyperspace. So the next phase of the experiment is to see how the point appears as mapped to some 3-D spaces; specifically our XYX, XYW, XZW and YZW spaces. That is the purpose of the secondary analysis conducted in Steps 2 through 5.

The general procedure for conducting a secondary analysis of a point in hyperspace is to specify only the meaningful components in each case. For the XYZ mapping, the only relevant components of the true coordinates are the X-, Y- and Z-axis components; the W-axis component is virtually meaningless from that space. So Step 2 cites just those three components for the reference and translated versions of the point.

Step 3 deals with the XYW spatial view, so it cites only the components of those axes. Follow that same idea through Steps 4 and 5 for their respective spaces.

By way of an exercise, make up some coordinates for points immersed in hyperspace and apply some 4-D translation terms of your own. Conduct both the primary and secondary analyses, using Example 19-1 as a guide.

19-2 TRANSLATING LINES IN 4-D SPACE

You know by now that you can define a line in 4-D space in terms of the coordinates of its two endpoints. You have just seen the equation for translating points in 4 dimensions, so it figures that you should also be able to translate lines in 4-D space. Simply translate the endpoints individually, then draw the line between them in their new position. Indeed, the matter of translating lines in 4-D space follows quite naturally from the procedures for translating lines in any lower-dimensioned space.

Applying Equation 19-1 to lines in 4-D space is a simple and straightforward procedure; and if you are reasonably careful about it, there is little chance of making any serious errors. On the other hand, doing an analysis of the reference and translated versions of the lines and making 3-D perspective views of both can be a time-consuming process. But you can save yourself some time, paper and pencil lead by keeping in mind these general principles for the translation of lines:

☐ The length of a line is an invariant under translations of any number of dimensions.

☐ The direction cosines of a line are invariant under translations of any dimension.

☐ The angle between two lines is an invariant under translations of any dimension.

As far as the translated version of a line is concerned, recalling those principles can save you the trouble of having to do a complete analysis from scratch. Of course it is still necessary to do the complete analysis of the original reference line; and you will find that it is necessary to designate the new endpoint coordinates for

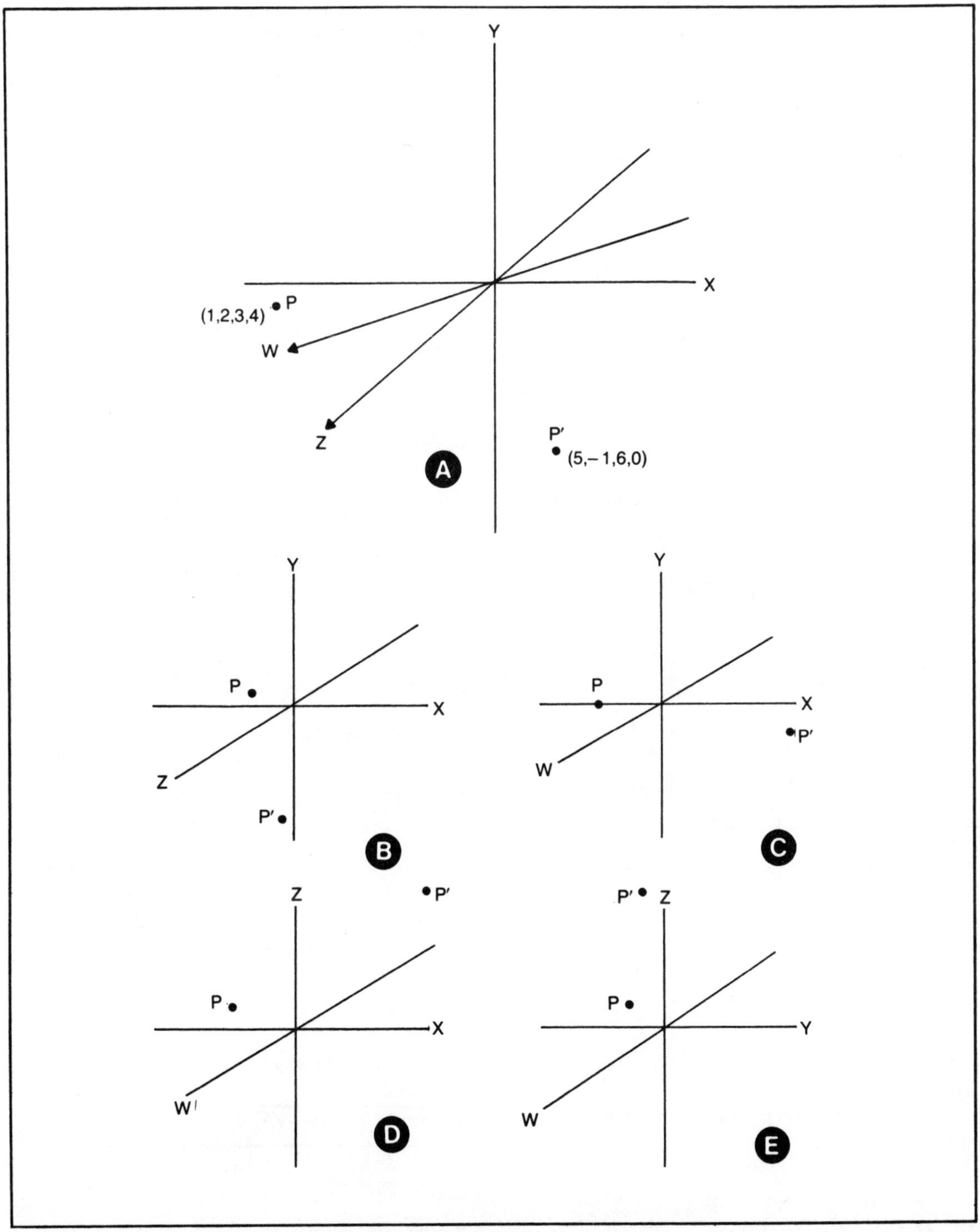

Fig. 19-1. Translating a point P to P′ in hyperspace as suggested in Example 19-1. (A) A general hyperspace view. (B) XYZ perspective view. (C) XYW perspective view. (D) XZW perspective view. (E) YZW perspective view.

the translated line. But there is no need to do the work all over again as far as the length, direction cosines and angles between lines in the translated version are concerned. They will be identical to their counterparts in the reference line.

The following example specifies a reference line that is imbedded in hyperspace, cites a particular 4-D translation, and asks for a complete primary and secondary analysis of the situation.

Example 19-2

Here are the specifications for a line imbedded in hyperspace:

$$L = \overline{P_1P_2} \qquad P_1 = (-1,-1,-1,-1)$$
$$P_2 = (1,1,1,1)$$

Subject the line to the following 4-D translation and conduct primary and secondary analyses of the entire situation.

$$T_x = 2, \quad T_y = -2, \quad T_z = 1, \quad T_w = -1$$

1. Gathering the components of the coordinates:

$$x_1 = -1 \quad y_1 = -1 \quad z_1 = -1 \quad w_1 = -1$$
$$x_2 = 1 \quad y_2 = 1 \quad z_2 = 1 \quad w_2 = 1$$

2. Translating endpoint P_1:

$$x_1' = x_1 + T_x \quad y_1' = y_1 + T_y$$
$$x_1' = -1 + 2 \quad y_1' = -1 + (-2)$$
$$x_1' = 1 \quad y_1' = -3$$

$$z_1' = z_1 + T_z \quad w_1' = w_1 + T_w$$
$$z_1' = -1 + 1 \quad w_1' = -1 + (-1)$$
$$z_1' = 0 \quad w_1' = -2$$

So point $P_1 = (-1,-1,-1,-1)$ is translated to $P_1' = (1,-3,0,-2)$.

3. Translating endpoint P_2:

$$x_2' = x_2 + T_x \quad y_2' = y_2 + T_y$$
$$x_2' = 1 + 2 \quad y_2' = 1 + (-2)$$
$$x_2' = 3 \quad y_2' = -1$$

$$z_2' = z_2 + T_z \quad w_2' = w_2 + T_w$$
$$z_2' = 1 + 1 \quad w_2' = 1 + (-1)$$
$$z_2' = 2 \quad w_2' = 0$$

So point $P_2 = (1,1,1,1)$ is translated to $P_2' = (3,-1,2,0)$.

4. Conducting the primary, or true hyperspace, analysis of the reference line:

(A) Plotted Points

$$P_1 = (x_1,y_1,z_1,w_1) \qquad P_2 = (x_2,y_2,z_2,w_2)$$

(B) Length

$$s = \sqrt{(x_2-x_1)^2 + (y_2-y_1)^2 + (z_2-z_1)^2 + (w_2-w_1)^2}$$
$$s = \sqrt{[1-(-1)]^2 + [1-(-1)]^2 + [1-(-1)]^2 + [1-(-1)]^2}$$
$$s = \sqrt{2^2 + 2^2 + 2^2 + 2^2}$$
$$s = \sqrt{16}$$
$$s = 4$$

The reference version of the line has a hyperspace length of 4 units.

(C) Direction cosines:

$$\cos\alpha = (x_2-x_1)/s \qquad \cos\beta = (y_2-y_1)/s$$
$$\cos\alpha = [1-(-1)]/4 \qquad \cos\beta = [1-(-1)]/4$$
$$\cos\alpha = 1/2 \qquad \cos\beta = 1/2$$

$$\cos\gamma = (z_2-z_1)/s \qquad \cos\delta = (w_2-w_1)/s$$
$$\cos\gamma = [1-(-1)]/4 \qquad \cos\delta = [1-(-1)]/4$$
$$\cos\gamma = 1/2 \qquad \cos\delta = 1/2$$

5. Conducting the primary, or true hyperspace, analysis of the translated version of the line:

(A) Plotted Points

$$P_1' = (x_1',y_1',z_1',w_1') \qquad P_2' = (x_2',y_2',z_2',w_2')$$

(B) Length

$$s' = \sqrt{(x_2'-x_1')^2 + (y_2'-y_1')^2 + (z_2'-z_1')^2 + (w_2'-w_1')^2}$$
$$s' = \sqrt{[3-(1)]^2 + [-1-(-3)]^2 + [2-(0)]^2 + [0-(-2)]^2}$$
$$s' = \sqrt{2^2 + 2^2 + 2^2 + 2^2}$$
$$s' = \sqrt{16}$$
$$s' = 4$$

The translated version of the line has a hyperspace length of 4 units.

(C) Direction cosines:

$$\cos\alpha' = (x_2'-x_1')/s' \qquad \cos\beta' = (y_2'-y_1')/s'$$
$$\cos\alpha' = [3-(1)]/4 \qquad \cos\beta' = [-1-(-3)]/4$$
$$\cos\alpha' = 1/2 \qquad \cos\beta' = 1/2$$

$$\cos\gamma' = (z_2-z_1)/s \qquad \cos\delta' = (w_2-w_1)/s$$
$$\cos\gamma' = [2-(0)]/4 \qquad \cos\delta' = [0-(-2)]/4$$
$$\cos\gamma' = 1/2 \qquad \cos\delta' = 1/2$$

6. Conducting a secondary analysis of the reference line:

(A) XYZ space

Plotted Points:

$$P_1 = (x_1,y_1,z_1) \qquad P_2 = (x_2,y_2,z_2)$$

Length:

$$s = \sqrt{(x_2-x_1)^2 + (y_2-y_1)^2 + (z_2-z_1)^2}$$
$$s = \sqrt{[1-(-1)]^2 + [1-(-1)]^2 + [1-(-1)]^2}$$
$$s = \sqrt{2^2 + 2^2 + 2^2}$$
$$s = \sqrt{12}$$
$$s = 2\sqrt{3}$$

Direction cosines:

$\cos\alpha = (x_2 - x_1)/s$ $\quad$ $\cos\beta = (y_2 - y_1)/s$
$\cos\alpha = [1-(-1)]/[2\sqrt{3}]$ $\quad$ $\cos\beta = [1-(-1)]/[2\sqrt{3}]$
$\cos\alpha = \sqrt{3}/3$ $\quad$ $\cos\beta = \sqrt{3}/3$

$\cos\gamma = (z_2 - z_1)/s$
$\cos\gamma = [1-(-1)]/[2\sqrt{3}]$
$\cos\gamma = \sqrt{3}/3$

(B) XYW Space

Plotted Points:

$P_1 = (x_1, y_1, w_1)$ $\qquad$ $P_2 = (x_2, y_2, w_2)$

Length:

$s = \sqrt{(x_2 - x_1)^2 + (y_2 - y_1)^2 + (w_2 - w_1)^2}$
$s = \sqrt{[1-(-1)]^2 + [1-(-1)]^2 + [1-(-1)]^2}$
$s = \sqrt{2^2 + 2^2 + 2^2}$
$s = \sqrt{12}$
$s = 2\sqrt{3}$

Direction cosines:

$\cos\alpha = (x_2 - x_1)/s$ $\quad$ $\cos\beta = (y_2 - y_1)/s$
$\cos\alpha = [1-(-1)]/[2\sqrt{3}]$ $\quad$ $\cos\beta = [1-(-1)]/[2\sqrt{3}]$
$\cos\alpha = \sqrt{3}/3$ $\quad$ $\cos\beta = \sqrt{3}/3$

$\cos\delta = (w_2 - w_1)/s$
$\cos\delta = [1-(-1)]/[2\sqrt{3}]$
$\cos\delta = \sqrt{3}/3$

(C) XZW Space

Plotted Points:

$P_1 = (x_1, z_1, w_1)$ $\qquad$ $P_2 = (x_2, z_2, w_2)$

Length:

$s = \sqrt{(x_2 - x_1)^2 + (z_2 - z_1)^2 + (w_2 - w_1)^2}$
$s = \sqrt{[1-(-1)]^2 + [1-(-1)]^2 + [1-(-1)]^2}$
$s = \sqrt{2^2 + 2^2 + 2^2}$
$s = \sqrt{12}$
$s = 2\sqrt{3}$

Direction cosines:

$\cos\alpha = (x_2 - x_1)/s$ $\quad$ $\cos\gamma = (z_2 - z_1)/s$
$\cos\alpha = [1-(-1)]/[2\sqrt{3}]$ $\quad$ $\cos\gamma = [1-(-1)]/[2\sqrt{3}]$
$\cos\alpha = \sqrt{3}/3$ $\quad$ $\cos\gamma = \sqrt{3}/3$

$\cos\delta = (w_2 - w_1)/s$
$\cos\delta = [1-(-1)]/[2\sqrt{3}]$
$\cos\delta = \sqrt{3}/3$

(D) YZW Space

Plotted Points:

$P_1 = (y_1, z_1, w_1)$ $\qquad$ $P_2 = (y_2, z_2, w_2)$

Length:

$s = \sqrt{(y_2 - y_1)^2 + (z_2 - z_1)^2 + (w_2 - 2_1)^2}$
$s = \sqrt{[1-(-1)]^2 + [1-(-1)]^2 + [1-(-1)]^2}$
$s = \sqrt{2^2 + 2^2 + 2^2}$
$s = \sqrt{12}$
$s = 2\sqrt{3}$

Direction cosines:

$\cos\beta = (y_2 - y_1)/s$ $\quad$ $\cos\gamma = (z_2 - z_1)/s$
$\cos\beta = [1-(-1)]/[2\sqrt{3}]$ $\quad$ $\cos\gamma = [1-(-1)]/[2\sqrt{3}]$
$\cos\beta = \sqrt{3}/3$ $\quad$ $\cos\gamma = \sqrt{3}/3$

$\cos\delta = (w_2 - w_1)/s$
$\cos\delta = [1-(-1)]/[2\sqrt{3}]$
$\cos\delta = \sqrt{3}/3$

7. Conducting a secondary analysis of the translated line:

(A) X'Y'Z' Space

Plotted Points:

$P_1' = (x_1', y_1', z_1')$ $\qquad$ $P_2' = (x_2', y_2', z_2')$

Length:

$s' = \sqrt{(x_2' - x_1')^2 + (y_2' - y_1')^2 + (z_2' - z_1')^2}$
$s' = \sqrt{(3-1)^2 + [-1-(-3)]^2 + [0-(-2)]^2}$
$s' = \sqrt{2^2 + 2^2 + 2^2}$
$s' = \sqrt{12}$
$s' = 2\sqrt{3}$

Direction cosines:

$\cos\alpha' = (x'_2 - x'_1)/s$ $\quad$ $\cos\beta' = (y_2' - y_1')/s$
$\cos\alpha' = (3-1)/(2\sqrt{3})$ $\quad$ $\cos\beta' = [-1-(-3)]/(2\sqrt{3})$
$\cos\alpha' = \sqrt{3}/3$ $\quad$ $\cos\beta' = \sqrt{3}/3$

$\cos\gamma' = (z'_2 - z'_1)/s$
$\cos\gamma' = (0-2)/(2\sqrt{3})$
$\cos\gamma' = \sqrt{3}/3$

(B) X'Y'W' Space

Plotted Points:

$P_1' = (x_1', y_1', w_1')$ $\qquad$ $P_2' = (x_2', y_2', w_2')$

Length:

$s' = \sqrt{(x_2' - x_1')^2 + (y_2' - y_1')^2 + (w_2' - w_1')^2}$
$s' = \sqrt{(3-1)^2 + [-1-(-3)]^2 + [0-(-2)]^2}$
$s' = \sqrt{2^2 + 2^2 + 2^2}$
$s' = \sqrt{12}$
$s' = 2\sqrt{3}$

Direction cosines:

$\cos\alpha' = (x_2' - x_1')/s$ $\quad$ $\cos\beta' = (y_2' - y_1')/s$
$\cos\alpha' = (3-1)/(2\sqrt{3})$ $\quad$ $\cos\beta' = [-1-(-3)]/(2\sqrt{3})$
$\cos\alpha' = \sqrt{3}/3$ $\quad$ $\cos\beta' = \sqrt{3}/3$

$\cos\delta' = (w'_2 - w'_1)/s$
$\cos\delta' = [0-(-2)]/(2\sqrt{3})$
$\cos\delta' = \sqrt{3}/3$

(C) X'Z'W' Space

Plotted Points:

$P_1' = (x_1', z_1', w_1')$ $P_2' = (x_2', z_2', w_2')$

Length:

$s' = \sqrt{(x_2'-x_1')^2 + (z_2'-z_1')^2 + (w_2'-w_1')^2}$
$s' = \sqrt{(3-1)^2 + (2-0)^2 + [0-(-2)]^2}$
$s' = \sqrt{2^2 + 2^2 + 2^2}$
$s' = \sqrt{12}$
$s' = 2\sqrt{3}$

Direction cosines:

$\cos\alpha' = (x_2'-x_1')/s$
$\cos\alpha' = (3-1)/(2\sqrt{3})$
$\cos\alpha' = \sqrt{3}/3$

$\cos\gamma' = (z_2'-z_1')/s$
$\cos\gamma' = (2-0)/(2\sqrt{3})$
$\cos\gamma' = \sqrt{3}/3$

$\cos\delta' = (w_2'-w_1')/s$
$\cos\delta' = [0-(-2)]/(2\sqrt{3})$
$\cos\delta' = \sqrt{3}/3$

(D) Y'Z'W' Space

Plotted Points:

$P_1' = (y_1', z_1', w_1')$ $P_2' = (y_2', z_2', w_2')$

Length:

$s' = \sqrt{(y_2'-y_1')^2 + (z_2'-z_1')^2 + (w_2'-w_1')^2}$
$s' = \sqrt{[-1-(-3)]^2 + (2-0)^2 + [0-(-2)]^2}$
$s' = \sqrt{2^2 + 2^2 + 2^2}$
$s' = \sqrt{12}$
$s' = 2\sqrt{3}$

Direction cosines:

$\cos\beta' = (y_2'-y_1')/s$
$\cos\beta' = [-1-(-3)]/(2\sqrt{3})$
$\cos\beta' = \sqrt{3}/3$

$\cos\gamma' = (z_2'-z_1')/s$
$\cos\gamma' = (2-0)/(2\sqrt{3})$
$\cos\gamma' = \sqrt{3}/3$

$\cos\delta' = (w'_2 - w'_1)/s$
$\cos\delta' = [0-(-2)]/(2\sqrt{3})$
$\cos\delta' = \sqrt{3}/3$

The experiment is divided into three basic phases:

☐ Perform the designated 4-D translation.

☐ Conduct a primary analysis of both the reference and translated lines, including:
a. the true, hyperspace lengths
b. the true, hyperspace direction cosines

☐ Conduct a secondary analysis of both the reference and translated lines, including:
a. the lengths and direction cosines in XYZ space
b. the lengths and direction cosines in XYW space
c. the lengths and direction cosines in XZW space
d. the lengths and direction cosines in YZW space

After citing the specifications for the reference line and designating a particular 4-D translation in the opening part of the exercise, Step 1 gathers the components of the reference line into a form that makes them compatible with the basic 4-D translation equation, Equation 9-1.

Steps 2 and 3 then perform the actual translation, ultimately providing the coordinates of the two endpoints of the translated version of the line.

Step 4 represents the complete primary analysis of the reference line as it appears in hyperspace. That includes both its length and its four direction cosines. See that line plotted as line L in the 4-D perspective drawing in Fig. 19-2A.

Step 5 is the complete primary analysis of the translated version of the line, L'. You can see that, in spite of its different position in hyperspace, its length and direction cosines are equal to those of the reference line.

Step 6 returns to the reference line, but this time the analysis is on the secondary level. It shows the length and direction cosines of the reference line as observed from four possible 3-D spaces. Clearly the line is shorter than its true, hyperspace version; and you can see from the direction cosines in the secondary analyses that the orientation of the line is also different from its hyperspace version.

Step 7 carries out the same sort of secondary analysis, but referring to the translated version of the line.

You can see the 3-D mappings of the entire translation situation in Figs. 19-2B through 19-2E. What you are seeing there is a line that actually exists in hyperspace and is being subjected to a 4-D translation, but as viewed from the four possible 3-D views. We live in a 3-D space, and you are free to select any one of the four as our own. That is how the hyperspace translation would appear to us.

Try your hand at the procedure by working the following exercises.

Exercise 19-1

Perform the 4-D translations that are specified for the reference lines in the following instances. Check your results by calculating and comparing the lengths of both the reference and translated versions. They should be the same.

1. $L_1 = \overline{P_1P_2}$ $P_1 = (0,1,1,0)$
$P_2 = (1,0,0,1)$

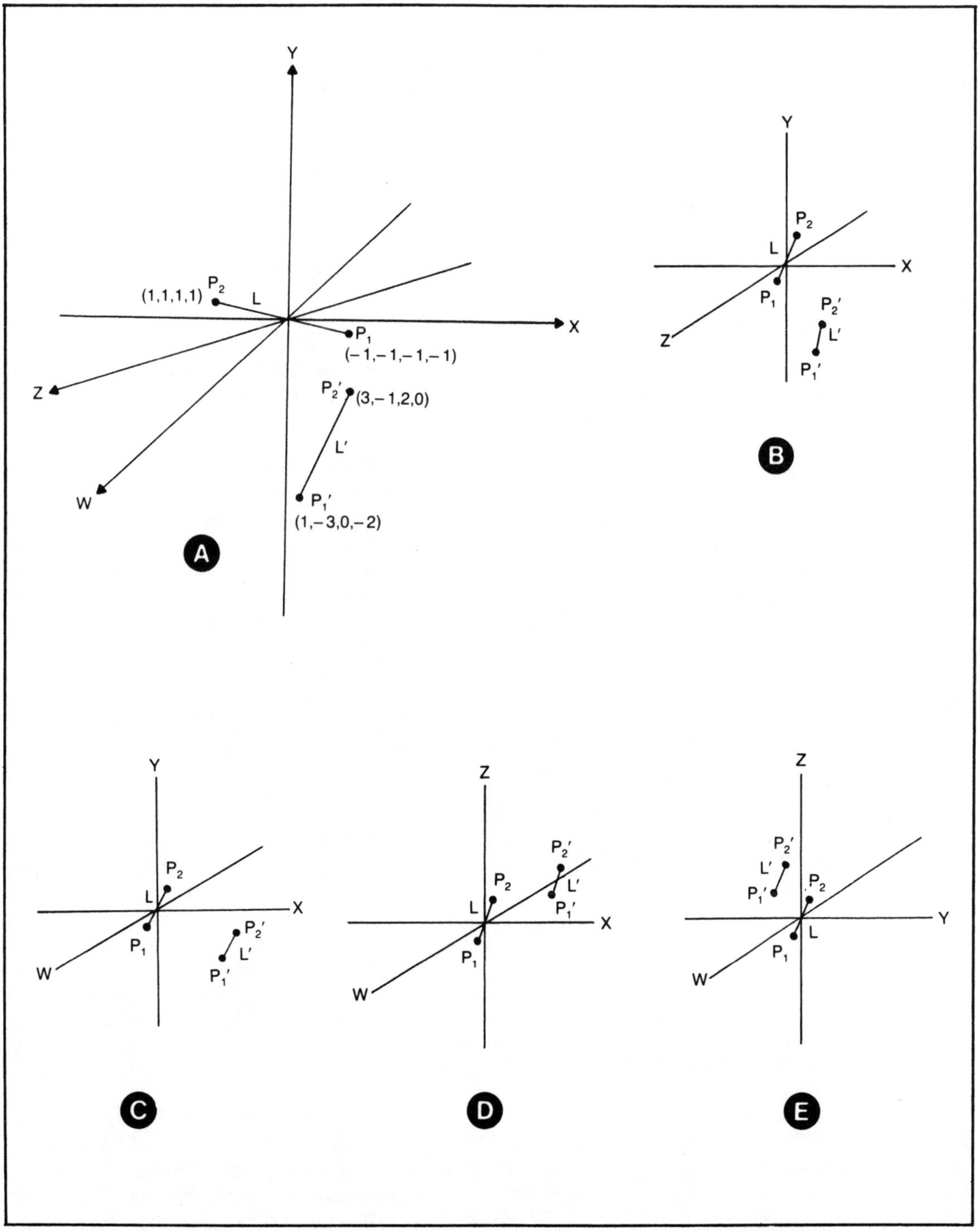

Fig. 19-2. Translating a line L to L′ hyperspace as suggested in Example 19-2. (A) A general hyperspace view. (B) XYZ perspective view. (C) XYW perspective view. (D) XZW perspective view. (E) YZW perspective view.

$$T_x = -1, \quad T_y = 2, \quad T_z = -1, \quad T_w = 0$$

2. $L_1 = \overline{P_1P_2}$ $\quad P_1 = (-2,2,3,-1)$
$\quad P_2 = (0,-1,1,1)$

$$T_x = 0, \quad T_y = -2, \quad T_z = 1, \quad T_w = 1$$

3. $L_1 = \overline{P_1P_2}$ $\quad P_1 = (2,2,-1,-1)$
$\quad P_2 = (0,0,1,1)$

$$T_x = -2, \quad T_y = -2, \quad T_z = 0, \quad T_w = 1$$

4. $L_1 = \overline{P_1P_2}$ $\quad P_1 = (0,-1,-2,-1)$
$\quad P_2 = (0,-1,-2,1)$

$$T_x = 0, \quad T_y = 1, \quad T_z = 2, \quad T_w = -3$$

19-3 TRANSLATING PLANE FIGURES IN 4-D SPACE

Just as you can translate a line in 4-D space by translating its two endpoints, so you can translate a plane figure that is immersed in 4-D space by translating its individual endpoints. And, again, little changes in the analysis but the position of the figure; the analysis of its lines—length, direction cosines and angles between lines—turns up identical results.

Work your way through Exercise 19-2. It cites a triangle and a particular 4-D translation that is to be applied to it. Recall that a complete analysis of a plane figure includes the lengths of its lines, the direction cosines for all of the lines, and the angles between the lines.

Exercise 19-2

Apply the suggested 4-D translation to the triangle specified below. Conduct analyses of both versions, and check your results by comparing the lengths of lines and the angles for both the reference and translated versions; they should be the same.

$F = \overline{L_1L_2L_3}$	$L_1 = \overline{P_1P_2}$	$P_1 = (-1,0,0,1)$
	$L_2 = \overline{P_2P_3}$	$P_2 = (1,0,0,1)$
	$L_3 = \overline{P_3P_1}$	$P_3 = (0,1,0,0)$

$$T_x = 0, \quad T_y = -1, \quad T_z = 1, \quad T_w = 1$$

19-4 TRANSLATING SPACE OBJECTS IN 4-D SPACE

There is little more to be said about translating space objects in 4-D space than about translating a plane figure in that sort of higher-dimensioned space. Basically, the transformation, itself, is an exceedingly simple one—apply the translation equation to each of the points in the space object, then specify the same S-F-L-P parameters. Nothing changes but the coordinates of the points. The real excitement lays in the analysis of the object as it appears in the four different 3-D perspective views of 4-D space.

If you do care to carry out complete analyses of both the reference and translated versions of space objects, you will occasionally encounter situations where the direction cosines for a given 3-D perspective view appear different for the reference and translated version. That is a simple mathematical anomaly, however; and if you calculate the angle between the lines, you will find that it is 0 degrees—the lines are, in fact, parallel to one another and, therefore, unchanged with respect to their orientation in space.

Exercise 19-3 cites a reference space object that is immersed in 4-D space. Work through the analysis, perform the suggested translation, and work out enough of the analysis of the translated version to make some good plots of the situation.

Exercise 19-3

Given the space object specified below, apply a 4-D translation of:

$$T_x = 1, \quad T_y = -1, \quad T_z = 1, \quad T_w = -1$$

Conduct analyses of both the reference and translated versions, and check your results by comparing the lengths of lines and the angles.

$S = \overline{F_1F_2F_3F_4}$	$L_1 = \overline{P_1P_2}$	$P_1 = (0,0,1,1)$
$F_1 = \overline{L_1L_2L_3}$	$L_2 = \overline{P_2P_3}$	$P_2 = (1,0,0,1)$
$F_2 = \overline{L_1L_4L_6}$	$L_3 = \overline{P_1P_3}$	$P_3 = (0,0,0,1)$
$F_3 = \overline{L_2L_4L_5}$	$L_4 = \overline{P_2P_4}$	$P_4 = (0,1,0,1)$
$F_4 = \overline{L_3L_5L_6}$	$L_5 = \overline{P_3P_4}$	
	$L_6 = \overline{P_1P_4}$	

19-5 TRANSLATING HYPERSPACE OBJECTS

Experiments involving the translation of hyperspace objects can be more exciting than the previous ones by virtue of the fact that they are hyperspace objects. The translation operation, itself, is almost trivial. The idea is to apply the translation equation, Equation 19-1, to some specified hyperspace object. The real task lays in doing the analyses and perspective-view drawings as viewed from the four alternative 3-D spaces. Such experiments show how a hyperspace object being translated in 4-D space appears to we lesser beings.

In spite of the fact that you are considering a relatively sophisticated hyperspace object, the general principles of translations hold firm: the lengths of the lines, the direction cosines and the angles between lines remain unaffected. Those are invariants for any kind of geometric entity that is undergoing a simple translation in any number of dimensions.

Exercise 19-4 cites a hyperspace object and designates a particular 4-D translation to be applied to it. Since it isn't always clear what angles are relevant to such objects, I've specified them for you. Attempt a complete set of primary and secondary analyses for the translation situation. It is a lot of work, but that is a small price to pay for doing something few people have ever done before. And don't be shy about applying some translations of your own choosing to some hyperspace objects of your own design; you might even discover something new.

Exercise 19-4

Given the hyperspace object specified below, apply a 4-D translation of:

$$T_x = -1, \quad T_y = 2, \quad T_z = 0, \quad T_w = -2$$

Conduct the analyses of both the reference and translated versions of the space object, and check your overall results by comparing these relevant angles:

$\phi_{1,2}$	$\phi_{1,3}$	$\phi_{1,4}$	$\phi_{1,7}$	$\phi_{1,8}$
$\phi_{1,10}$	$\phi_{2,3}$	$\phi_{2,5}$	$\phi_{2,7}$	$\phi_{2,9}$
$\phi_{2,10}$	$\phi_{3,6}$	$\phi_{3,8}$	$\phi_{3,9}$	$\phi_{3,10}$
$\phi_{4,5}$	$\phi_{4,6}$	$\phi_{4,7}$	$\phi_{4,8}$	$\phi_{4,10}$
$\phi_{5,6}$	$\phi_{5,7}$	$\phi_{5,9}$	$\phi_{5,10}$	$\phi_{6,8}$
$\phi_{6,9}$	$\phi_{6,10}$	$\phi_{7,8}$	$\phi_{7,9}$	$\phi_{8,9}$

$H = \overline{S_1S_2S_3S_4S_5}$

$S_1 = \overline{F_1F_2F_3F_4}$
$S_2 = \overline{F_4F_5F_6F_7}$
$S_3 = \overline{F_1F_5F_8F_9}$
$S_4 = \overline{F_2F_6F_8F_{10}}$
$S_5 = \overline{F_3F_7F_9F_{10}}$

$F_1 = \overline{L_1L_2L_3}$	$L_1 = \overline{P_1P_3}$	$P_1 = (0,0,0,0)$
$F_2 = \overline{L_1L_3L_8}$	$L_2 = \overline{P_1P_4}$	$P_2 = (0,0,0,1)$
$F_3 = \overline{L_2L_3L_9}$	$L_3 = \overline{P_1P_5}$	$P_3 = (1,0,0,0)$
$F_4 = \overline{L_7L_8L_9}$	$L_4 = \overline{P_2P_3}$	$P_4 = (0,1,0,0)$
$F_5 = \overline{L_4L_5L_7}$	$L_5 = \overline{P_2P_4}$	$P_5 = (0,0,1,0)$
$F_6 = \overline{L_4L_6L_8}$	$L_6 = \overline{P_2P_5}$	
$F_7 = \overline{L_5L_6L_9}$	$L_7 = \overline{P_3P_4}$	
$F_8 = \overline{L_1L_4L_{10}}$	$L_8 = \overline{P_3P_5}$	
$F_9 = \overline{L_2L_5L_{10}}$	$L_9 = \overline{P_4P_5}$	
$F_{10} = \overline{L_3L_6L_{10}}$	$L_{10} = \overline{P_1P_2}$	

19-6 ALTERNATIVE VERSIONS OF THE GEOMETRIC EQUATION

Equation 19-2 shows the basic equation for geometric translation reorganized to solve for the translation terms. Given the coordinate of a reference point and a translated version of that point, the expressions lead directly to the translation that exists between them.

$$\begin{aligned} T_x &= x_n' - x_n \\ T_y &= y_n' - y_n \\ T_z &= z_n' - z_n \\ T_w &= w_n' - w_n \end{aligned} \qquad \textbf{Equation 19-2}$$

where:

T_x = the X-axis translation term
T_y = the Y-axis translation term
T_z = the Z-axis translation term
T_w = the W-axis translation term
x_n', y_n', z_n', w_n' = the coordinate of the translated version of point *n* relative to the frame of reference
x_n, y_n, z_n, w_n = the coordinate of point *n* relative to the frame of reference

Earlier discussions cited lower-dimensioned versions of this equation as a means for determining the translation that exists between two coordinate systems. That can be done in the case of geometric entities that are separated by a translation in 4-D space as well, but I am leaving applications of that sort to your own sense of ambition and desire to try it. What is of more practical importance to an experimenter working in a hyperspace environment is the ability to solve for the translation terms that are necessary for adjusting the position of a point, line, figure, space object or hyperspace object in 4-D space.

Here is the basic idea. You have seen from previous work that it is far easier to derive the specifications for entities in 4-D space when they are situated in a particular way with respect to the origin of the coordinate system. Many interesting kinds of experiments, on the other hand, call for working with the entities as situated elsewhere on the coordinate system. Equation 19-2 provides a means for relocating the points to some other desired position.

Using that technique, begin with the specifications for the entity as designated in the simplest location on the coordinate system. Then cite the new, desired coordinate for one of the critical points in the entity. Regarding the original coordinate as the reference coordinate and the desired coordinate as the translated version of the point, Equation 19-2 provides the necessary translation terms for moving that point and all others in the figure or object at hand. Calculate the translation terms for the critical point you have chosen, then use those terms as a basis for translating the remaining points by Equation 19-1.

Select the specifications for a hyperspace object, or a lower-dimensioned object that is immersed in hyperspace, and then attempt to translate it to a particular place in hyperspace. Use any one of the previous examples and exercises as a source of specifications.

Equation 19-3 is the inverse form of the basic equation for geometric translations in 4-D space. Taken in its most literal form, it lets you calculate the reference-point coordinates, given the translated version of the coordinates and the translation terms. You may use it that way if you choose, but I am showing it here only as a means for leading into the relativistic version of 4-D translations.

$$\begin{aligned} x_n &= x_n' - T_x \\ y_n &= y_n' - T_y \\ z_n &= z_n' - T_z \\ w_n &= w_n' - T_w \end{aligned} \qquad \textbf{Equation 19-3}$$

where:

x_n, y_n, z_n, w_n = the coordinate of point *n* relative to the frame of reference

x_n', y_n', z_n', w_n' = the coordinate of the translated version of point *n* relative to the frame of reference

T_x = the X-axis translation term

T_y = the Y-axis translation term

T_z = the Z-axis translation term

T_w = the W-axis translation term

Note: This is the inverse version of Equation 19-1

19-7 RELATIVISTIC TRANSLATIONS IN 4-D

From the tone of the foregoing discussions, it is easy to lose sight of the fact that a translation—a translation in any number of dimensions—really represents a translation that exists between two coordinate systems. Specificially, a geometric translation in four dimensions actually represents a translation-type offset between a 4-D reference coordinate system and another 4-D system that is otherwise identical to it. The discussion to this point in this chapter regard a 4-D translation as a particular means for moving geometric entities from one place to another in 4-D space. Both views are quite valid, and they serve particular purposes.

In the context of relativistic translations, however, the idea of moving entities from place to place is not very meaningful. Here it is better to view a translation as an expression of the displacement between two otherwise identical 4-D coordinate systems.

$$\begin{aligned} x_n^* &= x_n - T_x \\ y_n^* &= y_n - T_y \\ z_n^* &= z_n - T_z \\ w_n^* &= w_n - T_w \end{aligned} \qquad \textbf{Equation 19-4}$$

where:

$x_n^*, y_n^*, z_n^*, w_n^*$ = the coordinate of reference point *n* relative to the translated frame of reference

x_n, y_n, z_n, w_n = the coordinate of reference point *n* relative to the observer's frame of reference

T_x = the X-axis translation term relative to the observer's frame of reference

T_y = the Y-axis translation term relative to the observer's frame of reference

T_z = the Z-axis translation term relative to the observer's frame of reference

T_w = the W-axis translation term relative to the observer's frame of reference

Equation 19-4 represents the basic relativistic version of a 4-dimensional translation that exists between two 4-D coordinate systems. Used in its most direct fashion, it turns up the coordinates of some point in 4-D space as regarded from a translated frame of reference. Generally the application begins by specifying some geometric entity as it appears from the viewpoint of the observer's 4-D frame of reference. Then establishing the translation terms for the displacement of a translated frame of reference, relative to the observer's frame of reference, the equations show how the entity appears relative to that second frame of reference.

Example 19-3 illustrates the situation with a specific example.

Example 19-3

The following specifications represent a space object that is immersed in hyperspace and viewed from an observer's 4-D frame of reference:

$S = \overline{F_1F_2F_3F_4F_5}$

$F_1 = \overline{L_1L_2L_3L_4}$
$F_2 = \overline{L_1L_5L_8}$
$F_3 = \overline{L_2L_5L_6}$
$F_4 = \overline{L_3L_6L_7}$
$F_5 = \overline{L_4L_7L_8}$

$L_1 = \overline{P_1P_2}$
$L_2 = \overline{P_2P_3}$
$L_3 = \overline{P_3P_4}$
$L_4 = \overline{P_1P_4}$
$L_5 = \overline{P_2P_5}$
$L_6 = \overline{P_3P_5}$
$L_7 = \overline{P_4P_5}$
$L_8 = \overline{P_1P_5}$

$P_1 = (-1,0,1,1)$
$P_2 = (1,0,1,1)$
$P_3 = (1,0,-1,-1)$
$P_4 = (-1,0,-1,-1)$
$P_5 = (0,1,0,0)$

Assuming that the following 4-D translation exists between the observer's and translated frames of reference, determine the specifications of the object as viewed from the translated frame of reference.

$$T_x = -1, \quad T_y = 2, \quad T_z = 4, \quad T_w = 1$$

1. Gathering the given point coordinates for the observer's view of the matter:

$x_1 = -1 \quad y_1 = 0 \quad z_1 = 1 \quad w_1 = 1$
$x_2 = 1 \quad y_2 = 0 \quad z_2 = 1 \quad w_2 = 1$
$x_3 = 1 \quad y_3 = 0 \quad z_3 = -1 \quad w_3 = -1$
$x_4 = -1 \quad y_4 = 0 \quad z_4 = -1 \quad w_4 = -1$
$x_5 = 0 \quad y_5 = 1 \quad z_5 = 0 \quad w_5 = 0$

2. Applying Equation 9-4 to each of the points:

$$x_1^* = x_1 - T_x \qquad y_1^* = y_1 - T_y$$
$$x_1^* = -1-(-1) \qquad y_1^* = 0-2$$
$$x_1^* = 0 \qquad y_1^* = -2$$

$$z_1^* = z_1 - T_z \qquad w_1^* = w_1 - T_w$$
$$z_1^* = 1-4 \qquad w_1^* = 1-1$$
$$z_1^* = -3 \qquad w_1^* = 0$$

Thus $P_1 = (-1,0,1,1)$ appears as $P_1^* = (0,-2,-3,0)$

$$x_2^* = x_2 - T_x \qquad y_2^* = y_2 - T_y$$
$$x_2^* = 1-(-1) \qquad y_2^* = 0-2$$
$$x_2^* = 2 \qquad y_2^* = -2$$

$$z_2^* = z_2 - T_z \qquad w_2^* = w_2 - T_w$$
$$z_2^* = 1-4 \qquad w_2^* = 1-1$$
$$z_2^* = -3 \qquad w_2^* = 0$$

Thus $P_2 = (-1,0,1,1)$ appears as $P_2^* = (2,-2,-3,0)$

$$x_3^* = x_3 - T_x \qquad y_3^* = y_3 - T_y$$
$$x_3^* = 1-(-1) \qquad y_3^* = 0-2$$
$$x_3^* = 2 \qquad y_3^* = -2$$

$$z_3^* = z_3 - T_z \qquad w_3^* = w_3 - T_w$$
$$z_3^* = -1-4 \qquad w_3^* = -1-1$$
$$z_3^* = -5 \qquad w_3^* = -2$$

Thus $P_3 = (1,0,1,1)$ appears as $P_3^* = (2,-2,-5,-2)$

$$x_4^* = x_4 - T_x \qquad y_4^* = y_4 - T_y$$
$$x_4^* = -1-(-1) \qquad y_4^* = 0-2$$
$$x_4^* = 0 \qquad y_4^* = -2$$

$$z_4^* = z_4 - T_z \qquad w_4^* = w_4 - T_4$$
$$z_4^* = -1-4 \qquad w_4^* = -1-1$$
$$z_4^* = -5 \qquad w_4^* = -2$$

Thus $P_4 = (-1,0,-1,-1)$ appears as $P_4^* = (0,-2,-4,-2)$

$$x_5^* = x_5 - T_x \qquad y_5^* = y_5 - T_y$$
$$x_5^* = 0-(-1) \qquad y_5^* = 1-2$$
$$x_5^* = 1 \qquad y_5^* = -1$$

$$z_5^* = z_5 - T_z \qquad w_5^* = w_5 - T_w$$
$$z_5^* = 0-4 \qquad w_5^* = 0-1$$
$$z_5^* = -4 \qquad w_5^* = -1$$

Thus $P_5 = (0,1,0,0)$ appears as $P_5^* = (1,-1,-4,-1)$

3. Summarizing the results, the reference object appears to have these specifications from the translated frame of reference:

$$S^* = \overline{F_1^* F_2^* F_3^* F_4^* F_5^*}$$

$$F_1^* = \overline{L_1^* L_2^* L_3^* L_4^*}$$
$$F_2^* = \overline{L_1^* L_5^* L_8^*}$$
$$F_3^* = \overline{L_2^* L_5^* L_6^*}$$
$$F_4^* = \overline{L_3^* L_6^* L_7^*}$$
$$F_5^* = \overline{L_4^* L_7^* L_8^*}$$

$$L_1^* = \overline{P_1^* P_2^*}$$
$$L_2^* = \overline{P_2^* P_3^*}$$
$$L_3^* = \overline{P_3^* P_4^*}$$
$$L_4^* = \overline{P_1^* P_4^*}$$
$$L_5^* = \overline{P_2^* P_5^*}$$
$$L_6^* = \overline{P_3^* P_5^*}$$
$$L_7^* = \overline{P_4^* P_5^*}$$
$$L_8^* = \overline{P_1^* P_5^*}$$

$$P_1^* = (0,-2,-3,0)$$
$$P_2^* = (2,-2,-3,0)$$
$$P_3^* = (2,-2,-5,-2)$$
$$P_4^* = (0,-2,-5,-2)$$
$$P_5^* = (1,-1,-4,-1)$$

That particular example happens to cite a space object that is immersed in hyperspace; and you can tell that it is a space object by the fact that it has a valid S specification, but no H specification.

Step 1, as usual, denotes the components of the reference points in their equation-type format, and then Step 2 applies the given 4-D relativistic translation equation, Equation 19-4, to those points.

Step 3 summarizes the results by specifying the object as it appears from the translated frame of reference. It is left to you to extend the example to include plots of the results and conduct full primary and secondary analyses of the situation.

$$T_x = x_n - x_n^*$$
$$T_y = y_n - y_n^*$$
$$T_z = z_n - z_n^*$$
$$T_w = w_n - w_n^*$$

Equation 19-5

where:

T_x = the X-axis translation that exists between the observer's and translated frames of reference, relative to the observer's

T_y = the Y-axis translation that exists between the observer's and translated frames of reference, relative to the observer's

T_z = the Z-axis translation that exists between the

observer's and translated frames of reference, relative to the observer's

T_w = the W-axis translation that exists between the observer's and translated frames of reference, relative to the observer's

x_n, y_n, z_n, w_n = the coordinate of reference point n relative to the observer's frame of reference

$x_n^*, y_n^*, z_n^*, w_n^*$ = the coordinate of reference point n relative to the translated frame of reference

Equation 19-5 lets you determine the 4-D translation that exists between two 4-dimensional frames of reference, given the coordinates of a single point from both the observer's and translated frame of reference.

The inverse form of the 4-D relativistic translation equation is shown here as Equation 19-6. Its application answers the question: How does an object specified from a translated frame of reference appear to that of the observer? It follows the same general pattern as the inverse relativistic equations for lower-dimension spaces, and I am leaving it up to you to build some interesting experiments around it.

$$\begin{aligned} x_n &= x_n^* + T_x \\ y_n &= y_n^* + T_y \\ z_n &= z_n^* + T_z \\ w_n &= w_n^* + T_w \end{aligned} \qquad \textbf{Equation 19-6}$$

where:

x_n, y_n, z_n, w_n = the coordinate of reference point n relative to the observer's frame of reference

$x_n^*, y_n^*, z_n^*, w_n^*$ = the coordinate of reference point n relative to the translated frame of reference

T_x = the X-axis translation that exists between the observer's and translated frames of reference, relative to the observer's

T_y = the Y-axis translation that exists between the observer's and translated frames of reference, relative to the observer's

T_z = the Z-axis translation that exists between the observer's and translated frames of reference, relative to the observer's

T_w = the W-axis translation that exists between the observer's and translated frames of reference, relative to the observer's

Note: This is the inverse form of Equation 19-4.

19-8 COMBINATIONS OF 4-D TRANSLATIONS

Multiple translations in 4-D space are commutative. That is, you can apply a given series of translations in any desired order and still end up with the same overall translation. This, of course, applies to geometric translations as well as their relativistic versions.

Scalings in 4-D Space

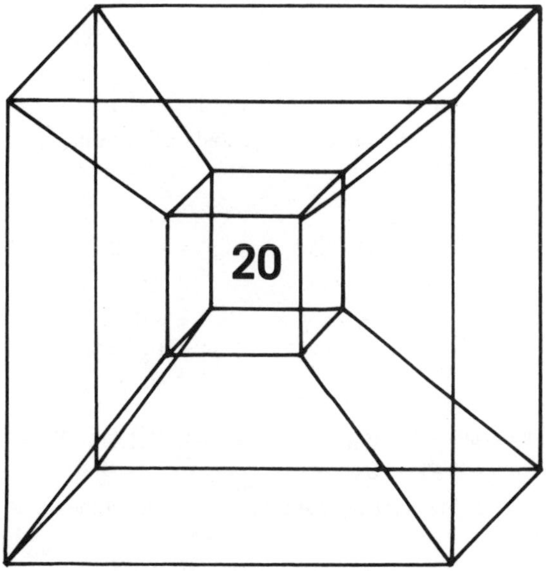

Translating points that are specified in 4-D space is an extension of the procedure for scaling in 3-D space. In the 4-D case, however, there are four scaling terms—one for each of the 4-D coordinate axes. Multiplying each of those scaling factors by their corresponding components in the coordinate of a point carries out the scaling.

Four-dimensional scalings have the same general characteristics as those described for lower dimensions; and you can take advantage of your previous experience to save a lot of time and effort when doing complete analyses of scaling effects in 4-D space.

Furthermore, there are useful alternative forms of the geometric scaling equation of 4 dimensions as well as the usual family of relativistic forms.

Reflection-type scalings in 4-D space are especially interesting because they take place about a 3-dimensional space. It is difficult to visualize the situation, but it follows quite naturally from the same ideas in lower dimensions:

☐ In 1-D space, reflections take place about a point (0-D space).

☐ In 2-D space, reflections take place about a line (1-D space).

☐ In 3-D space, reflections take place about a plane (2-D space).

☐ In 4-D space, reflections take place about a 3-D space.

20-1 THE BASIC SCALING EQUATION FOR 4-D SPACE

As shown in Equation 20-1, the fundamental equation for geometric scalings in 4-D space is an extension of the scaling equation for lower spaces. It enables you to scale the components of a reference point by multiplying the individual reference components by their respective scaling factors.

$$\begin{aligned} x_n' &= x_n K_x \\ y_n' &= y_n K_y \\ z_n' &= z_n K_z \\ w_n' &= w_n K_w \end{aligned} \qquad \textbf{Equation 20-1}$$

where:

x_n', y_n', z_n', w_n' = the coordinate of the scaled version of reference point *n*, relative to the chosen frame of reference

x_n, y_n, z_n, w_n = the coordinate of reference point *n*, relative to the chosen frame of reference.

K_x, K_y, K_z, and K_w = the scaling factors for the coordinate axes, relative to the chosen frame of reference.

Example 20-1 demonstrates the procedure for scaling a point within a 4-dimensional coordinate system.

Example 20-1

Given a reference point P = (−1,−1,1,1) in hyperspace, subject it to the following 4-D scaling:

$$K_x = 0.5,\ K_y = 2,\ K_z = -1,\ K_w = -0.5$$

1. Gathering the components of the reference point:

$$x = -1 \quad y = -1 \quad z = 1 \quad w = 1$$

2. Carrying out the designated scaling by means of Equation 20-4:

$$x' = xK_x \qquad y' = yK_y$$
$$x' = -1\,(0.5) \qquad y' = -1\,(2)$$
$$x' = -0.5 \qquad y' = -2$$

$$z' = zK_z \qquad w' = wK_w$$
$$z' = 1(-1) \qquad w' = 1\,(-0.5)$$
$$z' = -1 \qquad w' = -0.5$$

The scaling thus transforms P = (−1,−1,1,1) to P′ = (−0.5,−2,−1,−0.5).

3. Summarizing the primary analysis of the situation:

$$P = (-1,-1,1,1)$$
$$P' = (-0.5,-2,-1,-0.5)$$

See the primary plot in Fig. 20-1A.

4. Setting up the secondary analysis for XYZ space:

$$P = (x,y,z) = (-1,-1,1)$$
$$P' = (x',y',z') = (-0.5,-2,-1)$$

See the plot in Fig. 20-1B.

5. Setting up the secondary analysis for XYW space:

$$P = (x,y,w) = (-1,-1,1)$$
$$P' = (x',y',w') = (-0.5,-2,-0.5)$$

See the plot in Fig. 20-1C.

6. Setting up the secondary analysis for XZW space:

$$P = (x,z,w) = (-1,1,1)$$
$$P' = (x',z',w') = (-0.5,-1,-0.5)$$

See the plot in Fig. 20-1D.

7. Setting up the secondary analysis for YZW space:

$$P = (y,z,w) = (-1,1,1)$$
$$P' = (y',z',w') = (-2,-1,-0.5)$$

See the plot in Fig. 20-1E.

The example simply cites a point that is immersed in hyperspace and then suggests a 4-D scaling to be applied to it. The basic scaling procedure outlined for you in Steps 1 and 2 is not a difficult one; and it probably isn't very exciting, either. See the brief summary of the results in Step 3.

Now, 4-D scalings, even for simple points, becomes more interesting when regarding them from the four alternative 3-D spaces—from 3-D views that creatures such as ourselves might be forced to adopt. Step 4 views the situation from an XYZ space, Step 5 looks at the same hyperspace situation from the perspective of XYW space, and Steps 6 and 7 regard it from the XZW and YZW spaces, respectively. See the plots in Fig. 20-1 of that example.

By way of an exercise, make up some coordinates for points in 4 dimensions and apply some 4-D scaling terms of your own.

20-2 SCALING LINES IN 4-D SPACE

A line in 4-D space can be defined in terms of the coordinates of its two endpoints, so scaling a line is a matter of scaling the coordinates of its endpoints. That applies to scaling lines in any number of dimensions, and using Equation 20-1 with lines in 4-D space is a relatively simple extension of what you've done before.

Doing a complete primary and secondary analysis of a scaling in four dimensions can be a time-consuming process, however, because none of the line parameters is invariant. Everything can change under a scaling in 4-D space—point positions, lengths of lines, direction cosines, and the angle between lines.

The effects that various values of scaling factors have on points and lines in hyperspace are identical to the effects achieved in the lower dimensions:

1. Any scaling factor greater than 1 will expand the scale of its axis.
2. A scaling factor of 1 causes no change in the scale of its axis.
3. A scaling factor that is greater than zero but less than 1 will reduce the scaling of its axis.
4. A scaling fact that is equal to zero will reduce its axis to a point.
5. A scaling factor that is greater than −1 but less than zero will reduce and reflect the scale of its axis.
6. A scaling factor of −1 will reflect its axis but not change its scale in any other way.
7. A scaling factor less than −1 will both expand and reflect the scaling of its axis.

Table 20-1 summarizes the special nature of reflection-type scalings in four dimensions. Notice, par-

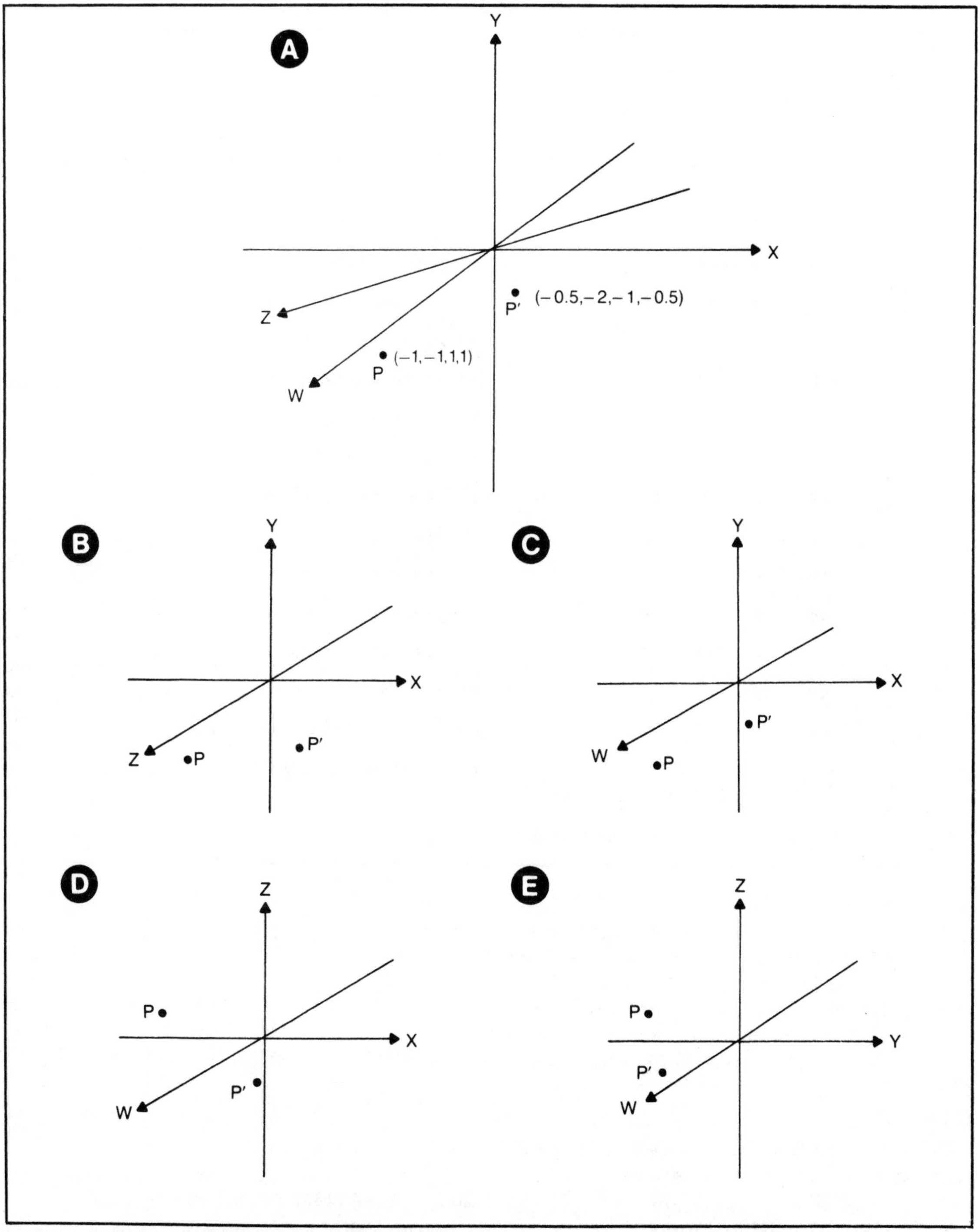

Fig. 20-1. Scaling a point P to P′ in hyperspace as suggested in Example 20-1. (A) A general hyperspace view. (B) XYZ perspective view. (C) XYW perspective view. (D) XZW perspective view. (E) YZW perspective view.

Table 20-1. A Summary of 4-D, Reflection-Type Scalings.

K_x	K_y	K_z	K_w	
>0	>0	>0	>0	No reflection
>0	>0	>0	<0	XYZ-space reflection
>0	>0	<0	>0	XYW-space reflection
>0	>0	<0	<0	XY-plane reflection
>0	<0	>0	>0	XZW-space reflection
>0	<0	>0	<0	XW-plane reflection
>0	<0	<0	>0	XZ-plane reflection
>0	<0	<0	<0	X-axis reflection
<0	>0	>0	>0	YZW-space reflection
<0	>0	>0	<0	YZ-plane reflection
<0	>0	<0	>0	YW-plane reflection
<0	>0	<0	<0	Y-axis reflection
<0	<0	>0	>0	ZW-plane reflection
<0	<0	>0	<0	Z-axis reflection
<0	<0	<0	>0	W-axis reflection
<0	<0	<0	<0	Undefined (5-dimensional) reflection

Note: <="less than"
>="greater than"

ticularly, the new idea of a reflection about a 3-D space.

Experiment with those reflection-type scalings. Cite any off-origin point in hyperspace, apply some reflection-type scalings (using Equation 20-1), and compare the plots of the reference and scaled versions as seen from the four alternative 3-D spaces. Also try your hand at scaling lines in hyperspace by working Exercise 20-1.

Exercise 20-1

Perform the 4-D scalings that are specified for the reference lines in the following instances.

1. $L_1 = \overline{P_1P_2}$ $\quad P_1 = (0,1,1,0)$, $P_2 = (1,0,0,1)$
   $K_x = -1,\ K_y = 0.5,\ K_z = 1,\ K_w = 2$
2. $L_1 = \overline{P_1P_2}$ $\quad P_1 = (-2,2,3,-1)$, $P_2 = (0,-1,1,1)$
   $K_x = 1,\ K_y = -2,\ K_z = -1,\ K_w = 1$
3. $L_1 = \overline{P_1P_2}$ $\quad P_1 = (2,2,-1,-1)$, $P_2 = (0,0,1,1)$
   $K_x = -1,\ K_y = 0.5,\ K_z = -0.5,\ K_w = 1$
4. $L_1 = \overline{P_1P_2}$ $\quad P_1 = (0,-1,-2,-1)$, $P_2 = (0,-1,-2,-1)$
   $K_x = 0,\ K_y = 1,\ K_z = 0.5,\ K_w = -2$

20-3 SCALING PLANE FIGURES IN 4-D SPACE

Scaling lines and scaling plane figures is much the same sort of process: scale the individual points, and then construct the lines according to the original specifications. That's all there is to the scaling part of the job. But unless the scaling happens to be a perfectly symmetrical one, a complete analysis can be a lengthy one.

Try Exercise 20-2. It suggests a complete analysis of a triangle in 4-D space, a certain 4-dimensional scaling, and a complete analysis of the resulting scaled version of the figure.

Exercise 20-2

Given the quadralateral specified below, apply a 4-D scaling of:

$$K_z = 2, \quad K_y = 1, \quad K_z = 1, \quad K_w = 0.5$$

Conduct analyses of both versions.

$$F = \overline{L_1L_2L_3L_4}$$

$$L_1 = \overline{P_1P_2} \qquad P_1 = (-2,0,1,1)$$
$$L_2 = \overline{P_2P_3} \qquad P_2 = (1,0,0,-1)$$
$$L_3 = \overline{P_3P_4} \qquad P_3 = (2,1,1,-1)$$
$$L_4 = \overline{P_1P_4} \qquad P_4 = (-1,1,1,1)$$

20-4 SCALING SPACE OBJECTS IN 4-D SPACE

The procedure for scaling a space object that is immersed in hyperspace is really no different from scaling a plane figure in a higher-dimensioned space. The specifications are a bit more involved and the analysis calls for a lot more steps; but the overall idea is no different. The scaling is simple, the analysis is a bit rough. The payoff for all the analytical work is the excitement that is inherent in seeing a scaled space object as it appears from the four different 3-D perspective views of 4-D space.

Bear in mind, though, that a 3-D mapping of an entity immersed in hyperspace often degenerates the entity to one of a lower dimension—a 5-sided pyramid in hyperspace often degenerates to a simple, plane triangle or rectangle in one or more of the alternative 3-D spaces. You know that degeneration is taking place when your secondary analyses begin turning up lines of zero length and undefinable direction cosines and angles.

Exercise 20-3 cites a reference space object that is immersed in 4-D space. Work through the analysis of the reference object, perform the suggested scaling, and conduct a complete analysis of the scaled version.

Exercise 20-3

Given the space object specified below, apply a 4-D scaling of:

$$K_x = 1, K_y = -1, K_z = 0.5, K_w = -0.5$$

Conduct analyses of both the reference and scaled versions.

$$S = \overline{F_1F_2F_3F_4}$$

$$F_1 = \overline{L_1L_2L_3} \qquad L_1 = \overline{P_1P_2} \qquad P_1 = (1,1,0,0)$$
$$F_2 = \overline{L_1L_4L_6} \qquad L_2 = \overline{P_2P_3} \qquad P_2 = (2,1,-1,0)$$
$$F_3 = \overline{L_2L_4L_5} \qquad L_3 = \overline{P_1P_3} \qquad P_3 = (1,1,-1,0)$$
$$F_4 = \overline{L_3L_5L_6} \qquad L_4 = \overline{P_2P_4} \qquad P_4 = (1,2,-1,0)$$
$$L_5 = \overline{P_3P_4}$$
$$L_6 = \overline{P_1P_4}$$

20-5 SCALING HYPERSPACE OBJECTS

Scaling hyperspace objects can be more exciting than scaling space objects by virtue of the fact that the hyperspace objects are quite unusual entities. The scaling operation, itself, is almost trivial; the idea is to apply the scaling equation, Equation 20-1, to the point coordinates some specified hyperspace object. The difficult part is doing the analyses and making the perspective-view drawings.

Example 20-2 cites a hyperspace object and a set of 4-D scaling factors. Figure 20-2 plots the reference version of the object, while Figure 20-3 shows its scaled version.

Example 20-2

Here are the specifications for a "standard" hypercube:

$$H = \overline{S_1S_2S_3S_4S_5S_6S_7S_8}$$

$$S_1 = \overline{F_1F_2F_3F_4F_5F_6} \qquad S_2 = \overline{F_5F_{11}F_{15}F_{19}F_{21}F_{22}}$$
$$S_3 = \overline{F_7F_8F_9F_{10}F_{11}F_{12}} \qquad S_4 = \overline{F_6F_{12}F_{16}F_{20}F_{23}F_{24}}$$
$$S_5 = \overline{F_1F_7F_{13}F_{14}F_{15}F_{16}} \qquad S_6 = \overline{F_4F_{10}F_{14}F_{18}F_{22}F_{24}}$$
$$S_7 = \overline{F_3F_9F_{17}F_{18}F_{19}F_{20}} \qquad S_8 = \overline{F_2F_8F_{13}F_{17}F_{21}F_{23}}$$

$$F_1 = \overline{L_1L_3L_5L_7} \qquad F_2 = \overline{L_3L_{11}L_{19}L_{21}}$$
$$F_3 = \overline{L_9L_{11}L_{13}L_{15}} \qquad F_4 = \overline{L_7L_{15}L_{17}L_{23}}$$
$$F_5 = \overline{L_1L_9L_{17}L_{19}} \qquad F_6 = \overline{L_5L_{13}L_{21}L_{23}}$$
$$F_7 = \overline{L_2L_4L_6L_8} \qquad F_8 = \overline{L_4L_{12}L_{20}L_{22}}$$
$$F_9 = \overline{L_{10}L_{12}L_{14}L_{16}} \qquad F_{10} = \overline{L_8L_{16}L_{18}L_{24}}$$
$$F_{11} = \overline{L_2L_{10}L_{18}L_{20}} \qquad F_{12} = \overline{L_6L_{14}L_{22}L_{24}}$$
$$F_{13} = \overline{L_3L_4L_{26}L_{27}} \qquad F_{14} = \overline{L_7L_8L_{25}L_{28}}$$
$$F_{15} = \overline{L_1L_2L_{25}L_{26}} \qquad F_{16} = \overline{L_5L_6L_{27}L_{28}}$$

$F_{17} = \overline{L_{11}L_{12}L_{30}L_{31}}$ $F_{18} = \overline{L_{15}L_{16}L_{29}L_{30}}$ $L_1 = \overline{P_1P_3}$ $L_2 = \overline{P_2P_4}$ $L_3 = \overline{P_3P_5}$ $L_4 = \overline{P_4P_6}$

$F_{19} = \overline{L_9L_{10}L_{29}L_{30}}$ $F_{20} = \overline{L_{13}L_{14}L_{31}L_{32}}$ $L_5 = \overline{P_5P_7}$ $L_6 = \overline{P_6P_8}$ $L_7 = \overline{P_1P_7}$ $L_8 = \overline{P_2P_8}$

$F_{21} = \overline{L_{19}L_{20}L_{26}L_{30}}$ $F_{22} = \overline{L_{17}L_{18}L_{25}L_{29}}$ $L_9 = \overline{P_9P_{11}}$ $L_{10} = \overline{P_{10}P_{12}}$ $L_{11} = \overline{P_{11}P_{13}}$ $L_{12} = \overline{P_{12}P_{14}}$

$F_{23} = \overline{L_{21}L_{22}L_{27}L_{31}}$ $F_{24} = \overline{L_{23}L_{24}L_{28}L_{32}}$ $L_{13} = \overline{P_{13}P_{15}}$ $L_{14} = \overline{P_{14}P_{16}}$ $L_{15} = \overline{P_9P_{15}}$ $L_{16} = \overline{P_{10}P_{16}}$

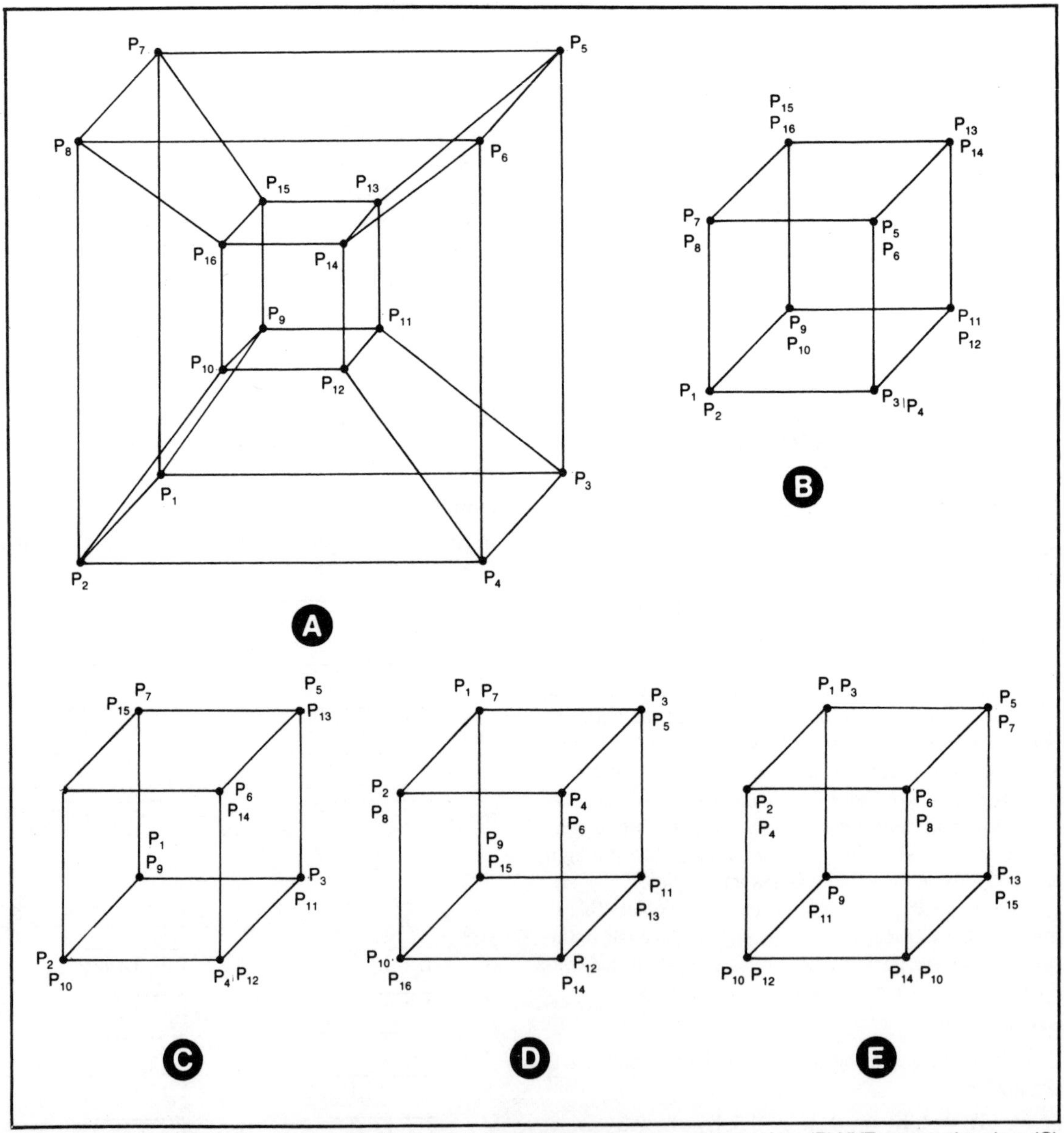

Fig. 20-2. The reference hypercube specified in Example 20-2. (A) A general hyperspace view. (B) XYZ perspective view. (C) XYW perspective view. (D) XZW perspective view. (E) YZW perspective view.

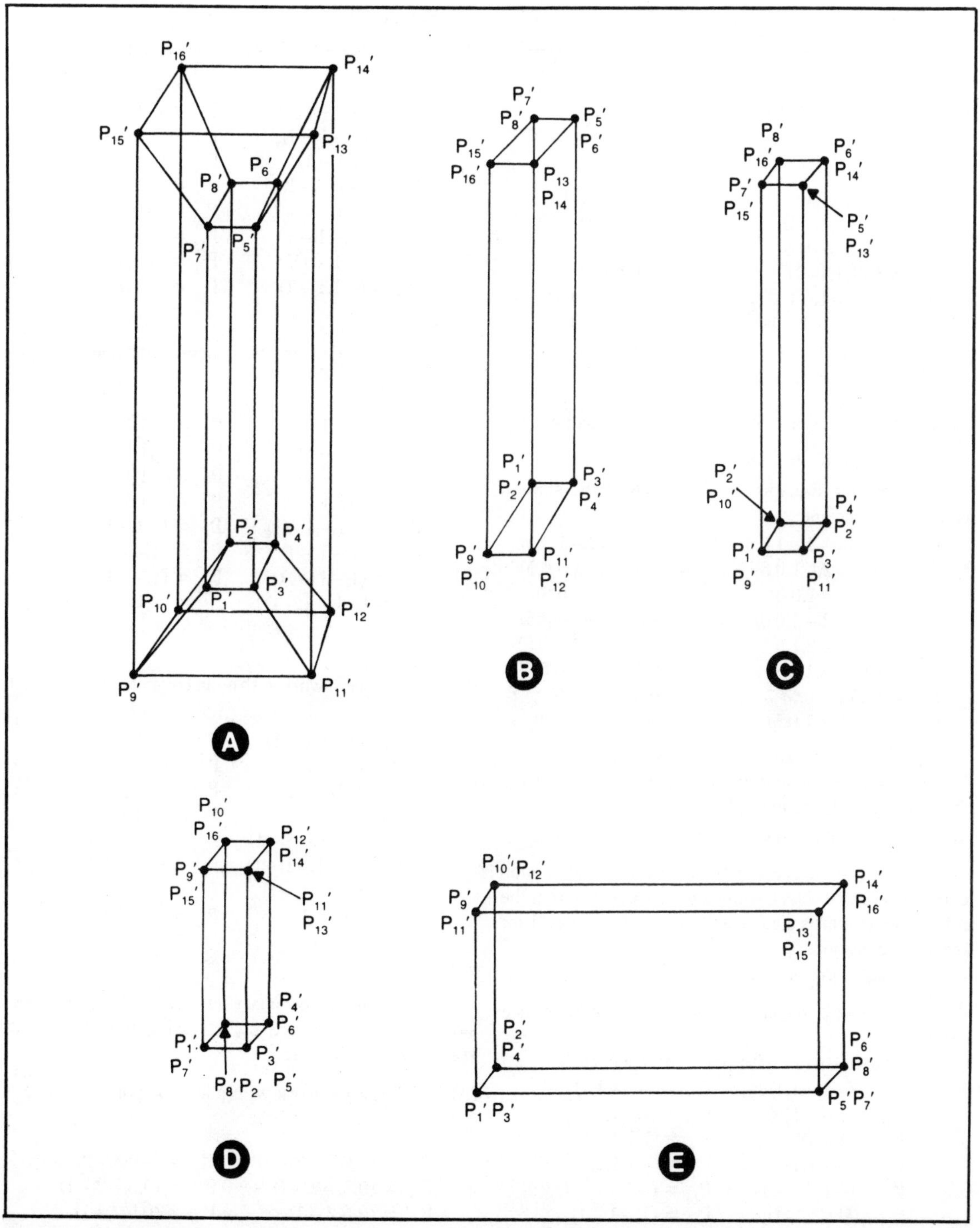

Fig. 20-3. A hypercube that is scaled as cited in Example 20-2. (A) A general hyperspace view. (B) XYZ perspective view. (C) XYW perspective view. (D) XZW perspective view. (E) YZW perspective view.

$L_{17} = \overline{P_1P_9}$ $L_{18} = \overline{P_2P_{10}}$ $L_{19} = \overline{P_3P_{11}}$ $L_{20} = \overline{P_4P_{12}}$
$L_{21} = \overline{P_5P_{13}}$ $L_{22} = \overline{P_6P_{14}}$ $L_{23} = \overline{P_7P_{15}}$ $L_{24} = \overline{P_8P_{16}}$
$L_{25} = \overline{P_1P_2}$ $L_{26} = \overline{P_3P_4}$ $L_{27} = \overline{P_5P_6}$ $L_{28} = \overline{P_7P_8}$
$L_{29} = \overline{P_9P_{10}}$ $L_{30} = \overline{P_{11}P_{12}}$ $L_{31} = \overline{P_{13}P_{14}}$ $L_{32} = \overline{P_{15}P_{16}}$

$P_1 = (-1,-1,1,-1)$ $P_2 = (-1,-1,1,1)$
$P_3 = (1,-1,1,-1)$ $P_4 = (1,-1,1,1)$
$P_5 = (1,1,1,-1)$ $P_6 = (1,1,1,1)$
$P_7 = (-1,1,1-1)$ $P_8 = (-1,1,1,1)$
$P_9 = (-1,-1,-1,-1)$ $P_{10} = (-1,-1,-1,1)$
$P_{11} = (1,-1,-1,-1)$ $P_{12} = (1,-1,-1,1)$
$P_{13} = (1,1,-1,-1)$ $P_{14} = (1,1,-1,1)$
$P_{15} = (-1,1,-1,-1)$ $P_{16} = (-1,1,-1,1)$

Subject that hyperspace object to this 4-D scaling:

$K_x = 0.5,\ K_y = 2,\ K_z = -1,\ K_w = -0.5$

1. Applying Equation 20-1 to each of the sixteen points, they become:

$P_1' = (-0.5,-2,-1,0.5)$ $P_2' = (-0.5,-2,-1,0.5)$
$P_3' = (0.5,-2,-1,0.5)$ $P_4' = (0.5,-2,-1,-0.5)$
$P_5' = (0.5,2,-1,0.5)$ $P_6' = (0.5,2,-1,-0.5)$
$P_7' = (-0.5,2,-1,0.5)$ $P_8' = (-0.5,2,-1,-0.5)$
$P_9' = (-0.5,-2,1,0.5)$ $P_{10}' = (-0.5,-2,1,-0.5)$
$P_{11}' = (0.5,-2,1,0.5)$ $P_{12}' = (0.5,-2,1,-0.5)$
$P_{13}' = (0.5,2,1,0.5)$ $P_{14}' = (0.5,2,1,-0.5)$
$P_{15}' = (-0.5,2,1,0.5)$ $P_{16}' = (-0.5,2,1,-0.5)$

2. Conduct a primary analysis of the reference hyperspace object, using the specifications as shown at the beginning of this example.

See the plot in Fig. 20-2A.

3. Rewrite the specifications for the scaled version of the hyperspace object, using the L-F-S-H data from the reference version as a guide and the results in Step 1 for the P specifications.

See the plot in Fig. 20-3A.

4. Conduct a secondary analysis of the reference version:

(A) XYZ Space—work with point components (x_n,y_n,z_n)

$P_1 = (-1,-1,1)$ $P_2 = (-1,-1,1)$
$P_3 = (1,-1,1)$ $P_4 = (1,-1,1)$
$P_5 = (1,1,1)$ $P_6 = (1,1,1)$
$P_7 = (-1,1,1)$ $P_8 = (-1,1,1)$
$P_9 = (-1,-1,-1)$ $P_{10} = (-1,-1,-1)$
$P_{11} = (1,-1,-1)$ $P_{12} = (1,-1,-1)$
$P_{13} = (1,1,-1)$ $P_{14} = (1,1,-1)$
$P_{15} = (-1,1,-1)$ $P_{16} = (-1,1,-1)$

See Fig. 20-2B.

(B) XYW Space—work with point components (x_n,y_n,w_n)

$P_1 = (-1,-1,-1)$ $P_2 = (-1,-1,1)$
$P_3 = (1,-1,-1)$ $P_4 = (1,-1,1)$
$P_5 = (1,1,-1)$ $P_6 = (1,1,1)$
$P_7 = (-1,1,-1)$ $P_8 = (-1,1,1)$
$P_9 = (-1,-1,-1)$ $P_{10} = (-1,-1,1)$
$P_{11} = (1,-1,-1)$ $P_{12} = (1,-1,1)$
$P_{13} = (1,1,-1)$ $P_{14} = (1,1,1)$
$P_{15} = (-1,1,-1)$ $P_{16} = (-1,1,1)$

See Fig. 20-2C.

(C) XZW Space—work with point components (x_n,z_n,w_n)

$P_1 = (-1,1,-1)$ $P_2 = (-1,1,1)$
$P_3 = (1,1,-1)$ $P_4 = (1,1,1)$
$P_5 = (1,1,-1)$ $P_6 = (1,1,1)$
$P_7 = (-1,1,-1)$ $P_8 = (-1,1,1)$
$P_9 = (-1,-1,-1)$ $P_{10} = (-1,-1,1)$
$P_{11} = (1,-1,-1)$ $P_{12} = (1,-1,1)$
$P_{13} = (1,-1,-1)$ $P_{14} = (1,-1,1)$
$P_{15} = (-1,-1,-1)$ $P_{16} = (-1,-1,1)$

See Fig. 20-2D.

(D) YZW Space—work with point components (y_n,z_n,w_n)

$P_1 = (-1,1,-1)$ $P_2 = (-1,1,1)$
$P_3 = (-1,1,-1)$ $P_4 = (-1,1,1)$
$P_5 = (1,1,-1)$ $P_6 = (1,1,1)$
$P_7 = (1,1,-1)$ $P_8 = (1,1,1)$
$P_9 = (-1,-1,-1)$ $P_{10} = (-1,-1,1)$
$P_{11} = (-1,-1,-1)$ $P_{12} = (-1,-1,1)$
$P_{13} = (1,-1,-1)$ $P_{14} = (1,-1,1)$
$P_{15} = (1,-1,-1)$ $P_{16} = (1,-1,1)$

See Fig. 20-2E.

5. Conduct a secondary analysis of the scaled version of the hyperspace object, using the 3-D points specified in each case:

(A) X'Y'Z' Space—work with point components (x_n',y_n',z_n')

$P_1' = (-0.5,-2,-1)$ $P_2' = (-0.5,-2,-1)$
$P_3' = (0.5,-2,-1)$ $P_4' = (0.5,-2,-1)$
$P_5' = (0.5,2,-1)$ $P_6' = (0.5,2,-1)$
$P_7' = (-0.5,2,-1)$ $P_8' = (-0.5,2,-1)$
$P_9' = (-0.5,-2,1)$ $P_{10}' = (-0.5,-2,1)$

$P_{11}' = (0.5,-2,1)$ $P_{12}' = (0.5,-2,1)$
$P_{13}' = (0.5,2,1)$ $P_{14}' = (0.5,2,1)$
$P_{15}' = (-0.5,2,1)$ $P_{16}' = (-0.5,2,1)$

See Fig. 20-3B.

(B) X′Y′W′ Space—work with point components (x_n', y_n' w_n')

$P_1' = (-0.5,-2,0.5)$ $P_2' = (-0.5,-2,0.5)$
$P_3' = (0.5,-2,0.5)$ $P_4' = (0.5,-2,-0.5)$
$P_5' = (0.5,2,0.5)$ $P_6' = (0.5,2,-0.5)$
$P_7' = (-0.5,2,0.5)$ $P_8' = (-0.5,2,-0.5)$
$P_9' = (-0.5,-2,0.5)$ $P_{10}' = (-0.5,-2,-0.5)$
$P_{11}' = (0.5,-2,0.5)$ $P_{12}' = (0.5,-2,-0.5)$
$P_{13}' = (0.5,2,0.5)$ $P_{14}' = (0.5,2,-0.5)$
$P_{15}' = (-0.5,2,0.5)$ $P_{16}' = (-0.5,2,-0.5)$

See Fig. 20-3C.

(C) X′Z′W′ Space—work with point components (x_n' z_n',w_n')

$P_1' = (-0.5,-1,0.5)$ $P_2' = (-0.5,-1,0.5)$
$P_3' = (0.5,-1,0.5)$ $P_4' = (0.5,-1,-0.5)$
$P_5' = (0.5,-1,0.5)$ $P_6' = (0.5,-1,-0.5)$
$P_7' = (-0.5,-1,0.5)$ $P_8' = (-0.5,-1,-0.5)$
$P_9' = (-0.5,1,0.5)$ $P_{10}' = (-0.5,1,-0.5)$
$P_{11}' = (0.5,1,0.5)$ $P_{12}' = (0.5,1,-0.5)$
$P_{13}' = (0.5,1,0.5)$ $P_{14}' = (0.5,1,-0.5)$
$P_{15}' = (-0.5,1,0.5)$ $P_{16}' = (-0.5,1,-0.5)$

See Fig. 20-3D.

(D) Y′Z′W′ Space—work with point components (y_n', z_n',w_n')

$P_1' = (-2,-1,0.5)$ $P_2' = (-2,-1,-0.5)$
$P_3' = (-2,-1,0.5)$ $P_4' = (-2,-1,-0.5)$
$P_5' = (2,-1,0.5)$ $P_6' = (2,-1,-0.5)$
$P_7' = (2,-1,0.5)$ $P_8' = (2,-1,-0.5)$
$P_9' = (-2,1,0.5)$ $P_{10}' = (-2,1,-0.5)$
$P_{11}' = (-2,1,0.5)$ $P_{12}' = (-2,1,-0.5)$
$P_{13}' = (2,1,0.5)$ $P_{14}' = (2,1,-0.5)$
$P_{15}' = (2,1,0.5)$ $P_{16}' = (2,1,-0.5)$

See Fig. 20-3E.

Step 1 in that Example summarizes the results of applying the given scaling factors to the points specified for the reference version of the hyperspace object. Those two sets of points—the points for the reference object and the points for the scaled version—are the main working elements of the entire experiment. The remaining specifications simply show how the points are to be tied together to form the lines, planes and spaces of the hyperspace objects.

The remainder of the work in that example concern the development of the primary and secondary analyses of the situation. Steps 2 and 3, for instance, suggest a primary analysis; an analysis of the reference and scaled objects from their true, hyperspace perspectives. I have not completed the analysis in detail, so it is left to you to calculate the lengths, direction cosines and angles for both versions.

Step 4 outlines the procedure and cites the plotted points for conducting a secondary analysis of the reference version of the hyperspace object. The general idea is to use only those components of each coordinate that are relevant to the space at hand. Those coordinates are the stuff of the mathematical analysis, and you should use them for determining the lengths, direction cosines and angles as they appear mapped to a 3-D space.

Step 5 outlines the same sort of secondary analysis, but beginning with the scaled version of the hyperspace object.

20-6 ALTERNATIVE VERSIONS OF THE GEOMETRIC EQUATION

Equation 20-2 is the basic equation for geometric scaling reorganized to solve for the scaling terms. Given the coordinate of a reference point and a scaled version of that point, the expressions lead directly to the scaling that exists between them.

$$K_x = x_n/x_n' \qquad K_z = z_n/z_n'$$
$$K_y = y_n/y_n' \qquad K_w = w_n/w_n'$$

Equation 20-2

where:

x_n',y_n',z_n',w_n' = the coordinate of the scaled version of reference point n, relative to the chosen frame of reference

x_n,y_n,z_n,w_n = the coordinate of reference point n, relative to the chosen frame of reference

K_x, K_y, K_z, and K_w = the scaling factors for the coordinate axes, relative to the chosen frame of reference

Earlier discussions of scalings in lower-dimensioned spaces cited appropriate versions of this equation as a means for determining the scaling that exists between two coordinate systems. That can be done in the case of coordinate systems that are separated by a scaling in 4-D space as well.

The equation can be of great practical importance when using it to change one or more of the dimensions of a geometric entity that happens to be specified in some convenient, but less desirable form. The general idea is to make a pyramid taller, for example, by applying a positive scaling greater than 1 to the vertical axis of its coordinate system. The scaled coordinates can then become the basis for a newly specified pyramid.

Equation 20-3 is the inverse form of the basic equation for geometric scalings in 4-D space. Taken in its most literal form, it lets you calculate the reference-point coordinates, given the scaled version of the coordinates and the scaling terms. It leads quite naturally to the equation for relativistic scalings.

$$x_n = x_n'/K_x \qquad z_n = z_n'/K_z$$
$$y_n = y_n'/K_y \qquad w_n = w_n'/K_w \qquad \textbf{Equation 20-3}$$

where:

x_n', y_n', z_n', w_n' = the coordinate of the scaled version of reference point *n*, relative to the chosen frame of reference

x_n, y_n, z_n, w_n = the coordinate of reference point *n*, relative to the chosen frame of reference

K_x, K_y, K_z, and K_w = the scaling factors for the coordinate axes, relative to the chosen frame of reference

Note: This is the inverse form of Equation 20-1

20-7 RELATIVISTIC SCALINGS IN 4-D

It is easy to lose sight of the fact that a scaling—a scaling in any number of dimensions—really represents a scaling that exists between two coordinate systems. Specifically, a geometric scaling in 4 dimensions actually represents a scaling-type offset between a 4-D reference coordinate system and another 4-D system that is otherwise identical to it. A couple of discussions earlier in this chapter regard a 4-D scaling as a particular means for changing the size of a geometric entity in one or more dimensions in 4-D space. Both views are quite valid, and they serve particular purposes.

In the context of relativistic scalings, however, the idea of changing the size of an object is not very meaningful. Here it is better to view a scaling as an expression of the size difference, or reflective change in the senses of the axes, between two otherwise identical 4-D coordinate systems.

$$x_n{}^* = x_n/K_x \qquad z_n{}^* = z_n/K_z$$
$$y_n{}^* = y_n/K_y \qquad w_n{}^* = w_n/K_w \qquad \textbf{Equation 20-4}$$

where:

x_n, y_n, z_n, w_n = the coordinate of reference point *n* relative to the observer's frame of reference

$x_n{}^*, y_n{}^*, z_n{}^*, w_n{}^*$ = the coordinate of reference point *n* relative to the scaled frame of reference

K_x = the X-axis scaling that exists between the observer's and scaled frames of reference, relative to the observer's

K_y = the Y-axis scaling that exists between the observer's and scaled frames of reference, relative to the observer's

K_z = the Z-axis scaling that exists between the observer's and scaled frames of reference, relative to the observer's

K_w = the W-axis scaling that exists between the observer's and scaled frames of reference, relative to the observer's

Equation 20-4 represents the basic relativistic version of a 4-dimensional scaling that exists between two 4-D coordinate systems. Used in its most direct fashion, it turns up the coordinates of some point in 4-D space as regarded from a scaled frame of reference. Generally the application begins by specifying some geometric entity as it appears from the viewpoint of the observer's 4-D frame of reference. The next step is to establish the scaling factors for the scaling of a second frame of reference, relative to the observer's frame of reference. The equation then shows how the entity appears relative to the second frame of reference. Example 20-3 illustrates the situation with a specific example.

Example 20-3

Here are the specifications for a hyperspace version of a simple prism:

$$H = \overline{S_1S_2S_3S_4S_5S_6S_7S_8}$$

$$S_1 = \overline{F_1F_2F_3F_4F_5} \qquad S_2 = \overline{F_6F_7F_8F_9F_{10}}$$
$$S_3 = \overline{F_1F_6F_{11}F_{12}F_{13}} \qquad S_4 = \overline{F_{14}F_{15}F_{16}F_{17}F_{18}}$$
$$S_5 = \overline{F_2F_7F_{11}F_{15}F_{19}F_{20}} \qquad S_6 = \overline{F_4F_9F_{13}F_{18}F_{20}F_{21}}$$
$$S_7 = \overline{F_3F_8F_{12}F_{14}F_{19}F_{21}} \qquad S_8 = \overline{F_4F_9F_{13}F_{18}F_{20}F_{21}}$$

$$F_1 = \overline{L_2L_3L_9} \qquad F_2 = \overline{L_2L_4L_{10}L_{18}} \qquad F_3 = \overline{L_3L_4L_{14}L_{17}}$$
$$F_4 = \overline{L_9L_{10}L_{14}L_{22}} \qquad F_5 = \overline{L_{17}L_{18}L_{22}} \qquad F_6 = \overline{L_5L_6L_{11}}$$
$$F_7 = \overline{L_5L_7L_{12}L_{20}} \qquad F_8 = \overline{L_6L_7L_{15}L_{19}} \qquad F_9 = \overline{L_{11}L_{12}L_{15}L_{23}}$$
$$F_{10} = \overline{L_{19}L_{20}L_{23}} \qquad F_{11} = \overline{L_1L_2L_5L_8} \qquad F_{12} = \overline{L_1L_3L_6L_{19}}$$
$$F_{13} = \overline{L_8L_9L_{11}} \qquad F_{14} = \overline{L_{16}L_{17}L_{19}L_{21}} \qquad F_{15} = \overline{L_{19}L_{20}L_{23}}$$
$$F_{16} = \overline{L_{21}L_{22}L_{23}L_{24}} \qquad F_{17} = \overline{L_1L_4L_7L_{16}} \qquad F_{18} = \overline{L_8L_{10}L_{12}L_{24}}$$
$$F_{19} = \overline{L_{13}L_{14}L_{15}L_{21}}$$

$L_1 = \overline{P_1P_2}$ $L_2 = \overline{P_1P_3}$ $L_3 = \overline{P_1P_5}$
$L_4 = \overline{P_1P_7}$ $L_5 = \overline{P_2P_4}$ $L_6 = \overline{P_2P_6}$
$L_7 = \overline{P_2P_8}$ $L_8 = \overline{P_3P_4}$ $L_9 = \overline{P_3P_5}$
$L_{10} = \overline{P_3P_{11}}$ $L_{11} = \overline{P_4P_6}$ $L_{12} = \overline{P_4P_{12}}$
$L_{13} = \overline{P_5P_6}$ $L_{14} = \overline{P_5P_9}$ $L_{15} = \overline{P_6P_{10}}$
$L_{16} = \overline{P_7P_8}$ $L_{17} = \overline{P_7P_9}$ $L_{18} = \overline{P_7P_{11}}$
$L_{19} = \overline{P_8P_{10}}$ $L_{20} = \overline{P_8P_{12}}$ $L_{21} = \overline{P_9P_{10}}$
$L_{22} = \overline{P_9P_{11}}$ $L_{23} = \overline{P_{10}P_{12}}$ $L_{24} = \overline{P_{11}P_{12}}$

$P_1 = (0,0,0,0)$ $P_2 = (0,0,0,1)$ $P_3 = (1,0,0,0)$
$P_4 = (1,0,0,1)$ $P_5 = (0,1,0,0)$ $P_6 = (0,1,0,1)$
$P_7 = (0,0,1,0)$ $P_8 = (0,0,1,1)$ $P_9 = (0,1,1,0)$
$P_{10} = (0,1,1,1)$ $P_{11} = (1,0,1,0)$ $P_{12} = (1,0,1,1)$

Assuming that those specifications are taken from an observer's frame of reference, what will they become as viewed from an alternative frame of reference that is scaled this way with respect to the observer's frame of reference?

$$K_x = 1/2,\ K_y = -1,\ K_z = -1/2,\ K_w = 2$$

1. Applying the basic 4-D relativistic scaling equation, Equation 20-4, to the twelve points of the hyperspace object:

$P_1^* = (0,0,0,0)$ $P_2^* = (0,0,0,1/2)$
$P_3^* = (2,0,0,0)$ $P_4^* = (2,0,0,1/2)$
$P_5^* = (0,-1,0,0)$ $P_6^* = (0,-1,0,1/2)$
$P_7^* = (0,0,-2,0)$ $P_8^* = (0,0,-2,1/2)$
$P_9^* = (0,-1,-2,0)$ $P_{10}^* = (0,-1,-2,1/2)$
$P_{11}^* = (2,0,-2,0)$ $P_{12}^* = (2,0,-2,1/2)$

2. Conduct a primary analysis of the hyperspace object as viewed from the observer's frame of reference by plotting and using the hyperspace points originally specified. See the plot in Fig. 20-4A.

3. Conduct a primary analysis of the hyperspace object as viewed from the scaled frame of reference by plotting the P* points that result in Step 1. See the plot in Fig. 20-5A.

4. Conduct a full secondary analysis of the reference object as viewed from the four different 3-D observer's frames of reference:

(A) XYZ Space—plot and work with points (x_n,y_n,z_n)

$P_1 = (0,0,0)$ $P_2 = (0,0,0)$ $P_3 = (1,0,0)$
$P_4 = (1,0,0)$ $P_5 = (0,1,0)$ $P_6 = (0,1,0)$
$P_7 = (0,0,1)$ $P_8 = (0,0,1)$ $P_9 = (0,1,1)$
$P_{10} = (0,1,1)$ $P_{11} = (1,0,1)$ $P_{12} = (1,0,1)$

See Fig. 20-4B.

(B) XYW Space—plot and work with points (x_n,y_n,w_n)

$P_1 = (0,0,0)$ $P_2 = (0,0,1)$ $P_3 = (1,0,0)$
$P_4 = (1,0,1)$ $P_5 = (0,1,0)$ $P_6 = (0,1,1)$
$P_7 = (0,0,0)$ $P_8 = (0,0,1)$ $P_9 = (0,1,0)$
$P_{10} = (0,1,1)$ $P_{11} = (1,0,0)$ $P_{12} = (1,0,1)$

See Fig. 20-4C.

(C) XZW Space—plot and work with points (x_n,z_n,w_n)

$P_1 = (0,0,0)$ $P_2 = (0,0,1)$ $P_3 = (1,0,0)$
$P_4 = (1,0,1)$ $P_5 = (0,0,0)$ $P_6 = (0,0,1)$
$P_7 = (0,1,0)$ $P_8 = (0,1,1)$ $P_9 = (0,1,0)$
$P_{10} = (0,1,1)$ $P_{11} = (1,1,0)$ $P_{12} = (1,1,1)$

See Fig. 20-4D.

(D) YZW Space—plot and work with points (y_n,z_n,w_n)

$P_1 = (0,0,0)$ $P_2 = (0,0,1)$ $P_3 = (0,0,0)$
$P_4 = (0,0,1)$ $P_5 = (1,0,0)$ $P_6 = (1,0,1)$
$P_7 = (0,1,0)$ $P_8 = (0,1,1)$ $P_9 = (1,1,0)$
$P_{10} = (1,1,1)$ $P_{11} = (0,1,0)$ $P_{12} = (0,1,1)$

See Fig. 20-4E.

5. Conduct a full secondary analysis of the reference object as viewed from the four alternative, 3-D scaled spaces:

(A) X*Y*Z* Space—plot and work with point (x_n^*, y_n^*, z_n^*)

$P_1^* = (0,0,0)$ $P_2^* = (0,0,0)$
$P_3^* = (2,0,0)$ $P_4^* = (2,0,0)$
$P_5^* = (0,-1,0)$ $P_6^* = (0,-1,0)$
$P_7^* = (0,0,-2)$ $P_8^* = (0,0,-2)$
$P_9^* = (0,-1,-2)$ $P_{10}^* = (0,-1,-2)$
$P_{11}^* = (2,0,-2)$ $P_{12}^* = (2,0,-2)$

See Fig. 20-5B.

(B) X*Y*W* Space—plot and work with point (x_n^*, y_n^*, w_n^*)

$P_1^* = (0,0,0)$ $P_2^* = (0,0,1/2)$
$P_3^* = (2,0,0)$ $P_4^* = (2,0,1/2)$
$P_5^* = (0,-1,0)$ $P_6^* = (0,-1,1/2)$
$P_7^* = (0,0,0)$ $P_8^* = (0,0,1/2)$
$P_9^* = (0,-1,0)$ $P_{10}^* = (0,-1,1/2)$
$P_{11}^* = (2,0,0)$ $P_{12}^* = (2,0,1/2)$

See Fig. 20-5C.

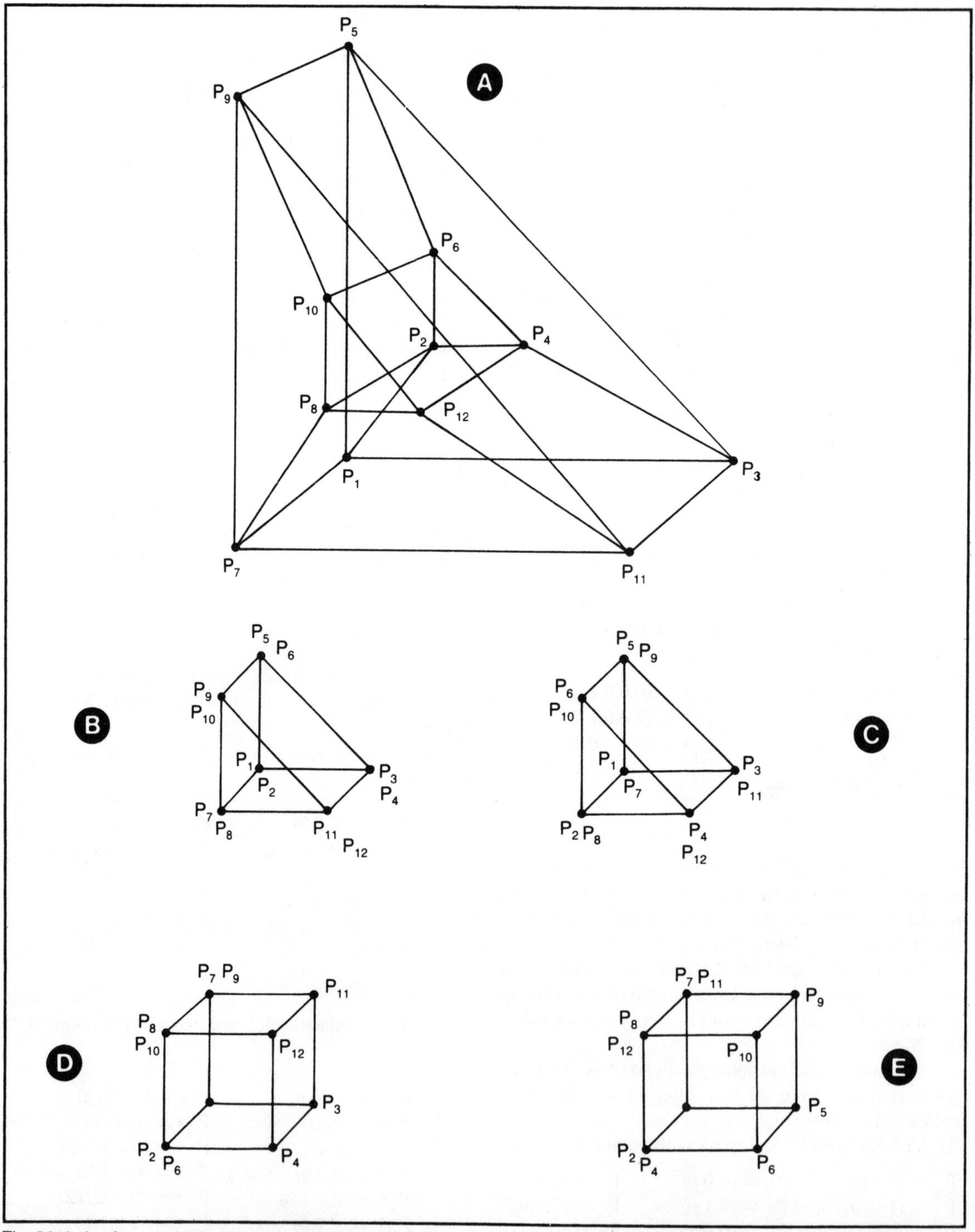

Fig. 20-4. A reference hyperprism as viewed from the observer's frame of reference and cited in Example 20-3. (A) A general hyperspace view. (B) XYZ perspective view. (C) XYW perspective view. (D) XZW perspective view. (E) YZW perspective view.

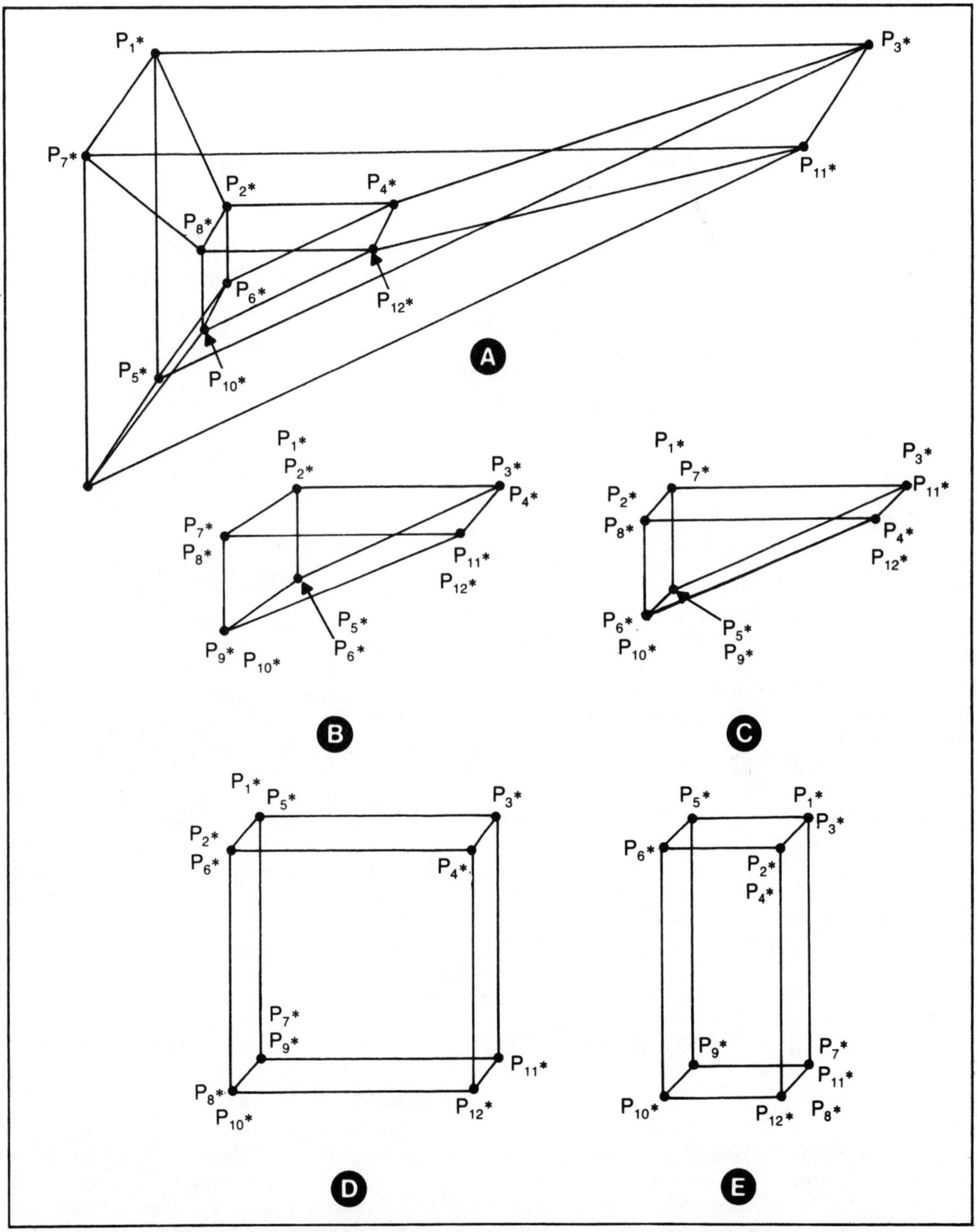

Fig. 20-5. A reference hyperprism as viewed from a scaled frame of reference—see Example 20-3. (A) A general hyperspace view. (B) XYZ perspective view. (C) XYW perspective view. (D) XZW perspective view. (E) YZW perspective view.

(C) X*Z*W* Space—plot and work with point (x_n*, z_n*,w_n*)

$P_1* = (0,0,0)$ $P_2* = (0,0,1/2)$
$P_3* = (2,0,0)$ $P_4* = (2,0,1/2)$
$P_5* = (0,0,0)$ $P_6* = (0,0,1/2)$
$P_7* = (0,-2,0)$ $P_8* = (0,-2,1/2)$
$P_9* = (0,-2,0)$ $P_{10}* = (0,-2,1/2)$
$P_{11}* = (2,-2,0)$ $P_{12}* = (2,-2,1/2)$

See Fig. 20-5D.

(D) Y*Z*W* Space—plot and work with point (y_n*, z_n*,w_n*)

$P_1* = (0,0,0)$ $P_2* = (0,0,1/2)$
$P_3* = (0,0,0)$ $P_4* = (0,0,1/2)$
$P_5* = (-1,0,0)$ $P_6* = (-1,0,1/2)$
$P_7* = (0,-2,0)$ $P_8* = (0,-2,1/2)$
$P_9* = (-1,-2,0)$ $P_{10}* = (-1,-2,1/2)$
$P_{11}* = (0,-2,0)$ $P_{12}* = (0,-2,1/2)$

See Fig. 20-5E.

The example opens by citing a particular hyperspace object and a scaling that is to exist between the observer's and scaled frames of reference. The original specifications represent the viewpoint of the observer, and the objective of the experiment is to apply Equation 20-4 to that data to generate a hyperspace view of the object as seen from the scaled frame of reference. See the drawings in Figures 20-4 and 20-5.

You will find that the example shows only a partial primary and secondary analysis of the situation. It is up to you to generate the lengths, direction cosines and angles for both versions of the object.

Equation 20-5 lets you determine the 4-D scaling that exists between two 4-dimensional frames of reference, given the coordinates of a pair of points from both the observer's and translated frame of reference. See Example 20-6.

$$K_x = x_n/x_n*$$
$$K_y = y_n/y_n*$$
$$K_z = z_n/z_n*$$
$$K_w = w_n/w_n*$$

Equation 20-5

where:

x_n,y_n,z_n,w_n = the coordinate of reference point *n* relative to the observer's frame of reference

x_n*,y_n*,z_n*,w_n* = the coordinate of reference point *n* relative to the scaled frame of reference

K_x = the X-axis scaling that exists between the observer's and scaled frames of reference, relative to the observer's

K_y = the Y-axis scaling that exists between the observer's and scaled frames of reference, relative to the observer's

K_z = the Z-axis scaling that exists between the observer's and scaled frames of reference, relative to the observer's

K_w = the W-axis scaling that exists between the observer's and scaled frames of reference, relative to the observer's

I am including the inverse form of the relativistic 4-D scaling equation only as a matter of record. It follows the same general pattern as its lower-dimension versions.

$$x_n = x_n*/K_x$$
$$y_n = y_n*/K_y$$
$$z_n = z_n*/K_z$$
$$w_n = w_n*/K_w$$

Equation 20-6

where:

x_n,y_n,z_n,w_n = the coordinate of reference point *n* relative to the observer's frame of reference

x_n*,y_n*,z_n*,w_n* = the coordinate of reference point *n* relative to the scaled frame of reference

K_x = the X-axis scaling that exists between the observer's and scaled frames of reference, relative to the observer's

K_y = the Y-axis scaling that exists between the observer's and scaled frames of reference, relative to the observer's

K_z = the Z-axis scaling that exists between the observer's and scaled frames of reference, relative to the observer's

K_w = the W-axis scaling that exists between the observer's and scaled frames of reference, relative to the observer's

Note: This is the inverse form of Equation 20-4.

20-8 COMBINATIONS OF 4-D SCALINGS

Multiple scalings in 4-D space are commutative. That is, you can apply a given series of scalings in any desired order and still end up with the same overall scaling. This, of course, applies to geometric scalings as well as their relativistic counterparts.

Rotations in 4-D Space

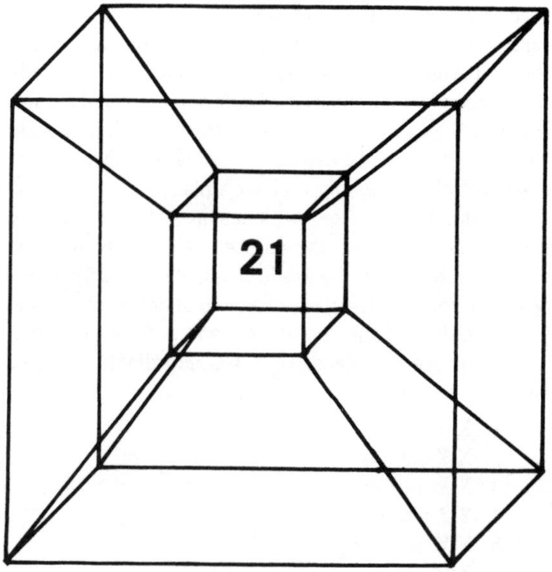

The rotation equations for points in 4-D space aren't much more difficult to solve than those for 3-D space. If you had no real difficulty rotating objects in 3-D space, it is unlikely that you will encounter any special difficulties here. Visualizing the situation is a different matter, however.

It is difficult to imagine what a 4-D rotation might look like from a 4-D point of view, and we are forced to work with it in 3-dimensional bits and pieces. Here is a step-by-step account of the evolution of rotations:

1. Rotations are not defined for 1-D space.
2. Rotations in 2-D space take place about a point of rotation.
3. Rotations in 3-D space take place about a line, or axis, of rotation.
4. Rotations in 4-D space take place about a plane of rotation.

Perhaps you can appreciate the fact that a rotation in a space of a given number of dimensions takes place about a geometric element of a space that is two dimensions lower. So rotations are impossible in 1-D space because we have not defined a space of -1 dimensions. Rotations in 2-D space are possible because $2-1 = 0$; and a 0-D element is a point—2-D rotations take place about a point. By the same token, rotations in 3-D space take place about a line; a line is the basic element of $3-2 = 1$, or 1-D space. And breaking new ground by extending that line of reasoning, it follows that rotations in 4-D space take place about the basic element of 2-D space—about a plane.

It is difficult enough to visualize 4-D space, but the notion of rotating a 4-dimensional coordinate system about a plane is the most abstract idea offered in this book. (If you think you can visualize it, you probably don't understand it properly.) The only way to get the barest hold on the subject is through the tools of analytic geometry; without that, the entire subject of 4-D rotation is nothing but idle speculation.

21-1 THE EQUATIONS OF 4-D ROTATION

Recall from Chapter 14 that simple rotations in 3-D space take place about one of the three axes of the coordinate system—about the X axis, the Y axis, or the Z axis. Furthermore, there are three different 3-D rotation equations: one for each of the possible axes of rotation.

The three equations appear quite similar in two respects. First, they use the same pattern of sine and cosine functions of the angle of rotation. And second, the components of points corresponding to the chosen axis of rotation are not affected by the rotation equations. Doing an X-axis rotation, for example, does not affect the X-axis components of any point in the space.

It so happens that the equations for rotations in 4-D space work much the same way. The equations use the same pattern of sine and cosine functions for the angle of rotation. You will soon see what I mean by that. The only difference is that there are two different axes that are not affected by a given rotation equation in 4 dimensions. Two axes define a plane, thus it is proper to conclude that a 4-D rotation takes place about a plane.

Count the number of planes in a 4-D coordinate system, and you have the number of different rotation equations that must be made available. Identify those planes in terms of the pairs of axes that define them, and you have defined the rotation equations for 4-D space.

What are the planes of a 4-D coordinate system? Here is the list:

1. X-Y plane
2. X-Z plane
3. X-W plane
4. Y-Z plane
5. Y-W plane
6. Z-W plane

There are six different planes in a 4-D coordinate system, so from the foregoing discussion, it follows that there must be six different rotation equations for 4-D space—one for each of the possible planes of rotation. See them defined completely in Equations 21-1 through 21-6.

$$\begin{aligned} x_n' &= x_n\cos\theta - y_n\sin\theta \\ y_n' &= x_n\sin\theta + y_n\cos\theta \\ z_n' &= z_n \\ w_n' &= w_n \end{aligned}$$

Equation 21-1

where:

x_n', y_n', z_n', w_n' = the coordinate of the rotated version of point *n* relative to the frame of reference

x_n, y_n, z_n, w_n = the coordinate of point *n* relative to the frame of reference

θ = the angle of rotation about the Z-W plane

Note: This set of expressions applies only to rotations about the Z-W plane of the frame of reference.

$$\begin{aligned} x_n' &= x_n\cos\theta - z_n\sin\theta \\ y_n' &= y_n \\ z_n' &= x_n\sin\theta + z_n\cos\theta \\ w_n' &= w_n \end{aligned}$$

Equation 21-2

where:

x_n', y_n', z_n', w_n' = the coordinate of the rotated version of point *n* relative to the frame of reference

x_n, y_n, z_n, w_n = the coordinate of point *n* relative to the frame of reference

θ = the angle of rotation about the Y-W plane

Note: These expressions apply only to rotations about the Y-W plane of the frame of reference.

$$\begin{aligned} x_n' &= x_n \\ y_n' &= y_n\cos\theta - z_n\sin\theta \\ z_n' &= y_n\sin\theta + z_n\cos\theta \\ w_n' &= w_n \end{aligned}$$

Equation 21-3

where:

x_n', y_n', z_n', w_n' = the coordinate of the rotated version of point *n* relative to the frame of reference

x_n, y_n, z_n, w_n = the coordinate of point *n* relative to the frame of reference

θ = the angle of rotation about the X-W plane

Note: These expressions apply only to rotations about the X-W plane of the frame of reference.

$$\begin{aligned} x_n' &= x_n\cos\theta - w_n\sin\theta \\ y_n' &= y_n \\ z_n' &= z_n \\ w_n' &= x_n\sin\theta + w_n\cos\theta \end{aligned}$$

Equation 21-4

where:

x_n', y_n', z_n', w_n' = the coordinate of the rotated version of point *n* relative to the frame of reference

x_n, y_n, z_n, w_n = the coordinate of point *n* relative to the frame of reference

θ = the angle of rotation about the Y-Z plane

Note: These expressions apply only to rotations about the Y-Z plane of the frame of reference.

$$\begin{aligned} x_n' &= x_n \\ y_n' &= y_n\cos\theta - w_n\sin\theta \\ z_n' &= z_n \\ w_n' &= y_n\sin\theta + w_n\cos\theta \end{aligned}$$

Equation 21-5

where:

x_n', y_n', z_n', w_n' = the coordinate of the rotated version of point *n* relative to the frame of reference

x_n, y_n, z_n, w_n = the coordinate of point *n* relative to the frame of reference

θ = the angle of rotation about the X-Z plane

Note: These expressions apply only to rotations about the X-Z plane of the frame of reference.

$$\begin{aligned} x_n' &= x_n \\ y_n' &= y_n \\ z_n' &= z_n\cos\theta - w_n\sin\theta \\ w_n' &= z_n\sin\theta + w_n\cos\theta \end{aligned}$$

Equation 21-6

where:

x_n', y_n', z_n', w_n' = the coordinate of rotated point *n* relative to the frame of reference

x_n, y_n, z_n, w_n = the coordinate of point *n* relative to the frame of reference.

θ = the angle of rotation about the X-Y plane

Note: These expressions apply only to rotations about the X-Y plane of the frame of reference.

Compare those equations with the 3-D versions in Chapter 14, and you will see the same pattern of sine and cosine functions of the angle of rotation. The big difference is that each 4-D rotation has two unaffected axes in each case, instead of one. That, in an analytical sense, is the only difference between 3-D and 4-D rotations. Generally speaking 4-D space and rotations in that space are simple extensions of the same effects in our more familiar 3-D space. The fourth dimension is analytically simple, but it is perceptually difficult.

Example 21-1 demonstrates how easy it is to rotate a single point in 4-D space about a plane of the coordinate system.

Example 21-1

Let P = (1,–1,1,–1) be a reference point that is specified in hyperspace. Beginning with that reference point in each case, calculate and plot its rotated coordinate under:

(1) A Z-W plane rotation of 30 degrees
(2) A Y-W plane rotation of 30 degrees
(3) An XW plane rotation of 30 degrees
(4) A Y-Z plane rotation of 30 degrees
(5) An X-Z plane rotation of 30 degrees
(6) An X-Y plane rotation of 30 degrees

In all of those instances:

$x = 1 \quad y = -1 \quad z = 1 \quad w = -1$

$\cos 30° = \sqrt{3}/2 \qquad \sin 30° = 1/2$

1. Using Equation 21-1 to perform the Z-W planar rotation:

$x' = x\cos\theta - \sin\theta$ | $y' = x\sin\theta + y\cos\theta$
$x' = 1\,(\sqrt{3}/2) - (-1)\,(1/2)$ | $y' = 1\,(1/2) + (-1)(\sqrt{3}/2)$
$x' = (\sqrt{3} + 1)/2$ | $y' = (1 - \sqrt{3})/2$

$z' = z$ | $w' = w$
$z' = 1$ | $w' = -1$

(A) For a true, primary-level hyperspace analysis:
P = (1,–1,1,–1) is rotated to P′ = (1.37,–0.37,1,–1)
(B) For a secondary analysis in XYZ space:
P = (1,–1,1) is rotated to P′ = (1.37,–0.37,1)
(C) For a secondary analysis in XYW space:
P = (1,–1,–1) is rotated to P′ = (1.37,–0.37,–1)
(D) For a secondary analysis in XZW space:
P = (1,1,–1) is rotated to P′ = (1.37,1,–1)
(E) For a secondary analysis in YZW space:
P = (–1,1,–1) is rotated to P′ = (–0.37,1,–1)

2. Using Equation 21-2 to perform the Y-W planar rotation:

$x' = x\cos\theta - z\sin\theta$ | $y' = y$
$x' = 1(\sqrt{3}/2) - 1(1/2)$ | $y' = -1$
$x' = (\sqrt{3} - 1)/2$

$z' = x\sin + z\cos$ | $w' = w$
$z' = 1(1/2) + 1(\sqrt{3}/2)$ | $w' = -1$
$z' = (1 + \sqrt{3}/2$

(A) For a true, primary-level hyperspace analysis:
P = (1,–1,1,–1) is rotated to P′ = (0.37,–1,1.37,–1)
(B) For a secondary analysis in XYZ space:
P = (1,–1,1) is rotated to P′ = (0.37,–1,1.37)
(C) For a secondary analysis in XYW space:
P = (1,–1,–1) is rotated to P′ = (0.37,–1,–1)
(D) For a secondary analysis in XZW space:
P = (1,–1,1,–1) is rotated to P′ = (0.37,1.37,–1)
(E) For a secondary analysis in YZW space:
P = (–1,1,–1) is rotated to P′ = (–1,1.37,–1)

3. Using Equation 21-3 to perform the X-W planar rotation:

$x' = x$ | $y' = y\cos\theta - z\sin\theta$
$x' = 1$ | $y' = -1(\sqrt{3}/2) - 1(1/2)$
| $y' = -(\sqrt{3} + 1)/2$

$z' = y\sin + z\cos$ | $w' = w$
$z' = -1(1/2) + 1(\sqrt{3}/2)$ | $w' = -1$
$z' = (\sqrt{3} - 1)/2$

(A) For a true, primary-level hyperspace analysis:
Point P=(1,–1,1,–1) is rotated to P′=(1,–1.37,0.37,–1)
(B) For a secondary analysis in XYZ space:
Point P = (1,–1,1) is rotated to P′ = (1,–1.37,0.37)
(C) For a secondary analysis in XYW space:
Point P = (1,–1,–1) is rotated to P′ = (1,–1.37,–1)
(D) For a secondary analysis in XZW space:
Point P = 1,1,–1) is rotated to P′ = (1,0.37,1)
(E) For a secondary analysis in YZW space:
Point P = (–1,1,–1) is rotated to P′ = (–1.37,0.37,–1)

4. Using Equation 21-4 to perform the Y-Z planar rotation:

$x' = x\cos\theta - w\sin\theta$ | $y' = y$
$x' = 1(\sqrt{3}/2) - (-1)(1/2)$ | $y' = -1$
$x' = (\sqrt{3} + 1)/2$

$x' = z$ | $w' = x\sin\theta + w\cos\theta$
$z' = 1$ | $w' = 1(1/2) + (-1)(\sqrt{3}/2)$
| $w' = (1 - \sqrt{3})/2$

(A) For a true, primary-level hyperspace analysis:
Point P = (1,–1,1,–1) is rotated to P′ = (1.37,–1,1,–0.37)
(B) For a secondary analysis in XYZ space:
Point P = (1,–1,1) is rotated to P′ = (1.37,–1,1)
(C) For a secondary analysis in XYW space:
Point P = (1,–1,–1) is rotated to P′ = (1.37,–1,–0.37)
(D) For a secondary analysis in XZW space:
Point P = (1,1,–1) is rotated to P′ = (1.37,1,–0.37)
(E) For a secondary analysis in YZW space:
Point P P = (–1,1–) is rotated to P′ = (–1,1,–0.37)

5. Using Equation 21-5 to perform the X-Z planar rotation:

$$x' = x \qquad y' = y\cos\theta - w\sin\theta$$
$$x' = 1 \qquad y' = -1(\sqrt{3}/2) - 1(-1)(1/2)$$
$$y' = (1 - \sqrt{3})/2$$

$$z' = z \qquad w' = y\sin\theta + w\cos\theta$$
$$z' = 1 \qquad w' = -1(1/2) + (-1)(\sqrt{3}/2)$$
$$w' = -(1 + \sqrt{3})/2$$

(A) For a true, primary-level hyperspace analysis:
Point P = (1,–1,1,–1) is rotated to P′ = (1,–0.37,1,1.37)
(B) For a secondary analysis in XYZ space:
Point P = (1,–1,1) is rotated to P′ = (1,–0.37,1)
(C) For a secondary analysis in XYW space:
Point P = (1,–1,–1) is rotated to P′ = (1,–0.37,1.37)
(D) For a secondary analysis in XZW space:
Point P = (1,1,–1) is rotated to P′ = (1,1,1.37)
(E) For a secondary analysis in YZW space:
Point P = (–1,1,–1) is rotated to P′ = (–0.37,1,1.37)

6. Using Equation 21-6 to perform the X-Y planar rotation:

$$x' = x \qquad y' = y$$
$$x' = 1 \qquad y' = -1$$

$$z' = z\cos\theta - w\sin\theta \qquad w' = z\sin\theta + w\cos\theta$$
$$z' = 1(\sqrt{3}/2) - (-1)(1/2) \qquad w' = 1(1/2) + (-1)(\sqrt{3}/2)$$
$$z' = (\sqrt{3} + 1)/2 \qquad w' = (1 - \sqrt{3})/2$$

(A) For a true, primary-level hyperspace analysis:
Point P = (1,–1,1,–1) is rotated to P′ = (1,–1,1.37,0.37)
(B) For a secondary analysis in XYZ space:
Point P = (1,–1,1) is rotated to P′ = (1,–1,1.37)
(C) For a secondary analysis in XYW space:
Point P = (1,–1,–1) is rotated to P′ = (1,–1.0.37)
(D) For a secondary analysis in XZW space:
Point P = (1,1,–1) is rotated to P′ = (1,1.37,0.37)
(E) For a secondary analysis in YZW space:
Point P = (–1,1,–1) is rotated to P′ = (–1,1.37,0.37)

Try specifying some points in 4-D space of your own. Rotate them around the six different planes of rotation and conduct a complete primary and secondary analysis in each case.

21-2 ROTATING LINES IN 4-D SPACE

Rotating a line that is immersed in 4-D space is a simple matter of rotating its two endpoints through the same angle and about the same plane. And as is the case for lower-dimensioned rotations, 4-D rotations:

1. Length, or the distance between two points, is an invariant under 4-D rotations.

2. Direction cosines are variant under 4-D rotations.

3. The angle between two lines is invariant under 4-D rotations.

Those principles, demonstrated to some extent in the analysis offered in the previous example, represent further support of the notion that 4-D space is a straight forward extension of lower-dimensioned, more familiar spaces. The effects of rotating lines in 4-D space, however, can be unusual and even quite dramatic at times; and that, alone, justifies all of the extra analytical work that goes into setting up and conducting the experiments.

Get involved in rotating lines in 4-D space by working through Exercise 21-1.

Exercise 21-1

Given this line in hyperspace:

$$L = \overline{P_1P_2} \qquad P_1 = (-1,-1,1,1)$$
$$P_2 = (1,1,0,-1)$$

Rotate it 30, 45, 60 and 90 degrees about each of the six spaces of 4-D rotation. Plot the line and conduct primary and secondary analyses of the results in each case. The analyses, of course, include the lengths of lines and their direction cosines as well as their endpoint coordinates.

21-3 ROTATING PLANE FIGURES IN 4-D SPACE

Again, there is nothing new here in principle. The general procedure for rotating plane figures immersed in 4-D space is not really different from rotating such figures in 3-D space. Rotation in the 4-dimensional case takes place about a chosen plane of rotation, though, and of course the analysis is a bit more involved. But it's nothing to fear if you have taken the time and trouble to master the experiments for 3-D rotations.

See if you agree after working out the experiments suggested in Exercise 21-2.

Exercise 21-2

Given the following triangle immersed in hyperspace, rotate it 30, 45, 60 and 90 degrees about each of the six spaces of 4-D rotation. Plot and analyze the results in each case, remembering that the analyses include the lengths of the lines, direction cosines of each line, and the three angles in the figure.

$$F = \overline{L_1L_2L_3} \qquad L_1 = \overline{P_1P_2} \qquad P_1 = (0,0,1,0)$$
$$L_2 = \overline{P_2P_3} \qquad P_2 = (1,0,0,0)$$
$$L_3 = \overline{P_3P_1} \qquad P_3 = (0,1,0,1)$$

21-4 ROTATING SPACE OBJECTS IN 4-D SPACE

By this time, you are probably fully prepared to deal with the rotation of a 3-dimensional space object that is imbedded in hyperspace. You have seen that virtually every principle from 3-D rotations carries directly over to the 4-D situations.

Try your hand at the experiments in Exercise 21-3.

Exercise 21-3

Here are the specifications for a 5-sided pyramid that happens to be resting in hyperspace:

$$S = \overline{F_1F_2F_3F_4F_5}$$
$$F_1 = \overline{L_1L_2L_3L_4} \qquad L_1 = \overline{P_1P_2} \qquad P_1 = (-1,0,1,1)$$
$$F_2 = \overline{L_1L_5L_6} \qquad L_2 = \overline{P_2P_3} \qquad P_2 = (1,0,1,1)$$
$$F_3 = \overline{L_2L_6L_7} \qquad L_3 = \overline{P_3P_4} \qquad P_3 = (1,0,-1,1)$$
$$F_4 = \overline{L_3L_7L_8} \qquad L_4 = \overline{P_1P_4} \qquad P_4 = (-1,0,-1,1)$$
$$F_5 = \overline{L_4L_5L_8} \qquad L_5 = \overline{P_1P_5} \qquad P_5 = (0,1,0,1)$$
$$L_6 = \overline{P_2P_7}$$
$$L_7 = \overline{P_3P_5}$$
$$L_8 = \overline{P_4P_5}$$

1. Conduct a complete analysis of this object

A. Primary analysis
   (1) Hyperspace perspective plot
   (2) Lengths of the eight lines
   (3) Direction cosines for the eight lines
   (4) Relevant angles:

$\phi_{1,2}$	$\phi_{1,4}$	$\phi_{1,5}$	$\phi_{1,6}$
$\phi_{2,3}$	$\phi_{2,6}$	$\phi_{2,7}$	$\phi_{3,4}$
$\phi_{3,7}$	$\phi_{3,8}$	$\phi_{4,5}$	$\phi_{4,8}$
$\phi_{5,6}$	$\phi_{5,8}$	$\phi_{6,7}$	$\phi_{7,8}$

B. Secondary analysis in XYZ space
   (1) XYZ perspective plot
   (2) Lengths of lines
   (3) Direction cosines for the lines
   (4) Relevant angles between lines

C. Secondary analysis in XYW space
   (1) XYW perspective plot
   (2) Lengths of lines
   (3) Direction cosines for the lines
   (4) Relevant angles between lines

Special Note: The object degenerates to a triangle in this instance. Two lines are reduced to points, thereby creating some undefined direction cosines and angles.

D. Secondary analysis in XZW space
   (1) XZW perspective plot
   (2) Lengths of lines
   (3) Direction cosines for the lines
   (4) Relevant angles between lines

Special Note: The object degenerates to a square having a critical point in its center. All lines, direction cosines and angles remain well-defined, however.

E. Secondary analysis in YZW space
   (1) YZW perspective plot
   (2) Lengths of lines
   (3) Direction cosines for the lines
   (4) Relevant angles between lines

Special Note: The object degenerates to a triangle and two lines become points. There are thus some undefined direction cosines and angles.

2. Rotate the reference object 45 degrees about the Y-Z plane.

3. Repeat the analyses in Step 1, using the new point coordinates generated in Step 2.

21-5 ROTATING HYPERSPACE OBJECTS IN 4-D SPACE

Now here is something especially tricky, but equally intriguing. Not only is the space a very abstract one, not only is the notion of rotating something about a coordinate plane an abstract one, but so is the object in that space. The visual essence of the situation is lost on the limited perceptual mechanisms of the human brain. The mathematical analyses and limited 3-D renderings of the situations are the only points of contact we can have through such experiments.

The overall procedure is troublesome only in that it involves a lot of different points. The analytic task, however, can be almost overwhelming at times. It's worth the trouble, though; especially when you consider how few human beings through history have even attempted, let alone completed, such a task in any responsible fashion.

I am going to conduct one such experiment for you. Further work is left to you. See Example 21-2.

Table 21-1. Summary of the Analysis of the Reference Hyperspace Object Cited in Example 21-2. (A) Primary Analysis. (B) XYZ-Space Secondary Analysis.

(A, B, C, and D correspond to α, β, γ, and δ.)

SPECIFICATIONS

$H=\overline{S_1S_2S_3S_4S_5S_6S_7}$

$S_1=\overline{F_1F_2F_4F_7F_{10}F_{14}}$

$S_2=\overline{F_1F_3F_5F_8F_{11}}$

$S_3=\overline{F_2F_3F_6F_9F_{15}}$

$S_4=\overline{F_4F_5F_6F_{12}F_{16}}$

$S_5=\overline{F_7F_8F_9F_{13}F_{17}}$

$S_6=\overline{F_{10}F_{11}F_{12}F_{13}F_{18}}$

$S_7=\overline{F_{14}F_{15}F_{16}F_{17}F_{18}}$

$F_{10}=\overline{L_8L_9L_{11}L_{18}}$

$F_{11}=\overline{L_8L_{10}L_{12}}$

$F_{12}=\overline{L_9L_{10}L_{19}}$

$F_{13}=\overline{L_{11}L_{12}L_{20}}$

$F_{14}=\overline{L_{13}L_{14}L_{16}L_{18}}$

$F_{15}=\overline{L_{13}L_{15}L_{17}}$

$F_{16}=\overline{L_{14}L_{15}L_{19}}$

$F_{17}=\overline{L_{16}L_{17}L_{20}}$

$F_{18}=\overline{L_{18}L_{19}L_{20}}$

$L_{10}=\overline{P_3P_9}$

$L_{11}=\overline{P_4P_8}$

$L_{12}=\overline{P_4P_9}$

$L_{13}=\overline{P_5P_6}$

$L_{14}=\overline{P_5P_7}$

$L_{15}=\overline{P_5P_9}$

$L_{16}=\overline{P_6P_8}$

$L_{17}=\overline{P_6P_9}$

$L_{18}=\overline{P_7P_8}$

$L_{19}=\overline{P_7P_9}$

$L_{20}=\overline{P_8P_9}$

$F_1=\overline{L_1L_2L_5L_8}$

$F_2=\overline{L_1L_3L_6L_{13}}$

$F_3=\overline{L_1L_4L_7}$

$F_4=\overline{L_2L_3L_9L_{14}}$

$F_5=\overline{L_1L_4L_{10}}$

$F_6=\overline{L_3L_4L_{15}}$

$F_7=\overline{L_5L_6L_{11}L_{16}}$

$F_8=\overline{L_5L_7L_{12}}$

$F_9=\overline{L_6L_7L_{17}}$

$L_1=\overline{P_1P_2}$

$L_2=\overline{P_1P_3}$

$L_3=\overline{P_1P_5}$

$L_4=\overline{P_1P_9}$

$L_5=\overline{P_2P_4}$

$L_6=\overline{P_2P_6}$

$L_7=\overline{P_2P_9}$

$L_8=\overline{P_3P_4}$

$L_9=\overline{P_3P_7}$

$P_1=(-1,0,-1,-1)$ $P_2=(-1,0,-1,1)$

$P_3=(1,0,-1,-1)$ $P_4=(1,0,-1,1)$

$P_5=(-1,0,1,-1)$ $P_6=(-1,0,1,1)$

$P_7=(1,0,1,-1)$ $P_8=(1,0,1,1)$

$P_9=(0,1,0,0)$

LENGTHS

$s_1=2$	$s_2=2$	$s_3=2$	$s_4=2$
$s_5=2$	$s_6=2$	$s_7=2$	$s_8=2$
$s_9=2$	$s_{10}=2$	$s_{11}=2$	$s_{12}=2$
$s_{13}=2$	$s_{14}=2$	$s_{15}=2$	$s_{16}=2$
$s_{17}=2$	$s_{18}=2$	$s_{19}=2$	$s_{20}=2$

A

DIRECTION COSINES

Line	A	B	C	D	Line	A	B	C	D
1	0	0	0	1	11	0	0	1	0
2	1	0	0	0	12	−1/2	1/2	1/2	1/2
3	0	0	1	0	13	0	0	0	1
4	1/2	1/2	1/2	1/2	14	1	0	0	0
5	1	0	0	0	15	1/2	1/2	−1/2	1/2
6	0	0	1	0	16	1	0	0	0
7	1/2	1/2	1/2	−1/2	17	1/2	1/2	−1/2	−1/2
8	0	0	0	1	18	0	0	0	1
9	0	0	1	0	19	−1/2	1/2	−1/2	1/2
10	−1/2	1/2	1/2	1/2	20	−1/2	1/2	−1/2	−1/2

A

ANGLES

$\phi_{1,2}=90^\circ$ $\phi_{1,3}=90^\circ$ $\phi_{1,4}=60^\circ$
$\phi_{1,5}=90^\circ$ $\phi_{1,6}=90^\circ$ $\phi_{1,7}=60^\circ$
$\phi_{2,3}=90^\circ$ $\phi_{2,4}=60^\circ$ $\phi_{2,8}=90^\circ$
$\phi_{2,9}=90^\circ$ $\phi_{2,10}=60^\circ$ $\phi_{3,4}=60^\circ$
$\phi_{3,13}=90^\circ$ $\phi_{3,14}=90^\circ$ $\phi_{3,15}=60^\circ$
$\phi_{4,7}=60^\circ$ $\phi_{4,10}=60^\circ$ $\phi_{4,15}=60^\circ$
$\phi_{5,6}=90^\circ$ $\phi_{5,7}=60^\circ$ $\phi_{5,8}=90^\circ$
$\phi_{5,11}=90^\circ$ $\phi_{5,12}=60^\circ$ $\phi_{6,7}=60^\circ$
$\phi_{6,13}=90^\circ$ $\phi_{6,16}=90^\circ$ $\phi_{6,17}=60^\circ$
$\phi_{7,12}=60^\circ$ $\phi_{7,17}=60^\circ$ $\phi_{8,9}=90^\circ$

$\phi_{8,10}=60^\circ$ $\phi_{8,11}=90^\circ$ $\phi_{8,12}=60^\circ$
$\phi_{9,10}=60^\circ$ $\phi_{9,14}=90^\circ$ $\phi_{9,18}=90^\circ$
$\phi_{9,19}=60^\circ$ $\phi_{10,12}=60^\circ$ $\phi_{10,19}=60^\circ$
$\phi_{11,12}=60^\circ$ $\phi_{11,16}=90^\circ$ $\phi_{11,18}=90^\circ$
$\phi_{11,20}=60^\circ$ $\phi_{12,20}=60^\circ$ $\phi_{13,14}=90^\circ$
$\phi_{13,15}=60^\circ$ $\phi_{13,16}=90^\circ$ $\phi_{13,17}=60^\circ$
$\phi_{14,15}=60^\circ$ $\phi_{14,18}=90^\circ$ $\phi_{14,19}=60^\circ$
$\phi_{15,17}=60^\circ$ $\phi_{15,19}=60^\circ$ $\phi_{16,17}=60^\circ$
$\phi_{16,18}=90^\circ$ $\phi_{16,20}=60^\circ$ $\phi_{17,20}=60^\circ$
$\phi_{18,19}=60^\circ$ $\phi_{18,20}=60^\circ$ $\phi_{19,20}=60^\circ$

B

Table 21-1. Summary of the Analysis of the Reference Hyperspace Object Cited in Example 21-2. (A) Primary Analysis. (B) XYZ-Space Secondary Analysis. (Continued from 327.)

PLOTTED POINTS

P_1=(-1,Ø,-1)	P_2=(-1,Ø,-1)
P_3=(1,Ø,-1)	P_4=(1,Ø,-1)
P_5=(-1,Ø,1)	P_6=(-1,Ø,1)
P_7=(1,Ø,1)	P_8=(1,Ø,1)
P_9=(Ø,1,Ø)	

LENGTHS

s_1=Ø	s_2=2	s_3=2	$s_4=\sqrt{3}$
s_5=2	s_6=2	$s_7=\sqrt{3}$	s_8=Ø
s_9=2	$s_{1Ø}=\sqrt{3}$	s_{11}=2	$s_{12}=\sqrt{3}$
s_{13}=Ø	s_{14}=2	$s_{15}=\sqrt{3}$	s_{16}=2
$s_{17}=\sqrt{3}$	s_{18}=Ø	$s_{19}=\sqrt{3}$	$s_{2Ø}=\sqrt{3}$

DIRECTION COSINES

Line	A	B	C
1		undefined	
2	1	Ø	Ø
3	Ø	Ø	1
4	$\sqrt{3}/3$	$\sqrt{3}/3$	$\sqrt{3}/3$
5	1	Ø	Ø
6	Ø	Ø	1
7	$\sqrt{3}/3$	$\sqrt{3}/3$	$\sqrt{3}/3$
8		undefined	
9	Ø	Ø	1
1Ø	$-\sqrt{3}/3$	$\sqrt{3}/3$	$\sqrt{3}/3$
11	Ø	Ø	1
12	$-\sqrt{3}/3$	$\sqrt{3}/3$	$\sqrt{3}/3$
13		undefined	
14	1	Ø	Ø
15	$\sqrt{3}/3$	$\sqrt{3}/3$	3/3
16	1	Ø	Ø
17	$\sqrt{3}/3$	$\sqrt{3}/3$	$\sqrt{3}/3$
18		undefined	
19	$-\sqrt{3}/3$	$\sqrt{3}/3$	$-\sqrt{3}/3$
2Ø	$-\sqrt{3}/3$	$\sqrt{3}/3$	$-\sqrt{3}/3$

ANGLES

$\phi_{1,2}$=undef	$\phi_{1,3}$=undef	$\phi_{1,4}$=undef
$\phi_{1,5}$=undef	$\phi_{1,6}$=undef	$\phi_{1,7}$=undef
$\phi_{2,3}$=9Ø°	$\phi_{2,4}$=54° 28'	$\phi_{2,8}$=undef
$\phi_{2,9}$=9Ø°	$\phi_{2,1Ø}$=54° 28'	$\phi_{3,4}$=54° 28'
$\phi_{3,13}$=undef	$\phi_{3,14}$=9Ø°	$\phi_{3,15}$=54° 28'
$\phi_{4,7}$=Ø°	$\phi_{4,1Ø}$=54° 28'	$\phi_{4,15}$=28° 32'
$\phi_{5,6}$=9Ø°	$\phi_{5,7}$=54° 28'	$\phi_{5,8}$=undef
$\phi_{5,11}$=9Ø°	$\phi_{5,12}$=54° 28'	$\phi_{6,7}$=54° 28'
$\phi_{6,13}$=undef	$\phi_{6,16}$=9Ø°	$\phi_{6,17}$=54° 28'
$\phi_{7,12}$=54° 28'	$\phi_{7,17}$=7Ø° 32'	$\phi_{8,9}$=undef
$\phi_{8,1Ø}$=undef	$\phi_{8,11}$=undef	$\phi_{8,12}$=undef
$\phi_{9,1Ø}$=54° 28'	$\phi_{9,14}$=9Ø°	$\phi_{9,18}$=undef
$\phi_{9,19}$=54° 28'	$\phi_{1Ø,12}$=7Ø° 32'	$\phi_{1Ø,19}$=54° 28'
$\phi_{11,12}$=54° 28	$\phi_{11,16}$=9Ø°	$\phi_{11,18}$=undef
$\phi_{11,2Ø}$=54° 28'	$\phi_{12,2Ø}$=54° 28'	$\phi_{13,14}$=undef
$\phi_{13,15}$=undef	$\phi_{13,16}$=undef	$\phi_{13,17}$=undef
$\phi_{14,15}$=54° 28'	$\phi_{14,18}$=undef	$\phi_{14,19}$=54° 28'
$\phi_{15,17}$=7Ø° 32'	$\phi_{15,19}$=54° 28'	$\phi_{16,17}$=54° 28'
$\phi_{16,18}$=undef	$\phi_{16,2Ø}$=54° 28'	$\phi_{17,2Ø}$=54° 28'
$\phi_{18,19}$=undef	$\phi_{18,2Ø}$=undef	$\phi_{19,2Ø}$=Ø°

(B)

Example 21-2

Given the specifications for a "standard" hyperpyramid (see Table 21-1A), apply a 4-D rotation of 45 degrees about the X-Y plane. Conduct a primary analysis and an XYZ-space secondary analysis of both versions of the object.

1. Conduct a primary analysis of the reference version of the hyperspace object. See Table 21-1A and Fig. 21-1A.

2. Conduct an XYZ-space secondary analysis of the reference object. See Table 21-1B and Fig. 21-1B.

3. Apply the X-Y planar rotation (Equation 21-6) to the points of the reference object. Compare the P and P′ specifications in Tables 21-1A and 21-2A, respectively.

4. Conduct a primary analysis of the rotated version of the hyperspace object. See Table 21-2A and Fig. 21-2A.

5. Conduct an XYZ-space secondary analysis of the rotated object. See Table 21-2B and 21-8B.

Table 21-1A cites the specifications and a complete primary analysis of the hyperspace object that is used as the reference object in that example. Those specifications and the general hyperspace drawing in Fig. 21-1A show that it is a 4-D version of an object we normally view as a 5-sided pyramid.

The XYZ-space secondary analysis of that reference object (Table 21-1B) shows that it appears as a 5-sided pyramid as mapped to that 3-dimensional space. That's how we would see that hyperspace object in our own world. Virtually all of the W-axis information is lost—becoming undefined—but enough remains to make sense of it.

Table 21-2A specifies the rotated version of the same hyperspace object. In this instance, it is rotated 45 degrees about the X-Y plane. A comparison of the data shows that only the direction cosines change at the primary-analysis level. Matter is quite different, however, when comparing the data at the XYZ-space, secondary-analysis level.

Experiments involving the rotation and analysis of hyperspace object are perhaps the most important of all. Rotating a hyperspace object and looking at its various 3-D mappings amounts to turning that unfamiliar object around in your hands and looking at it from different angles. It is, indeed, a time-consuming procedure, but it represents the most meaningful and dramatic way to experiment with hyperspace and hyperspace object.

Select some other hyperspace objects specified elsewhere in this book, apply some 4-D rotations of your own choosing, and conduct complete analyses of the results. Be sure to keep track of all your information for future studies.

21-6 INVERSE FORMS OF THE GEOMETRIC ROTATION EQUATIONS

Equations 21-7 through 21-12 are the inverse forms of the rotation equations for 4-D space. You can use them for determining the coordinate of reference points relative to the frame of reference, given the angle of rotation, the plane of rotation, and the coordinate of the rotated version of the point. Their real significance however, is the fact that they lead directly to the relativistic versions of the 4-D rotation equations.

$$\begin{aligned} x_n &= x_n'\cos\theta + y_n'\sin\theta \\ y_n &= x_n'\sin\theta + y_n'\cos\theta \\ z_n &= z_n' \\ w_n &= w_n' \end{aligned}$$

Equation 21-7

where:

x_n', y_n', z_n', w_n' = the coordinate of the rotated version of point *n* relative to the frame of reference

x_n', y_n', z_n', w_n' = the coordinate of the rotated version of point *n* relative to the frame of reference

θ = the angle of rotation about the Z-W plane

Note: This is the inverse form of Equation 21-1.

$$\begin{aligned} x_n &= x_n'\cos\theta + z_n'\sin\theta \\ y_n &= y_n' \\ z_n &= -x_n'\sin\theta + z_n'\cos\theta \\ w_n &= w_n' \end{aligned}$$

Equation 21-8

where:

x_n, y_n, z_n, w_n = the coordinate of point *n* relative to the frame of reference

x_n', y_n', z_n', w_n' = the coordinate of the rotated version of point *n* relative to the frame of reference

θ = the angle of rotation about the Y-W plane of the frame of reference

Note: This is the inverse version of Equation 21-2.

$$\begin{aligned} x_n &= x_n' \\ y_n &= y_n'\cos\theta + z_n'\sin\theta \\ z_n &= -y_n'\sin\theta + z_n'\cos\theta \\ w_n &= w_n' \end{aligned}$$

Equation 21-9

where:

x_n, y_n, z_n, w_n = the coordinate of point *n* relative to the frame of reference

x_n', y_n', z_n', w_n' = the coordinate of the rotated version of point *n* relative to the frame of reference

θ = the angle of rotation about the X-W plane relative to the frame of reference

Note: This is the inverse version of Equation 21-3.

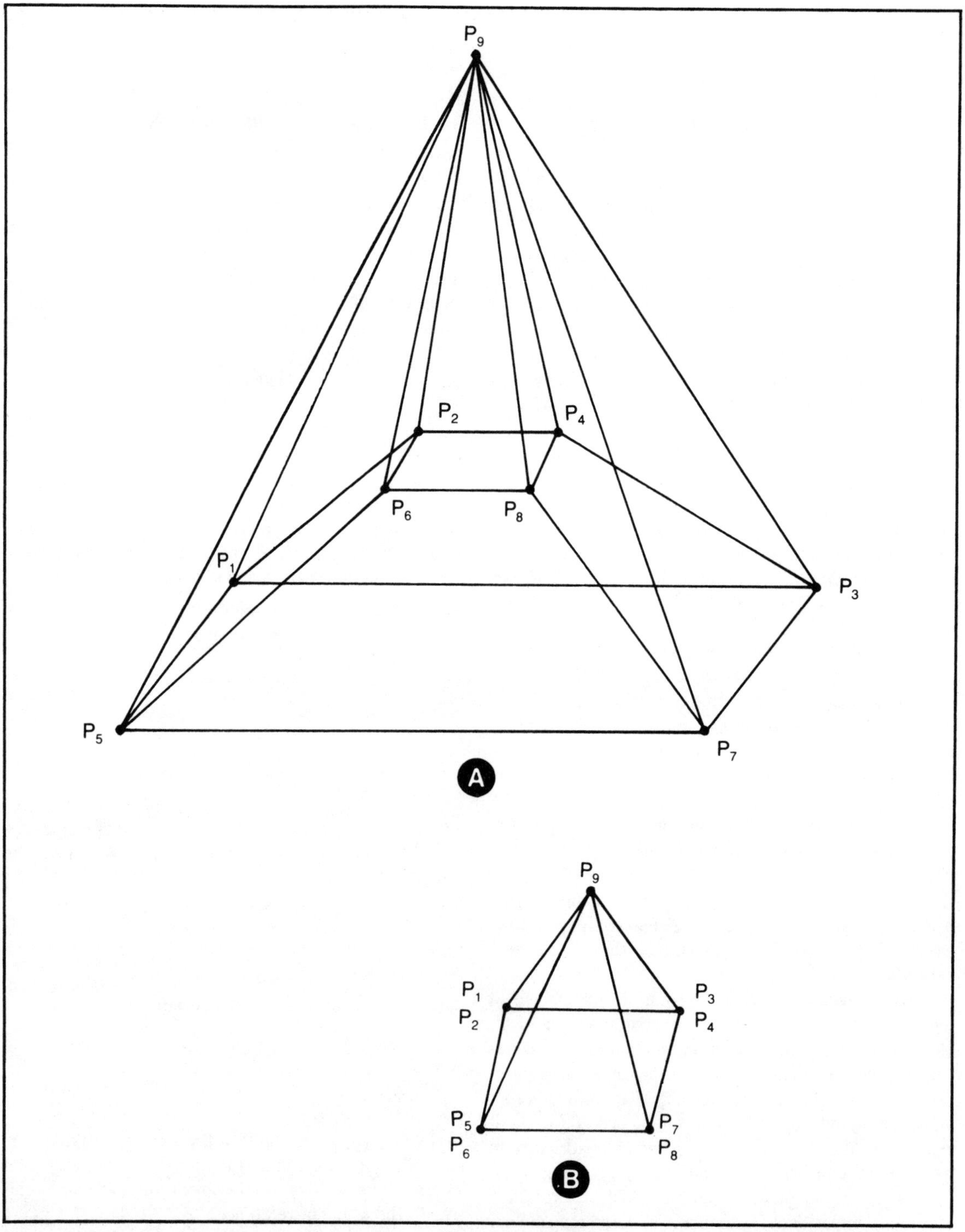

Fig. 21-1. The reference hyperpyramid cited in Example 21-2. (A) A general hyperspace view. (B) XYZ perspective view.

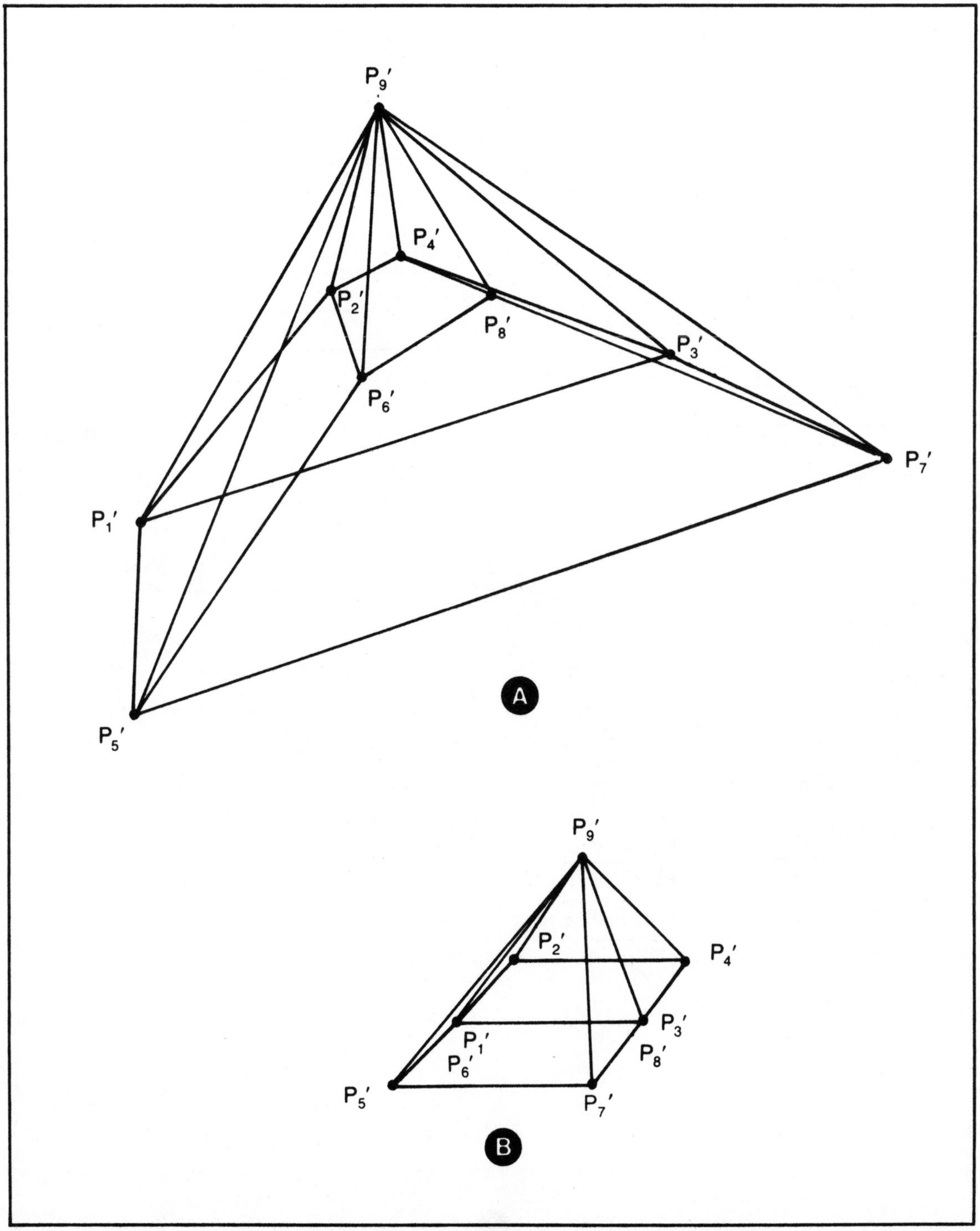

Fig. 21-2. The rotated version of the hyperpyramid cited in Example 21-2. (A) A general hyperspace view. (B) XYZ perspective view.

Table 21-2. Summary of the Analysis of the Rotated Version of the Hyperspace Object Cited in Example 21-2. (A) Primary Analysis. (B) XYZ-Space Secondary Analysis.

(A, B, C, and D correspond to α, β, γ, and δ.)

SPECIFICATIONS

$H'=\overline{S'_1S'_2S'_3S'_4S'_5S'_6S'_7}$

$S'_1=\overline{F'_1F'_2F'_4F'_7F'_{10}F'_{14}}$

$S'_2=\overline{F'_1F'_3F'_5F'_8F'_{11}}$

$S'_3=\overline{F'_2F'_3F'_6F'_9F'_{15}}$

$S'_4=\overline{F'_4F'_5F'_6F'_{12}F'_{16}}$

$S'_5=\overline{F'_7F'_8F'_9F'_{13}F'_{17}}$

$S'_6=\overline{F'_{10}F'_{11}F'_{12}F'_{13}F'_{18}}$

$S'_7=\overline{F'_{14}F'_{15}F'_{16}F'_{17}F'_{18}}$

$F'_{10}=\overline{L'_8L'_9L'_{11}L'_{18}}$

$F'_{11}=\overline{L'_8L'_{10}L'_{12}}$

$F'_{12}=\overline{L'_9L'_{10}L'_{19}}$

$F'_{13}=\overline{L'_{11}L'_{12}L'_{20}}$

$F'_{14}=\overline{L'_{13}L'_{14}L'_{16}L'_{18}}$

$F'_{15}=\overline{L'_{13}L'_{15}L'_{17}}$

$F'_{16}=\overline{L'_{14}L'_{15}L'_{19}}$

$F'_{17}=\overline{L'_{16}L'_{17}L'_{20}}$

$F'_{18}=\overline{L'_{18}L'_{19}L'_{20}}$

$L'_{10}=\overline{P'_3P'_9}$

$L'_{11}=\overline{P'_4P'_8}$

$L'_{12}=\overline{P'_4P'_9}$

$L'_{13}=\overline{P'_5P'_6}$

$L'_{14}=\overline{P'_5P'_7}$

$L'_{15}=\overline{P'_5P'_9}$

$L'_{16}=\overline{P'_6P'_8}$

$L'_{17}=\overline{P'_6P'_9}$

$L'_{18}=\overline{P'_7P'_8}$

$L'_{19}=\overline{P'_7P'_9}$

$L'_{20}=\overline{P'_8P'_9}$

$F'_1=\overline{L'_1L'_2L'_5L'_8}$

$F'_2=\overline{L'_1L'_3L'_6L'_{13}}$

$F'_3=\overline{L'_1L'_4L'_7}$

$F'_4=\overline{L'_2L'_3L'_9L'_{14}}$

$F'_5=\overline{L'_1L'_4L'_{10}}$

$F'_6=\overline{L'_3L'_4L'_{15}}$

$F'_7=\overline{L'_5L'_6L'_{11}L'_{16}}$

$F'_8=\overline{L'_5L'_7L'_{12}}$

$F'_9=\overline{L'_6L'_7L'_{17}}$

$L'_1=\overline{P'_1P'_2}$

$L'_2=\overline{P'_1P'_3}$

$L'_3=\overline{P'_1P'_5}$

$L'_4=\overline{P'_1P'_9}$

$L'_5=\overline{P'_2P'_4}$

$L'_6=\overline{P'_2P'_6}$

$L'_7=\overline{P'_2P'_9}$

$L'_8=\overline{P'_3P'_4}$

$L'_9=\overline{P'_3P'_7}$

$P'_1=(-1,0,0,-2)$ $P'_2=(-1,0,-2,0)$

$P'_3=(1,0,0,-2)$ $P'_4=(1,0,-2,0)$

$P'_5=(-1,0,2,0)$ $P'_6=(-1,0,0,2)$

$P'_7=(1,0,2,0)$ $P'_8=(1,0,0,2)$

$P'_9=(0,1,0,0)$

LENGTHS

$s'_1=2$	$s'_2=2$	$s'_3=2$	$s'_4=2$
$s'_5=2$	$s'_6=2$	$s'_7=2$	$s'_8=2$
$s'_9=2$	$s'_{10}=2$	$s'_{11}=2$	$s'_{12}=2$
$s'_{13}=2$	$s'_{14}=2$	$s'_{15}=2$	$s'_{16}=2$
$s'_{17}=2$	$s'_{18}=2$	$s'_{19}=2$	$s'_{20}=2$

A

DIRECTION COSINES

Line'	A'	B'	C'	D'
1	0	0	$-\sqrt{2}/2$	$\sqrt{2}/2$
2	1	0	0	0
3	0	0	$\sqrt{2}/2$	$\sqrt{2}/2$
4	1/2	1/2	0	$-\sqrt{2}/2$
5	1	0	0	0
6	0	0	$\sqrt{2}/2$	$\sqrt{2}/2$
7	1/2	1/2	$\sqrt{2}/2$	0
8	0	0	$-\sqrt{2}/2$	$\sqrt{2}/2$
9	0	0	$\sqrt{2}/2$	$\sqrt{2}/2$
10	-1/2	1/2	0	$\sqrt{2}/2$

DIRECTION COSINES

Line'	A'	B'	C'	D'
11	0	0	$\sqrt{2}/2$	$\sqrt{2}/2$
12	-1/2	1/2	$\sqrt{2}/2$	0
13	0	0	$-\sqrt{2}/2$	$\sqrt{2}/2$
14	1	0	0	0
15	1/2	1/2	$-\sqrt{2}/2$	0
16	1	0	0	0
17	1/2	1/2	0	$-\sqrt{2}/2$
18	0	0	$-\sqrt{2}/2$	$\sqrt{2}/2$
19	-1/2	1/2	$-\sqrt{2}/2$	0
20	-1/2	1/2	0	$-\sqrt{2}/2$

ANGLES

$\phi'_{1,2}=90^\circ$	$\phi'_{1,3}=90^\circ$	$\phi'_{1,4}=60^\circ$	$\phi'_{8,10}=60^\circ$	$\phi'_{8,11}=90^\circ$	$\phi'_{8,12}=60^\circ$
$\phi'_{1,5}=90^\circ$	$\phi'_{1,6}=90^\circ$	$\phi'_{1,7}=60^\circ$	$\phi'_{9,10}=60^\circ$	$\phi'_{9,14}=90^\circ$	$\phi'_{9,18}=90^\circ$
$\phi'_{2,3}=90^\circ$	$\phi'_{2,4}=60^\circ$	$\phi'_{2,8}=90^\circ$	$\phi'_{9,19}=60^\circ$	$\phi'_{10,12}=60^\circ$	$\phi'_{10,19}=60^\circ$
$\phi'_{2,9}=90^\circ$	$\phi'_{2,10}=60^\circ$	$\phi'_{3,4}=60^\circ$	$\phi'_{11,12}=60^\circ$	$\phi'_{11,16}=90^\circ$	$\phi'_{11,18}=90^\circ$
$\phi'_{3,13}=90^\circ$	$\phi'_{3,14}=90^\circ$	$\phi'_{3,15}=60^\circ$	$\phi'_{11,20}=60^\circ$	$\phi'_{12,20}=60^\circ$	$\phi'_{13,14}=90^\circ$
$\phi'_{4,7}=60^\circ$	$\phi'_{4,10}=60^\circ$	$\phi'_{4,15}=60^\circ$	$\phi'_{13,15}=60^\circ$	$\phi'_{13,16}=90^\circ$	$\phi'_{13,17}=60^\circ$
$\phi'_{5,6}=90^\circ$	$\phi'_{5,7}=60^\circ$	$\phi'_{5,8}=90^\circ$	$\phi'_{14,15}=60^\circ$	$\phi'_{14,18}=90^\circ$	$\phi'_{14,19}=60^\circ$
$\phi'_{5,11}=90^\circ$	$\phi'_{5,12}=60^\circ$	$\phi'_{6,7}=60^\circ$	$\phi'_{15,17}=60^\circ$	$\phi'_{15,19}=60^\circ$	$\phi'_{16,17}=60^\circ$
$\phi'_{6,13}=90^\circ$	$\phi'_{6,16}=90^\circ$	$\phi'_{6,17}=60^\circ$	$\phi'_{16,18}=90^\circ$	$\phi'_{16,20}=60^\circ$	$\phi'_{17,20}=60^\circ$
$\phi'_{7,12}=60^\circ$	$\phi'_{7,17}=60^\circ$	$\phi'_{8,9}=90^\circ$	$\phi'_{18,19}=60^\circ$	$\phi'_{18,20}=60^\circ$	$\phi'_{19,20}=60^\circ$

A

Table 21-2. Summary of the Analysis of the Rotated Version of the Hyperspace Object Cited in Example 21-2. (A) Primary Analysis. (B) XYZ-Space Secondary Analysis. (Continued from page 333.

PLOTTED POINTS

P'_{1}=(-1,Ø,Ø)	P'_{2}=(-1,Ø,- 2)
P'_{3}=(1,Ø,Ø)	P'_{4}=(1,Ø,- 2)
P'_{5}=(-1,Ø, 2)	P'_{6}=(-1,Ø,Ø)
P'_{7}=(1,Ø, 2)	P'_{8}=(1,Ø,Ø)
P'_{9}=(Ø,1,Ø)	

LENGTHS

$s'_{1}=\sqrt{2}$	$s'_{2}=2$	$s'_{3}=\sqrt{2}$	$s'_{4}=\sqrt{2}$
$s'_{5}=2$	$s'_{6}=\sqrt{2}$	$s'_{7}=2$	$s'_{8}=\sqrt{2}$
$s'_{9}=\sqrt{2}$	$s'_{1Ø}=\sqrt{2}$	$s'_{11}=\sqrt{2}$	$s'_{12}=2$
$s'_{13}=\sqrt{2}$	$s'_{14}=2$	$s'_{15}=2$	$s'_{16}=2$
$s'_{17}=\sqrt{2}$	$s'_{18}=\sqrt{2}$	$s'_{19}=2$	$s'_{2Ø}=\sqrt{2}$

DIRECTION COSINES

Line'	A'	B'	C'
1	Ø	Ø	-1
2	1	Ø	Ø
3	Ø	Ø	1
4	$\sqrt{2}/2$	$\sqrt{2}/2$	Ø
5	1	Ø	Ø
6	Ø	Ø	1
7	1/2	1/2	$\sqrt{2}/2$
8	Ø	Ø	1
9	Ø	Ø	1
1Ø	$-\sqrt{2}/2$	$\sqrt{2}/2$	Ø
11	Ø	Ø	1
12	-1/2	1/2	$\sqrt{2}/2$
13	Ø	Ø	-1
14	1	Ø	Ø
15	1/2	1/2	$-\sqrt{2}/2$
16	1	Ø	Ø
17	$\sqrt{2}/2$	$\sqrt{2}/2$	Ø
18	Ø	Ø	-1
19	-1/2	1/2	$-\sqrt{2}/2$
2Ø	$-\sqrt{2}/2$	$\sqrt{2}/2$	Ø

ANGLES

$\phi'_{1,2}=9Ø^\circ$	$\phi'_{1,3}=Ø^\circ$	$\phi'_{1,4}=9Ø^\circ$
$\phi'_{1,5}=9Ø^\circ$	$\phi'_{1,6}=Ø^\circ$	$\phi'_{1,7}=45^\circ$
$\phi'_{2,3}=9Ø^\circ$	$\phi'_{2,4}=45^\circ$	$\phi'_{2,8}=9Ø^\circ$
$\phi'_{2,9}=9Ø^\circ$	$\phi'_{2,1Ø}=45^\circ$	$\phi'_{3,4}=9Ø^\circ$
$\phi'_{3,13}=Ø^\circ$	$\phi'_{3,14}=9Ø^\circ$	$\phi'_{3,15}=45^\circ$
$\phi'_{4,7}=45^\circ$	$\phi'_{4,1Ø}=9Ø^\circ$	$\phi'_{4,15}=45^\circ$
$\phi'_{5,6}=9Ø^\circ$	$\phi'_{5,7}=6Ø^\circ$	$\phi'_{5,8}=9Ø^\circ$
$\phi'_{5,11}=9Ø^\circ$	$\phi'_{5,12}=6Ø^\circ$	$\phi'_{6,7}=45^\circ$
$\phi'_{6,13}=Ø^\circ$	$\phi'_{6,16}=9Ø^\circ$	$\phi'_{6,17}=9Ø^\circ$
$\phi'_{7,12}=6Ø^\circ$	$\phi'_{7,17}=45^\circ$	$\phi'_{8,9}=Ø^\circ$
$\phi'_{8,1Ø}=9Ø^\circ$	$\phi'_{8,11}=Ø^\circ$	$\phi'_{8,12}=45^\circ$
$\phi'_{9,1Ø}=9Ø^\circ$	$\phi'_{9,14}=9Ø^\circ$	$\phi'_{9,18}=Ø^\circ$
$\phi'_{9,19}=45^\circ$	$\phi'_{1Ø,12}=45^\circ$	$\phi'_{1Ø,19}=45^\circ$
$\phi'_{11,12}=45^\circ$	$\phi'_{11,16}=9Ø^\circ$	$\phi'_{11,18}=Ø^\circ$
$\phi'_{11,2Ø}=9Ø^\circ$	$\phi'_{12,2Ø}=45^\circ$	$\phi'_{13,14}=9Ø^\circ$
$\phi'_{13,15}=45^\circ$	$\phi'_{13,16}=9Ø^\circ$	$\phi'_{13,17}=9Ø^\circ$
$\phi'_{14,15}=6Ø^\circ$	$\phi'_{14,18}=9Ø^\circ$	$\phi'_{14,19}=6Ø^\circ$
$\phi'_{15,17}=45^\circ$	$\phi'_{15,19}=6Ø^\circ$	$\phi'_{16,17}=45^\circ$
$\phi'_{16,18}=9Ø^\circ$	$\phi'_{16,2Ø}=45^\circ$	$\phi'_{17,2Ø}=9Ø^\circ$
$\phi'_{18,19}=45^\circ$	$\phi'_{18,2Ø}=9Ø^\circ$	$\phi'_{19,2Ø}=45^\circ$

B

$$
\begin{aligned}
x_n &= x_n'\cos\theta + w_n'\sin\theta \\
y_n &= y_n' \\
z_n &= z_n' \\
w_n &= -x_n'\sin\theta + w_n'\cos\theta
\end{aligned}
\qquad \textbf{Equation 21-10}
$$

where:

x_n, y_n, z_n, w_n = the coordinate of point *n* relative to the frame of reference

x_n', y_n', z_n', w_n' = the coordinate of the rotated version of point *n* relative to the frame of reference

θ = the angle of rotation about the Y-Z plane relative to the frame of reference

Note: This is the inverse version of Equation 21-4.

$$
\begin{aligned}
x_n &= x_n' \\
y_n &= y_n'\cos\theta + w_n'\sin\theta \\
z_n &= z_n' \\
w_n &= -y_n'\sin\theta + w_n'\cos\theta
\end{aligned}
\qquad \textbf{Equation 21-11}
$$

where:

x_n, y_n, z_n, w_n = the coordinate of point *n* relative to the frame of reference

x_n', y_n', z_n', w_n' = the coordinate of the rotated version of point *n* relative to the frame of reference

θ = the angle of rotation about the X-Z plane relative to the frame of reference

Note: This is the inverse version of Equation 21-5.

$$
\begin{aligned}
x_n &= x_n' \\
y_n &= y_n' \\
z_n &= z_n'\cos\theta + w_n'\sin\theta \\
w_n &= -z_n'\sin\theta + w_n'\cos\theta
\end{aligned}
\qquad \textbf{Equation 21-12}
$$

where:

x_n, y_n, z_n, w_n = the coordinate of point *n* relative to the frame of reference

x_n', y_n', z_n', w_n' = the coordinate of the rotated version of point *n* relative to the frame of reference

θ = the angle of rotation about the X-Y plane relative to the frame of reference.

Note: This is the inverse version of Equation 21-6.

21-7 RELATIVISTIC 4-D ROTATIONS

The relativistic versions of the 4-D rotation equations deal with two different hyperspace frames of reference: the observer's and the rotated frames of reference. After specifying some geometric entity as regarded in the observer's hyperspace, select the angle and plane of rotation that represents the orientation of the rotated hyperspace frame of reference. Applying the appropriate relativistic equation then leads directly to the coordinates of the entity's points as viewed from the rotated frame of reference. There is nothing new; it is simply a 4-dimensional analogue of the same procedure in 3-D space.

$$
\begin{aligned}
x_n^* &= x_n\cos\theta + y_n\sin\theta \\
y_n^* &= -x_n\sin\theta + y_n\sin\theta \\
z_n^* &= z_n \\
w_n^* &= w_n
\end{aligned}
\qquad \textbf{Equation 21-13}
$$

where:

$x_n^*, y_n^*, z_n^*, w_n^*$ = the coordinate of point *n* relative to the rotated frame of reference

x_n, y_n, z_n, w_n = the coordinate of point *n* relative to the observer's frame of reference

θ = the angle of rotation about the Z-W plane that exists between the observer's and the rotated frames of reference, relative to the observer's frame of reference

Note: These expressions apply only where there is a Z-W plane rotation.

$$
\begin{aligned}
x_n^* &= x_n\cos\theta + z_n\sin\theta \\
y_n^* &= y_n \\
z_n^* &= -x_n\sin\theta + z_n\cos\theta \\
w_n^* &= w_n
\end{aligned}
\qquad \textbf{Equation 21-14}
$$

where:

$x_n^*, y_n^*, z_n^*, w_n^*$ = the coordinate of point *n* relative to the rotated frame of reference

x_n, y_n, z_n, w_n = the coordinate of point *n* relative to the observer's frame of reference

θ = the angle of rotation about the Y-W plane that exists between the observer's and the rotated frame of reference, relative to the observer's frame of reference

Note: These expressions apply only where there is a Y-W plane rotation.

$$
\begin{aligned}
x_n^* &= x_n \\
y_n^* &= y_n\cos\theta + z_n\sin\theta \\
z_n^* &= -y_n\sin\theta + z_n\cos\theta \\
w_n^* &= w_n
\end{aligned}
\qquad \textbf{Equation 21-15}
$$

where:

$x_n^*, y_n^*, z_n^*, w_n^*$ = the coordinate of point *n* relative to the rotated frame of reference

x_n, y_n, z_n, w_n = the coordinate of point *n* relative to the observer's frame of reference

θ = the angle of rotation about the X-W plane that exists between the observer's and the rotated frame of reference, relative to the observer's frame of reference

Note: The expressions apply only where there is an X-W plane rotation.

$$x_n^* = x_n\cos\theta + w_n\sin\theta$$
$$y_n^* = y_n$$
$$z_n^* = z_n \qquad \textbf{Equation 21-16}$$
$$w_n^* = -x_n\sin\theta + w_n\cos\theta$$

where:

$x_n^*, y_n^*, x_n^*, w_n^*$ = the coordinate of point n relative to the rotated frame of reference

x_n, y_n, z_n, w_n = the coordinate of point n relative to the observer's frame of reference

θ = the angle of rotation about the Y-Z plane that exists between the observer's and the rotated frames of reference, relative to the observer's frame of reference

Note: These expressions apply only where there is a Y-Z plane rotation.

$$x_n^* = x_n$$
$$y_n^* = y_n\cos\theta + w_n\sin\theta \qquad \textbf{Equation 21-17}$$
$$z_n^* = z_n$$
$$w_n^* = -y_n\sin\theta + w_n\cos\theta$$

where:

$x_n^*, y_n^*, z_n^*, w_n^*$ = the coordinate of point n relative to the rotated frame of reference

x_n, y_n, z_n, w_n = the coordinate of point n relative to the observer's frame of reference

θ = the angle of rotation about the X-Z axis that exists between the observer's and the rotated frames of reference, relative to the observer's frame of reference

Note: These expressions apply only where there is an X-Z plane rotation.

$$x_n^* = x_n$$
$$y_n^* = y_n \qquad \textbf{Equation 21-18}$$
$$z_n^* = z_n\cos\theta + w_n\sin\theta$$
$$w_n^* = -z_n\sin\theta + w_n\cos\theta$$

where:

$x_n^*, y_n^*, z_n^*, w_n^*$ = the coordinate of point n relative to the rotated frame of reference

x_n, y_n, z_n, w_n = the coordinate of point n relative to the observer's frame of reference

θ = the angle of rotation about the X-Y axis that exists between the observer's and the rotated frames of reference, relative to the observer's frame of reference

Note: These expressions apply only where there is an X-Y plane rotation.

The whole idea takes an intriguing turn when you play the role of an intelligent creature that can perceive only a 3-D space. Pick a 3-D space, such as the XYZ space, and specify a space object within it. That is your view—an observer's view—of an object.

But then suppose that you know that you are really living in a 4-D universe. You cannot directly perceive elements of that fourth dimension, but you have access to the mathematics for working with it.

Next, imagine that there is another universe that differs from your own in just one respect: it is rotated in hyperspace—rotated by some angle about a plane—relative to your version of the universe. The question is this: How will your 3-D space and the object specified within it appear to another intelligent being of your calibre? The 4-D rotation equations supply the mathematical technique for answering the question, and the resulting view of the object in X*Y*Z* space shows how it appears to your counterpart in that alternative universe.

By way of further speculation, we might well live in a 4-dimensional universe; but with our senses limited to a 3-dimensional perspective, there could be other intelligent beings existing in alternative spaces that are only rotated about a plane with respect to our own. We cannot perceive one another directly, but it is possible to exert some influence, in a hyperspace fashion, on one another.

The notion of having intelligent creatures living in alternative, hyperspace-separated worlds is a far-fetched one. Remove the notion of intelligent beings, however, and the idea has the full support of modern mathematics.

Try some experiments with relativistic hyperspace rotations. Let your imagination roam as you do the work, and I guarantee that you will have a lot of fun.

21-8 SUCCESSIVE 4-D ROTATIONS

Successive rotations in 4-D space about the same plane are commutative. You can perform a certain series of rotations about the X-W plane, for instance, and come up with the same overall rotation, regardless of the order to choose to apply those rotations.

Successive rotations about different planes, however, are not commutative. Suppose that you execute a rotation about the X-W plane, followed by one about the Y-Z plane. That will yield a particular result. But you are likely to get an entirely different result if you perform the Y-Z rotation first. You saw those same principles illustrated in the 3-D case in Chapter 14, and they apply equally well here.

Experiments in Four Dimensions

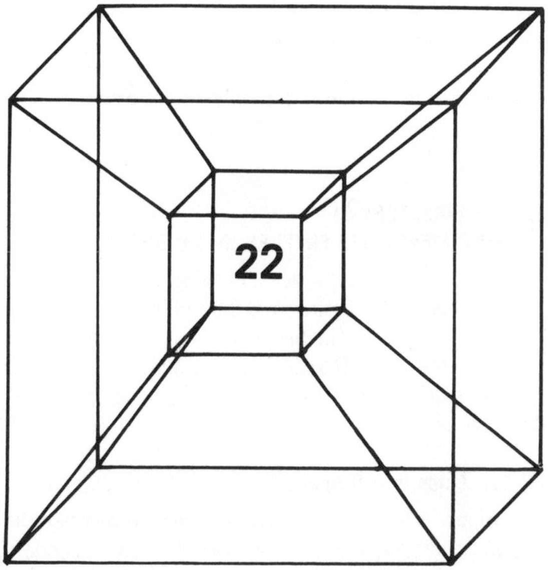

The fact that this chapter carries the title of the entire book implies that it represents a culmination of everything presented in the earlier chapters. If I could have assumed that every reader had a good background and a lot of experience in the mathematics of lower dimensions, I could have begun the book here. Rather, it is my hope that you have used all of the prior work to get yourself prepared for the finale of the present work. Perhaps one day we can resume the work in such a way that this chapter will be the first in a series of new adventures in space.

As defined earlier, four-dimensional space supports has five different kinds of geometric entities: points, lines, plane figures, space objects, and hyperspace objects. Certainly you can now appreciate that there is a clear sequence of entities that represents an evolutionary development from the most primitive to the most sophisticated kinds of entities in 4-D space: hyperspace objects are specified in terms of the space objects that bound them, space objects are defined in terms of the plane figures that bound them, plane figures are specified in terms of the lines that bound them and, finally, lines are defined in terms of the points that bound them. Every geometric entity is ultimately defined in terms of the coordinates of its points. Manipulate just the points, and you end up manipulating the entire geometric entity that they define; no matter how simple or complex it might be.

The H-S-F-L-P specifications for objects in 4-D space reflect the evolution of geometric entities. The P specifications cite the coordinates of points, the L specifications cite the points that bound lines, the F specifications cite the lines that bound a plane figure, the S specifications cite the planes that bound the space object, and the H specifications cite the space object that bound the hyperspace object under study at the moment.

And in keeping with the idea that the character of a geometric entity, especially when it is considered the object of a geometric or relativistic transformation, rests solely with the manipulation of the coordinates of its points; the H, S, F and L specifications remain unchanged.

The following discussions emphasize these ideas, and extend them to include the formal analysis of the entities. The format is identical to that used for 2-D space in Chapter 9 and for 3-D space in Chapter 15. As in those earlier presentations, the main purpose is to demonstrate the logical simplicity and consistency of both the specifications and analyses of geometric entities in 4-D space. And by now, you are also in good position to appreciate the consistency of ideas from lower-dimensioned spaces to the higher-dimensioned ones.

You will find here a catalog of lower-dimension objects as specified in 4-D space; and although you have already studied some procedures for generating hyperspace objects of your own (Chapter 18), I am going to include a brief catalog of standard hyperspace objects as well. Each case will include the general, primary specifications that can remain fixed under any sort of transformation, a guide to the analysis of the entities, and some suggested coordinates for several different reference versions.

22-1 A DIRECTORY OF LOWER-DIMENSIONED ENTITIES IN 4-D SPACE

Four-dimensional space can be an environment for points, lines, plane figures and space objects as well as hyperspace objects. The lower-dimensioned entities can be immersed in a 4-D space and manipulated in a much more dynamic fashion than is possible in their "home" dimensions.

22-1.1 Lines in 4-D Space

Figure 22-1 and Table 22-1 plot and summarize the specifications and both the primary and the secondary analyses of any line in 4-D space. The line is defined in terms of two general endpoints, and the endpoints are defined in terms of their 4-D coordinates. That completes the specifications, the L-P specifications for any line that is imbedded in a 4-D space.

The analyses of a line in 4-D space ought to include its length and its set of four direction cosines: alpha, beta, gamma and delta. As far as the primary analysis is concerned, those analytic line parameters represent the lines in their true form. The four separate secondary analyses, however, cite those parameters as they are mapped (and generally distorted) to the four 3-D perspective views: XYZ, XYW, XZW, and YZW views.

In all of the analyses, the length of the line and its direction cosines are all calculated from the primitive endpoint coordinates. The endpoint coordinates for the primary analysis are the 4-D versions of the coordinates stated in the original specifications. For the XYZ perspective analysis, however, the endpoints are specified in terms of their X, Y, and Z components; for the XYW perspective analysis, the endpoints are taken from the X, Y, and W components; for the XZW perspective analysis, the endpoints are taken from the X, Z and W components; and, finally, for the YZW perspective analysis, the components of the endpoints are taken from the original Y, X and W components.

So when it comes to specifying and analyzing a line in 4-D space, all you have to do is supply some specific values for the coordinates in the original specifications. Select some 4-D endpoint coordinates, plot the line on a 4-D coordinate system, and complete the primary analysis by working out the length and direction cosines. Repeat the procedure for the four 3-D perspective mappings to complete the secondary analyses. In a manner of speaking, those secondary analyses show how the line immersed in 4-D space appears to creatures, such as ourselves, that are confined to one of those individual 3-D spaces.

Don't be surprised whenever the line reduces to a single point as plotted and analyzed from one of the 3-D perspective views. As you have probably seen many times before, that can happen under certain circumstances.

If you care to extend your understanding of direction cosines for lines to include those in 4-D space, you will find that:

1. A line is parallel to the X axis whenever its alpha direction cosine is equal to 1 or -1, and it is perpendicular to that axis whenever its alpha direction cosine is equal to zero. That means there are angles of zero or 90 degrees, respectively, between the line and the X axis.
2. A line is parallel to the Y axis whenever its beta direction cosine is equal to 1 or -1, and it is perpendicular to the Y axis whenever its beta direction cosine is equal to zero.
3. A line is parallel to the Z axis whenever its gamma direction cosine is equal to 1 or -1, and it is perpendicular to that axis whenever its gamma direction cosine is equal to zero.
4. A line is parallel to the W axis whenever its delta direction cosine is equal to 1 or -1, and it is perpendicular to that axis whenever its delta direction cosine is equal to zero.

Generally speaking, you can always determine the angle between a line and a particular axis of the coordinate system by figuring the arc cosine of the corresponding direction cosine: alpha for the X axis, beta for the Y axis, gamma for the Z axis, and delta for the W axis. Those principles apply to the secondary perspective analyses as well as the true primary analysis.

22-1.2 Triangles in 4-D Space

Figure 22-2 and Table 22-2 outline the specifications and procedure of analysis for any triangle in 4-D space. The F specification in the table shows that every triangle is composed of three lines. That specification never changes for triangles. Then the specifications for the

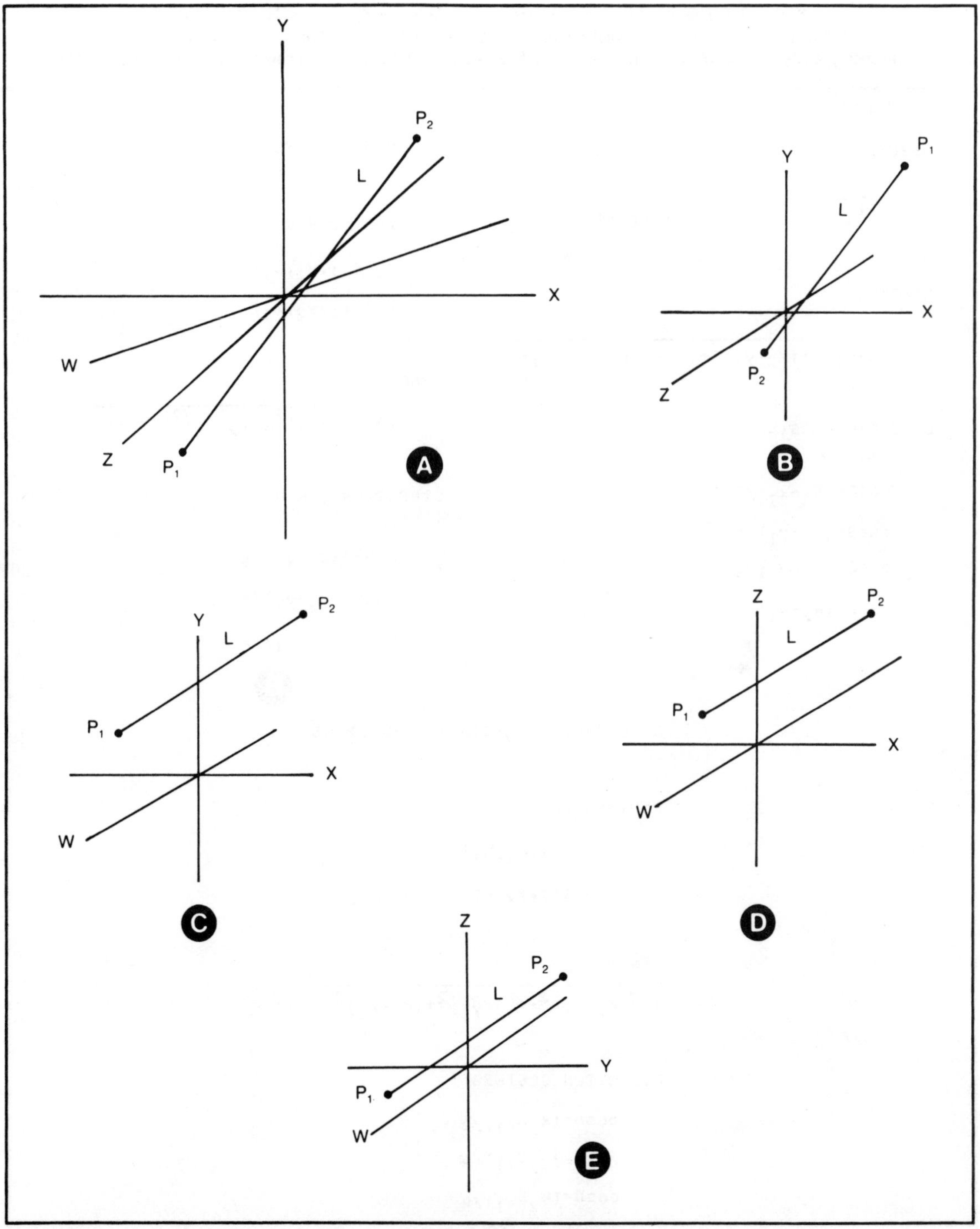

Fig. 22-1. A line specified in 4-D space. (A) General hyperspace view. (B) XYZ view. (C) XYW perspective view. (D) XZW perspective view. (E) YZW perspective view. See outlines of analyses in Table 22-1.

Table 22-1. Specifications and Analyses of a Line in 4-D Space. See Fig. 22-1. (A) Specifications and Primary Analysis in Hyperspace. (B) Secondary Analysis in XYZ Space. (C) Secondary Analysis in XYW Space. (D) Secondary Analysis in XZW Space. (E) Secondary Analysis in YZW Space.

(A, B, C, and D correspond to α, β, γ, and δ.)

SPECIFICATIONS

$L=\overline{P_1P_2}$

$P_1=(x_1,y_1,z_1,w_1)$

$P_2=(x_2,y_2,z_1,w_1)$

LENGTH

$s=\sqrt{(x_2-x_1)^2+(y_2-y_1)^2+(z_2-z_1)^2+(w_2-w_1)^2}$

DIRECTION COSINES

$\cos A=(x_2-x_1)/s$

$\cos B=(y_2-y_1)/s$

$\cos C=(z_2-z_1)/s$

$\cos D=(w_2-w_1)/s$

A

SECONARY ANALYSIS -- XYZ SPACE

PLOTTED POINTS

$P_1=(x_1,y_1,z_1)$

$P_2=(x_2,y_2,z_1)$

LENGTH

$s=\sqrt{(x_2-x_1)^2+(y_2-y_1)^2+(z_2-z_1)^2}$

DIRECTION COSINES

$\cos A=(x_2-x_1)/s$

$\cos B=(y_2-y_1)/s$

$\cos C=(z_2-z_1)/s$

B

SECONDARY ANALYSIS -- XYW SPACE

PLOTTED POINTS

$P_1=(x_1,y_1,w_1)$

$P_2=(x_2,y_2,w_1)$

LENGTH

$s=\sqrt{(x_2-x_1)^2+(y_2-y_1)^2+(w_2-w_1)^2}$

DIRECTION COSINES

$\cos A=(x_2-x_1)/s$

$\cos B=(y_2-y_1)/s$

$\cos D=(w_2-w_1)/s$

C

SECONDARY ANALYSIS -- XZW SPACE

PLOTTED POINTS

$P_1=(x_1,z_1,w_1)$

$P_2=(x_2,z_1,w_1)$

LENGTH

$$s=\sqrt{(x_2-x_1)^2+(z_2-z_1)^2+(w_2-w_1)^2}$$

DIRECTION COSINES

$\cos A=(x_2-x_1)/s$

$\cos C=(z_2-z_1)/s$

$\cos D=(w_2-w_1)/s$

D

SECONDARY ANALYSIS -- YZW SPACE

PLOTTED POINTS

$P_1=(y_1,z_1,w_1)$

$P_2=(y_2,z_1,w_1)$

LENGTH

$$s=\sqrt{(y_2-y_1)^2+(z_2-z_1)^2+(w_2-w_1)^2}$$

DIRECTION COSINES

$\cos B=(y_2-y_1)/s$

$\cos C=(z_2-z_1)/s$

$\cos D=(w_2-w_1)/s$

E

three lines show that the figure is composed of just three points. That particular configuration of lines and point specifications represents the closure of the three lines to create a complete, unbroken, triangular figure in 4-D space.

Everything that is expressed in the specifications and both levels of analysis rests ultimately with your choice of point coordinates; the form of the specifications and the equations for working out the analyses are virtually identical for every triangular figure in 4-D space.

You can generate any number of different triangles in 4-D space by assigning some carefully selected, specific numbers to the components of the coordinates in the P specifications. See some examples in Table 22-2F.

While you are conducting formal analyses of some triangles in 4-D space, bear in mind the character of triangles that are common to such figures in two or more dimensions: the sum of the three angles in a triangle is 180 degrees; if one of the angles is supposed to be greater than 90 degrees, subtract the calculated value from 180 degrees to determine that angle; if the angle between any two lines happens to be 90 degrees, you have specified a right triangle; and if an angle between two lines calculates to be zero degrees, you've made a mistake in specifying the original coordinates of carrying out the arithmetic.

22-1.3 Quadralaterals in 4-D Space

The F specification for the quadralateral specified in Table 22-3A shows the four lines that bound it, and the L specifications define those lines in terms of their 4-D endpoints. This particular set of specifications ensure complete closure of the figure. Refer to some of the sets of coordinates suggested in Table 22-3F, and the general specifications will always lead you to a valid 4-sided plane figure immersed in 4-D space (Fig. 22-3).

The primary and secondary analyses includes the lengths of the four lines, the direction cosines, and values for the four angles between them.

While working with the suggested coordinates, plotting the figures and conducting complete analyses of them, it will be helpful to bear in mind some special ideas that have been outlined for this sort of figure in Chapter 9: the sum of the four angles in a quadralateral is 360 degrees two lines in the figure are parallel to one another if the sum of the products of their corresponding direction cosines is 1 or −1, and two lines in the figure are perpen-

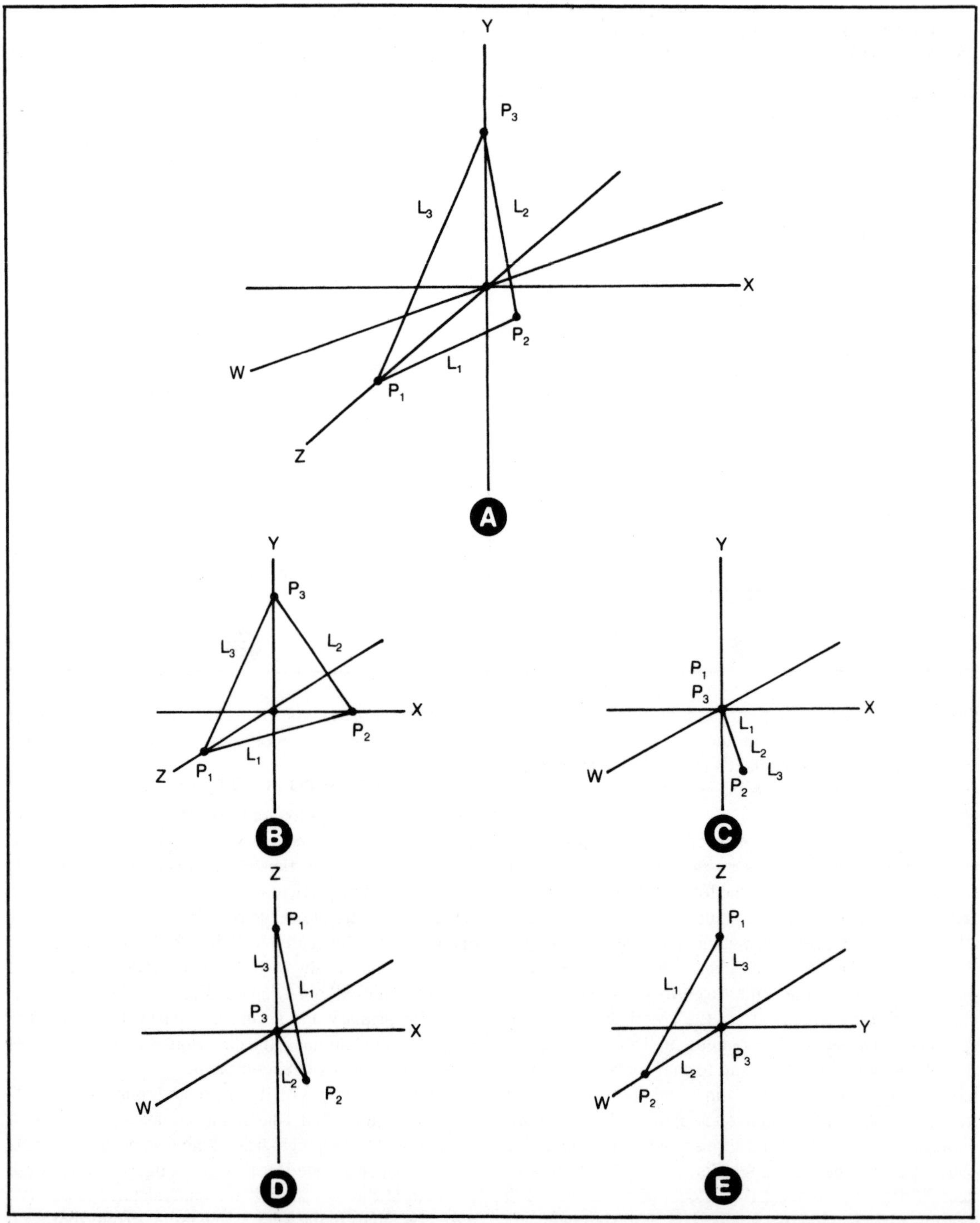

Fig. 22-2. A triangle specified in 4-D space. (A) General hyperspace view. (B) XYZ perspective view. (C) XYW perspective view. (D) XZW perspective view. (E) YZW perspective view. See outlines of analyses in Table 22-2.

Table 22-2. Specifications and Analyses of a Triangle in 4-D Space. See Fig. 22-2. (A) Specifications and Primary Analysis in Hyperspace. (B) Secondary Analysis in XYZ Space. (C) Secondary Analysis in XYW Space. (D) Secondary Analysis in XZW Space. (E) Secondary Analysis in YZW Space. (F) Suggested Coordinates.

(A, B, C, and D correspond to α, β, γ, and δ.)

SPECIFICATIONS

$F=\overline{L_1L_2L_3}$

$L_1=\overline{P_1P_2}$ $\quad P_1=(x_1,y_1,z_1,w_1)$

$L_2=\overline{P_2P_3}$ $\quad P_2=(x_2,y_2,z_2,w_2)$

$L_3=\overline{P_3P_1}$ $\quad P_3=(x_3,y_3,z_3,w_3)$

LENGTH

$$s_1=\sqrt{(x_2-x_1)^2+(y_2-y_1)^2+(z_2-z_1)^2+(w_2-w_1)^2}$$

$$s_2=\sqrt{(x_3-x_2)^2+(y_3-y_2)^2+(z_3-z_2)^2+(w_3-w_2)^2}$$

$$s_3=\sqrt{(x_1-x_3)^2+(y_1-y_3)^2+(z_1-z_3)^2+(w_1-w_3)^2}$$

DIRECTION COSINES

$\cos A_1=(x_2-x_1)/s_1$ $\quad \cos B_1=(y_2-y_1)/s_1$

$\cos C_1=(z_2-z_1)/s_1$ $\quad \cos D_1=(w_2-w_1)/s_1$

$\cos A_2=(x_3-x_2)/s_2$ $\quad \cos B_2=(y_3-y_2)/s_2$

$\cos C_2=(z_3-z_2)/s_2$ $\quad \cos D_2=(w_3-w_2)/s_2$

$\cos A_3=(x_1-x_3)/s_3$ $\quad \cos B_3=(y_1-y_3)/s_3$

$\cos C_3=(z_1-z_2)/s_3$ $\quad \cos D_3=(w_1-w_2)/s_3$

ANGLES

$$\cos\phi_{1,2}=\cos A_1\cos A_2+\cos B_1\cos B_2+\cos C_1\cos C_2+\cos D_1\cos D_2$$

$$\cos\phi_{1,3}=\cos A_1\cos A_3+\cos B_1\cos B_3+\cos C_1\cos C_3+\cos D_1\cos D_3$$

$$\cos\phi_{2,3}=\cos A_2\cos A_3+\cos B_2\cos B_3+\cos C_2\cos C_3+\cos D_2\cos D_3$$

CHECK

$$\phi_{1,2}+\phi_{1,3}+\phi_{2,3}=180^\circ$$

A

Table 22-2. Specifications and Analyses of a Triangle in 4-D Space. See Fig. 22-2. (A) Specifications and Primary Analysis in Hyperspace. (B) Secondary Analysis in XYZ Space. (C) Secondary Analysis in XYW Space. (D) Secondary Analysis in YZW Space. (E) Secondary Analysis in YZW Space. (F) Suggested Coordinates. (Continued from page 343.)

SECONDARY ANALYSIS -- XYZ SPACE

PLOTTED POINTS

$P_1=(x_1,y_1,z_1)$

$P_2=(x_2,y_2,z_2)$

$P_3=(x_3,y_3,z_3)$

LENGTH

$s_1=\sqrt{(x_2-x_1)^2+(y_2-y_1)^2+(z_2-z_1)^2}$

$s_2=\sqrt{(x_3-x_2)^2+(y_3-y_2)^2+(z_3-z_2)^2}$

$s_3=\sqrt{(x_1-x_3)^2+(y_1-y_3)^2+(z_1-z_3)^2}$

DIRECTION COSINES

$\cos A_1=(x_2-x_1)/s_1$ $\cos B_1=(y_2-y_1)/s_1$ $\cos C_1=(z_2-z_1)/s_1$

$\cos A_2=(x_3-x_2)/s_2$ $\cos B_2=(y_3-y_2)/s_2$ $\cos C_2=(z_3-z_2)/s_2$

$\cos A_3=(x_1-x_3)/s_3$ $\cos B_3=(y_1-y_3)/s_3$ $\cos C_3=(z_1-z_2)/s_3$

ANGLES

$\cos\phi_{1,2}=\cos A_1\cos A_2+\cos B_1\cos B_2+\cos C_1\cos C_2$

$\cos\phi_{1,3}=\cos A_1\cos A_3+\cos B_1\cos B_3+\cos C_1\cos C_3$

$\cos\phi_{2,3}=\cos A_2\cos A_3+\cos B_2\cos B_3+\cos C_2\cos C_3$

CHECK

$\phi_{1,2}+\phi_{1,3}+\phi_{2,3}=180^\circ$

B

SECONDARY ANALYSIS -- XYW SPACE

PLOTTED POINTS

$P_1=(x_1,y_1,w_1)$

$P_2=(x_2,y_2,w_2)$

$P_3=(x_3,y_3,w_3)$

LENGTH

$s_1=\sqrt{(x_2-x_1)^2+(y_2-y_1)^2+(w_2-w_1)^2}$

$s_2=\sqrt{(x_3-x_2)^2+(y_3-y_2)^2+(w_3-w_2)^2}$

$s_3=\sqrt{(x_1-x_3)^2+(y_1-y_3)^2+(w_1-w_3)^2}$

DIRECTION COSINES

$\cos A_1=(x_2-x_1)/s_1$ $\cos B_1=(y_2-y_1)/s_1$

$\cos D_1=(w_2-w_1)/s_1$

$\cos A_2=(x_3-x_2)/s_2$ $\cos B_2=(y_3-y_2)/s_2$

$\cos D_2=(w_3-w_2)/s_2$

$\cos A_3=(x_1-x_3)/s_3$ $\cos B_3=(y_1-y_3)/s_3$

$\cos D_3=(w_1-w_2)/s_3$

ANGLES

$\cos\phi_{1,2}=\cos A_1\cos A_2+\cos B_1\cos B_2+\cos D_1\cos D_2$

$\cos\phi_{1,3}=\cos A_1\cos A_3+\cos B_1\cos B_3+\cos D_1\cos D_3$

$\cos\phi_{2,3}=\cos A_2\cos A_3+\cos B_2\cos B_3+\cos D_2\cos D_3$

CHECK

$\phi_{1,2}+\phi_{1,3}+\phi_{2,3}=180^\circ$

C

SECONDARY ANALYSIS -- XZW SPACE

PLOTTED POINTS

$P_1=(x_1,z_1,w_1)$

$P_2=(x_2,z_2,w_2)$

$P_3=(x_3,z_3,w_3)$

LENGTH

$s_1=\sqrt{(x_2-x_1)^2+(z_2-z_1)^2+(w_2-w_1)^2}$

$s_2=\sqrt{(x_3-x_2)^2+(z_3-z_2)^2+(w_3-w_2)^2}$

$s_3=\sqrt{(x_1-x_3)^2+(z_1-z_3)^2+(w_1-w_3)^2}$

DIRECTION COSINES

$\cos A_1=(x_2-x_1)/s_1$ $\cos C_1=(z_2-z_1)/s_1$

$\cos D_1=(w_2-w_1)/s_1$

$\cos A_2=(x_3-x_2)/s_2$ $\cos C_2=(z_3-z_2)/s_2$

$\cos D_2=(w_3-w_2)/s_2$

$\cos A_3=(x_1-x_3)/s_3$ $\cos C_3=(z_1-z_2)/s_3$

$\cos D_3=(w_1-w_2)/s_3$

ANGLES

$\cos\phi_{1,2}=\cos A_1\cos A_2+\cos C_1\cos C_2+\cos D_1\cos D_2$

$\cos\phi_{1,3}=\cos A_1\cos A_3+\cos C_1\cos C_3+\cos D_1\cos D_3$

$\cos\phi_{2,3}=\cos A_2\cos A_3+\cos C_2\cos C_3+\cos D_2\cos D_3$

CHECK

$\phi_{1,2}+\phi_{1,3}+\phi_{2,3}=180^\circ$

D

Table 22-2. Specifications and Analyses of a Triangle in 4-D Space. See Fig. 22-2. (A) Specifications and Primary Analysis in Hyperspace. (B) Secondary Analysis in XYZ Space. (C) Secondary Analysis in XYW Space. (D) Secondary Analysis in YZW Space. (E) Secondary Analysis in YZW Space. (F) Suggested Coordinates. (Continued from page 345.)

SECONDARY ANALYSIS -- YZW SPACE

PLOTTED POINTS

$P_1=(y_1,z_1,w_1)$

$P_2=(y_2,z_2,w_2)$

$P_3=(y_3,z_3,w_3)$

LENGTH

$s_1=\sqrt{(y_2-y_1)^2+(z_2-z_1)^2+(w_2-w_1)^2}$

$s_2=\sqrt{(y_3-y_2)^2+(z_3-z_2)^2+(w_3-w_2)^2}$

$s_3=\sqrt{(y_1-y_3)^2+(z_1-z_3)^2+(w_1-w_3)^2}$

DIRECTION COSINES

$\cos B_1=(y_2-y_1)/s_1$ $\cos C_1=(z_2-z_1)/s_1$

$\cos D_1=(w_2-w_1)/s_1$

$\cos B_2=(y_3-y_2)/s_2$ $\cos C_2=(z_3-z_2)/s_2$

$\cos D_2=(w_3-w_2)/s_2$

$\cos B_3=(y_1-y_3)/s_3$ $\cos C_3=(z_1-z_2)/s_3$

$\cos D_3=(w_1-w_2)/s_3$

ANGLES

$\cos\emptyset_{1,2}=\cos B_1\cos B_2+\cos C_1\cos C_2+\cos D_1\cos D_2$

$\cos\emptyset_{1,3}=\cos B_1\cos B_3+\cos C_1\cos C_3+\cos D_1\cos D_3$

$\cos\emptyset_{2,3}=\cos B_2\cos B_3+\cos C_2\cos C_3+\cos D_2\cos D_3$

CHECK

$\emptyset_{1,2}+\emptyset_{1,3}+\emptyset_{2,3}=180^\circ$

E

FIGURE VERSION 1:

$P_1=(0,0,0,0)$ $P_2=(1,0,0,1)$ $P_3=(0,1,1,0)$

FIGURE VERSION 2:

$P_1=(-1,0,0,-1)$ $P_2=(1,0,0,1)$ $P_3=(0,1,1,0)$

FIGURE VERSION 3:

$P_1=(-1,-1,-1,-1)$ $P_2=(2,1,1,2)$ $P_3=(3,3,3,3)$

F

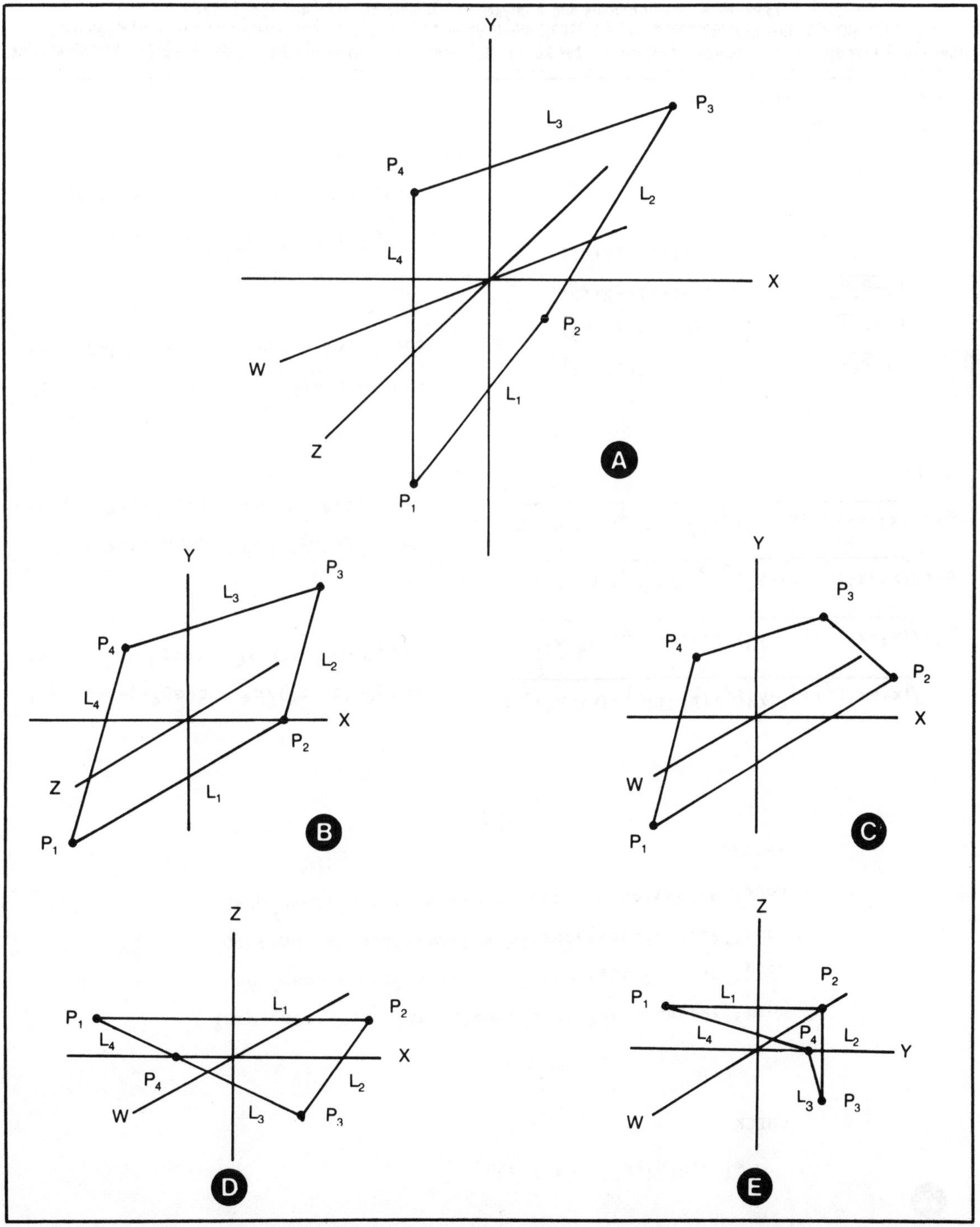

Fig. 22-3. A quadralateral specified in 4-D space. (A) General hyperspace perspective view. (B) XYZ perspective view. (C) XYW perspective view. (D) XZW perspective view. (E) YZW perspective view. See outlines of analyses in Table 22-3.

Table 22-3. Specifications and Analyses of a Quadralateral in 4-D Space. See Fig. 22-3. (A) Specifications and Primary Analysis in Hyperspace. (B) Secondary Analysis in XYZ Space. (C) Secondary Analysis in XYW Space. (D) Secondary Analysis in XZW Space. (E) Secondary Analysis in YZW Space. (F) Suggested Coordinates.

(A, B, C, and D correspond to α, β, γ, and δ.)

SPECIFICATIONS

$F=\overline{L_1L_2L_3L_4}$

$L_1=\overline{P_1P_2}$ $\quad P_1=(x_1,y_1,z_1,w_1)$

$L_2=\overline{P_2P_3}$ $\quad P_2=(x_2,y_2,z_2,w_2)$

$L_3=\overline{P_3P_4}$ $\quad P_3=(x_3,y_3,z_3,w_3)$

$L_4=\overline{P_4P_1}$ $\quad P_4=(x_4,y_4,z_4,w_4)$

LENGTH

$s_1=\sqrt{(x_2-x_1)^2+(y_2-y_1)^2+(z_2-z_1)^2+(w_2-w_1)^2}$

$s_2=\sqrt{(x_3-x_2)^2+(y_3-y_2)^2+(z_3-z_2)^2+(w_3-w_2)^2}$

$s_3=\sqrt{(x_4-x_3)^2+(y_4-y_3)^2+(z_4-z_3)^2+(w_4-w_3)^2}$

$s_4=\sqrt{(x_1-x_4)^2+(y_1-y_4)^2+(z_1-z_4)^2+(w_1-w_4)^2}$

DIRECTION COSINES

$\cos A_1=(x_2-x_1)/s_1$ $\quad \cos B_1=(y_2-y_1)/s_1$

$\cos C_1=(z_2-z_1)/s_1$ $\quad \cos D_1=(w_2-w_1)/s_1$

$\cos A_2=(x_3-x_2)/s_2$ $\quad \cos B_2=(y_3-y_2)/s_2$

$\cos C_2=(z_3-z_2)/s_2$ $\quad \cos D_2=(w_3-w_2)/s_2$

$\cos A_3=(x_4-x_3)/s_3$ $\quad \cos B_3=(y_4-y_3)/s_3$

$\cos C_3=(z_4-z_3)/s_3$ $\quad \cos D_3=(w_4-w_3)/s_3$

$\cos A_4=(x_1-x_4)/s_4$ $\quad \cos B_4=(y_1-y_4)/s_4$

$\cos C_4=(z_1-z_4)/s_4$ $\quad \cos D_4=(w_1-w_4)/s_4$

ANGLES

$\cos\phi_{1,2}=\cos A_1\cos A_2+\cos B_1\cos B_2+\cos C_1\cos C_2+\cos D_1\cos D_2$

$\cos\phi_{1,4}=\cos A_1\cos A_4+\cos B_1\cos B_4+\cos C_1\cos C_4+\cos D_1\cos D_4$

$\cos\phi_{2,3}=\cos A_2\cos A_3+\cos B_2\cos B_3+\cos C_2\cos C_3+\cos D_2\cos D_3$

$\cos\phi_{3,4}=\cos A_3\cos A_4+\cos B_3\cos B_4+\cos C_3\cos C_4+\cos D_3\cos D_4$

CHECK

$\phi_{1,2}+\phi_{1,4}+\phi_{2,3}+\phi_{3,4}=360^\circ$

A

SECONDARY ANALYSIS -- XYZ SPACE

PLOTTED POINTS

$P_1=(x_1,y_1,z_1)$

$P_2=(x_2,y_2,z_2)$

$P_3=(x_3,y_3,z_3)$

$P_4=(x_4,y_4,z_4)$

LENGTH

$s_1=\sqrt{(x_2-x_1)^2+(y_2-y_1)^2+(z_2-z_1)^2}$

$s_2=\sqrt{(x_3-x_2)^2+(y_3-y_2)^2+(z_3-z_2)^2}$

$s_3=\sqrt{(x_4-x_3)^2+(y_4-y_3)^2+(z_4-z_3)^2}$

$s_4=\sqrt{(x_1-x_4)^2+(y_1-y_4)^2+(z_1-z_4)^2}$

DIRECTION COSINES

$\cos A_1=(x_2-x_1)/s_1$ $\cos B_1=(y_2-y_1)/s_1$ $\cos C_1=(z_2-z_1)/s_1$

$\cos A_2=(x_3-x_2)/s_2$ $\cos B_2=(y_3-y_2)/s_2$ $\cos C_2=(z_3-z_2)/s_2$

$\cos A_3=(x_4-x_3)/s_3$ $\cos B_3=(y_4-y_3)/s_3$ $\cos C_3=(z_4-z_3)/s_3$

$\cos A_4=(x_1-x_4)/s_4$ $\cos B_4=(y_1-y_4)/s_4$ $\cos C_4=(z_1-z_4)/s_4$

ANGLES

$\cos\phi_{1,2}=\cos A_1\cos A_2+\cos B_1\cos B_2+\cos C_1\cos C_2$

$\cos\phi_{1,4}=\cos A_1\cos A_4+\cos B_1\cos B_4+\cos C_1\cos C_4$

$\cos\phi_{2,3}=\cos A_2\cos A_3+\cos B_2\cos B_3+\cos C_2\cos C_3$

$\cos\phi_{3,4}=\cos A_3\cos A_4+\cos B_3\cos B_4+\cos C_3\cos C_4$

CHECK

B

$\phi_{1,2}+\phi_{1,4}+\phi_{2,3}+\phi_{3,4}=360^\circ$

Table 22-3. Specifications and Analyses of a Quadralateral in 4-D Space. See Fig. 22-3. (A) Specifications and Primary Analysis in Hyperspace. (B) Secondary Analysis in XYZ Space. (C) Secondary Analysis in XYW Space. (D) Secondary Analysis in XZW Space. (E) Secondary Analysis in YZW Space. (F) Suggested Coordinates. (Continued from page 349.)

SECONDARY ANALYSIS -- XYW SPACE

PLOTTED POINTS

$P_1=(x_1,y_1,w_1)$

$P_2=(x_2,y_2,w_2)$

$P_3=(x_3,y_3,w_3)$

$P_4=(x_4,y_4,w_4)$

LENGTH

$s_1=\sqrt{(x_2-x_1)^2+(y_2-y_1)^2+(w_2-w_1)^2}$

$s_2=\sqrt{(x_3-x_2)^2+(y_3-y_2)^2+(w_3-w_2)^2}$

$s_3=\sqrt{(x_4-x_3)^2+(y_4-y_3)^2+(w_4-w_3)^2}$

$s_4=\sqrt{(x_1-x_4)^2+(y_1-y_4)^2+(w_1-w_4)^2}$

DIRECTION COSINES

$\cos A_1=(x_2-x_1)/s_1$ $\cos B_1=(y_2-y_1)/s_1$ $\cos D_1=(w_2-w_1)/s_1$

$\cos A_2=(x_3-x_2)/s_2$ $\cos B_2=(y_3-y_2)/s_2$ $\cos D_2=(w_3-w_2)/s_2$

$\cos A_3=(x_4-x_3)/s_3$ $\cos B_3=(y_4-y_3)/s_3$ $\cos D_3=(w_4-w_3)/s_3$

$\cos A_4=(x_1-x_4)/s_4$ $\cos B_4=(y_1-y_4)/s_4$ $\cos D_4=(w_1-w_4)/s_4$

ANGLES

$\cos\phi_{1,2}=\cos A_1\cos A_2+\cos B_1\cos B_2+\cos D_1\cos D_2$

$\cos\phi_{1,4}=\cos A_1\cos A_4+\cos B_1\cos B_4+\cos D_1\cos D_4$

$\cos\phi_{2,3}=\cos A_2\cos A_3+\cos B_2\cos B_3+\cos D_2\cos D_3$

$\cos\phi_{3,4}=\cos A_3\cos A_4+\cos B_3\cos B_4+\cos D_3\cos D_4$

CHECK

C

$\phi_{1,2}+\phi_{1,4}+\phi_{2,3}+\phi_{3,4}=360^\circ$

SECONDARY ANALYSIS -- XZW SPACE

PLOTTED POINTS

$P_1=(x_1,z_1,w_1)$

$P_2=(x_2,z_2,w_2)$

$P_3=(x_3,z_3,w_3)$

$P_4=(x_4,z_4,w_4)$

LENGTH

$$s_1=\sqrt{(x_2-x_1)^2+(z_2-z_1)^2+(w_2-w_1)^2}$$

$$s_2=\sqrt{(x_3-x_2)^2+(z_3-z_2)^2+(w_3-w_2)^2}$$

$$s_3=\sqrt{(x_4-x_3)^2+(z_4-z_3)^2+(w_4-w_3)^2}$$

$$s_4=\sqrt{(x_1-x_4)^2+(z_1-z_4)^2+(w_1-w_4)^2}$$

DIRECTION COSINES

$\cos A_1=(x_2-x_1)/s_1$ $\cos C_1=(z_2-z_1)/s_1$ $\cos D_1=(w_2-w_1)/s_1$

$\cos A_2=(x_3-x_2)/s_2$ $\cos C_2=(z_3-z_2)/s_2$ $\cos D_2=(w_3-w_2)/s_2$

$\cos A_3=(x_4-x_3)/s_3$ $\cos C_3=(z_4-z_3)/s_3$ $\cos D_3=(w_4-w_3)/s_3$

$\cos A_4=(x_1-x_4)/s_4$ $\cos C_4=(z_1-z_4)/s_4$ $\cos D_4=(w_1-w_4)/s_4$

ANGLES

$\cos\phi_{1,2}=\cos A_1\cos A_2+\cos C_1\cos C_2+\cos D_1\cos D_2$

$\cos\phi_{1,4}=\cos A_1\cos A_4+\cos C_1\cos C_4+\cos D_1\cos D_4$

$\cos\phi_{2,3}=\cos A_2\cos A_3+\cos C_2\cos C_3+\cos D_2\cos D_3$

$\cos\phi_{3,4}=\cos A_3\cos A_4+\cos C_3\cos C_4+\cos D_3\cos D_4$

CHECK

$\phi_{1,2}+\phi_{1,4}+\phi_{2,3}+\phi_{3,4}=360^\circ$

D

Table 22-3. Specifications and Analyses of a Quadralateral in 4-D Space. See Fig. 22-3. (A) Specifications and Primary Analysis in Hyperspace. (B) Secondary Analysis in XYZ Space. (C) Secondary Analysis in XYW Space. (D) Secondary Analysis in XZW Space. (E) Secondary Analysis in YZW Space. (F) Suggested Coordinates. (Continued from page 351.)

SECONDARY ANALYSIS -- YZW SPACE

PLOTTED POINTS

$P_1=(y_1,z_1,w_1)$

$P_2=(y_2,z_2,w_2)$

$P_3=(y_3,z_3,w_3)$

$P_4=(y_4,z_4,w_4)$

LENGTH

$$s_1=\sqrt{(y_2-y_1)^2+(z_2-z_1)^2+(w_2-w_1)^2}$$

$$s_2=\sqrt{(y_3-y_2)^2+(z_3-z_2)^2+(w_3-w_2)^2}$$

$$s_3=\sqrt{(y_4-y_3)^2+(z_4-z_3)^2+(w_4-w_3)^2}$$

$$s_4=\sqrt{(y_1-y_4)^2+(z_1-z_4)^2+(w_1-w_4)^2}$$

DIRECTION COSINES

$\cos B_1=(y_2-y_1)/s_1$	$\cos C_1=(z_2-z_1)/s_1$	$\cos D_1=(w_2-w_1)/s_1$
$\cos B_2=(y_3-y_2)/s_2$	$\cos C_2=(z_3-z_2)/s_2$	$\cos D_2=(w_3-w_2)/s_2$
$\cos B_3=(y_4-y_3)/s_3$	$\cos C_3=(z_4-z_3)/s_3$	$\cos D_3=(w_4-w_3)/s_3$
$\cos B_4=(y_1-y_4)/s_4$	$\cos C_4=(z_1-z_4)/s_4$	$\cos D_4=(w_1-w_4)/s_4$

ANGLES

$$\cos\phi_{1,2}=\cos B_1\cos B_2+\cos C_1\cos C_2+\cos D_1\cos D_2$$

$$\cos\phi_{1,4}=\cos B_1\cos B_4+\cos C_1\cos C_4+\cos D_1\cos D_4$$

$$\cos\phi_{2,3}=\cos B_2\cos B_3+\cos C_2\cos C_3+\cos D_2\cos D_3$$

$$\cos\phi_{3,4}=\cos B_3\cos B_4+\cos C_3\cos C_4+\cos D_3\cos D_4$$

E

CHECK

$$\phi_{1,2}+\phi_{1,4}+\phi_{2,3}+\phi_{3,4}=360^\circ$$

FIGURE VERSION 1:

$P_1=(-1,-1,-1,-1)$ $P_2=(1,-1,-1,1)$

$P_3=(1,1,1,1)$ $P_4=(-1,1,1,-1)$

FIGURE VERSION 2:

$P_1=(-2,-1,-1,-2)$ $P_2=(2,-1,-1,2)$

$P_3=(2,1,1,2)$ $P_4=(-2,1,1,-2)$

FIGURE VERSION 3:

$P_1=(-3,-2,-2,-3)$ $P_2=(1,-2,-2,1)$

$P_3=(1,2,2,1)$ $P_4=(-1,2,2,-1)$

FIGURE VERSION 4:

$P_1=(-1,0,0,-1)$ $P_2=(0,-1,-1,0)$

$P_3=(1,0,0,1)$ $P_4=(0,1,1,0)$

F

dicular to one another if the sum of the products of their corresponding direction cosines is zero.

Like triangles that are mapped to 3-D space, quadralaterals often degenerate to simple lines.

22-1.4 Pentagons in 4-D Space

The drawings in Fig. 22-4 and the specifications cited in Table 22-4 show that a general pentagon is composed of five bounding lines and five different points. The specifications are identical to those for a pentagon in 2-D space except at one level—the point coordinates here require four components instead of just two.

Use the general specifications and the suggested sets of coordinates in Table 22-4F to generate a file of pentagons. Check your work by using the rule that says the sum of the angle of the pentagon is 540 degrees.

22-1.5 The General 4-Sided Pyramid

Figure 22-5 shows the four perspective views of a 4-sided pyramid that is immersed in 4-D space. Table 22-5 shows the general specifications and general outlines for conducting complete primary and secondary analyses of it.

You can use those specifications to generate any number of variations of this 4-sided pyramid. All you have to do is substitute valid and specific component values for the point coordinates (see some examples in Table 22-5F).

After specifying some coordinates and conducting the primary analysis, you can check the results by working with the angles. Recalling that the sum of the angles in a triangle is 180 degrees, you can determine which angles appear in one of the sides of the space object and add them together. If the sum is close to 180 degrees, you can be reasonably sure that your analysis of that side is correct. Doing the same for the three remaining sides completes your check of the work. This sort of doublechecking is especially important because space objects that are immersed in 4-D space are about impossible to perceive in anything but a totally analytic fashion.

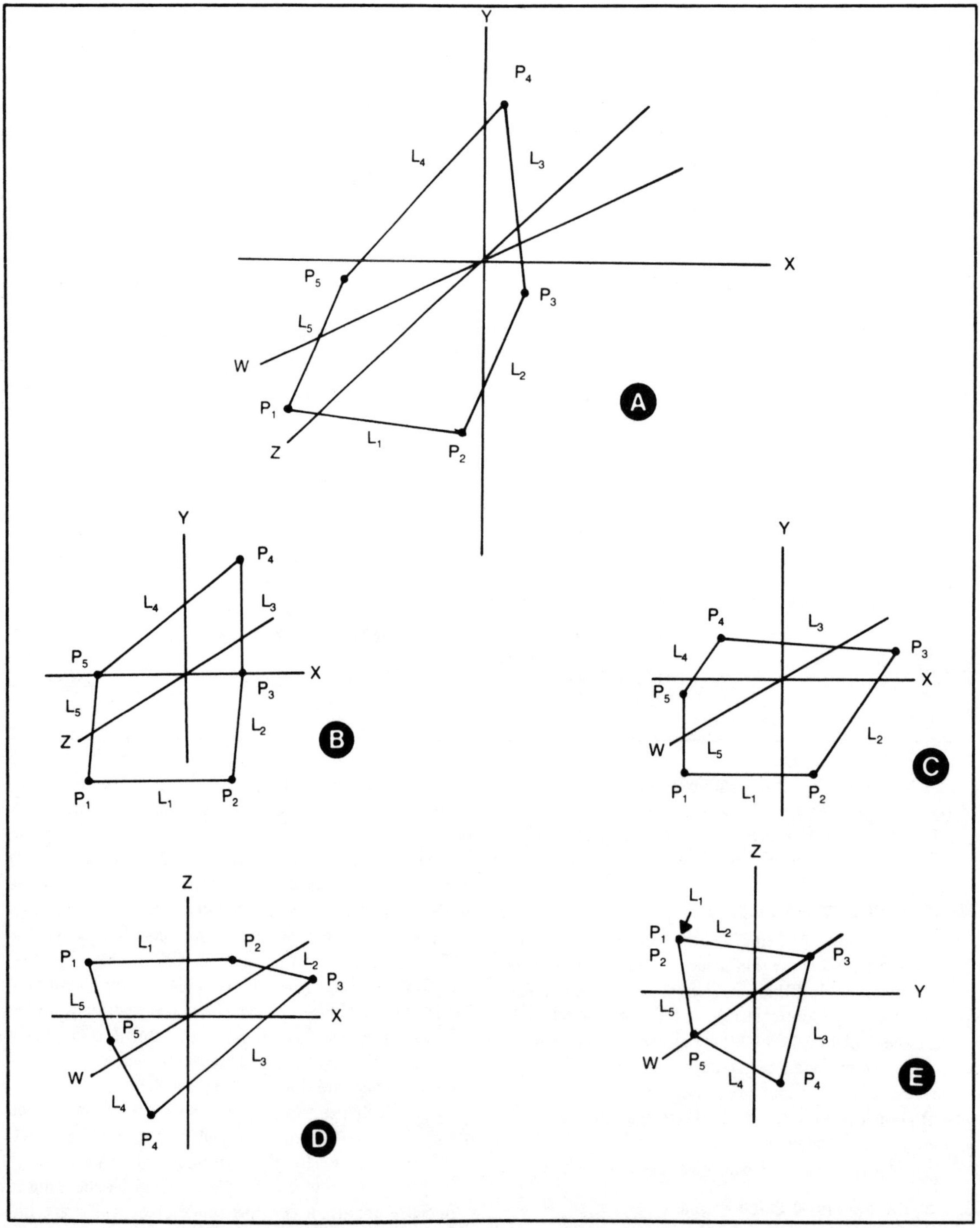

Fig. 22-4. A pentagon specified in 4-D space. (A) General hyperspace view. (B) XYZ perspective view. (C) XYW perspective view. (D) XZW perspective view. (E) YZW perspective view. See outlines of analyses in Table 22-4.

Table 22-4. Specifications and Analyses of a Pentagon in 4-D Space. See Fig. 22-4. (A) Specifications and Primary Analysis in Hyperspace. (B) Secondary Analysis in XYZ Space. (C) Secondary Analysis in XYW Space. (D) Secondary Analysis in XZW Space. (E) Secondary Analysis in YZW Space. (F) Suggested Coordinates.

(A, B, C, and D correspond to α, β, γ, and δ.)

SPECIFICATIONS

$$F=\overline{L_1L_2L_3L_4L_5}$$

$L_1=\overline{P_1P_2}$ $\quad P_1=(x_1,y_1,z_1,w_1)$

$L_2=\overline{P_2P_3}$ $\quad P_2=(x_2,y_2,z_2,w_2)$

$L_3=\overline{P_3P_4}$ $\quad P_3=(x_3,y_3,z_3,w_3)$

$L_4=\overline{P_4P_5}$ $\quad P_4=(x_4,y_4,z_4,w_4)$

$L_5=\overline{P_5P_1}$ $\quad P_5=(x_5,y_5,z_5,w_5)$

LENGTH

$$s_1=\sqrt{(x_2-x_1)^2+(y_2-y_1)^2+(z_2-z_1)^2+(w_2-w_1)^2}$$

$$s_2=\sqrt{(x_3-x_2)^2+(y_3-y_2)^2+(z_3-z_2)^2+(w_3-w_2)^2}$$

$$s_3=\sqrt{(x_4-x_3)^2+(y_4-y_3)^2+(z_4-z_3)^2+(w_4-w_3)^2}$$

$$s_4=\sqrt{(x_5-x_4)^2+(y_5-y_4)^2+(z_5-z_4)^2+(w_5-w_4)^2}$$

$$s_5=\sqrt{(x_1-x_5)^2+(y_1-y_5)^2+(z_1-z_5)^2+(w_1-w_5)^2}$$

DIRECTION COSINES

$\cos A_1=(x_2-x_1)/s_1$ $\quad \cos B_1=(y_2-y_1)/s_1$

$\cos C_1=(z_2-z_1)/s_1$ $\quad \cos D_1=(w_2-w_1)/s_1$

$\cos A_2=(x_3-x_2)/s_2$ $\quad \cos B_2=(y_3-y_2)/s_2$

$\cos C_2=(z_3-z_2)/s_2$ $\quad \cos D_2=(w_3-w_2)/s_2$

$\cos A_3=(x_4-x_3)/s_3$ $\quad \cos B_3=(y_4-y_3)/s_3$

$\cos C_3=(z_4-z_3)/s_3$ $\quad \cos D_3=(w_4-w_3)/s_3$

$\cos A_4=(x_5-x_4)/s_4$ $\quad \cos B_4=(y_5-y_4)/s_4$

$\cos C_4=(z_5-z_4)/s_4$ $\quad \cos D_4=(w_5-w_4)/s_4$

$\cos A_5=(x_1-x_5)/s_5$ $\quad \cos B_5=(y_1-y_5)/s_5$

$\cos C_5=(z_1-z_5)/s_5$ $\quad \cos D_5=(w_1-w_5)/s_5$

ANGLES

$$\cos\phi_{1,2}=\cos A_1\cos A_2+\cos B_1\cos B_2+\cos C_1\cos C_2+\cos D_1\cos D_2$$

$$\cos\phi_{2,3}=\cos A_2\cos A_3+\cos B_2\cos B_3+\cos C_2\cos C_3+\cos D_2\cos D_3$$

$$\cos\phi_{3,4}=\cos A_3\cos A_4+\cos B_3\cos B_4+\cos C_3\cos C_4+\cos D_3\cos D_4$$

$$\cos\phi_{4,5}=\cos A_4\cos A_5+\cos B_4\cos B_5+\cos C_4\cos C_5+\cos D_4\cos D_5$$

$$\cos\phi_{5,1}=\cos A_5\cos A_1+\cos B_5\cos B_1+\cos C_5\cos C_1+\cos D_5\cos D_1$$

A

CHECK

$$\phi_{1,2}+\phi_{2,3}+\phi_{3,4}+\phi_{4,5}+\phi_{5,1}=540^\circ$$

Table 22-4. Specifications and Analyses of a Pentagon in 4-D Space. See Fig. 22-4. (A) Specifications and Primary Analysis in Hyperspace. (B) Secondary Analysis in XYZ Space. (C) Secondary Analysis in XYW Space. (D) Secondary Analysis in XZW Space. (E) Secondary Analysis in YZW Space. (F) Suggested Coordinates. (Continued from page 355.)

SECONDARY ANALYSIS -- XYZ SPACE

PLOTTED POINTS

$P_1=(x_1,y_1,z_1)$

$P_2=(x_2,y_2,z_2)$

$P_3=(x_3,y_3,z_3)$

$P_4=(x_4,y_4,z_4)$

$P_5=(x_5,y_5,z_5)$

LENGTH

$s_1=\sqrt{(x_2-x_1)^2+(y_2-y_1)^2+(z_2-z_1)^2}$

$s_2=\sqrt{(x_3-x_2)^2+(y_3-y_2)^2+(z_3-z_2)^2}$

$s_3=\sqrt{(x_4-x_3)^2+(y_4-y_3)^2+(z_4-z_3)^2}$

$s_4=\sqrt{(x_5-x_4)^2+(y_5-y_4)^2+(z_5-z_4)^2}$

$s_5=\sqrt{(x_1-x_5)^2+(y_1-y_5)^2+(z_1-z_5)^2}$

DIRECTION COSINES

$\cos A_1=(x_2-x_1)/s_1$	$\cos B_1=(y_2-y_1)/s_1$	$\cos C_1=(z_2-z_1)/s_1$
$\cos A_2=(x_3-x_2)/s_2$	$\cos B_2=(y_3-y_2)/s_2$	$\cos C_2=(z_3-z_2)/s_2$
$\cos A_3=(x_4-x_3)/s_3$	$\cos B_3=(y_4-y_3)/s_3$	$\cos C_3=(z_4-z_3)/s_3$
$\cos A_4=(x_5-x_4)/s_4$	$\cos B_4=(y_5-y_4)/s_4$	$\cos C_4=(z_5-z_4)/s_4$
$\cos A_5=(x_1-x_5)/s_5$	$\cos B_5=(y_1-y_5)/s_5$	$\cos C_5=(z_1-z_5)/s_5$

ANGLES

$\cos\phi_{1,2}=\cos A_1\cos A_2+\cos B_1\cos B_2+\cos C_1\cos C_2$

$\cos\phi_{2,3}=\cos A_2\cos A_3+\cos B_2\cos B_3+\cos C_2\cos C_3$

$\cos\phi_{3,4}=\cos A_3\cos A_4+\cos B_3\cos B_4+\cos C_3\cos C_4$

$\cos\phi_{4,5}=\cos A_4\cos A_5+\cos B_4\cos B_5+\cos C_4\cos C_5$

$\cos\phi_{5,1}=\cos A_5\cos A_1+\cos B_5\cos B_1+\cos C_5\cos C_1$

CHECK

$\phi_{1,2}+\phi_{2,3}+\phi_{3,4}+\phi_{4,5}+\phi_{5,1}=540^\circ$

B

SECONDARY ANALYSIS -- XYW SPACE

PLOTTED POINTS

$P_1=(x_1,y_1,w_1)$

$P_2=(x_2,y_2,w_2)$

$P_3=(x_3,y_3,w_3)$

$P_4=(x_4,y_4,w_4)$

$P_5=(x_5,y_5,w_5)$

LENGTH

$s_1=\sqrt{(x_2-x_1)^2+(y_2-y_1)^2+(w_2-w_1)^2}$

$s_2=\sqrt{(x_3-x_2)^2+(y_3-y_2)^2+(w_3-w_2)^2}$

$s_3=\sqrt{(x_4-x_3)^2+(y_4-y_3)^2+(w_4-w_3)^2}$

$s_4=\sqrt{(x_5-x_4)^2+(y_5-y_4)^2+(w_5-w_4)^2}$

$s_5=\sqrt{(x_1-x_5)^2+(y_1-y_5)^2+(w_1-w_5)^2}$

DIRECTION COSINES

$\cos A_1=(x_2-x_1)/s_1$ $\cos B_1=(y_2-y_1)/s_1$

$\cos D_1=(w_2-w_1)/s_1$

$\cos A_2=(x_3-x_2)/s_2$ $\cos B_2=(y_3-y_2)/s_2$

$\cos D_2=(w_3-w_2)/s_2$

$\cos A_3=(x_4-x_3)/s_3$ $\cos B_3=(y_4-y_3)/s_3$

$\cos D_3=(w_4-w_3)/s_3$

$\cos A_4=(x_5-x_4)/s_4$ $\cos B_4=(y_5-y_4)/s_4$

$\cos D_4=(w_5-w_4)/s_4$

$\cos A_5=(x_1-x_5)/s_5$ $\cos B_5=(y_1-y_5)/s_5$

$\cos D_5=(w_1-w_5)/s_5$

ANGLES

$\cos\phi_{1,2}=\cos A_1\cos A_2+\cos B_1\cos B_2+\cos D_1\cos D_2$

$\cos\phi_{2,3}=\cos A_2\cos A_3+\cos B_2\cos B_3+\cos D_2\cos D_3$

$\cos\phi_{3,4}=\cos A_3\cos A_4+\cos B_3\cos B_4+\cos D_3\cos D_4$

$\cos\phi_{4,5}=\cos A_4\cos A_5+\cos B_4\cos B_5+\cos D_4\cos D_5$

$\cos\phi_{5,1}=\cos A_5\cos A_1+\cos B_5\cos B_1+\cos D_5\cos D_1$

CHECK

$\phi_{1,2}+\phi_{2,3}+\phi_{3,4}+\phi_{4,5}+\phi_{5,1}=540^\circ$

C

Table 22-4. Specifications and Analyses of a Pentagon in 4-D Space. See Fig. 22-4. (A) Specifications and Primary Analysis in Hyperspace. (B) Secondary Analysis in XYZ Space. (C) Secondary Analysis in XYW Space. (D) Secondary Analysis in XZW Space. (E) Secondary Analysis in YZW Space. (F) Suggested Coordinates. (Continued from page 357.)

SECONDARY ANALYSIS -- XZW SPACE

PLOTTED POINTS

$P_1=(x_1,z_1,w_1)$

$P_2=(x_2,z_2,w_2)$

$P_3=(x_3,z_3,w_3)$

$P_4=(x_4,z_4,w_4)$

$P_5=(x_5,z_5,w_5)$

DIRECTION COSINES

$\cos A_1=(x_2-x_1)/s_1$ $\cos C_1=(z_2-z_1)/s_1$

$\cos D_1=(w_2-w_1)/s_1$

$\cos A_2=(x_3-x_2)/s_2$ $\cos C_2=(z_3-z_2)/s_2$

$\cos D_2=(w_3-w_2)/s_2$

$\cos A_3=(x_4-x_3)/s_3$ $\cos C_3=(z_4-z_3)/s_3$

$\cos D_3=(w_4-w_3)/s_3$

$\cos A_4=(x_5-x_4)/s_4$ $\cos C_4=(z_5-z_4)/s_4$

$\cos D_4=(w_5-w_4)/s_4$

$\cos A_5=(x_1-x_5)/s_5$ $\cos C_5=(z_1-z_5)/s_5$

$\cos D_5=(w_1-w_5)/s_5$

LENGTH

$s_1=\sqrt{(x_2-x_1)^2+(z_2-z_1)^2+(w_2-w_1)^2}$

$s_2=\sqrt{(x_3-x_2)^2+(z_3-z_2)^2+(w_3-w_2)^2}$

$s_3=\sqrt{(x_4-x_3)^2+(z_4-z_3)^2+(w_4-w_3)^2}$

$s_4=\sqrt{(x_5-x_4)^2+(z_5-z_4)^2+(w_5-w_4)^2}$

$s_5=\sqrt{(x_1-x_5)^2+(z_1-z_5)^2+(w_1-w_5)^2}$

ANGLES

$\cos\phi_{1,2}=\cos A_1\cos A_2+\cos C_1\cos C_2+\cos D_1\cos D_2$

$\cos\phi_{2,3}=\cos A_2\cos A_3+\cos C_2\cos C_3+\cos D_2\cos D_3$

$\cos\phi_{3,4}=\cos A_3\cos A_4+\cos C_3\cos C_4+\cos D_3\cos D_4$

$\cos\phi_{4,5}=\cos A_4\cos A_5+\cos C_4\cos C_5+\cos D_4\cos D_5$

$\cos\phi_{5,1}=\cos A_5\cos A_1+\cos C_5\cos C_1+\cos D_5\cos D_1$

CHECK

D

$\phi_{1,2}+\phi_{2,3}+\phi_{3,4}+\phi_{4,5}+\phi_{5,1}=540^\circ$

SECONDARY ANALYSIS -- YZW SPACE

PLOTTED POINTS

$P_1=(y_1,z_1,w_1)$

$P_2=(y_2,z_2,w_2)$

$P_3=(y_3,z_3,w_3)$

$P_4=(y_4,z_4,w_4)$

$P_5=(y_5,z_5,w_5)$

DIRECTION COSINES

$\cos B_1=(y_2-y_1)/s_1 \quad \cos C_1=(z_2-z_1)/s_1$

$\cos D_1=(w_2-w_1)/s_1$

$\cos B_2=(y_3-y_2)/s_2 \quad \cos C_2=(z_3-z_2)/s_2$

$\cos D_2=(w_3-w_2)/s_2$

$\cos B_3=(y_4-y_3)/s_3 \quad \cos C_3=(z_4-z_3)/s_3$

$\cos D_3=(w_4-w_3)/s_3$

$\cos B_4=(y_5-y_4)/s_4 \quad \cos C_4=(z_5-z_4)/s_4$

$\cos D_4=(w_5-w_4)/s_4$

$\cos B_5=(y_1-y_5)/s_5 \quad \cos C_5=(z_1-z_5)/s_5$

$\cos D_5=(w_1-w_5)/s_5$

LENGTH

$s_1=\sqrt{(y_2-y_1)^2+(z_2-z_1)^2+(w_2-w_1)^2}$

$s_2=\sqrt{(y_3-y_2)^2+(z_3-z_2)^2+(w_3-w_2)^2}$

$s_3=\sqrt{(y_4-y_3)^2+(z_4-z_3)^2+(w_4-w_3)^2}$

$s_4=\sqrt{(y_5-y_4)^2+(z_5-z_4)^2+(w_5-w_4)^2}$

$s_5=\sqrt{(y_1-y_5)^2+(z_1-z_5)^2+(w_1-w_5)^2}$

ANGLES

$\cos\phi_{1,2}=\cos B_1\cos B_2+\cos C_1\cos C_2+\cos D_1\cos D_2$

$\cos\phi_{2,3}=\cos B_2\cos B_3+\cos C_2\cos C_3+\cos D_2\cos D_3$

$\cos\phi_{3,4}=\cos B_3\cos B_4+\cos C_3\cos C_4+\cos C_3\cos D_4$

$\cos\phi_{4,5}=\cos B_4\cos B_5+\cos C_4\cos C_5+\cos D_4\cos D_5$

$\cos\phi_{5,1}=\cos B_5\cos B_1+\cos C_5\cos C_1+\cos D_5\cos D_1$

CHECK

$\phi_{1,2}+\phi_{2,3}+\phi_{3,4}+\phi_{4,5}+\phi_{5,1}=540^\circ$

E

FIGURE VERSION 1:

$P_1=(-1,0,0,-1) \quad P_2=(1,0,0,1)$

$P_3=(1,1,1,1) \quad P_4=(0,2,2,0)$

$P_5=(-1,1,1,-1)$

FIGURE VERSION 2:

$P_1=(-2,0,0,-2) \quad P_2=(2,0,0,2)$

$P_3=(3,4,4,3) \quad P_4=(0,6,6,0)$

$P_5=(-3,4,4,-3)$

F

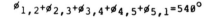

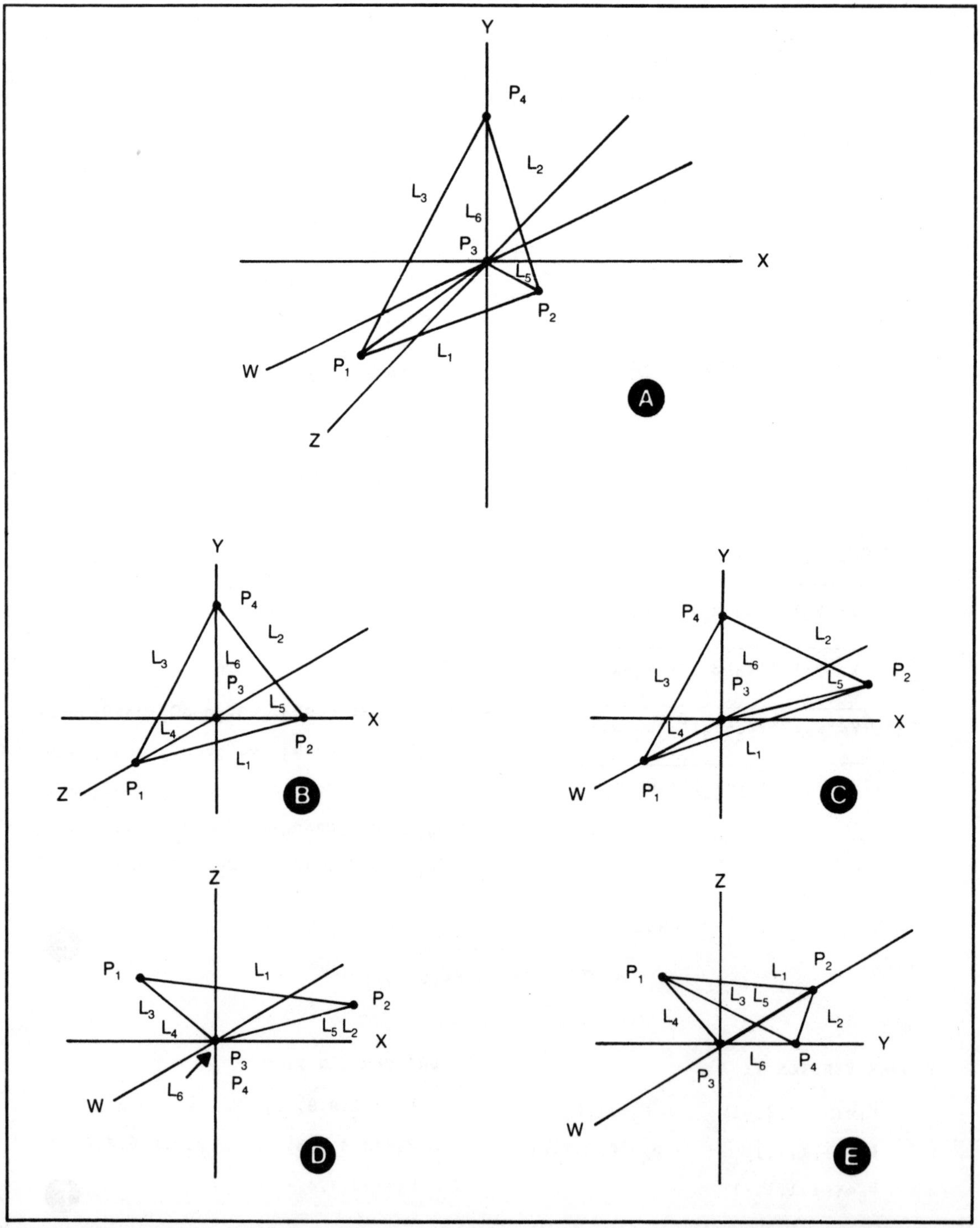

Fig. 22-5. A 4-sided pyramid specified in 4-D space. (A) General hyperspace view. (B) XYZ perspective view. (C) XYW perspective view. (D) XZW perspective view. (E) YZW perspective view. See outlines of analyses in Table 22-5.

Table 22-5. Specifications and Analyses of a 4-Sided Pyramid in 4-D Space. See Fig. 22-5. (A) Specifications and Primary Analysis in Hyperspace. (B) Secondary Analysis in XYZ Space. (C) Secondary Analysis in XYW Space. (D) Secondary Analysis in XZW Space. (E) Secondary Analysis in YZW Space. (F) Suggested Coordinates.

(A, B, C, and D correspond to α, β, γ, and δ.)

A

SPECIFICATIONS

$S=\overline{F_1F_2F_3F_4}$

$F_1=\overline{L_1L_2L_3}$

$F_2=\overline{L_1L_4L_5}$

$F_3=\overline{L_2L_5L_6}$

$F_4=\overline{L_3L_4L_6}$

$L_1=\overline{P_1P_2}$

$L_2=\overline{P_2P_3}$

$L_3=\overline{P_3P_1}$

$L_4=\overline{P_1P_4}$

$L_5=\overline{P_2P_4}$

$L_6=\overline{P_3P_4}$

$P_1=(x_1,y_1,z_1,w_1)$

$P_2=(x_2,y_2,z_2,w_2)$

$P_3=(x_3,y_3,z_3,w_3)$

$P_4=(x_4,y_4,z_4,w_4)$

LENGTHS

$$s_1=\overline{(x_2-x_1)^2+(y_2-y_1)^2+(z_2-z_1)^2+(w_2-w_1)^2}$$

$$s_2=\overline{(x_3-x_2)^2+(y_3-y_2)^2+(z_3-z_2)^2+(w_3-w_2)^2}$$

$$s_3=\overline{(x_1-x_3)^2+(y_1-y_3)^2+(z_1-z_3)^2+(w_1-w_3)^2}$$

$$s_4=\overline{(x_4-x_1)^2+(y_4-y_1)^2+(z_4-z_1)^2+(w_4-w_1)^2}$$

$$s_5=\overline{(x_4-x_2)^2+(y_4-y_2)^2+(z_4-z_2)^2+(w_4-w_2)^2}$$

$$s_6=\overline{(x_4-x_3)^2+(y_4-y_3)^2+(z_4-z_3)^2+(w_4-w_3)^2}$$

DIRECTION COSINES

$\cos A_1=(x_2-x_1)/s_1$ $\cos B_1=(y_2-y_1)/s_1$

$\cos C_1=(z_2-z_1)/s_1$ $\cos D_1=(w_2-w_1)/s$

$\cos A_2=(x_3-x_2)/s_2$ $\cos B_2=(y_3-y_2)/s_2$

$\cos C_2=(z_3-z_2)/s_2$ $\cos D_2=(w_3-w_2)/s$

Table 22-5. Specifications and Analyses of a 4-Sided Pyramid in 4-D Space. See Fig. 22-5. (A) Specifications and Primary Analysis in Hyperspace. (B) Secondary Analysis in XYZ Space. (C) Secondary Analysis in XYW Space. (D) Secondary Analysis in XZW Space. (E) Secondary Analysis in YZW Space. (F) Suggested Coordinates. (Continued from page 361.)

DIRECTION COSINES

$\cos A_3=(x_1-x_3)/s_3$ $\cos B_3=(y_1-y_3)/s_3$

$\cos C_3=(z_1-z_3)/s_3$ $\cos D_3=(w_1-w_3)/s$

$\cos A_4=(x_4-x_1)/s_4$ $\cos B_4=(y_4-y_1)/s_4$

$\cos C_4=(z_4-z_1)/s_4$ $\cos D_4=(w_4-w_1)/s$

$\cos A_5=(x_4-x_2)/s_5$ $\cos B_5=(y_4-y_2)/s_5$

$\cos C_5=(z_4-z_2)/s_5$ $\cos D_5=(w_4-w_2)/s$

$\cos A_6=(x_4-x_3)/s_6$ $\cos B_6=(y_4-y_3)/s_6$

$\cos C_6=(z_4-z_3)/s_6$ $\cos D_6=(w_4-w_3)/s$

ANGLES

$\cos\phi_{1,2}=\cos A_1\cos A_2+\cos B_1\cos B_2+\cos C_1\cos C_2+\cos D_1\cos D_2$

$\cos\phi_{1,3}=\cos A_1\cos A_3+\cos B_1\cos B_3+\cos C_1\cos C_3+\cos D_1\cos D_3$

$\cos\phi_{1,4}=\cos A_1\cos A_4+\cos B_1\cos B_4+\cos C_1\cos C_4+\cos D_1\cos D_4$

$\cos\phi_{2,3}=\cos A_2\cos A_3+\cos B_2\cos B_3+\cos C_2\cos C_3+\cos D_2\cos D_3$

$\cos\phi_{2,5}=\cos A_2\cos A_5+\cos B_2\cos B_5+\cos C_2\cos C_5+\cos D_2\cos D_5$

$\cos\phi_{2,6}=\cos A_2\cos A_6+\cos B_2\cos B_6+\cos C_2\cos C_6+\cos D_2\cos D_6$

$\cos\phi_{3,4}=\cos A_3\cos A_4+\cos B_3\cos B_4+\cos C_3\cos C_4+\cos D_3\cos D_4$

$\cos\phi_{3,6}=\cos A_3\cos A_6+\cos B_3\cos B_6+\cos C_3\cos C_6+\cos D_3\cos D_6$

$\cos\phi_{4,5}=\cos A_4\cos A_5+\cos B_4\cos B_5+\cos C_4\cos C_5+\cos D_4\cos D_5$

$\cos\phi_{4,6}=\cos A_4\cos A_6+\cos B_4\cos B_6+\cos C_4\cos C_6+\cos D_4\cos D_6$

$\cos\phi_{5,6}=\cos A_5\cos A_6+\cos B_5\cos B_6+\cos C_5\cos C_6+\cos D_5\cos D_6$

CHECK

F_1: $\phi_{1,2}+\phi_{1,3}+\phi_{2,3}=180^\circ$

F_2: $\phi_{1,4}+\phi_{1,5}+\phi_{4,5}=180$

F_3: $\phi_{2,5}+\phi_{2,6}+\phi_{5,6}=180^\circ$

F_4: $\phi_{3,4}+\phi_{3,6}+\phi_{4,6}=180^\circ$

A

SECONDARY ANALYSIS -- XYZ SPACE

PLOTTED POINTS

$P_1=(x_1,y_1,z_1)$

$P_2=(x_2,y_2,z_2)$

$P_3=(x_3,y_3,z_3)$

$P_4=(x_4,y_4,z_4)$

LENGTHS

$$s_1=\sqrt{(x_2-x_1)^2+(y_2-y_1)^2+(z_2-z_1)^2}$$

$$s_2=\sqrt{(x_3-x_2)^2+(y_3-y_2)^2+(z_3-z_2)^2}$$

$$s_3=\sqrt{(x_1-x_3)^2+(y_1-y_3)^2+(z_1-z_3)^2}$$

$$s_4=\sqrt{(x_4-x_1)^2+(y_4-y_1)^2+(z_4-z_1)^2}$$

$$s_5=\sqrt{(x_4-x_2)^2+(y_4-y_2)^2+(z_4-z_2)^2}$$

$$s_6=\sqrt{(x_4-x_3)^2+(y_4-y_3)^2+(z_4-z_3)^2}$$

DIRECTION COSINES

$\cos A_1=(x_2-x_1)/s_1$ $\cos B_1=(y_2-y_1)/s_1$

$\cos C_1=(z_2-z_1)/s_1$

$\cos A_2=(x_3-x_2)/s_2$ $\cos B_2=(y_3-y_2)/s_2$

$\cos C_2=(z_3-z_2)/s_2$

$\cos A_3=(x_1-x_3)/s_3$ $\cos B_3=(y_1-y_3)/s_3$

$\cos C_3=(z_1-z_3)/s_3$

$\cos A_4=(x_4-x_1)/s_4$ $\cos B_4=(y_4-y_1)/s_4$

$\cos C_4=(z_4-z_1)/s_4$

$\cos A_5=(x_4-x_2)/s_5$ $\cos B_5=(y_4-y_2)/s_5$

$\cos C_5=(z_4-z_2)/s_5$

$\cos A_6=(x_4-x_3)/s_6$ $\cos B_6=(y_4-y_3)/s_6$

$\cos C_6=(z_4-z_3)/s_6$

ANGLES

$\cos\phi_{1,2}=\cos A_1\cos A_2+\cos B_1\cos B_2+\cos C_1\cos C_2$

$\cos\phi_{1,3}=\cos A_1\cos A_3+\cos B_1\cos B_3+\cos C_1\cos C_3$

$\cos\phi_{1,4}=\cos A_1\cos A_4+\cos B_1\cos B_4+\cos C_1\cos C_4$

$\cos\phi_{2,3}=\cos A_2\cos A_3+\cos B_2\cos B_3+\cos C_2\cos C_3$

$\cos\phi_{2,5}=\cos A_2\cos A_5+\cos B_2\cos B_5+\cos C_2\cos C_5$

$\cos\phi_{2,6}=\cos A_2\cos A_6+\cos B_2\cos B_6+\cos C_2\cos C_6$

$\cos\phi_{3,4}=\cos A_3\cos A_4+\cos B_3\cos B_4+\cos C_3\cos C_4$

$\cos\phi_{3,6}=\cos A_3\cos A_6+\cos B_3\cos B_6+\cos C_3\cos C_6$

$\cos\phi_{4,5}=\cos A_4\cos A_5+\cos B_4\cos B_5+\cos C_4\cos C_5$

$\cos\phi_{4,6}=\cos A_4\cos A_6+\cos B_4\cos B_6+\cos C_4\cos C_6$

$\cos\phi_{5,6}=\cos A_5\cos A_6+\cos B_5\cos B_6+\cos C_5\cos C_6$

CHECK

F_1: $\phi_{1,2}+\phi_{1,3}+\phi_{2,3}=180^\circ$

F_2: $\phi_{1,4}+\phi_{1,5}+\phi_{4,5}=180^\circ$

F_3: $\phi_{2,5}+\phi_{2,6}+\phi_{5,6}=180^\circ$

F_4: $\phi_{3,4}+\phi_{3,6}+\phi_{4,6}=180^\circ$

B

Table 22-5. Specifications and Analysis of a 4-Sided Pyramid in 4-D Space. See Fig. 22-5. (A) Specifications and Primary Analysis in Hyperspace. (B) Secondary Analysis in XYZ Space. (C) Secondary Analysis in XYW Space. (D) Secondary Analysis in XZW Space. (E) Secondary Analysis in YZW Space. (F) Suggested Coordinates. (Continued from page 363.)

SECONDARY ANALYSIS -- XYW SPACE

PLOTTED POINTS

$P_1=(x_1,y_1,w_1)$

$P_2=(x_2,y_2,w_2)$

$P_3=(x_3,y_3,w_3)$

$P_4=(x_4,y_4,w_4)$

LENGTHS

$s_1=\sqrt{(x_2-x_1)^2+(y_2-y_1)^2+(w_2-w_1)^2}$

$s_2=\sqrt{(x_3-x_2)^2+(y_3-y_2)^2+(w_3-w_2)^2}$

$s_3=\sqrt{(x_1-x_3)^2+(y_1-y_3)^2+(w_1-w_3)^2}$

$s_4=\sqrt{(x_4-x_1)^2+(y_4-y_1)^2+(w_4-w_1)^2}$

$s_5=\sqrt{(x_4-x_2)^2+(y_4-y_2)^2+(w_4-w_2)^2}$

$s_6=\sqrt{(x_4-x_3)^2+(y_4-y_3)^2+(w_4-w_3)^2}$

DIRECTION COSINES

$\cos A_1=(x_2-x_1)/s_1$ $\cos B_1=(y_2-y_1)/s_1$
$\cos D_1=(w_2-w_1)/s_1$

$\cos A_2=(x_3-x_2)/s_2$ $\cos B_2=(y_3-y_2)/s_2$
$\cos D_2=(w_3-w_2)/s_2$

$\cos A_3=(x_1-x_3)/s_3$ $\cos B_3=(y_1-y_3)/s_3$
$\cos D_3=(w_1-w_3)/s_3$

$\cos A_4=(x_4-x_1)/s_4$ $\cos B_4=(y_4-y_1)/s_4$
$\cos D_4=(w_4-w_1)/s_4$

$\cos A_5=(x_4-x_2)/s_5$ $\cos B_5=(y_4-y_2)/s_5$
$\cos D_5=(w_4-w_2)/s_5$

$\cos A_6=(x_4-x_3)/s_6$ $\cos B_6=(y_4-y_3)/s_6$
$\cos D_6=(w_4-w_3)/s_6$

ANGLES

$\cos\phi_{1,2}=\cos A_1\cos A_2+\cos B_1\cos B_2+\cos D_1\cos D_2$

$\cos\phi_{1,3}=\cos A_1\cos A_3+\cos B_1\cos B_3+\cos D_1\cos D_3$

$\cos\phi_{1,4}=\cos A_1\cos A_4+\cos B_1\cos B_4+\cos D_1\cos D_4$

$\cos\phi_{2,3}=\cos A_2\cos A_3+\cos B_2\cos B_3+\cos D_2\cos D_3$

$\cos\phi_{2,5}=\cos A_2\cos A_5+\cos B_2\cos B_5+\cos D_2\cos D_5$

$\cos\phi_{2,6}=\cos A_2\cos A_6+\cos B_2\cos B_6+\cos D_2\cos D_6$

$\cos\phi_{3,4}=\cos A_3\cos A_4+\cos B_3\cos B_4+\cos D_3\cos D_4$

$\cos\phi_{3,6}=\cos A_3\cos A_6+\cos B_3\cos B_6+\cos D_3\cos D_6$

$\cos\phi_{4,5}=\cos A_4\cos A_5+\cos B_4\cos B_5+\cos D_4\cos D_5$

$\cos\phi_{4,6}=\cos A_4\cos A_6+\cos B_4\cos B_6+\cos D_4\cos D_6$

C $\cos\phi_{5,6}=\cos A_5\cos A_6+\cos B_5\cos B_6+\cos D_5\cos D_6$

CHECK

F_1: $\phi_{1,2}+\phi_{1,3}+\phi_{2,3}=180^\circ$

F_2: $\phi_{1,4}+\phi_{1,5}+\phi_{4,5}=180^\circ$

F_3: $\phi_{2,5}+\phi_{2,6}+\phi_{5,6}=180^\circ$

F_4: $\phi_{3,4}+\phi_{3,6}+\phi_{4,6}=180^\circ$

SECONDARY ANALYSIS -- XZW SPACE

PLOTTED POINTS

$P_1=(x_1,z_1,w_1)$

$P_2=(x_2,z_2,w_2)$

$P_3=(x_3,z_3,w_3)$

$P_4=(x_4,z_4,w_4)$

LENGTHS

$s_1=\sqrt{(x_2-x_1)^2+(z_2-z_1)^2+(w_2-w_1)^2}$

$s_2=\sqrt{(x_3-x_2)^2+(z_3-z_2)^2+(w_3-w_2)^2}$

$s_3=\sqrt{(x_1-x_3)^2+(z_1-z_3)^2+(w_1-w_3)^2}$

$s_4=\sqrt{(x_4-x_1)^2+(z_4-z_1)^2+(w_4-w_1)^2}$

$s_5=\sqrt{(x_4-x_2)^2+(z_4-z_2)^2+(w_4-w_2)^2}$

$s_6=\sqrt{(x_4-x_3)^2+(z_4-z_3)^2+(w_4-w_3)^2}$

DIRECTION COSINES

$\cos A_1=(x_2-x_1)/s_1$ $\cos C_1=(z_2-z_1)/s_1$

$\cos D_1=(w_2-w_1)/s_1$

$\cos A_2=(x_3-x_2)/s_2$ $\cos C_2=(z_3-z_2)/s_2$

$\cos D_2=(w_3-w_2)/s_2$

$\cos A_3=(x_1-x_3)/s_3$ $\cos C_3=(z_1-z_3)/s_3$

$\cos D_3=(w_1-w_3)/s_3$

$\cos A_4=(x_4-x_1)/s_4$ $\cos C_4=(z_4-z_1)/s_4$

$\cos D_4=(w_4-w_1)/s_4$

$\cos A_5=(x_4-x_2)/s_5$ $\cos C_5=(z_4-z_2)/s_5$

$\cos D_5=(w_4-w_2)/s_5$

$\cos A_6=(x_4-x_3)/s_6$ $\cos C_6=(z_4-z_3)/s_6$

$\cos D_6=(w_4-w_3)/s_6$

ANGLES

$\cos\phi_{1,2}=\cos A_1\cos A_2+\cos C_1\cos C_2+\cos D_1\cos D_2$

$\cos\phi_{1,3}=\cos A_1\cos A_3+\cos C_1\cos C_3+\cos D_1\cos D_3$

$\cos\phi_{1,4}=\cos A_1\cos A_4+\cos C_1\cos C_4+\cos D_1\cos D_4$

$\cos\phi_{2,3}=\cos A_2\cos A_3+\cos C_2\cos C_3+\cos D_2\cos D_3$

$\cos\phi_{2,5}=\cos A_2\cos A_5+\cos C_2\cos C_5+\cos D_2\cos D_5$

$\cos\phi_{2,6}=\cos A_2\cos A_6+\cos C_2\cos C_6+\cos D_2\cos D_6$

$\cos\phi_{3,4}=\cos A_3\cos A_4+\cos C_3\cos C_4+\cos D_3\cos D_4$

$\cos\phi_{3,6}=\cos A_3\cos A_6+\cos C_3\cos C_6+\cos D_3\cos D_6$

$\cos\phi_{4,5}=\cos A_4\cos A_5+\cos C_4\cos C_5+\cos D_4\cos D_5$

$\cos\phi_{4,6}=\cos A_4\cos A_6+\cos C_4\cos C_6+\cos D_4\cos D_6$

$\cos\phi_{5,6}=\cos A_5\cos A_6+\cos C_5\cos C_6+\cos D_5\cos D_6$

CHECK

F_1: $\phi_{1,2}+\phi_{1,3}+\phi_{2,3}=180^\circ$

F_2: $\phi_{1,4}+\phi_{1,5}+\phi_{4,5}=180^\circ$

F_3: $\phi_{2,5}+\phi_{2,6}+\phi_{5,6}=180^\circ$

F_4: $\phi_{3,4}+\phi_{3,6}+\phi_{4,6}=180^\circ$

D

Table 22-5. Specifications and Analyses of a 4-Sided Pyramid in 4-D Space. See Fig. 22-5. (A) Specifications and Primary Analysis in Hyperspace. (B) Secondary Analysis in XYZ Space. (C) Secondary Analysis in XYW Space. (D) Secondary Analysis in XZW Space. (E) Secondary Analysis in YZW Space. (F) Suggested Coordinates. (Continued from page 365.)

SECONDARY ANALYSIS -- YZW SPACE

PLOTTED POINTS

$P_1=(y_1,z_1,w_1)$

$P_2=(y_2,z_2,w_2)$

$P_3=(y_3,z_3,w_3)$

$P_4=(y_4,z_4,w_4)$

LENGTHS

$$s_1=\sqrt{(y_2-y_1)^2+(z_2-z_1)^2+(w_2-w_1)^2}$$

$$s_2=\sqrt{(y_3-y_2)^2+(z_3-z_2)^2+(w_3-w_2)^2}$$

$$s_3=\sqrt{(y_1-y_3)^2+(z_1-z_3)^2+(w_1-w_3)^2}$$

$$s_4=\sqrt{(y_4-y_1)^2+(z_4-z_1)^2+(w_4-w_1)^2}$$

$$s_5=\sqrt{(y_4-y_2)^2+(z_4-z_2)^2+(w_4-w_2)^2}$$

$$s_6=\sqrt{(y_4-y_3)^2+(z_4-z_3)^2+(w_4-w_3)^2}$$

DIRECTION COSINES

$\cos B_1=(y_2-y_1)/s_1$ $\quad \cos C_1=(z_2-z_1)/s_1$

$\cos D_1=(w_2-w_1)/s_1$

$\cos B_2=(y_3-y_2)/s_2$ $\quad \cos C_2=(z_3-z_2)/s_2$

$\cos D_2=(w_3-w_2)/s_2$

$\cos B_3=(y_1-y_3)/s_3$ $\quad \cos C_3=(z_1-z_3)/s_3$

$\cos D_3=(w_1-w_3)/s_3$

$\cos B_4=(y_4-y_1)/s_4$ $\quad \cos C_4=(z_4-z_1)/s_4$

$\cos D_4=(w_4-w_1)/s_4$

$\cos B_5=(y_4-y_2)/s_5$ $\quad \cos C_5=(z_4-z_2)/s_5$

$\cos D_5=(w_4-w_2)/s_5$

$\cos B_6=(y_4-y_3)/s_6$ $\quad \cos C_6=(z_4-z_3)/s_6$

$\cos D_6=(w_4-w_3)/s_6$

ANGLES

$\cos\phi_{1,2}=\cos B_1\cos B_2+\cos C_1\cos C_2+\cos D_1\cos D_2$

$\cos\phi_{1,3}=\cos B_1\cos B_3+\cos C_1\cos C_3+\cos D_1\cos D_3$

$\cos\phi_{1,4}=\cos B_1\cos B_4+\cos C_1\cos C_4+\cos D_1\cos D_4$

$\cos\phi_{2,3}=\cos B_2\cos B_3+\cos C_2\cos C_3+\cos D_2\cos D_3$

$\cos\phi_{2,5}=\cos B_2\cos B_5+\cos C_2\cos C_5+\cos D_2\cos D_5$

$\cos\phi_{2,6}=\cos B_2\cos B_6+\cos C_2\cos C_6+\cos D_2\cos D_6$

$\cos\phi_{3,4}=\cos B_3\cos B_4+\cos C_3\cos C_4+\cos D_3\cos D_4$

$\cos\phi_{3,6}=\cos B_3\cos B_6+\cos C_3\cos C_6+\cos D_3\cos D_6$

$\cos\phi_{4,5}=\cos B_4\cos B_5+\cos C_4\cos C_5+\cos D_4\cos D_5$

$\cos\phi_{4,6}=\cos B_4\cos B_6+\cos C_4\cos C_6+\cos D_4\cos D_6$

$\cos\phi_{5,6}=\cos B_5\cos B_6+\cos C_5\cos C_6+\cos D_5\cos D_6$

E

CHECK

F_1: $\phi_{1,2}+\phi_{1,3}+\phi_{2,3}=180^\circ$

F_2: $\phi_{1,4}+\phi_{1,5}+\phi_{4,5}=180^\circ$

F_3: $\phi_{2,5}+\phi_{2,6}+\phi_{5,6}=180^\circ$

F_4: $\phi_{3,4}+\phi_{3,6}+\phi_{4,6}=180^\circ$

```
HYPERSPACE OBJECT VERSION 1

P1=(Ø,Ø,1,Ø)          P2=(1,Ø,Ø,Ø)          P3=(Ø,Ø,Ø,Ø)
P4=(Ø,1,Ø,1)

HYPERSPACE OBJECT VERSION 2

P1=(Ø,Ø,1,Ø)          P2=(1,Ø,Ø,Ø)          P3=(-1,Ø,-1,Ø)
P4=(Ø,1,Ø,1)

HYPERSPACE OBJECT VERSION 3

P1=(Ø,Ø,1,Ø)          P2=(1,1,1,1)          P3=(Ø,-2,-1,-2)
P4=(2,2,Ø,2)
```

F

22-1.6 An Object Bounded by 4 Triangles and 1 Rectangle

Figure 22-6 and Table 22-6 represent a 5-sided space object that is bounded by four triangles and a single rectangle. The 5-sided pyramid cited in earlier projects in 3-D space is one version of it. The specifications, however, are written in a very general form, thereby giving you the opportunity to generate any number of different versions by simply plugging in a set of five carefully selected coordinates (see Table 22-6F).

22-1.7 A Space Object of 2 Triangles and 3 Rectangles

The 5-sided pyramid described in the previous discussion is bounded by four triangles and a single rectangle. The object featured here is also a 5-sided space object, but it is bounded by two triangles and three rectangles. It can be generally described as a prism object. See Fig. 22-7 and Table 22-7A.

Substitute the suggested specific coordinates (Table 22-7F) into the original specifications, and you will find a family of 5-sided objects of this type. Conduct complete primary and secondary analyses of several of the suggested version to build up your file of experimental data.

22-1.8 A Space Object of 6 Rectangles

Figure 22-8 and Table 22-8 show the plots and analyses of 6-sided objects where all sides happen to be quadralateral figures. The sets of coordinates in Table 22-8F are intended to provide you with a selection of these space objects. Conduct complete analyses of them, including the plots, and save them for future work.

Check your work at the primary-analysis level by summing the four angles in each side. Since they are quadralaterals, the sum in each case should come out to about 180 degrees. Also recall that pairs of lines are parallel when the sum of the produces of their corresponding direction cosines is 1 or -1; and they are perpendicular when the sum of the products of their corresponding direction cosines is zero. Don't count on "eyeball" views as a technique for checking any work in 4-D space.

22-1.9 A 7-Sided Space Object

Figure 22-9 and Table 22-9 represent a 7-sided space object that is bounded by two pentagons and five rectangles. The whole works, however, is immersed in 4-D space as indicated by the presence of four components in each point coordinate.

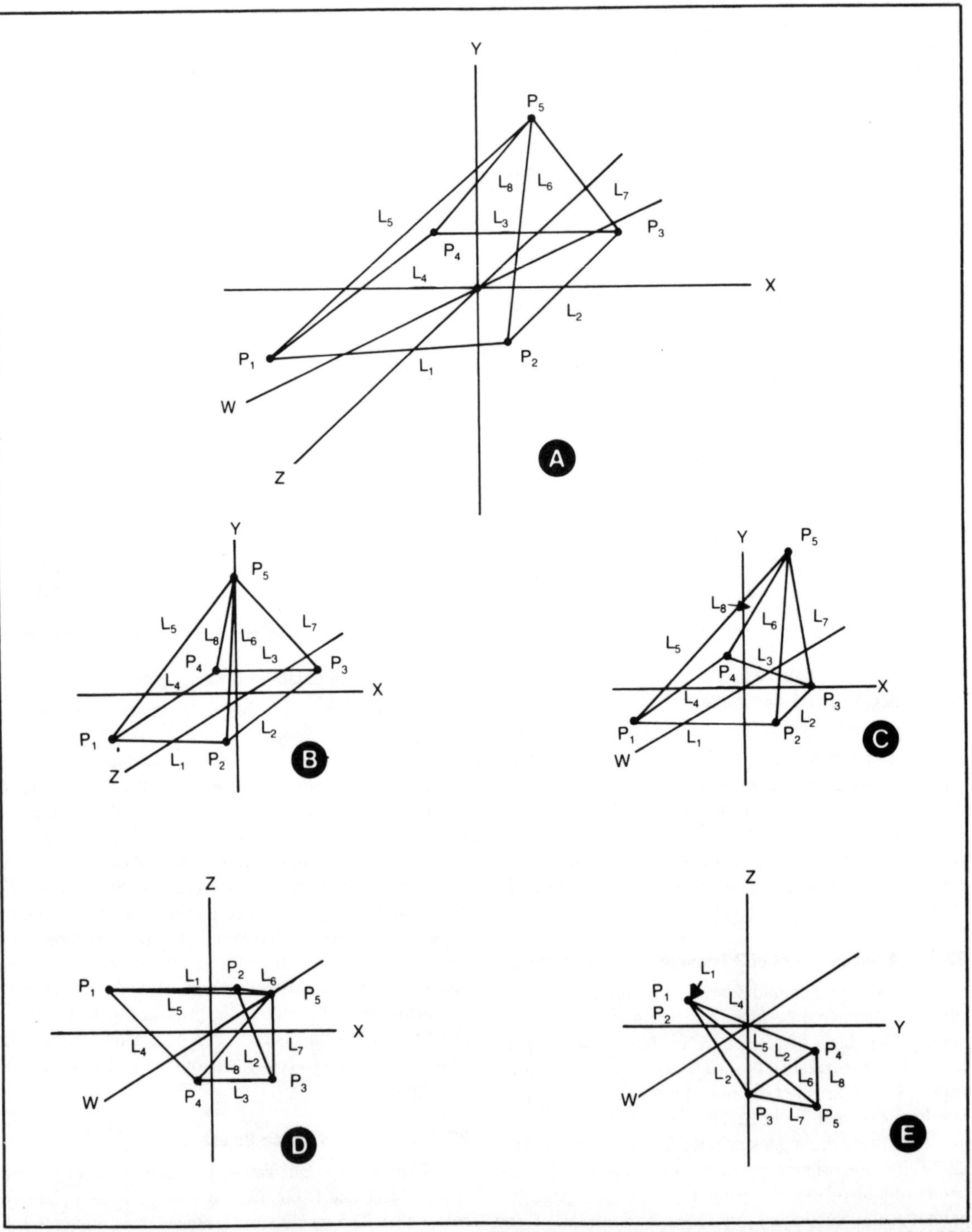

Fig. 22-6. A 5-sided pyramid specified in 4-D space. (A) General hyperspace view. (B) XYZ perspective view. (C) XYW perspective view. (D) XZW perspective view. (E) YZW perspective view. See outlines of analyses in Table 22-6.

Table 22-6. Specifications and Analyses of a 5-Sided Pyramid in 4-D Space. See Fig. 22-6. (A) Specifications and Primary Analysis in Hyperspace. (B) Secondary Analysis in XYZ Space. (C) Secondary Analysis in XYW Space. (D) Secondary Analysis in XZW Space. (E) Secondary Analysis in YZW Space. (F) Suggested Coordinates.

(A, B, C, and D correspond to α, β, γ, and δ.)

SPECIFICATIONS

$S=\overline{F_1F_2F_3F_4F_5}$

$F_1=\overline{L_1L_2L_3L_4}$

$F_2=\overline{L_1L_5L_6}$

$F_3=\overline{L_2L_6L_7}$

$F_4=\overline{L_3L_7L_8}$

$F_5=\overline{L_4L_5L_8}$

$L_1=\overline{P_1P_2}$

$L_2=\overline{P_2P_3}$

$L_3=\overline{P_3P_4}$

$L_4=\overline{P_4P_1}$

$L_5=\overline{P_1P_5}$

$L_6=\overline{P_2P_5}$

$L_7=\overline{P_3P_5}$

$L_8=\overline{P_4P_5}$

$P_1=(x_1,y_1,z_1,w_1)$

$P_2=(x_2,y_2,z_2,w_2)$

$P_3=(x_3,y_3,z_3,w_3)$

$P_4=(x_4,y_4,z_4,w_4)$

$P_5=(x_5,y_5,z_5,w_5)$

LENGTHS

$s_1=\sqrt{(x_2-x_1)^2+(y_2-y_1)^2+(z_2-z_1)^2+(w_2-w_1)^2}$

$s_2=\sqrt{(x_3-x_2)^2+(y_3-y_2)^2+(z_3-z_2)^2+(w_3-w_2)^2}$

$s_3=\sqrt{(x_4-x_3)^2+(y_4-y_3)^2+(z_4-z_3)^2+(w_4-w_3)^2}$

$s_4=\sqrt{(x_1-x_4)^2+(y_1-y_4)^2+(z_1-z_4)^2+(w_1-w_4)^2}$

$s_5=\sqrt{(x_1-x_5)^2+(y_1-y_5)^2+(z_1-z_5)^2+(w_1-w_5)^2}$

$s_6=\sqrt{(x_5-x_2)^2+(y_5-y_2)^2+(z_5-z_2)^2+(w_5-w_2)^2}$

$s_7=\sqrt{(x_5-x_3)^2+(y_5-y_3)^2+(z_5-z_1)^2+(w_5-w_3)^2}$

$s_8=\sqrt{(x_5-x_4)^2+(y_5-y_4)^2+(z_5-z_4)^2+(w_5-w_4)^2}$

DIRECTION COSINES

$\cos A_5=(x_1-x_5)/s_5$ $\quad\cos B_5=(y_1-y_5)/s_5$

$\cos C_5=(z_1-z_5)/s_5$ $\quad\cos D_5=(w_1-w_5)/s$

$\cos A_6=(x_5-x_2)/s_6$ $\quad\cos B_6=(y_5-y_2)/s_6$

$\cos C_6=(z_5-z_2)/s_6$ $\quad\cos D_6=(w_5-w_2)/s$

$\cos A_7=(x_5-x_3)/s_7$ $\quad\cos B_7=(y_5-y_3)/s_7$

$\cos C_7=(z_5-z_1)/s_7$ $\quad\cos D_7=(w_5-w_3)/s$

$\cos A_8=(x_5-x_4)/s_8$ $\quad\cos B_8=(y_5-y_4)/s_8$

$\cos C_8=(z_5-z_4)/s_8$ $\quad\cos D_8=(w_5-w_4)/s$

$\cos A_1=(x_2-x_1)/s_1$ $\quad\cos B_1=(y_2-y_1)/s_1$

$\cos C_1=(z_2-z_1)/s_1$ $\quad\cos D_1=(w_2-w_1)/s$

$\cos A_2=(x_3-x_2)/s_2$ $\quad\cos B_2=(y_3-y_2)/s_2$

$\cos C_2=(z_3-z_2)/s_2$ $\quad\cos D_2=(w_3-w_2)/s$

$\cos A_3=(x_4-x_3)/s_3$ $\quad\cos B_3=(y_4-y_3)/s_3$

$\cos C_3=(z_4-z_3)/s_3$ $\quad\cos D_3=(w_4-w_3)/s$

$\cos A_4=(x_1-x_4)/s_4$ $\quad\cos B_4=(y_1-y_4)/s_4$

$\cos C_4=(z_1-z_4)/s_4$ $\quad\cos D_4=(w_1-w_4)/s$

ANGLES

$\cos\phi_{1,2}=\cos A_1\cos A_2+\cos B_1\cos B_2+\cos C_1\cos C_2+\cos D_1\cos D_2$

$\cos\phi_{1,4}=\cos A_1\cos A_4+\cos B_1\cos B_4+\cos C_1\cos C_4+\cos D_1\cos D_4$

$\cos\phi_{1,5}=\cos A_1\cos A_5+\cos B_1\cos B_5+\cos C_1\cos C_5+\cos D_1\cos D_5$

$\cos\phi_{1,6}=\cos A_1\cos A_6+\cos B_1\cos B_6+\cos C_1\cos C_6+\cos D_1\cos D_6$

$\cos\phi_{2,3}=\cos A_2\cos A_3+\cos B_2\cos B_3+\cos C_2\cos C_3+\cos D_2\cos D_3$

$\cos\phi_{2,6}=\cos A_2\cos A_6+\cos B_2\cos B_6+\cos C_2\cos C_6+\cos D_2\cos D_6$

$\cos\phi_{2,7}=\cos A_2\cos A_7+\cos B_2\cos B_7+\cos C_2\cos C_7+\cos D_2\cos D_7$

$\cos\phi_{3,7}=\cos A_3\cos A_7+\cos B_3\cos B_7+\cos C_3\cos C_7+\cos D_3\cos D_7$

$\cos\phi_{3,8}=\cos A_3\cos A_8+\cos B_3\cos B_8+\cos C_3\cos C_8+\cos D_3\cos D_8$

$\cos\phi_{4,5}=\cos A_4\cos A_5+\cos B_4\cos B_5+\cos C_4\cos C_5+\cos D_4\cos D_5$

$\cos\phi_{4,8}=\cos A_4\cos A_8+\cos B_4\cos B_8+\cos C_4\cos C_8+\cos D_4\cos D_8$

$\cos\phi_{5,6}=\cos A_5\cos A_6+\cos B_5\cos B_6+\cos C_5\cos C_6+\cos D_5\cos D_6$

$\cos\phi_{5,7}=\cos A_5\cos A_7+\cos B_5\cos B_7+\cos C_5\cos C_7+\cos D_5\cos D_7$

$\cos\phi_{5,8}=\cos A_5\cos A_8+\cos B_5\cos B_8+\cos C_5\cos C_8+\cos D_5\cos D_8$

$\cos\phi_{6,7}=\cos A_6\cos A_7+\cos B_6\cos B_7+\cos C_6\cos C_7+\cos D_6\cos D_7$

$\cos\phi_{7,8}=\cos A_7\cos A_8+\cos B_7\cos B_8+\cos C_7\cos C_8+\cos D_7\cos D_8$

CHECK

F_1: $\phi_{1,2}+\phi_{2,3}+\phi_{3,4}+\phi_{1,4}=360^\circ$

F_2: $\phi_{1,5}+\phi_{1,6}+\phi_{5,6}=180^\circ$

F_3: $\phi_{2,6}+\phi_{2,7}+\phi_{6,7}=180^\circ$

F_4: $\phi_{3,7}+\phi_{3,8}+\phi_{7,8}=180^\circ$

F_5: $\phi_{4,5}+\phi_{4,8}+\phi_{5,8}=180^\circ$

A

Table 22-6. Specifications and Analyses of a 5-Sided Pyramid in 4-D Space. See Fig. 22-6. (A) Specifications and Primary Analysis in Hyperspace. (B) Secondary Analysis in XYZ Space. (C) Secondary Analysis in XYW Space. (D) Secondary Analysis in XZW Space. (E) Secondary Analysis in YZW Space. (F) Suggested Coordinates. (Continued from page 369.)

SECONDARY ANALYSIS -- XYZ SPACE

PLOTTED POINTS

$P_1=(x_1,y_1,z_1)$

$P_2=(x_2,y_2,z_2)$

$P_3=(x_3,y_3,z_3)$

$P_4=(x_4,y_4,z_4)$

$P_5=(x_5,y_5,z_5)$

LENGTHS

$s_1=\sqrt{(x_2-x_1)^2+(y_2-y_1)^2+(z_2-z_1)^2}$

$s_2=\sqrt{(x_3-x_2)^2+(y_3-y_2)^2+(z_3-z_2)^2}$

$s_3=\sqrt{(x_4-x_3)^2+(y_4-y_3)^2+(z_4-z_3)^2}$

$s_4=\sqrt{(x_1-x_4)^2+(y_1-y_4)^2+(z_1-z_4)^2}$

$s_5=\sqrt{(x_1-x_5)^2+(y_1-y_5)^2+(z_1-z_5)^2}$

$s_6=\sqrt{(x_5-x_2)^2+(y_5-y_2)^2+(z_5-z_2)^2}$

$s_7=\sqrt{(x_5-x_3)^2+(y_5-y_3)^2+(z_5-z_1)^2}$

$s_8=\sqrt{(x_5-x_4)^2+(y_5-y_4)^2+(z_5-z_4)^2}$

DIRECTION COSINES

$\cos A_1=(x_2-x_1)/s_1$ $\cos B_1=(y_2-y_1)/s_1$
$\cos C_1=(z_2-z_1)/s_1$

$\cos A_2=(x_3-x_2)/s_2$ $\cos B_2=(y_3-y_2)/s_2$
$\cos C_2=(z_3-z_2)/s_2$

$\cos A_3=(x_4-x_3)/s_3$ $\cos B_3=(y_4-y_3)/s_3$
$\cos C_3=(z_4-z_3)/s_3$

$\cos A_4=(x_1-x_4)/s_4$ $\cos B_4=(y_1-y_4)/s_4$
$\cos C_4=(z_1-z_4)/s_4$

$\cos A_5=(x_1-x_5)/s_5$ $\cos B_5=(y_1-y_5)/s_5$
$\cos C_5=(z_1-z_5)/s_5$

$\cos A_6=(x_5-x_2)/s_6$ $\cos B_6=(y_5-y_2)/s_6$
$\cos C_6=(z_5-z_2)/s_6$

$\cos A_7=(x_5-x_3)/s_7$ $\cos B_7=(y_5-y_3)/s_7$
$\cos C_7=(z_5-z_1)/s_7$

$\cos A_8=(x_5-x_4)/s_8$ $\cos B_8=(y_5-y_4)/s_8$
$\cos C_8=(z_5-z_4)/s_8$

ANGLES

$\cos\phi_{1,2}=\cos A_1\cos A_2+\cos B_1\cos B_2+\cos C_1\cos C_2$

$\cos\phi_{1,4}=\cos A_1\cos A_4+\cos B_1\cos B_4+\cos C_1\cos C_4$

$\cos\phi_{1,5}=\cos A_1\cos A_5+\cos B_1\cos B_5+\cos C_1\cos C_5$

$\cos\phi_{1,6}=\cos A_1\cos A_6+\cos B_1\cos B_6+\cos C_1\cos C_6$

$\cos\phi_{2,3}=\cos A_2\cos A_3+\cos B_2\cos B_3+\cos C_2\cos C_3$

$\cos\phi_{2,6}=\cos A_2\cos A_6+\cos B_2\cos B_6+\cos C_2\cos C_6$

$\cos\phi_{2,7}=\cos A_2\cos A_7+\cos B_2\cos B_7+\cos C_2\cos C_7$

$\cos\phi_{3,7}=\cos A_3\cos A_7+\cos B_3\cos B_7+\cos C_3\cos C_7$

$\cos\phi_{3,8}=\cos A_3\cos A_8+\cos B_3\cos B_8+\cos C_3\cos C_8$

$\cos\phi_{4,5}=\cos A_4\cos A_5+\cos B_4\cos B_5+\cos C_4\cos C_5$

$\cos\phi_{4,8}=\cos A_4\cos A_8+\cos B_4\cos B_8+\cos C_4\cos C_8$

$\cos\phi_{5,6}=\cos A_5\cos A_6+\cos B_5\cos B_6+\cos C_5\cos C_6$

$\cos\phi_{5,7}=\cos A_5\cos A_7+\cos B_5\cos B_7+\cos C_5\cos C_7$

$\cos\phi_{5,8}=\cos A_5\cos A_8+\cos B_5\cos B_8+\cos C_5\cos C_8$

$\cos\phi_{6,7}=\cos A_6\cos A_7+\cos B_6\cos B_7+\cos C_6\cos C_7$

$\cos\phi_{7,8}=\cos A_7\cos A_8+\cos B_7\cos B_8+\cos C_7\cos C_8$

CHECK

F_1: $\phi_{1,2}+\phi_{2,3}+\phi_{3,4}+\phi_{1,4}=360^\circ$

F_2: $\phi_{1,5}+\phi_{1,6}+\phi_{5,6}=180^\circ$

F_3: $\phi_{2,6}+\phi_{2,7}+\phi_{6,7}=180^\circ$

F_4: $\phi_{3,7}+\phi_{3,8}+\phi_{7,8}=180^\circ$

F_5: $\phi_{4,5}+\phi_{4,8}+\phi_{5,8}=180^\circ$

B

SECONDARY ANALYSIS -- XYW SPACE

PLOTTED POINTS

$P_1=(x_1,y_1,w_1)$

$P_2=(x_2,y_2,w_2)$

$P_3=(x_3,y_3,w_3)$

$P_4=(x_4,y_4,w_4)$

$P_5=(x_5,y_5,w_5)$

LENGTHS

$s_1=\sqrt{(x_2-x_1)^2+(y_2-y_1)^2+(w_2-w_1)^2}$

$s_2=\sqrt{(x_3-x_2)^2+(y_3-y_2)^2+(w_3-w_2)^2}$

$s_3=\sqrt{(x_4-x_3)^2+(y_4-y_3)^2+(w_4-w_3)^2}$

$s_4=\sqrt{(x_1-x_4)^2+(y_1-y_4)^2+(w_1-w_4)^2}$

$s_5=\sqrt{(x_1-x_5)^2+(y_1-y_5)^2+(w_1-w_5)^2}$

$s_6=\sqrt{(x_5-x_2)^2+(y_5-y_2)^2+(w_5-w_2)^2}$

$s_7=\sqrt{(x_5-x_3)^2+(y_5-y_3)^2+(w_5-w_1)^2}$

$s_8=\sqrt{(x_5-x_4)^2+(y_5-y_4)^2+(w_5-w_4)^2}$

DIRECTION COSINES

$\cos A_1=(x_2-x_1)/s_1$ $\cos B_1=(y_2-y_1)/s_1$
$\cos D_1=(w_2-w_1)/s_1$

$\cos A_2=(x_3-x_2)/s_2$ $\cos B_2=(y_3-y_2)/s_2$
$\cos D_2=(w_3-w_2)/s_2$

$\cos A_3=(x_4-x_3)/s_3$ $\cos B_3=(y_4-y_3)/s_3$
$\cos D_3=(w_4-w_3)/s_3$

$\cos A_4=(x_1-x_4)/s_4$ $\cos B_4=(y_1-y_4)/s_4$
$\cos D_4=(w_1-w_4)/s_4$

$\cos A_5=(x_1-x_5)/s_5$ $\cos B_5=(y_1-y_5)/s_5$
$\cos D_5=(w_1-w_5)/s_5$

$\cos A_6=(x_5-x_2)/s_6$ $\cos B_6=(y_5-y_2)/s_6$
$\cos D_6=(w_5-w_2)/s_6$

$\cos A_7=(x_5-x_3)/s_7$ $\cos B_7=(y_5-y_3)/s_7$
$\cos D_7=(w_5-w_1)/s_7$

$\cos A_8=(x_5-x_4)/s_8$ $\cos B_8=(y_5-y_4)/s_8$
$\cos D_8=(w_5-w_4)/s_8$

ANGLES

$\cos\phi_{1,2}=\cos A_1\cos A_2+\cos B_1\cos B_2+\cos D_1\cos D_2$

$\cos\phi_{1,4}=\cos A_1\cos A_4+\cos B_1\cos B_4+\cos D_1\cos D_4$

$\cos\phi_{1,5}=\cos A_1\cos A_5+\cos B_1\cos B_5+\cos D_1\cos D_5$

$\cos\phi_{1,6}=\cos A_1\cos A_6+\cos B_1\cos B_6+\cos D_1\cos D_6$

$\cos\phi_{2,3}=\cos A_2\cos A_3+\cos B_2\cos B_3+\cos D_2\cos D_3$

$\cos\phi_{2,6}=\cos A_2\cos A_6+\cos B_2\cos B_6+\cos D_2\cos D_6$

$\cos\phi_{2,7}=\cos A_2\cos A_7+\cos B_2\cos B_7+\cos D_2\cos D_7$

$\cos\phi_{3,7}=\cos A_3\cos A_7+\cos B_3\cos B_7+\cos D_3\cos D_7$

$\cos\phi_{3,8}=\cos A_3\cos A_8+\cos B_3\cos B_8+\cos D_3\cos D_8$

$\cos\phi_{4,5}=\cos A_4\cos A_5+\cos B_4\cos B_5+\cos D_4\cos D_5$

$\cos\phi_{4,8}=\cos A_4\cos A_8+\cos B_4\cos B_8+\cos D_4\cos D_8$

$\cos\phi_{5,6}=\cos A_5\cos A_6+\cos B_5\cos B_6+\cos D_5\cos D_6$

$\cos\phi_{5,7}=\cos A_5\cos A_7+\cos B_5\cos B_7+\cos D_5\cos D_7$

$\cos\phi_{5,8}=\cos A_5\cos A_8+\cos B_5\cos B_8+\cos D_5\cos D_8$

$\cos\phi_{6,7}=\cos A_6\cos A_7+\cos B_6\cos B_7+\cos D_6\cos D_7$

$\cos\phi_{7,8}=\cos A_7\cos A_8+\cos B_7\cos B_8+\cos D_7\cos D_8$

CHECK

F_1: $\phi_{1,2}+\phi_{2,3}+\phi_{3,4}+\phi_{1,4}=360^\circ$

F_2: $\phi_{1,5}+\phi_{1,6}+\phi_{5,6}=180^\circ$

F_3: $\phi_{2,6}+\phi_{2,7}+\phi_{6,7}=180^\circ$

F_4: $\phi_{3,7}+\phi_{3,8}+\phi_{7,8}=180^\circ$

F_5: $\phi_{4,5}+\phi_{4,8}+\phi_{5,8}=180^\circ$

C

Table 22-6. Specifications and Analyses of a 5-Sided Pyramid in 4-D Space. See Fig. 22-6. (A) Specifications and Primary Analysis in Hyperspace. (B) Secondary Analysis in XYZ Space. (C) Secondary Analysis in XYW Space. (D) Secondary Analysis in XZW Space. (E) Secondary Analysis in YZW Space. (F) Suggested Coordinates. (Continued from page 371.)

SECONDARY ANALYSIS -- XZW SPACE

PLOTTED POINTS

$P_1=(x_1,z_1,w_1)$

$P_2=(x_2,z_2,w_2)$

$P_3=(x_3,z_3,w_3)$

$P_4=(x_4,z_4,w_4)$

$P_5=(x_5,z_5,w_5)$

LENGTHS

$s_1=\sqrt{(x_2-x_1)^2+(z_2-z_1)^2+(w_2-w_1)^2}$

$s_2=\sqrt{(x_3-x_2)^2+(z_3-z_2)^2+(w_3-w_2)^2}$

$s_3=\sqrt{(x_4-x_3)^2+(z_4-z_3)^2+(w_4-w_3)^2}$

$s_4=\sqrt{(x_1-x_4)^2+(z_1-z_4)^2+(w_1-w_4)^2}$

$s_5=\sqrt{(x_1-x_5)^2+(z_1-z_5)^2+(w_1-w_5)^2}$

$s_6=\sqrt{(x_5-x_2)^2+(z_5-z_2)^2+(w_5-w_2)^2}$

$s_7=\sqrt{(x_5-x_3)^2+(z_5-z_3)^2+(w_5-w_1)^2}$

$s_8=\sqrt{(x_5-x_4)^2+(z_5-z_4)^2+(w_5-w_4)^2}$

DIRECTION COSINES

$\cos A_1=(x_2-x_1)/s_1$ $\cos C_1=(z_2-z_1)/s_1$

$\cos D_1=(w_2-w_1)/s_1$

$\cos A_2=(x_3-x_2)/s_2$ $\cos C_2=(z_3-z_2)/s_2$

$\cos D_2=(w_3-w_2)/s_2$

$\cos A_3=(x_4-x_3)/s_3$ $\cos C_3=(z_4-z_3)/s_3$

$\cos D_3=(w_4-w_3)/s_3$

$\cos A_4=(x_1-x_4)/s_4$ $\cos C_4=(z_1-z_4)/s_4$

$\cos D_4=(w_1-w_4)/s_4$

$\cos A_5=(x_1-x_5)/s_5$ $\cos C_5=(z_1-z_5)/s_5$

$\cos D_5=(w_1-w_5)/s_5$

$\cos A_6=(x_5-x_2)/s_6$ $\cos C_6=(z_5-z_2)/s_6$

$\cos D_6=(w_5-w_2)/s_6$

$\cos A_7=(x_5-x_3)/s_7$ $\cos C_7=(z_5-z_3)/s_7$

$\cos D_7=(w_5-w_1)/s_7$

$\cos A_8=(x_5-x_4)/s_8$ $\cos C_8=(z_5-z_4)/s_8$

$\cos D_8=(w_5-w_4)/s_8$

ANGLES

$\cos\phi_{1,2}=\cos A_1\cos A_2+\cos C_1\cos C_2+\cos D_1\cos D_2$

$\cos\phi_{1,4}=\cos A_1\cos A_4+\cos C_1\cos C_4+\cos D_1\cos D_4$

$\cos\phi_{1,5}=\cos A_1\cos A_5+\cos C_1\cos C_5+\cos D_1\cos D_5$

$\cos\phi_{1,6}=\cos A_1\cos A_6+\cos C_1\cos C_6+\cos D_1\cos D_6$

$\cos\phi_{2,3}=\cos A_2\cos A_3+\cos C_2\cos C_3+\cos D_2\cos D_3$

$\cos\phi_{2,6}=\cos A_2\cos A_6+\cos C_2\cos C_6+\cos D_2\cos D_6$

$\cos\phi_{2,7}=\cos A_2\cos A_7+\cos C_2\cos C_7+\cos D_2\cos D_7$

$\cos\phi_{3,7}=\cos A_3\cos A_7+\cos C_3\cos C_7+\cos D_3\cos D_7$

$\cos\phi_{3,8}=\cos A_3\cos A_8+\cos C_3\cos C_8+\cos D_3\cos D_8$

$\cos\phi_{4,5}=\cos A_4\cos A_5+\cos C_4\cos C_5+\cos D_4\cos D_5$

$\cos\phi_{4,8}=\cos A_4\cos A_8+\cos C_4\cos C_8+\cos D_4\cos D_8$

$\cos\phi_{5,6}=\cos A_5\cos A_6+\cos C_5\cos C_6+\cos D_5\cos D_6$

$\cos\phi_{5,7}=\cos A_5\cos A_7+\cos C_5\cos C_7+\cos D_5\cos D_7$

$\cos\phi_{5,8}=\cos A_5\cos A_8+\cos C_5\cos C_8+\cos D_5\cos D_8$

$\cos\phi_{6,7}=\cos A_6\cos A_7+\cos C_6\cos C_7+\cos D_6\cos D_7$

$\cos\phi_{7,8}=\cos A_7\cos A_8+\cos C_7\cos C_8+\cos D_7\cos D_8$

CHECK

F_1: $\phi_{1,2}+\phi_{2,3}+\phi_{3,4}+\phi_{1,4}=360^\circ$

F_2: $\phi_{1,5}+\phi_{1,6}+\phi_{5,6}=180^\circ$

F_3: $\phi_{2,6}+\phi_{2,7}+\phi_{6,7}=180^\circ$

F_4: $\phi_{3,7}+\phi_{3,8}+\phi_{7,8}=180^\circ$

F_5: $\phi_{4,5}+\phi_{4,8}+\phi_{5,8}=180^\circ$

D

SECONDARY ANALYSIS -- YZW SPACE

PLOTTED POINTS

$P_1=(y_1,z_1,w_1)$

$P_2=(y_2,z_2,w_2)$

$P_3=(y_3,z_3,w_3)$

$P_4=(y_4,z_4,w_4)$

$P_5=(y_5,z_5,w_5)$

LENGTHS

$s_1=\sqrt{(y_2-y_1)^2+(z_2-z_1)^2+(w_2-w_1)^2}$

$s_2=\sqrt{(y_3-y_2)^2+(z_3-z_2)^2+(w_3-w_2)^2}$

$s_3=\sqrt{(y_4-y_3)^2+(z_4-z_3)^2+(w_4-w_3)^2}$

$s_4=\sqrt{(y_1-y_4)^2+(z_1-z_4)^2+(w_1-w_4)^2}$

$s_5=\sqrt{(y_1-y_5)^2+(z_1-z_5)^2+(w_1-w_5)^2}$

$s_6=\sqrt{(y_5-y_2)^2+(z_5-z_2)^2+(w_5-w_2)^2}$

$s_7=\sqrt{(y_5-y_3)^2+(z_5-z_3)^2+(w_5-w_1)^2}$

$s_8=\sqrt{(y_5-y_4)^2+(z_5-z_4)^2+(w_5-w_4)^2}$

DIRECTION COSINES

$\cos B_1=(y_2-y_1)/s_1$ $\quad \cos C_1=(z_2-z_1)/s_1$

$\cos D_1=(w_2-w_1)/s_1$

$\cos B_2=(y_3-y_2)/s_2$ $\quad \cos C_2=(z_3-z_2)/s_2$

$\cos D_2=(w_3-w_2)/s_2$

$\cos B_3=(y_4-y_3)/s_3$ $\quad \cos C_3=(z_4-z_3)/s_3$

$\cos D_3=(w_4-w_3)/s_3$

$\cos B_4=(y_1-y_4)/s_4$ $\quad \cos C_4=(z_1-z_4)/s_4$

$\cos D_4=(w_1-w_4)/s_4$

$\cos B_5=(y_1-y_5)/s_5$ $\quad \cos C_5=(z_1-z_5)/s_5$

$\cos D_5=(w_1-w_5)/s_5$

$\cos B_6=(y_5-y_2)/s_6$ $\quad \cos C_6=(z_5-z_2)/s_6$

$\cos D_6=(w_5-w_2)/s_6$

$\cos B_7=(y_5-y_3)/s_7$ $\quad \cos C_7=(z_5-z_3)/s_7$

$\cos D_7=(w_5-w_1)/s_7$

$\cos B_8=(y_5-y_4)/s_8$ $\quad \cos C_8=(z_5-z_4)/s_8$

$\cos D_8=(w_5-w_4)/s_8$

ANGLES

$\cos\phi_{1,2}=\cos B_1\cos B_2+\cos C_1\cos C_2+\cos D_1\cos D_2$

$\cos\phi_{1,4}=\cos B_1\cos B_4+\cos C_1\cos C_4+\cos D_1\cos D_4$

$\cos\phi_{1,5}=\cos B_1\cos B_5+\cos C_1\cos C_5+\cos D_1\cos D_5$

$\cos\phi_{1,6}=\cos B_1\cos B_6+\cos C_1\cos C_6+\cos D_1\cos D_6$

$\cos\phi_{2,3}=\cos B_2\cos B_3+\cos C_2\cos C_3+\cos D_2\cos D_3$

$\cos\phi_{2,6}=\cos B_2\cos B_6+\cos C_2\cos C_6+\cos D_2\cos D_6$

$\cos\phi_{2,7}=\cos B_2\cos B_7+\cos C_2\cos C_7+\cos D_2\cos D_7$

$\cos\phi_{3,7}=\cos B_3\cos B_7+\cos C_3\cos C_7+\cos D_3\cos D_7$

$\cos\phi_{3,8}=\cos B_3\cos B_8+\cos C_3\cos C_8+\cos D_3\cos D_8$

$\cos\phi_{4,5}=\cos B_4\cos B_5+\cos C_4\cos C_5+\cos D_4\cos D_5$

$\cos\phi_{4,8}=\cos B_4\cos B_8+\cos C_4\cos C_8+\cos D_4\cos D_8$

$\cos\phi_{5,6}=\cos B_5\cos B_6+\cos C_5\cos C_6+\cos D_5\cos D_6$

$\cos\phi_{5,7}=\cos B_5\cos B_7+\cos C_5\cos C_7+\cos D_5\cos D_7$

$\cos\phi_{5,8}=\cos B_5\cos B_8+\cos C_5\cos C_8+\cos D_5\cos D_8$

$\cos\phi_{6,7}=\cos B_6\cos B_7+\cos C_6\cos C_7+\cos D_6\cos D_7$

$\cos\phi_{7,8}=\cos B_7\cos B_8+\cos C_7\cos C_8+\cos D_7\cos D_8$

CHECK

F_1: $\phi_{1,2}+\phi_{2,3}+\phi_{3,4}+\phi_{1,4}=360^\circ$

F_2: $\phi_{1,5}+\phi_{1,6}+\phi_{5,6}=180^\circ$

F_3: $\phi_{2,6}+\phi_{2,7}+\phi_{6,7}=180^\circ$

F_4: $\phi_{3,7}+\phi_{3,8}+\phi_{7,8}=180^\circ$

F_5: $\phi_{4,5}+\phi_{4,8}+\phi_{5,8}=180^\circ$

E

Table 22-6.Specifications and Analyses of a 5-Sided Pyramid in 4-D Space. See Fig. 22-6. (A) Specifications and Primary Analysis in Hyperspace. (B) Secondary Analysis in XYZ Space. (C) Secondary Analysis in XYW Space. (D) Secondary Analysis in XZW Space. (E) Secondary Analysis in YZW Space. (F) Suggested Coordinates. **(Continued from page 373.)**

```
HYPERSPACE OBJECT VERSION 1

P1=(-1,Ø,1,Ø)          P2=(1,Ø,1,Ø)          P3=(Ø,1,-1,1)
P4=(-1,Ø,-1,Ø)         P5=(Ø,1,Ø,1)


HYPERSPACE OBJECT VERSION 2

P1=(-1,Ø,1,Ø)          P2=(Ø,Ø,1,Ø)          P3=(1,Ø,-1,Ø)
P4=(Ø,Ø,-1,Ø)          P5=(Ø,1,Ø,1)


HYPERSPACE OBJECT VERSION 3

P1=(-1,Ø,1,Ø)          P2=(Ø,Ø,1,Ø)          P3=(Ø,Ø,-1,Ø)
P4=(-1,Ø,-1,Ø)         P5=(1,1,Ø,1)
```

F

Conduct complete analyses of the object, using some of the suggested coordinates, recalling that the sum of the angles in the pentagons should be close to 540 degrees, while the sum of the angles in the rectangular sides has to be close to 360 degrees.

21-2 A CATALOG OF HYPERSPACE OBJECTS

In the context of this book, hyperspace objects represent some of the most interesting and complex types of geometric entities we ever study. This brief catalog of them is intended to show some general specifications and, more important, the step-by-step procedures for conducting both primary and secondary analyses of them. The latter procedures, incidentally, apply to any of the valid custom hyperspace object you might generate on your own (see Chapter 18).

22-2.1 A Hyperspace Object of All Triangles

Figure 22-10 and Table 22-10 specify the simplest sort of hyperspace object. As with objects and figures of lower dimensions, you can generate a wide variety of objects in this class by substituting appropriate values for the coordinates of the points. Of course it is easy to make up some coordinate values and plug them into the P specifications, but that doesn't guarantee a valid space object. You must be careful about the selection of those coordinates for the reference version of the figure—and you most likely know by now, that is not an easy task. I've supplied some valid coordinates for several different versions of this particular hyperspace object; and unless you feel you are especially confident at generating coordinates of your own, I suggest using these as reference points for your experiments.

You can always doublecheck the results of the primary analysis by observing the sums of the angles bounded by the specified plane figures. The sums of angles in a triangle—even those included in a hyperspace object—must be close to 180 degrees.

The most intriguing phase of the experiments, I think, is generating the 3-D perspective drawings and analyses. Those show how the hyperspace object appears as mapped to 3-D spaces. You will find that such objects often map as plane figures in those lower-dimensioned spaces.

22-2.2 The Standard Hyperpyramid

If you are one of those individuals who believe that 5-sided pyramids have unusual psychic powers (calming the nerves and sharpening razar blades, for instance) you will get something special to wonder about here. If an

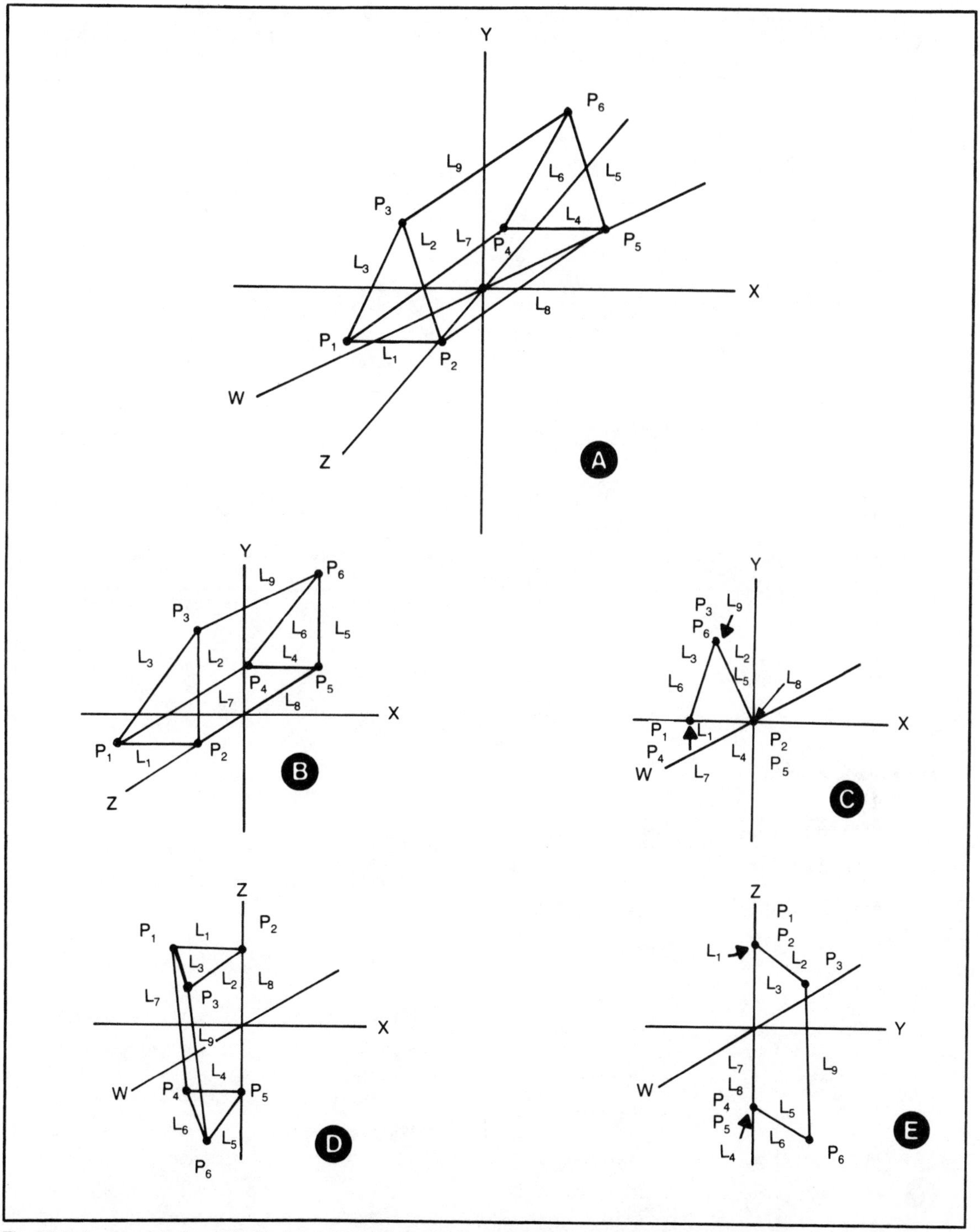

Fig. 22-7. A prism object specified in 4-D space. (A) General hyperspace view. (B) XYZ perspective view. (C) XYW perspective view. (D) XZW perspective view. (E) YZW perspective view. See outlines of analyses in Table 22-7.

Table 22-7. Specifications and Analyses of a Prism in 4-D Space. See Fig. 22-7.
(A) Specifications and Primary Analysis in Hyperspace. (B) Secondary Analysis in XYZ Space. (C) Secondary Analysis in XYW Space. (D) Secondary Analysis in XZW Space. (E) Secondary Analysis in YZW Space. (F) Suggested Coordinates.

(A, B, C, and D correspond to α, β, γ, and δ.)

SPECIFICATIONS

$S=\overline{F_1F_2F_3F_4F_5}$

$F_1=\overline{L_1L_2L_3}$

$F_2=\overline{L_1L_4L_7L_8}$

$F_3=\overline{L_2L_5L_8L_9}$

$F_4=\overline{L_3L_6L_7L_9}$

$F_5=\overline{L_4L_5L_6}$

$L_1=\overline{P_1P_2}$

$L_2=\overline{P_2P_3}$

$L_3=\overline{P_3P_1}$

$L_4=\overline{P_4P_5}$

$L_5=\overline{P_5P_6}$

$L_6=\overline{P_6P_4}$

$L_7=\overline{P_1P_4}$

$L_8=\overline{P_2P_5}$

$L_9=\overline{P_3P_6}$

$P_1=(x_1,y_1,z_1,w_1)$

$P_2=(x_2,y_2,z_2,w_2)$

$P_3=(x_3,y_3,z_3,w_3)$

$P_4=(x_4,y_4,z_4,w_4)$

$P_5=(x_5,y_5,z_5,w_5)$

LENGTHS

$s_1=\sqrt{(x_2-x_1)^2+(y_2-y_1)^2+(z_2-z_1)^2+(w_2-w_1)^2}$

$s_2=\sqrt{(x_3-x_2)^2+(y_3-y_2)^2+(z_3-z_2)^2+(w_3-w_2)^2}$

$s_3=\sqrt{(x_1-x_3)^2+(y_1-y_3)^2+(z_1-z_3)^2+(w_1-w_3)^2}$

$s_4=\sqrt{(x_5-x_4)^2+(y_5-y_4)^2+(z_5-z_4)^2+(w_5-w_4)^2}$

$s_5=\sqrt{(x_6-x_5)^2+(y_6-y_5)^2+(z_6-z_5)^2+(w_6-w_5)^2}$

$s_6=\sqrt{(x_4-x_6)^2+(y_4-y_6)^2+(z_4-z_6)^2+(w_4-w_6)^2}$

$s_7=\sqrt{(x_4-x_1)^2+(y_4-y_1)^2(z_4-z_1)^2+(w_4-w_1)^2}$

$s_8=\sqrt{(x_5-x_2)^2+(y_5-y_2)^2+(z_5-z_2)^2+(w_5-w_2)^2}$

$s_9=\sqrt{(x_6-x_3)^2+(y_6-y_3)^2+(z_6-z_3)^2+(w_6-w_3)^2}$

DIRECTION COSINES

$\cos A_1=(x_2-x_1)/s_1$ $\qquad \cos B_1=(y_2-y_1)/s_1$

$\cos C_1=(z_2-z_1)/s_1$ $\qquad \cos D_1=(w_2-w_1)/s_1$

$\cos A_2=(x_3-x_2)/s_2$ $\qquad \cos B_2=(y_3-y_2)/s_2$

$\cos C_2=(z_3-z_2)/s_2$ $\qquad \cos D_2=(w_3-w_2)/s_2$

$\cos A_3=(x_1-x_3)/s_3$ $\qquad \cos B_3=(y_1-y_3)/s_3$

$\cos C_3=(z_1-z_3)/s_3$ $\qquad \cos D_3=(w_1-w_3)/s_3$

$\cos A_4=(x_5-x_4)/s_4$ $\qquad \cos B_4=(y_5-y_4)/s_4$

$\cos C_4=(z_5-z_4)/s_4$ $\qquad \cos D_4=(w_5-w_4)/s_4$

$\cos A_5=(x_6-x_5)/s_5$ $\qquad \cos B_5=(y_6-y_5)/s_5$

$\cos C_5=(z_6-z_5)/s_5$ $\qquad \cos D_5=(w_6-w_5)/s_5$

$\cos A_6=(x_4-x_6)/s_6$ $\qquad \cos B_6=(y_4-y_6)/s_6$

$\cos C_6=(z_4-z_6)/s_6$ $\qquad \cos D_6=(w_4-w_6)/s_6$

$\cos A_7=(x_4-x_1)/s_7$ $\qquad \cos B_7=(y_4-y_1)/s_7$

$\cos C_7=(z_4-z_1)/s_7$ $\qquad \cos D_7=(w_4-w_1)/s_7$

$\cos A_8=(x_5-x_2)/s_8$ $\qquad \cos B_8=(y_5-y_2)/s_8$

$\cos C_8=(z_5-z_2)/s_8$ $\qquad \cos D_8=(w_5-w_2)/s_8$

$\cos A_9=(x_6-x_3)/s_9$ $\qquad \cos B_9=(y_6-y_3)/s_9$

$\cos C_9=(z_6-z_3)/s_9$ $\qquad \cos D_9=(w_6-w_3)/s_9$

A

ANGLES

$$\cos\phi_{1,2}=\cos A_1\cos A_2+\cos B_1\cos B_2+\cos C_1\cos C_2+\cos D_1\cos D_2$$

$$\cos\phi_{1,3}=\cos A_1\cos A_3+\cos B_1\cos B_3+\cos C_1\cos C_3+\cos D_1\cos D_3$$

$$\cos\phi_{1,7}=\cos A_1\cos A_7+\cos B_1\cos B_7+\cos C_1\cos C_7+\cos D_1\cos D_7$$

$$\cos\phi_{1,8}=\cos A_1\cos A_8+\cos B_1\cos B_8+\cos C_1\cos C_8+\cos D_1\cos D_8$$

$$\cos\phi_{2,3}=\cos A_2\cos A_3+\cos B_2\cos B_3+\cos C_2\cos C_3+\cos D_2\cos D_3$$

$$\cos\phi_{2,8}=\cos A_2\cos A_8+\cos B_2\cos B_8+\cos C_2\cos C_8+\cos D_2\cos D_8$$

$$\cos\phi_{2,9}=\cos A_2\cos A_9+\cos B_2\cos B_9+\cos C_2\cos C_9+\cos D_2\cos D_9$$

$$\cos\phi_{3,7}=\cos A_3\cos A_7+\cos B_3\cos B_7+\cos C_3\cos C_7+\cos D_3\cos D_7$$

$$\cos\phi_{3,9}=\cos A_3\cos A_9+\cos B_3\cos B_9+\cos C_3\cos C_9+\cos D_3\cos D_9$$

$$\cos\phi_{4,5}=\cos A_4\cos A_5+\cos B_4\cos B_5+\cos C_4\cos C_5+\cos D_4\cos D_5$$

$$\cos\phi_{4,6}=\cos A_4\cos A_6+\cos B_4\cos B_6+\cos C_4\cos C_6+\cos D_4\cos D_6$$

$$\cos\phi_{4,7}=\cos A_4\cos A_7+\cos B_4\cos B_7+\cos C_4\cos C_7+\cos D_4\cos D_7$$

$$\cos\phi_{4,8}=\cos A_4\cos A_8+\cos B_4\cos B_8+\cos C_4\cos C_8+\cos D_4\cos D_8$$

$$\cos\phi_{5,6}=\cos A_5\cos A_6+\cos B_5\cos B_6+\cos C_5\cos C_6+\cos D_4\cos D_8$$

$$\cos\phi_{5,8}=\cos A_5\cos A_8+\cos B_5\cos B_8+\cos C_5\cos C_8+\cos D_5\cos D_8$$

$$\cos\phi_{5,9}=\cos A_5\cos A_9+\cos B_5\cos B_9+\cos C_5\cos C_9+\cos D_5\cos D_9$$

$$\cos\phi_{6,7}=\cos A_6\cos A_7+\cos B_6\cos B_7+\cos C_6\cos C_7+\cos D_6\cos D_7$$

$$\cos\phi_{6,9}=\cos A_6\cos A_9+\cos B_6\cos B_9+\cos C_6\cos C_9+\cos D_6\cos D_9$$

CHECK

F_1: $\phi_{1,2}+\phi_{1,3}+\phi_{2,3}=180^\circ$

F_2: $\phi_{1,7}+\phi_{1,8}+\phi_{4,7}+\phi_{4,8}=360^\circ$

F_3: $\phi_{2,8}+\phi_{2,9}+\phi_{5,8}+\phi_{5,9}=360^\circ$

F_4: $\phi_{3,7}+\phi_{3,9}+\phi_{6,7}+\phi_{6,9}=360^\circ$

F_5: $\phi_{4,5}+\phi_{4,6}+\phi_{5,6}=180^\circ$

A

SECONDARY ANALYSIS -- XYZ SPACE

PLOTTED POINTS

$P_1=(x_1,y_1,z_1)$

$P_2=(x_2,y_2,z_2)$

$P_3=(x_3,y_3,z_3)$

$P_4=(x_4,y_4,z_4)$

$P_5=(x_5,y_5,z_5)$

LENGTHS

$$s_1=\sqrt{(x_2-x_1)^2+(y_2-y_1)^2+(z_2-z_1)^2}$$

$$s_2=\sqrt{(x_3-x_2)^2+(y_3-y_2)^2+(z_3-z_2)^2}$$

$$s_3=\sqrt{(x_1-x_3)^2+(y_1-y_3)^2+(z_1-z_3)^2}$$

$$s_4=\sqrt{(x_5-x_4)^2+(y_5-y_4)^2+(z_5-z_4)^2}$$

$$s_5=\sqrt{(x_6-x_5)^2+(y_6-y_5)^2+(z_6-z_5)^2}$$

$$s_6=\sqrt{(x_4-x_6)^2+(y_4-y_6)^2+(z_4-z_6)^2}$$

$$s_7=\sqrt{(x_4-x_1)^2+(y_4-y_1)^2+(z_4-z_1)^2}$$

$$s_8=\sqrt{(x_5-x_2)^2+(y_5-y_2)^2+(z_5-z_2)^2}$$

$$s_9=\sqrt{(x_6-x_3)^2+(y_6-y_3)^2+(z_6-z_3)^2}$$

B

Table 22-7. Specifications and Analyses of a Prism in 4-D Space. See Fig. 22-7. (A) Specifications and Primary Analysis in Hyperspace. (B) Secondary Analysis in XYZ Space. (C) Secondary Analysis in XYW Space. (D) Secondary Analysis in XZW Space. (E) Secondary Analysis in YZW Space. (F) Suggested Coordinates. **(Continued from page 377.)**

DIRECTION COSINES

$\cos A_1=(x_2-x_1)/s_1$ $\quad \cos B_1=(y_2-y_1)/s_1$
$\cos C_1=(z_2-z_1)/s_1$

$\cos A_2=(x_3-x_2)/s_2$ $\quad \cos B_2=(y_3-y_2)/s_2$
$\cos C_2=(z_3-z_2)/s_2$

$\cos A_3=(x_1-x_3)/s_3$ $\quad \cos B_3=(y_1-y_3)/s_3$
$\cos C_3=(z_1-z_3)/s_3$

$\cos A_4=(x_5-x_4)/s_4$ $\quad \cos B_4=(y_5-y_4)/s_4$
$\cos C_4=(z_5-z_4)/s_4$

$\cos A_5=(x_6-x_5)/s_5$ $\quad \cos B_5=(y_6-y_5)/s_5$
$\cos C_5=(z_6-z_5)/s_5$

$\cos A_6=(x_4-x_6)/s_6$ $\quad \cos B_6=(y_4-y_6)/s_6$
$\cos C_6=(z_4-z_6)/s_6$

$\cos A_7=(x_4-x_1)/s_7$ $\quad \cos B_7=(y_4-y_1)/s_7$
$\cos C_7=(z_4-z_1)/s_7$

$\cos A_8=(x_5-x_2)/s_8$ $\quad \cos B_8=(y_5-y_2)/s_8$
$\cos C_8=(z_5-z_2)/s_8$

$\cos A_9=(x_6-x_3)/s_9$ $\quad \cos B_9=(y_6-y_3)/s_9$
$\cos C_9=(z_6-z_3)/s_9$

ANGLES

$\cos\phi_{1,2}=\cos A_1\cos A_2+\cos B_1\cos B_2+\cos C_1\cos C_2$

$\cos\phi_{1,3}=\cos A_1\cos A_3+\cos B_1\cos B_3+\cos C_1\cos C_3$

$\cos\phi_{1,7}=\cos A_1\cos A_7+\cos B_1\cos B_7+\cos C_1\cos C_7$

$\cos\phi_{1,8}=\cos A_1\cos A_8+\cos B_1\cos B_8+\cos C_1\cos C_8$

$\cos\phi_{2,3}=\cos A_2\cos A_3+\cos B_2\cos B_3+\cos C_2\cos C_3$

$\cos\phi_{2,8}=\cos A_2\cos A_8+\cos B_2\cos B_8+\cos C_2\cos C_8$

$\cos\phi_{2,9}=\cos A_2\cos A_9+\cos B_2\cos B_9+\cos C_2\cos C_9$

$\cos\phi_{3,7}=\cos A_3\cos A_7+\cos B_3\cos B_7+\cos C_3\cos C_7$

$\cos\phi_{3,9}=\cos A_3\cos A_9+\cos B_3\cos B_9+\cos C_3\cos C_9$

$\cos\phi_{4,5}=\cos A_4\cos A_5+\cos B_4\cos B_5+\cos C_4\cos C_5$

$\cos\phi_{4,6}=\cos A_4\cos A_6+\cos B_4\cos B_6+\cos C_4\cos C_6$

$\cos\phi_{4,7}=\cos A_4\cos A_7+\cos B_4\cos B_7+\cos C_4\cos C_7$

$\cos\phi_{4,8}=\cos A_4\cos A_8+\cos B_4\cos B_8+\cos C_4\cos C_8$

$\cos\phi_{5,6}=\cos A_5\cos A_6+\cos B_5\cos B_6+\cos C_5\cos C_6$

$\cos\phi_{5,8}=\cos A_5\cos A_8+\cos B_5\cos B_8+\cos C_5\cos C_8$

$\cos\phi_{5,9}=\cos A_5\cos A_9+\cos B_5\cos B_9+\cos C_5\cos C_9$

$\cos\phi_{6,7}=\cos A_6\cos A_7+\cos B_6\cos B_7+\cos C_6\cos C_7$

$\cos\phi_{6,9}=\cos A_6\cos A_9+\cos B_6\cos B_9+\cos C_6\cos C_9$

CHECK

F_1: $\phi_{1,2}+\phi_{1,3}+\phi_{2,3}=180^\circ$

F_2: $\phi_{1,7}+\phi_{1,8}+\phi_{4,7}+\phi_{4,8}=360^\circ$

F_3: $\phi_{2,8}+\phi_{2,9}+\phi_{5,8}+\phi_{5,9}=360^\circ$

F_4: $\phi_{3,7}+\phi_{3,9}+\phi_{6,7}+\phi_{6,9}=360^\circ$

F_5: $\phi_{4,5}+\phi_{4,6}+\phi_{5,6}=180^\circ$

B

SECONDARY ANALYSIS -- XYW SPACE

PLOTTED POINTS

$P_1=(x_1,y_1,w_1)$

$P_2=(x_2,y_2,w_2)$

$P_3=(x_3,y_3,w_3)$

$P_4=(x_4,y_4,w_4)$

$P_5=(x_5,y_5,w_5)$

LENGTHS

$s_1=\sqrt{(x_2-x_1)^2+(y_2-y_1)^2+(w_2-w_1)^2}$

$s_2=\sqrt{(x_3-x_2)^2+(y_3-y_2)^2+(w_3-w_2)^2}$

$s_3=\sqrt{(x_1-x_3)^2+(y_1-y_3)^2+(w_1-w_3)^2}$

$s_4=\sqrt{(x_5-x_4)^2+(y_5-y_4)^2+(w_5-w_4)^2}$

$s_5=\sqrt{(x_6-x_5)^2+(y_6-y_5)^2+(w_6-w_5)^2}$

$s_6=\sqrt{(x_4-x_6)^2+(y_4-y_6)^2+(w_4-w_6)^2}$

$s_7=\sqrt{(x_4-x_1)^2+(y_4-y_1)^2+(w_4-w_1)^2}$

$s_8=\sqrt{(x_5-x_2)^2+(y_5-y_2)^2+(w_5-w_2)^2}$

$s_9=\sqrt{(x_6-x_3)^2+(y_6-y_3)^2+(w_6-w_3)^2}$

DIRECTION COSINES

$\cos A_1=(x_2-x_1)/s_1$ $\cos B_1=(y_2-y_1)/s_1$

$\cos D_1=(w_2-w_1)/s_1$

$\cos A_2=(x_3-x_2)/s_2$ $\cos B_2=(y_3-y_2)/s_2$

$\cos D_2=(w_3-w_2)/s_2$

$\cos A_3=(x_1-x_3)/s_3$ $\cos B_3=(y_1-y_3)/s_3$

$\cos D_3=(w_1-w_3)/s_3$

$\cos A_4=(x_5-x_4)/s_4$ $\cos B_4=(y_5-y_4)/s_4$

$\cos D_4=(w_5-w_4)/s_4$

$\cos A_5=(x_6-x_5)/s_5$ $\cos B_5=(y_6-y_5)/s_5$

$\cos D_5=(w_6-w_5)/s_5$

$\cos A_6=(x_4-x_6)/s_6$ $\cos B_6=(y_4-y_6)/s_6$

$\cos D_6=(w_4-w_6)/s_6$

$\cos A_7=(x_4-x_1)/s_7$ $\cos B_7=(y_4-y_1)/s_7$

$\cos D_7=(w_4-w_1)/s_7$

$\cos A_8=(x_5-x_2)/s_8$ $\cos B_8=(y_5-y_2)/s_8$

$\cos D_8=(w_5-w_2)/s_8$

$\cos A_9=(x_6-x_3)/s_9$ $\cos B_9=(y_6-y_3)/s_9$

$\cos D_9=(w_6-w_3)/s_9$

ANGLES

$\cos\phi_{1,2}=\cos A_1\cos A_2+\cos B_1\cos B_2+\cos D_1\cos D_2$

$\cos\phi_{1,3}=\cos A_1\cos A_3+\cos B_1\cos B_3+\cos D_1\cos D_3$

$\cos\phi_{1,7}=\cos A_1\cos A_7+\cos B_1\cos B_7+\cos D_1\cos D_7$

$\cos\phi_{1,8}=\cos A_1\cos A_8+\cos B_1\cos B_8+\cos D_1\cos D_8$

$\cos\phi_{2,3}=\cos A_2\cos A_3+\cos B_2\cos B_3+\cos D_2\cos D_3$

$\cos\phi_{2,8}=\cos A_2\cos A_8+\cos B_2\cos B_8+\cos D_2\cos D_8$

$\cos\phi_{2,9}=\cos A_2\cos A_9+\cos B_2\cos B_9+\cos D_2\cos D_9$

$\cos\phi_{3,7}=\cos A_3\cos A_7+\cos B_3\cos B_7+\cos D_3\cos D_7$

$\cos\phi_{3,9}=\cos A_3\cos A_9+\cos B_3\cos B_9+\cos D_3\cos D_9$

$\cos\phi_{4,5}=\cos A_4\cos A_5+\cos B_4\cos B_5+\cos D_4\cos D_5$

$\cos\phi_{4,6}=\cos A_4\cos A_6+\cos B_4\cos B_6+\cos D_4\cos D_6$

$\cos\phi_{4,7}=\cos A_4\cos A_7+\cos B_4\cos B_7+\cos D_4\cos D_7$

$\cos\phi_{4,8}=\cos A_4\cos A_8+\cos B_4\cos B_8+\cos D_4\cos D_8$

$\cos\phi_{5,6}=\cos A_5\cos A_6+\cos B_5\cos B_6+\cos D_5\cos D_6$

$\cos\phi_{5,8}=\cos A_5\cos A_8+\cos B_5\cos B_8+\cos D_5\cos D_8$

$\cos\phi_{5,9}=\cos A_5\cos A_9+\cos B_5\cos B_9+\cos D_5\cos D_9$

$\cos\phi_{6,7}=\cos A_6\cos A_7+\cos B_6\cos B_7+\cos D_6\cos D_7$

$\cos\phi_{6,9}=\cos A_6\cos A_9+\cos B_6\cos B_9+\cos D_6\cos D_9$

CHECK

F_1: $\phi_{1,2}+\phi_{1,3}+\phi_{2,3}=180^\circ$

F_2: $\phi_{1,7}+\phi_{1,8}+\phi_{4,7}+\phi_{4,8}=360^\circ$

F_3: $\phi_{2,8}+\phi_{2,9}+\phi_{5,8}+\phi_{5,9}=360^\circ$

F_4: $\phi_{3,7}+\phi_{3,9}+\phi_{6,7}+\phi_{6,9}=360^\circ$

F_5: $\phi_{4,5}+\phi_{4,6}+\phi_{5,6}=180^\circ$

C

Table 22-7. Specifications and Analyses of a Prism in 4-D Space. See Fig. 22-7. (A) Specifications and Primary Analysis in Hyperspace. (B) Secondary Analysis in XYZ Space. (C) Secondary Analysis in XYW Space. (D) Secondary Analysis in XZW Space. (E) Secondary Analysis in YZW Space. (F) Suggested Coordinates. (Continued from page 379.)

SECONDARY ANALYSIS -- XZW SPACE

PLOTTED POINTS

$P_1=(x_1,z_1,w_1)$ $P_2=(x_2,z_2,w_2)$ $P_3=(x_3,z_3,w_3)$ $P_4=(x_4,z_4,w_4)$ $P_5=(x_5,z_5,w_5)$

LENGTHS

$s_1=\sqrt{(x_2-x_1)^2+(z_2-z_1)^2+(w_2-w_1)^2}$

$s_2=\sqrt{(x_3-x_2)^2+(z_3-z_2)^2+(w_3-w_2)^2}$

$s_3=\sqrt{(x_1-x_3)^2+(z_1-z_3)^2+(w_1-w_3)^2}$

$s_4=\sqrt{(x_5-x_4)^2+(z_5-z_4)^2+(w_5-w_4)^2}$

$s_5=\sqrt{(x_6-x_5)^2+(z_6-z_5)^2+(w_6-w_5)^2}$

$s_6=\sqrt{(x_4-x_6)^2+(z_4-z_6)^2+(w_4-w_6)^2}$

$s_7=\sqrt{(x_4-x_1)^2+(z_4-z_1)^2+(w_4-w_1)^2}$

$s_8=\sqrt{(x_5-x_2)^2+(z_5-z_2)^2+(w_5-w_2)^2}$

$s_9=\sqrt{(x_6-x_3)^2+(z_6-z_3)^2+(w_6-w_3)^2}$

ANGLES

$\cos\phi_{1,2}=\cos A_1\cos A_2+\cos C_1\cos C_2+\cos D_1\cos D_2$

$\cos\phi_{1,3}=\cos A_1\cos A_3+\cos C_1\cos C_3+\cos D_1\cos D_3$

$\cos\phi_{1,7}=\cos A_1\cos A_7+\cos C_1\cos C_7+\cos D_1\cos D_7$

$\cos\phi_{1,8}=\cos A_1\cos A_8+\cos C_1\cos C_8+\cos D_1\cos D_8$

$\cos\phi_{2,3}=\cos A_2\cos A_3+\cos C_2\cos C_3+\cos D_2\cos D_3$

$\cos\phi_{2,8}=\cos A_2\cos A_8+\cos C_2\cos C_8+\cos D_2\cos D_8$

$\cos\phi_{2,9}=\cos A_2\cos A_9+\cos C_2\cos C_9+\cos D_2\cos D_9$

$\cos\phi_{3,7}=\cos A_3\cos A_7+\cos C_3\cos C_7+\cos D_3\cos D_7$

$\cos\phi_{3,9}=\cos A_3\cos A_9+\cos C_3\cos C_9+\cos D_3\cos D_9$

$\cos\phi_{4,5}=\cos A_4\cos A_5+\cos C_4\cos C_5+\cos D_4\cos D_5$

$\cos\phi_{4,6}=\cos A_4\cos A_6+\cos C_4\cos C_6+\cos D_4\cos D_6$

$\cos\phi_{4,7}=\cos A_4\cos A_7+\cos C_4\cos C_7+\cos D_4\cos D_7$

D $\cos\phi_{4,8}=\cos A_4\cos A_8+\cos C_4\cos C_8+\cos D_4\cos D_8$

$\cos\phi_{5,6}=\cos A_5\cos A_6+\cos C_5\cos C_6+\cos D_5\cos D_6$

DIRECTION COSINES

$\cos A_1=(x_2-x_1)/s_1$ $\cos C_1=(z_2-z_1)/s_1$

$\cos D_1=(w_2-w_1)/s_1$

$\cos A_2=(x_3-x_2)/s_2$ $\cos C_2=(z_3-z_2)/s_2$

$\cos D_2=(w_3-w_2)/s_2$

$\cos A_3=(x_1-x_3)/s_3$ $\cos C_3=(z_1-z_3)/s_3$

$\cos D_3=(w_1-w_3)/s_3$

$\cos A_4=(x_5-x_4)/s_4$ $\cos C_4=(z_5-z_4)/s_4$

$\cos D_4=(w_5-w_4)/s_4$

$\cos A_5=(x_6-x_5)/s_5$ $\cos C_5=(z_6-z_5)/s_5$

$\cos D_5=(w_6-w_5)/s_5$

$\cos A_6=(x_4-x_6)/s_6$ $\cos C_6=(z_4-z_6)/s_6$

$\cos D_6=(w_4-w_6)/s_6$

$\cos A_7=(x_4-x_1)/s_7$ $\cos C_7=(z_4-z_1)/s_7$

$\cos D_7=(w_4-w_1)/s_7$

$\cos A_8=(x_5-x_2)/s_8$ $\cos C_8=(z_5-z_2)/s_8$

$\cos D_8=(w_5-w_2)/s_8$

$\cos A_9=(x_6-x_3)/s_9$ $\cos C_9=(z_6-z_3)/s_9$

$\cos D_9=(w_6-w_3)/s_9$

$\cos\phi_{5,8}=\cos A_5\cos A_8+\cos C_5\cos C_8+\cos D_5\cos D_8$

$\cos\phi_{5,9}=\cos A_5\cos A_9+\cos C_5\cos C_9+\cos D_5\cos D_9$

$\cos\phi_{6,7}=\cos A_6\cos A_7+\cos C_6\cos C_7+\cos D_6\cos D_7$

$\cos\phi_{6,9}=\cos A_6\cos A_9+\cos C_6\cos C_9+\cos D_6\cos D_9$

CHECK

F_1: $\phi_{1,2}+\phi_{1,3}+\phi_{2,3}=180^\circ$

F_2: $\phi_{1,7}+\phi_{1,8}+\phi_{4,7}+\phi_{4,8}=360^\circ$

F_3: $\phi_{2,8}+\phi_{2,9}+\phi_{5,8}+\phi_{5,9}=360^\circ$

F_4: $\phi_{3,7}+\phi_{3,9}+\phi_{6,7}+\phi_{6,9}=360^\circ$

F_5: $\phi_{4,5}+\phi_{4,6}+\phi_{5,6}=180^\circ$

SECONDARY ANALYSIS -- YZW SPACE

PLOTTED POINTS

$P_1=(y_1,z_1,w_1)$

$P_2=(y_2,z_2,w_2)$

$P_3=(y_3,z_3,w_3)$

$P_4=(y_4,z_4,w_4)$

$P_5=(y_5,z_5,w_5)$

LENGTHS

$$s_1=\sqrt{(y_2-y_1)^2+(z_2-z_1)^2+(w_2-w_1)^2}$$

$$s_2=\sqrt{(y_3-y_2)^2+(z_3-z_2)^2+(w_3-w_2)^2}$$

$$s_3=\sqrt{(y_1-y_3)^2+(z_1-z_3)^2+(w_1-w_3)^2}$$

$$s_4=\sqrt{(y_5-y_4)^2+(z_5-z_4)^2+(w_5-w_4)^2}$$

$$s_5=\sqrt{(y_6-y_5)^2+(z_6-z_5)^2+(w_6-w_5)^2}$$

$$s_6=\sqrt{(y_4-y_6)^2+(z_4-z_6)^2+(w_4-w_6)^2}$$

$$s_7=\sqrt{(y_4-y_1)^2+(z_4-z_1)^2+(w_4-w_1)^2}$$

$$s_8=\sqrt{(y_5-y_2)^2+(z_5-z_2)^2+(w_5-w_2)^2}$$

$$s_9=\sqrt{(y_6-y_3)^2+(z_6-z_3)^2+(w_6-w_3)^2}$$

ANGLES

$\cos\phi_{1,2}=\cos B_1\cos B_2+\cos C_1\cos C_2+\cos D_1\cos D_2$

$\cos\phi_{1,3}=\cos B_1\cos B_3+\cos C_1\cos C_3+\cos D_1\cos D_3$

$\cos\phi_{1,7}=\cos B_1\cos B_7+\cos C_1\cos C_7+\cos D_1\cos D_7$

$\cos\phi_{1,8}=\cos B_1\cos B_8+\cos C_1\cos C_8+\cos D_1\cos D_8$

$\cos\phi_{2,3}=\cos B_2\cos B_3+\cos C_2\cos C_3+\cos D_2\cos D_3$

$\cos\phi_{2,8}=\cos B_2\cos B_8+\cos C_2\cos C_8+\cos D_2\cos D_8$

$\cos\phi_{2,9}=\cos B_2\cos B_9+\cos C_2\cos C_9+\cos D_2\cos D_9$

$\cos\phi_{3,7}=\cos B_3\cos B_7+\cos C_3\cos C_7+\cos D_3\cos D_7$

$\cos\phi_{3,9}=\cos B_3\cos B_9+\cos C_3\cos C_9+\cos D_3\cos D_9$

$\cos\phi_{4,5}=\cos B_4\cos B_5+\cos C_4\cos C_5+\cos D_4\cos D_5$

$\cos\phi_{4,6}=\cos B_4\cos B_6+\cos C_4\cos C_6+\cos D_4\cos D_6$

$\cos\phi_{4,7}=\cos B_4\cos B_7+\cos C_4\cos C_7+\cos D_4\cos D_7$

$\cos\phi_{4,8}=\cos B_4\cos B_8+\cos C_4\cos C_8+\cos D_4\cos D_8$

$\cos\phi_{5,6}=\cos B_5\cos B_6+\cos C_5\cos C_6+\cos D_5\cos D_6$

DIRECTION COSINES

$\cos B_1=(y_2-y_1)/s_1$ $\quad\cos C_1=(z_2-z_1)/s_1$

$\cos D_1=(w_2-w_1)/s_1$

$\cos B_2=(y_3-y_2)/s_2$ $\quad\cos C_2=(z_3-z_2)/s_2$

$\cos D_2=(w_3-w_2)/s_2$

$\cos B_3=(y_1-y_3)/s_3$ $\quad\cos C_3=(z_1-z_3)/s_3$

$\cos D_3=(w_1-w_3)/s_3$

$\cos B_4=(y_5-y_4)/s_4$ $\quad\cos C_4=(z_5-z_4)/s_4$

$\cos D_4=(w_5-w_4)/s_4$

$\cos B_5=(y_6-y_5)/s_5$ $\quad\cos C_5=(z_6-z_5)/s_5$

$\cos D_5=(w_6-w_5)/s_5$

$\cos B_6=(y_4-y_6)/s_6$ $\quad\cos C_6=(z_4-z_6)/s_6$

$\cos D_6=(w_4-w_6)/s_6$

$\cos B_7=(y_4-y_1)/s_7$ $\quad\cos C_7=(z_4-z_1)/s_7$

$\cos D_7=(w_4-w_1)/s_7$

$\cos B_8=(y_5-y_2)/s_8$ $\quad\cos C_8=(z_5-z_2)/s_8$

$\cos D_8=(w_5-w_2)/s_8$

$\cos B_9=(y_6-y_3)/s_9$ $\quad\cos C_9=(z_6-z_3)/s_9$

$\cos D_9=(w_6-w_3)/s_9$

$\cos\phi_{5,8}=\cos B_5\cos B_8+\cos C_5\cos C_8+\cos D_5\cos D_8$

$\cos\phi_{5,9}=\cos B_5\cos B_9+\cos C_5\cos C_9+\cos D_5\cos D_9$

$\cos\phi_{6,7}=\cos B_6\cos B_7+\cos C_6\cos C_7+\cos D_6\cos D_7$

$\cos\phi_{6,9}=\cos B_6\cos B_9+\cos C_6\cos C_9+\cos D_6\cos D_9$

CHECK

F_1: $\quad\phi_{1,2}+\phi_{1,3}+\phi_{2,3}=180^\circ$

F_2: $\quad\phi_{1,7}+\phi_{1,8}+\phi_{4,7}+\phi_{4,8}=360^\circ$

F_3: $\quad\phi_{2,8}+\phi_{2,9}+\phi_{5,8}+\phi_{5,9}=360^\circ$

F_4: $\quad\phi_{3,7}+\phi_{3,9}+\phi_{6,7}+\phi_{6,9}=360^\circ$

F_5: $\quad\phi_{4,5}+\phi_{4,6}+\phi_{5,6}=180^\circ$

Table 22-7. Specifications and Analyses of a Prism in 4-D Space. See Fig. 22-7. (A) Specifications and Primary Analysis in Hyperspace. (B) Secondary Analysis in XYZ Space. (C) Secondary Analysis in XYW Space. (D) Secondary Analysis in XZW Space. (E) Secondary Analysis in YZW Space. (F) Suggested Coordinates. **(Continued from page 381.)**

F

HYPERSPACE OBJECT VERSION 1

P_1=(-1,Ø,1,Ø)	P_2=(Ø,Ø,1,Ø)	P_3=(Ø,1,1,1)
P_4=(-1,Ø,-1,Ø)	P_5=(Ø,Ø,-1,Ø)	P_6=(Ø,1,-1,1)

HYPERSPACE OBJECT VERSION 2

P_1=(-1,1,1,1)	P_2=(Ø,1,1,1)	P_3=(Ø,2,1,2)
P_4=(Ø,-1,-1,-1)	P_5=(1,-1,-1,-1)	P_6=(1,-1,-1,-1)

HYPERSPACE OBJECT VERSION 3

P_1=(-1,Ø,1,Ø)	P_2=(Ø,Ø,1,Ø)	P_3=(1,1,1,1)
P_4=(-1,Ø,-1,Ø)	P_5=(Ø,Ø,-1,Ø)	P_6=(-1,1,Ø,1)

ordinary, 3-dimensional pyramid is capable of such feats, imagine what a 4-dimensional version of it can do!

At any rate, Fig. 22-11 shows the pyramid and Table 22-11 outlines the specifications, primary analysis and secondary analysis. Table 22-11F is important, too, because it suggests some point coordinates for a couple of different variations.

Doublecheck the results of your work by referring to the F specifications and angles. The F specifications indicate which lines bound a plane figure; and given that information, you should be able to determine the angles included in it. From there, you can check the work by summing the angles in each plane figure. In this particular case, you are working with triangles and quadralaterals, and the sums of the angles ought to be close to 180 and 360 degrees, respectively.

22-2.3 A Hyperprism

I am specifying a general hyperprism for you in Fig. 22-12 and Table 22-12. Use the suggested coordinates in Table 22-12F as a starting point for conducting complete analyses of this figure.

Use the guide to summing angles to check your results.

22-2.4 A "Hyperhouse" Object

Just to be somewhat consistent with some previous work, I am including here the specifications and guides for complete analyses of a 4-D version of the "Monopoly house." It's a strange little figure in the literature of hyperspace, but worthy of study. Use my suggested coordinates in Table 22-13F as a guide for getting started. Also see Fig. 22-13.

22-3 A SUMMARY OF SIMPLE TRANSFORMATIONS IN 4-D SPACE

The simple transformations—those based on a single type of transformation—in 4-D space are the same as those in lower-dimensioned spaces:

- ☐ Translations or combinations of translations.
- ☐ Scalings or combinations of scalings.
- ☐ Rotations or combinations of rotations.

Here is a brief summary of the transformations and their effects on geometric entities that they influence:

☐ Under simple translations: length is an invariant, direction cosines are invariant, angles between lines are invariant, and all simple translations are commutative.

☐ Under simple scalings: length is a variant, direction cosines are variant, angles between lines are variant, and combinations of simple scalings are commutative.

☐ Under simple rotations: length is an invariant, direction cosines are variant, angles between lines are invariant, and simple rotations about one particular axis

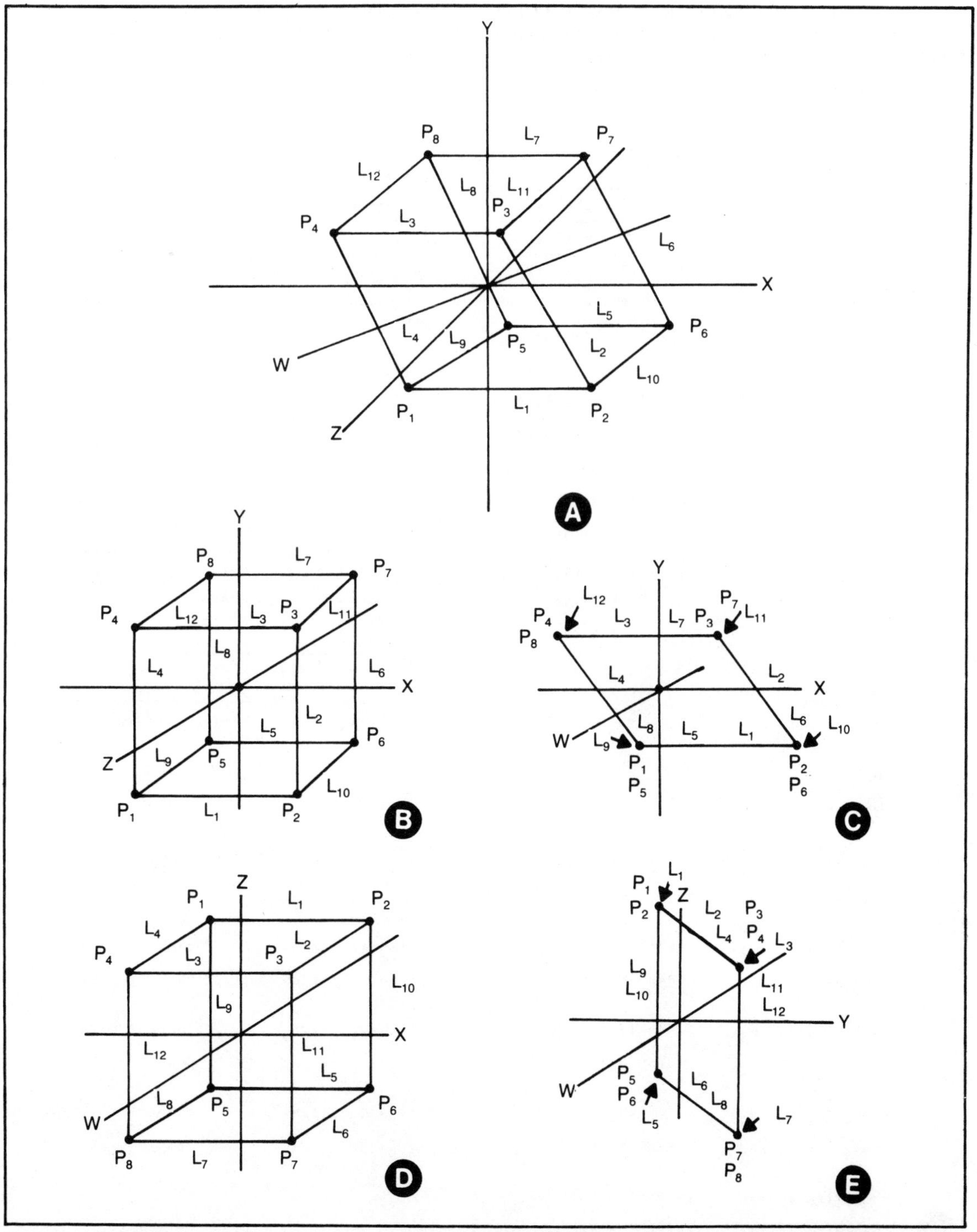

Fig. 22-8. A 6-sided object specified in 4-D space. (A) General hyperspace view. (B) XYZ perspective view. (C) XYW perspective view. (D) XZW perspective view. (E) YZW perspective view. See outlines of analyses in Table 22-8.

Table 22-8. Specifications and Analyses of a 6-Sided Object in 4-D Space. See Fig. 22-8. (A) Specifications and Analysis in Hyperspace. (B) Secondary Analysis in XYZ Space. (C) Secondary Analysis in XYW Space. (D) Secondary Analysis in XZW Space. (E) Secondary Analysis in YZW Space. (F) Suggested Coordinates.

(A, B, C, and D correspond to α, β, γ, and δ.)

SPECIFICATIONS

$S=\overline{F_1F_2F_3F_4F_5F_6}$

$F_1=\overline{L_1L_2L_3L_4}$

$F_2=\overline{L_1L_5L_9L_{10}}$

$F_3=\overline{L_2L_6L_{10}L_{11}}$

$F_4=\overline{L_3L_7L_{11}L_{12}}$

$F_5=\overline{L_4L_8L_9L_{12}}$

$F_6=\overline{L_5L_6L_7L_8}$

$L_1=\overline{P_1P_2}$

$L_2=\overline{P_2P_3}$

$L_3=\overline{P_3P_4}$

$L_4=\overline{P_4P_1}$

$L_5=\overline{P_5P_6}$

$L_6=\overline{P_6P_7}$

$L_7=\overline{P_7P_8}$

$L_8=\overline{P_8P_5}$

$L_9=\overline{P_1P_5}$

$L_{10}=\overline{P_2P_6}$

$L_{11}=\overline{P_3P_7}$

$L_{12}=\overline{P_4P_8}$

$P_1=(x_1,y_1,z_1,w_1)$

$P_2=(x_2,y_2,z_2,w_2)$

$P_3=(x_3,y_3,z_3,w_3)$

$P_4=(x_4,y_4,z_4,w_4)$

$P_5=(x_5,y_5,z_5,w_5)$

$P_6=(x_6,y_6,z_6,w_6)$

$P_7=(x_7,y_7,z_7,w_7)$

$P_8=(x_8,y_8,z_8,w_8)$

LENGTHS

$s_1=\sqrt{(x_2-x_1)^2+(y_2-y_1)^2+(z_2-z_1)^2+(w_2-w_1)^2}$

$s_2=\sqrt{(x_3-x_2)^2+(y_3-y_2)^2+(z_3-z_2)^2+(w_3-w_2)^2}$

$s_3=\sqrt{(x_4-x_3)^2+(y_4-y_3)^2+(z_4-z_3)^2+(w_4-w_3)^2}$

$s_4=\sqrt{(x_1-x_4)^2+(y_1-y_4)^2+(z_1-z_4)^2+(w_1-w_4)^2}$

$s_5=\sqrt{(x_6-x_5)^2+(y_6-y_5)^2+(z_6-z_5)^2+(w_6-w_5)^2}$

$s_6=\sqrt{(x_7-x_6)^2+(y_7-y_6)^2+(z_7-z_6)^2+(w_7-w_6)^2}$

$s_7=\sqrt{(x_8-x_7)^2+(y_8-y_7)^2+(z_8-z_7)^2+(w_8-w_7)^2}$

$s_8=\sqrt{(x_5-x_8)^2+(y_5-y_8)^2+(z_5-z_8)^2+(w_5-w_8)^2}$

$s_9=\sqrt{(x_5-x_1)^2+(y_5-y_1)^2+(z_5-z_1)^2+(w_5-w_1)^2}$

$s_{10}=\sqrt{(x_6-x_2)^2+(y_6-y_2)^2+(z_6-z_2)^2+(w_6-w_2)^2}$

$s_{11}=\sqrt{(x_7-x_3)^2+(y_7-y_3)^2+(z_7-z_3)^2+(w_7-w_3)^2}$

$s_{12}=\sqrt{(x_8-x_4)^2+(y_8-y_4)^2+(z_8-z_4)^2+(w_8-w_4)^2}$

DIRECTION COSINES

$\cos A_1=(x_2-x_1)/s_1$ $\cos B_1=(y_2-y_1)/s_1$
$\cos C_1=(z_2-z_1)/s_1$ $\cos D_1=(w_2-w_1)/s_1$

$\cos A_2=(x_3-x_2)/s_2$ $\cos B_2=(y_3-y_2)/s_2$
$\cos C_2=(z_3-z_2)/s_2$ $\cos D_2=(w_3-w_2)/s_2$

$\cos A_3=(x_4-x_3)/s_3$ $\cos B_3=(y_4-y_3)/s_3$
$\cos C_3=(z_4-z_3)/s_3$ $\cos D_3=(w_4-w_3)/s_3$

$\cos A_4=(x_1-x_4)/s_4$ $\cos B_4=(y_1-y_4)/s_4$
$\cos C_4=(z_1-z_4)/s_4$ $\cos D_4=(w_1-w_4)/s_4$

$\cos A_5=(x_6-x_5)/s_5$ $\cos B_5=(y_6-y_5)/s_5$
$\cos C_5=(z_6-z_5)/s_5$ $\cos D_5=(w_6-w_5)/s_5$

$\cos A_6=(x_7-x_6)/s_6$ $\cos B_6=(y_7-y_6)/s_6$
$\cos C_6=(z_7-z_6)/s_6$ $\cos D_6=(w_7-w_6)/s_6$

$\cos A_7=(x_8-x_7)/s_7$ $\cos B_7=(y_8-y_7)/s_7$
$\cos C_7=(z_8-z_7)/s_7$ $\cos D_7=(w_8-w_7)/s_7$

$\cos A_8=(x_5-x_8)/s_8$ $\cos B_8=(y_5-y_8)/s_8$
$\cos C_8=(z_5-z_8)/s_8$ $\cos D_8=(w_5-w_8)/s_8$

$\cos A_9=(x_5-x_1)/s_9$ $\cos B_9=(y_5-y_1)/s_9$
$\cos C_9=(z_5-z_1)/s_9$ $\cos D_9=(w_5-w_1)/s_9$

A

DIRECTION COSINES (Cont'd)

$\cos A_{10}=(x_6-x_2)/s_{10}$ $\quad$ $\cos B_{10}=(y_6-y_2)/s_{10}$

$\cos C_{10}=(z_6-z_2)/s_{10}$ $\quad$ $\cos D_{10}=(w_6-w_2)/s_{10}$

$\cos A_{11}=(x_7-x_3)/s_{11}$ $\quad$ $\cos B_{11}=(y_7-y_3)/s_{11}$

$\cos C_{11}=(z_7-z_3)/s_{11}$ $\quad$ $\cos D_{11}=(w_7-x_3)/s_{11}$

$\cos A_{12}=(x_8-x_4)/s_{12}$ $\quad$ $\cos B_{12}=(y_8-y_4)/s_{12}$

$\cos C_{12}=(z_8-z_4)/s_{12}$ $\quad$ $\cos D_{12}=(w_8-w_4)/s_{12}$

CHECK

F_1: $\phi_{1,2}+\phi_{1,4}+\phi_{2,3}+\phi_{3,4}=360^\circ$

F_2: $\phi_{1,9}+\phi_{1,10}+\phi_{5,9}+\phi_{5,10}=360^\circ$

F_3: $\phi_{2,10}+\phi_{2,11}+\phi_{6,10}+\phi_{6,11}=360^\circ$

F_4: $\phi_{3,11}+\phi_{3,12}+\phi_{7,11}+\phi_{7,12}=360^\circ$

F_5: $\phi_{4,9}+\phi_{4,12}+\phi_{8,9}+\phi_{8,12}=360^\circ$

F_6: $\phi_{5,6}+\phi_{5,8}+\phi_{6,7}+\phi_{7,8}=360^\circ$

ANGLES

$\cos\phi_{1,2}=\cos A_1\cos A_2+\cos B_1\cos B_2+\cos C_1\cos C_2+\cos D_1\cos D_2$

$\cos\phi_{1,4}=\cos A_1\cos A_4+\cos B_1\cos B_4+\cos C_1\cos C_4+\cos D_1\cos D_4$

$\cos\phi_{1,9}=\cos A_1\cos A_9+\cos B_1\cos B_9+\cos C_1\cos C_9+\cos D_1\cos D_9$

$\cos\phi_{1,10}=\cos A_1\cos A_{10}+\cos B_1\cos B_{10}+\cos C_1\cos C_{10}+\cos D_1\cos D_{10}$

$\cos\phi_{2,3}=\cos A_2\cos A_3+\cos B_2\cos B_3+\cos C_2\cos C_3+\cos D_2\cos D_3$

$\cos\phi_{2,10}=\cos A_2\cos A_{10}+\cos B_2\cos B_{10}+\cos C_2\cos C_{10}+\cos D_2\cos D_{10}$

$\cos\phi_{2,11}=\cos A_2\cos A_{11}+\cos B_2\cos B_{11}+\cos C_2\cos C_{11}+\cos D_2\cos D_{11}$

$\cos\phi_{3,4}=\cos A_3\cos A_4+\cos B_3\cos B_4+\cos C_3\cos C_4+\cos D_3\cos D_4$

$\cos\phi_{3,11}=\cos A_3\cos A_{11}+\cos B_3\cos B_{11}+\cos C_3\cos C_{11}+\cos D_3\cos D_{11}$

$\cos\phi_{3,11}=\cos A_3\cos A_{11}+\cos B_3\cos B_{11}+\cos C_3\cos C_{11}+\cos D_3\cos D_{11}$

$\cos\phi_{4,9}=\cos A_4\cos A_9+\cos B_4\cos B_9+\cos C_4\cos C_9+\cos D_4\cos D_9$

$\cos\phi_{4,12}=\cos A_4\cos A_{12}+\cos B_4\cos B_{12}+\cos C_4\cos C_{12}+\cos D_4\cos D_{12}$

$\cos\phi_{5,6}=\cos A_5\cos A_6+\cos B_5\cos B_6+\cos C_5\cos C_6+\cos D_5\cos D_6$

$\cos\phi_{5,8}=\cos A_5\cos A_8+\cos B_5\cos B_8+\cos C_5\cos C_8+\cos D_5\cos D_8$

$\cos\phi_{5,9}=\cos A_5\cos A_9+\cos B_5\cos B_9+\cos C_5\cos C_9+\cos D_5\cos D_9$

$\cos\phi_{5,10}=\cos A_5\cos A_{10}+\cos B_5\cos B_{10}+\cos C_5\cos C_{10}+\cos D_5\cos D_{10}$

$\cos\phi_{6,7}=\cos A_6\cos A_7+\cos B_6\cos B_7+\cos C_6\cos C_7+\cos D_6\cos D_7$

$\cos\phi_{6,10}=\cos A_6\cos A_{10}+\cos B_6\cos B_{10}+\cos C_6\cos C_{10}+\cos A_6\cos D_{10}$

$\cos\phi_{6,11}=\cos A_6\cos A_{11}+\cos B_6\cos B_{11}+\cos C_6\cos C_{11}+\cos D_6\cos D_{11}$

$\cos\phi_{7,8}=\cos A_7\cos A_8+\cos B_7\cos B_8+\cos C_7\cos C_8+\cos D_7\cos D_8$

$\cos\phi_{7,11}=\cos A_7\cos A_{11}+\cos B_7\cos B_{11}+\cos C_7\cos C_{11}+\cos D_7\cos D_{11}$

$\cos\phi_{7,12}=\cos A_7\cos A_{12}+\cos B_7\cos B_{12}+\cos C_7\cos C_{12}+\cos D_7\cos D_{12}$

$\cos\phi_{8,9}=\cos A_8\cos A_9+\cos B_8\cos B_9+\cos C_8\cos C_9+\cos D_8\cos D_9$

$\cos\phi_{8,12}=\cos A_8\cos A_{12}+\cos B_8\cos B_{12}+\cos C_8\cos C_{12}+\cos D_8\cos D_{12}$

Table 22-8. Specifications and Analyses of a 6-Sided Object in 4-D Space. See Fig. 22-8. (A) Specifications and Analysisi in Hyperspace. (B) Secondary Analysis in XYZ Space. (C) Secondary Analysis in XYW Space. (D) Secondary Analysis in XZW Space. (E) Secondary Analysis in YZW Space. (F) Suggested Coordinates. (Continued from page 385.)

SECONDARY ANALYSIS -- XYZ SPACE

PLOTTED POINTS

$P_1=(x_1,y_1,z_1)$ $P_2=(x_2,y_2,z_2)$

$P_3=(x_3,y_3,z_3)$ $P_4=(x_4,y_4,z_4)$

$P_5=(x_5,y_5,z_5)$ $P_6=(x_6,y_6,z_6)$

$P_7=(x_7,y_7,z_7)$ $P_8=(x_8,y_8,z_8)$

LENGTHS

$$s_1=\sqrt{(x_2-x_1)^2+(y_2-y_1)^2+(z_2-z_1)^2}$$

$$s_2=\sqrt{(x_3-x_2)^2+(y_3-y_2)^2+(z_3-z_2)^2}$$

$$s_3=\sqrt{(x_4-x_3)^2+(y_4-y_3)^2+(z_4-z_3)^2}$$

$$s_4=\sqrt{(x_1-x_4)^2+(y_1-y_4)^2+(z_1-z_4)^2}$$

$$s_5=\sqrt{(x_6-x_5)^2+(y_6-y_5)^2+(z_6-z_5)^2}$$

$$s_6=\sqrt{(x_7-x_6)^2+(y_7-y_6)^2+(z_7-z_6)^2}$$

$$s_7=\sqrt{(x_8-x_7)^2+(y_8-y_7)^2+(z_8-z_7)^2}$$

$$s_8=\sqrt{(x_5-x_8)^2+(y_5-y_8)^2+(z_5-z_8)^2}$$

$$s_9=\sqrt{(x_5-x_1)^2+(y_5-y_1)^2+(z_5-z_1)^2}$$

$$s_{10}=\sqrt{(x_6-x_2)^2+(y_6-y_2)^2+(z_6-z_2)^2}$$

$$s_{11}=\sqrt{(x_7-x_3)^2+(y_7-y_3)^2+(z_7-z_3)^2}$$

$$s_{12}=\sqrt{(x_8-x_4)^2+(y_8-y_4)^2+(z_8-z_4)^2}$$

DIRECTION COSINES

$\cos A_1=(x_2-x_1)/s_1$ $\cos B_1=(y_2-y_1)/s_1$

$\cos C_1=(z_2-z_1)/s_1$

$\cos A_2=(x_3-x_2)/s_2$ $\cos B_2=(y_3-y_2)/s_2$

$\cos C_2=(z_3-z_2)/s_2$

$\cos A_3=(x_4-x_3)/s_3$ $\cos B_3=(y_4-y_3)/s_3$

$\cos C_3=(z_4-z_3)/s_3$

$\cos A_4=(x_1-x_4)/s_4$ $\cos B_4=(y_1-y_4)/s_4$

$\cos C_4=(z_1-z_4)/s_4$

$\cos A_5=(x_6-x_5)/s_5$ $\cos B_5=(y_6-y_5)/s_5$

$\cos C_5=(z_6-z_5)/s_5$

$\cos A_6=(x_7-x_6)/s_6$ $\cos B_6=(y_7-y_6)/s_6$

$\cos C_6=(z_7-z_6)/s_6$

$\cos A_7=(x_8-x_7)/s_7$ $\cos B_7=(y_8-y_7)/s_7$

$\cos C_7=(z_8-z_7)/s_7$

$\cos A_8=(x_5-x_8)/s_8$ $\cos B_8=(y_5-y_8)/s_8$

$\cos C_8=(z_5-z_8)/s_8$

$\cos A_9=(x_5-x_1)/s_9$ $\cos B_9=(y_5-y_1)/s_9$

$\cos C_9=(z_5-z_1)/s_9$

$\cos A_{10}=(x_6-x_2)/s_{10}$ $\cos B_{10}=(y_6-y_2)/s_{10}$

$\cos C_{10}=(z_6-z_2)/s_{10}$

$\cos A_{11}=(x_7-x_3)/s_{11}$ $\cos B_{11}=(y_7-y_3)/s_{11}$

$\cos C_{11}=(z_7-z_3)/s_{11}$

$\cos A_{12}=(x_8-x_4)/s_{12}$ $\cos B_{12}=(y_8-y_4)/s_{12}$

$\cos B_{12}=(z_8-z_4)/s_{12}$

B

ANGLES

$$\cos\phi_{1,2}=\cos A_1\cos A_2+\cos B_1\cos B_2+\cos C_1\cos C_2$$

$$\cos\phi_{1,4}=\cos A_1\cos A_4+\cos B_1\cos B_4+\cos C_1\cos C_4$$

$$\cos\phi_{1,9}=\cos A_1\cos A_9+\cos B_1\cos B_9+\cos C_1\cos C_9$$

$$\cos\phi_{1,10}=\cos A_1\cos A_{10}+\cos B_1\cos B_{10}+\cos C_1\cos C_{10}$$

$$\cos\phi_{2,3}=\cos A_2\cos A_3+\cos B_2\cos B_3+\cos C_2\cos C_3$$

$$\cos\phi_{2,10}=\cos A_2\cos A_{10}+\cos B_2\cos B_{10}+\cos C_2\cos C_{10}$$

$$\cos\phi_{2,11}=\cos A_2\cos A_{11}+\cos B_2\cos B_{11}+\cos C_2\cos C_{11}$$

$$\cos\phi_{3,4}=\cos A_3\cos A_4+\cos B_3\cos B_4+\cos C_3\cos C_4$$

$$\cos\phi_{3,11}=\cos A_3\cos A_{11}+\cos B_3\cos B_{11}+\cos C_3\cos C_{11}$$

$$\cos\phi_{3,11}=\cos A_3\cos A_{11}+\cos B_3\cos B_{11}+\cos C_3\cos C_{11}$$

$$\cos\phi_{4,9}=\cos A_4\cos A_9+\cos B_4\cos B_9+\cos C_4\cos C_9$$

$$\cos\phi_{4,12}=\cos A_4\cos A_{12}+\cos B_4\cos B_{12}+\cos C_4\cos C_{12}$$

$$\cos\phi_{5,6}=\cos A_5\cos A_6+\cos B_5\cos B_6+\cos C_5\cos C_6$$

$$\cos\phi_{5,8}=\cos A_5\cos A_8+\cos B_5\cos B_8+\cos C_5\cos C_8$$

$$\cos\phi_{5,9}=\cos A_5\cos A_9+\cos B_5\cos B_9+\cos C_5\cos C_9$$

$$\cos\phi_{5,10}=\cos A_5\cos A_{10}+\cos B_5\cos B_{10}+\cos C_5\cos C_{10}$$

$$\cos\phi_{6,7}=\cos A_6\cos A_7+\cos B_6\cos B_7+\cos C_6\cos C_7$$

$$\cos\phi_{6,10}=\cos A_6\cos A_{10}+\cos B_6\cos B_{10}+\cos C_6\cos C_{10}$$

$$\cos\phi_{6,11}=\cos A_6\cos A_{11}+\cos B_6\cos B_{11}+\cos C_6\cos C_{11}$$

$$\cos\phi_{7,8}=\cos A_7\cos A_8+\cos B_7\cos B_8+\cos C_7\cos C_8$$

$$\cos\phi_{7,11}=\cos A_7\cos A_{11}+\cos B_7\cos B_{11}+\cos C_7\cos C_{11}$$

$$\cos\phi_{7,12}=\cos A_7\cos A_{12}+\cos B_7\cos B_{12}+\cos C_7\cos C_{12}$$

$$\cos\phi_{8,9}=\cos A_8\cos A_9+\cos B_8\cos B_9+\cos C_8\cos C_9$$

$$\cos\phi_{8,12}=\cos A_8\cos A_{12}+\cos B_8\cos B_{12}+\cos C_8\cos C_{12}$$

CHECK

F_1: $\phi_{1,2}+\phi_{1,4}+\phi_{2,3}+\phi_{3,4}=360^\circ$

F_2: $\phi_{1,9}+\phi_{1,10}+\phi_{5,9}+\phi_{5,10}=360^\circ$

F_3: $\phi_{2,10}+\phi_{2,11}+\phi_{6,10}+\phi_{6,11}=360^\circ$

F_4: $\phi_{3,11}+\phi_{3,12}+\phi_{7,11}+\phi_{7,12}=360^\circ$

F_5: $\phi_{4,9}+\phi_{4,12}+\phi_{8,9}+\phi_{8,12}=360^\circ$

F_6: $\phi_{5,6}+\phi_{5,8}+\phi_{6,7}+\phi_{7,8}=360^\circ$

Table 22-8. Specifications and Analyses of a 6-Sided Object in 4-D Space. See Fig. 22-8. (A) Specifications and Analysis in Hyperspace. (B) Secondary Analysis in XYZ Space. (C) Secondary Analysis in XYW Space. (D) Secondary Analysis in XZW Space. (E) Secondary Analysis in YZW Space. (F) Suggested Coordinates. (Continued from page 387.)

SECONDARY ANALYSIS -- XYW SPACE

PLOTTED POINTS

$P_1=(x_1,y_1,w_1)$ $P_2=(x_2,y_2,w_2)$

$P_3=(x_3,y_3,w_3)$ $P_4=(x_4,y_4,w_4)$

$P_5=(x_5,y_5,w_5)$ $P_6=(x_6,y_6,w_6)$

$P_7=(x_7,y_7,w_7)$ $P_8=(x_8,y_8,w_8)$

LENGTHS

$$s_1=\sqrt{(x_2-x_1)^2+(y_2-y_1)^2+(w_2-w_1)^2}$$

$$s_2=\sqrt{(x_3-x_2)^2+(y_3-y_2)^2+(w_3-w_2)^2}$$

$$s_3=\sqrt{(x_4-x_3)^2+(y_4-y_3)^2+(w_4-w_3)^2}$$

$$s_4=\sqrt{(x_1-x_4)^2+(y_1-y_4)^2+(w_1-w_4)^2}$$

$$s_5=\sqrt{(x_6-x_5)^2+(y_6-y_5)^2+(w_6-w_5)^2}$$

$$s_6=\sqrt{(x_7-x_6)^2+(y_7-y_6)^2+(w_7-w_6)^2}$$

$$s_7=\sqrt{(x_8-x_7)^2+(y_8-y_7)^2+(w_8-w_7)^2}$$

$$s_8=\sqrt{(x_5-x_8)^2+(y_5-y_8)^2+(w_5-w_8)^2}$$

$$s_9=\sqrt{(x_5-x_1)^2+(y_5-y_1)^2+(w_5-w_1)^2}$$

LENGTHS (Cont'd)

$$s_{10}=\sqrt{(x_6-x_2)^2+(y_6-y_2)^2+(w_6-w_2)^2}$$

$$s_{11}=\sqrt{(x_7-x_3)^2+(y_7-y_3)^2+(w_7-w_3)^2}$$

$$s_{12}=\sqrt{(x_8-x_4)^2+(y_8-y_4)^2+(w_8-w_4)^2}$$

DIRECTION COSINES

$\cos A_1=(x_2-x_1)/s_1$ $\cos B_1=(y_2-y_1)/s_1$
$\cos D_1=(w_2-w_1)/s_1$

$\cos A_2=(x_3-x_2)/s_2$ $\cos B_2=(y_3-y_2)/s_2$
$\cos D_2=(w_3-w_2)/s_2$

$\cos A_3=(x_4-x_3)/s_3$ $\cos B_3=(y_4-y_3)/s_3$
$\cos D_3=(w_4-w_3)/s_3$

$\cos A_4=(x_1-x_4)/s_4$ $\cos B_4=(y_1-y_4)/s_4$
$\cos D_4=(w_1-w_4)/s_4$

$\cos A_5=(x_6-x_5)/s_5$ $\cos B_5=(y_6-y_5)/s_5$
$\cos D_5=(w_6-w_5)/s_5$

$\cos A_6=(x_7-x_6)/s_6$ $\cos B_6=(y_7-y_6)/s_6$
$\cos D_6=(w_7-w_6)/s_6$

$\cos A_7=(x_8-x_7)/s_7$ $\cos B_7=(y_8-y_7)/s_7$
$\cos D_7=(w_8-w_7)/s_7$

$\cos A_8=(x_5-x_8)/s_8$ $\cos B_8=(y_5-y_8)/s_8$
$\cos D_8=(w_5-w_8)/s_8$

$\cos A_9=(x_5-x_1)/s_9$ $\cos B_9=(y_5-y_1)/s_9$
$\cos D_9=(w_5-w_1)/s_9$

$\cos A_{10}=(x_6-x_2)/s_{10}$ $\cos B_{10}=(y_6-y_2)/s_{10}$
$\cos D_{10}=(w_6-w_2)/s_{10}$

$\cos A_{11}=(x_7-x_3)/s_{11}$ $\cos B_{11}=(y_7-y_3)/s_{11}$
$\cos D_{11}=(w_7-w_3)/s_{11}$

$\cos A_{12}=(x_8-x_4)/s_{12}$ $\cos B_{12}=(y_8-y_4)/s_{12}$
$\cos D_{12}=(w_8-w_4)/s_{12}$

C

ANGLES

$\cos\phi_{1,2}=\cos A_1\cos A_2+\cos B_1\cos B_2+\cos D_1\cos D_2$

$\cos\phi_{1,4}=\cos A_1\cos A_4+\cos B_1\cos B_4+\cos D_1\cos D_4$

$\cos\phi_{1,9}=\cos A_1\cos A_9+\cos A_1\cos B_9+\cos B_1\cos D_9$

$\cos\phi_{1,10}=\cos A_1\cos A_{10}+\cos B_1\cos B_{10}+\cos D_1\cos D_{10}$

$\cos\phi_{2,3}=\cos A_2\cos A_3+\cos B_2\cos B_3+\cos D_2\cos D_3$

$\cos\phi_{2,10}=\cos A_2\cos A_{10}+\cos B_2\cos B_{10}+\cos D_2\cos D_{10}$

$\cos\phi_{2,11}=\cos A_2\cos A_{11}+\cos B_2\cos B_{11}+\cos D_2\cos D_{11}$

$\cos\phi_{3,4}=\cos A_3\cos A_4+\cos B_3\cos B_4+\cos D_3\cos D_4$

$\cos\phi_{3,11}=\cos A_3\cos A_{11}+\cos B_3\cos B_{11}+\cos D_3\cos D_{11}$

$\cos\phi_{3,11}=\cos A_3\cos A_{11}+\cos B_3\cos B_{11}+\cos D_3\cos D_{11}$

$\cos\phi_{4,9}=\cos A_4\cos A_9+\cos B_4\cos B_9+\cos D_4\cos D_9$

$\cos\phi_{4,12}=\cos A_4\cos A_{12}+\cos B_4\cos B_{12}+\cos D_4\cos D_{12}$

$\cos\phi_{5,6}=\cos A_5\cos A_6+\cos B_5\cos B_6+\cos D_5\cos D_6$

$\cos\phi_{5,8}=\cos A_5\cos A_8+\cos B_5\cos B_8+\cos D_5\cos D_8$

$\cos\phi_{5,9}=\cos A_5\cos A_9+\cos B_5\cos B_9+\cos D_5\cos D_9$

$\cos\phi_{5,10}=\cos A_5\cos A_{10}+\cos B_5\cos B_{10}+\cos D_5\cos D_{10}$

$\cos\phi_{6,7}=\cos A_6\cos A_7+\cos B_6\cos B_7+\cos D_6\cos D_7$

$\cos\phi_{6,10}=\cos A_6\cos A_{10}+\cos B_6\cos B_{10}+\cos D_6\cos D_{10}$

$\cos\phi_{6,11}=\cos A_6\cos A_{11}+\cos B_6\cos B_{11}+\cos D_6\cos D_{11}$

$\cos\phi_{7,8}=\cos A_7\cos A_8+\cos B_7\cos B_8+\cos D_7\cos D_8$

$\cos\phi_{7,11}=\cos A_7\cos A_{11}+\cos B_7\cos B_{11}+\cos D_7\cos D_{11}$

$\cos\phi_{7,12}=\cos A_7\cos A_{12}+\cos B_7\cos B_{12}+\cos D_7\cos D_{12}$

$\cos\phi_{8,9}=\cos A_8\cos A_9+\cos B_8\cos B_9+\cos D_8\cos D_9$

$\cos\phi_{8,12}=\cos A_8\cos A_{12}+\cos B_8\cos B_{12}+\cos D_8\cos D_{12}$

CHECK

F_1: $\phi_{1,2}+\phi_{1,4}+\phi_{2,3}+\phi_{3,4}=360^\circ$

F_2: $\phi_{1,9}+\phi_{1,10}+\phi_{5,9}+\phi_{5,10}=360^\circ$

F_3: $\phi_{2,10}+\phi_{2,11}+\phi_{6,10}+\phi_{6,11}=360^\circ$

F_4: $\phi_{3,11}+\phi_{3,12}+\phi_{7,11}+\phi_{7,12}=360^\circ$

F_5: $\phi_{4,9}+\phi_{4,12}+\phi_{8,9}+\phi_{8,12}=360^\circ$

F_6: $\phi_{5,6}+\phi_{5,8}+\phi_{6,7}+\phi_{7,8}=360^\circ$

C

SECONDARY ANALYSIS -- XZW SPACE

PLOTTED POINTS

$P_1=(x_1,z_1,w_1)$ $P_2=(x_2,z_2,w_2)$

$P_3=(x_3,z_3,w_3)$ $P_4=(x_4,z_4,w_4)$

$P_5=(x_5,z_5,w_5)$ $P_6=(x_6,z_6,w_6)$

$P_7=(x_7,z_7,w_7)$ $P_8=(x_8,z_8,w_8)$

LENGTHS

$s_1=\sqrt{(x_2-x_1)^2+(z_2-z_1)^2+(w_2-w_1)^2}$

$s_2=\sqrt{(x_3-x_2)^2+(z_3-z_2)^2+(w_3-w_2)^2}$

$s_3=\sqrt{(x_4-x_3)^2+(z_4-z_3)^2+(w_4-w_3)^2}$

$s_4=\sqrt{(x_1-x_4)^2+(z_1-z_4)^2+(w_1-w_4)^2}$

LENGTHS (Cont'd)

$s_5=\sqrt{(x_6-x_5)^2+(z_6-z_5)^2+(w_6-w_5)^2}$

$s_6=\sqrt{(x_7-x_6)^2+(z_7-z_6)^2+(w_7-w_6)^2}$

$s_7=\sqrt{(x_8-x_7)^2+(z_8-z_7)^2+(w_8-w_7)^2}$

$s_8=\sqrt{(x_5-x_8)^2+(z_5-z_8)^2+(w_5-w_8)^2}$

$s_9=\sqrt{(x_5-x_1)^2+(z_5-z_1)^2+(w_5-w_1)^2}$

$s_{10}=\sqrt{(x_6-x_2)^2+(z_6-z_2)^2+(w_6-w_2)^2}$

$s_{11}=\sqrt{(x_7-x_3)^2+(z_7-z_3)^2+(w_7-w_3)^2}$

$s_{12}=\sqrt{(x_8-x_4)^2+(z_8-z_4)^2+(w_8-w_4)^2}$

D

Table 22-8. Specifications and Analyses of a 6-Sided Object in 4-D Space. See Fig. 22-8. (A) Specifications and Analysis in Hyperspace. (B) Secondary Analysis in XYZ Space. (C) Secondry Analysis in XYW Space. (D) Secondary Analysis in XZW Space. (E) Secondary Analysis in YZW Space. (F) Suggested Coordinates. (Continued from page 389.)

DIRECTION COSINES

$\cos A_1=(x_2-x_1)/s_1$ $\quad \cos C_1=(z_2-z_1)/s_1$

$\cos D_1=(w_2-w_1)/s_1$

$\cos A_2=(x_3-x_2)/s_2$ $\quad \cos C_2=(z_3-z_2)/s_2$

$\cos D_2=(w_3-w_2)/s_2$

$\cos A_3=(x_4-x_3)/s_3$ $\quad \cos C_3=(z_4-z_3)/s_3$

$\cos D_3=(w_4-w_3)/s_3$

$\cos A_4=(x_1-x_4)/s_4$ $\quad \cos C_4=(z_1-z_4)/s_4$

$\cos D_4=(w_1-w_4)/s_4$

$\cos A_5=(x_6-x_5)/s_5$ $\quad \cos C_5=(z_6-z_5)/s_5$

$\cos D_5=(w_6-w_5)/s_5$

$\cos A_6=(x_7-x_6)/s_6$ $\quad \cos C_6=(z_7-z_6)/s_6$

$\cos D_6=(w_7-w_6)/s_6$

$\cos A_7=(x_8-x_7)/s_7$ $\quad \cos C_7=(z_8-z_7)/s_7$

$\cos D_7=(w_8-w_7)/s_7$

$\cos A_8=(x_5-x_8)/s_8$ $\quad \cos C_8=(z_5-z_8)/s_8$

$\cos D_8=(w_5-w_8)/s_8$

$\cos A_9=(x_5-x_1)/s_9$ $\quad \cos C_9=(z_5-z_1)/s_9$

$\cos D_9=(w_5-w_1)/s_9$

$\cos A_{10}=(x_6-x_2)/s_{10}$ $\quad \cos C_{10}=(z_6-z_2)/s_{10}$

$\cos D_{10}=(w_6-w_2)/s_{10}$

$\cos A_{11}=(x_7-x_3)/s_{11}$ $\quad \cos C_{11}=(z_7-z_3)/s_{11}$

$\cos D_{11}=(w_7-w_3)/s_{11}$

$\cos A_{12}=(x_8-x_4)/s_{12}$ $\quad \cos C_{12}=(z_8-z_4)/s_{12}$

$\cos D_{12}=(w_8-w_4)/s_{12}$

ANGLES

$\cos\phi_{1,2}=\cos A_1\cos A_2+\cos C_1\cos C_2+\cos D_1\cos D_2$

$\cos\phi_{1,4}=\cos A_1\cos A_4+\cos C_1\cos C_4+\cos D_1\cos D_4$

$\cos\phi_{1,9}=\cos A_1\cos A_9+\cos C_1\cos C_9+\cos D_1\cos D_9$

$\cos\phi_{1,10}=\cos A_1\cos A_{10}+\cos C_1\cos C_{10}+\cos D_1\cos D_{10}$

$\cos\phi_{2,3}=\cos A_2\cos A_3+\cos C_2\cos C_3+\cos D_2\cos D_3$

$\cos\phi_{2,10}=\cos A_2\cos A_{10}+\cos C_2\cos C_{10}+\cos D_2\cos D_{10}$

$\cos\phi_{2,11}=\cos A_2\cos A_{11}+\cos C_2\cos C_{11}+\cos D_2\cos D_{11}$

$\cos\phi_{3,4}=\cos A_3\cos A_4+\cos C_3\cos C_4+\cos D_3\cos D_4$

$\cos\phi_{3,11}=\cos A_3\cos A_{11}+\cos C_3\cos C_{11}+\cos D_3\cos D_{11}$

$\cos\phi_{3,11}=\cos A_3\cos A_{11}+\cos C_3\cos C_{11}+\cos D_3\cos D_{11}$

$\cos\phi_{4,9}=\cos A_4\cos A_9+\cos C_4\cos C_9+\cos D_4\cos D_9$

$\cos\phi_{4,12}=\cos A_4\cos A_{12}+\cos C_4\cos C_{12}+\cos D_4\cos D_{12}$

$\cos\phi_{5,6}=\cos A_5\cos A_6+\cos C_5\cos C_6+\cos D_5\cos D_6$

$\cos\phi_{5,8}=\cos A_5\cos A_8+\cos C_5\cos C_8+\cos D_5\cos D_8$

$\cos\phi_{5,9}=\cos A_5\cos A_9+\cos C_5\cos C_9+\cos D_5\cos D_9$

$\cos\phi_{5,10}=\cos A_5\cos A_{10}+\cos C_5\cos C_{10}+\cos D_5\cos D_{10}$

$\cos\phi_{6,7}=\cos A_6\cos A_7+\cos C_6\cos C_7+\cos D_6\cos D_7$

$\cos\phi_{6,10}=\cos A_6\cos A_{10}+\cos C_6\cos C_{10}+\cos D_6\cos D_{10}$

$\cos\phi_{6,11}=\cos A_6\cos A_{11}+\cos C_6\cos C_{11}+\cos D_6\cos D_{11}$

$\cos\phi_{7,8}=\cos A_7\cos A_8+\cos C_7\cos C_8+\cos D_7\cos D_8$

$\cos\phi_{7,11}=\cos A_7\cos A_{11}+\cos C_7\cos C_{11}+\cos D_7\cos D_{11}$

$\cos\phi_{7,12}=\cos A_7\cos A_{12}+\cos C_7\cos C_{12}+\cos D_7\cos D_{12}$

$\cos\phi_{8,9}=\cos A_8\cos A_9+\cos C_8\cos C_9+\cos D_8\cos D_9$

$\cos\phi_{8,12}=\cos A_8\cos A_{12}+\cos C_8\cos C_{12}+\cos D_8\cos D_{12}$

CHECK

F_1: $\phi_{1,2}+\phi_{1,4}+\phi_{2,3}+\phi_{3,4}=360^\circ$

F_2: $\phi_{1,9}+\phi_{1,10}+\phi_{5,9}+\phi_{5,10}=360^\circ$

F_3: $\phi_{2,10}+\phi_{2,11}+\phi_{6,10}+\phi_{6,11}=360^\circ$

F_4: $\phi_{3,11}+\phi_{3,12}+\phi_{7,11}+\phi_{7,12}=360^\circ$

F_5: $\phi_{4,9}+\phi_{4,12}+\phi_{8,9}+\phi_{8,12}=360^\circ$

F_6: $\phi_{5,6}+\phi_{5,8}+\phi_{6,7}+\phi_{7,8}=360^\circ$

D

SECONDARY ANALYSIS -- YZW SPACE

PLOTTED POINTS

$P_1=(y_1,z_1,w_1)$ $P_2=(y_2,z_2,w_2)$

$P_3=(y_3,z_3,w_3)$ $P_4=(y_4,z_4,w_4)$

$P_5=(y_5,z_5,w_5)$ $P_6=(y_6,z_6,w_6)$

$P_7=(y_7,z_7,w_7)$ $P_8=(y_8,z_8,w_8)$

LENGTHS

$$s_1=\sqrt{(y_2-y_1)^2+(z_2-z_1)^2+(w_2-w_1)^2}$$

$$s_2=\sqrt{(y_3-y_2)^2+(z_3-z_2)^2+(w_3-w_2)^2}$$

$$s_3=\sqrt{(y_4-y_3)^2+(z_4-z_3)^2+(w_4-w_3)^2}$$

$$s_4=\sqrt{(y_1-y_4)^2+(z_1-z_4)^2+(w_1-w_4)^2}$$

$$s_5=\sqrt{(y_6-y_5)^2+(z_6-z_5)^2+(w_6-w_5)^2}$$

$$s_6=\sqrt{(y_7-y_6)^2+(z_7-z_6)^2+(w_7-w_6)^2}$$

$$s_7=\sqrt{(y_8-y_7)^2+(z_8-z_7)^2+(w_8-w_7)^2}$$

$$s_8=\sqrt{(y_5-y_8)^2+(z_5-z_8)^2+(w_5-w_8)^2}$$

$$s_9=\sqrt{(y_5-y_1)^2+(z_5-z_1)^2+(w_5-w_1)^2}$$

$$s_{10}=\sqrt{(y_6-y_2)^2+(z_6-z_2)^2+(w_6-w_2)^2}$$

$$s_{11}=\sqrt{(y_7-y_3)^2+(z_7-z_3)^2+(w_7-w_3)^2}$$

$$s_{12}=\sqrt{(y_8-y_4)^2+(z_8-z_4)^2+(w_8-w_4)^2}$$

DIRECTION COSINES

$\cos B_1=(y_2-y_1)/s_1$ $\cos C_1=(z_2-z_1)/s_1$

$\cos D_1=(w_2-w_1)/s_1$

$\cos B_2=(y_3-y_2)/s_2$ $\cos C_2=(z_3-z_2)/s_2$

$\cos D_2=(w_3-w_2)/s_2$

DIRECTION COSINES (Cont'd)

$\cos B_3=(y_4-y_3)/s_3$ $\cos C_3=(z_4-z_3)/s_3$

$\cos D_3=(w_4-w_3)/s_3$

$\cos B_4=(y_1-y_4)/s_4$ $\cos C_4=(z_1-z_4)/s_4$

$\cos D_4=(w_1-w_4)/s_4$

$\cos B_5=(y_6-y_5)/s_5$ $\cos C_5=(z_6-z_5)/s_5$

$\cos D_5=(w_6-w_5)/s_5$

$\cos B_6=(y_7-y_6)/s_6$ $\cos C_6=(z_7-z_6)/s_6$

$\cos D_6=(w_7-w_6)/s_6$

$\cos B_7=(y_8-y_7)/s_7$ $\cos C_7=(z_8-z_7)/s_7$

$\cos D_7=(w_8-w_7)/s_7$

$\cos B_8=(y_5-y_8)/s_8$ $\cos C_8=(z_5-z_8)/s_8$

$\cos D_8=(w_5-w_8)/s_8$

$\cos B_9=(y_5-y_1)/s_9$ $\cos C_9=(z_5-z_1)/s_9$

$\cos D_9=(w_5-w_1)/s_9$

$\cos B_{10}=(y_6-y_2)/s_{10}$ $\cos C_{10}=(z_6-z_2)/s_{10}$

$\cos D_{10}=(w_6-w_2)/s_{10}$

$\cos B_{11}=(y_7-y_3)/s_{11}$ $\cos C_{11}=(z_7-z_3)/s_{11}$

$\cos D_{11}=(w_7-w_3)/s_{11}$

$\cos B_{12}=(y_8-y_4)/s_{12}$ $\cos C_{12}=(z_8-z_4)/s_{12}$

$\cos D_{12}=(w_8-w_4)/s_{12}$

ANGLES

$\cos\phi_{1,2}=\cos B_1\cos B_2+\cos C_1\cos C_2+\cos D_1\cos D_2$

$\cos\phi_{1,4}=\cos B_1\cos B_4+\cos C_1\cos C_4+\cos D_1\cos D_4$

$\cos\phi_{1,9}=\cos B_1\cos B_9+\cos C_1\cos C_9+\cos D_1\cos D_9$

$\cos\phi_{1,10}=\cos B_1\cos B_{10}+\cos C_1\cos C_{10}+\cos D_1\cos D_{10}$

$\cos\phi_{2,3}=\cos B_2\cos B_3+\cos C_2\cos C_3+\cos D_2\cos D_3$

$\cos\phi_{2,10}=\cos B_2\cos B_{10}+\cos C_2\cos C_{10}+\cos D_2\cos D_{10}$

E

Table 22-8. Specifications and Analyses of a 6-Sided Object in 4-D Space. See Fig. 22-8. (A) Specifications and Analysis in Hyperspace. (B) Secondary Analysis in XYZ Space. (C) Secondary Analysis in XYW Space. (D) Secondary Analysis in XZW Space. (E) Secondary Analysis in YZW Space. (F) Suggested Coordinates. (Continued from page 391.)

ANGLES (Cont'd)

$\cos\phi_{2,11}=\cos B_2\cos B_{11}+\cos C_2\cos C_{11}+\cos D_2\cos D_{11}$

$\cos\phi_{3,4}=\cos B_3\cos B_4+\cos C_3\cos C_4+\cos D_3\cos D_4$

$\cos\phi_{3,11}=\cos B_3\cos B_{11}+\cos C_3\cos C_{11}+\cos D_3\cos D_{11}$

$\cos\phi_{3,11}=\cos B_3\cos B_{11}+\cos C_3\cos C_{11}+\cos D_3\cos D_{11}$

$\cos\phi_{4,9}=\cos B_4\cos B_9+\cos C_4\cos C_9+\cos D_4\cos D_9$

$\cos\phi_{4,12}=\cos B_4\cos B_{12}+\cos C_4\cos C_{12}+\cos D_4\cos D_{12}$

$\cos\phi_{5,6}=\cos B_5\cos B_6+\cos C_5\cos C_6+\cos D_5\cos D_6$

$\cos\phi_{5,8}=\cos B_5\cos B_8+\cos C_5\cos C_8+\cos D_5\cos D_8$

$\cos\phi_{5,9}=\cos B_5\cos B_9+\cos C_5\cos C_9+\cos D_5\cos D_9$

$\cos\phi_{5,10}=\cos B_5\cos B_{10}+\cos C_5\cos C_{10}+\cos D_5\cos D_{10}$

$\cos\phi_{6,7}=\cos B_6\cos B_7+\cos C_6\cos C_7+\cos D_6\cos D_7$

$\cos\phi_{6,10}=\cos B_6\cos C_{10}+\cos C_6\cos D_{10}+\cos D_6\cos D_{10}$

$\cos\phi_{6,11}=\cos B_6\cos B_{11}+\cos C_6\cos C_{11}+\cos D_6\cos D_{11}$

$\cos\phi_{7,8}=\cos B_7\cos B_8+\cos C_7\cos C_8+\cos D_7\cos D_8$

$\cos\phi_{7,11}=\cos B_7\cos B_{11}+\cos C_7\cos C_{11}+\cos D_7\cos D_{11}$

$\cos\phi_{7,12}=\cos B_7\cos B_{12}+\cos C_7\cos C_{12}+\cos D_7\cos D_{12}$

$\cos\phi_{8,9}=\cos B_8\cos B_9+\cos C_8\cos C_9+\cos D_8\cos D_9$

E $\cos\phi_{8,12}=\cos B_8\cos B_{12}+\cos C_8\cos C_{12}+\cos D_8\cos D_{12}$

CHECK

F_1: $\phi_{1,2}+\phi_{1,4}+\phi_{2,3}+\phi_{3,4}=360^\circ$

F_2: $\phi_{1,9}+\phi_{1,10}+\phi_{5,9}+\phi_{5,10}=360^\circ$

F_3: $\phi_{2,10}+\phi_{2,11}+\phi_{6,10}+\phi_{6,11}=360^\circ$

F_4: $\phi_{3,11}+\phi_{3,12}+\phi_{7,11}+\phi_{7,12}=360^\circ$

F_5: $\phi_{4,9}+\phi_{4,12}+\phi_{8,9}+\phi_{8,12}=360^\circ$

F_6: $\phi_{5,6}+\phi_{5,8}+\phi_{6,7}+\phi_{7,8}=360^\circ$

HYPERSPACE OBJECT VERSION 1

$P_1=(-1,-1,1,-1)$ $P_2=(1,-1,1,-1)$ $P_3=(1,1,1,1)$

$P_4=(-1,1,1,1)$ $P_5=(-1,-1,-1,-1)$ $P_6=(1,-1,-1,-1)$

$P_7=(1,1,-1,1)$ $P_8=(-1,1,-1,1)$

HYPERSPACE OBJECT VERSION 2

$P_1=(-1,-1,1,-1)$ $P_2=(1,-1,1,-1)$ $P_3=(2,1,1,1)$

$P_4=(0,1,1,1)$ $P_5=(-1,-1,-1,-1)$ $P_6=(1,-1,-1,-1)$

$P_7=(2,1,-1,1)$ $P_8=(0,1,-1,1)$

HYPERSPACE OBJECT VERSION 3

$P_1=(-1,-2,0,-2)$ $P_2=(1,-2,1,-2)$ $P_3=(2,0,1,0)$

$P_4=(0,0,1,0)$ $P_5=(-1,0,-1,0)$ $P_6=(1,0,-1,0)$

$P_7=(2,2,0,2)$ $P_8=(0,2,-1,2)$

F

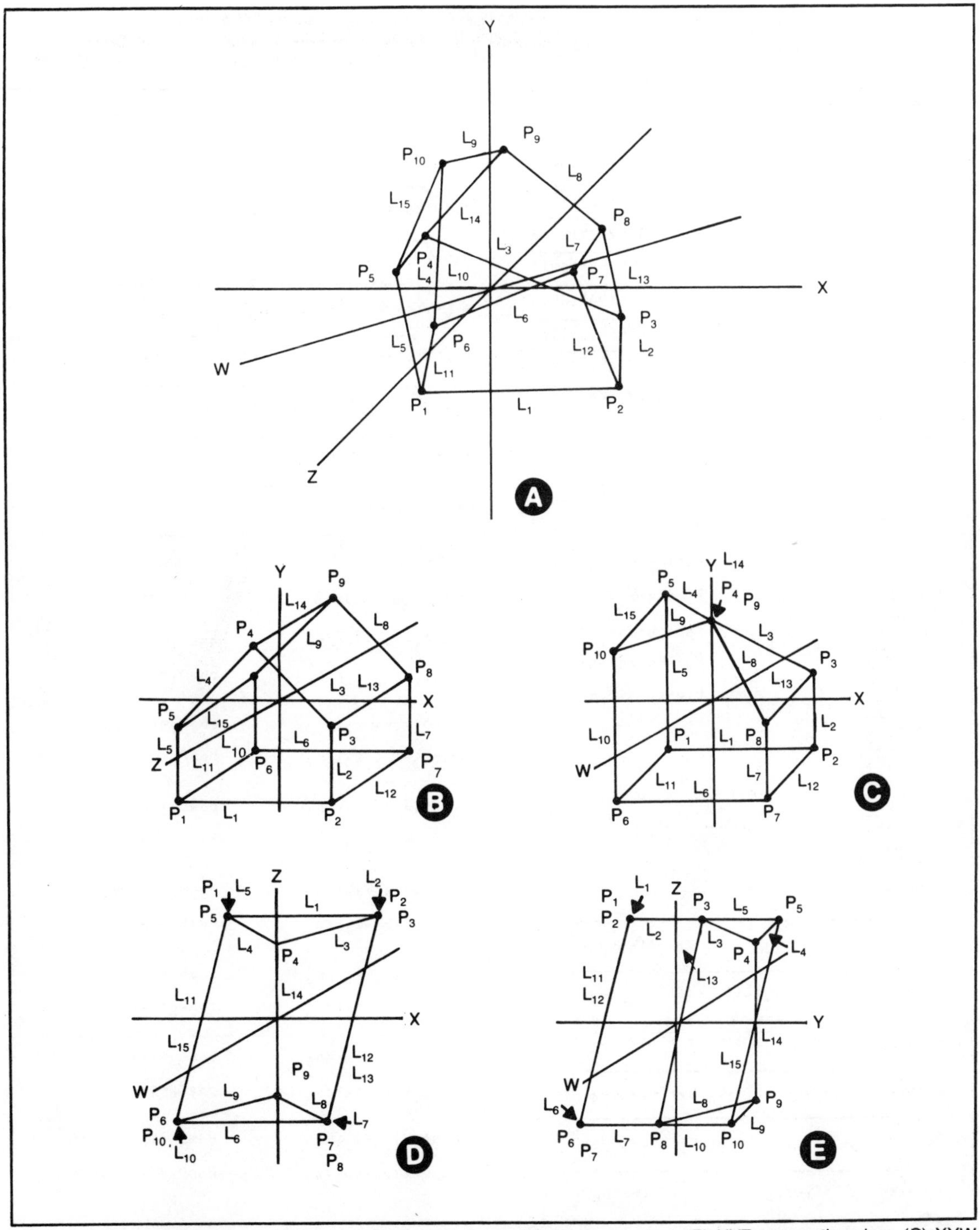

Fig. 22-9. A 7-sided object specified in 4-D space. (A) General hyperspace view. (B) XYZ perspective view. (C) XYW perspective view. (D) XZW perspective view. (E) YZW perspective view. See outlines of analyses in Table 22-9.

Table 22-9. Specifications and Analyses of a 7-Sided Object in 4-D Space. See Fig. 22-9. (A) Specifications and Analysis in Hyperspace. (B) Secondary Analysis in XYZ Space. (C) Secondary Analysis in XYW Space. (D) Secondary Analysis in XZW Space. (E) Secondary Analysis in YZW Space. (F) Suggested Coordinates.

(A, B, C, and D correspond to α, β, γ, and δ.)

SPECIFICATIONS

$S=\overline{F_1F_2F_3F_4F_5F_6F_7}$

$F_1=\overline{L_1L_2L_3L_4L_5}$

$F_2=\overline{L_1L_6L_{11}L_{12}}$

$F_3=\overline{L_2L_7L_{12}L_{13}}$

$F_4=\overline{L_3L_8L_{13}L_{14}}$

$F_5=\overline{L_4L_9L_{14}L_{15}}$

$F_6=\overline{L_5L_{10}L_{11}L_{15}}$

$F_7=\overline{L_6L_7L_8L_9L_{10}}$

$L_1=\overline{P_1P_2}$ $P_1=(x_1,y_1,z_1,w_1)$

$L_2=\overline{P_2P_3}$ $P_2=(x_2,y_2,z_2,w_2)$

$L_3=\overline{P_3P_4}$ $P_3=(x_3,y_3,z_3,w_3)$

$L_4=\overline{P_4P_5}$ $P_4=(x_4,y_4,z_4,w_4)$

$L_5=\overline{P_5P_1}$ $P_5=(x_5,y_5,z_5,w_5)$

$L_6=\overline{P_6P_7}$ $P_6=(x_6,y_6,z_6,w_6)$

$L_7=\overline{P_7P_8}$ $P_7=(x_7,y_7,z_7,w_7)$

$L_8=\overline{P_8P_9}$ $P_8=(x_8,y_8,z_8,w_8)$

$L_9=\overline{P_9P_{10}}$ $P_9=(x_9,y_9,z_9,w_9)$

$L_{10}=\overline{P_{10}P_6}$ $P_{10}=(x_{10},y_{10},z_{10},w_{10})$

$L_{11}=\overline{P_1P_6}$

$L_{12}=\overline{P_2P_7}$

$L_{13}=\overline{P_3P_8}$

$L_{14}=\overline{P_4P_9}$

$L_{15}=\overline{P_5P_{10}}$

LENGTHS

$$s_1=\sqrt{(x_2-x_1)^2+(y_2-y_1)^2+(z_2-z_1)^2+(w_2-w_1)^2}$$

$$s_2=\sqrt{(x_3-x_2)^2+(y_3-y_2)^2+(z_3-z_2)^2+(w_3-w_2)^2}$$

$$s_3=\sqrt{(x_4-x_3)^2+(y_4-y_3)^2+(z_4-z_3)^2+(w_4-w_3)^2}$$

$$s_4=\sqrt{(x_5-x_4)^2+(y_5-y_4)^2+(z_5-z_4)^2+(w_5-w_4)^2}$$

$$s_5=\sqrt{(x_1-x_5)^2+(y_1-y_5)^2+(z_1-z_5)^2+(w_1-w_5)^2}$$

$$s_6=\sqrt{(x_7-x_6)^2+(y_7-y_6)^2+(z_7-z_6)^2+(w_7-w_6)^2}$$

A

LENGTHS (Cont'd)

$$s_7=\sqrt{(x_8-x_7)^2+(y_8-y_7)^2+(z_8-z_7)^2+(w_8-w_7)^2}$$

$$s_8=\sqrt{(x_9-x_8)^2+(y_9-y_8)^2+(z_9-z_8)^2+(w_9-w_8)^2}$$

$$s_9=\sqrt{(x_{10}-x_9)^2+(y_{10}-y_9)^2+(z_{10}-z_9)^2+(w_{10}-w_9)^2}$$

$$s_{10}=\sqrt{(x_6-x_{10})^2+(y_6-y_{10})^2+(z_6-z_{10})^2+(w_6-w_{10})^2}$$

$$s_{11}=\sqrt{(x_6-x_1)^2+(y_6-y_1)^2+(z_6-z_1)^2+(w_6-w_1)^2}$$

$$s_{12}=\sqrt{(x_7-x_2)^2+(y_7-y_2)^2+(z_7-z_2)^2+(w_7-w_2)^2}$$

$$s_{13}=\sqrt{(x_8-x_3)^2+(y_8-y_3)^2+(z_8-z_3)^2+(w_8-w_3)^2}$$

$$s_{14}=\sqrt{(x_9-x_4)^2+(y_9-y_4)^2+(z_9-z_4)^2+(w_9-w_4)^2}$$

$$s_{15}=\sqrt{(x_{10}-x_5)^2+(y_{10}-y_5)^2+(z_{10}-z_5)^2+(w_{10}-w_5)^2}$$

DIRECTION COSINES

$\cos A_1=(x_2-x_1)/s_1$ $\cos B_1=(y_2-y_1)/s_1$

$\cos C_1=(z_2-z_1)/s_1$ $\cos D_1=(w_2-w_1)/s_1$

$\cos A_2=(x_3-x_2)/s_2$ $\cos B_2=(y_3-y_2)/s_2$

$\cos C_2=(z_3-z_2)/s_2$ $\cos D_2=(w_3-w_2)/s_2$

$\cos A_3=(x_4-x_3)/s_3$ $\cos B_3=(y_4-y_3)/s_3$

$\cos C_3=(z_4-z_3)/s_3$ $\cos D_3=(w_4-w_3)/s_3$

$\cos A_4=(x_5-x_4)/s_4$ $\cos B_4=(y_5-y_4)/s_4$

$\cos C_4=(z_5-z_4)/s_4$ $\cos D_4=(w_5-w_4)/s_4$

$\cos A_5=(x_1-x_5)/s_5$ $\cos B_5=(y_1-y_5)/s_5$

$\cos C_5=(z_1-z_5)/s_5$ $\cos D_5=(w_1-w_5)/s_5$

$\cos A_6=(x_7-x_6)/s_6$ $\cos B_6=(y_7-y_6)/s_6$

$\cos C_6=(z_7-z_6)/s_6$ $\cos D_6=(w_7-w_6)/s_6$

$\cos A_7=(x_8-x_7)/s_7$ $\cos B_7=(y_8-y_7)/s_7$

$\cos C_7=(z_8-z_7)/s_7$ $\cos D_7=(w_8-w_7)/s_7$

DIRECTION COSINES (Cont'd)

$\cos A_8=(x_9-x_8)/s_8$ $\cos B_8=(y_9-y_8)/s_8$

$\cos C_8=(z_9-z_8)/s_8$ $\cos D_8=(w_9-w_8)/s_8$

$\cos A_9=(x_{10}-x_9)/s_9$ $\cos B_9=(y_9-y_9)/s_9$

$\cos C_9=(z_{10}-z_9)/s_9$ $\cos D_9=(w_9-w_9)/s_9$

$\cos A_{10}=(x_6-x_{10})/s_{10}$ $\cos B_{10}=(y_6-y_{10})/s_{10}$

$\cos C_{10}=(z_6-z_{10})/s_{10}$ $\cos D_{10}=(w_6-w_{10})/s_{10}$

$\cos A_{11}=(x_6-x_1)/s_{11}$ $\cos B_{11}=(y_6-y_1)/s_{11}$

$\cos C_{11}=(z_6-z_1)/s_{11}$ $\cos D_{11}=(w_6-w_1)/s_{11}$

$\cos A_{12}=(x_7-x_2)/s_{12}$ $\cos B_{12}=(y_7-y_2)/s_{12}$

$\cos C_{12}=(z_7-z_2)/s_{12}$ $\cos D_{12}=(w_7-w_2)/s_{12}$

$\cos A_{13}=(x_8-x_3)/s_{13}$ $\cos B_{13}=(y_8-y_3)/s_{13}$

$\cos C_{13}=(z_8-z_3)/s_{13}$ $\cos D_{13}=(w_8-w_3)/s_{13}$

$\cos A_{14}=(x_9-x_4)/s_{14}$ $\cos B_{14}=(y_9-y_4)/s_{14}$

$\cos B_{14}=(z_9-z_4)/s_{14}$ $\cos C_{14}=(w_9-w_4)/s_{14}$

$\cos A_{15}=(x_{10}-x_5)/s_{15}$ $\cos B_{15}=(y_{10}-y_5)/s_{15}$

$\cos C_{15}=(z_{10}-z_5)/s_{15}$ $\cos D_{15}=(w_{10}-w_5)/s_{15}$

ANGLES

$\cos\phi_{1,2}=\cos A_1\cos A_2+\cos B_1\cos B_2+\cos C_1\cos C_2+\cos D_1\cos D_2$

$\cos\phi_{1,5}=\cos A_1\cos A_5+\cos B_1\cos B_5+\cos C_1\cos C_5+\cos D_1\cos D_5$

$\cos\phi_{1,11}=\cos A_1\cos A_{11}+\cos B_1\cos B_{11}+\cos C_1\cos C_{11}+\cos D_1\cos D_{11}$

$\cos\phi_{1,12}=\cos A_1\cos A_{12}+\cos B_1\cos B_{12}+\cos C_1\cos C_{12}+\cos D_1\cos D_{12}$

$\cos\phi_{2,3}=\cos A_2\cos A_3+\cos B_2\cos B_3+\cos C_2\cos C_3+\cos D_2\cos D_3$

$\cos\phi_{2,12}=\cos A_2\cos A_{12}+\cos B_2\cos B_{12}+\cos C_2\cos C_{12}+\cos D_2\cos D_{12}$

$\cos\phi_{2,13}=\cos A_2\cos A_{13}+\cos B_2\cos B_{13}+\cos C_2\cos C_{13}+\cos D_2\cos D_{13}$

$\cos\phi_{3,4}=\cos A_3\cos A_4+\cos B_3\cos B_4+\cos C_3\cos C_4+\cos D_3\cos D_4$

$\cos\phi_{3,13}=\cos A_3\cos A_{13}+\cos B_3\cos B_{13}+\cos C_3\cos C_{13}+\cos D_3\cos D_{13}$

Table 22-9. Specifications and Analyses of a 7-Sided Object in 4-D Space. See Fig. 22-9. (A) Specifications and Analysis in Hyperspace. (B) Secondary Analysis in XYZ Space. (C) Secondary Analysis in XYW Space. (D) Secondary Analysis in XZW Space. (E) Secondary Analysis in YZW Space. (F) Suggested in Coordinates. (Continued from page 395.)

ANGLES (Con't'd)

$$\cos\phi_{3,14}=\cos A_3\cos A_{14}+\cos B_3\cos B_{14}+\cos C_3\cos C_{14}+\cos D_3\cos D_{14}$$

$$\cos\phi_{4,5}=\cos A_4\cos A_5+\cos B_4\cos B_5+\cos C_4\cos C_5+\cos D_4\cos D_5$$

$$\cos\phi_{4,14}=\cos A_4\cos A_{14}+\cos B_4\cos B_{14}+\cos C_4\cos C_{14}+\cos D_4\cos D_{14}$$

$$\cos\phi_{4,15}=\cos A_4\cos A_{15}+\cos B_4\cos B_{15}+\cos C_4\cos C_{15}+\cos D_4\cos D_{15}$$

$$\cos\phi_{5,11}=\cos A_5\cos A_{11}+\cos B_5\cos B_{11}+\cos C_5\cos C_{11}+\cos D_5\cos D_{15}$$

$$\cos\phi_{5,15}=\cos A_5\cos A_{15}+\cos B_5\cos B_{15}+\cos C_5\cos C_{15}+\cos D_5\cos D_{15}$$

$$\cos\phi_{6,7}=\cos A_6\cos A_7+\cos B_6\cos B_7+\cos C_6\cos C_7+\cos D_6\cos D_7$$

$$\cos\phi_{6,10}=\cos A_6\cos A_{10}+\cos B_6\cos B_{10}+\cos C_6\cos C_{10}+\cos D_6\cos D_{10}$$

$$\cos\phi_{6,11}=\cos A_6\cos A_{11}+\cos B_6\cos B_{11}+\cos C_6\cos C_{11}+\cos D_6\cos D_{11}$$

$$\cos\phi_{6,12}=\cos A_6\cos A_{12}+\cos B_6\cos B_{12}+\cos C_6\cos C_{12}+\cos D_6\cos D_{12}$$

$$\cos\phi_{7,8}=\cos A_7\cos A_8+\cos B_7\cos B_8+\cos C_7\cos C_8+\cos D_7\cos D_8$$

$$\cos\phi_{7,12}=\cos A_7\cos A_{12}+\cos B_7\cos B_{12}+\cos C_7\cos C_{12}+\cos D_7\cos D_{12}$$

$$\cos\phi_{7,13}=\cos A_7\cos A_{13}+\cos B_7\cos B_{13}+\cos C_7\cos C_{13}+\cos D_7\cos D_{13}$$

$$\cos\phi_{8,9}=\cos A_8\cos A_9+\cos A_8\cos B_9+\cos B_8\cos C_9+\cos C_8\cos D_9$$

$$\cos\phi_{8,13}=\cos A_8\cos A_{13}+\cos B_8\cos B_{13}+\cos C_8\cos C_{13}+\cos D_8\cos D_{13}$$

$$\cos\phi_{8,14}=\cos A_8\cos A_{14}+\cos B_8\cos B_{14}+\cos C_8\cos C_{14}+\cos D_8\cos D_{14}$$

$$\cos\phi_{9,10}=\cos A_9\cos A_{10}+\cos B_9\cos B_{10}+\cos C_9\cos C_{10}+\cos D_9\cos D_{14}$$

$$\cos\phi_{9,14}=\cos A_9\cos A_{14}+\cos B_9\cos B_{14}+\cos C_9\cos C_{14}+\cos D_9\cos D_{14}$$

$$\cos\phi_{9,15}=\cos A_9\cos A_{15}+\cos B_9\cos B_{15}+\cos C_9\cos C_{15}+\cos D_9\cos D_{15}$$

$$\cos\phi_{10,11}=\cos A_{10}\cos A_{11}+\cos B_{10}\cos B_{11}+\cos C_{10}\cos C_{11}+\cos D_{10}\cos D_{11}$$

$$\cos\phi_{10,15}=\cos A_{10}\cos A_{15}+\cos B_{10}\cos B_{15}+\cos C_{10}\cos C_{15}+\cos D_{10}\cos D_{15}$$

CHECK

F_1: $\phi_{1,2}+\phi_{1,5}+\phi_{2,3}+\phi_{3,4}+\phi_{4,5}=540^\circ$

F_2: $\phi_{1,12}+\phi_{1,11}+\phi_{6,11}+\phi_{6,12}=360^\circ$

F_3: $\phi_{2,12}+\phi_{2,13}+\phi_{7,12}+\phi_{7,13}=360^\circ$

F_4: $\phi_{3,13}+\phi_{3,14}+\phi_{8,13}+\phi_{8,14}=360^\circ$

F_5: $\phi_{4,14}+\phi_{4,15}+\phi_{9,14}+\phi_{9,15}=360^\circ$

F_6: $\phi_{5,11}+\phi_{5,15}+\phi_{10,11}+\phi_{10,15}=360^\circ$

F_7: $\phi_{6,7}+\phi_{6,10}+\phi_{7,8}+\phi_{8,9}+\phi_{9,10}=540^\circ$

A

SECONDARY ANALYSIS -- XYZ SPACE

PLOTTED POINTS

$P_1=(x_1,y_1,z_1)$ $P_2=(x_2,y_2,z_2)$ $P_3=(x_3,y_3,z_3)$

$P_4=(x_4,y_4,z_4)$ $P_5=(x_5,y_5,z_5)$ $P_6=(x_6,y_6,z_6)$

$P_7=(x_7,y_7,z_7)$ $P_8=(x_8,y_8,z_8)$ $P_9=(x_9,y_9,z_9)$

$P_{10}=(x_{10},y_{10},z_{10})$

LENGTHS

$$s_1=\sqrt{(x_2-x_1)^2+(y_2-y_1)^2+(z_2-z_1)^2}$$

$$s_2=\sqrt{(x_3-x_2)^2+(y_3-y_2)^2+(z_3-z_2)^2}$$

$$s_3=\sqrt{(x_4-x_3)^2+(y_4-y_3)^2+(z_4-z_3)^2}$$

$$s_4=\sqrt{(x_5-x_4)^2+(y_5-y_4)^2+(z_5-z_4)^2}$$

$$s_5=\sqrt{(x_1-x_5)^2+(y_1-y_5)^2+(z_1-z_5)^2}$$

$$s_6=\sqrt{(x_7-x_6)^2+(y_7-y_6)^2+(z_7-z_6)^2}$$

$$s_7=\sqrt{(x_8-x_7)^2+(y_8-y_7)^2+(z_8-z_7)^2}$$

$$s_8=\sqrt{(x_9-x_8)^2+(y_9-y_8)^2+(z_9-z_8)^2}$$

$$s_9=\sqrt{(x_{10}-x_9)^2+(y_{10}-y_9)^2+(z_{10}-z_9)^2}$$

$$s_{10}=\sqrt{(x_6-x_{10})^2+(y_6-y_{10})^2+(z_6-z_{10})^2}$$

$$s_{11}=\sqrt{(x_6-x_1)^2+(y_6-y_1)^2+(z_6-z_1)^2}$$

$$s_{12}=\sqrt{(x_7-x_2)^2+(y_7-y_2)^2+(z_7-z_2)^2}$$

$$s_{13}=\sqrt{(x_8-x_3)^2+(y_8-y_3)^2+(z_8-z_3)^2}$$

$$s_{14}=\sqrt{(x_9-x_4)^2+(y_9-y_4)^2+(z_9-z_4)^2}$$

$$s_{15}=\sqrt{(x_{10}-x_5)^2+(y_{10}-y_5)^2+(z_{10}-z_5)^2}$$

DIRECTION COSINES

$\cos A_1=(x_2-x_1)/s_1$ $\cos B_1=(y_2-y_1)/s_1$

$\cos C_1=(z_2-z_1)/s_1$

$\cos A_2=(x_3-x_2)/s_2$ $\cos B_2=(y_3-y_2)/s_2$

$\cos C_2=(z_3-z_2)/s_2$

$\cos A_3=(x_4-x_3)/s_3$ $\cos B_3=(y_4-y_3)/s_3$

$\cos C_3=(z_4-z_3)/s_3$

$\cos A_4=(x_5-x_4)/s_4$ $\cos B_4=(y_5-y_4)/s_4$

$\cos C_4=(z_5-z_4)/s_4$

$\cos A_5=(x_1-x_5)/s_5$ $\cos B_5=(y_1-y_5)/s_5$

$\cos C_5=(z_1-z_5)/s_5$

$\cos A_6=(x_7-x_6)/s_6$ $\cos B_6=(y_7-y_6)/s_6$

$\cos C_6=(z_7-z_6)/s_6$

$\cos A_7=(x_8-x_7)/s_7$ $\cos B_7=(y_8-y_7)/s_7$

$\cos C_7=(z_8-z_7)/s_7$

$\cos A_8=(x_9-x_8)/s_8$ $\cos B_8=(y_9-y_8)/s_8$

$\cos C_8=(z_9-z_8)/s_8$

$\cos A_9=(x_{10}-x_9)/s_9$ $\cos B_9=(y_{10}-y_9)/s_9$

$\cos C_9=(z_{10}-z_9)/s_9$

$\cos A_{10}=(x_6-x_{10})/s_{10}$ $\cos B_{10}=(y_6-y_{10})/s_{10}$

$\cos C_{10}=(z_6-z_{10})/s_{10}$

$\cos A_{11}=(x_6-x_1)/s_{11}$ $\cos B_{11}=(y_6-y_1)/s_{11}$

$\cos C_{11}=(z_6-z_1)/s_{11}$

$\cos A_{12}=(x_7-x_2)/s_{12}$ $\cos B_{12}=(y_7-y_2)/s_{12}$

$\cos C_{12}=(z_7-z_2)/s_{12}$

$\cos A_{13}=(x_8-x_3)/s_{13}$ $\cos B_{13}=(y_8-y_3)/s_{13}$

$\cos C_{13}=(z_8-z_3)/s_{13}$

B

Table 22-9. Specifications and Analyses of a 7-Sided Object in 4-D Space. See Fig. 22-9. (A) Specifications and Analysis in Hyperspace. (B) Secondary Analysis in XYZ Space. (C) Secondary Analysis in XYW Space. (D) Secondary Analysis in ZXW Space. (E) Secondary Analysis in YZW Space. (F) Suggested in Coordinates. (Continued from page 397.)

DIRECTION COSINES (Cont'd)

$\cos A_{14}=(x_9-x_4)/s_{14}$ $\cos B_{14}=(y_9-y_4)/s_{14}$

$\cos C_{14}=(z_9-z_4)/s_{14}$

$\cos A_{15}=(x_{10}-x_5)/s_{15}$ $\cos B_{15}=(y_{10}-y_5)/s_{15}$

$\cos C_{15}=(z_{10}-z_5)/s_{15}$

ANGLES

$\cos\phi_{1,2}=\cos A_1\cos A_2+\cos B_1\cos B_2+\cos C_1\cos C_2$

$\cos\phi_{1,5}=\cos A_1\cos A_5+\cos B_1\cos B_5+\cos C_1\cos C_5$

$\cos\phi_{1,11}=\cos A_1\cos A_{11}+\cos B_1\cos B_{11}+\cos C_1\cos C_{11}$

$\cos\phi_{1,12}=\cos A_1\cos A_{12}+\cos B_1\cos B_{12}+\cos C_1\cos C_{12}$

$\cos\phi_{2,3}=\cos A_2\cos A_3+\cos B_2\cos B_3+\cos C_2\cos C_3$

$\cos\phi_{2,12}=\cos A_2\cos A_{12}+\cos B_2\cos B_{12}+\cos C_2\cos C_{12}$

$\cos\phi_{2,13}=\cos A_2\cos A_{13}+\cos B_2\cos B_{13}+\cos C_2\cos C_{13}$

$\cos\phi_{3,4}=\cos A_3\cos A_4+\cos B_3\cos B_4+\cos C_3\cos C_4$

$\cos\phi_{3,13}=\cos A_3\cos A_{13}+\cos B_3\cos B_{13}+\cos C_3\cos C_{13}$

$\cos\phi_{3,14}=\cos A_3\cos A_{14}+\cos B_3\cos B_{14}+\cos C_3\cos C_{14}$

$\cos\phi_{4,5}=\cos A_4\cos A_5+\cos B_4\cos B_5+\cos C_4\cos C_5$

$\cos\phi_{4,14}=\cos A_4\cos A_{14}+\cos B_4\cos B_{14}+\cos C_4\cos C_{14}$

$\cos\phi_{4,15}=\cos A_4\cos A_{15}+\cos B_4\cos B_{15}+\cos C_4\cos C_{15}$

$\cos\phi_{5,11}=\cos A_5\cos A_{11}+\cos B_5\cos B_{11}+\cos C_5\cos C_{11}$

$\cos\phi_{5,15}=\cos A_5\cos A_{15}+\cos B_5\cos B_{15}+\cos C_5\cos C_{15}$

$\cos\phi_{6,7}=\cos A_6\cos A_7+\cos B_6\cos B_7+\cos C_6\cos C_7$

$\cos\phi_{6,10}=\cos A_6\cos A_{10}+\cos B_6\cos B_{10}+\cos C_6\cos C_{10}$

$\cos\phi_{6,11}=\cos A_6\cos A_{11}+\cos B_6\cos B_{11}+\cos C_6\cos C_{11}$

$\cos\phi_{6,12}=\cos A_6\cos A_{12}+\cos B_6\cos B_{12}+\cos C_6\cos C_{12}$

$\cos\phi_{7,8}=\cos A_7\cos A_8+\cos B_7\cos B_8+\cos C_7\cos C_8$

$\cos\phi_{7,12}=\cos A_7\cos A_{12}+\cos B_7\cos B_{12}+\cos C_7\cos C_{12}$

$\cos\phi_{7,13}=\cos A_7\cos A_{13}+\cos B_7\cos B_{13}+\cos C_7\cos C_{13}$

ANGLES (Cont'd)

$\cos\phi_{8,9}=\cos A_8\cos A_9+\cos B_8\cos B_9+\cos C_8\cos C_9$

$\cos\phi_{8,13}=\cos A_8\cos A_{13}+\cos B_8\cos B_{13}+\cos C_8\cos C_{13}$

$\cos\phi_{8,14}=\cos A_8\cos A_{14}+\cos B_8\cos B_{14}+\cos C_8\cos C_{14}$

$\cos\phi_{9,10}=\cos A_9\cos A_{10}+\cos B_9\cos B_{10}+\cos C_9\cos C_{10}$

$\cos\phi_{9,14}=\cos A_9\cos A_{14}+\cos B_9\cos B_{14}+\cos C_9\cos C_{14}$

$\cos\phi_{9,15}=\cos A_9\cos A_{15}+\cos B_9\cos B_{15}+\cos C_9\cos C_{15}$

$\cos\phi_{10,11}=\cos A_{10}\cos A_{11}+\cos B_{10}\cos B_{11}+\cos C_{10}\cos C_{11}$

$\cos\phi_{10,15}=\cos A_{10}\cos A_{15}+\cos B_{10}\cos B_{15}+\cos C_{10}\cos C_{15}$

CHECK

F_1: $\phi_{1,2}+\phi_{1,5}+\phi_{2,3}+\phi_{3,4}+\phi_{4,5}=540^\circ$

F_2: $\phi_{1,12}+\phi_{1,11}+\phi_{6,11}+\phi_{6,12}=360^\circ$

F_3: $\phi_{2,12}+\phi_{2,13}+\phi_{7,12}+\phi_{7,13}=360^\circ$

F_4: $\phi_{3,13}+\phi_{3,14}+\phi_{8,13}+\phi_{8,14}=360^\circ$

F_5: $\phi_{4,14}+\phi_{4,15}+\phi_{9,14}+\phi_{9,15}=360^\circ$

F_6: $\phi_{5,11}+\phi_{5,15}+\phi_{10,11}+\phi_{10,15}=360^\circ$

F_7: $\phi_{6,7}+\phi_{6,10}+\phi_{7,8}+\phi_{8,9}+\phi_{9,10}=540^\circ$

B

SECONDARY ANALYSIS -- XYW SPACE

PLOTTED POINTS

$P_1=(x_1,y_1,w_1)$ $P_2=(x_2,y_2,w_2)$ $P_3=(x_3,y_3,w_3)$

$P_4=(x_4,y_4,w_4)$ $P_5=(x_5,y_5,w_5)$ $P_6=(x_6,y_6,w_6)$

$P_7=(x_7,y_7,w_7)$ $P_8=(x_8,y_8,w_8)$ $P_9=(x_9,y_9,w_9)$

$P_{10}=(x_{10},y_{10},w_{10})$

LENGTHS

$$s_1=\sqrt{(x_2-x_1)^2+(y_2-y_1)^2+(w_2-w_1)^2}$$

$$s_2=\sqrt{(x_3-x_2)^2+(y_3-y_2)^2+(w_3-w_2)^2}$$

$$s_3=\sqrt{(x_4-x_3)^2+(y_4-y_3)^2+(w_4-w_3)^2}$$

$$s_4=\sqrt{(x_5-x_4)^2+(y_5-y_4)^2+(w_5-w_4)^2}$$

$$s_5=\sqrt{(x_1-x_5)^2+(y_1-y_5)^2+(w_1-w_5)^2}$$

$$s_6=\sqrt{(x_7-x_6)^2+(y_7-y_6)^2+(w_7-w_6)^2}$$

$$s_7=\sqrt{(x_8-x_7)^2+(y_8-y_7)^2+(w_8-w_7)^2}$$

$$s_8=\sqrt{(x_9-x_8)^2+(y_9-y_8)^2+(w_9-w_8)^2}$$

$$s_9=\sqrt{(x_{10}-x_9)^2+(y_{10}-y_9)^2+(w_{10}-w_9)^2}$$

$$s_{10}=\sqrt{(x_6-x_{10})^2+(y_6-y_{10})^2+(w_6-w_{10})^2}$$

$$s_{11}=\sqrt{(x_6-x_1)^2+(y_6-y_1)^2+(w_6-w_1)^2}$$

$$s_{12}=\sqrt{(x_7-x_2)^2+(y_7-y_2)^2+(w_7-w_2)^2}$$

$$s_{13}=\sqrt{(x_8-x_3)^2+(y_8-y_3)^2+(w_8-w_3)^2}$$

$$s_{14}=\sqrt{(x_9-x_4)^2+(y_9-y_4)^2+(w_9-w_4)^2}$$

$$s_{15}=\sqrt{(x_{10}-x_5)^2+(y_{10}-y_5)^2+(w_{10}-w_5)^2}$$

DIRECTION COSINES

$\cos A_1=(x_2-x_1)/s_1$ $\cos B_1=(y_2-y_1)/s_1$

$\cos D_1=(w_2-w_1)/s_1$

$\cos A_2=(x_3-x_2)/s_2$ $\cos B_2=(y_3-y_2)/s_2$

$\cos D_2=(w_3-w_2)/s_2$

$\cos A_3=(x_4-x_3)/s_3$ $\cos B_3=(y_4-y_3)/s_3$

$\cos D_3=(w_4-w_3)/s_3$

$\cos A_4=(x_5-x_4)/s_4$ $\cos B_4=(y_5-y_4)/s_4$

$\cos D_4=(w_5-w_4)/s_4$

$\cos A_5=(x_1-x_5)/s_5$ $\cos B_5=(y_1-y_5)/s_5$

$\cos D_5=(w_1-w_5)/s_5$

$\cos A_6=(x_7-x_6)/s_6$ $\cos B_6=(y_7-y_6)/s_6$

$\cos D_6=(w_7-w_6)/s_6$

$\cos A_7=(x_8-x_7)/s_7$ $\cos B_7=(y_8-y_7)/s_7$

$\cos D_7=(w_8-w_7)/s_7$

$\cos A_8=(x_9-x_8)/s_8$ $\cos B_8=(y_9-y_8)/s_8$

$\cos D_8=(w_9-w_8)/s_8$

$\cos A_9=(x_{10}-x_9)/s_9$ $\cos B_9=(y_{10}-y_9)/s_9$

$\cos D_9=(w_{10}-w_9)/s_9$

$\cos A_{10}=(x_6-x_{10})/s_{10}$ $\cos B_{10}=(y_6-y_{10})/s_{10}$

$\cos D_{10}=(w_6-w_{10})/s_{10}$

$\cos A_{11}=(x_6-x_1)/s_{11}$ $\cos B_{11}=(y_6-y_1)/s_{11}$

$\cos D_{11}=(w_6-w_1)/s_{11}$

$\cos A_{12}=(x_7-x_2)/s_{12}$ $\cos B_{12}=(y_7-y_2)/s_{12}$

$\cos D_{12}=(w_7-w_2)/s_{12}$

$\cos A_{13}=(x_8-x_3)/s_{13}$ $\cos B_{13}=(y_8-y_3)/s_{13}$

$\cos D_{13}=(w_8-w_3)/s_{13}$

Table 22-9. Specifications and Analyses of a 7-Sided Object in 4-D Space. See Fig. 22-9. (A) Specifications and Analysis in Hyperspace. (B) Secondary Analysis in XYZ Space. (C) Secondary Analysis in XYW Space. (D) Secondary Analysis in XZW Space. (E) Secondary Analysis in YZW Space. (F) Suggested in Coordinates. (Continued from page 399.)

DIRECTION COSINES (Cont'd)

$\cos A_{14}=(x_9-x_4)/s_{14}$ $\cos B_{14}=(y_9-y_4)/s_{14}$

$\cos D_{14}=(w_9-w_4)/s_{14}$

$\cos A_{15}=(x_{10}-x_5)/s_{15}$ $\cos B_{15}=(y_{10}-y_5)/s_{15}$

$\cos D_{15}=(w_{10}-w_5)/s_{15}$

ANGLES

$\cos\phi_{1,2}=\cos A_1\cos A_2+\cos B_1\cos B_2+\cos D_1\cos D_2$

$\cos\phi_{1,5}=\cos A_1\cos A_5+\cos B_1\cos B_5+\cos D_1\cos D_5$

$\cos\phi_{1,11}=\cos A_1\cos A_{11}+\cos B_1\cos B_{11}+\cos D_1\cos D_{11}$

$\cos\phi_{1,12}=\cos A_1\cos A_{12}+\cos B_1\cos B_{12}+\cos D_1\cos D_{12}$

$\cos\phi_{2,3}=\cos A_2\cos A_3+\cos B_2\cos B_3+\cos D_2\cos D_3$

$\cos\phi_{2,12}=\cos A_2\cos A_{12}+\cos B_2\cos B_{12}+\cos D_2\cos D_{12}$

$\cos\phi_{2,13}=\cos A_2\cos A_{13}+\cos B_2\cos B_{13}+\cos D_2\cos D_{13}$

$\cos\phi_{3,4}=\cos A_3\cos A_4+\cos B_3\cos B_4+\cos D_3\cos D_4$

$\cos\phi_{3,13}=\cos A_3\cos A_{13}+\cos B_3\cos B_{13}+\cos D_3\cos D_{13}$

$\cos\phi_{3,14}=\cos A_3\cos A_{14}+\cos B_3\cos B_{14}+\cos D_3\cos D_{14}$

$\cos\phi_{4,5}=\cos A_4\cos A_5+\cos B_4\cos B_5+\cos D_4\cos D_5$

$\cos\phi_{4,14}=\cos A_4\cos A_{14}+\cos B_4\cos B_{14}+\cos D_4\cos D_{14}$

$\cos\phi_{4,15}=\cos A_4\cos A_{15}+\cos B_4\cos B_{15}+\cos D_4\cos D_{15}$

$\cos\phi_{5,11}=\cos A_5\cos A_{11}+\cos B_5\cos B_{11}+\cos D_5\cos D_{11}$

$\cos\phi_{5,15}=\cos A_5\cos A_{15}+\cos B_5\cos B_{15}+\cos D_5\cos D_{15}$

$\cos\phi_{6,7}=\cos A_6\cos A_7+\cos B_6\cos B_7+\cos D_6\cos D_7$

$\cos\phi_{6,10}=\cos A_6\cos A_{10}+\cos B_6\cos B_{10}+\cos D_6\cos D_{10}$

$\cos\phi_{6,11}=\cos A_6\cos A_{11}+\cos B_6\cos B_{11}+\cos D_6\cos D_{11}$

$\cos\phi_{6,12}=\cos A_6\cos A_{12}+\cos B_6\cos B_{12}+\cos D_6\cos D_{12}$

$\cos\phi_{7,8}=\cos A_7\cos A_8+\cos B_7\cos B_8+\cos D_7\cos D_8$

$\cos\phi_{7,12}=\cos A_7\cos A_{12}+\cos B_7\cos B_{12}+\cos D_7\cos D_{12}$

$\cos\phi_{7,13}=\cos A_7\cos A_{13}+\cos B_7\cos B_{13}+\cos D_7\cos D_{13}$

ANGLES (Cont'd)

$\cos\phi_{8,9}=\cos A_8\cos A_9+\cos B_8\cos B_9+\cos D_8\cos D_9$

$\cos\phi_{8,13}=\cos A_8\cos A_{13}+\cos B_8\cos B_{13}+\cos D_8\cos D_{13}$

$\cos\phi_{8,14}=\cos A_8\cos A_{14}+\cos B_8\cos B_{14}+\cos D_8\cos D_{14}$

$\cos\phi_{9,10}=\cos A_9\cos A_{10}+\cos B_9\cos B_{10}+\cos D_9\cos D_{10}$

$\cos\phi_{9,14}=\cos A_9\cos A_{14}+\cos B_9\cos B_{14}+\cos D_9\cos D_{14}$

$\cos\phi_{9,15}=\cos A_9\cos A_{15}+\cos B_9\cos B_{15}+\cos D_9\cos D_{15}$

$\cos\phi_{10,11}=\cos A_{10}\cos A_{11}+\cos B_{10}\cos B_{11}+\cos D_{10}\cos D_{11}$

$\cos\phi_{10,15}=\cos A_{10}\cos A_{15}+\cos B_{10}\cos B_{15}+\cos D_{10}\cos D_{15}$

CHECK

F_1: $\phi_{1,2}+\phi_{1,5}+\phi_{2,3}+\phi_{3,4}+\phi_{4,5}=540^\circ$

F_2: $\phi_{1,12}+\phi_{1,11}+\phi_{6,11}+\phi_{6,12}=360^\circ$

F_3: $\phi_{2,12}+\phi_{2,13}+\phi_{7,12}+\phi_{7,13}=360^\circ$

F_4: $\phi_{3,13}+\phi_{3,14}+\phi_{8,13}+\phi_{8,14}=360^\circ$

F_5: $\phi_{4,14}+\phi_{4,15}+\phi_{9,14}+\phi_{9,15}=360^\circ$

F_6: $\phi_{5,11}+\phi_{5,15}+\phi_{10,11}+\phi_{10,15}=360^\circ$

F_7: $\phi_{6,7}+\phi_{6,10}+\phi_{7,8}+\phi_{8,9}+\phi_{9,10}=540^\circ$

C

SECONDARY ANALYSIS -- XZW SPACE

PLOTTED POINTS

$P_1=(x_1,z_1,w_1)$ $P_2=(x_2,z_2,w_2)$ $P_3=(x_3,z_3,w_3)$

$P_4=(x_4,z_4,w_4)$ $P_5=(x_5,z_5,w_5)$ $P_6=(x_6,z_6,w_6)$

$P_7=(x_7,z_7,w_7)$ $P_8=(x_8,z_8,w_8)$ $P_9=(x_9,z_9,w_9)$

$P_{10}=(x_{10},z_{10},w_{10})$

LENGTHS

$s_1=\sqrt{(x_2-x_1)^2+(z_2-z_1)^2+(w_2-w_1)^2}$

$s_2=\sqrt{(x_3-x_2)^2+(z_3-z_2)^2+(w_3-w_2)^2}$

$s_3=\sqrt{(x_4-x_3)^2+(z_4-z_3)^2+(w_4-w_3)^2}$

$s_4=\sqrt{(x_5-x_4)^2+(z_5-z_4)^2+(w_5-w_4)^2}$

$s_5=\sqrt{(x_1-x_5)^2+(z_1-z_5)^2+(w_1-w_5)^2}$

$s_6=\sqrt{(x_7-x_6)^2+(z_7-z_6)^2+(w_7-w_6)^2}$

$s_7=\sqrt{(x_8-x_7)^2+(z_8-z_7)^2+(w_8-w_7)^2}$

$s_8=\sqrt{(x_9-x_8)^2+(z_9-z_8)^2+(w_9-w_8)^2}$

$s_9=\sqrt{(x_{10}-x_9)^2+(z_{10}-z_9)^2+(w_{10}-w_9)^2}$

$s_{10}=\sqrt{(x_6-x_{10})^2+(z_6-z_{10})^2+(w_6-w_{10})^2}$

$s_{11}=\sqrt{(x_6-x_1)^2+(z_6-z_1)^2+(w_6-w_1)^2}$

$s_{12}=\sqrt{(x_7-x_2)^2+(z_7-z_2)^2+(w_7-w_2)^2}$

$s_{13}=\sqrt{(x_8-x_3)^2+(z_8-z_3)^2+(w_8-w_3)^2}$

$s_{14}=\sqrt{(x_9-x_4)^2+(z_9-z_4)^2+(w_9-w_4)^2}$

$s_{15}=\sqrt{(x_{10}-x_5)^2+(z_{10}-z_5)^2+(w_{10}-w_5)^2}$

DIRECTION COSINES

$\cos A_1=(x_2-x_1)/s_1$ $\cos C_1=(z_2-z_1)/s_1$

$\cos D_1=(w_2-w_1)/s_1$

$\cos A_2=(x_3-x_2)/s_2$ $\cos C_2=(z_3-z_2)/s_2$

$\cos D_2=(w_3-w_2)/s_2$

$\cos A_3=(x_4-x_3)/s_3$ $\cos C_3=(z_4-z_3)/s_3$

$\cos D_3=(w_4-w_3)/s_3$

$\cos A_4=(x_5-x_4)/s_4$ $\cos C_4=(z_5-z_4)/s_4$

$\cos D_4=(w_5-w_4)/s_4$

$\cos A_5=(x_1-x_5)/s_5$ $\cos C_5=(z_1-z_5)/s_5$

$\cos D_5=(w_1-w_5)/s_5$

$\cos A_6=(x_7-x_6)/s_6$ $\cos C_6=(z_7-z_6)/s_6$

$\cos D_6=(w_7-w_6)/s_6$

$\cos A_7=(x_8-x_7)/s_7$ $\cos C_7=(z_8-z_7)/s_7$

$\cos D_7=(w_8-w_7)/s_7$

$\cos A_8=(x_9-x_8)/s_8$ $\cos C_8=(z_9-z_8)/s_8$

$\cos D_8=(w_9-w_8)/s_8$

$\cos A_9=(x_{10}-x_9)/s_9$ $\cos C_9=(z_{10}-z_9)/s_9$

$\cos D_9=(w_{10}-w_9)/s_9$

$\cos A_{10}=(x_6-x_{10})/s_{10}$ $\cos C_{10}=(z_6-z_{10})/s_{10}$

$\cos D_{10}=(w_6-w_{10})/s_{10}$

$\cos A_{11}=(x_6-x_1)/s_{11}$ $\cos C_{11}=(z_6-z_1)/s_{11}$

$\cos D_{11}=(w_6-w_1)/s_{11}$

$\cos A_{12}=(x_7-x_2)/s_{12}$ $\cos C_{12}=(z_7-z_2)/s_{12}$

$\cos D_{12}=(w_7-w_2)/s_{12}$

$\cos A_{13}=(x_8-x_3)/s_{13}$ $\cos C_{13}=(z_8-z_3)/s_{13}$

$\cos D_{13}=(w_8-w_3)/s_{13}$

D

Table 22-9. Specifications and Analyses of a 7-Sided Object in 4-D Space. See Fig. 22-9. (A) Specifications and Analysis in Hyperspace. (B) Secondary Analysis in XYZ Space. (C) Secondary Analysis in XYW Space. (D) Secondary Analysis in XZW Space. (E) Secondary Analysis in YZW Space. (F) Suggested in Coordinates. **(Continued from page 401.)**

DIRECTION COSINES (Cont'd)

$\cos A_{14}=(x_9-x_4)/s_{14}$ $\quad\cos C_{14}=(z_9-z_4)/s_{14}$

$\cos D_{14}=(w_9-w_4)/s_{14}$

$\cos A_{15}=(x_{10}-x_5)/s_{15}$ $\quad\cos C_{15}=(z_{10}-z_5)/s_{15}$

$\cos D_{15}=(w_{10}-w_5)/s_{15}$

ANGLES

$\cos\phi_{1,2}=\cos A_1\cos A_2+\cos C_1\cos C_2+\cos D_1\cos D_2$

$\cos\phi_{1,5}=\cos A_1\cos A_5+\cos C_1\cos C_5+\cos D_1\cos D_5$

$\cos\phi_{1,11}=\cos A_1\cos A_{11}+\cos C_1\cos C_{11}+\cos D_1\cos D_{11}$

$\cos\phi_{1,12}=\cos A_1\cos A_{12}+\cos C_1\cos C_{12}+\cos D_1\cos D_{12}$

$\cos\phi_{2,3}=\cos A_2\cos A_3+\cos C_2\cos C_3+\cos D_2\cos D_3$

$\cos\phi_{2,12}=\cos A_2\cos A_{12}+\cos C_2\cos C_{12}+\cos D_2\cos D_{12}$

$\cos\phi_{2,13}=\cos A_2\cos A_{13}+\cos C_2\cos C_{13}+\cos D_2\cos D_{13}$

$\cos\phi_{3,4}=\cos A_3\cos A_4+\cos C_3\cos C_4+\cos D_3\cos D_4$

$\cos\phi_{3,13}=\cos A_3\cos A_{13}+\cos C_3\cos C_{13}+\cos D_3\cos D_{13}$

$\cos\phi_{3,14}=\cos A_3\cos A_{14}+\cos C_3\cos C_{14}+\cos D_3\cos D_{14}$

$\cos\phi_{4,5}=\cos A_4\cos A_5+\cos C_4\cos C_5+\cos D_4\cos D_5$

$\cos\phi_{4,14}=\cos A_4\cos A_{14}+\cos C_4\cos C_{14}+\cos D_4\cos D_{14}$

$\cos\phi_{4,15}=\cos A_4\cos A_{15}+\cos C_4\cos C_{15}+\cos D_4\cos D_{15}$

$\cos\phi_{5,11}=\cos A_5\cos A_{11}+\cos C_5\cos C_{11}+\cos D_5\cos D_{11}$

$\cos\phi_{5,15}=\cos A_5\cos A_{15}+\cos C_5\cos C_{15}+\cos D_5\cos D_{15}$

$\cos\phi_{6,7}=\cos A_6\cos A_7+\cos C_6\cos C_7+\cos D_6\cos D_7$

$\cos\phi_{6,10}=\cos A_6\cos A_{10}+\cos C_6\cos C_{10}+\cos D_6\cos D_{10}$

$\cos\phi_{6,11}=\cos A_6\cos A_{11}+\cos C_6\cos C_{11}+\cos D_6\cos D_{11}$

$\cos\phi_{6,12}=\cos A_6\cos A_{12}+\cos C_6\cos C_{12}+\cos D_6\cos D_{12}$

$\cos\phi_{7,8}=\cos A_7\cos A_8+\cos C_7\cos C_8+\cos D_7\cos D_8$

$\cos\phi_{7,12}=\cos A_7\cos A_{12}+\cos C_7\cos C_{12}+\cos D_7\cos D_{12}$

$\cos\phi_{7,13}=\cos A_7\cos A_{13}+\cos C_7\cos C_{13}+\cos D_7\cos D_{13}$

ANGLES (Cont'd)

$\cos\phi_{8,9}=\cos A_8\cos A_9+\cos C_8\cos C_9+\cos D_8\cos D_9$

$\cos\phi_{8,13}=\cos A_8\cos A_{13}+\cos C_8\cos C_{13}+\cos D_8\cos D_{13}$

$\cos\phi_{8,14}=\cos A_8\cos A_{14}+\cos C_8\cos C_{14}+\cos D_8\cos D_{14}$

$\cos\phi_{9,10}=\cos A_9\cos A_{10}+\cos C_9\cos C_{10}+\cos D_9\cos D_{10}$

$\cos\phi_{9,14}=\cos A_9\cos A_{14}+\cos C_9\cos C_{14}+\cos D_9\cos D_{14}$

$\cos\phi_{9,15}=\cos A_9\cos A_{15}+\cos C_9\cos C_{15}+\cos D_9\cos D_{15}$

$\cos\phi_{10,11}=\cos A_{10}\cos A_{11}+\cos C_{10}\cos C_{11}+\cos D_{10}\cos D_{11}$

$\cos\phi_{10,15}=\cos A_{10}\cos A_{15}+\cos C_{10}\cos C_{15}+\cos D_{10}\cos D_{15}$

CHECK

F_1: $\quad\phi_{1,2}+\phi_{1,5}+\phi_{2,3}+\phi_{3,4}+\phi_{4,5}=540^\circ$

F_2: $\quad\phi_{1,12}+\phi_{1,11}+\phi_{6,11}+\phi_{6,12}=360^\circ$

F_3: $\quad\phi_{2,12}+\phi_{2,13}+\phi_{7,12}+\phi_{7,13}=360^\circ$

F_4: $\quad\phi_{3,13}+\phi_{3,14}+\phi_{8,13}+\phi_{8,14}=360^\circ$

F_5: $\quad\phi_{4,14}+\phi_{4,15}+\phi_{9,14}+\phi_{9,15}=360^\circ$

F_6: $\quad\phi_{5,11}+\phi_{5,15}+\phi_{10,11}+\phi_{10,15}=360^\circ$

F_7: $\quad\phi_{6,7}+\phi_{6,10}+\phi_{7,8}+\phi_{8,9}+\phi_{9,10}=540^\circ$

D

SECONDARY ANALYSIS -- YZW SPACE

PLOTTED POINTS

$P_1=(y_1,z_1,w_1)$ $P_2=(y_2,z_2,w_2)$ $P_3=(y_3,z_3,w_3)$

$P_4=(y_4,z_4,w_4)$ $P_5=(y_5,z_5,w_5)$ $P_6=(y_6,z_6,w_6)$

$P_7=(y_7,z_7,w_7)$ $P_8=(y_8,z_8,w_8)$ $P_9=(y_9,z_9,w_9)$

$P_{10}=(y_{10},z_{10},w_{10})$

LENGTHS

$s_1=\sqrt{(y_2-y_1)^2+(z_2-z_1)^2+(w_2-w_1)^2}$

$s_2=\sqrt{(y_3-y_2)^2+(z_3-z_2)^2+(w_3-w_2)^2}$

$s_3=\sqrt{(y_4-y_3)^2+(z_4-z_3)^2+(w_4-w_3)^2}$

$s_4=\sqrt{(y_5-y_4)^2+(z_5-z_4)^2+(w_5-w_4)^2}$

$s_5=\sqrt{(y_1-y_5)^2+(z_1-z_5)^2+(w_1-w_5)^2}$

$s_6=\sqrt{(y_7-y_6)^2+(z_7-z_6)^2+(w_7-w_6)^2}$

$s_7=\sqrt{(y_8-y_7)^2+(z_8-z_7)^2+(w_8-w_7)^2}$

$s_8=\sqrt{(y_9-y_8)^2+(z_9-z_8)^2+(w_9-w_8)^2}$

$s_9=\sqrt{(y_{10}-y_9)^2+(z_{10}-z_9)^2+(w_{10}-w_9)^2}$

$s_{10}=\sqrt{(y_6-y_{10})^2+(z_6-z_{10})^2+(w_6-w_{10})^2}$

$s_{11}=\sqrt{(y_6-y_1)^2+(z_6-z_1)^2+(w_6-w_1)^2}$

$s_{12}=\sqrt{(y_7-y_2)^2+(z_7-z_2)^2+(w_7-w_2)^2}$

$s_{13}=\sqrt{(y_8-y_3)^2+(z_8-z_3)^2+(w_8-w_3)^2}$

$s_{14}=\sqrt{(y_9-y_4)^2+(z_9-z_4)^2+(w_9-w_4)^2}$

$s_{15}=\sqrt{(y_{10}-y_5)^2+(z_{10}-z_5)^2+(w_{10}-w_5)^2}$

DIRECTION COSINES

$\cos B_1=(y_2-y_1)/s_1$ $\cos C_1=(z_2-z_1)/s_1$

$\cos D_1=(w_2-w_1)/s_1$

$\cos B_2=(y_3-y_2)/s_2$ $\cos C_2=(z_3-z_2)/s_2$

$\cos D_2=(w_3-w_2)/s_2$

$\cos B_3=(y_4-y_3)/s_3$ $\cos C_3=(z_4-z_3)/s_3$

$\cos D_3=(w_4-w_3)/s_3$

$\cos B_4=(y_5-y_4)/s_4$ $\cos C_4=(z_5-z_4)/s_4$

$\cos D_4=(w_5-w_4)/s_4$

$\cos B_5=(y_1-y_5)/s_5$ $\cos C_5=(z_1-z_5)/s_5$

$\cos D_5=(w_1-w_5)/s_5$

$\cos B_6=(y_7-y_6)/s_6$ $\cos C_6=(z_7-z_6)/s_6$

$\cos D_6=(w_7-w_6)/s_6$

$\cos B_7=(y_8-y_7)/s_7$ $\cos C_7=(z_8-z_7)/s_7$

$\cos D_7=(w_8-w_7)/s_7$

$\cos B_8=(y_9-y_8)/s_8$ $\cos C_8=(z_9-z_8)/s_8$

$\cos D_8=(w_9-w_8)/s_8$

$\cos B_9=(y_{10}-y_9)/s_9$ $\cos C_9=(z_{10}-z_9)/s_9$

$\cos D_9=(w_{10}-w_9)/s_9$

$\cos B_{10}=(y_6-y_{10})/s_{10}$ $\cos C_{10}=(z_6-z_{10})/s_{10}$

$\cos D_{10}=(w_6-w_{10})/s_{10}$

$\cos B_{11}=(y_6-y_1)/s_{11}$ $\cos C_{11}=(z_6-z_1)/s_{11}$

$\cos D_{11}=(w_6-w_1)/s_{11}$

$\cos B_{12}=(y_7-y_2)/s_{12}$ $\cos C_{12}=(z_7-z_2)/s_{12}$

$\cos D_{12}=(w_7-w_2)/s_{12}$

$\cos B_{13}=(y_8-y_3)/s_{13}$ $\cos C_{13}=(z_8-z_3)/s_{13}$

$\cos D_{13}=(w_8-w_3)/s_{13}$

E

Table 22-9. Specifications and Analyses of a 7-Sided Object in 4-D Space. See Fig. 22-9. (A) Specifications and Analysis in Hyperspace. (B) Secondary Analysis in XYZ Space. (C) Secondary Analysis in XYW Space. (D) Secondary Analysis in XZW Space. (E) Secondary Analysis in YZW Space. (F) Suggested in Coordinates. (Continued from page 403.)

DIRECTION COSINES (Cont'd)

$\cos B_{14}=(y_9-y_4)/s_{14}$ $\quad \cos C_{14}=(z_9-z_4)/s_{14}$

$\cos D_{14}=(w_9-w_4)/s_{14}$

$\cos B_{15}=(y_{10}-y_5)/s_{15}$ $\quad \cos C_{15}=(z_{10}-z_5)/s_{15}$

$\cos D_{15}=(w_{10}-w_5)/s_{15}$

ANGLES

$\cos\phi_{1,2}=\cos B_1\cos B_2+\cos C_1\cos C_2+\cos D_1\cos D_2$

$\cos\phi_{1,5}=\cos B_1\cos B_5+\cos C_1\cos C_5+\cos D_1\cos D_5$

$\cos\phi_{1,11}=\cos B_1\cos B_{11}+\cos C_1\cos C_{11}+\cos D_1\cos D_{11}$

$\cos\phi_{1,12}=\cos B_1\cos B_{12}+\cos C_1\cos C_{12}+\cos D_1\cos D_{12}$

$\cos\phi_{2,3}=\cos B_2\cos B_3+\cos C_2\cos C_3+\cos D_2\cos D_3$

$\cos\phi_{2,12}=\cos B_2\cos B_{12}+\cos C_2\cos C_{12}+\cos D_2\cos D_{12}$

$\cos\phi_{2,13}=\cos B_2\cos B_{13}+\cos C_2\cos C_{13}+\cos D_2\cos D_{13}$

$\cos\phi_{3,4}=\cos B_3\cos B_4+\cos C_3\cos C_4+\cos D_3\cos D_4$

$\cos\phi_{3,13}=\cos B_3\cos B_{13}+\cos C_3\cos C_{13}+\cos D_3\cos D_{13}$

$\cos\phi_{3,14}=\cos B_3\cos B_{14}+\cos C_3\cos C_{14}+\cos D_3\cos D_{14}$

$\cos\phi_{4,5}=\cos B_4\cos B_5+\cos C_4\cos C_5+\cos D_4\cos D_5$

$\cos\phi_{4,14}=\cos B_4\cos B_{14}+\cos C_4\cos C_{14}+\cos D_4\cos D_{14}$

$\cos\phi_{4,15}=\cos B_4\cos B_{15}+\cos C_4\cos C_{15}+\cos D_4\cos D_{15}$

$\cos\phi_{5,11}=\cos B_5\cos B_{11}+\cos C_5\cos C_{11}+\cos D_5\cos D_{11}$

$\cos\phi_{5,15}=\cos B_5\cos B_{15}+\cos C_5\cos C_{15}+\cos D_5\cos D_{15}$

$\cos\phi_{6,7}=\cos B_6\cos B_7+\cos C_6\cos C_7+\cos D_6\cos D_7$

$\cos\phi_{6,10}=\cos B_6\cos B_{10}+\cos C_6\cos C_{10}+\cos D_6\cos D_{10}$

$\cos\phi_{6,11}=\cos B_6\cos B_{11}+\cos C_6\cos C_{11}+\cos D_6\cos D_{11}$

$\cos\phi_{6,12}=\cos B_6\cos B_{12}+\cos C_6\cos C_{12}+\cos D_6\cos D_{12}$

$\cos\phi_{7,8}=\cos B_7\cos B_8+\cos C_7\cos C_8+\cos D_7\cos D_8$

$\cos\phi_{7,12}=\cos B_7\cos B_{12}+\cos C_7\cos C_{12}+\cos D_7\cos D_{12}$

$\cos\phi_{7,13}=\cos B_7\cos B_{13}+\cos C_7\cos C_{13}+\cos D_7\cos D_{13}$

ANGLES (Cont'd)

$\cos\phi_{8,9}=\cos B_8\cos B_9+\cos C_8\cos C_9+\cos D_8\cos D_9$

$\cos\phi_{8,13}=\cos B_8\cos B_{13}+\cos C_8\cos C_{13}+\cos D_8\cos D_{13}$

$\cos\phi_{8,14}=\cos B_8\cos B_{14}+\cos C_8\cos C_{14}+\cos D_8\cos D_{14}$

$\cos\phi_{9,10}=\cos B_9\cos B_{10}+\cos C_9\cos C_{10}+\cos D_9\cos D_{10}$

$\cos\phi_{9,14}=\cos B_9\cos B_{14}+\cos C_9\cos C_{14}+\cos D_9\cos D_{14}$

$\cos\phi_{9,15}=\cos B_9\cos B_{15}+\cos C_9\cos C_{15}+\cos D_9\cos D_{15}$

$\cos\phi_{10,11}=\cos B_{10}\cos B_{11}+\cos C_{10}\cos C_{11}+\cos D_{10}\cos D_{11}$

$\cos\phi_{10,15}=\cos B_{10}\cos B_{15}+\cos C_{10}\cos C_{15}+\cos D_{10}\cos D_{15}$

CHECK

F_1: $\phi_{1,2}+\phi_{1,5}+\phi_{2,3}+\phi_{3,4}+\phi_{4,5}=540^\circ$

F_2: $\phi_{1,12}+\phi_{1,11}+\phi_{6,11}+\phi_{6,12}=360^\circ$

F_3: $\phi_{2,12}+\phi_{2,13}+\phi_{7,12}+\phi_{7,13}=360^\circ$

F_4: $\phi_{3,13}+\phi_{3,14}+\phi_{8,13}+\phi_{8,14}=360^\circ$

F_5: $\phi_{4,14}+\phi_{4,15}+\phi_{9,14}+\phi_{9,15}=360^\circ$

F_6: $\phi_{5,11}+\phi_{5,15}+\phi_{10,11}+\phi_{10,15}=360^\circ$

F_7: $\phi_{6,7}+\phi_{6,10}+\phi_{7,8}+\phi_{8,9}+\phi_{9,10}=540^\circ$

E

HYPERSPACE OBJECT VERSION 1

$P_1=(-1,-1,1,-1)$	$P_2=(1,-1,1,-1)$	$P_3=(1,1,1,1)$
$P_4=(0,2,1,2)$	$P_5=(-1,1,1,1)$	$P_6=(-1,-1,-1,-1)$
$P_7=(1,-1,-1,-1)$	$P_8=(1,1,-1,1)$	$P_9=(0,2,-1,2)$
$P_{10}=(-1,1,-1,1)$		

HYPERSPACE OBJECT VERSION 2

$P_1=(-2,-2,1,-2)$	$P_2=(0,-2,1,-2)$	$P_3=(0,0,1,0)$
$P_4=(-1,1,1,1)$	$P_5=(-2,0,1,0)$	$P_6=(0,0,-1,0)$
$P_7=(2,0,-1,0)$	$P_8=(2,2,-1,2)$	$P_9=(1,4,-1,4)$
$P_{10}=(0,2,-1,2)$		

HYPERSPACE OBJECT VERSION 3

$P_1=(-1,-1,1,-1)$	$P_2=(1,-1,1,-1)$	$P_3=(1,1,1,1)$
$P_4=(0,0.5,1,0.5)$	$P_5=(-1,1,1,1)$	$P_6=(-1,-1,-1,-1)$
$P_7=(1,-1,-1,-1)$	$P_8=(1,1,-1,1)$	$P_9=(0,0.5,-1,0.5)$
$P_{10}=(-1,1,-1,1)$		

are commutative. (Combinations of simple rotations about different axes are not commutative.)

I hope you can see that the principles of simple transformations and their effects are virtually identical in any number of dimensions.

22-4 SIMPLE TRANSLATIONS AND SUGGESTED EXPERIMENTS

Simple translations in any number of dimensions express a spatial displacement between a reference entity and a translated version of it. Recall that the two versions are identical in every respect except their position with respect to the frame-of-reference coordinate system. The principle holds for translations in 4-D space as well as the more familiar, lower-dimensioned spaces.

22-4.1 Suggested Experiments

The most straightforward sort of experiment with simple translations in 4-D space follows this general outline:

☐ Specify a reference figure, space object, or hyperspace object in 4-D space (see Sections 22-1 and 22-2).

☐ Conduct complete primary and secondary analyses of the reference entity.

☐ Specify and carry out a chosen simple 4-D translation.

☐ Set up the formal specifications for the translated version of the entity, and conduct complete primary and secondary analyses of it.

☐ Plot both the reference and translated version of the geometric entity.

Running that sort of experiment with a variety of space and hyperspace objects not only demonstrates the linear displacement of entities in hyperspace; it also helps build files of complete specifications and analyses that can be invaluable for later experiments.

Simple hyperspace translations also let you position a line, figure, space or hyperspace object at some desired place in a 4-D coordinate system. The procedure in this

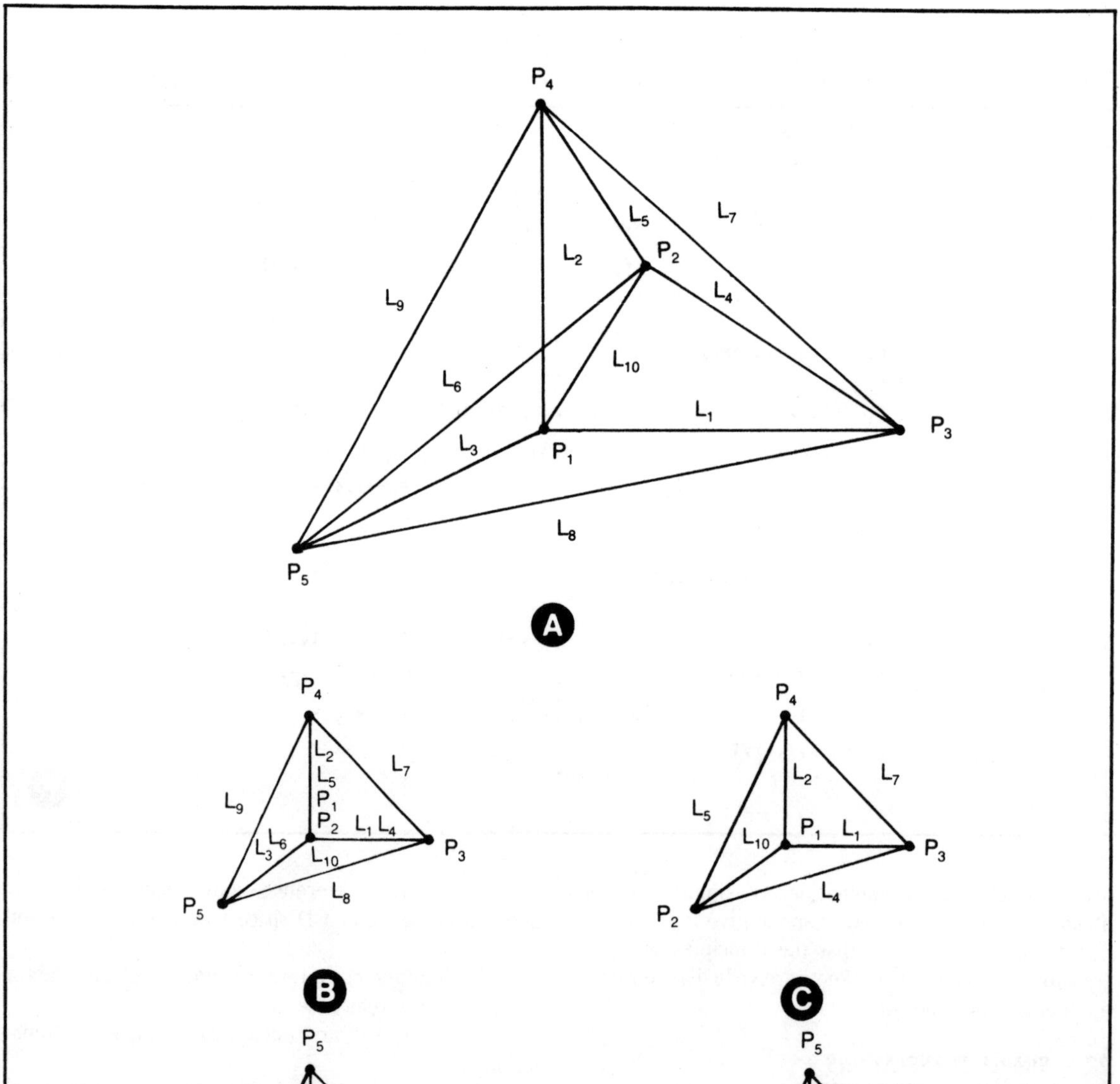

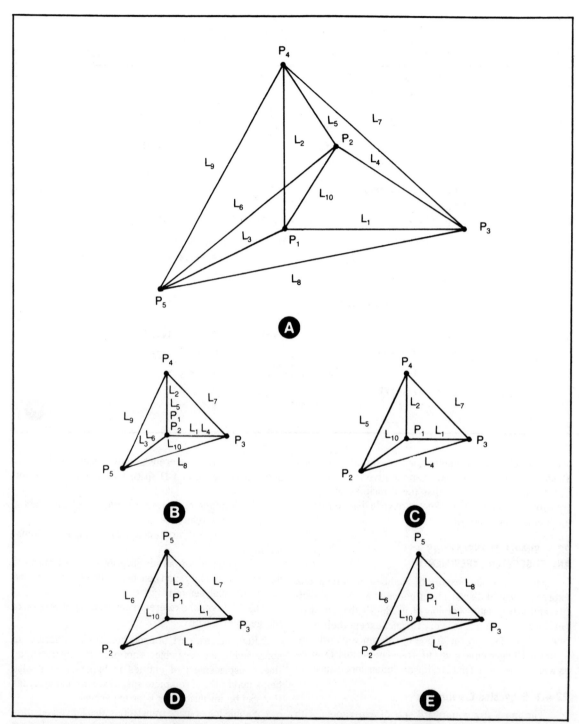

Fig. 22-10. Plots of the hyperspace object specified in Table 22-10A. (A) General hyperspace view. (B) XYZ perspective view. (C) XYW perspective view. (D) XZW perspective view. (E) YZW perspective view. See outlines of analyses of Table 22-10.

Table 22-10. Specifications and Analyses of the Hyperspace Object Shown in Fig. 22-10. (A) Specifications and Analyses in Hyperspace. (B) Secondary Analysis in XYZ Space. (C) Secondary Analysis in XYW Space. (D) Secondary Analysis in XZW Space. (E) Secondary Analysis in YZW Space. (F) Suggested Coordinates.

(A, B, C, and D correspond to α, β, γ, and δ.)

SPECIFICATIONS

$H=\overline{S_1S_2S_3S_4S_5}$

$S_1=\overline{F_1F_2F_3F_4}$

$S_2=\overline{F_4F_5F_6F_7}$

$S_3=\overline{F_1F_5F_8F_9}$

$S_4=\overline{F_2F_6F_8F_{10}}$

$S_5=\overline{F_3F_7F_9F_{10}}$

$F_1=\overline{L_1L_2L_7}$	$L_1=\overline{P_1P_3}$	$P_1=(x_1,y_1,z_1,w_1)$
$F_2=\overline{L_1L_3L_8}$	$L_2=\overline{P_1P_4}$	$P_2=(x_2,y_2,z_2,w_2)$
$F_3=\overline{L_2L_3L_9}$	$L_3=\overline{P_1P_5}$	$P_3=(x_3,y_3,z_3,w_3)$
$F_4=\overline{L_7L_8L_9}$	$L_4=\overline{P_2P_3}$	$P_4=(x_4,y_4,z_4,w_4)$
$F_5=\overline{L_4L_5L_7}$	$L_5=\overline{P_2P_4}$	$P_5=(x_5,y_5,z_5,w_5)$
$F_6=\overline{L_4L_6L_8}$	$L_6=\overline{P_2P_5}$	
$F_7=\overline{L_5L_6L_9}$	$L_7=\overline{P_3P_4}$	
$F_8=\overline{L_1L_4L_{10}}$	$L_8=\overline{P_3P_5}$	
$F_9=\overline{L_2L_5L_{10}}$	$L_9=\overline{P_4P_5}$	
$F_{10}=\overline{L_3L_6L_{10}}$	$L_{10}=\overline{P_1P_2}$	

PRIMARY ANALYSIS

LENGTHS

$s_1=\sqrt{(x_3-x_1)^2+(y_3-y_1)^2+(z_3-z_1)^2+(w_3-w_1)^2}$

$s_2=\sqrt{(x_4-x_1)^2+(y_4-y_1)^2+(z_4-z_1)^2+(w_4-w_1)^2}$

$s_3=\sqrt{(x_5-x_1)^2+(y_5-y_1)^2+(z_5-z_1)^2+(w_5-w_1)^2}$

$s_4=\sqrt{(x_3-x_2)^2+(y_3-y_2)^2+(z_3-z_2)^2+(w_3-w_2)^2}$

$s_5=\sqrt{(x_4-x_2)^2+(y_4-y_2)^2+(z_4-z_2)^2+(w_4-w_2)^2}$

LENGTHS (Cont'd)

$s_6=\sqrt{(x_5-x_2)^2+(y_5-y_2)^2+(z_5-z_2)^2+(w_5-w_2)^2}$

$s_7=\sqrt{(x_4-x_3)^2+(y_4-y_3)^2+(z_4-z_3)^2+(w_4-w_3)^2}$

$s_8=\sqrt{(x_5-x_3)^2+(y_5-y_3)^2+(z_5-z_3)^2+(w_5-w_3)^2}$

$s_9=\sqrt{(x_5-x_4)^2+(y_5-y_4)^2+(z_5-z_4)^2+(w_5-w_4)^2}$

$s_{10}=\sqrt{(x_2-x_1)^2+(y_2-y_1)^2+(z_2-z_1)^2+(w_2-w_1)^2}$

DIRECTION COSINES

$\cos A_1=(x_3-x_1/s_1$	$\cos B_1=(y_3-y_1)/s_1$
$\cos C_1=(z_3-z_1)/s_1$	$\cos D_1=(w_3-w_1)/s_1$
$\cos A_2=(x_4-x_1)/s_2$	$\cos B_2=(y_4-y_1)/s_2$
$\cos C_2=(z_4-z_1)/s_2$	$\cos D_2=(w_4-w_1)/s_2$
$\cos A_3=(x_5-x_1)/s_2$	$\cos B_3=(y_5-y_1)/s_3$
$\cos C_3=(z_5-z_1)/s_3$	$\cos D_3=(w_5-w_1)/s_3$
$\cos A_4=(x_3-x_2)/s_4$	$\cos B_4=(y_3-y_2)/s_4$
$\cos C_4=(z_3-z_2)/s_4$	$\cos D_4=(w_3-w_2)/s_4$
$\cos A_5=(x_4-x_2)/s_5$	$\cos B_5=(y_4-y_2)/s_5$
$\cos C_5=(z_4-z_2)/s_5$	$\cos D_5=(w_4-w_2/s_5$
$\cos A_6=(x_5-x_2)/s_6$	$\cos B_6=(y_5-y_2)/s_6$
$\cos C_6=(z_5-z_2)/s_6$	$\cos D_6=(w_5-w_2)/s_6$
$\cos A_7=(x_4-x_3)/s_7$	$\cos B_7=(y_4-y_3)/s_7$
$\cos C_7=(z_4-z_3)/s_7$	$\cos D_7=(w_4-w_3)/s_7$
$\cos A_8=(x_5-x_3)/s_8$	$\cos B_8=(y_5-y_3)/s_8$
$\cos C_8=(z_5-z_3)/s_8$	$\cos D_8=(w_5-w_3)/s_8$
$\cos A_9=(x_5-x_4)/s_9$	$\cos B_9(y_5-y_4)/s_9$
$\cos C_9=(z_5-z_4)/s_9$	$\cos D_9=(w_5-w_4)/s_9$
$\cos A_{10}=(x_2-x_1)/s_{10}$	$\cos B_{10}=(y_2-y_1)/s_{10}$
$\cos C_{10}=(z_2-z_1)/s_{10}$	$\cos D_{10}=(w_2-w_1)/s_{10}$

A

Table 22-10. Specifications and Analyses of the Hyperspace Object Shown in Fig. 22-10. (A) Specifications and Analysis in Hyperspace. (B) Secondary Analysis in XYZ Space. (C) Secondary Analysis in XYW Space. (D) Secondary Analysis in XZW Space. (E) Secondary Analysis in YZW Space. (F) Suggested Coordinates. **(Continued from page 407.)**

ANGLES

$\cos\phi_{1,2}=\cos A_1\cos A_2+\cos B_1\cos B_2+\cos C_1\cos C_2+\cos D_1\cos D_2$

$\cos\phi_{1,3}=\cos A_1\cos A_3+\cos B_1\cos B_3+\cos C_1\cos C_3+\cos D_1\cos D_3$

$\cos\phi_{1,4}=\cos A_1\cos A_4+\cos B_1\cos B_4+\cos C_1\cos C_4+\cos D_1\cos D_4$

$\cos\phi_{1,7}=\cos A_1\cos A_7+\cos B_1\cos B_7+\cos C_1\cos C_7+\cos D_1\cos D_7$

$\cos\phi_{1,8}=\cos A_1\cos A_8+\cos B_1\cos B_8+\cos C_1\cos C_8+\cos D_1\cos D_8$

$\cos\phi_{1,10}=\cos A_1\cos A_{10}+\cos B_1\cos B_{10}+\cos C_1\cos C_{10}+\cos D_1\cos D_{10}$

$\cos\phi_{2,3}=\cos A_2\cos A_3+\cos B_2\cos B_3+\cos C_2\cos C_3+\cos D_2\cos D_3$

$\cos\phi_{2,5}=\cos A_2\cos A_5+\cos B_2\cos B_5+\cos C_2\cos C_5+\cos D_2\cos D_5$

$\cos\phi_{2,7}=\cos A_2\cos A_7+\cos B_2\cos B_7+\cos C_2\cos C_7+\cos D_2\cos D_7$

$\cos\phi_{2,10}=\cos A_2\cos A_{10}+\cos B_2\cos B_{10}+\cos C_2\cos C_{10}+\cos D_2\cos D_{10}$

$\cos\phi_{3,6}=\cos A_3\cos A_6+\cos B_3\cos B_6+\cos C_3\cos C_6+\cos D_3\cos D_6$

$\cos\phi_{3,8}=\cos A_3\cos A_8+\cos B_3\cos B_8+\cos C_3\cos C_8+\cos D_3\cos D_8$

$\cos\phi_{3,9}=\cos A_3\cos A_9+\cos B_3\cos B_9+\cos C_3\cos C_9+\cos D_3\cos D_9$

$\cos\phi_{3,10}=\cos A_3\cos A_{10}+\cos B_3\cos B_{10}+\cos C_3\cos C_{10}+\cos D_3\cos D_{10}$

$\cos\phi_{4,5}=\cos A_4\cos A_5+\cos B_4\cos B_5+\cos C_4\cos C_5+\cos D_4\cos D_5$

$\cos\phi_{4,6}=\cos A_4\cos A_6+\cos B_4\cos B_6+\cos C_4\cos C_6+\cos D_4\cos D_6$

$\cos\phi_{4,7}=\cos A_4\cos A_7+\cos B_4\cos B_7+\cos C_4\cos C_7+\cos D_4\cos D_7$

$\cos\phi_{4,8}=\cos A_4\cos A_8+\cos B_4\cos B_8+\cos C_4\cos C_8+\cos D_4\cos D_8$

$\cos\phi_{4,10}=\cos A_4\cos A_{10}+\cos B_4\cos B_{10}+\cos C_4\cos C_{10}+\cos D_4\cos D_{10}$

$\cos\phi_{5,6}=\cos A_5\cos A_6+\cos B_5\cos B_6+\cos C_5\cos C_6+\cos D_5\cos D_6$

$\cos\phi_{5,7}=\cos A_5\cos A_7+\cos B_5\cos B_7+\cos C_5\cos C_7+\cos D_5\cos D_7$

$\cos\phi_{5,9}=\cos A_5\cos A_9+\cos B_5\cos B_9+\cos C_5\cos C_9+\cos D_5\cos D_9$

$\cos\phi_{5,10}=\cos A_5\cos A_{10}+\cos B_5\cos B_{10}+\cos C_5\cos C_{10}+\cos D_5\cos D_{10}$

$\cos\phi_{6,8}=\cos A_6\cos A_8+\cos B_6\cos B_8+\cos C_6\cos C_8+\cos D_6\cos D_8$

$\cos\phi_{6,9}=\cos A_6\cos A_9+\cos B_6\cos B_9+\cos C_6\cos C_9+\cos D_6\cos D_9$

$\cos\phi_{6,10}=\cos A_6\cos A_{10}+\cos B_6\cos B_{10}+\cos C_6\cos C_{10}+\cos D_6\cos D_{10}$

$\cos\phi_{7,8}=\cos A_7\cos A_8+\cos B_7\cos B_8+\cos C_7\cos C_8+\cos D_7\cos D_8$

$\cos\phi_{7,9}=\cos A_7\cos A_9+\cos B_7\cos B_9+\cos C_7\cos C_9+\cos D_7\cos D_9$

$\cos\phi_{8,9}=\cos A_8\cos A_9+\cos B_8\cos B_9+\cos C_8\cos C_9+\cos D_8\cos D_9$

CHECK

F_1: $\phi_{1,2}+\phi_{1,7}+\phi_{2,7}=180^\circ$

F_2: $\phi_{1,3}+\phi_{1,8}+\phi_{3,8}=180^\circ$

F_3: $\phi_{2,3}+\phi_{2,9}+\phi_{3,9}=180^\circ$

F_4: $\phi_{7,8}+\phi_{7,9}+\phi_{8,9}=180^\circ$

F_5: $\phi_{4,5}+\phi_{4,7}+\phi_{5,7}=180^\circ$

F_6: $\phi_{4,6}+\phi_{4,8}+\phi_{6,8}=180^\circ$

F_7: $\phi_{5,6}+\phi_{5,9}+\phi_{6,9}=180^\circ$

F_8: $\phi_{1,4}+\phi_{1,10}+\phi_{4,10}=180^\circ$

F_9: $\phi_{2,5}+\phi_{2,10}+\phi_{5,10}=180^\circ$

F_{10}: $\phi_{3,6}+\phi_{3,10}+\phi_{6,10}=180^\circ$

A

SECONDARY ANALYSIS -- XYZ SPACE

PLOTTED POINTS

$P_1=(x_1,y_1,z_1)$

$P_2=(x_2,y_2,z_2)$

$P_3=(x_3,y_3,z_3)$

$P_4=(x_4,y_4,z_4)$

$P_5=(x_5,y_5,z_5)$

LENGTHS

$s_1=\sqrt{(x_3-x_1)^2+(y_3-y_1)^2+(z_3-z_1)^2}$

$s_2=\sqrt{(x_4-x_1)^2+(y_4-y_1)^2+(z_4-z_1)^2}$

$s_3=\sqrt{(x_5-x_1)^2+(y_5-y_1)^2+(z_5-z_1)^2}$

$s_4=\sqrt{(x_3-x_2)^2+(y_3-y_2)^2+(z_3-z_2)^2}$

$s_5=\sqrt{(x_4-x_2)^2+(y_4-y_2)^2+(z_4-z_2)^2}$

$s_6=\sqrt{(x_5-x_2)^2+(y_5-y_2)^2+(z_5-z_2)^2}$

$s_7=\sqrt{(x_4-x_3)^2+(y_4-y_3)^2+(z_4-z_3)^2}$

$s_8=\sqrt{(x_5-x_3)^2+(y_5-y_3)^2+(z_5-z_3)^2}$

$s_9=\sqrt{(x_5-x_4)^2+(y_5-y_4)^2+(z_5-z_4)^2}$

$s_{1\emptyset}=\sqrt{(x_2-x_1)^2+(y_2-y_1)^2+(z_2-z_1)^2}$

DIRECTION COSINES

$\cos A_1=(x_3-x_1/s_1$ $\quad$ $\cos B_1=(y_3-y_1)/s_1$

$\cos C_1=(z_3-z_1)/s_1$

$\cos A_2=(x_4-x_1)/s_2$ $\quad$ $\cos B_2=(y_4-y_1)/s_2$

$\cos C_2=(z_4-z_1)/s_2$

$\cos A_3=(x_5-x_1)/s_2$ $\quad$ $\cos B_3=(y_5-y_1)/s_3$

$\cos C_3=(z_5-z_1)/s_3$

DIRECTION COSINES (Cont'd)

$\cos A_4=(x_3-x_2)/s_4$ $\quad$ $\cos B_4=(y_3-y_2)/s_4$

$\cos C_4=(z_3-z_2)/s_4$

$\cos A_5=(x_4-x_2)/s_5$ $\quad$ $\cos B_5=(y_4-y_2)/s_5$

$\cos C_5=(z_4-z_2)/s_5$

$\cos A_6=(x_5-x_2)/s_6$ $\quad$ $\cos B_6=(y_5-y_2)/s_6$

$\cos C_6=(z_5-z_2)/s_6$

$\cos A_7=(x_4-x_3)/s_7$ $\quad$ $\cos B_7=(y_4-y_3)/s_7$

$\cos C_7=(z_4-z_3)/s_7$

$\cos A_8=(x_5-x_3)/s_8$ $\quad$ $\cos B_8=(y_5-y_3)/s_8$

$\cos C_8=(z_5-z_3)/s_8$

$\cos A_9=(x_5-x_4)/s_9$ $\quad$ $\cos B_9(y_5-y_4)/s_9$

$\cos C_9=(z_5-z_4)/s_9$

$\cos A_{1\emptyset}=(x_2-x_1)/s_{1\emptyset}$ $\quad$ $\cos B_{1\emptyset}=(y_2-y_1)/s_{1\emptyset}$

$\cos C_{1\emptyset}=(z_2-z_1)/s_{1\emptyset}$

ANGLES

$\cos\emptyset_{1,2}=\cos A_1\cos A_2+\cos B_1\cos B_2+\cos C_1\cos C_2$

$\cos\emptyset_{1,3}=\cos A_1\cos A_3+\cos B_1\cos B_3+\cos C_1\cos C_3$

$\cos\emptyset_{1,4}=\cos A_1\cos A_4+\cos B_1\cos B_4+\cos C_1\cos C_4$

$\cos\emptyset_{1,7}=\cos A_1\cos A_7+\cos B_1\cos B_7+\cos C_1\cos C_7$

$\cos\emptyset_{1,8}=\cos A_1\cos A_8+\cos B_1\cos B_8+\cos C_1\cos C_8$

$\cos\emptyset_{1,1\emptyset}=\cos A_1\cos A_{1\emptyset}+\cos B_1\cos B_{1\emptyset}+\cos C_1\cos C_{1\emptyset}$

$\cos\emptyset_{2,3}=\cos A_2\cos A_3+\cos B_2\cos B_3+\cos C_2\cos C_3$

$\cos\emptyset_{2,5}=\cos A_2\cos A_5+\cos B_2\cos B_5+\cos C_2\cos C_5$

$\cos\emptyset_{2,7}=\cos A_2\cos A_7+\cos B_2\cos B_7+\cos C_2\cos C_7$

$\cos\emptyset_{2,1\emptyset}=\cos A_2\cos A_{1\emptyset}+\cos B_2\cos B_{1\emptyset}+\cos C_2\cos C_{1\emptyset}$

$\cos\emptyset_{3,6}=\cos A_3\cos A_6+\cos B_3\cos B_6+\cos C_3\cos C_6$

$\cos\emptyset_{3,8}=\cos A_3\cos A_8+\cos B_3\cos B_8+\cos C_3\cos C_8$

$\cos\emptyset_{3,9}=\cos A_3\cos A_9+\cos B_3\cos B_9+\cos C_3\cos C_9$

B

Table 22-10. Specifications and Analyses of the Hyperspace Object Shown in Fig. 22-10. (A) Specifications and Analysis in Hyperspace. (B) Secondary Analysis in XYZ Space. (C) Secondary in XYW Space. (D) Secondary Analysis in XZW Space. (E) Secondary Analysis in YZW Space. (F) Suggested Coordinates. (Continued from page 409.)

ANGLES (Cont'd)

$\cos\phi_{3,10}=\cos A_3\cos A_{10}+\cos B_3\cos B_{10}+\cos C_3\cos C_{10}$

$\cos\phi_{4,5}=\cos A_4\cos A_5+\cos B_4\cos B_5+\cos C_4\cos C_5$

$\cos\phi_{4,6}=\cos A_4\cos A_6+\cos B_4\cos B_6+\cos C_4\cos C_6$

$\cos\phi_{4,7}=\cos A_4\cos A_7+\cos B_4\cos B_7+\cos C_4\cos C_7$

$\cos\phi_{4,8}=\cos A_4\cos A_8+\cos B_4\cos B_8+\cos C_4\cos C_8$

$\cos\phi_{4,10}=\cos A_4\cos A_{10}+\cos B_4\cos B_{10}+\cos C_4\cos C_{10}$

$\cos\phi_{5,6}=\cos A_5\cos A_6+\cos B_5\cos B_6+\cos C_5\cos C_6$

$\cos\phi_{5,7}=\cos A_5\cos A_7+\cos B_5\cos B_7+\cos C_5\cos C_7$

$\cos\phi_{5,9}=\cos A_5\cos A_9+\cos B_5\cos B_9+\cos C_5\cos C_9$

$\cos\phi_{5,10}=\cos A_5\cos A_{10}+\cos B_5\cos B_{10}+\cos C_5\cos C_{10}$

$\cos\phi_{6,8}=\cos A_6\cos A_8+\cos B_6\cos B_8+\cos C_6\cos C_8$

$\cos\phi_{6,9}=\cos A_6\cos A_9+\cos B_6\cos B_9+\cos C_6\cos C_9$

$\cos\phi_{6,10}=\cos A_6\cos A_{10}+\cos B_6\cos B_{10}+\cos C_6\cos C_{10}$

$\cos\phi_{7,8}=\cos A_7\cos A_8+\cos B_7\cos B_8+\cos C_7\cos C_8$

$\cos\phi_{7,9}=\cos A_7\cos A_9+\cos B_7\cos B_9+\cos C_7\cos C_9$

$\cos\phi_{8,9}=\cos A_8\cos A_9+\cos B_8\cos B_9+\cos C_8\cos C_9$

CHECK

F_1:	$\phi_{1,2}+\phi_{1,7}+\phi_{2,7}=180^\circ$
F_2:	$\phi_{1,3}+\phi_{1,8}+\phi_{3,8}=180^\circ$
F_3:	$\phi_{2,3}+\phi_{2,9}+\phi_{3,9}=180^\circ$
F_4:	$\phi_{7,8}+\phi_{7,9}+\phi_{8,9}=180^\circ$
F_5:	$\phi_{4,5}+\phi_{4,7}+\phi_{5,7}=180^\circ$
F_6:	$\phi_{4,6}+\phi_{4,8}+\phi_{6,8}=180^\circ$
F_7:	$\phi_{5,6}+\phi_{5,9}+\phi_{6,9}=180^\circ$
F_8:	$\phi_{1,4}+\phi_{1,10}+\phi_{4,10}=180^\circ$
F_9:	$\phi_{2,5}+\phi_{2,10}+\phi_{5,10}=180^\circ$
F_{10}:	$\phi_{3,6}+\phi_{3,10}+\phi_{6,10}=180^\circ$

B

SECONDARY ANALYSIS -- XYW SPACE

PLOTTED POINTS

$P_1=(x_1,y_1,w_1)$

$P_2=(x_2,y_2,w_2)$

$P_3=(x_3,y_3,w_3)$

$P_4=(x_4,y_4,w_4)$

$P_5=(x_5,y_5,w_5)$

LENGTHS

$s_1=\sqrt{(x_3-x_1)^2+(y_3-y_1)^2+(w_3-w_1)^2}$

$s_2=\sqrt{(x_4-x_1)^2+(y_4-y_1)^2+(w_4-w_1)^2}$

$s_3=\sqrt{(x_5-x_1)^2+(y_5-y_1)^2+(w_5-w_1)^2}$

$s_4=\sqrt{(x_3-x_2)^2+(y_3-y_2)^2+(w_3-w_2)^2}$

$s_5=\sqrt{(x_4-x_2)^2+(y_4-y_2)^2+(w_4-w_2)^2}$

$s_6=\sqrt{(x_5-x_2)^2+(y_5-y_2)^2+(w_5-w_2)^2}$

$s_7=\sqrt{(x_4-x_3)^2+(y_4-y_3)^2+(w_4-w_3)^2}$

$s_8=\sqrt{(x_5-x_3)^2+(y_5-y_3)^2+(w_5-w_3)^2}$

$s_9=\sqrt{(x_5-x_4)^2+(y_5-y_4)^2+(w_5-w_4)^2}$

$s_{10}=\sqrt{(x_2-x_1)^2+(y_2-y_1)^2+(w_2-w_1)^2}$

C

DIRECTION COSINES

$cosA_1=(x_3-x_1/s_1$ $\quad$ $cosB_1=(y_3-y_1)/s_1$

$cosD_1=(w_3-w_1)/s_1$

$cosA_2=(x_4-x_1)/s_2$ $\quad$ $cosB_2=(y_4-y_1)/s_2$

$cosD_2=(w_4-w_1)/s_2$

$cosA_3=(x_5-x_1)/s_2$ $\quad$ $cosB_3=(y_5-y_1)/s_3$

$cosD_3=(w_5-w_1)/s_3$

$cosA_4=(x_3-x_2)/s_4$ $\quad$ $cosB_4=(y_3-y_2)/s_4$

$cosD_4=(w_3-w_2)/s_4$

$cosA_5=(x_4-x_2)/s_5$ $\quad$ $cosB_5=(y_4-y_2)/s_5$

$cosD_5=(w_4-w_2/s_5$

$cosA_6=(x_5-x_2)/s_6$ $\quad$ $cosB_6=(y_5-y_2)/s_6$

$cosD_6=(w_5-w_2)/s_6$

$cosA_7=(x_4-x_3)/s_7$ $\quad$ $cosB_7=(y_4-y_3)/s_7$

$cosD_7=(w_4-w_3)/s_7$

$cosA_8=(x_5-x_3)/s_8$ $\quad$ $cosB_8=(y_5-y_3)/s_8$

$cosD_8=(w_5-w_3)/s_8$

$cosA_9=(x_5-x_4)/s_9$ $\quad$ $cosB_9(y_5-y_4)/s_9$

$cosD_9=(w_5-w_4)/s_9$

$cosA_{10}=(x_2-x_1)/s_{10}$ $\quad$ $cosB_{10}=(y_2-y_1)/s_{10}$

$cosD_{10}=(w_2-w_1)/s_{10}$

ANGLES

$cos\phi_{1,2}=cosA_1cosA_2+cosB_1cosB_2+cosD_1cosD_2$

$cos\phi_{1,3}=cosA_1cosA_3+cosB_1cosB_3+cosD_1cosD_3$

Table 22-10. Specifications and Analyses of the Hyperspace Object Shown in Fig. 22-10. (A) Specifications and Analysis in Hyperspace. (B) Secondary Analysis in XYZ Space. (C) Secondary Analysis in XYW Space. (D) Secondary Analysis in XZW Space. (E) Secondary Analysis in YZW Space. (F) Suggested Coordinates. **(Continued from page 411)**

ANGLES (Cont'd)

$\cos\phi_{1,4}=\cos A_1\cos A_4+\cos B_1\cos B_4+\cos D_1\cos D_4$

$\cos\phi_{1,7}=\cos A_1\cos A_7+\cos B_1\cos B_7+\cos D_1\cos D_7$

$\cos\phi_{1,8}=\cos A_1\cos A_8+\cos B_1\cos B_8+\cos D_1\cos D_8$

$\cos\phi_{1,10}=\cos A_1\cos A_{10}+\cos B_1\cos B_{10}+\cos D_1\cos D_{10}$

$\cos\phi_{2,3}=\cos A_2\cos A_3+\cos B_2\cos B_3+\cos D_2\cos D_3$

$\cos\phi_{2,5}=\cos A_2\cos A_5+\cos B_2\cos B_5+\cos D_2\cos D_5$

$\cos\phi_{2,7}=\cos A_2\cos A_7+\cos B_2\cos B_7+\cos D_2\cos D_7$

$\cos\phi_{2,10}=\cos A_2\cos A_{10}+\cos B_2\cos B_{10}+\cos D_2\cos D_{10}$

$\cos\phi_{3,6}=\cos A_3\cos A_6+\cos B_3\cos B_6+\cos D_3\cos D_6$

$\cos\phi_{3,8}=\cos A_3\cos A_8+\cos B_3\cos B_8+\cos D_3\cos D_8$

$\cos\phi_{3,9}=\cos A_3\cos A_9+\cos B_3\cos B_9+\cos D_3\cos D_9$

$\cos\phi_{3,10}=\cos A_3\cos A_{10}+\cos B_3\cos B_{10}+\cos D_3\cos D_{10}$

$\cos\phi_{4,5}=\cos A_4\cos A_5+\cos B_4\cos B_5+\cos D_4\cos D_5$

$\cos\phi_{4,6}=\cos A_4\cos A_6+\cos B_4\cos B_6+\cos D_4\cos D_6$

$\cos\phi_{4,7}=\cos A_4\cos A_7+\cos B_4\cos B_7+\cos D_4\cos D_7$

$\cos\phi_{4,8}=\cos A_4\cos A_8+\cos B_4\cos B_8+\cos D_4\cos D_8$

$\cos\phi_{4,10}=\cos A_4\cos A_{10}+\cos B_4\cos B_{10}+\cos D_4\cos D_{10}$

$\cos\phi_{5,6}=\cos A_5\cos A_6+\cos B_5\cos B_6+\cos D_5\cos D_6$

$\cos\phi_{5,7}=\cos A_5\cos A_7+\cos B_5\cos B_7+\cos D_5\cos D_7$

$\cos\phi_{5,9}=\cos A_5\cos B_9+\cos B_5\cos B_9+\cos D_5\cos D_9$

$\cos\phi_{5,10}=\cos A_5\cos A_{10}+\cos B_5\cos B_{10}+\cos D_5\cos D_{10}$

$\cos\phi_{6,8}=\cos A_6\cos A_8+\cos B_6\cos B_8+\cos D_6\cos D_8$

$\cos\phi_{6,9}=\cos A_6\cos A_9+\cos B_6\cos B_9+\cos D_6\cos D_9$

$\cos\phi_{6,10}=\cos A_6\cos A_{10}+\cos B_6\cos B_{10}+\cos D_6\cos D_{10}$

$\cos\phi_{7,8}=\cos A_7\cos A_8+\cos B_7\cos B_8+\cos D_7\cos D_8$

$\cos\phi_{7,9}=\cos A_7\cos A_9+\cos B_7\cos B_9+\cos D_7\cos D_9$

$\cos\phi_{8,9}=\cos A_8\cos A_9+\cos B_8\cos B_9+\cos D_8\cos D_9$

CHECK

F_1:	$\phi_{1,2}+\phi_{1,7}+\phi_{2,7}=180^\circ$
F_2:	$\phi_{1,3}+\phi_{1,8}+\phi_{3,8}=180^\circ$
F_3:	$\phi_{2,3}+\phi_{2,9}+\phi_{3,9}=180^\circ$
F_4:	$\phi_{7,8}+\phi_{7,9}+\phi_{8,9}=180^\circ$
F_5:	$\phi_{4,5}+\phi_{4,7}+\phi_{5,7}=180^\circ$
F_6:	$\phi_{4,6}+\phi_{4,8}+\phi_{6,8}=180^\circ$
F_7:	$\phi_{5,6}+\phi_{5,9}+\phi_{6,9}=180^\circ$
F_8:	$\phi_{1,4}+\phi_{1,10}+\phi_{4,10}=180^\circ$
F_9:	$\phi_{2,5}+\phi_{2,10}+\phi_{5,10}=180^\circ$
F_{10}:	$\phi_{3,6}+\phi_{3,10}+\phi_{6,10}=180^\circ$

C

SECONDARY ANALYSIS -- XZW SPACE

PLOTTED POINTS

$P_1=(x_1,z_1,w_1)$

$P_2=(x_2,z_2,w_2)$

$P_3=(x_3,z_3,w_3)$

$P_4=(x_4,z_4,w_4)$

$P_5=(x_5,z_5,w_5)$

LENGTHS

$s_1=\sqrt{(x_3-x_1)^2+(z_3-z_1)^2+(w_3-w_1)^2}$

$s_2=\sqrt{(x_4-x_1)^2+(z_4-z_1)^2+(w_4-w_1)^2}$

$s_3=\sqrt{(x_5-x_1)^2+(z_5-z_1)^2+(w_5-w_1)^2}$

$s_4=\sqrt{(x_3-x_2)^2+(z_3-z_2)^2+(w_3-w_2)^2}$

$s_5=\sqrt{(x_4-x_2)^2+(z_4-z_2)^2+(w_4-w_2)^2}$

$s_6=\sqrt{(x_5-x_2)^2+(z_5-z_2)^2+(w_5-w_2)^2}$

$s_7=\sqrt{(x_4-x_3)^2+(z_4-z_3)^2+(w_4-w_3)^2}$

$s_8=\sqrt{(x_5-x_3)^2+(z_5-z_3)^2+(w_5-w_3)^2}$

$s_9=\sqrt{(x_5-x_4)^2+(z_5-z_4)^2+(w_5-w_4)^2}$

$s_{10}=\sqrt{(x_2-x_1)^2+(z_2-z_1)^2+(w_2-w_1)^2}$

DIRECTION COSINES

$\cos A_1=(x_3-x_1/s_1$ $\qquad \cos C_1=(z_3-z_1)/s_1$

$\cos D_1=(w_3-w_1)/s_1$

$\cos A_2=(x_4-x_1)/s_2$ $\qquad \cos C_2=(z_4-z_1)/s_2$

$\cos D_2=(w_4-w_1)/s_2$

$\cos A_3=(x_5-x_1)/s_2$ $\qquad \cos C_3=(z_5-z_1)/s_3$

$\cos D_3=(w_5-w_1)/s_3$

$\cos A_4=(x_3-x_2)/s_4$ $\qquad \cos C_4=(z_3-z_2)/s_4$

$\cos D_4=(w_3-w_2)/s_4$

$\cos A_5=(x_4-x_2)/s_5$ $\qquad \cos C_5=(z_4-z_2)/s_5$

$\cos D_5=(w_4-w_2/s_5$

$\cos A_6=(x_5-x_2)/s_6$ $\qquad \cos C_6=(z_5-z_2)/s_6$

$\cos D_6=(w_5-w_2)/s_6$

$\cos A_7=(x_4-x_3)/s_7$ $\qquad \cos C_7=(z_4-z_3)/s_7$

$\cos D_7=(w_4-w_3)/s_7$

$\cos A_8=(x_5-x_3)/s_8$ $\qquad \cos C_8=(z_5-z_3)/s_8$

$\cos D_8=(w_5-w_3)/s_8$

$\cos A_9=(x_5-x_4)/s_9$ $\qquad \cos C_9=(z_5-z_4)/s_9$

$\cos D_9=(w_5-w_4)/s_9$

$\cos A_{10}=(x_2-x_1)/s_{10}$ $\qquad \cos C_{10}=(z_2-z_1)/s_{10}$

$\cos D_{10}=(w_2-w_1)/s_{10}$

D

Table 22-10. Specifications and Analyses of the Hyperspace Object Shown in Fig. 22-10. (A) Specifications and Analysis in Hyperspace. (B) Secondary Analysis in XYZ Space. (C) Secondary Analysis in XYW Space. (D) Secondary Analysis in XZW Space. (E) Secondary Analysis in YZW Space. (F) Suggested Coordinates. (Continued from page 413.)

ANGLES

$\cos\phi_{1,2}=\cos A_1\cos A_2+\cos C_1\cos C_2+\cos D_1\cos D_2$

$\cos\phi_{1,3}=\cos A_1\cos A_3+\cos C_1\cos C_3+\cos D_1\cos D_3$

$\cos\phi_{1,4}=\cos A_1\cos A_4+\cos C_1\cos C_4+\cos D_1\cos D_4$

$\cos\phi_{1,7}=\cos A_1\cos A_7+\cos C_1\cos C_7+\cos D_1\cos D_7$

$\cos\phi_{1,8}=\cos A_1\cos A_8+\cos C_1\cos C_8+\cos D_1\cos D_8$

$\cos\phi_{1,10}=\cos A_1\cos A_{10}+\cos C_1\cos C_{10}+\cos D_1\cos D_{10}$

$\cos\phi_{2,3}=\cos A_2\cos A_3+\cos C_2\cos C_3+\cos D_2\cos D_3$

$\cos\phi_{2,5}=\cos A_2\cos A_5+\cos C_2\cos C_5+\cos D_2\cos D_5$

$\cos\phi_{2,7}=\cos A_2\cos A_7+\cos C_2\cos C_7+\cos D_2\cos D_7$

$\cos\phi_{2,10}=\cos A_2\cos A_{10}+\cos C_2\cos C_{10}+\cos D_2\cos D_{10}$

$\cos\phi_{3,6}=\cos A_3\cos A_6+\cos C_3\cos C_6+\cos D_3\cos D_6$

$\cos\phi_{3,8}=\cos A_3\cos A_8+\cos C_3\cos C_8+\cos D_3\cos D_8$

$\cos\phi_{3,9}=\cos A_3\cos A_9+\cos C_3\cos C_9+\cos D_3\cos D_9$

$\cos\phi_{3,10}=\cos A_3\cos A_{10}+\cos C_3\cos C_{10}+\cos D_3\cos D_{10}$

$\cos\phi_{4,5}=\cos A_4\cos A_5+\cos C_4\cos C_5+\cos D_4\cos D_5$

$\cos\phi_{4,6}=\cos A_4\cos A_6+\cos C_4\cos C_6+\cos D_4\cos D_6$

$\cos\phi_{4,7}=\cos A_4\cos A_7+\cos C_4\cos C_7+\cos D_4\cos D_7$

$\cos\phi_{4,8}=\cos A_4\cos A_8+\cos C_4\cos C_8+\cos D_4\cos D_8$

$\cos\phi_{4,10}=\cos A_4\cos A_{10}+\cos C_4\cos C_{10}+\cos D_4\cos D_{10}$

$\cos\phi_{5,6}=\cos A_5\cos A_6+\cos C_5\cos C_6+\cos D_5\cos D_6$

$\cos\phi_{5,7}=\cos A_5\cos A_7+\cos C_5\cos C_7+\cos D_5\cos D_7$

$\cos\phi_{5,9}=\cos A_5\cos A_9+\cos C_5\cos C_9+\cos D_5\cos D_9$

$\cos\phi_{5,10}=\cos A_5\cos A_{10}+\cos C_5\cos C_{10}+\cos D_5\cos D_{10}$

$\cos\phi_{6,8}=\cos A_6\cos A_8+\cos C_6\cos C_8+\cos D_6\cos D_8$

$\cos\phi_{6,9}=\cos A_6\cos A_9+\cos C_6\cos C_9+\cos D_6\cos D_9$

$\cos\phi_{6,10}=\cos A_6\cos A_{10}+\cos C_6\cos C_{10}+\cos D_6\cos D_{10}$

$\cos\phi_{7,8}=\cos A_7\cos A_8+\cos C_7\cos C_8+\cos D_7\cos D_8$

$\cos\phi_{7,9}=\cos A_7\cos A_9+\cos C_7\cos C_9+\cos D_7\cos D_9$

D

$\cos\phi_{8,9}=\cos A_8\cos A_9+\cos C_8\cos C_9+\cos D_8\cos D_9$

CHECK

F_1:	$\phi_{1,2}+\phi_{1,7}+\phi_{2,7}=180^\circ$
F_2:	$\phi_{1,3}+\phi_{1,8}+\phi_{3,8}=180^\circ$
F_3:	$\phi_{2,3}+\phi_{2,9}+\phi_{3,9}=180^\circ$
F_4:	$\phi_{7,8}+\phi_{7,9}+\phi_{8,9}=180^\circ$
F_5:	$\phi_{4,5}+\phi_{4,7}+\phi_{5,7}=180^\circ$
F_6:	$\phi_{4,6}+\phi_{4,8}+\phi_{6,8}=180^\circ$
F_7:	$\phi_{5,6}+\phi_{5,9}+\phi_{6,9}=180^\circ$
F_8:	$\phi_{1,4}+\phi_{1,10}+\phi_{4,10}=180^\circ$
F_9:	$\phi_{2,5}+\phi_{2,10}+\phi_{5,10}=180^\circ$
F_{10}:	$\phi_{3,6}+\phi_{3,10}+\phi_{6,10}=180^\circ$

SECONDARY ANALYSIS -- YZW SPACE

PLOTTED POINTS

$P_1=(y_1,z_1,w_1)$

$P_2=(y_2,z_2,w_2)$

$P_3=(y_3,z_3,w_3)$

$P_4=(y_4,z_4,w_4)$

$P_5=(y_5,z_5,w_5)$

LENGTHS

$s_1=\sqrt{(y_3-y_1)^2+(z_3-z_1)^2+(w_3-w_1)^2}$

$s_2=\sqrt{(y_4-y_1)^2+(z_4-z_1)^2+(w_4-w_1)^2}$

$s_3=\sqrt{(y_5-y_1)^2+(z_5-z_1)^2+(w_5-w_1^2}$

$s_4=\sqrt{(y_3-y_2)^2+(z_3-z_2)^2+(w_3-w_2)^2}$

$s_5=\sqrt{(y_4-y_2)^2+(z_4-z_2)^2+(w_4-w_2)^2}$

$s_6=\sqrt{(y_5-y_2)^2+(z_5-z_2)^2+(w_5-w_2)^2}$

$s_7=\sqrt{(y_4-y_3)^2+(z_4-z_3)^2+(w_4-w_3)^2}$

$s_8=\sqrt{(y_5-y_3)^2+(z_5-z_3)^2+(w_5-w_3)^2}$

$s_9=\sqrt{(y_5-y_4)^2+(z_5-z_4)^2+(w_5-w_4)^2}$

$s_{10}=\sqrt{(y_2-y_1)^2+(z_2-z_1)^2+(w_2-w_1)^2}$

DIRECTION COSINES

$\cos B_1=(y_3-y_1)/s_1$ $\cos C_1=(z_3-z_1)/s_1$

$\cos D_1=(w_3-w_1)/s_1$

$\cos B_2=(y_4-y_1)/s_2$ $\cos C_2=(z_4-z_1)/s_2$

$\cos D_2=(w_4-w_1)/s_2$

$\cos B_3=(y_5-y_1)/s_3$ $\cos C_3=(z_5-z_1)/s_3$

$\cos D_3=(w_5-w_1)/s_3$

$\cos B_4=(y_3-y_2)/s_4$ $\cos C_4=(z_3-z_2)/s_4$

$\cos D_4=(w_3-w_2)/s_4$

$\cos B_5=(y_4-y_2)/s_5$ $\cos C_5=(z_4-z_2)/s_5$

$\cos D_5=(w_4-w_2/s_5$

$\cos B_6=(y_5-y_2)/s_6$ $\cos C_6=(z_5-z_2)/s_6$

$\cos D_6=(w_5-w_2)/s_6$

$\cos B_7=(y_4-y_3)/s_7$ $\cos C_7=(z_4-z_3)/s_7$

$\cos D_7=(w_4-w_3)/s_7$

$\cos B_8=(y_5-y_3)/s_8$ $\cos C_8=(z_5-z_3)/s_8$

$\cos D_8=(w_5-w_3)/s_8$

$\cos B_9(y_5-y_4)/s_9$ $\cos C_9=(z_5-z_4)/s_9$

$\cos D_9=(w_5-w_4)/s_9$

$\cos B_{10}=(y_2-y_1)/s_{10}$ $\cos C_{10}=(z_2-z_1)/s_{10}$

$\cos D_{10}=(w_2-w_1)/s_{10}$

E

Table 22-10. Specifications and Analyses of the Hyperspace Object Shown in Fig. 22-10. (A) Specifications and Analysis in Hyperspace. (B) Secondary Analysis in XYZ Space. (C) Secondary Analysis in XYW Space. (D) Secondary Analysis in XZW Space. (E) Secondary Analysis in YZW Space. (F) Suggested Coordinates. (Continued from page 415.)

ANGLES

$\cos\phi_{1,2}=\cos B_1\cos B_2+\cos C_1\cos C_2+\cos D_1\cos D_2$

$\cos\phi_{1,3}=\cos B_1\cos B_3+\cos C_1\cos C_3+\cos D_1\cos D_3$

$\cos\phi_{1,4}=\cos B_1\cos B_4+\cos C_1\cos C_4+\cos D_1\cos D_4$

$\cos\phi_{1,7}=\cos B_1\cos B_7+\cos C_1\cos C_7+\cos D_1\cos D_7$

$\cos\phi_{1,8}=\cos B_1\cos B_8+\cos C_1\cos C_8+\cos D_1\cos D_8$

$\cos\phi_{1,10}=\cos B_1\cos B_{10}+\cos C_1\cos C_{10}+\cos D_1\cos D_{10}$

$\cos\phi_{2,3}=\cos B_2\cos B_3+\cos C_2\cos C_3+\cos D_2\cos D_3$

$\cos\phi_{2,5}=\cos B_2\cos B_5+\cos C_2\cos C_5+\cos D_2\cos D_5$

$\cos\phi_{2,7}=\cos B_2\cos B_7+\cos C_2\cos C_7+\cos D_2\cos D_7$

$\cos\phi_{2,10}=\cos B_2\cos B_{10}+\cos C_2\cos C_{10}+\cos D_2\cos D_{10}$

$\cos\phi_{3,6}=\cos B_3\cos B_6+\cos C_3\cos C_6+\cos D_3\cos D_6$

$\cos\phi_{3,8}=\cos B_3\cos B_8+\cos C_3\cos C_8+\cos D_3\cos D_8$

$\cos\phi_{3,9}=\cos B_3\cos B_9+\cos C_3\cos C_9+\cos D_3\cos D_9$

$\cos\phi_{3,10}=\cos B_3\cos B_{10}+\cos C_3\cos C_{10}+\cos D_3\cos D_{10}$

$\cos\phi_{4,5}=\cos B_4\cos B_5+\cos C_4\cos C_5+\cos D_4\cos D_5$

$\cos\phi_{4,6}=\cos B_4\cos B_6+\cos C_4\cos C_6+\cos D_4\cos D_6$

$\cos\phi_{4,7}=\cos B_4\cos B_7+\cos C_4\cos C_7+\cos D_4\cos D_7$

$\cos\phi_{4,8}=\cos B_4\cos B_8+\cos C_4\cos C_8+\cos D_4\cos D_8$

$\cos\phi_{4,10}=\cos B_4\cos B_{10}+\cos C_4\cos C_{10}+\cos D_4\cos D_{10}$

$\cos\phi_{5,6}=\cos B_5\cos B_6+\cos C_5\cos C_6+\cos D_5\cos D_6$

$\cos\phi_{5,7}=\cos B_5\cos B_7+\cos C_5\cos C_7+\cos D_5\cos D_7$

$\cos\phi_{5,9}=\cos B_5\cos B_9+\cos C_5\cos C_9+\cos D_5\cos D_9$

$\cos\phi_{5,10}=\cos B_5\cos B_{10}+\cos C_5\cos C_{10}+\cos D_5\cos D_{10}$

$\cos\phi_{6,8}=\cos B_6\cos B_8+\cos C_6\cos C_8+\cos D_6\cos D_8$

$\cos\phi_{6,9}=\cos B_6\cos B_9+\cos C_6\cos C_9+\cos D_6\cos D_9$

$\cos\phi_{6,10}=\cos B_6\cos B_{10}+\cos C_6\cos C_{10}+\cos D_6\cos D_{10}$

$\cos\phi_{7,8}=\cos B_7\cos B_8+\cos C_7\cos C_8+\cos D_7\cos D_8$

$\cos\phi_{7,9}=\cos B_7\cos B_9+\cos C_7\cos C_9+\cos D_7\cos D_9$

$\cos\phi_{8,9}=\cos B_8\cos B_9+\cos C_8\cos C_9+\cos D_8\cos D_9$

CHECK

F_1: $\phi_{1,2}+\phi_{1,7}+\phi_{2,7}=180^\circ$

F_2: $\phi_{1,3}+\phi_{1,8}+\phi_{3,8}=180^\circ$

F_3: $\phi_{2,3}+\phi_{2,9}+\phi_{3,9}=180^\circ$

F_4: $\phi_{7,8}+\phi_{7,9}+\phi_{8,9}=180^\circ$

F_5: $\phi_{4,5}+\phi_{4,7}+\phi_{5,7}=180^\circ$

F_6: $\phi_{4,6}+\phi_{4,8}+\phi_{6,8}=180^\circ$

F_7: $\phi_{5,6}+\phi_{5,9}+\phi_{6,9}=180^\circ$

F_8: $\phi_{1,4}+\phi_{1,10}+\phi_{4,10}=180^\circ$

F_9: $\phi_{2,5}+\phi_{2,10}+\phi_{5,10}=180^\circ$

F_{10}: $\phi_{3,6}+\phi_{3,10}+\phi_{6,10}=180^\circ$

E

HYPERSPACE VERSION 1

$P_1=(0,0,0,0)$

$P_2=(0,0,0,1)$

$P_3=(1,0,0,0)$

$P_4=(0,1,0,0)$

$P_5=(0,0,1,0)$

HYPERSPACE VERSION 2

$P_1=(-1,0,0,1)$

$P_2=(-1,0,0,2)$

$P_3=(0,0,0,1)$

$P_4=(-1,1,0,1)$

$P_5=(-1,0,1,1)$

HYPERSPACE VERSION 3

$P_1=(0,0,0,0)$

$P_2=(0,0,0,1)$

$P_3=(1,0,1,0)$

$P_4=(0,1,0,0)$

$P_5=(-1,0,1,0)$

F

case is identical to that suggested for doing the same job in lower-dimensioned spaces. It's a bit trickier in perceptual sense, however, because it is usually difficult to have anything but an intuitive notion of where you want the critical point to go (unless it happens to be a matter of translating some 4-D point to the origin of the coordinate system). In any event, here is the general outline:

☐ Specify a reference entity in 4-D space, and plot its general 4-D view and the four 3-D perspective views.

☐ Select a point in the entity that you want to displace to another position in the 4-D coordinate system. Specify the coordinate of that new position, too.

☐ Solve for the translation terms that are necessary for displacing that selected point in your reference entity to the new place in the coordinate system.

☐ Translate the entire reference entity to its new position by applying the translation terms to all of the points in it.

☐ Specify the entity in its new position and plot its general 4-D and 3-D perspective views.

Applying successive 4-D translations to some reference entity, using the point specifications for the previous version as the starting point for each translation, generates some meaningful and visually interesting experiments. It is a time-consuming affair; especially making the plots. But how often do you think anyone has ever had your opportunity to do such a thing? Not many, you can be sure. Try this plan:

☐ Select a combination of translation terms for 4-D space. Keep the values relatively small so that you can do at least four or five successive translations without running the plots off the page.

☐ Select an initial geometric entity. Hyperspace objects create the most interesting results, but of course space object and plane figures immersed in hyperspace are simpler to plot.

☐ Apply the selected 4-D translation to the entity and plot the reference and translated versions. Use the general 4-D plots if you wish, but you will find that one or more of the 3-D perspective plots are equally interesting. To achieve a more colorful result, plot each version with a different color of pencil or ink.

☐ Repeat the translation, using the results of the previous translation as a reference entity, until you either run near the edge of the paper or get tired of the work.

Experiments with relativistic 4-D translations can

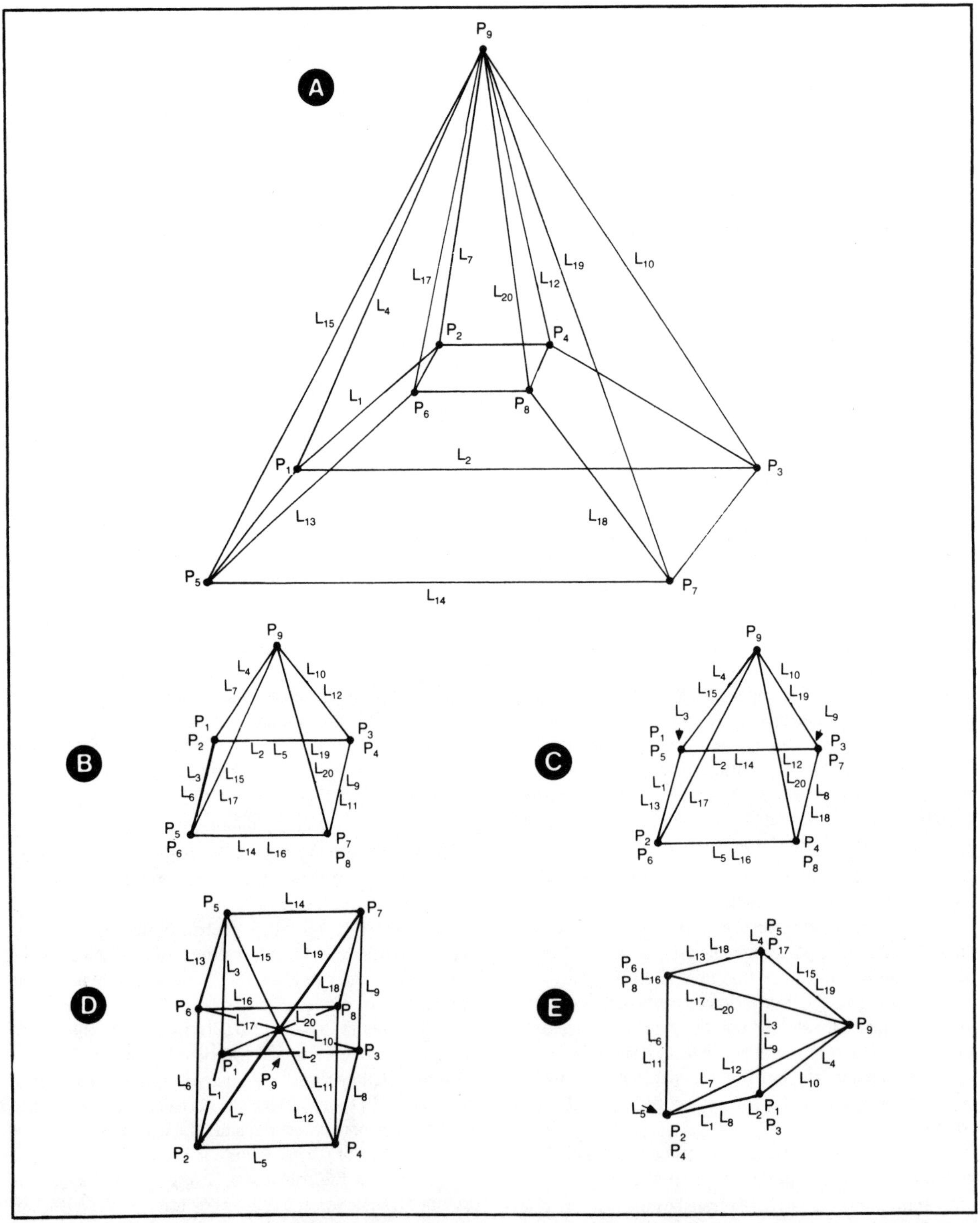

Fig. 22-11. Plots of the hyperspace object specified in Table 22-11A. (A) General hyperspace view. (B) XYZ perspective view. (C) XYW perspective view. (D) XZW perspective view. (E) YZW perspective view. See outlines of analyses in Table 22-11.

Table 22-11. Specifications and Analyses of the Hyperspace Object Shown in Fig. 22-11. (A) Specifications and Analysis in Hyperspace. (B) Secondary Analysis in XYZ Space. (C) Secondary Analysis in XYW Space. (D) Secondary Analysis in XZW Space. (E) Secondary Analysis in YZW Space. (F) Suggested Coordinates.

(A, B, C, and D correspond to α, β, γ, and δ.)

SPECIFICATIONS

$H=\overline{S_1S_2S_3S_4S_5S_6S_7}$

$S_1=\overline{F_1F_2F_4F_7F_{10}F_{14}}$

$S_2=\overline{F_1F_3F_5F_8F_{11}}$

$S_3=\overline{F_2F_3F_6F_9F_{15}}$

$S_4=\overline{F_4F_5F_6F_{12}F_{16}}$

$S_5=\overline{F_7F_8F_9F_{13}F_{17}}$

$S_6=\overline{F_{10}F_{11}F_{12}F_{13}F_{18}}$

$S_7=\overline{F_{14}F_{15}F_{16}F_{17}F_{18}}$

$F_1=\overline{L_1L_2L_5L_8}$

$F_2=\overline{L_1L_3L_6L_{13}}$

$F_3=\overline{L_1L_4L_7}$

$F_4=\overline{L_2L_3L_9L_{14}}$

$F_5=\overline{L_2L_4L_{10}}$

$F_6=\overline{L_3L_4L_{15}}$

$F_7=\overline{L_5L_6L_{11}L_{16}}$

$F_8=\overline{L_5L_7L_{12}}$

$F_9=\overline{L_6L_7L_{17}}$

$F_{10}=\overline{L_8L_9L_{11}L_{18}}$

$F_{11}=\overline{L_8L_{10}L_{12}}$

$F_{12}=\overline{L_9L_{10}L_{19}}$

$F_{13}=\overline{L_{11}L_{12}L_{20}}$

$F_{14}=\overline{L_{13}L_{14}L_{16}L_{18}}$

$F_{15}=\overline{L_{13}L_{15}L_{17}}$

$F_{16}=\overline{L_{14}L_{15}L_{19}}$

$F_{17}=\overline{L_{16}L_{17}L_{20}}$

$F_{18}=\overline{L_{18}L_{19}L_{20}}$

$L_1=P_1P_2$

$L_2=\overline{P_1P_3}$

$L_3=\overline{P_1P_5}$

$L_4=\overline{P_1P_9}$

$L_5=\overline{P_2P_4}$

$L_6=\overline{P_2P_6}$

$L_7=\overline{P_2P_9}$

$L_8=\overline{P_3P_4}$

$L_9=\overline{P_3P_7}$

$L_{10}=\overline{P_3P_9}$

$L_{11}=\overline{P_4P_8}$

$L_{12}=\overline{P_4P_9}$

$L_{13}=\overline{P_5P_6}$

$L_{14}=\overline{P_5P_7}$

$L_{15}=\overline{P_5P_9}$

$L_{16}=\overline{P_6P_8}$

$L_{17}=\overline{P_6P_9}$

$L_{18}=\overline{P_7P_8}$

$L_{19}=\overline{P_7P_9}$

$L_{20}=\overline{P_8P_9}$

SPECIFICATIONS (Cont'd)

$P_1=(x_1,y_1,z_1,w_1)$

$P_2=(x_2,y_2,z_2,w_2)$

$P_3=(x_3,y_3,z_3,w_3)$

$P_4=(x_4,y_4,z_4,w_4)$

$P_5=(x_5,y_5,z_5,w_5)$

$P_6=(x_6,y_6,z_6,w_6)$

$P_7=(x_7,y_7,z_7,w_7)$

$P_8=(x_8,y_8,z_8,w_8)$

$P_9=(x_9,y_9,z_9,w_9)$

LENGTHS

$$s_1=\sqrt{(x_2-x_1)^2+(y_2-y_1)^2+(z_2-z_1)^2+(w_2-w_1)^2}$$

$$s_2=\sqrt{(x_3-x_1)^2+(y_3-y_1)^2+(z_3-z_1)^2+(w_3-w_1)^2}$$

$$s_3=\sqrt{(x_5-x_1)^2+(y_5-y_1)^2+(z_5-z_1)^2+(w_5-w_1)^2}$$

$$s_4=\sqrt{(x_9-x_1)^2+(y_9-y_1)^2+(z_9-z_1)^2+(w_9-w_1)^2}$$

$$s_5=\sqrt{(x_4-x_2)^2+(y_4-y_2)^2+(z_4-z_2)^2+(w_4-w_2)^2}$$

$$s_6=\sqrt{(x_6-x_2)^2+(y_6-y_2)^2+(z_6-z_2)^2+(w_6-w_2)^2}$$

$$s_7=\sqrt{(x_9-x_2)^2+(y_9-y_2)^2+(z_9-z_2)^2+(w_9-w_2)^2}$$

$$s_8=\sqrt{(x_4-x_3)^2+(y_4-y_3)^2+(z_4-z_3)^2+(w_4-w_3)^2}$$

$$s_9=\sqrt{(x_7-x_3)^2+(y_7-y_3)^2+(z_7-z_3)^2+(w_7-w_3)^2}$$

$$s_{10}=\sqrt{(x_9-x_3)^2+(y_9-y_3)^2+(z_9-z_3)^2+(w_9-w_3)^2}$$

$$s_{11}=\sqrt{(x_8-x_4)^2+(y_8-y_4)^2+(z_8-z_4)^2+(w_8-w_4)^2}$$

$$s_{12}=\sqrt{(x_9-x_4)^2+(y_9-y_4)^2+(z_9-z_4)^2+(w_9-w_4)^2}$$

$$s_{13}=\sqrt{(x_6-x_5)^2+(y_6-y_5)^2+(z_6-z_5)^2+(w_6-w_5)^2}$$

$$s_{14}=\sqrt{(x_7-x_5)^2+(y_7-y_5)^2+(z_7-z_5)^2+(w_7-w_5)^2}$$

A

Table 22-11. Specifications and Analyses of the Hyperspace Object Shown in Fig. 22-11. (A) Specifications and Analysis in Hyperspace. (B) Secondary Analysis in XYZ Space. (C) Secondary Analysis in XYW Space. (D) Secondary Analysis in XZW Space. (E) Secondary Analysis in YZW Space. (F) Suggested Coordinates. **(Continued from page 419.)**

LENGTHS (Cont'd)

$$s_{15}=\sqrt{(x_9-x_5)^2+(y_9-y_5)^2+(z_9-z_5)^2+(w_9-w_5)^2}$$

$$s_{16}=\sqrt{(x_8-x_6)^2+(y_8-y_6)^2+(z_8-z_6)^2+(w_8-w_6)^2}$$

$$s_{17}=\sqrt{(x_9-x_6)^2+(y_9-y_6)^2+(z_9-z_6)^2+(w_9-w_6)^2}$$

$$s_{18}=\sqrt{(x_8-x_7)^2+(y_8-y_7)^2+(z_8-z_7)^2+(w_8-w_7)^2}$$

$$s_{19}=\sqrt{(x_9-x_7)^2+(y_9-y_7)^2+(z_9-z_7)^2+(w_9-w_7)^2}$$

$$s_{20}=\sqrt{(x_9-x_8)^2+(y_9-y_8)^2+(z_9-z_8)^2+(w_9-w_8)^2}$$

DIRECTION COSINES

$\cos A_1=(x_2-x_1)/s_1$ $\cos B_1=(y_2-y_1)/s_1$
$\cos C_1=(z_2-z_1)/s_1$ $\cos D_1=(w_2-w_1)/s_1$

$\cos A_2=(x_3-x_1)/s_2$ $\cos B_2=(y_3-y_1)/s_2$
$\cos C_2=(z_3-z_1)/s_2$ $\cos D_2=(w_3-w_1)/s_2$

$\cos A_3=(x_5-x_1)/s_3$ $\cos B_3=(y_5-y_1)/s_3$
$\cos C_3=(z_5-z_1)/s_3$ $\cos D_3=(w_5-w_1)/s_3$

$\cos A_4=(x_9-x_1)/s_4$ $\cos B_4=(y_9-y_1)/s_4$
$\cos C_4=(z_9-z_1)/s_4$ $\cos D_4=(w_9-w_1)/s_4$

$\cos A_5=(x_4-x_2)/s_5$ $\cos B_5=(y_4-y_2)/s_5$
$\cos C_5=(z_4-z_2)/s_5$ $\cos D_5=(w_4-w_2)/s_5$

$\cos A_6=(x_6-x_2)/s_6$ $\cos B_6=(y_6-y_2)/s_6$
$\cos C_6=(z_6-z_2)/s_6$ $\cos D_6=(w_6-w_2)/s_6$

$\cos A_7=(x_9-x_2)/s_7$ $\cos B_7=(y_9-y_2)/s_7$
$\cos C_7=(z_9-z_2)/s_7$ $\cos D_7=(w_9-w_2)/s_7$

$\cos A_8=(x_4-x_3)/s_8$ $\cos B_8=(y_4-y_3)/s_8$
$\cos C_8=(z_4-z_3)/s_8$ $\cos D_8=(w_4-w_3)/s_8$

$\cos A_9=(x_7-x_3)/s_9$ $\cos B_9=(y_7-y_3)/s_9$
$\cos C_9=(z_7-z_3)/s_9$ $\cos D_9=(w_7-w_3)/s_9$

DIRECTION COSINES (Cont'd)

$\cos A_{10}=(x_9-x_3)/s_{10}$ $\cos B_{10}=(y_9-y_3)/s_{10}$
$\cos C_{10}=(z_9-z_3)/s_{10}$ $\cos D_{10}=(w_9-w_3)/s_{10}$

$\cos A_{11}=(x_8-x_4)/s_{11}$ $\cos B_{11}=(y_8-y_4)/s_{11}$
$\cos C_{11}=(z_8-z_4)/s_{11}$ $\cos D_{11}=(w_8-w_4)/s_{11}$

$\cos A_{12}=(x_9-x_4)/s_{12}$ $\cos B_{12}=(y_9-y_4)/s_{12}$
$\cos C_{12}=(z_9-z_4)/s_{12}$ $\cos D_{12}=(w_9-w_4)/s_{12}$

$\cos A_{13}=(x_6-x_5)/s_{13}$ $\cos B_{13}=(y_6-y_5)/s_{13}$
$\cos C_{13}=(z_6-z_5)/s_{13}$ $\cos D_{13}=(w_6-w_5)/s_{13}$

$\cos A_{14}=(x_7-x_5)/s_{14}$ $\cos B_{14}=(y_7-y_5)/s_{14}$
$\cos C_{14}=(z_7-z_5)/s_{14}$ $\cos D_{14}=(w_7-w_5)/s_{14}$

$\cos A_{15}=(x_9-x_5)/s_{15}$ $\cos B_{15}=(y_9-y_5)/s_{15}$
$\cos C_{15}=(z_9-z_5)/s_{15}$ $\cos D_{15}=(w_9-w_5)/s_{15}$

$\cos A_{16}=(x_8-x_6)/s_{16}$ $\cos B_{16}=(y_8-y_6)/s_{16}$
$\cos C_{16}=(z_8-z_6)/s_{16}$ $\cos D_{16}=(w_8-w_6)/s_{16}$

$\cos A_{17}=(x_9-x_6)/s_{17}$ $\cos B_{17}=(y_9-y_6)/s_{17}$
$\cos C_{17}=(z_9-z_6)/s_{17}$ $\cos D_{17}=(w_9-w_6)/s_{17}$

$\cos A_{18}=(x_8-x_7)/s_{18}$ $\cos B_{18}=(y_8-y_7)/s_{18}$
$\cos C_{18}=(z_8-z_7)/s_{18}$ $\cos D_{18}=(w_8-w_7)/s_{18}$

$\cos A_{19}=(x_9-x_7)/s_{19}$ $\cos B_{19}=(y_9-y_7)/s_{19}$
$\cos C_{19}=(z_9-z_7)/s_{19}$ $\cos D_{19}=(w_9-w_7)/s_{19}$

$\cos A_{20}=(x_9-x_8)/s_{20}$ $\cos B_{20}=(y_9-y_8)/s_{20}$
$\cos C_{20}=(z_9-z_8)/s_{20}$ $\cos D_{20}=(w_9-w_8)/s_{20}$

A

ANGLES

$\cos\phi_{1,2}=\cos A_1\cos A_2+\cos B_1\cos B_2+\cos C_1\cos C_2+\cos D_1\cos D_2$

$\cos\phi_{1,3}=\cos A_1\cos A_3+\cos B_1\cos B_3+\cos C_1\cos C_3+\cos D_1\cos D_3$

$\cos\phi_{1,4}=\cos A_1\cos A_4+\cos B_1\cos B_4+\cos C_1\cos C_4+\cos D_1\cos D_4$

$\cos\phi_{1,5}=\cos A_1\cos A_5+\cos B_1\cos B_5+\cos C_1\cos C_5+\cos D_1\cos D_5$

$\cos\phi_{1,6}=\cos A_1\cos A_6+\cos B_1\cos B_6+\cos C_1\cos C_6+\cos D_1\cos D_6$

$\cos\phi_{1,7}=\cos A_1\cos A_7+\cos B_1\cos B_7+\cos C_1\cos C_7+\cos D_1\cos D_7$

$\cos\phi_{2,3}=\cos A_2\cos A_3+\cos B_2\cos B_3+\cos C_2\cos C_3+\cos D_2\cos D_3$

$\cos\phi_{2,4}=\cos A_2\cos A_4+\cos B_2\cos B_4+\cos C_2\cos C_4+\cos C_2\cos D_4$

$\cos\phi_{2,8}=\cos A_2\cos A_8+\cos B_2\cos B_8+\cos C_2\cos C_8+\cos D_2\cos D_8$

$\cos\phi_{2,9}=\cos A_2\cos A_9+\cos B_2\cos B_9+\cos C_2\cos C_9+\cos D_2\cos D_9$

$\cos\phi_{2,10}=\cos A_2\cos A_{10}+\cos B_2\cos B_{10}+\cos C_2\cos C_{10}+\cos D_2\cos D_{10}$

$\cos\phi_{3,4}=\cos A_3\cos A_4+\cos B_3\cos B_4+\cos C_3\cos C_4+\cos D_3\cos D_4$

$\cos\phi_{3,13}=\cos A_3\cos A_{13}+\cos B_3\cos B_{13}+\cos C_3\cos C_{13}+\cos D_3\cos D_{13}$

$\cos\phi_{3,14}=\cos A_3\cos A_{14}+\cos B_3\cos B_{14}+\cos C_3\cos C_{14}+\cos D_3\cos D_{14}$

$\cos\phi_{3,15}=\cos A_3\cos A_{15}+\cos B_3\cos B_{15}+\cos C_3\cos C_{15}+\cos D_3\cos D_{15}$

$\cos\phi_{4,7}=\cos A_4\cos A_7+\cos B_4\cos B_7+\cos C_4\cos C_7+\cos D_4\cos D_7$

$\cos\phi_{4,10}=\cos A_4\cos A_{10}+\cos B_4\cos B_{10}+\cos C_4\cos C_{10}+\cos D_4\cos D_{10}$

$\cos\phi_{4,15}=\cos A_4\cos A_{15}+\cos B_4\cos B_{15}+\cos C_4\cos C_{15}+\cos D_4\cos D_{15}$

$\cos\phi_{5,6}=\cos A_5\cos A_6+\cos B_5\cos B_6+\cos C_5\cos C_6+\cos D_5\cos D_6$

$\cos\phi_{5,7}=\cos A_5\cos A_7+\cos B_5\cos B_7+\cos C_5\cos C_7+\cos D_5\cos D_7$

$\cos\phi_{5,8}=\cos A_5\cos A_8+\cos B_5\cos B_8+\cos C_5\cos C_8+\cos D_5\cos D_8$

$\cos\phi_{5,11}=\cos A_5\cos A_{11}+\cos B_5\cos B_1+\cos C_5\cos C_{11}+\cos D_5\cos D_{11}$

$\cos\phi_{5,12}=\cos A_5\cos A_{12}+\cos B_5\cos B_{12}+\cos C_5\cos C_{12}+\cos D_5\cos D_{12}$

$\cos\phi_{6,7}=\cos A_6\cos A_7+\cos B_6\cos B_7+\cos C_6\cos C_7+\cos D_6\cos D_7$

$\cos\phi_{6,13}=\cos A_6\cos A_{13}+\cos B_6\cos B_{13}+\cos C_6\cos C_{13}+\cos D_6\cos D_{13}$

$\cos\phi_{6,16}=\cos A_6\cos A_{16}+\cos B_6\cos B_{16}+\cos C_6\cos C_{16}+\cos D_6\cos D_{16}$

$\cos\phi_{6,17}=\cos A_6\cos A_{17}+\cos B_6\cos B_{17}+\cos C_6\cos C_{17}+\cos D_6\cos D_{17}$

$\cos\phi_{7,12}=\cos A_7\cos A_{12}+\cos B_7\cos B_{12}+\cos C_7\cos C_{12}+\cos D_7\cos D_{12}$

$\cos\phi_{7,17}=\cos A_7\cos A_{17}+\cos B_7\cos B_{17}+\cos C_7\cos C_{17}+\cos D_7\cos D_{17}$

$\cos\phi_{8,9}=\cos A_8\cos A_9+\cos B_8\cos B_9+\cos C_8\cos C_9+\cos D_8\cos D_9$

$\cos\phi_{8,10}=\cos A_8\cos A_{10}+\cos B_8\cos B_{10}+\cos C_8\cos C_{10}+\cos D_8\cos D_{10}$

$\cos\phi_{8,11}=\cos A_8\cos A_{11}+\cos B_8\cos B_{11}+\cos C_8\cos C_{11}+\cos D_8\cos D_{11}$

$\cos\phi_{8,12}=\cos A_8\cos A_{12}+\cos B_8\cos B_{12}+\cos C_8\cos C_{12}+\cos D_8\cos D_{12}$ A

$\cos\phi_{9,10}=\cos A_9\cos A_{10}+\cos B_9\cos B_{10}+\cos C_9\cos C_{10}+\cos D_9\cos D_{10}$

$\cos\phi_{9,14}=\cos A_9\cos A_{14}+\cos B_9\cos B_{14}+\cos C_9\cos C_{14}+\cos D_9\cos D_{14}$

Table 22-11. Specifications and Analyses of the Hyperspace Object Shown in Fig. 22-11. (A) Specifications and Analysis in Hyperspace. (B) Secondary Analysis in XYZ Space. (C) Secondary Analysis in XYW Space. (D) Secondary Analysis in XYZ Space. (E) Secondary Analysis in YZW Space. (F) Suggested Coordinates. **(Continued from page 421.)**

ANGLES (Cont'd)

$$\cos\phi_{9,18}=\cos A_9\cos A_{18}+\cos B_9\cos B_{18}+\cos C_9\cos C_{18}+\cos D_9\cos D_{18}$$

$$\cos\phi_{9,19}=\cos A_9\cos A_{19}+\cos B_9\cos B_{19}+\cos C_9\cos C_{19}+\cos D_9\cos D_{19}$$

$$\cos\phi_{10,12}=\cos A_{10}\cos A_{12}+\cos B_{10}\cos B_{12}+\cos C_{10}\cos C_{12}+\cos D_{10}\cos D_{12}$$

$$\cos\phi_{10,19}=\cos A_{10}\cos A_{19}+\cos B_{10}\cos B_{19}+\cos C_{10}\cos C_{19}+\cos D_{10}\cos D_{19}$$

$$\cos\phi_{11,12}=\cos A_{11}\cos A_{12}+\cos B_{11}\cos B_{12}+\cos C_{11}\cos C_{12}+\cos D_{11}\cos D_{12}$$

$$\cos\phi_{11,16}=\cos A_{11}\cos A_{16}+\cos B_{11}\cos B_{16}+\cos C_{11}\cos C_{16}+\cos D_{11}\cos D_{16}$$

$$\cos\phi_{11,18}=\cos A_{11}\cos A_{18}+\cos B_{11}\cos B_{18}+\cos C_{11}\cos C_{18}+\cos D_{11}\cos D_{18}$$

$$\cos\phi_{11,20}=\cos A_{11}\cos A_{20}+\cos B_{11}\cos B_{20}+\cos C_{11}\cos C_{20}+\cos D_{11}\cos D_{20}$$

$$\cos\phi_{12,20}=\cos A_{12}\cos A_{20}+\cos B_{12}\cos B_{20}+\cos C_{12}\cos C_{20}+\cos D_{12}\cos D_{20}$$

$$\cos\phi_{13,14}=\cos A_{13}\cos A_{14}+\cos B_{13}\cos B_{14}+\cos C_{13}\cos C_{14}+\cos D_{13}\cos D_{14}$$

$$\cos\phi_{13,15}=\cos A_{13}\cos A_{15}+\cos B_{13}\cos B_{145}+\cos C_{13}\cos C_{15}+\cos D_{13}\cos D_{15}$$

$$\cos\phi_{13,16}=\cos A_{13}\cos A_{16}+\cos B_{13}\cos B_{16}+\cos C_{13}\cos C_{16}+\cos D_{13}\cos D_{16}$$

$$\cos\phi_{13,17}=\cos A_{13}\cos A_{17}+\cos B_{13}\cos B_{17}+\cos C_{13}\cos C_{17}+\cos D_{13}\cos D_{17}$$

$$\cos\phi_{14,15}=\cos A_{14}\cos A_{15}+\cos B_{14}\cos B_{15}+\cos C_{14}\cos C_{15}+\cos D_{14}\cos D_{15}$$

$$\cos\phi_{14,18}=\cos A_{14}\cos A_{18}+\cos B_{14}\cos B_{18}+\cos C_{14}\cos C_{18}+\cos D_{14}\cos D_{18}$$

$$\cos\phi_{14,19}=\cos A_{14}\cos A_{19}+\cos B_{14}\cos B_{19}+\cos C_{14}\cos C_{19}+\cos D_{14}\cos D_{19}$$

$$\cos\phi_{15,17}=\cos A_{15}\cos A_{17}+\cos B_{15}\cos B_{17}+\cos C_{15}\cos C_{17}+\cos D_{15}\cos D_{17}$$

$$\cos\phi_{15,19}=\cos A_{15}\cos A_{19}+\cos B_{15}\cos B_{19}+\cos C_{15}\cos C_{19}+\cos D_{15}\cos D_{19}$$

$$\cos\phi_{16,17}=\cos A_{16}\cos A_{17}+\cos B_{16}\cos B_{17}+\cos C_{16}\cos C_{17}+\cos D_{16}\cos D_{17}$$

$$\cos\phi_{16,18}=\cos A_{16}\cos A_{18}+\cos B_{16}\cos B_{18}+\cos C_{16}\cos C_{18}+\cos D_{16}\cos D_{18}$$

$$\cos\phi_{16,20}=\cos A_{16}\cos A_{20}+\cos B_{16}\cos B_{20}+\cos C_{16}\cos C_{20}+\cos D_{16}\cos D_{20}$$

$$\cos\phi_{17,20}=\cos A_{17}\cos A_{20}+\cos B_{17}\cos B_{20}+\cos C_{17}\cos C_{20}+\cos D_{17}\cos D_{20}$$

$$\cos\phi_{18,19}=\cos A_{18}\cos A_{19}+\cos B_{18}\cos B_{19}+\cos C_{18}\cos C_{19}+\cos D_{18}\cos D_{19}$$

$$\cos\phi_{18,20}=\cos A_{18}\cos A_{20}+\cos B_{18}\cos B_{20}+\cos C_{18}\cos C_{20}+\cos D_{18}\cos D_{20}$$

$$\cos\phi_{19,20}=\cos A_{19}\cos A_{20}+\cos B_{19}\cos B_{20}+\cos C_{19}\cos C_{20}+\cos D_{19}\cos D_{20}$$

CHECK

F_1:	$\phi_{1,2}+\phi_{1,5}+\phi_{2,8}+\phi_{5,8}=360^\circ$	F_{10}:	$\phi_{8,9}+\phi_{8,11}+\phi_{9,18}+\phi_{11,18}=360^\circ$
F_2:	$\phi_{1,3}+\phi_{1,6}+\phi_{3,13}+\phi_{6,13}=360^\circ$	F_{11}:	$\phi_{8,10}+\phi_{8,12}+\phi_{10,12}=180^\circ$
F_3:	$\phi_{1,4}+\phi_{1,7}+\phi_{4,7}=180^\circ$	F_{12}:	$\phi_{9,10}+\phi_{9,19}+\phi_{10,19}=180^\circ$
F_4:	$\phi_{2,3}+\phi_{2,9}+\phi_{3,14}+\phi_{9,14}=360^\circ$	F_{13}:	$\phi_{11,12}+\phi_{11,20}+\phi_{12,20}=180^\circ$
F_5:	$\phi_{2,4}+\phi_{2,10}+\phi_{4,10}=180^\circ$	F_{14}:	$\phi_{13,14}+\phi_{13,16}+\phi_{14,18}+\phi_{16,18}=360^\circ$
F_6:	$\phi_{3,4}+\phi_{3,15}+\phi_{4,15}=180^\circ$	F_{15}:	$\phi_{13,15}+\phi_{13,17}+\phi_{15,17}=180^\circ$
F_7:	$\phi_{5,6}+\phi_{5,11}+\phi_{6,16}+\phi_{11,16}=360^\circ$	F_{16}:	$\phi_{14,15}+\phi_{14,19}+\phi_{15,19}=180^\circ$
F_8:	$\phi_{5,7}+\phi_{5,12}+\phi_{7,12}=180^\circ$	F_{17}:	$\phi_{16,17}+\phi_{16,20}+\phi_{17,20}=180^\circ$
F_9:	$\phi_{6,7}+\phi_{6,17}+\phi_{7,17}=180^\circ$	F_{18}:	$\phi_{18,19}+\phi_{18,20}+\phi_{19,20}=180^\circ$

A

SECONDARY ANALYSIS -- XYZ SPACE

PLOTTED POINTS

$P_1=(x_1,y_1,z_1)$

$P_2=(x_2,y_2,z_2)$

$P_3=(x_3,y_3,z_3)$

$P_4=(x_4,y_4,z_4)$

$P_5=(x_5,y_5,z_5)$

$P_6=(x_6,y_6,z_6)$

$P_7=(x_7,y_7,z_7)$

$P_8=(x_8,y_8,z_8)$

$P_9=(x_9,y_9,z_9)$

LENGTHS

$s_1=\sqrt{(x_2-x_1)^2+(y_2-y_1)^2+(z_2-z_1)^2}$

$s_2=\sqrt{(x_3-x_1)^2+(y_3-y_1)^2+(z_3-z_1)^2}$

$s_3=\sqrt{(x_5-x_1)^2+(y_5-y_1)^2+(z_5-z_1)^2}$

$s_4=\sqrt{(x_9-x_1)^2+(y_9-y_1)^2+(z_9-z_1)^2}$

$s_5=\sqrt{(x_4-x_2)^2+(y_4-y_2)^2+(z_4-z_2)^2}$

$s_6=\sqrt{(x_6-x_2)^2+(y_6-y_2)^2+(z_6-z_2)^2}$

$s_7=\sqrt{(x_9-x_2)^2+(y_9-y_2)^2+(z_9-z_2)^2}$

$s_8=\sqrt{(x_4-x_3)^2+(y_4-y_3)^2+(z_4-z_3)^2}$

$s_9=\sqrt{(x_7-x_3)^2+(y_7-y_3)^2+(z_7-z_3)^2}$

$s_{10}=\sqrt{(x_9-x_3)^2+(y_9-y_3)^2+(z_9-z_3)^2}$

$s_{11}=\sqrt{(x_8-x_4)^2+(y_8-y_4)^2+(z_8-z_4)^2}$

$s_{12}=\sqrt{(x_9-x_4)^2+(y_9-y_4)^2+(z_9-z_4)^2}$

LENGTHS (Cont'd)

$s_{13}=\sqrt{(x_6-x_5)^2+(y_6-y_5)^2+(z_6-z_5)^2}$

$s_{14}=\sqrt{(x_7-x_5)^2+(y_7-y_5)^2+(z_7-z_5)^2}$

$s_{15}=\sqrt{(x_9-x_5)^2+(y_9-y_5)^2+(z_9-z_5)^2}$

$s_{16}=\sqrt{(x_8-x_6)^2+(y_8-y_6)^2+(z_8-z_6)^2}$

$s_{17}=\sqrt{(x_9-x_6)^2+(y_9-y_6)^2+(z_9-z_6)^2}$

$s_{18}=\sqrt{(x_8-x_7)^2+(y_8-y_7)^2+(z_8-z_7)^2}$

$s_{19}=\sqrt{(x_9-x_7)^2+(y_9-y_7)^2+(z_9-z_7)^2}$

$s_{20}=\sqrt{(x_9-x_8)^2+(y_9-y_8)^2+(z_9-z_8)^2}$

DIRECTION COSINES

$\cos A_1=(x_2-x_1)/s_1$ $\cos B_1=(y_2-y_1)/s_1$

$\cos C_1=(z_2-z_1)/s_1$

$\cos A_2=(x_3-x_1)/s_2$ $\cos B_2=(y_3-y_1)/s_2$

$\cos C_2=(z_3-z_1)/s_2$

$\cos A_3=(x_5-x_1)/s_3$ $\cos B_3=(y_5-y_1)/s_3$

$\cos C_3=(z_5-z_1)/s_3$

$\cos A_4=(x_9-x_1)/s_4$ $\cos B_4=(y_9-y_1)/s_4$

$\cos C_4=(z_9-z_1)/s_4$

$\cos A_5=(x_4-x_2)/s_5$ $\cos B_5=(y_4-y_2)/s_5$

$\cos C_5=(z_4-z_2)/s_5$

$\cos A_6=(x_6-x_2)/s_6$ $\cos B_6=(y_6-y_2)/s_6$

$\cos C_6=(z_6-z_2)/s_6$

$\cos A_7=(x_9-x_2)/s_7$ $\cos B_7=(y_9-y_2)/s_7$

$\cos C_7=(z_9-z_2)/s_7$

$\cos A_8=(x_4-x_3)/s_8$ $\cos B_8=(y_4-y_3)/s_8$

$\cos C_8=(z_4-z_3)/s_8$

B

Table 22-11. Specifications and Analyses of the Hyperspace Object Shown in Fig. 22-11. (A) Specifications and Analysis in Hyperspace. (B) Secondary Analysis in XYZ Space. (C) Secondary Analysis in XYW Space. (D) Secondary Analysis in XZW Space. (E) Secondary Analysis in YZW Space. (F) Suggested Coordinates. **(Continued from page 423.)**

DIRECTION COSINES (Cont'd)

$\cos A_9=(x_7-x_3)/s_9$ $\cos B_9=(y_7-y_3)/s_9$

$\cos C_9=(z_7-z_3)/s_9$

$\cos A_{10}=(x_9-x_3)/s_{10}$ $\cos B_{10}=(y_9-y_3)/s_{10}$

$\cos C_{10}=(z_9-z_3)/s_{10}$

$\cos A_{11}=(x_8-x_4)/s_{11}$ $\cos B_{11}=(y_8-y_4)/s_{11}$

$\cos C_{11}=(z_8-z_4)/s_{11}$

$\cos A_{12}=(x_9-x_4)/s_{12}$ $\cos B_{12}=(y_9-y_4)/s_{12}$

$\cos C_{12}=(z_9-z_4)/s_{12}$

$\cos A_{13}=(x_6-x_5)/s_{13}$ $\cos B_{13}=(y_6-y_5)/s_{13}$

$\cos C_{13}=(z_6-z_5)/s_{13}$

$\cos A_{14}=(x_7-x_5)/s_{14}$ $\cos B_{14}=(y_7-y_5)/s_{14}$

$\cos C_{14}=(z_7-z_5)/s_{14}$

$\cos A_{15}=(x_9-x_5)/s_{15}$ $\cos B_{15}=(y_9-y_5)/s_{15}$

$\cos C_{15}=(z_9-z_5)/s_{15}$

$\cos A_{16}=(x_8-x_6)/s_{16}$ $\cos B_{16}=(y_8-y_6)/s_{16}$

$\cos C_{16}=(z_8-z_6)/s_{16}$

$\cos A_{17}=(x_9-x_6)/s_{17}$ $\cos B_{17}=(y_9-y_6)/s_{17}$

$\cos C_{17}=(z_9-z_6)/s_{17}$

$\cos A_{18}=(x_8-x_7)/s_{18}$ $\cos B_{18}=(y_8-y_7)/s_{18}$

$\cos C_{18}=(z_8-z_7)/s_{18}$

$\cos A_{19}=(x_9-x_7)/s_{19}$ $\cos B_{19}=(y_9-y_7)/s_{19}$

$\cos C_{19}=(z_9-z_7)/s_{19}$

$\cos A_{20}=(x_9-x_8)/s_{20}$ $\cos B_{20}=(y_9-y_8)/s_{20}$

$\cos C_{20}=(z_9-z_8)/s_{20}$

ANGLES

$\cos\phi_{1,2}=\cos A_1\cos A_2+\cos B_1\cos B_2+\cos C_1\cos C_2$

$\cos\phi_{1,3}=\cos A_1\cos A_3+\cos B_1\cos B_3+\cos C_1\cos C_3$

$\cos\phi_{1,4}=\cos A_1\cos A_4+\cos B_1\cos B_4+\cos C_1\cos C_4$

$\cos\phi_{1,5}=\cos A_1\cos A_5+\cos B_1\cos B_5+\cos C_1\cos C_5$

$\cos\phi_{1,6}=\cos A_1\cos A_6+\cos B_1\cos B_6+\cos C_1\cos C_6$

$\cos\phi_{1,7}=\cos A_1\cos A_7+\cos B_1\cos B_7+\cos C_1\cos C_7$

$\cos\phi_{2,3}=\cos A_2\cos A_3+\cos B_2\cos B_3+\cos C_2\cos C_3$

$\cos\phi_{2,4}=\cos A_2\cos A_4+\cos B_2\cos B_4+\cos C_2\cos C_4$

$\cos\phi_{2,8}=\cos A_2\cos A_8+\cos B_2\cos B_8+\cos C_2\cos C_8$

$\cos\phi_{2,9}=\cos A_2\cos A_9+\cos B_2\cos B_9+\cos C_2\cos C_9$

$\cos\phi_{2,10}=\cos A_2\cos A_{10}+\cos B_2\cos B_{10}+\cos C_2\cos C_{10}$

$\cos\phi_{3,4}=\cos A_3\cos A_4+\cos B_3\cos B_4+\cos C_3\cos C_4$

$\cos\phi_{3,13}=\cos A_3\cos A_{13}+\cos B_3\cos B_{13}+\cos C_3\cos C_{13}$

$\cos\phi_{3,14}=\cos A_3\cos A_{14}+\cos B_3\cos B_{14}+\cos C_3\cos C_{14}$

$\cos\phi_{3,15}=\cos A_3\cos A_{15}+\cos B_3\cos B_{15}+\cos C_3\cos C_{15}$

$\cos\phi_{4,7}=\cos A_4\cos A_7+\cos B_4\cos B_7+\cos C_4\cos C_7$

$\cos\phi_{4,10}=\cos A_4\cos A_{10}+\cos B_4\cos B_{10}+\cos C_4\cos C_{10}$

$\cos\phi_{4,15}=\cos A_4\cos A_{15}+\cos B_4\cos B_{15}+\cos C_4\cos C_{15}$

$\cos\phi_{5,6}=\cos A_5\cos A_6+\cos B_5\cos B_6+\cos C_5\cos C_6$

$\cos\phi_{5,7}=\cos A_5\cos A_7+\cos B_5\cos B_7+\cos C_5\cos C_7$

$\cos\phi_{5,8}=\cos A_5\cos A_8+\cos B_5\cos B_8+\cos C_5\cos C_8$

$\cos\phi_{5,11}=\cos A_5\cos A_1+\cos B_5\cos B_{11}+\cos C_5\cos C_{11}$

$\cos\phi_{5,12}=\cos A_5\cos A_{12}+\cos B_5\cos B_{12}+\cos C_5\cos C_{12}$

$\cos\phi_{6,7}=\cos A_6\cos A_7+\cos B_6\cos B_7+\cos C_6\cos C_7$

$\cos\phi_{6,13}=\cos A_6\cos A_{13}+\cos B_6\cos B_{13}+\cos C_6\cos C_{13}$

$\cos\phi_{6,16}=\cos A_6\cos A_{16}+\cos B_6\cos B_{16}+\cos C_6\cos C_{16}$

$\cos\phi_{6,17}=\cos A_6\cos A_{17}+\cos B_6\cos B_{17}+\cos C_6\cos C_{17}$

$\cos\phi_{7,12}=\cos A_7\cos A_{12}+\cos B_7\cos B_{12}+\cos C_7\cos C_{12}$

$\cos\phi_{7,17}=\cos A_7\cos A_{17}+\cos B_7\cos B_{17}+\cos C_7\cos C_{17}$

$\cos\phi_{8,9}=\cos A_8\cos A_9+\cos B_8\cos B_9+\cos C_8\cos C_9$

$\cos\phi_{8,10}=\cos A_8\cos A_{10}+\cos B_8\cos B_{10}+\cos C_8\cos C_{10}$

B

ANGLES (Cont'd)

$\cos\phi_{8,11}=\cos A_8\cos A_{11}+\cos B_8\cos B_{11}+\cos C_8\cos C_{11}$

$\cos\phi_{8,12}=\cos A_8\cos A_{12}+\cos B_8\cos B_{12}+\cos C_8\cos C_{12}$

$\cos\phi_{9,10}=\cos A_9\cos A_{10}+\cos B_9\cos B_{10}+\cos C_9\cos C_{10}$

$\cos\phi_{9,14}=\cos A_9\cos A_{14}+\cos B_9\cos B_{14}+\cos C_9\cos C_{14}$

$\cos\phi_{9,18}=\cos A_9\cos A_{18}+\cos B_9\cos B_{18}+\cos C_9\cos C_{18}$

$\cos\phi_{9,19}=\cos A_9\cos A_{19}+\cos B_9\cos B_{19}+\cos C_9\cos C_{19}$

$\cos\phi_{10,12}=\cos A_{10}\cos A_{12}+\cos B_{10}\cos B_{12}+\cos C_{10}\cos C_{12}$

$\cos\phi_{10,19}=\cos A_{10}\cos A_{19}+\cos B_{10}\cos B_{19}+\cos C_{10}\cos C_{19}$

$\cos\phi_{11,12}=\cos A_{11}\cos A_{12}+\cos B_{11}\cos B_{12}+\cos C_{11}\cos C_{12}$

$\cos\phi_{11,16}=\cos A_{11}\cos A_{16}+\cos B_{11}\cos B_{16}+\cos C_{11}\cos C_{16}$

$\cos\phi_{11,18}=\cos A_{11}\cos A_{18}+\cos B_{11}\cos B_{18}+\cos C_{11}\cos C_{18}$

$\cos\phi_{11,20}=\cos A_{11}\cos A_{20}+\cos B_{11}\cos B_{20}+\cos C_{11}\cos C_{20}$

$\cos\phi_{12,20}=\cos A_{12}\cos A_{20}+\cos B_{12}\cos B_{20}+\cos C_{12}\cos C_{20}$

$\cos\phi_{13,14}=\cos A_{13}\cos A_{14}+\cos B_{13}\cos B_{14}+\cos C_{13}\cos C_{14}$

$\cos\phi_{13,15}=\cos A_{13}\cos A_{145}+\cos B_{13}\cos B_{15}+\cos C_{13}\cos C_{15}$

$\cos\phi_{13,16}=\cos A_{13}\cos A_{16}+\cos B_{13}\cos B_{16}+\cos C_{13}\cos C_{16}$

$\cos\phi_{13,17}=\cos A_{13}\cos A_{17}+\cos B_{13}\cos B_{17}+\cos C_{13}\cos C_{17}$

$\cos\phi_{14,15}=\cos A_{14}\cos A_{15}+\cos B_{14}\cos B_{15}+\cos C_{14}\cos C_{15}$

$\cos\phi_{14,18}=\cos A_{14}\cos A_{18}+\cos B_{14}\cos B_{18}+\cos C_{14}\cos C_{18}$

$\cos\phi_{14,19}=\cos A_{14}\cos A_{19}+\cos B_{14}\cos B_{19}+\cos C_{14}\cos C_{19}$

$\cos\phi_{15,17}=\cos A_{15}\cos A_{17}+\cos B_{15}\cos B_{17}+\cos C_{15}\cos C_{17}$

$\cos\phi_{15,19}=\cos A_{15}\cos A_{19}+\cos B_{15}\cos B_{19}+\cos C_{15}\cos C_{19}$

$\cos\phi_{16,17}=\cos A_{16}\cos A_{17}+\cos B_{16}\cos B_{17}+\cos C_{16}\cos C_{17}$

$\cos\phi_{16,18}=\cos A_{16}\cos A_{18}+\cos B_{16}\cos B_{18}+\cos C_{16}\cos C_{18}$

$\cos\phi_{16,20}=\cos A_{16}\cos A_{20}+\cos B_{16}\cos B_{20}+\cos C_{16}\cos C_{20}$

$\cos\phi_{17,20}=\cos A_{17}\cos A_{20}+\cos B_{17}\cos B_{20}+\cos C_{17}\cos C_{20}$

$\cos\phi_{18,19}=\cos A_{18}\cos A_{19}+\cos B_{18}\cos B_{19}+\cos C_{18}\cos C_{19}$

$\cos\phi_{18,20}=\cos A_{18}\cos A_{20}+\cos B_{18}\cos B_{20}+\cos C_{18}\cos C_{20}$

$\cos\phi_{19,20}=\cos A_{19}\cos A_{20}+\cos B_{19}\cos B_{20}+\cos C_{19}\cos C_{20}$

CHECK

F_1:	$\phi_{1,2}+\phi_{1,5}+\phi_{2,8}+\phi_{5,8}=360^\circ$
F_2:	$\phi_{1,3}+\phi_{1,6}+\phi_{3,13}+\phi_{6,13}=360^\circ$
F_3:	$\phi_{1,4}+\phi_{1,7}+\phi_{4,7}=180^\circ$
F_4:	$\phi_{2,3}+\phi_{2,9}+\phi_{3,14}+\phi_{9,14}=360^\circ$
F_5:	$\phi_{2,4}+\phi_{2,10}+\phi_{4,10}=180^\circ$
F_6:	$\phi_{3,4}+\phi_{3,15}+\phi_{4,15}=180^\circ$
F_7:	$\phi_{5,6}+\phi_{5,11}+\phi_{6,16}+\phi_{11,16}=360^\circ$
F_8:	$\phi_{5,7}+\phi_{5,12}+\phi_{7,12}=180^\circ$
F_9:	$\phi_{6,7}+\phi_{6,17}+\phi_{7,17}=180^\circ$
F_{10}:	$\phi_{8,9}+\phi_{8,11}+\phi_{9,18}+\phi_{11,18}=360^\circ$
F_{11}:	$\phi_{8,10}+\phi_{8,12}+\phi_{10,12}=180^\circ$
F_{12}:	$\phi_{9,10}+\phi_{9,19}+\phi_{10,19}=180^\circ$
F_{13}:	$\phi_{11,12}+\phi_{11,20}+\phi_{12,20}=180^\circ$
F_{14}:	$\phi_{13,14}+\phi_{13,16}+\phi_{14,18}+\phi_{16,18}=360^\circ$
F_{15}:	$\phi_{13,15}+\phi_{13,17}+\phi_{15,17}=180^\circ$
F_{16}:	$\phi_{14,15}+\phi_{14,19}+\phi_{15,19}=180^\circ$
F_{17}:	$\phi_{16,17}+\phi_{16,20}+\phi_{17,20}=180^\circ$
F_{18}:	$\phi_{18,19}+\phi_{18,20}+\phi_{19,20}=180^\circ$

B

Table 22-11. Specifications and Analyses of the Hyperspace Object Shown in Fig. 22-11. (A) Specifications and Analysis in Hyperspace. (B) Secondary Analysis in XYZ Space. (C) Secondary Analysis in XYW Space. (D) Secondary Analysis in XZW Space. (E) Secondary Analysis in YZW Space. (F) Suggested Coordinates. (Continued from page 425.)

SECONDARY ANALYSIS -- XYW SPACE

PLOTTED POINTS

$P_1=(x_1,y_1,w_1)$

$P_2=(x_2,y_2,w_2)$

$P_3=(x_3,y_3,w_3)$

$P_4=(x_4,y_4,w_4)$

$P_5=(x_5,y_5,w_5)$

$P_6=(x_6,y_6,w_6)$

$P_7=(x_7,y_7,w_7)$

$P_8=(x_8,y_8,w_8)$

$P_9=(x_9,y_9,w_9)$

LENGTHS

$s_1=\sqrt{(x_2-x_1)^2+(y_2-y_1)^2+(w_2-w_1)^2}$

$s_2=\sqrt{(x_3-x_1)^2+(y_3-y_1)^2+(w_3-w_1)^2}$

$s_3=\sqrt{(x_5-x_1)^2+(y_5-y_1)^2+(w_5-w_1)^2}$

$s_4=\sqrt{(x_9-x_1)^2+(y_9-y_1)^2+(w_9-w_1)^2}$

$s_5=\sqrt{(x_4-x_2)^2+(y_4-y_2)^2+(w_4-w_2)^2}$

$s_6=\sqrt{(x_6-x_2)^2+(y_6-y_2)^2+(w_6-w_2)^2}$

$s_7=\sqrt{(x_9-x_2)^2+(y_9-y_2)^2+(w_9-w_2)^2}$

$s_8=\sqrt{(x_4-x_3)^2+(y_4-y_3)^2+(w_4-w_3)^2}$

$s_9=\sqrt{(x_7-x_3)^2+(y_7-y_3)^2+(w_7-w_3)^2}$

$s_{10}=\sqrt{(x_9-x_3)^2+(y_9-y_3)^2+(w_9-w_3)^2}$

$s_{11}=\sqrt{(x_8-x_4)^2+(y_8-y_4)^2+(w_8-w_4)^2}$

$s_{12}=\sqrt{(x_9-x_4)^2+(y_9-y_4)^2+(w_9-w_4)^2}$

$s_{13}=\sqrt{(x_6-x_5)^2+(y_6-y_5)^2+(w_6-w_5)^2}$

C

$s_{14}=\sqrt{(x_7-x_5)^2+(y_7-y_5)^2+(w_7-w_5)^2}$

LENGTHS (Cont'd)

$s_{15}=\sqrt{(x_9-x_5)^2+(y_9-y_5)^2+(w_9-w_5)^2}$

$s_{16}=\sqrt{(x_8-x_6)^2+(y_8-y_6)^2+(w_8-w_6)^2}$

$s_{17}=\sqrt{(x_9-x_6)^2+(y_9-y_6)^2+(w_9-w_6)^2}$

$s_{18}=\sqrt{(x_8-x_7)^2+(y_8-y_7)^2+(w_8-w_7)^2}$

$s_{19}=\sqrt{(x_9-x_7)^2+(y_9-y_7)^2+(w_9-w_7)^2}$

$s_{20}=\sqrt{(x_9-x_8)^2+(y_9-y_8)^2+(w_9-w_8)^2}$

DIRECTION COSINES

$\cos A_1=(x_2-x_1)/s_1$ $\cos B_1=(y_2-y_1)/s_1$
$\cos D_1=(w_2-w_1)/s_1$

$\cos A_2=(x_3-x_1)/s_2$ $\cos B_2=(y_3-y_1)/s_2$
$\cos D_2=(w_3-w_1)/s_2$

$\cos A_3=(x_5-x_1)/s_3$ $\cos B_3=(y_5-y_1)/s_3$
$\cos D_3=(w_5-w_1)/s_3$

$\cos A_4=(x_9-x_1)/s_4$ $\cos B_4=(y_9-y_1)/s_4$
$\cos D_4=(w_9-w_1)/s_4$

$\cos A_5=(x_4-x_2)/s_5$ $\cos B_5=(y_4-y_2)/s_5$
$\cos D_5=(w_4-w_2)/s_5$

$\cos A_6=(x_6-x_2)/s_6$ $\cos B_6=(y_6-y_2)/s_6$
$\cos D_6=(w_6-w_2)/s_6$

$\cos A_7=(x_9-x_2)/s_7$ $\cos B_7=(y_9-y_2)/s_7$
$\cos D_7=(w_9-w_2)/s_7$

$\cos A_8=(x_4-x_3)/s_8$ $\cos B_8=(y_4-y_3)/s_8$
$\cos D_8=(w_4-w_3)/s_8$

$\cos A_9=(x_7-x_3)/s_9$ $\cos B_9=(y_7-y_3)/s_9$
$\cos D_9=(w_7-w_3)/s_9$

DIRECTION COSINES (Cont'd)

$\cos A_{10}=(x_9-x_3)/s_{10}$ $\cos B_{10}=(y_9-y_3)/s_{10}$

$\cos D_{10}=(w_9-w_3)/s_{10}$

$\cos A_{11}=(x_8-x_4)/s_{11}$ $\cos B_{11}=(y_8-y_4)/s_{11}$

$\cos D_{11}=(w_8-w_4)/s_{11}$

$\cos A_{12}=(x_9-x_4)/s_{12}$ $\cos B_{12}=(y_9-y_4)/s_{12}$

$\cos D_{12}=(w_9-w_4)/s_{12}$

$\cos A_{13}=(x_6-x_5)/s_{13}$ $\cos B_{13}=(y_6-y_5)/s_{13}$

$\cos D_{13}=(w_6-w_5)/s_{13}$

$\cos A_{14}=(x_7-x_5)/s_{14}$ $\cos B_{14}=(y_7-y_5)/s_{14}$

$\cos D_{14}=(w_7-w_5)/s_{14}$

$\cos A_{15}=(x_9-x_5)/s_{15}$ $\cos B_{15}=(y_9-y_5)/s_{15}$

$\cos D_{15}=(w_9-w_5)/s_{15}$

$\cos A_{16}=(x_8-x_6)/s_{16}$ $\cos B_{16}=(y_8-y_6)/s_{16}$

$\cos D_{16}=(w_8-w_6)/s_{16}$

$\cos A_{17}=(x_9-x_6)/s_{17}$ $\cos B_{17}=(y_9-y_6)/s_{17}$

$\cos D_{17}=(w_9-w_6)/s_{17}$

$\cos A_{18}=(x_8-x_7)/s_{18}$ $\cos B_{18}=(y_8-y_7)/s_{18}$

$\cos D_{18}=(w_8-w_7)/s_{18}$

$\cos A_{19}=(x_9-x_7)/s_{19}$ $\cos B_{19}=(y_9-y_7)/s_{19}$

$\cos D_{19}=(w_9-w_7)/s_{19}$

$\cos A_{20}=(x_9-x_8)/s_{20}$ $\cos B_{20}=(y_9-y_8)/s_{20}$

$\cos D_{20}=(w_9-w_8)/s_{20}$

ANGLES

$\cos\phi_{1,2}=\cos A_1\cos A_2+\cos B_1\cos B_2+\cos D_1\cos D_2$

$\cos\phi_{1,3}=\cos A_1\cos A_3+\cos B_1\cos B_3+\cos D_1\cos D_3$

$\cos\phi_{1,4}=\cos A_1\cos A_4+\cos B_1\cos B_4+\cos D_1\cos D_4$

$\cos\phi_{1,5}=\cos A_1\cos A_5+\cos B_1\cos B_5+\cos D_1\cos D_5$

$\cos\phi_{1,6}=\cos A_1\cos A_6+\cos B_1\cos B_6+\cos D_1\cos D_6$

ANGLES (Cont'd)

$\cos\phi_{1,7}=\cos A_1\cos A_7+\cos B_1\cos B_7+\cos D_1\cos D_7$

$\cos\phi_{2,3}=\cos A_2\cos A_3+\cos B_2\cos B_3+\cos D_2\cos D_3$

$\cos\phi_{2,4}=\cos A_2\cos A_4+\cos B_2\cos B_4+\cos D_2\cos D_4$

$\cos\phi_{2,8}=\cos A_2\cos A_8+\cos B_2\cos B_8+\cos D_2\cos D_8$

$\cos\phi_{2,9}=\cos A_2\cos A_9+\cos B_2\cos B_9+\cos D_2\cos D_9$

$\cos\phi_{2,10}=\cos A_2\cos A_{10}+\cos B_2\cos B_{10}+\cos D_2\cos D_{10}$

$\cos\phi_{3,4}=\cos A_3\cos A_4+\cos B_3\cos B_4+\cos D_3\cos D_4$

$\cos\phi_{3,13}=\cos A_3\cos A_{13}+\cos B_3\cos B_{13}+\cos D_3\cos D_{13}$

$\cos\phi_{3,14}=\cos A_3\cos A_{14}+\cos B_3\cos B_{14}+\cos D_3\cos D_{14}$

$\cos\phi_{3,15}=\cos A_3\cos A_{15}+\cos B_3\cos B_{15}+\cos D_3\cos D_{15}$

$\cos\phi_{4,7}=\cos A_4\cos A_7+\cos B_4\cos B_7+\cos D_4\cos D_7$

$\cos\phi_{4,10}=\cos A_4\cos A_{10}+\cos B_4\cos B_{10}+\cos D_4\cos D_{10}$

$\cos\phi_{4,15}=\cos A_4\cos A_{15}+\cos B_4\cos B_{15}+\cos D_4\cos D_{15}$

$\cos\phi_{5,6}=\cos A_5\cos A_6+\cos B_5\cos B_6+\cos D_5\cos D_6$

$\cos\phi_{5,7}=\cos A_5\cos A_7+\cos B_5\cos B_7+\cos D_5\cos D_7$

$\cos\phi_{5,8}=\cos A_5\cos A_8+\cos B_5\cos B_8+\cos D_5\cos D_8$

$\cos\phi_{5,11}=\cos A_5\cos A_1+\cos B_5\cos B_{11}+\cos D_5\cos D_{11}$

$\cos\phi_{5,12}=\cos A_5\cos A_{12}+\cos B_5\cos B_{12}+\cos D_5\cos D_{12}$

$\cos\phi_{6,7}=\cos A_6\cos A_7+\cos B_6\cos B_7+\cos D_6\cos D_7$

$\cos\phi_{6,13}=\cos A_6\cos A_{13}+\cos B_6\cos B_{13}+\cos D_6\cos D_{13}$

$\cos\phi_{6,16}=\cos A_6\cos A_{16}+\cos B_6\cos B_{16}+\cos D_6\cos D_{16}$

$\cos\phi_{6,17}=\cos A_6\cos A_{17}+\cos B_6\cos B_{17}+\cos D_6\cos D_{17}$

$\cos\phi_{7,12}=\cos A_7\cos A_{12}+\cos B_7\cos B_{12}+\cos D_7\cos D_{12}$

$\cos\phi_{7,17}=\cos A_7\cos A_{17}+\cos B_7\cos B_{17}+\cos D_7\cos D_{17}$

$\cos\phi_{8,9}=\cos A_8\cos A_9+\cos B_8\cos B_9+\cos D_8\cos D_9$

$\cos\phi_{8,10}=\cos A_8\cos A_{10}+\cos B_8\cos B_{10}+\cos D_8\cos D_{10}$

$\cos\phi_{8,11}=\cos A_8\cos A_{11}+\cos B_8\cos B_{11}+\cos D_8\cos D_{11}$

$\cos\phi_{8,12}=\cos A_8\cos A_{12}+\cos B_8\cos B_{12}+\cos D_8\cos D_{12}$

$\cos\phi_{9,10}=\cos A_9\cos A_{10}+\cos B_9\cos B_{10}+\cos D_9\cos D_{10}$

$\cos\phi_{9,14}=\cos A_9\cos A_{14}+\cos B_9\cos B_{14}+\cos D_9\cos D_{14}$

$\cos\phi_{9,18}=\cos A_9\cos A_{18}+\cos B_9\cos B_{18}+\cos D_9\cos D_{18}$

$\cos\phi_{9,19}=\cos A_9\cos A_{19}+\cos B_9\cos B_{19}+\cos D_9\cos D_{19}$

$\cos\phi_{10,12}=\cos A_{10}\cos A_{12}+\cos B_{10}\cos B_{12}+\cos D_{10}\cos D_{12}$

$\cos\phi_{10,19}=\cos A_{10}\cos A_{19}+\cos B_{10}\cos B_{19}+\cos D_{10}\cos D_{19}$

$\cos\phi_{11,12}=\cos A_{11}\cos A_{12}+\cos B_{11}\cos B_{12}+\cos D_{11}\cos D_{12}$

Table 22-11. Specifications and Analyses of the Hyperspace Object Shown in Fig. 22-11. (A) Specifications and Analysis in Hyperspace. (B) Secondary Analysis in XYZ Space. (C) Secondary Analysis in XYW Space. (D) Secondary Analysis in XZW Space. (E) Secondary Analysis in YZW Space. (F) Suggested Coordinates. **(Continued from page 427.)**

ANGLES (Cont'd)

$\cos\phi_{11,16}=\cos A_{11}\cos A_{16}+\cos B_{11}\cos B_{16}+\cos D_{11}\cos D_{16}$

$\cos\phi_{11,18}=\cos A_{11}\cos A_{18}+\cos B_{11}\cos B_{18}+\cos D_{11}\cos D_{18}$

$\cos\phi_{11,20}=\cos A_{11}\cos A_{20}+\cos B_{11}\cos B_{20}+\cos D_{11}\cos D_{20}$

$\cos\phi_{12,20}=\cos A_{12}\cos A_{20}+\cos B_{12}\cos B_{20}+\cos D_{12}\cos D_{20}$

$\cos\phi_{13,14}=\cos A_{13}\cos A_{14}+\cos B_{13}\cos B_{14}+\cos D_{13}\cos D_{14}$

$\cos\phi_{13,15}=\cos A_{13}\cos A_{145}+\cos B_{13}\cos B_{15}+\cos D_{13}\cos D_{15}$

$\cos\phi_{13,16}=\cos A_{13}\cos A_{16}+\cos B_{13}\cos B_{16}+\cos D_{13}\cos D_{16}$

$\cos\phi_{13,17}=\cos A_{13}\cos A_{17}+\cos B_{13}\cos B_{17}+\cos D_{13}\cos D_{17}$

$\cos\phi_{14,15}=\cos A_{14}\cos A_{15}+\cos B_{14}\cos B_{15}+\cos D_{14}\cos D_{15}$

$\cos\phi_{14,18}=\cos A_{14}\cos A_{18}+\cos B_{14}\cos B_{18}+\cos D_{14}\cos D_{18}$

$\cos\phi_{14,19}=\cos A_{14}\cos A_{19}+\cos B_{14}\cos B_{19}+\cos D_{14}\cos D_{19}$

$\cos\phi_{15,17}=\cos A_{15}\cos A_{17}+\cos B_{15}\cos B_{17}+\cos D_{15}\cos D_{17}$

$\cos\phi_{15,19}=\cos A_{15}\cos A_{19}+\cos B_{15}\cos B_{19}+\cos D_{15}\cos D_{19}$

$\cos\phi_{16,17}=\cos A_{16}\cos A_{17}+\cos B_{16}\cos B_{17}+\cos D_{16}\cos D_{17}$

$\cos\phi_{16,18}=\cos A_{16}\cos A_{18}+\cos B_{16}\cos B_{18}+\cos D_{16}\cos D_{18}$

$\cos\phi_{16,20}=\cos A_{16}\cos A_{20}+\cos B_{16}\cos B_{20}+\cos D_{16}\cos D_{20}$

$\cos\phi_{17,20}=\cos A_{17}\cos A_{20}+\cos B_{17}\cos B_{20}+\cos D_{17}\cos D_{20}$

$\cos\phi_{18,19}=\cos A_{18}\cos A_{19}+\cos B_{18}\cos B_{19}+\cos D_{18}\cos D_{19}$

$\cos\phi_{18,20}=\cos A_{18}\cos A_{20}+\cos B_{18}\cos B_{20}+\cos D_{18}\cos D_{20}$

$\cos\phi_{19,20}=\cos A_{19}\cos A_{20}+\cos B_{19}\cos B_{20}+\cos D_{19}\cos D_{20}$

CHECK

F_1: $\phi_{1,2}+\phi_{1,5}+\phi_{2,8}+\phi_{5,8}=360^\circ$

F_2: $\phi_{1,3}+\phi_{1,6}+\phi_{3,13}+\phi_{6,13}=360^\circ$

F_3: $\phi_{1,4}+\phi_{1,7}+\phi_{4,7}=180^\circ$

F_4: $\phi_{2,3}+\phi_{2,9}+\phi_{3,14}+\phi_{9,14}=360^\circ$

F_5: $\phi_{2,4}+\phi_{2,10}+\phi_{4,10}=180^\circ$

F_6: $\phi_{3,4}+\phi_{3,15}+\phi_{4,15}=180^\circ$

F_7: $\phi_{5,6}+\phi_{5,11}+\phi_{6,16}+\phi_{11,16}=360^\circ$

F_8: $\phi_{5,7}+\phi_{5,12}+\phi_{7,12}=180^\circ$

F_9: $\phi_{6,7}+\phi_{6,17}+\phi_{7,17}=180^\circ$

F_{10}: $\phi_{8,9}+\phi_{8,11}+\phi_{9,18}+\phi_{11,18}=360^\circ$

F_{11}: $\phi_{8,10}+\phi_{8,12}+\phi_{10,12}=180^\circ$

F_{12}: $\phi_{9,10}+\phi_{9,19}+\phi_{10,19}=180^\circ$

F_{13}: $\phi_{11,12}+\phi_{11,20}+\phi_{12,20}=180^\circ$

F_{14}: $\phi_{13,14}+\phi_{13,16}+\phi_{14,18}+\phi_{16,18}=360^\circ$

F_{15}: $\phi_{13,15}+\phi_{13,17}+\phi_{15,17}=180^\circ$

F_{16}: $\phi_{14,15}+\phi_{14,19}+\phi_{15,19}=180^\circ$

F_{17}: $\phi_{16,17}+\phi_{16,20}+\phi_{17,20}=180^\circ$

F_{18}: $\phi_{18,19}+\phi_{18,20}+\phi_{19,20}=180^\circ$

C

SECONDARY ANALYSIS -- XZW SPACE

PLOTTED POINTS

$P_1=(x_1,z_1,w_1)$

$P_2=(x_2,z_2,w_2)$

$P_3=(x_3,z_3,w_3)$

$P_4=(x_4,z_4,w_4)$

$P_5=(x_5,z_5,w_5)$

$P_6=(x_6,z_6,w_6)$

$P_7=(x_7,z_7,w_7)$

$P_8=(x_8,z_8,w_8)$

$P_9=(x_9,z_9,w_9)$

LENGTHS

$s_1=\sqrt{(x_2-x_1)^2+(z_2-z_1)^2+(w_2-w_1)^2}$

$s_2=\sqrt{(x_3-x_1)^2+(z_3-z_1)^2+(w_3-w_1)^2}$

$s_3=\sqrt{(x_5-x_1)^2+(z_5-z_1)^2+(w_5-w_1)^2}$

$s_4=\sqrt{(x_9-x_1)^2+(z_9-z_1)^2+(w_9-w_1)^2}$

$s_5=\sqrt{(x_4-x_2)^2+(z_4-z_2)^2+(w_4-w_2)^2}$

$s_6=\sqrt{(x_6-x_2)^2+(z_6-z_2)^2+(w_6-w_2)^2}$

$s_7=\sqrt{(x_9-x_2)^2+(z_9-z_2)^2+(w_9-w_2)^2}$

D

LENGTHS (Cont'd)

$s_8=\sqrt{(x_4-x_3)^2+(z_4-z_3)^2+(w_4-w_3)^2}$

$s_9=\sqrt{(x_7-x_3)^2+(z_7-z_3)^2+(w_7-w_3)^2}$

$s_{10}=\sqrt{(x_9-x_3)^2+(z_9-z_3)^2+(w_9-w_3)^2}$

$s_{11}=\sqrt{(x_8-x_4)^2+(z_8-z_4)^2+(w_8-w_4)^2}$

$s_{12}=\sqrt{(x_9-x_4)^2+(z_9-z_4)^2+(w_9-w_4)^2}$

$s_{13}=\sqrt{(x_6-x_5)^2+(z_6-z_5)^2+(w_6-w_5)^2}$

$s_{14}=\sqrt{(x_7-x_5)^2+(z_7-z_5)^2+(w_7-w_5)^2}$

$s_{15}=\sqrt{(x_9-x_5)^2+(z_9-z_5)^2+(w_9-w_5)^2}$

$s_{16}=\sqrt{(x_8-x_6)^2+(z_8-z_6)^2+(w_8-w_6)^2}$

$s_{17}=\sqrt{(x_9-x_6)^2+(z_9-z_6)^2+(w_9-w_6)^2}$

$s_{18}=\sqrt{(x_8-x_7)^2+(z_8-z_7)^2+(w_8-w_7)^2}$

$s_{19}=\sqrt{(x_9-x_7)^2+(z_9-z_7)^2+(w_9-w_7)^2}$

$s_{20}=\sqrt{(x_9-x_8)^2+(z_9-z_8)^2+(w_9-w_8)^2}$

DIRECTION COSINES

$\cos A_1=(x_2-x_1)/s_1$ $\cos C_1=(z_2-z_1)/s_1$

$\cos D_1=(w_2-w_1)/s_1$

$\cos A_2=(x_3-x_1)/s_2$ $\cos C_2=(z_3-z_1)/s_2$

$\cos D_2=(w_3-w_1)/s_2$

$\cos A_3=(x_5-x_1)/s_3$ $\cos C_3=(z_5-z_1)/s_3$

$\cos D_3=(w_5-w_1)/s_3$

$\cos A_4=(x_9-x_1)/s_4$ $\cos C_4=(z_9-z_1)/s_4$

$\cos D_4=(w_9-w_1)/s_4$

$\cos A_5=(x_4-x_2)/s_5$ $\cos C_5=(z_4-z_2)/s_5$

$\cos D_5=(w_4-w_2)/s_5$

$\cos A_6=(x_6-x_2)/s_6$ $\cos C_6=(z_6-z_2)/s_6$

$\cos D_6=(w_6-w_2)/s_6$

DIRECTION COSINES (Cont'd)

$\cos A_7=(x_9-x_2)/s_7$ $\cos C_7=(z_9-z_2)/s_7$

$\cos D_7=(w_9-w_2)/s_7$

$\cos A_8=(x_4-x_3)/s_8$ $\cos C_8=(z_4-z_3)/s_8$

$\cos D_8=(w_4-w_3)/s_8$

$\cos A_9=(x_7-x_3)/s_9$ $\cos C_9=(z_7-z_3)/s_9$

$\cos D_9=(w_7-w_3)/s_9$

$\cos A_{10}=(x_9-x_3)/s_{10}$ $\cos C_{10}=(z_9-z_3)/s_{10}$

$\cos D_{10}=(w_9-w_3)/s_{10}$

$\cos A_{11}=(x_8-x_4)/s_{11}$ $\cos C_{11}=(z_8-z_4)/s_{11}$

$\cos D_{11}=(w_8-w_4)/s_{11}$

$\cos A_{12}=(x_9-x_4)/s_{12}$ $\cos C_{12}=(z_9-z_4)/s_{12}$

$\cos D_{12}=(w_9-w_4)/s_{12}$

$\cos A_{13}=(x_6-x_5)/s_{13}$ $\cos C_{13}=(z_6-z_5)/s_{13}$

$\cos D_{13}=(w_6-w_5)/s_{13}$

$\cos A_{14}=(x_7-x_5)/s_{14}$ $\cos C_{14}=(z_7-z_5)/s_{14}$

$\cos D_{14}=(w_7-w_5)/s_{14}$

$\cos A_{15}=(x_9-x_5)/s_{15}$ $\cos C_{15}=(z_9-z_5)/s_{15}$

$\cos D_{15}=(w_9-w_5)/s_{15}$

$\cos A_{16}=(x_8-x_6)/s_{16}$ $\cos C_{16}=(z_8-z_6)/s_{16}$

$\cos D_{16}=(w_8-w_6)/s_{16}$

$\cos A_{17}=(x_9-x_6)/s_{17}$ $\cos C_{17}=(z_9-z_6)/s_{17}$

$\cos D_{17}=(w_9-w_6)/s_{17}$

$\cos A_{18}=(x_8-x_7)/s_{18}$ $\cos C_{18}=(z_8-z_7)/s_{18}$

$\cos D_{18}=(w_8-w_7)/s_{18}$

$\cos A_{19}=(x_9-x_7)/s_{19}$ $\cos C_{19}=(z_9-z_7)/s_{19}$

$\cos D_{19}=(w_9-w_7)/s_{19}$

$\cos A_{20}=(x_9-x_8)/s_{20}$ $\cos C_{20}=(z_9-z_8)/s_{20}$

$\cos D_{20}=(w_9-w_8)/s_{20}$

D

Table 22-11. Specifications and Analyses of the Hyperspace Object Shown in Fig. 22-11. (A) Specifications and Analysis in Hyperspace. (B) Secondary Analysis in XYZ Space. (C) Secondary Analysis in XYW Space. (D) Secondary Analysis in XZW Space. (E) Secondary Analysis in YZW Space. (F) Suggested Coordinates. (Continued from page 429.)

ANGLES

$\cos\phi_{1,2}=\cos A_1\cos A_2+\cos C_1\cos C_2+\cos D_1\cos D_2$

$\cos\phi_{1,3}=\cos A_1\cos A_3+\cos C_1\cos C_3+\cos D_1\cos D_3$

$\cos\phi_{1,4}=\cos A_1\cos A_4+\cos C_1\cos C_4+\cos D_1\cos D_4$

$\cos\phi_{1,5}=\cos A_1\cos A_5+\cos C_1\cos C_5+\cos D_1\cos D_5$

$\cos\phi_{1,6}=\cos A_1\cos A_6+\cos C_1\cos C_6+\cos D_1\cos D_6$

$\cos\phi_{1,7}=\cos A_1\cos A_7+\cos C_1\cos C_7+\cos D_1\cos D_7$

$\cos\phi_{2,3}=\cos A_2\cos A_3+\cos C_2\cos C_3+\cos D_2\cos D_3$

$\cos\phi_{2,4}=\cos A_2\cos A_4+\cos C_2\cos C_4+\cos D_2\cos D_4$

$\cos\phi_{2,8}=\cos A_2\cos A_8+\cos C_2\cos C_8+\cos D_2\cos D_8$

$\cos\phi_{2,9}=\cos A_2\cos A_9+\cos C_2\cos C_9+\cos D_2\cos D_9$

$\cos\phi_{2,10}=\cos A_2\cos A_{10}+\cos C_2\cos C_{10}+\cos D_2\cos D_{10}$

$\cos\phi_{3,4}=\cos A_3\cos A_4+\cos C_3\cos C_4+\cos D_3\cos D_4$

$\cos\phi_{3,13}=\cos A_3\cos A_{13}+\cos C_3\cos C_{13}+\cos D_3\cos D_{13}$

$\cos\phi_{3,14}=\cos A_3\cos A_{14}+\cos C_3\cos C_{14}+\cos D_3\cos D_{14}$

$\cos\phi_{3,15}=\cos A_3\cos A_{15}+\cos C_3\cos C_{15}+\cos D_3\cos D_{15}$

$\cos\phi_{4,7}=\cos A_4\cos A_7+\cos C_4\cos C_7+\cos D_4\cos D_7$

$\cos\phi_{4,10}=\cos A_4\cos A_{10}+\cos C_4\cos C_{10}+\cos D_4\cos D_{10}$

$\cos\phi_{4,15}=\cos A_4\cos A_{15}+\cos C_4\cos C_{15}+\cos D_4\cos D_{15}$

$\cos\phi_{5,6}=\cos A_5\cos A_6+\cos C_5\cos C_6+\cos D_5\cos D_6$

$\cos\phi_{5,7}=\cos A_5\cos A_7+\cos C_5\cos C_7+\cos D_5\cos D_7$

$\cos\phi_{5,8}=\cos A_5\cos A_8+\cos C_5\cos C_8+\cos D_5\cos D_8$

$\cos\phi_{5,11}=\cos A_5\cos A_1+\cos C_5\cos C_{11}+\cos D_5\cos D_{11}$

$\cos\phi_{5,12}=\cos A_5\cos A_{12}+\cos C_5\cos C_{12}+\cos D_5\cos D_{12}$

$\cos\phi_{6,7}=\cos A_6\cos A_7+\cos C_6\cos C_7+\cos D_6\cos D_7$

$\cos\phi_{6,13}=\cos A_6\cos A_{13}+\cos C_6\cos C_{13}+\cos D_6\cos D_{13}$

$\cos\phi_{6,16}=\cos A_6\cos A_{16}+\cos C_6\cos C_{16}+\cos D_6\cos D_{16}$

$\cos\phi_{6,17}=\cos A_6\cos A_{17}+\cos C_6\cos C_{17}+\cos D_6\cos D_{17}$

$\cos\phi_{7,12}=\cos A_7\cos A_{12}+\cos C_7\cos C_{12}+\cos D_7\cos D_{12}$

$\cos\phi_{7,17}=\cos A_7\cos A_{17}+\cos C_7\cos C_{17}+\cos D_7\cos D_{17}$

$\cos\phi_{8,9}=\cos A_8\cos A_9+\cos C_8\cos C_9+\cos D_8\cos D_9$

ANGLES (Cont'd)

$\cos\phi_{8,10}=\cos A_8\cos A_{10}+\cos C_8\cos C_{10}+\cos D_8\cos D_{10}$

$\cos\phi_{8,11}=\cos A_8\cos A_{11}+\cos C_8\cos C_{11}+\cos D_8\cos D_{11}$

$\cos\phi_{8,12}=\cos A_8\cos A_{12}+\cos C_8\cos C_{12}+\cos D_8\cos D_{12}$

$\cos\phi_{9,10}=\cos A_9\cos A_{10}+\cos C_9\cos C_{10}+\cos D_9\cos D_{10}$

$\cos\phi_{9,14}=\cos A_9\cos A_{14}+\cos C_9\cos C_{14}+\cos D_9\cos D_{14}$

$\cos\phi_{9,18}=\cos A_9\cos A_{18}+\cos C_9\cos C_{18}+\cos D_9\cos D_{18}$

$\cos\phi_{9,19}=\cos A_9\cos A_{19}+\cos C_9\cos C_{19}+\cos D_9\cos D_{19}$

$\cos\phi_{10,12}=\cos A_{10}\cos A_{12}+\cos C_{10}\cos C_{12}+\cos D_{10}\cos D_{12}$

$\cos\phi_{10,19}=\cos A_{10}\cos A_{19}+\cos C_{10}\cos C_{19}+\cos D_{10}\cos D_{19}$

$\cos\phi_{11,12}=\cos A_{11}\cos A_{12}+\cos C_{11}\cos C_{12}+\cos D_{11}\cos D_{12}$

$\cos\phi_{11,16}=\cos A_{11}\cos A_{16}+\cos C_{11}\cos C_{16}+\cos D_{11}\cos D_{16}$

$\cos\phi_{11,18}=\cos A_{11}\cos A_{18}+\cos C_{11}\cos C_{18}+\cos D_{11}\cos D_{18}$

$\cos\phi_{11,20}=\cos A_{11}\cos A_{20}+\cos C_{11}\cos C_{20}+\cos D_{11}\cos D_{20}$

$\cos\phi_{12,20}=\cos A_{12}\cos A_{20}+\cos C_{12}\cos C_{20}+\cos D_{12}\cos D_{20}$

$\cos\phi_{13,14}=\cos A_{13}\cos A_{14}+\cos C_{13}\cos C_{14}+\cos D_{13}\cos D_{14}$

$\cos\phi_{13,15}=\cos A_{13}\cos A_{145}+\cos C_{13}\cos C_{15}+\cos D_{13}\cos D_{15}$

$\cos\phi_{13,16}=\cos A_{13}\cos A_{16}+\cos C_{13}\cos C_{16}+\cos D_{13}\cos D_{16}$

$\cos\phi_{13,17}=\cos A_{13}\cos A_{17}+\cos C_{13}\cos C_{17}+\cos D_{13}\cos D_{17}$

$\cos\phi_{14,15}=\cos A_{14}\cos A_{15}+\cos C_{14}\cos C_{15}+\cos D_{14}\cos D_{15}$

$\cos\phi_{14,18}=\cos A_{14}\cos A_{18}+\cos C_{14}\cos C_{18}+\cos D_{14}\cos D_{18}$

$\cos\phi_{14,19}=\cos A_{14}\cos A_{19}+\cos C_{14}\cos C_{19}+\cos D_{14}\cos D_{19}$

$\cos\phi_{15,17}=\cos A_{15}\cos A_{17}+\cos C_{15}\cos C_{17}+\cos D_{15}\cos D_{17}$

$\cos\phi_{15,19}=\cos A_{15}\cos A_{19}+\cos C_{15}\cos C_{19}+\cos D_{15}\cos D_{19}$

$\cos\phi_{16,17}=\cos A_{15}\cos A_{17}+\cos C_{16}\cos C_{17}+\cos D_{16}\cos D_{17}$

$\cos\phi_{16,18}=\cos A_{16}\cos A_{18}+\cos C_{16}\cos C_{18}+\cos D_{16}\cos D_{18}$

$\cos\phi_{16,20}=\cos A_{16}\cos A_{20}+\cos C_{16}\cos C_{20}+\cos D_{16}\cos D_{20}$

$\cos\phi_{17,20}=\cos A_{17}\cos A_{20}+\cos C_{17}\cos C_{20}+\cos D_{17}\cos D_{20}$

$\cos\phi_{18,19}=\cos A_{18}\cos A_{19}+\cos C_{18}\cos C_{19}+\cos D_{18}\cos D_{19}$

$\cos\phi_{18,20}=\cos A_{18}\cos A_{20}+\cos C_{18}\cos C_{20}+\cos D_{18}\cos D_{20}$

$\cos\phi_{19,20}=\cos A_{19}\cos A_{20}+\cos C_{19}\cos C_{20}+\cos D_{19}\cos D_{20}$

D

CHECK

F_1: $\phi_{1,2}+\phi_{1,5}+\phi_{2,8}+\phi_{5,8}=360^\circ$

F_2: $\phi_{1,3}+\phi_{1,6}+\phi_{3,13}+\phi_{6,13}=360^\circ$

F_3: $\phi_{1,4}+\phi_{1,7}+\phi_{4,7}=180^\circ$

F_4: $\phi_{2,3}+\phi_{2,9}+\phi_{3,14}+\phi_{9,14}=360^\circ$

F_5: $\phi_{2,4}+\phi_{2,10}+\phi_{4,10}=180^\circ$

F_6: $\phi_{3,4}+\phi_{3,15}+\phi_{4,15}=180^\circ$

F_7: $\phi_{5,6}+\phi_{5,11}+\phi_{6,16}+\phi_{11,16}=360^\circ$

F_8: $\phi_{5,7}+\phi_{5,12}+\phi_{7,12}=180^\circ$

F_9: $\phi_{6,7}+\phi_{6,17}+\phi_{7,17}=180^\circ$

F_{10}: $\phi_{8,9}+\phi_{8,11}+\phi_{9,18}+\phi_{11,18}=360^\circ$

F_{11}: $\phi_{8,10}+\phi_{8,12}+\phi_{10,12}=180^\circ$

F_{12}: $\phi_{9,10}+\phi_{9,19}+\phi_{10,19}=180^\circ$

F_{13}: $\phi_{11,12}+\phi_{11,20}+\phi_{12,20}=180^\circ$

F_{14}: $\phi_{13,14}+\phi_{13,16}+\phi_{14,18}+\phi_{16,18}=360^\circ$

F_{15}: $\phi_{13,15}+\phi_{13,17}+\phi_{15,17}=180^\circ$

F_{16}: $\phi_{14,15}+\phi_{14,19}+\phi_{15,19}=180^\circ$

F_{17}: $\phi_{16,17}+\phi_{16,20}+\phi_{17,20}=180^\circ$

F_{18}: $\phi_{18,19}+\phi_{18,20}+\phi_{19,20}=180^\circ$

D

SECONDARY ANALYSIS -- YZW SPACE

PLOTTED POINTS

$P_1=(y_1,z_1,w_1)$

$P_2=(y_2,z_2,w_2)$

$P_3=(y_3,z_3,w_3)$

$P_4=(y_4,z_4,w_4)$

$P_5=(y_5,z_5,w_5)$

$P_6=(y_6,z_6,w_6)$

$P_7=(y_7,z_7,w_7)$

$P_8=(y_8,z_8,w_8)$

$P_9=(y_9,z_9,w_9)$

LENGTHS

$s_1=\sqrt{(y_2-y_1)^2+(z_2-z_1)^2+(w_2-w_1)^2}$

$s_2=\sqrt{(y_3-y_1)^2+(z_3-z_1)^2+(w_3-w_1)^2}$

$s_3=\sqrt{(y_5-y_1)^2+(z_5-z_1)^2+(w_5-w_1)^2}$

$s_4=\sqrt{(y_9-y_1)^2+(z_9-z_1)^2+(w_9-w_1)^2}$

$s_5=\sqrt{(y_4-y_2)^2+(z_4-z_2)^2+(w_4-w_2)^2}$

$s_6=\sqrt{(y_6-y_2)^2+(z_6-z_2)^2+(w_6-w_2)^2}$

LENGTHS (Cont'd)

$s_7=\sqrt{(y_9-y_2)^2+(z_9-z_2)^2+(w_9-w_2)^2}$

$s_8=\sqrt{(y_4-y_3)^2+(z_4-z_3)^2+(w_4-w_3)^2}$

$s_9=\sqrt{(y_7-y_3)^2+(z_7-z_3)^2+(w_7-w_3)^2}$

$s_{10}=\sqrt{(y_9-y_3)^2+(z_9-z_3)^2+(w_9-w_3)^2}$

$s_{11}=\sqrt{(y_8-y_4)^2+(z_8-z_4)^2+(w_8-w_4)^2}$

$s_{12}=\sqrt{(y_9-y_4)^2+(z_9-z_4)^2+(w_9-w_4)^2}$

$s_{13}=\sqrt{(y_6-y_5)^2+(z_6-z_5)^2+(w_6-w_5)^2}$

$s_{14}=\sqrt{(y_7-y_5)^2+(z_7-z_5)^2+(w_7-w_5)^2}$

$s_{15}=\sqrt{(y_9-y_5)^2+(z_9-z_5)^2+(w_9-w_5)^2}$

$s_{16}=\sqrt{(y_8-y_6)^2+(z_8-z_6)^2+(w_8-w_6)^2}$

$s_{17}=\sqrt{(y_9-y_6)^2+(z_9-z_6)^2+(w_9-w_6)^2}$

$s_{18}=\sqrt{(y_8-y_7)^2+(z_8-z_7)^2+(w_8-w_7)^2}$

$s_{19}=\sqrt{(y_9-y_7)^2+(z_9-z_7)^2+(w_9-w_7)^2}$

$s_{20}=\sqrt{(y_9-y_8)^2+(z_9-z_8)^2+(w_9-w_8)^2}$

E

Table 22-11. Specifications and Analyses of the Hyperspace Object Shown in Fig. 22-11. (A) Specifications and Analysis in Hyperspace. (B) Secondary Analysis in XYZ Space. (C) Secondary Analysis in XYW Space. (D) Secondary Analysis in XZW Space. (E) Secondary Analysis in YZW Space. (F) Suggested Coordinates. (Continued from page 431.)

DIRECTION COSINES

$\cos B_1=(y_2-y_1)/s_1$ $\cos C_1=(z_2-z_1)/s_1$
$\cos D_1=(w_2-w_1)/s_1$

$\cos B_2=(y_3-y_1)/s_2$ $\cos C_2=(z_3-z_1)/s_2$
$\cos D_2=(w_3-w_1)/s_2$

$\cos B_3=(y_5-y_1)/s_3$ $\cos C_3=(z_5-z_1)/s_3$
$\cos D_3=(w_5-w_1)/s_3$

$\cos B_4=(y_9-y_1)/s_4$ $\cos C_4=(z_9-z_1)/s_4$
$\cos D_4=(w_9-w_1)/s_4$

$\cos B_5=(y_4-y_2)/s_5$ $\cos C_5=(z_4-z_2)/s_5$
$\cos D_5=(w_4-w_2)/s_5$

$\cos B_6=(y_6-y_2)/s_6$ $\cos C_6=(z_6-z_2)/s_6$
$\cos D_6=(w_6-w_2)/s_6$

$\cos B_7=(y_9-y_2)/s_7$ $\cos C_7=(z_9-z_2)/s_7$
$\cos D_7=(w_9-w_2)/s_7$

$\cos B_8=(y_4-y_3)/s_8$ $\cos C_8=(z_4-z_3)/s_8$
$\cos D_8=(w_4-w_3)/s_8$

$\cos B_9=(y_7-y_3)/s_9$ $\cos C_9=(z_7-z_3)/s_9$
$\cos D_9=(w_7-w_3)/s_9$

$\cos B_{10}=(y_9-y_3)/s_{10}$ $\cos C_{10}=(z_9-z_3)/s_{10}$
$\cos D_{10}=(w_9-w_3)/s_{10}$

$\cos B_{11}=(y_8-y_4)/s_{11}$ $\cos C_{11}=(z_8-z_4)/s_{11}$
$\cos D_{11}=(w_8-w_4)/s_{11}$

$\cos B_{12}=(y_9-y_4)/s_{12}$ $\cos C_{12}=(z_9-z_4)/s_{12}$
$\cos D_{12}=(w_9-w_4)/s_{12}$

$\cos B_{13}=(y_6-y_5)/s_{13}$ $\cos C_{13}=(z_6-z_5)/s_{13}$
$\cos D_{13}=(w_6-w_5)/s_{13}$

E

DIRECTION COSINES (Cont'd)

$\cos B_{14}=(y_7-y_5)/s_{14}$ $\cos C_{14}=(z_7-z_5)/s_{14}$
$\cos D_{14}=(w_7-w_5)/s_{14}$

$\cos B_{15}=(y_9-y_5)/s_{15}$ $\cos C_{15}=(z_9-z_5)/s_{15}$
$\cos D_{15}=(w_9-w_5)/s_{15}$

$\cos B_{16}=(y_8-y_6)/s_{16}$ $\cos C_{16}=(z_8-z_6)/s_{16}$
$\cos D_{16}=(w_8-w_6)/s_{16}$

$\cos B_{17}=(y_9-y_6)/s_{17}$ $\cos C_{17}=(z_9-z_6)/s_{17}$
$\cos D_{17}=(w_9-w_6)/s_{17}$

$\cos B_{18}=(y_8-y_7)/s_{18}$ $\cos C_{18}=(z_8-z_7)/s_{18}$
$\cos D_{18}=(w_8-w_7)/s_{18}$

$\cos B_{19}=(y_9-y_7)/s_{19}$ $\cos C_{19}=(z_9-z_7)/s_{19}$
$\cos D_{19}=(w_9-w_7)/s_{19}$

$\cos B_{20}=(y_9-y_8)/s_{20}$ $\cos C_{20}=(z_9-z_8)/s_{20}$
$\cos D_{20}=(w_9-w_8)/s_{20}$

ANGLES

$\cos\phi_{1,2}=\cos B_1\cos B_2+\cos C_1\cos C_2+\cos D_1\cos D_2$
$\cos\phi_{1,3}=\cos B_1\cos B_3+\cos C_1\cos C_3+\cos D_1\cos D_3$
$\cos\phi_{1,4}=\cos B_1\cos B_4+\cos C_1\cos C_4+\cos D_1\cos D_4$
$\cos\phi_{1,5}=\cos B_1\cos B_5+\cos C_1\cos C_5+\cos D_1\cos D_5$
$\cos\phi_{1,6}=\cos B_1\cos B_6+\cos C_1\cos C_6+\cos D_1\cos D_6$
$\cos\phi_{1,7}=\cos B_1\cos B_7+\cos C_1\cos C_7+\cos D_1\cos D_7$
$\cos\phi_{2,3}=\cos B_2\cos B_3+\cos C_2\cos C_3+\cos D_2\cos D_3$
$\cos\phi_{2,4}=\cos B_2\cos B_4+\cos C_2\cos C_4+\cos D_2\cos D_4$
$\cos\phi_{2,8}=\cos B_2\cos B_8+\cos C_2\cos C_8+\cos D_2\cos D_8$
$\cos\phi_{2,9}=\cos B_2\cos B_9+\cos C_2\cos C_9+\cos D_2\cos D_9$
$\cos\phi_{2,10}=\cos B_2\cos B_{10}+\cos C_2\cos C_{10}+\cos D_2\cos D_{10}$
$\cos\phi_{3,4}=\cos B_3\cos B_4+\cos C_3\cos C_4+\cos D_3\cos D_4$
$\cos\phi_{3,13}=\cos B_3\cos B_{13}+\cos C_3\cos C_{13}+\cos D_3\cos D_{13}$
$\cos\phi_{3,14}=\cos B_3\cos B_{14}+\cos C_3\cos C_{14}+\cos D_3\cos D_{14}$
$\cos\phi_{3,15}=\cos B_3\cos B_{15}+\cos C_3\cos C_{15}+\cos D_3\cos D_{15}$

ANGLES (Cont'd)

$$\cos\phi_{4,7}=\cos B_4\cos B_7+\cos C_4\cos C_7+\cos D_4\cos D_7$$

$$\cos\phi_{4,10}=\cos B_4\cos B_{10}+\cos C_4\cos C_{10}+\cos D_4\cos D_{10}$$

$$\cos\phi_{4,15}=\cos B_4\cos B_{15}+\cos C_4\cos C_{15}+\cos D_4\cos D_{15}$$

$$\cos\phi_{5,6}=\cos B_5\cos B_6+\cos C_5\cos C_6+\cos D_5\cos D_6$$

$$\cos\phi_{5,7}=\cos B_5\cos B_7+\cos C_5\cos C_7+\cos D_5\cos D_7$$

$$\cos\phi_{5,8}=\cos B_5\cos B_8+\cos C_5\cos C_8+\cos D_5\cos D_8$$

$$\cos\phi_{5,11}=\cos B_5\cos B_1+\cos C_5\cos C_{11}+\cos D_5\cos D_{11}$$

$$\cos\phi_{5,12}=\cos B_5\cos B_{12}+\cos C_5\cos C_{12}+\cos D_5\cos D_{12}$$

$$\cos\phi_{6,7}=\cos B_6\cos B_7+\cos C_6\cos C_7+\cos D_6\cos D_7$$

$$\cos\phi_{6,13}=\cos B_6\cos B_{13}+\cos C_6\cos C_{13}+\cos D_6\cos D_{13}$$

$$\cos\phi_{6,16}=\cos B_6\cos B_{16}+\cos C_6\cos C_{16}+\cos D_6\cos D_{16}$$

$$\cos\phi_{6,17}=\cos B_6\cos B_{17}+\cos C_6\cos C_{17}+\cos D_6\cos D_{17}$$

$$\cos\phi_{7,12}=\cos B_7\cos B_{12}+\cos C_7\cos C_{12}+\cos D_7\cos D_{12}$$

$$\cos\phi_{7,17}=\cos B_7\cos B_{17}+\cos C_7\cos C_{17}+\cos D_7\cos D_{17}$$

$$\cos\phi_{8,9}=\cos B_8\cos B_9+\cos C_8\cos C_9+\cos D_8\cos D_9$$

$$\cos\phi_{8,10}=\cos B_8\cos B_{10}+\cos C_8\cos C_{10}+\cos D_8\cos D_{10}$$

$$\cos\phi_{8,11}=\cos B_8\cos B_{11}+\cos C_8\cos C_{11}+\cos D_8\cos D_{11}$$

$$\cos\phi_{8,12}=\cos B_8\cos B_{12}+\cos C_8\cos C_{12}+\cos D_8\cos D_{12}$$

$$\cos\phi_{9,10}=\cos B_9\cos B_{10}+\cos C_9\cos C_{10}+\cos D_9\cos D_{10}$$

$$\cos\phi_{9,14}=\cos B_9\cos B_{14}+\cos C_9\cos C_{14}+\cos D_9\cos D_{14}$$

$$\cos\phi_{9,18}=\cos B_9\cos B_{18}+\cos C_9\cos C_{18}+\cos D_9\cos D_{18}$$

$$\cos\phi_{9,19}=\cos B_9\cos B_{19}+\cos C_9\cos C_{19}+\cos D_9\cos D_{19}$$

$$\cos\phi_{10,12}=\cos B_{10}\cos B_{12}+\cos C_{10}\cos C_{12}+\cos D_{10}\cos D_{12}$$

$$\cos\phi_{10,19}=\cos B_{10}\cos B_{19}+\cos C_{10}\cos C_{19}+\cos D_{10}\cos D_{19}$$

$$\cos\phi_{11,12}=\cos B_{11}\cos B_{12}+\cos C_{11}\cos C_{12}+\cos D_{11}\cos D_{12}$$

$$\cos\phi_{11,16}=\cos B_{11}\cos B_{16}+\cos C_{11}\cos C_{16}+\cos D_{11}\cos D_{16}$$

$$\cos\phi_{11,18}=\cos B_{11}\cos B_{18}+\cos C_{11}\cos C_{18}+\cos D_{11}\cos D_{18}$$

$$\cos\phi_{11,20}=\cos B_{11}\cos B_{20}+\cos C_{11}\cos C_{20}+\cos D_{11}\cos D_{20}$$

$$\cos\phi_{12,20}=\cos B_{12}\cos B_{20}+\cos C_{12}\cos C_{20}+\cos D_{12}\cos D_{20}$$

$$\cos\phi_{13,14}=\cos B_{13}\cos B_{14}+\cos C_{13}\cos C_{14}+\cos D_{13}\cos D_{14}$$

$$\cos\phi_{13,15}=\cos B_{13}\cos B_{145}+\cos C_{13}\cos C_{15}+\cos D_{13}\cos D_{15}$$

$$\cos\phi_{13,16}=\cos B_{13}\cos B_{16}+\cos C_{13}\cos C_{16}+\cos D_{13}\cos D_{16}$$

ANGLES (Cont'd)

$$\cos\phi_{13,17}=\cos B_{13}\cos B_{17}+\cos C_{13}\cos C_{17}+\cos D_{13}\cos D_{17}$$

$$\cos\phi_{14,15}=\cos B_{14}\cos B_{15}+\cos C_{14}\cos C_{15}+\cos D_{14}\cos D_{15}$$

$$\cos\phi_{14,18}=\cos B_{14}\cos B_{18}+\cos C_{14}\cos C_{18}+\cos D_{14}\cos D_{18}$$

$$\cos\phi_{14,19}=\cos B_{14}\cos B_{19}+\cos C_{14}\cos C_{19}+\cos D_{14}\cos D_{19}$$

$$\cos\phi_{15,17}=\cos B_{15}\cos B_{17}+\cos C_{15}\cos C_{17}+\cos D_{15}\cos D_{17}$$

$$\cos\phi_{15,19}=\cos B_{15}\cos B_{19}+\cos C_{15}\cos C_{19}+\cos D_{15}\cos D_{19}$$

$$\cos\phi_{16,17}=\cos B_{16}\cos B_{17}+\cos C_{16}\cos C_{17}+\cos D_{16}\cos D_{17}$$

$$\cos\phi_{16,18}=\cos B_{16}\cos B_{18}+\cos C_{16}\cos C_{18}+\cos D_{16}\cos D_{18}$$

$$\cos\phi_{16,20}=\cos B_{16}\cos B_{20}+\cos C_{16}\cos C_{20}+\cos D_{16}\cos D_{20}$$

$$\cos\phi_{17,20}=\cos B_{17}\cos B_{20}+\cos C_{17}\cos C_{20}+\cos D_{17}\cos D_{20}$$

$$\cos\phi_{18,19}=\cos B_{18}\cos B_{19}+\cos C_{18}\cos C_{19}+\cos D_{18}\cos D_{19}$$

$$\cos\phi_{18,20}=\cos B_{18}\cos B_{20}+\cos C_{18}\cos C_{20}+\cos D_{18}\cos D_{20}$$

$$\cos\phi_{19,20}=\cos B_{19}\cos B_{20}+\cos C_{19}\cos C_{20}+\cos D_{19}\cos D_{20}$$

CHECK

F_1: $\phi_{1,2}+\phi_{1,5}+\phi_{2,8}+\phi_{5,8}=360^\circ$

F_2: $\phi_{1,3}+\phi_{1,6}+\phi_{3,13}+\phi_{6,13}=360^\circ$

F_3: $\phi_{1,4}+\phi_{1,7}+\phi_{4,7}=180^\circ$

F_4: $\phi_{2,3}+\phi_{2,9}+\phi_{3,14}+\phi_{9,14}=360^\circ$

F_5: $\phi_{2,4}+\phi_{2,10}+\phi_{4,10}=180^\circ$

F_6: $\phi_{3,4}+\phi_{3,15}+\phi_{4,15}=180^\circ$

F_7: $\phi_{5,6}+\phi_{5,11}+\phi_{6,16}+\phi_{11,16}=360^\circ$

F_8: $\phi_{5,7}+\phi_{5,12}+\phi_{7,12}=180^\circ$

F_9: $\phi_{6,7}+\phi_{6,17}+\phi_{7,17}=180^\circ$

F_{10}: $\phi_{8,9}+\phi_{8,11}+\phi_{9,18}+\phi_{11,18}=360^\circ$

F_{11}: $\phi_{8,10}+\phi_{8,12}+\phi_{10,12}=180^\circ$

F_{12}: $\phi_{9,10}+\phi_{9,19}+\phi_{10,19}=180^\circ$

F_{13}: $\phi_{11,12}+\phi_{11,20}+\phi_{12,20}=180^\circ$

F_{14}: $\phi_{13,14}+\phi_{13,16}+\phi_{14,18}+\phi_{16,18}=360^\circ$

F_{15}: $\phi_{13,15}+\phi_{13,17}+\phi_{15,17}=180^\circ$

F_{16}: $\phi_{14,15}+\phi_{14,19}+\phi_{15,19}=180^\circ$

F_{17}: $\phi_{16,17}+\phi_{16,20}+\phi_{17,20}=180^\circ$

F_{18}: $\phi_{18,19}+\phi_{18,20}+\phi_{19,20}=180^\circ$

E

Table 22-11. Specifications and Analyses of the Hyperspace Object Shown in Fig. 22-11. (A) Specifications and Analysis in Hyperspace. (B) Secondary Analysis in XYZ Space. (C) Secondary Analysis in XYW Space. (D) Secondary Analysis in XZW Space. (E) Secondary Analysis in YZW Space. (F) Suggested Coordinates. **(Continued from page 433.)**

HYPERSPACE VERSION 1

P_1=(-1,0,-1,-1)	P_2=(-1,0,-1,1)
P_3=(1,0,-1,-1)	P_4=(1,0,-1,1)
P_5=(-1,0,1,-1)	P_6=(-1,0,1,1)
P_7=(1,0,1,-1)	P_8=(1,0,1,1)
P_9=(0,1,0,0)	

HYPERSPACE VERSION 2

P_1=(0,0,-2,-1)	P_2=(0,0,0,1)
P_3=(2,0,-2,-1)	P_4=(2,0,0,1)
P_5=(-2,0,-2,-1)	P_6=(-2,0,0,1)
P_7=(0,0,-2,-1)	P_8=(0,0,0,2)
P_9=(0,1,0,0)	

HYPERSPACE VERSION 3

P_1=(0,0,-2,-1)	P_2=(0,0,-2,1)
P_3=(2,0,0,-1)	P_4=(2,0,0,1)
P_5=(-2,0,0,-1)	P_6=(-2,0,0,1)
P_7=(0,0,2,-1)	P_8=(0,0,2,1)
P_9=(0,1,0,0)	

F

be especially meaningful because they show how displacements in hyperspace appear to alternative 3-D spaces. Perhaps we do live in a 4-dimensional universe; and perhaps some of the puzzling features of space, as we currently perceive it, is caused by relativistic displacements in hyperspace. Think about that while you set up some experiments along this line:

☐ Specify a reference space or hyperspace object in the 4-D space of the observer's frame of reference.

☐ Conduct complete primary and secondary analyses of the object as viewed from the observer's frame of reference. Select a 3-D perspective that is to represent your known space, and plot the entity as it appears within it.

☐ Specify a translation that is to exist between the observer's and translated frames of reference, then use the relativistic forms of the 4-D translation equations to determine the specifications for the space object as it appears from the translated frame of reference.

☐ Conduct complete primary and secondary analyses of the entity as it appears from the translated frame of reference, and compare the results with the analysis of the space object in the observer's frame of reference. The comparison is especially meaningful in the 3-D perspective space that you have chosen to represent the perceptible universe.

An interesting variation of that experiment combines the simple geometric and relativistic versions of the translation equations. The objective is to deal with the geometric 4-D translation of a space or hyperspace object as viewed from two different coordinate systems which are, themselves, separated by a relativistic translation. It is a matter of dealing with 4-D relativistic views of a displacement of some object in hyperspace. Give the idea some serious thought on a philosophical level; it may hold more real truth than we can imagine at this time. Here is a general procedure:

☐ Establish a fixed, relativistic translation that is to exist between the two 4-D frames of reference.

☐ Specify the initial space or hyperspace object as it is to appear from the observer's frame of reference, and then conduct at least a primary analysis of it.

☐ Apply the relativistic translation equation to get the specifications for the object as it appears from the translated frame of reference. Do at least a primary analysis of the results.

☐ Apply some geometric translation to the initial object in the observer's frame of reference, and then apply the relativistic equation to find out the specifications of that translated version as regarded from the translated frame of reference.

☐ Repeat the above step as desired to create the impression of a body moving in hyperspace.

The ultimate hyperspace translation experiment is

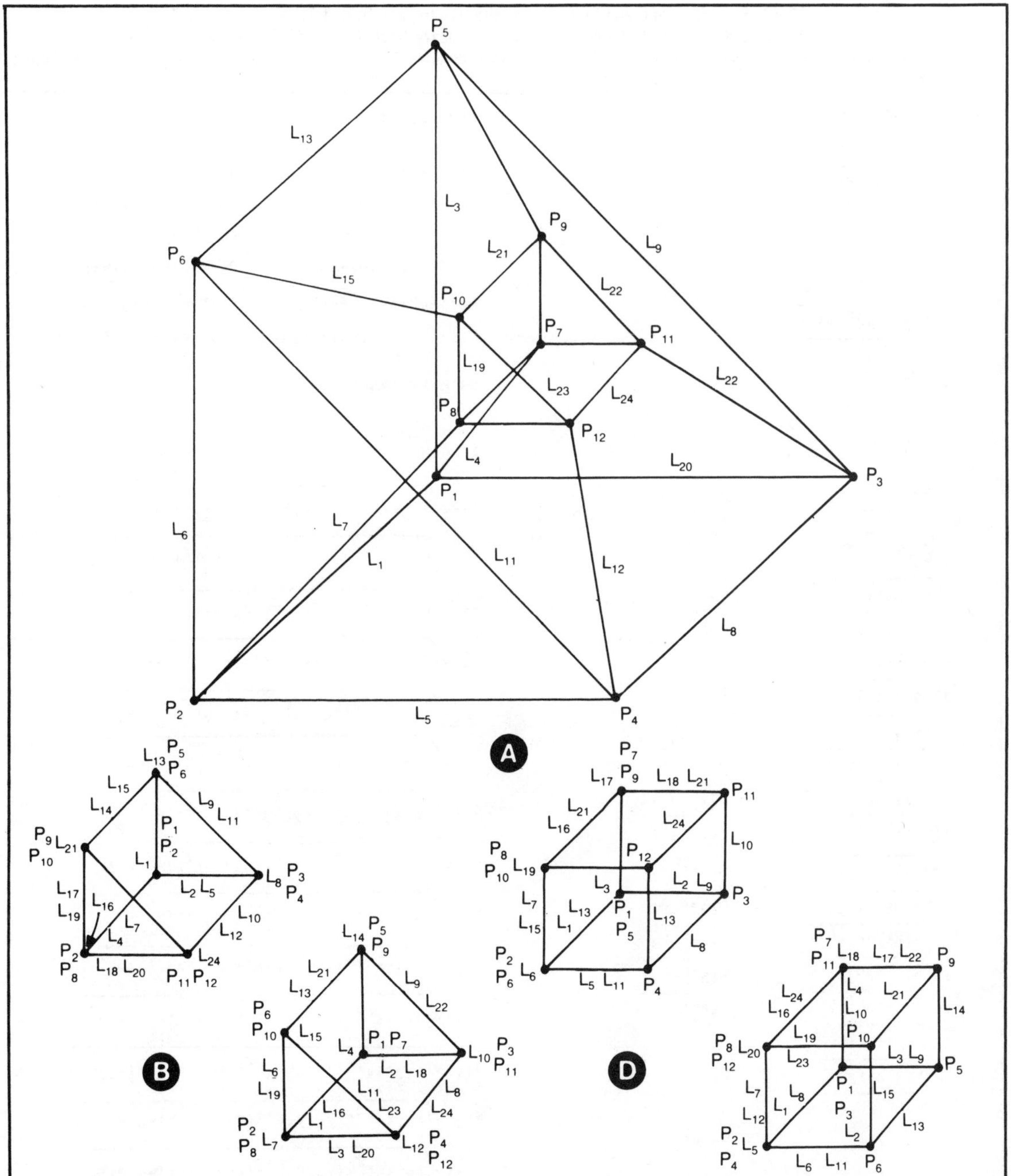

Fig. 22-12. Plots of the hyperspace object specified in Table 22-12A. (A) General hyperspace view. (B) XYZ perspective view. (C) XYW perspective view. (D) XZW perspective view. (E) YZW perspective view. See outlines of analyses in Table 22-12.

Table 22-12. Specifications and Analyses for the Hyperspace Object Shown in Fig. 22-12. (A) Specifications and Analysis in Hyperspace. (B) Secondary Analysis in XYZ Space. (C) Secondary Analysis in XYW Space. (D) Secondary Analysis in XZW Space. (E) Secondary Analysis in YZW Space. (F) Suggested Coordinates.

(A, B, C, and D correspond to α, β, γ, and δ.)

SPECIFICATIONS

$H=\overline{S_1S_2S_3S_4S_5S_6S_7S_8}$

$S_1=\overline{F_1F_2F_3F_4F_5}$

$S_2=\overline{F_6F_7F_8F_9F_{10}}$

$S_3=\overline{F_1F_6F_{11}F_{12}F_{13}}$

$S_4=\overline{F_1F_{15}F_{16}F_{17}F_{18}}$

$S_5=\overline{F_2F_7F_{11}F_{15}F_{19}F_{20}}$

$S_6=\overline{F_4F_9F_{13}F_{18}F_{20}F_{21}}$

$S_7=\overline{F_3F_8F_{12}F_{14}F_{19}F_{21}}$

$S_8=\overline{F_4F_9F_{13}F_{18}F_{20}F_{21}}$

$F_1=\overline{L_2L_3L_9}$	$L_1=\overline{P_1P_2}$
$F_2=\overline{L_2L_4L_{10}L_{18}}$	$L_2=\overline{P_1P_3}$
$F_3=\overline{L_3L_4L_{14}L_{17}}$	$L_3=\overline{P_1P_5}$
$F_4=\overline{L_9L_{10}L_{14}L_{22}}$	$L_4=\overline{P_1P_7}$
$F_5=\overline{L_{17}L_{18}L_{22}}$	$L_5=\overline{P_2P_4}$
$F_6=\overline{L_5L_6L_{11}}$	$L_6=\overline{P_2P_6}$
$F_7=\overline{L_5L_7L_{12}L_{20}}$	$L_7=\overline{P_2P_8}$
$F_8=\overline{L_6L_7L_{15}L_{19}}$	$L_8=\overline{P_3P_4}$
$F_9=\overline{L_{11}L_{12}L_{15}L_{23}}$	$L_9=\overline{P_3P_5}$
$F_{10}=\overline{L_{19}L_{20}L_{23}}$	$L_{10}=\overline{P_4P_{11}}$
$F_{11}=\overline{L_1L_2L_5L_8}$	$L_{11}=\overline{P_4P_6}$
$F_{12}=\overline{L_1L_3L_6L_{13}}$	$L_{12}=\overline{P_4P_{12}}$
$F_{13}=\overline{L_8L_9L_{11}L_{13}}$	$L_{13}=\overline{P_5P_6}$
$F_{14}=\overline{L_{16}L_{17}L_{19}L_{21}}$	$L_{14}=\overline{P_5P_9}$
$F_{15}=\overline{L_{19}L_{20}L_{23}}$	$L_{15}=\overline{P_6P_{10}}$
$F_{16}=\overline{L_{21}L_{22}L_{23}L_{24}}$	$L_{16}=\overline{P_7P_8}$
$F_{17}=\overline{L_1L_4L_7L_{16}}$	$L_{17}=\overline{P_7P_9}$
$F_{18}=\overline{L_8L_{10}L_{12}L_{24}}$	$L_{18}=\overline{P_7P_{11}}$
$F_{19}=\overline{L_{13}L_{14}L_{15}L_{21}}$	$L_{19}=\overline{P_8P_{10}}$
	$L_{20}=\overline{P_8P_{12}}$
	$L_{21}=\overline{P_9P_{10}}$
	$L_{22}=\overline{P_9P_{11}}$
	$L_{23}=\overline{P_{10}P_{12}}$
	$L_{24}=\overline{P_{11}P_{12}}$

SPECIFICATIONS (Cont'd)

$P_1=(x_1,y_1,z_1,w_1)$	$P_2=(x_2,y_2,z_2,w_2)$
$P_3=(x_3,y_3,z_3,w_3)$	$P_4=(x_4,y_4,z_4,w_4)$
$P_5=(x_5,y_5,z_5,w_5)$	$P_6=(x_6,y_6,z_6,w_6)$
$P_7=(x_7,y_7,z_7,w_7)$	$P_8=(x_8,y_8,z_8,w_8)$
$P_9=(x_9,y_9,z_9,w_9)$	$P_{10}=(x_{10},y_{10},z_{10},w_{10})$
$P_{11}=(x_{11},y_{11},z_{11},w_{11})$	$P_{12}=(x_{12},y_{12},z_{12},w_{12})$

PRIMARY ANALYSIS

LENGTHS

$$s_1=\sqrt{(x_2-x_1)^2+(y_2-y_1)^2+(z_2-z_1)^2+(w_2-w_1)^2}$$

$$s_2=\sqrt{(x_3-x_1)^2+(y_3-y_1)^2+(z_3-z_1)^2+(w_3-w_1)^2}$$

$$s_3=\sqrt{(x_5-x_1)^2+(y_5-y_1)^2+(z_5-z_1)^2+(w_5-w_1)^2}$$

$$s_4=\sqrt{(x_7-x_1)^2+(y_7-y_1)^2+(z_7-z_1)^2+(w_7-w_1)^2}$$

$$s_5=\sqrt{(x_4-x_2)^2+(y_4-y_2)^2+(z_4-z_2)^2+(w_4-w_2)^2}$$

$$s_6=\sqrt{(x_6-x_2)^2+(y_6-y_2)^2+(z_6-z_2)^2+(w_6-w_2)^2}$$

$$s_7=\sqrt{(x_8-x_2)^2+(y_8-y_2)^2+(z_8-z_2)^2+(w_8-w_2)^2}$$

$$s_8=\sqrt{(x_4-x_3)^2+(y_4-y_3)^2+(z_4-z_3)^2+(w_4-w_3)^2}$$

$$s_9=\sqrt{(x_5-x_3)^2+(y_5-y_3)^2+(z_5-z_3)^2+(w_5-w_3)^2}$$

$$s_{10}=\sqrt{(x_{11}-x_3)^2+(y_{11}-y_3)^2+(z_{11}-z_3)^2+(w_{11}-w_3)^2}$$

$$s_{11}=\sqrt{(x_6-x_4)^2+(y_6-y_4)^2+(z_6-z_4)^2+(w_6-w_4)^2}$$

$$s_{12}=\sqrt{(x_{12}-x_4)^2+(y_{12}-y_4)^2+(z_{12}-z_4)^2+(w_{12}-w_4)^2}$$

$$s_{13}=\sqrt{(x_6-x_5)^2+(y_6-y_5)^2+(z_6-z_5)^2+(w_6-w_5)^2}$$

$$s_{14}=\sqrt{(x_9-x_5)^2+(y_9-y_5)^2+(z_9-z_5)^2+(w_9-w_5)^2}$$

$$s_{15}=\sqrt{(x_{10}-x_6)^2+(y_{10}-y_6)^2+(z_{10}-z_6)^2+(w_{10}-w_6)^2}$$

A

$$s_{16}=\sqrt{(x_8-x_7)^2+(y_8-y_7)^2+(z_8-z_7)^2+(w_8-w_7)^2}$$

LENGTHS (Cont'd)

$$s_{17}=\sqrt{(x_9-x_7)^2+(y_9-y_7)^2+(z_9-z_7)^2+(w_9-w_7)^2}$$

$$s_{18}=\sqrt{(x_{11}-x_7)^2+(y_{11}-y_7)^2+(z_{11}-z_7)^2+(w_{11}-w_7)^2}$$

$$s_{19}=\sqrt{(x_{10}-x_8)^2+(y_{10}-y_8)^2+(z_{10}-z_8)^2+(w_{10}-w_8)^2}$$

$$s_{20}=\sqrt{(x_{12}-x_8)^2+(y_{12}-y_8)^2+(z_{12}-z_8)^2+(w_{12}-w_8)^2}$$

$$s_{21}=\sqrt{(x_{10}-x_9)^2+(y_{10}-y_9)^2+(z_{10}-z_9)^2+(w_{10}-w_9)^2}$$

$$s_{22}=\sqrt{(x_{11}-x_9)^2+(y_{11}-y_9)^2+(z_{11}-z_9)^2+(w_{11}-w_9)^2}$$

$$s_{23}=\sqrt{(x_{12}-x_{10})^2+(y_{12}-y_{10})^2+(z_{12}-z_{10})^2+(w_{12}-w_{10})^2}$$

$$s_{24}=\sqrt{(x_{12}-x_{11})^2+(y_{12}-y_{11})^2+(z_{12}-z_{11})^2+(w_{12}-w_{11})^2}$$

DIRECTION COSINES

$\cos A_1=(x_2-x_1)/s_1$ $\cos B_1=(y_2-y_1)/s_1$
$\cos C_1=(z_2-z_1)/s_1$ $\cos D_1=(w_2-w_1)/s_1$

$\cos A_2=(x_3-x_1)/s_2$ $\cos B_2=(y_3-y_1)/s_2$
$\cos C_2=(z_3-z_1)/s_2$ $\cos D_2=(w_3-w_1)/s_2$

$\cos A_3=(x_5-x_1)/s_3$ $\cos B_3=(y_5-y_1)/s_3$
$\cos C_3=(z_5-z_1)/s_3$ $\cos D_3=(w_5-w_1)/s_3$

$\cos A_4=(x_7-x_1)/s_4$ $\cos B_4=(y_7-y_1)/s_4$
$\cos C_4=(z_7-z_1)/s_4$ $\cos D_4=(w_7-w_1)/s_4$

$\cos A_5=(x_4-x_2)/s_5$ $\cos B_5=(y_4-y_2)/s_5$
$\cos C_5=(z_4-z_2)/s_5$ $\cos D_5=(w_4-w_2)/s_5$

$\cos A_6=(x_6-x_2)/s_6$ $\cos B_6=(y_6-y_2)/s_6$
$\cos C_6=(z_6-z_2)/s_6$ $\cos D_6=(w_6-w_2)/s_6$

$\cos A_7=(x_8-x_2)/s_7$ $\cos B_7=(y_8-y_2)/s_7$
$\cos C_7=(z_8-z_2)/s_7$ $\cos D_7=(w_8-w_2)/s_7$

$\cos A_8=(x_4-x_3)/s_8$ $\cos B_8=(y_4-y_3)/s_8$
$\cos C_8=(z_4-z_3)/s_8$ $\cos D_8=(w_4-w_3)/s_8$

A

DIRECTION COSINES (Cont'd)

$\cos A_9=(x_5-x_3)/s_9$ $\cos B_9=(y_5-y_3)/s_9$
$\cos C_9=(z_5-z_3)/s_9$ $\cos D_9=(w_5-w_3)/s_9$

$\cos A_{10}=(x_{11}-x_3)/s_{10}$ $\cos B_{10}=(y_{11}-y_3)/s_{10}$
$\cos C_{10}=(z_{11}-z_3)/s_{10}$ $\cos D_{10}=(w_{11}-w_3)/s_{10}$

$\cos A_{11}=(x_6-x_4)/s_{11}$ $\cos B_{11}=(y_6-y_4)/s_{11}$
$\cos C_{11}=(z_6-z_4)/s_{11}$ $\cos D_{11}=(w_6-w_4)/s_{11}$

$\cos A_{12}=(x_{12}-x_4)/s_{12}$ $\cos B_{12}=(y_{12}-y_4)/s_{12}$
$\cos C_{12}=(z_{12}-z_4)/s_{12}$ $\cos D_{12}=(w_{12}-w_4)/s_{12}$

$\cos A_{13}=(x_6-x_5)/s_{13}$ $\cos B_{13}=(y_6-y_5)/s_{13}$
$\cos C_{13}=(z_6-z_5)/s_{13}$ $\cos D_{13}=(w_6-w_5)/s_{13}$

$\cos A_{14}=(x_9-x_5)/s_{14}$ $\cos B_{14}=(y_9-y_5)/s_{14}$
$\cos C_{14}=(z_9-z_5)/s_{14}$ $\cos D_{14}=(w_9-w_5)/s_{14}$

$\cos A_{15}=(x_{10}-x_6)/s_{15}$ $\cos B_{15}=(y_{10}-y_6)/s_{15}$
$\cos C_{15}=(z_{10}-z_6)/s_{15}$ $\cos D_{15}=(w_{10}-w_6)/s_{15}$

$\cos A_{16}=(x_8-x_7)/s_{16}$ $\cos B_{16}=(y_8-y_7)/s_{16}$
$\cos C_{16}=(z_8-z_7)/s_{16}$ $\cos D_{16}=(w_8-w_7)/s_{16}$

$\cos A_{17}=(x_9-x_7)/s_{17}$ $\cos B_{17}=(y_9-y_7)/s_{17}$
$\cos C_{17}=(z_9-z_7)/s_{17}$ $\cos D_{17}=(w_9-w_7)/s_{17}$

$\cos A_{18}=(x_{11}-x_7)/s_{18}$ $\cos B_{18}=(y_{11}-y_7)/s_{18}$
$\cos C_{18}=(z_{11}-z_7)/s_{18}$ $\cos D_{18}=(w_{11}-w_7)/s_{18}$

$\cos A_{19}=(x_{10}-x_8)/s_{19}$ $\cos B_{19}=(y_{10}-y_8)/s_{19}$
$\cos C_{19}=(z_{10}-z_8)/s_{19}$ $\cos D_{19}=(w_{10}-w_8)/s_{19}$

$\cos A_{20}=(x_{12}-x_8)/s_{20}$ $\cos B_{20}=(y_{12}-y_8)/s_{20}$
$\cos C_{20}=(z_{12}-z_8)/s_{20}$ $\cos D_{20}=(w_{12}-w_8)/s_{20}$

$\cos A_{21}=(x_{10}-x_9)/s_{21}$ $\cos B_{21}=(y_{10}-y_9)/s_{21}$
$\cos C_{21}=(z_{10}-z_9)/s_{21}$ $\cos D_{21}=(w_{10}-w_9)/s_{21}$

$\cos A_{22}=(x_{11}-x_9)/s_{22}$ $\cos B_{22}=(y_{11}-y_9)/s_{22}$
$\cos C_{22}=(z_{11}-z_9)/s_{22}$ $\cos D_{22}=(w_{11}-w_9)/s_{22}$

Table 22-12. Specifications and Analyses for the Hyperspace Object Shown in Fig. 22-12. (A) Specifications and Analysis in Hyperspace. (B) Secondary Analysis in XYZ Space. (C) Secondary Analysis in XYW Space. (D) Secondary Analyis in XZW Space. (E) Secondary Analysis in YZW Space. (F) Suggested Coordinates. **(Continued from page 437.)**

DIRECTION COSINES (Cont'd)

$\cos A_{23}=(x_{12}-x_{10})/s_{23}$ $\cos B_{23}=(y_{12}-y_{10})/s_{23}$

$\cos C_{23}=(z_{12}-z_{10})/s_{23}$ $\cos D_{23}=(w_{12}-w_{10})/s_{23}$

$\cos A_{24}=(x_{12}-x_{11})/s_{24}$ $\cos B_{24}=(y_{12}-y_{11})/s_{24}$

$\cos C_{24}=(z_{12}-z_{11})/s_{24}$ $\cos D_{24}=(w_{12}-w_{11})/s_{24}$

ANGLES

$\cos\phi_{1,2}=\cos A_1\cos A_2+\cos B_1\cos B_2+\cos C_1\cos C_2+\cos D_1\cos D_2$

$\cos\phi_{1,3}=\cos A_1\cos A_3+\cos B_1\cos B_3+\cos C_1\cos C_3+\cos D_1\cos D_3$

$\cos\phi_{1,4}=\cos A_1\cos A_4+\cos B_1\cos B_4+\cos C_1\cos C_4+\cos D_1\cos D_4$

$\cos\phi_{1,5}=\cos A_1\cos A_5+\cos B_1\cos B_5+\cos C_1\cos C_5+\cos D_1\cos D_5$

$\cos\phi_{1,6}=\cos A_1\cos A_6+\cos B_1\cos B_6+\cos C_1\cos C_6+\cos D_1\cos D_6$

$\cos\phi_{1,7}=\cos A_1\cos A_7+\cos B_1\cos B_7+\cos C_1\cos C_7+\cos D_1\cos D_7$

$\cos\phi_{2,3}=\cos A_2\cos A_3+\cos B_2\cos B_3+\cos C_2\cos C_3+\cos D_2\cos D_3$

$\cos\phi_{2,4}=\cos A_2\cos A_4+\cos B_2\cos B_4+\cos C_2\cos C_4+\cos D_2\cos D_4$

$\cos\phi_{2,8}=\cos A_2\cos A_8+\cos B_2\cos B_8+\cos C_2\cos C_8+\cos D_2\cos D_8$

$\cos\phi_{2,9}=\cos A_2\cos A_9+\cos B_2\cos B_9+\cos C_2\cos C_9+\cos D_2\cos D_9$

$\cos\phi_{2,10}=\cos A_2\cos A_{10}+\cos B_2\cos B_{10}+\cos C_2\cos C_{10}+\cos D_2\cos D_{10}$

$\cos\phi_{3,4}=\cos A_3\cos A_4+\cos B_3\cos B_4+\cos C_3\cos C_4+\cos D_3\cos D_4$

$\cos\phi_{3,9}=\cos A_3\cos A_9+\cos B_3\cos B_9+\cos C_3\cos C_9+\cos D_3\cos D_9$

$\cos\phi_{3,13}=\cos A_3\cos A_{13}+\cos B_3\cos B_{13}+\cos C_3\cos C_{13}+\cos D_3\cos D_{13}$

$\cos\phi_{3,14}=\cos A_3\cos A_{14}+\cos B_3\cos B_{14}+\cos C_3\cos C_{14}+\cos D_3\cos D_{14}$

$\cos\phi_{4,16}=\cos A_4\cos A_{16}+\cos B_4\cos B_{16}+\cos C_4\cos C_{16}+\cos D_4\cos D_{16}$

$\cos\phi_{4,17}=\cos A_4\cos A_{17}+\cos B_4\cos B_{17}+\cos C_4\cos C_{17}+\cos D_4\cos D_{17}$

$\cos\phi_{4,18}=\cos A_4\cos A_{18}+\cos B_4\cos B_{18}+\cos C_4\cos C_{18}+\cos D_4\cos D_{18}$

$\cos\phi_{5,6}=\cos A_5\cos A_6+\cos B_5\cos B_6+\cos C_5\cos C_6+\cos D_5\cos D_6$

$\cos\phi_{5,7}=\cos A_5\cos A_7+\cos B_5\cos B_7+\cos C_5\cos C_7+\cos D_5\cos D_7$

$\cos\phi_{5,8}=\cos A_5\cos A_8+\cos B_5\cos B_8+\cos C_5\cos C_8+\cos D_5\cos D_8$

$\cos\phi_{5,11}=\cos A_5\cos A_{11}+\cos B_5\cos B_{11}+\cos C_5\cos C_{11}+\cos D_5\cos D_{11}$

$\cos\phi_{5,12}=\cos A_5\cos A_{12}+\cos B_5\cos B_{12}+\cos C_5\cos C_{12}+\cos D_5\cos D_{12}$

$\cos\phi_{6,7}=\cos A_6\cos A_7+\cos B_6\cos B_7+\cos C_6\cos C_7+\cos D_6\cos D_7$

$\cos\phi_{6,11}=\cos A_6\cos A_{11}+\cos B_6\cos B_{11}+\cos C_6\cos C_{11}+\cos D_6\cos D_{11}$

$\cos\phi_{6,13}=\cos A_6\cos A_{13}+\cos B_6\cos B_{13}+\cos C_6\cos C_{13}+\cos D_6\cos D_{13}$

$\cos\phi_{6,15}=\cos A_6\cos A_{15}+\cos B_6\cos B_{15}+\cos C_6\cos C_{15}+\cos D_6\cos D_{15}$

A

ANGLES (Cont'd)

$\cos\phi_{7,16}=\cos A_7\cos A_{16}+\cos B_7\cos B_{16}+\cos C_7\cos C_{16}+\cos D_7\cos D_{16}$

$\cos\phi_{7,19}=\cos A_7\cos A_{19}+\cos B_7\cos B_{19}+\cos C_7\cos C_{19}+\cos D_7\cos D_{19}$

$\cos\phi_{7,20}=\cos A_7\cos A_{20}+\cos B_7\cos B_{20}+\cos C_7\cos C_{20}+\cos D_7\cos D_{20}$

$\cos\phi_{8,9}=\cos A_8\cos A_9+\cos B_8\cos B_9+\cos C_8\cos C_9+\cos D_8\cos D_9$

$\cos\phi_{8,10}=\cos A_8\cos A_{10}+\cos B_8\cos B_{10}+\cos C_8\cos C_{10}+\cos D_8\cos D_{10}$

$\cos\phi_{8,11}=\cos A_8\cos A_{11}+\cos B_8\cos B_{11}+\cos C_8\cos C_{11}+\cos D_8\cos D_{11}$

$\cos\phi_{8,12}=\cos A_8\cos A_{12}+\cos B_8\cos B_{12}+\cos C_8\cos C_{12}+\cos D_8\cos D_{12}$

$\cos\phi_{9,10}=\cos A_9\cos A_{10}+\cos B_9\cos B_{10}+\cos C_9\cos C_{10}+\cos D_9\cos D_{10}$

$\cos\phi_{9,13}=\cos A_9\cos A_{13}+\cos B_9\cos B_{13}+\cos C_9\cos C_{13}+\cos D_9\cos D_{13}$

$\cos\phi_{9,14}=\cos A_9\cos A_{14}+\cos B_9\cos B_{14}+\cos C_9\cos C_{14}+\cos D_9\cos D_{14}$

$\cos\phi_{10,18}=\cos A_{10}\cos A_{18}+\cos B_{10}\cos B_{18}+\cos C_{10}\cos C_{18}+\cos D_{10}\cos D_{18}$

$\cos A_{10,22}=\cos A_{10}\cos A_{22}+\cos B_{10}\cos B_{22}+\cos C_{10}\cos C_{22}+\cos D_{10}\cos D_{22}$

$\cos\phi_{10,24}=\cos A_{10}\cos A_{24}+\cos B_{10}\cos B_{24}+\cos C_{10}\cos C_{24}+\cos D_{10}\cos D_{24}$

$\cos\phi_{11,12}=\cos A_{11}\cos A_{12}+\cos B_{11}\cos B_{12}+\cos C_{11}\cos C_{12}+\cos D_{11}\cos D_{12}$

$\cos\phi_{11,13}=\cos A_{11}\cos A_{13}+\cos B_{11}\cos B_{13}+\cos C_{11}\cos C_{13}+\cos D_{11}\cos D_{13}$

$\cos\phi_{11,15}=\cos A_{11}\cos A_{15}+\cos B_{11}\cos B_{15}+\cos C_{11}\cos C_{15}+\cos C_{11}\cos D_{15}$

$\cos\phi_{12,20}=\cos A_{12}\cos A_{20}+\cos B_{12}\cos B_{20}+\cos C_{12}\cos C_{20}+\cos D_{12}\cos D_{20}$

$\cos\phi_{12,23}=\cos A_{12}\cos A_{23}+\cos B_{12}\cos B_{23}+\cos C_{12}\cos C_{23}+\cos D_{12}\cos D_{23}$

$\cos\phi_{12,24}=\cos A_{12}\cos A_{24}+\cos B_{12}\cos B_{24}+\cos C_{12}\cos C_{24}+\cos D_{12}\cos D_{24}$

$\cos\phi_{13,14}=\cos A_{13}\cos A_{14}+\cos B_{13}\cos B_{14}+\cos C_{13}\cos C_{14}+\cos D_{13}\cos D_{14}$

$\cos\phi_{13,15}=\cos A_{13}\cos A_{15}+\cos B_{13}\cos B_{15}+\cos C_{13}\cos C_{15}+\cos D_{13}\cos D_{15}$

$\cos\phi_{14,17}=\cos A_{14}\cos A_{17}+\cos B_{14}\cos B_{17}+\cos C_{14}\cos C_{17}+\cos D_{14}\cos D_{17}$

$\cos\phi_{14,21}=\cos A_{14}\cos A_{21}+\cos B_{14}\cos B_{21}+\cos C_{14}\cos C_{21}+\cos D_{14}\cos D_{21}$

$\cos\phi_{14,22}=\cos A_{14}\cos A_{22}+\cos B_{14}\cos B_{22}+\cos C_{14}\cos C_{22}+\cos D_{14}\cos D_{22}$

$\cos\phi_{15,19}=\cos A_{15}\cos A_{19}+\cos B_{15}\cos B_{19}+\cos C_{15}\cos C_{19}+\cos D_{15}\cos D_{19}$

$\cos\phi_{15,21}=\cos A_{15}\cos A_{21}+\cos B_{15}\cos B_{21}+\cos C_{15}\cos C_{21}+\cos D_{15}\cos D_{21}$

$\cos\phi_{15,23}=\cos A_{15}\cos A_{23}+\cos B_{15}\cos B_{23}+\cos C_{15}\cos C_{23}+\cos D_{15}\cos D_{23}$

$\cos\phi_{16,17}=\cos A_{16}\cos A_{17}+\cos B_{16}\cos B_{17}+\cos C_{16}\cos C_{17}+\cos D_{16}\cos D_{17}$

$\cos\phi_{16,18}=\cos A_{16}\cos A_{18}+\cos B_{16}\cos B_{18}+\cos C_{16}\cos C_{18}+\cos D_{16}\cos D_{18}$

$\cos\phi_{16,19}=\cos A_{16}\cos A_{19}+\cos B_{16}\cos B_{19}+\cos C_{16}\cos C_{19}+\cos D_{16}\cos D_{19}$

$\cos\phi_{16,20}=\cos A_{16}\cos A_{20}+\cos B_{16}\cos B_{20}+\cos C_{16}\cos C_{20}+\cos D_{16}\cos D_{20}$

$\cos\phi_{17,18}=\cos A_{17}\cos A_{18}+\cos B_{17}\cos B_{18}+\cos C_{17}\cos C_{18}+\cos D_{17}\cos D_{18}$

$\cos\phi_{17,21}=\cos A_{17}\cos A_{21}+\cos B_{17}\cos B_{21}+\cos C_{17}\cos C_{21}+\cos D_{17}\cos D_{21}$

$\cos\phi_{17,22}=\cos A_{17}\cos A_{22}+\cos B_{17}\cos B_{22}+\cos C_{17}\cos C_{22}+\cos D_{17}\cos D_{22}$

Table 22-12. Specifications and Analyses for the Hyperspace Object Shown in Fig. 22-12. (A) Specifications and Analysis in Hyperspace. (B) Secondary Analysis in XYZ Space. (C) Secondary Analysis in XYW Space. (D) Secondary Analysis in XZW Space. (E) Secondary Analysis in YZW Space. (F) Suggested Coordinates. (Continued from page 439.)

ANGLES (Cont'd)

$\cos\phi_{18,22}=\cos A_{18}\cos A_{22}+\cos B_{18}\cos B_{22}+\cos C_{18}\cos C_{22}+\cos D_{18}\cos D_{22}$

$\cos\phi_{18,24}=\cos A_{18}\cos A_{24}+\cos B_{18}\cos B_{24}+\cos C_{18}\cos C_{24}+\cos D_{18}\cos D_{24}$

$\cos\phi_{19,20}=\cos A_{19}\cos A_{20}+\cos B_{19}\cos B_{20}+\cos C_{19}\cos C_{20}+\cos D_{19}\cos D_{20}$

$\cos\phi_{19,21}=\cos A_{19}\cos A_{21}+\cos B_{19}\cos B_{21}+\cos C_{19}\cos C_{21}+\cos D_{19}\cos D_{21}$

$\cos\phi_{19,23}=\cos A_{19}\cos A_{23}+\cos B_{19}\cos B_{23}+\cos C_{19}\cos C_{23}+\cos D_{19}\cos D_{23}$

$\cos\phi_{20,23}=\cos A_{20}\cos A_{23}+\cos B_{20}\cos B_{23}+\cos C_{20}\cos C_{23}+\cos D_{20}\cos D_{23}$

$\cos\phi_{20,24}=\cos A_{20}\cos A_{24}+\cos B_{20}\cos B_{24}+\cos C_{20}\cos C_{24}+\cos D_{20}\cos D_{24}$

$\cos\phi_{21,22}=\cos A_{21}\cos A_{22}+\cos B_{21}\cos B_{22}+\cos C_{21}\cos C_{22}+\cos D_{21}\cos D_{22}$

$\cos\phi_{21,23}=\cos A_{21}\cos A_{23}+\cos B_{21}\cos B_{23}+\cos C_{21}\cos C_{23}+\cos D_{21}\cos D_{23}$

$\cos\phi_{22,24}=\cos A_{22}\cos A_{24}+\cos B_{22}\cos B_{24}+\cos C_{22}\cos C_{24}+\cos D_{22}\cos D_{24}$

$\cos\phi_{23,24}=\cos A_{23}\cos A_{24}+\cos B_{23}\cos B_{24}+\cos C_{23}\cos C_{24}+\cos D_{23}\cos D_{24}$

CHECK

F_1: $\phi_{2,3}+\phi_{2,9}+\phi_{3,9}=180^\circ$

F_2: $\phi_{2,4}+\phi_{2,10}+\phi_{4,18}+\phi_{10,18}=360^\circ$

F_3: $\phi_{3,4}+\phi_{3,14}+\phi_{4,17}+\phi_{14,17}=360^\circ$

F_4: $\phi_{9,10}+\phi_{9,14}+\phi_{10,22}+\phi_{14,22}=360^\circ$

F_5: $\phi_{17,18}+\phi_{17,22}+\phi_{18,22}=180^\circ$

F_6: $\phi_{5,6}+\phi_{5,11}+\phi_{6,11}=180^\circ$

F_7: $\phi_{5,7}+\phi_{5,12}+\phi_{7,20}+\phi_{12,20}=360^\circ$

F_8: $\phi_{6,7}+\phi_{6,15}+\phi_{7,19}+\phi_{15,19}=360^\circ$

F_9: $\phi_{11,12}+\phi_{11,15}+\phi_{12,23}+\phi_{15,23}=360^\circ$

F_{10}: $\phi_{19,20}+\phi_{19,23}+\phi_{20,23}=180^\circ$

F_{11}: $\phi_{1,2}+\phi_{1,5}+\phi_{2,8}+\phi_{5,8}=360^\circ$

F_{12}: $\phi_{1,3}+\phi_{1,6}+\phi_{3,13}+\phi_{6,13}=360^\circ$

F_{13}: $\phi_{8,9}+\phi_{8,11}+\phi_{9,13}+\phi_{11,13}=360^\circ$

F_{14}: $\phi_{16,17}+\phi_{16,19}+\phi_{17,21}+\phi_{19,21}=360^\circ$

F_{15}: $\phi_{21,22}+\phi_{21,23}+\phi_{22,24}+\phi_{23,24}=360^\circ$

F_{16}: $\phi_{1,3}+\phi_{1,7}+\phi_{4,16}+\phi_{7,16}=360^\circ$

F_{17}: $\phi_{8,10}+\phi_{8,12}+\phi_{10,24}+\phi_{12,24}=360^\circ$

F_{18}: $\phi_{13,14}+\phi_{13,15}+\phi_{14,21}+\phi_{15,21}=360^\circ$

A

SECONDARY ANALYSIS -- XYZ SPACE
(See Fig. 22-12B)

PLOTTED POINTS

$P_1=(x_1,y_1,z_1)$ $P_2=(x_2,y_2,z_2)$

$P_3=(x_3,y_3,z_3)$ $P_4=(x_4,y_4,z_4)$

$P_5=(x_5,y_5,z_5)$ $P_6=(x_6,y_6,z_6)$

$P_7=(x_7,y_7,z_7)$ $P_8=(x_8,y_8,z_8)$

$P_9=(x_9,y_9,z_9)$ $P_{10}=(x_{10},y_{10},z_{10})$

$P_{11}=(x_{11},y_{11},z_{11})$ $P_{12}=(x_{12},y_{12},z_{12})$

DIRECTION COSINES

$s_1=\sqrt{(x_2-x_1)^2+(y_2-y_1)^2+(z_2-z_1)^2}$

$s_2=\sqrt{(x_3-x_1)^2+(y_3-y_1)^2+(z_3-z_1)^2}$

$s_3=\sqrt{(x_5-x_1)^2+(y_5-y_1)^2+(z_5-z_1)^2}$

$s_4=\sqrt{(x_7-x_1)^2+(y_7-y_1)^2+(z_7-z_1)^2}$

$s_5=\sqrt{(x_4-x_2)^2+(y_4-y_2)^2+(z_4-z_2)^2}$

$s_6=\sqrt{(x_6-x_2)^2+(y_6-y_2)^2+(z_6-z_2)^2}$

$s_7=\sqrt{(x_8-x_2)^2+(y_8-y_2)^2+(z_8-z_2)^2}$

$s_8=\sqrt{(x_4-x_3)^2+(y_4-y_3)^2+(z_4-z_3)^2}$

$s_9=\sqrt{(x_5-x_3)^2+(y_5-y_3)^2+(z_5-z_3)^2}$

$s_{10}=\sqrt{(x_{11}-x_3)^2+(y_{11}-y_3)^2+(z_{11}-z_3)^2}$

$s_{11}=\sqrt{(x_6-x_4)^2+(y_6-y_4)^2+(z_6-z_4)^2}$

$s_{12}=\sqrt{(x_{12}-x_4)^2+(y_{12}-y_4)^2+(z_{12}-z_4)^2}$

$s_{13}=\sqrt{(x_6-x_5)^2+(y_6-y_5)^2+(z_6-z_5)^2}$

$s_{14}=\sqrt{(x_9-x_5)^2+(y_9-y_5)^2+(z_9-z_5)^2}$

$s_{15}=\sqrt{(x_{10}-x_6)^2+(y_{10}-y_6)^2+(z_{10}-z_6)^2}$

LENGTHS (Cont'd)

$s_{16}=\sqrt{(x_8-x_7)^2+(y_8-y_7)^2+(z_8-z_7)^2}$

$s_{17}=\sqrt{(x_9-x_7)^2+(y_9-y_7)^2+(z_9-z_7)^2}$

$s_{18}=\sqrt{(x_{11}-x_7)^2+(y_{11}-y_7)^2+(z_{11}-z_7)^2}$

$s_{19}=\sqrt{(x_{10}-x_8)^2+(y_{10}-y_8)^2+(z_{10}-z_8)^2}$

$s_{20}=\sqrt{(x_{12}-x_8)^2+(y_{12}-y_8)^2+(z_{12}-z_8)^2}$

$s_{21}=\sqrt{(x_{10}-x_9)^2+(y_{10}-y_9)^2+(z_{10}-z_9)^2}$

$s_{22}=\sqrt{(x_{11}-x_9)^2+(y_{11}-y_9)^2+(z_{11}-z_9)^2}$

$s_{23}=\sqrt{(x_{12}-x_{10})^2+(y_{12}-y_{10})^2+(z_{12}-z_{10})^2}$

$s_{24}=\sqrt{(x_{12}-x_{11})^2+(y_{12}-y_{11})^2+(z_{12}-z_{11})^2}$

DIRECTION COSINES

$\cos A_1=(x_2-x_1)/s_1$ $\cos B_1=(y_2-y_1)/s_1$
$\cos C_1=(z_2-z_1)/s_1$

$\cos A_2=(x_3-x_1)/s_2$ $\cos B_2=(y_3-y_1)/s_2$
$\cos C_2=(z_3-z_1)/s_2$

$\cos A_3=(x_5-x_1)/s_3$ $\cos B_3=(y_5-y_1)/s_3$
$\cos C_3=(z_5-z_1)/s_3$

$\cos A_4=(x_7-x_1)/s_4$ $\cos B_4=(y_7-y_1)/s_4$
$\cos C_4=(z_7-z_1)/s_4$

$\cos A_5=(x_4-x_2)/s_5$ $\cos B_5=(y_4-y_2)/s_5$
$\cos C_5=(z_4-z_2)/s_5$

$\cos A_6=(x_6-x_2)/s_6$ $\cos B_6=(y_6-y_2)/s_6$
$\cos C_6=(z_6-z_2)/s_6$

$\cos A_7=(x_8-x_2)/s_7$ $\cos B_7=(y_8-y_2)/s_7$
$\cos C_7=(z_8-z_2)/s_7$

B

Table 22-12. Specifications and Analysis for the Hyperspace Object Shown in Fig. 22-12. (A) Specifications and Analysis in Hyperspace. (B) Secondary Analysis in XYZ Space. (C) Secondary Analysis in XYW Space. (D) Secondary Analysis in XZW Space. (E) Secondary Analysis in YZW Space. (F) Suggested Coordinates. (Continued from page 441.)

DIRECTION COSINES (Cont'd)

$\cos A_8=(x_4-x_3)/s_8$ $\cos B_8=(y_4-y_3)/s_8$
$\cos C_8=(z_4-z_3)/s_8$

$\cos A_9=(x_5-x_3)/s_9$ $\cos B_9=(y_5-y_3)/s_9$
$\cos C_9=(z_5-z_3)/s_9$

$\cos A_{10}=(x_{11}-x_3)/s_{10}$ $\cos B_{10}=(y_{11}-y_3)/s_{10}$
$\cos C_{10}=(z_{11}-z_3)/s_{10}$

$\cos A_{11}=(x_6-x_4)/s_{11}$ $\cos B_{11}=(y_6-y_4)/s_{11}$
$\cos C_{11}=(z_6-z_4)/s_{11}$

$\cos A_{12}=(x_{12}-x_4)/s_{12}$ $\cos B_{12}=(y_{12}-y_4)/s_{12}$
$\cos C_{12}=(z_{12}-z_4)/s_{12}$

$\cos A_{13}=(x_6-x_5)/s_{13}$ $\cos B_{13}=(y_6-y_5)/s_{13}$
$\cos C_{13}=(z_6-z_5)/s_{13}$

$\cos A_{14}=(x_9-x_5)/s_{14}$ $\cos B_{14}=(y_9-y_5)/s_{14}$
$\cos C_{14}=(z_9-z_5)/s_{14}$

$\cos A_{15}=(x_{10}-x_6)/s_{15}$ $\cos B_{15}=(y_{10}-y_6)/s_{15}$
$\cos C_{15}=(z_{10}-z_6)/s_{15}$

$\cos A_{16}=(x_8-x_7)/s_{16}$ $\cos B_{16}=(y_8-y_7)/s_{16}$
$\cos C_{16}=(z_8-z_7)/s_{16}$

$\cos A_{17}=(x_9-x_7)/s_{17}$ $\cos B_{17}=(y_9-y_7)/s_{17}$
$\cos C_{17}=(z_9-z_7)/s_{17}$

$\cos A_{18}=(x_{11}-x_7)/s_{18}$ $\cos B_{18}=(y_{11}-y_7)/s_{18}$
$\cos C_{18}=(z_{11}-z_7)/s_{18}$

$\cos A_{19}=(x_{10}-x_8)/s_{19}$ $\cos B_{19}=(y_{10}-y_8)/s_{19}$
$\cos C_{19}=(z_{10}-z_8)/s_{19}$

$\cos A_{20}=(x_{12}-x_8)/s_{20}$ $\cos B_{20}=(y_{12}-y_8)/s_{20}$
$\cos C_{20}=(z_{12}-z_8)/s_{20}$

$\cos A_{21}=(x_{10}-x_9)/s_{21}$ $\cos B_{21}=(y_{10}-y_9)/s_{21}$
$\cos C_{21}=(z_{10}-z_9)/s_{21}$

DIRECTION COSINES (Cont'd)

$\cos A_{22}=(x_{11}-x_9)/s_{22}$ $\cos B_{22}=(y_{11}-y_9)/s_{22}$
$\cos C_{22}=(z_{11}-z_9)/s_{22}$

$\cos A_{23}=(x_{12}-x_{10})/s_{23}$ $\cos B_{23}=(y_{12}-y_{10})/s_{23}$
$\cos C_{23}=(z_{12}-z_{10})/s_{23}$

$\cos A_{24}=(x_{12}-x_{11})/s_{24}$ $\cos B_{24}=(y_{12}-y_{11})/s_{24}$
$\cos C_{24}=(z_{12}-z_{11})/s_{24}$

ANGLES

$\cos\phi_{1,2}=\cos A_1\cos A_2+\cos B_1\cos B_2+\cos C_1\cos C_2$
$\cos\phi_{1,3}=\cos A_1\cos A_3+\cos B_1\cos B_3+\cos C_1\cos C_3$
$\cos\phi_{1,4}=\cos A_1\cos A_4+\cos B_1\cos B_4+\cos C_1\cos C_4$
$\cos\phi_{1,5}=\cos A_1\cos A_5+\cos B_1\cos B_5+\cos C_1\cos C_5$
$\cos\phi_{1,6}=\cos A_1\cos A_6+\cos B_1\cos B_6+\cos C_1\cos C_6$
$\cos\phi_{1,7}=\cos A_1\cos A_7+\cos B_1\cos B_7+\cos C_1\cos C_7$
$\cos\phi_{2,3}=\cos A_2\cos A_3+\cos B_2\cos B_3+\cos C_2\cos C_3$
$\cos\phi_{2,4}=\cos A_2\cos A_4+\cos B_2\cos B_4+\cos C_2\cos C_4$
$\cos\phi_{2,8}=\cos A_2\cos A_8+\cos B_2\cos B_8+\cos C_2\cos C_8$
$\cos\phi_{2,9}=\cos A_2\cos A_9+\cos B_2\cos B_9+\cos C_2\cos C_9$
$\cos\phi_{2,10}=\cos A_2\cos A_{10}+\cos B_2\cos B_{10}+\cos C_2\cos C_{10}$
$\cos\phi_{3,4}=\cos A_3\cos A_4+\cos B_3\cos B_4+\cos C_3\cos C_4$
$\cos\phi_{3,9}=\cos A_3\cos A_9+\cos B_3\cos B_9+\cos C_3\cos C_9$
$\cos\phi_{3,13}=\cos A_3\cos A_{13}+\cos B_3\cos B_{13}+\cos C_3\cos C_{13}$
$\cos\phi_{3,14}=\cos A_3\cos A_{14}+\cos B_3\cos B_{14}+\cos C_3\cos C_{14}$
$\cos\phi_{4,16}=\cos A_4\cos A_{16}+\cos B_4\cos B_{16}+\cos C_4\cos C_{16}$
$\cos\phi_{4,17}=\cos A_4\cos A_{17}+\cos B_4\cos B_{17}+\cos C_4\cos C_{17}$
$\cos\phi_{4,18}=\cos A_4\cos A_{18}+\cos B_4\cos B_{18}+\cos C_4\cos C_{18}$
$\cos\phi_{5,6}=\cos A_5\cos A_6+\cos B_5\cos B_6+\cos C_5\cos C_6$
$\cos\phi_{5,7}=\cos A_5\cos A_7+\cos B_5\cos B_7+\cos C_5\cos C_7$
$\cos\phi_{5,8}=\cos A_5\cos A_8+\cos B_5\cos B_8+\cos C_5\cos C_8$
$\cos\phi_{5,11}=\cos A_5\cos A_{11}+\cos B_5\cos B_{11}+\cos C_5\cos C_{11}$
$\cos\phi_{5,12}=\cos A_5\cos A_{12}+\cos B_5\cos B_{12}+\cos C_5\cos C_{12}$

B

$\cos\phi_{6,7}=\cos A_6\cos A_7+\cos B_6\cos B_7+\cos C_6\cos C_7$

ANGLES (Cont'd)

$\cos\phi_{6,11}=\cos A_6\cos A_{11}+\cos B_6\cos B_{11}+\cos C_6\cos C_{11}$

$\cos\phi_{6,13}=\cos A_6\cos A_{13}+\cos B_6\cos B_{13}+\cos C_6\cos C_{13}$

$\cos\phi_{6,15}=\cos A_6\cos A_{15}+\cos B_6\cos B_{15}+\cos C_6\cos C_{15}$

$\cos\phi_{7,16}=\cos A_7\cos A_{16}+\cos B_7\cos B_{16}+\cos C_7\cos C_{16}$

$\cos\phi_{7,19}=\cos A_7\cos A_{19}+\cos B_7\cos B_{19}+\cos C_7\cos C_{19}$

$\cos\phi_{7,20}=\cos A_7\cos A_{20}+\cos B_7\cos B_{20}+\cos C_7\cos C_{20}$

$\cos\phi_{8,9}=\cos A_8\cos A_9+\cos B_8\cos B_9+\cos C_8\cos C_9$

$\cos\phi_{8,10}=\cos A_8\cos A_{10}+\cos B_8\cos B_{10}+\cos C_8\cos C_{10}$

$\cos\phi_{8,11}=\cos A_8\cos A_{11}+\cos B_8\cos B_{11}+\cos C_8\cos C_{11}$

$\cos\phi_{8,12}=\cos A_8\cos A_{12}+\cos B_8\cos B_{12}+\cos C_8\cos C_{12}$

$\cos\phi_{9,10}=\cos A_9\cos A_{10}+\cos B_9\cos B_{10}+\cos C_9\cos C_{10}$

$\cos\phi_{9,13}=\cos A_9\cos A_{13}+\cos B_9\cos B_{13}+\cos C_9\cos C_{13}$

$\cos\phi_{9,14}=\cos A_9\cos A_{14}+\cos B_9\cos B_{14}+\cos C_9\cos C_{14}$

$\cos\phi_{10,18}=\cos A_{10}\cos A_{18}+\cos B_{10}\cos B_{18}+\cos C_{10}\cos C_{18}$

$\cos A_{10,22}=\cos A_{10}\cos A_{22}+\cos B_{10}\cos B_{22}+\cos C_{10}\cos C_{22}$

$\cos\phi_{10,24}=\cos A_{10}\cos A_{24}+\cos B_{10}\cos B_{24}+\cos C_{10}\cos C_{24}$

$\cos\phi_{11,12}=\cos A_{11}\cos A_{12}+\cos B_{11}\cos B_{12}+\cos C_{11}\cos C_{12}$

$\cos\phi_{11,13}=\cos A_{11}\cos A_{13}+\cos B_{11}\cos B_{13}+\cos C_{11}\cos C_{13}$

$\cos\phi_{11,15}=\cos A_{11}\cos A_{15}+\cos B_{11}\cos B_{15}+\cos C_{11}\cos C_{15}$

$\cos\phi_{12,20}=\cos A_{12}\cos A_{20}+\cos B_{12}\cos B_{20}+\cos C_{12}\cos C_{20}$

$\cos\phi_{12,23}=\cos A_{12}\cos A_{23}+\cos B_{12}\cos B_{23}+\cos C_{12}\cos C_{23}$

$\cos\phi_{12,24}=\cos A_{12}\cos A_{24}+\cos B_{12}\cos B_{24}+\cos C_{12}\cos C_{24}$

$\cos\phi_{13,14}=\cos A_{13}\cos A_{14}+\cos B_{13}\cos B_{14}+\cos C_{13}\cos C_{14}$

$\cos\phi_{13,15}=\cos A_{13}\cos A_{15}+\cos B_{13}\cos B_{15}+\cos C_{13}\cos C_{15}$

$\cos\phi_{14,17}=\cos A_{14}\cos A_{17}+\cos B_{14}\cos B_{17}+\cos C_{14}\cos C_{17}$

$\cos\phi_{14,21}=\cos A_{14}\cos A_{21}+\cos B_{14}\cos B_{21}+\cos C_{14}\cos C_{21}$

$\cos\phi_{14,22}=\cos A_{14}\cos A_{22}+\cos B_{14}\cos B_{22}+\cos C_{14}\cos C_{22}$

$\cos\phi_{15,19}=\cos A_{15}\cos A_{19}+\cos B_{15}\cos B_{19}+\cos C_{15}\cos C_{19}$

$\cos\phi_{15,21}=\cos A_{15}\cos A_{21}+\cos B_{15}\cos B_{21}+\cos C_{15}\cos C_{21}$

$\cos\phi_{15,23}=\cos A_{15}\cos A_{23}+\cos B_{15}\cos B_{23}+\cos C_{15}\cos C_{23}$

$\cos\phi_{16,17}=\cos A_{16}\cos A_{17}+\cos B_{16}\cos B_{17}+\cos C_{16}\cos C_{17}$

$\cos\phi_{16,18}=\cos A_{16}\cos A_{18}+\cos B_{16}\cos B_{18}+\cos C_{16}\cos C_{18}$

$\cos\phi_{16,19}=\cos A_{16}\cos A_{19}+\cos B_{16}\cos B_{19}+\cos C_{16}\cos C_{19}$

B

ANGLES (Cont'd)

$\cos\phi_{16,20}=\cos A_{16}\cos A_{20}+\cos B_{16}\cos B_{20}+\cos C_{16}\cos C_{20}$

$\cos\phi_{17,18}=\cos A_{17}\cos A_{18}+\cos B_{17}\cos B_{18}+\cos C_{17}\cos C_{18}$

$\cos\phi_{17,21}=\cos A_{17}\cos A_{21}+\cos B_{17}\cos B_{21}+\cos C_{17}\cos C_{21}$

$\cos\phi_{17,22}=\cos A_{17}\cos A_{22}+\cos B_{17}\cos B_{22}+\cos C_{17}\cos C_{22}$

$\cos\phi_{18,22}=\cos A_{18}\cos A_{22}+\cos B_{18}\cos B_{22}+\cos C_{18}\cos C_{22}$

$\cos\phi_{18,24}=\cos A_{18}\cos A_{24}+\cos B_{18}\cos B_{24}+\cos C_{18}\cos C_{24}$

$\cos\phi_{19,20}=\cos A_{19}\cos A_{20}+\cos B_{19}\cos B_{20}+\cos C_{19}\cos C_{20}$

$\cos\phi_{19,21}=\cos A_{19}\cos A_{21}+\cos B_{19}\cos B_{21}+\cos C_{19}\cos C_{21}$

$\cos\phi_{19,23}=\cos A_{19}\cos A_{23}+\cos B_{19}\cos B_{23}+\cos C_{19}\cos C_{23}$

$\cos\phi_{20,23}=\cos A_{20}\cos A_{23}+\cos B_{20}\cos B_{23}+\cos C_{20}\cos C_{23}$

$\cos\phi_{20,24}=\cos A_{20}\cos A_{24}+\cos B_{20}\cos B_{24}+\cos C_{20}\cos C_{24}$

$\cos\phi_{21,22}=\cos A_{21}\cos A_{22}+\cos B_{21}\cos B_{22}+\cos C_{21}\cos C_{22}$

$\cos\phi_{21,23}=\cos A_{21}\cos A_{23}+\cos B_{21}\cos B_{23}+\cos C_{21}\cos C_{23}$

$\cos\phi_{22,24}=\cos A_{22}\cos A_{24}+\cos B_{22}\cos B_{24}+\cos C_{22}\cos C_{24}$

$\cos\phi_{23,24}=\cos A_{23}\cos A_{24}+\cos B_{23}\cos B_{24}+\cos C_{23}\cos C_{24}$

CHECK

F_1: $\phi_{2,3}+\phi_{2,9}+\phi_{3,9}=180^\circ$

F_2: $\phi_{2,4}+\phi_{2,10}+\phi_{4,18}+\phi_{10,18}=360^\circ$

F_3: $\phi_{3,4}+\phi_{3,14}+\phi_{4,17}+\phi_{14,17}=360^\circ$

F_4: $\phi_{9,10}+\phi_{9,14}+\phi_{10,22}+\phi_{14,22}=360^\circ$

F_5: $\phi_{17,18}+\phi_{17,22}+\phi_{18,22}=180^\circ$

F_6: $\phi_{5,6}+\phi_{5,11}+\phi_{6,11}=180^\circ$

F_7: $\phi_{5,7}+\phi_{5,12}+\phi_{7,20}+\phi_{12,20}=360^\circ$

F_8: $\phi_{6,7}+\phi_{6,15}+\phi_{7,19}+\phi_{15,19}=360^\circ$

F_9: $\phi_{11,12}+\phi_{11,15}+\phi_{12,23}+\phi_{15,23}=360^\circ$

F_{10}: $\phi_{19,20}+\phi_{19,23}+\phi_{20,23}=180^\circ$

F_{11}: $\phi_{1,2}+\phi_{1,5}+\phi_{2,8}+\phi_{5,8}=360^\circ$

F_{12}: $\phi_{1,3}+\phi_{1,6}+\phi_{3,13}+\phi_{6,13}=360^\circ$

F_{13}: $\phi_{8,9}+\phi_{8,11}+\phi_{9,13}+\phi_{11,13}=360^\circ$

F_{14}: $\phi_{16,17}+\phi_{16,19}+\phi_{17,21}+\phi_{19,21}=360^\circ$

F_{15}: $\phi_{21,22}+\phi_{21,23}+\phi_{22,24}+\phi_{23,24}=360^\circ$

F_{16}: $\phi_{1,3}+\phi_{1,7}+\phi_{4,16}+\phi_{7,16}=360^\circ$

F_{17}: $\phi_{8,10}+\phi_{8,12}+\phi_{10,24}+\phi_{12,24}=360^\circ$

F_{18}: $\phi_{13,14}+\phi_{13,15}+\phi_{14,21}+\phi_{15,21}=360^\circ$

Table 22-12. Specifications and Analyses for the Hyperspace Object Shown in Fig. 22-12. (A) Specifications and Analysis in Hyperspace. (B) Secondary Analysis in XYZ Space. (C) Secondary Analysis in XYW Space. (D) Secondary Analysis in XZW Space. (E) Secondary Analysis in YZW Space. (F) Suggested Coordinates. (Continued from page 443.)

SECONDARY ANALYSIS -- XYW SPACE
(See Fig. 22-12C)

PLOTTED POINTS

$P_1=(x_1,y_1,w_1)$	$P_2=(x_2,y_2,w_2)$
$P_3=(x_3,y_3,w_3)$	$P_4=(x_4,y_4,w_4)$
$P_5=(x_5,y_5,w_5)$	$P_6=(x_6,y_6,w_6)$
$P_7=(x_7,y_7,w_7)$	$P_8=(x_8,y_8,w_8)$
$P_9=(x_9,y_9,w_9)$	$P_{10}=(x_{10},y_{10},w_{10})$
$P_{11}=(x_{11},y_{11},w_{11})$	$P_{12}=(x_{12},y_{12},w_{12})$

LENGTHS

$$s_1=\sqrt{(x_2-x_1)^2+(y_2-y_1)^2+(w_2-w_1)^2}$$

$$s_2=\sqrt{(x_3-x_1)^2+(y_3-y_1)^2+(w_3-w_1)^2}$$

$$s_3=\sqrt{(x_5-x_1)^2+(y_5-y_1)^2+(w_5-w_1)^2}$$

$$s_4=\sqrt{(x_7-x_1)^2+(y_7-y_1)^2+(w_7-w_1)^2}$$

$$s_5=\sqrt{(x_4-x_2)^2+(y_4-y_2)^2+(w_4-w_2)^2}$$

$$s_6=\sqrt{(x_6-x_2)^2+(y_6-y_2)^2+(w_6-w_2)^2}$$

$$s_7=\sqrt{(x_8-x_2)^2+(y_8-y_2)^2+(w_8-w_2)^2}$$

$$s_8=\sqrt{(x_4-x_3)^2+(y_4-y_3)^2+(w_4-w_3)^2}$$

$$s_9=\sqrt{(x_5-x_3)^2+(y_5-y_3)^2+(w_5-w_3)^2}$$

$$s_{10}=\sqrt{(x_{11}-x_3)^2+(y_{11}-y_3)^2+(w_{11}-w_3)^2}$$

$$s_{11}=\sqrt{(x_6-x_4)^2+(y_6-y_4)^2+(w_6-w_4)^2}$$

$$s_{12}=\sqrt{(x_{12}-x_4)^2+(y_{12}-y_4)^2+(w_{12}-w_4)^2}$$

$$s_{13}=\sqrt{(x_6-x_5)^2+(y_6-y_5)^2+(w_6-w_5)^2}$$

$$s_{14}=\sqrt{(x_9-x_5)^2+(y_9-y_5)^2+(w_9-w_5)^2}$$

$$s_{15}=\sqrt{(x_{10}-x_6)^2+(y_{10}-y_6)^2+(w_{10}-w_6)^2}$$

LENGTHS (Cont'd)

$$s_{16}=\sqrt{(x_8-x_7)^2+(y_8-y_7)^2+(w_8-w_7)^2}$$

$$s_{17}=\sqrt{(x_9-x_7)^2+(y_9-y_7)^2+(w_9-w_7)^2}$$

$$s_{18}=\sqrt{(x_{11}-x_7)^2+(y_{11}-y_7)^2+(w_{11}-w_7)^2}$$

$$s_{19}=\sqrt{(x_{10}-x_8)^2+(y_{10}-y_8)^2+(w_{10}-w_8)^2}$$

$$s_{20}=\sqrt{(x_{12}-x_8)^2+(y_{12}-y_8)^2+(w_{12}-w_8)^2}$$

$$s_{21}=\sqrt{(x_{10}-x_9)^2+(y_{10}-y_9)^2+(w_{10}-w_9)^2}$$

$$s_{22}=\sqrt{(x_{11}-x_9)^2+(y_{11}-y_9)^2+(w_{11}-w_9)^2}$$

$$s_{23}=\sqrt{(x_{12}-x_{10})^2+(y_{12}-y_{10})^2+(w_{12}-w_{10})^2}$$

$$s_{24}=\sqrt{(x_{12}-x_{11})^2+(y_{12}-y_{11})^2+(w_{12}-w_{11})^2}$$

DIRECTION COSINES

$\cos A_1=(x_2-x_1)/s_1$ $\quad$ $\cos B_1=(y_2-y_1)/s_1$

$\cos D_1=(w_2-w_1)/s_1$

$\cos A_2=(x_3-x_1)/s_2$ $\quad$ $\cos B_2=(y_3-y_1)/s_2$

$\cos D_2=(w_3-w_1)/s_2$

$\cos A_3=(x_5-x_1)/s_3$ $\quad$ $\cos B_3=(y_5-y_1)/s_3$

$\cos D_3=(w_5-w_1)/s_3$

$\cos A_4=(x_7-x_1)/s_4$ $\quad$ $\cos B_4=(y_7-y_1)/s_4$

$\cos D_4=(w_7-w_1)/s_4$

$\cos A_5=(x_4-x_2)/s_5$ $\quad$ $\cos B_5=(y_4-y_2)/s_5$

$\cos D_5=(w_4-w_2)/s_5$

$\cos A_6=(x_6-x_2)/s_6$ $\quad$ $\cos B_6=(y_6-y_2)/s_6$

$\cos D_6=(w_6-w_2)/s_6$

$\cos A_7=(x_8-x_2)/s_7$ $\quad$ $\cos B_7=(y_8-y_2)/s_7$

$\cos D_7=(w_8-w_2)/s_7$

C

DIRECTION COSINES (Cont'd)

$\cos A_8=(x_4-x_3)/s_8$ $\cos B_8=(y_4-y_3)/s_8$

$\cos D_8=(w_4-w_3)/s_8$

$\cos A_9=(x_5-x_3)/s_9$ $\cos B_9=(y_5-y_3)/s_9$

$\cos D_9=(w_5-w_3)/s_9$

$\cos A_{10}=(x_{11}-x_3)/s_{10}$ $\cos B_{10}=(y_{11}-y_3)/s_{10}$

$\cos D_{10}=(w_{11}-w_3)/s_{10}$

$\cos A_{11}=(x_6-x_4)/s_{11}$ $\cos B_{11}=(y_6-y_4)/s_{11}$

$\cos D_{11}=(w_6-w_4)/s_{11}$

$\cos A_{12}=(x_{12}-x_4)/s_{12}$ $\cos B_{12}=(y_{12}-y_4)/s_{12}$

$\cos D_{12}=(w_{12}-w_4)/s_{12}$

$\cos A_{13}=(x_6-x_5)/s_{13}$ $\cos B_{13}=(y_6-y_5)/s_{13}$

$\cos D_{13}=(w_6-w_5)/s_{13}$

$\cos A_{14}=(x_9-x_5)/s_{14}$ $\cos B_{14}=(y_9-y_5)/s_{14}$

$\cos D_{14}=(w_9-w_5)/s_{14}$

$\cos A_{15}=(x_{10}-x_6)/s_{15}$ $\cos B_{15}=(y_{10}-y_6)/s_{15}$

$\cos D_{15}=(w_{10}-w_6)/s_{15}$

$\cos A_{16}=(x_8-x_7)/s_{16}$ $\cos B_{16}=(y_8-y_7)/s_{16}$

$\cos D_{16}=(w_8-w_7)/s_{16}$

$\cos A_{17}=(x_9-x_7)/s_{17}$ $\cos B_{17}=(y_9-y_7)/s_{17}$

$\cos D_{17}=(w_9-w_7)/s_{17}$

$\cos A_{18}=(x_{11}-x_7)/s_{18}$ $\cos B_{18}=(y_{11}-y_7)/s_{18}$

$\cos D_{18}=(w_{11}-w_7)/s_{18}$

$\cos A_{19}=(x_{10}-x_8)/s_{19}$ $\cos B_{19}=(y_{10}-y_8)/s_{19}$

$\cos D_{19}=(w_{10}-w_8)/s_{19}$

$\cos A_{20}=(x_{12}-x_8)/s_{20}$ $\cos B_{20}=(y_{12}-y_8)/s_{20}$

$\cos D_{20}=(w_{12}-w_8)/s_{20}$

DIRECTION COSINES (Cont'd)

$\cos A_{21}=(x_{10}-x_9)/s_{21}$ $\cos B_{21}=(y_{10}-y_9)/s_{21}$

$\cos D_{21}=(w_{10}-w_9)/s_{21}$

$\cos A_{22}=(x_{11}-x_9)/s_{22}$ $\cos B_{22}=(y_{11}-y_9)/s_{22}$

$\cos D_{22}=(w_{11}-w_9)/s_{22}$

$\cos A_{23}=(x_{12}-x_{10})/s_{23}$ $\cos B_{23}=(y_{12}-y_{10})/s_{23}$

$\cos D_{23}=(w_{12}-w_{10})/s_{23}$

$\cos A_{24}=(x_{12}-x_{11})/s_{24}$ $\cos B_{24}=(y_{12}-y_{11})/s_{24}$

$\cos D_{24}=(w_{12}-w_{11})/s_{24}$

ANGLES

$\cos\phi_{1,2}=\cos A_1\cos A_2+\cos B_1\cos B_2+\cos D_1\cos D_2$

$\cos\phi_{1,3}=\cos A_1\cos A_3+\cos B_1\cos B_3+\cos D_1\cos D_3$

$\cos\phi_{1,4}=\cos A_1\cos A_4+\cos B_1\cos B_4+\cos D_1\cos D_4$

$\cos\phi_{1,5}=\cos A_1\cos A_5+\cos B_1\cos B_5+\cos D_1\cos D_5$

$\cos\phi_{1,6}=\cos A_1\cos A_6+\cos B_1\cos B_6+\cos D_1\cos D_6$

$\cos\phi_{1,7}=\cos A_1\cos A_7+\cos B_1\cos B_7+\cos D_1\cos D_7$

$\cos\phi_{2,3}=\cos A_2\cos A_3+\cos B_2\cos B_3+\cos D_2\cos D_3$

$\cos\phi_{2,4}=\cos A_2\cos A_4+\cos B_2\cos B_4+\cos D_2\cos D_4$

$\cos\phi_{2,8}=\cos A_2\cos A_8+\cos B_2\cos B_8+\cos D_2\cos D_3$

$\cos\phi_{2,9}=\cos A_2\cos A_9+\cos B_2\cos B_9+\cos D_2\cos D_9$

$\cos\phi_{2,10}=\cos A_2\cos A_{10}+\cos B_2\cos B_{10}+\cos D_2\cos D_{10}$

$\cos\phi_{3,4}=\cos A_3\cos A_4+\cos B_3\cos B_4+\cos D_3\cos D_4$

$\cos\phi_{3,9}=\cos A_3\cos A_9+\cos B_3\cos B_9+\cos D_3\cos D_9$

$\cos\phi_{3,13}=\cos A_3\cos A_{13}+\cos B_3\cos B_{13}+\cos D_3\cos D_{13}$

$\cos\phi_{3,14}=\cos A_3\cos A_{14}+\cos B_3\cos B_{14}+\cos D_3\cos D_{14}$

$\cos\phi_{4,16}=\cos A_4\cos A_{16}+\cos B_4\cos B_{16}+\cos D_4\cos D_{16}$

$\cos\phi_{4,17}=\cos A_4\cos A_{17}+\cos B_4\cos B_{17}+\cos D_4\cos D_{17}$

$\cos\phi_{4,18}=\cos A_4\cos A_{18}+\cos B_4\cos B_{18}+\cos D_4\cos D_{18}$

$\cos\phi_{5,6}=\cos A_5\cos A_6+\cos B_5\cos B_6+\cos D_5\cos D_6$

$\cos\phi_{5,7}=\cos A_5\cos A_7+\cos B_5\cos B_7+\cos D_5\cos D_7$

$\cos\phi_{5,8}=\cos A_5\cos A_8+\cos B_5\cos B_8+\cos D_5\cos D_8$

$\cos\phi_{5,11}=\cos A_5\cos A_{11}+\cos B_5\cos B_{11}+\cos D_5\cos D_{11}$

C

Table 22-12. Specifications and Analyses for the Hyperspace Object Shown in Fig. 22-12. (A) Specifications and Analysis in Hyperspace. (B) Secondary Analysis in XYZ Space. (C) Secondary Analysis in XYW Space. (D) Secondary Analysis in XZW Space. (E) Secondary Analysis in YZW Space. (F) Suggested Coordinates. (Continued from page 445.)

ANGLES (Cont'd)

$\cos\phi_{5,12}=\cos A_5\cos A_{12}+\cos B_5\cos B_{12}+\cos D_5\cos D_{12}$

$\cos\phi_{6,7}=\cos A_6\cos A_7+\cos B_6\cos B_7+\cos D_6\cos D_7$

$\cos\phi_{6,11}=\cos A_6\cos A_{11}+\cos B_6\cos B_{11}+\cos D_6\cos D_{11}$

$\cos\phi_{6,13}=\cos A_6\cos A_{13}+\cos B_6\cos B_{13}+\cos D_6\cos D_{13}$

$\cos\phi_{6,15}=\cos A_6\cos A_{15}+\cos B_6\cos B_{15}+\cos D_6\cos D_{15}$

$\cos\phi_{7,16}=\cos A_7\cos A_{16}+\cos B_7\cos B_{16}+\cos D_7\cos D_{16}$

$\cos\phi_{7,19}=\cos A_7\cos A_{19}+\cos B_7\cos B_{19}+\cos D_7\cos D_{19}$

$\cos\phi_{7,20}=\cos A_7\cos A_{20}+\cos B_7\cos B_{20}+\cos D_7\cos D_{20}$

$\cos\phi_{8,9}=\cos A_8\cos A_9+\cos B_8\cos B_9+\cos D_8\cos D_9$

$\cos\phi_{8,10}=\cos A_8\cos A_{10}+\cos B_8\cos B_{10}+\cos D_8\cos D_{10}$

$\cos\phi_{8,11}=\cos A_8\cos A_{11}+\cos B_8\cos B_{11}+\cos D_8\cos D_{11}$

$\cos\phi_{8,12}=\cos A_8\cos A_{12}+\cos B_8\cos B_{12}+\cos D_8\cos D_{12}$

$\cos\phi_{9,10}=\cos A_9\cos A_{10}+\cos B_9\cos B_{10}+\cos D_9\cos D_{10}$

$\cos\phi_{9,13}=\cos A_9\cos A_{13}+\cos B_9\cos B_{13}+\cos D_9\cos D_{13}$

$\cos\phi_{9,14}=\cos A_9\cos A_{14}+\cos B_9\cos B_{14}+\cos D_9\cos D_{14}$

$\cos\phi_{10,18}=\cos A_{10}\cos A_{18}+\cos B_{10}\cos B_{18}+\cos D_{10}\cos D_{18}$

$\cos A_{10,22}=\cos A_{10}\cos A_{22}+\cos B_{10}\cos B_{22}+\cos D_{10}\cos D_{22}$

$\cos\phi_{10,24}=\cos A_{10}\cos A_{24}+\cos B_{10}\cos B_{24}+\cos D_{10}\cos D_{24}$

$\cos\phi_{11,12}=\cos A_{11}\cos A_{12}+\cos B_{11}\cos B_{12}+\cos D_{11}\cos D_{12}$

$\cos\phi_{11,13}=\cos A_{11}\cos A_{13}+\cos B_{11}\cos B_{13}+\cos D_{11}\cos D_{13}$

$\cos\phi_{11,15}=\cos A_{11}\cos A_{15}+\cos B_{11}\cos B_{15}+\cos D_{11}\cos D_{15}$

$\cos\phi_{12,20}=\cos A_{12}\cos A_{20}+\cos B_{12}\cos B_{20}+\cos D_{12}\cos D_{20}$

$\cos\phi_{12,23}=\cos A_{12}\cos A_{23}+\cos B_{12}\cos B_{23}+\cos D_{12}\cos D_{23}$

$\cos\phi_{12,24}=\cos A_{12}\cos A_{24}+\cos B_{12}\cos B_{24}+\cos D_{12}\cos D_{24}$

$\cos\phi_{13,14}=\cos A_{13}\cos A_{14}+\cos B_{13}\cos B_{14}+\cos D_{13}\cos D_{14}$

$\cos\phi_{13,15}=\cos A_{13}\cos A_{15}+\cos B_{13}\cos B_{15}+\cos D_{13}\cos D_{15}$

$\cos\phi_{14,17}=\cos A_{14}\cos A_{17}+\cos B_{14}\cos B_{17}+\cos D_{14}\cos D_{17}$

$\cos\phi_{14,21}=\cos A_{14}\cos A_{21}+\cos B_{14}\cos B_{21}+\cos D_{14}\cos D_{21}$

$\cos\phi_{14,22}=\cos A_{14}\cos A_{22}+\cos B_{14}\cos B_{22}+\cos D_{14}\cos D_{22}$

$\cos\phi_{15,19}=\cos A_{15}\cos A_{19}+\cos B_{15}\cos B_{19}+\cos D_{15}\cos D_{19}$

$\cos\phi_{15,21}=\cos A_{15}\cos A_{21}+\cos B_{15}\cos B_{21}+\cos D_{15}\cos D_{21}$

$\cos\phi_{15,23}=\cos A_{15}\cos A_{23}+\cos B_{15}\cos B_{23}+\cos D_{15}\cos D_{23}$

$\cos\phi_{16,17}=\cos A_{16}\cos A_{17}+\cos B_{16}\cos B_{17}+\cos D_{16}\cos D_{17}$

$\cos\phi_{16,18}=\cos A_{16}\cos A_{18}+\cos B_{16}\cos B_{18}+\cos D_{16}\cos D_{18}$

C

ANGLES (Cont'd)

$\cos\phi_{16,19}=\cos A_{16}\cos A_{19}+\cos B_{16}\cos B_{19}+\cos D_{16}\cos D_{19}$

$\cos\phi_{16,20}=\cos A_{16}\cos A_{20}+\cos B_{16}\cos B_{20}+\cos D_{16}\cos D_{20}$

$\cos\phi_{17,18}=\cos A_{17}\cos A_{18}+\cos B_{17}\cos B_{18}+\cos D_{17}\cos D_{18}$

$\cos\phi_{17,21}=\cos A_{17}\cos A_{21}+\cos B_{17}\cos B_{21}+\cos D_{17}\cos D_{21}$

$\cos\phi_{17,22}=\cos A_{17}\cos A_{22}+\cos B_{17}\cos B_{22}+\cos D_{17}\cos D_{22}$

$\cos\phi_{18,22}=\cos A_{18}\cos A_{22}+\cos B_{18}\cos B_{22}+\cos D_{18}\cos D_{22}$

$\cos\phi_{18,24}=\cos A_{18}\cos A_{24}+\cos B_{18}\cos B_{24}+\cos D_{18}\cos D_{24}$

$\cos\phi_{19,20}=\cos A_{19}\cos A_{20}+\cos B_{19}\cos B_{20}+\cos D_{19}\cos D_{20}$

$\cos\phi_{19,21}=\cos A_{19}\cos A_{21}+\cos B_{19}\cos B_{21}+\cos D_{19}\cos D_{21}$

$\cos\phi_{19,23}=\cos A_{19}\cos A_{23}+\cos B_{19}\cos B_{23}+\cos D_{19}\cos D_{23}$

$\cos\phi_{20,23}=\cos A_{20}\cos A_{23}+\cos B_{20}\cos B_{23}+\cos D_{20}\cos D_{23}$

$\cos\phi_{20,24}=\cos A_{20}\cos A_{24}+\cos B_{20}\cos B_{24}+\cos D_{20}\cos D_{24}$

$\cos\phi_{21,22}=\cos A_{21}\cos A_{22}+\cos B_{21}\cos B_{22}+\cos D_{21}\cos D_{22}$

$\cos\phi_{21,23}=\cos A_{21}\cos A_{23}+\cos B_{21}\cos B_{23}+\cos D_{21}\cos D_{23}$

$\cos\phi_{22,24}=\cos A_{22}\cos A_{24}+\cos B_{22}\cos B_{24}+\cos D_{22}\cos D_{24}$

$\cos\phi_{23,24}=\cos A_{23}\cos A_{24}+\cos B_{23}\cos B_{24}+\cos D_{23}\cos D_{24}$

CHECK

F_1: $\phi_{2,3}+\phi_{2,9}+\phi_{3,9}=180^\circ$

F_2: $\phi_{2,4}+\phi_{2,10}+\phi_{4,18}+\phi_{10,18}=360^\circ$

F_3: $\phi_{3,4}+\phi_{3,14}+\phi_{4,17}+\phi_{14,17}=360^\circ$

F_4: $\phi_{9,10}+\phi_{9,14}+\phi_{10,22}+\phi_{14,22}=360^\circ$

F_5: $\phi_{17,18}+\phi_{17,22}+\phi_{18,22}=180^\circ$

F_6: $\phi_{5,6}+\phi_{5,11}+\phi_{6,11}=180^\circ$

F_7: $\phi_{5,7}+\phi_{5,12}+\phi_{7,20}+\phi_{12,20}=360^\circ$

F_8: $\phi_{6,7}+\phi_{6,15}+\phi_{7,19}+\phi_{15,19}=360^\circ$

F_9: $\phi_{11,12}+\phi_{11,15}+\phi_{12,23}+\phi_{15,23}=360^\circ$

F_{10}: $\phi_{19,20}+\phi_{19,23}+\phi_{20,23}=180^\circ$

F_{11}: $\phi_{1,2}+\phi_{1,5}+\phi_{2,8}+\phi_{5,8}=360^\circ$

F_{12}: $\phi_{1,3}+\phi_{1,6}+\phi_{3,13}+\phi_{6,13}=360^\circ$

F_{13}: $\phi_{8,9}+\phi_{8,11}+\phi_{9,13}+\phi_{11,13}=360^\circ$

F_{14}: $\phi_{16,17}+\phi_{16,19}+\phi_{17,21}+\phi_{19,21}=360^\circ$

F_{15}: $\phi_{21,22}+\phi_{21,23}+\phi_{22,24}+\phi_{23,24}=360^\circ$

F_{16}: $\phi_{1,3}+\phi_{1,7}+\phi_{4,16}+\phi_{7,16}=360^\circ$

F_{17}: $\phi_{8,10}+\phi_{8,12}+\phi_{10,24}+\phi_{12,24}=360^\circ$

F_{18}: $\phi_{13,14}+\phi_{13,15}+\phi_{14,21}+\phi_{15,21}=360^\circ$

SECONDARY ANALYSIS -- XZW SPACE

(See Fig. 22-12D)

PLOTTED POINTS

$P_1=(x_1,z_1,w_1)$ $P_2=(x_2,z_2,w_2)$

$P_3=(x_3,z_3,w_3)$ $P_4=(x_4,z_4,w_4)$

$P_5=(x_5,z_5,w_5)$ $P_6=(x_6,z_6,w_6)$

$P_7=(x_7,z_7,w_7)$ $P_8=(x_8,z_8,w_8)$

$P_9=(x_9,z_9,w_9)$ $P_{10}=(x_{10},z_{10},w_{10})$

$P_{11}=(x_{11},z_{11},w_{11})$ $P_{12}=(x_{12},z_{12},w_{12})$

LENGTHS

$$s_1=\sqrt{(x_2-x_1)^2+(z_2-z_1)^2+(w_2-w_1)^2}$$

$$s_2=\sqrt{(x_3-x_1)^2+(z_3-z_1)^2+(w_3-w_1)^2}$$

$$s_3=\sqrt{(x_5-x_1)^2+(z_5-z_1)^2+(w_5-w_1)^2}$$

$$s_4=\sqrt{(x_7-x_1)^2+(z_7-z_1)^2+(w_7-w_1)^2}$$

$$s_5=\sqrt{(x_4-x_2)^2+(z_4-z_2)^2+(w_4-w_2)^2}$$

$$s_6=\sqrt{(x_6-x_2)^2+(z_6-z_2)^2+(w_6-w_2)^2}$$

$$s_7=\sqrt{(x_8-x_2)^2+(z_8-z_2)^2+(w_8-w_2)^2}$$

$$s_8=\sqrt{(x_4-x_3)^2+(z_4-z_3)^2+(w_4-w_3)^2}$$

$$s_9=\sqrt{(x_5-x_3)^2+(z_5-z_3)^2+(w_5-w_3)^2}$$

$$s_{10}=\sqrt{(x_{11}-x_3)^2+(z_{11}-z_3)^2+(w_{11}-w_3)^2}$$

$$s_{11}=\sqrt{(x_6-x_4)^2+(z_6-z_4)^2+(w_6-w_4)^2}$$

$$s_{12}=\sqrt{(x_{12}-x_4)^2+(z_{12}-z_4)^2+(w_{12}-w_4)^2}$$

$$s_{13}=\sqrt{(x_6-x_5)^2+(z_6-z_5)^2+(w_6-w_5)^2}$$

$$s_{14}=\sqrt{(x_9-x_5)^2+(z_9-z_5)^2+(w_9-w_5)^2}$$

$$s_{15}=\sqrt{(x_{10}-x_6)^2+(z_{10}-z_6)^2+(w_{10}-w_6)^2}$$

LENGTHS (Cont'd)

$$s_{16}=\sqrt{(x_8-x_7)^2+(z_8-z_7)^2+(w_8-w_7)^2}$$

$$s_{17}=\sqrt{(x_9-x_7)^2+(z_9-z_7)^2+(w_9-w_7)^2}$$

$$s_{18}=\sqrt{(x_{11}-x_7)^2+(z_{11}-z_7)^2+(w_{11}-w_7)^2}$$

$$s_{19}=\sqrt{(x_{10}-x_8)^2+(z_{10}-z_8)^2+(w_{10}-w_8)^2}$$

$$s_{20}=\sqrt{(x_{12}-x_8)^2+(z_{12}-z_8)^2+(w_{12}-w_8)^2}$$

$$s_{21}=\sqrt{(x_{10}-x_9)^2+(z_{10}-z_9)^2+(w_{10}-w_9)^2}$$

$$s_{22}=\sqrt{(x_{11}-x_9)^2+(z_{11}-z_9)^2+(w_{11}-w_9)^2}$$

$$s_{23}=\sqrt{(x_{12}-x_{10})^2+(z_{12}-z_{10})^2+(w_{12}-w_{10})^2}$$

$$s_{24}=\sqrt{(x_{12}-x_{11})^2+(z_{12}-z_{11})^2+(w_{12}-w_{11})^2}$$

DIRECTION COSINES

$\cos A_1=(x_2-x_1)/s_1$ $\cos C_1=(z_2-z_1)/s_1$

$\cos D_1=(w_2-w_1)/s_1$

$\cos A_2=(x_3-x_1)/s_2$ $\cos C_2=(z_3-z_1)/s_2$

$\cos D_2=(w_3-w_1)/s_2$

$\cos A_3=(x_5-x_1)/s_3$ $\cos C_3=(z_5-z_1)/s_3$

$\cos D_3=(w_5-w_1)/s_3$

$\cos A_4=(x_7-x_1)/s_4$ $\cos C_4=(z_7-z_1)/s_4$

$\cos D_4=(w_7-w_1)/s_4$

$\cos A_5=(x_4-x_2)/s_5$ $\cos C_5=(z_4-z_2)/s_5$

$\cos D_5=(w_4-w_2)/s_5$

$\cos A_6=(x_6-x_2)/s_6$ $\cos C_6=(z_6-z_2)/s_6$

$\cos D_6=(w_6-w_2)/s_6$

$\cos A_7=(x_8-x_2)/s_7$ $\cos C_7=(z_8-z_2)/s_7$

$\cos D_7=(w_8-w_2)/s_7$

D

Table 22-12. Specifications and Analyses for the Hyperspace Object Shown in Fig. 22-12. (A) Specifications and Analysis in Hyperspace. (B) Secondary Analysis in XYZ Space. (C) Secondary Analysis in XYW Space. (D) Secondary Analysis in XZW Space. (E) Secondary Analysis in YZW Space. (F) Suggested Coordinates. (Continued from page 447.)

DIRECTION COSINES (Cont'd)

$\cos A_8=(x_4-x_3)/s_8$ $\cos C_8=(z_4-z_3)/s_8$
$\cos D_8=(w_4-w_3)/s_8$

$\cos A_9=(x_5-x_3)/s_9$ $\cos C_9=(z_5-z_3)/s_9$
$\cos D_9=(w_5-w_3)/s_9$

$\cos A_{10}=(x_{11}-x_3)/s_{10}$ $\cos C_{10}=(z_{11}-z_3)/s_{10}$
$\cos D_{10}=(w_{11}-w_3)/s_{10}$

$\cos A_{11}=(x_6-x_4)/s_{11}$ $\cos C_{11}=(z_6-z_4)/s_{11}$
$\cos D_{11}=(w_6-w_4)/s_{11}$

$\cos A_{12}=(x_{12}-x_4)/s_{12}$ $\cos C_{12}=(z_{12}-z_4)/s_{12}$
$\cos D_{12}=(w_{12}-w_4)/s_{12}$

$\cos A_{13}=(x_6-x_5)/s_{13}$ $\cos C_{13}=(z_6-z_5)/s_{13}$
$\cos D_{13}=(w_6-w_5)/s_{13}$

$\cos A_{14}=(x_9-x_5)/s_{14}$ $\cos C_{14}=(z_9-z_5)/s_{14}$
$\cos D_{14}=(w_9-w_5)/s_{14}$

$\cos A_{15}=(x_{10}-x_6)/s_{15}$ $\cos C_{15}=(z_{10}-z_6)/s_{15}$
$\cos D_{15}=(w_{10}-w_6)/s_{15}$

$\cos A_{16}=(x_8-x_7)/s_{16}$ $\cos C_{16}=(z_8-z_7)/s_{16}$
$\cos D_{16}=(w_8-w_7)/s_{16}$

$\cos A_{17}=(x_9-x_7)/s_{17}$ $\cos C_{17}=(z_9-z_7)/s_{17}$
$\cos D_{17}=(w_9-w_7)/s_{17}$

$\cos A_{18}=(x_{11}-x_7)/s_{18}$ $\cos C_{18}=(z_{11}-z_7)/s_{18}$
$\cos D_{18}=(w_{11}-w_7)/s_{18}$

$\cos A_{19}=(x_{10}-x_8)/s_{19}$ $\cos C_{19}=(z_{10}-z_8)/s_{19}$
$\cos D_{19}=(w_{10}-w_8)/s_{19}$

$\cos A_{20}=(x_{12}-x_8)/s_{20}$ $\cos C_{20}=(z_{12}-z_8)/s_{20}$
$\cos D_{20}=(w_{12}-w_8)/s_{20}$

$\cos A_{21}=(x_{10}-x_9)/s_{21}$ $\cos C_{21}=(z_{10}-z_9)/s_{21}$
$\cos D_{21}=(w_{10}-w_9)/s_{21}$

D

DIRECTION COSINES (Cont'd)

$\cos A_{22}=(x_{11}-x_9)/s_{22}$ $\cos C_{22}=(z_{11}-z_9)/s_{22}$
$\cos D_{22}=(w_{11}-w_9)/s_{22}$

$\cos A_{23}=(x_{12}-x_{10})/s_{23}$ $\cos C_{23}=(z_{12}-z_{10})/s_{23}$
$\cos D_{23}=(w_{12}-w_{10})/s_{23}$

$\cos A_{24}=(x_{12}-x_{11})/s_{24}$ $\cos C_{24}=(z_{12}-z_{11})/s_{24}$
$\cos D_{24}=(w_{12}-w_{11})/s_{24}$

ANGLES

$\cos\phi_{1,2}=\cos A_1\cos A_2+\cos C_1\cos C_2+\cos D_1\cos D_2$
$\cos\phi_{1,3}=\cos A_1\cos A_3+\cos C_1\cos C_3+\cos D_1\cos D_3$
$\cos\phi_{1,4}=\cos A_1\cos A_4+\cos C_1\cos C_4+\cos D_1\cos D_4$
$\cos\phi_{1,5}=\cos A_1\cos A_5+\cos C_1\cos C_5+\cos D_1\cos D_5$
$\cos\phi_{1,6}=\cos A_1\cos A_6+\cos C_1\cos C_6+\cos D_1\cos D_6$
$\cos\phi_{1,7}=\cos A_1\cos A_7+\cos C_1\cos C_7+\cos D_1\cos D_7$
$\cos\phi_{2,3}=\cos A_2\cos A_3+\cos C_2\cos C_3+\cos D_2\cos D_3$
$\cos\phi_{2,4}=\cos A_2\cos A_4+\cos C_2\cos C_4+\cos D_2\cos D_4$
$\cos\phi_{2,8}=\cos A_2\cos A_8+\cos C_2\cos C_8+\cos D_2\cos D_8$
$\cos\phi_{2,9}=\cos A_2\cos A_9+\cos C_2\cos C_9+\cos D_2\cos D_9$
$\cos\phi_{2,10}=\cos A_2\cos A_{10}+\cos C_2\cos C_{10}+\cos D_2\cos D_{10}$
$\cos\phi_{3,4}=\cos A_3\cos A_4+\cos C_3\cos C_4+\cos D_3\cos D_4$
$\cos\phi_{3,9}=\cos A_3\cos A_9+\cos C_3\cos C_9+\cos D_3\cos D_9$
$\cos\phi_{3,13}=\cos A_3\cos A_{13}+\cos C_3\cos C_{13}+\cos D_3\cos D_{13}$
$\cos\phi_{3,14}=\cos A_3\cos A_{14}+\cos C_3\cos C_{14}+\cos D_3\cos D_{14}$
$\cos\phi_{4,16}=\cos A_4\cos A_{16}+\cos C_4\cos C_{16}+\cos D_4\cos D_{16}$
$\cos\phi_{4,17}=\cos A_4\cos A_{17}+\cos C_4\cos C_{17}+\cos D_4\cos D_{17}$
$\cos\phi_{4,18}=\cos A_4\cos A_{18}+\cos C_4\cos C_{18}+\cos D_4\cos D_{18}$
$\cos\phi_{5,6}=\cos A_5\cos A_6+\cos C_5\cos C_6+\cos D_5\cos D_6$
$\cos\phi_{5,7}=\cos A_5\cos A_7+\cos C_5\cos C_7+\cos D_5\cos D_7$
$\cos\phi_{5,8}=\cos A_5\cos A_8+\cos C_5\cos C_8+\cos D_5\cos D_8$
$\cos\phi_{5,11}=\cos A_5\cos A_{11}+\cos C_5\cos C_{11}+\cos D_5\cos D_{11}$
$\cos\phi_{5,12}=\cos A_5\cos A_{12}+\cos C_5\cos C_{12}+\cos D_5\cos D_{12}$
$\cos\phi_{6,7}=\cos A_6\cos A_7+\cos C_6\cos C_7+\cos D_6\cos D_7$
$\cos\phi_{6,11}=\cos A_6\cos A_{11}+\cos C_6\cos C_{11}+\cos D_6\cos D_{11}$

ANGLES (Cont'd)

$\cos\phi_{6,13}=\cos A_6\cos A_{13}+\cos C_6\cos C_{13}+\cos D_6\cos D_{13}$

$\cos\phi_{6,15}=\cos A_6\cos A_{15}+\cos C_6\cos C_{15}+\cos D_6\cos D_{15}$

$\cos\phi_{7,16}=\cos A_7\cos A_{16}+\cos C_7\cos C_{16}+\cos D_7\cos D_{16}$

$\cos\phi_{7,19}=\cos A_7\cos A_{19}+\cos C_7\cos C_{19}+\cos D_7\cos D_{19}$

$\cos\phi_{7,20}=\cos A_7\cos A_{20}+\cos C_7\cos C_{20}+\cos D_7\cos D_{20}$

$\cos\phi_{8,9}=\cos A_8\cos A_9+\cos C_8\cos C_9+\cos D_8\cos D_9$

$\cos\phi_{8,10}=\cos A_8\cos A_{10}+\cos C_8\cos C_{10}+\cos D_8\cos D_{10}$

$\cos\phi_{8,11}=\cos A_8\cos A_{11}+\cos C_8\cos C_{11}+\cos D_8\cos D_{11}$

$\cos\phi_{8,12}=\cos A_8\cos A_{12}+\cos C_8\cos C_{12}+\cos D_8\cos D_{12}$

$\cos\phi_{9,10}=\cos A_9\cos A_{10}+\cos C_9\cos C_{10}+\cos D_9\cos D_{10}$

$\cos\phi_{9,13}=\cos A_9\cos A_{13}+\cos C_9\cos C_{13}+\cos D_9\cos D_{13}$

$\cos\phi_{9,14}=\cos A_9\cos A_{14}+\cos C_9\cos C_{14}+\cos D_9\cos D_{14}$

$\cos\phi_{10,18}=\cos A_{10}\cos A_{18}+\cos C_{10}\cos C_{18}+\cos D_{10}\cos D_{18}$

$\cos A_{10,22}=\cos A_{10}\cos A_{22}+\cos C_{10}\cos C_{22}+\cos D_{10}\cos D_{22}$

$\cos\phi_{10,24}=\cos A_{10}\cos A_{24}+\cos C_{10}\cos C_{24}+\cos D_{10}\cos D_{24}$

$\cos\phi_{11,12}=\cos A_{11}\cos A_{12}+\cos C_{11}\cos C_{12}+\cos D_{11}\cos D_{12}$

$\cos\phi_{11,13}=\cos A_{11}\cos A_{13}+\cos C_{11}\cos C_{13}+\cos D_{11}\cos D_{13}$

$\cos\phi_{11,15}=\cos A_{11}\cos A_{15}+\cos C_{11}\cos C_{15}+\cos D_{11}\cos D_{15}$

$\cos\phi_{12,20}=\cos A_{12}\cos A_{20}+\cos C_{12}\cos C_{20}+\cos D_{12}\cos D_{20}$

$\cos\phi_{12,23}=\cos A_{12}\cos A_{23}+\cos C_{12}\cos C_{23}+\cos D_{12}\cos D_{23}$

$\cos\phi_{12,24}=\cos A_{12}\cos A_{24}+\cos C_{12}\cos C_{24}+\cos D_{12}\cos D_{24}$

$\cos\phi_{13,14}=\cos A_{13}\cos A_{14}+\cos C_{13}\cos C_{14}+\cos D_{13}\cos D_{14}$

$\cos\phi_{13,15}=\cos A_{13}\cos A_{15}+\cos C_{13}\cos C_{15}+\cos D_{13}\cos D_{15}$

$\cos\phi_{14,17}=\cos A_{14}\cos A_{17}+\cos C_{14}\cos C_{17}+\cos D_{14}\cos D_{17}$

$\cos\phi_{14,21}=\cos A_{14}\cos A_{21}+\cos C_{14}\cos C_{21}+\cos D_{14}\cos D_{21}$

$\cos\phi_{14,22}=\cos A_{14}\cos A_{22}+\cos C_{14}\cos C_{22}+\cos D_{14}\cos D_{22}$

$\cos\phi_{15,19}=\cos A_{15}\cos A_{19}+\cos C_{15}\cos C_{19}+\cos D_{15}\cos D_{19}$

$\cos\phi_{15,21}=\cos A_{15}\cos A_{21}+\cos C_{15}\cos C_{21}+\cos D_{15}\cos D_{21}$

$\cos\phi_{15,23}=\cos A_{15}\cos A_{23}+\cos C_{15}\cos C_{23}+\cos D_{15}\cos D_{23}$

$\cos\phi_{16,17}=\cos A_{16}\cos A_{17}+\cos C_{16}\cos C_{17}+\cos D_{16}\cos D_{17}$

$\cos\phi_{16,18}=\cos A_{16}\cos A_{18}+\cos C_{16}\cos C_{18}+\cos D_{16}\cos D_{18}$

$\cos\phi_{16,19}=\cos A_{16}\cos A_{19}+\cos C_{16}\cos C_{19}+\cos D_{16}\cos D_{19}$

$\cos\phi_{16,20}=\cos A_{16}\cos A_{20}+\cos C_{16}\cos C_{20}+\cos D_{16}\cos D_{20}$

$\cos\phi_{17,18}=\cos A_{17}\cos A_{18}+\cos C_{17}\cos C_{18}+\cos D_{17}\cos D_{18}$ D

$\cos\phi_{17,21}=\cos A_{17}\cos A_{21}+\cos C_{17}\cos C_{21}+\cos D_{17}\cos D_{21}$

ANGLES (Cont'd)

$\cos\phi_{17,22}=\cos A_{17}\cos A_{22}+\cos C_{17}\cos C_{22}+\cos D_{17}\cos D_{22}$

$\cos\phi_{18,22}=\cos A_{18}\cos A_{22}+\cos C_{18}\cos C_{22}+\cos D_{18}\cos D_{22}$

$\cos\phi_{18,24}=\cos A_{18}\cos A_{24}+\cos C_{18}\cos C_{24}+\cos D_{18}\cos D_{24}$

$\cos\phi_{19,20}=\cos A_{19}\cos A_{20}+\cos C_{19}\cos C_{20}+\cos D_{19}\cos D_{20}$

$\cos\phi_{19,21}=\cos A_{19}\cos A_{21}+\cos C_{19}\cos C_{21}+\cos D_{19}\cos D_{21}$

$\cos\phi_{19,23}=\cos A_{19}\cos A_{23}+\cos C_{19}\cos C_{23}+\cos D_{19}\cos D_{23}$

$\cos\phi_{20,23}=\cos A_{20}\cos A_{23}+\cos C_{20}\cos C_{23}+\cos D_{20}\cos D_{23}$

$\cos\phi_{20,24}=\cos A_{20}\cos A_{24}+\cos C_{20}\cos C_{24}+\cos D_{20}\cos D_{24}$

$\cos\phi_{21,22}=\cos A_{21}\cos A_{22}+\cos C_{21}\cos C_{22}+\cos D_{21}\cos D_{22}$

$\cos\phi_{21,23}=\cos A_{21}\cos A_{23}+\cos C_{21}\cos C_{23}+\cos D_{21}\cos D_{23}$

$\cos\phi_{22,24}=\cos A_{22}\cos A_{24}+\cos C_{22}\cos C_{24}+\cos D_{22}\cos D_{24}$

$\cos\phi_{23,24}=\cos A_{23}\cos A_{24}+\cos C_{23}\cos C_{24}+\cos D_{23}\cos D_{24}$

CHECK

F_1: $\phi_{2,3}+\phi_{2,9}+\phi_{3,9}=180^\circ$

F_2: $\phi_{2,4}+\phi_{2,10}+\phi_{4,18}+\phi_{10,18}=360^\circ$

F_3: $\phi_{3,4}+\phi_{3,14}+\phi_{4,17}+\phi_{14,17}=360^\circ$

F_4: $\phi_{9,10}+\phi_{9,14}+\phi_{10,22}+\phi_{14,22}=360^\circ$

F_5: $\phi_{17,18}+\phi_{17,22}+\phi_{18,22}=180^\circ$

F_6: $\phi_{5,6}+\phi_{5,11}+\phi_{6,11}=180^\circ$

F_7: $\phi_{5,7}+\phi_{5,12}+\phi_{7,20}+\phi_{12,20}=360^\circ$

F_8: $\phi_{6,7}+\phi_{6,15}+\phi_{7,19}+\phi_{15,19}=360^\circ$

F_9: $\phi_{11,12}+\phi_{11,15}+\phi_{12,23}+\phi_{15,23}=360^\circ$

F_{10}: $\phi_{19,20}+\phi_{19,23}+\phi_{20,23}=180^\circ$

F_{11}: $\phi_{1,2}+\phi_{1,5}+\phi_{2,8}+\phi_{5,8}=360^\circ$

F_{12}: $\phi_{1,3}+\phi_{1,6}+\phi_{3,13}+\phi_{6,13}=360^\circ$

F_{13}: $\phi_{8,9}+\phi_{8,11}+\phi_{9,13}+\phi_{11,13}=360^\circ$

F_{14}: $\phi_{16,17}+\phi_{16,19}+\phi_{17,21}+\phi_{19,21}=360^\circ$

F_{15}: $\phi_{21,22}+\phi_{21,23}+\phi_{22,24}+\phi_{23,24}=360^\circ$

F_{16}: $\phi_{1,3}+\phi_{1,7}+\phi_{4,16}+\phi_{7,16}=360^\circ$

F_{17}: $\phi_{8,10}+\phi_{8,12}+\phi_{10,24}+\phi_{12,24}=360^\circ$

F_{18}: $\phi_{13,14}+\phi_{13,15}+\phi_{14,21}+\phi_{15,21}=360^\circ$

Table 22-12. Specifications and Analyses for the Hyperspace Object Shown in Fig. 22-12. (A) Specifications and Analysis in Hyperspace. (B) Secondary Analysis in XYZ Space. (C) Secondary Analysis in XYW Space. (D) Secondary Analysis in XZW Space. (E) Secondary Analysis in YZW Space. (F) Suggested Coordinates. (Continued from page 449.)

SECONDARY ANALYSIS -- YZW SPACE (See Fig. 22-12E)

PLOTTED POINTS

$P_1=(y_1,z_1,w_1)$	$P_2=(y_2,z_2,w_2)$
$P_3=(y_3,z_3,w_3)$	$P_4=(y_4,z_4,w_4)$
$P_5=(y_5,z_5,w_5)$	$P_6=(y_6,z_6,w_6)$
$P_7=(y_7,z_7,w_7)$	$P_8=(y_8,z_8,w_8)$
$P_9=(y_9,z_9,w_9)$	$P_{10}=(y_{10},z_{10},w_{10})$
$P_{11}=(y_{11},z_{11},w_{11})$	$P_{12}=(y_{12},z_{12},w_{12})$

LENGTHS

$$s_1=\sqrt{(y_2-y_1)^2+(z_2-z_1)^2+(w_2-w_1)^2}$$

$$s_2=\sqrt{(y_3-y_1)^2+(z_3-z_1)^2+(w_3-w_1)^2}$$

$$s_3=\sqrt{(y_5-y_1)^2+(z_5-z_1)^2+(w_5-w_1)^2}$$

$$s_4=\sqrt{(y_7-y_1)^2+(z_7-z_1)^2+(w_7-w_1)^2}$$

$$s_5=\sqrt{(y_4-y_2)^2+(z_4-z_2)^2+(w_4-w_2)^2}$$

$$s_6=\sqrt{(y_6-y_2)^2+(z_6-z_2)^2+(w_6-w_2)^2}$$

$$s_7=\sqrt{(y_8-y_2)^2+(z_8-z_2)^2+(w_8-w_2)^2}$$

$$s_8=\sqrt{(y_4-y_3)^2+(z_4-z_3)^2+(w_4-w_3)^2}$$

$$s_9=\sqrt{(y_5-y_3)^2+(z_5-z_3)^2+(w_5-w_3)^2}$$

$$s_{10}=\sqrt{(y_{11}-y_3)^2+(z_{11}-z_3)^2+(w_{11}-w_3)^2}$$

$$s_{11}=\sqrt{(y_6-y_4)^2+(z_6-z_4)^2+(w_6-w_4)^2}$$

$$s_{12}=\sqrt{(y_{12}-y_4)^2+(z_{12}-z_4)^2+(w_{12}-w_4)^2}$$

$$s_{13}=\sqrt{(y_6-y_5)^2+(z_6-z_5)^2+(w_6-w_5)^2}$$

$$s_{14}=\sqrt{(y_9-y_5)^2+(z_9-z_5)^2+(w_9-w_5)^2}$$

$$s_{15}=\sqrt{(y_{10}-y_6)^2+(z_{10}-z_6)^2+(w_{10}-w_6)^2}$$

LENGTHS (Cont'd)

$$s_{16}=\sqrt{(y_8-y_7)^2+(z_8-z_7)^2+(w_8-w_7)^2}$$

$$s_{17}=\sqrt{(y_9-y_7)^2+(z_9-z_7)^2+(w_9-w_7)^2}$$

$$s_{18}=\sqrt{(y_{11}-y_7)^2+(z_{11}-z_7)^2+(w_{11}-w_7)^2}$$

$$s_{19}=\sqrt{(y_{10}-y_8)^2+(z_{10}-z_8)^2+(w_{10}-w_8)^2}$$

$$s_{20}=\sqrt{(y_{12}-y_8)^2+(z_{12}-z_8)^2+(w_{12}-w_8)^2}$$

$$s_{21}=\sqrt{(y_{10}-y_9)^2+(z_{10}-z_9)^2+(w_{10}-w_9)^2}$$

$$s_{22}=\sqrt{(y_{11}-y_9)^2+(z_{11}-z_9)^2+(w_{11}-w_9)^2}$$

$$s_{23}=\sqrt{(y_{12}-y_{10})^2+(z_{12}-z_{10})^2+(w_{12}-w_{10})^2}$$

$$s_{24}=\sqrt{(y_{12}-y_{11})^2+(z_{12}-z_{11})^2+(w_{12}-w_{11})^2}$$

DIRECTION COSINES

$\cos B_1=(y_2-y_1)/s_1$ $\cos C_1=(z_2-z_1)/s_1$
$\cos D_1=(w_2-w_1)/s_1$

$\cos B_2=(y_3-y_1)/s_2$ $\cos C_2=(z_3-z_1)/s_2$
$\cos D_2=(w_3-w_1)/s_2$

$\cos B_3=(y_5-y_1)/s_3$ $\cos C_3=(z_5-z_1)/s_3$
$\cos D_3=(w_5-w_1)/s_3$

$\cos B_4=(y_7-y_1)/s_4$ $\cos C_4=(z_7-z_1)/s_4$
$\cos D_4=(w_7-w_1)/s_4$

$\cos B_5=(y_4-y_2)/s_5$ $\cos C_5=(z_4-z_2)/s_5$
$\cos D_5=(w_4-w_2)/s_5$

$\cos B_6=(y_6-y_2)/s_6$ $\cos C_6=(z_6-z_2)/s_6$
$\cos D_6=(w_6-w_2)/s_6$

$\cos B_7=(y_8-y_2)/s_7$ $\cos C_7=(z_8-z_2)/s_7$
$\cos D_7=(w_8-w_2)/s_7$

E

DIRECTION COSINES (Cont'd)

$\cos B_8=(y_4-y_3)/s_8$ $\cos C_8=(z_4-z_3)/s_8$

$\cos D_8=(w_4-w_3)/s_8$

$\cos B_9=(y_5-y_3)/s_9$ $\cos C_9=(z_5-z_3)/s_9$

$\cos D_9=(w_5-w_3)/s_9$

$\cos B_{10}=(y_{11}-y_3)/s_{10}$ $\cos C_{10}=(z_{11}-z_3)/s_{10}$

$\cos D_{10}=(w_{11}-w_3)/s_{10}$

$\cos B_{11}=(y_6-y_4)/s_{11}$ $\cos C_{11}=(z_6-z_4)/s_{11}$

$\cos D_{11}=(w_6-w_4)/s_{11}$

$\cos B_{12}=(y_{12}-y_4)/s_{12}$ $\cos C_{12}=(z_{12}-z_4)/s_{12}$

$\cos D_{12}=(w_{12}-w_4)/s_{12}$

$\cos B_{13}=(y_6-y_5)/s_{13}$ $\cos C_{13}=(z_6-z_5)/s_{13}$

$\cos D_{13}=(w_6-w_5)/s_{13}$

$\cos B_{14}=(y_9-y_5)/s_{14}$ $\cos C_{14}=(z_9-z_5)/s_{14}$

$\cos D_{14}=(w_9-w_5)/s_{14}$

$\cos B_{15}=(y_{10}-y_6)/s_{15}$ $\cos C_{15}=(z_{10}-z_6)/s_{15}$

$\cos D_{15}=(w_{10}-w_6)/s_{15}$

$\cos B_{16}=(y_8-y_7)/s_{16}$ $\cos C_{16}=(z_8-z_7)/s_{16}$

$\cos D_{16}=(w_8-w_7)/s_{16}$

$\cos B_{17}=(y_9-y_7)/s_{17}$ $\cos C_{17}=(z_9-z_7)/s_{17}$

$\cos D_{17}=(w_9-w_7)/s_{17}$

$\cos B_{18}=(y_{11}-y_7)/s_{18}$ $\cos C_{18}=(z_{11}-z_7)/s_{18}$

$\cos D_{18}=(w_{11}-w_7)/s_{18}$

$\cos B_{19}=(y_{10}-y_8)/s_{19}$ $\cos C_{19}=(z_{10}-z_8)/s_{19}$

$\cos D_{19}=(w_{10}-w_8)/s_{19}$

$\cos B_{20}=(y_{12}-y_8)/s_{20}$ $\cos C_{20}=(z_{12}-z_8)/s_{20}$

$\cos D_{20}=(w_{12}-w_8)/s_{20}$

$\cos B_{21}=(y_{10}-y_9)/s_{21}$ $\cos C_{21}=(z_{10}-z_9)/s_{21}$

$\cos D_{21}=(w_{10}-w_9)/s_{21}$

$\cos B_{22}=(y_{11}-y_9)/s_{22}$ $\cos C_{22}=(z_{11}-z_9)/s_{22}$

$\cos D_{22}=(w_{11}-w_9)/s_{22}$

$\cos B_{23}=(y_{12}-y_{10})/s_{23}$ $\cos C_{23}=(z_{12}-z_{10})/s_{23}$

$\cos D_{23}=(w_{12}-w_{10})/s_{23}$

$\cos B_{24}=(y_{12}-y_{11})/s_{24}$ $\cos C_{24}=(z_{12}-z_{11})/s_{24}$

$\cos D_{24}=(w_{12}-w_{11})/s_{24}$

ANGLES

$\cos\phi_{1,2}=\cos B_1\cos B_2+\cos C_1\cos C_2+\cos D_1\cos D_2$

$\cos\phi_{1,3}=\cos B_1\cos B_3+\cos C_1\cos C_3+\cos D_1\cos D_3$

$\cos\phi_{1,4}=\cos B_1\cos B_4+\cos C_1\cos C_4+\cos D_1\cos D_4$

$\cos\phi_{1,5}=\cos B_1\cos B_5+\cos C_1\cos C_5+\cos D_1\cos D_5$

$\cos\phi_{1,6}=\cos B_1\cos B_6+\cos C_1\cos C_6+\cos D_1\cos D_6$

$\cos\phi_{1,7}=\cos B_1\cos B_7+\cos C_1\cos C_7+\cos D_1\cos D_7$

$\cos\phi_{2,3}=\cos B_2\cos B_3+\cos C_2\cos C_3+\cos D_2\cos D_3$

$\cos\phi_{2,4}=\cos B_2\cos B_4+\cos C_2\cos C_4+\cos D_2\cos D_4$

$\cos\phi_{2,8}=\cos B_2\cos B_8+\cos C_2\cos C_8+\cos D_2\cos D_8$

$\cos\phi_{2,9}=\cos B_2\cos B_9+\cos C_2\cos C_9+\cos D_2\cos D_9$

$\cos\phi_{2,10}=\cos B_2\cos B_{10}+\cos C_2\cos C_{10}+\cos D_2\cos D_{10}$

$\cos\phi_{3,4}=\cos B_3\cos B_4+\cos C_3\cos C_4+\cos D_3\cos D_4$

$\cos\phi_{3,9}=\cos B_3\cos B_9+\cos C_3\cos C_9+\cos D_3\cos D_9$

$\cos\phi_{3,13}=\cos B_3\cos B_{13}+\cos C_3\cos C_{13}+\cos D_3\cos D_{13}$

$\cos\phi_{3,14}=\cos B_3\cos B_{14}+\cos C_3\cos C_{14}+\cos D_3\cos D_{14}$

$\cos\phi_{4,16}=\cos B_4\cos B_{16}+\cos C_4\cos C_{16}+\cos D_4\cos D_{16}$

$\cos\phi_{4,17}=\cos B_4\cos B_{17}+\cos C_4\cos C_{17}+\cos D_4\cos D_{17}$

$\cos\phi_{4,18}=\cos B_4\cos B_{18}+\cos C_4\cos C_{18}+\cos D_4\cos D_{18}$

$\cos\phi_{5,6}=\cos B_5\cos B_6+\cos C_5\cos C_6+\cos D_5\cos D_6$

$\cos\phi_{5,7}=\cos B_5\cos B_7+\cos C_5\cos C_7+\cos D_5\cos D_7$

$\cos\phi_{5,8}=\cos B_5\cos B_8+\cos C_5\cos C_8+\cos D_5\cos D_8$

$\cos\phi_{5,11}=\cos B_5\cos B_{11}+\cos C_5\cos C_{11}+\cos D_5\cos D_{11}$

E

Table 22-12. Specifications and Analyses for the Hyperspace Object Shown in Fig. 22-12. (A) Specifications and Analysis in Hyperspace. (B) Secondary Analysis in XYZ Space, (C) Secondary Analysis in XYW Space. (D) Secondary Analysis in XZW Space. (E) Secondary Analysis in YZW Space. (F) Suggested Coordinates. (Continued from page 451.)

ANGLES (Cont'd)

$\cos\phi_{5,12}=\cos B_5\cos B_{12}+\cos C_5\cos C_{12}+\cos D_5\cos D_{12}$

$\cos\phi_{6,7}=\cos B_6\cos B_7+\cos C_6\cos C_7+\cos D_6\cos D_7$

$\cos\phi_{6,11}=\cos B_6\cos B_{11}+\cos C_6\cos C_{11}+\cos D_6\cos D_{11}$

$\cos\phi_{6,13}=\cos B_6\cos B_{13}+\cos C_6\cos C_{13}+\cos D_6\cos D_{13}$

$\cos\phi_{6,15}=\cos B_6\cos B_{15}+\cos C_6\cos C_{15}+\cos D_6\cos D_{15}$

$\cos\phi_{7,16}=\cos B_7\cos B_{16}+\cos C_7\cos C_{16}+\cos D_7\cos D_{16}$

$\cos\phi_{7,19}=\cos B_7\cos B_{19}+\cos C_7\cos C_{19}+\cos D_7\cos D_{19}$

$\cos\phi_{7,20}=\cos B_7\cos B_{20}+\cos C_7\cos C_{20}+\cos D_7\cos D_{20}$

$\cos\phi_{8,9}=\cos B_8\cos B_9+\cos C_8\cos C_9+\cos D_8\cos D_9$

$\cos\phi_{8,10}=\cos B_8\cos B_{10}+\cos C_8\cos C_{10}+\cos D_8\cos D_{10}$

$\cos\phi_{8,11}=\cos B_8\cos B_{11}+\cos C_8\cos C_{11}+\cos D_8\cos D_{11}$

$\cos\phi_{8,12}=\cos B_8\cos B_{12}+\cos C_8\cos C_{12}+\cos D_8\cos D_{12}$

$\cos\phi_{9,10}=\cos B_9\cos B_{10}+\cos C_9\cos C_{10}+\cos D_9\cos D_{10}$

$\cos\phi_{9,13}=\cos B_9\cos B_{13}+\cos C_9\cos C_{13}+\cos D_9\cos D_{13}$

$\cos\phi_{9,14}=\cos B_9\cos B_{14}+\cos C_9\cos C_{14}+\cos D_9\cos D_{14}$

$\cos\phi_{10,18}=\cos B_{10}\cos B_{18}+\cos C_{10}\cos C_{18}+\cos D_{10}\cos D_{18}$

$\cos B_{10,22}=\cos B_{10}\cos B_{22}+\cos C_{10}\cos C_{22}+\cos D_{10}\cos D_{22}$

$\cos\phi_{10,24}=\cos B_{10}\cos B_{24}+\cos C_{10}\cos C_{24}+\cos D_{10}\cos D_{24}$

$\cos\phi_{11,12}=\cos B_{11}\cos B_{12}+\cos C_{11}\cos C_{12}+\cos D_{11}\cos D_{12}$

$\cos\phi_{11,13}=\cos B_{11}\cos B_{13}+\cos C_{11}\cos C_{13}+\cos D_{11}\cos D_{13}$

$\cos\phi_{11,15}=\cos B_{11}\cos B_{15}+\cos C_{11}\cos C_{15}+\cos D_{11}\cos D_{15}$

$\cos\phi_{12,20}=\cos B_{12}\cos B_{20}+\cos C_{12}\cos C_{20}+\cos D_{12}\cos D_{20}$

$\cos\phi_{12,23}=\cos B_{12}\cos B_{23}+\cos C_{12}\cos C_{23}+\cos D_{12}\cos D_{23}$

$\cos\phi_{12,24}=\cos B_{12}\cos B_{24}+\cos C_{12}\cos C_{24}+\cos D_{12}\cos D_{24}$

$\cos\phi_{13,14}=\cos B_{13}\cos B_{14}+\cos C_{13}\cos C_{14}+\cos D_{13}\cos D_{14}$

$\cos\phi_{13,15}=\cos B_{13}\cos B_{15}+\cos C_{13}\cos C_{15}+\cos D_{13}\cos D_{15}$

$\cos\phi_{14,17}=\cos B_{14}\cos B_{17}+\cos C_{14}\cos C_{17}+\cos D_{14}\cos D_{17}$

$\cos\phi_{14,21}=\cos B_{14}\cos B_{21}+\cos C_{14}\cos C_{21}+\cos D_{14}\cos D_{21}$

$\cos\phi_{14,22}=\cos B_{14}\cos B_{22}+\cos C_{14}\cos C_{22}+\cos D_{14}\cos D_{22}$

$\cos\phi_{15,19}=\cos B_{15}\cos B_{19}+\cos C_{15}\cos C_{19}+\cos D_{15}\cos D_{19}$

$\cos\phi_{15,21}=\cos B_{15}\cos B_{21}+\cos C_{15}\cos C_{21}+\cos D_{15}\cos D_{21}$

$\cos\phi_{15,23}=\cos B_{15}\cos B_{23}+\cos C_{15}\cos C_{23}+\cos D_{15}\cos D_{23}$

$\cos\phi_{16,17}=\cos B_{16}\cos B_{17}+\cos C_{16}\cos C_{17}+\cos D_{16}\cos D_{17}$

$\cos\phi_{16,18}=\cos B_{16}\cos B_{18}+\cos C_{16}\cos C_{18}+\cos D_{16}\cos D_{18}$

$\cos\phi_{16,19}=\cos B_{16}\cos B_{19}+\cos C_{16}\cos C_{19}+\cos D_{16}\cos D_{19}$

ANGLES (Cont'd)

$\cos\phi_{16,20}=\cos B_{16}\cos B_{20}+\cos C_{16}\cos C_{20}+\cos D_{16}\cos D_{20}$

$\cos\phi_{17,18}=\cos B_{17}\cos B_{18}+\cos C_{17}\cos C_{18}+\cos D_{17}\cos D_{18}$

$\cos\phi_{17,21}=\cos B_{17}\cos B_{21}+\cos C_{17}\cos C_{21}+\cos D_{17}\cos D_{21}$

$\cos\phi_{17,22}=\cos B_{17}\cos B_{22}+\cos C_{17}\cos C_{22}+\cos D_{17}\cos D_{22}$

$\cos\phi_{18,22}=\cos B_{18}\cos B_{22}+\cos C_{18}\cos C_{22}+\cos D_{18}\cos D_{22}$

$\cos\phi_{18,24}=\cos B_{18}\cos B_{24}+\cos C_{18}\cos C_{24}+\cos D_{18}\cos D_{24}$

$\cos\phi_{19,20}=\cos B_{19}\cos B_{20}+\cos C_{19}\cos C_{20}+\cos D_{19}\cos D_{20}$

$\cos\phi_{19,21}=\cos B_{19}\cos B_{21}+\cos C_{19}\cos C_{21}+\cos D_{19}\cos D_{21}$

$\cos\phi_{19,23}=\cos B_{19}\cos B_{23}+\cos C_{19}\cos C_{23}+\cos D_{19}\cos D_{23}$

$\cos\phi_{20,23}=\cos B_{20}\cos B_{23}+\cos C_{20}\cos C_{23}+\cos D_{20}\cos D_{23}$

$\cos\phi_{20,24}=\cos B_{20}\cos B_{24}+\cos C_{20}\cos C_{24}+\cos D_{20}\cos D_{24}$

$\cos\phi_{21,22}=\cos B_{21}\cos B_{22}+\cos C_{21}\cos C_{22}+\cos D_{21}\cos D_{22}$

$\cos\phi_{21,23}=\cos B_{21}\cos B_{23}+\cos C_{21}\cos C_{23}+\cos D_{21}\cos D_{23}$

$\cos\phi_{22,24}=\cos B_{22}\cos B_{24}+\cos C_{22}\cos C_{24}+\cos D_{22}\cos D_{24}$

$\cos\phi_{23,24}=\cos B_{23}\cos B_{24}+\cos C_{23}\cos C_{24}+\cos D_{23}\cos D_{24}$

E

CHECK

F_1:	$\phi_{2,3}+\phi_{2,9}+\phi_{3,9}=180^\circ$
F_2:	$\phi_{2,4}+\phi_{2,10}+\phi_{4,18}+\phi_{10,18}=360^\circ$
F_3:	$\phi_{3,4}+\phi_{3,14}+\phi_{4,17}+\phi_{14,17}=360^\circ$
F_4:	$\phi_{9,10}+\phi_{9,14}+\phi_{10,22}+\phi_{14,22}=360^\circ$
F_5:	$\phi_{17,18}+\phi_{17,22}+\phi_{18,22}=180^\circ$
F_6:	$\phi_{5,6}+\phi_{5,11}+\phi_{6,11}=180^\circ$
F_7:	$\phi_{5,7}+\phi_{5,12}+\phi_{7,20}+\phi_{12,20}=360^\circ$
F_8:	$\phi_{6,7}+\phi_{6,15}+\phi_{7,19}+\phi_{15,19}=360^\circ$
F_9:	$\phi_{11,12}+\phi_{11,15}+\phi_{12,23}+\phi_{15,23}=360^\circ$
F_{10}:	$\phi_{19,20}+\phi_{19,23}+\phi_{20,23}=180^\circ$
F_{11}:	$\phi_{1,2}+\phi_{1,5}+\phi_{2,8}+\phi_{5,8}=360^\circ$
F_{12}:	$\phi_{1,3}+\phi_{1,6}+\phi_{3,13}+\phi_{6,13}=360^\circ$
F_{13}:	$\phi_{8,9}+\phi_{8,11}+\phi_{9,13}+\phi_{11,13}=360^\circ$
F_{14}:	$\phi_{16,17}+\phi_{16,19}+\phi_{17,21}+\phi_{19,21}=360^\circ$
F_{15}:	$\phi_{21,22}+\phi_{21,23}+\phi_{22,24}+\phi_{23,24}=360^\circ$
F_{16}:	$\phi_{1,3}+\phi_{1,7}+\phi_{4,16}+\phi_{7,16}=360^\circ$
F_{17}:	$\phi_{8,10}+\phi_{8,12}+\phi_{10,24}+\phi_{12,24}=360^\circ$
F_{18}:	$\phi_{13,14}+\phi_{13,15}+\phi_{14,21}+\phi_{15,21}=360^\circ$

HYPERSPACE VERSION 1

$P_1=(0,0,0,0)$	$P_2=(0,0,0,1)$
$P_3=(1,0,0,0)$	$P_4=(1,0,0,1)$
$P_5=(0,1,0,0)$	$P_6=(0,1,0,1)$
$P_7=(0,0,1,0)$	$P_8=(0,0,1,1)$
$P_9=(0,1,1,0)$	$P_{10}=(0,1,1,1)$
$P_{11}=(1,0,1,0)$	$P_{12}=(1,0,1,1)$

HYPERSPACE VERSION 2

$P_1=(0,0,-1,-1)$	$P_2=(0,0,-1,1)$
$P_3=(1,0,-1,-1)$	$P_4=(1,0,-1,1)$
$P_5=(0,1,-1,-1)$	$P_6=(0,1,-1,1)$
$P_7=(0,0,1,-1)$	$P_8=(0,0,1,1)$
$P_9=(0,1,1,-1)$	$P_{10}=(0,1,1,1)$
$P_{11}=(1,0,1,-1)$	$P_{12}=(1,0,1,1)$

HYPERSPACE VERSION 3

$P_1=(0,0,0,0)$	$P_2=(0,0,0,1)$
$P_3=(1,0,1,0)$	$P_4=(1,0,1,1)$
$P_5=(0,1,0,0)$	$P_6=(0,1,0,1)$
$P_7=(-1,0,1,0)$	$P_8=(-1,0,1,1)$
$P_9=(-1,1,1,0)$	$P_{10}=(-1,1,1,1)$
$P_{11}=(0,0,2,0)$	$P_{12}=(0,0,2,1)$

F

one where you force both the relativistic and geometric translation terms to change. The general experimental outline is the same as the one just described except that the relativisitic translation terms change, too. The frames of reference are displaced in a relativistic sense, and the observed geometric entity is moving through hyperspace at the same time. It's an odd idea, but possibly a very real one. At any rate, you have an opportunity to deal with it in both an analytic and descriptive fashion. That beats simply reading about it in some popularized book on relativity.

22-4.2 Some Special Studies

There are some important lines of study for investigators who are interested in the most general views of geometric and relativistic translations in hyperspace. Please don't regard the following suggestions as mere exercises. Such suggestions offered in earlier chapters dealing with lower-dimensioned spaces might rightly be regarded as exercises. These aren't; particularly when you attempt to interpret the work in both a physical and philosophical way.

☐ Conduct formal analytic proofs of the invariance of length, direction cosines and angles under both geometric and relativistic translations in 4-D space.

☐ See if you can discover a general relationship between the direction cosines of line in 4-D space and its length as mapped to the four different 3-D spaces.

☐ Regard time as the W-axis translation term in a 4-D relativistic translation situation, then see if you can generate some equations that express the geometric "velocity" of a moving body as regarded from two 3-D spaces separated by hyperspace time. If we were working with nonlinear translations, the interpretation of this study would line up nicely with those of modern relativity.

☐ Research other text sources for discussions of the volumes of space objects from an analytic viewpoint, and then see if you can develop a technique for determining the volume of a hyperspace object, given its primary specifications.

☐ Use the results of the above to prove analytically that the volume of a hyperspace object as mapped to 3-D space remains fixed under simple translations in 4-D space.

22-5 SIMPLE SCALINGS AND SUGGESTED EXPERIMENTS

Simple scaling transformations in 4-D space, both symmetrical and asymmetrical, can expand, contract and

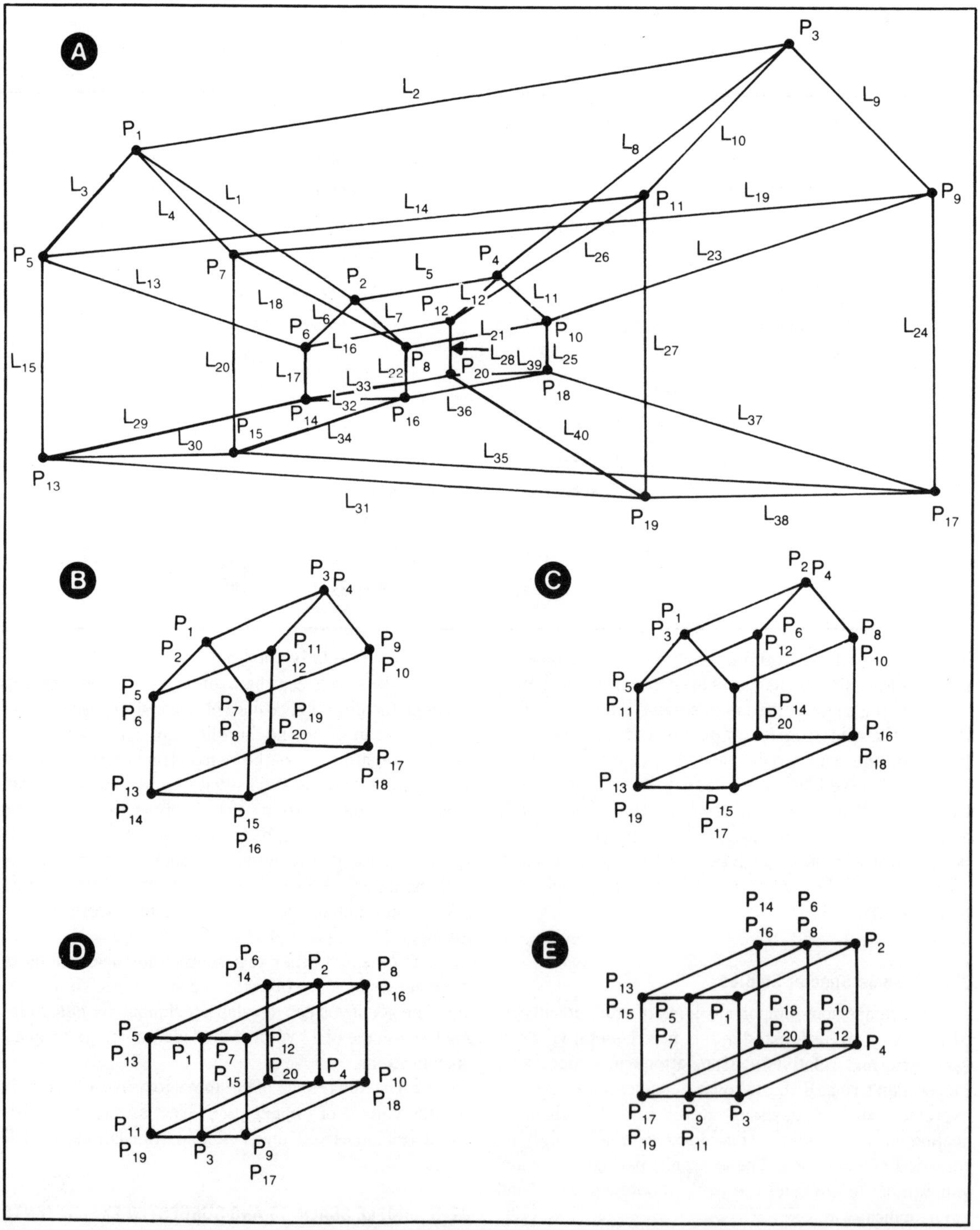

Fig. 22-13. Plots of the hyperspace object specified in Table 22-13A. (A) General hyperspace view. (B) XYZ perspective view. (C) XYW perspective view. (D) XZW perspective view. (E) YZW perspective view. See outlines of analyses in Table 22-13.

Table 22-13. Specifications and Analyses for the Hyperspace Object Shown in Fig. 22-13. (A) Specifications and Analysis in Hyperspace. (B) Secondary Analysis in XYZ Space. (C) Secondary Analysis in XYW Space. (D) Secondary Analysis is XZW Space. (E) Secondary Analysis in YZW Space. (F) Suggested Coordinates.

(A, B, C, and D correspond to α, β, γ, and δ.)

SPECIFICATIONS

$H=\overline{S_1S_2S_3S_4S_5S_6S_7S_8S_9}$

$S_1=\overline{F_4F_5F_6F_{12}F_{16}F_{20}F_{26}}$ $S_2=\overline{F_7F_8F_9F_{13}F_{17}F_{21}F_{27}}$

$S_3=\overline{F_2F_3F_6F_9F_{15}F_{19}F_{24}}$ $S_4=\overline{F_{10}F_{11}F_{12}F_{13}F_{22}F_{23}F_{29}}$

$S_5=\overline{F_{24}F_{25}F_{26}F_{27}F_{28}F_{29}}$ $S_6=\overline{F_1F_2F_4F_7F_{11}F_{14}}$

$S_7=\overline{F_1F_3F_5F_8F_{10}F_{18}}$ $S_8=\overline{F_{14}F_{15}F_{16}F_{17}F_{23}F_{25}}$

$S_9=\overline{F_{18}F_{19}F_{21}F_{22}F_{26}F_{28}}$

$F_1=\overline{L_1L_2L_5L_8}$ $F_2=\overline{L_1L_3L_6L_{13}}$

$F_3=\overline{L_1L_4L_7L_{18}}$ $F_4=\overline{L_2L_3L_{10}L_{14}}$

$F_5=\overline{L_2L_4L_9L_{19}}$ $F_6=\overline{L_3L_4L_{15}L_{20}L_{30}}$

$F_7=\overline{L_5L_6L_{12}L_{16}}$ $F_8=\overline{L_5L_7L_{11}L_{21}}$

$F_9=\overline{L_6L_7L_{17}L_{22}L_{32}}$ $F_{10}=\overline{L_8L_9L_{11}L_{23}}$

$F_{11}=\overline{L_8L_{10}L_{12}L_{26}}$ $F_{12}=\overline{L_9L_{10}L_{24}L_{27}L_{38}}$

$F_{13}=\overline{L_{11}L_{12}L_{25}L_{28}L_{39}}$ $F_{14}=\overline{L_{13}L_{14}L_{16}L_{26}}$

$F_{15}=\overline{L_{13}L_{15}L_{17}L_{29}}$ $F_{16}=\overline{L_{14}L_{15}L_{27}L_{31}}$

$F_{17}=\overline{L_{16}L_{17}L_{28}L_{33}}$ $F_{18}=\overline{L_{18}L_{19}L_{21}L_{23}}$

$F_{19}=\overline{L_{18}L_{20}L_{22}L_{34}}$ $F_{20}=\overline{L_{19}L_{20}L_{24}L_{35}}$

$F_{21}=\overline{L_{21}L_{22}L_{25}L_{26}}$ $F_{22}=\overline{L_{23}L_{24}L_{25}L_{37}}$

$F_{23}=\overline{L_{26}L_{27}L_{28}L_{40}}$ $F_{24}=\overline{L_{29}L_{30}L_{32}L_{34}}$

$F_{25}=\overline{L_{29}L_{31}L_{33}L_{34}}$ $F_{26}=\overline{L_{30}L_{31}L_{35}L_{38}}$

$F_{27}=\overline{L_{32}L_{33}L_{36}L_{39}}$ $F_{28}=\overline{L_{34}L_{35}L_{36}L_{37}}$

$F_{29}=\overline{L_{37}L_{38}L_{39}L_{40}}$

SPECIFICATIONS (Cont'd)

$L_1=\overline{P_1P_2}$ $L_2=\overline{P_1P_3}$

$L_3=\overline{P_1P_5}$ $L_4=\overline{P_1P_7}$

$L_5=\overline{P_2P_4}$ $L_6=\overline{P_2P_6}$

$L_7=\overline{P_2P_8}$ $L_8=\overline{P_3P_4}$

$L_9=\overline{P_3P_9}$ $L_{10}=\overline{P_3P_{11}}$

$L_{11}=\overline{P_4P_{10}}$ $L_{12}=\overline{P_4P_{12}}$

$L_{13}=\overline{P_5P_6}$ $L_{14}=\overline{P_5P_{11}}$

$L_{15}=\overline{P_5P_{13}}$ $L_{16}=\overline{P_6P_{12}}$

SPECIFICATIONS (Cont'd)

$L_{17}=\overline{P_6P_{14}}$ $L_{18}=\overline{P_7P_8}$

$L_{19}=\overline{P_7P_9}$ $L_{20}=\overline{P_7P_{15}}$

$L_{21}=\overline{P_8P_{10}}$ $L_{22}=\overline{P_8P_{16}}$

$L_{23}=\overline{P_9P_{10}}$ $L_{24}=\overline{P_9P_{17}}$

$L_{25}=\overline{P_{10}P_{18}}$ $L_{26}=\overline{P_{11}P_{12}}$

$L_{27}=\overline{P_{11}P_{19}}$ $L_{28}=\overline{P_{12}P_{20}}$

$L_{29}=\overline{P_{13}P_{14}}$ $L_{30}=\overline{P_{13}P_{15}}$

$L_{31}=\overline{P_{13}P_{19}}$ $L_{32}=\overline{P_{14}P_{16}}$

$L_{33}=\overline{P_{14}P_{20}}$ $L_{34}=\overline{P_{15}P_{16}}$

$L_{35}=\overline{P_{15}P_{17}}$ $L_{36}=\overline{P_{16}P_{18}}$

$L_{37}=\overline{P_{17}P_{18}}$ $L_{38}=\overline{P_{17}P_{19}}$

$L_{39}=\overline{P_{18}P_{20}}$ $L_{40}=\overline{P_{19}P_{20}}$

$P_1=(x_1,y_1,z_1,w_1)$ $P_2=(x_2,y_2,z_2,w_2)$

$P_3=(x_3,y_3,z_3,w_3)$ $P_4=(x_4,y_4,z_4,w_4)$

$P_5=(x_5,y_5,z_5,w_5)$ $P_6=(x_6,y_6,z_6,w_6)$

$P_7=(x_7,y_7,z_7,w_7)$ $P_8=(x_8,y_8,z_8,w_8)$

$P_9=(x_9,y_9,z_9,w_9)$ $P_{10}=(x_{10},y_{10},z_{10},w_{10})$

$P_{11}=(x_{11},y_{11},z_{11},w_{11})$ $P_{12}=(x_{12},y_{12},z_{12},w_{12})$

$P_{13}=(x_{13},y_{13},z_{13},w_{13})$ $P_{14}=(x_{14},y_{14},z_{14},w_{14})$

$P_{15}=(x_{15},y_{15},z_{15},w_{15})$ $P_{16}=(x_{16},y_{16},z_{16},w_{16})$

$P_{17}=(x_{17},y_{17},z_{17},w_{17})$ $P_{18}=(x_{18},y_{18},z_{18},w_{18})$

$P_{19}=(x_{18},y_{19},z_{19},w_{19})$ $P_{20}=(x_{20},y_{20},z_{20},w_{20})$

PRIMARY ANALYSIS (See Fig. 22-13A)

LENGTHS

$s_1=\sqrt{(x_2-x_1)^2+(y_2-y_1)^2+(z_2-z_1)^2+(w_2-w_1)^2}$

$s_2=\sqrt{(x_3-x_1)^2+(y_3-y_1)^2+(z_3-z_1)^2+(w_3-w_1)^2}$

$s_3=\sqrt{(x_5-x_1)^2+(y_5-y_1)^2+(z_5-z_1)^2+(w_5-w_1)^2}$

$s_4=\sqrt{(x_7-x_1)^2+(y_7-y_1)^2+(z_7-z_1)^2+(w_7-w_1)^2}$

$s_5=\sqrt{(x_4-x_2)^2+(y_4-y_2)^2+(z_4-z_2)^2+(w_4-w_2)^2}$

A

Table 22-13. Specifications and Analyses for the Hyperspace Object Shown in Fig. 22-13. (A) Specifications and Analysis in Hyperspace. (B) Secondary Analysis in XYZ Space. (C) Secondary Analysis in XYW Space. (D) Secondary Analysis in XZW Space. (E) Secondary Analysis in YZW Space. (F) Suggested Coordinates. **(Continued from page 455.)**

LENGTHS (Cont'd)

$$s_6=\sqrt{(x_6-x_2)^2+(y_6-y_2)^2+(z_6-z_2)^2+(w_6-w_2)^2}$$

$$s_7=\sqrt{(x_8-x_2)^2+(y_8-y_2)^2+(z_8-z_2)^2+(w_3-w_2)^2}$$

$$s_8=\sqrt{(x_4-x_3)^2+(y_4-y_3)^2+(z_4-z_3)^2+(w_4-w_3)^2}$$

$$s_9=\sqrt{(x_9-x_3)^2+(y_9-y_3)^2+(z_9-z_3)^2+(w_9-w_3)^2}$$

$$s_{10}=\sqrt{(x_{11}-x_3)^2+(y_{11}-y_3)^2+(z_{11}-z_3)^2+(w_{11}-w_3)^2}$$

$$s_{11}=\sqrt{(x_{10}-x_4)^2+(y_{10}-y_4)^2+(z_{10}-z_4)^2+(w_{10}-w_4)^2}$$

$$s_{12}=\sqrt{(x_{12}-x_4)^2+(y_{12}-y_4)^2+(z_{12}-z_4)^2+(w_{12}-w_4)^2}$$

$$s_{13}=\sqrt{(x_6-x_5)^2+(y_6-y_5)^2+(z_6-z_5)^2+(w_6-w_5)^2}$$

$$s_{14}=\sqrt{(x_{11}-x_5)^2+(y_{11}-y_5)^2+(z_{11}-z_5)^2+(w_{11}-w_5)^2}$$

$$s_{15}=\sqrt{(x_{13}-x_5)^2+(y_{13}-y_5)^2+(z_{13}-z_5)^2+(w_{13}-w_5)^2}$$

$$s_{16}=\sqrt{(x_{12}-x_6)^2+(y_{12}-y_6)^2+(z_{12}-z_6)^2+(w_{12}-w_6)^2}$$

$$s_{17}=\sqrt{(x_{14}-x_6)^2+(y_{14}-y_6)^2+(z_{14}-z_6)^2+(w_{14}-w_6)^2}$$

$$s_{18}=\sqrt{(x_8-x_7)^2+(y_8-y_7)^2+(z_8-z_7)^2+(w_8-w_7)^2}$$

$$s_{19}=\sqrt{(x_9-x_7)^2+(y_9-y_7)^2+(z_9-z_7)^2+(w_9-w_7)^2}$$

$$s_{20}=\sqrt{(x_{15}-x_7)^2+(y_{15}-y_7)^2+(z_{15}-z_7)^2+(w_{15}-w_7)^2}$$

$$s_{21}=\sqrt{(x_{10}-x_8)^2+(y_{10}-y_8)^2+(z_{10}-z_8)^2+(w_{10}-w_8)^2}$$

$$s_{22}=\sqrt{(x_{16}-x_8)^2+(y_{16}-y_8)^2+(z_{16}-z_8)^2+(w_{16}-w_8)^2}$$

$$s_{23}=\sqrt{(x_{10}-x_9)^2+(y_{10}-y_9)^2+(z_{10}-z_9)^2+(w_{10}-w_9)^2}$$

$$s_{24}=\sqrt{(x_{17}-x_9)^2+(y_{17}-y_9)^2+(z_{17}-z_9)^2+(w_{17}-w_9)^2}$$

$$s_{25}=\sqrt{(x_{18}-x_{10})^2+(y_{18}-y_{10})^2+(z_{18}-z_{10})^2+(w_{18}-w_{10})^2}$$

A

LENGTHS (Cont'd)

$$s_{26}=\sqrt{(x_{12}-x_{11})^2+(y_{12}-y_{11})^2+(z_{12}-z_{11})^2+(w_{12}-w_{11})^2}$$

$$s_{27}=\sqrt{(x_{19}-x_{11})^2+(y_{19}-y_{11})^2+(z_{19}-z_{11})^2+(w_{19}-w_{11})^2}$$

$$s_{28}=\sqrt{(x_{20}-x_{12})^2+(y_{20}-y_{12})^2+(z_{20}-z_{12})^2+(w_{20}-w_{12})^2}$$

$$s_{29}=\sqrt{(x_{14}-x_{13})^2+(y_{14}-y_{13})^2+(z_{14}-z_{13})^2+(w_{14}-w_{13})^2}$$

$$s_{30}=\sqrt{(x_{15}-x_{13})^2+(y_{15}-y_{13})^2+(z_{15}-z_{13})^2+(w_{15}-w_{13})^2}$$

$$s_{31}=\sqrt{(x_{19}-x_{13})^2+(y_{19}-y_{13})^2+(z_{19}-z_{13})^2+(w_{19}-w_{13})^2}$$

$$s_{32}=\sqrt{(x_{16}-x_{14})^2+(y_{16}-y_{14})^2+(z_{16}-z_{14})^2+(w_{16}-w_{14})^2}$$

$$s_{33}=\sqrt{(x_{20}-x_{14})^2+(y_{20}-y_{14})^2+(z_{20}-z_{14})^2+(w_{20}-w_{14})^2}$$

$$s_{34}=\sqrt{(x_{16}-x_{15})^2+(y_{16}-y_{15})^2+(z_{16}-z_{15})^2+(w_{16}-w_{15})^2}$$

$$s_{35}=\sqrt{(x_{17}-x_{15})^2+(y_{17}-y_{15})^2+(z_{17}-z_{15})^2+(w_{17}-w_{15})^2}$$

$$s_{36}=\sqrt{(x_{18}-x_{16})^2+(y_{18}-y_{16})^2+(z_{18}-z_{16})^2+(w_{18}-w_{16})^2}$$

$$s_{37}=\sqrt{(x_{18}-x_{17})^2+(y_{18}-y_{17})^2+(z_{18}-z_{17})^2+(w_{18}-w_{17})^2}$$

$$s_{38}=\sqrt{(x_{19}-x_{17})^2+(y_{19}-y_{17})^2+(z_{19}-z_{17})^2+(w_{19}-w_{17})^2}$$

$$s_{39}=\sqrt{(x_{20}-x_{18})^2+(y_{20}-y_{18})^2+(z_{20}-z_{18})^2+(w_{20}-w_{18})^2}$$

$$s_{40}=\sqrt{(x_{20}-x_{19})^2+(y_{20}-y_{19})^2+(z_{20}-z_{19})^2+(w_{20}-w_{19})^2}$$

DIRECTION COSINES

$\cos A_1=(x_2-x_1)$ $\cos B_1=(y_2-y_1)$

$\cos C_1=(z_2-z_1)$ $\cos D_1=(w_2-w_1)$

$\cos A_2=(x_3-x_1)$ $\cos B_2=(y_3-y_1)$

$\cos C_2=(z_3-z_1)$ $\cos D_2=(w_3-w_1)$

$\cos A_3=(x_5-x_1)$ $\cos B_3=(y_5-y_1)$

$\cos C_3=(z_5-z_1)$ $\cos D_3=(w_5-w_1)$

$\cos A_4=(x_7-x_1)$ $\cos B_4=(y_7-y_1)$

$\cos C_4=(z_7-z_1)$ $\cos D_4=(w_7-w_1)$

DIRECTION COSINES (Cont'd)

$\cos A_5=(x_4-x_2)$	$\cos B_5=(y_4-y_2)$
$\cos C_5=(z_4-z_2)$	$\cos D_5=(w_4-w_2)$
$\cos A_6=(x_6-x_2)$	$\cos B_6=(y_6-y_2)$
$\cos C_6=(z_6-z_2)$	$\cos D_6=(w_6-w_2)$
$\cos A_7=(x_8-x_2)$	$\cos B_7=(y_8-y_2)$
$\cos C_7=(z_8-z_2)$	$\cos D_7=(w_8-w_2)$
$\cos A_8=(x_4-x_3)$	$\cos B_8=(y_4-y_3)$
$\cos C_8=(z_4-z_3)$	$\cos D_8=(w_4-w_3)$
$\cos A_9=(x_9-x_3)$	$\cos B_9=(y_9-y_3)$
$\cos C_9=(z_9-z_3)$	$\cos D_9=(w_9-w_3)$
$\cos A_{10}=(x_{11}-x_3)$	$\cos B_{10}=(y_{11}-y_3)$
$\cos C_{10}=(z_{11}-z_3)$	$\cos D_{10}=(w_{11}-w_3)$
$\cos A_{11}=(x_{10}-x_4)$	$\cos B_{11}=(y_{10}-y_4)$
$\cos C_{11}=(z_{10}-z_4)$	$\cos D_{11}=(w_{10}-w_4)$
$\cos A_{12}=(x_{12}-x_4)$	$\cos B_{12}=(y_{12}-y_4)$
$\cos C_{12}=(z_{12}-z_4)$	$\cos D_{12}=(w_{12}-w_4)$
$\cos A_{13}=(x_6-x_5)$	$\cos B_{13}=(y_6-y_5)$
$\cos C_{13}=(z_6-z_5)$	$\cos D_{13}=(w_6-w_5)$
$\cos A_{14}=(x_{11}-x_5)$	$\cos B_{14}=(y_{11}-y_5)$
$\cos C_{14}=(z_{11}-z_5)$	$\cos D_{14}=(w_{11}-w_5)$
$\cos A_{15}=(x_{13}-x_5)$	$\cos B_{15}=(y_{13}-y_5)$
$\cos C_{15}=(z_{13}-z_5)$	$\cos D_{15}=(w_{13}-w_5)$
$\cos A_{16}=(x_{12}-x_6)$	$\cos B_{16}=(y_{12}-y_6)$
$\cos C_{16}=(z_{12}-z_6)$	$\cos D_{16}=(w_{12}-w_6)$
$\cos A_{17}=(x_{14}-x_6)$	$\cos B_{17}=(y_{14}-y_6)$
$\cos C_{17}=(z_{14}-z_6)$	$\cos D_{17}=(w_{14}-w_6)$
$\cos A_{18}=(x_8-x_7)$	$\cos B_{18}=(y_8-y_7)$
$\cos C_{18}=(z_8-z_7)$	$\cos D_{18}=(w_8-w_7)$

DIRECTION COSINES (Cont'd)

$\cos A_{19}=(x_9-x_7)$	$\cos B_{19}=(y_9-y_7)$
$\cos C_{19}=(z_9-z_7)$	$\cos D_{19}=(w_9-w_7)$
$\cos A_{20}=(x_{15}-x_7)$	$\cos B_{20}=(y_{15}-y_7)$
$\cos C_{20}=(z_{15}-z_7)$	$\cos D_{20}=(w_{15}-w_7)$
$\cos A_{21}=(x_{10}-x_8)$	$\cos B_{21}=(y_{10}-y_8)$
$\cos C_{21}=(z_{10}-z_8)$	$\cos D_{21}=(w_{10}-w_8)$
$\cos A_{22}=(x_{16}-x_8)$	$\cos B_{22}=(y_{16}-y_8)$
$\cos C_{22}=(z_{16}-z_8)$	$\cos D_{22}=(w_{16}-w_8)$
$\cos A_{23}=(x_{10}-x_9)$	$\cos B_{23}=(y_{10}-y_9)$
$\cos C_{23}=(z_{10}-z_9)$	$\cos D_{23}=(w_{10}-w_9)$
$\cos A_{24}=(x_{17}-x_9)$	$\cos B_{24}=(y_{17}-y_9)$
$\cos C_{24}=(z_{17}-z_9)$	$\cos D_{24}=(w_{17}-w_9)$
$\cos A_{25}=(x_{18}-x_{10})$	$\cos B_{25}=(y_{18}-y_{10})$
$\cos C_{25}=(z_{18}-z_{10})$	$\cos D_{25}=(w_{18}-w_{10})$
$\cos C_{26}=(x_{12}-x_{11})$	$\cos B_{26}=(y_{12}-y_{11})$
$\cos C_{26}=(z_{12}-z_{11})$	$\cos D_{26}=(w_{12}-w_{11})$
$\cos A_{27}=(x_{19}-x_{11})$	$\cos B_{27}=(y_{19}-y_{11})$
$\cos C_{27}=(z_{19}-z_{11})$	$\cos D_{27}=(w_{19}-w_{11})$
$\cos A_{28}=(x_{20}-x_{12})$	$\cos B_{28}=(y_{20}-y_{12})$
$\cos C_{28}=(z_{20}-z_{12})$	$\cos D_{28}=(w_{20}-w_{12})$
$\cos A_{29}=(x_{14}-x_{13})$	$\cos B_{29}=(y_{14}-y_{13})$
$\cos C_{29}=(z_{14}-z_{13})$	$\cos D_{29}=(w_{14}-w_{13})$
$\cos A_{30}=(x_{15}-x_{13})$	$\cos B_{30}=(y_{15}-y_{13})$
$\cos C_{30}=(z_{15}-z_{13})$	$\cos D_{30}=(w_{15}-w_{13})$
$\cos A_{31}=(x_{19}-x_{13})$	$\cos B_{31}=(y_{19}-y_{13})$
$\cos C_{31}=(z_{19}-z_{13})$	$\cos D_{31}=(w_{19}-w_{13})$

A

Table 22-13. Specifications and Analyses for the Hyperspace Object Shown in Fig. 22-13. (A) Specifications and Analysis in Hyperspace. (B) Secondary Analysis in XYZ Space. (C) Secondary Analysis in XYW Space. (D) Secondary Analysis in XZW Space. (E) Secondary Analysis in YZW Space. (F) Suggested Coordinates. (Continued from page 457.)

DIRECTION COSINES (Cont'd)

$\cos A_{32}=(x_{16}-x_{14})$ $\cos B_{32}=(y_{16}-y_{14})$
$\cos C_{32}=(z_{16}-z_{14})$ $\cos D_{32}=(w_{16}-w_{14})$

$\cos A_{33}=(x_{20}-x_{14})$ $\cos B_{33}=(y_{20}-y_{14})$
$\cos C_{33}=(z_{20}-z_{14})$ $\cos D_{33}=(w_{20}-w_{14})$

$\cos A_{34}=(x_{16}-x_{15})$ $\cos B_{34}=(y_{16}-y_{15})$
$\cos C_{34}=(z_{16}-z_{15})$ $\cos D_{34}=(w_{16}-w_{15})$

$\cos A_{35}=(x_{17}-x_{15})$ $\cos B_{35}=(y_{17}-y_{15})$
$\cos C_{35}=(z_{17}-z_{15})$ $\cos D_{35}=(w_{17}-w_{15})$

$\cos A_{36}=(x_{18}-x_{16})$ $\cos B_{36}=(y_{18}-y_{16})$
$\cos C_{36}=(z_{18}-z_{16})$ $\cos D_{36}=(w_{18}-w_{16})$

DIRECTION COSINES (Cont'd)

$\cos A_{37}=(x_{18}-x_{17})$ $\cos B_{37}=(y_{18}-y_{17})$
$\cos C_{37}=(z_{18}-z_{17})$ $\cos D_{37}=(w_{18}-w_{17})$

$\cos A_{38}=(x_{19}-x_{17})$ $\cos B_{38}=(y_{19}-y_{17})$
$\cos C_{38}=(z_{19}-z_{17})$ $\cos D_{38}=(w_{19}-w_{17})$

$\cos A_{39}=(x_{20}-x_{18})$ $\cos B_{39}=(y_{20}-y_{18})$
$\cos C_{39}=(z_{20}-z_{18})$ $\cos D_{39}=(w_{20}-w_{18})$

$\cos A_{40}=(x_{20}-x_{19})$ $\cos B_{40}=(y_{20}-y_{19})$
$\cos C_{40}=(z_{20}-z_{19})$ $\cos D_{40}=(w_{20}-w_{19})$

ANGLES

$\cos\phi_{1,2}=\cos A_1\cos A_2+\cos B_1\cos B_2+\cos C_1\cos C_2+\cos D_1\cos D_2$

$\cos\phi_{1,3}=\cos A_1\cos A_3+\cos B_1\cos B_3+\cos C_1\cos C_3+\cos D_1\cos D_3$

$\cos\phi_{1,4}=\cos A_1\cos A_4+\cos B_1\cos B_4+\cos C_1\cos C_4+\cos D_1\cos D_4$

$\cos\phi_{1,5}=\cos A_1\cos A_5+\cos B_1\cos B_5+\cos C_1\cos C_5+\cos D_1\cos D_5$

$\cos\phi_{1,6}=\cos A_1\cos A_6+\cos B_1\cos B_6+\cos C_1\cos C_6+\cos D_1\cos D_6$

$\cos\phi_{1,7}=\cos A_1\cos A_7+\cos B_1\cos B_7+\cos C_1\cos C_7+\cos D_1\cos D_7$

$\cos\phi_{2,3}=\cos A_2\cos A_3+\cos B_2\cos B_3+\cos C_2\cos C_3+\cos D_2\cos D_3$

$\cos\phi_{2,4}=\cos A_2\cos A_4+\cos B_2\cos B_4+\cos C_2\cos C_4+\cos D_2\cos D_4$

$\cos\phi_{2,8}=\cos A_2\cos A_8+\cos B_2\cos B_8+\cos C_2\cos C_8+\cos D_2\cos D_8$

$\cos\phi_{2,9}=\cos A_2\cos A_9+\cos B_2\cos B_9+\cos C_2\cos C_9+\cos D_2\cos D_9$

$\cos\phi_{2,10}=\cos A_2\cos A_{10}+\cos B_2\cos B_{10}+\cos C_2\cos C_{10}+\cos D_2\cos D_{10}$

$\cos\phi_{3,4}=\cos A_3\cos A_4+\cos B_3\cos B_4+\cos C_3\cos C_4+\cos D_3\cos D_4$

$\cos\phi_{3,13}=\cos A_3\cos A_{13}+\cos B_3\cos B_{13}+\cos C_3\cos C_{13}+\cos D_3\cos D_{13}$

$\cos\phi_{3,14}=\cos A_3\cos A_{14}+\cos B_3\cos B_{14}+\cos C_3\cos C_{14}+\cos D_3\cos D_{14}$

$\cos\phi_{3,15}=\cos A_3\cos A_{15}+\cos B_3\cos B_{15}+\cos C_3\cos C_{15}+\cos D_3\cos D_{15}$

$\cos\phi_{4,18}=\cos A_4\cos A_{18}+\cos B_4\cos B_{18}+\cos C_4\cos C_{18}+\cos D_4\cos D_{18}$

$\cos\phi_{4,19}=\cos A_4\cos A_{19}+\cos B_4\cos B_{19}+\cos C_4\cos C_{19}+\cos D_4\cos D_{19}$

$\cos\phi_{4,20}=\cos A_4\cos A_{20}+\cos B_4\cos B_{20}+\cos C_4\cos C_{20}+\cos D_4\cos D_{20}$

A

ANGLES (Cont'd)

$\cos\phi_{5,6}=\cos A_5\cos A_6+\cos B_5\cos B_6+\cos C_5\cos C_6+\cos D_5\cos D_6$

$\cos\phi_{5,7}=\cos A_5\cos A_7+\cos B_5\cos B_7+\cos C_5\cos C_7+\cos D_5\cos D_7$

$\cos\phi_{5,8}=\cos A_5\cos A_8+\cos B_5\cos B_8+\cos C_5\cos C_8+\cos D_5\cos D_8$

$\cos\phi_{5,11}=\cos A_5\cos A_{11}+\cos B_5\cos B_{11}+\cos C_5\cos C_{11}+\cos D_5\cos D_{11}$

$\cos\phi_{5,12}=\cos A_5\cos A_{12}+\cos B_5\cos B_{12}+\cos C_5\cos C_{12}+\cos D_5\cos D_{12}$

$\cos\phi_{6,7}=\cos A_6\cos A_7+\cos B_6\cos B_7+\cos C_6\cos C_7+\cos D_6\cos D_7$

$\cos\phi_{6,13}=\cos A_6\cos A_{13}+\cos B_6\cos B_{13}+\cos C_6\cos C_{13}+\cos D_6\cos D_{13}$

$\cos\phi_{6,16}=\cos A_6\cos A_{16}+\cos B_6\cos B_{16}+\cos C_6\cos C_{16}+\cos D_6\cos D_{16}$

$\cos\phi_{6,17}=\cos A_6\cos A_{17}+\cos B_6\cos B_{17}+\cos C_6\cos C_{17}+\cos D_6\cos D_{17}$

$\cos\phi_{7,18}=\cos A_7\cos A_{18}+\cos B_7\cos B_{18}+\cos C_7\cos C_{18}+\cos D_7\cos D_{18}$

$\cos\phi_{7,21}=\cos A_7\cos A_{21}+\cos B_7\cos B_{21}+\cos C_7\cos C_{21}+\cos D_7\cos D_{21}$

$\cos\phi_{7,22}=\cos A_7\cos A_{22}+\cos B_7\cos B_{22}+\cos C_7\cos C_{22}+\cos D_7\cos D_{22}$

$\cos\phi_{8,9}=\cos A_8\cos A_9+\cos B_8\cos B_9+\cos C_8\cos C_9+\cos D_8\cos D_9$

$\cos\phi_{8,10}=\cos A_8\cos A_{10}+\cos B_8\cos B_{190}+\cos C_8\cos C_{10}+\cos D_8\cos D_{10}$

$\cos\phi_{8,11}=\cos A_8\cos A_{11}+\cos B_8\cos B_{11}+\cos C_8\cos C_{11}+\cos D_8\cos D_{11}$

$\cos\phi_{8,12}=\cos A_8\cos A_{12}+\cos B_8\cos B_{12}+\cos C_8\cos C_{12}+\cos D_8\cos D_{12}$

$\cos\phi_{9,10}=\cos A_9\cos A_{10}+\cos B_9\cos B_{10}+\cos C_9\cos C_{10}+\cos D_9\cos D_{10}$

$\cos\phi_{9,19}=\cos A_9\cos A_{19}+\cos B_9\cos B_{19}+\cos C_9\cos C_{19}+\cos D_9\cos D_{19}$

$\cos\phi_{9,23}=\cos A_9\cos A_{23}+\cos B_9\cos B_{23}+\cos C_9\cos C_{23}+\cos D_9\cos D_{23}$

$\cos\phi_{9,24}=\cos A_9\cos A_{24}+\cos B_9\cos B_{24}+\cos C_9\cos C_{24}+\cos D_9\cos D_{24}$

$\cos\phi_{10,14}=\cos A_{10}\cos A_{14}+\cos B_{10}\cos B_{14}+\cos C_{10}\cos C_{14}+\cos D_{10}\cos D_{14}$

$\cos\phi_{10,26}=\cos A_{10}\cos A_{26}+\cos B_{10}\cos B_{26}+\cos C_{10}\cos C_{26}+\cos D_{10}\cos D_{26}$

$\cos\phi_{10,27}=\cos A_{10}\cos A_{27}+\cos B_{10}\cos B_{27}+\cos C_{10}\cos C_{27}+\cos D_{10}\cos D_{27}$

$\cos\phi_{11,12}=\cos A_{11}\cos A_{12}+\cos B_{11}\cos B_{12}+\cos C_{11}\cos C_{12}+\cos D_{11}\cos D_{12}$

$\cos\phi_{11,21}=\cos A_{11}\cos A_{21}+\cos B_{11}\cos B_{21}+\cos C_{11}\cos C_{21}+\cos D_{11}\cos D_{21}$

$\cos\phi_{11,23}=\cos A_{11}\cos A_{23}+\cos B_{11}\cos B_{23}+\cos C_{11}\cos C_{23}+\cos D_{11}\cos D_{23}$

$\cos\phi_{11,25}=\cos A_{11}\cos A_{25}+\cos B_{11}\cos B_{25}+\cos C_{11}\cos C_{25}+\cos D_{11}\cos D_{25}$

$\cos\phi_{12,16}=\cos A_{12}\cos A_{16}+\cos B_{12}\cos B_{16}+\cos C_{12}\cos C_{16}+\cos D_{12}\cos D_{16}$

$\cos\phi_{12,26}=\cos A_{12}\cos A_{26}+\cos B_{12}\cos B_{26}+\cos C_{12}\cos C_{26}+\cos D_{12}\cos D_{26}$

$\cos\phi_{12,28}=\cos A_{12}\cos A_{28}+\cos B_{12}\cos B_{28}+\cos C_{12}\cos C_{28}+\cos D_{12}\cos D_{28}$

$\cos\phi_{13,14}=\cos A_{13}\cos A_{14}+\cos B_{13}\cos B_{14}+\cos C_{13}\cos C_{14}+\cos D_{13}\cos D_{14}$

$\cos\phi_{13,15}=\cos A_{13}\cos A_{15}+\cos B_{13}\cos B_{15}+\cos C_{13}\cos C_{15}+\cos D_{13}\cos D_{15}$

$\cos\phi_{13,16}=\cos A_{13}\cos A_{16}+\cos B_{13}\cos B_{16}+\cos C_{13}\cos C_{16}+\cos D_{13}\cos D_{16}$

$\cos\phi_{13,17}=\cos A_{13}\cos A_{17}+\cos B_{13}\cos B_{17}+\cos C_{13}\cos C_{17}+\cos D_{13}\cos D_{17}$

A

Table 22-13. Specifications and Analyses for the Hyperspace Object Shown in Fig. 22-13. (A) Specifications and Analysis in Hyperspace. (B) Secondary Analysis in XYZ Space. (C) Secondary Analysis in XYW Space. (D) Secondary Analysis in XZW Space. (E) Secondary Analysis in YZW Space. (F) Suggested Coordinates. (Continued from page 459.)

A

ANGLES (Cont'd)

$$\cos\phi_{14,15}=\cos A_{14}\cos A_{15}+\cos B_{14}\cos B_{15}+\cos C_{14}\cos C_{15}+\cos D_{14}\cos D_{15}$$

$$\cos\phi_{14,26}=\cos A_{14}\cos A_{26}+\cos B_{14}\cos B_{26}+\cos C_{14}\cos C_{26}+\cos D_{14}\cos D_{26}$$

$$\cos\phi_{14,27}=\cos A_{14}\cos A_{27}+\cos B_{14}\cos B_{27}+\cos C_{14}\cos C_{27}+\cos D_{14}\cos D_{27}$$

$$\cos\phi_{15,29}=\cos A_{15}\cos A_{29}+\cos B_{15}\cos B_{29}+\cos C_{15}\cos C_{29}+\cos D_{15}\cos D_{29}$$

$$\cos\phi_{15,30}=\cos A_{15}\cos A_{30}+\cos B_{15}\cos B_{30}+\cos C_{15}\cos C_{30}+\cos D_{15}\cos D_{30}$$

$$\cos\phi_{15,31}=\cos A_{15}\cos A_{31}+\cos B_{15}\cos B_{31}+\cos C_{15}\cos C_{31}+\cos D_{15}\cos D_{31}$$

$$\cos\phi_{16,17}=\cos A_{16}\cos A_{17}+\cos B_{16}\cos B_{17}+\cos C_{16}\cos C_{17}+\cos D_{16}\cos D_{32}$$

$$\cos\phi_{16,26}=\cos A_{16}\cos A_{26}+\cos B_{16}\cos B_{26}+\cos C_{16}\cos C_{26}+\cos D_{16}\cos D_{26}$$

$$\cos\phi_{16,28}=\cos A_{16}\cos A_{28}+\cos B_{16}\cos B_{28}+\cos C_{16}\cos C_{28}+\cos D_{16}\cos D_{28}$$

$$\cos\phi_{17,29}=\cos A_{17}\cos A_{29}+\cos B_{17}\cos B_{29}+\cos C_{17}\cos C_{29}+\cos D_{17}\cos D_{29}$$

$$\cos\phi_{17,32}=\cos A_{17}\cos A_{32}+\cos B_{17}\cos B_{32}+\cos C_{17}\cos C_{32}+\cos D_{17}\cos D_{32}$$

$$\cos\phi_{17,33}=\cos A_{17}\cos A_{33}+\cos B_{17}\cos B_{33}+\cos C_{17}\cos C_{33}+\cos D_{17}\cos D_{33}$$

$$\cos\phi_{18,19}=\cos A_{18}\cos A_{19}+\cos B_{18}\cos B_{19}+\cos C_{18}\cos C_{19}+\cos D_{18}\cos D_{19}$$

$$\cos\phi_{18,20}=\cos A_{18}\cos A_{20}+\cos B_{18}\cos B_{20}+\cos C_{178}\cos C_{20}+\cos D_{18}\cos D_{20}$$

$$\cos\phi_{18,21}=\cos A_{18}\cos A_{21}+\cos B_{18}\cos B_{21}+\cos C_{18}\cos C_{21}+\cos D_{18}\cos D_{21}$$

$$\cos\phi_{18,22}=\cos A_{18}\cos A_{22}+\cos B_{18}\cos B_{22}+\cos C_{18}\cos C_{22}+\cos D_{18}\cos D_{22}$$

$$\cos\phi_{19,20}=\cos A_{19}\cos A_{20}+\cos B_{19}\cos B_{20}+\cos C_{19}\cos C_{20}+\cos D_{19}\cos D_{20}$$

$$\cos\phi_{19,23}=\cos A_{19}\cos A_{23}+\cos B_{19}\cos B_{23}+\cos C_{19}\cos C_{23}+\cos D_{19}\cos D_{23}$$

$$\cos\phi_{19,24}=\cos A_{19}\cos A_{24}+\cos B_{19}\cos B_{24}+\cos C_{19}\cos C_{24}+\cos D_{19}\cos D_{24}$$

$$\cos\phi_{20,30}=\cos A_{20}\cos A_{30}+\cos B_{20}\cos B_{30}+\cos C_{20}\cos C_{30}+\cos D_{20}\cos D_{30}$$

$$\cos\phi_{20,34}=\cos A_{20}\cos A_{34}+\cos B_{20}\cos B_{34}+\cos C_{20}\cos C_{34}+\cos D_{20}\cos D_{34}$$

$$\cos\phi_{20,35}=\cos A_{20}\cos A_{35}+\cos B_{20}\cos B_{35}+\cos C_{20}\cos C_{35}+\cos D_{20}\cos D_{35}$$

$$\cos\phi_{21,22}=\cos A_{21}\cos A_{22}+\cos B_{212}\cos B_{22}+\cos C_{21}\cos C_{22}+\cos D_{21}\cos D_{22}$$

$$\cos\phi_{21,23}=\cos A_{21}\cos A_{23}+\cos B_{21}\cos B_{23}+\cos C_{21}\cos C_{23}+\cos D_{21}\cos D_{23}$$

$$\cos\phi_{21,25}=\cos A_{21}\cos A_{25}+\cos B_{21}\cos B_{25}+\cos C_{21}\cos C_{25}+\cos D_{21}\cos D_{25}$$

$$\cos\phi_{22,32}=\cos A_{22}\cos A_{32}+\cos B_{22}\cos B_{32}+\cos C_{22}\cos C_{32}+\cos D_{22}\cos D_{32}$$

$$\cos\phi_{22,34}=\cos A_{22}\cos A_{34}+\cos B_{22}\cos B_{34}+\cos C_{22}\cos C_{34}+\cos D_{22}\cos D_{34}$$

$$\cos\phi_{22,36}=\cos A_{22}\cos A_{36}+\cos B_{22}\cos B_{36}+\cos C_{22}\cos C_{36}+\cos D_{22}\cos D_{36}$$

$$\cos\phi_{23,24}=\cos A_{23}\cos A_{24}+\cos B_{23}\cos B_{24}+\cos C_{23}\cos C_{24}+\cos D_{23}\cos D_{24}$$

$$\cos\phi_{23,25}=\cos A_{23}\cos A_{25}+\cos B_{23}\cos B_{25}+\cos C_{23}\cos C_{25}+\cos D_{23}\cos D_{25}$$

$$\cos\phi_{24,35}=\cos A_{24}\cos A_{35}+\cos B_{24}\cos B_{35}+\cos C_{24}\cos C_{35}+\cos D_{24}\cos D_{35}$$

$$\cos\phi_{24,37}=\cos A_{24}\cos A_{37}+\cos B_{24}\cos B_{37}+\cos C_{24}\cos C_{37}+\cos D_{24}\cos D_{37}$$

$$\cos\phi_{24,37}=\cos A_{24}\cos A_{37}+\cos B_{24}\cos B_{37}+\cos C_{24}\cos C_{37}+\cos D_{24}\cos D_{37}$$

$$\cos\phi_{25,36}=\cos A_{25}\cos A_{36}+\cos B_{25}\cos B_{36}+\cos C_{25}\cos C_{36}+\cos D_{25}\cos D_{36}$$

ANGLES (Cont'd)

$\cos\phi_{25,37}=\cos A_{25}\cos A_{37}+\cos B_{25}\cos B_{37}+\cos C_{25}\cos C_{37}+\cos D_{25}\cos D_{37}$

$\cos\phi_{25,39}=\cos A_{25}\cos A_{39}+\cos B_{25}\cos B_{39}+\cos C_{25}\cos C_{39}+\cos D_{25}\cos D_{39}$

$\cos\phi_{26,27}=\cos A_{26}\cos A_{27}+\cos B_{26}\cos B_{27}+\cos C_{26}\cos C_{27}+\cos D_{26}\cos D_{27}$

$\cos\phi_{26,28}=\cos A_{26}\cos A_{28}+\cos B_{26}\cos B_{28}+\cos C_{26}\cos C_{28}+\cos D_{26}\cos D_{28}$

$\cos\phi_{27,31}=\cos A_{27}\cos A_{31}+\cos B_{27}\cos B_{31}+\cos C_{27}\cos C_{31}+\cos D_{27}\cos D_{31}$

$\cos\phi_{27,38}=\cos A_{27}\cos A_{38}+\cos B_{27}\cos B_{38}+\cos C_{27}\cos C_{38}+\cos D_{27}\cos D_{38}$

$\cos\phi_{27,40}=\cos A_{27}\cos A_{40}+\cos B_{27}\cos B_{40}+\cos C_{27}\cos C_{40}+\cos D_{27}\cos D_{40}$

$\cos\phi_{28,33}=\cos A_{28}\cos A_{33}+\cos B_{28}\cos C_{33}+\cos C_{28}\cos D_{33}+\cos D_{28}\cos D_{33}$

$\cos\phi_{28,39}=\cos A_{28}\cos A_{39}+\cos B_{28}\cos B_{39}+\cos C_{28}\cos C_{39}+\cos D_{28}\cos D_{39}$

$\cos\phi_{28,40}=\cos A_{28}\cos A_{40}+\cos B_{28}\cos B_{40}+\cos C_{28}\cos C_{40}+\cos D_{28}\cos D_{40}$

$\cos\phi_{29,31}=\cos A_{29}\cos A_{31}+\cos B_{29}\cos B_{31}+\cos C_{29}\cos C_{31}+\cos D_{29}\cos D_{31}$

$\cos\phi_{29,31}=\cos A_{29}\cos A_{31}+\cos B_{29}\cos B_{31}+\cos C_{29}\cos C_{31}+\cos D_{29}\cos D_{31}$

$\cos\phi_{29,32}=\cos A_{29}\cos A_{32}+\cos B_{29}\cos B_{32}+\cos C_{29}\cos C_{32}+\cos D_{29}\cos D_{32}$

$\cos\phi_{29,33}=\cos A_{29}\cos A_{33}+\cos B_{29}\cos B_{33}+\cos C_{29}\cos C_{33}+\cos D_{29}\cos D_{33}$

$\cos\phi_{30,31}=\cos A_{30}\cos A_{31}+\cos B_{30}\cos B_{31}+\cos C_{30}\cos C_{31}+\cos D_{30}\cos D_{31}$

$\cos\phi_{30,34}=\cos A_{30}\cos A_{34}+\cos B_{30}\cos B_{34}+\cos C_{30}\cos C_{34}+\cos D_{30}\cos D_{34}$

$\cos\phi_{30,35}=\cos A_{30}\cos A_{35}+\cos B_{30}\cos B_{35}+\cos C_{30}\cos C_{35}+\cos D_{30}\cos D_{35}$

$\cos\phi_{31,38}=\cos A_{31}\cos A_{38}+\cos B_{31}\cos B_{38}+\cos C_{31}\cos C_{38}+\cos D_{31}\cos D_{38}$

$\cos\phi_{31,40}=\cos A_{31}\cos A_{40}+\cos B_{31}\cos B_{40}+\cos C_{31}\cos C_{40}+\cos D_{31}\cos D_{40}$

$\cos\phi_{32,33}=\cos A_{32}\cos A_{33}+\cos B_{32}\cos B_{33}+\cos C_{32}\cos C_{33}+\cos D_{32}\cos D_{33}$

$\cos\phi_{32,34}=\cos A_{32}\cos A_{34}+\cos B_{32}\cos B_{34}+\cos C_{32}\cos C_{34}+\cos D_{32}\cos D_{34}$

$\cos\phi_{32,36}=\cos A_{32}\cos A_{36}+\cos B_{32}\cos B_{36}+\cos C_{32}\cos C_{36}+\cos D_{32}\cos D_{36}$

$\cos\phi_{33,39}=\cos A_{33}\cos A_{39}+\cos B_{33}\cos B_{39}+\cos C_{33}\cos C_{39}+\cos D_{33}\cos D_{39}$

$\cos\phi_{33,40}=\cos A_{33}\cos A_{40}+\cos B_{33}\cos B_{40}+\cos C_{33}\cos C_{40}+\cos D_{33}\cos D_{40}$

$\cos\phi_{34,35}=\cos A_{34}\cos A_{35}+\cos B_{34}\cos B_{35}+\cos C_{34}\cos C_{35}+\cos D_{34}\cos D_{35}$

$\cos\phi_{34,36}=\cos A_{34}\cos A_{36}+\cos B_{34}\cos B_{36}+\cos C_{34}\cos C_{36}+\cos D_{34}\cos D_{36}$

$\cos\phi_{35,37}=\cos A_{35}\cos A_{37}+\cos B_{35}\cos B_{37}+\cos C_{35}\cos C_{37}+\cos D_{35}\cos D_{37}$

$\cos\phi_{35,38}=\cos A_{35}\cos A_{38}+\cos B_{35}\cos B_{38}+\cos C_{35}\cos C_{38}+\cos D_{35}\cos D_{38}$

$\cos\phi_{36,37}=\cos A_{36}\cos A_{37}+\cos B_{36}\cos B_{37}+\cos C_{36}\cos C_{37}+\cos D_{36}\cos D_{37}$

$\cos\phi_{36,39}=\cos A_{36}\cos A_{39}+\cos B_{36}\cos B_{39}+\cos C_{36}\cos C_{39}+\cos D_{36}\cos D_{39}$

$\cos\phi_{37,38}=\cos A_{37}\cos A_{38}+\cos B_{37}\cos B_{38}+\cos C_{37}\cos C_{38}+\cos D_{37}\cos D_{38}$

$\cos\phi_{37,39}=\cos A_{37}\cos A_{39}+\cos B_{37}\cos B_{39}+\cos C_{37}\cos C_{39}+\cos D_{37}\cos D_{39}$

$\cos\phi_{38,40}=\cos A_{38}\cos A_{40}+\cos B_{38}\cos B_{40}+\cos C_{38}\cos C_{40}+\cos D_{38}\cos D_{40}$

$\cos\phi_{39,40}=\cos A_{39}\cos A_{40}+\cos B_{39}\cos B_{40}+\cos C_{39}\cos C_{40}+\cos D_{39}\cos D_{40}$

A

Table 22-13. Specifications and Analyses for the Hyperspace Object Shown in Fig. 22-13. (A) Specifications and Analysis in Hyperspace. (B) Secondary Analysis in XYZ Space. (C) Secondary Analysis in XYW Space. (D) Secondary Analysis in XZW Space. (E) Secondary Analysis in YZW Space. (F) Suggested Coordinates. (Continued from page 461.)

A

CHECK

F_{1}: $\phi_{1,2}+\phi_{1,5}+\phi_{2,8}+\phi_{5,8}=360^{\circ}$

F_{2}: $\phi_{1,3}+\phi_{1,6}+\phi_{3,13}+\phi_{6,13}=360^{\circ}$

F_{3}: $\phi_{1,4}+\phi_{1,7}+\phi_{4,18}+\phi_{7,18}=360^{\circ}$

F_{4}: $\phi_{2,3}+\phi_{2,10}+\phi_{3,14}+\phi_{10,14}=360^{\circ}$

F_{5}: $\phi_{2,4}+\phi_{2,9}+\phi_{4,19}+\phi_{9,19}=360^{\circ}$

F_{6}: $\phi_{3,4}+\phi_{3,15}+\phi_{4,20}+\phi_{15,30}+\phi_{20,30}=550^{\circ}$

F_{7}: $\phi_{5,6}+\phi_{5,12}+\phi_{6,16}+\phi_{12,16}=360^{\circ}$

F_{8}: $\phi_{5,7}+\phi_{5,11}+\phi_{7,21}+\phi_{11,21}=360^{\circ}$

F_{9}: $\phi_{6,7}+\phi_{6,17}+\phi_{7,22}+\phi_{17,32}+\phi_{22,32}=550^{\circ}$

F_{10}: $\phi_{8,9}+\phi_{8,11}+\phi_{9,23}+\phi_{11,23}=360^{\circ}$

F_{11}: $\phi_{8,10}+\phi_{8,12}+\phi_{10,26}+\phi_{12,26}=360^{\circ}$

F_{12}: $\phi_{9,10}+\phi_{9,24}+\phi_{10,27}+\phi_{24,38}+\phi_{27,38}=550^{\circ}$

F_{13}: $\phi_{11,12}+\phi_{11,25}+\phi_{12,28}+\phi_{25,39}+\phi_{28,39}=550^{\circ}$

F_{14}: $\phi_{13,14}+\phi_{13,16}+\phi_{14,26}+\phi_{16,26}=360^{\circ}$

F_{15}: $\phi_{13,15}+\phi_{13,17}+\phi_{15,29}+\phi_{17,29}=360^{\circ}$

F_{16}: $\phi_{14,15}+\phi_{14,27}+\phi_{15,31}+\phi_{27,31}=360^{\circ}$

F_{17}: $\phi_{16,17}+\phi_{16,28}+\phi_{17,33}+\phi_{28,33}=360^{\circ}$

F_{18}: $\phi_{18,19}+\phi_{18,21}+\phi_{19,23}+\phi_{21,23}=360^{\circ}$

F_{19}: $\phi_{18,20}+\phi_{18,22}+\phi_{20,34}+\phi_{22,34}=360^{\circ}$

F_{20}: $\phi_{19,20}+\phi_{19,24}+\phi_{20,35}+\phi_{24,35}=360^{\circ}$

F_{21}: $\phi_{21,22}+\phi_{21,25}+\phi_{22,36}+\phi_{25,36}=360^{\circ}$

F_{22}: $\phi_{23,24}+\phi_{23,25}+\phi_{24,37}+\phi_{25,37}=360^{\circ}$

F_{23}: $\phi_{26,27}+\phi_{26,28}+\phi_{27,40}+\phi_{28,40}=360^{\circ}$

F_{24}: $\phi_{29,30}+\phi_{29,32}+\phi_{30,34}+\phi_{32,34}=360^{\circ}$

F_{25}: $\phi_{29,31}+\phi_{29,33}+\phi_{31,34}+\phi_{33,34}=360^{\circ}$

F_{26}: $\phi_{30,31}+\phi_{30,35}+\phi_{31,38}+\phi_{35,38}=360^{\circ}$

F_{27}: $\phi_{32,33}+\phi_{32,36}+\phi_{33,39}+\phi_{36,39}=360^{\circ}$

F_{28}: $\phi_{34,35}+\phi_{34,36}+\phi_{35,37}+\phi_{36,36}=360^{\circ}$

F_{29}: $\phi_{37,38}+\phi_{37,39}+\phi_{38,40}+\phi_{39,40}=360^{\circ}$

B

SECONDARY ANALYSIS -- XYZ SPACE
(See Fig. 22-13B)

PLOTTED POINTS

$P_{1}=(x_{1},y_{1},z_{1})$	$P_{2}=(x_{2},y_{2},z_{2})$
$P_{3}=(x_{3},y_{3},z_{3})$	$P_{4}=(x_{4},y_{4},z_{4})$
$P_{5}=(x_{5},y_{5},z_{5})$	$P_{6}=(x_{6},y_{6},z_{6})$
$P_{7}=(x_{7},y_{7},z_{7})$	$P_{8}=(x_{8},y_{8},z_{8})$
$P_{9}=(x_{9},y_{9},z_{9})$	$P_{10}=(x_{10},y_{10},z_{10})$
$P_{11}=(x_{11},y_{11},z_{11})$	$P_{12}=(x_{12},y_{12},z_{12})$
$P_{13}=(x_{13},y_{13},z_{13})$	$P_{14}=(x_{14},y_{14},z_{14})$
$P_{15}=(x_{15},y_{15},z_{15})$	$P_{16}=(x_{16},y_{16},z_{16})$
$P_{17}=(x_{17},y_{17},z_{17})$	$P_{18}=(x_{18},y_{18},z_{18})$
$P_{19}=(x_{18},y_{19},z_{19})$	$P_{20}=(x_{20},y_{20},z_{20})$

LENGTHS

$$s_{1}=\sqrt{(x_{2}-x_{1})^{2}+(y_{2}-y_{1})^{2}+(z_{2}-z_{1})^{2}}$$

$$s_{2}=\sqrt{(x_{3}-x_{1})^{2}+(y_{3}-y_{1})^{2}+(z_{3}-z_{1})^{2}}$$

$$s_{3}=\sqrt{(x_{5}-x_{1})^{2}+(y_{5}-y_{1})^{2}+(z_{5}-z_{1})^{2}}$$

$$s_{4}=\sqrt{(x_{7}-x_{1})^{2}+(y_{7}-y_{1})^{2}+(z_{7}-z_{1})^{2}}$$

$$s_{5}=\sqrt{(x_{4}-x_{2})^{2}+(y_{4}-y_{2})^{2}+(z_{4}-z_{2})^{2}}$$

$$s_{6}=\sqrt{(x_{6}-x_{2})^{2}+(y_{6}-y_{2})^{2}+(z_{6}-z_{2})^{2}}$$

$$s_{7}=\sqrt{(x_{8}-x_{2})^{2}+(y_{8}-y_{2})^{2}+(z_{8}-z_{2})^{2}}$$

$$s_{8}=\sqrt{(x_{4}-x_{3})^{2}+(y_{4}-y_{3})^{2}+(z_{4}-z_{3})^{2}}$$

$$s_{9}=\sqrt{(x_{9}-x_{3})^{2}+(y_{9}-y_{3})^{2}+(z_{9}-z_{3})^{2}}$$

$$s_{10}=\sqrt{(x_{11}-x_{3})^{2}+(y_{11}-y_{3})^{2}+(z_{11}-z_{3})^{2}}$$

$$s_{11}=\sqrt{(x_{10}-x_{4})^{2}+(y_{10}-y_{4})^{2}+(z_{10}-z_{4})^{2}}$$

B

LENGTHS (Cont'd)

$$s_{12}=\sqrt{(x_{12}-x_4)^2+(y_{12}-y_4)^2+(z_{12}-z_4)^2}$$

$$s_{13}=\sqrt{(x_6-x_5)^2+(y_6-y_5)^2+(z_6-z_5)^2}$$

$$s_{14}=\sqrt{(x_{11}-x_5)^2+(y_{11}-y_5)^2+(z_{11}-z_5)^2}$$

$$s_{15}=\sqrt{(x_{13}-x_5)^2+(y_{13}-y_5)^2+(z_{13}-z_5)^2}$$

$$s_{16}=\sqrt{(x_{12}-x_6)^2+(y_{12}-y_6)^2+(z_{12}-z_6)^2}$$

$$s_{17}=\sqrt{(x_{14}-x_6)^2+(y_{14}-y_6)^2+(z_{14}-z_6)^2}$$

$$s_{18}=\sqrt{(x_8-x_7)^2+(y_8-y_7)^2+(z_8-z_7)^2}$$

$$s_{19}=\sqrt{(x_9-x_7)^2+(y_9-y_7)^2+(z_9-z_7)^2}$$

$$s_{20}=\sqrt{(x_{15}-x_7)^2+(y_{15}-y_7)^2+(z_{15}-z_7)^2}$$

$$s_{21}=\sqrt{(x_{10}-x_8)^2+(y_{10}-y_8)^2+(z_{10}-z_8)^2}$$

$$s_{22}=\sqrt{(x_{16}-x_8)^2+(y_{16}-y_8)^2+(z_{16}-z_8)^2}$$

$$s_{23}=\sqrt{(x_{10}-x_9)^2+(y_{10}-y_9)^2+(z_{10}-z_9)^2}$$

$$s_{24}=\sqrt{(x_{17}-x_9)^2+(y_{17}-y_9)^2+(z_{17}-z_9)^2}$$

$$s_{25}=\sqrt{(x_{18}-x_{10})^2+(y_{18}-y_{10})^2+(z_{18}-z_{10})^2}$$

$$s_{26}=\sqrt{(x_{12}-x_{11})^2+(y_{12}-y_{11})^2+(z_{12}-z_{11})^2}$$

$$s_{27}=\sqrt{(x_{19}-x_{11})^2+(y_{19}-y_{11})^2+(z_{19}-z_{11})^2}$$

$$s_{28}=\sqrt{(x_{20}-x_{12})^2+(y_{20}-y_{12})^2+(z_{20}-z_{12})^2}$$

$$s_{29}=\sqrt{(x_{14}-x_{13})^2+(y_{14}-y_{13})^2+(z_{14}-z_{13})^2}$$

$$s_{30}=\sqrt{(x_{15}-x_{13})^2+(y_{15}-y_{13})^2+(z_{15}-z_{13})^2}$$

$$s_{31}=\sqrt{(x_{19}-x_{13})^2+(y_{19}-y_{13})^2+(z_{19}-z_{13})^2}$$

$$s_{32}=\sqrt{(x_{16}-x_{14})^2+(y_{16}-y_{14})^2+(z_{16}-z_{14})^2}$$

$$s_{33}=\sqrt{(x_{20}-x_{14})^2+(y_{20}-y_{14})^2+(z_{20}-z_{14})^2}$$

LENGTHS (Cont'd)

$$s_{34}=\sqrt{(x_{16}-x_{15})^2+(y_{16}-y_{15})^2+(z_{16}-z_{15})^2}$$

$$s_{35}=\sqrt{(x_{17}-x_{15})^2+(y_{17}-y_{15})^2+(z_{17}-z_{15})^2}$$

$$s_{36}=\sqrt{(x_{18}-x_{16})^2+(y_{18}-y_{16})^2+(z_{18}-z_{16})^2}$$

$$s_{37}=\sqrt{(x_{18}-x_{17})^2+(y_{18}-y_{17})^2+(z_{18}-z_{17})^2}$$

$$s_{38}=\sqrt{(x_{19}-x_{17})^2+(y_{19}-y_{17})^2+(z_{19}-z_{17})^2}$$

$$s_{39}=\sqrt{(x_{20}-x_{18})^2+(y_{20}-y_{18})^2+(z_{20}-z_{18})^2}$$

$$s_{40}=\sqrt{(x_{20}-x_{19})^2+(y_{20}-y_{19})^2+(z_{20}-z_{19})^2}$$

DIRECTION COSINES

$\cos A_1=(x_2-x_1)/s_1$ $\cos B_1=(y_2-y_1)/s_1$
$\cos C_1=(z_2-z_1)/s_1$

$\cos A_2=(x_3-x_1)/s_2$ $\cos B_2=(y_3-y_1)/s_2$
$\cos C_2=(z_3-z_1)/s_2$

$\cos A_3=(x_5-x_1)/s_3$ $\cos B_3=(y_5-y_1)/s_3$
$\cos C_3=(z_5-z_1)/s_3$

$\cos A_4=(x_7-x_1)/s_4$ $\cos B_4=(y_7-y_1)/s_4$
$\cos C_4=(z_7-z_1)/s_4$

$\cos A_5=(x_4-x_2)/s_5$ $\cos B_5=(y_4-y_2)/s_5$
$\cos C_5=(z_4-z_2)/s_5$

$\cos A_6=(x_6-x_2)/s_6$ $\cos B_6=(y_6-y_2)/s_6$
$\cos C_6=(z_6-z_2)/s_6$

$\cos A_7=(x_8-x_2)/s_7$ $\cos B_7=(y_8-y_2)/s_7$
$\cos C_7=(z_8-z_2)/s_7$

$\cos A_8=(x_4-x_3)/s_8$ $\cos B_8=(y_4-y_3)/s_8$
$\cos C_8=(z_4-z_3)/s_8$

Table 22-13. Specifications and Analyses for the Hyperspace Object Shown in Fig. 22-13. (A) Specifications and Analysis in Hyperspace. (B) Secondary Analysis in XYZ Space. (C) Secondary Analysis in XYW Space. (D) Secondary Analysis in XZW Space. (E) Secondary Analysis in YZW Space. (F) Suggested Coordinates. (Continued from page 463.)

B

DIRECTION COSINES (Cont'd)

$\cos A_{9}=(x_{9}-x_{3})/s_{9}$ $\cos B_{9}=(y_{9}-y_{3})/s_{9}$
$\cos C_{9}=(z_{9}-z_{3})/s_{9}$

$\cos A_{10}=(x_{11}-x_{3})/s_{10}$ $\cos B_{10}=(y_{11}-y_{3})/s_{10}$
$\cos C_{10}=(z_{11}-z_{3})/s_{10}$

$\cos A_{11}=(x_{10}-x_{4})/s_{11}$ $\cos B_{11}=(y_{10}-y_{4})/s_{11}$
$\cos C_{11}=(z_{10}-z_{4})/s_{11}$

$\cos A_{12}=(x_{12}-x_{4})/s_{12}$ $\cos B_{12}=(y_{12}-y_{4})/s_{12}$
$\cos C_{12}=(z_{12}-z_{4})/s_{12}$

$\cos A_{13}=(x_{6}-x_{5})/s_{13}$ $\cos B_{13}=(y_{6}-y_{5})/s_{13}$
$\cos C_{13}=(z_{6}-z_{5})/s_{13}$

$\cos A_{14}=(x_{11}-x_{5})/s_{14}$ $\cos B_{14}=(y_{11}-y_{5})/s_{14}$
$\cos C_{14}=(z_{11}-z_{5})/s$

$\cos A_{15}=(x_{13}-x_{5})/s_{15}$ $\cos B_{15}=(y_{13}-y_{5})/s_{15}$
$\cos C_{15}=(z_{13}-z_{5})/s_{15}$

$\cos A_{16}=(x_{12}-x_{6})/s_{16}$ $\cos B_{16}=(y_{12}-y_{6})/s_{16}$
$\cos C_{16}=(z_{12}-z_{6})/s_{16}$

$\cos A_{17}=(x_{14}-x_{6})/s_{17}$ $\cos B_{17}=(y_{14}-y_{6})/s_{17}$
$\cos C_{17}=(z_{14}-z_{6})/s_{17}$

$\cos A_{18}=(x_{8}-x_{7})/s_{18}$ $\cos B_{18}=(y_{8}-y_{7})/s_{18}$
$\cos C_{18}=(z_{8}-z_{7})/s_{18}$

$\cos A_{19}=(x_{9}-x_{7})/s_{19}$ $\cos B_{19}=(y_{9}-y_{7})/s_{19}$
$\cos C_{19}=(z_{9}-z_{7})/s_{19}$

$\cos A_{20}=(x_{15}-x_{7})/s_{20}$ $\cos B_{20}=(y_{15}-y_{7})/s_{20}$
$\cos C_{20}=(z_{15}-z_{7})/s_{20}$

$\cos A_{21}=(x_{10}-x_{8})/s_{21}$ $\cos B_{21}=(y_{10}-y_{8})/s_{21}$
$\cos C_{21}=(z_{10}-z_{8})/s_{21}$

$\cos A_{22}=(x_{16}-x_{8})/s_{22}$ $\cos B_{22}=(y_{16}-y_{8})/s_{22}$
$\cos C_{22}=(z_{16}-z_{8})/s_{22}$

DIRECTION COSINES (Cont'd)

$\cos A_{23}=(x_{10}-x_{9})/s_{23}$ $\cos B_{23}=(y_{10}-y_{9})/s_{23}$
$\cos C_{23}=(z_{10}-z_{9})/s_{23}$

$\cos A_{24}=(x_{17}-x_{9})/s_{24}$ $\cos B_{24}=(y_{17}-y_{9})/s_{24}$
$\cos C_{24}=(z_{17}-z_{9})/s_{24}$

$\cos A_{25}=(x_{18}-x_{10})/s_{25}$ $\cos B_{25}=(y_{18}-y_{10})/s_{25}$
$\cos C_{25}=(z_{18}-z_{10})/s_{25}$

$\cos A_{26}=(x_{12}-x_{11})/s_{26}$ $\cos B_{26}=(y_{12}-y_{11})/s_{26}$
$\cos C_{26}=(z_{12}-z_{11})/s_{26}$

$\cos A_{27}=(x_{19}-x_{11})/s_{27}$ $\cos B_{27}=(y_{19}-y_{11})/s_{27}$
$\cos C_{27}=(z_{19}-z_{11})/s_{27}$

$\cos A_{28}=(x_{20}-x_{12})/s_{28}$ $\cos B_{28}=(y_{20}-y_{12})/s_{28}$
$\cos C_{28}=(z_{20}-z_{12})/s_{28}$

$\cos A_{29}=(x_{14}-x_{13})/s_{29}$ $\cos B_{29}=(y_{14}-y_{13})/s_{29}$
$\cos C_{29}=(z_{14}-z_{13})/s_{29}$

$\cos A_{30}=(x_{15}-x_{13})/s_{30}$ $\cos B_{30}=(y_{15}-y_{13})/s_{30}$
$\cos C_{30}=(z_{15}-z_{13})/s_{30}$

$\cos A_{31}=(x_{19}-x_{13})/s_{31}$ $\cos B_{31}=(y_{19}-y_{13})/s_{31}$
$\cos C_{31}=(z_{19}-z_{13})/s_{31}$

$\cos A_{32}=(x_{16}-x_{14})/s_{32}$ $\cos B_{32}=(y_{16}-y_{14})/s_{32}$
$\cos C_{32}=(z_{16}-z_{14})/s_{32}$

$\cos A_{33}=(x_{20}-x_{14})/s_{33}$ $\cos B_{33}=(y_{20}-y_{14})/s_{33}$
$\cos C_{33}=(z_{20}-z_{14})/s_{33}$

$\cos A_{34}=(x_{16}-x_{15})/s_{34}$ $\cos B_{34}=(y_{16}-y_{15})/s_{34}$
$\cos C_{34}=(z_{16}-z_{15})/s_{34}$

$\cos A_{35}=(x_{17}-x_{15})/s_{35}$ $\cos B_{35}=(y_{17}-y_{15})/s_{35}$
$\cos C_{35}=(z_{17}-z_{15})/s_{35}$

DIRECTION COSINES (Cont'd)

$\cos A_{36}=(x_{18}-x_{16})/s_{36}$ $\cos B_{36}=(y_{18}-y_{16})/s_{36}$

$\cos C_{36}=(z_{18}-z_{16})/s_{36}$

$\cos A_{37}=(x_{18}-x_{17})/s_{37}$ $\cos B_{37}=(y_{18}-y_{17})/s_{37}$

$\cos C_{37}=(z_{18}-z_{17})/s_{37}$

$\cos A_{38}=(x_{19}-x_{17})/s_{38}$ $\cos B_{38}=(y_{19}-y_{17})/s_{38}$

$\cos C_{38}=(z_{19}-z_{17})/s_{38}$

$\cos A_{39}=(x_{20}-x_{18})/s_{39}$ $\cos B_{39}=(y_{20}-y_{18})/s_{39}$

$\cos C_{39}=(z_{20}-z_{18})/s_{39}$

$\cos A_{40}=(x_{20}-x_{19})/s_{40}$ $\cos B_{40}=(y_{20}-y_{19})/s_{40}$

$\cos C_{40}=(z_{20}-z_{19})/s_{40}$

ANGLES

$\cos\phi_{1,2}=\cos A_1\cos A_2+\cos B_1\cos B_2+\cos C_1\cos C_2$

$\cos\phi_{1,3}=\cos A_1\cos A_3+\cos B_1\cos B_3+\cos C_1\cos C_3$

$\cos\phi_{1,4}=\cos A_1\cos A_4+\cos B_1\cos B_4+\cos C_1\cos C_4$

$\cos\phi_{1,5}=\cos A_1\cos A_5+\cos B_1\cos B_5+\cos C_1\cos C_5$

$\cos\phi_{1,6}=\cos A_1\cos A_6+\cos B_1\cos B_6+\cos C_1\cos C_6$

$\cos\phi_{1,7}=\cos A_1\cos A_7+\cos B_1\cos B_7+\cos C_1\cos C_7$

$\cos\phi_{2,3}=\cos A_2\cos A_3+\cos B_2\cos B_3+\cos C_2\cos C_3$

$\cos\phi_{2,4}=\cos A_2\cos A_4+\cos B_2\cos B_4+\cos C_2\cos C_4$

$\cos\phi_{2,8}=\cos A_2\cos A_8+\cos B_2\cos B_8+\cos C_2\cos C_8$

$\cos\phi_{2,9}=\cos A_2\cos A_9+\cos B_2\cos B_9+\cos C_2\cos C_9$

$\cos\phi_{2,10}=\cos A_2\cos A_{10}+\cos B_2\cos B_{10}+\cos C_2\cos C_{10}$

$\cos\phi_{3,4}=\cos A_3\cos A_4+\cos B_3\cos B_4+\cos C_3\cos C_4$

$\cos\phi_{3,13}=\cos A_3\cos A_{13}+\cos B_3\cos B_{13}+\cos C_3\cos C_{13}$

$\cos\phi_{3,14}=\cos A_3\cos A_{14}+\cos B_3\cos B_{14}+\cos C_3\cos C_{14}$

$\cos\phi_{3,15}=\cos A_3\cos A_{15}+\cos B_3\cos B_{15}+\cos C_3\cos C_{15}$

$\cos\phi_{4,18}=\cos A_4\cos A_{18}+\cos B_4\cos B_{18}+\cos C_4\cos C_{18}$

$\cos\phi_{4,19}=\cos A_4\cos A_{19}+\cos B_4\cos B_{19}+\cos C_4\cos C_{19}$

$\cos\phi_{4,20}=\cos A_4\cos A_{20}+\cos B_4\cos B_{20}+\cos C_4\cos C_{20}$

$\cos\phi_{5,6}=\cos A_5\cos A_6+\cos B_5\cos B_6+\cos C_5\cos C_6$

$\cos\phi_{5,7}=\cos A_5\cos A_7+\cos B_5\cos B_7+\cos C_5\cos C_7$

B

ANGLES (Cont'd)

$\cos\phi_{5,8}=\cos A_5\cos A_8+\cos B_5\cos B_8+\cos C_5\cos C_8$

$\cos\phi_{5,11}=\cos A_5\cos A_{11}+\cos B_5\cos B_{11}+\cos C_5\cos C_{11}$

$\cos\phi_{5,12}=\cos A_5\cos A_{12}+\cos B_5\cos B_{12}+\cos C_5\cos C_{12}$

$\cos\phi_{6,7}=\cos A_6\cos A_7+\cos B_6\cos B_7+\cos C_6\cos C_7$

$\cos\phi_{6,13}=\cos A_6\cos A_{13}+\cos B_6\cos B_{13}+\cos C_6\cos C_{13}$

$\cos\phi_{6,16}=\cos A_6\cos A_{16}+\cos B_6\cos B_{16}+\cos C_6\cos C_{16}$

$\cos\phi_{6,17}=\cos A_6\cos A_{17}+\cos B_6\cos B_{17}+\cos C_6\cos C_{17}$

$\cos\phi_{7,18}=\cos A_7\cos A_{18}+\cos B_7\cos B_{18}+\cos C_7\cos C_{18}$

$\cos\phi_{7,21}=\cos A_7\cos A_{21}+\cos B_7\cos B_{21}+\cos C_7\cos C_{21}$

$\cos\phi_{7,22}=\cos A_7\cos A_{22}+\cos B_7\cos B_{22}+\cos C_7\cos C_{22}$

$\cos\phi_{8,9}=\cos A_8\cos A_9+\cos B_8\cos B_9+\cos C_8\cos C_9$

$\cos\phi_{8,10}=\cos A_8\cos A_{10}+\cos B_8\cos B_{190}+\cos C_8\cos C_{10}$

$\cos\phi_{8,11}=\cos A_8\cos A_{11}+\cos B_8\cos B_{11}+\cos C_8\cos C_{11}$

$\cos\phi_{8,12}=\cos A_8\cos A_{12}+\cos B_8\cos B_{12}+\cos C_8\cos C_{12}$

$\cos\phi_{9,10}=\cos A_9\cos A_{10}+\cos B_9\cos B_{10}+\cos C_9\cos C_{10}$

$\cos\phi_{9,19}=\cos A_9\cos A_{19}+\cos B_9\cos B_{19}+\cos C_9\cos C_{19}$

$\cos\phi_{9,23}=\cos A_9\cos A_{23}+\cos B_9\cos B_{23}+\cos C_9\cos C_{23}$

$\cos\phi_{9,24}=\cos A_9\cos A_{24}+\cos B_9\cos B_{24}+\cos C_9\cos C_{24}$

$\cos\phi_{10,14}=\cos A_{10}\cos A_{14}+\cos B_{10}\cos B_{14}+\cos C_{10}\cos C_{14}$

$\cos\phi_{10,26}=\cos A_{10}\cos A_{26}+\cos B_{10}\cos B_{26}+\cos C_{10}\cos C_{26}$

$\cos\phi_{10,27}=\cos A_{10}\cos A_{27}+\cos B_{10}\cos B_{27}+\cos C_{10}\cos C_{27}$

$\cos\phi_{11,12}=\cos A_{11}\cos A_{12}+\cos B_{11}\cos B_{12}+\cos C_{11}\cos C_{12}$

$\cos\phi_{11,21}=\cos A_{11}\cos A_{21}+\cos B_{11}\cos B_{21}+\cos C_{11}\cos C_{21}$

$\cos\phi_{11,23}=\cos A_{11}\cos A_{23}+\cos B_{11}\cos B_{23}+\cos C_{11}\cos C_{23}$

$\cos\phi_{11,25}=\cos A_{11}\cos A_{25}+\cos B_{11}\cos B_{25}+\cos C_{11}\cos C_{25}$

$\cos\phi_{12,16}=\cos A_{12}\cos A_{16}+\cos B_{12}\cos B_{16}+\cos C_{12}\cos C_{16}$

$\cos\phi_{12,26}=\cos A_{12}\cos A_{26}+\cos B_{12}\cos B_{26}+\cos C_{12}\cos C_{26}$

$\cos\phi_{12,28}=\cos A_{12}\cos A_{28}+\cos B_{12}\cos B_{28}+\cos C_{12}\cos C_{28}$

$\cos\phi_{13,14}=\cos A_{13}\cos A_{14}+\cos B_{13}\cos B_{14}+\cos C_{13}\cos C_{14}$

$\cos\phi_{13,15}=\cos A_{13}\cos A_{15}+\cos B_{13}\cos B_{15}+\cos C_{13}\cos C_{15}$

$\cos\phi_{13,16}=\cos A_{13}\cos A_{16}+\cos B_{13}\cos B_{16}+\cos C_{13}\cos C_{16}$

$\cos\phi_{13,17}=\cos A_{13}\cos A_{17}+\cos B_{13}\cos B_{17}+\cos C_{13}\cos C_{17}$

$\cos\phi_{14,15}=\cos A_{14}\cos A_{15}+\cos B_{14}\cos B_{15}+\cos C_{14}\cos C_{15}$

$\cos\phi_{14,26}=\cos A_{14}\cos A_{26}+\cos B_{14}\cos B_{26}+\cos C_{14}\cos C_{26}$

Table 22-13. Specifications and Analyses for the Hyperspace Object Shown in Fig. 22-13. (A) Specifications and Analysis in Hyperspace. (B) Secondary Analysis in XYZ Space. (C) Secondary Analysis in XYW Space. (D) Secondary Analysis in XZW Space. (E) Secondary Analysis in YZW Space. (F) Suggested Coordinates. **(Continued from page 465.)**

B

ANGLES (Cont'd)

$\cos\phi_{14,27}=\cos A_{14}\cos A_{27}+\cos B_{14}\cos B_{27}+\cos C_{14}\cos C_{27}$

$\cos\phi_{15,29}=\cos A_{15}\cos A_{29}+\cos B_{15}\cos B_{29}+\cos C_{15}\cos C_{29}$

$\cos\phi_{15,30}=\cos A_{15}\cos A_{30}+\cos B_{15}\cos B_{30}+\cos C_{15}\cos C_{30}$

$\cos\phi_{15,31}=\cos A_{15}\cos A_{31}+\cos B_{15}\cos B_{31}+\cos C_{15}\cos C_{31}$

$\cos\phi_{16,17}=\cos A_{16}\cos A_{17}+\cos B_{16}\cos B_{17}+\cos C_{16}\cos C_{17}$

$\cos\phi_{16,26}=\cos A_{16}\cos A_{26}+\cos B_{16}\cos B_{26}+\cos C_{16}\cos C_{26}$

$\cos\phi_{16,28}=\cos A_{16}\cos A_{28}+\cos B_{16}\cos B_{28}+\cos C_{16}\cos C_{28}$

$\cos\phi_{17,29}=\cos A_{17}\cos A_{29}+\cos B_{17}\cos B_{29}+\cos C_{17}\cos C_{29}$

$\cos\phi_{17,32}=\cos A_{17}\cos A_{32}+\cos B_{17}\cos B_{32}+\cos C_{17}\cos C_{32}$

$\cos\phi_{17,33}=\cos A_{17}\cos A_{33}+\cos B_{17}\cos B_{33}+\cos C_{17}\cos C_{33}$

$\cos\phi_{18,19}=\cos A_{18}\cos A_{19}+\cos B_{18}\cos B_{19}+\cos C_{18}\cos C_{19}$

$\cos\phi_{18,20}=\cos A_{18}\cos A_{20}+\cos B_{18}\cos B_{20}+\cos C_{178}\cos C_{20}$

$\cos\phi_{18,21}=\cos A_{18}\cos A_{21}+\cos B_{18}\cos B_{21}+\cos C_{18}\cos C_{21}$

$\cos\phi_{18,22}=\cos A_{18}\cos A_{22}+\cos B_{18}\cos B_{22}+\cos C_{18}\cos C_{22}$

$\cos\phi_{19,20}=\cos A_{19}\cos A_{20}+\cos B_{19}\cos B_{20}+\cos C_{19}\cos C_{20}$

$\cos\phi_{19,23}=\cos A_{19}\cos A_{23}+\cos B_{19}\cos B_{23}+\cos C_{19}\cos C_{23}$

$\cos\phi_{19,24}=\cos A_{19}\cos A_{24}+\cos B_{19}\cos B_{24}+\cos C_{19}\cos C_{24}$

$\cos\phi_{20,30}=\cos A_{20}\cos A_{30}+\cos B_{20}\cos B_{30}+\cos C_{20}\cos C_{30}$

$\cos\phi_{20,34}=\cos A_{20}\cos A_{34}+\cos B_{20}\cos B_{34}+\cos C_{20}\cos C_{34}$

$\cos\phi_{20,35}=\cos A_{20}\cos A_{35}+\cos B_{20}\cos B_{35}+\cos C_{20}\cos C_{35}$

$\cos\phi_{21,22}=\cos A_{21}\cos A_{22}+\cos B_{212}\cos B_{22}+\cos C_{21}\cos C_{22}$

$\cos\phi_{21,23}=\cos A_{21}\cos A_{23}+\cos B_{21}\cos B_{23}+\cos C_{21}\cos C_{23}$

$\cos\phi_{21,25}=\cos A_{21}\cos A_{25}+\cos B_{21}\cos B_{25}+\cos C_{21}\cos C_{25}$

$\cos\phi_{22,32}=\cos A_{22}\cos A_{32}+\cos B_{22}\cos B_{32}+\cos C_{22}\cos C_{32}$

$\cos\phi_{22,34}=\cos A_{22}\cos A_{34}+\cos B_{22}\cos B_{34}+\cos C_{22}\cos C_{34}$

$\cos\phi_{22,36}=\cos A_{22}\cos A_{36}+\cos B_{22}\cos B_{36}+\cos C_{22}\cos C_{36}$

$\cos\phi_{23,24}=\cos A_{23}\cos A_{24}+\cos B_{23}\cos B_{24}+\cos C_{23}\cos C_{24}$

$\cos\phi_{23,25}=\cos A_{23}\cos A_{25}+\cos B_{23}\cos B_{25}+\cos C_{23}\cos C_{25}$

$\cos\phi_{24,35}=\cos A_{24}\cos A_{35}+\cos B_{24}\cos B_{35}+\cos C_{24}\cos C_{35}$

$\cos\phi_{24,37}=\cos A_{24}\cos A_{37}+\cos B_{24}\cos B_{37}+\cos C_{24}\cos C_{37}$

$\cos\phi_{24,37}=\cos A_{24}\cos A_{37}+\cos B_{24}\cos B_{37}+\cos C_{24}\cos C_{37}$

$\cos\phi_{25,36}=\cos A_{25}\cos A_{36}+\cos B_{25}\cos B_{36}+\cos C_{25}\cos C_{36}$

$\cos\phi_{25,37}=\cos A_{25}\cos A_{37}+\cos B_{25}\cos B_{37}+\cos C_{25}\cos C_{37}$

ANGLES (Cont'd)

$\cos\phi_{25,39}=\cos A_{25}\cos A_{39}+\cos B_{25}\cos B_{39}+\cos C_{25}\cos C_{39}$

$\cos\phi_{26,27}=\cos A_{26}\cos A_{27}+\cos B_{26}\cos B_{27}+\cos C_{26}\cos C_{27}$

$\cos\phi_{26,28}=\cos A_{26}\cos A_{28}+\cos B_{26}\cos B_{28}+\cos C_{26}\cos C_{28}$

$\cos\phi_{27,31}=\cos A_{27}\cos A_{31}+\cos B_{27}\cos B_{31}+\cos C_{27}\cos C_{31}$

$\cos\phi_{27,38}=\cos A_{27}\cos A_{38}+\cos B_{27}\cos B_{38}+\cos C_{27}\cos C_{38}$

$\cos\phi_{27,40}=\cos A_{27}\cos A_{40}+\cos B_{27}\cos B_{40}+\cos C_{27}\cos C_{40}$

$\cos\phi_{28,33}=\cos A_{28}\cos A_{33}+\cos B_{28}\cos C_{33}+\cos C_{28}\cos D_{33}$

$\cos\phi_{28,39}=\cos A_{28}\cos A_{39}+\cos B_{28}\cos B_{39}+\cos C_{28}\cos C_{39}$

$\cos\phi_{28,40}=\cos A_{28}\cos A_{40}+\cos B_{28}\cos B_{40}+\cos C_{28}\cos C_{40}$

$\cos\phi_{29,31}=\cos A_{29}\cos A_{31}+\cos B_{29}\cos B_{31}+\cos C_{29}\cos C_{31}$

$\cos\phi_{29,31}=\cos A_{29}\cos A_{31}+\cos B_{29}\cos B_{31}+\cos C_{29}\cos C_{31}$

$\cos\phi_{29,32}=\cos A_{29}\cos A_{32}+\cos B_{29}\cos B_{32}+\cos C_{29}\cos C_{32}$

$\cos\phi_{29,33}=\cos A_{29}\cos A_{33}+\cos B_{29}\cos B_{33}+\cos C_{29}\cos C_{33}$

$\cos\phi_{30,31}=\cos A_{30}\cos A_{31}+\cos B_{30}\cos B_{31}+\cos C_{30}\cos C_{31}$

$\cos\phi_{30,34}=\cos A_{30}\cos A_{34}+\cos B_{30}\cos B_{34}+\cos C_{30}\cos C_{34}$

$\cos\phi_{30,35}=\cos A_{30}\cos A_{35}+\cos B_{30}\cos B_{35}+\cos C_{30}\cos C_{35}$

$\cos\phi_{31,38}=\cos A_{31}\cos A_{38}+\cos B_{31}\cos B_{38}+\cos C_{31}\cos C_{38}$

$\cos\phi_{31,40}=\cos A_{31}\cos A_{40}+\cos B_{31}\cos B_{40}+\cos C_{31}\cos C_{40}$

$\cos\phi_{32,33}=\cos A_{32}\cos A_{33}+\cos B_{32}\cos B_{33}+\cos C_{32}\cos C_{33}$

$\cos\phi_{32,34}=\cos A_{32}\cos A_{34}+\cos B_{32}\cos B_{34}+\cos C_{32}\cos C_{34}$

$\cos\phi_{32,36}=\cos A_{32}\cos A_{36}+\cos B_{32}\cos B_{36}+\cos C_{32}\cos C_{36}$

$\cos\phi_{33,39}=\cos A_{33}\cos A_{39}+\cos B_{33}\cos B_{39}+\cos C_{33}\cos C_{39}$

$\cos\phi_{33,40}=\cos A_{33}\cos A_{40}+\cos B_{33}\cos B_{40}+\cos C_{33}\cos C_{40}$

$\cos\phi_{34,35}=\cos A_{34}\cos A_{35}+\cos B_{34}\cos B_{35}+\cos C_{34}\cos C_{35}$

$\cos\phi_{34,36}=\cos A_{34}\cos A_{36}+\cos B_{34}\cos B_{36}+\cos C_{34}\cos C_{36}$

$\cos\phi_{35,37}=\cos A_{35}\cos A_{37}+\cos B_{35}\cos B_{37}+\cos C_{35}\cos C_{37}$

$\cos\phi_{35,38}=\cos A_{35}\cos A_{38}+\cos B_{35}\cos B_{38}+\cos C_{35}\cos C_{38}$

$\cos\phi_{36,37}=\cos A_{36}\cos A_{37}+\cos B_{36}\cos B_{37}+\cos C_{36}\cos C_{37}$

$\cos\phi_{36,39}=\cos A_{36}\cos A_{39}+\cos B_{36}\cos B_{39}+\cos C_{36}\cos C_{39}$

$\cos\phi_{37,38}=\cos A_{37}\cos A_{38}+\cos B_{37}\cos B_{38}+\cos C_{37}\cos C_{38}$

$\cos\phi_{37,39}=\cos A_{37}\cos A_{39}+\cos B_{37}\cos B_{39}+\cos C_{37}\cos C_{39}$

$\cos\phi_{38,40}=\cos A_{38}\cos A_{40}+\cos B_{38}\cos B_{40}+\cos C_{38}\cos C_{40}$

$\cos\phi_{39,40}=\cos A_{39}\cos A_{40}+\cos B_{39}\cos B_{40}+\cos C_{39}\cos C_{40}$

CHECK

B

F_1: $\phi_{1,2}+\phi_{1,5}+\phi_{2,8}+\phi_{5,8}=360^\circ$

F_2: $\phi_{1,3}+\phi_{1,6}+\phi_{3,13}+\phi_{6,13}=360^\circ$

F_3: $\phi_{1,4}+\phi_{1,7}+\phi_{4,18}+\phi_{7,18}=360^\circ$

F_4: $\phi_{2,3}+\phi_{2,10}+\phi_{3,14}+\phi_{10,14}=360^\circ$

F_5: $\phi_{2,4}+\phi_{2,9}+\phi_{4,19}+\phi_{9,19}=360^\circ$

F_6: $\phi_{3,4}+\phi_{3,15}+\phi_{4,20}+\phi_{15,30}+\phi_{20,30}=550^\circ$

F_7: $\phi_{5,6}+\phi_{5,12}+\phi_{6,16}+\phi_{12,16}=360^\circ$

F_8: $\phi_{5,7}+\phi_{5,11}+\phi_{7,21}+\phi_{11,21}=360^\circ$

F_9: $\phi_{6,7}+\phi_{6,17}+\phi_{7,22}+\phi_{17,32}+\phi_{22,32}=550^\circ$

F_{10}: $\phi_{8,9}+\phi_{8,11}+\phi_{9,23}+\phi_{11,23}=360^\circ$

F_{11}: $\phi_{8,10}+\phi_{8,12}+\phi_{10,26}+\phi_{12,26}=360^\circ$

F_{12}: $\phi_{9,10}+\phi_{9,24}+\phi_{10,27}+\phi_{24,38}+\phi_{27,38}=550^\circ$

F_{13}: $\phi_{11,12}+\phi_{11,25}+\phi_{12,28}+\phi_{25,39}+\phi_{28,39}=550^\circ$

F_{14}: $\phi_{13,14}+\phi_{13,16}+\phi_{14,26}+\phi_{16,26}=360^\circ$

F_{15}: $\phi_{13,15}+\phi_{13,17}+\phi_{15,29}+\phi_{17,29}=360^\circ$

F_{16}: $\phi_{14,15}+\phi_{14,27}+\phi_{15,31}+\phi_{27,31}=360^\circ$

F_{17}: $\phi_{16,17}+\phi_{16,28}+\phi_{17,33}+\phi_{28,33}=360^\circ$

F_{18}: $\phi_{18,19}+\phi_{18,21}+\phi_{19,23}+\phi_{21,23}=360^\circ$

F_{19}: $\phi_{18,20}+\phi_{18,22}+\phi_{20,34}+\phi_{22,34}=360^\circ$

F_{20}: $\phi_{19,20}+\phi_{19,24}+\phi_{20,35}+\phi_{24,35}=360^\circ$

F_{21}: $\phi_{21,22}+\phi_{21,25}+\phi_{22,36}+\phi_{25,36}=360^\circ$

F_{22}: $\phi_{23,24}+\phi_{23,25}+\phi_{24,37}+\phi_{25,37}=360^\circ$

F_{23}: $\phi_{26,27}+\phi_{26,28}+\phi_{27,40}+\phi_{28,40}=360^\circ$

F_{24}: $\phi_{29,30}+\phi_{29,32}+\phi_{30,34}+\phi_{32,34}=360^\circ$

F_{25}: $\phi_{29,31}+\phi_{29,33}+\phi_{31,34}+\phi_{33,34}=360^\circ$

F_{26}: $\phi_{30,31}+\phi_{30,35}+\phi_{31,38}+\phi_{35,38}=360^\circ$

F_{27}: $\phi_{32,33}+\phi_{32,36}+\phi_{33,39}+\phi_{36,39}=360^\circ$

F_{28}: $\phi_{34,35}+\phi_{34,36}+\phi_{35,37}+\phi_{36,36}=360^\circ$

F_{29}: $\phi_{37,38}+\phi_{37,39}+\phi_{38,40}+\phi_{39,40}=360^\circ$

SECONDARY ANALYSIS -- XYW SPACE (See Fig. 22-13C) **C**

PLOTTED POINTS

$P_1=(x_1,y_1,w_1)$	$P_2=(x_2,y_2,w_2)$
$P_3=(x_3,y_3,w_3)$	$P_4=(x_4,y_4,w_4)$
$P_5=(x_5,y_5,w_5)$	$P_6=(x_6,y_6,w_6)$
$P_7=(x_7,y_7,w_7)$	$P_8=(x_8,y_8,w_8)$
$P_9=(x_9,y_9,w_9)$	$P_{10}=(x_{10},y_{10},w_{10})$
$P_{11}=(x_{11},y_{11},w_{11})$	$P_{12}=(x_{12},y_{12},w_{12})$
$P_{13}=(x_{13},y_{13},w_{13})$	$P_{14}=(x_{14},y_{14},w_{14})$
$P_{15}=(x_{15},y_{15},w_{15})$	$P_{16}=(x_{16},y_{16},w_{16})$
$P_{17}=(x_{17},y_{17},w_{17})$	$P_{18}=(x_{18},y_{18},w_{18})$
$P_{19}=(x_{18},y_{19},w_{19})$	$P_{20}=(x_{20},y_{20},w_{20})$

LENGTHS

$$s_1=\sqrt{(x_2-x_1)^2+(y_2-y_1)^2+(w_2-w_1)^2}$$

$$s_2=\sqrt{(x_3-x_1)^2+(y_3-y_1)^2+(w_3-w_1)^2}$$

LENGTHS (Cont'd)

$$s_3=\sqrt{(x_5-x_1)^2+(y_5-y_1)^2+(w_5-w_1)^2}$$

$$s_4=\sqrt{(x_7-x_1)^2+(y_7-y_1)^2+(w_7-w_1)^2}$$

$$s_5=\sqrt{(x_4-x_2)^2+(y_4-y_2)^2+(w_4-w_2)^2}$$

$$s_6=\sqrt{(x_6-x_2)^2+(y_6-y_2)^2+(w_6-w_2)^2}$$

$$s_7=\sqrt{(x_8-x_2)^2+(y_8-y_2)^2+(w_8-w_2)^2}$$

$$s_8=\sqrt{(x_4-x_3)^2+(y_4-y_3)^2+(w_4-w_3)^2}$$

$$s_9=\sqrt{(x_9-x_3)^2+(y_9-y_3)^2+(w_9-w_3)^2}$$

$$s_{10}=\sqrt{(x_{11}-x_3)^2+(y_{11}-y_3)^2+(w_{11}-w_3)^2}$$

$$s_{11}=\sqrt{(x_{10}-x_4)^2+(y_{10}-y_4)^2+(w_{10}-w_4)^2}$$

$$s_{12}=\sqrt{(x_{12}-x_4)^2+(y_{12}-y_4)^2+(w_{12}-w_4)^2}$$

$$s_{13}=\sqrt{(x_6-x_5)^2+(y_6-y_5)^2+(w_6-w_5)^2}$$

Table 22-13. Specifications and Analyses for the Hyperspace Object Shown in Fig. 22-13. (A) Specifications and Analysis in Hyperspace. (B) Secondary Analysis in XYZ Space. (C) Secondary Analysis in XYW Space. (D) Secondary Analysis in XZW Space. (E) Secondary Analysis in YZW Space. (F) Suggested Coordinates. **(Continued from page 467.)**

C

LENGTHS (Cont'd)

$$s_{14}=\sqrt{(x_{11}-x_5)^2+(y_{11}-y_5)^2+(w_{11}-w_5)^2}$$

$$s_{15}=\sqrt{(x_{13}-x_5)^2+(y_{13}-y_5)^2+(w_{13}-w_5)^2}$$

$$s_{16}=\sqrt{(x_{12}-x_6)^2+(y_{12}-y_6)^2+(w_{12}-w_6)^2}$$

$$s_{17}=\sqrt{(x_{14}-x_6)^2+(y_{14}-y_6)^2+(w_{14}-w_6)^2}$$

$$s_{18}=\sqrt{(x_8-x_7)^2+(y_8-y_7)^2+(w_8-w_7)^2}$$

$$s_{19}=\sqrt{(x_9-x_7)^2+(y_9-y_7)^2+(w_9-w_7)^2}$$

$$s_{20}=\sqrt{(x_{15}-x_7)^2+(y_{15}-y_7)^2+(w_{15}-w_7)^2}$$

$$s_{21}=\sqrt{(x_{10}-x_8)^2+(y_{10}-y_8)^2+(w_{10}-w_8)^2}$$

$$s_{22}=\sqrt{(x_{16}-x_8)^2+(y_{16}-y_8)^2+(w_{16}-w_8)^2}$$

$$s_{23}=\sqrt{(x_{10}-x_9)^2+(y_{10}-y_9)^2+(w_{10}-w_9)^2}$$

$$s_{24}=\sqrt{(x_{17}-x_9)^2+(y_{17}-y_9)^2+(w_{17}-w_9)^2}$$

$$s_{25}=\sqrt{(x_{18}-x_{10})^2+(y_{18}-y_{10})^2+(w_{18}-w_{10})^2}$$

$$s_{26}=\sqrt{(x_{12}-x_{11})^2+(y_{12}-y_{11})^2+(w_{12}-w_{11})^2}$$

$$s_{27}=\sqrt{(x_{19}-x_{11})^2+(y_{19}-y_{11})^2+(w_{19}-w_{11})^2}$$

$$s_{28}=\sqrt{(x_{20}-x_{12})^2+(y_{20}-y_{12})^2+(w_{20}-w_{12})^2}$$

$$s_{29}=\sqrt{(x_{14}-x_{13})^2+(y_{14}-y_{13})^2+(w_{14}-w_{13})^2}$$

$$s_{30}=\sqrt{(x_{15}-x_{13})^2+(y_{15}-y_{13})^2+(w_{15}-w_{13})^2}$$

$$s_{31}=\sqrt{(x_{19}-x_{13})^2+(y_{19}-y_{13})^2+(w_{19}-w_{13})^2}$$

$$s_{32}=\sqrt{(x_{16}-x_{14})^2+(y_{16}-y_{14})^2+(w_{16}-w_{14})^2}$$

$$s_{33}=\sqrt{(x_{20}-x_{14})^2+(y_{20}-y_{14})^2+(w_{20}-w_{14})^2}$$

$$s_{34}=\sqrt{(x_{16}-x_{15})^2+(y_{16}-y_{15})^2+(w_{16}-w_{15})^2}$$

$$s_{35}=\sqrt{(x_{17}-x_{15})^2+(y_{17}-y_{15})^2+(w_{17}-w_{15})^2}$$

$$s_{36}=\sqrt{(x_{18}-x_{16})^2+(y_{18}-y_{16})^2+(w_{18}-w_{16})^2}$$

$$s_{37}=\sqrt{(x_{18}-x_{17})^2+(y_{18}-y_{17})^2+(w_{18}-w_{17})^2}$$

$$s_{38}=\sqrt{(x_{19}-x_{17})^2+(y_{19}-y_{17})^2+(w_{19}-w_{17})^2}$$

$$s_{39}=\sqrt{(x_{20}-x_{18})^2+(y_{20}-y_{18})^2+(w_{20}-w_{18})^2}$$

$$s_{40}=\sqrt{(x_{20}-x_{19})^2+(y_{20}-y_{19})^2+(w_{20}-w_{19})^2}$$

DIRECTION COSINES

$\cos A_1=(x_2-x_1)/s_1$ $\cos B_1=(y_2-y_1)/s_1$
$\cos D_1=(w_2-w_1)/s_1$

$\cos A_2=(x_3-x_1)/s_2$ $\cos B_2=(y_3-y_1)/s_2$
$\cos D_2=(w_3-w_1)/s_2$

$\cos A_3=(x_5-x_1)/s_3$ $\cos B_3=(y_5-y_1)/s_3$
$\cos D_3=(w_5-w_1)/s_3$

$\cos A_4=(x_7-x_1)/s_4$ $\cos B_4=(y_7-y_1)/s_4$
$\cos D_4=(w_7-w_1)/s_4$

$\cos A_5=(x_4-x_2)/s_5$ $\cos B_5=(y_4-y_2)/s_5$
$\cos D_5=(w_4-w_2)/s_5$

$\cos A_6=(x_6-x_2)/s_6$ $\cos B_6=(y_6-y_2)/s_6$
$\cos D_6=(w_6-w_2)/s_6$

$\cos A_7=(x_8-x_2)/s_7$ $\cos B_7=(y_8-y_2)/s_7$
$\cos D_7=(w_8-w_2)/s_7$

$\cos A_8=(x_4-x_3)/s_8$ $\cos B_8=(y_4-y_3)/s_8$
$\cos D_8=(w_4-w_3)/s_8$

$\cos A_9=(x_9-x_3)/s_9$ $\cos B_9=(y_9-y_3)/s_9$
$\cos D_9=(w_9-w_3)/s_9$

$\cos A_{10}=(x_{11}-x_3)/s_{10}$ $\cos B_{10}=(y_{11}-y_3)/s_{10}$
$\cos D_{10}=(w_{11}-w_3)/s_{10}$

DIRECTION COSINES (Cont'd)

$\cos A_{11}=(x_{10}-x_4)/s_{11}$ $\cos B_{11}=(y_{10}-y_4)/s_{11}$

$\cos D_{11}=(w_{10}-w_4)/s_{11}$

$\cos A_{12}=(x_{12}-x_4)/s_{12}$ $\cos B_{12}=(y_{12}-y_4)/s_{12}$

$\cos D_{12}=(w_{12}-w_4)/s_{12}$

$\cos A_{13}=(x_6-x_5)/s_{13}$ $\cos B_{13}=(y_6-y_5)/s_{13}$

$\cos D_{13}=(w_6-w_5)/s_{13}$

$\cos A_{14}=(x_{11}-x_5)/s_{14}$ $\cos B_{14}=(y_{11}-y_5)/s_{14}$

$\cos D_{14}=(w_{11}-w_5)/s$

$\cos A_{15}=(x_{13}-x_5)/s_{15}$ $\cos B_{15}=(y_{13}-y_5)/s_{15}$

$\cos D_{15}=(w_{13}-w_5)/s_{15}$

$\cos A_{16}=(x_{12}-x_6)/s_{16}$ $\cos B_{16}=(y_{12}-y_6)/s_{16}$

$\cos D_{16}=(w_{12}-w_6)/s_{16}$

$\cos A_{17}=(x_{14}-x_6)/s_{17}$ $\cos B_{17}=(y_{14}-y_6)/s_{17}$

$\cos D_{17}=(w_{14}-w_6)/s_{17}$

$\cos A_{18}=(x_8-x_7)/s_{18}$ $\cos B_{18}=(y_8-y_7)/s_{18}$

$\cos D_{18}=(w_8-w_7)/s_{18}$

$\cos A_{19}=(x_9-x_7)/s_{19}$ $\cos B_{19}=(y_9-y_7)/s_{19}$

$\cos D_{19}=(w_9-w_7)/s_{19}$

$\cos A_{20}=(x_{15}-x_7)/s_{20}$ $\cos B_{20}=(y_{15}-y_7)/s_{20}$

$\cos D_{20}=(w_{15}-w_7)/s_{20}$

$\cos A_{21}=(x_{10}-x_8)/s_{21}$ $\cos B_{21}=(y_{10}-y_8)/s_{21}$

$\cos D_{21}=(w_{10}-w_8)/s_{21}$

$\cos A_{22}=(x_{16}-x_8)/s_{22}$ $\cos B_{22}=(y_{16}-y_8)/s_{22}$

$\cos D_{22}=(w_{16}-w_8)/s_{22}$

$\cos A_{23}=(x_{10}-x_9)/s_{23}$ $\cos B_{23}=(y_{10}-y_9)/s_{23}$

$\cos D_{23}=(w_{10}-w_9)/s_{23}$

C

DIRECTION COSINES (Cont'd)

$\cos A_{24}=(x_{17}-x_9)/s_{24}$ $\cos B_{24}=(y_{17}-y_9)/s_{24}$

$\cos D_{24}=(w_{17}-w_9)/s_{24}$

$\cos A_{25}=(x_{18}-x_{10})/s_{25}$ $\cos B_{25}=(y_{18}-y_{10})/s_{25}$

$\cos D_{25}=(w_{18}-w_{10})/s_{25}$

$\cos A_{26}=(x_{12}-x_{11})/s_{26}$ $\cos B_{26}=(y_{12}-y_{11})/s_{26}$

$\cos D_{26}=(w_{12}-w_{11})/s_{26}$

$\cos A_{27}=(x_{19}-x_{11})/s_{27}$ $\cos B_{27}=(y_{19}-y_{11})/s_{27}$

$\cos D_{27}=(w_{19}-w_{11})/s_{27}$

$\cos A_{28}=(x_{20}-x_{12})/s_{28}$ $\cos B_{28}=(y_{20}-y_{12})/s_{28}$

$\cos D_{28}=(w_{20}-w_{12})/s_{28}$

$\cos A_{29}=(x_{14}-x_{13})/s_{29}$ $\cos B_{29}=(y_{14}-y_{13})/s_{29}$

$\cos D_{29}=(w_{14}-w_{13})/s_{29}$

$\cos A_{30}=(x_{15}-x_{13})/s_{30}$ $\cos B_{30}=(y_{15}-y_{13})/s_{30}$

$\cos D_{30}=(w_{15}-w_{13})/s_{30}$

$\cos A_{31}=(x_{19}-x_{13})/s_{31}$ $\cos B_{31}=(y_{19}-y_{13})/s_{31}$

$\cos D_{31}=(w_{19}-w_{13})/s_{31}$

$\cos A_{32}=(x_{16}-x_{14})/s_{32}$ $\cos B_{32}=(y_{16}-y_{14})/s_{32}$

$\cos D_{32}=(w_{16}-w_{14})/s_{32}$

$\cos A_{33}=(x_{20}-x_{14})/s_{33}$ $\cos B_{33}=(y_{20}-y_{14})/s_{33}$

$\cos D_{33}=(w_{20}-w_{14})/s_{33}$

$\cos A_{34}=(x_{16}-x_{15})/s_{34}$ $\cos B_{34}=(y_{16}-y_{15})/s_{34}$

$\cos D_{34}=(w_{16}-w_{15})/s_{34}$

$\cos A_{35}=(x_{17}-x_{15})/s_{35}$ $\cos B_{35}=(y_{17}-y_{15})/s_{35}$

$\cos D_{35}=(w_{17}-w_{15})/s_{35}$

$\cos A_{36}=(x_{18}-x_{16})/s_{36}$ $\cos B_{36}=(y_{18}-y_{16})/s_{36}$

$\cos D_{36}=(w_{18}-w_{16})/s_{36}$

Table 22-13. Specifications and Analyses for the Hyperspace Object Shown in Fig. 22-13. (A) Specifications and Analysis in Hyperspace. (B) Secondary Analysis in XYZ Space. (C) Secondary Analysis in XYW Space. (D) Secondary Analysis in XZW Space. (E) Secondary Analysis in YZW Space. (F) Suggested Coordinates. (Continued from page 469.)

DIRECTION COSINES (Cont'd)

$\cos A_{37}=(x_{18}-x_{17})/s_{37}$ $\cos B_{37}=(y_{18}-y_{17})/s_{37}$

$\cos D_{37}=(w_{18}-w_{17})/s_{37}$

$\cos A_{38}=(x_{19}-x_{17})/s_{38}$ $\cos B_{38}=(y_{19}-y_{17})/s_{38}$

$\cos D_{38}=(w_{19}-w_{17})/s_{38}$

$\cos A_{39}=(x_{20}-x_{18})/s_{39}$ $\cos B_{39}=(y_{20}-y_{18})/s_{39}$

$\cos D_{39}=(w_{20}-w_{18})/s_{39}$

$\cos A_{40}=(x_{20}-x_{19})/s_{40}$ $\cos B_{40}=(y_{20}-y_{19})/s_{40}$

$\cos D_{40}=(w_{20}-w_{19})/s_{40}$

ANGLES

$\cos\phi_{1,2}=\cos A_1\cos A_2+\cos B_1\cos B_2+\cos D_1\cos D_2$

$\cos\phi_{1,3}=\cos A_1\cos A_3+\cos B_1\cos B_3+\cos D_1\cos D_3$

$\cos\phi_{1,4}=\cos A_1\cos A_4+\cos B_1\cos B_4+\cos D_1\cos D_4$

$\cos\phi_{1,5}=\cos A_1\cos A_5+\cos B_1\cos B_5+\cos D_1\cos D_5$

$\cos\phi_{1,6}=\cos A_1\cos A_6+\cos B_1\cos B_6+\cos D_1\cos D_6$

$\cos\phi_{1,7}=\cos A_1\cos A_7+\cos B_1\cos B_7+\cos D_1\cos D_7$

$\cos\phi_{2,3}=\cos A_2\cos A_3+\cos B_2\cos B_3+\cos D_2\cos D_3$

$\cos\phi_{2,4}=\cos A_2\cos A_4+\cos B_2\cos B_4+\cos D_2\cos D_4$

$\cos\phi_{2,8}=\cos A_2\cos A_8+\cos B_2\cos B_8+\cos D_2\cos D_8$

$\cos\phi_{2,9}=\cos A_2\cos A_9+\cos B_2\cos B_9+\cos D_2\cos D_9$

$\cos\phi_{2,10}=\cos A_2\cos A_{10}+\cos B_2\cos B_{10}+\cos D_2\cos D_{10}$

$\cos\phi_{3,4}=\cos A_3\cos A_4+\cos B_3\cos B_4+\cos D_3\cos D_4$

$\cos\phi_{3,13}=\cos A_3\cos A_{13}+\cos B_3\cos B_{13}+\cos D_3\cos D_{13}$

$\cos\phi_{3,14}=\cos A_3\cos A_{14}+\cos B_3\cos B_{14}+\cos D_3\cos D_{14}$

$\cos\phi_{3,15}=\cos A_3\cos A_{15}+\cos B_3\cos B_{15}+\cos D_3\cos D_{15}$

$\cos\phi_{4,18}=\cos A_4\cos A_{18}+\cos B_4\cos B_{18}+\cos D_4\cos D_{18}$

$\cos\phi_{4,19}=\cos A_4\cos A_{19}+\cos B_4\cos B_{19}+\cos D_4\cos D_{19}$

$\cos\phi_{4,20}=\cos A_4\cos A_{20}+\cos B_4\cos B_{20}+\cos D_4\cos D_{20}$

$\cos\phi_{5,6}=\cos A_5\cos A_6+\cos B_5\cos B_6+\cos D_5\cos D_6$

$\cos\phi_{5,7}=\cos A_5\cos A_7+\cos B_5\cos B_7+\cos D_5\cos D_7$

$\cos\phi_{5,8}=\cos A_5\cos A_8+\cos B_5\cos B_8+\cos D_5\cos D_8$

$\cos\phi_{5,11}=\cos A_5\cos A_{11}+\cos B_5\cos B_{11}+\cos D_5\cos D_{11}$

C

ANGLES (Cont'd)

$\cos\phi_{5,12}=\cos A_5\cos A_{12}+\cos B_5\cos B_{12}+\cos D_5\cos D_{12}$

$\cos\phi_{6,7}=\cos A_6\cos A_7+\cos B_6\cos B_7+\cos D_6\cos D_7$

$\cos\phi_{6,13}=\cos A_6\cos A_{13}+\cos B_6\cos B_{13}+\cos D_6\cos D_{13}$

$\cos\phi_{6,16}=\cos A_6\cos A_{16}+\cos B_6\cos B_{16}+\cos D_6\cos D_{16}$

$\cos\phi_{6,17}=\cos A_6\cos A_{17}+\cos B_6\cos B_{17}+\cos D_6\cos D_{17}$

$\cos\phi_{7,18}=\cos A_7\cos A_{18}+\cos B_7\cos B_{18}+\cos D_7\cos D_{18}$

$\cos\phi_{7,21}=\cos A_7\cos A_{21}+\cos B_7\cos B_{21}+\cos D_7\cos D_{21}$

$\cos\phi_{7,22}=\cos A_7\cos A_{22}+\cos B_7\cos B_{22}+\cos D_7\cos D_{22}$

$\cos\phi_{8,9}=\cos A_8\cos A_9+\cos B_8\cos B_9+\cos D_8\cos D_9$

$\cos\phi_{8,10}=\cos A_8\cos A_{10}+\cos B_8\cos B_{190}+\cos D_8\cos D_{10}$

$\cos\phi_{8,11}=\cos A_8\cos A_{11}+\cos B_8\cos B_{11}+\cos D_8\cos D_{11}$

$\cos\phi_{8,12}=\cos A_8\cos A_{12}+\cos B_8\cos B_{12}+\cos D_8\cos D_{12}$

$\cos\phi_{9,10}=\cos A_9\cos A_{10}+\cos B_9\cos B_{10}+\cos D_9\cos D_{10}$

$\cos\phi_{9,19}=\cos A_9\cos A_{19}+\cos B_9\cos B_{19}+\cos D_9\cos D_{19}$

$\cos\phi_{9,23}=\cos A_9\cos A_{23}+\cos B_9\cos B_{23}+\cos D_9\cos D_{23}$

$\cos\phi_{9,24}=\cos A_9\cos A_{24}+\cos B_9\cos B_{24}+\cos D_9\cos D_{24}$

$\cos\phi_{10,14}=\cos A_{10}\cos A_{14}+\cos B_{10}\cos B_{14}+\cos D_{10}\cos D_{14}$

$\cos\phi_{10,26}=\cos A_{10}\cos A_{26}+\cos B_{10}\cos B_{26}+\cos D_{10}\cos D_{26}$

$\cos\phi_{10,27}=\cos A_{10}\cos A_{27}+\cos B_{10}\cos B_{27}+\cos D_{10}\cos D_{27}$

$\cos\phi_{11,12}=\cos A_{11}\cos A_{12}+\cos B_{11}\cos B_{12}+\cos D_{11}\cos D_{12}$

$\cos\phi_{11,21}=\cos A_{11}\cos A_{21}+\cos B_{11}\cos B_{21}+\cos D_{11}\cos D_{21}$

$\cos\phi_{11,23}=\cos A_{11}\cos A_{23}+\cos B_{11}\cos B_{23}+\cos D_{11}\cos D_{23}$

$\cos\phi_{11,25}=\cos A_{11}\cos A_{25}+\cos B_{11}\cos B_{25}+\cos D_{11}\cos D_{25}$

$\cos\phi_{12,16}=\cos A_{12}\cos A_{16}+\cos B_{12}\cos B_{16}+\cos D_{12}\cos D_{16}$

$\cos\phi_{12,26}=\cos A_{12}\cos A_{26}+\cos B_{12}\cos B_{26}+\cos D_{12}\cos D_{26}$

$\cos\phi_{12,28}=\cos A_{12}\cos A_{28}+\cos B_{12}\cos B_{28}+\cos D_{12}\cos D_{28}$

$\cos\phi_{13,14}=\cos A_{13}\cos A_{14}+\cos B_{13}\cos B_{14}+\cos D_{13}\cos D_{14}$

$\cos\phi_{13,15}=\cos A_{13}\cos A_{15}+\cos B_{13}\cos B_{15}+\cos D_{13}\cos D_{15}$

$\cos\phi_{13,16}=\cos A_{13}\cos A_{16}+\cos B_{13}\cos B_{16}+\cos D_{13}\cos D_{16}$

$\cos\phi_{13,17}=\cos A_{13}\cos A_{17}+\cos B_{13}\cos B_{17}+\cos D_{13}\cos D_{17}$

$\cos\phi_{14,15}=\cos A_{14}\cos A_{15}+\cos B_{14}\cos B_{15}+\cos D_{14}\cos D_{15}$

$\cos\phi_{14,26}=\cos A_{14}\cos A_{26}+\cos B_{14}\cos B_{26}+\cos D_{14}\cos D_{26}$

$\cos\phi_{14,27}=\cos A_{14}\cos A_{27}+\cos B_{14}\cos B_{27}+\cos D_{14}\cos D_{27}$

ANGLES (Cont'd)

$\cos\phi_{15,29}=\cos A_{15}\cos A_{29}+\cos B_{15}\cos B_{29}+\cos D_{15}\cos D_{29}$

$\cos\phi_{15,30}=\cos A_{15}\cos A_{30}+\cos B_{15}\cos B_{30}+\cos D_{15}\cos D_{30}$

$\cos\phi_{15,31}=\cos A_{15}\cos A_{31}+\cos B_{15}\cos B_{31}+\cos D_{15}\cos D_{31}$

$\cos\phi_{16,17}=\cos A_{16}\cos A_{17}+\cos B_{16}\cos B_{17}+\cos D_{16}\cos D_{17}$

$\cos\phi_{16,26}=\cos A_{16}\cos A_{26}+\cos B_{16}\cos B_{26}+\cos D_{16}\cos D_{26}$

$\cos\phi_{16,28}=\cos A_{16}\cos A_{28}+\cos B_{16}\cos B_{28}+\cos D_{16}\cos D_{28}$

$\cos\phi_{17,29}=\cos A_{17}\cos A_{29}+\cos B_{17}\cos B_{29}+\cos D_{17}\cos D_{29}$

$\cos\phi_{17,32}=\cos A_{17}\cos A_{32}+\cos B_{17}\cos B_{32}+\cos D_{17}\cos D_{32}$

$\cos\phi_{17,33}=\cos A_{17}\cos A_{33}+\cos B_{17}\cos B_{33}+\cos D_{17}\cos D_{33}$

$\cos\phi_{18,19}=\cos A_{18}\cos A_{19}+\cos B_{18}\cos B_{19}+\cos D_{18}\cos D_{19}$

$\cos\phi_{18,20}=\cos A_{18}\cos A_{20}+\cos B_{18}\cos B_{20}+\cos D_{178}\cos D_{20}$

$\cos\phi_{18,21}=\cos A_{18}\cos A_{21}+\cos B_{18}\cos B_{21}+\cos D_{18}\cos D_{21}$

$\cos\phi_{18,22}=\cos A_{18}\cos A_{22}+\cos B_{18}\cos B_{22}+\cos D_{18}\cos D_{22}$

$\cos\phi_{19,20}=\cos A_{19}\cos A_{20}+\cos B_{19}\cos B_{20}+\cos D_{19}\cos D_{20}$

$\cos\phi_{19,23}=\cos A_{19}\cos A_{23}+\cos B_{19}\cos B_{23}+\cos D_{19}\cos D_{23}$

$\cos\phi_{19,24}=\cos A_{19}\cos A_{24}+\cos B_{19}\cos B_{24}+\cos D_{19}\cos D_{24}$

$\cos\phi_{20,30}=\cos A_{20}\cos A_{30}+\cos B_{20}\cos B_{30}+\cos D_{20}\cos D_{30}$

$\cos\phi_{20,34}=\cos A_{20}\cos A_{34}+\cos B_{20}\cos B_{34}+\cos D_{20}\cos D_{34}$

$\cos\phi_{20,35}=\cos A_{20}\cos A_{35}+\cos B_{20}\cos B_{35}+\cos D_{20}\cos D_{35}$

$\cos\phi_{21,22}=\cos A_{21}\cos A_{22}+\cos B_{212}\cos B_{22}+\cos D_{21}\cos D_{22}$

$\cos\phi_{21,23}=\cos A_{21}\cos A_{23}+\cos B_{21}\cos B_{23}+\cos D_{21}\cos D_{23}$

$\cos\phi_{21,25}=\cos A_{21}\cos A_{25}+\cos B_{21}\cos B_{25}+\cos D_{21}\cos D_{25}$

$\cos\phi_{22,32}=\cos A_{22}\cos A_{32}+\cos B_{22}\cos B_{32}+\cos D_{22}\cos D_{32}$

$\cos\phi_{22,34}=\cos A_{22}\cos A_{34}+\cos B_{22}\cos B_{34}+\cos D_{22}\cos D_{34}$

$\cos\phi_{22,36}=\cos A_{22}\cos A_{36}+\cos B_{22}\cos B_{36}+\cos D_{22}\cos D_{36}$

$\cos\phi_{23,24}=\cos A_{23}\cos A_{24}+\cos B_{23}\cos B_{24}+\cos D_{23}\cos D_{24}$

$\cos\phi_{23,25}=\cos A_{23}\cos A_{25}+\cos B_{23}\cos B_{25}+\cos D_{23}\cos D_{25}$

$\cos\phi_{24,35}=\cos A_{24}\cos A_{35}+\cos B_{24}\cos B_{35}+\cos D_{24}\cos D_{35}$

$\cos\phi_{24,37}=\cos A_{24}\cos A_{37}+\cos B_{24}\cos B_{37}+\cos D_{24}\cos D_{37}$

$\cos\phi_{24,37}=\cos A_{24}\cos A_{37}+\cos B_{24}\cos B_{37}+\cos D_{24}\cos D_{37}$

$\cos\phi_{25,36}=\cos A_{25}\cos A_{36}+\cos B_{25}\cos B_{36}+\cos D_{25}\cos D_{36}$

$\cos\phi_{25,37}=\cos A_{25}\cos A_{37}+\cos B_{25}\cos B_{37}+\cos D_{25}\cos D_{37}$

$\cos\phi_{25,39}=\cos A_{25}\cos A_{39}+\cos B_{25}\cos B_{39}+\cos D_{25}\cos D_{39}$

ANGLES (Cont'd)

$\cos\phi_{26,27}=\cos A_{26}\cos A_{27}+\cos B_{26}\cos B_{27}+\cos D_{26}\cos D_{27}$

$\cos\phi_{26,28}=\cos A_{26}\cos A_{28}+\cos B_{26}\cos B_{28}+\cos D_{26}\cos D_{28}$

$\cos\phi_{27,31}=\cos A_{27}\cos A_{31}+\cos B_{27}\cos B_{31}+\cos D_{27}\cos D_{31}$

$\cos\phi_{27,38}=\cos A_{27}\cos A_{38}+\cos B_{27}\cos B_{38}+\cos D_{27}\cos D_{38}$

$\cos\phi_{27,40}=\cos A_{27}\cos A_{40}+\cos B_{27}\cos B_{40}+\cos D_{27}\cos D_{40}$

$\cos\phi_{28,33}=\cos A_{28}\cos A_{33}+\cos B_{28}\cos D_{33}+\cos D_{28}\cos D_{33}$

$\cos\phi_{28,39}=\cos A_{28}\cos A_{39}+\cos B_{28}\cos B_{39}+\cos D_{28}\cos D_{39}$

$\cos\phi_{28,40}=\cos A_{28}\cos A_{40}+\cos B_{28}\cos B_{40}+\cos D_{28}\cos D_{40}$

$\cos\phi_{29,31}=\cos A_{29}\cos A_{31}+\cos B_{29}\cos B_{31}+\cos D_{29}\cos D_{31}$

$\cos\phi_{29,31}=\cos A_{29}\cos A_{31}+\cos B_{29}\cos B_{31}+\cos D_{29}\cos D_{31}$

$\cos\phi_{29,32}=\cos A_{29}\cos A_{32}+\cos B_{29}\cos B_{32}+\cos D_{29}\cos D_{32}$

$\cos\phi_{29,33}=\cos A_{29}\cos A_{33}+\cos B_{29}\cos B_{33}+\cos D_{29}\cos D_{33}$

$\cos\phi_{30,31}=\cos A_{30}\cos A_{31}+\cos B_{30}\cos B_{31}+\cos D_{30}\cos D_{31}$

$\cos\phi_{30,34}=\cos A_{30}\cos A_{34}+\cos B_{30}\cos B_{34}+\cos D_{30}\cos D_{34}$

$\cos\phi_{30,35}=\cos A_{30}\cos A_{35}+\cos B_{30}\cos B_{35}+\cos D_{30}\cos D_{35}$

$\cos\phi_{31,38}=\cos A_{31}\cos A_{38}+\cos B_{31}\cos B_{38}+\cos D_{31}\cos D_{38}$

$\cos\phi_{31,40}=\cos A_{31}\cos A_{40}+\cos B_{31}\cos B_{40}+\cos D_{31}\cos D_{40}$

$\cos\phi_{32,33}=\cos A_{32}\cos A_{33}+\cos B_{32}\cos B_{33}+\cos D_{32}\cos D_{33}$

$\cos\phi_{32,34}=\cos A_{32}\cos A_{34}+\cos B_{32}\cos B_{34}+\cos D_{32}\cos D_{34}$

$\cos\phi_{32,36}=\cos A_{32}\cos A_{36}+\cos B_{32}\cos B_{36}+\cos D_{32}\cos D_{36}$

$\cos\phi_{33,39}=\cos A_{33}\cos A_{39}+\cos B_{33}\cos B_{39}+\cos D_{33}\cos D_{39}$

$\cos\phi_{33,40}=\cos A_{33}\cos A_{40}+\cos B_{33}\cos B_{40}+\cos D_{33}\cos D_{40}$

$\cos\phi_{34,35}=\cos A_{34}\cos A_{35}+\cos B_{34}\cos B_{35}+\cos D_{34}\cos D_{35}$

$\cos\phi_{34,36}=\cos A_{34}\cos A_{36}+\cos B_{34}\cos B_{36}+\cos D_{34}\cos D_{36}$

$\cos\phi_{35,37}=\cos A_{35}\cos A_{37}+\cos B_{35}\cos B_{37}+\cos D_{35}\cos D_{37}$

$\cos\phi_{35,38}=\cos A_{35}\cos A_{38}+\cos B_{35}\cos B_{38}+\cos D_{35}\cos D_{38}$

$\cos\phi_{36,37}=\cos A_{36}\cos A_{37}+\cos B_{36}\cos B_{37}+\cos D_{36}\cos D_{37}$

$\cos\phi_{36,39}=\cos A_{36}\cos A_{39}+\cos B_{36}\cos B_{39}+\cos D_{36}\cos D_{39}$

$\cos\phi_{37,38}=\cos A_{37}\cos A_{38}+\cos B_{37}\cos B_{38}+\cos D_{37}\cos D_{38}$

$\cos\phi_{37,39}=\cos A_{37}\cos A_{39}+\cos B_{37}\cos B_{39}+\cos D_{37}\cos D_{39}$

$\cos\phi_{38,40}=\cos A_{38}\cos A_{40}+\cos B_{38}\cos B_{40}+\cos D_{38}\cos D_{40}$

$\cos\phi_{39,40}=\cos A_{39}\cos A_{40}+\cos B_{39}\cos B_{40}+\cos D_{39}\cos D_{40}$

Table 22-13. Specifications and Analyses for the Hyperspace Object Shown in Fig. 22-13. (A) Specifications and Analysis in Hyperspace. (B) Secondary Analysis in XYZ Space. (C) Secondary Analysis in XYW Space. (D) Secondary Analysis in XZW Space. (E) Secondary Analysis in YZW Space. (F) Suggested Coordinates. **(Continued from page 471.)**

C

CHECK

F_1: $\phi_{1,2}+\phi_{1,5}+\phi_{2,8}+\phi_{5,8}=360^{\circ}$

F_2: $\phi_{1,3}+\phi_{1,6}+\phi_{3,13}+\phi_{6,13}=360^{\circ}$

F_3: $\phi_{1,4}+\phi_{1,7}+\phi_{4,18}+\phi_{7,18}=360^{\circ}$

F_4: $\phi_{2,3}+\phi_{2,10}+\phi_{3,14}+\phi_{10,14}=360^{\circ}$

F_5: $\phi_{2,4}+\phi_{2,9}+\phi_{4,19}+\phi_{9,19}=360^{\circ}$

F_6: $\phi_{3,4}+\phi_{3,15}+\phi_{4,20}+\phi_{15,30}+\phi_{20,30}=550^{\circ}$

F_7: $\phi_{5,6}+\phi_{5,12}+\phi_{6,16}+\phi_{12,16}=360^{\circ}$

F_8: $\phi_{5,7}+\phi_{5,11}+\phi_{7,21}+\phi_{11,21}=360^{\circ}$

F_9: $\phi_{6,7}+\phi_{6,17}+\phi_{7,22}+\phi_{17,32}+\phi_{22,32}=550^{\circ}$

F_{10}: $\phi_{8,9}+\phi_{8,11}+\phi_{9,23}+\phi_{11,23}=360^{\circ}$

F_{11}: $\phi_{8,10}+\phi_{8,12}+\phi_{10,26}+\phi_{12,26}=360^{\circ}$

F_{12}: $\phi_{9,10}+\phi_{9,24}+\phi_{10,27}+\phi_{24,38}+\phi_{27,38}=550^{\circ}$

F_{13}: $\phi_{11,12}+\phi_{11,25}+\phi_{12,28}+\phi_{25,39}+\phi_{28,39}=550^{\circ}$

F_{14}: $\phi_{13,14}+\phi_{13,16}+\phi_{14,26}+\phi_{16,26}=360^{\circ}$

F_{15}: $\phi_{13,15}+\phi_{13,17}+\phi_{15,29}+\phi_{17,29}=360^{\circ}$

F_{16}: $\phi_{14,15}+\phi_{14,27}+\phi_{15,31}+\phi_{27,31}=360^{\circ}$

F_{17}: $\phi_{16,17}+\phi_{16,28}+\phi_{17,33}+\phi_{28,33}=360^{\circ}$

F_{18}: $\phi_{18,19}+\phi_{18,21}+\phi_{19,23}+\phi_{21,23}=360^{\circ}$

F_{19}: $\phi_{18,20}+\phi_{18,22}+\phi_{20,34}+\phi_{22,34}=360^{\circ}$

F_{20}: $\phi_{19,20}+\phi_{19,24}+\phi_{20,35}+\phi_{24,35}=360^{\circ}$

F_{21}: $\phi_{21,22}+\phi_{21,25}+\phi_{22,36}+\phi_{25,36}=360^{\circ}$

F_{22}: $\phi_{23,24}+\phi_{23,25}+\phi_{24,37}+\phi_{25,37}=360^{\circ}$

F_{23}: $\phi_{26,27}+\phi_{26,28}+\phi_{27,40}+\phi_{28,40}=360^{\circ}$

F_{24}: $\phi_{29,30}+\phi_{29,32}+\phi_{30,34}+\phi_{32,34}=360^{\circ}$

F_{25}: $\phi_{29,31}+\phi_{29,33}+\phi_{31,34}+\phi_{33,34}=360^{\circ}$

F_{26}: $\phi_{30,31}+\phi_{30,35}+\phi_{31,38}+\phi_{35,38}=360^{\circ}$

F_{27}: $\phi_{32,33}+\phi_{32,36}+\phi_{33,39}+\phi_{36,39}=360^{\circ}$

F_{28}: $\phi_{34,35}+\phi_{34,36}+\phi_{35,37}+\phi_{36,36}=360^{\circ}$

F_{29}: $\phi_{37,38}+\phi_{37,39}+\phi_{38,40}+\phi_{39,40}=360^{\circ}$

D

SECONDARY ANALYSIS -- XZW SPACE (See Fig. 22-13D)

PLOTTED POINTS

$P_1=(x_1,z_1,w_1)$	$P_2=(x_2,z_2,w_2)$
$P_3=(x_3,z_3,w_3)$	$P_4=(x_4,z_4,w_4)$
$P_5=(x_5,z_5,w_5)$	$P_6=(x_6,z_6,w_6)$
$P_7=(x_7,z_7,w_7)$	$P_8=(x_8,z_8,w_8)$
$P_9=(x_9,z_9,w_9)$	$P_{10}=(x_{10},z_{10},w_{10})$
$P_{11}=(x_{11},z_{11},w_{11})$	$P_{12}=(x_{12},z_{12},w_{12})$
$P_{13}=(x_{13},z_{13},w_{13})$	$P_{14}=(x_{14},z_{14},w_{14})$
$P_{15}=(x_{15},z_{15},w_{15})$	$P_{16}=(x_{16},z_{16},w_{16})$
$P_{17}=(x_{17},z_{17},w_{17})$	$P_{18}=(x_{18},z_{18},w_{18})$
$P_{19}=(x_{18},z_{19},w_{19})$	$P_{20}=(x_{20},z_{20},w_{20})$

LENGTHS

$$s_1=\sqrt{(x_2-x_1)^2+(z_2-z_1)^2+(w_2-w_1)^2}$$

$$s_2=\sqrt{(x_3-x_1)^2+(z_3-z_1)^2+(w_3-w_1)^2}$$

$$s_3=\sqrt{(x_5-x_1)^2+(z_5-z_1)^2+(w_5-w_1)^2}$$

$$s_4=\sqrt{(x_7-x_1)^2+(z_7-z_1)^2+(w_7-w_1)^2}$$

$$s_5=\sqrt{(x_4-x_2)^2+(z_4-z_2)^2+(w_4-w_2)^2}$$

$$s_6=\sqrt{(x_6-x_2)^2+(z_6-z_2)^2+(w_6-w_2)^2}$$

$$s_7=\sqrt{(x_8-x_2)^2+(z_8-z_2)^2+(w_8-w_2)^2}$$

$$s_8=\sqrt{(x_4-x_3)^2+(z_4-z_3)^2+(w_4-w_3)^2}$$

$$s_9=\sqrt{(x_9-x_3)^2+(z_9-z_3)^2+(w_9-w_3)^2}$$

$$s_{10}=\sqrt{(x_{11}-x_3)^2+(z_{11}-z_3)^2+(w_{11}-w_3)^2}$$

$$s_{11}=\sqrt{(x_{10}-x_4)^2+(z_{10}-z_4)^2+(w_{10}-w_4)^2}$$

LENGTHS (Cont'd)

$$s_{12}=\sqrt{(x_{12}-x_4)^2+(z_{12}-z_4)^2+(w_{12}-w_4)^2}$$

$$s_{13}=\sqrt{(x_6-x_5)^2+(z_6-z_5)^2+(w_6-w_5)^2}$$

$$s_{14}=\sqrt{(x_{11}-x_5)^2+(z_{11}-z_5)^2+(w_{11}-w_5)^2}$$

$$s_{15}=\sqrt{(x_{13}-x_5)^2+(z_{13}-z_5)^2+(w_{13}-w_5)^2}$$

$$s_{16}=\sqrt{(x_{12}-x_6)^2+(z_{12}-z_6)^2+(w_{12}-w_6)^2}$$

$$s_{17}=\sqrt{(x_{14}-x_6)^2+(z_{14}-z_6)^2+(w_{14}-w_6)^2}$$

$$s_{18}=\sqrt{(x_8-x_7)^2+(z_8-z_7)^2+(w_8-w_7)^2}$$

$$s_{19}=\sqrt{(x_9-x_7)^2+(z_9-z_7)^2+(w_9-w_7)^2}$$

$$s_{20}=\sqrt{(x_{15}-x_7)^2+(z_{15}-z_7)^2+(w_{15}-w_7)^2}$$

$$s_{21}=\sqrt{(x_{10}-x_8)^2+(z_{10}-z_8)^2+(w_{10}-w_8)^2}$$

$$s_{22}=\sqrt{(x_{16}-x_8)^2+(z_{16}-z_8)^2+(w_{16}-w_8)^2}$$

$$s_{23}=\sqrt{(x_{10}-x_9)^2+(z_{10}-z_9)^2+(w_{10}-w_9)^2}$$

$$s_{24}=\sqrt{(x_{17}-x_9)^2+(z_{17}-z_9)^2+(w_{17}-w_9)^2}$$

$$s_{25}=\sqrt{(x_{18}-x_{10})^2+(z_{18}-z_{10})^2+(w_{18}-w_{10})^2}$$

$$s_{26}=\sqrt{(x_{12}-x_{11})^2+(z_{12}-z_{11})^2+(w_{12}-w_{11})^2}$$

$$s_{27}=\sqrt{(x_{19}-x_{11})^2+(z_{19}-z_{11})^2+(w_{19}-w_{11})^2}$$

$$s_{28}=\sqrt{(x_{20}-x_{12})^2+(z_{20}-z_{12})^2+(w_{20}-w_{12})^2}$$

$$s_{29}=\sqrt{(x_{14}-x_{13})^2+(z_{14}-z_{13})^2+(w_{14}-w_{13})^2}$$

$$s_{30}=\sqrt{(x_{15}-x_{13})^2+(z_{15}-z_{13})^2+(w_{15}-w_{13})^2}$$

$$s_{31}=\sqrt{(x_{19}-x_{13})^2+(z_{19}-z_{13})^2+(w_{19}-w_{13})^2}$$

$$s_{32}=\sqrt{(x_{16}-x_{14})^2+(z_{16}-z_{14})^2+(w_{16}-w_{14})^2}$$

$$s_{33}=\sqrt{(x_{20}-x_{14})^2+(z_{20}-z_{14})^2+(w_{20}-w_{14})^2}$$

LENGTHS (Cont'd)

$$s_{34}=\sqrt{(x_{16}-x_{15})^2+(z_{16}-z_{15})^2+(w_{16}-w_{15})^2}$$

$$s_{35}=\sqrt{(x_{17}-x_{15})^2+(z_{17}-z_{15})^2+(w_{17}-w_{15})^2}$$

$$s_{36}=\sqrt{(x_{18}-x_{16})^2+(z_{18}-z_{16})^2+(w_{18}-w_{16})^2}$$

$$s_{37}=\sqrt{(x_{18}-x_{17})^2+(z_{18}-z_{17})^2+(w_{18}-w_{17})^2}$$

$$s_{38}=\sqrt{(x_{19}-x_{17})^2+(z_{19}-z_{17})^2+(w_{19}-w_{17})^2}$$

$$s_{39}=\sqrt{(x_{20}-x_{18})^2+(z_{20}-z_{18})^2+(w_{20}-w_{18})^2}$$

$$s_{40}=\sqrt{(x_{20}-x_{19})^2+(z_{20}-z_{19})^2+(w_{20}-w_{19})^2}$$

DIRECTION COSINES

$\cos A_1=(x_2-x_1)/s_1$ $\cos C_1=(z_2-z_1)/s_1$
$\cos D_1=(w_2-w_1)/s_1$

$\cos A_2=(x_3-x_1)/s_2$ $\cos C_2=(z_3-z_1)/s_2$
$\cos D_2=(w_3-w_1)/s_2$

$\cos A_3=(x_5-x_1)/s_3$ $\cos C_3=(z_5-z_1)/s_3$
$\cos D_3=(w_5-w_1)/s_3$

$\cos A_4=(x_7-x_1)/s_4$ $\cos C_4=(z_7-z_1)/s_4$
$\cos D_4=(w_7-w_1)/s_4$

$\cos A_5=(x_4-x_2)/s_5$ $\cos C_5=(z_4-z_2)/s_5$
$\cos D_5=(w_4-w_2)/s_5$

$\cos A_6=(x_6-x_2)/s_6$ $\cos C_6=(z_6-z_2)/s_6$
$\cos D_6=(w_6-w_2)/s_6$

$\cos A_7=(x_8-x_2)/s_7$ $\cos C_7=(z_8-z_2)/s_7$
$\cos D_7=(w_8-w_2)/s_7$

$\cos A_8=(x_4-x_3)/s_8$ $\cos C_8=(z_4-z_3)/s_8$
$\cos D_8=(w_4-w_3)/s_8$

Table 22-13. Specifications and Analyses for the Hyperspace Object Shown in Fig. 22-13. (A) Specifications and Analysis in Hyperspace. (B) Secondary Analysis in XYZ Space. (C) Secondary Analysis in XYW Space. (D) Secondary Analysis in XZW Space. (E) Secondary Analysis in YZW Space. (F) Suggested Coordinates. (Continued from page 473.)

DIRECTION COSINES (Cont'd)

$\cos A_9=(x_9-x_3)/s_9$ $\cos C_9=(z_9-z_3)/s_9$

$\cos D_9=(w_9-w_3)/s_9$

$\cos A_{10}=(x_{11}-x_3)/s_{10}$ $\cos C_{10}=(z_{11}-z_3)/s_{10}$

$\cos D_{10}=(w_{11}-w_3)/s_{10}$

$\cos A_{11}=(x_{10}-x_4)/s_{11}$ $\cos C_{11}=(z_{10}-z_4)/s_{11}$

$\cos D_{11}=(w_{10}-w_4)/s_{11}$

$\cos A_{12}=(x_{12}-x_4)/s_{12}$ $\cos C_{12}=(z_{12}-z_4)/s_{12}$

$\cos D_{12}=(w_{12}-w_4)/s_{12}$

$\cos A_{13}=(x_6-x_5)/s_{13}$ $\cos C_{13}=(z_6-z_5)/s_{13}$

$\cos D_{13}=(w_6-w_5)/s_{13}$

$\cos A_{14}=(x_{11}-x_5)/s_{14}$ $\cos C_{14}=(z_{11}-z_5)/s_{14}$

$\cos D_{14}=(w_{11}-w_5)/s$

$\cos A_{15}=(x_{13}-x_5)/s_{15}$ $\cos C_{15}=(z_{13}-z_5)/s_{15}$

$\cos D_{15}=(w_{13}-w_5)/s_{15}$

$\cos A_{16}=(x_{12}-x_6)/s_{16}$ $\cos C_{16}=(z_{12}-z_6)/s_{16}$

$\cos D_{16}=(w_{12}-w_6)/s_{16}$

$\cos A_{17}=(x_{14}-x_6)/s_{17}$ $\cos C_{17}=(z_{14}-z_6)/s_{17}$

$\cos D_{17}=(w_{14}-w_6)/s_{17}$

$\cos A_{18}=(x_8-x_7)/s_{18}$ $\cos C_{18}=(z_8-z_7)/s_{18}$

$\cos D_{18}=(w_8-w_7)/s_{18}$

$\cos A_{19}=(x_9-x_7)/s_{19}$ $\cos C_{19}=(z_9-z_7)/s_{19}$

$\cos D_{19}=(w_9-w_7)/s_{19}$

$\cos A_{20}=(x_{15}-x_7)/s_{20}$ $\cos C_{20}=(z_{15}-z_7)/s_{20}$

$\cos D_{20}=(w_{15}-w_7)/s_{20}$

$\cos A_{21}=(x_{10}-x_8)/s_{21}$ $\cos C_{21}=(z_{10}-z_8)/s_{21}$

$\cos D_{21}=(w_{10}-w_8)/s_{21}$

DIRECTION COSINES (Cont'd)

$\cos A_{22}=(x_{16}-x_8)/s_{22}$ $\cos C_{22}=(z_{16}-z_8)/s_{22}$

$\cos D_{22}=(w_{16}-w_8)/s_{22}$

$\cos A_{23}=(x_{10}-x_9)/s_{23}$ $\cos C_{23}=(z_{10}-z_9)/s_{23}$

$\cos D_{23}=(w_{10}-w_9)/s_{23}$

$\cos A_{24}=(x_{17}-x_9)/s_{24}$ $\cos C_{24}=(z_{17}-z_9)/s_{24}$

$\cos D_{24}=(w_{17}-w_9)/s_{24}$

$\cos A_{25}=(x_{18}-x_{10})/s_{25}$ $\cos C_{25}=(z_{18}-z_{10})/s_{25}$

$\cos D_{25}=(w_{18}-w_{10})/s_{25}$

$\cos A_{26}=(x_{12}-x_{11})/s_{26}$ $\cos C_{26}=(z_{12}-z_{11})/s_{26}$

$\cos D_{26}=(w_{12}-w_{11})/s_{26}$

$\cos A_{27}=(x_{19}-x_{11})/s_{27}$ $\cos C_{27}=(z_{19}-z_{11})/s_{27}$

$\cos D_{27}=(w_{19}-w_{11})/s_{27}$

$\cos A_{28}=(x_{20}-x_{12})/s_{28}$ $\cos C_{28}=(z_{20}-z_{12})/s_{28}$

$\cos D_{28}=(w_{20}-w_{12})/s_{28}$

$\cos A_{29}=(x_{14}-x_{13})/s_{29}$ $\cos C_{29}=(z_{14}-z_{13})/s_{29}$

$\cos D_{29}=(w_{14}-w_{13})/s_{29}$

$\cos A_{30}=(x_{15}-x_{13})/s_{30}$ $\cos C_{30}=(z_{15}-z_{13})/s_{30}$

$\cos D_{30}=(w_{15}-w_{13})/s_{30}$

$\cos A_{31}=(x_{19}-x_{13})/s_{31}$ $\cos C_{31}=(z_{19}-z_{13})/s_{31}$

$\cos D_{31}=(w_{19}-w_{13})/s_{31}$

$\cos A_{32}=(x_{16}-x_{14})/s_{32}$ $\cos C_{32}=(z_{16}-z_{14})/s_{32}$

$\cos D_{32}=(w_{16}-w_{14})/s_{32}$

$\cos A_{33}=(x_{20}-x_{14})/s_{33}$ $\cos C_{33}=(z_{20}-z_{14})/s_{33}$

$\cos D_{33}=(w_{20}-w_{14})/s_{33}$

$\cos A_{34}=(x_{16}-x_{15})/s_{34}$ $\cos C_{34}=(z_{16}-z_{15})/s_{34}$

$\cos D_{34}=(w_{16}-w_{15})/s_{34}$

D

DIRECTION COSINES (Cont'd)

$\cos A_{35} = (x_{17}-x_{15})/s_{35}$ $\cos C_{35} = (z_{17}-z_{15})/s_{35}$

$\cos D_{35} = (w_{17}-w_{15})/s_{35}$

$\cos A_{36} = (x_{18}-x_{16})/s_{36}$ $\cos C_{36} = (z_{18}-z_{16})/s_{36}$

$\cos D_{36} = (w_{18}-w_{16})/s_{36}$

$\cos A_{37} = (x_{18}-x_{17})/s_{37}$ $\cos C_{37} = (z_{18}-z_{17})/s_{37}$

$\cos D_{37} = (w_{18}-w_{17})/s_{37}$

$\cos A_{38} = (x_{19}-x_{17})/s_{38}$ $\cos C_{38} = (z_{19}-z_{17})/s_{38}$

$\cos D_{38} = (w_{19}-w_{17})/s_{38}$

$\cos A_{39} = (x_{20}-x_{18})/s_{39}$ $\cos C_{39} = (z_{20}-z_{18})/s_{39}$

$\cos D_{39} = (w_{20}-w_{18})/s_{39}$

$\cos A_{40} = (x_{20}-x_{19})/s_{40}$ $\cos C_{40} = (z_{20}-z_{19})/s_{40}$

$\cos D_{40} = (w_{20}-w_{19})/s_{40}$

ANGLES

$\cos\phi_{1,2} = \cos A_1 \cos A_2 + \cos C_1 \cos C_2 + \cos D_1 \cos D_2$

$\cos\phi_{1,3} = \cos A_1 \cos A_3 + \cos C_1 \cos C_3 + \cos D_1 \cos D_3$

$\cos\phi_{1,4} = \cos A_1 \cos A_4 + \cos C_1 \cos C_4 + \cos D_1 \cos D_4$

$\cos\phi_{1,5} = \cos A_1 \cos A_5 + \cos C_1 \cos C_5 + \cos D_1 \cos D_5$

$\cos\phi_{1,6} = \cos A_1 \cos A_6 + \cos C_1 \cos C_6 + \cos D_1 \cos D_6$

$\cos\phi_{1,7} = \cos A_1 \cos A_7 + \cos C_1 \cos C_7 + \cos D_1 \cos D_7$

$\cos\phi_{2,3} = \cos A_2 \cos A_3 + \cos C_2 \cos C_3 + \cos D_2 \cos D_3$

$\cos\phi_{2,4} = \cos A_2 \cos A_4 + \cos C_2 \cos C_4 + \cos D_2 \cos D_4$

$\cos\phi_{2,8} = \cos A_2 \cos A_8 + \cos C_2 \cos C_8 + \cos D_2 \cos D_8$

$\cos\phi_{2,9} = \cos A_2 \cos A_9 + \cos C_2 \cos C_9 + \cos D_2 \cos D_9$

$\cos\phi_{2,10} = \cos A_2 \cos A_{10} + \cos C_2 \cos C_{10} + \cos D_2 \cos D_{10}$

$\cos\phi_{3,4} = \cos A_3 \cos A_4 + \cos C_3 \cos C_4 + \cos D_3 \cos D_4$

$\cos\phi_{3,13} = \cos A_3 \cos A_{13} + \cos C_3 \cos C_{13} + \cos D_3 \cos D_{13}$

$\cos\phi_{3,14} = \cos A_3 \cos A_{14} + \cos C_3 \cos C_{14} + \cos D_3 \cos D_{14}$

$\cos\phi_{3,15} = \cos A_3 \cos A_{15} + \cos C_3 \cos C_{15} + \cos D_3 \cos D_{15}$

$\cos\phi_{4,18} = \cos A_4 \cos A_{18} + \cos C_4 \cos C_{18} + \cos D_4 \cos D_{18}$

$\cos\phi_{4,19} = \cos A_4 \cos A_{19} + \cos C_4 \cos C_{19} + \cos D_4 \cos D_{19}$

$\cos\phi_{4,20} = \cos A_4 \cos A_{20} + \cos C_4 \cos C_{20} + \cos D_4 \cos D_{20}$

ANGLES (Cont'd)

$\cos\phi_{5,6} = \cos A_5 \cos A_6 + \cos C_5 \cos C_6 + \cos D_5 \cos D_6$

$\cos\phi_{5,7} = \cos A_5 \cos A_7 + \cos C_5 \cos C_7 + \cos D_5 \cos D_7$

$\cos\phi_{5,8} = \cos A_5 \cos A_8 + \cos C_5 \cos C_8 + \cos D_5 \cos D_8$

$\cos\phi_{5,11} = \cos A_5 \cos A_{11} + \cos C_5 \cos C_{11} + \cos D_5 \cos D_{11}$

$\cos\phi_{5,12} = \cos A_5 \cos A_{12} + \cos C_5 \cos C_{12} + \cos D_5 \cos D_{12}$

$\cos\phi_{6,7} = \cos A_6 \cos A_7 + \cos C_6 \cos C_7 + \cos D_6 \cos D_7$

$\cos\phi_{6,13} = \cos A_6 \cos A_{13} + \cos C_6 \cos C_{13} + \cos D_6 \cos D_{13}$

$\cos\phi_{6,16} = \cos A_6 \cos A_{16} + \cos C_6 \cos C_{16} + \cos D_6 \cos D_{16}$

$\cos\phi_{6,17} = \cos A_6 \cos A_{17} + \cos C_6 \cos C_{17} + \cos D_6 \cos D_{17}$

$\cos\phi_{7,18} = \cos A_7 \cos A_{18} + \cos C_7 \cos C_{18} + \cos D_7 \cos D_{18}$

$\cos\phi_{7,21} = \cos A_7 \cos A_{21} + \cos C_7 \cos C_{21} + \cos D_7 \cos D_{21}$

$\cos\phi_{7,22} = \cos A_7 \cos A_{22} + \cos C_7 \cos C_{22} + \cos D_7 \cos D_{22}$

$\cos\phi_{8,9} = \cos A_8 \cos A_9 + \cos C_8 \cos C_9 + \cos D_8 \cos D_9$

$\cos\phi_{8,10} = \cos A_8 \cos A_{10} + \cos C_8 \cos C_{190} + \cos D_8 \cos D_{10}$

$\cos\phi_{8,11} = \cos A_8 \cos A_{11} + \cos C_8 \cos C_{11} + \cos D_8 \cos D_{11}$

$\cos\phi_{8,12} = \cos A_8 \cos A_{12} + \cos C_8 \cos C_{12} + \cos D_8 \cos D_{12}$

$\cos\phi_{9,10} = \cos A_9 \cos A_{10} + \cos C_9 \cos C_{10} + \cos D_9 \cos D_{10}$

$\cos\phi_{9,19} = \cos A_9 \cos A_{19} + \cos C_9 \cos C_{19} + \cos D_9 \cos D_{19}$

$\cos\phi_{9,23} = \cos A_9 \cos A_{23} + \cos C_9 \cos C_{23} + \cos D_9 \cos D_{23}$

$\cos\phi_{9,24} = \cos A_9 \cos A_{24} + \cos C_9 \cos C_{24} + \cos D_9 \cos D_{24}$

$\cos\phi_{10,14} = \cos A_{10} \cos A_{14} + \cos C_{10} \cos C_{14} + \cos D_{10} \cos D_{14}$

$\cos\phi_{10,26} = \cos A_{10} \cos A_{26} + \cos C_{10} \cos C_{26} + \cos D_{10} \cos D_{26}$

$\cos\phi_{10,27} = \cos A_{10} \cos A_{27} + \cos C_{10} \cos C_{27} + \cos D_{10} \cos D_{27}$

$\cos\phi_{11,12} = \cos A_{11} \cos A_{12} + \cos C_{11} \cos C_{12} + \cos D_{11} \cos D_{12}$

$\cos\phi_{11,21} = \cos A_{11} \cos A_{21} + \cos C_{11} \cos C_{21} + \cos D_{11} \cos D_{21}$

$\cos\phi_{11,23} = \cos A_{11} \cos A_{23} + \cos C_{11} \cos C_{23} + \cos D_{11} \cos D_{23}$

$\cos\phi_{11,25} = \cos A_{11} \cos A_{25} + \cos C_{11} \cos C_{25} + \cos D_{11} \cos D_{25}$

$\cos\phi_{12,16} = \cos A_{12} \cos A_{16} + \cos C_{12} \cos C_{16} + \cos D_{12} \cos D_{16}$

$\cos\phi_{12,26} = \cos A_{12} \cos A_{26} + \cos C_{12} \cos C_{26} + \cos D_{12} \cos D_{26}$

$\cos\phi_{12,28} = \cos A_{12} \cos A_{28} + \cos C_{12} \cos C_{28} + \cos D_{12} \cos D_{28}$

$\cos\phi_{13,14} = \cos A_{13} \cos A_{14} + \cos C_{13} \cos C_{14} + \cos D_{13} \cos D_{14}$

$\cos\phi_{13,15} = \cos A_{13} \cos A_{15} + \cos C_{13} \cos C_{15} + \cos D_{13} \cos D_{15}$

$\cos\phi_{13,16} = \cos A_{13} \cos A_{16} + \cos C_{13} \cos C_{16} + \cos D_{13} \cos D_{16}$

$\cos\phi_{13,17} = \cos A_{13} \cos A_{17} + \cos C_{13} \cos C_{17} + \cos D_{13} \cos D_{17}$

Table 22-13. Specifications and Analyses for the Hyperspace Object Shown in Fig. 22-13. (A) Specifications and Analysis in Hyperspace. (B) Secondary Analysis in XYZ Space. (C) Secondary Analysis in XYW Space. (D) Secondary Analysis in XZW Space. (E) Secondary Analysis in YZW Space. (F) Suggested Coordinates. (Continued from page 475.)

ANGLES (Cont'd)

$\cos\phi_{14,15}=\cos A_{14}\cos A_{15}+\cos C_{14}\cos C_{15}+\cos D_{14}\cos D_{15}$

$\cos\phi_{14,26}=\cos A_{14}\cos A_{26}+\cos C_{14}\cos C_{26}+\cos D_{14}\cos D_{26}$

$\cos\phi_{14,27}=\cos A_{14}\cos A_{27}+\cos C_{14}\cos C_{27}+\cos D_{14}\cos D_{27}$

$\cos\phi_{15,29}=\cos A_{15}\cos A_{29}+\cos C_{15}\cos C_{29}+\cos D_{15}\cos D_{29}$

$\cos\phi_{15,30}=\cos A_{15}\cos A_{30}+\cos C_{15}\cos C_{30}+\cos D_{15}\cos D_{30}$

$\cos\phi_{15,31}=\cos A_{15}\cos A_{31}+\cos C_{15}\cos C_{31}+\cos D_{15}\cos D_{31}$

$\cos\phi_{16,17}=\cos A_{16}\cos A_{17}+\cos C_{16}\cos C_{17}+\cos D_{16}\cos D_{17}$

$\cos\phi_{16,26}=\cos A_{16}\cos A_{26}+\cos C_{16}\cos C_{26}+\cos D_{16}\cos D_{26}$

$\cos\phi_{16,28}=\cos A_{16}\cos A_{28}+\cos C_{16}\cos C_{28}+\cos D_{16}\cos D_{28}$

$\cos\phi_{17,29}=\cos A_{17}\cos A_{29}+\cos C_{17}\cos C_{29}+\cos D_{17}\cos D_{29}$

$\cos\phi_{17,32}=\cos A_{17}\cos A_{32}+\cos C_{17}\cos C_{32}+\cos D_{17}\cos D_{32}$

$\cos\phi_{17,33}=\cos A_{17}\cos A_{33}+\cos C_{17}\cos C_{33}+\cos D_{17}\cos D_{33}$

$\cos\phi_{18,19}=\cos A_{18}\cos A_{19}+\cos C_{18}\cos C_{19}+\cos D_{18}\cos D_{19}$

$\cos\phi_{18,20}=\cos A_{18}\cos A_{20}+\cos C_{18}\cos C_{20}+\cos D_{178}\cos D_{20}$

$\cos\phi_{18,21}=\cos A_{18}\cos A_{21}+\cos C_{18}\cos C_{21}+\cos D_{18}\cos D_{21}$

$\cos\phi_{18,22}=\cos A_{18}\cos A_{22}+\cos C_{18}\cos C_{22}+\cos D_{18}\cos D_{22}$

$\cos\phi_{19,20}=\cos A_{19}\cos A_{20}+\cos C_{19}\cos C_{20}+\cos D_{19}\cos D_{20}$

$\cos\phi_{19,23}=\cos A_{19}\cos A_{23}+\cos C_{19}\cos C_{23}+\cos D_{19}\cos D_{23}$

$\cos\phi_{19,24}=\cos A_{19}\cos A_{24}+\cos C_{19}\cos C_{24}+\cos D_{19}\cos D_{24}$

$\cos\phi_{20,30}=\cos A_{20}\cos A_{30}+\cos C_{20}\cos C_{30}+\cos D_{20}\cos D_{30}$

$\cos\phi_{20,34}=\cos A_{20}\cos A_{34}+\cos C_{20}\cos C_{34}+\cos D_{20}\cos D_{34}$

$\cos\phi_{20,35}=\cos A_{20}\cos A_{35}+\cos C_{20}\cos C_{35}+\cos D_{20}\cos D_{35}$

$\cos\phi_{21,22}=\cos A_{21}\cos A_{22}+\cos C_{212}\cos C_{22}+\cos D_{21}\cos D_{22}$

$\cos\phi_{21,23}=\cos A_{21}\cos A_{23}+\cos C_{21}\cos C_{23}+\cos D_{21}\cos D_{23}$

$\cos\phi_{21,25}=\cos A_{21}\cos A_{25}+\cos C_{21}\cos C_{25}+\cos D_{21}\cos D_{25}$

$\cos\phi_{22,32}=\cos A_{22}\cos A_{32}+\cos C_{22}\cos C_{32}+\cos D_{22}\cos D_{32}$

$\cos\phi_{22,34}=\cos A_{22}\cos A_{34}+\cos C_{22}\cos C_{34}+\cos D_{22}\cos D_{34}$

$\cos\phi_{22,36}=\cos A_{22}\cos A_{36}+\cos C_{22}\cos C_{36}+\cos D_{22}\cos D_{36}$

$\cos\phi_{23,24}=\cos A_{23}\cos A_{24}+\cos C_{23}\cos C_{24}+\cos D_{23}\cos D_{24}$

$\cos\phi_{23,25}=\cos A_{23}\cos A_{25}+\cos C_{23}\cos C_{25}+\cos D_{23}\cos D_{25}$

$\cos\phi_{24,35}=\cos A_{24}\cos A_{35}+\cos C_{24}\cos C_{35}+\cos D_{24}\cos D_{35}$

$\cos\phi_{24,37}=\cos A_{24}\cos A_{37}+\cos C_{24}\cos C_{37}+\cos D_{24}\cos D_{37}$

$\cos\phi_{24,37}=\cos A_{24}\cos A_{37}+\cos C_{24}\cos C_{37}+\cos D_{24}\cos D_{37}$

$\cos\phi_{25,36}=\cos A_{25}\cos A_{36}+\cos C_{25}\cos C_{36}+\cos D_{25}\cos D_{36}$

D

ANGLES (Cont'd)

$\cos\phi_{25,37}=\cos A_{25}\cos A_{37}+\cos C_{25}\cos C_{37}+\cos D_{25}\cos D_{37}$

$\cos\phi_{25,39}=\cos A_{25}\cos A_{39}+\cos C_{25}\cos C_{39}+\cos D_{25}\cos D_{39}$

$\cos\phi_{26,27}=\cos A_{26}\cos A_{27}+\cos C_{26}\cos C_{27}+\cos D_{26}\cos D_{27}$

$\cos\phi_{26,28}=\cos A_{26}\cos A_{28}+\cos C_{26}\cos C_{28}+\cos D_{26}\cos D_{28}$

$\cos\phi_{27,31}=\cos A_{27}\cos A_{31}+\cos C_{27}\cos C_{31}+\cos D_{27}\cos D_{31}$

$\cos\phi_{27,38}=\cos A_{27}\cos A_{38}+\cos C_{27}\cos C_{38}+\cos D_{27}\cos D_{38}$

$\cos\phi_{27,40}=\cos A_{27}\cos A_{40}+\cos C_{27}\cos C_{40}+\cos D_{27}\cos D_{40}$

$\cos\phi_{28,33}=\cos A_{28}\cos A_{33}+\cos C_{28}\cos D_{33}+\cos D_{28}\cos D_{33}$

$\cos\phi_{28,39}=\cos A_{28}\cos A_{39}+\cos C_{28}\cos C_{39}+\cos D_{28}\cos D_{39}$

$\cos\phi_{28,40}=\cos A_{28}\cos A_{40}+\cos C_{28}\cos C_{40}+\cos D_{28}\cos D_{40}$

$\cos\phi_{29,31}=\cos A_{29}\cos A_{31}+\cos C_{29}\cos C_{31}+\cos D_{29}\cos D_{31}$

$\cos\phi_{29,31}=\cos A_{29}\cos A_{31}+\cos C_{29}\cos C_{31}+\cos D_{29}\cos D_{31}$

$\cos\phi_{29,32}=\cos A_{29}\cos A_{32}+\cos C_{29}\cos C_{32}+\cos D_{29}\cos D_{32}$

$\cos\phi_{29,33}=\cos A_{29}\cos A_{33}+\cos C_{29}\cos C_{33}+\cos D_{29}\cos D_{33}$

$\cos\phi_{30,31}=\cos A_{30}\cos A_{31}+\cos C_{30}\cos C_{31}+\cos D_{30}\cos D_{31}$

$\cos\phi_{30,34}=\cos A_{30}\cos A_{34}+\cos C_{30}\cos C_{34}+\cos D_{30}\cos D_{34}$

$\cos\phi_{30,35}=\cos A_{30}\cos A_{35}+\cos C_{30}\cos C_{35}+\cos D_{30}\cos D_{35}$

$\cos\phi_{31,38}=\cos A_{31}\cos A_{38}+\cos C_{31}\cos C_{38}+\cos D_{31}\cos D_{38}$

$\cos\phi_{31,40}=\cos A_{31}\cos A_{40}+\cos C_{31}\cos C_{40}+\cos D_{31}\cos D_{40}$

$\cos\phi_{32,33}=\cos A_{32}\cos A_{33}+\cos C_{32}\cos C_{33}+\cos D_{32}\cos D_{33}$

$\cos\phi_{32,34}=\cos A_{32}\cos A_{34}+\cos C_{32}\cos C_{34}+\cos D_{32}\cos D_{34}$

$\cos\phi_{32,36}=\cos A_{32}\cos A_{36}+\cos C_{32}\cos C_{36}+\cos D_{32}\cos D_{36}$

$\cos\phi_{33,39}=\cos A_{33}\cos A_{39}+\cos C_{33}\cos C_{39}+\cos D_{33}\cos D_{39}$

$\cos\phi_{33,40}=\cos A_{33}\cos A_{40}+\cos C_{33}\cos C_{40}+\cos D_{33}\cos D_{40}$

$\cos\phi_{34,35}=\cos A_{34}\cos A_{35}+\cos C_{34}\cos C_{35}+\cos D_{34}\cos D_{35}$

$\cos\phi_{34,36}=\cos A_{34}\cos A_{36}+\cos C_{34}\cos C_{36}+\cos D_{34}\cos D_{36}$

$\cos\phi_{35,37}=\cos A_{35}\cos A_{37}+\cos C_{35}\cos C_{37}+\cos D_{35}\cos D_{37}$

$\cos\phi_{35,38}=\cos A_{35}\cos A_{38}+\cos C_{35}\cos C_{38}+\cos D_{35}\cos D_{38}$

$\cos\phi_{36,37}=\cos A_{36}\cos A_{37}+\cos C_{36}\cos C_{37}+\cos D_{36}\cos D_{37}$

$\cos\phi_{36,39}=\cos A_{36}\cos A_{39}+\cos C_{36}\cos C_{39}+\cos D_{36}\cos D_{39}$

$\cos\phi_{37,38}=\cos A_{37}\cos A_{38}+\cos C_{37}\cos C_{38}+\cos D_{37}\cos D_{38}$

$\cos\phi_{37,39}=\cos A_{37}\cos A_{39}+\cos C_{37}\cos C_{39}+\cos D_{37}\cos D_{39}$

$\cos\phi_{38,40}=\cos A_{38}\cos A_{40}+\cos C_{38}\cos C_{40}+\cos D_{38}\cos D_{40}$

$\cos\phi_{39,40}=\cos A_{39}\cos A_{40}+\cos C_{39}\cos C_{40}+\cos D_{39}\cos D_{40}$

CHECK

F_1: $\phi_{1,2}+\phi_{1,5}+\phi_{2,8}+\phi_{5,8}=360^\circ$

F_2: $\phi_{1,3}+\phi_{1,6}+\phi_{3,13}+\phi_{6,13}=360^\circ$

F_3: $\phi_{1,4}+\phi_{1,7}+\phi_{4,18}+\phi_{7,18}=360^\circ$

F_4: $\phi_{2,3}+\phi_{2,10}+\phi_{3,14}+\phi_{10,14}=360^\circ$

F_5: $\phi_{2,4}+\phi_{2,9}+\phi_{4,19}+\phi_{9,19}=360^\circ$

F_6: $\phi_{3,4}+\phi_{3,15}+\phi_{4,20}+\phi_{15,30}+\phi_{20,30}=550^\circ$

F_7: $\phi_{5,6}+\phi_{5,12}+\phi_{6,16}+\phi_{12,16}=360^\circ$

F_8: $\phi_{5,7}+\phi_{5,11}+\phi_{7,21}+\phi_{11,21}=360^\circ$

F_9: $\phi_{6,7}+\phi_{6,17}+\phi_{7,22}+\phi_{17,32}+\phi_{22,32}=550^\circ$

F_{10}: $\phi_{8,9}+\phi_{8,11}+\phi_{9,23}+\phi_{11,23}=360^\circ$

F_{11}: $\phi_{8,10}+\phi_{8,12}+\phi_{10,26}+\phi_{12,26}=360^\circ$

F_{12}: $\phi_{9,10}+\phi_{9,24}+\phi_{10,27}+\phi_{24,38}+\phi_{27,38}=550^\circ$

F_{13}: $\phi_{11,12}+\phi_{11,25}+\phi_{12,28}+\phi_{25,39}+\phi_{28,39}=550^\circ$

F_{14}: $\phi_{13,14}+\phi_{13,16}+\phi_{14,26}+\phi_{16,26}=360^\circ$

F_{15}: $\phi_{13,15}+\phi_{13,17}+\phi_{15,29}+\phi_{17,29}=360^\circ$

F_{16}: $\phi_{14,15}+\phi_{14,27}+\phi_{15,31}+\phi_{27,31}=360^\circ$

F_{17}: $\phi_{16,17}+\phi_{16,28}+\phi_{17,33}+\phi_{28,33}=360^\circ$

F_{18}: $\phi_{18,19}+\phi_{18,21}+\phi_{19,23}+\phi_{21,23}=360^\circ$

F_{19}: $\phi_{18,20}+\phi_{18,22}+\phi_{20,34}+\phi_{22,34}=360^\circ$

F_{20}: $\phi_{19,20}+\phi_{19,24}+\phi_{20,35}+\phi_{24,35}=360^\circ$

F_{21}: $\phi_{21,22}+\phi_{21,25}+\phi_{22,36}+\phi_{25,36}=360^\circ$

F_{22}: $\phi_{23,24}+\phi_{23,25}+\phi_{24,37}+\phi_{25,37}=360^\circ$

F_{23}: $\phi_{26,27}+\phi_{26,28}+\phi_{27,40}+\phi_{28,40}=360^\circ$

F_{24}: $\phi_{29,30}+\phi_{29,32}+\phi_{30,34}+\phi_{32,34}=360^\circ$

F_{25}: $\phi_{29,31}+\phi_{29,33}+\phi_{31,34}+\phi_{33,34}=360^\circ$

F_{26}: $\phi_{30,31}+\phi_{30,35}+\phi_{31,38}+\phi_{35,38}=360^\circ$

F_{27}: $\phi_{32,33}+\phi_{32,36}+\phi_{33,39}+\phi_{36,39}=360^\circ$

F_{28}: $\phi_{34,35}+\phi_{34,36}+\phi_{35,37}+\phi_{36,36}=360^\circ$

F_{29}: $\phi_{37,38}+\phi_{37,39}+\phi_{38,40}+\phi_{39,40}=360^\circ$

SECONDARY ANALYSIS -- YZW SPACE (See Fig. 22-13E)

PLOTTED POINTS

$P_1=(y_1,z_1,w_1)$ $P_2=(y_2,z_2,w_2)$

$P_3=(y_3,z_3,w_3)$ $P_4=(y_4,z_4,w_4)$

$P_5=(y_5,z_5,w_5)$ $P_6=(y_6,z_6,w_6)$

$P_7=(y_7,z_7,w_7)$ $P_8=(y_8,z_8,w_8)$

$P_9=(y_9,z_9,w_9)$ $P_{10}=(y_{10},z_{10},w_{10})$

$P_{11}=(y_{11},z_{11},w_{11})$ $P_{12}=(y_{12},z_{12},w_{12})$

$P_{13}=(y_{13},z_{13},w_{13})$ $P_{14}=(y_{14},z_{14},w_{14})$

$P_{15}=(y_{15},z_{15},w_{15})$ $P_{16}=(y_{16},z_{16},w_{16})$

$P_{17}=(y_{17},z_{17},w_{17})$ $P_{18}=(y_{18},z_{18},w_{18})$

$P_{19}=(y_{18},z_{19},w_{19})$ $P_{20}=(y_{20},z_{20},w_{20})$

LENGTHS

$s_1=\sqrt{(y_2-y_1)^2+(z_2-z_1)^2+(w_2-w_1)^2}$

$s_2=\sqrt{(y_3-y_1)^2+(z_3-z_1)^2+(w_3-w_1)^2}$

$s_3=\sqrt{(y_5-y_1)^2+(z_5-z_1)^2+(w_5-w_1)^2}$

$s_4=\sqrt{(y_7-y_1)^2+(z_7-z_1)^2+(w_7-w_1)^2}$

$s_5=\sqrt{(y_4-y_2)^2+(z_4-z_2)^2+(w_4-w_2)^2}$

$s_6=\sqrt{(y_6-y_2)^2+(z_6-z_2)^2+(w_6-w_2)^2}$

$s_7=\sqrt{(y_8-y_2)^2+(z_8-z_2)^2+(w_8-w_2)^2}$

$s_8=\sqrt{(y_4-y_3)^2+(z_4-z_3)^2+(w_4-w_3)^2}$

$s_9=\sqrt{(y_9-y_3)^2+(z_9-z_3)^2+(w_9-w_3)^2}$

$s_{10}=\sqrt{(y_{11}-y_3)^2+(z_{11}-z_3)^2+(w_{11}-w_3)^2}$

$s_{11}=\sqrt{(y_{10}-y_4)^2+(z_{10}-z_4)^2+(w_{10}-w_4)^2}$

E

Table 22-13. Specifications and Analyses for the Hyperspace Object Shown in Fig. 22-13. (A) Specifications and Analysis in Hyperspace. (B) Secondary Analysis in XYZ Space. (C) Secondary Analysis in XYW Space. (D) Secondary Analysis in XZW Space. (E) Secondary Analysis in YZW Space. (F) Suggested Coordinates. (Continued from page 477.)

LENGTHS (Cont'd)

$$s_{12}=\sqrt{(y_{12}-y_4)^2+(z_{12}-z_4)^2+(w_{12}-w_4)^2}$$

$$s_{13}=\sqrt{(y_6-y_5)^2+(z_6-z_5)^2+(w_6-w_5)^2}$$

$$s_{14}=\sqrt{(y_{11}-y_5)^2+(z_{11}-z_5)^2+(w_{11}-w_5)^2}$$

$$s_{15}=\sqrt{(y_{13}-y_5)^2+(z_{13}-z_5)^2+(w_{13}-w_5)^2}$$

$$s_{16}=\sqrt{(y_{12}-y_6)^2+(z_{12}-z_6)^2+(w_{12}-w_6)^2}$$

$$s_{17}=\sqrt{(y_{14}-y_6)^2+(z_{14}-z_6)^2+(w_{14}-w_6)^2}$$

$$s_{18}=\sqrt{(y_8-y_7)^2+(z_8-z_7)^2+(w_8-w_7)^2}$$

$$s_{19}=\sqrt{(y_9-y_7)^2+(z_9-z_7)^2+(w_9-w_7)^2}$$

$$s_{20}=\sqrt{(y_{15}-y_7)^2+(z_{15}-z_7)^2+(w_{15}-w_7)^2}$$

$$s_{21}=\sqrt{(y_{10}-y_8)^2+(z_{10}-z_8)^2+(w_{10}-w_8)^2}$$

$$s_{22}=\sqrt{(y_{16}-y_8)^2+(z_{16}-z_8)^2+(w_{16}-w_8)^2}$$

$$s_{23}=\sqrt{(y_{10}-y_9)^2+(z_{10}-z_9)^2+(w_{10}-w_9)^2}$$

$$s_{24}=\sqrt{(y_{17}-y_9)^2+(z_{17}-z_9)^2+(w_{17}-w_9)^2}$$

$$s_{25}=\sqrt{(y_{18}-y_{10})^2+(z_{18}-z_{10})^2+(w_{18}-w_{10})^2}$$

$$s_{26}=\sqrt{(y_{12}-y_{11})^2+(z_{12}-z_{11})^2+(w_{12}-w_{11})^2}$$

$$s_{27}=\sqrt{(y_{19}-y_{11})^2+(z_{19}-z_{11})^2+(w_{19}-w_{11})^2}$$

$$s_{28}=\sqrt{(y_{20}-y_{12})^2+(z_{20}-z_{12})^2+(w_{20}-w_{12})^2}$$

$$s_{29}=\sqrt{(y_{14}-y_{13})^2+(z_{14}-z_{13})^2+(w_{14}-w_{13})^2}$$

$$s_{30}=\sqrt{(y_{15}-y_{13})^2+(z_{15}-z_{13})^2+(w_{15}-w_{13})^2}$$

$$s_{31}=\sqrt{(y_{19}-y_{13})^2+(z_{19}-z_{13})^2+(w_{19}-w_{13})^2}$$

$$s_{32}=\sqrt{(y_{16}-y_{14})^2+(z_{16}-z_{14})^2+(w_{16}-w_{14})^2}$$

$$s_{33}=\sqrt{(y_{20}-y_{14})^2+(z_{20}-z_{14})^2+(w_{20}-w_{14})^2}$$

LENGTHS (Cont'd)

$$s_{34}=\sqrt{(y_{16}-y_{15})^2+(z_{16}-z_{15})^2+(w_{16}-w_{15})^2}$$

$$s_{35}=\sqrt{(y_{17}-y_{15})^2+(z_{17}-z_{15})^2+(w_{17}-w_{15})^2}$$

$$s_{36}=\sqrt{(y_{18}-y_{16})^2+(z_{18}-z_{16})^2+(w_{18}-w_{16})^2}$$

$$s_{37}=\sqrt{(y_{18}-y_{17})^2+(z_{18}-z_{17})^2+(w_{18}-w_{17})^2}$$

$$s_{38}=\sqrt{(y_{19}-y_{17})^2+(z_{19}-z_{17})^2+(w_{19}-w_{17})^2}$$

$$s_{39}=\sqrt{(y_{20}-y_{18})^2+(z_{20}-z_{18})^2+(w_{20}-w_{18})^2}$$

$$s_{40}=\sqrt{(y_{20}-y_{19})^2+(z_{20}-z_{19})^2+(w_{20}-w_{19})^2}$$

DIRECTION COSINES

$\cos B_1=(y_2-y_1)/s_1$ $\cos C_1=(z_2-z_1)/s_1$

$\cos D_1=(w_2-w_1)/s_1$

$\cos B_2=(y_3-y_1)/s_2$ $\cos C_2=(z_3-z_1)/s_2$

$\cos D_2=(w_3-w_1)/s_2$

$\cos B_3=(y_5-y_1)/s_3$ $\cos C_3=(z_5-z_1)/s_3$

$\cos D_3=(w_5-w_1)/s_3$

$\cos B_4=(y_7-y_1)/s_4$ $\cos C_4=(z_7-z_1)/s_4$

$\cos D_4=(w_7-w_1)/s_4$

$\cos B_5=(y_4-y_2)/s_5$ $\cos C_5=(z_4-z_2)/s_5$

$\cos D_5=(w_4-w_2)/s_5$

$\cos B_6=(y_6-y_2)/s_6$ $\cos C_6=(z_6-z_2)/s_6$

$\cos D_6=(w_6-w_2)/s_6$

$\cos B_7=(y_8-y_2)/s_7$ $\cos C_7=(z_8-z_2)/s_7$

$\cos D_7=(w_8-w_2)/s_7$

$\cos B_8=(y_4-y_3)/s_8$ $\cos C_8=(z_4-z_3)/s_8$

$\cos D_8=(w_4-w_3)/s_8$

E

$\cos B_9=(y_9-y_3)/s_9$ $\cos C_9=(z_9-z_3)/s_9$

$\cos D_9=(w_9-w_3)/s_9$

DIRECTION COSINES (Cont'd)

$\cos B_{10}=(y_{11}-y_3)/s_{10}$ $\cos C_{10}=(z_{11}-z_3)/s_{10}$

$\cos D_{10}=(w_{11}-w_3)/s_{10}$

$\cos B_{11}=(y_{10}-y_4)/s_{11}$ $\cos C_{11}=(z_{10}-z_4)/s_{11}$

$\cos D_{11}=(w_{10}-w_4)/s_{11}$

$\cos B_{12}=(y_{12}-y_4)/s_{12}$ $\cos C_{12}=(z_{12}-z_4)/s_{12}$

$\cos D_{12}=(w_{12}-w_4)/s_{12}$

$\cos B_{13}=(y_6-y_5)/s_{13}$ $\cos C_{13}=(z_6-z_5)/s_{13}$

$\cos D_{13}=(w_6-w_5)/s_{13}$

$\cos B_{14}=(y_{11}-y_5)/s_{14}$ $\cos C_{14}=(z_{11}-z_5)/s_{14}$

$\cos D_{14}=(w_{11}-w_5)/s$

$\cos B_{15}=(y_{13}-y_5)/s_{15}$ $\cos C_{15}=(z_{13}-z_5)/s_{15}$

$\cos D_{15}=(w_{13}-w_5)/s_{15}$

$\cos B_{16}=(y_{12}-y_6)/s_{16}$ $\cos C_{16}=(z_{12}-z_6)/s_{16}$

$\cos D_{16}=(w_{12}-w_6)/s_{16}$

$\cos B_{17}=(y_{14}-y_6)/s_{17}$ $\cos C_{17}=(z_{14}-z_6)/s_{17}$

$\cos D_{17}=(w_{14}-w_6)/s_{17}$

$\cos B_{18}=(y_8-y_7)/s_{18}$ $\cos C_{18}=(z_8-z_7)/s_{18}$

$\cos D_{18}=(w_8-w_7)/s_{18}$

$\cos B_{19}=(y_9-y_7)/s_{19}$ $\cos C_{19}=(z_9-z_7)/s_{19}$

$\cos D_{19}=(w_9-w_7)/s_{19}$

$\cos B_{20}=(y_{15}-y_7)/s_{20}$ $\cos C_{20}=(z_{15}-z_7)/s_{20}$

$\cos D_{20}=(w_{15}-w_7)/s_{20}$

$\cos B_{21}= (y_{10}-y_8)/s_{21}$ $\cos C_{21}=(z_{10}-z_8)/s_{21}$

$\cos D_{21}=(w_{10}-w_8)/s_{21}$

$\cos B_{22}= (y_{16}-y_8)/s_{22}$ $\cos C_{22}=(z_{16}-z_8)/s_{22}$

$\cos D_{22}=(w_{16}-w_8)/s_{22}$

DIRECTION COSINES (Cont'd)

$\cos B_{23}= (y_{10}-y_9)/s_{23}$ $\cos C_{23}=(z_{10}-z_9)/s_{23}$

$\cos D_{23}=(w_{10}-w_9)/s_{23}$

$\cos B_{24}= (y_{17}-y_9)/s_{24}$ $\cos C_{24}=(z_{17}-z_9)/s_{24}$

$\cos D_{24}=(w_{17}-w_9)/s_{24}$

$\cos B_{25}= (y_{18}-y_{10})/s_{25}$ $\cos C_{25}=(z_{18}-z_{10})/s_{25}$

$\cos D_{25}=(w_{18}-w_{10})/s_{25}$

$\cos B_{26}= (y_{12}-y_{11})/s_{26}$ $\cos C_{26}=(z_{12}-z_{11})/s_{26}$

$\cos D_{26}=(w_{12}-w_{11})/s_{26}$

$\cos B_{27}= (y_{19}-y_{11})/s_{27}$ $\cos C_{27}=(z_{19}-z_{11})/s_{27}$

$\cos D_{27}=(w_{19}-w_{11})/s_{27}$

$\cos B_{28}= (y_{20}-y_{12})/s_{28}$ $\cos C_{28}=(z_{20}-z_{12})/s_{28}$

$\cos D_{28}=(w_{20}-w_{12})/s_{28}$

$\cos B_{29}= (y_{14}-y_{13})/s_{29}$ $\cos C_{29}=(z_{14}-z_{13})/s_{29}$

$\cos D_{29}=(w_{14}-w_{13})/s_{29}$

$\cos B_{30}= (y_{15}-y_{13})/s_{30}$ $\cos C_{30}=(z_{15}-z_{13})/s_{30}$

$\cos D_{30}=(w_{15}-w_{13})/s_{30}$

$\cos B_{31}= (y_{19}-y_{13})/s_{31}$ $\cos C_{31}=(z_{19}-z_{13})/s_{31}$

$\cos D_{31}=(w_{19}-w_{13})/s_{31}$

$\cos B_{32}= (y_{16}-y_{14})/s_{32}$ $\cos C_{32}=(z_{16}-z_{14})/s_{32}$

$\cos D_{32}=(w_{16}-w_{14})/s_{32}$

$\cos B_{33}= (y_{20}-y_{14})/s_{33}$ $\cos C_{33}=(z_{20}-z_{14})/s_{33}$

$\cos D_{33}=(w_{20}-w_{14})/s_{33}$

$\cos B_{34}= (y_{16}-y_{15})/s_{34}$ $\cos C_{34}=(z_{16}-z_{15})/s_{34}$

$\cos D_{34}=(w_{16}-w_{15})/s_{34}$

$\cos B_{35}= (y_{17}-y_{15})/s_{35}$ $\cos C_{35}=(z_{17}-z_{15})/s_{35}$

$\cos D_{35}=(w_{17}-w_{15})/s_{35}$

Table 22-13. Specifications and Analyses for the Hyperspace Object Shown in Fig. 22-13. (A) Specifications and Analysis in Hyperspace. (B) Secondary Analysis in XYZ Space. (C) Secondary Analysis in XYW Space. (D) Secondary Analysis in XZW Space. (E) Secondary Analysis in YZW Space. (F) Suggested Coordinates. (Continued from page 479.)

DIRECTION COSINES (Cont'd)

$\cos B_{36} = (y_{18}-y_{16})/s_{36}$ $\cos C_{36} = (z_{18}-z_{16})/s_{36}$
$\cos D_{36} = (w_{18}-w_{16})/s_{36}$

$\cos B_{37} = (y_{18}-y_{17})/s_{37}$ $\cos C_{37} = (z_{18}-z_{17})/s_{37}$
$\cos D_{37} = (w_{18}-w_{17})/s_{37}$

$\cos B_{38} = (y_{19}-y_{17})/s_{38}$ $\cos C_{38} = (z_{19}-z_{17})/s_{38}$
$\cos D_{38} = (w_{19}-w_{17})/s_{38}$

$\cos B_{39} = (y_{20}-y_{18})/s_{39}$ $\cos C_{39} = (z_{20}-z_{18})/s_{39}$
$\cos D_{39} = (w_{20}-w_{18})/s_{39}$

$\cos B_{40} = (y_{20}-y_{19})/s_{40}$ $\cos C_{40} = (z_{20}-z_{19})/s_{40}$
$\cos D_{40} = (w_{20}-w_{19})/s_{40}$

ANGLES

$\cos\phi_{1,2} = \cos B_1 \cos B_2 + \cos C_1 \cos C_2 + \cos D_1 \cos D_2$
$\cos\phi_{1,3} = \cos B_1 \cos B_3 + \cos C_1 \cos C_3 + \cos D_1 \cos D_3$
$\cos\phi_{1,4} = \cos B_1 \cos B_4 + \cos C_1 \cos C_4 + \cos D_1 \cos D_4$
$\cos\phi_{1,5} = \cos B_1 \cos B_5 + \cos C_1 \cos C_5 + \cos D_1 \cos D_5$
$\cos\phi_{1,6} = \cos B_1 \cos B_6 + \cos C_1 \cos C_6 + \cos D_1 \cos D_6$
$\cos\phi_{1,7} = \cos B_1 \cos B_7 + \cos C_1 \cos C_7 + \cos D_1 \cos D_7$
$\cos\phi_{2,3} = \cos B_2 \cos B_3 + \cos C_2 \cos C_3 + \cos D_2 \cos D_3$
$\cos\phi_{2,4} = \cos B_2 \cos B_4 + \cos C_2 \cos C_4 + \cos D_2 \cos D_4$
$\cos\phi_{2,8} = \cos B_2 \cos B_8 + \cos C_2 \cos C_8 + \cos D_2 \cos D_8$
$\cos\phi_{2,9} = \cos B_2 \cos B_9 + \cos C_2 \cos C_9 + \cos D_2 \cos D_9$
$\cos\phi_{2,10} = \cos B_2 \cos B_{10} + \cos C_2 \cos C_{10} + \cos D_2 \cos D_{10}$
$\cos\phi_{3,4} = \cos B_3 \cos B_4 + \cos C_3 \cos C_4 + \cos D_3 \cos D_4$
$\cos\phi_{3,13} = \cos B_3 \cos B_{13} + \cos C_3 \cos C_{13} + \cos D_3 \cos D_{13}$
$\cos\phi_{3,14} = \cos B_3 \cos B_{14} + \cos C_3 \cos C_{14} + \cos D_3 \cos D_{14}$
$\cos\phi_{3,15} = \cos B_3 \cos B_{15} + \cos C_3 \cos C_{15} + \cos D_3 \cos D_{15}$
$\cos\phi_{4,18} = \cos B_4 \cos B_{18} + \cos C_4 \cos C_{18} + \cos D_4 \cos D_{18}$
$\cos\phi_{4,19} = \cos B_4 \cos B_{19} + \cos C_4 \cos C_{19} + \cos D_4 \cos D_{19}$
$\cos\phi_{4,20} = \cos B_4 \cos B_{20} + \cos C_4 \cos C_{20} + \cos D_4 \cos D_{20}$
$\cos\phi_{5,6} = \cos B_5 \cos B_6 + \cos C_5 \cos C_6 + \cos D_5 \cos D_6$
$\cos\phi_{5,7} = \cos B_5 \cos B_7 + \cos C_5 \cos C_7 + \cos D_5 \cos D_7$

E

ANGLES (Cont'd)

$\cos\phi_{5,8} = \cos B_5 \cos B_8 + \cos C_5 \cos C_8 + \cos D_5 \cos D_8$
$\cos\phi_{5,11} = \cos B_5 \cos B_{11} + \cos C_5 \cos C_{11} + \cos D_5 \cos D_{11}$
$\cos\phi_{5,12} = \cos B_5 \cos B_{12} + \cos C_5 \cos C_{12} + \cos D_5 \cos D_{12}$
$\cos\phi_{6,7} = \cos B_6 \cos B_7 + \cos C_6 \cos C_7 + \cos D_6 \cos D_7$
$\cos\phi_{6,13} = \cos B_6 \cos B_{13} + \cos C_6 \cos C_{13} + \cos D_6 \cos D_{13}$
$\cos\phi_{6,16} = \cos B_6 \cos B_{16} + \cos C_6 \cos C_{16} + \cos D_6 \cos D_{16}$
$\cos\phi_{6,17} = \cos B_6 \cos B_{17} + \cos C_6 \cos C_{17} + \cos D_6 \cos D_{17}$
$\cos\phi_{7,18} = \cos B_7 \cos B_{18} + \cos C_7 \cos C_{18} + \cos D_7 \cos D_{18}$
$\cos\phi_{7,21} = \cos B_7 \cos B_{21} + \cos C_7 \cos C_{21} + \cos D_7 \cos D_{21}$
$\cos\phi_{7,22} = \cos B_7 \cos B_{22} + \cos C_7 \cos C_{22} + \cos D_7 \cos D_{22}$
$\cos\phi_{8,9} = \cos B_8 \cos B_9 + \cos C_8 \cos C_9 + \cos D_8 \cos D_9$
$\cos\phi_{8,10} = \cos B_8 \cos B_{10} + \cos C_8 \cos C_{190} + \cos D_8 \cos D_{10}$
$\cos\phi_{8,11} = \cos B_8 \cos B_{11} + \cos C_8 \cos C_{11} + \cos D_8 \cos D_{11}$
$\cos\phi_{8,12} = \cos B_8 \cos B_{12} + \cos C_8 \cos C_{12} + \cos D_8 \cos D_{12}$
$\cos\phi_{9,10} = \cos B_9 \cos B_{10} + \cos C_9 \cos C_{10} + \cos D_9 \cos D_{10}$
$\cos\phi_{9,19} = \cos B_9 \cos B_{19} + \cos C_9 \cos C_{19} + \cos D_9 \cos D_{19}$
$\cos\phi_{9,23} = \cos B_9 \cos B_{23} + \cos C_9 \cos C_{23} + \cos D_9 \cos D_{23}$
$\cos\phi_{9,24} = \cos B_9 \cos B_{24} + \cos C_9 \cos C_{24} + \cos D_9 \cos D_{24}$
$\cos\phi_{10,14} = \cos B_{10} \cos B_{14} + \cos C_{10} \cos C_{14} + \cos D_{10} \cos D_{14}$
$\cos\phi_{10,26} = \cos B_{10} \cos B_{26} + \cos C_{10} \cos C_{26} + \cos D_{10} \cos D_{26}$
$\cos\phi_{10,27} = \cos B_{10} \cos B_{27} + \cos C_{10} \cos C_{27} + \cos D_{10} \cos D_{27}$
$\cos\phi_{11,12} = \cos B_{11} \cos B_{12} + \cos C_{11} \cos C_{12} + \cos D_{11} \cos D_{12}$
$\cos\phi_{11,21} = \cos B_{11} \cos B_{21} + \cos C_{11} \cos C_{21} + \cos D_{11} \cos D_{21}$
$\cos\phi_{11,23} = \cos B_{11} \cos B_{23} + \cos C_{11} \cos C_{23} + \cos D_{11} \cos D_{23}$
$\cos\phi_{11,25} = \cos B_{11} \cos B_{25} + \cos C_{11} \cos C_{25} + \cos D_{11} \cos D_{25}$
$\cos\phi_{12,16} = \cos B_{12} \cos B_{16} + \cos C_{12} \cos C_{16} + \cos D_{12} \cos D_{16}$
$\cos\phi_{12,26} = \cos B_{12} \cos B_{26} + \cos C_{12} \cos C_{26} + \cos D_{12} \cos D_{26}$
$\cos\phi_{12,28} = \cos B_{12} \cos B_{28} + \cos C_{12} \cos C_{28} + \cos D_{12} \cos D_{28}$
$\cos\phi_{13,14} = \cos B_{13} \cos B_{14} + \cos C_{13} \cos C_{14} + \cos D_{13} \cos D_{14}$
$\cos\phi_{13,15} = \cos B_{13} \cos B_{15} + \cos C_{13} \cos C_{15} + \cos D_{13} \cos D_{15}$
$\cos\phi_{13,16} = \cos B_{13} \cos B_{16} + \cos C_{13} \cos C_{16} + \cos D_{13} \cos D_{16}$
$\cos\phi_{13,17} = \cos B_{13} \cos B_{17} + \cos C_{13} \cos C_{17} + \cos D_{13} \cos D_{17}$
$\cos\phi_{14,15} = \cos B_{14} \cos B_{15} + \cos C_{14} \cos C_{15} + \cos D_{14} \cos D_{15}$
$\cos\phi_{14,26} = \cos B_{14} \cos B_{26} + \cos C_{14} \cos C_{26} + \cos D_{14} \cos D_{26}$

ANGLES (Cont'd)

$\cos\phi_{14,27}=\cos B_{14}\cos B_{27}+\cos C_{14}\cos C_{27}+\cos D_{14}\cos D_{27}$

$\cos\phi_{15,29}=\cos B_{15}\cos B_{29}+\cos C_{15}\cos C_{29}+\cos D_{15}\cos D_{29}$

$\cos\phi_{15,30}=\cos B_{15}\cos B_{30}+\cos C_{15}\cos C_{30}+\cos D_{15}\cos D_{30}$

$\cos\phi_{15,31}=\cos B_{15}\cos B_{31}+\cos C_{15}\cos C_{31}+\cos D_{15}\cos D_{31}$

$\cos\phi_{16,17}=\cos B_{16}\cos B_{17}+\cos C_{16}\cos C_{17}+\cos D_{16}\cos D_{17}$

$\cos\phi_{16,26}=\cos B_{16}\cos B_{26}+\cos C_{16}\cos C_{26}+\cos D_{16}\cos D_{26}$

$\cos\phi_{16,28}=\cos B_{16}\cos B_{28}+\cos C_{16}\cos C_{28}+\cos D_{16}\cos D_{28}$

$\cos\phi_{17,29}=\cos B_{17}\cos B_{29}+\cos C_{17}\cos C_{29}+\cos D_{17}\cos D_{29}$

$\cos\phi_{17,32}=\cos B_{17}\cos B_{32}+\cos C_{17}\cos C_{32}+\cos D_{17}\cos D_{32}$

$\cos\phi_{17,33}=\cos B_{17}\cos B_{33}+\cos C_{17}\cos C_{33}+\cos D_{17}\cos D_{33}$

$\cos\phi_{18,19}=\cos B_{18}\cos B_{19}+\cos C_{18}\cos C_{19}+\cos D_{18}\cos D_{19}$

$\cos\phi_{18,20}=\cos B_{18}\cos B_{20}+\cos C_{18}\cos C_{20}+\cos D_{178}\cos D_{20}$

$\cos\phi_{18,21}=\cos B_{18}\cos B_{21}+\cos C_{18}\cos C_{21}+\cos D_{18}\cos D_{21}$

$\cos\phi_{18,22}=\cos B_{18}\cos B_{22}+\cos C_{18}\cos C_{22}+\cos D_{18}\cos D_{22}$

$\cos\phi_{19,20}=\cos B_{19}\cos B_{20}+\cos C_{19}\cos C_{20}+\cos D_{19}\cos D_{20}$

$\cos\phi_{19,23}=\cos B_{19}\cos B_{23}+\cos C_{19}\cos C_{23}+\cos D_{19}\cos D_{23}$

$\cos\phi_{19,24}=\cos B_{19}\cos B_{24}+\cos C_{19}\cos C_{24}+\cos D_{19}\cos D_{24}$

$\cos\phi_{20,30}=\cos B_{20}\cos B_{30}+\cos C_{20}\cos C_{30}+\cos D_{20}\cos D_{30}$

$\cos\phi_{20,34}=\cos B_{20}\cos B_{34}+\cos C_{20}\cos C_{34}+\cos D_{20}\cos D_{34}$

$\cos\phi_{20,35}=\cos B_{20}\cos B_{35}+\cos C_{20}\cos C_{35}+\cos D_{20}\cos D_{35}$

$\cos\phi_{21,22}=\cos B_{21}\cos B_{22}+\cos C_{212}\cos C_{22}+\cos D_{21}\cos D_{22}$

$\cos\phi_{21,23}=\cos B_{21}\cos B_{23}+\cos C_{21}\cos C_{23}+\cos D_{21}\cos D_{23}$

$\cos\phi_{21,25}=\cos B_{21}\cos B_{25}+\cos C_{21}\cos C_{25}+\cos D_{21}\cos D_{25}$

$\cos\phi_{22,32}=\cos B_{22}\cos B_{32}+\cos C_{22}\cos C_{32}+\cos D_{22}\cos D_{32}$

$\cos\phi_{22,34}=\cos B_{22}\cos B_{34}+\cos C_{22}\cos C_{34}+\cos D_{22}\cos D_{34}$

$\cos\phi_{22,36}=\cos B_{22}\cos B_{36}+\cos C_{22}\cos C_{36}+\cos D_{22}\cos D_{36}$

$\cos\phi_{23,24}=\cos B_{23}\cos B_{24}+\cos C_{23}\cos C_{24}+\cos D_{23}\cos D_{24}$

$\cos\phi_{23,25}=\cos B_{23}\cos B_{25}+\cos C_{23}\cos C_{25}+\cos D_{23}\cos D_{25}$

$\cos\phi_{24,35}=\cos B_{24}\cos B_{35}+\cos C_{24}\cos C_{35}+\cos D_{24}\cos D_{35}$

$\cos\phi_{24,37}=\cos B_{24}\cos B_{37}+\cos C_{24}\cos C_{37}+\cos D_{24}\cos D_{37}$

$\cos\phi_{24,37}=\cos B_{24}\cos B_{37}+\cos C_{24}\cos C_{37}+\cos D_{24}\cos D_{37}$

$\cos\phi_{25,36}=\cos B_{25}\cos B_{36}+\cos C_{25}\cos C_{36}+\cos D_{25}\cos D_{36}$

$\cos\phi_{25,37}=\cos B_{25}\cos B_{37}+\cos C_{25}\cos C_{37}+\cos D_{25}\cos D_{37}$

$\cos\phi_{25,39}=\cos B_{25}\cos B_{39}+\cos C_{25}\cos C_{39}+\cos D_{25}\cos D_{39}$

ANGLES (Cont'd)

$\cos\phi_{26,27}=\cos B_{26}\cos B_{27}+\cos C_{26}\cos C_{27}+\cos D_{26}\cos D_{27}$

$\cos\phi_{26,28}=\cos B_{26}\cos B_{28}+\cos C_{26}\cos C_{28}+\cos D_{26}\cos D_{28}$

$\cos\phi_{27,31}=\cos B_{27}\cos B_{31}+\cos C_{27}\cos C_{31}+\cos D_{27}\cos D_{31}$

$\cos\phi_{27,38}=\cos B_{27}\cos B_{38}+\cos C_{27}\cos C_{38}+\cos D_{27}\cos D_{38}$

$\cos\phi_{27,40}=\cos B_{27}\cos B_{40}+\cos C_{27}\cos C_{40}+\cos D_{27}\cos D_{40}$

$\cos\phi_{28,33}=\cos B_{28}\cos B_{33}+\cos C_{28}\cos D_{33}+\cos D_{28}\cos D_{33}$

$\cos\phi_{28,39}=\cos B_{28}\cos B_{39}+\cos C_{28}\cos C_{39}+\cos D_{28}\cos D_{39}$

$\cos\phi_{28,40}=\cos B_{28}\cos B_{40}+\cos C_{28}\cos C_{40}+\cos D_{28}\cos D_{40}$

$\cos\phi_{29,31}=\cos B_{29}\cos B_{31}+\cos C_{29}\cos C_{31}+\cos D_{29}\cos D_{31}$

$\cos\phi_{29,31}=\cos B_{29}\cos B_{31}+\cos C_{29}\cos C_{31}+\cos D_{29}\cos D_{31}$

$\cos\phi_{29,32}=\cos B_{29}\cos B_{32}+\cos C_{29}\cos C_{32}+\cos D_{29}\cos D_{32}$

$\cos\phi_{29,33}=\cos B_{29}\cos B_{33}+\cos C_{29}\cos C_{33}+\cos D_{29}\cos D_{33}$

$\cos\phi_{30,31}=\cos B_{30}\cos B_{31}+\cos C_{30}\cos C_{31}+\cos D_{30}\cos D_{31}$

$\cos\phi_{30,34}=\cos B_{30}\cos B_{34}+\cos C_{30}\cos C_{34}+\cos D_{30}\cos D_{34}$

$\cos\phi_{30,35}=\cos B_{30}\cos B_{35}+\cos C_{30}\cos C_{35}+\cos D_{30}\cos D_{35}$

$\cos\phi_{31,38}=\cos B_{31}\cos B_{38}+\cos C_{31}\cos C_{38}+\cos D_{31}\cos D_{38}$

$\cos\phi_{31,40}=\cos B_{31}\cos B_{40}+\cos C_{31}\cos C_{40}+\cos D_{31}\cos D_{40}$

$\cos\phi_{32,33}=\cos B_{32}\cos B_{33}+\cos C_{32}\cos C_{33}+\cos D_{32}\cos D_{33}$

$\cos\phi_{32,34}=\cos B_{32}\cos B_{34}+\cos C_{32}\cos C_{34}+\cos D_{32}\cos D_{34}$

$\cos\phi_{32,36}=\cos B_{32}\cos B_{36}+\cos C_{32}\cos C_{36}+\cos D_{32}\cos D_{36}$

$\cos\phi_{33,39}=\cos B_{33}\cos B_{39}+\cos C_{33}\cos C_{39}+\cos D_{33}\cos D_{39}$

$\cos\phi_{33,40}=\cos B_{33}\cos B_{40}+\cos C_{33}\cos C_{40}+\cos D_{33}\cos D_{40}$

$\cos\phi_{34,35}=\cos B_{34}\cos B_{35}+\cos C_{34}\cos C_{35}+\cos D_{34}\cos D_{35}$

$\cos\phi_{34,36}=\cos B_{34}\cos B_{36}+\cos C_{34}\cos C_{36}+\cos D_{34}\cos D_{36}$

$\cos\phi_{35,37}=\cos B_{35}\cos B_{37}+\cos C_{35}\cos C_{37}+\cos D_{35}\cos D_{37}$

$\cos\phi_{35,38}=\cos B_{35}\cos B_{38}+\cos C_{35}\cos C_{38}+\cos D_{35}\cos D_{38}$

$\cos\phi_{36,37}=\cos B_{36}\cos B_{37}+\cos C_{36}\cos C_{37}+\cos D_{36}\cos D_{37}$

$\cos\phi_{36,39}=\cos B_{36}\cos B_{39}+\cos C_{36}\cos C_{39}+\cos D_{36}\cos D_{39}$

$\cos\phi_{37,38}=\cos B_{37}\cos B_{38}+\cos C_{37}\cos C_{38}+\cos D_{37}\cos D_{38}$

$\cos\phi_{37,39}=\cos B_{37}\cos B_{39}+\cos C_{37}\cos C_{39}+\cos D_{37}\cos D_{39}$

$\cos\phi_{38,40}=\cos B_{38}\cos B_{40}+\cos C_{38}\cos C_{40}+\cos D_{38}\cos D_{40}$

$\cos\phi_{39,40}=\cos B_{39}\cos B_{40}+\cos C_{39}\cos C_{40}+\cos D_{39}\cos D_{40}$

Table 22-13. Specifications and Analyses for the Hyperspace Object Shown in Fig. 22-13. (A) Specifications and Analysis in Hyperspace. (B) Secondary Analysis in XYZ Space. (C) Secondary Analysis in XYW Space. (D) Secondary Analysis in XZW Space. (E) Secondary Analysis in YZW Space. (F) Suggested Coordinates. (Continued from page 481.)

CHECK

F_1:	$\phi_{1,2}+\phi_{1,5}+\phi_{2,8}+\phi_{5,8}=360^\circ$
F_2:	$\phi_{1,3}+\phi_{1,6}+\phi_{3,13}+\phi_{6,13}=360^\circ$
F_3:	$\phi_{1,4}+\phi_{1,7}+\phi_{4,18}+\phi_{7,18}=360^\circ$
F_4:	$\phi_{2,3}+\phi_{2,10}+\phi_{3,14}+\phi_{10,14}=360^\circ$
F_5:	$\phi_{2,4}+\phi_{2,9}+\phi_{4,19}+\phi_{9,19}=360^\circ$
F_6:	$\phi_{3,4}+\phi_{3,15}+\phi_{4,20}+\phi_{15,30}+\phi_{20,30}=550^\circ$
F_7:	$\phi_{5,6}+\phi_{5,12}+\phi_{6,16}+\phi_{12,16}=360^\circ$
F_8:	$\phi_{5,7}+\phi_{5,11}+\phi_{7,21}+\phi_{11,21}=360^\circ$
F_9:	$\phi_{6,7}+\phi_{6,17}+\phi_{7,22}+\phi_{17,32}+\phi_{22,32}=550^\circ$
F_{10}:	$\phi_{8,9}+\phi_{8,11}+\phi_{9,23}+\phi_{11,23}=360^\circ$
F_{11}:	$\phi_{8,10}+\phi_{8,12}+\phi_{10,26}+\phi_{12,26}=360^\circ$
F_{12}:	$\phi_{9,10}+\phi_{9,24}+\phi_{10,27}+\phi_{24,38}+\phi_{27,38}=550^\circ$
F_{13}:	$\phi_{11,12}+\phi_{11,25}+\phi_{12,28}+\phi_{25,39}+\phi_{28,39}=550^\circ$
F_{14}:	$\phi_{13,14}+\phi_{13,16}+\phi_{14,26}+\phi_{16,26}=360^\circ$
F_{15}:	$\phi_{13,15}+\phi_{13,17}+\phi_{15,29}+\phi_{17,29}=360^\circ$
F_{16}:	$\phi_{14,15}+\phi_{14,27}+\phi_{15,31}+\phi_{27,31}=360^\circ$
F_{17}:	$\phi_{16,17}+\phi_{16,28}+\phi_{17,33}+\phi_{28,33}=360^\circ$
F_{18}:	$\phi_{18,19}+\phi_{18,21}+\phi_{19,23}+\phi_{21,23}=360^\circ$
F_{19}:	$\phi_{18,20}+\phi_{18,22}+\phi_{20,34}+\phi_{22,34}=360^\circ$
F_{20}:	$\phi_{19,20}+\phi_{19,24}+\phi_{20,35}+\phi_{24,35}=360^\circ$
F_{21}:	$\phi_{21,22}+\phi_{21,25}+\phi_{22,36}+\phi_{25,36}=360^\circ$
F_{22}:	$\phi_{23,24}+\phi_{23,25}+\phi_{24,37}+\phi_{25,37}=360^\circ$
F_{23}:	$\phi_{26,27}+\phi_{26,28}+\phi_{27,40}+\phi_{28,40}=360^\circ$
F_{24}:	$\phi_{29,30}+\phi_{29,32}+\phi_{30,34}+\phi_{32,34}=360^\circ$
F_{25}:	$\phi_{29,31}+\phi_{29,33}+\phi_{31,34}+\phi_{33,34}=360^\circ$
F_{26}:	$\phi_{30,31}+\phi_{30,35}+\phi_{31,38}+\phi_{35,38}=360^\circ$
F_{27}:	$\phi_{32,33}+\phi_{32,36}+\phi_{33,39}+\phi_{36,39}=360^\circ$
F_{28}:	$\phi_{34,35}+\phi_{34,36}+\phi_{35,37}+\phi_{36,36}=360^\circ$
F_{29}:	$\phi_{37,38}+\phi_{37,39}+\phi_{38,40}+\phi_{39,40}=360^\circ$

HYPERSPACE VERSION 1

$P_1=(0,1,1,1)$	$P_2=(0,1,1,-1)$	$P_3=(0,1,-1,1)$
$P_4=(0,1,-1,-1)$	$P_5=(-1,0,1,1)$	$P_6=(-1,0,1,-1)$
$P_7=(1,0,1,1)$	$P_8=(1,0,1,-1)$	$P_9=(1,0,-1,0)$
$P_{10}=(1,0,-1,-1)$	$P_{11}=(-1,0,-1,1)$	$P_{12}=(-1,0,-1,-1)$
$P_{13}=(-1,-1,1,1)$	$P_{14}=(-1,-1,1,-1)$	$P_{15}=(1,-1,1,1)$
$P_{16}=(1,-1,1,-1)$	$P_{17}=(1,-1,-1,1)$	$P_{18}=(1,-1,-1,-1)$
$P_{19}=(-1,-1,-1,1)$	$P_{20}=(-1,-1,-1,-1)$	

HYPERSPACE VERSION 2

$P_1=(0,4,1,1)$	$P_2=(0,4,1,-1)$	$P_3=(0,4,-1,1)$
$P_4=(0,4,-1,-1)$	$P_5=(-1,2,1,1)$	$P_6=(-1,2,1,-1)$
$P_7=(1,2,1,1)$	$P_8=(1,2,1,-1)$	$P_9=(1,2,-1,1)$
$P_{10}=(1,2,-1,-1)$	$P_{11}=(-1,2,-1,1)$	$P_{12}=(-1,2,-1,-1)$
$P_{13}=(-1,0,1,1)$	$P_{14}=(-1,0,1,-1)$	$P_{15}=(1,0,1,1)$
$P_{16}=(1,0,1,-1)$	$P_{17}=(1,0,-1,1)$	$P_{18}=(1,0,-1,-1)$
$P_{19}=(-1,0,-1,1)$	$P_{20}=(-1,0,-1,-1)$	

F

```
HYPERSPACE VERSION 3

P1=(1,1,1,1)         P2=(1,1,1,-1)        P3=(-1,1,-1,1)
P4=(-1,1,-1,-1)      P5=(0,0,2,1)         P6=(0,0,2,-1)
P7=(2,0,0,1)         P8=(2,0,0,-1)        P9=(0,0,-2,1)
P10=(0,0,-2,-1)      P11=(-2,0,0,1)       P12=(-2,0,0,-1)
P13=(0,-1,-2,1)      P14=(0,-1,-2,-1)     P15=(2,-1,0,1)
P16=(2,-1,0,-1)      P17=(-2,-1,-2,1)     P18=(-2,-1,-2,-1)
P19=(-2,-1,0,1)      P20=(-2,-1,0,-1)
```

reflect (reverse) the scaling of a coordinate system. The critical parameters of a line in hyperspace—length, direction cosines and angles between lines—are variant except under certain symmetrical scalings. A good rule-of-thumb is to not count on any line parameter being unchanged by any sort of geometric or relativistic scaling transformation in 4-D space.

22-5.1 Some Suggested Experiments

You have already seen that geometric scalings of a figure, space or hyperspace object offers a chance to mold the entity into some other desired form. Consider this type of experiment:

☐ Specify a reference entity according to the methods outlined in Sections 22-1 and 22-2.

☐ Conduct complete primary and secondary analyses of that reference entity.

☐ Specify some desired scaling factors and carry out the scaling operation.

☐ Write out the specifications for the scaled version of the entity, and conduct complete analyses of it.

☐ Plot both the reference and scaled versions of the entity, using the general 4-D views and the alternative 3-D perspective views.

Conduct such experiments for a number of different figures, space and hyperspace objects specified in 4-D space. Be sure to select both symmetrical and asymmetrical scalings at one time or another. Selecting scaling factors to accomplish the effect that you want can't be difficult at first because the notion of a 4th dimension is so terribly abstract. Given some experience—first-hand experience—you will find that you can get pretty good at it. You might not be able to "see" hyperspace, but you can certainly learn to manipulate it.

Then try applying successive hyperspace scalings to some interesting geometric entities, making them expand, shrink or reflect about an axis, plane or 3-D space. Plotting the results to 3-D perspective views can be a time-consuming process, but the results can be striking.

Finally, try some experiments built around the 4-D relativistic scaling equations, plotting a space object as it appears from two different frames of reference—frames of reference that differ only in the scaling that exists between them. Asymmetrical and reflection-type relativistic scaling effects are particularly intriguing, because they can represent strange, but possibly very real, hyperspace scalings. (Set up some experiments that combine relativistic and geometric scalings in four dimensions.)

22-5.2 Some Suggested Special Studies

Experimenters who feel prepared to deal with 4-D scalings in a more general fashion might find the following special studies pointing the way to some original research:

☐ Derive a general equation for the length of a scaled line in four dimensions as a function of the coordinates of the points in the reference version of the figure and the scaling factors.

☐ Derive a general equation for the length of a scaled line in 4-D space as a function of its reference length and the scaling factors.

☐ Derive a general equation for calculating the angle between two lines in 4-D space as a function of the corresponding angle in the reference figure and the given scaling factors.

22-6 SIMPLE ROTATIONS AND SUGGESTED EXPERIMENTS

Simple rotations in 4-D space turn a coordinate system about a selected space of rotation. The reference and rotated versions of any geometric entities specified within the coordinate systems are identical except for their position and angular orientation with respect to the coordinate planes. The lengths of lines and angles between them are invariant.

22-6.1 Taking Advantage of the Invariants

Only the direction cosines of lines are variant under simple rotations in 4-D space. The fact that the lengths of corresponding lines and the corresponding angles are invariant makes it possible to doublecheck the results of hyperspace rotation by comparing those values for the reference and rotated versions.

As mentioned earlier, hyperspace rotations represent the most sophisticated and potentially useful transformations described in this book. If you have been working diligently along these lines, you doubtless agree. Doing the mathematics of the rotations is a moderately difficult task; one that is perhaps eclipsed only by the time and effort involved in analyzing and plotting the results. But it's all worth it in the long run.

22-6.2 Suggested Experiments and Studies

The most straightforward kinds of experiments with simple rotations in 4-D space take this general form:

☐ Specify a reference figure, space or hyperspace object in four dimensions.

☐ If you haven't already done so in earlier work, conduct complete analyses of the reference entity.

☐ Carry out a desired rotation about a selected plane.

☐ Establish the formal specifications for the rotated version of the entity and conduct complete primary and secondary analyses of it.

☐ Generate a complete series of plots for both the reference and rotated versions of the entity.

Try applying a succession of some 4-D rotations of relatively small angle such as 15 degrees to an interesting reference entity, and then applying another 15-degree rotation to the result. Also try different planes of rotation.

Then, of course, there is the matter of relativistic rotations. The objective is to specify a relativistic rotation that exists between an observer's and rotated coordinate system, specify a space or hyperspace object as it would appear from the observer's point of view, and then use the relativistic rotation equations to find out how that figure appears from the rotated frame of reference.

Combining relativistic and geometric rotations into the same experiment is the most complex sort of task you can carry out with the information provided in this book.

☐ Specify a fixed relativistic 4-D rotation that is to exist between the observer's and the rotated frames of reference.

☐ Specify a reference object relative to the observer's frame of reference.

☐ Apply the relativistic rotation equation to see how the reference object appears from the rotated frame of reference; especially from one of the alternative 3-D spaces.

☐ Apply some geometric angle of rotation to the space object.

☐ Repeat the two steps above any number of times, plotting the results in each case. Such work can be made visually interesting by using different colors of pencil or ink.

An interpretation of that last experiment is in order. What you are doing is considering a body that is rotating in hyperspace; considering it from two different 4-D frames of reference that are, themselves, separated by a relativistic rotation in hyperspace. The 3-D mappings and secondary analyses of the results strongly suggest something about the nature of bodies and the gravitation that surrounds any mass in space. It's Isaac Newton updated.

As far as special studies are concerned, consider these:

☐ Conduct a formal analytic proof of the fact that length is an invariant under 4-D rotations, regardless of the chosen space of rotation.

☐ Conduct a formal analytic proof that the internal angles of a space objects in 4-D space are invariant under simple 4-D rotations, regardless of the chosen axis of rotation. Extend the proof to include the angles of a hyperspace object.

☐ Derive a set of equations that determine the direction cosines of a line that is being rotated in 4-D space as a function of the angles and spaces of rotation.

☐ Derive a set of equations that express directly the coordinate of a point that is transformed both geometrically and in a relativistic sense.

☐ Consult a physics text for the general equations expressing force in terms of angular velocity and mass, and then see if you can derive some equations for expressing a force generated by relativistic hyperspace rotations.

22-7 COMBINATIONS OF TRANSLATIONS, SCALINGS AND ROTATIONS

You saw in Chapters 9 and 15 that combinations of transformations are not commutative in 2-D and 3-D spaces. Of course they are not commutative in 4-D space, either, and you can use the derivation technique outlined in Chapter 9 to derive equations that lead directly to the coordinates of points under 4-D versions of various combinations of two or more hyperspace transformations.

Combinations of translations and rotations line up with some of the work you might find in an advanced course in physics; combinations of all three major transformations are generally left to space-time ideas of a modern variety.

This isn't grocery-list arithmetic you are dealing with here. It is mathematics in its powerful form—a form that calls for the application of a lot of imagination, discipline and philosophical interpretation. Somewhere in all of this; somewhere among the suggested experiments and roughly sketched ideas; are gems of truth just waiting to be uncovered and entered into the record of human intellectual achievement. Perhaps the overall feeling I hope you have gained from your study is a clear understanding of the fact that no one has to be cloistered within the walls of a famous university or think tank in order to deal with such matters in a responsible way.

Work hard, think clearly, play around with the mathematics, study, let your imagination and intuition roam, harness your enthusiasm with a creative application of sound principles. What happens after that is up to you.

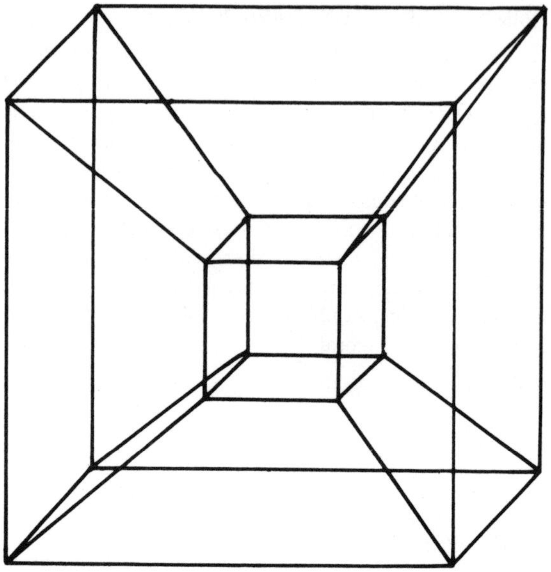

Appendix
Computer Programs

Experimenters who have access to a personal computer can save some time, paper, and pencil lead by using the BASIC program suggested in this appendix. Essentially the program takes care of calculating coordinates for geometric translations, scalings and rotations in dimensions 1 through 4; and it can determine the lengths and direction cosines for lines. The notation and terminiology appearing on the screen follows that of the book. So if you have been studying the book carefully, you will be able to begin using the program with very little practice.

The program always begins by requesting the number of dimensions for the experiment you are about to begin. After that, you will see a menu that gives you an opportunity to select what you want to do next. You can always return to that menu after doing some operations; and that means you have complete control over the experiment.

The main menu has this general form:

1—Specify all new coordinates
2—Display current coordinates
3—Edit current coordinates
4—Translate the coordinates
5—Scale the coordinates
6—Rotate the coordinates
7—Calculate lengths of lines
8—Quit the program

SPECIFY ALL NEW COORDINATES

This operation gives you the opportunity to specify the coordinates for as many as 32 different points in the number of dimensions you specified in the opening phase of the program. Of course you must use it first when setting up a new experiment, but you can return to it at any later time to start a new experiment in the same number of dimensions.

The routine first requests the number of points you want to use, then it prompts you to enter the appropriate coordinate for each point. After you have completed the coordinate-entry operation, the program summarizes your coordinates on the screen and returns to the main menu.

DISPLAY CURRENT COORDINATES

That menu selection lets you view a summary of the current coordinates at any time through the experiment. The program returns to the main menu after you've had a chance to study those coordinates.

EDIT CURRENT COORDINATES

This option gives you a chance to change selected coordinates. It is most useful in instances where you have perhaps made an error in specifying some of the original coordinates; but you can also use it to make changes in the coordinates during the course of an experiment. After you have edited a point, the program summarizes the revised list of coordinates, then gives you a chance to edit another or return to the main menu.

TRANSLATE THE COORDINATES

This operation begins by requesting an appropriate selection of translation terms, then the program executes the translation and displays the results. The program returns to the main menu from that point.

SCALE THE COORDINATES

The program requests an appropriate number of scaling factors, executes the scaling transformation, displays the results and returns to the main menu.

ROTATE THE COORDINATES

This part of the program can be executed only for dimensions 2 through 4. It first requests an angle of rotation and then asks you to specify an axis of rotation (if you are working in 3-D space) or a plane of rotation (if you are working in 4-D) space. Rotations in 2-D space are automatically set for rotations about the origin. The routine then executes the translation on all of your current point coordinates, displays the resulting coordinates and returns to the main menu.

CALCULATE LENGTHS OF LINES

This menu selection begins by displaying the current coordinates and requesting the two that are to define the endpoints of a line. The program then calculates and displays the length of the line and its direction cosines. After that, you have the choice of specifying another line or returning to the main menu.

RESTORE ORIGINAL COORDINATES

This option should be used only when you want to restart an experiment, using the coordinates you originally entered under the *Specify all new coordinates.* The results of any previous transformations will be lost. The program returns to the main menu after executing this operation.

QUIT THE PROGRAM

Use this menu selection only to end your work or change the number of dimensions for the next experiment.

There are two different listings here for the same program. One is written in Microsoft BASIC® (for the IBM Personal Computer and TRS-80 Models I and III). The other is written in Applesoft BASIC® for all Apple machines equipped with floating point BASIC. Both require at least 16K of RAM.

TRS-80 users need not be concerned with lowercase letters in the messages (unless, of course, the machine has that capability), but they must make the following modifications of the Microsoft version:

160 PRINT "EXPERIMENTS IN FOUR DIMENSIONS";:PRINT TAB(40) "Setup Routine"

170 PRINT STRING$(64,"-"):PRINT

270 MESS$="EXPERIMENTS IN "+X$+DCHR(10)+STRING$(64,"-")+CHR$(10)

3050 IF POS(X)>50 THEN PRINT

Slight modifications of the Applesoft version make it suitable for ATARI computers.

MICROSOFT VERSION

```
10 '
20 '             EXPERIMENTS IN FOUR DIMENSIONS
30 '
40 ' Based on the book EXPERIMENTS IN FOUR DIMENSIONS
50 '             BY DAVID L. HEISERMAN
60 '
70 'Copyright 1983
80 'Tab Books Inc.
90 'Blue Ridge Summit  PA
100 '
110 '            Setup routine
```

```
120 '
130 DIM C(2,32,4),CP$(2,32,4)
140 DEFINT N
150 KEY OFF:CLS
160 PRINT "EXPERIMENTS IN FOUR DIMENSIONS";:PRINT TAB(50) "Setup Routine
170 PRINT STRING$(80,"~"):PRINT
180 INPUT "How many dimensions (1-4)";ND
190 IF ND>=1 AND ND<=4 THEN 210
200 PRINT:PRINT "Entry error. Try again ...":PRINT:GOTO 180
210 D$=" DIMENSIONS"
220 ON ND GOTO 230,240,250,260
230 X$="ONE":D$=" DIMENSION":GOTO 270
240 X$="TWO":GOTO 270
250 X$="THREE":GOTO 270
260 X$="FOUR"
270 MESS$="EXPERIMENTS IN "+X$+D$+CHR$(10)+STRING$(80,"~")+CHR$(10)
280 '
1000 '
1010 '          Main Menu
1020 '
1030 CLS:PRINT MESS$:PRINT:PRINT "Main Menu:":PRINT
1040 PRINT TAB(5) "1 -- Specify all new coordinates
1050 PRINT TAB(5) "2 -- Display current coordinates
1060 PRINT TAB(5) "3 -- Edit current coordinates
1070 PRINT TAB(5) "4 -- Translate the coordinates
1080 PRINT TAB(5) "5 -- Scale the coordinates
1090 PRINT TAB(5) "6 -- Rotate the coordinates
1100 PRINT TAB(5) "7 -- Calculate lengths of lines
1110 PRINT TAB(5) "8 -- Restore original coordinates
1120 PRINT TAB(5) "9 -- Quit the program
1130 PRINT:GOSUB 1300
1140 IF ASC(K$)>=49 AND ASC(K$)<=57 THEN 1160
1150 PRINT:PRINT "Entry error.  Try again ...":GOTO 1130
1160 ON VAL(K$) GOTO 1170,1180,1190,1200,1210,1220,1230,1240,1250
1170 T=1:GOSUB 2000:GOTO 1000
1180 GOSUB 3000:GOSUB 1400:GOTO 1000
1190 GOSUB 4000:GOTO 1000
1200 T=2:GOSUB 5000:GOTO 1000
1210 T=2:GOSUB 6000:GOTO 1000
1220 T=2:GOSUB 7000:GOTO 1000
1230 GOSUB 8000:GOTO 1000
1240 T=1:GOSUB 9000:GOTO 1000
1250 CLS:KEY ON:END
1300 '
1310 '          Select one
1320 '
1330 PRINT:PRINT "Select one ..."
1340 GOSUB 1440:RETURN
1350 '
1400 '
1410 '           Strike any key
1420 '
1430 PRINT "Strike any key to return to Main Menu ..."
1440 K$=INKEY$:IF K$="" THEN 1440
1450 RETURN
1460 '
2000 '
```

```
2010 '           New coordinates
2020 '
2030 CLS:PRINT MESS$:PRINT
2040 INPUT "How many points (1-32)";NP
2050 IF NP>=1 AND NP<=32 THEN 2070
2060 PRINT:PRINT "Entry error. Try again ...":PRINT:GOTO 2040
2070 CLS:PRINT MESS$:PRINT
2080 PRINT "Enter your"NP"point coordinate";
2090 IF NP>1 THEN PRINT "s";
2100 PRINT " as x";
2110 IF ND>1 THEN PRINT ",y";
2120 IF ND>2 THEN PRINT ",z";
2130 IF ND>3 THEN PRINT ",w"
2140 PRINT:PRINT
2150 FOR N=1 TO NP
2160 PRINT TAB(20) "P"MID$(STR$(N),2);
2170 ON ND GOTO 2180,2190,2200,2210
2180 INPUT "=",C(1,N,1):GOTO 2220
2190 INPUT "=",C(1,N,1),C(1,N,2):GOTO 2220
2200 INPUT "=",C(1,N,1),C(1,N,2),C(1,N,3):GOTO 2220
2210 INPUT "=",C(1,N,1),C(1,N,2),C(1,N,3),C(1,N,4)
2220 FOR X=1 TO ND:C(2,N,X)=C(1,N,X):NEXT X
2230 NEXT N
2240 GOSUB 3000:GOSUB 1400:GOTO 1000
2250 '
3000 '
3010 '           Pack and print coordinates
3020 '
3030 CLS:PRINT MESS$:PRINT "Summary of current points:":PRINT
3040 FOR N=1 TO NP
3050 IF POS(X)>60 THEN PRINT
3060 PRINT "P";:IF T>1 THEN PRINT "'";
3070 PRINT MID$(STR$(N),2)"=(";
3080 FOR X=1 TO ND
3090 CP$(1,N,X)=STR$(INT(1000*C(1,N,X))/1000)
3100 IF ASC(LEFT$(CP$(1,N,X),1))=32 THEN CP$(1,N,X)=MID$(CP$(1,N,X),2)
3110 PRINT CP$(1,N,X)",";
3120 NEXT X
3130 PRINT CHR$(29)+")",
3140 NEXT N
3150 PRINT:PRINT:RETURN
3160 '
4000 '
4010 '           Edit
4020 '
4030 GOSUB 3000
4040 PRINT:PRINT "Which point to you want to edit?"
4050 INPUT "(Strike Return to return to Main Menu) ",PE$
4060 IF PE$="" THEN RETURN
4070 IF LEFT$(PE$,1)="P" OR LEFT$(PE$,1)="p" THEN PE$=MID$(PE$,2)
4080 PE=VAL(PE$)
4090 IF PE>=1 AND PE<=NP THEN 4110
4100 PRINT:PRINT "Entry error. Try again ...":GOTO 4040
4110 PRINT:PRINT TAB(5) "P"PE$"=";
4120 ON ND GOTO 4130,4140,4150,4160
4130 INPUT C(1,PE,1):GOTO 4170
4140 INPUT C(1,PE,1),C(1,PE,2):GOTO 4170
```

```
4150 INPUT C(1,PE,1),C(1,PE,2),C(1,PE,3):GOTO 4170
4160 INPUT C(1,PE,1),C(1,PE,2),C(1,PE,3),C(1,PE,4)
4170 GOTO 4000
4180 '
5000 '
5010 '          Translation
5020 '
5030 CLS:PRINT MESS$
5040 PRINT "Specify the translation term";
5050 IF ND>1 THEN PRINT "s";
5060 PRINT ":":PRINT
5070 PRINT TAB(30);:INPUT "Tx=",T(1)
5080 IF ND>1 THEN PRINT TAB(30);:INPUT "Ty=",T(2)
5090 IF ND>2 THEN PRINT TAB(30);:INPUT "Tz=",T(3)
5100 IF ND>3 THEN PRINT TAB(30);:INPUT "Tw=",T(4)
5110 FOR N=1 TO NP:FOR X=1 TO ND
5120 C(1,N,X)=C(1,N,X)+T(X)
5130 NEXT X:NEXT N
5140 GOSUB 3000:GOTO 1400:RETURN
5150 '
6000 '
6010 '          Scaling
6020 '
6030 CLS:PRINT MESS$
6040 PRINT "Specify the scaling factor";
6050 IF ND>1 THEN PRINT "s";
6060 PRINT ":":PRINT
6070 PRINT TAB(30);:INPUT "Kx=",T(1)
6080 IF ND>1 THEN PRINT TAB(30);:INPUT "Ky=",T(2)
6090 IF ND>2 THEN PRINT TAB(30);:INPUT "Kz=",T(3)
6100 IF ND>3 THEN PRINT TAB(30);:INPUT "Kw=",T(4)
6110 FOR N=1 TO NP:FOR X=1 TO ND
6120 C(1,N,X)=C(1,N,X)*T(X)
6130 NEXT X:NEXT N
6140 GOSUB 3000:GOTO 1400:RETURN
6150 '
7000 '
7010 '          Rotation
7020 '
7030 CLS:PRINT MESS$:PRINT
7040 IF ND>1 THEN 7070
7050 PRINT "You cannot rotate in just one dimension."
7060 GOSUB 1400:RETURN
7070 INPUT "Specify an angle of rotation (degrees): ";ANG
7080 RAD=ANG*(3.141593/180)
7090 ON ND-1 GOTO 7100,7110,7220
7100 T(1)=1:T(2)=2:GOTO 7390
7110 CLS:PRINT MESS$:PRINT
7120 PRINT "Axes of rotation:":PRINT
7130 PRINT TAB(5) "1 -- X axis
7140 PRINT TAB(5) "2 -- Y axis
7150 PRINT TAB(5) "3 -- Z axis
7160 GOSUB 1300
7170 IF ASC(K$)<49 OR ASC(K$)>51 THEN GOSUB 1440:GOTO 7170
7180 ON VAL(K$) GOTO 7190,7200,7210
7190 T(1)=2:T(2)=3:GOTO 7390
7200 T(1)=1:T(2)=3:GOTO 7390
```

```
7210 T(1)=1:T(2)=2:GOTO 7390
7220 CLS:PRINT MESS$:PRINT
7230 PRINT "Planes of rotation:":PRINT
7240 PRINT TAB(5) "1 -- X-Y plane
7250 PRINT TAB(5) "2 -- X-Z plane
7260 PRINT TAB(5) "3 -- X-W plane
7270 PRINT TAB(5) "4 -- Y-Z plane
7280 PRINT TAB(5) "5 -- Y-W plane
7290 PRINT TAB(5) "6 -- Z-W plane
7300 GOSUB 1300
7310 IF ASC(K$)<49 OR ASC(K$)>54 THEN GOSUB 1440:GOTO 7310
7320 ON VAL(K$) GOTO 7330,7340,7350,7360,7370,7380
7330 T(1)=3:T(2)=4:GOTO 7390
7340 T(1)=2:T(2)=4:GOTO 7390
7350 T(1)=2:T(2)=3:GOTO 7390
7360 T(1)=1:T(2)=4:GOTO 7390
7370 T(1)=1:T(2)=3:GOTO 7390
7380 T(1)=1:T(2)=2
7390 PRINT:PRINT "Ok"
7400 FOR N=1 TO NP
7410 R1=C(1,N,T(1))*COS(RAD)-C(1,N,T(2))*SIN(RAD)
7420 R2=C(1,N,T(1))*SIN(RAD)+C(1,N,T(2))*COS(RAD)
7430 C(1,N,T(1))=R1:C(1,N,T(2))=R2
7440 NEXT N
7450 GOSUB 3000:GOSUB 1400:RETURN
8000 '
8010 '          Length
8020 '
8030 GOSUB 3000
8040 PRINT:INPUT "Specify one of the endpoints: P",E1
8050 IF E1>=1 AND E1<=NP THEN 8070
8060 PRINT:PRINT "Entry error.  Try again ...":GOTO 8040
8070 INPUT "Specify the second endpoint:  P",E2
8080 IF E2>=1 AND E2<=NP THEN 8100
8090 PRINT:PRINT "Entry error.  Try again ...":GOTO 8070
8100 CLS:PRINT MESS$:PRINT
8110 PRINT "The length of a line between points P";
8120 PRINT MID$(STR$(E1),2)" and P";
8130 PRINT MID$(STR$(E2),2)" is: ";
8140 T(0)=0
8150 FOR X=1 TO ND
8160 D(X)=C(1,E2,X)-C(1,E1,X)
8170 T(0)=T(0)+D(X)^2
8180 NEXT X
8190 T(0)=SQR(T(0)):PRINT INT(1000*T(0))/1000
8200 PRINT:PRINT "The direction cosine";
8210 IF ND>1 THEN PRINT "s";
8220 PRINT " for that line";
8230 IF ND>1 THEN PRINT " are:"; ELSE PRINT " is:";
8240 PRINT:PRINT
8250 PRINT TAB(35) "A="INT(1000*D(1)/T(0))/1000
8260 IF ND>1 THEN PRINT TAB(35) "B="INT(1000*D(2)/T(0))/1000
8270 IF ND>2 THEN PRINT TAB(35) "C="INT(1000*D(3)/T(0))/1000
8280 IF ND>3 THEN PRINT TAB(35) "D="INT(1000*D(4)/T(0))/1000
8290 PRINT:PRINT:PRINT "Do you want to find another length (Y/N)?"
8300 GOSUB 1440
8310 IF K$="Y" OR K$="y" THEN 8000
```

```
8320 IF K$="N" OR K$="n" THEN RETURN
8330 GOTO 8300
9000 '
9010 '          Restore
9020 '
9030 PRINT:PRINT "Ok"
9040 FOR N=1 TO NP:FOR X=1 TO ND
9050 C(1,N,X)=C(2,N,X)
9060 NEXT X:NEXT N
9070 GOSUB 3000:GOSUB 1400
9080 RETURN
```

APPLESOFT VERSION

```
10   REM   ** EXPERIMENTS IN FOUR DIMENSIONS **
20   REM
30   REM
40   REM  -- BASED ON THE BOOK "EXPERIMENTS IN FOUR DIMENSIONS"
50   REM     BY DAVID L. HEISERMAN
60   REM
70   REM  COPYRIGHT 1983
80   REM     TAB BOOKS INC.
90   REM     BLUE RIDGE SUMMIT  PA
100  REM
110  REM      -- SETUP ROUTINE --
120  REM
130  DIM C(2,32,4): DIM CP$(2,32,4)
140 L$ = "": FOR N = 1 TO 39:L$ = L$ + "=": NEXT N
150  TEXT : HOME
160  PRINT "EXPERIMENTS IN FOUR DIMENSIONS"
170  PRINT : PRINT "-- SETUP ROUTINE --"
180  PRINT L$: PRINT : PRINT
190  INPUT "HOW MANY DIMENSIONS? ";ND
200 ND =  INT (ND)
210  IF ND >  = 1 AND ND <  = 4 THEN 240
220  PRINT : PRINT : PRINT
230  PRINT "ENTRY ERROR.  TRY AGAIN ...": PRINT : GOTO 190
240 D$ = " DIMENSIONS"
250  ON ND GOTO 260,270,280,290
260 X$ = "ONE":D$ = " DIMENSION": GOTO 300
270 X$ = "TWO": GOTO 300
280 X$ = "THREE": GOTO 300
290 X$ = "FOUR"
300 MESS$ = "EXPERIMENTS IN " + X$ + D$
310 MESS$ = MESS$ +  CHR$ (13) + L$ +  CHR$ (13)
320  REM
1000  REM
1010  REM       MAIN MENU
1020  REM
1030  HOME : PRINT MESS$: PRINT
1040  PRINT "MAIN MENU": PRINT
1050  PRINT  TAB( 5)"1 -- SPECIFY ALL NEW COORDINATES"
1060  PRINT  TAB( 5)"2 -- DISPLAY CURRENT COORDINATES"
1070  PRINT  TAB( 5)"3 -- EDIT CURRENT COORDINATES"
1080  PRINT  TAB( 5)"4 -- TRANSLATE THE COORDINATES"
1090  PRINT  TAB( 5)"5 -- SCALE THE COORDINATES"
1100  PRINT  TAB( 5)"6 -- ROTATE THE COORDINATES"
```

```
1110  PRINT  TAB( 5)"7 -- CALCULATE LENGTHS OF LINES"
1120  PRINT  TAB( 5)"8 -- RESTORE ORIGINAL COORDINATES"
1130  PRINT  TAB( 5)"9 -- QUIT THE PROGRAM"
1140  PRINT : GOSUB 1300
1150  IF  ASC (K$) >  = 49 AND  ASC (K$) <  = 57 THEN 1170
1160  PRINT : PRINT "ENTRY ERROR. TRY AGAIN ...": GOTO 1140
1170  ON  VAL (K$) GOTO 1180,1190,1200,1210,1220,1230,1240,1250,1260
1180 T = 1: GOSUB 2000: GOTO 1000
1190  GOSUB 3000: GOSUB 1400: GOTO 1000
1200  GOSUB 4000: GOTO 1000
1210 T = 2: GOSUB 5000: GOTO 1000
1220 T = 2: GOSUB 6000: GOTO 1000
1230 T = 2: GOSUB 7000: GOTO 1000
1240  GOSUB 8000: GOTO 1000
1250 T = 1: GOSUB 9000: GOTO 1000
1260  HOME : END
1300  REM
1310  REM   -- SELECT ONE --
1320  REM
1330  PRINT : PRINT "SELECT ONE ... ";
1340  GOSUB 1440: RETURN
1350  REM
1400  REM
1410  REM   -- STRIKE ANY KEY --
1420  REM
1430  PRINT "STRIKE ANY KEY TO RETURN TO MAIN MENU"
1440  GET K$
1450  RETURN
1460  REM
2000  REM
2010  REM   -- NEW COORDINATES --
2020  REM
2030  HOME : PRINT MESS$: PRINT
2040  INPUT "HOW MANY POINTS (1-32)? ";NP
2050 NP =  INT (NP)
2060  IF NP >  = 1 AND NP <  = 32 THEN 2080
2070  PRINT : PRINT "ENTRY ERROR. TRY AGAIN ...": PRINT : GOTO 2040
2080  HOME : PRINT MESS$: PRINT
2090  PRINT "ENTER YOUR "NP" POINT COORDINATE";
2100  IF NP > 1 THEN  PRINT "S";
2110  PRINT : PRINT "AS X";
2120  IF ND > 1 THEN  PRINT ",Y";
2130  IF ND > 2 THEN  PRINT ",Z";
2140  IF ND > 3 THEN  PRINT ",W"
2150  PRINT : PRINT
2160  FOR N = 1 TO NP
2170  PRINT  TAB( 5)"P"N;
2180  ON ND GOTO 2190,2200,2210,2220
2190  INPUT "=";C(1,N,1): GOTO 2230
2200  INPUT "=";C(1,N,1),C(1,N,2): GOTO 2230
2210  INPUT "=";C(1,N,1),C(1,N,2),C(1,N,3): GOTO 2230
2220  INPUT "=";C(1,N,1),C(1,N,2),C(1,N,3),C(1,N,4)
2230  FOR X = 1 TO ND:C(2,N,X) = C(1,N,X): NEXT X
2240  NEXT N
2250  GOSUB 3000: GOSUB 1400: GOTO 1000
2260  REM
3000  REM
```

```
3010 REM    -- PACK AND PRINT --
3020 REM
3030 HOME : PRINT MESS$
3040 PRINT : PRINT "SUMMARY OF CURRENT POINTS:": PRINT
3050 FOR N = 1 TO NP
3060 IF  POS (X) > 28 THEN  PRINT
3070 PRINT "P";: IF T > 1 THEN  PRINT "'";
3080 PRINT N"=(";
3090 FOR X = 1 TO ND
3100 CP$(1,N,X) =  STR$ ( INT (1000 * C(1,N,X)) / 1000)
3110 PRINT CP$(1,N,X)",";
3120 NEXT X
3130 PRINT  CHR$ (8) + ")",
3140 NEXT N
3150 PRINT : PRINT : RETURN
3160 REM
4000 REM
4010 REM    -- EDIT --
4020 REM
4030 GOSUB 3000
4040 PRINT : PRINT "WHICH POINT DO YOU WANT TO EDIT?"
4050 INPUT "(STRIKE 'RETURN' TO RETURN TO MAIN MENU)";PE$
4060 IF PE$ = "" THEN  RETURN
4070 IF  LEFT$ (PE$,1) = "P" THEN PE$ =  MID$ (PE$,2)
4080 PE =  VAL (PE$)
4090 IF PE >  = 1 AND PE <  = NP THEN 4110
4100 PRINT : PRINT "ENTRY ERROR. TRY AGAIN ...": GOTO 4040
4110 PRINT : PRINT  TAB( 5)"P"PE$"=";
4120 ON ND GOTO 4130,4140,4150,4160
4130 INPUT C(1,PE,1): GOTO 4170
4140 INPUT C(1,PE,1),C(1,PE,2): GOTO 4170
4150 INPUT C(1,PE,1),C(1,PE,2),C(1,PE,3): GOTO 4170
4160 INPUT C(1,PE,1),C(1,PE,2),C(1,PE,3),C(1,PE,4)
4170 GOTO 4000
4180 REM
5000 REM
5010 REM    -- TRANSLATION --
5030 HOME : PRINT MESS$
5040 PRINT "SPECIFY THE TRANSLATION TERM";
5050 IF NP > 1 THEN  PRINT "S";
5060 PRINT ":": PRINT
5070 PRINT  TAB( 10);: INPUT "TX=";T(1)
5080 IF ND > 1 THEN  PRINT  TAB( 10);: INPUT "TY=";T(2)
5090 IF ND > 2 THEN  PRINT  TAB( 10);: INPUT "TX=";T(3)
5100 IF ND > 3 THEN  PRINT  TAB( 10);: INT PUT"TW=";T(4)
5110 FOR N = 1 TO NP: FOR X = 1 TO ND
5120 C(1,N,X) = C(1,N,X) + T(X)
5130 NEXT X: NEXT N
5140 GOSUB 3000: GOTO 1400: RETURN
5150 REM
6000 REM
6010 REM    -- SCALING --
6020 REM
6030 HOME : PRINT MESS$
6040 PRINT "SPECIFY THE SCALING FACTOR";
6050 IF ND > 1 THEN  PRINT "S";
6060 PRINT ":": PRINT
```

```
6070  PRINT  TAB( 10);: INPUT "KX=";T(1)
6080  IF ND > 1 THEN  PRINT  TAB( 10);: INPUT "KY=";T(2)
6090  IF ND > 2 THEN  PRINT  TAB( 10);: INPUT "KZ=";T(3)
6100  IF ND > 3 THEN  PRINT  TAB( 10);: INPUT "KW=";T(4)
6110  FOR N = 1 TO NP: FOR X = 1 TO ND
6120 C(1,N,X) = C(1,N,X) * T(X)
6130  NEXT X: NEXT N
6140  GOSUB 3000: GOTO 1400: RETURN
6150  REM
7000  REM
7010  REM   -- ROTATION --
7020  REM
7030  HOME : PRINT MESS$: PRINT
7040  IF ND > 1 THEN 7070
7050  PRINT "YOU CANNOT ROTATE IN JUST ONE DIMENSION."
7060  GOSUB 1400: RETURN
7070  PRINT "SPECIFY AN ANGLE OF ROTATION"
7080  INPUT ANG
7090 RAD = ANG * (3.141593 / 180)
7100  ON ND - 1 GOTO 7110,7120,7230
7110 T(1) = 1:T(2) = 2: GOTO 7400
7120  HOME : PRINT MESS$: PRINT
7130  PRINT "AXES OF ROTATION:": PRINT
7140  PRINT  TAB( 5)"1 -- X AXIS"
7150  PRINT  TAB( 5)"2 -- Y AXIS"
7160  PRINT  TAB( 5)"3 -- Z AXIS"
7170  GOSUB 1300
7180  IF  ASC (K$) < 49 OR  ASC (K$) > 51 THEN  GOSUB 1440: GOTO 7180
7190  ON  VAL (K$) GOTO 7200,7210,7220
7200 T(1) = 2:T(2) = 3: GOTO 7400
7210 T(1) = 1:T(2) = 3: GOTO 7400
7220 T(1) = 1:T(2) = 2: GOTO 7400
7230  HOME : PRINT MESS$: PRINT
7240  PRINT "PLANES OF ROTATI
7250  PRINT  TAB( 5)"1 -- X-Y PLANE"
7260  PRINT  TAB( 5)"2 -- X-Z PLANE"
7270  PRINT  TAB( 5)"3 -- X-W PLANE"
7280  PRINT  TAB( 5)"4 -- Y-Z PLANE"
7290  PRINT  TAB( 5)"5 -- Y-W PLANE"
7300  PRINT  TAB( 5)"6 -- Z-W PLANE"
7310  GOSUB 1300
7320  IF  ASC (K$) < 49 OR  ASC (K$) > 54 THEN  GOSUB 1440: GOTO 7320
7330  ON  VAL (K$) GOTO 7340,7350,7360,7370,7380,7390
7340 T(1) = 3:T(2) = 4: GOTO 7400
7350 T(1) = 2:T(2) = 4: GOTO 7400
7360 T(1) = 2:T(2) = 3: GOTO 7400
7370 T(1) = 1:T(2) = 4: GOTO 7400
7380 T(1) = 1:T(2) = 3: GOTO 7400
7390 T(1) = 1:T(2) = 2
7400  PRINT : PRINT "OK"
7410  FOR N = 1 TO NP
7420 R1 = C(1,N,T(1)) *  COS (RAD) - C(1,N,T(2)) *  SIN (RAD)
7430 R2 = C(1,N,T(1)) *  SIN (RAD) + C(1,N,T(2)) *  COS (RAD)
7440 C(1,N,T(1)) = R1:C(1,N,T(2)) = R2
7450  NEXT N
7460  GOSUB 3000: GOSUB 1400: RETURN
7470  REM
```

```
8000  REM
8010  REM   -- LENGTH --
8020  REM
8030  GOSUB 3000
8040  PRINT : INPUT "SPECIFY ONE OF THE ENDPOINTS: P";E1
8050  IF E1 >  = 1 AND E1 <  = NP THEN 8070
8060  PRINT : PRINT "ENTRY ERROR. TRY AGAIN ...": GOTO 8040
8070  INPUT "SPECIFY THE SECOND ENDPOINT:  P";E2
8080  IF E2 >  = 1 AND E2 <  = NP THEN 8100
8090  PRINT : PRINT "ENTRY ERROR. TRY AGAIN ...": GOTO 8070
8100  HOME : PRINT MESS$: PRINT
8110  PRINT "THE LENGTH OF A LINE BETWEEN POINTS"
8120  PRINT "P"E1" AND P"E2" IS: ";
8130 T(0) = 0
8140  FOR X = 1 TO ND
8150 D(X) = C(1,E2,X) - C(1,E1,X)
8160 T(0) = T(0) + D(X) ↑ 2
8170  NEXT X
8180 T(0) =  SQR (T(0))
8190  PRINT  INT (1000 * T(0)) / 1000
8200  PRINT : PRINT "THE DIRECTION COSINE";
8210  IF ND > 1 THEN  PRINT "S";
8220  PRINT " FOR THAT LINE";
8230  IF ND > 1 THEN  PRINT " ARE:";: GOTO 8250
8240  PRINT " IS:";
8250  PRINT : PRINT
8260  PRINT  TAB( 10)"A=" INT (1000 * D(1) / T(0)) / 1000
8270  IF ND > 1 THEN  PRINT  TAB( 10)"B=" INT (1000 * D(2) / T(0)) / 1000
8280  IF ND > 2 THEN  PRINT  TAB( 10)"C=" INT (1000 * D(3) / T(0)) / 1000
8290  IF ND > 3 THEN  PRINT  TAB( 10)"D=" INT (1000 * D(4) / T(0)) / 1000
8300  PRINT : PRINT
8310  PRINT "DO YOU WANT TO FIND ANOTHER LENGTH (Y/N)?"
8320  GOSUB 1440
8330  IF K$ = "Y" THEN 8000
8340  IF K$ = "N" THEN  RETURN
8350  GOTO 8320
9000  REM
9010  REM   -- RESTORE --
9040  FOR N = 1 TO NP: FOR X = 1 TO ND
9050 C(1,N,X) = C(2,N,X)
9060  NEXT X: NEXT N
9070  GOSUB 3000: GOSUB 1400
9080  RETURN
9120  REM
9130  PRINT : PRINT "OK"
```

Index

Index

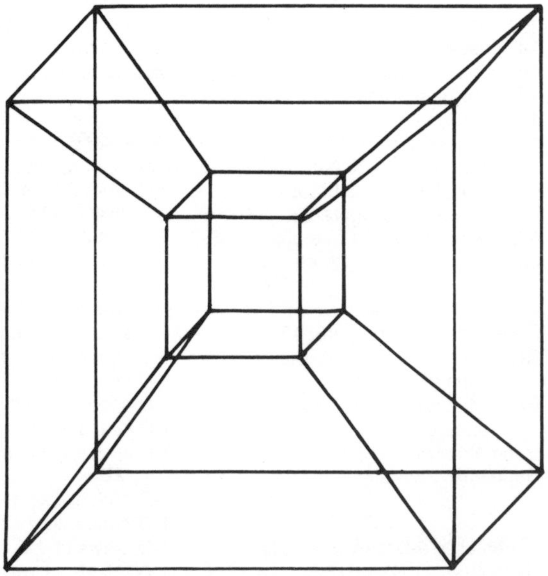